Thrust Tectonics and Hydrocarbon Systems

Edited by
K. R. McClay

AAPG MEMOIR 82

Published by
The American Association of Petroleum Geologists
Tulsa, Oklahoma, U.S.A.

ISBN: 0-89181-363-2

AAPG Editor: Ernest A. Mancini
Geoscience Director: J. B. "Jack" Thomas

This publication is available from:

The AAPG Bookstore
P.O. Box 979
Tulsa, OK U.S.A. 74101-0979
Phone: 1-918-584-2555 or 1-800-364-AAPG (U.S.A. only)
Fax: 1-918-560-2652 or 1-800-898-2274 (U.S.A. only)
E-mail: bookstore@aapg.org
www.aapg.org

Dedication

This volume is dedicated to the late Peter Coney (1930–1999), Professor, at the Department of Geosciences, University of Arizona, and Visiting Professor in Tectonics, Royal Holloway, University of London. Peter was fundamental in shaping many aspects of the study of tectonics, including suspect terranes, metamorphic core complexes, orogenic collapse and extension, and the geodynamic evolution of the Pyrenees. He compiled terrane analyses of the western North American Cordillera, Alaska, and Australia, and had just completed a terrane analysis of the world before his untimely death. He had the uncanny ability to cut through the tectonic noise and ask the fundamental questions, which inevitably lead to new theories and discoveries. The photograph shows him at his best, in the field, in the Pyrenees, Spain, pondering the geodynamics of the thrust belt and its uplift and exhumation.

Table of Contents

SECTION 3—FAULT-RELATED FOLDS IN THRUST SYSTEMS

SECTION 4—CASE STUDIES

Foreword

This volume resulted from the International Conference on Thrust Tectonics, convened by the editor and the Fault Dynamics Research Group, Department of Geology, Royal Holloway, University of London, in April 1999. The papers presented here are a selection of key presentations on thrust tectonics, as well as invited submissions from other authors who were not able to attend the meeting. The papers presented in this volume reflect the current research on thrust tectonics and its applications to hydrocarbon systems. The chapters are subdivided into four broad categories, each of which covers the most important aspects of thrust-belt studies. The categories are

(1) geodynamics of thrust systems,
(2) analog modeling of thrust systems,
(3) fault-related folds in thrust systems, and
(4) case studies.

Thrust Tectonics and Hydrocarbon Systems covers a wide and diverse range of topics, from individual case studies of petroleum systems in thrust belts to the geodynamics of the Andes. It aims to reflect the current key aspects of thrust-belt research and the application of this research to exploring for and exploiting hydrocarbon resources.

I would like to thank my colleagues in the Department of Geology at Royal Holloway, and in particular Tim Dooley, Paul Whitehouse, and Kevin D'Souza, for all of their help both pre- and postconference. Last, but not least, I wish to acknowledge the tremendous support provided by April Harper, both in helping organize the conference and in helping organize and arrange this volume. The editorial staff at the AAPG are thanked for taking the final manuscripts and producing the completed publication.

Ken McClay, Editor
2004

SPONSORS

The conference Thrust Tectonics '99 was generously sponsored by the following companies and organizations:

AGIP (U.K.)
Amerada Hess
Anadarko
ARCO British
BP Exploration
BG Exploration
Canadian Petroleum
Chevron
Conoco
Enterprise
Fault Dynamics Research Group RHUL
Lasmo
Monument Oil & Gas
Premier
Sipetrol
Texaco
Union Texas
Unocal

The American Association of Petroleum Geologists thanks the following for their generous contributions to this publication:

Conoco Phillips
EnCana Corporation
Talisman Energy
Fold-Fault Research Project—University of Calgary and Queen's University
Tectonics and Structural Geology Research—Science Alliance Center of Excellence—University of Tennessee

REVIEWERS

After the conference, many manuscripts were submitted for consideration for this volume. The following people are sincerely thanked for their time and effort in reviewing one or more of the manuscripts, and in some cases rereviewing revised manuscripts. The review and correction process has been long and exhausting, and the editor is deeply grateful both to authors and reviewers for their efforts and patience.

Richard Allmendinger
Joaquina Alvarez Marron
Andrea Argnani
Alistair Beach
Denis Brown
Mayte Bulnes
Jean-Paul Burg
Martin Burkhard
Rob Butler
Peter Cobbold
Bernard Colletta
Mark Cooper
Mike Coward
Jonathan Craig
Ian Davison
Peter DeCelles
Tarik Djebbar
Tim Dooley
Chris Elders
Peter Ellis
Eric Erslev
Mark Fischer
Thomas Flottmann
Mary Ford
Dick Glen
Rick Groshong
Robert Hall
Stuart Hardy
Bob Hatcher
Tony Haywood
Kevin Hill
Martin Insley
Chris Izatt
Jamie Jamison
Bob Jennings
Chuck Kluth
Hemin Koyi
Philip Maclaurin
Niell Manktelow
Jacques Malavieille
Alex Maltman
Steve Marshak
Buffy McClelland
Shankar Mitra
Chris Morley
Eric Mountjoy
Jean-Louis Mugnier
Josep-Anton Muñoz
Josep Poblet
Ray Price
Geoff Rait
Robert Ratcliff
Pedro Restrepo-Pace
Dave Richards
David Roberts
David G. Roberts
Dietrich Roeder
François Roure
Mark Rowan
Francesco Salvini
John Shaw
Rick Sibson
Jorge Skarmeta
Debbie Spratt
Fabrizio Storti
John Suppe
Peter Talling
Antonio Teixell
Bob Thompson
Phaedra Upton
Wesley Wallace
Dave Waltham
Stan White
Scott Wilkerson
Graham Williams
Yasuhiro Yamada
Tomas Zapata

McClay, K. R., 2004, Thrust tectonics and hydrocarbon systems; Introduction, *in* K. R. McClay, ed., Thrust tectonics and hydrocarbon systems: AAPG Memoir 82, p. ix–xx.

Introduction

Thrust Tectonics and Hydrocarbon Systems; Introduction

K. R. McClay

Fault Dynamics Research Group, Geology Department, Royal Holloway University of London, United Kingdom

INTRODUCTION

The tectonic settings and internal characteristics of fold-and-thrust belts are remarkably varied and diverse. They include

- classic thin-skinned Canadian Rocky Mountain–style foreland fold-and-thrust belts (Bally et al., 1966; Price, 1981; Barclay and Smith, 1992);
- basement-involved Laramide uplifts of the western U.S.A. (Schmidt et al., 1993);
- subduction-related thrust systems of the Andes (Ramos et al., 2004 [this volume]), the Zagros (Beydoun et al., 1992), and Papua New Guinea (Cole et al., 2000; Hill et al., 2000; and Hill et al., 2004 [this volume]);
- collision terranes of the Himalayas (e.g., Mugnier et al., 2004 [this volume]), and the Italian Apennines (Coward et al., 1999; Turrini and Rennison, 2004 [this volume]); and
- deep-water, gravitationally driven fold-and-thrust belts, such as those offshore West Africa or in the Gulf of Mexico (Rowan et al., 2004 [this volume]).

Many of these fold-and-thrust belts contain major hydrocarbon accumulations in complex structural traps. Hydrocarbon exploration in many onshore fold-and-thrust belts commonly is very difficult because of rugged and severe terrain, complex structures, and poor-quality seismic data.

Key to understanding the 4-D (heighth, width, depth, time) evolution of fold-and-thrust-belt systems are an appreciation of the driving mechanisms for thrust-belt development, and knowledge of the timing of thrust movements, their burial and thermal evolution, and their uplift and exhumation histories. Understanding both the spatial and temporal evolution of fold-and-thrust belts is vital for viable hydrocarbon exploration and development in these complex terranes. This volume presents key papers that address current issues in understanding fold-and-thrust-belt structures. These include (1) fold-and-thrust-belt geodynamics, (2) 3-D thrust geometries, (3) scaled analog modeling of thrust belts as a means of developing new kinematic models for their evolution, (4) new models of thrust-fault-related folds and synkinematic growth strata, and (5) case studies of particular thrust terranes.

Several key topics are addressed in this volume.

THE CRITICAL COULOMB WEDGE MODEL

Following the benchmark paper by Davis et al. (1983), the Critical Coulomb Wedge hypothesis has become widely accepted as a general model for the development of foreland fold-and-thrust belts (Dahlen, 1990; Brown, 2004 [this volume]; Mugnier et al., 2004 [this volume]; McClay and Whitehouse, 2004 [this volume]). This simple wedge model was further improved by Koons (1990) and Willett et al., (1993) to explain doubly vergent fold-and-thrust belts, such as the Pyrenees (Figure 1) and other collisional orogens. Doubly vergent orogenic wedges (Figure 1) have been simulated successfully in the laboratory (cf., Storti et al., 2000; McClay and Whitehouse, 2004 [this volume]) and provide important constraints for the kinematic evolution of thrust systems. In particular, they demonstrate multiple thrust activities and the dramatic feedback effects of syntectonic erosion and sedimentation on thrust-belt evolution.

DYNAMIC OROGENS—INTERACTION WITH SURFICIAL PROCESSES

The dynamic interaction between thrust-belt evolution and surficial processes—synkinematic uplift,

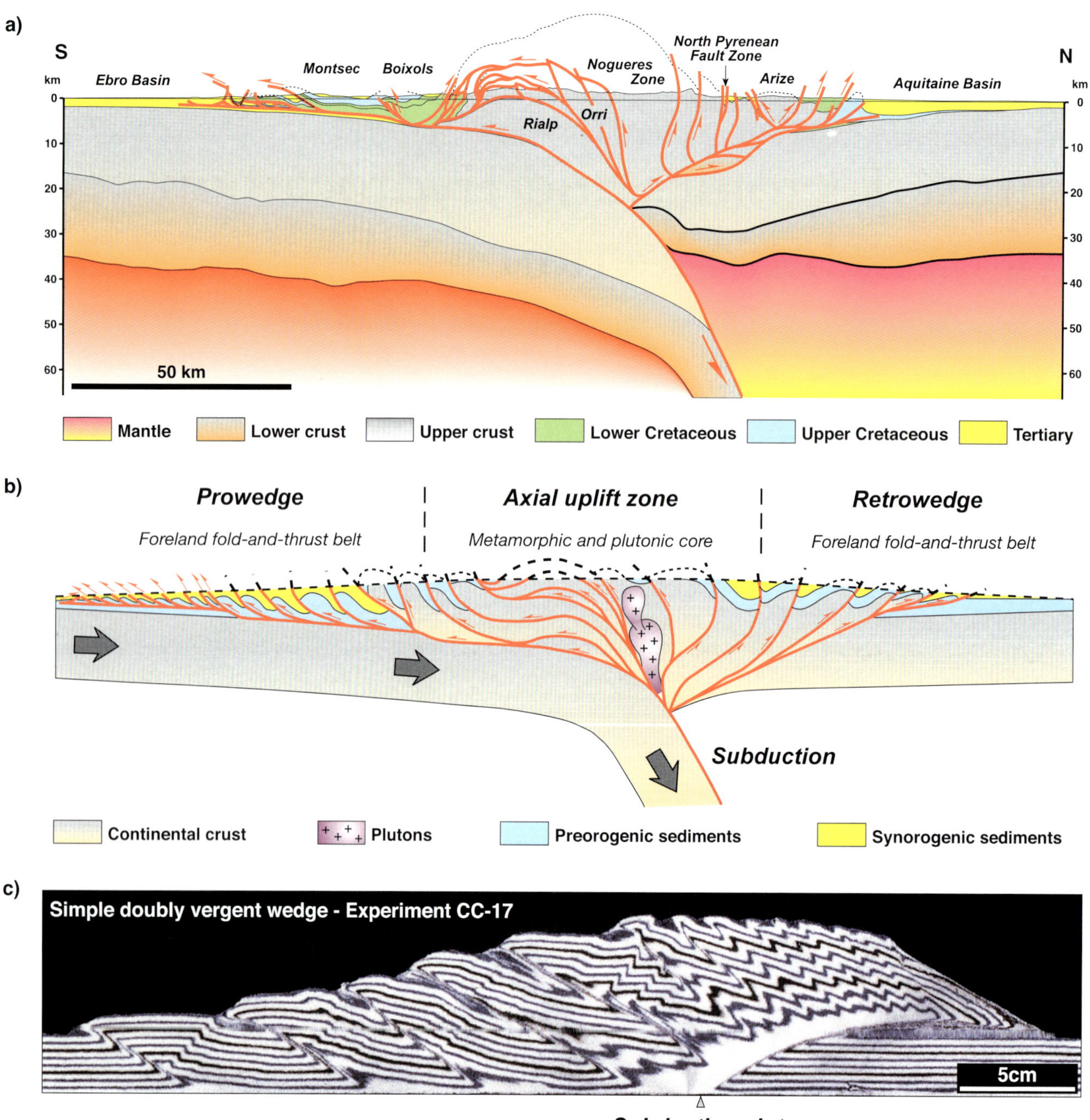

FIGURE 1. Doubly vergent orogens. (a) Cross section through the Pyrenees (modified from Muñoz, 1992). (b) Conceptual model of a doubly vergent orogen (from McClay and Whitehouse, 2004). (c) Analog model of a simple, doubly vergent orogen (from McClay and Whitehouse, 2004).

erosion, and sedimentation— has been the focus of much recent research (Beaumont et al., 1992; Storti and McClay, 1995; Meigs and Burbank, 1997; Hardy et al., 1998; Beaumont et al., 2000a, b; Burbank and Anderson, 2001; McClay and Whitehouse, 2004 [this volume]; Mugnier et al., 2004 [this volume]; and others). Clearly, the dynamic feedback between thrusting, uplift, erosion, and sedimentation exerts dramatic controls and effects on thrust-belt morphologies and evolution (Figure 2). Field studies combined with numerical modeling (Beaumont et al., 2000a) and scaled analog modeling (McClay and Whitehouse, 2004 [this volume]) clearly demonstrate that this dynamic feedback affects thrust morphologies, thrust sequences, thrust activities, and thrust rates. Developing new dynamic models and critically testing them with field data from natural fold-thrust belts will be essential if we are to understand these dynamic feedback effects.

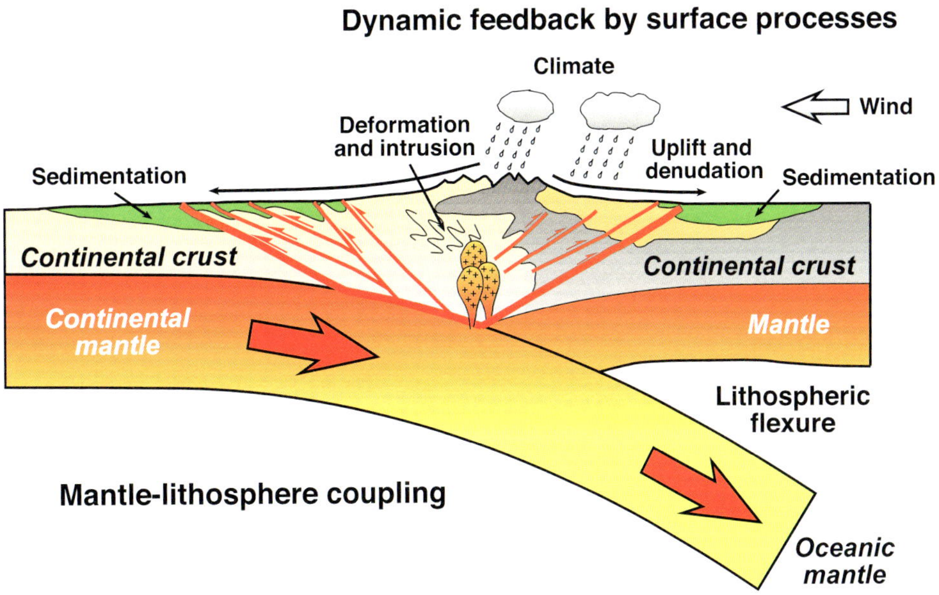

FIGURE 2. Conceptual model of a collisional orogen, showing the interaction of surficial processes (modified from Beaumont et al., 2000a).

INVERSION AND BASEMENT INVOLVEMENT IN THRUST BELTS

An important topic at the Thrust Tectonics '99 conference as well as in many of the chapters in this volume is the recognition that basement is involved in many thrust belts, whether it is incorporated in thrust sheets, or inversion and preexisting basement fabrics control subsequent thrust architectures. Contractional inversion of older extensional faults has now been recognized widely in many fold-and-thrust belts, e.g., in the Alps (Schmid et al., 1996; Pfiffner et al., 2000); the Apennines (Coward et al., 1999); the Pyrenees (Figure 1a) (Muñoz, 1992); Papua New Guinea (Figure 3) (Buchanan and Warburton, 1996; Cole et al., 2000; Hill et al., 2004 [this volume]); the sub-Andean basins, such as the Neuquen (Figure 4) (Rojas et al., 1999); the Andes (Ramos et al., 2004 [this volume]); and in the Atlas Mountains of Morocco (Beauchamp, 2004 [this volume]). Many fold-and-thrust belts appear to contain elements of both thin-skinned thrusting and basement-involved, thick-skinned thrusting. As a result, section construction and restoration needs to include both these thick- and thin-skinned elements. Basement involvement also has major implications for distribution of reservoirs and for hydrocarbon systems in fold-and-thrust belts.

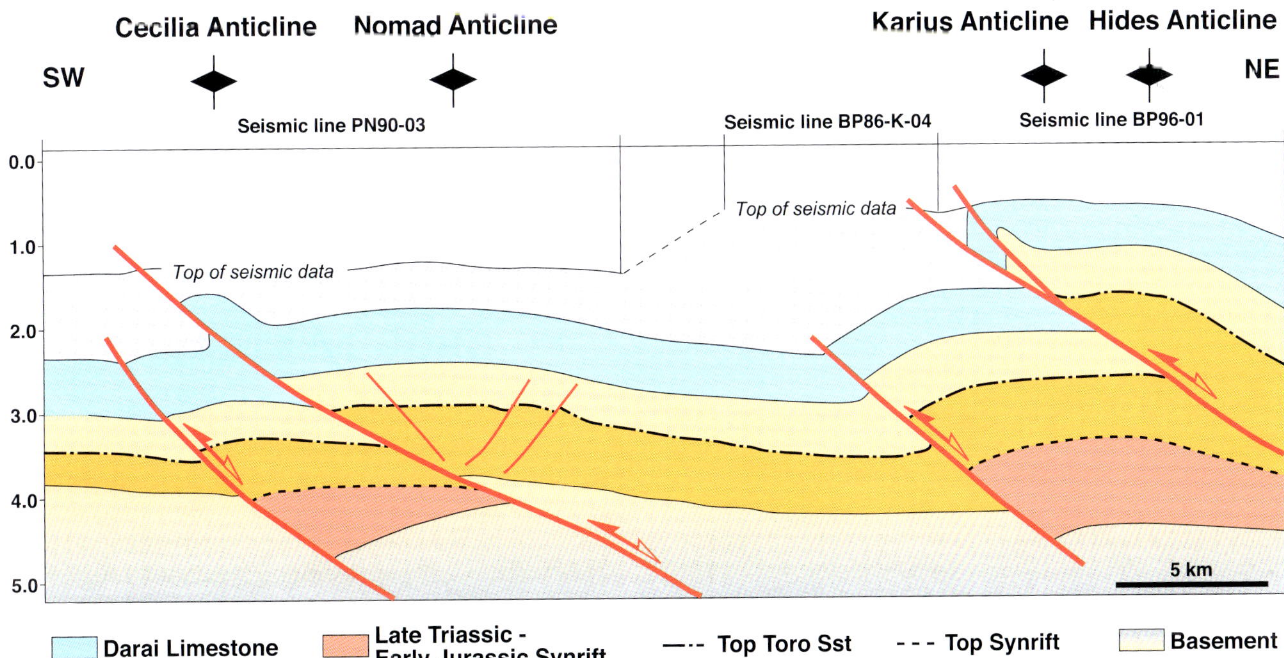

FIGURE 3. Line drawing of a seismic section through the Papua New Guinea Fold Belt, showing inversion of Late Jurassic–Early Jurassic rift faults (adapted from Cole et al., 2000).

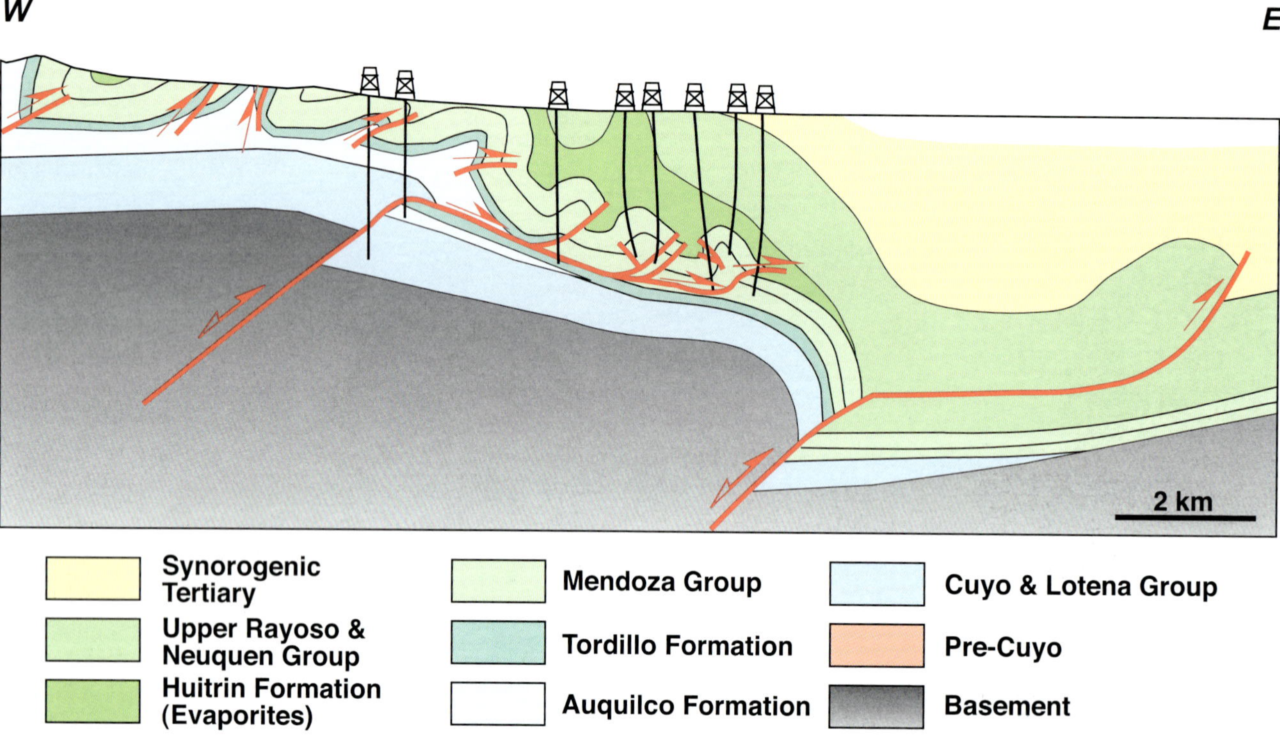

FIGURE 4. Cross section through the Rio Salado–Puesto Rojas oil field, Malargue Fold Belt, northern Neuquen Basin, Argentina (modified from Rojas et al., 1999).

THRUST-FAULT-RELATED FOLD SYSTEMS

Geometric and kinematic models for fault-related folding in thrust belts have evolved dramatically from the kink-band fault-bend-fold models published by John Suppe in 1983. We now recognize that many folds in thrust belts commonly are hybrid structures that have undergone a complex evolution (e.g., Figure 5) (Wallace and Homza, 2004 [this volume]). New kinematic models have been proposed, including those that include progressive limb rotation (e.g., Poblet and McClay, 1996; Poblet et al., 2004 [this volume]) and trishear models (Erslev, 1991; Allmendinger, 1998; Allmendinger et al., 2004 [this volume]). Trishear modeling appears to be a powerful tool for both forward and reverse modeling of thrust-related folds (Allmendinger, 1998) and has the added attraction of modeling strain distribution (and hence, possible fracture distribution) in complex fault-fold systems (Allmendinger et al., 2004 [this volume]). Detachment folding also has been recognized as an important fold style in many thrust belts (Poblet and McClay, 1996; Poblet et al., 2004 [this volume]) and is probably more common than hitherto recognized.The geometric fold models described by Suppe (1983, 1985; Mosar and Suppe, 1992; Mitra, 1990) and the trishear models (Allmendinger, 1998; Allmendinger et al., 2004 [this volume]), however, do not take into account variations of physical properties in the layers that are being folded— that is, the mechanical stratigraphy. It is clear from field and laboratory studies presented in this volume (Dixon, 2004 [this volume]; Spratt et al., 2004 [this volume]) that variations in mechanical properties both along strike and across the stratigraphy have dramatic effects on the folds produced by fault-related folding. Many fold structures are clearly hybrids and have evolved in a complex fashion, perhaps initiating as detachment folds and evolving through fault-propagation folds to transported fault-propagation folds (cf. Figure 6). Analysis of thrust-fault-related folds needs to consider more complex deformation paths that reflect this type of evolution.

GROWTH STRATA

Following the benchmark paper by Suppe et al. (1992), considerable attention has focused on the development of growth strata (Figure 7) in fold-thrust belts. As Suppe et al. (1992) stated, growth strata act as a tectonic tape recorder that tracks the spatial and temporal evolution of thrust structures. Models for growth-strata geometries have become more sophisticated and include fault-propagation folds (Figure 7) and detachment folds (Poblet et al., 1997), as well as trishear models (Figure 8) (Ford et al., 1997; Allmendinger, 1998). Field studies such as those by Poblet and Hardy (1995)

FIGURE 5. Hybrid fault-propagation fold in Mississippian limestones, southern Canadian Rocky Mountains (photograph by K. McClay).

and Poblet et al. (1998) in the Spanish Pyrenees show that detachment folds in these terranes mainly grow by progressive limb rotation. These models clearly also have great applicability to growth folds that develop in deep-water fold belts (cf. Rowan et al., 2004 [this volume]; Shaw et al., 2004 [this volume]). It is important to note that most of the models for growth-strata development are two-dimensional, and more research is needed on 3-D growth-strata systems.

This volume is divided into four broad categories: geodynamics of thrust systems, analog modeling of thrust systems, fault-related folds in thrust systems, and case studies.

The section on **Geodynamics of Thrust Systems** covers a wide variety of topics and reflects current research on developing an understanding of how thrust belts work— from the plate-driving mechanisms, to understanding the frictional constraints on individual thrust-sheet motion. In particular, this section has a number of contributions that discuss how basement thrust sheets evolve and how they relate to foreland fold-and-thrust-belt systems that typically are thin-skinned in character. Other chapters analyze the interrelationships between thrust-belt dynamics, syntectonic erosion and sedimentation, synchronous and out of sequence thrust motion, and thrust-belt curvatures.

The section devoted to **Analog Modeling of Thrust Systems** presents a series of six chapters that show how scaled physical modeling provides important insights into how thrust belts work. In particular, the chapters illustrate the dynamic feedback between surface processes (erosion and sedimentation) and thrust dynamics, as well as the kinematics of fault-related folding in thrust systems. Physical models provide important constraints for the sequences and activities on faults within thrust systems, as well as for the dynamics of thrust-related fold development. The results of analog model studies may be used as templates for section balancing and for understanding the progressive evolution of thrust systems.

The section on **Fault-Related Folds in Thrust Systems** focuses on the kinematic and geometric evolution of thrust-related folds and involves theoretical models, numerical models, and case studies. Eight chapters in this section include theoretical models of fault-bend folding, detachment-fold systems, numerical modeling of fault-related fold systems, and growth-strata geometries in thrust-related folds.

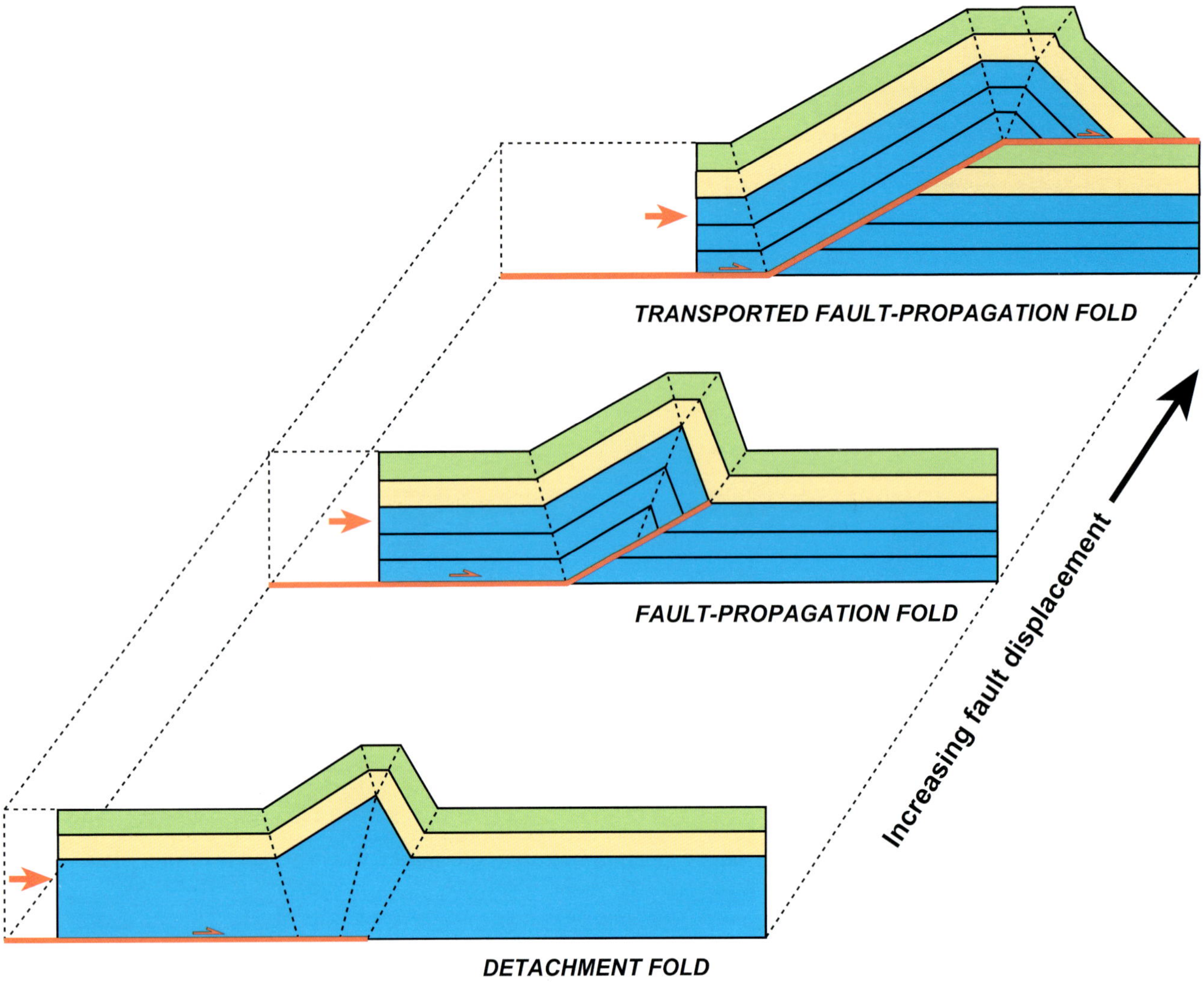

FIGURE 6. Conceptual model of the progressive evolution, with increasing displacement, of a detachment folds to a fault-propagation fold and then to a transported fault-propagation fold (modified from Dobson, 1991).

The final section of this volume comprises **Case Studies** and contains 10 chapters that focus on hydrocarbon-bearing thrust systems, in locations from the Albanides to Kalimantan. In these contributions, the analysis and characterization of the thrust-belt systems is linked to the understanding of the hydrocarbon systems, trap formation, and migration.

The chapters presented in each section are briefly reviewed next, and the key aspects of each contribution are highlighted.

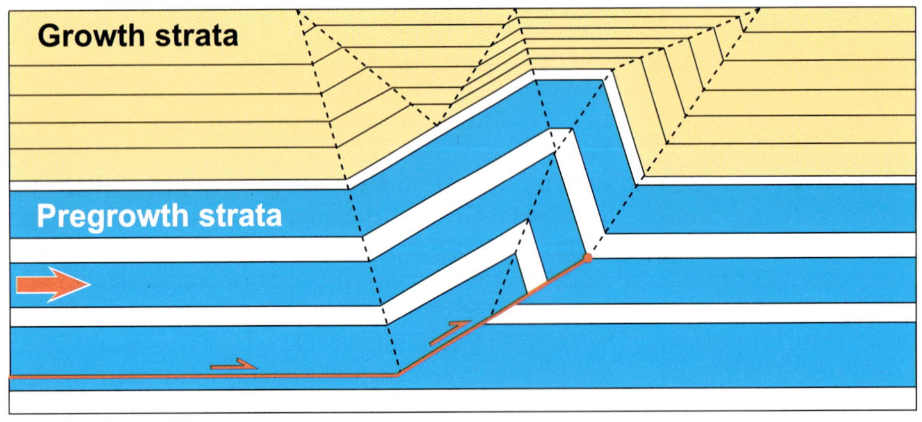

FIGURE 7. Simple kink-band style fault-propagation fold with growth strata (modified from Suppe et al., 1992).

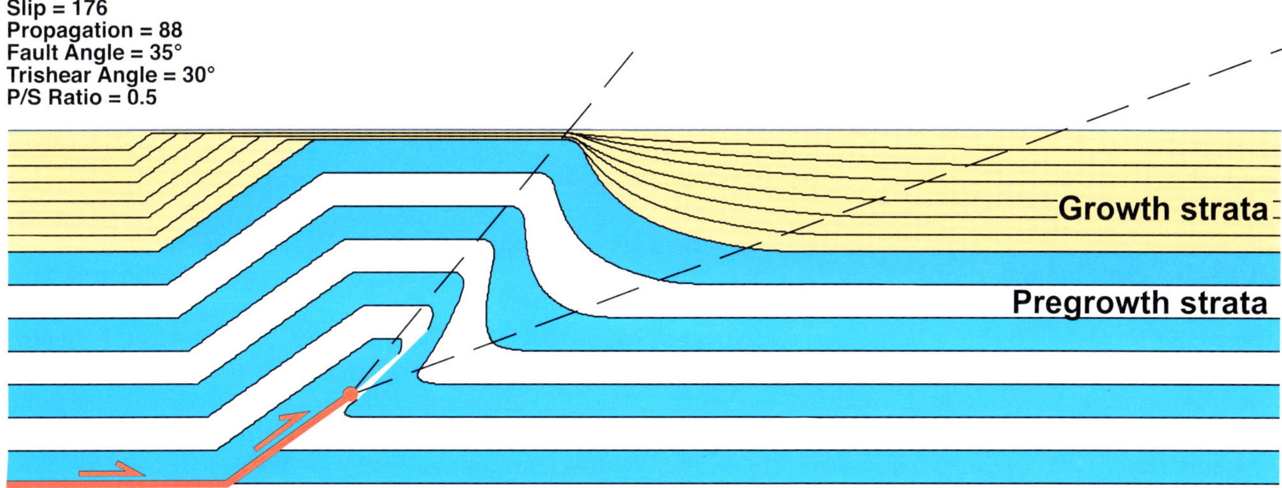

FIGURE 8. Numerical trishear model of a fault-propagation fold with synkinematic growth strata (constructed with the trishear modeling program kindly provided by R. Allmendinger).

GEODYNAMICS OF THRUST SYSTEMS

The nine chapters in this section cover the broad topic of thrustbelt geodynamics, from frictional mechanics, hinterland deformation, rates of thrust belt evolution, and the development of recesses and salients in thrust belt systems, to the formation of deep-water, gravity-driven fold-and-thrust belts. The diversity of topics in this section attests to the many fundamental problems that remain in our understanding of the geodynamics of fold-and-thrust belts.

In Chapter 1, Sibson discusses the role of frictional mechanics and pore-fluid systems in the mechanics of seismogenic thrust systems. He argues that elevated pore-fluid pressures play a pivotal role in the reactivation and continued slip on most thrust faults, and, in particular, for low-angle (<10°) and high-angle (>50°) thrust faults. Where preexisting steep faults (>50°) are reactivated in inversion, Sibson proposes that elevated pore-fluid pressures are essential for thrust motion. He concludes that fault valve action involves the redistribution of large volumes of crustal fluids and in particular may play a role in hydrocarbon migration.

Hatcher (Chapter 2) analyzes the mechanical properties of thrust sheets, from those involving crystalline basement, to classic foreland fold-and-thrust belts. Using displacement/length plots for different types of thrust sheets, Hatcher demonstrates that there are three fundamental types of thrust belt systems: (1) classic foreland fold-and-thrust belts,(2) internal crystalline thrust sheets that involve basement, and (3) fold-related thrust sheets formed by the plastic shearing of the common limb between an anticline-syncline pair (these form below the brittle-plastic transition). These three types have fundamentally different length/displacement characteristics that depend on the structural/lithological strengths of the units in the thrust system.

The Andes typically are taken as the classic ocean-continent subduction system with the development of hinterland-vergent foreland fold-and-thrust belt systems. **Ramos et al. (Chapter 3)** review the geodynamics of three transects across the Andes, from 22°S to 40°S. They demonstrate that there is a close relationship between topography, crustal shortening, and structural styles. In particular, they show that the sub-Andean fold-thrust belt is a mixture of both thin-skinned and thick-skinned thrust systems and that the amount of shortening decreases progressively from north to south. They also show that in some sections, several thrusts were moving simultaneously and that out of sequence thrusting has also been recognized.

The theme of basement-involved thrusting, penetrative plastic deformation, and tectonic wedging in the southern Canadian Cordillera is discussed by **Brown (Chapter 4)**. This chapter uses a critical taper wedge model and shows that the driving mechanisms for foreland thrusting are tectonic wedging and underplating in the metamorphic hinterland.

The seismic expression and analysis of basement-involved thrusting in the Sierras Pampeanas of western Argentina are addressed in **Chapter 5, by Cristallini et al**. Using reprocessed seismic data they propose a doubly vergent wedge model for the Sierra Aconquija in the Sierras Pampeanas. They also conclude that some of the thrust faults that bound the Sierras Pampeanas are inverted extensional faults. These Andean examples of basement uplifts show many similarities to the Laramide uplifts of the western U.S.A.

Chapter 6, by Mugnier et al., analyzes thrust-sheet motion in the Himalayas of western Nepal. The authors integrate field, GPS, and earthquake data to demonstrate that shortening along the Main Frontal

Fault of the Himalayas varies along strike and is influenced by surficial processes (erosion) and episodic, out-of-sequence thrust reactivation internally within the belt. They conclude that the overall total shortening rate for the Himalayas of western Nepal remained relatively constant at 19 mm/yr for the last 11 m.y. Their model of episodic thrust slip influenced by surficial processes, fault-valve activity, and out-of-sequence reactivation may be applicable to other thrust belts.

Chapter 7, by Husson et al., continues the Himalayan theme with a study on the sediment cycle within the synorogenic sediment wedge in the sub-Himalayan zone. Using balanced cross sections and structural and morphological data, the authors conclude that synorogenic sediments accreted to the frontal part of the fold thrust belt have a recycling time of 6.5 m.y. (i.e., 6.5 m.y. for incorporation into the wedge, uplift in the thrust sheets, and subsequent erosion). They also estimate that 21% of the accreted material is subducted in duplexes beneath the Lesser Himalayas. They estimate erosion rates of this region to be between 1.8 mm and 2.1 mm per year.

In Chapter 8, Marshak reviews the various models for the development of curvature in fold-thrust belts. He concludes that, in general, curvatures in fold-thrust belts, such as salients and recesses, are controlled by local features such as basin geometry, the impact of indenters, interaction with foreland obstacles, interaction with strike-slip faults, buckling of the down-going slab, and overlapping of nonparallel thrust belts. He concludes that most thrust-belt curvatures are controlled by variations in basin geometries— thicknesses in the thrust sequences and/or facies control on rheological properties of the detachment units.

Chapter 9, by Rowan et al., addresses the important topic of deep-water fold-and-thrust belts developed on passive margins such as the Gulf of Mexico and offshore West Africa. The authors illustrate the various structural styles associated with these deep-water fold belts and distinguish between shale- and salt-dominated systems. Unlike orogenic fold-and-thrust belts, these systems are driven by gravitational forces and synsedimentary loading and therefore may exhibit different controls and structural styles when compared with typical orogenic fold-and-thrust belts.

ANALOG MODELING OF THRUST SYSTEMS

Scaled physical modeling has proved to be a powerful tool for simulating the progressive evolution of thrust-and-fold systems. Since the end of the nineteenth century, it has been widely used, although not always correctly scaled, to simulate thin-skinned thrust systems. Following Davis et al.'s (1983) benchmark paper on Coulomb wedge thrust tectonics, scaled analog models of thrust systems have focused primarily on the detailed understanding of wedge tectonics in both 2-D and 3-D. These analog models provide insights into the fourth dimension— time— and into the evolution of fold-and-thrust belts. They also provide templates for understanding both the geometric and kinematic development of thrust systems. In particular, the dynamic interactions between surface processes (erosion and sedimentation) have been demonstrated by a number of authors (cf. Storti and McClay, 1995; Storti et al., 2000).

McClay and Whitehouse (Chapter 10) show that 2-D analog modeling of doubly vergent orogenic wedges successfully simulates the formation of fold-thrust belts that have a Coulomb wedge geometry. Synchronous thrust activities were observed, and when syntectonic sedimentation and/or erosion was introduced, out-of-sequence thrusting and reactivation of hinterland thrusts occurred. These results have important implications for the geodynamic evolution of fold-thrust belts, for the sequences of thrust faults, and for section construction and restoration in fold-thrust belts.

Koyi et al. (Chapter 11) use sandbox models of thrust belts to study the development of layer-parallel strain within thrust systems and its use as a guide to strain distributions in natural thrust belts. They demonstrate that differences in mechanical stratigraphy produce dramatically different strain patterns within a thrust belt. These results have important implications for section reconstruction and restoration in fold-thrust belts.

In Chapter 12, Dixon uses very elegant centrifuge models to illustrate the different behavior of basin facies and platform sequences during thrusting. The strong mechanical contrast between platform and basin units affects the geometries and sequences of thrusting, with the facies boundary between them being a significant structural heterogeneity. This boundary controls the type and orientation of structures that develop during thrusting.

Chapter 13, by Dixon and Spratt, uses scaled physical models to analyze the development of lateral ramps and tear faults in thrust belts. The models show how thrust sheets accommodate displacement over frontal ramps bounded either by lateral tear faults or lateral ramps. These results may be used as templates for interpreting these complex structures, which may not be well-imaged on seismic sections.

Chapter 14, by Spratt et al., analyzes changes in structural style across transverse lithofacies boundaries in the Canadian Rocky Mountains and compares these with scaled analog models. The authors show that carbonate bank margins strongly control the structural style. In particular, they demonstrate facies-controlled along-strike changes from fault-cored folds to detachment folds, within the same thrust sheet. These examples and models give important clues to the significance of stratigraphic control on structural style within thrust belts.

Yamada and McClay (Chapter 15) show that inversion of complex 3-D listric faults during compression produces steep reverse faults with doubly plunging anticlinal culminations in their hanging walls. Their models illustrate the complexities of inversion thrust systems in which preexisting basement-involved extensional faults exert a strong control on the contractional geometries.

FAULT-RELATED FOLDING IN THRUST SYSTEMS

The section on fault-related folding in thrust belts covers a wide range of structures, from shear fault-bend folds to detachment folds and trishear folding.

Chapter 16, by Suppe et al., describes a new model for shear fault-bend folds. This new model accounts for internal shear, within the sequence, as it undergoes folding; and as a result, the backlimb dips less steeply than the underlying ramp. Backlimbs may rotate and increase in dip as folding occurs. The authors illustrate this new model of fault-related folds with exquisite seismic examples for the Nankai and Cascadia accretionary prisms.

Chapter 17, by Wallace and Homza, examines the differences between fault-propagation folds, detachment folds, and thrust-truncated folds. The authors present geometric models for these different fold types and illustrate the problems of distinguishing between them, using examples from the Brooks Range, northern Alaska. They emphasize that breakthrough of existing folds may be more common than was hitherto recognized.

Allmendinger et al. (Chapter 18) use trishear kinematic modeling to produce forward models of complex thrust folds and extensional folds. They use this technique to examine strains and strain paths, fractures, and fault-slip magnitudes. They demonstrate the power of this method of forward modeling with examples from the Neuquen Basin in Argentina, where many hydrocarbon accumulations occur in basement-cored anticlines.

Chapter 19, by Poblet et al., presents a new geometric method for analyzing detachment folds to determine the fold history, amplification, and associated fault slip. This technique is independent of the actual kinematic model of folding. The authors demonstrate that natural folds show a variety of uplift rates and do not fit simple kink-band-fold kinematic models. These results can be applied to the analysis of fold systems in a variety of hydrocarbon-bearing thrust structures.

Chapter 20, by Shaw et al., addresses the structural controls on growth stratigraphy in fault-related folds, using well-documented seismic examples. This chapter presents new examples of growth-strata architectures in thrust-related folds that are inferred to form by kink-band migration, as well as other examples in which fold amplification by limb rotation has occurred. The authors discuss the implications for both structural and stratigraphic traps in these types of fold systems.

Chapter 21, by Strayer et al., uses distinct-element numerical models to study the influence of growth strata on the evolution of fault-related folds. They demonstrate that the presence of growth strata that are deposited while a fault-related fold develops strongly influences the resultant fold geometries. Gentle ramp anticlines occur where there are few or no growth strata. In contrast, tight to overturned fault-propagation type folds occur where there are thick growth sequences. The numerical models produce growth-strata architectures that compare well with natural examples.

Erickson et al. (Chapter 22) use numerical models of hinge migration in fault-bend folds to examine the internal deformation within layers as they migrate through a fold hinge and up a thrust ramp. The authors develop predictive models for the distribution of shear bands within the layer, depending on the mechanical properties of the layers. These results have important implications for fracture patterns in thrust-related folds.

Salvini and Storti (Chapter 23) address problems similar to those examined by Erickson et al. (2004 [this volume]), but they use a novel numerical modeling technique: Hybrid Cellular Automata (HCA). This powerful new technique has the potential to generate numerical models that allow prediction of internal deformation features (fractures) within complex fault-bend fold systems. Their models permit the prediction of fracture panels within thrust-fault-related folds and show close similarities to natural examples.

CASE STUDIES

The chapters in this section present specific case studies of hydrocarbon systems in thrust belts, from the Albanides, to Papua New Guinea; in eastern and central Australia; in Kalimantan, Indonesia; the Apennines; Morocco; western Canada; and Colombia. A theme common to most of these papers is the recognition of basement-involved thrust systems intimately linked with thin-skinned thrust systems.

Chapter 24, by Roure et al., uses a combination of balanced sections together with forward kinematic and thermal modeling to analyze the petroleum systems of the Outer Albanides. The authors demonstrate that variations in structural styles and décollement levels are related to the architecture of the prethrusting basins in the northern Outer Albanides and inversion in the southern Albanides. Kinematic and thermal modeling are used to predict hydrocarbon systems in various parts of the thrust belt.

Chapter 25, by Hill et al., uses the Tertiary fold belt in Papua New Guinea to demonstrate that previous thin-skinned models for this system were inappropriate because of a lack of data and because they ignored the mechanical stratigraphy. Hill et al. demonstrate that the Papuan Fold Belt consists of inverted basement-involved extensional faults together with asymmetric detachment fold systems. The authors demonstrate that along-strike changes in the structural styles and hydrocarbon systems are a result of the position of the deformation front relative to the mechanically strong Australian lithosphere in front of it.

Chapter 26, by Korsch, describes the evolution of the Permian-Triassic retroforeland thrust system of the New England orogen, eastern Australia. Korsch demonstrates that both thick- and thin-skinned thrusting operated simultaneously. The interior of the thrust system was dominated by planar, thick-skinned thrusts, whereas the foreland basin deformation was dominated by thin-skinned thrusting. Reactivation of Early Permian extensional faults produced inversion structures that commonly contain commercial hydrocarbons.

In Chapter 27, Flottmann et al. describe the tectonic styles of the intracratonic fold-thrust belts that flank the Amadeus Basin in central Australia. They demonstrate that both the south and north flanks of the basin are deformed by basement-involved thrust complexes in which complex triangle-zone-like structures were developed in the footwalls to major basement thrust complexes.

Chapter 28, by Turrini and Rennison, describes the 3-D structural styles of the Campano-Lucano Arch in the southern Apennine hydrocarbon province. The authors use a combination of sandbox models and 3-D reconstructions to develop a new model of a younger, underlying autochthonous thrust system that deforms an overlying allochthonous thrust system such that a dome-and-basin interference pattern results. They relate the geometries of the thrust system to known and potential hydrocarbon accumulations in the region.

Chapter 29, by Cooper et al., describes the geometries of thrust-related folds in the subsurface of the Monkman area of the Rocky Mountain Foothills of British Columbia, western Canada. Cooper et al. show that the main traps in this area are fault-propagation and detachment folds in which the fracturing in the hinges and forelimbs enhances the reservoir properties.

Chapter 30, by Restrepo-Pace et al., analyzes the evolution of the fold-and-thrust belt along the west flank of the Eastern Colombian Cordillera. The authors propose that the doubly vergent character of the Eastern Cordillera was the result of superposition of the east-vergent, late Miocene to Pliocene Andean deformation on an earlier, west-vergent, late Paleocene to Eocene fold-and-thrust belt. They propose that late Oligocene to early Miocene intercutaneous thrust wedges will be the targets of future hydrocarbon exploration in this belt.

Chapter 31, by Chambers et al., examines thick- and thin-skinned inversion-related thrusting in the prolific Kutai Basin, Kalimantan, Indonesia. Chambers et al. propose that compressionally reactivated basement extensional faults inverted the Paleogene depocenters as anticlines, whereas the overlying Neogene sequences were detached near the top of an overpressured zone, thereby producing a thin-skinned thrust belt. In this chapter, the importance of basement reactivation is emphasized, even when it occurs in a progradational delta environment.

Chapter 32, by Beauchamp, describes the complex superposed folding generated by inversion of a preexisting rift accommodation zone in the Atlas Mountains of Morocco. Beauchamp proposes that disharmonic superposed folding may be characteristic of inverted accommodation zones.

Thick-skinned thrust models are proposed by **Butler et al. in Chapter 33** on the Apennine Thrust Belt of Italy. The authors use seismic and field data to show that, in parts of the Apennines, basement-involved inversion models are most applicable to describe the structural style. However, in other areas, classic thin-skinned thrust models are more viable. Butler et al. show that the Apennines display different structural styles and hence different displacements along strike.

CONCLUSIONS

Although the chapters in this volume do not cover all aspects of thrust tectonics and hydrocarbon systems, they do summarize key aspects of modern thrust-belt studies. Three main themes are dominant in the current research on fold-and-thrust belts: (1) the dynamic interaction between surface processes and fold-thrust-belt evolution, (2) the role of basement involvement and the contractional reactivation (inversion) of preexisting extensional faults in many thrust belts, and (3) the mechanisms of fault-related folding and growth-strata architectures as indicators of timing and mechanisms of folding. It is apparent that many thrust-and-fold belts are combinations of both thin-skinned thrusting and basement-involved thrusting as a result of reactivation of preexisting extensional faults. Physical modeling studies of fold-and-thrust-belt systems have provided valuable insights into possible evolutionary patterns for fold-and-thrust belts. In particular, these studies demonstrate the dynamic interaction between surface processes, thrust architectures, and thrust-fault activities. Complex and sophisticated fault-related fold models have been developed over the last decade. Many fault-related folds appear to be hybrid structures involving changes in their deformation mechanisms through time and changes in their kinematic evolution. In particular, recognition of the importance of detachment folding, limb rotation, and growth-strata geometries has allowed significant advances in our understanding of thrust-belt folds. Trishear modeling has provided a new tool for understanding complex hybrid fold systems. However, much research still needs to be carried out, particularly using 3-D.

The case studies presented in this volume strongly demonstrate that there now is widespread recognition of basement involvement and inversion in many thrust-and-fold belts. Many thrust belts appear to be combinations of thick- and thin-skinned thrusting. This has major implications for the development of reservoir sections in thrust belts and for the migration and entrapment of hydrocarbons.

REFERENCES CITED

Allmendinger, R. W., 1998, Inverse and forward numerical modeling of trishear fault-propagation folds: Tectonics, v. 17, no. 4, p. 640–656.

Allmendinger, R. W., T. Zapata, R. Manceda, and F. Dzelalija, 2004, Trishear kinematic modeling of structures, with examples from the Neuquén Basin, Argentina, *in* K. R. McClay, ed., Thrust tectonics and hydrocarbon systems: AAPG Memoir 82, p. 356–371.

Bally, A. W., P. L. Gordy, and G. A. Stewart, 1966, Structure, seismic data and orogenic evolution of the southern Canadian Rockies: Bulletin of Canadian Petroleum Geology, v. 14, p. 337–381.

Barclay, J. E., and D. G. Smith, 1992, Western Canada foreland basin oil and gas plays, *in* R. W. MacQueen and D. A. Leckie, eds., Foreland basins and fold belts: AAPG Memoir 55, p. 81–105.

Beauchamp, W., 2004, Superposed folding resulting from inversion of a synrift accommodation zone, Atlas Mountains, Morocco, *in* K. R. McClay, ed., Thrust tectonics and hydrocarbon systems: AAPG Memoir 82, p. 635–646.

Beaumont, C., P. Fullsack, and J. H. Hamilton, 1992, Erosional control of active compressional orogens, *in* K. R. McClay, ed., Thrust tectonics: London, Chapman & Hall, p. 377–390.

Beaumont, C., H. Kooi, and S. Willett, 2000a, Progress in coupled tectonic-surface process models with applications to rifted margins and collisional orogens, *in* M. A. Summerfield, ed., Geomorphology and global tectonics: New York, John Wiley, p. 29–55.

Beaumont, C., J. A. Muñoz, J. Hamilton, and P. Fullsack, 2000b, Factors controlling the Alpine evolution of the central Pyrenees inferred from a comparison of observations and geodynamical models: Journal of Geophysical Research, v. 105, p. 8121–8145.

Beydoun, Z. R., M. W. Clarke, G. W. Hughes, and R. Stoneley, 1992, Petroleum in the Zagros Basin; a late Tertiary foreland basin overprinted onto the outer edge of a vast hydrocarbon-rich Paleozoic-Mesozoic passive-margin shelf, *in* R. W. MacQueen and D. A. Leckie, eds., Foreland basins and fold belts: AAPG Memoir 55, p. 309–339.

Brown, R. L., 2004, Thrust-belt accretion and hinterland underplating of orogenic wedges— An example from the Canadian Cordillera, *in* K. R. McClay, ed., Thrust tectonics and hydrocarbon systems: AAPG Memoir 82, p. 51–65.

Buchanan, P. G., and J. Warburton, 1996, The influence of pre-existing basin architecture in the development of the Papuan fold-and-thrust belt: implications for petroleum prospectivity, *in* P. G. Buchanan, ed., Petroleum exploration, development and production in Papua New Guinea, Proceedings of the third PNG Petroleum Convention, Port Moresby, September 1996, p. 89–109.

Burbank, D. W., and R. S. Anderson, 2001, Tectonic geomorphology: Oxon, U.K., Blackwell Science, 274 p.

Cole, J. P., M. Parish, and D. Schmidt, 2000, Sub-thrust plays in the Papuan fold belt: The next generation of exploration targets, *in* P. G. Buchanan, A. M. Grainge, and R. C. N. Thornton, eds., Papua New Guineas petroleum industry in the 21st century: Proceedings of the Fourth PNG Petroleum Conference, Port Moresby, 29 to 31 May, 2000, p. 87–100.

Coward, M. P., M. De Donatis, S. Mazzoli, W. Paltrinieri, and F. C. Wezel, 1999, Frontal part of the Northern Apennines fold-and-thrust belt in the Romagna-Marche area (Italy): Shallow and deep structural styles: Tectonics, v. 18, p. 559–574.

Dahlen, F. A., 1990, Critical taper model of fold-and-thrust belts and accretionary wedges: Annual Review of Earth and Planetary Sciences, v. 18, p. 55–99.

Davis, D., J. Suppe, and F. A. Dahlen, 1983, Mechanics of fold/thrust belts and accretionary wedges: Journal of Geophysical Research, v. 88, p. 1153–1172.

Dixon, J. M., 2004, Physical (centrifuge) modeling of fold-thrust shortening across carbonate bank margins— Timing, vergence, and style of deformation, *in* K. R. McClay, ed., Thrust tectonics and hydrocarbon systems: AAPG Memoir 82, p. 223–238.

Dobson, J. M. M., 1991, The dynamics of foreland fold-and-thrust belts, Southern Canadian Rocky Mountains, SW Alberta, Canada: unpublished PhD thesis, Royal Holloway, University of London, 316 p

Erickson, S. G., L. S. Strayer, and J. Suppe, 2004, Numerical modeling of hinge-zone migration in fault-bend folds, *in* K. R. McClay, ed., Thrust tectonics and hydrocarbon systems: AAPG Memoir 82, p. 438–452.

Erslev, E. A., 1991, Trishear fault-propagation folding: Geology, v. 19, no. 6, p. 617–620.

Ford, M., E. A. Williams, A. Artoni, J. Vergés, and S. Hardy, 1997, Progressive evolution of a fault-related fold pair from growth strata geometries, Saint Llorenç de Morunys, SE Pyrenees: Journal of Structural Geology, v. 19, no. 3–4, p. 413–441.

Hardy, S., C. Duncan, J. Masek, and D. Brown, 1998, Minimum work, fault activity and the growth of critical wedges in fold-and-thrust belts: Basin Research, v. 10, p. 365–373.

Hill, K. C., M. S. Norvick, J. T. Keetley, and A. Adams, 2000, Structural and stratigraphic shelf-edge hydrocarbon plays in the Papuan fold belt, *in* P. G. Buchanan, A. M. Grainge, and R. C. N. Thornton, eds., Papua New Guineas petroleum industry in the 21st century: Proceedings of the Fourth PNG Petroleum Conference, Port Moresby, 29 to 31 May, 2000, p. 67–85.

Hill, K. C., J. T. Keetley, R. Dan Kendrick, and E. Sutriyono,

2004, Structure and hydrocarbon potential of the New Guinea fold belt, *in* K. R. McClay, ed., Thrust tectonics and hydrocarbon systems: AAPG Memoir 82, p. 494–514.

Koons, P. O., 1990, Two-sided orogens: Collision and erosion from sandbox to the Southern Alps, New Zealand: Geology, v. 18, p. 679–682.

McClay, K. R., and P. S. Whitehouse, 2004, Analog modeling of doubly vergent thrust wedges, *in* K. R. McClay, ed., Thrust tectonics and hydrocarbon systems: AAPG Memoir 82, p. 184–206.

Meigs, A. J., and D. W. Burbank, 1997, Growth of the South Pyrenean orogenic wedge: Tectonics, v. 16, p. 239–258.

Mitra S., 1990, Fault-propagation folds: Geometry, kinematic evolution, and hydrocarbon traps: AAPG Bulletin, v. 74, p. 921–945.

Mosar, J., and J. Suppe, 1992, Role of shear in fault-propagation folding, *in* K. R. McClay ed., Thrust tectonics: London, Chapman & Hall, p. 377–390.

Mugnier, J.-L., P. Huyghe, P. Leturmy, and F. Jouanne, 2004, Episodicity and rates of thrust-sheet motion in the Himalayas (Western Nepal), *in* K. R. McClay, ed., Thrust tectonics and hydrocarbon systems: AAPG Memoir 82, p. 91–114.

Muñoz, J. A., 1992, Evolution of a continental collision belt: ECORS-Pyrenees crustal balanced cross-section, *in* K. R. McClay, ed., Thrust tectonics: London, Chapman & Hall, p. 235–246.

Pfiffner, O. A., S. Ellis, and C. Beaumont, 2000, Collision tectonics in the Swiss Alps: Insight from geodynamic modeling: Tectonics, v. 19, p. 1065–1094.

Poblet, J., M. Bulnes, K. McClay, and S. Hardy, 2004, Plots of crestal structural relief and fold area versus shortening— A graphical technique to unravel the kinematics of thrust-related folds, *in* K. R. McClay, ed., Thrust tectonics and hydrocarbon systems: AAPG Memoir 82, p. 372–399.

Poblet, J., and S. Hardy, 1995, Reverse modeling of detachment folds— Application to the Pico-del-Aguila Anticline in the south central Pyrenees (Spain): Journal of Structural Geology, v. 17, p. 1707–1724.

Poblet, J., and K. McClay, 1996, Geometry and kinematics of single-layer detachment folds: AAPG Bulletin, v. 80, p. 1085–1109.

Poblet, J., K. McClay, F. Storti, and J. A. Muñoz, 1997, Geometries of syntectonic sediments associated with single-layer detachment folds: Journal of Structural Geology, v. 19, p. 369–381.

Poblet, J., J. A. Muñoz, A. Travé, and J. Serra-Keil, 1998, Quantifying the kinematics of detachment folds using 3D geometry: Application to the Mediano anticline (Pyrenees, Spain): Geological Society of America Bulletin, v. 110, p. 111–125.

Price, R. A., 1981, The Cordilleran foreland thrust and fold belt in the southern Canadian Rock Mountains, *in* K. R. McClay and N. J. Price, eds., Thrust and nappe tectonics: Geological Society of London Special Publication 9, p. 427–448.

Ramos, V. A., T. Zapata, E. Cristallini, and A. Introcaso, 2004, The Andean thrust system— Latitudinal variations in structural styles and orogenic shortening, *in* K. R. McClay, ed., Thrust tectonics and hydrocarbon systems: AAPG Memoir 82, p. 30–50.

Rojas, L., N. Muñoz, J. P. Radic, and K. McClay, 1999, The stratigraphic controls in the transference of displacement from basement thrusts to the sedimentary cover in the Malargüe fold-thrust belt, Neuquén basin, Argentina: The Puesto Rojas oil field example, *in* K. R. McClay, ed.: Thrust Tectonics 99, Royal Holloway University of London, April 26–29, Abstracts, p. 119–121.

Rowan, M. G., F. J. Peel, and B. C. Vendeville, 2004, Gravity-driven fold belts on passive margins, *in* K. R. McClay, ed., Thrust tectonics and hydrocarbon systems: AAPG Memoir 82, p. 157–182.

Schmid, S. M., O. A. Pfiffner, N. Froitzheim, G. Schönborn, and E. Kissling, 1996, Geophysical-geological transect and tectonic evolution of the Swiss-Italian Alps: Tectonics, v. 15, p. 1036–1064.

Schmidt, C. J., P. W. Genovese, and R. B. Chase, 1993, Role of basement fabric and cover-rock lithology on the geometry and kinematics of twelve folds in the Rocky Mountain foreland, *in* C. J. Schmidt, R. B. Chase, and E. A. Erslev, eds., Laramide basement deformation in the Rocky Mountain Foreland of the western United States: Boulder, Colorado, GSA Special Paper 280, p. 1–44.

Spratt, D. A., J. M. Dixon, and E. T. Beattie, 2004, Changes in structural style controlled by lithofacies contrast across transverse carbonate bank margins— Canadian Rocky Mountains and scaled physical models, *in* K. R. McClay, ed., Thrust tectonics and hydrocarbon systems: AAPG Memoir 82, p. 259–275.

Shaw, J. H., E. Novoa, and C. D. Connors, 2004, Structural controls on growth stratigraphy in contractional fault-related folds, *in* K. R. McClay, ed., Thrust tectonics and hydrocarbon systems: AAPG Memoir 82, p. 400–412.

Storti, F., and K. R. McClay, 1995, Influence of syntectonic sedimentation on thrust wedges in analog models: Geology, v. 23, p. 999–1002.

Storti, F., F. Salvini, and K. R. McClay, 2000, Velocity-partitioned synchronous thrusting and thrust polarity reversal in experimental doubly vergent thrust wedges accreted at different syntectonic sedimentation rates: implications for natural orogens: Tectonics, v. 19, p. 378–396.

Suppe, J., 1983, Geometry and kinematics of fault-bend folding: American Journal of Science, v. 283, p. 684–721.

Suppe, J., 1985, Principles of structural geology: Englewood Cliffs, N.J., Prentice-Hall Inc., 537 p.

Suppe, J., G. T. Chou, and S. C. Hook, 1992, Rates of folding and faulting determined from growth strata, *in* K. R. McClay ed., Thrust tectonics: London, Chapman & Hall, p. 105–121.

Turrini, C., and P. Rennison, 2004, Structural style from the southern Apennines' hydrocarbon province— An integrated view, *in* K. R. McClay, ed., Thrust tectonics and hydrocarbon systems: AAPG Memoir 82, p. 558–578.

Wallace, W. K., and T. X. Homza, 2004, Detachment folds versus fault-propagation folds, and their truncation by thrust faults, *in* K. R. McClay, ed., Thrust tectonics and hydrocarbon systems: AAPG Memoir 82, p. 324–355.

Willett, S., C. Beaumont, and P. Fullsack, 1993, Mechanical model for the tectonics of doubly vergent compressional orogens: Geology, v. 21, p. 371–374.

Geodynamics of Thrust Systems

Sibson, R. H., 2004, Frictional mechanics of seismogenic thrust systems in the upper continental crust—Implications for fluid overpressures and redistribution, *in* K. R. McClay, ed., Thrust tectonics and hydrocarbon systems: AAPG Memoir 82, p. 1–17.

Frictional Mechanics of Seismogenic Thrust Systems in the Upper Continental Crust—Implications for Fluid Overpressures and Redistribution

R. H. Sibson

Department of Geology, University of Otago, Dunedin, New Zealand

ABSTRACT

Intracontinental reverse-fault ruptures (magnitude, M > 5.5) in the upper, seismogenic crust have dips, δ, that lie in the range 10–60° and are broadly compatible with expectations from laboratory measurements of rock friction. Two distinct peaks occur, at $\delta = 30 \pm 5°$ and $\delta = 50 \pm 5°$. Although the lower peak, at $\delta \approx 30°$, corresponds to expected optimal orientation for frictional reactivation under horizontal compression and may be attributed to ramp failure, the peak at $\delta \approx 50°$ is attributed to the compressional reactivation of inherited normal faults during positive inversion. A notable contrast with active subduction earthquakes and a data set from the Himalayan frontal thrust system is a general scarcity of low-angle thrusts with $\delta \leq 10°$. Frictional-mechanics analysis of reverse-fault reactivation endorses long-standing suggestions that active thrust systems are likely to be fluid overpressured. However, such analysis emphasizes that both steeper reverse faults ($\delta = 50 \pm 5°$) and very low angle thrusts ($\delta \leq 10°$) must be overpressured locally with respect to their surroundings (with $P_f \rightarrow \sigma_3$) to become reactivated in preference to the formation or reactivation of more favorably oriented faults. Maximum sustainable overpressures are, however, limited to sublithostatic values by the presence of faults that are oriented at angles that are less than those required for frictional lockup (i.e., $\delta < 60°$) in the prevailing stress field.

Dynamic processes of fluid redistribution tied to the earthquake stress cycle include upward migration of overpressured fluids through fault-valve action, especially on steeper reverse faults ($\delta > 60°$), and episodic to-and-fro migration of fluids, along strike, induced by mean-stress cycling coupled with strong σ_2 directional permeability. Remarkable similarities between steep reverse-slip structures hosting mesozonal gold-quartz vein systems and comparable structural assemblages developed at much higher structural levels in sedimentary basins suggest that extreme valving action (involving redistribution of large volumes of overpressured fluids) may also play a role in hydrocarbon migration, especially in regions undergoing positive tectonic inversion.

INTRODUCTION

The mapped complexity of fold-thrust belts contrasts with simple expectations from frictional Andersonian fault theory where, in a triaxial stress field (principal compressive stresses $\sigma_1 > \sigma_2 > \sigma_3 = \sigma_v$), planar thrust faults should initiate with dips of approximately 25–30° (Anderson, 1951). These structural complexities in the finite deformation state arise principally from heterogeneity in the rock mass, particularly subhorizontal layering with marked competence contrasts. More competent beds may serve as σ_1 stress guides at various stages of the evolution of a thrust complex (Figure 1), their presence tending to deflect fault slip into flat-ramp-flat assemblages and to promote buckling instabilities. Inherited discontinuities, such as normal faults derived from previous phases of crustal extension, may also contribute to finite complexity. Segmentation along strike may involve thrusts splaying and merging at branch points, soft linkage between en échelon strands, or hard linkage involving lateral ramps and transfer faults (Boyer and Elliot, 1982). As is the case with other seismically active fault systems, earthquakes on reverse faults are affected by such structural complexity. Seismological analyses are beginning to demonstrate good correlation between rupture complexity and structural segmentation inferred from geometrical irregularities (e.g., bends, transfer faults, etc.) in surface rupture traces (Nabelek, 1985; Haessler et al., 1992).

Related issues concern the strength and sliding mechanics of thrust faults in the upper crust: whether slip principally involves pressure-sensitive frictional slip, perhaps aided by fluid overpressure (Hubbert and Rubey, 1959), or uses other nonfrictional mechanisms such as pressure-solution sliding (Elliot, 1976) and various forms of distributed deformation (Wojtal, 1992). This, in turn, is coupled to the question of the proportion of fault slip that is accomodated by seismic fault slip, which is generally interpreted as resulting from pressure-sensitive frictional instability (Scholz, 1990). Of special relevance is the role of elevated fluid pressure, P_f, usefully defined in relation to the vertical stress, σ_v, by the pore-fluid factor:

$$\lambda_v = P_f/\sigma_v \tag{1}$$

so that hydrostatic fluid pressures are represented by $\lambda_v \approx 0.4$, suprahydrostatic conditions by $\lambda_v > 0.4$, and lithostatic levels of fluid pressure by $\lambda_v = 1$. Effective principal compressive stresses in fluid-saturated crust are then $\sigma_1' = (\sigma_1 - P_f) > \sigma_2' = (\sigma_2 - P_f) > \sigma_3' = (\sigma_3 - P_f)$ and the effective overburden pressure at depth z becomes

$$\sigma_v' = (\sigma_v - P_f) = \rho g z(1 - \lambda_v) \tag{2}$$

where ρ is the average rock density and g is the gravitational acceleration.

A key issue is whether fault slip increments that lead to a complex finite deformation approximate to "Andersonian" behavior or are themselves inherently more complex. Earthquakes within reverse-fault assemblages arise from slip increments and provide a means for exploring the progressive evolution of such assemblages. Using the dip distribution of active reverse-fault ruptures as a basis for discussion, this chapter addresses the extent to which simple frictional mechanics is applicable to thrust faulting in the upper, seismogenic crust. Although limited by its two-dimensional character, the analysis has implications for maximum sustainable fluid overpressure and processes of dynamic fluid redistribution in seismically active fold-thrust belts.

DIP DISTRIBUTION FOR INTRACONTINENTAL REVERSE-FAULT RUPTURES

Dips of active reverse faults may be estimated from *P*-wave focal mechanisms, from centroid moment tensor (CMT) solutions, and also from waveform and geodetic modeling techniques. The first method likely yields the best estimate of fault dip at rupture initiation and is the principal technique employed, although some means

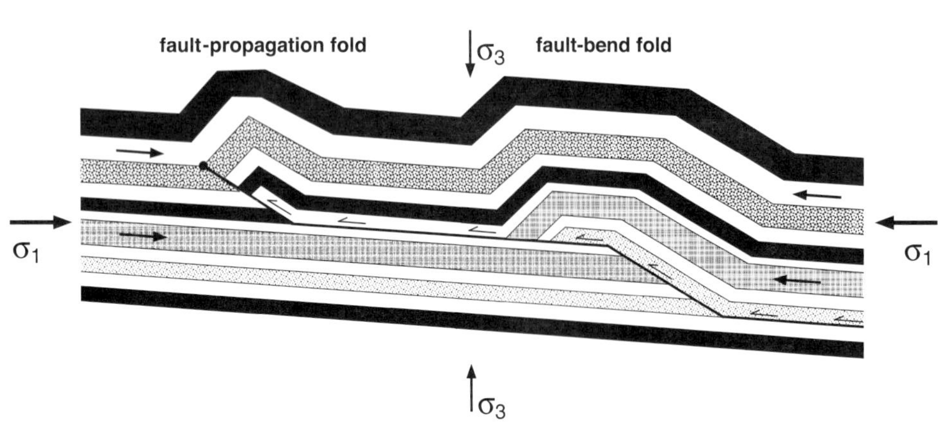

FIGURE 1. Competent beds acting as σ_1 stress guides in flat-ramp thrust assemblages associated with fault-bend and fault-propagation folds developed in gently inclined strata.

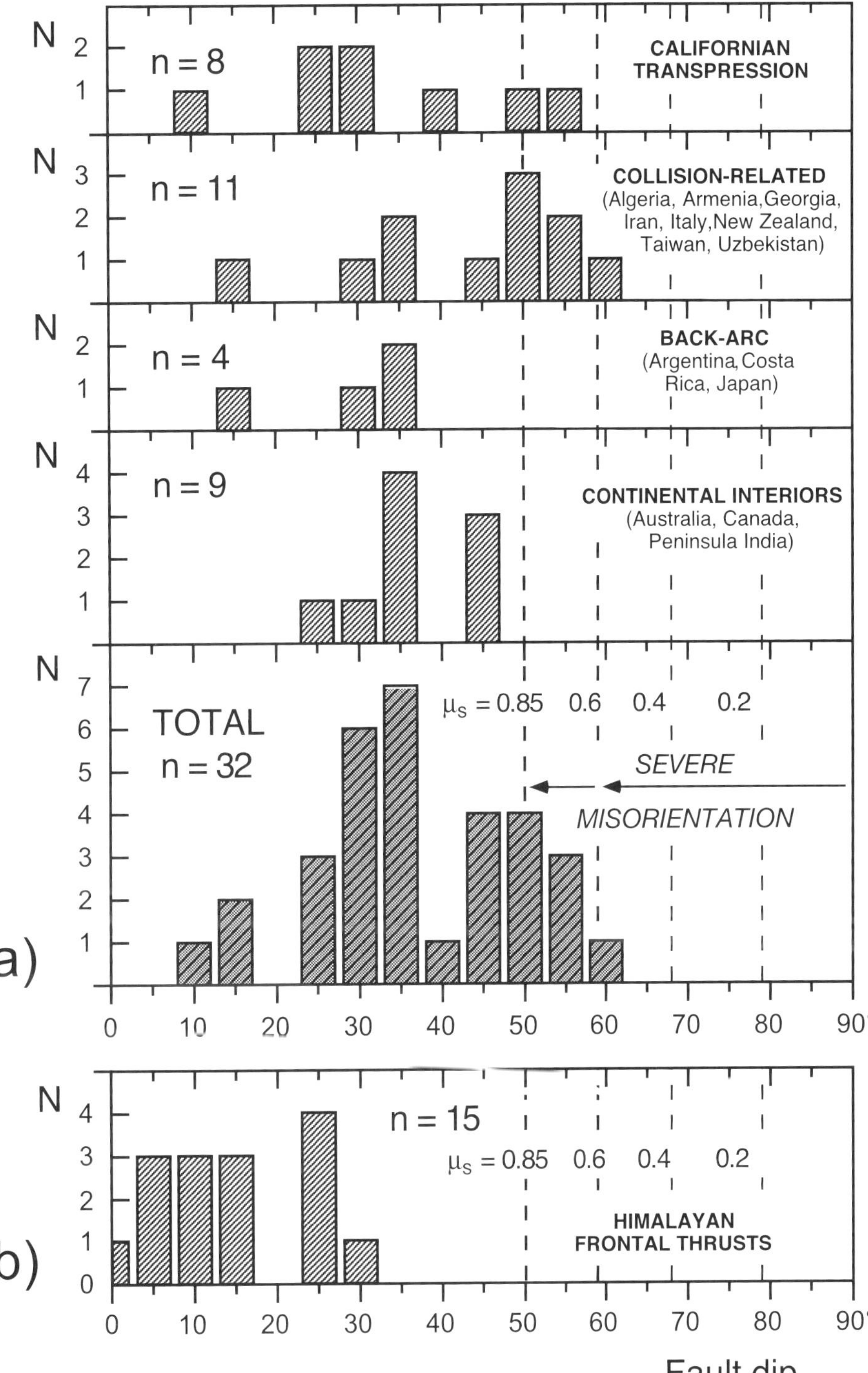

FIGURE 2. (a) Dip distribution for 32 intracontinental reverse-fault ruptures (M > 5.5; slip vector raking within 30° of dip direction) (data from Sibson and Xie, 1998, plus the USGS National Earthquake Information Center [NEIC] CMT dip estimate of 29° for the 20/09/99 M7.6 Chi-chi, Taiwan, earthquake). Vertical dashed lines are expected lockup angles for various friction coefficients, μ_s, assuming horizontal σ_1. (b) Dip distribution of 15 ruptures in the Himalayan thrust front, where the rupture plane has been distinguished on topographic grounds (data from the compilation of Molnar and Lyon-Caen (1989), plus the likely dip for the 28/03/99 M6.6 thrust rupture obtained from the NEIC CMT catalog).

must be used to distinguish the true rupture plane from the auxilliary plane in the focal mechanism (Sibson and Xie, 1998). Two data sets are presented, both for reverse-fault ruptures in which the slip vector rakes within ±30° of the dip direction (Figure 2). Such dip estimates are uncertain by at least ±5°; however, the data have been plotted in 5° bins centered at 5°, 10°, 15°, etc., because estimated dips are frequently rounded to the nearest 5° value. It should be borne in mind that the estimates are mostly for fault dip at the initiation of a thrust rupture; evidence exists in some cases for a change in fault dip as the rupture evolves.

Figure 2a compiles estimates of rupture dip, δ, for 32 intracontinental reverse-fault ruptures having magnitude, M > 5.5, in which the fault plane is clearly discriminated by the presence of a surface-rupture trace or from aftershock distributions. Earthquakes of this size typically have along-strike rupture dimensions greater than about 5 km. Nine of the earthquakes are from continental shields, four are from back-arc settings, seven are from transpressional portions of the San Andreas system in California (plus one possible subduction event in northern California), and eleven are from areas that can be described broadly as being related to zones of active continental collision. Estimated focal depths range from ~3 km to 20 km, but the majority lie within the 5–15 km depth interval. In many instances, therefore, the ruptures probably originated within crystalline basement. Observed reverse-fault dips range from 10–60°, with two distinct peaks at

30 ±5° and 50 ±5°. Only one of these ruptures has a dip, $\delta \leq 10°$.

Figure 2b is a set of dip estimates for reverse-slip earthquakes of comparable magnitude along the Himalayan frontal thrust system, the majority at 15 ±5 km depth. Throughout this data set (largely derived from the focal mechanism compilation of Molnar and Lyon-Caen [1989]), rupture planes have been discriminated on topographic grounds that assume the mountains to occupy the faults' hanging walls. In marked contrast to Figure 2a, nearly 50% of the ruptures in this compilation have $\delta \leq 10°$. In fact, the distribution of thrust dips along the Himalayan front is comparable to that found for subduction thrust ruptures (Molnar and Chen, 1982). Note that the average range-front slope in the vicinity of the frontal thrusts is approximately 8°, thus allowing σ_3 trajectories in this region to deviate significantly from the vertical and perhaps contributing to the lower than usual thrust dips (see also Dahlen, 1990).

MECHANICS OF FRICTIONAL REACTIVATION

Earthquake ruptures generally involve the frictional reactivation of part or all of an existing fault surface (Scholz, 1990). In fluid-saturated crust with pore fluid pressure, P_f, the criterion for frictional reactivation (reshear) of an existing cohesionless fault with static coefficient of rock friction, μ_s, is simply expressed as Amontons' Law:

$$\tau = \mu_s \sigma_n' = \mu_s(\sigma_n - P_f) \tag{3}$$

where τ and σ_n are, respectively, the shear and normal stresses acting on the fault. Laboratory experiments yield a surprisingly restricted range for rock friction ($0.6 < \mu_s < 0.85$) that is almost independent of lithology (Byerlee, 1978). However, the range is supported by structural relationships in natural fault systems (Sibson, 1994a). When the fault contains the intermediate stress, σ_2, so that the problem becomes two-dimensional, the reactivation criterion may be rewritten in terms of the ratio of effective principal stresses as:

$$\frac{\sigma_1'}{\sigma_3'} = \frac{(\sigma_1 - P_f)}{(\sigma_3 - P_f)} = \frac{(1 + \mu_s \cot \theta_r)}{(1 - \mu_s \tan \theta_r)} \tag{4}$$

where θ_r is the reactivation angle between σ_1 and the fault (Sibson, 1985). Ease of frictional reactivation, therefore, depends on the coefficient of static friction and the orientation of the fault with respect to the prevailing stress field (Figure 3). The stress ratio for reactivation reaches a positive minimum when the fault is optimally oriented for reactivation at $\theta_r^* = 0.5 \tan^{-1}(1/\mu_s)$. As θ_r decreases or increases

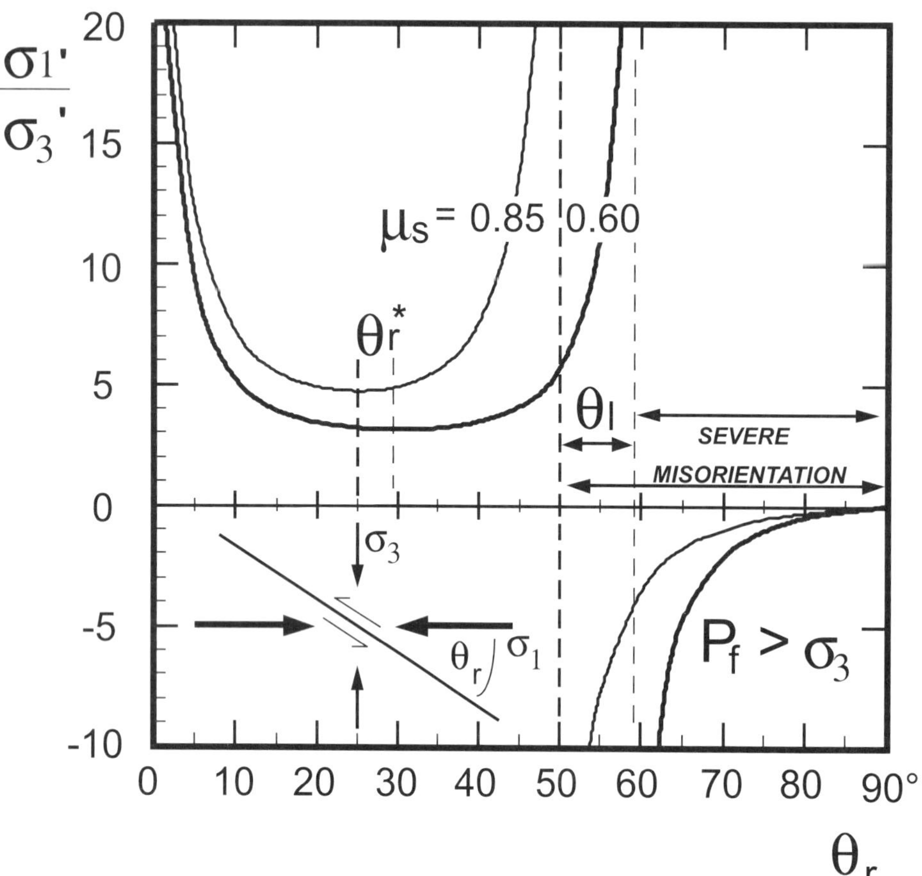

FIGURE 3. Stress conditions for 2-D frictional reactivation. The stress ratio (σ_1'/σ_3') required for reactivation of a cohesionless fault (with pole lying in the σ_1/σ_3 plane) is plotted against the reactivation angle, θ_r, for $\mu_s = 0.6$ (bold line) and $\mu_s = 0.85$ (light line) (after Sibson, 1985). θ_r^* and θ_l are, respectively, the optimal angle for frictional reactivation and the angle of frictional lockup. Reactivation of severely misoriented faults requires the tensile overpressure condition, $P_f > \sigma_3$, to be met.

away from this optimal orientation, the stress ratio needed for reactivation rises, tending towards infinity as $\theta_r \to 0$ or $\theta_r \to 2\theta_r^*$, the angle of frictional lockup. Reactivation of severely misoriented faults with $\theta_r > 2\theta_r^*$ is possible only if $\sigma_3' = (\sigma_3 - P_f) < 0$, requiring the tensile overpressure condition, $P_f > \sigma_3$, to be met. For the Byerlee (1978) range of friction coefficients, optimal reactivation occurs with $25° < \theta_r < 30°$ and frictional lockup occurs at $50° < \theta_r < 60°$.

A key issue in this analysis of reverse-fault dips is the extent to which σ_1 stress trajectories in the seismogenic upper crust depart from horizontal. Topographic effects can perturb trajectories significantly near the surface, but most of the dip estimates comes from depths of approximately 10 km and are likely to be affected only by longer wavelength topography. Analytical and experimental work on accretionary wedges suggests departures by up to a few degrees (Dahlen, 1990), while Westaway (1998) argues that significant deviations of tens of degrees may occur in special circumstances as the result of lower crustal flow. Throughout this analysis, we assume pure Andersonian reverse faulting, with $\sigma_v = \sigma_3$, but the possibility of significant σ_1 departures from the horizontal must be kept in mind.

Given this assumption, the reactivation angle corresponds directly to the fault dip so that for Byerlee friction, optimal reactivation is expected at dips of 25–30° and frictional lockup at dips of 50–60°. Allowing for the uncertainties in dip estimates, the observed distribution of 10–60° for reverse-fault dips (Figure 2a) is broadly compatible with the lower bound of the Byerlee friction range, with $\mu_s \approx 0.6$, although lower coefficients remain allowable. In addition, the well-defined peak at 30 ±5° corresponds to expected optimal reactivation for $\mu_s \approx 0.6$ and likely corresponds to ramp failure in flat-ramp-flat thrust assemblages (Figure 1). The subsidiary peak at 50 ±5° probably arises largely from compressional reactivation in collision zones of inherited normal faults (positive tectonic inversion) (Sibson and Xie, 1998). Throughout the rest of this chapter, $\mu_s \approx 0.6$ is adopted as a likely lower bound for fault friction in consolidated rock.

To explore likely stress levels associated with reverse faulting with $\sigma_v = \sigma_3$, the reactivation criterion (Equation 4) may be rewritten in terms of the differential stress required for reactivation at depth, z:

$$(\sigma_1 - \sigma_3) = \frac{\mu_s(\cot\theta_r + \tan\theta_r)}{(1 - \mu_s\tan\theta_r)}\rho g z(1 - \lambda_v) \qquad (5)$$

The resulting plot for $\mu_s \approx 0.6$ at a 10-km depth (Figure 4) places constraints on the differential stresses needed for reverse-fault reactivation at different dips and fluid pressure values, on the assumption of horizontal σ_1. Peaks in the dip distributions at 30 ±5° and 50 ±5° are shaded. The depth chosen is a typical focal depth for the thrust earthquakes under consideration, but inferred stress levels are readily scaled to other depths. At a 10-km depth under hydrostatic fluid pressure ($\lambda_v = 0.4$), even an optimally oriented fault requires a differential stress of approximately 350 MPa (3.5 kbar) for reshear. Differential stress for optimal reactivation can only be reduced below 100 MPa (1 kbar) for $\lambda_v > 0.8$. The plot also illustrates how, again for horizontal σ_1, misoriented faults with dips in the 50 ±5° peak require much higher λ_v values if they are to be reactivated at the same levels of differential stress as are optimally oriented structures. The same applies to very low angle thrusts ($\delta < 10°$) driven by subhorizontal σ_1.

This point is emphasized in Figure 5 (also derived from Equation 5), which shows the required increase in λ_v for a nonoptimally oriented fault to be reactivated at the same differential stress level as one that is optimally oriented. Reactivation of faults with dips in the 50 ±5° range, or < 10°, require a substantial localized increase in λ_v. Note also that reactivation to slightly beyond the angle of frictional lockup is possible if supralithostatic fluid pressures ($\lambda_v > 1.0$) can be sustained.

BRITTLE-FAILURE-MODE PLOT FOR A COMPRESSIONAL REGIME

Another factor to be considered in the reactivation of nonoptimally oriented faults is the likelihood that, with increasing misorientation, new, favorably oriented faults may form at lower differential stress levels than those required for reshear. This possibility may be explored on a brittle failure mode plot constructed for a compressional regime with $\sigma_v = \sigma_3$, using Sibson's (1998) procedure.

Composite failure envelopes for intact rock with different tensile strengths, T, all with a representative average coefficient of internal friction, $\mu_i = 0.75$ (Jaeger and Cook, 1979), are transcribed from a standard Mohr diagram of τ versus σ_n' (Figure 6a) to a plot of differential stress versus effective least stress, $\sigma_3' = \sigma_v' = (\sigma_v - P_f)$ (Figure 6b). Solid lines on the plot define failure criteria for intact rocks with different tensile strengths. Sedimentary rocks generally have tensile strengths in the range $1 < T < 10$ MPa, but tensile strength of crystalline rocks may reach 20 MPa or more (Lockner, 1995). Note first that only compressional shear failure can occur for positive σ_3'. Formation of hydraulic extension fractures and of extensional shears both require $\sigma_3' < 0$, the former occurring when

$$P_f = \sigma_3 + T \qquad (6)$$

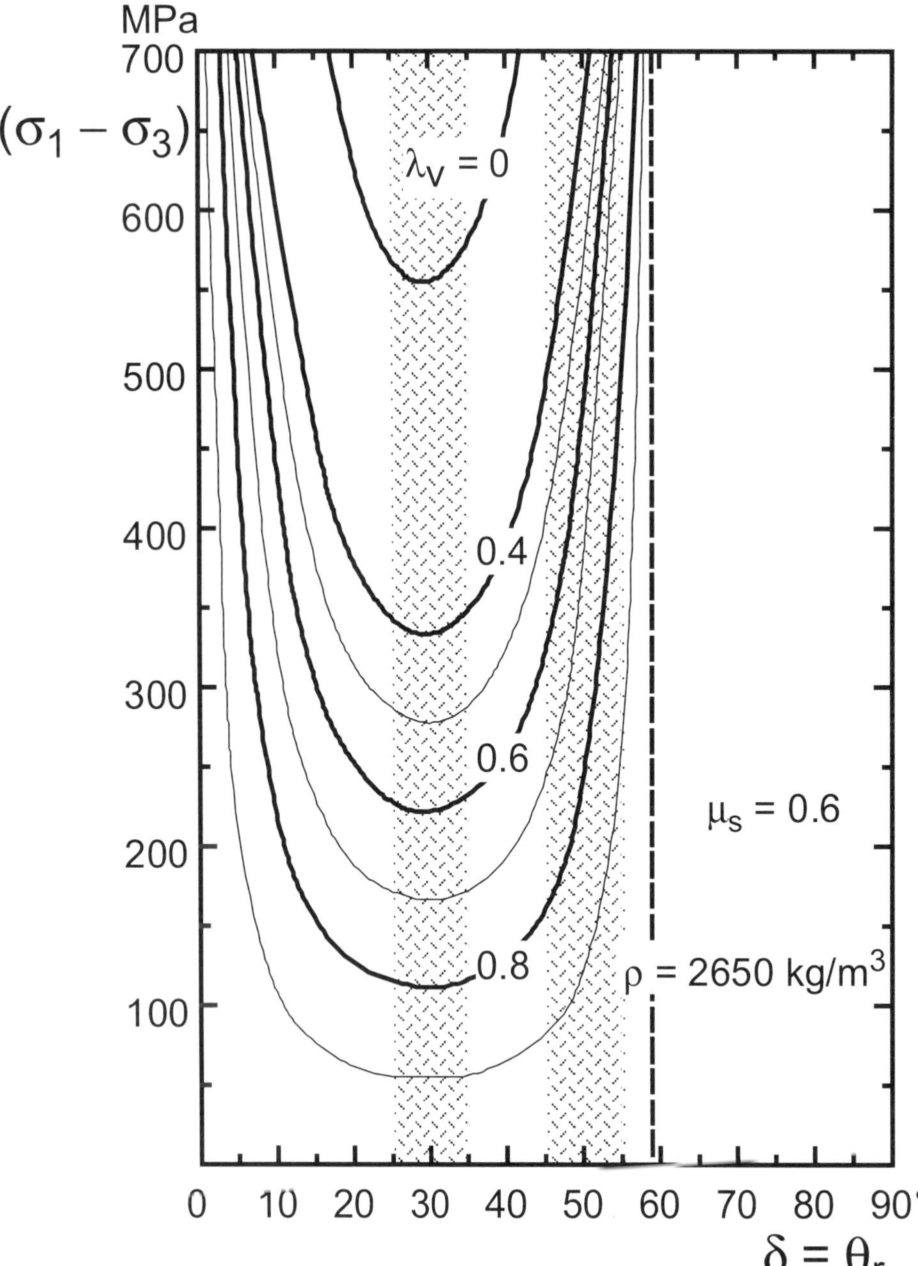

Figure 4. Differential stress required for frictional reactivation of a cohesionless thrust fault, versus dip angle (= reactivation angle), at a 10-km depth, for $\sigma_v = \sigma_3$, $\mu_s = 0.6$, and varying values of the pore-fluid factor, λ_v. Shaded areas represent the two peaks in the dip distribution of reverse-fault ruptures.

provided the differential stress $(\sigma_1 - \sigma_3) < 4T$. This, in turn, requires either $\sigma_3 < 0$ through some form of stress heterogeneity, or the tensile overpressure condition $P_f > \sigma_3$ to be maintained.

The dashed lines represent the reshear condition derived from Equation 5 with $\mu_s = 0.6$ for different values of the reactivation angle (or fault dip). Reshear criteria appropriate to the two peaks on the dip histogram, at 30° and 50°, are emboldened. Consider the intercept between the $\theta_r = 50°$ reshear condition and the intact failure criterion for T = 5 MPa in Figure 6b. At $\sigma_3' <$ 37 MPa, reshear will occur at a lower value of differential stress than will formation of a new fault, whereas for $\sigma_3' >$ 37 MPa the reverse is the case, and a new fault will form in preference to reshear. Combinations of depth and fluid pressure conditions where this changeover occurs can then be estimated from Equation 2. Overall, it is clear that the more an existing fault departs from optimal orientation, the lower must be the value of σ_3' (i.e., $P_f \rightarrow \sigma_3$) for reactivation to occur in preference to formation of a new, favorably oriented fault. Moreover, to prevent new faults from forming, the overpressure must be localized in the vicinity of the existing misoriented fault (see also Hill, 1993). For severe misorientation, with $\theta_r > 60°$ (beyond frictional lockup), reshear of cohesionless faults requires the tensile overpressure condition $P_f > \sigma_3$ to be met.

Figure 6 also shows that the presence of an existing cohesionless fault that is oriented at reactivation angles less than ~65° to σ_1 prevents formation of hydraulic extension and extensional-shear fractures because, with increasing fluid pressure, the fault will always reactivate to provide a permeable drainage path before the appropriate extensional failure conditions can be achieved.

STRUCTURAL PERMEABILITY OF ACTIVE REVERSE-FAULT SYSTEMS

At shallow depths in sedimentary basins where rocks may have high initial porosity and permeability, faults may function either as relatively permeable conduits

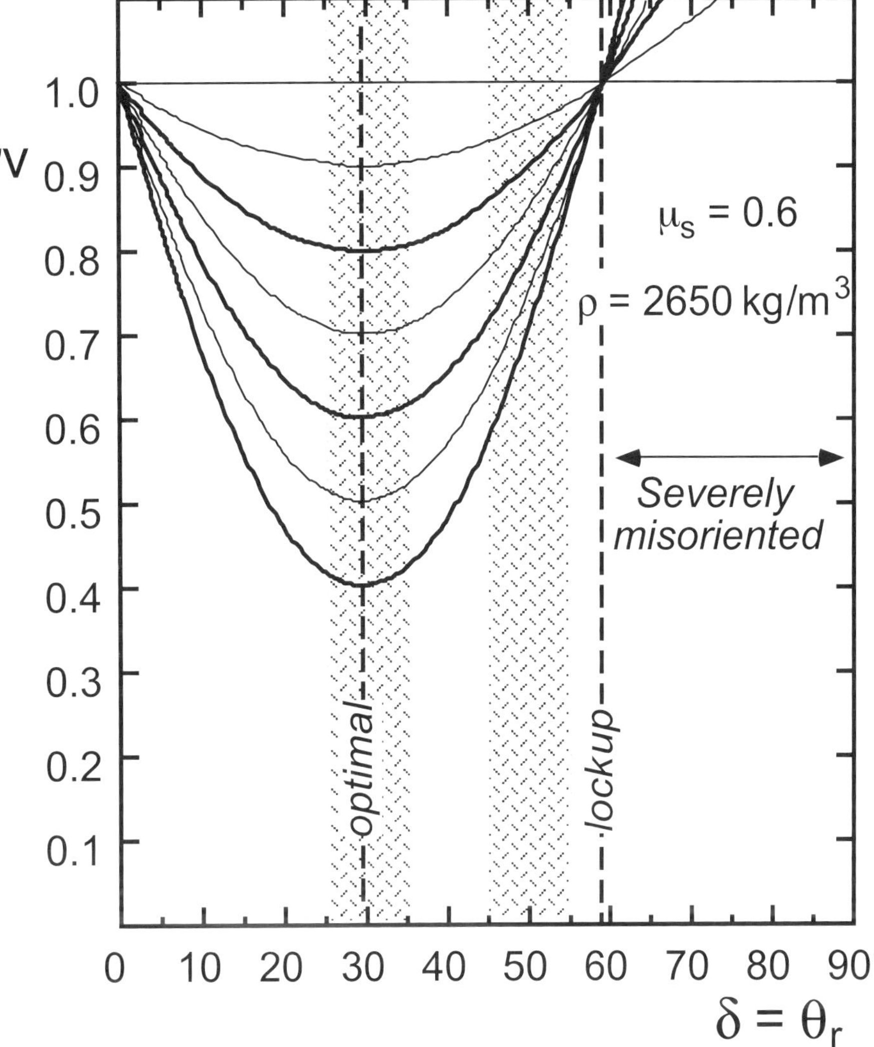

FIGURE 5. Plot of pore-fluid factor, λ_v, against reverse-fault dip, $\delta = \theta_r$, for $\sigma_v = \sigma_3$ and $\mu_s = 0.6$, demonstrating the localized λ_v needed for reshear of a misoriented fault to occur at the same differential stress as for an optimally oriented reverse fault. Shaded areas again represent the two peaks in the dip distribution of reverse-fault ruptures.

or as impermeable seals, depending on the character of the host rock, juxtaposition effects, and total fault slip (e.g., Knipe et al., 1998). In lower porosity host rocks at deeper levels, evidence of dynamic, probably transitory fault and fracture permeability that was active during the evolution of fold-thrust belts is often preserved in the form of hydrothermal vein fills, along with other stress-controlled features affecting rock-mass permeability, such as stylolites (e.g., Cox et al., 1991; Sibson, 1994b). Gold-bearing quartz-vein systems reveal details of high-flux conduits; a volumetric scaling ratio in excess of 1000:1 is likely for high-grade veins carrying more than 10 ppm. Au (Seward, 1993). In this regard, mesozonal vein systems that developed in sub- to low-greenschist environments toward the base of the seismogenic zone have proved especially valuable for the evidence they provide on structural permeability at depths where larger ruptures typically nucleate (Sibson et al., 1988; Cox, 1995; Robert et al., 1995; Sibson and Scott, 1998; Nguyen et al., 1998).

Figure 7 illustrates orientations of different stress-controlled components of structural permeability in a compressional regime with $\sigma_v = \sigma_3$. Components that may form in the prevailing stress field include macroscopic extension and extensional-shear fractures opening in the σ_3 direction, faults generally initiating as Coulomb shears at approximately ±25 to 30° to σ_1 but locally deflected to lower dips by bedding anisotropy, and stylolitic solution surfaces containing seams of insoluble minerals forming perpendicular to σ_1. Under appropriate fluid-pressure conditions (Equation 5), existing faults and other potential slip surfaces (e.g., bedding) may also be reactivated, at other than optimal orientations, to create permeable channelways following failure. The basic components may combine together into more complex forms of structural permeability, for example, dilational jogs allowing slip transfer between en échelon faults, and various types of mesh structure that incorporate low-displacement faults interlinked by extensional and extensional-shear fractures (Sibson, 1996). An important characteristic is that the mutual intersection of many of these features, as well as the orientation of pipelike features such as dilational jogs and saddle reefs, all lie in the σ_2 direction. This raises the possibility of enhanced directional permeability along strike, reinforced by macroscopic fold structures (Figure 7d).

Activation of many of these components depends strongly on fluid pressure. Flat-lying macroscopic extensional and extensional-shear fractures, which may enormously enhance the bulk permeability of the rock mass, can only develop when the tensile overpressure condition ($P_f > \sigma_3$) is met (Figure 6). In general, this

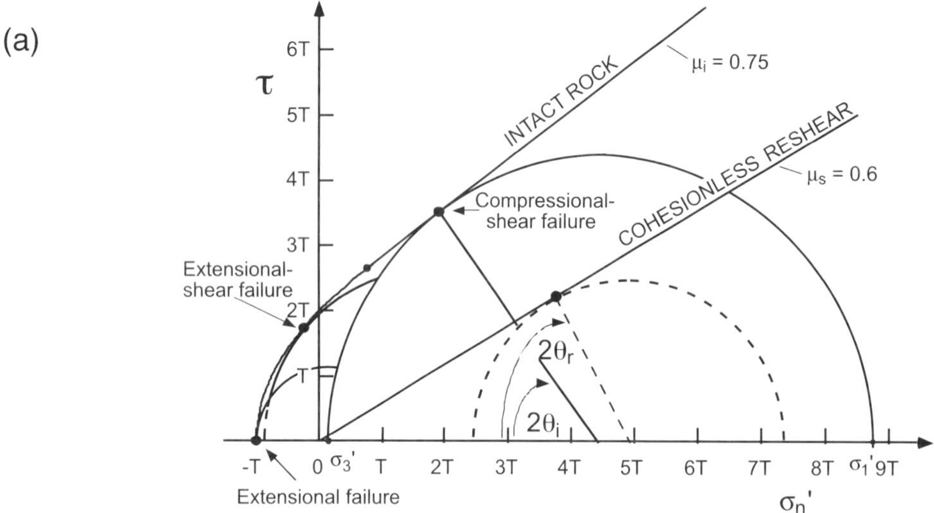

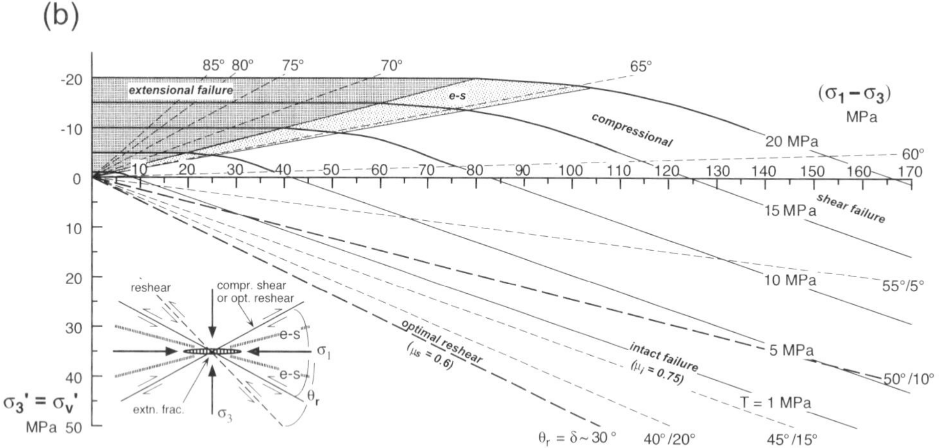

FIGURE 6. (a) Mohr diagram in τ/σ_n' space representing failure stress states for intact rock (μ_i = 0.75) leading to the formation of extensional, extensional-shear, and shear fractures, normalized to tensile strength, T, plus the reshear condition for a cohesionless fault with $\mu_s = 0.6$. (b) Brittle-failure-mode plot of differential stress, $(\sigma_1 - \sigma_3)$, versus σ_3', for $\sigma_v = \sigma_3$, defining stress/fluid-pressure conditions for different modes of brittle failure in intact rock characterized by various values of tensile strength, T, with μ_i = 0.75 (bold lines). Dashed lines are the reshear criteria for existing cohesionless faults, with $\mu_s = 0.6$ oriented at θ_r to σ_1 and containing the σ_2 direction (bold dashed lines represent the two peaks on the dip histogram at 30° and 50°).

requires supralithostatic overpressures ($\lambda_v > 1.0$), unless heterogeneity in a compressional stress regime exists such that $\sigma_3 < \sigma_v$, for example, at fault tips or in dilational stepovers (Segall and Pollard, 1980; Ohlmacher and Aydin, 1997). However, as previously discussed, the tensile overpressure condition can only be achieved in the absence of throughgoing cohesionless faults that are oriented at less than frictional lockup to the σ_1 direction. Widespread vein development on extension and extensional-shear fractures near a major throughgoing reverse fault is, therefore, to be expected only at the time of the fault's initial formation, or when components of the thrust system have become severely misoriented for reactivation in the prevailing stress field, or when the fault has regained cohesive strength through hydrothermal cementation (Nguyen et al., 1998). Absence of veining does not, by itself, reflect absence of fluid overpressures; it merely shows that the tensile overpressure condition has not been achieved.

An association between steep, severely misoriented reverse faults and extensive development of subsidiary extensional vein sets, as portrayed in Figure 7b and c, has in fact been widely reported (Robert and Brown, 1986; Cox et al., 1991; Sibson, 1990, 1995; Robert et al., 1995; Windh, 1995). Misorientation of slip surfaces in prevailing compressional stress fields may arise variously through fault inheritance from former tectonic regimes (for example, positive tectonic inversion), from progressive domino steepening of original imbricate thrusts, or from rotation of bedding slip surfaces during growth of fault-related folds with the development of associated saddle-reef structures in fold hinges.

OVERPRESSURING IN THRUST REGIMES

Following the seminal work of Hubbert and Rubey (1959), it is now readily apparent that many thrust systems, often with associated flexural-slip folding, were active in highly overpressured crust (Price, 1975; Cello and Nur, 1988; Henderson et al., 1990). Development of fluid overpressures is, in fact, especially likely in compressional regimes, because many of the structural components enhancing structural permeability are horizontal or flat-lying (Figure 7a), thereby inhibiting vertical transport of fluids. Evidence of overpressuring in active

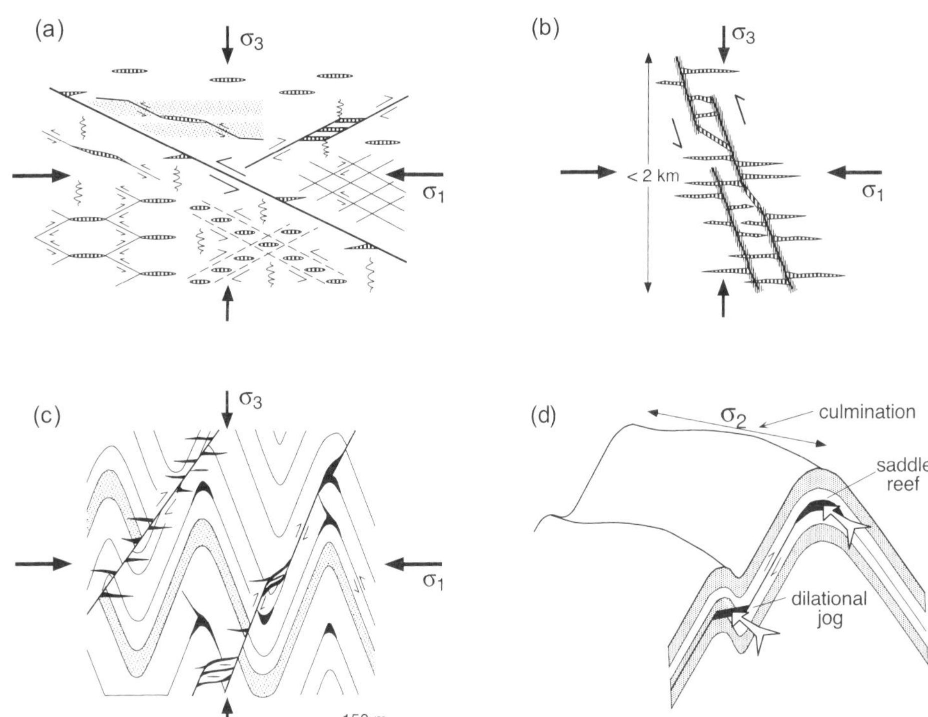

FIGURE 7. Structural permeability in compressional stress regimes, with $\sigma_v = \sigma_3$. (a) Components of stress-controlled structural permeability (faults = straight lines, extension and extensional-shear veins = hachured, stylolites = squiggles). (b) Mesh complex of fault veins and flat-lying extensional veins associated with steep reverse faults, based on mesozonal lode gold systems (after Robert et al., 1995). (c) Cartoon representation of gold-quartz-vein systems associated with upright fold structures in the Victorian gold field, Australia (after Cox et al., 1991). (d) 3-D characteristics of structural permeability in Victorian gold field fold structures (after Cox et al., 1991).

thrust systems comes from borehole and geophysical measurements in sedimentary basins and accretionary wedges, as well as through modeling of critical wedge tapers (Davis et al., 1983; Dahlen, 1990; Westbrook, 1991; Moore et al., 1995).

Additional evidence for overpressuring to near-lithostatic values comes from sets of extension veins sometimes associated with exhumed reverse-fault systems; tapering of veins suggests that the faults themselves were the principal conduits for the overpressured fluids (Figure 7b and c). Such assemblages have been recorded not only from sedimentary and metasedimentary successions (Cox, 1995; Sibson, 1995), but also from high-grade metamorphic-plutonic complexes (Robert et al., 1995), which indicates that fluid overpressuring may also occur within crystalline basement. In this regard, it is interesting to note, from seismic profiling, the recent demonstration of flat-lying bright reflective zones at depths of 18–23 km beneath the crystalline San Gabriel Mountains, California. Ryberg and Fuis (1998) attribute these bright reflectors, underlying the seismogenic regime and proximal to the seismically active thrust faults bounding the rangefront to the south, to near-lithostatic overpressures developed in the mid-crust.

Controls on Maximum Sustainable Overpressure

Brittle structures of different kinds, either newly formed or reactivated in the prevailing stress field, impose constraints on the maximum overpressure that can be sustained. Hydraulic-extension fracturing defines the lowest sustainable σ_3' for a particular rock type (Figure 6) and thus provides the upper limit for fluid overpressure in intact crust, requiring $\lambda_v > 1$ for a compressional regime, unless $\sigma_3 < \sigma_v$ through local stress heterogeneity. Maximum sustainable overpressure is then calculated simply from Equation 6 and can be represented on a plot of λ_v versus depth for different values of tensile strength (Secor, 1965) (Figure 8). However, this situation is only achievable under low differential stress [where $(\sigma_1 - \sigma_3) < 4T$] and in the absence of existing faults oriented at angles less than frictional lockup. This is because reactivation of such faults creates permeable discharge paths that allow overpressures to bleed off, postfailure, through fault-valve action.

Another situation of interest arises when the crust is pervaded by cohesionless thrust faults that are optimally oriented for reactivation. The reactivation criterion for an optimally oriented thrust (Equation 5) then reduces to

$$(\sigma_1 - \sigma_3) = 2.12\rho g z(1 - \lambda_v) \quad (7)$$

for $\mu_s = 0.6$, allowing maximum sustainable overpressure to be estimated for a given level of differential stress and depth (Figure 8). Note that the higher the differential stress, the lower the sustainable λ_v at a particular depth. Curves defining maximum sustainable overpressure for non-optimally oriented reverse faults and for faults retaining some cohesive strength lie between these end-member scenarios. Measured and inferred overpressures ($0.7 < \lambda_v < 0.97$) at depths up to a few kilometers in accretionary wedges, both onshore and offshore (Davis et al., 1983), are broadly compatible with the curves in Figure 8, provided that differential stress is less than a few tens of megapascals.

FLUID REDISTRIBUTION COUPLED TO THE EARTHQUAKE STRESS CYCLE

Fluid interactions with reverse-fault systems may be divided into those that are essentially *static,* with faults behaving either as passive barriers to flow or as fluid conduits, and *dynamic* interactions in which fluid redistribution is modulated by fault-slip increments coupled to the earthquake stress cycle (Sibson, 1994b). Static interactions between passive reverse faults and fluid flow depend critically on the relative permeability between the fault zone and the host rocks (Knipe et al., 1998) and, being site specific, are not discussed further; we focus our attention here on dynamic processes of redistribution that are likely to be associated with seismogenic reverse-fault systems.

Two interlinked factors, episodic changes in the tectonic stress state and in fault zone permeability, need to be considered. First, reverse faults are load-strengthening: the vertical stress, $\sigma_v = \sigma_3$, stays constant while the horizontal greatest stress, $\sigma_h = \sigma_1$, increases through the interseismic period during loading to failure (Sibson, 1991). Thus shear stress on the fault, τ, mean stress, $\bar{\sigma}$, and fault normal stress, σ_n (and hence

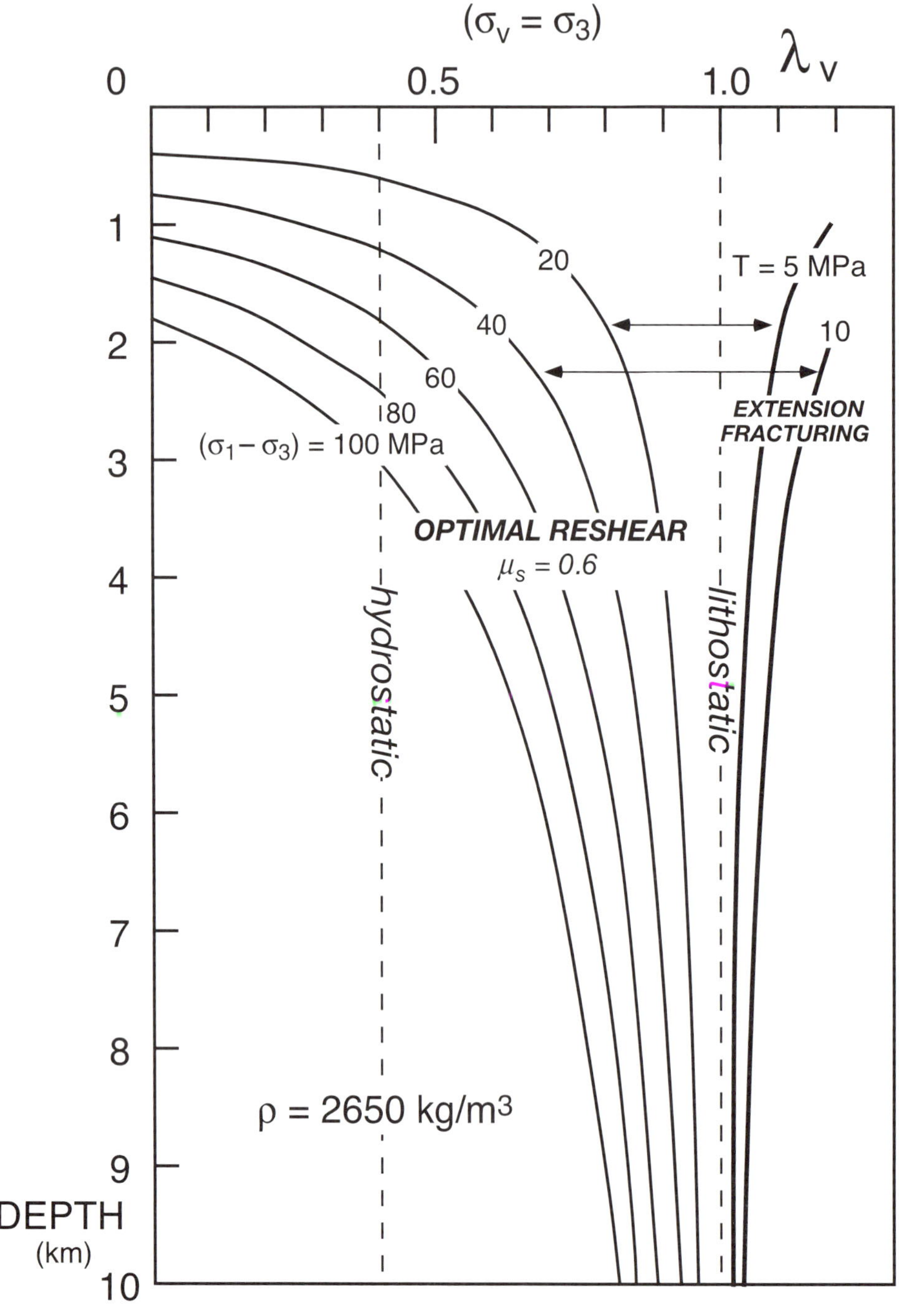

FIGURE 8. Maximum sustainable overpressure in a compressional regime ($\sigma_v = \sigma_3$) for intact crust in which fluid pressure is limited by the formation of flat hydraulic extension fractures, and for crust penetrated by an array of optimally oriented thrusts, with $\mu_s = 0.6$ at various values of differential stress.

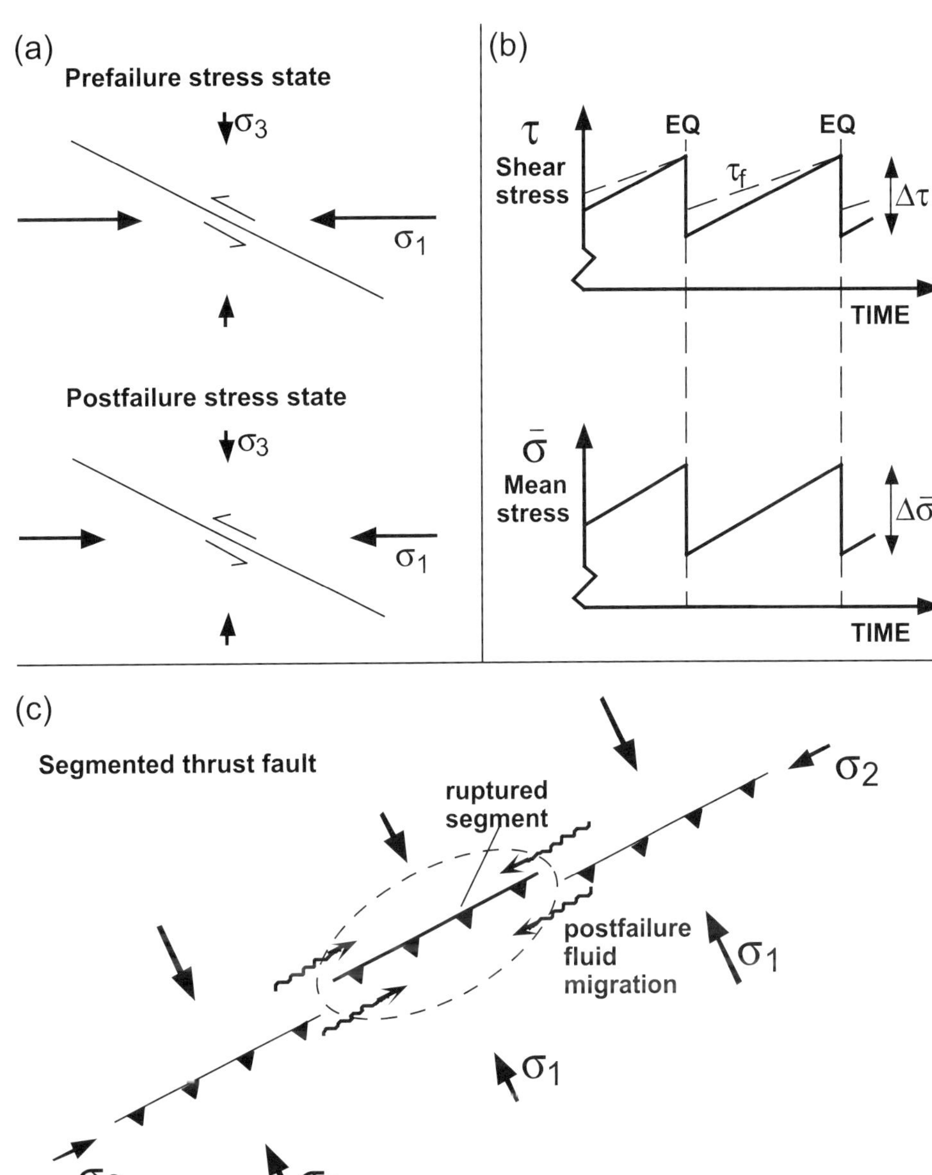

FIGURE 9. Expected fluid redistribution from mean-stress cycling around reverse faults. (a) Variations in far-field stress state before and after failure. (b) Coupled mean and shear stress cycling around an optimally oriented thrust fault (τ_f is frictional strength of fault). (c) Expected along-strike fluid migration, after failure of one segment in a reverse-fault system.

fault frictional strength), all increase during progressive loading, only to decrease at failure (Figure 9a). Second, fault permeability is likely to be highest in the postseismic period immediately following rupture (Brown and Bruhn, 1996), but it may also vary throughout the loading cycle, with solution transfer in fine-grained gouge and hydrothermal cementation reducing permeability through the interseismic period. High postfailure permeability on reverse faults is also enhanced by the decrease in fault normal stress.

Given these considerations, two dynamic processes, acting singly or in combination, seem likely to dominate fluid redistribution around reverse faults.

Mean-stress Cycling

Cyclical changes in mean stress, $\bar{\sigma}$ ("the grasp on the sponge"), tied to the sawtooth cycle of shear stress accumulation and release on reverse faults (Figure 9b), may induce substantial fluid redistribution, especially in high-porosity rocks near the earth's surface where background fluid pressures are hydrostatic. However, these changes may also contribute to fluid redistribution in overpressured crust. Amplitude of mean-stress cycling, $\Delta\bar{\sigma}$, commonly will be of the same order as that for shear stress cycling, $\Delta\tau$. From seismology, $0.1 < \Delta\tau < 10$ MPa (Kanamori and Anderson, 1975), implying potential differences in hydraulic head of 10–1000 m.

Consider a segmented reverse-fault system under progressive loading, with individual segments failing in isolation. In terms of bulk effects for the whole system, progressive increases in σ_1 and $\bar{\sigma}$ throughout the interseismic period should lead to negative dilatation expelling fluid from the vicinity of the faults. Subhorizontal σ_2 directional permeability associated with the system dictates that much of the fluid redistribution should occur along strike, though stress heterogeneities at rupture tips, combined with fault overlaps, may complicate the pattern locally. Sudden decreases in σ_1 and $\bar{\sigma}$ accompanying rupture of an individual reverse-fault segment then lead to a transitory local decrease in fluid pressure, thereby drawing fluid back in to the vicinity of the fault, again predominantly from along strike (Figure 9c). Such fluid loss strengthens adjacent fault segments, with fluid influx into the area of reduced mean stress around the failed segment perhaps contributing to time-dependent aftershock activity. Mean-stress cycling associated with failure of different segments within a reverse-fault system therefore should cause fluid to oscillate to and fro along strike.

Fault-valve Action

It is now also clear that seismic activity, including large-scale rupturing, may occur in areas of overpressured crust under compression. For instance, there are grounds for believing that major thrust ruptures with along-strike dimensions of tens of kilometers, such as the 1931 M7.8 Napier, New Zealand, the 1983 M6.7 Coalinga, California, and the 1999 M7.6 Chi-chi, Taiwan, earthquakes, all occurred within crust that, at least locally, was strongly overpressured (Davis et al., 1983; Yerkes et al., 1990; Allis et al., 1998). Fault-valve action (postfailure discharge of overpressured fluids along a fault) potentially arises wherever a seismically active fault transects an impermeable barrier bounding a suprahydrostatically pressured portion of the crust. Overpressuring may be laterally extensive or restricted to the vicinity of the fault itself, as suggested by fault-localized vein systems (Figure 10). Low-permeability barriers acting as seals through the interseismic period may be purely stratigraphic features (e.g., shale horizons) or they may result from hydrothermal cementation (e.g., Tigert and Al-Shaieb, 1990).

Strong valving action is especially likely in compressional regimes because overpressuring more commonly develops in such settings (Osborne and Swarbrick, 1997); and because the load-strengthening character of reverse faults means that fault-normal stresses increase through interseismic periods as faults are loaded to failure, thereby reducing permeability, but decrease postfailure to enhance fault permeability. However, as discharge proceeds, valving action should lead to postfailure strengthening of the fault through the drop in fluid pressure (Figure 10a). In this fault-valve model, successive failure episodes leading to postseismic discharge of overpressured fluid are linked, through fluctuations in fluid pressure, to the coupled cycling of both shear stress and fault strength (Sibson, 1992).

Extreme Valving Action

Minor valving activity involving small fluid volumes and low-amplitude fluid-pressure cycling is probably widespread wherever seismically active reverse faults occur in overpressured crust (Figure 10b). However, a case can be made, on mechanical grounds and on the association of mesozonal gold-quartz lodes with high-angle reverse faults, that extreme fault-valve behavior that redistributes large fluid volumes is associated

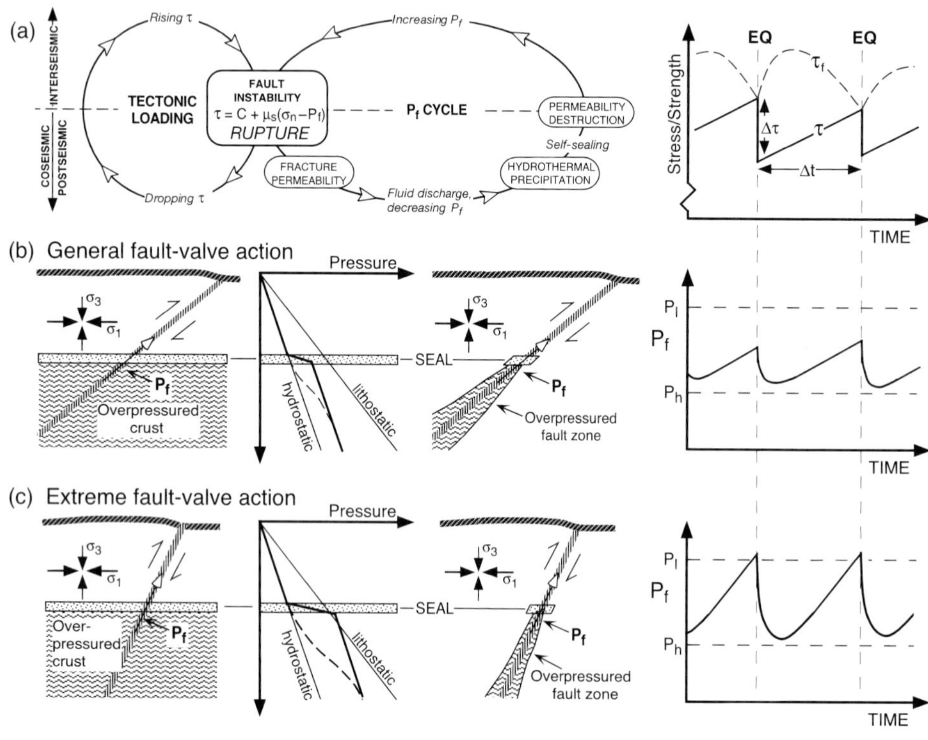

FIGURE 10. Fluid redistribution from valving action on reverse faults. (a) Coupled cycling of shear stress, τ, fluid-pressure, P_f, and fault strength, τ_f, during fault-valve action, showing how the recurrence interval between successive slip events (Δt) is affected by the cycling of both shear stress and fault strength. (b) Valving action and fluid-pressure cycling on reverse faults oriented at reactivation angles less than lockup, transecting impermeable seals capping uniformly overpressured crust or a locally overpressured portion of the fault zone. (c) Extreme valve action involving large-amplitude fluid-pressure cycling on severely misoriented steep reverse faults for the two cases as in (b). Bold lines on the pressure/depth plots represent the prefailure fluid-pressure state, dashed lines the postfailure state.

with steep reverse faults (Sibson et al., 1988; Cox, 1995; Robert et al., 1995). High-angle reverse faults that are severely misoriented for frictional reactivation (i.e., that are oriented beyond frictional lockup) make particularly effective fault valves that discharge large fluid volumes postfailure (Figure 10c) for three reasons. (1) The requirement for attaining the tensile overpressure condition prefailure allows substantial fluid volumes to accumulate, within a dilatant mesh structure, to maximum sustainable supralithostatic fluid overpressure, with ready access to the fault postfailure. (2) Steep faults make an efficient drainage channel across the steepest hydraulic gradients. (3) Reduction of normal stress enhances postfailure fault permeability (Sibson, 1990; Sibson and Scott, 1998). Support for the development of extreme valve action on steep reverse faults comes from fluid-inclusion studies of associated gold-quartz vein systems, which demonstrate repeated fluid-pressure cycling between lithostatic and hydrostatic levels at a depth of approximately 10 km (Cox, 1995; Robert et al., 1995). In most cases, extreme valving action is likely to be short-lived and involves comparatively minor reverse-fault displacement, because it can only be sustained as long as fluid pressures can reaccumulate back to lithostatic levels.

Tectonic Settings for Extreme Valve Action

Tectonic settings in which extreme fault-valve action is likely to develop on steep reverse faults include accretionary wedges and collisional zones, where domino steepening of original imbricate thrust stacks results from progressive regional contraction; and areas of positive tectonic inversion, where inherited normal faults with moderate-to-steep dips are reactivated in compression. This latter situation is of special relevance because extensional basins form under comparatively low mean stress ($\bar{\sigma} < \sigma_V$), which allows porosity to be retained to greater depth, and thus they have an inherently greater capacity for storing fluids. The switch to a compressional regime ($\bar{\sigma} > \sigma_V$), especially if it is sudden, may induce a general increase in fluid pressure throughout the basin as a consequence of increased tectonic mean stress (under constant fluid pressure, mean stress can be expected to increase by about an order of magnitude, according to Sibson, 1995). This concomitant boosting of fluid pressure accompanying positive inversion promotes compressional reactivation of former normal faults that are not well oriented for reactivation, which helps to account for the 50° subsidiary peak in the dip histogram for active reverse faults (Figure 2). Accompanying valve action may help redistribute fluids into hanging-wall antiforms developed during inversion. Extreme valving sets in when the inherited faults are sufficiently steep to be severely misoriented for reactivation under compression.

SANTA BARBARA CHANNEL–VENTURA BASIN: AN ACTIVE FAULT-VALVE PROVINCE?

The eastern Santa Barbara Channel–Ventura Basin region in the western Transverse Ranges of California is an area of active tectonic inversion where inherited normal and/or strike-slip faults that developed during Miocene extension have been reactivated since the Pliocene, within an actively contracting fold-thrust belt characterized by seismically active reverse faults and rapidly amplifying folds (Yeats et al., 1988; Shaw and Suppe, 1994). Historical seismicity includes earthquakes up to at least M6.3, but the area is also characterized by intermittent earthquake swarm activity, which is unusual for a compressional tectonic regime (Sylvester et al., 1970; Henyey and Teng, 1985). With the Ventura-Rincon anticlinorium ranking as a supergiant oil field on a world scale, a great deal of subsurface information is available from exploration and production drilling and from seismic profiling.

The region has the hallmarks of an active fault-valve province. Overpressuring is widepread, with λ_v values approaching 0.9 at depths of 3 km, notably in the core of the young (<0.2 Ma) and actively amplifying Ventura Anticline where the crestal region is uplifting differentially at approximately 10 mm/yr (Yeats, 1983). Local production from strata as young as Pleistocene age suggests that hydrocarbon migration is still ongoing. Surface discharges of hydrocarbon and aqueous fluids have been recorded following major historical earthquakes within the western Transverse Ranges (Nunn, 1925; Hamilton et al., 1969). Major reverse faults onshore, such as the opposite-dipping Oak Ridge and Red Mountain Faults, remain microseismically active though steeply dipping, which is consistent with near-lithostatic overpressuring at depth (Sibson, 1995). For example, the near-surface dip of approximately 60° for the Red Mountain Fault, as defined by oil-field drilling, can be extrapolated, through microearthquake clusters, to a 12-km depth, with focal mechanisms defining almost pure reverse dip-slip (Yeats et al., 1987).

Suggestions that microearthquake swarms denote active migration of overpressured fluids make their occurrence in this area of active compressional tectonics especially interesting (Hill, 1977; Sibson, 1996). Noteworthy in this respect is the 1984 earthquake swarm (>400 events) that occurred in the eastern Santa Barbara Channel, elongated east–west along the oil-producing Oak Ridge trend (Henyey and Teng, 1985; Shaw and

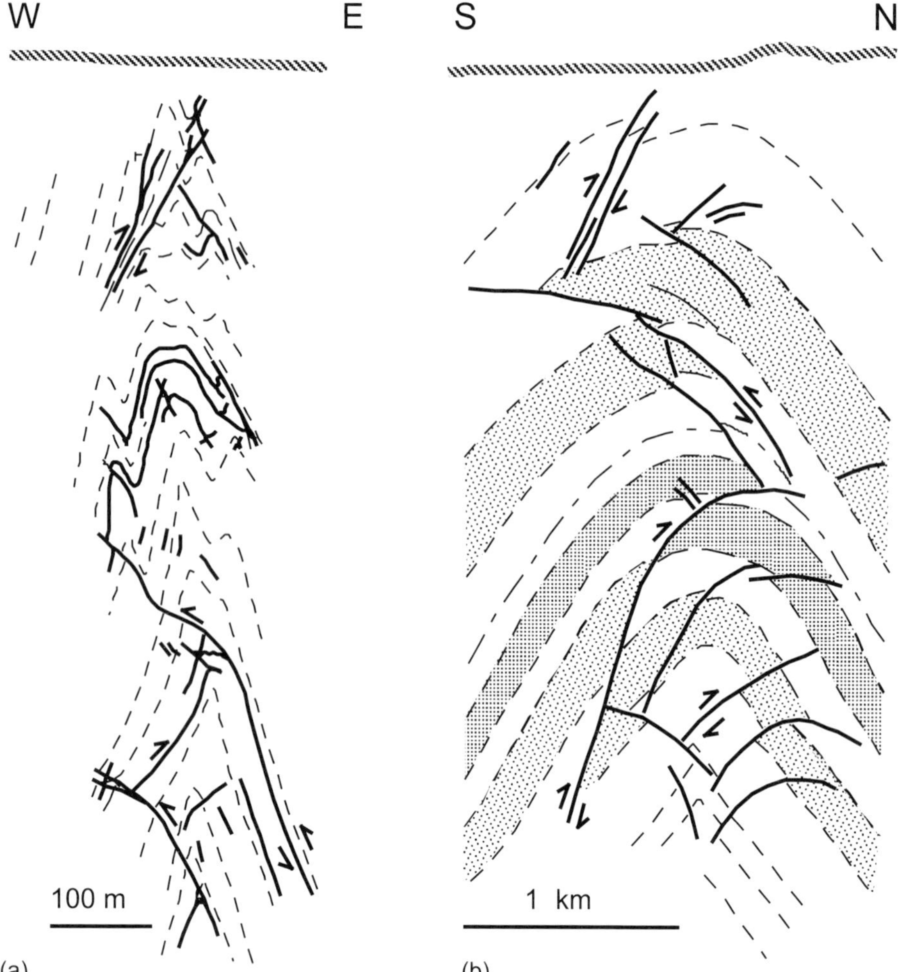

FIGURE 11. (a) Gold-quartz veins hosted in saddle reef structures and reverse faults, many rooted in bedding shears within the Garden Gully Anticline in deformed Ordovician turbidites, Confidence Extended Mine, Bendigo, Australia (after Cox et al., 1991, from Whitelaw, 1918). (b) Reverse faulting, often rooted in bedding shears, defined by drilling in the core of the growing, fluid-overpressured Ventura anticline of southern California, where oil is produced from actively deforming Plio-cene-Pleistocene strata (after Yeats, 1983).

Suppe, 1994). In north-south cross sections, foci from the swarm define a steeply south-dipping plane over a depth range of 7–15 km, with predominantly reverse-slip focal mechanisms suggesting either slip on planes dipping 0–20°N or on planes dipping 70–90°S. Despite the steep alignment of foci, Shaw and Suppe (1994) interpreted the swarm to be the result of shallow north-dipping bedding-plane slip associated with an active fold axial surface. However, an alternative interpretation, allowable from seismic and subsurface data, involves reverse reactivation of a steeply south-dipping fault driven by upward migration of lithostatically overpressured fluids. The swarm thus would represent an episode of fault-valve behavior.

Hydrocarbon Versus Hydrothermal Flow Paths

Figure 11 draws attention to the remarkable similarities between the internal structure of the actively growing and strongly overpressured Ventura Anticline in California (Yeats, 1983) and the system of gold-quartz veins hosted by a folded and cleaved sequence of Ordovician turbidites within the Garden Gully Anticline (Whitelaw, 1918), near Bendigo in the state of Victoria, Australia. Structural permeability in both fold closures is dominated by reverse-slip faults and bedding planes with predominantly moderate to steep dips. Gold-quartz veining on these structures in the core of the Garden Gully Anticline provides compelling evidence for episodic large-volume flow along the discontinuities (Cox et al., 1991). A reasonable inference is that fluid redistribution within the actively growing Ventura Anticline is likewise episodic and concentrated on reverse-slip surfaces. Bruhn et al. (2000) have recently drawn attention to the migration of overpressured fluids into somewhat comparable reverse-fault cored anticlines in Cook Inlet, Alaska. Although it occurs on different scales and represents very different structural levels in the crust, massive fluid redistribution in all these cases seems attributable to extreme valving action on reverse-slip discontinuities.

CONCLUSIONS

1) The observed dip distribution for M > 5.5 reverse-slip ruptures in the upper continental crust (mostly in the 5–15 km depth range) is broadly consistent with the lower bound of Byerlee's (1978) rock-friction range ($0.6 < \mu_s < 0.85$), with optimal reactivation occurring at dips of about 30° and frictional lockup at $\delta \approx 60°$. Although the interpretation depends strongly on the assumption of subhorizontal

σ_1 trajectories, rupture dimensions (>5 km) are such that small-scale stress heterogeneities are unlikely to be important. General scarcity of very low-angle thrust ruptures ($\delta \leq 10°$), apart from in the Himalayan frontal thrust system, suggests either that slip in flat-ramp-flat assemblages occurs seismically on the ramps but aseismically on the flats, as suggested by Bombolakis (1992) and Shaw and Suppe (1994), or that active onshore development of foreland fold-thrust belts is restricted to major zones of continental collision.

2) The existence of a subsidiary peak at $\delta \approx 50°$ on the histogram suggests that positive tectonic inversion, involving compressional reactivation of normal faults inherited from earlier crustal extension, is widespread in collision zones. Frictional-mechanics analysis shows that reverse reactivation at such dips implies fluid overpressuring that is localized to the vicinity of the faults to prevent the formation of new, favorably oriented thrusts. However, maximum sustainable overpressures are limited to sublithostatic values by the presence of throughgoing cohesionless faults oriented for reactivation at angles less than frictional lockup in the prevailing stress field. Overall, the implication is that overpressuring, with $P_f \rightarrow \sigma_3$, is widespread in the vicinity of reverse faults but that the tensile overpressure condition, $P_f > \sigma_3$, which is where extensional veining can be expected to develop, is generally attained only at the time of thrust initiation or in the vicinity of severely misoriented reverse faults ($\delta > 60°$).

3) Dynamic processes of fluid redistribution around seismically active reverse faults depend on the load-strengthening character of reverse-fault systems, the containing effect of compressional stress fields on overpressured fluids, and the likelihood of strong σ_2 directional permeability along strike. Oscillatory redistribution to and fro along strike may arise from mean-stress cycling, while fault-valve action in overpressured crust promotes episodic upward migration of overpressured fluids focused along reverse faults. Extreme valving action involving substantial fluid volumes tends to be associated with steep reverse faults that are severely misoriented for reactivation ($\delta > 60°$). It is especially likely to occur in shortening accretionary wedges and in areas of positive tectonic inversion. In this latter situation, fluid-rich sedimentary basins that originally developed in extension undergo contraction, with increased mean stress boosting fluid pressure and promoting compressional reactivation of steep former normal faults.

4) Similarities between steep reverse-slip structures that host mesozonal gold-quartz-vein systems and those that developed at higher structural levels in sedimentary basins undergoing positive tectonic inversion suggest that extreme valving action may also play an important role in hydrocarbon migration.

ACKNOWLEDGMENTS

Thanks to Stephen Cox, Howard Poulsen, and François Robert, who, over many years, have focused my attention on these issues, and to Ken McClay for kindly allowing late submission of this manuscript. Helpful reviews by Alex Maltman, Stan White, and Francesca Ghisetti are gratefully acknowledged. This work was funded by the NZ Public Good Science Fund through FRST Contract #C05611.

REFERENCES CITED

Allis, R. G., R. Funnell, and X. Zhan, 1998, From basins to mountains and back again: N.Z. basin evolution since 10 Ma, *in* G. B. Arehart and J. R. Hulston, eds., Proceedings of the 9th International Symposium on Water-Rock Interaction, Taupo, New Zealand, March 30–April 3, 1998, p. 3–9.

Anderson, E. M., 1951, The dynamics of faulting and dyke formation with application to Britain (2nd ed.): Edinburgh, Oliver and Boyd, 206 p.

Bombolakis, E. G., 1992, A developmental stage of a foreland belt, *in* K. R. McClay, ed., Thrust tectonics: London, Chapman & Hall, p. 33–40.

Boyer, S. E., and D. Elliot, 1982, Thrust systems: AAPG Bulletin, v. 66, p. 1196–1230.

Brown, S. R., and R. L. Bruhn, 1996, Formation of voids and veins during faulting: Journal of Structural Geology, v. 18, p. 657–671.

Bruhn, R. L., R. L. Parry, and M. P. Bunds, 2000, Tectonics, fluid migration, and fluid pressure in a deformed fore-arc basin: Geological Society of America Bulletin, v. 112, p. 550–563.

Byerlee, J. D., 1978, Friction of rocks: Pure and Applied Geophysics, v. 116, p. 615–626.

Cello, G., and A. Nur, 1988, Emplacement of foreland thrust systems: Tectonics, v. 7, p. 261–271.

Cox, S. F., 1995, Faulting processes at high fluid pressures: an example from the Wattle Gully Fault, Victoria, Australia: Journal of Geophysical Research, v. 100, p. 12,841–12,859.

Cox, S. F., V. J. Wall, M. A. Etheridge, and T. F. Potter, 1991, Deformation and metamorphic processes in the formation of mesothermal vein-hosted gold deposits—examples from the Lachlan Fold Belt in central Victoria, Australia: Ore Geology Reviews, v. 6, p. 391–423.

Dahlen, F. A., 1990, Critical taper model of fold-and-thrust belts and accretionary wedges: Annual Reviews of Earth and Planetary Science, v. 18, p. 55–99.

Davis, D. J., J. Suppe, and F. A. Dahlen, 1983, The mechanics of fold-and-thrust belts and accretionary wedges: Journal of Geophysical Research, v. 88, p. 1153–1172.

Elliot, D., 1976, The energy balance and deformation mechanisms of thrust sheets: Philosphical Transactions of the Royal Society, v. A 283, p. 289–312.

Haessler, H., A. Deschamps, H. Dufumier, H. Fuenzalida, and A. Cisternas, 1992, The rupture process of the Armenian earthquake from broad-band teleseismic body waves: Geophysical Journal International, v. 109, p. 151–161.

Hamilton, R. M., R. F. Yerkes, R. D. Brown, R. O. Burford, and J. M. DeNoyer, 1969, Seismicity and associated effects, Santa Barbara region: U.S. Geological Survey Professional Paper, v. 679, p. 47–72.

Henderson, J. R., M. N. Henderson, and T. O. Wright, 1990, Water-sill hypothesis for the origin of certain veins in the Meguma Group, Nova Scotia, Canada: Geology, v. 18, p. 654–657.

Henyey, T. L., and T. Teng, 1985, Seismic studies of the Dos Cuadras and Beta offshore oil fields, southern California OCS: Final Technical Report submitted to the Department of the Interior, Minerals Management Service, Center for Earth Sciences, University of Southern California, Los Angeles, California, 40 p.

Hill, D. P., 1977, A model for earthquake swarms: Journal of Geophysical Research, v. 82, p. 347–352.

Hill, D. P., 1993, A note on ambient pore pressure, fault-confined pore pressure, and apparent friction: Seismological Society of America Bulletin, v. 83, p. 583–586.

Hubbert, M. K., and W. W. Rubey, 1959, Role of fluid pressure in mechanics of overthrust faulting: Geological Society of America Bulletin, v. 70, p. 115–166.

Jaeger, J. C., and N. G. W. Cook, 1979, Fundamentals of rock mechanics (3rd ed.): London, Chapman & Hall, 593 p.

Kanamori, H., and D. L. Anderson, 1975, Theoretical basis of some empirical relations in seismology: Seismological Society of America Bulletin, v. 65, p. 1073–1095.

Knipe, R. J., G. Jones, and Q. J. Fisher, 1998, Faulting, fault sealing and fluid flow in hydrocarbon reservoirs: an introduction, *in* G. Jones, Q. J. Fisher, and R. J. Knipe, eds., Faulting, fault sealing and fluid flow in hydrocarbon reservoirs: Geological Society of London Special Publication 147, p. vii–xxi.

Lockner, D. A., 1995, Rock failure, *in* Rock physics and phase relations: a handbook of physical constants: American Geophysical Union Reference Shelf, v. 3, p. 127–147.

Molnar, P., and W-P. Chen, 1982, Seismicity and mountain building, *in* K. J. Hsu, ed., Mountain building processes: London, Academic Press, p. 41–57.

Molnar, P., and H. Lyon-Caen, 1989, Fault plane solutions and active tectonics of the Tibetan Plateau and its margins: Geophysical Journal International, v. 99, p. 123–153.

Moore, J. C., and 27 others, 1995, Abnormal fluid pressures and fault zone dilation in the Barbados accretionary prism: evidence from logging while drilling: Geology, v. 23, p. 605–608.

Nabelek, J., 1985, Geometry and mechanism of faulting of the 1980 El Asnam, Algeria, earthquake from inversion of teleseismic body waves and comparison with field observations: Journal of Geophysical Research, v. 90, p. 12,713–12,728.

Nguyen, P. T., S. F. Cox, L. B. Harris, and C. McA. Powell, 1998, Fault-valve behaviour in optimally oriented shear zones: an example at the Revenge gold mine, Kambalda, Western Australia: Journal of Structural Geology, v. 20, p. 1625–1640.

Nunn, H., 1925, Municipal problems of Santa Barbara: Seismological Society of America Bulletin, v. 15, p. 308–319.

Ohlmacher, G. C., and A. Aydin, 1997, Mechanics of vein, fault, and solution surface formation in the Appalachian Valley and Ridge, northeastern Tennessee, U.S.A.: implications for fault friction, state of stress and fluid pressure: Journal of Structural Geology, v. 19, p. 927–944.

Osborne, M. J., and R. E. Swarbrick, 1997, Mechanisms for generating overpressure in sedimentary basins: a reevaluation: AAPG Bulletin, v. 81, p. 1023–1041.

Price, N. J., 1975, Fluids in the crust of the Earth: Science Progress, v. 62, p. 59–87.

Robert, F., and A. C. Brown, 1986, Archean gold-bearing quartz veins at the Sigma Mine, Abitibi greenstone belt, Quebec: Part I. Geologic relations and formation of the vein system: Economic Geology, v. 81, p. 578–592.

Robert, F., A-M. Boullier, and K. Firdaous, 1995, Gold-quartz veins in metamorphic terrains and their bearing on the role of fluids in faulting: Journal of Geophysical Research 100, 12,861–12,879.

Ryberg, T., and G. S. Fuis, 1998, The San Gabriel Mountains' bright reflective zone: possible evidence of young mid-crustal thrust faulting in southern California: Tectonophysics, v. 286, p. 31–46.

Scholz, C. H., 1990, The mechanics of earthquakes and faulting: Cambridge, Cambridge University Press, 493 p.

Secor, D. T., 1965, Role of fluid pressure in jointing: American Journal of Science, v. 263, p. 633–646.

Segall, P., and D. D. Pollard, 1980, Mechanics of discontinuous faults: Journal of Geophysical Research, v. 85, p. 4337–4350.

Seward, T. M., 1993, The hydrothermal geochemistry of gold, *in* R. P. Foster, ed., Gold metallogeny and exploration: London, Chapman & Hall, p. 37–62.

Shaw, J. H., and J. Suppe, 1994, Active faulting and fold growth in the eastern Santa Barbara Channel, California: Geological Society of America Bulletin, v. 106, p. 607–626.

Sibson, R. H., 1985, A note on fault reactivation: Journal of Structural Geology, v. 7, p. 751–754.

Sibson, R. H., 1990, Conditions for fault-valve behaviour, *in* R. J. Knipe and E. H. Rutter, eds., Deformation mechanisms, rheology, and tectonics: Geological Society of London Special Publication 54, p. 15–28.

Sibson, R. H., 1991, Loading of faults to failure: Seismological Society of America Bulletin, v. 81, p. 2493–2497.

Sibson, R. H., 1992, Implications of fault-valve behaviour for rupture nucleation and recurrence: Tectonophysics, v. 211, p. 283–293.

Sibson, R. H., 1994a, An assessment of field evidence for

"Byerlee" friction: Pure and Applied Geophysics, v. 142, p. 645–662.

Sibson, R. H., 1994b, Crustal stress, faulting and fluid flow, *in* J. Parnell, ed., Geofluids: origin, migration and evolution of fluids in sedimentary basins: Geological Society of London Special Publication 78, p. 69–84.

Sibson, R. H., 1995, Selective fault reactivation during basin inversion: potential for fluid redistribution through fault-valve action, *in* J. G. Buchanan and P. G. Buchanan, eds., Basin inversion: Geological Society of London Special Publication 88, p. 3–19.

Sibson, R. H., 1996, Structural permeability of fluid-driven fault-fracture meshes: Journal of Structural Geology, v. 18, p. 1031–1042.

Sibson, R. H., 1998, Brittle failure mode plots for compressional and extensional tectonic regimes: Journal of Structural Geology, v. 20, p. 655–660.

Sibson, R. H., and J. Scott, 1998, Stress/fault controls on the containment and release of overpressured fluids: examples from gold-quartz vein systems in Juneau, Alaska; Victoria, Australia and Otago, New Zealand: Ore Geology Reviews, v. 13, p. 293–306.

Sibson, R. H., and G. Xie, 1998, Dip range for intracontinental reverse fault ruptures: truth not stranger than friction?: Seismological Society of America Bulletin, v. 88, p. 1014–1022.

Sibson, R. H., F. Robert, and K. H. Poulsen, 1988, High-angle reverse faults, fluid-pressure cycling, and mesothermal gold-quartz deposits: Geology, v. 16, p. 551–555.

Sylvester, A. G., S. W. Stewart, and C. H. Scholz, 1970, Earthquake swarm in the Santa Barbara Channel, California, 1968: Seismological Society of America Bulletin, v. 60, p. 1047–1060.

Tigert, V., and Z. Al-Shaieb, 1990, Pressure seals: their diagenetic banding patterns: Earth Science Reviews, v. 29, p. 227–240.

Westaway, R., 1998, Dependence of active normal fault dips on lower crustal flow regimes: Journal of the Geological Society, London, v. 155, p. 233–254.

Westbrook, G. K., 1991, Geophysical evidence for the role of fluids in accretionary wedge tectonics: Philosphical Transactions of the Royal Society, v. A 335, p. 227–242.

Whitelaw, H. S., 1918, The Confidence group of mines: Geological Survey of Victoria Bulletin, v. 30, 32 p.

Windh, J., 1995, Saddle reef and related gold mineralization, Hill End gold field, Australia: evolution of an auriferous vein system during progressive deformation: Economic Geology, v. 90, p. 1764–1775.

Wojtal, S., 1992, One-dimensional models for plane and non-plane power-law flow in shortening and elongating thrust zones, *in* K. R. McClay, ed., Thrust tectonics: London, Chapman & Hall, p. 41–52.

Yeats, R. S., 1983, Large-scale Quaternary detachments in Ventura Basin, southern California: Journal of Geophysical Research, v. 88, p. 569–583.

Yeats, R. S., W. H. K. Lee, and R. F. Yerkes, 1987, Geology and seismicity of the eastern Red Mountain fault, Ventura County: U.S. Geological Survey Professional Paper, v. 1339, p. 161–167.

Yeats, R. S., G. J. Huftile, and F. B. Grigsby, 1988, Oak Ridge fault, Ventura fold belt, and the Sisar décollement: Geology, v. 16, p. 1112–1116.

Yerkes, R. F., P. Levine, and C. M. Wentworth, 1990, Abnormally high fluid pressures in the region of the Coalinga earthquake sequence and their significance: U.S. Geological Survey Professional Paper, v. 1487, p. 235–257.

2

Hatcher, R. D. Jr., 2004, Properties of thrusts and upper bounds for the size of thrust sheets, *in* K. R. McClay, ed., Thrust tectonics and hydrocarbon systems: AAPG Memoir 82, p. 18–29.

Properties of Thrusts and Upper Bounds for the Size of Thrust Sheets

Robert D. Hatcher Jr.

Department of Earth and Planetary Sciences, University of Tennessee, Knoxville, Tennessee, U.S.A.

ABSTRACT

Basic types of thrusts recognizable in many orogens include: (1) foreland fold-thrust belt (and accretionary complex) thrusts, (2) Type C crystalline thrust sheets that detach along the ductile- (plastic) brittle transition in either continental or oceanic crust, and (3) Type F (fold-related) crystalline thrust sheets that are generated beneath the ductile-brittle transition by plastically shearing the common limb between antiforms and synforms.

Spacing of thrusts in a foreland fold-thrust belt is determined by the thickness and shear strength of the dominant structural-lithic unit and by basement irregularities, facies changes, and the development of stress maxima that localize buckling instabilities.

Displacement-length plots for the Canadian Rockies and Alberta Foothills, along with the Wyoming-Idaho-Utah thrust belt thrusts, plot on a linear curve. Appalachian foreland fold-thrust belt thrusts, however, plot on three curves that separate triangle zone thrusts from thrusts of small displacement from thrusts of large displacement. Type C sheets plot on a curve of positive slope separate from Alpine plastic Type F sheets, which plot on a curve of negative slope, because pure Type F sheets may involve a greater component of constrictional flow.

Upper bounds for the size of thrust sheets are delimited by the thickness and strength of their strong components, the size of the orogen of which a thrust sheet is a part, and by other inherent properties of a deforming wedge. Foreland fold-thrust belts contain thrust sheets that reach their maximum size as a function of the thickness and along-strike continuity of structural-lithic units. The maximum size of Type C sheets may be attained by the constancy of the geothermal gradient over a wide area during continent–continent collision.

INTRODUCTION

Thrust fault properties and the processes that form thrusts have been debated for more than 150 years. Today we are able to apply seismic imaging and computer technologies to better understand them in three dimensions, but several attributes and processes related to the formation of thrust sheets remain obscure.

The purpose of this chapter is to identify and evaluate the parameters that may control the upper bounds for the size of thrust sheets. Size is regarded here as the overall dimensions of a thrust sheet: along-strike length, across-strike width, and thickness. Intuitively, properties like thickness of the dominant structural-lithic unit, nature of the detachment, and the overall strength of the sheet should provide limits on the size of thrust sheets. If these were the only variables, however, thrust sheet size and upper size limits would be very predictable. In addition, the size of thrust sheets may also be related proportionally to the size of the orogen in which they form. For example, the Alps are a relatively small orogen and the largest thrust sheet (the Silvretta composite crystalline sheet) should be smaller than the largest thrust sheet in the medium-size Appalachians (the Blue Ridge–Piedmont megathrust sheet), and those should ideally be smaller than the largest thrust sheets in a large orogen, the North American Cordillera (Yukon-Tanana crystalline sheet, and Monashee detachment and Selkirk allochthon). All three, however, contain foreland fold-thrust belts with thrust sheets of about the same maximum size and displacement, but the Appalachian Blue Ridge–Piedmont megathrust sheet is anomalously much larger relative to the size of the orogen. Thus, either other variables are involved or the suggested direct proportionality between thrust size and orogen size is an oversimplification.

Studies of thrusts in most orogens have been confined to foreland fold-thrust belts, at least partly because they contain economic accumulations of hydrocarbons. As a result, we know more about the emplacement mechanisms and geometry of foreland fold-thrust-belt sheets. Foreland thrust sheets and certain types of internides crystalline thrust sheets, however, exhibit similar geometry and kinematics and deserve to be considered together for analysis, despite important material differences.

SIZE OF THRUST SHEETS

Thrust sheet size may be delimited partly by the thickness and strength of strong components as well as by the mechanical properties of a deforming wedge (Davis et al., 1983; Dahlen, 1990; Mandal et al., 1997). All foreland fold-thrust belt thrusts (with a few exceptions like the Muddy Mountains thrust in Nevada; Brock and Engelder, 1977), crustal slab Type C (composite crystalline) thrust sheets (Hatcher and Hooper, 1992), and ophiolites, regardless of size, have a weak basal unit (Figure 1). Local irregularities in the basement (or on the basement surface), facies changes that alter mechanical properties, and other factors that affect mechanical stratigraphy may also influence the size of thrust sheets. It is difficult to understand the linear along-strike continuity and across-strike width or regular spacing of successive thrust sheets, however, without concluding that some independent, overriding mechanical properties dominate the size parameter.

Willis (1893) recognized the relationship between the size of foreland fold-thrust-belt structures and the relative strength and thickness of structural-lithic units

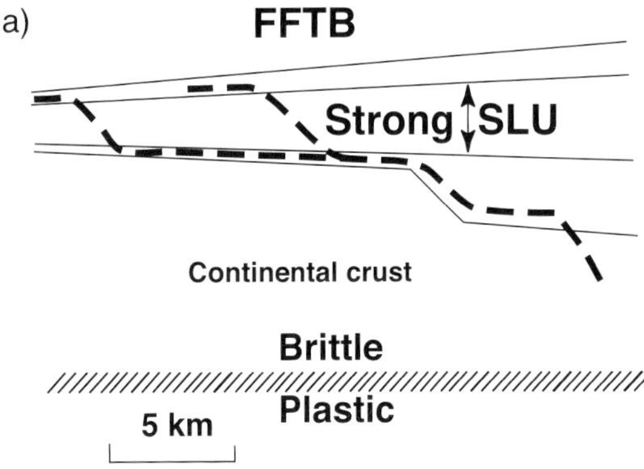

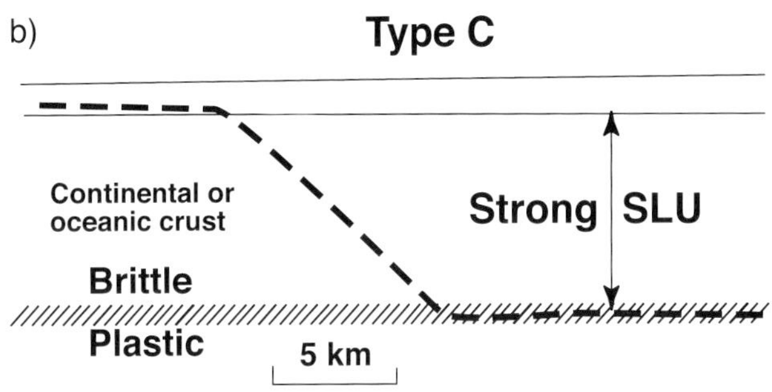

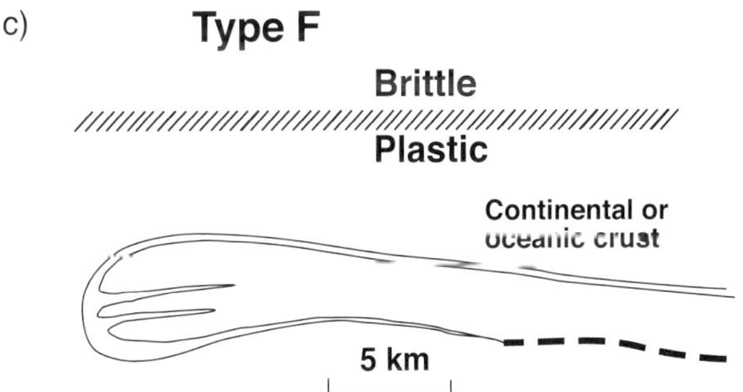

FIGURE 1. Three environments of thrust faulting. (a) Foreland fold-thrust belt (FFTB), where thrusts are generated by an indentor by propagation into a wedge-shaped assemblage of sedimentary rocks with a weak basal unit (local ductile-brittle transition) and a strong structural-lithic unit (SLU) that influences the size of thrust imbricates (after Fermor, 1999). (b) Type C crustal-slab thrust sheet that detaches along the ductile-brittle transition, transporting all of the crust above the transition in the sheet. These thrust sheets serve as the indentors that push the foreland fold-thrust belt in front of them, generally in continent–continent collision zones. (c) Plastic, Type F—fold-related—thrust sheets that form with a lobate shape below the ductile-brittle transition, or cool and evolve (as Type T, transitional) into smaller Type C sheets.

that compose them. A direct relationship probably exists between the thickness of strong structural-lithic units and size of thrust sheets; this relationship defines an upper limit relating the size of thrust sheets to the thickness of strong structural-lithic units (e.g., Woodward et al., 1988). These factors appear related to the locations of structural-lithic units in a foreland fold-thrust belt and of individual thrusts in foreland fold-thrust belts. Strong structural-lithic units may range from a section of strong sedimentary rocks, such as carbonate or massive sandstone; to a slab of continental crust that encompasses a substantial proportion of the crust; to an ophiolite—a slab of oceanic crust (or volcanic island arc) that may include upper mantle rocks (Figure 1). Weak structural-lithic units include evaporites, shale, and coal or thin-bedded carbonate rocks in sedimentary sequences ("effective ductile-brittle transitions"), and the ductile-brittle transition in the crust or mantle. Thicknesses of structural-lithic units range from a few millimeters to meters in mesoscopic thrusts, to 1–3 km in foreland fold-thrust belts, to 10 km or more in Type C crystalline slabs.

The three types of thrusts depicted in Figure 1 have contrasting attributes. The foreland fold-thrust-belt thrusts (Figure 1a) characteristically propagate into a strongly anisotropic foreland-thinning wedge of layered sedimentary rocks that contains weak and strong structural-lithic units (Chapple, 1978; Davis et al., 1983). Thrusts in foreland fold-thrust belts ideally propagate parallel to bedding in weak structural-lithic units (material of low yield strength, <10 MPa; Ranalli, 1987), and ramp (refract) at 18 to 30° across strong structural-lithic units. Detachment in strong structural-lithic units in foreland fold-thrust belts has always been an enigma, but transient ductile-brittle transitions that result from local, anomalously high fluid pressures that produce strain-induced weak zones may provide an explanation. The basal weak structural-lithic unit in foreland fold-thrust belts exhibits plastic or low-strength frictional-brittle thixotropic behavior, thus serving as a local ductile-brittle transition. Faults move incrementally or continuously, depending on continuity of the mechanically weak properties of the weak structural-lithic unit, as the fault propagates or is reactivated. Where stratigraphic thinning, facies changes, and topographic irregularities interrupt a weak zone, a thrust will refract to a higher weak structural-lithic unit, or ultimately, with sufficient displacement, to the surface. The largest sheets in foreland fold-thrust belts—the Lewis, McConnell, and Purcell in the Canadian Cordillera, and the Saltville and Whiteoak Mountain in the Appalachians—have displacements on the order of 100 km (determined from published and unpublished cross sections) and strike lengths on the order of 700 km, including linked systems (Figure 1). All are dominated by thick carbonate structural-lithic units. These may approach a critical mechanical upper bound for size of foreland fold-thrust-belt thrusts. Foreland fold-thrust-belt thrust sheet sizes range from mesoscopic to thrust sheets with strike lengths of >700 km, displacements of >100 km, and dominant structural-lithic unit thicknesses that are >3 km.

The largest thrust sheets, by almost an order of magnitude, occur in the internides of mountain chains and consist of composite (Type C) crystalline megathrust sheets (Hatcher and Hooper, 1992), no doubt largely because they have greater strength and cohesiveness than the largest foreland sheets. The larger size of Type C crystalline sheets is no doubt also related to their temperature (and heat flow) at the time they formed: a strong structural-lithic unit of continental basement would detach on the ductile-brittle transition 10 or more kilometers deep in the crust, and will be stronger and thus larger than the largest thrust sheet in the foreland fold-thrust belt in the same orogen (Hatcher and Williams, 1986) (Figure 1b). Type C sheets commonly serve as the indentors that drive the thin-skinned foreland fold-thrust belt deformation in front of them (Blue Ridge–Piedmont; Yukon-Tanana and Monashee-Selkirk; and Jotun). The behavior of Type C sheets is geometrically similar to that of large foreland fold-thrust-belt sheets composed of a strong structural-lithic unit and a weak basal unit, with the principal important mechanical differences being size and composition of strong structural-lithic units, and the fact that they propagate along the ductile-brittle transition in either continental or oceanic crust (and mantle) (Figure 1b). The thickness of these sheets is determined largely by the geothermal gradient where they form. Sheet thickness ranges from a few kilometers in thin, warm oceanic crust to 10 km or more in continental crust (Armstrong and Dick, 1974). These thrusts form in the internides of mountain chains and propagate until the ductile-brittle transition changes level and the thrust is forced to propagate (ramp) toward the surface. These thrusts frequently ramp into a higher, weak structural-lithic unit in the sedimentary wedge in the platform margin that becomes the master detachment for the foreland fold-thrust belt. The Type C sheet, because of its size and energy, serves as an indentor and pushes the foreland fold-thrust belt like a snowplow in front of it (it actually *is* the snowplow), thus mechanically linking foreland fold-thrust belt and Type C thrusts. The size of Type C thrusts ranges from the size of moderate to large foreland fold-thrust-belt thrusts to the megathrust sheets like the Blue Ridge–Piedmont in the Appalachians, the Silvretta in the eastern Alps, and the Monashee detachment–Selkirk allochthon in British Columbia. The latter has a displacement of more than 200 km and a strike length greater than 600 km (Brown et al., 1986, 1993; Wheeler and McFeeley, 1991).

Smaller crystalline sheets form in the internides as products of folding at moderate to high temperatures (Types F: fold-related, and T: transitional sheets), but exhibit purely plastic behavior throughout with only a strain or velocity gradient at the base. Type F sheets (Figure 1c) contrast markedly in geometry and mechanics with the other two classes in Figure 1. They form entirely below the ductile-brittle transition, are a byproduct of folding, and are created by plastic shearing of the common limb between an antiform and synform (Hatcher and Hooper, 1992). These thrusts move by directed plastic flow at moderate to high temperatures in the crust and mantle or in salt or ice at surface temperatures. The dominant mechanism may be flexural to passive flow (passive amplification), or involve ductility contrast between layers of different material properties (e.g., amphibolite or quartzite with pelitic interlayers). Layering and thus buckling and flexural flow still play a role in the thrusting process here. Useful natural analogs may be seen in glacial ice (e.g., Hudleston, 1976) and salt (e.g., Muehlberger, 1968; Jackson et al., 1990) and also experimentally in artificial analogs (e.g., Bucher, 1956). Cooling during emplacement of Type F thrusts may lower the ductile-brittle transition and cause Type F thrusts to gain strength and be transformed into Type C sheets, thus increasing their stability and potentially their lateral extent and displacement. All Type F sheets do not evolve into Type C sheets. Eastern Pennine Alps Type F sheets (Adula, Tambo, Suretta) have a lobate shape, with the long axis extended in the direction of transport, whereas western Pennine Type F sheets (St. Bernard, Dente Blanche, Monte Rosa) probably originated as Type F sheets, but may have cooled during emplacement and transformed into Type T (transitional) or C sheets. Type C sheets do not necessarily originate as Type F sheets; most Type C sheets probably form independently, as slabs detached along the horizontal ductile-brittle transition. Interestingly, displacement on the Appalachian Blue Ridge–Piedmont megathrust sheet decreases northeastward where the thrust becomes blind beneath the Blue Ridge anticlinorium, suggesting it is a crustal analog of the Powell Valley anticline that ramped from the ductile-brittle transition into the sedimentary section (Harris, 1979). The Blue Ridge anticlinorium may, however, be an artifact of the early Paleozoic Taconian orogeny (Geiser in Hatcher et al., 1989).

Rheology of internides Type F and C sheets ranges from plastic to elastic-plastic, whereas that of foreland fold-thrust-belt sheets ranges from elastic (Coulomb, on the mesoscale) to elastic-plastic (or even plastic) on the scale of the orogen. The foreland undergoes permanent, unrecoverable body deformation and thus may be considered a plastically deformed mass (Chapple, 1978). Proportionally smaller segments of foreland fold-thrust belts (e.g., in the Alberta Foothills, Appalachian Plateau, and Jura Mountains) and subduction complex sheets may be smaller because of the dimensions and strength of their strong structural-lithic units. Smaller thrust sheets form in the outer parts of foreland fold-thrust belts, by a combination of stratigraphy—thinner structural-lithic units in the distally tapered part of the deforming wedge, and the small amount of available energy remaining for deformation. Willis (1893) may have been the first to recognize the importance of structural-lithic units in determining the properties of thrust sheets. Woodward et al. (1988) also attempted to apply this concept to the Appalachian foreland fold-thrust belt, where they addressed the importance of structural-lithic units in controlling thrust sheet thickness and of facies in controlling the areal distribution of structural-lithic units. The variable properties of structural-lithic units may ultimately be the factor that controls the size and displacement of thrust sheets.

Large Type C sheets formed in the internides of orogens like the Appalachians (Blue Ridge–Piedmont), Canadian–Alaska Cordillera (Monashee–Selkirk allochthon, Yukon-Tanana) (Figure 2a), and Scandinavian Caledonides (Jotun, Seve, Särv). Type C sheets proportionally larger than those in the corresponding foreland fold-thrust belt also exist in the Alps (Austroalpine sheet) (Figure 1). Strong structural-lithic units in Type C sheets are an order of magnitude thicker than those in foreland fold-thrust-belt sheets, partly accounting for the much greater size of these sheets. The dimensions of these sheets may thus define the upper bound for the size of thrust sheets composed of quartz-rich crust. The upper bound for ophiolite sheets, and also for smaller Type C sheets, is more difficult to estimate, because it requires reconstruction of larger, formerly more continuous ophiolite sheets and identification of the source, which may have been partially consumed during subduction or eroded following obduction. One of the largest preserved ophiolites is the Samail in Oman (at least 700 km long, with greater than 200 km of displacement), yet it is only a remnant of a larger sheet of unknown original dimensions (Coleman, 1981).

SPACING OF THRUSTS

Attempts have been made to predict the spacing of thrust sheets, which is partly a function of size, thickness of structural-lithic units, location of basement irregularities and facies boundaries, and other factors. Such determinations are made difficult by contrasting relationships of thrust sheet spacing that increases toward the foreland in the southern Appalachians (Figure 3a) but decreases toward the foreland in the Canadian Cordillera (Figure 2a). Wiltschko and Eastman (1983), Bombolakis (1986), and Mandal et al. (1997) concluded that basement topographic surface irregularities can focus thrust nucleation and influence thrust spacing. Dixon (1982) pointed out, however, that the regular spacing of thrusts in the Idaho-Wyoming-Utah thrust

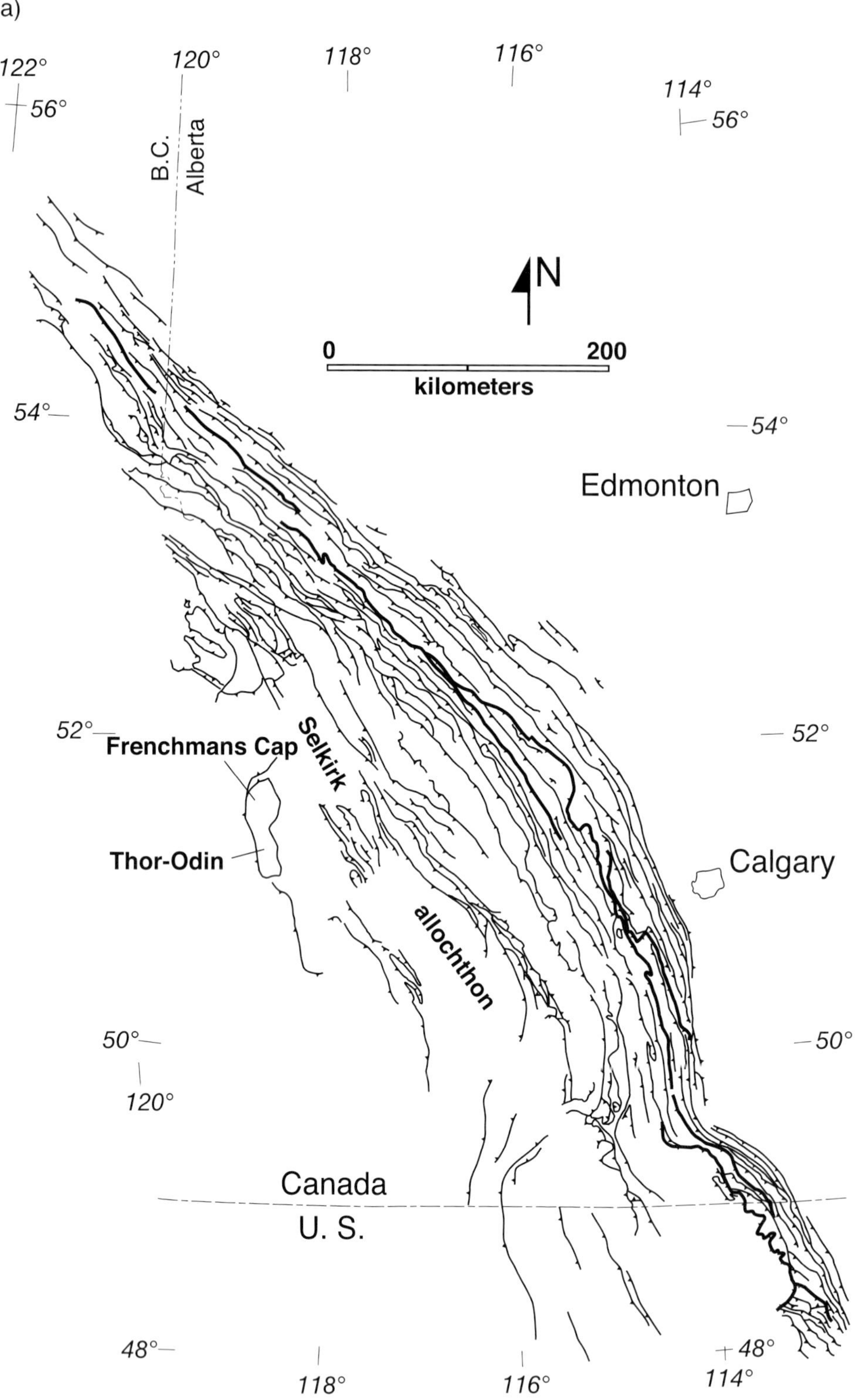

FIGURE 2. (a) Map of the Alberta–British Columbia foreland fold-thrust belt. Bold lines represent the approximate boundary between the Foothills and Rocky Mountains (from Wheeler and McFeely, 1991). (b) Thrusts (teethed lines) and major normal faults (not ornamented) in the Wyoming-Idaho-Utah thrust belt (after Royse et al., 1975). (c) Displacement-length plot for thrusts in the Alberta–British Columbia and Wyoming thrust belts (modified from Fermor, 1999). Names of Alberta Foothills thrusts and triangle zones are shaded. Wyoming-Utah-Idaho thrusts are indicated as open circles. [Additional data on Canadian Rockies thrusts from Price and Mountjoy (1970) and Price (1981).]

belt developed over a featureless basement surface, so other factors must be involved.

Liu and Dixon (1995) suggested that spacing of thrust ramps in duplexes is linearly related to the thickness of strata involved in the duplex. They also concluded that buckling instability in layered assemblages can produce stress concentrators that result in regularly spaced thrust ramps. Thickness of the strong structural-lithic unit has also been shown experimentally to be a factor in determining size and spacing (Mandal et al., 1997).

Regular thrust spacing exists in parts of the southern Appalachian foreland fold-thrust belt, in Tennessee, northern Georgia, and south western Virginia. Regional facies changes (e.g., changes from carbonate to fine siliciclastics) clearly influence the spacing and behavior of thrusts by providing mechanical interfaces and variations in structural-lithic unit thickness. Basement topography produced by Neoproterozoic extensional faults related to the rifting of Laurentia doubtlessly also influenced spacing of thrust sheets in the southern Appalachians (Hatcher et al., 1998). Mandl and Shippam (1981) were able to experimentally estimate critical shear strength within a rectangular thrust sheet and predict how imbrication would occur within a sheet of specified dimensions and mechanical properties (e.g., competent thrust sheet, plastic substratum). They were able to predict

Figure 2. (cont.).

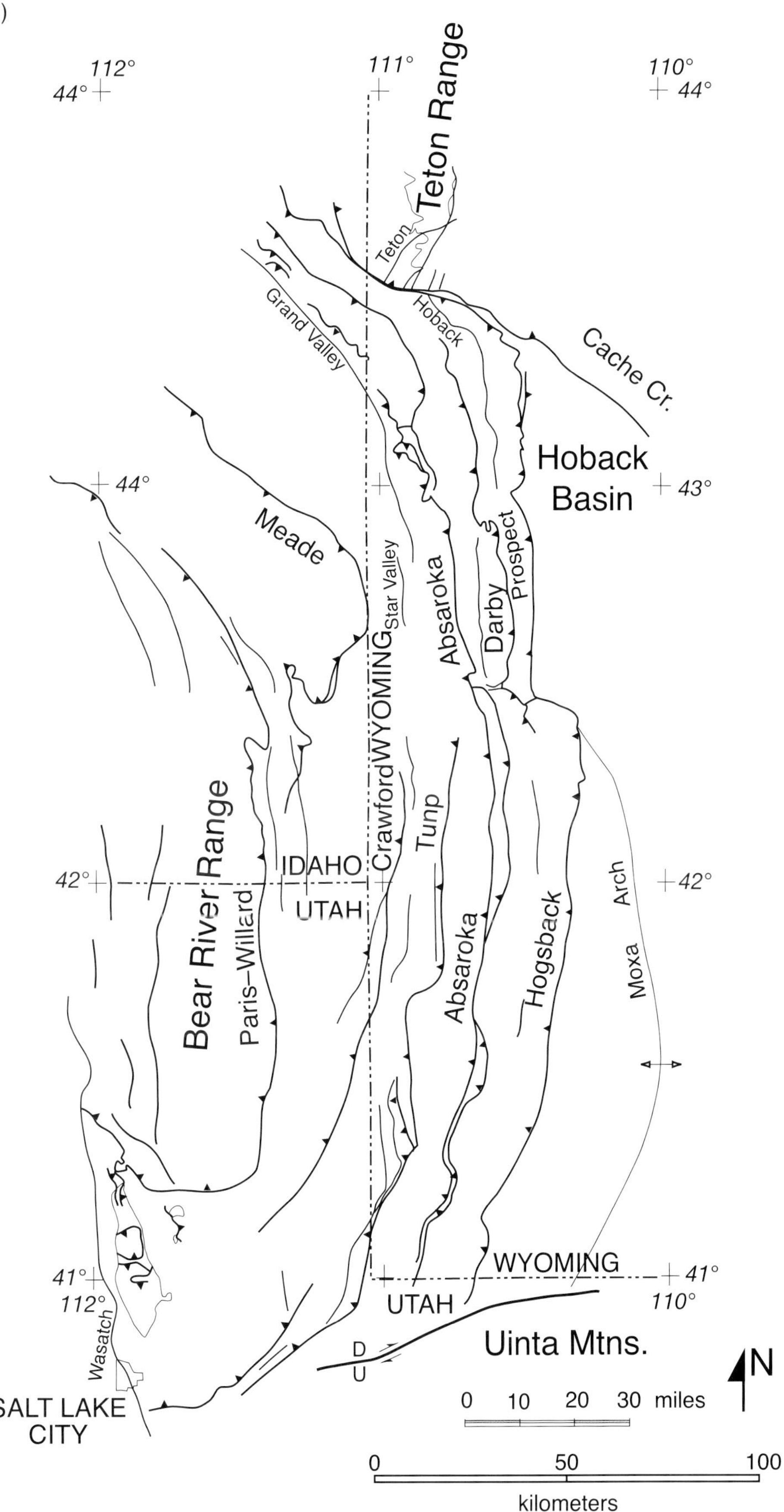

that a rectangular block that was bounded on two ends by thrusts would be imbricated from the end closest to the push in blocks of decreasing spacing toward the foreland, but with a lengthy gap from the foreland-propagating imbricates to the most foreland-ward thrust. This latter is one of the bounding thrusts and would have formed prior to initiation of imbrication, in contrast with the way most foreland fold-thrust belts are thought to form. Interestingly, the pattern of imbrication and spacing of thrust sheets predicted by Mandl and Shippam (1981) is almost exactly that in the Tennessee Valley and Ridge foreland fold-thrust belt (Figure 3a and 3b). A series of thrusts formed here that is more closely spaced toward the northwest, except for the westernmost thrusts (Cumberland Plateau and Pine Mountain) that are spaced well to the west of the more hinterlandward thrusts. Except for the predicted spacing of the most forelandward thrust, the imbrication pattern could conceivably develop without the older thrust already there. The spacing between the imbricate fan and the older thrust, however, lends credence to the Mandl and Shippam (1981) hypothesis that the older thrust should already be there, making the imbricate fan an initially out-of-sequence array that then develops in piggyback (forward-breaking) fashion as it evolves. Alternatively, the entire array and spacing could have developed because of changes in thickness of the

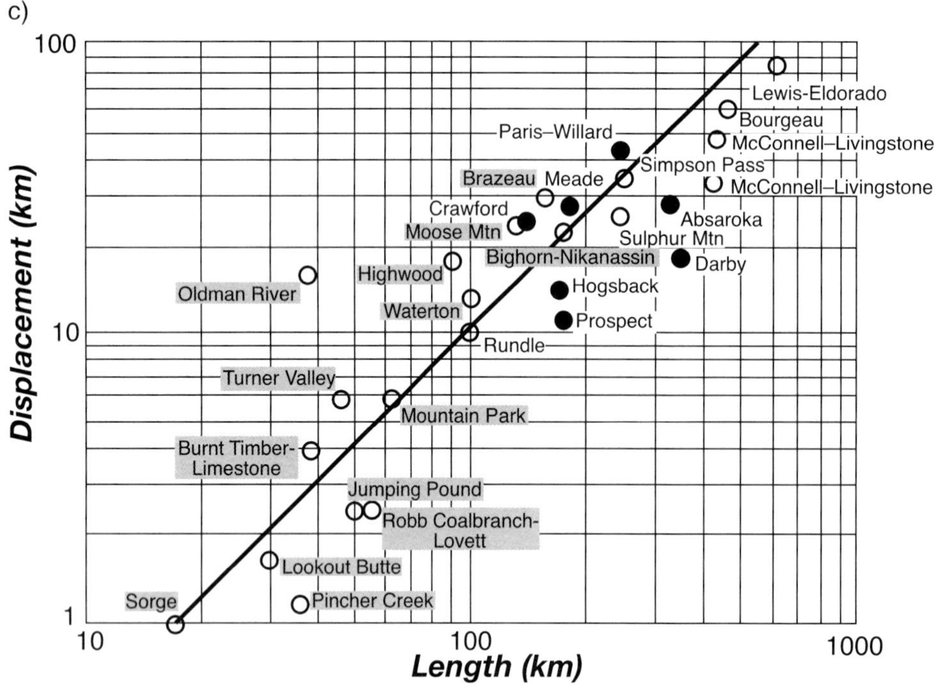

FIGURE 2. (cont.).

dominant structural-lithic unit, which occurs in the southern and central Appalachian foreland fold-thrust belt (Rich, 1934; Harris and Milici, 1977; Mitra, 1988). The existence of the westernmost thrusts is related to westward pinching of the Cambrian detachment that forced the basal thrust to ramp into either the weak Devonian–Mississippian shale in the Pine Mountain block or the Pennsylvanian coal and shale in the Cumberland Plateau overthrust to the southwest.

DISPLACEMENT-LENGTH PLOTS

Elliott (1976) showed that a direct relationship exists between displacement and strike length of thrust sheets in the Alberta Foothills and Front Ranges of the Canadian Rockies (Figure 2). Fermor (1999) reconfirmed this relationship for the same area. Wyoming thrust belt and other Canadian Rockies thrusts also plot on Fermor's log-log displacement-length curve (Figure 2c). Ironically, the relationship is not as clear for Appalachian foreland fold-thrust-belt sheets (Figure 3). Small Appalachian foreland fold-thrust-belt thrust sheets—largely triangle zones and small-displacement thrust sheets in the Cumberland–Allegheny Plateau—plot on a separate curve from that of moderate-displacement Appalachian thrusts and a third curve of lower slope is possible for large-displacement Appalachian foreland fold-thrust belt curves (Figure 3c).

Type C crystalline slab sheets (Figure 4) plot on a displacement-length curve that is not unlike the curve that Fermor (1999) generated or that generated here for the Appalachian foreland fold-thrust belt (Figure 3c), but none of these lines are colinear. The Type C thrust sheets plot on a line (Figure 4c) that has a lesser slope than the foreland fold-thrust-belt thrust curves. This may reflect the greater variation in the size of Type C sheets or perhaps more uniform thickness. Megathrust (large Type C) sheets (Figure 1b) plot on a length-displacement curve (Figure 4c) that has a lower slope than that of foreland fold-thrust-belt sheets (Figures 2c and 3c). The Blue Ridge–Piedmont sheet, for example, is more than 1100 km long and has a maximum displacement greater than 350 km (Figure 3a). Several Scandinavian Caledonides composite sheets are even larger, with strike lengths of greater than 1400 km and displacements of more than 600 km (Gee, 1978; Gee et al., 1985) (Figure 4c).

Some Type F sheets (e.g., Walhalla) have strike lengths and displacements similar to those of thrusts in foreland fold-thrust belts, and plot on similar curves of different slope in displacement-length plots (Figure 4c). Dominantly plastic sheets, like the Suretta, Tambo, and others in the Pennine Alps, have greater displacements than along-strike length and thus plot on a curve with negative slope (Figure 4c). These may have been the least cohesive and may have flowed into place. Alternatively, they could have been laterally and vertically confined into a state of northwest-directed, constricted flow during emplacement. Some of the sheets in the Appalachian Inner Piedmont (e.g., Sugarloaf Mountain–Six Mile–Alto), and the Pennine Great St. Bernard nappe, which formed as Type F sheets, however, cooled sufficiently to become coherent and to move as intact sheets (evolving from F to T or C) rather than as flowing lobate masses (Figure 4b).

UPPER BOUNDS FOR THRUST-SHEET SIZE

The largest thrust sheets in foreland fold-thrust belts approach lengths of 700 km and have displacements on the order of 100 to 150 km. These include the Lewis and Purcell thrusts in the southern Alberta-Montana Rockies and the Saltville and Whiteoak Mountain thrusts

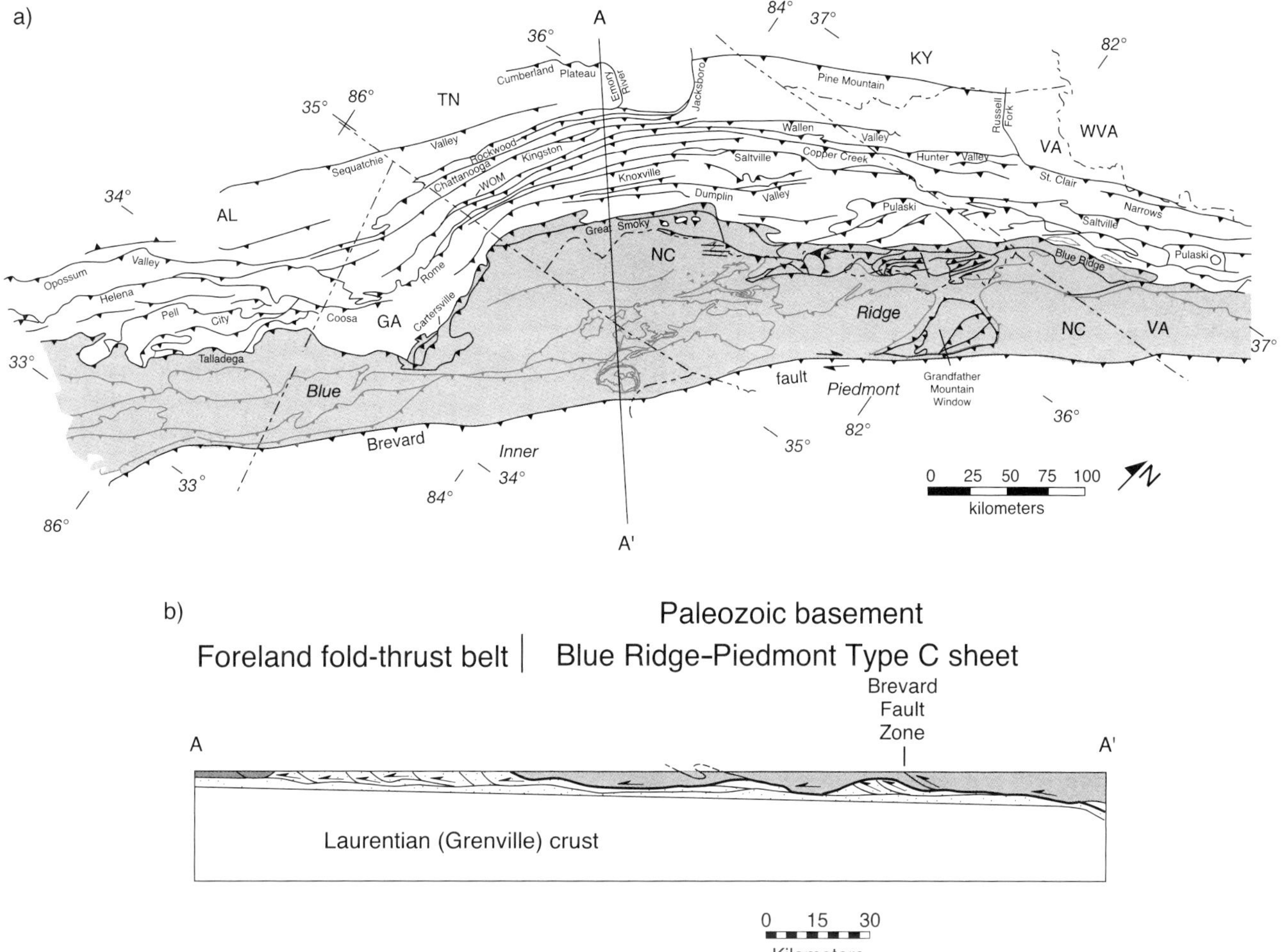

Figure 3. (a) Map of Alleghanian (Carboniferous-Permian) thrust faults in the Appalachian Valley and Ridge foreland fold-thrust belt from Virginia to Alabama. The shaded region represents part of the Blue Ridge–Piedmont Type C megathrust sheet indentor. Faults shown in solid black are Alleghanian thrusts that formed during emplacement of the Blue Ridge–Piedmont sheet that pushed the Valley and Ridge in front of it. Light teethed lines are early to middle Paleozoic thrusts. WOM = Whiteoak Mountain fault. Sources of data include Gwinn (1964), Hardeman (1966), Johnson (1993), and Osborne et al. (1988). (b) Cross section through the Valley and Ridge of Tennessee and the Blue Ridge–Piedmont megathrust sheet in Tennessee, North Carolina, and Georgia. Note the spacing of thrusts from southeast to northwest in both (a) and (b). Rocks of the foreland fold-thrust belt are stippled in the cross section; the Blue Ridge–Piedmont Type C crystalline slab is shaded. Location of line A–A′ indicated in (a). (c) Displacement-length plot for Appalachian foreland fold-thrust-belt thrusts. Note that small-displacement and triangle-zone (names shaded) thrusts mostly plot on a different, more steeply sloping curve, and very large thrusts plot on a third, more gently sloping curve (gray). The dashed line is Fermor's (1999) Canadian Rockies curve.

in the southern Appalachians. Each contains strong structural-lithic units that range up to 2 km thick and are composed mostly of carbonate rocks. Interestingly, thrusts immediately adjacent on either side of the Appalachian examples have relatively small displacements of 10 km or less (with subsidiary imbricates having even less than 1 km of displacement), yet the structural-lithic units for these are approximately the same thickness. The upper bound for the size of foreland fold-thrust-belt thrusts may thus be determined by the strength, thickness, and lateral extent of the dominant structural-lithic units in the system, and may dictate that maximum displacement is on the order of 100 km before subsidiary imbricate thrusts begin to form (both in- and out-of-sequence). An additional factor favoring generation of larger thrusts in foreland fold-thrust belts could be the lack of initiation of buckling instabilities or fewer mechanical obstacles to detachment propagation within the wedge, so that larger thrust sheets formed.

The largest thrust sheets in any orogen are crystalline Type C crustal slab megathrust sheets (Figure 1b). They are always larger than the largest foreland fold-thrust-belt thrusts in the same orogen (Hatcher and

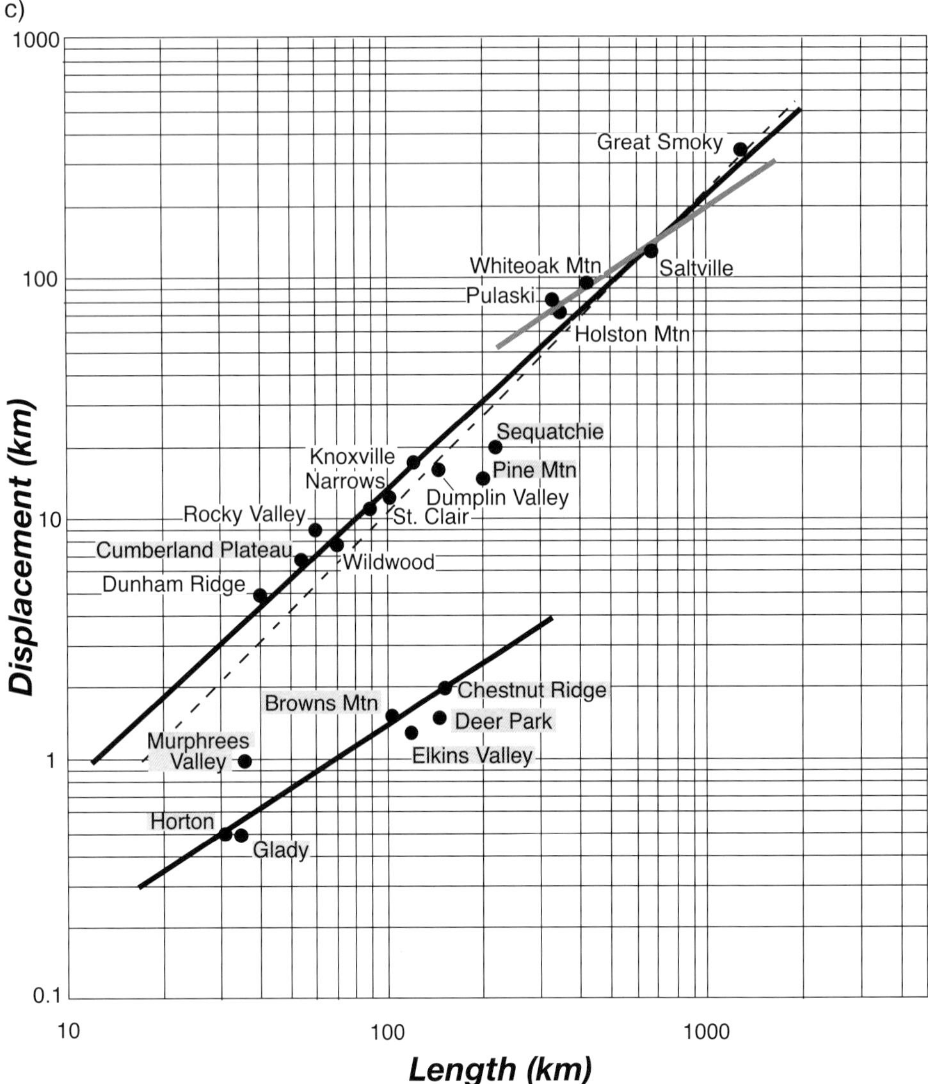

FIGURE 3. (cont.).

Williams, 1986). As might be expected because of the involvement of crystalline rocks and a detachment that is some 10 km deep in the crust, they also plot on a displacement-length curve that differs from those for foreland fold-thrust-belt thrusts or triangle zones (Figure 4c). The upper bound for these huge thrusts is difficult to estimate, because the largest sheets known, the Blue Ridge–Piedmont megathrust sheet and several of the composite allochthons in the Scandinavian Caledonides (Gee, 1978; Gee et al., 1985), have displacements of hundreds of kilometers, but their absolute upper displacement limits remain unknown. The crustal uniformity of the geothermal gradient where the Type C sheets are being formed may be a factor that delimits their size. More precise statements can, however, be made about the minimum displacements on Type C sheets, and this remains the basis for interpreting the points plotted in Figure 4c.

CONCLUSIONS

1) Three fundamental types of thrust sheets can be defined: (1) foreland fold-thrust-belt (accretionary-complex) thrusts, (2) Type C crystalline thrust (and megathrust) sheets that are detached along the ductile- (plastic) brittle transition in either continental or oceanic crust, and (3) Type F (fold-related) crystalline thrust sheets that are generated beneath the ductile-brittle transition by the plastic shearing of the common limb between antiforms and synforms. A transitional (Type T) class probably also exists where Type F sheets cool, gain strength, and evolve toward Type C sheets but do not attain the maximum size of Type C megathrust sheets.
2) Foreland fold-thrust-belt thrust sheets are smaller analogs of Type C sheets.
3) Displacement-length plots help distinguish families of thrusts that form by similar or different mechanisms.
4) Thrust-sheet size may be delimited by thickness and strength of strong structural-lithic units, but other factors are probably also involved.
5) Imbrication of a thrust-bounded block may be predicted experimentally and matches the thrust pattern in part of the southern Appalachian foreland fold-thrust belt.
6) Localization of ramps by preexisting inhomogeneities (e.g., basement topography, facies boundaries) may affect the size of thrust sheets.
7) The upper bounds for thrust-sheet size are probably related to the maximum size and lateral continuity for the structural-lithic units in the thrust system. This maximum appears to be 100 km of displacement in foreland fold-thrust belts and between 500

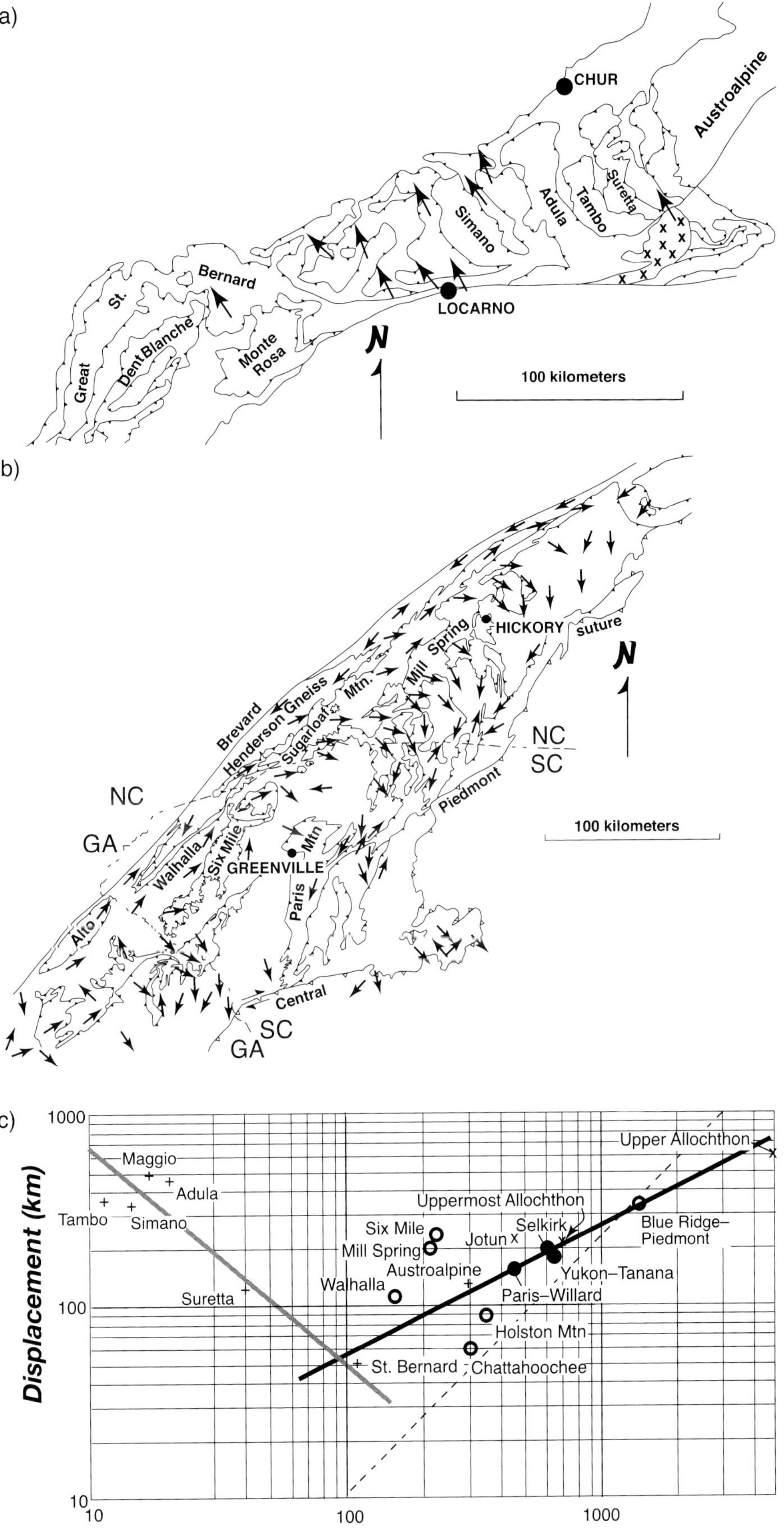

FIGURE 4. (a) Part of the Pennine Alps showing major Type F thrust sheets, along with part of the Type C Austroalpine sheet to the east. Arrows are orientations of mineral stretching lineations. The area with the x pattern is the Bergell pluton. (b) Map of the northern part of the southern Appalachian Inner Piedmont showing the configuration of major Type F and C thrust sheets. Arrows indicate orientations of shallow-plunging mineral stretching lineations that probably trace flow paths of thrust sheets. (c) Displacement-length plot for crystalline Types F and C thrust sheets from several orogens. The curve with negative slope for Pennine Alps thrusts indicates these lobate thrusts have displacements that are much greater than their along-strike length. The others, including the St. Bernard sheet, probably cooled enough that the sheet gained enough strength to evolve into Type C thrusts. Alpine thrusts are indicated by **+**; Appalachian thrusts are indicated by open circles; Canadian Rockies thrusts and the Paris–Willard thrust in the Wyoming thrust belt are indicated by solid circles; and Scandinavian Caledonides thrusts are indicated by an **x**. Data for the Alpine thrusts are derived from Spicher (1980). The dashed line is Fermor's (1999) Canadian Rockies curve. Data for Canadian Cordillera thrust sheets were derived from Wheeler and McFeely (1991). The Paris-Willard thrust sheet dimensions in Wyoming are from Royse et al. (1975). Data for Scandinavian Caledonides thrusts are from Gee et al. (1985). Data for Appalachian thrusts (open circles) are from Hatcher et al. (1990), state geologic maps, and unpublished maps and sections.

and 1000 km of displacement in large crustal-slab Type C megathrust sheets in orogens formed by continent-continent collision.

ACKNOWLEDGMENTS

Support for this research has been provided by the U.S. National Science Foundation and more recently by the U.S. Department of Energy National Petroleum Technology Office and the National Cooperative Geologic Mapping Program, Educational Component (administered by the U.S. Geological Survey). This paper was improved following reviews by Mark Cooper and Robert I. Thompson, and additionally by the editing of Anne H. Thomas.

REFERENCES CITED

Armstrong, R. L., and H. J. B. Dick, 1974, A model for the development of thin overthrust sheets of crystalline rock: Geology, v. 1, p. 35–40.

Bombolakis, E. G., 1986, Thrust-fault mechanics and the origin of a frontal ramp: Journal of Structural Geology, v. 8, p. 281–290.

Brock, W. G., and T. Engelder, 1977, Deformation associated with the movement of the Muddy Mountains overthrust in the Buffington window, southeastern Nevada: Geological Society of America Bulletin, v. 88, p. 1667–1677.

Brown, R. L., C. Beaumont, and S. D. Willett, 1993, Comparison of the Selkirk fan structure with mechanical models: implications for interpretation of the southern Canadian Cordillera: Geology, v. 21, p. 1015–1018.

Brown, R. L., J. M. Journeay, L. S. Lane, D. C. Murphy, and C. J. Rees, 1986, Obduction, backfolding and piggyback thrusting in the metamorphic hinterland of the southeastern Canadian Cordillera: Journal of Structural Geology, v. 8, p. 255–268.

Bucher, W. H., 1956, Role of gravity in orogenesis: Geological Society of America Bulletin, v. 76, p. 1295–1318.

Chapple, W. M., 1978, Mechanics of thin-skinned fold-and-thrust belts: Geological Society of America Bulletin, v. 89, p. 1189–1198.

Coleman, R. B., 1981, Tectonic setting for ophiolite obduction in Oman: Journal of Geophysical Research, v. 86, p. 2497–2508.

Dahlen, F. A., 1990, Critical taper model of fold-and-thrust belts and accretionary wedges: Annual Review of Earth and Planetary Science, v. 18, p. 55–99.

Davis, D., J. Suppe, and F. A. Dahlen, 1983, Mechanics of fold-and-thrust belts and accretionary wedges: Journal of Geophysical Research, v. 88, p. 1153–1172.

Dixon, J. S., 1982, Regional structural synthesis, Wyoming salient of western overthrust belt: American Association of Petroleum Geologists Bulletin, v. 66, p. 1560–1580.

Elliott, D., 1976, The motion of thrust sheets: Journal of Geophysical Research, v. 81, p. 949–963.

Fermor, P., 1999, Aspects of the three-dimensional structure of the Alberta Foothills and Front Ranges: Geological Society of America Bulletin, v. 111, p. 317–346.

Gee, D. G., 1978, Nappe displacement in the Scandinavian Caledonides: Tectonophysics, v. 47, p. 393–419.

Gee, D. G., R. Kumpulainen, D. Roberts, M. B. Stephens, A. Thon, and E. Zachrisson, 1985, Scandinavian Caledonides tectonostratigraphic map: Geological Survey of Sweden, scale 1:2,000,000.

Gwinn, V. J., 1964, Thin-skinned tectonics in the Plateau and northwestern Valley and Ridge provinces of the central Appalachians: Geological Society of America Bulletin, v. 75, p. 863–900.

Hardeman, W. D., 1966, Geologic map of Tennessee: Tennessee Division of Geology, scale 1:250,000.

Harris, L. D., 1979, Similarities between the thick-skinned Blue Ridge anticlinorium and the thin-skinned Powell Valley anticline: Geological Society of America Bulletin, v. 90, p. 525–539.

Harris, L. D., and R. C. Milici, 1977, Characteristics of thin-skinned style of deformation in the southern Appalachians, and potential hydrocarbon traps: U.S. Geological Survey Professional Paper 1018, 40 p.

Hatcher, R. D. Jr., and R. J. Hooper, 1992, Evolution of crystalline thrust sheets in the internal parts of mountain chains, *in* K. R. McClay, ed., Thrust tectonics: London, Chapman & Hall, p. 217–233.

Hatcher, R. D. Jr., and R. T. Williams, 1986, Mechanical model for single thrust sheets Part 1: Crystalline thrust sheets and their relationships to the mechanical/thermal behavior of orogenic belts: Geological Society of America Bulletin, v. 97, p. 975–985.

Hatcher, R. D. Jr., P. H. Osberg, P. Robinson, and W. A. Thomas, 1990, Tectonic map of the U.S. Appalachians: Geological Society of America, The Geology of North America, v. F-2, Plate 1, scale 1:2,500,000.

Hatcher, R. D. Jr., W. A. Thomas, P. A. Geiser, A. W. Snoke, S. Mosher, and D. V. Wiltschko, 1989, Alleghanian orogen, Chapter 5, *in* R. D. Hatcher Jr., W. A. Thomas, and G. W. Viele, eds., The Appalachian–Ouachita orogen in the United States: Boulder, Colorado: Geological Society of America, The Geology of North America, v. F–2, p. 233–318.

Hatcher, R. D. Jr., P. J. Lemiszki, and C. Montes, 1998, Boundary conditions of foreland fold-thrust-belt deformation from 3–D reconstruction of a curved segment of the southern Appalachian basement surface and base of the Blue Ridge–Piedmont thrust sheet: Progress report and initial results: American Association of Petroleum Geologists Extended Abstracts, v. 1, p. A281–A284.

Hudleston, P. J., 1976, Recumbent folding in the base of the Barnes Ice Cap, Northwest Territories, Canada: Geological Society of America Bulletin, v. 87, p. 1684–1692.

Jackson, M. P. A., R. R. Cornelius, J. H. Craig, A. Gansser, J. Stocklin, and C. J. Talbot, 1990, Salt diapirs of the Great Kavir, central Iran: Geological Society of America Memoir 177, 139 p.

Johnson, S. S., 1993, Geologic map of Virginia: Virginia Division of Mineral Resources, scale 1:500,000.

Liu, S., and J. M. Dixon, 1995, Localization of duplex thrust-ramps by buckling: Analog and numerical modeling: Journal of Structural Geology, v. 17, p. 875–886.

Mandl, G., and G. K. Shippam, 1981, Mechanical model of thrust sheet gliding and imbrication, *in* K. R. McClay and N. J. Price, eds., Thrust and nappe tectonics: London, The Geological Society of London Special Publication 9, p. 79–98.

Mandal, N., A. Chattopadhyay, and S. Bose, 1997, Imbricate thrust spacing: Experimental and theoretical analyses, *in* S. Sengupta, ed., Evolution of Geological Structures in Micro- to Macro-scales: London, Chapman & Hall, p. 143–165.

Mitra, S., 1988, Three-dimensional geometry and kinematic evolution of the Pine Mountain thrust system, southern Appalachians: Geological Society of America Bulletin, v. 100, p. 72–95.

Muehlberger, W. R., 1968, Internal structures and mode of uplift of Texas and Louisiana salt domes: Geological Society of America Special Paper 88, p. 359–364.

Osborne, W. E., M. W. Szabo, T. L. Neathery, and C. W. Copeland, Jr., 1988, Geologic map of Alabama: Geological Survey of Alabama, scale 1:250,000.

Price, R. A., 1981, The Cordilleran foreland thrust and fold belt in the southern Canadian Rockies, *in* K. R. McClay and N. J. Price, eds., Thrust and nappe tectonics: Geological Society of London Special Publication 9, p. 427–448.

Price, R. A., and E. W. Mountjoy, 1970, Geologic structure of the Canadian Rocky Mountains between Bow and Athabaska Rivers—A progress report: Geological Association of Canada Special Paper 6, p. 7–25.

Ranalli, G., 1987, Rheology of the Earth: Boston, Allen and Unwin, 366 p.

Rich, J. L., 1934, Mechanics of low-angle overthrust faulting illustrated by the Cumberland thrust block, Virginia, Kentucky, and Tennessee: AAPG Bulletin, v. 18, p. 1584–1596.

Royse, F. Jr., M. A. Warner, and D. L. Reese, 1975, Thrust belt structural geometry and related structural problems, Wyoming–Idaho–northern Utah, *in* D. W. Bolyard, ed., Symposium on drilling frontiers of the Central Rocky Mountains: Denver, Colorado, Rocky Mountain Association of Geologists, p. 41–54.

Spicher, A., 1980, Tektonische Karte der Schweiz: Swiss Geological Commission, scale 1:500,000.

Wheeler, J. O., and P. McFeely, 1991, Tectonic assemblage map of the Canadian Cordillera and adjacent parts of the United States of America: Geological Survey of Canada Map 1712A, scale 1:2,000,000.

Willis, B., 1893, The mechanics of Appalachian structure: U.S. Geological Survey 13th Annual Report 1891–1892, Part 2, p. 212–281.

Wiltschko, D. V., and D. J. Eastman, 1983, Role of basement warps and faults in localizing thrust fault ramps, *in* R. D. Hatcher Jr., H. Williams, and I. Zietz, eds., Contributions to the tectonics and geophysics of mountain chains: Geological Society of America Memoir 158, p. 177–190.

Woodward, N. B., K. R. Walker, and C. T. Lutz, 1988, Relationships between early Paleozoic facies patterns and structural trends in the Saltville thrust family, Tennessee Valley and Ridge, southern Appalachians: Geological Society of America Bulletin, v. 100, p. 1758–1769.

3

Ramos, V. A., T. Zapata, E. Cristallini, and A. Introcaso, 2004, The Andean thrust system— Latitudinal variations in structural styles and orogenic shortening, *in* K. R. McClay, ed., Thrust tectonics and hydrocarbon systems: AAPG Memoir 82, p. 30–50.

The Andean Thrust System— Latitudinal Variations in Structural Styles and Orogenic Shortening

Victor A. Ramos

Laboratorio de Tectónica Andina, Departamento de Ciencias Geológicas, Facultad de Ciencias Exactas y Naturales, Universidad de Buenos Aires, Buenos Aires, Argentina

Tomás Zapata

REPSOL-YPF S.A., Exploración y Desarrollo en Faja Plegada, Buenos Aires, Argentina

Ernesto Cristallini

Laboratorio de Tectónica Andina, Departamento de Ciencias Geológicas, Facultad de Ciencias Exactas y Naturales, Universidad de Buenos Aires, Buenos Aires, Argentina; CONICET Consejo Nacional de Investigaciones Cientificas y Técnicas

Antonio Introcaso

Instituto de Física de Rosario, Universidad Nacional de Rosario, Rosario, Argentina; CONICET Consejo Nacional de Investigaciones Cientificas y Técnicas

ABSTRACT

The different segments of the Andean thrust system have distinctive topography and inferred crustal roots. These two characteristics both depend upon crustal shortening, and on this basis they provide independent constraints for evaluating estimates of Cenozoic shortening obtained by balanced structural cross-sections of different segments of the fold-and-thrust systems. Three transects in the Central Andes are analyzed: a northern (22–23°S), a central (32–33°S), and a southern segment (37–39°S). Each segment shows different amounts of orogenic shortening, generated through a complex combination of thin- and thick-skinned thrusting. Based on known age constraints, different shortening rates are calculated for each segment. Estimates of crustal shortening derived from gravity and seismic-refraction data are used to evaluate interpretations of the structural style. In some segments, where alternative styles were proposed, the crustal-shortening estimates are used to identify the more realistic models.

Crustal shortening, shortening rates, and the resulting topography decrease progressively from north to south. These variations cannot be fully explained by differential fore-arc rotation, as in the Bolivian orocline model. Instead, a close correlation is suggested between the age of oceanic crust being subducted and the amount of shortening and propagation of the orogenic front toward the foreland. This fact becomes more important than fore-arc rotation farther south of the Bolivian orocline. On this basis, the present topography of the Andes, along the Nazca plate boundary, can be correlated with the age of adjacent oceanic crust.

INTRODUCTION

Several structural styles and kinematic models have been proposed to explain the Cenozoic deformation of the antithetic fold-and-thrust belt that has developed in the eastern Andes, adjacent to the foreland region. Thin-skinned deformation, basement uplifts, and recently, rift inversion, have all been proposed for the same areas. However, several of these proposed models are incompatible with the orogenic shortening that is implied by the Andean crustal roots, as reflected in the present topography. Recently released digital topography of the Andean mountain belt provides two important facts. First, it illustrates the amount and shape of uplift, and thus it gives some hints about the amount and style of shortening. Second, it demonstrates pronounced longitudinal variation along the strike of the chain. This variation, as illustrated in Figure 1, points out the different kinds of processes that took place along strike in distinct segments of the Andes.

The objective of this chapter is to characterize the Andean thrust system using the digital topography, the inferred crustal roots, and the fold-and-thrust belts, which are recognized in a series of representative segments. These observations demonstrate a continuous decrease in orogenic shortening to the north and south of the Central Andes of Bolivia, despite the different and complex structural styles and kinematics. Isacks (1988) explained this variation in shortening, with reference to bending of the Bolivian orocline, and the differential rotation of the fore arc, as documented by paleomagnetic data. Several paleomagnetic studies have reinforced the importance of fore-arc block rotations at both sides of the Bolivian orocline during Cenozoic time (Roperch and Carlier, 1992; Butler et al., 1995; Somoza et al., 1996). However, a combination of oroclinal bending and post-mid-Cretaceous shear-driven rotation by oblique subduction is required to explain the rotations found (Beck, 1998).

The Isacks model was successful in explaining the variation in the amount of shortening at both sides of the Bolivian orocline. Nevertheless, the trend of decreasing shortening to the north and south of the orocline, as shown by Kley et al. (1999), exceeds by far the area of influence denoted by fore-arc rotation. This amount of shortening decreases southward to the triple junction among the Nazca, South America, and Antarctica plates at 46°30′S. South of the triple junction, a collision of an oceanic ridge segment against the subduction zone is associated with a significant increase of deformation, uplift, and shortening in the southern Patagonian Cordillera (Ramos, 1989; Kraemer et al., 1996).

Three segments will be analyzed to address this problem and characterize the Andean thrust system: a segment across the Puna high plateau and the Sub-Andean Ranges (22–23°S), another across the highest Andes of San Juan and Mendoza (32°–33°S), and a segment across the Neuquén Basin between 37° and 39°S.

CRUSTAL ROOTS AND TOPOGRAPHY

There is a remarkable correlation between crustal structure and topography along the entire western margin of South America (Isacks, 1988; Dewey and Lamb, 1992). The topography clearly expresses the width of the plate-boundary zone, where deformation, accommodated through different mechanisms along strike, produced different sets of structures. Gravity studies performed between 30° and 35°S, combined with scarce seismic-refraction data on both sides of the Andes, have demonstrated that wavelength and amplitude of the Bouguer anomalies beneath the Andes are mainly controlled by crustal roots (Introcaso et al., 1992). Seismic-refraction surveys and gravimetric studies between 20° and 26°S have also made it possible to infer the shape and depth of crustal roots (Götze et al., 1994). Recent deep seismic reflection and refraction data reported by the ANCORP working group (1999) shows that crust-mantle boundary has a transitional character, with a width of several hundreds of meters to a few kilometers. However, these results are compatible with a crustal thickness beneath the Altiplano and Puna of less than 70 km at this latitude (Giese et al., 1999; Romanyuk et al., 1999). Farther north at 20°S, broadband seismologic studies indicate a 70- to 74-km-thick crust (Beck et al., 1996).

On the other hand, seismologic studies by Whitman et al. (1992, 1996) have shown that the lithosphere in the Puna is substantially thinned, which explains the higher topography of the Puna in comparison with the Bolivian Altiplano (Allmendinger et al., 1997). This confirms that the present topography of the Puna-Altiplano region required, besides orogenic shortening, thermal uplift, as proposed by Isacks (1988). The region of thermal uplift corresponds to areas of lithospheric thinning, as depicted by Allmendinger et al. (1997).

Figure 2 summarizes the relationships between crustal thickness and topography between 23° and 39°S,

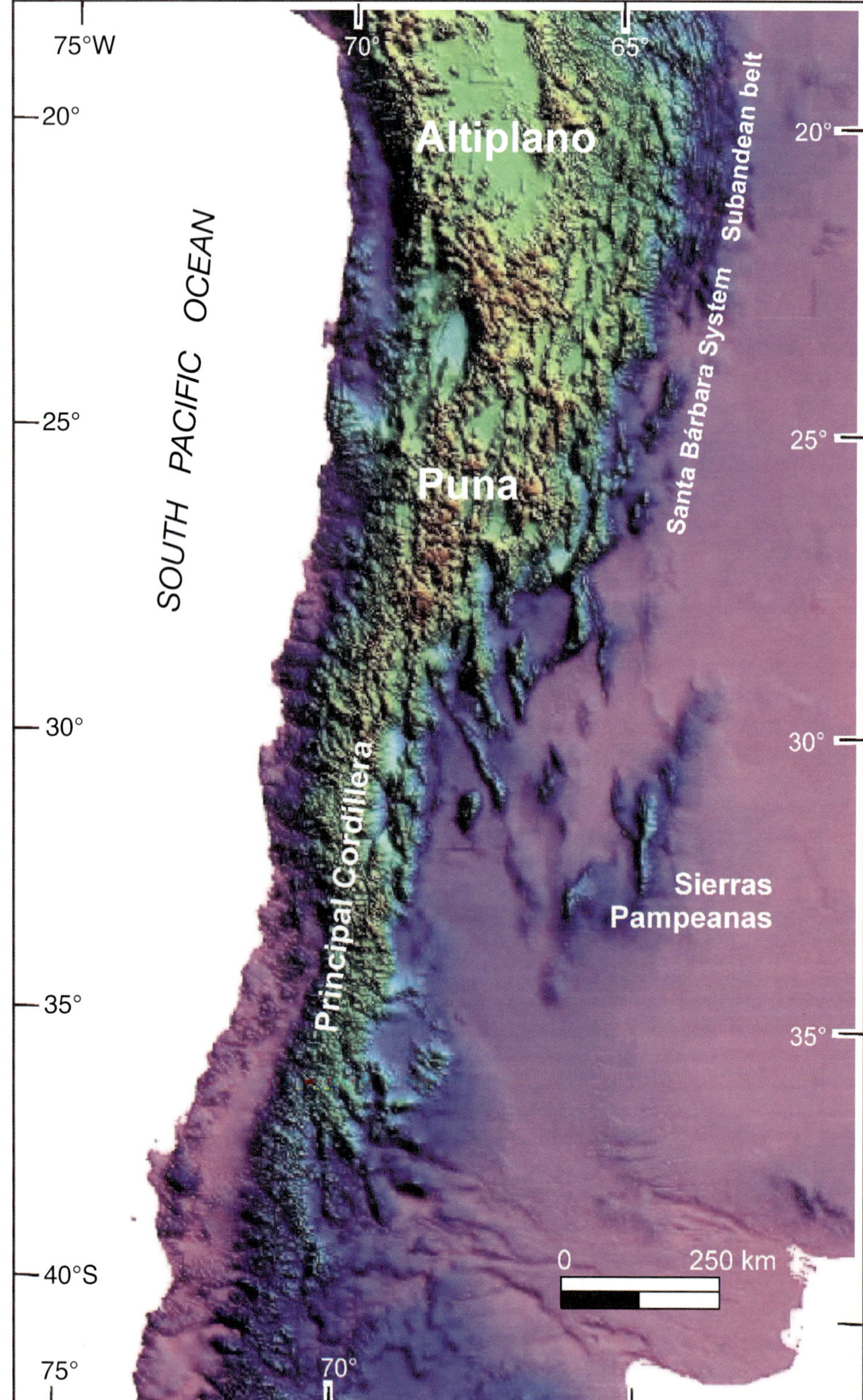

FIGURE 1. Digital topography of the Central Andes, based on U.S.G.S. data. Note the variation along strike and the close correlation of the flat-slab segment with the Sierras Pampeanas between 27°S and 33°S.

based on geophysical data mainly derived from Romanyuk et al. (1999), Götze et al. (1994), Introcaso et al. (1992), Martínez et al. (1997), and Couch et al. (1981). It is evident that the cross-sectional area of the crustal root decreases steadily from north to south, as does the crustal thickness. Based on the premises that magmatic addition and losses by erosion are negligible, as Isacks (1988), Allmendinger et al. (1990), and Introcaso et al. (1992) proposed, the crustal root area makes it possible to compute the orogenic shortening in each segment, assuming a starting crustal thickness. This starting crustal thickness is constrained by the fact that the orogen was covered by marine transgressions prior to the Andean deformation. In order to be below sea level, the maximum thickness of the crust should be less than normal.

Crustal root areas have been used to estimate the amount of shortening along each traverse, as previously proposed by Isacks (1988), Roeder (1988), Introcaso et al. (1992), and Allmendinger et al. (1990). This crustal shortening can be correlated with the amount of shortening across the Andes obtained from analysis of the shallow crust structure in the different fold-and-thrust-belt segments, which is strongly model dependent. This is done first in a northern segment, where more geologic and geophysical information is available. These results are then compared with central and southern segments, where geophysical information is scarcer.

NORTHERN SEGMENT

The northern segment includes the morphostructural provinces of Puna, Eastern Cordillera, and Sub-Andean Ranges at 22–23°S latitudes (Figure 1). The crustal balance of this area, based on seismic-refraction

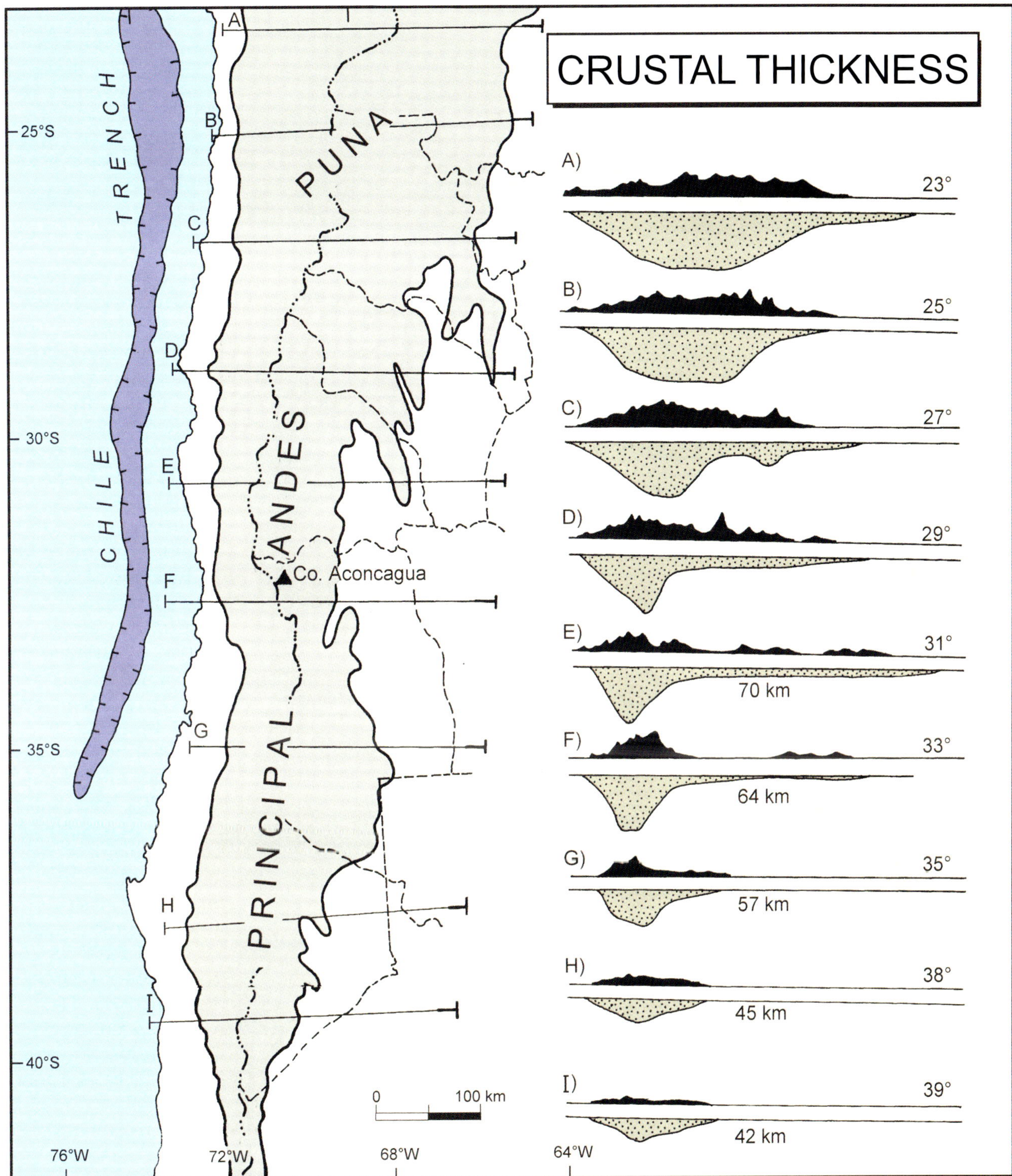

FIGURE 2. Crustal thickness and topography of the Andes between 23° and 40°S. Shaded area corresponds to elevations above the 3000 m.a.s.l. contour level. Crustal sections assume 33 km initial crust, topography has 10 times vertical exaggeration, and stippled pattern shows crustal roots with maximum thickness in kilometers (based, from north to south, on Götze et al., 1994; Introcaso et al., 1992; Martínez et al., 1997, and Couch et al., 1981).

data, indicates a shortening of 320 km, according to Schmitz (1993, 1994). This value correlates with the thin-skinned fold-and-thrust belt of the Sub-Andean Ranges and with the thick-skinned deformation of the Eastern and Western Precordillera and the Puna (Figure 3a).

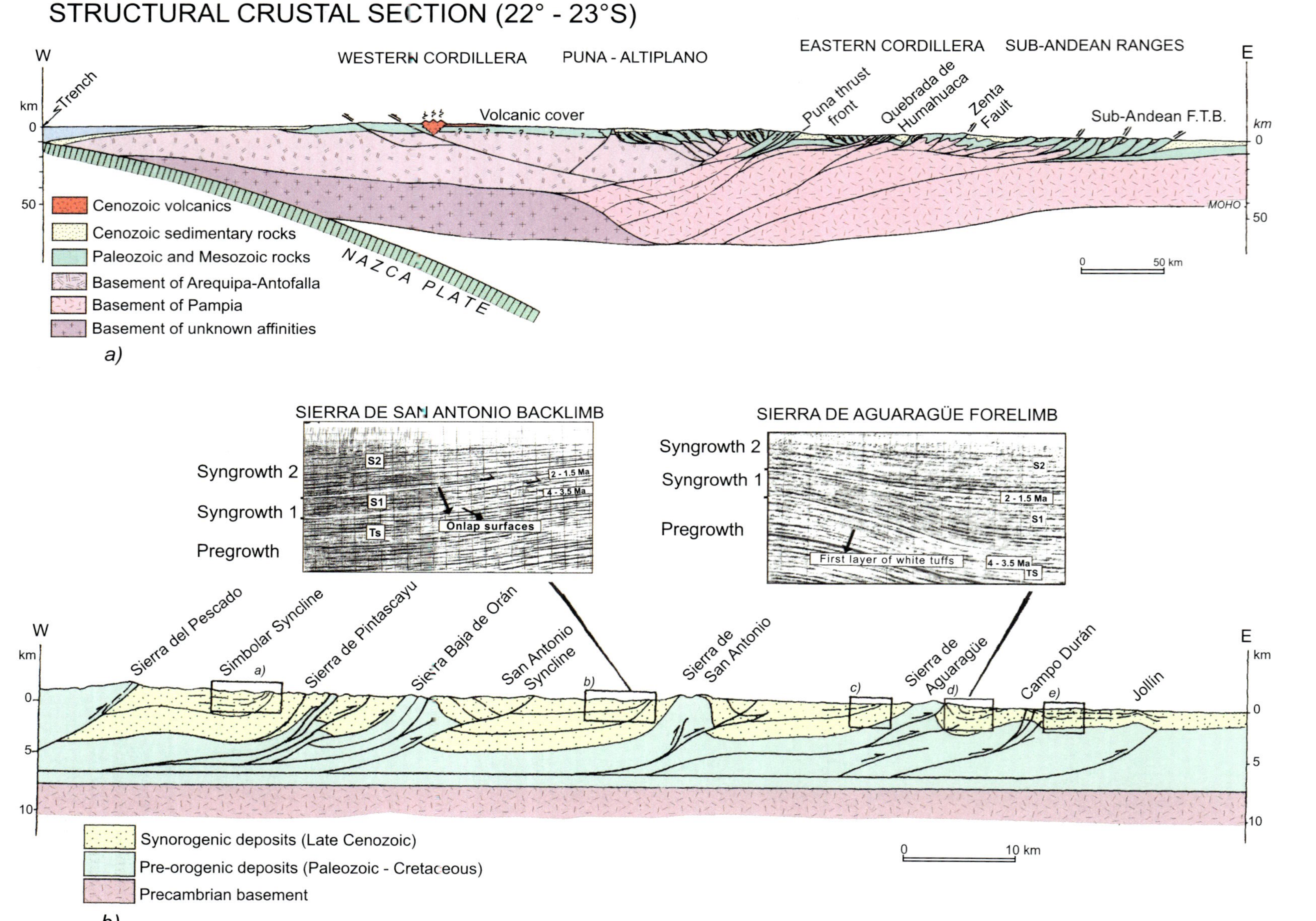

FIGURE 3. Structural crustal sections between 22° and 23°S. (a) Crustal-scale section based on Schmitz (1993), Inter-Andean and Sub-Andean belt from Kley (1996), eastern Cordillera from Heredia et al. (1999), and Puna from Gangui (1998). (b) The Sub-Andean belt section, immediately south of Figure 3a, based on Aramayo Flores (1989), and growth strata stratigraphy after Hernández et al. (1996), Mojica and Zorzín (1996), and Mosquera (1999).

Several authors have studied the Sub-Andean belt, from which most of the seismic information is available (Mingramm et al., 1979; Allmendinger et al., 1983; Roeder, 1988; Aramayo Flores, 1989; Baby et al., 1992, 1993; Dunn et al., 1995; Allmendinger and Zapata, 1996). Most of these authors accept a shortening between 132 and 75 km in a thin-skinned belt. The discrepancies are due mainly to the precise location of the section, because a southward decrease in shortening is evident within the southern Sub-Andean belt, from Santa Cruz de la Sierra (Bolivia, 18°S) up to Orán (northern Argentina, 23°S). Recent evaluation of the structure of this area, based on new seismic analyses, indicates an average thin-skinned shortening of 100 km (55%) for this belt (Giraudo et al., 1999). However, gravity data combined with seismic information require that the western part of the belt (Figure 3a) must include basement participation (Kley et al., 1996). This western part, named the Inter-Andean belt, has, together with the Sub-Andean belt, a shortening of 140–150 km.

Most authors have assumed a foreland breaking sequence of thrust-fault development in the Sub-Andean belt thrusts (Kley et al., 1996; Giraudo et al., 1999), but synorogenic sequences, mainly derived from the study of the growth strata, indicate a more complex development (Hernández et al., 1996). There is a general pattern of a foreland breaking thrust sequence that uplifted the Eastern Cordillera between 17 and 12 Ma, and, in a piggyback order, the Sub-Andean belt after 8.5 Ma (Hernández et al., 1996). However, as Figure 3b shows, following an initial foreland breaking sequence, most of the Sub-Andean belt displays evidence of simultaneous displacements. Mosquera (1999), using the lowest level of white tuffs as a marker-bed for correlation dated at 4–3.5 Ma, has identified at least two younger syngrowth sequences. These syngrowth sequences are associated with pulses of out-of-sequence thrusting and were dated at 2–1.5 Ma (S_1) and 1.5–0.25 (S_2) in the Simbolar syncline (section a in Figure 3b). Correlative syngrowth sequences were recognized in the San Antonio Syncline (section b), on both limbs of Sierra de Aguaragüe (sections c and d), and in Campo Durán Anticline backlimb (section e) (Mosquera, 1999). Even farther east, at Jollín, a new, active structure was developing within this time interval (Mojica and Zorzín, 1996).

These facts demonstrate a partitioning of the displacement across the thrust belt similar to those described by Jordan et al. (1993) in the Argentine Precordillera and Butler et al. (1999) in the Central Sicilian Thrust Belt. There are periods when all thrusts are simultaneously active in the Sub-Andean belt, thereby defining a mobile area that corresponds to what DeCelles and Gilest (1996) called a wedge top of the foreland system. This area shifts to the foreland as the foreland basin is cannibalized at the thrust front.

Another constraint from the synorogenic stratigraphy is the shortening rate of the Sub-Andean belt. Magnetostratigraphic data, which are only available in the southern part of the belt (Hernández et al. 1996), were used to determine a shortening rate of between 7 and 9 mm/yr— higher than the rates obtained by Butler et al. (1999) in Central Sicilia. The rate of shortening in the Sub-Andean belt is on the same order as the 7.65 mm/yr rate obtained by Cegarra and Ramos (1996) farther south, in the highest Andes, but is less than the peak rates documented for the Precordillera by Jordan et al. (1993).

Heredia et al. (1999) have recently studied the structure of the Eastern Cordillera thrust belt at these latitudes. These authors found evidence of inversion of Cretaceous normal faults at both sides of Quebrada de Humahuaca. The palinspastic restoration of their structural cross sections indicates a 55% shortening, which is responsible for a total minimum shortening for the Eastern Cordillera of 60 km. Heredia et al. (1999) assumed that the time of deformation for the Eastern Cordillera was between 17 and 11.4 Ma, based on the Marrett et al. (1994) data. This time interval gives a shortening rate of 9.3 mm/yr. It is necessary to remark that this rate could be lower, because there is evidence of neotectonic reactivation affecting Pliocene and Quaternary deposits in the central part of Eastern Cordillera near Tilcara. This confirms that at least part of the structures have been reactivated in an out-of-sequence mode, like the Tilcara thrust described by Salfity et al. (1984).

No regional structural cross sections are available farther west, in the Puna high plateau, but local studies by Gangui (1998) have documented a complex structure. Interpretation of a few available seismic lines shows a structure partially controlled by tectonic inversion of Cretaceous normal faults, as well as by new thrusts developed within the Puna Precambrian basement. Because most of western Puna is covered by Cenozoic volcanics, the data presented in Figure 3a, which are based on one of Gangui's (1998) local cross sections, indicate the kind of structure that may exist beneath the Puna. The thick-skinned deformation in the Puna was facilitated by the thermal structure (Springer and Förster, 1998). High heat flow (80–100 mW/m^2), favored the development of several potential ductile-brittle transitions within the middle and upper crust that might be the sites of the décollement levels. The minimum shortening in eastern Puna, although not as well constrained as in the Sub-Andean belt, is about 60 km and was bracketed between 26 and 18 Ma (Gangui, 1998); thus the shortening rate is about 7.5 mm/yr. Based on the surficial structure of the Cenozoic deposits, a conservative estimate of total shortening in the Puna is more than 100 km. The western sector of the Puna in the Chilean side has been deformed since about 45 Ma (late

Eocene), as indicated by recent dating of the synorogenic deposits of the Salar de Atacama (older than 43.8 ± 0.5 Ma, according to Mpodozis et al., 1999). Kraemer et al.'s (1999) recent studies demonstrated that the synorogenic deposits are older than 38 Ma west of Antofalla, which indicates the beginning of the foreland basin subsidence in the Argentine side of the Puna.

In the section analyzed, the shortening is 140–150 km in the Inter-Andean and Sub-Andean belt, more than 60 km in the Eastern Cordillera, and about 100 km in the Puna. The total shortening is 300–310 km at these latitudes, without taking into account the shortening of the Western Cordillera and adjacent areas to the west. These values are similar to the crustal shortening obtained by Schmitz (1994) and on the order of the estimates by Kley (1999). This shortening occurred in the last 45 million years, which gives an average shortening rate of 6.7–6.9mm/yr, but probably with periods of high activity during late Miocene and Pliocene–Quaternary times.

CENTRAL SEGMENT

The central segment is located between 32° and 33°S. This area includes the highest Andes of the states of San Juan and Mendoza, with elevations of nearly 7000 m above sea level in Mount Aconcagua. This region comprises the Principal and Frontal Cordilleras and the Precordillera (see Figure 4). The Andes at these latitudes are much narrower but higher compared with their northern segment (see Figure 1).

The changes along strike within this central segment are shown by the crustal cross sections of Figure 5. This figure illustrates two different tectonic styles related to local development of early extensional structures produced during Late Triassic and Early Jurassic rifting in the Principal Cordillera (Ramos et al., 1996a, b).

First-order Nazca plate segmentation, with the development of a flat subduction segment between 27° and 33°30'S (Jordan et al., 1983a), has obliterated other significant differences in structural styles along strike of the Principal and Frontal Cordilleras (Figure 6). Important changes in structural styles along the Principal Andes, north and south of 33°S, are controlled by the local inception of early Mesozoic rifting. The La Ramada fold-and-thrust belt in the north is the result of tectonic inversion of Permian–Triassic Choiyoi normal faulting and Late Triassic rift extension (Alvarez, 1996; Alvarez and Ramos, 1999; Rodríguez Fernández et al., 1999) (Figure 7). The main compressional deformation during Andean times has produced a complete inversion of rift structures, with a few exceptions. The normal fault on the western side of Cordón del Espinacito still shows normal throw, in spite of kinematic indicators that demonstrate top-to-the-east displacement (Ragona, 1993), because in this case the reverse displacement is less than normal displacement.

The Aconcagua fold-and-thrust belt in the Principal Cordillera at 33°S is dominated by thin-skinned deformation (Cegarra and Ramos, 1996) that was facilitated by the presence of thick gypsum of the Jurassic Auquilco Formation, which localized the detachment of the post-Jurassic cover (Figure 8). Triassic rifting at these latitudes occurred east of the foothills of the Frontal Cordillera within the Cuyo Basin (Dellapé and Hegedus, 1993). Because the present orogenic front is located in that basin, inversion is still incipient.

Based on the present topography, gravity data, and limited seismic refraction lines from both slopes of the Andes, the Andean crustal structure contains deep and narrow crustal roots varying from 70 to 64 km in depth (Introcaso et al., 1992). Based on the crustal structure in the La Ramada segment, the shortening is about 150 km, while in the Aconcagua segment it is only 130 km (Introcaso et al., 1992).

In the northern section, at 32°S, the shortening is less than 30 km in the La Ramada fold-and-thrust belt of Principal Cordillera (Cristallini, 1996); it is 85 km in the thin-skinned Precordillera fold-and-thrust belt at this latitude (Ramos et al., 1997); it is 10–20 km in the Sierras Pampeanas; but it is poorly constrained in the foothills of the Frontal Cordillera. However, based on surface geology and a detachment level at about 15 to 20 km farther north in western Precordillera, as proposed by Allmendinger et al. (1990), a minimum shortening of 20 km in the Frontal Cordillera seems reliable. Andean shortening on the Chilean side is minimal at these latitudes. Thus, crustal shortening across the 32°S section is between 135 and 155 km, which is comparable to the 150 km estimated by Introcaso et al. (1992) based on the crustal roots.

In the 33°S section, the largest shortening occurs in the Principal Cordillera, with 55 to 60 km, including the Chilean side, at the Aconcagua fold-and-thrust belt (Kozlowski et al., 1993); the thick-skinned Frontal Cordillera records about 18 km (Ramos et al., 1996a, b); and the remaining shortening has taken place either in the Precordillera (less than 30 km at these latitudes) or by inversion in the Cuyo Basin, and is less than 6.6 km in the Sierras Pampeanas of San Luis (Costa, 1992). Thus, total Andean shortening across the 33°S section is on the order of 110–115 km, slightly less than the 135 km estimated by Introcaso et al. (1992).

In both sections, the fold-and-thrust belts have developed in a foreland-breaking sequence, although out-of-sequence thrusts have been identified in both sections. The wedge top, as defined by DeCelles and Gilest (1996), is very narrow (less than 10 km) at these latitudes. This is substantially different from the development of the large wedge top observed in the Sub-Andean belt (Figure 3b).

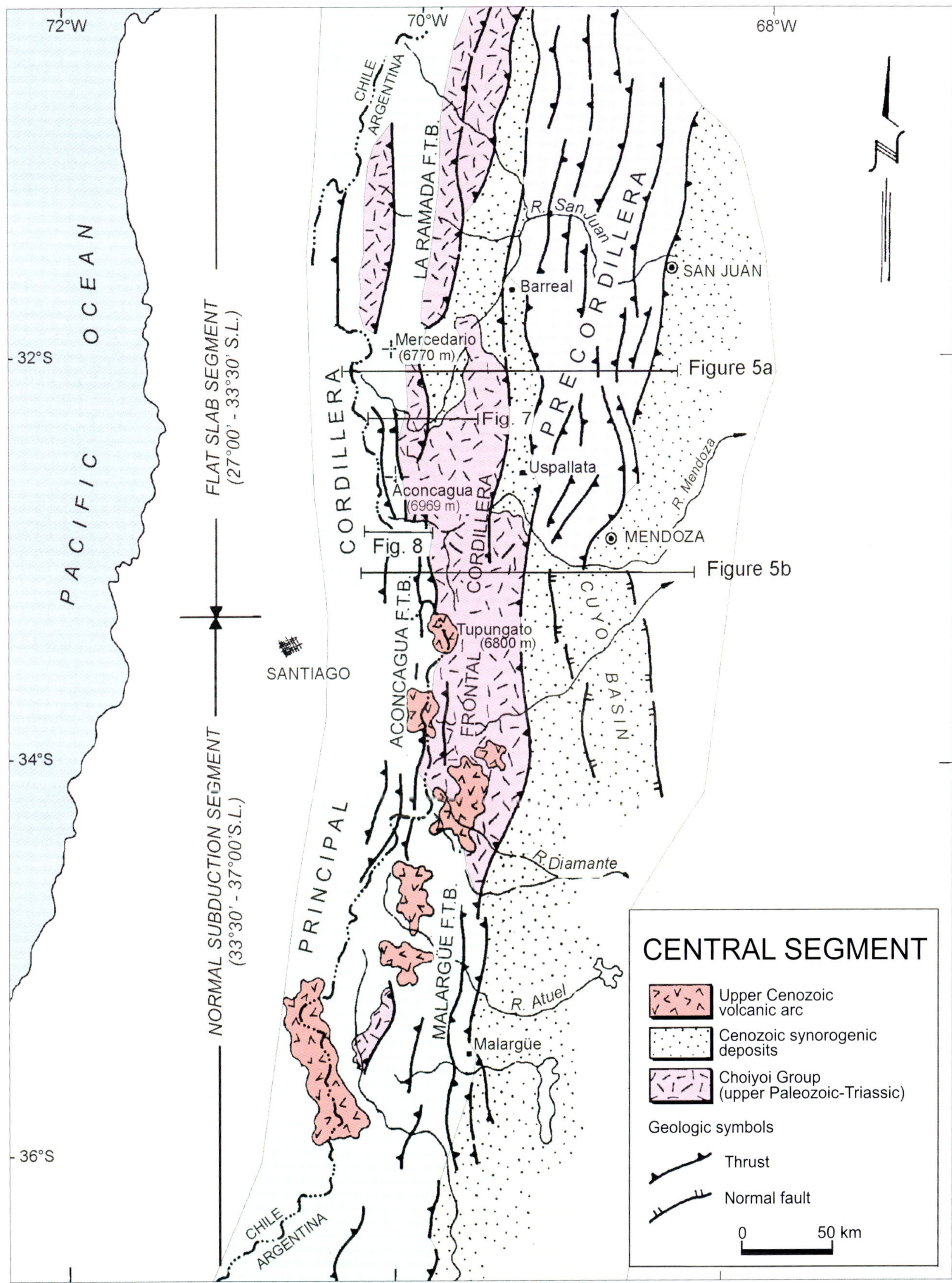

FIGURE 4. Regional structural units of the southern central Andes, with locations of the different crustal and structural cross sections of the central segment. In the Principal Cordillera the La Ramada, Aconcagua, and Malargüe belts are indicated (Based on Ramos et al., 1996b).

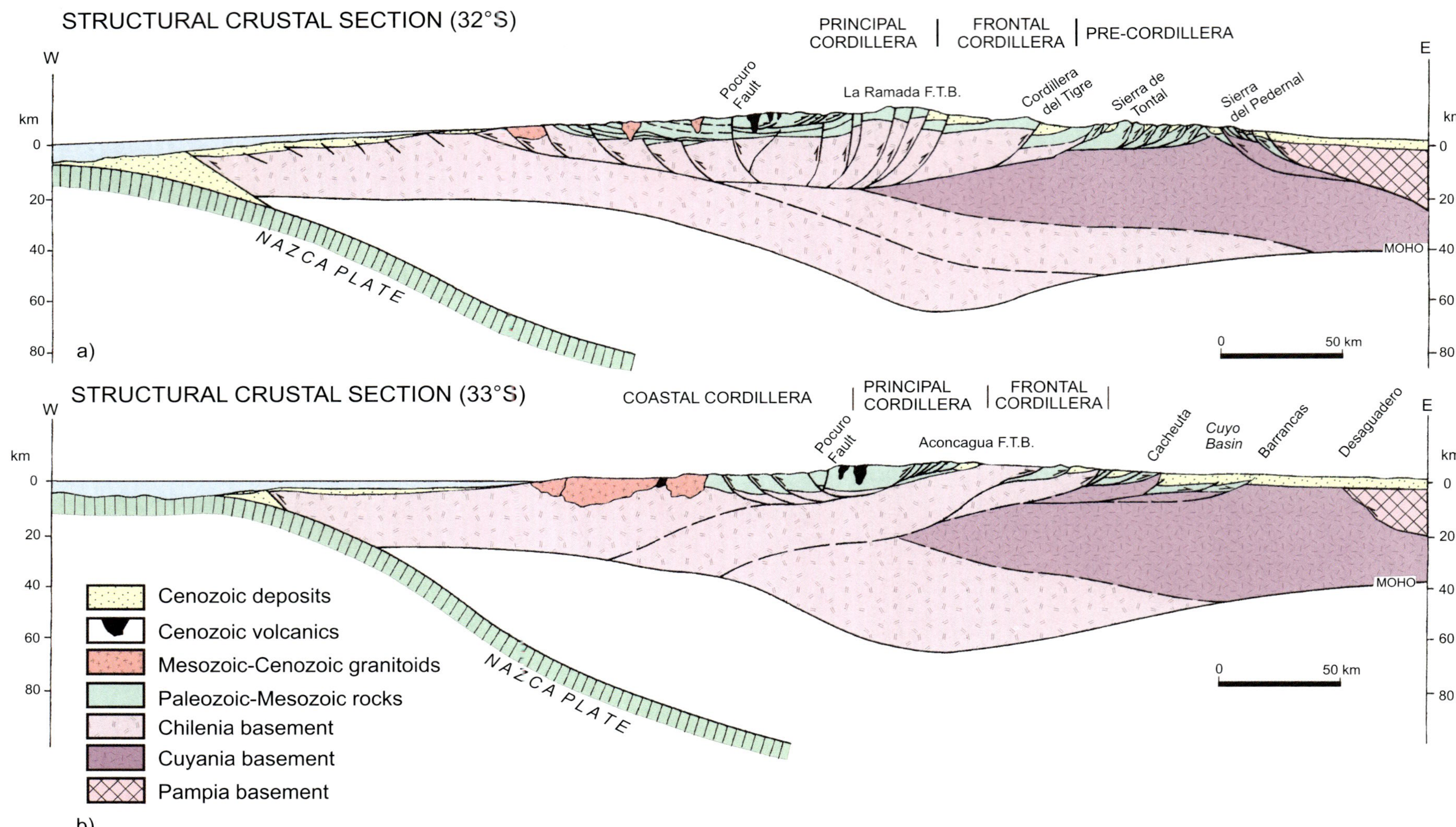

Figure 5. Structural crustal section of the central segment. The boundaries of the basement terranes are based on Ramos (1994). (a) Crustal section of the La Ramada belt at 32°S. Note that the Principal Cordillera is controlled by tectonic inversion of Triassic normal faults (according to Cristallini and Ramos, 1997). (b) Crustal section of the Aconcagua belt at 33°S. Note that the Principal Cordillera is dominated by thin-skinned deformation (according to Introcaso and Ramos, 1992).

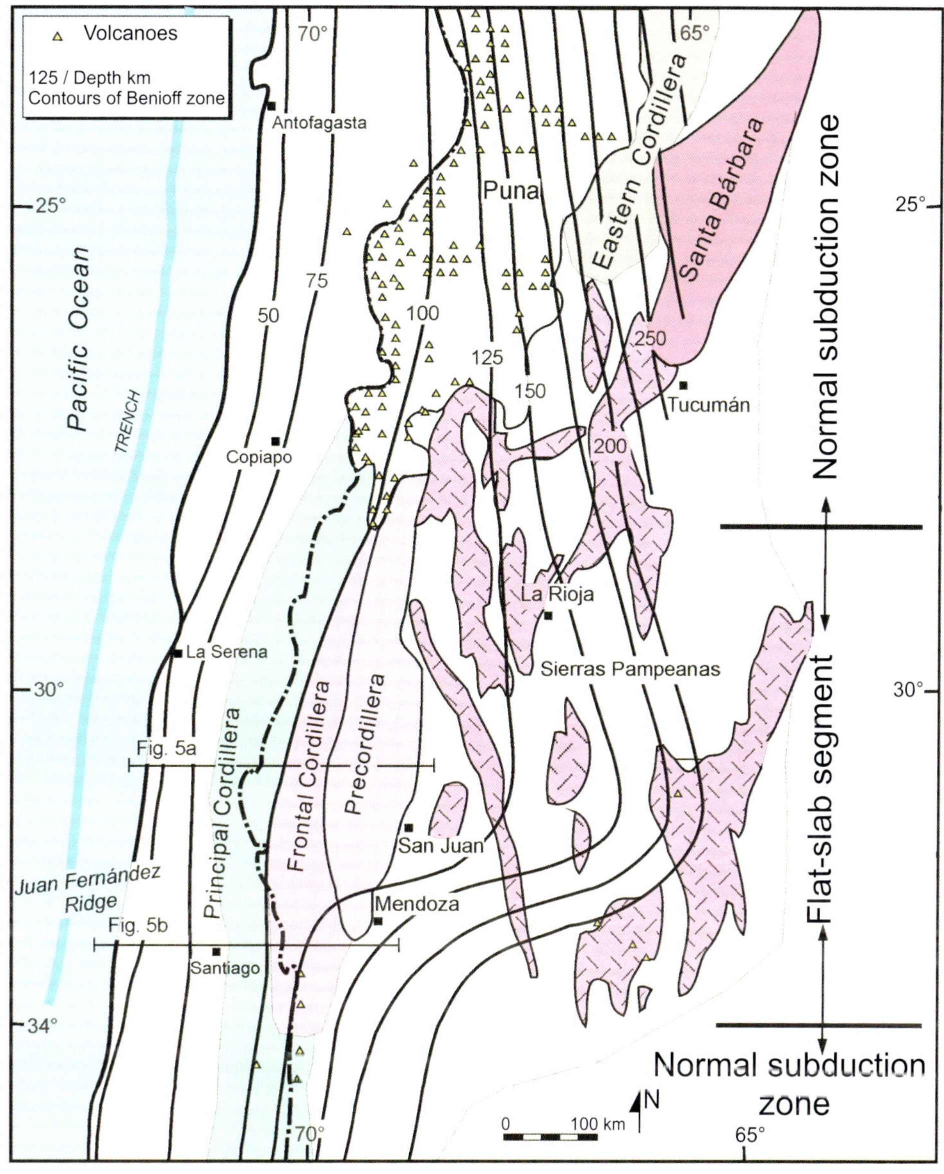

FIGURE 6. First-order Nazca plate segmentation related to the Wadati-Benioff zone geometry. The crustal sections of the central segment are in the middle of the flat-slab segment (Figure 5a) and in its southern boundary (Figure 5b). Based on Isacks (1988) and Cahill and Isacks (1992).

The time constraints along both sections indicate that the present structures developed in the last 20 million years (Ramos, 1996). Based on that estimate, the shortening rates at these latitudes vary, from 7.3–7.7 mm/yr along the La Ramada northern section, to 5.5–5.7 mm/yr in the Aconcagua southern section.

SOUTHERN SEGMENT

The southern segment crosses the Principal Cordillera and the Neuquén embayment, thereby encompassing the Agrio fold-and-thrust belt. The interpretation of the crustal structure shown in Figure 9 is less constrained than that of the other two segments and comes mainly from the work of Couch et al. (1981), Pacino (1993), and Martínez et al. (1997). It is based on gravity data, and features a small crustal root, less than 42–43 km deep (Figure 9). This fact poses important constraints on the structural style of the Agrio fold-and-thrust belt.

The structure of the Neuquén embayment has received considerable attention in recent years because of the basin's hydrocarbon potential. Several papers describing the Agrio fold-and-thrust belt have presented contrasting models and interpretations; some involve a thin-skinned style, based mainly on detachment folds controlled by the Jurassic Auquilco gypsum (Vásquez, 1979; Allen et al., 1984; Ploszkiewicz, 1987; Viñes, 1989; Eisner, 1991). In contrast, other models display a conspicuous basement involvement (Groeber, 1929; Kozlowski, 1991; Kozlowski et al., 1993, 1996; Uliana and Legarreta, 1993; Legarreta and Uliana, 1996).

Based on the surface geology and the basement control of the structures, there is no doubt that the inner sector of the Agrio fold-and-thrust belt has been produced by tectonic inversion of pre-Jurassic normal faults (Ramos, 1998). The outer sector of the belt, on the other hand, has a localized, thin-skinned deformation. Displacement transfer between the two sectors has been explained by different mechanisms. Deep basement faults with considerable displacement were recognized by Kozlowski et al. (1996, 1993); a combination of basement thrusts with surface detachment folds was proposed by Chaveau et al. (1996); and tectonic inversion was suggested by Booth and Coward (1996).

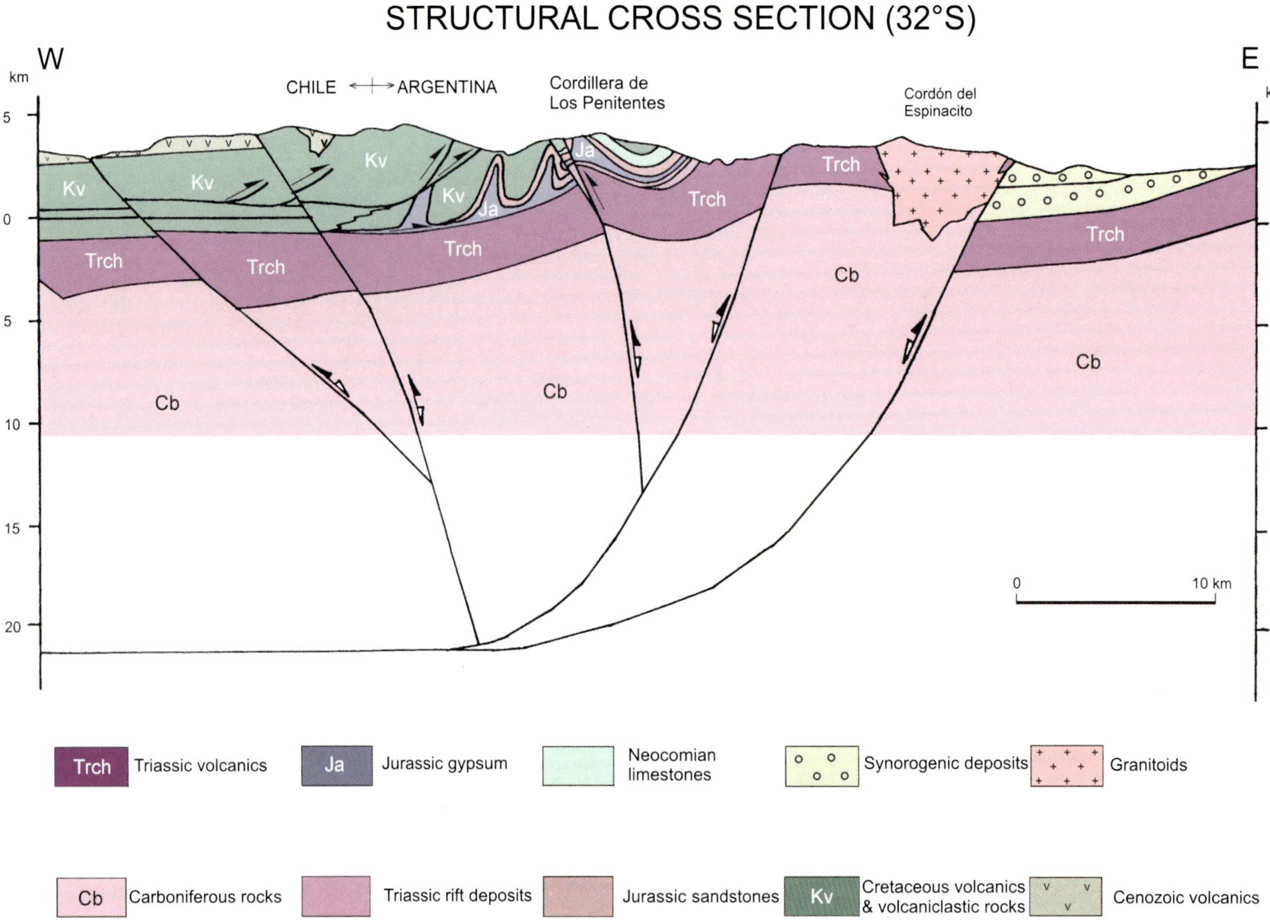

STRUCTURAL CROSS SECTION (32°S)

W E

CHILE ARGENTINA

Cordillera de Los Penitentes

Cordón del Espinacito

km

Kv Kv Kv Kv Ja Ja Trch Trch Trch Trch Trch Cb Cb Cb Cb

0 10 km

Trch Triassic volcanics

Ja Jurassic gypsum

Neocomian limestones

Synorogenic deposits

Granitoids

Cb Carboniferous rocks

Triassic rift deposits

Jurassic sandstones

Kv Cretaceous volcanics & volcaniclastic rocks

Cenozoic volcanics

FIGURE 7. Structural cross-section of the Principal Cordillera at the La Ramada belt (based on Cristallini, 1996). Note that first-order structures are controlled by Choiyoi Group geometry, reactivated during Late Triassic rifting.

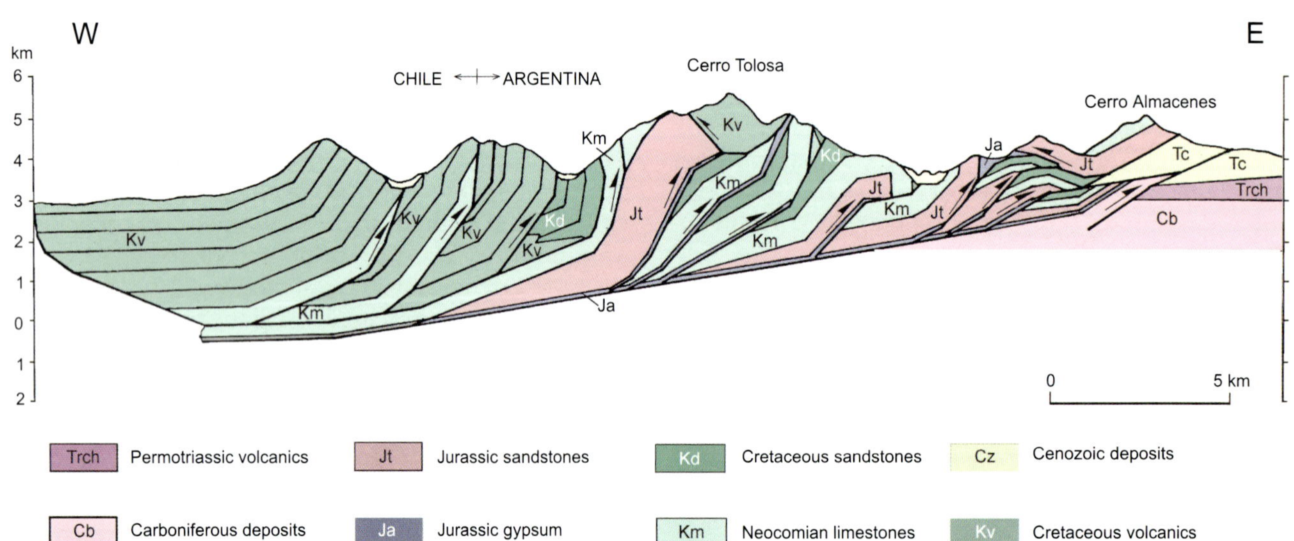

FIGURE 8. The thin-skinned Aconcagua belt across the principal Andes at 33°S (based on Cegarra and Ramos, 1996).

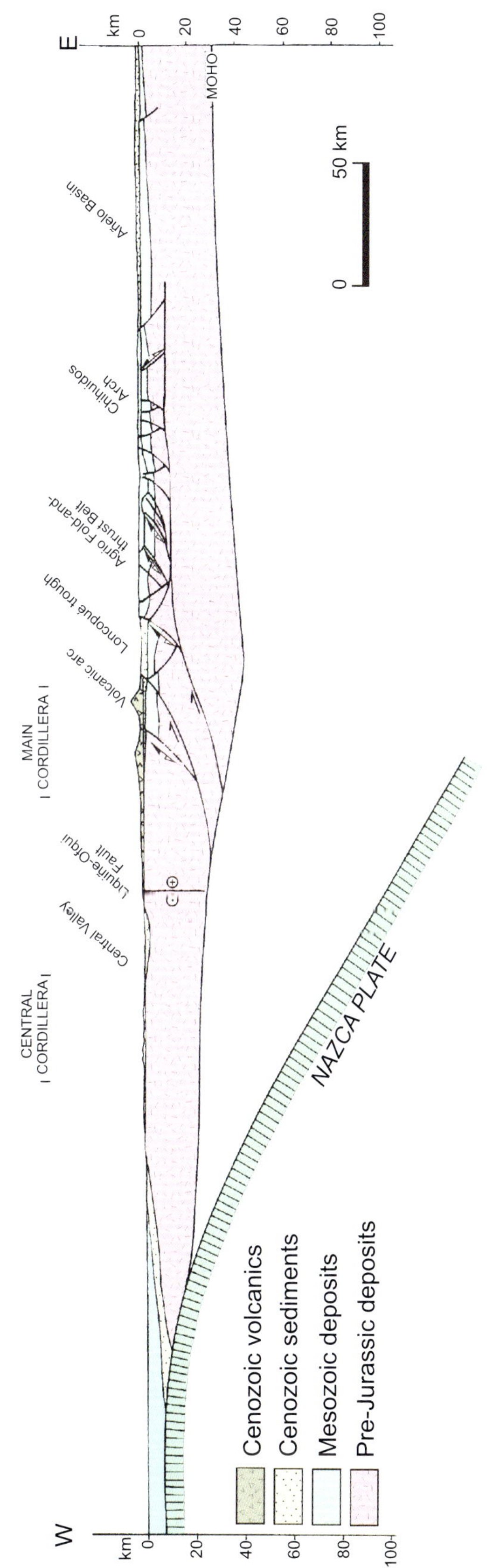

FIGURE 9. Structural crustal section of the southern segment. The upper crustal structures are based on Zapata et al. (1999). Crustal roots are from Couch et al. (1981) and Martínez et al. (1997).

The crustal structure imposes important constraints on previous models because it implies that the crustal shortening cannot exceed 44 km at 37°S and 20 km at the 39°S (Martínez et al., 1997). These crustal models assume that during the Pacific marine transgression that covered the entire orogen in the Early Cretaceous with shallow-water deposition, the crustal thickness should be slightly less than normal. Several of the proposed models, such as those of Vásquez (1979) and Allen et al. (1984), can be ruled out because of excessive shortening. The structural cross sections at 37° and 39°S (Zapata et al., 1999) have a strong basement control, as illustrated in Figure 10. The minimum shortening at 37°S is 36 km, and it is 19 km at 39°S, in agreement with what is inferred from the crustal structure.

Both structural sections of Figure 10 show a simple basement structure dominated by two or three important basement uplifts, such as the Cordillera del Viento, Volcán Tromen, and Sierra de Reyes blocks. The Cordillera del Viento and adjacent blocks plunge to the south, where they are covered by a Mesozoic sequence. A thicker Mesozoic sequence favored the development of the detachment folds and faults in the upper part of the section that characterize the surface deformation of the Agrio fold-and-thrust belt.

A common feature of both sections is the Loncopué Trough, a possible Jurassic depression that was reactivated during Cenozoic times. The Cenozoic Cura Mallín Basin, a volcaniclastic trough characterized by a northwest-trending half-graben system, was developed in the late Oligocene and early Miocene along the Chilean-Argentine border (Vergara et al., 1997). The extension and formation of this basin coincide with trenchward shifting of the magmatic arc, probably produced as a consequence of the steepening of the subduction zone at the end of the Eocene or the early Oligocene (Ramos and Folguera, 1999). Subsequent extensional reactivation at the end of the Miocene produced the present structure of the Loncopué Trough, which is linked to an important volcanic field having numerous monogenic basaltic cones. Late Cenozoic strike-slip displacements were responsible for the localized extension and compression observed in the Copahue region, along the volcanic axis. Partitioning of the stress through strike-slip displacements between the fore arc and the back arc explains the presently active tectonism as being a product of the oblique convergence. Diraison et al. (1998) have described a similar setting farther south that has the opposite sense of displacement controlled by a different basement fabric.

The timing of the main deformation in the Neuquén embayment is poorly known. The pre-Liassic half grabens were partially inverted during Cretaceous compressional deformation. The structure was mildly reactivated during the early Miocene, with final compressional deformation taking place in the middle late

STRUCTURAL CROSS SECTION (37°S)

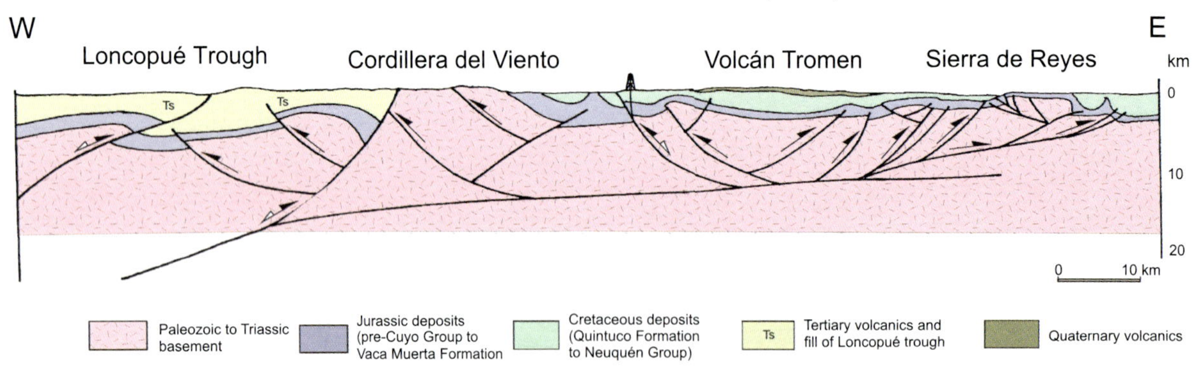

STRUCTURAL CROSS SECTION (39°S)

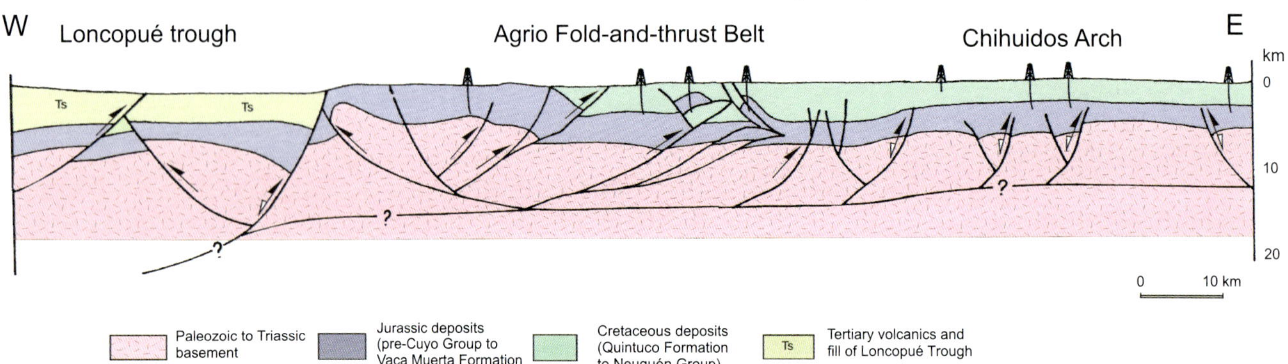

FIGURE 10. Structural cross sections of the Neuquén embayment, based on Zapata et al. (1999). (a) Structural cross section at 37°S, showing the basement uplift of Cordillera del Viento and the Sierra de Reyes, as controlled by tectonic inversion of previous normal faults. (b) Structural cross section at 39°S showing the tectonic inversion of normal faults in the inner sector of the Agrio belt and in the Chihuidos Arch.

Miocene, as indicated by the synorogenic sediments and associated volcaniclastic deposits. The Cenozoic deformation was a combination of extension, mild compression, and strike-slip displacements, as established by Lavenu and Cembrano (1999), and was very different from the active late Cenozoic deformation of previous segments. The average shortening rate has been less than 1 mm/yr since the early Miocene.

DISCUSSION

The differences in the thrusting among the three segments of this part of the Andes are strongly controlled by a series of factors related to the nature and physical state of the continental crust. The most important of these factors are the thermal gradient in the crust, the lithologic nature and structure of the basement, the former geologic history of the crust, and the previous existence of a sedimentary wedge. Other important factors are related to the kinematics and the properties of the subducting oceanic slab.

Basement Influence

The nature of the Andean basement is of primary importance in controlling compressional deformation. An example of this control can be seen in the northern segment. The old and strong Amazonia cratonic terrane underlies the Sub-Andean belt. Its rigid nature favored the detachment of the sedimentary cover from the basement, resulting in greater than 100-km shortening in the thin-skinned fold-thrust belt. On the other hand, farther west, the suture between Amazonia and the composite Arequipa-Antofalla terrane coincides with important changes in structural style (Figure 3). This composite basement terrane is clearly depicted in gravimetric maps (Götze et al., 1994). A Cretaceous graben system that developed in the hanging wall of the suture has controlled the style and structure of some sectors of the Eastern Cordillera and the adjacent Puna (Salfity et al., 1993; Kley and Monaldi, 1998; Cristallini et al., 1997).

The central segment shows similar controls that were exerted, by the nature of the basement, on the deformation. The thin-skinned Precordillera thrust belt (von Gosen, 1992) formed in the thick-layered strata of the early Paleozoic passive margin that were deposited on the ancient, highly metamorphosed and therefore presumably strong basement of the Cuyania terrane (Ramos et al., 1996a, b). In contrast, the adjacent Pampia and Chilenia terranes (see Figure 11) were rheologically weak and had no passive margin sedimentary wedge. Triassic rifting, which subsequently developed in the hanging wall of the suture, induced changes in the style of compressional deformation even within the Precordillera fold-and-thrust belt. Sectors of thin-skinned deformation alternate with others having inversion tectonics, as shown in the sections of Figure 5. The composite and relatively young nature of the Pampia terrane favored the basement thrusts of Sierras Pampeanas, as seen in the Sierras de Córdoba and farther west (Cristallini et al., 2004).

The southern segment developed in young, thin continental crust, which probably was accreted to Gondwana during Paleozoic times. The rheologically weak basement and the segmented nature of the substratum resulting from the Mesozoic rifting favored the basement's involvement in the deformation, as depicted in Figures 9 and 10.

Oceanic Slab Influence

The continuous trend of decreasing crustal roots observed at both sides of the Bolivian orocline can be only partially explained by fore-arc rotation and differences in orogenic shortening of the foreland fold-and-thrust belts, as proposed by Isacks (1988). The decrease in shortening continues as far south as 46°30′S, almost 3000 km away, in regions not affected by fore-arc rotation.

Jordan et al. (1983a, b) described the control exerted by the geometry of the oceanic Nazca plate beneath the Andes. The strong coupling between the Nazca and South America plates in the Pampean subhorizontal subduction segment controls the development of the Sierras Pampeanas in the adjacent broken foreland.

Kley et al. (1999) recently questioned the relationship between flat-slab segments and broken foreland and basement thrusts. A contrast exists between the Peruvian and the Pampean segments, despite their similar subduction geometry: In the Pampean segment there are well-developed basement thrust blocks in the foreland that are similar to the Colorado-Wyoming Rockies (Jordan and Allmendinger, 1986). As shown by Mathalone and Montoya (1995), although the basement in the Peruvian segment is involved in the deformation, there are no basement thrust blocks like those in the Sierras Pampeanas. Kley et al. (1999) took this fact as evidence that the existence of old basement highs, without significant sedimentary cover, was the main control of these basement thrusts. However, as Jordan et al. (1983b) and Pilger (1984) pointed out, the shallowing of the subduction zone occurred 15 million years ago in the Pampean segment and much more recently (approximately 5 million years ago) in the Peruvian segment). Therefore, the Peruvian segment has only an incipiently developed broken foreland. On the other hand, another flat-slab segment, the one developed in the Bucaramanga segment (Ramos, 1999) in the Northern Andes (north of 4°30′N) and described by Pennington (1981), is closely related to the basement uplift by tectonic inversion of the Mérida Andes that was described by Colletta et al. (1997).

The transition between the tectonic inversion of the Santa Bárbara fold-thrust belt and the Sierras Pampeanas basement thrusts is imperceptible (Figure 1); there is a smooth transition from a normal-dipping to a flat-slab subduction zone between 26° and 28°S.

However, in spite of the importance of the geometry of the Nazca plate, there are no significant variations, either in the crustal roots or in the amount of upper-plate shortening, immediately to the north and south of the flat-slab segment. Variations in structural styles in this area are mainly controlled by basement rheology and previous tectonic history (Allmendinger et al., 1983; Ramos et al., 1996a, b; Kley et al., 1999). To explain the continuous trend of decreasing shortening at both sides of the Bolivian orocline, far from the rotated fore-arc areas, it is necessary to examine other characteristics of the oceanic floor.

Golovchencko et al. (1981) present a magnetic lineation map of the adjacent oceanic floor, west of South America, that shows that the oldest and coldest oceanic crust reaching the trench is the early Eocene crust that is being subducted at both sides of the Bolivian orocline, along the central segment of the Central Andes. In this segment the Andes are widest, the crustal roots are largest, and the orogenic shortening is greatest. The present thrust front extends as much as 800 km away from the trench.

On the other hand, the crust being subducted south of the Gulf of Guayaquil (3°S latitude) is as young as early Oligocene, and the crustal shortening and crustal roots are smaller than in the Bolivian segment (Cabassi and Introcaso, 1999). The difference is more striking in the Southern Andes. The age of oceanic crust that is being subducted there varies from early Eocene in the north at the Bolivian segment to Quaternary at the triple junction of the Nazca, South America, and

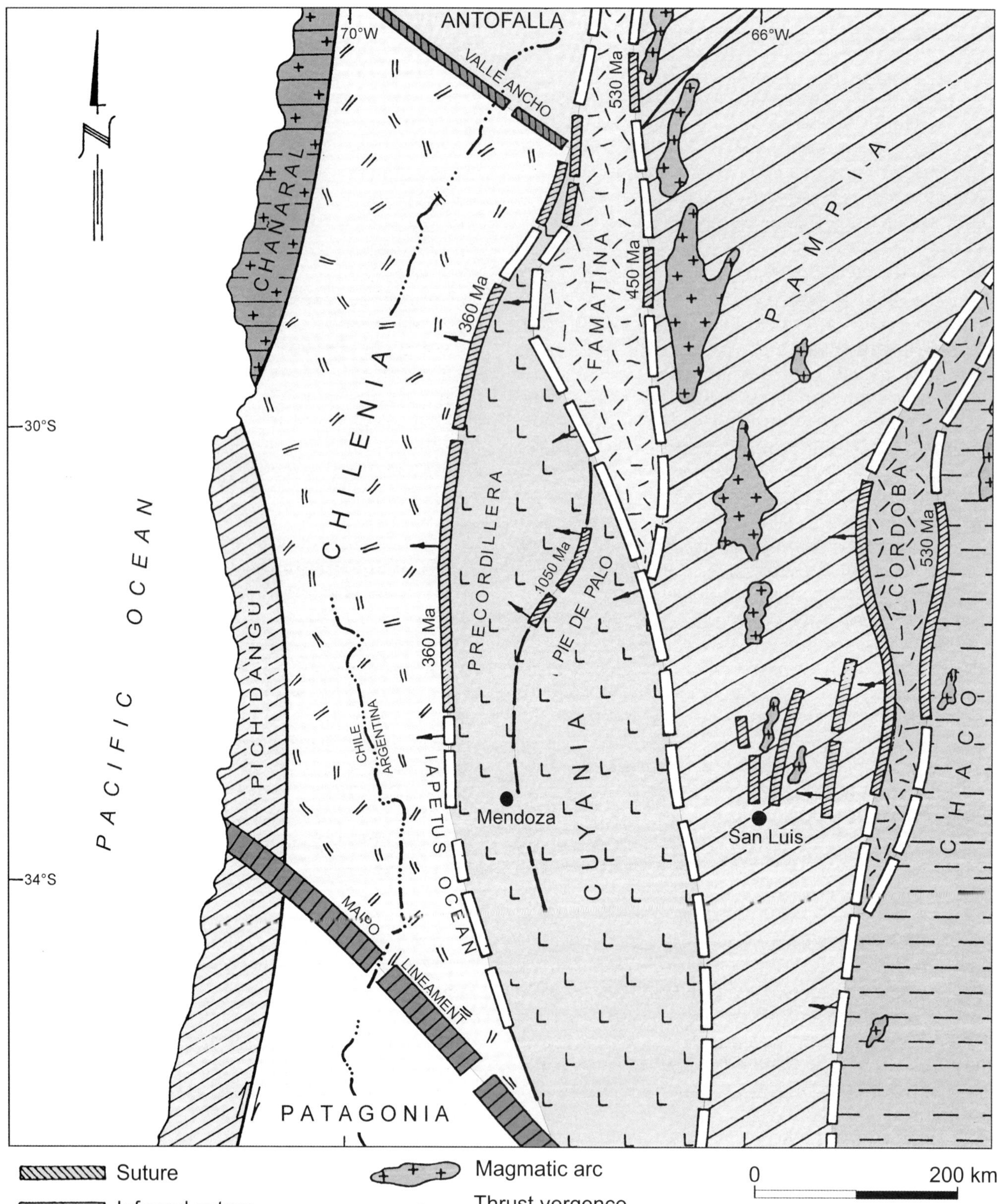

FIGURE 11. Terrane map of the central segment of the Andes, indicating the sutures and inferred sutures, magmatic arcs, and basement thrust vergence; based on Ramos (1996).

Antarctica plates, where the Chilean spreading ridge collides with the trench (Tebbens et al., 1997). This age decrease occurs in discrete segments, which show smaller crustal roots and orogenic shortening that matches the trend of decreasing age of oceanic-crust subduction.

Crustal Roots and Shortening

The amount of crustal shortening inferred from geophysical data, on the basis of crustal thickness derived from seismic-refraction and gravity data and other methods, generally matches the orogenic shortening derived form structural cross sections. The few exceptions found along the Andes (Kley and Monaldi, 1998) may be related either to lack of information, which locally is the result of the extensive cover by the late Cenozoic volcanic arc, or to the mass transfer by strike-slip faults to adjacent areas. The fit is adequate in those areas where there is good surface information.

On the other hand, the crustal shortening derived from the topography, which is used to constrain the interpretation of the crustal roots, gives an independent way to control the structural style of the surface deformation. Several previously proposed models seem to be inadequate when compared with the available information on size of the crustal roots. In some areas, such as the Puna or sectors of the Eastern Cordillera, models that advocate high-angle thrusting do not produce enough shortening to account for the crustal roots.

In the southern segment described here, some authors have proposed thin-skinned structural geometries with large amounts of shortening (Vásquez, 1979; Allen et al., 1984). However, the Agrio fold-and-thrust belt sections of Figure 10, constrained by the available seismic-reflection data, suggest substantial basement involvement and only about 20 km of crustal shortening. If the crustal roots are as small as suggested by the relatively subdued topography and gravity data of this region, it will be difficult to accommodate large magnitude shortening without significant tectonic erosion at the continental margin. The lack of foreland shifting of the volcanic arc between the Jurassic and the Paleogene demonstrates that no important tectonic erosion took place (Ramos, 1988).

There is a close relationship between orogenic shortening and propagation of the thrust front to the foreland. In the areas of greatest shortening, such as in the northern segment, the thrust front extends 750 to 800 km away from the trench. The separation decreases to the south, and at the latitude of the Neuquén Basin, the distance from the orogenic front in the foreland to the trench is about 280 km. The separation between the orogenic front and trench also correlates with the age of the subducting oceanic crust. The inboard limit of the Wadati-Benioff zone seismicity coincides, to a first order, with the Andean deformation front. It seems reasonable to suggest that younger, hotter oceanic lithosphere will be resorbed back into the mantle faster and therefore has a less profound inboard extent. Old, cold oceanic lithosphere occupies a larger downdip extent of the subducted plate and therefore has a larger asthenospheric wedge. As a result, deformation is more efficiently transmitted to the foreland and extends farther away from the trench.

Newly acquired space geodetic data record rates and direction of motion across the Andes, mainly between the continental margin, which is affected by the Nazca plate convergence, and stable South America (Norabuena et al., 1998). Recent data presented by Kendrick et al. (1999) show significant shortening rates between the western slope of the Andes and the average position of stations located in cratonic stable areas of Argentina and Brazil. This total relative motion is the result of several components, such as (1) transient elastic deformation on the locked portion of the plate interface that can be released during large thrust earthquakes, and (2) permanent deformation through crustal shortening and mountain uplift. The assessment of this permanent deformation is important for interpreting the Andean rates of shortening. The changes in displacement rates among Santiago, Chile (19.4 ± 0.2 mm/yr); San Juan (7.3 ± 0.3 mm/yr), Argentina; and La Plata, near Buenos Aires (1.9 ± 0.5 mm/yr) make it possible to evaluate the active shortening within the Principal Cordillera-Precordillera and Sierras Pampeanas in the Pampean flat-slab segment. These rates indicate a shortening between both slopes of the Andes of 12 ± 0.5 mm/yr, and within Sierras Pampeanas of 5.4 ± 0.5 mm/yr. If we compare these figures with those derived from crustal balance of Andean roots (7.65 mm/yr) or from structural cross sections (5.25 mm/yr) in the Principal Cordillera at these latitudes, the G.P.S. results are higher. The rates obtained by Zapata and Allmendinger (1996) for the Precordillera are within this range. This fact may indicate either a concentration of elastic deformation along the continental margin or an increase, in recent years, of the average Neogene shortening. On the other hand, figures obtained from the structural shortening computed for one of the most active areas of Sierras Pampeanas, mainly in the Pie de Palo area (5 mm/yr in the last 3 m.y., according to Ramos and Vujovich, 1995), give similar rates when compared with the G.P.S. data (5.4 mm/yr). The similar values may indicate that any elastic deformation accumulated in this region was minimal, and, if it existed, was released by the large Pie de Palo earthquake in 1977 (Smalley et al., 1993).

Although these values are still preliminary, they illustrate that space-based geodesy is opening a new era for the evaluation, in an independent way, of the present orogenic shortening in the Andes.

CONCLUSIONS

This overview attempts to relate topography to crustal roots and, on this basis, to evaluate the crustal shortening in three segments of the Andes. The correlation is good between crustal shortening data derived from the crustal roots and topography and orogenic shortening data derived from balanced cross sections of fold-and-thrust belts.

In the northern segment of the Andes, the structural style is a thick-skinned belt in the Puna, controlled mainly by the high heat flow and a shallow ductile-brittle transition in the basement but with subordinate control by tectonic inversion of older normal faults. The Eastern Cordillera is mainly a thick-skinned belt controlled by tectonic inversion, while the Sub-Andean belt is entirely dominated by thin-skinned tectonics. The Sub-Andean belt is developing as the wedge top of the foreland system, where shortening is partitioned in four or five thrusts that are moving simultaneously. Average shortening rates for the entire segment vary from 6.7 to 6.9 mm/yr, although locally they may reach 9.3 mm/yr.

The central segment shows contrasting structural styles, with alternating thin- and thick-skinned belts. Tectonic inversion dominates the thick-skinned belts, and the wedge-top is relatively reduced. In spite of the different heat flows and the distinct basement with the northern segment, the shortening rate varies, from 7.3–7.7 mm/yr in the north, to 5.5–5.7 mm/yr in the south.

The southern segment has a very low rate of shortening— less than 1 mm/yr— and probably is regulated by important strike-slip partitioning of the finite oblique stress in the fore arc.

The size of the crustal roots, the shortening within the fold-and-thrust belts, and the estimated shortening rates show a progressive decrease that correlates with the age of the subducting oceanic crust. This latter factor seems to control the amount of shortening detected at different latitudes, with larger shortening associated with the older, subducting oceanic crust.

ACKNOWLEDGMENTS

The authors are indebted to several colleagues of the Laboratorio de Tectónica Andina of the Universidad de Buenos Aires, as well as to YPF S.A. for logistic support. The SECYT PICT 06729 and CONICET PIP 4162 financed this work. The authors want to express their gratitude to reviewers Richard Allmendinger, Raymond Price, and Jorge Skarmeta for their comments and suggestions.

REFERENCES CITED

Allen, R., E. García, and J. Feehan, 1984, Low angle thrusting in the Neuquén basin, south of Chos Malal, northwest Neuquén province, Argentina: IX° Congreso Geológico Argentino, Actas, v. 2, p. 137–146.

Allmendinger, R. W., and T. R. Zapata, 1996, Imaging the Andean structure of the Eastern Cordillera on reprocessed YPF seismic reflection data: XIII° Congreso Geológico Argentino (Buenos Aires), Actas, v. 2, p. 125–134.

Allmendinger, R. W., V. A. Ramos, T. E. Jordan, M. A. Palma, and B. I. Isacks, 1983, Paleogeography and Andean structural geometry, Northwest Argentina: Tectonics, v. 2, p. 1–16.

Allmendinger, R. W., D. Figueroa, D. Snyder, J. Beer, C. Mpodozis, and B. L. Isacks, 1990, Foreland shortening and crustal balancing in the Andes at 30°S latitude: Tectonics, v. 9, p. 789–809.

Allmendinger, R. W., T. Jordan, S. Kay, and B. L. Isacks, 1997, The evolution of the Altiplano-Puna Plateau of the Central Andes: Annual Reviews of Earth and Planetary Sciences, v. 25, p. 139–174.

Alvarez, P. P., 1996, Los depósitos triásicos y jurásicos de la Alta Cordillera de San Juan, *in* Geología de la Región del Aconcagua, Provincias de San Juan y Mendoza: Dirección Nacional del Servicio Geológico Anales, v. 24, p. 59–137.

Alvarez, P. P., and V. A. Ramos, 1999, The Mercedario Rift System in the Principal Cordillera of Argentina and Chile (32°SL): Journal of South American Earth Sciences, v. 12, no. 1, p. 17–31.

ANCORP Working Group, 1999, Seismic reflection image revealing offset of Andean subduction-zone earthquake locations into oceanic mantle: Nature, v. 397, p. 341–344.

Aramayo Flores, R. F., 1989, El cinturón plegado y sobrecorrido del norte argentino: Boletín de Informaciones Petroleras, Tercera Epoca, v. VI, no. 17, p. 2–16.

Baby, P., G. Heráil, R. Salinas, and T. Sempere, 1992, Geometric and cinematic evolution of passive roof duplexes deduced from cross section balancing: example from the foreland thrust system of the southern Bolivian Subandean zone: Tectonics, v. 11, p. 523–536.

Baby, P., B. Guiller, J. Oller, and G. Montemurro, 1993, Modèle cinématique de la Zona Subandine du Coude de Santa Cruz (entre 16°S et 19°S, Bolivie) déduit de la construction de cartes équilibrées: Comptes Rendus Academie des Sciences de Paris, Série II, v. 317, p. 1477–1483.

Beck, M. E. Jr., 1998, On the mechanism of crustal block rotations in the Central Andes: Tectonophysics, v. 299, p. 75–92.

Beck, S. L., G. Zandt, S. C. Myers, T. C. Wallace, P. G. Silver, and L. Drake, 1996, Crustal-thickness variations in the central Andes: Geology, v. 24, p. 407–410.

Booth, J. L. M., and M. P. Coward, 1996, Basement faulting and inversion of the NW Neuquén basin, Argentina: III° International Symposium on Andean Geodynamics (Saint Maló), Abstracts, p. 295–298.

Butler, R. F., D. Richards, T. Sempere, and L. Marshall, 1995, Paleomagnetic determinations of vertical axis rotations from late Cretaceous and Paleocene of Bolivia: Geology, v. 23, p. 799–802.

Butler, R., H. Lickorish, and M. Casey, 1999, Displacement and deformation rates in thrust belts integrating strata with structure to examine fold and thrust processes, *in* K. McClay, ed., Thrust tectonics 99 (London), Abstracts, p. 10–13.

Cabassi, I. R., and A. Introcaso, 1999, Los Andes peruanos y argentino-chilenos: un estudio cortical comparativo: XIV° Congreso Geológico Argentino (Salta), Actas, v. 1, p. 295–297.

Cahill, T., and B. L. Isacks, 1992, Seismicity and the shape of the subducted Nazca plate: Journal of Geophysical Research, v. 97, p. 17503–17529.

Cegarra, M. I., and V. A. Ramos, 1996, La Faja Plegada y Corrida del Aconcagua, *in* Geología de la región del Aconcagua, Provincias de San Juan y Mendoza: Dirección Nacional del Servicio Geológico Anales, v. 24, p. 387–422.

Chaveau, V., B. Niviere, P. R. Cobbold, E. A. Rossello, J-F. Ballard, and H. T. Eichenseer, 1996, Structure of the Andean foothills, Chos Malal region, Neuquén basin, Argentina: III° International Symposium on Andean Geodynamics (Saint Maló), Abstracts, p. 315–318.

Colletta, B., F. Roure, B. de Toni, D. Loureiro, H. Passalacqua, and Y. Gou, 1997, Tectonic inheritance, crustal architecture and contrasting structural styles in the Venezuela Andes: Tectonics, v. 16, p. 777–794.

Costa, C. H., 1992, Neotectónica del sur de la Sierra de San Luis: Universidad Nacional de San Luis, Ph.D. Thesis (unpublished), 390 p.

Couch, R., R. Whitsett, B. Huehn, and L. Briceno-Guarupe, 1981, Structures of the continental margin of Peru and Chile: Geological Society of America Memoir, v. 154, p. 703–726.

Cristallini, E., 1996, La faja plegada y corrida de La Ramada, *in* Geología de la región del Aconcagua, provincias de San Juan y Mendoza: Dirección Nacional del Servicio Geológico Anales, v. 24, p. 349–385.

Cristallini, E. O., and V. A. Ramos, 1997, Estructura profunda de los Andes a los 32° de latitud sur (Argentina y Chile): VIII° Congreso Geológico Chileno (Antofagasta), Actas, v. 3, p. 1622–1625.

Cristallini, E., A. H. Comínguez, and V. A. Ramos, 1997, Deep structure of the Metan-Guachipas region: Tectonic inversion in northwestern Argentina: Journal of South American Earth Sciences, v. 10, p. 403–421.

Cristallini, E. O., A. Cominguez, V. A. Ramos, and E. D. Mercerat, 2004, Basement double-wedge thrusting in the northern Sierras Pampeanas of Argentina (27°S)—Constraints from deep-seismic reflection, *in* K. R. McClay, ed., Thrust tectonics and hydrocarbon systems, AAPG Memoir 82, p. 65–90.

DeCelles, P., and K. Gilest, 1996, Foreland basin systems: Basin Research, v. 8, p. 105–123.

Dellapé, D., and A. Hegedus, 1993, Inversión estructural de la cuenca Cuyana y su relación con las acumulaciones de hidrocarburos: XII° Congreso Geológico Argentino y II° Congreso de Exploración de Hidrocarburos (Mendoza) Actas, v. 3, p. 211–218.

Dewey, J. F., and S. H. Lamb, 1992, Active tectonics of the Andes: Tectonophysics, v. 205, p. 79–95.

Diraison, M., P. R. Cobbold, E. A. Rossello, and A. J. Amos, 1998, Neogene dextral transpression due to oblique convergence across the Andes of northwestern Patagonia, Argentina: Journal of South American Earth Sciences, v. 11, p. 519–532.

Dunn, J. F., K. G. Hartshorn, and P. W. Hartshorn, 1995, Structural styles and hydrocarbon potential of the Sub-Andean thrust belt of southern Bolivia, *in* A. J. Tankard, R. Suárez S., and H. J. Welsink, eds., Petroleum basins of South America: AAPG Memoir 62, p. 523–543.

Eisner, P., 1991, Tectonostratigraphic evolution of Neuquén Basin, Argentina: Houston, Texas, Rice University, Master's Thesis, 56 p.

Gangui, A., 1998, A combined structural interpretation based on seismic data and 3-D gravity modeling in the northern Puna/Eastern Cordillera: Berlín, Freie Universität, Ph.D. Dissertation, v. B(27), p. 1–176.

Giese, P., E. Scheuber, F. Schilling, M. Schmitz, and P. Wigger, 1999, Crustal thickening processes in the Central Andes and the different natures of the Moho-discontinuity: Journal of South American Earth Sciences, v. 12, no. 2, p. 201–220.

Giraudo, R., R. Limachi, E. Requena, and H. Guerra, 1999, Geología estructural de las regiones subandina y piedemonte entre los 18° y 22°30', Bolivia: un nuevo modelo de deformación: IV° Congreso de Exploración y Desarrollo de Hidrocarburos (Mar del Plata) Actas, v. 1, p. 405–426.

Golovchenko, X., R. L. Larson, and W. C. Pitman III, 1981, Plate tectonic map of the Circumpacific region, southeast quadrant, magnetic lineations: Circum-Pacific Council for Energy and Mineral Resources, A.A.P.G., Scale 1:10,000,000.

Götze, H. J., B. Lahmeyer, S. Schmidt, and S. Strunk, 1994, The lithospheric structure of the central Andes (20°–25°S) as inferred from quantitative interpretation of regional gravity, *in* K. J. Reutter, E. Scheuber, and P. J. Wigger, eds., Tectonics of Southern Central Andes: Structure and evolution of an active continental margin: Berlin, Springer-Verlag, p. 23–48.

Groeber, P., 1929, Líneas fundamentales de la geología del Neuquén, sur de Mendoza y regiones adyacentes: Dirección Nacional de Geología y Minería Publicación, v. 58, p. 1–110.

Heredia, N., L. R. Rodríguez-Fernández, R. E. Seggiaro, and M. A. González, 1999, Estructuras de inversión tectónica en la Cordillera Oriental de los Andes entre 23° y 24°S, provincia de Jujuy, NO de Argentina, *in* F. Colombo et al., eds., I Seminario Iberoamericano de Cuencas de Antepaís en los Andes, los Pirineos y los Varíscides, Universidad de Barcelona, Comunicaciones, p. 20–24.

Hernández, R. M., J. Reynolds, and A. Disalvo, 1996, Análisis tectosedimentario y ubicación geocronológica del Grupo Orán en el Río Iruya: Boletín de Informaciones Petroleras, Tercera Serie, v. XII(45), p. 80–93.

Introcaso, A., and V. A. Ramos, 1992, Geotransecta Valparaíso-Punta del Este; Parte I, perfil geofísico: Global Geotransect Program, I.C.L., Buenos Aires.

Introcaso, A., M. C. Pacino, and H. Fraga, 1992, Gravity, isostasy and Andean crustal shortening between latitudes 30° and 35°S: Tectonophysics, v. 205, p. 31–48.

Isacks, B., 1988, Uplift of the Central Andean plateau and

bending of the Bolivian orocline: Journal of Geophysical Research, v. 93, p. 3211–3231.

Jordan, T., and R. Allmendinger, 1986, The Sierras Pampeanas of Argentina: a modern analogue of Laramide deformation: American Journal of Science, v. 286, p. 737–764.

Jordan, T. E., B. Isacks, V. A. Ramos, and R. W. Allmendinger, 1983a, Mountain building in the Central Andes: Episodes, v. 1983(3), p. 20–26.

Jordan, T. E., B. L. Isacks, R. W. Allmendinger, J. A. Brewer, V. A. Ramos, and C. J. Ando, 1983b, Andean tectonics related to geometry of subducted Nazca plate: Geological Society of America Bulletin, v. 94, p. 341–361.

Jordan, T. E., R. W. Allmendinger, J. F. Damanti, and R. E. Drake, 1993, Chronology of motion in a complete thrust belt: the Precordillera, 30–31°S, Andes Mountains: Journal of Geology, v. 101, p. 137–158.

Kendrick, E. C., M. Bevis, R. F. Smalley Jr., O. Cifuentes, and F. Galban, 1999, Current rates of convergence across the Central Andes: estimates fron continuous GPS observations: Geophysical Research Letters, v. 26, p. 541–544.

Kley, J., 1996, Transition from basement involved to thin skinned thrusting in the Cordillera Oriental of southern Bolivia: Tectonics, v. 15, no. 4, p. 763–775.

Kley, J., 1999, Geologic and geometric constraints on a kinematic model of the Bolivian orocline: Journal of South American Earth Sciences, v. 12, no. 2, p. 221–235.

Kley, J., and C. R. Monaldi, 1998, Tectonic shortening and crustal thickness in the Central Andes: How good is the correlation?: Geology, v. 26, p. 723–726.

Kley, J., C. R. Monaldi, and J. A. Salfity, 1996, Along-strike segmentation of the Andean foreland: Third International Symposium on Andean Geodynamics, Orstom Editions, Collection Colloques et Seminairés, p. 403–406.

Kley, J., C. R. Monaldi, and J. A. Salfity, 1999, Along-strike segmentation of the Andean foreland: causes and consequences: Tectonophysics, v. 301, p. 75–94.

Kozlowski, E., 1991, Structural geology of the northwest Neuquina basin, Argentina: Cuarto Simposio Bolivariano "Exploración Petrolera de las Cuencas Subandinas," v. 1, p. 5–13, Bogotá.

Kozlowski, E., R. Manceda, and V. A. Ramos, 1993, Estructura, *in* V. A. Ramos, ed., Geología y Recursos Naturales de Mendoza, XII° Congreso Geológico Argentino y II° Congreso de Exploración de Hidrocarburos (Mendoza) Relatorio, p. 235–256.

Kozlowski, E. E., C. E. Cruz, and C. A. Sylwan, 1996, Geología estructural de la zona de Chos Malal, Cuenca Neuquina, Argentina: XIII° Congreso Geológico Argentino y III° Congreso de Exploración de Hidrocarburos (Buenos Aires), Actas, v. 1, p. 15–26.

Kraemer, P. E., A. Introcaso, and A. Robles, A. 1996, Perfil geológico-gravimétrico regional a los 50°20′ Latitud Sur; estructura crustal y acortamiento andino. Cuenca Austral y Cordillera Patagónica: 13° Congreso Geológico Argentino y 3° Congreso de Exploración de Hidrocarburos (Buenos Aires) Actas, v. 2, p. 423–432.

Kraemer, B., D. Adelmann, M. Alten, W. Schnurr, K. Erpenstein, E. Kiefer, P. van den Bogaard, and K. Görler, 1999, Incorporation of the Paleogene foreland into the Neogene Puna plateau: the Salar de Antofalla area, NW Argentina: Journal of South American Earth Sciences, v. 12, no. 2, p. 157–182.

Lavenu, A., and J. Cembrano, 1999, Compressional- and transpressional-stress pattern for Pliocene and Quaternary brittle deformation in fore arc and intra-arc zones (Andes of Central and Southern Chile): Journal of Structural Geology, v. 21, p. 1669–1691.

Legarreta, L., and M. A. Uliana, 1996, The Jurassic succession in west-central Argentina: stratal patterns, sequences and paleogeographic evolution: Palaeogeography, Palaeoclimatology, Palaeoecology, v. 120, p. 303–330.

Marrett, R., R. W. Allmendinger, R. N. Alonso, and R. E. Drake, 1994, Late Cenozoic tectonic evolution of the Puna Plateau and adjacent foreland, northwestern Argentine Andes: Journal of South American Earth Sciences, v. 7, no. 2, p. 179–207.

Martínez, M. P., M. E. Giménez, A. Introcaso, and J. A. Robles, 1997, Excesos de espesores corticales y acortamientos andinos en tres secciones ubicadas en 36°, 37° y 39° de latitud sur: 7° Congreso Geológico Chileno (Antofagasta) Actas, v. 1, p. 101–105.

Mathalone, J. M. P., and M. Montoya, 1995, Petroleum geology of the Subandean basins of Peru, *in* A. J. Tankard, R. Suárez, and H. J. Welsink, eds., Petroleum basins of South America: AAPG Memoir 62, p. 423–444.

Mingramm, A., A. Russo, A. Pozzo, and L. Cazau, 1979, Sierras Subandinas, *in* J. C. M. Turner, ed., Geologia regional Argentina: Cordoba, Academia Nacional de Ciencias, I, p. 95–138.

Mosquera, A., 1999, Evolución temporal de la deformación en las Sierras Subandinas Orientales: IV° Congreso de Exploración y Desarrollo de Hidrocarburos (Mar del Plata) Actas, v. 2, p. 563–578

Mojica, H. J., and V. Zorzin, 1996, Extensión oriental de la tectónica andina en la llanura salteña, Provincia de Salta: XIII° Congreso Geológico Argentino y III° Congreso Exploración de Hidrocarburos (Buenos Aires): Actas, v. 1, p. 77–83.

Mpodozis, C., C. Arriagada, and P. Roperch, 1999, Cretaceous to Paleogene geology of the Salar de Atacama basin, northern Chile: a reappraisal of the Purilactis Group stratigraphy: Fourth International Symposium on Andean Geodynamics (Göettingen), Abstracts, p. 523–526.

Norabuena, E., L. Leffler-Griffin, A. Mao, T. Dixon, S. Stein, I. S. Sacks, L. Ocola, and M. Ellis, 1998, Space geodetic observations of Nazca-South America convergence across the Andes: Science, v. 279, p. 358–362.

Pacino, M., 1993, Perfil gravimétrico transcontinental en la latitud 39°Sur: XII° Congreso Geológico Argentino y II° Congreso de Exploración de Hidrocarburos (Mendoza), Actas, v. 3, p. 282–285.

Pennington, W. D., 1981, Subduction of the Eastern Panama Basin and seismotectonics of Northwest South America: Journal of Geophysical Research, v. 86, p. 10753–10770.

Pilger, R. H., 1984, Cenozoic plate kinematics, subduction and magmatism: South American Andes: Journal Geological Society of London, v. 141, p. 793802.

Ploszkiewicz, J. V., 1987, Las zonas triangulares de la faja fallada y plegada de la Cuenca Neuquina. Argentina: X° Congreso Geológico Argentino (Tucumán) Actas, v. 1, p. 177–180.

Ragona, D., 1993, Estudio Geológico de la Ciénaga del Gaucho, Alta Cordillera de San Juan: Buenos Aires, Universidad de Buenos Aires, Trabajo Final de Licenciatura, (unpublished), 113 p.

Ramos, V. A., 1988, The tectonics of the Central Andes: 30° to 33° S latitude, *in* S. Clark and D. Burchfiel, eds., Processes in continental lithospheric deformation: Geological Society of America Special Paper 218, p. 31–54.

Ramos, V. A., 1989, Foothills structure in northern Magallanes Basin, Argentina: AAPG Bulletin, v. 73, p. 887–903.

Ramos, V. A., 1994, Terranes of southern Gondwanaland and their control in the Andean structure (30–33°S lat.), *in* K. J. Reutter, E. Scheuber, and P. J. Wigger, eds., Tectonics of the southern central Andes, structure and evolution of an active continental margin: Berlin, Springer Verlag, p. 249–261.

Ramos, V. A., 1996, Interpretación tectónica: Geología de la región del Aconcagua, provincias de San Juan y Mendoza, Dirección Nacional del Servicio Geológico Anales, v. 24, p. 447–460.

Ramos, V. A., 1998, Estructura del sector occidental de la faja plegada y corrida del Agrio, cuenca Neuquina, Argentina: X° Congreso Latinoamericano de Geología (Buenos Aires), Actas, v. 2, p. 105–110.

Ramos, V. A., 1999, Plate tectonic setting of the Andean Cordillera: Episodes, v. 22, no. 3, p. 183–190.

Ramos, V. A., and A. Folguera, 1999, Extensión cenozoica en la Cordillera Neuquina: IV Congreso de Exploración y Desarrollo de Hidrocarburos (Mar del Plata) Actas, v. 2, p. 661–664.

Ramos, V. A., and G. I. Vujovich, 1995, Descripción Geológica de la Hoja San Juan, provincias de San Juan y Mendoza: Servicio Geológico y Minero de Argentina, p. 1–230.

Ramos, V. A., M. B. Aguirre-Urreta, P. P. Alvarez, M. Cegarra, E. O. Cristallini, S. M. Kay, G. L. Lo Forte, F. Pereyra, and D. Pérez, 1996a, Geología de la región del Aconcagua, provincias de San Juan y Mendoza: Dirección Nacional del Servicio Geológico Anales, v. 24, p. 1–510.

Ramos, V. A., M. Cegarra, and E. Cristallini, 1996b, Cenozoic tectonics of the High Andes of west-central Argentina, (30°– 36°S latitude): Tectonophysics, v. 259, p. 185–200.

Ramos, V. A., M. Cegarra, G. Lo Forte, and A. Comínguez, 1997, El frente orogénico en la Sierra de Pedernal (San Juan Argentina): su migración a través de los depósitos orogénicos: VIII° Congreso Geológico Chileno (Antofagasta) Actas, v. 3, p. 1709–1713.

Rodríguez Fernández, L. R., N. Heredia, R. G. Espina, and M. I. Cegarra, 1999, Estratigrafía y estructura de los Andes Centrales Argentinos entre los 30° y 31° de latitud sur: Acta Geológica Hispánica, v. 32, no. 1–2, (1997), p. 51–75.

Roeder, D. H., 1988, Andean-age structure of Eastern Cordillera (Province of La Paz, Bolivia): Tectonics, v. 7, p. 23–40.

Romanyuk, T. V., H.-J. Götze, and P. F. Halvorson, 1999, A density model for the Andean Subduction zone: The Leading Edge (February), p. 264–268.

Roperch, P., and G. Carlier, 1992, Paleomagnetism of Mesozoic rocks from the central Andes of southern Peru: importance of rotations in the development of the Bolivian orocline: Journal of Geophysical Research, v. 97, p. 17233–17249.

Salfity, J., S. A. Gorustovich, M. C. Moya, and R. Amengual, 1984, Marco tectónico de la sedimentación y efusividad cenozoicas en la Puna argentina: IX° Congreso Geológico Argentino (Tucumán) Actas, v. 1, p. 539–554.

Salfity, J. A., C. R. Monaldi, R. A. Marquillas, and R. E. González, 1993, La inversión tectónica del Umbral de los Gallos en la cuenca del Grupo Salta durante la fase incaica: XII° Congreso Geológico Argentino y II° Congreso de Exploración de Hidrocarburos (Mendoza) Actas, v. 3, p. 200–210.

Schmitz, M., 1993, Kollisionsstrukturen in den Zentralen Anden: Ergebnisse refraktionsseismischer Messungen und Modellierung kurstaler Deformation: Berliner Geowissenschaftliche Abhandlunge B, v. 20, p. 1–127.

Schmitz, M., 1994, A balanced model of the southern Central Andes, Tectonics, v. 13, p. 484–492.

Smalley, R., J. Pujol, M. Regnier, J. Ming Chiu, J. L. Chatelain, B. Isacks, M. Araujo, and N. Puebla, 1993, Basement seismicity beneath the Andean Precordillera thin-skinned thrust belt and implications for crustal and lithospheric behaviour: Tectonics, v. 12, p. 63–76.

Somoza, R., S. Singer, and B. L. Coira, 1996, Paleomagnetism of upper Miocene ignimbrites at the Puna: an analysis of vertical-axis rotations in the Central Andes: Journal of Geophysical Research, v. 101(B5), p. 11387–11400.

Springer, M., and A. Förster, 1998, Heat-flow density across the Central Andean subduction zone: Tectonophysics, v. 291, p. 123–139.

Tebbens, S. F., S. Cande, L. Kovacs, J. C. Parra, J. L. LaBreque, and H. Vergara, 1997, The Chile ridge: a tectonic framework: Journal Geophysical Research, v. 102(B6), p. 12035–12059.

Uliana, M. A., and L. Legarreta, 1993, Hydrocarbons habitat in a Triassic-to-Cretaceous sub-Andean setting: Neuquén basin, Argentina: Journal of Petroleum Geologists, v. 16, p. 397–420.

Vergara, M., J. Moraga, and M. Zentilli, 1997, Evolución termotectónica de la cuenca terciaria entre Parral y Chillán: análisis por trazas de fisión en apatitas: VIII° Congreso Geológico Chileno (Antofagasta) Actas, v. 2, p. 1574–1578.

Vásquez, J. R., 1979, Informe preliminar sobre la tectónica del sur de la provincia de Mendoza: Buenos Aires,YPF S.A. (unpublished report).

Viñes, R. F., 1989, Interpretación de la estructura de Filo Morado: I° Congreso Nacional de Exploración de Hidrocarburos (Mar del Plata) Actas, v. 2, p. 1107–1124.

Von Gosen, W., 1992, Structural evolution of the Argentine Precordillera: the Río San Juan Section: Journal of Structural Geology, v. 14, p. 643–667.

Whitman, D., B. L. Isacks, and S. M. Kay, 1996, Lithospheric structure and along-stike segmentation of the central Andean Plateau: seismic Q, magmatism, flexure, topography and tectonics: Tectonophysics, v. 259, p. 29–40.

Whitman, D., B. L. Isacks, J-L. Chatelain, J-M. Chiu, and A. Perez, 1992, Attenuation of high frequencey seismic waves beneath the central Andes plateau: Journal of Geophysical Research, v. 97, p. 19929–19947.

Zapata, T. R., and R. W. Allmendinger, 1996, Growth stratal record of instantaneous and progressive limb rotation, Precordillera thrust belt and Bermejo basin, Argentina: Tectonics, v. 15, p. 1065–1083.

Zapata, T. R., I. Brissón, and F. Dzelalija, 1999, The role of basement in the Andean fold and thrust belt of the Neuquén basin, *in* K. McClay, ed., Thrust tectonics 99 (London): Abstracts, p. 122–124.

Brown, R. L., 2004, Thrust-belt accretion and hinterland underplating of orogenic wedges — An example from the Canadian Cordillera, *in* K. R. McClay, ed., Thrust tectonics and hydrocarbon systems: AAPG Memoir 82, p. 51–64.

Thrust-belt Accretion and Hinterland Underplating of Orogenic Wedges—An Example From the Canadian Cordillera

Richard L. Brown

Department of Earth Sciences and Ottawa-Carleton Geoscience Centre, Carleton University, Ottawa, Ontario, Canada

ABSTRACT

Geochronologic, structural, and metamorphic data obtained from the hinterland of the Canadian Cordillera indicate that middle-crust ductile deformation was diachronous and coincided temporally with thrust propagation in the Rocky Mountain Thrust-and-fold Belt. Rocks currently at the highest structural level were deeply buried and then exhumed to upper-crust levels during the Middle Jurassic, when the orogenic wedge was in its early stages of evolution. However, structurally deeper rocks were buried and then exhumed at progressively more recent times, such that the deepest exposed level was not buried until the Eocene. This late stage of burial coincided with the final stages of accretion at the toe of the wedge. These observations are interpreted in terms of critical-taper theory. This theory suggests that, at the rear of the wedge, rocks were displaced from a basal high-strain zone to higher levels, where they subsequently cooled as they were exhumed from beneath the extending and/or rapidly eroding upper part of the wedge. A general conclusion of this analysis is that the localized shear zone that underlies the base of the wedge in the frontal part of the orogen becomes a transient feature toward the rear of the wedge. As underplating developed, rocks that were deformed within the basal shear zone were displaced upward to become incorporated into the overlying wedge and ultimately exhumed.

INTRODUCTION

The Rocky Mountain Thrust-and-fold Belt of the southeastern Canadian Cordillera (Figure 1) is a classical thin-skinned deformation wedge. Listric thrust faults have shortened the miogeoclinal cover rocks and foredeep sedimentary rocks by approximately 50% (~300 km). This shortening has occurred above the uninvolved Precambrian basement, and all fault displacement and associated ductile shortening have been accommodated by shearing along the near basement dêcollement (Bally et al., 1966; Price and Mountjoy, 1970; Cook et al., 1992). Within the Rocky Mountain Thrust-and-fold Belt, the most distal and westerly part of the wedge includes miogeoclinal rocks that originally were deposited on the western continental margin of North America. Accordingly, these continental-margin rocks originally must have extended at least 300 km farther to the west, into the hinterland of the orogen. Direct evidence for the presence of Precambrian basement beneath the hinterland is found in the southern Omineca Belt, where a tectonic window exposes Precambrian crystalline rocks and an unconformably overlying cover sequence, collectively known as the Monashee complex (Figures 1 and 2). This complex is now thought to be autochthonous, and its crystalline basement is

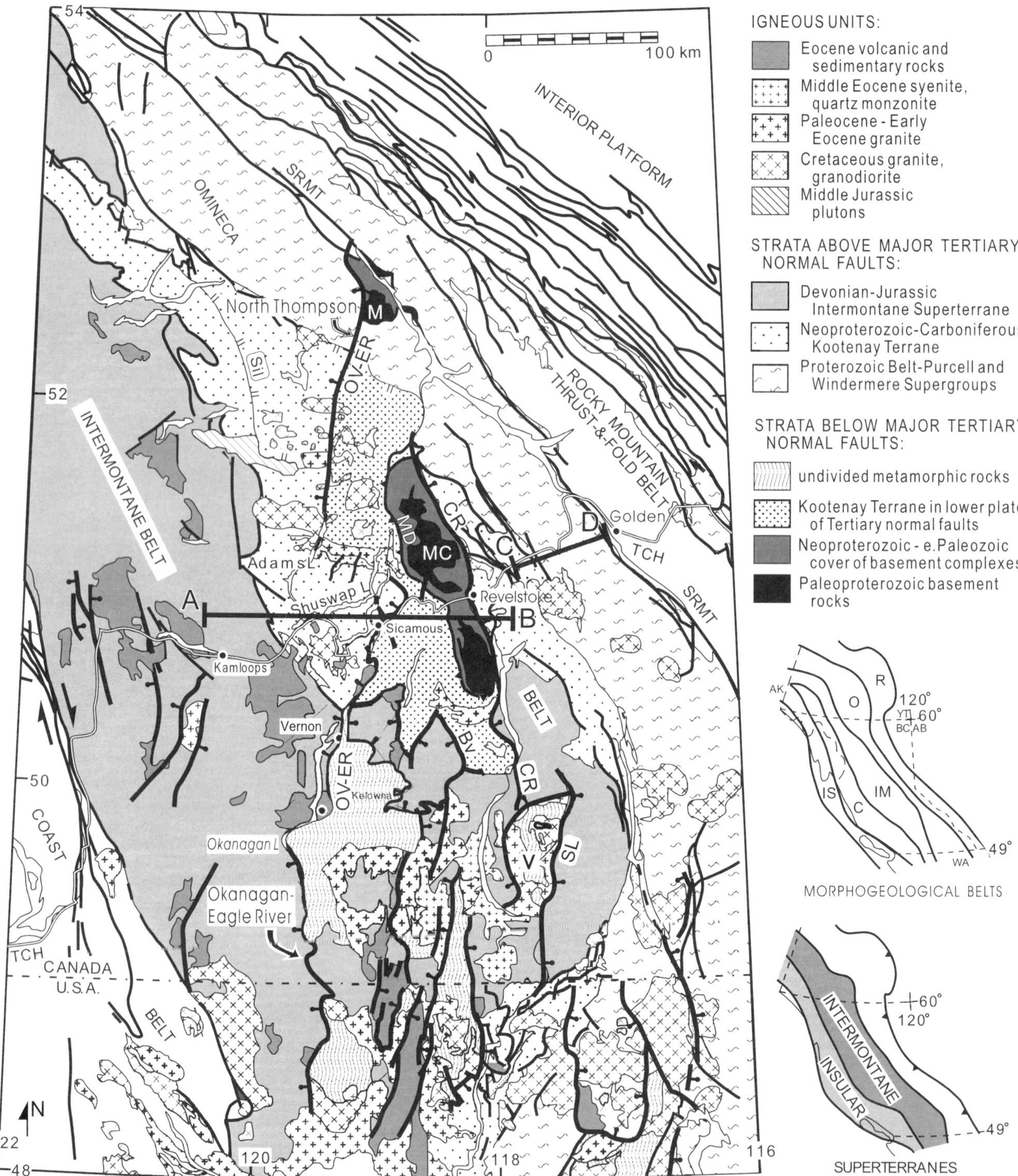

FIGURE 1. Tectonic assemblage map, simplified after Wheeler and McFeely (1991). Insets locate the Insular and Intermontane superterranes and the morphogeological belts: C = Coast Belt, CR = Columbia River Fault, O = Omineca Belt, OV–ER = Okanagan Valley–Eagle River normal fault system, IM = Intermontane Belt, IS = Insular Belt, M = Malton Gneiss, MC = Monashee complex, MD = Monashee dêcollement, R = Rocky Mountain Thrust-and-fold Belt, SiL = sillimanite isograd, SL = Slocan Lake fault, SRMT = Southern Rocky Mountain Trench, TCH = Trans-Canada highway, V = Valhalla complex. A-B, C-D = location of Figure 5 cross section.

correlated, on the basis of geochronologic signature and metamorphic-plutonic history, with rocks of the North American Precambrian basement that underlie the Rocky Mountain Thrust-and-fold Belt (Armstrong et al., 1991; Parrish, 1995; Crowley, 1997b, 1999). The structural boundary between the Monashee complex

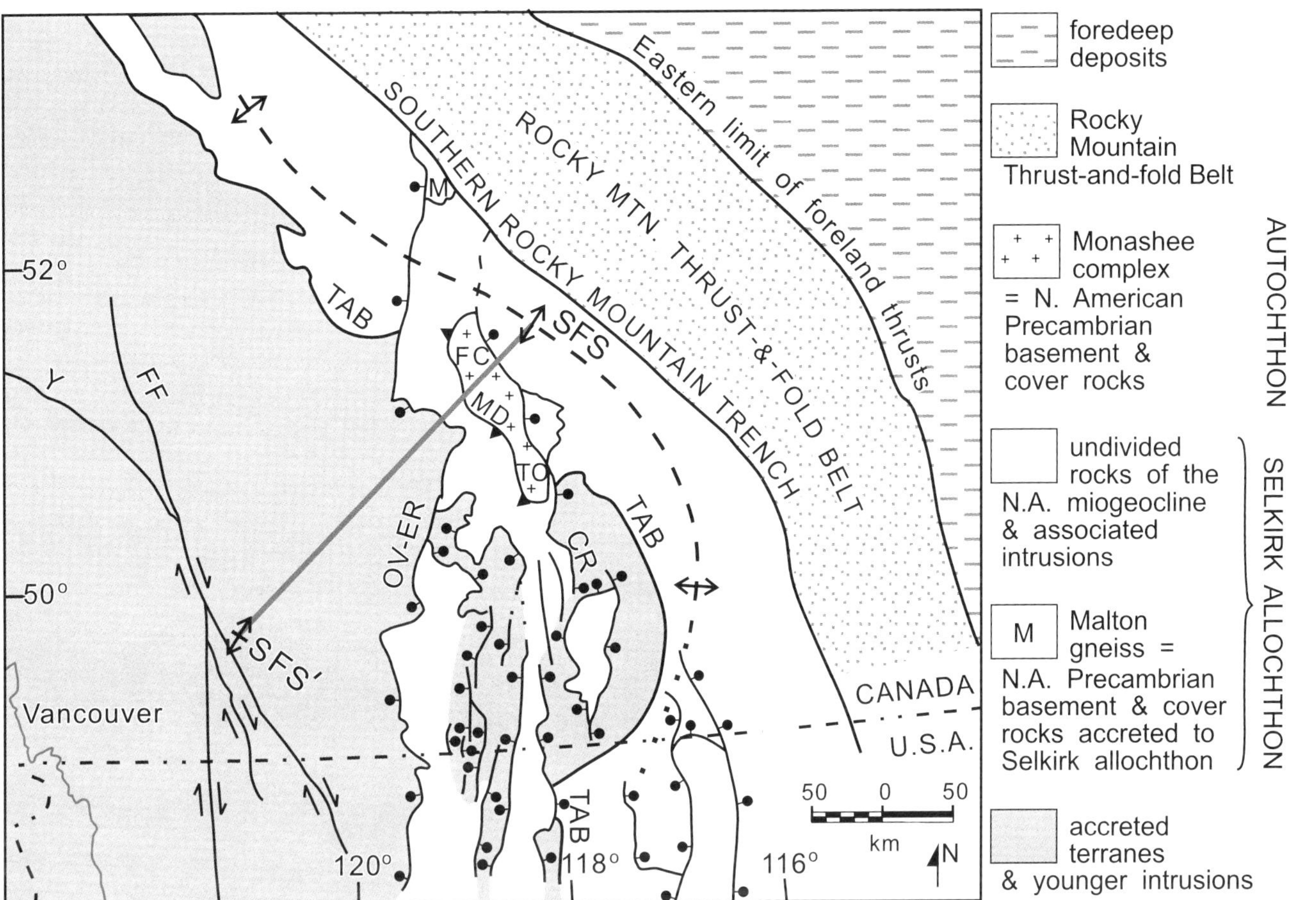

FIGURE 2. Simplified tectonic map of the southern Canadian Cordillera. SFS = Selkirk fan structure, TAB = terrane accretion boundary, OV–ER = Okanagan Valley–Eagle River normal fault system, CR = Columbia River Fault, FF = Fraser River Fault, Y = Yalakom Fault, MD = Monashee décollement, FC and TO = Frenchman Cap and Thor-Odin culminations of the Monashee complex, M = Malton Gneiss. Note that the structural style of the southern Rocky Mountain Thrust-and-fold Belt locally extends southwestward across the southern Rocky Mountain Trench into the eastern margin of the Omineca Belt. Also the eastern part of the Selkirk allochthon extends northeastward across the southern Rocky Mountain Trench into the Rocky Mountain Thrust-and-fold Belt.

and overlying allochthonous rocks, the Monashee décollement, is thought to be an extension of the basal décollement of the Rocky Mountain Thrust-and-fold Belt (Brown et al., 1992). Isotopic signatures of Eocene magmatic rocks (Armstrong, 1988) and crustal-scale cross sections constrained by geologic data and seismic-reflection data suggest that the North American crustal rocks extend as far west as the eastern edge of the Coast Belt, approximately 350 km to the west of the Rocky Mountain Thrust-and-fold Belt (Figures 1 and 2). From these observations it has been inferred that a near-basement shear zone also extends this far west beneath the hinterland (Brown et al., 1992; Cook et al., 1992). However, it is recognized that, within the hinterland, basement rocks are at least locally involved in the deformation of the orogenic wedge (cf. Simony et al., 1980).

This chapter presents an analysis of the deformation and thermal history of the orogen's hinterland. From the analysis, it is concluded that the discrete basal shear zone that characterizes the frontal part of the orogenic wedge in the Rocky Mountain Thrust-and-fold Belt does not extend to middle-crust depths beneath the hinterland. Rather, the ductile shear zone at the base of the wedge beneath the hinterland is a transient feature that was subjected to tectonic underplating and exhumation throughout the deformation history of the orogen. Ductile shearing, in association with underplating, gave rise to transposition of stratigraphy, diachronous generation of fold nappes, inverted metamorphic field gradients, and rapid exhumation.

ALLOCHTHONOUS ROCKS OF THE HINTERLAND

The western margin of the Rocky Mountain Thrust-and-fold Belt in the southern Canadian Cordillera is

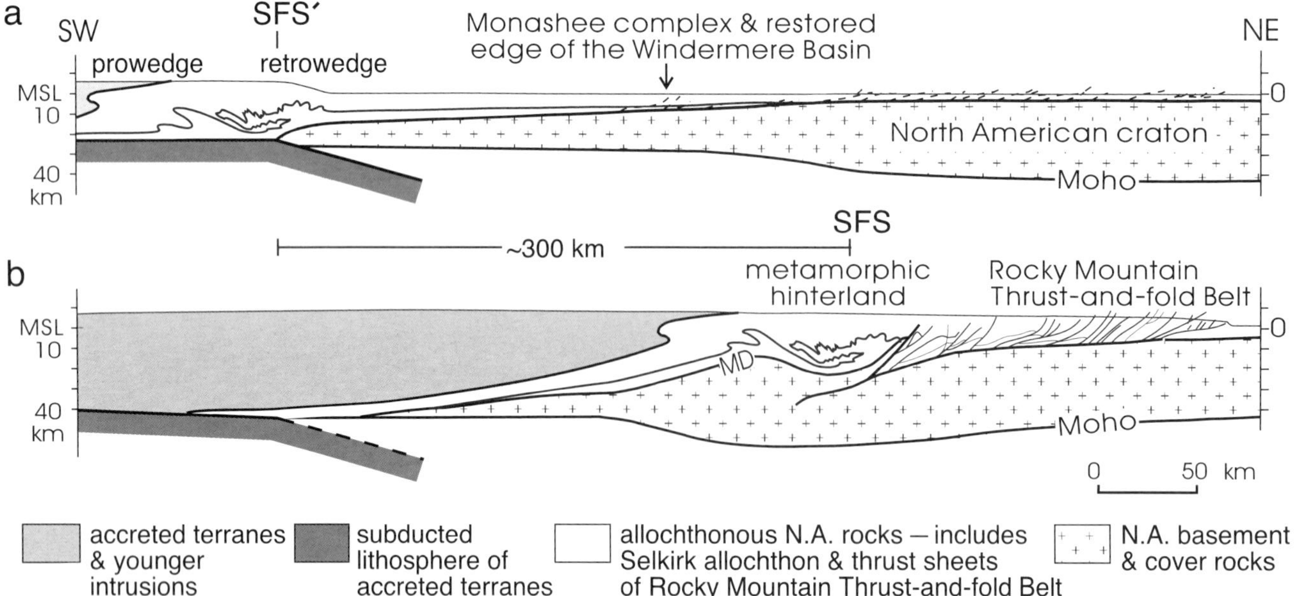

FIGURE 3. A: Middle Jurassic palinspastic restoration of SFS to SFS′. B: Tectonic cross section prior to Eocene extension. SFS = Selkirk fan structure, MD = Monashee dêcollement. Modified after Brown et al. (1993).

marked by the southern Rocky Mountain Trench (Figures 1 and 2), a distinct topographic depression that exhibits normal and reverse faulting as well as minor dextral shear. West of the trench, where exhumation has exposed deeper-level rocks, the classical Rocky Mountain style of thin-skinned thrusting gives way to a style of ductile deformation involving complex folding and transposition of stratigraphy. This part of the orogen, known as the Omineca Belt, is the exhumed hinterland of the Rocky Mountain Thrust-and-fold Belt (Figures 1 and 2).

The western margin of the Omineca Belt includes terranes of oceanic, marginal-basin, and island-arc affinities that have been thrust onto the North American miogeocline (Figures 1 and 2). Orogenic events associated with terrane accretion and subsequent migration of deformation from the hinterland to the foreland in the Mesozoic and Tertiary have been discussed in numerous publications (see Gabrielse and Yorath, 1991, and Cook et al., 1992, for comprehensive reviews). It is generally agreed that pervasive deformation of the North American continental margin began with terrane accretion along its western extremity and that the deformation front migrated from southwest to northeast, ultimately involving rocks of the Rocky Mountain Thrust-and-fold Belt. These allochthonous rocks of the hinterland are collectively known as the Selkirk allochthon (Figure 2). Geoscientists recognize that rocks involved in the allochthon extend eastward into the Rocky Mountain Thrust-and-fold Belt, and the nomenclature change at the Rocky Mountain Trench has no tectonic significance.

Structural Style of the Selkirk Allochthon

The Selkirk allochthon, with its incorporated accreted terranes, exhibits two distinct structural styles that reflect (1) post-accretion, southwesterly directed folding and thrusting and (2) superimposed northeasterly directed structures related to northeastward advance of the orogenic front. This superposition gave rise to the Selkirk fan structure, which is preserved at high structural levels within the central Omineca Belt (Figure 2) (Brown and Tippett, 1978; Brown et al., 1993; Colpron et al., 1998). Geochronologic and thermobarometric constraints confirm that the structural fan and associated fabrics originated during the Middle Jurassic at middle-crust depths and that the fan was subsequently rapidly exhumed (Colpron et al., 1998). Brown et al. (1993) have proposed that the southwestward-vergent structures originated in a prowedge setting at the distal edge of the North American plate and that the superimposed northeastward-vergent structures formed in a retrowedge setting (Figure 3) (see Willett et al., 1993, for mechanical models and nomenclature). Colpron et al. (1998) have recently challenged this interpretation; they suggest that a northeasterly driven crustal wedge caused the development of the southwestward-vergent structures. It appears to the present author that the existence of a crustal-scale tectonic wedge has yet to be demonstrated in the field. The fault that Colpron et al. consider to form the top of the wedge (Standfast Creek Fault, Figure 4 of Colpron et al., 1998, p.1064) has been shown to have very limited movement and

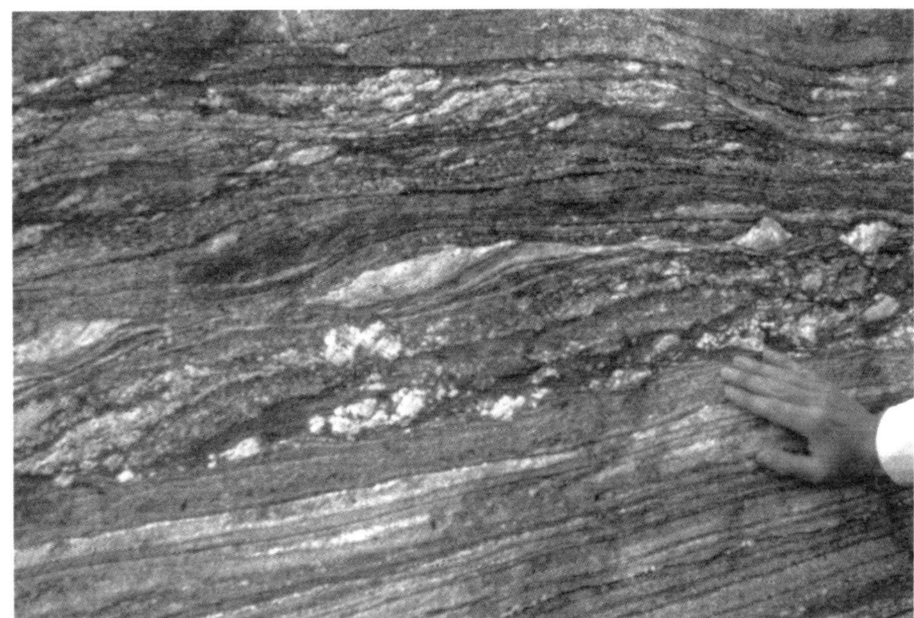

FIGURE 4. (a) Photo of transposed strata at the base of the Selkirk allochthon. View looking northwest, located at the northwest margin of Frenchman Cap dome (see Figure 1 and text for explanation). (b) Folded strata within the Monashee complex; fold intensity and degree of transposition increases up-section toward the highly transposed strata at the base of the Selkirk allochthon. View looking northwest in a location immediately to the southeast of (a).

to be primarily a zone of fold-limb attenuation. It has also been demonstrated that the structures beneath this fault are similar in style and vergence to the southwesterly verging folds that lie above the fault (Crowley and Brown, 1994). Colpron et al.(1998) consider the paucity of plutonic or volcanic rocks within the retrowedge, which would be attributed to eastward subduction, to rule out a prowedge setting for the Jurassic structures. However, this could be explained by either limited or shallow subduction. It is agreed that these southwest-verging structures originated during the Jurassic in close association with obduction of the accreted terranes. There may well be a role for tectonic wedging in the evolution of the orogen, but it is clear that the early southwestward-vergent structures developed above the underthrusted lithosphere of the accreted terranes, and as such, these structures evolved in a prowedge setting.

Rocks that now structurally underlie the Selkirk fan do not contain evidence of deformation during the Jurassic and are presumed to have been inboard of the orogenic front at that time. These deeper level strata are characteristically highly transposed with relict fold hinges and attenuated fold limbs preserved as boudins within the transposition foliation (Figure 4a, b) (see also Johnston et al., 2000). The transposed section is locally more than 10 km thick and extends down to the underlying Precambrian basement rocks of the Monashee complex (Figure 5). This highly strained panel exhibits kinematic evidence of predominantly northeasterly vergent deformation (see Johnson and Brown, 1996; Johnston et al., 2000).

Diachronous Thermal-mechanical History of the Transposition Zone

Several authors have studied the plutonic, metamorphic, and deformation history of parts of the transposed zone of the Selkirk allochthon and, on the basis of U-Pb geochronologic data, have concluded that the zone evolved diachronously (Brown and Carr, 1990; Carr, 1991; Scammell, 1993; Parrish, 1995). The uppermost

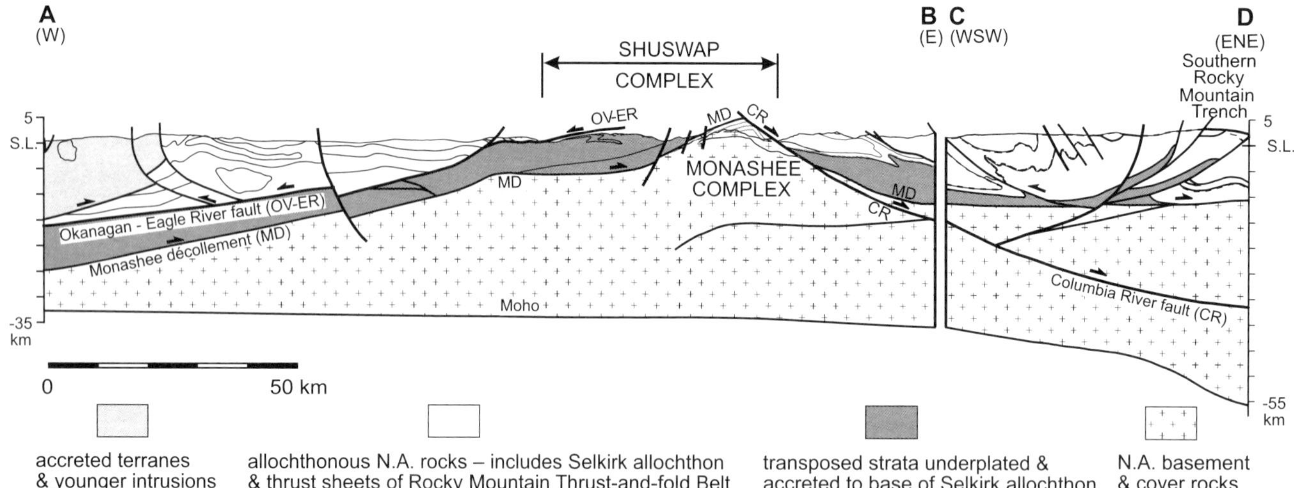

FIGURE 5. Crustal cross sections through the Omineca Belt (simplified after Johnson and Brown, 1996). Added to these cross sections is the exposed and inferred subsurface location of the transposed strata, herein interpreted to have been underplated during growth of the orogenic wedge. The cross sections are located in Figure 1. See Johnson and Brown (1996) for details related to construction of the cross sections and stratigraphic assemblages.

part of the zone exhibits high-temperature deformation that occurred in the Early Cretaceous, but at deeper levels there is no indication of this early history. There appears to be a downward decrease in age of high-temperature metamorphism, plutonism, and transposition fabrics. At the deepest exposed level, adjacent to the transposed contact with rocks of the Monashee complex, deep burial and associated ductile deformation did not occur until the latest Cretaceous (Parrish, 1995). The depths of burial at the time of high-temperature transposition are not precisely constrained, but interpretation of peak metamorphic mineral assemblages suggests pressures generally in the 6–7 kbar range (Scammell, 1993). The metamorphic grade of the highly transposed section is generally at or above the second sillimanite isograd. The absence of very high pressure rocks indicates that lower-crust rocks have not been incorporated into the transposed zone. Furthermore, the uniform pressure measurements throughout the more than 10-km-thick transposed section suggests that all of the high-temperature deformation occurred at approximately the same depth within the middle crust. For this to occur, overlying rocks must have been removed by erosion and/or tectonic denudation during progressive development of the transposition zone.

AUTOCHTHONOUS ROCKS OF THE HINTERLAND

The Monashee complex, a Precambrian basement massif with an unconformably overlying cover sequence, is exposed through a tectonic window in the Selkirk allochthon (Figure 2) (see also Brown et al.,1992). The Precambrian terrane exhibits a plutonic and metamorphic history that indicates an affinity with the Paleoproterozoic Wopmay orogenic belt of northern Canada, which at this latitude extends beneath the Rocky Mountain Thrust-and-fold Belt; such alignment appears to rule out major strike-slip motion with respect to North America (Crowley, 1999). Furthermore, contrary to earlier ideas on the origin of the complex (e.g., Brown et al., 1986), there is no evidence of deep burial by the overriding Cordilleran orogen prior to the latest Cretaceous; thus, its lateral motion is also restricted (cf. Parrish, 1995). The cover sequence, which unconformably overlies the crystalline basement, ranges from Paleoproterozoic to possibly as young as early Paleozoic (Crowley, 1999). The strata were deposited on a stable platform (Scammell and Brown, 1990), and the absence of deep-water Neoproterozoic Windermere Basin sediments supports the contention that the core of the Monashee complex is part of the crystalline basement, which lay beneath the North American miogeocline inboard of the Windermere Basin. These relationships are satisfied in palinspastic reconstruction if the Monashee complex is considered to be autochthonous with respect to the North American craton (Figure 3) (see also Price and Mountjoy, 1970). Nevertheless, the exposed upper part of the complex is characterized by high-grade metamorphism, polyphase folding, and transposition of stratigraphic units (Fyles, 1970; Reesor and Moore, 1971; McMillan, 1973; Brown, 1980; Hoy and Brown, 1980; Read and Klepacki, 1981; Journeay, 1986; Johnston et al., 2000).

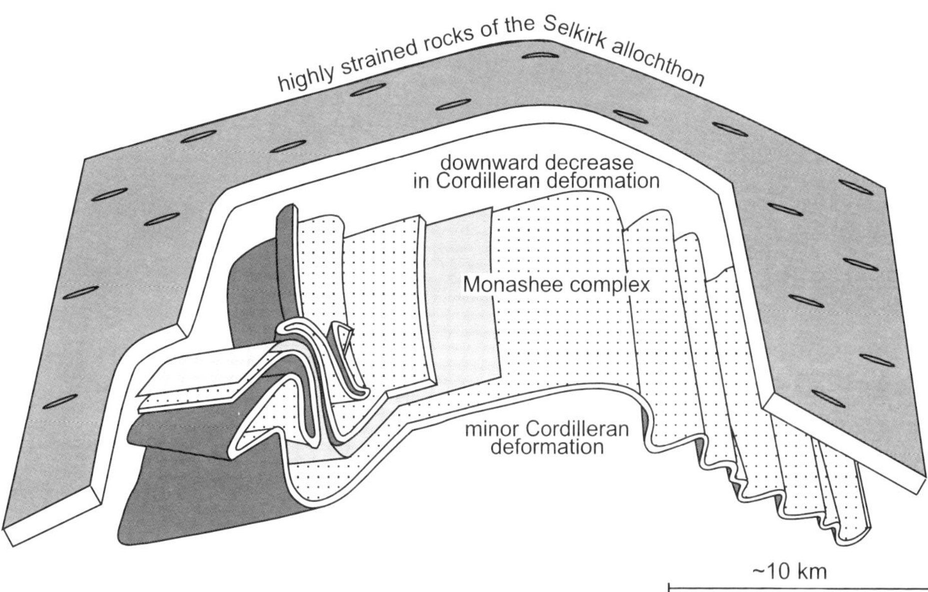

FIGURE 6. Schematic diagram illustrating the geometry of the northern end of Frenchman Cap dome and overlying transposed strata at the base of the Selkirk allochthon (looking north-northwest) (after Journeay, 1986). The boundary between the folded rocks of the Monashee complex and the diachronously transposed rocks of the Selkirk allochthon is the Monashee dêcollement.

Frenchman Cap and Thor Odin Domes

The Monashee complex has two major domal culminations: a northern dome called Frenchman Cap and a southern dome called Thor Odin. The intervening structural depression lies in the topographic valley traversed by the Trans-Canada highway (Figure 1). Major isoclinal folds have deformed the basement and cover rocks of the complex, and these structures exhibit complicated geometry as a result of progressive deformation and superposition of smaller scale folds (Figure 6) (Hoy and Brown, 1980, Journeay, 1986). It is important, for the purposes of this paper, to recognize in Frenchman Cap a downward decrease in intensity of Cordilleran deformation, from the highly strained zone of the overlying allochthonous rocks to the deepest exposed levels of the basement (Figures 4, 5, and 6) (see also Crowley, 1997a; Crowley et al., 2001). Strata of the Selkirk allochthon, which overlie the complex, are completely transposed. Isoclinal folding and associated limb attenuation has obliterated the stratigraphic succession; mylonitic fabrics, extensive boudinage of competent units, isolated hinges of rootless folds, and lenses of highly strained pegmatite characterize this structural level. Structurally down-section, the intensity of transposition decreases and stratigraphic markers become continuous features for several kilometers along strike. At the deepest exposed levels, the basement rocks of the complex preserve Precambrian fabrics, which have been weakly deformed by the Cordilleran overprint (Crowley, 1997a; Crowley et al., 2001). The recognition of this strain gradient further supports the contention that the Monashee complex is rooted in the autochthonous North American continental margin. The gradient is accordingly interpreted to be a deep-level equivalent of the discrete basal dêcollement that underlies the Rocky Mountain Thrust-and-fold Belt.

The strain gradient within the basement rocks of Frenchman Cap was accompanied by a downward decrease in temperature at the time of deformation (Crowley, 1997a, 1999). To the south, in Thor Odin dome, this gradient has not been recognized. Transposition of cover stratigraphy is extreme compared with that of the northern exposures, and Cordilleran deformation appears to have affected the deepest levels of exposed basement. It may turn out that the basement in Thor Odin has been more strongly influenced by deformation associated with Tertiary extension than is the case to the north in Frenchman Cap. Vanderhaeghe et al. (1999) have proposed a model of tectonic denudation that envisages a partially molten, highly mobile lower and middle crust beneath the Thor Odin region. They consider the evolution of Thor Odin to resemble a diapiric model that is invoked to explain the tectonic denudation of the French Massif Central. This interpretation is incompatible with the data cited previously from Frenchman Cap dome to the north, which indicate a downward decrease in temperature and mobility at the time of denudation.

Diachronous Thermal-mechanical History

Kilometer-scale northeasterly vergent isoclinal folds, which deform both the basement and cover rocks, characterize the upper part of the complex immediately beneath the highly transposed rocks of the overlying Selkirk allochthon (McNicoll and Brown, 1995). Thermal-chronologic studies in the Frenchman Cap dome have

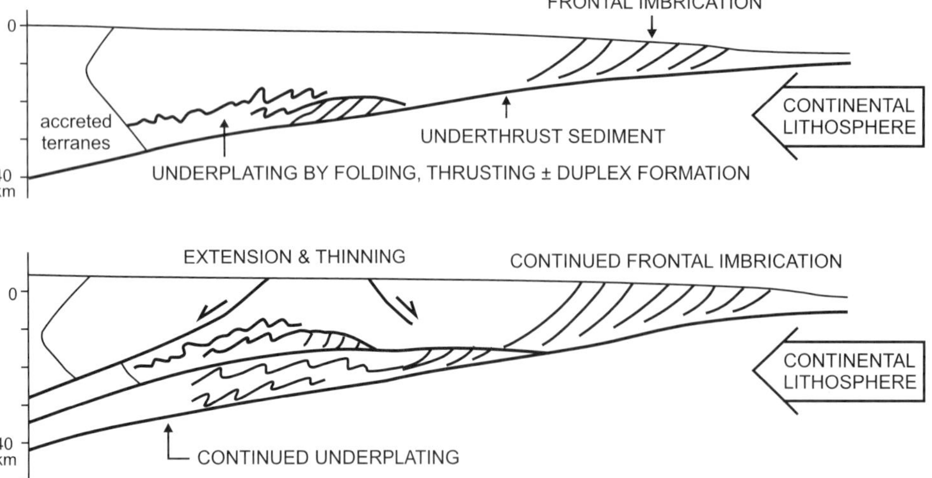

FIGURE 7. Generic and unscaled diagram of underplating illustrating the transient nature of a basal shear zone beneath the hinterland of an orogenic wedge. (See text for further explanation; diagram simplified after Platt, 1986.)

demonstrated that folding developed diachronously during progressive burial of the complex (Gibson et al., 1999; Crowley et al., 2001). At the highest structural level beneath the allochthon, folding and associated high-temperature metamorphism were underway by the latest Cretaceous; underlying folds have developed at progressively more recent times, and at the deepest exposed levels, high-temperature metamorphism and associated deformation were not initiated until the Eocene. Gibson et al. (1999) and Crowley et al. (2001) have concluded that the folds now at high structural levels within the complex originated to the southwest relative to underlying structures and that the current stacking of folds is a result of progressive deformation during northeastward advance of the orogenic front. Acceptance of this model implies that, with growth of the orogen, strata were carried southwestward on the North American basement, were buried to middle-crust levels, and ultimately, were ductilely deformed as they were underplated beneath the advancing Selkirk allochthon.

TOWARD AN UNDERPLATING MODEL

Any proposed tectonic model of the Cordilleran hinterland must consider the results of the above analysis. In summary, models must account for the following observations. (1) The uppermost level of the Selkirk allochthon was buried to middle-crust depths, deformed, metamorphosed, and rapidly exhumed in the Middle Jurassic. (2) The highly deformed strata beneath the upper part of the allochthon were progressively overridden and incorporated into the base of the Selkirk allochthon while it advanced northeastward toward the foreland. (3) This progressive incorporation and transposition occurred at middle-crust depths. (4) Previously deformed strata were exhumed as additional strata were underplated. (5) Timing of the advance of the allochthon and associated underplating coincided with progressive growth of the Rocky Mountain Thrust-and-fold Belt.

IMPLICATIONS OF UNDERPLATING

The Concept of a Discrete Basal Shear Zone

The idea that the Selkirk allochthon was emplaced, throughout the history of the orogen, by displacement along a single discrete basal shear zone is not compatible with underplating. Such underplating would inevitably lead to upward displacement of rocks within the basal shear zone as new material was accreted to the base of the allochthon (Figure 7) (cf. Platt, 1986). The boundary between the Selkirk allochthon and the underlying Monashee complex marks the position of the base of allochthonous rocks in the latest Cretaceous. As previously pointed out, the basement rocks of the complex were caught up in the deformation during the early Tertiary; one can presume that if the orogen had continued to grow, these basement-cored fold nappes would have been stripped from the underlying basement, accreted to the base of the allochthon, and transported northeastward above a new underlying shear zone. In this sense, the base of the allochthon can only be defined as a snapshot in time, because its location migrated downward with respect to newly underplated strata (Figure 8).

Inverted Metamorphic Field Gradients

Pelitic rocks of the transposition zone were heated to the point of generating sillimanite and potassium-feldspar mineral assemblages and melt. At the boundary

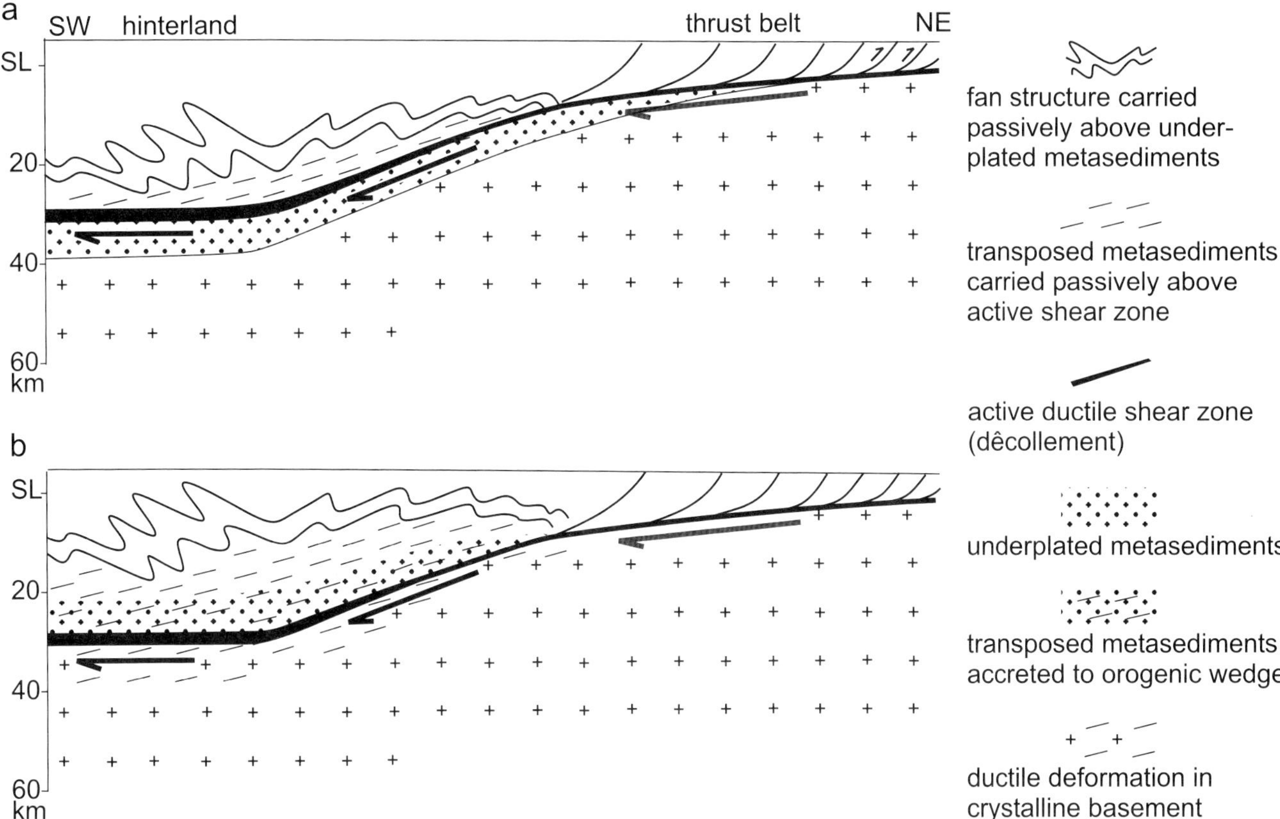

FIGURE 8. Unscaled diagram illustrating progressive underplating beneath the Selkirk fan structure. See text for further explanation. (a) Active shear zone at base of orogenic wedge is underlain by underplated metasediments. (b) Underplated metasediments accreted to orogenic wedge, transposed and incorporated into active shear zone. Deformed crystalline basement beneath active shear zone in forward models would be accreted to the base of the orogenic wedge. Note exhumation of the wedge is induced by underplating.

between the base of the Selkirk allochthon and the Monashee complex, there is an inverted metamorphic field gradient; underlying rocks of the complex were metamorphosed at a lower grade and presumably lower temperature than were the overlying rocks of the transposition zone. These observations have been interpreted by Gibson et al. (1999) and Crowley et al. (2001) to be a result of progressive northeastward emplacement of deeply buried hot rocks onto cooler higher-level rocks of the complex. In this interpretation, there is no requirement for a thermal inversion involving heat transfer from the allochthon to its underlying footwall.

Exhumation

There is clear evidence that the hinterland was exhumed during growth of the orogen; this was accomplished by a combination of erosion and extensional tectonics. A Jurassic record of significant exhumation is found in the upper structural level of the Selkirk allochthon. As discussed previously, these strata were deformed and metamorphosed at middle-crust depths and then rapidly exhumed. Rapid erosion may account for the high rates of exhumation, because as yet no extensional structures have been identified that are known to have been active during this time period. However, upper-crust extension faults may have been completely removed by subsequent erosion. Whether it was by erosion or tectonic denudation, this part of the allochthon had lost at least 10 km of its cover by the end of the Jurassic (Colpron et al., 1998). During the Cretaceous, the transposition zone must have been evolving while upper levels of the allochthon were being removed, because all structural levels within the transposition zone evolved diachronously at approximately the same crustal depth. Scammell's work (1993) indicates that there was a pulse of rapid exhumation at approximately 100 Ma that removed at least an additional 10 km of cover. Again, no extensional faults of that age have been identified. However, Scammell (1993) and Scammell et al. (1999) have suggested that the ductile deformation fabrics in the transposition zone reflect a component of thrust-parallel extension. Such middle-crust ductile thinning could account for decreased pressure in the underlying strata (Ring et al., 1999). Additional evidence of tectonic crustal thinning appears in early Tertiary. Johnston et al. (2000) have documented a west-dipping extensional shear zone along the west flank of Thor Odin dome that was active during the Paleocene. At that time, deeper levels within the Monashee complex were deforming in compression (Crowley, 1997a; Crowley et al., 2001), suggesting that crustal thinning was occurring as a result of northeastward extrusion of part of the middle-crust transposition zone. During the early Eocene, major crustal-scale extension affected the hinterland (Ewing, 1980; Price and Carmichael, 1986; Tempelman-Kluit and Parkinson, 1986; Brown and Journeay, 1987; Carr et al., 1987; Parrish et al., 1988; Carr, 1992). Much of this extension appears to postdate growth of the orogen and can be attributed to orogenic collapse (Coney and Harmes, 1984; Ranalli et al., 1989). However, U-Pb dating of high-temperature deformation within the Monashee complex indicates that folding was ongoing in the complex during the early stages of the development of normal-sense shear zones, such as the Okanagan Valley fault system (Figure 1) (Crowley et al., 2001; Johnston et al., 2000). Evidence of extensional tectonics overlapping in time with compressional tectonics requires a model that includes local extension during orogenic growth. Pulses of rapid erosion and/or extension are predicted in critical-taper models that include underplating (Figure 7) (cf. Platt, 1986).

Interpretation of Seismic-reflection Data

Seismic profiles of the southern Canadian Cordillera, generated by Lithoprobe, have been interpreted in combination with surface geologic data (e.g., Cook et al., 1992). These interpreted cross sections are presented as accurate representations of the current geometry of the orogen. Complete palinspastic restoration of these sections has yet to be attempted but would be an essential step in understanding how the orogen has evolved. Before the advent of Lithoprobe, Brown et al. (1986) created a restored section that now requires revision in light of Lithoprobe data and new geologic data outlined in this paper and elsewhere. A particular concern is that faults and ductile shear zones recorded in seismic profiles may have only transient significance. If an underplating model accurately describes the growth of the hinterland of the orogenic wedge, then a "basal shear zone" is only the position of the zone during the final stages of orogenic growth. Therefore this feature cannot be used as a template for palinspastic restoration. Furthermore, restoration of distributed ductile strain within allochthonous sheets may be far more significant than restoration of any bounding faults or shear zones displayed in seismic profiles.

Allochthonous Crystalline Thrust Sheets

In this chapter, the Monashee complex is interpreted to be an autochthonous mass that is rooted in the lower crust, although it has been shown that its upper part is significantly deformed and that this deformation includes fold nappes with crystalline basement cores. The point has been made that, with additional growth of the orogen, these basement nappes might well be accreted to the Selkirk allochthon and transported northeastward toward the foreland. Elsewhere in the hinterland, underplating appears to have reached deeper levels and created allochthonous basement sheets. One example is the Malton Gneiss, which occurs in the hanging wall of an out-of-sequence thrust fault and is capped by a mylonitic zone that separates it from overlying supracrustal rocks. These allochthonous rocks are considered to be a Precambrian basement nappe that has been displaced northeastward from the interior of the hinterland (Figure 1) (Simony et al., 1980; McDonough and Simony, 1988).

DISCUSSION

Relationships between the Hinterland and the Rocky Mountain Thrust-and-Fold Belt

The orogenic wedge in the Rocky Mountain Thrust-and-fold Belt developed during the interval from the Late Jurassic to the Paleocene (Price and Mountjoy, 1970). These timing constraints, together with U-Pb dating of high-temperature progressive deformation that occurred beneath the hinterland (Parrish, 1995), support the notion that middle-crust ductile deformation was kinematically linked to upper-crust thrust faulting. Another important consideration is the relationship between terrane accretion and shortening in the Rocky Mountain Thrust-and-fold Belt. As pointed out previously, oceanic and island arc terranes obducted onto the continental margin of North America during the Early Jurassic. This event is closely related in time with generation of southwestward-directed folds and thrusts in the distal part of the hinterland and the ultimate formation and exhumation of the Selkirk fan structure. This early growth of the orogen appears to predate significant deformation in the Rocky Mountain Thrust-and-fold Belt, which at that time was still part of the undeformed miogeocline (Brown and Tippett, 1978; Brown et al., 1986; Colpron et al., 1998). However, underplating of the fan structure and hinterland growth of the retrowedge during the Cretaceous and early Tertiary did overlap in time with thrusting in the Rocky Mountain Thrust-and-fold Belt (Figure 3). Future work may demonstrate that activity on specific faults within the Rocky Mountain Thrust-and-fold Belt correlates with a specific zone of transposition and underplating within the hinterland.

Was the Lower Crust Subducted?

This paper proposes that underplating occurred at middle-crust depths beneath the hinterland of the Cordilleran orogen. Such a model of orogen growth leaves open the question of what happened at lower-crust and subcrust depths. It would seem that because there is no evidence that these deeper level rocks have been incorporated into the orogen, they must have been displaced elsewhere. Ranalli et al. (1989) have suggested that the orogen may have had a subcrustal lithospheric root that developed during contraction of the orogen and that this root was delaminated and incorporated into the upper mantle during the Eocene, thus triggering orogenic collapse. Alternatively, the lithosphere below middle-crust depths may have been subducted at the plate boundary during contractional growth of the orogen. This mechanism is implied in recent interpretations of the Lithoprobe transect (cf. Clowes et al., 1998). A third option is that there may have been subduction rollback during orogenic growth. This third explanation would account for a thin lower crust that may have completely decoupled from the thickening middle crust. Deciding the relative validity of these or other theories is a topic for future research.

Relevance to Other Orogenic Belts

This paper concludes that in the Canadian Cordillera, penetrative ductile flow is the dominant mode of deformation at middle-crust depths beneath the hinterland. This is hardly a new discovery and should be obvious from observation of surface geology. Yet in this orogen and in many others, recent cross sections drawn with an eye on seismic reflectors have tended to overemphasize brittle faults and/or ductile shear zones. If the Cordillera is in any way a template for other orogens, important deformation in high-grade metamorphic environments of hinterland belts is likely to involve penetrative transposition of strata rather than rigid transport above discrete faults or ductile shear zones.

Much less obvious is recognition of the diachronous nature of folding and fabric development within crustal-scale zones of high strain. Shear zones commonly are assumed to have been active at approximately the same time across their entire width. The geochronologic data indicate that this is not the case within the infrastructure of the southern Canadian Cordillera, and similar diachronous relationships should be anticipated in other orogenic belts. These observations also have implications for interpreting the timing of a metamorphic thermal culmination. If the metamorphism is syntectonic, then peak metamorphism will generally be diachronous and an age constraint obtained at one outcrop may not have validity beyond the scale of that particular locality (Gibson et al., 1999).

Tectonic underplating has long been recognized as an important mechanism within subduction channels (e.g., Karig, 1974) but has not been generally applied to crustal thickening processes within the hinterland of evolving orogenic belts. However, the formation of duplex structures in the upper crust is a form of underplating that is well understood in foreland thrust belts. Although Platt (1986) approached the problem from the point of view of high-pressure rocks being rapidly exhumed from beneath an accretionary prism, he recognized that the hinterland of orogenic belts could also be exhumed by underplating and extension. In orogens such as the North American Cordillera, the level of underplating appears to lie within the middle crust, but the general aspects of Platt's model have been adopted in this paper. It appears most likely, to the present author, that underplating drives crustal thickening and that the resultant rapid erosion and/or normal faulting and ductile thinning of the supercritical wedge contribute to exhumation in most if not all orogenic belts (see Ring et al., 1999, for a recent review of exhumation processes).

Implications for Petroleum Exploration

Knowledge of the timing of tectonic activity in the foreland area of orogenic belts is an essential tool for petroleum exploration and development. Much information can be obtained directly from the stratigraphic record in the foredeep; however, frequently it is difficult to constrain models of progressive growth of thrust systems at the frontal region of the orogenic wedge. To understand the driving mechanism of foreland thrusting, one must establish the kinematic relationships between the evolving foreland and its hinterland. For example, renewed (out-of-sequence) thrusting or development of normal faults within a previously thrusted section may depend largely on a change in activity within the hinterland of the orogen. In the southern Canadian Cordillera, it appears that thrusting in the Rocky Mountain Thrust-and-fold Belt was driven by a combination of plate-boundary convergence and gravitational spreading of the hinterland and that deformation within the hinterland part of the orogenic wedge was kinematically linked to growth in the foreland.

ACKNOWLEDGMENTS

The author has benefited from collaboration over the years with several colleagues. In particular, I would like to thank Paul Williams for his constructive criticism of earlier ideas on the structure of the Monashee complex and for taking the lead in placing more emphasis on the role of transposition during progressive deformation. Discussions with Sharon Carr, James Crowley, Dan Gibson, Dennis Johnston, Murray Journeay, Randy Parrish, and Rob Scammell are gratefully acknowledged. Ray Price, Bob Hatcher, and an anonymous referee are thanked for their constructive reviews. A Natural Sciences and Engineering Research Council of Canada grant to RLB funded the research.

REFERENCES CITED

Armstrong, R. L., 1988, Mesozoic and early Cenozoic magmatic evolution of the Canadian Cordillera: Geological Society of America Special Paper 218, p. 55–91.

Armstrong, R. L., R. R. Parrish, P. van der Heyden, K. A. Scott, D. A. Runkle, and R. L. Brown, 1991, Early Proterozoic basement exposures in the southern Canadian Cordillera: core gneiss of Frenchman Cap, Unit I of the Grand Forks gneiss, and the Vaseaux Formation: Canadian Journal of Earth Sciences, v. 28, p. 1169–1201.

Bally, A. W., P. L. Gordy, and G. A. Stewart, 1966, Structure, seismic data and orogenic evolution of southern Canadian Rocky Mountains: Bulletin of Canadian Petroleum Geology, v. 14, p. 337–381.

Brown, R. L., 1980, Frenchman Cap dome, Shuswap complex, British Columbia: a progress report: Current Research, Part A, Geological Survey of Canada, Paper 80-1A, p. 47–51.

Brown, R. L., and S. D. Carr, 1990, Lithospheric thickening and orogenic collapse within the Canadian Cordillera: Proceedings of Pacific Rim '90 Congress, Brisbane, Australia, Australasian Institute of Mining and Metallurgy, v. 2, p. 1–10.

Brown, R. L., and J. M. Journeay, 1987, Tectonic denudation of the Shuswap metamorphic terrane of southeastern British Columbia: Geology, v. 15, p. 142–146.

Brown, R. L., and C. R. Tippett, 1978, The Selkirk fan structure of the southeastern Canadian Cordillera: Geological Society of America Bulletin, v. 89, p. 548–558.

Brown, R. L., J. M. Journeay, L. S. Lane, D. C. Murphy, and C. J. Rees, 1986, Obduction, backfolding and piggyback thrusting in the metamorphic hinterland of the southeastern Canadian Cordillera: Journal of Structural Geology, v. 8, p. 255–268.

Brown, R. L., S. D. Carr, B. J. Johnson, V. J. Coleman, F. A. Cook, and J. L. Varsek, 1992, The Monashee décollement of the southern Canadian Cordillera: a crustal-scale shear zone linking the Rocky Mountain Foreland Belt to lower crust beneath accreted terranes, *in* K. R. McClay, ed., Thrust tectonics: Chapman and Hall, London, p. 357–364.

Brown, R. L., C. Beaumont, and S. Willett, 1993, Comparison of the Selkirk fan structure with mechanical models: implications for interpretation of the southern Canadian Cordillera: Geology, v. 21, p. 1015–1018.

Carr, S. D., 1991, Three crustal zones in the Thor-Odin-Pinnacles area, southern Omineca Belt, British Columbia: Canadian Journal of Earth Sciences, v. 28, p. 2003–2023.

Carr, S. D., 1992, Tectonic setting and U-Pb geochronology of the early Tertiary Ladybird leucogranite suite, Thor-Odin-Pinnacles area, southern Omineca Belt, British Columbia: Tectonics, v. 11, p. 258–278.

Carr, S. D., R. R. Parrish, and R. L. Brown, 1987, Eocene structural development of the Valhalla complex, southeastern British Columbia: Tectonics, v. 6, p. 175–196.

Clowes, R. M., F. A. Cook, and J. N. Ludden, 1998, Lithoprobe leads to new perspectives on continental evolution: GSA Today, v. 8, p. 1–7.

Colpron, M., J. M. Warren, and R. A. Price, 1998, Selkirk fan structure, southeastern Canadian Cordillera: Tectonic wedging against an inherited basement ramp: Geological Society of America Bulletin, v. 110, p. 1060–1074.

Coney, P., and T. Harmes, 1984, Cordilleran metamorphic core complexes: Cenozoic extensional relics of Mesozoic compression: Geology, v. 12, p. 550–554.

Cook, F. A., J. L. Varsek, R. M. Clowes, E. R. Kanasewich, C. S. Spencer, R. Parrish, R. L. Brown, S. D. Carr, B. J. Johnson, and R. A. Price, 1992, LITHOPROBE crustal reflection cross section of the southern Canadian Cordillera, 1, Foreland Thrust and Fold Belt to Fraser River fault: Tectonics, v. 11, p. 12–35.

Crowley, J. L., 1997b, U-Pb geochronologic constraints on the cover sequence of the Monashee complex, Canadian cordillera: Paleoproterozoic deposition on basement: Canadian Journal of Earth Sciences, v. 34, p. 1008–1022.

Crowley, J. L., 1997a, U-Pb geochronology in Frenchman Cap dome of the Monashee complex, southern Canadian Cordillera: Early Tertiary tectonic overprint of a Proterozoic history: PhD thesis, Carleton University, Ottawa, Ontario.

Crowley, J. L., 1999, U-Pb geochronologic constraints on Paleoproterozoic tectonics in the Monashee complex, Canadian Cordillera: Elucidating an overprinted geologic history: Geological Society of America Bulletin, v. 111, p. 560–577.

Crowley, J. L., and R. L. Brown, 1994, Tectonic links between the Clachnacudainn terrane and Selkirk allochthon, southern Omineca Belt, Canadian Cordillera: Tectonics, v. 13, p. 1035–1051.

Crowley, J. L., R. L. Brown, and R. R. Parrish, 2001, Diachronous deformation and a strain gradient beneath the Selkirk allochthon, northern Monashee complex, southeastern Canadian Cordillera: Journal of Structural Geology, v. 23 (Special issue), Evolution of structures in deforming rocks, p. 1103–1121.

Ewing, T. E., 1980, Paleogene tectonic evolution of the Pacific northwest: Journal of Geology, v. 88, p. 619–638.

Fyles, J. T., 1970, The Jordan River area near Revelstoke, British Columbia: British Columbia Department of Mines and Petroleum Resources, Bulletin 57, 64 p.

Gabrielse, H., and C. J. Yorath, eds., 1991, Geology of the Cordilleran orogen in Canada: Geological Survey of Canada, Geology of Canada, No. 4 (*also* Geological Society of America, The Geology of North America, v. G–2), 844 p.

Gibson, D. H., R. L. Brown, and R. R. Parrish, 1999, Deformation-induced inverted metamorphic field gradients: an example from the southern Canadian Cordillera: Journal of Structural Geology v. 21, p. 751–767.

Hoy, T., and R. L. Brown, 1980, Geology of eastern margin of Shuswap complex, Frenchman Cap area: British Columbia Ministry of Energy, Mines and Petroleum Resources, Preliminary Map 43.

Johnson, B. J., and R. L. Brown, 1996, Crustal structure and early Tertiary extensional tectonics of the Omineca Belt at 51°N latitude, southern Canadian Cordillera: Canadian Journal of Earth Sciences, v. 33, p. 1596–1611.

Johnston, D. H., P. F. Williams, R. L. Brown, J. L. Crowley, and S. D. Carr, 2000, Northeastward extrusion and extensional exhumation of crystalline rocks of the Monashee complex, southeastern Canadian Cordillera: Journal of Structural Geology, v. 22, p. 603–626.

Journeay, M. J., 1986, Stratigraphy, internal strain and thermo-tectonic evolution of northern Frenchman Cap dome: an exhumed duplex structure, Omineca hinterland, S.E. Canadian Cordillera: Ph.D. thesis, Queen's University, Kingston, Ontario, 401 p.

Karig, D. E., 1974, Evolution of arc systems in the western Pacific: Annual Review of Earth and Planetary Sciences, v. 2, p. 51–75.

McDonough, M. R., and P. S. Simony, 1988, Structural evolution of basement gneisses and Hadrynian cover, Bulldog Creek area, Rocky Mountains, British Columbia: Canadian Journal of Earth Sciences, v. 25, p. 1687–1702.

McMillan, W. J., 1973, Petrology and structure of the west flank, Frenchman's Cap dome, near Revelstoke, British Columbia: Geological Survey of Canada, Paper 71–29.

McNicoll, V. J., and R. L. Brown, 1995, The Monashee décollement at Cariboo Alp, southern flank of the Monashee complex, southern British Columbia, Canada: Journal of Structural Geology, v. 17, p. 17–30.

Parrish, R. R., 1995, Thermal evolution of the southeastern Canadian Cordillera: Canadian Journal of Earth Sciences, v. 32, p. 1618–1642.

Parrish, R. R., S. D. Carr, and D. L. Parkinson, 1988, Eocene extensional tectonics of the southern Omineca Belt, British Columbia and Washington: Tectonics, v. 7, p. 181–212.

Platt, J. P., 1986, Dynamics of orogenic wedges and the uplift of high-pressure metamorphic rocks: Geological Society of America Bulletin, v. 97, p. 1037–1053.

Price, R. A., and D. M. Carmichael, 1986, Geometric test for Late Cretaceous-Paleogene intracontinental transform faulting in the Canadian Cordillera: Geology, v. 14, p. 468–471.

Price, R. A., and E. W. Mountjoy, 1970, Geologic structure of the Canadian Rocky Mountains between Bow and Athabasca rivers — a progress report, *in* J. O. Wheeler, ed., Structure of the Southern Canadian Cordillera: The Geological Association of Canada, Special Paper 6, p. 7–25.

Ranalli, G., R. L. Brown, and R. Bosdachin, 1989, A geodynamic model for extension in the Shuswap core complex, southeastern Canadian Cordillera: Canadian Journal of Earth Sciences, v. 26, p. 1647–1653.

Read, P. B., and D. W. Klepacki, 1981, Stratigraphy and structure: northern half of Thor-Odin nappe, Vernon east-half map area, southern British Columbia: Current Research, Part A, Geological Survey of Canada, Paper 81–1A, p. 169–173.

Reesor, J. E., and J. M. Moore, 1971, Petrology and structure of Thor-Odin gneiss dome, Shuswap metamorphic complex, British Columbia: Bulletin of the Geological Survey of Canada, v. 95, 149 p.

Ring, U., M. T. Brandon, S. D. Willett, and G. S. Lister, 1999, Exhumation processes, *in* U. Ring, M. T. Brandon, G. S. Lister, and S. D. Willett, eds., Exhumation processes: normal faulting, ductile flow and erosion: Geological Society of London Special Publication 154, p. 1–27.

Scammell, R. J., 1993, Mid-Cretaceous to Tertiary thermotectonic history of former mid-crustal rocks, southern Omineca Belt, Canadian Cordillera: Ph.D. thesis, Queen's University, Kingston, Ontario.

Scammell, R. J., and R. L. Brown, 1990, Cover gneisses of the Monashee Terrane: a record of synsedimentary rifting in the North American Cordillera: Canadian Journal of Earth Sciences, v. 27, p. 712–726.

Scammell, R. J., J. M. Dixon, and R. L. Brown, 1999, Thrust-parallel extension of a hinterland thrust sheet: an example from the Cordilleran thrust system of western North America: Abstract, Proceedings of Thrust Tectonics '99, Royal Holloway, University of London, U.K.

Simony, P. S., E. D. Ghent, D. Craw, W. Mitchell, and D. B. Robbins, 1980, Structural and metamorphic evolution of northeast flank of Shuswap complex in southern Canoe River area, British Columbia, *in* M. D. Crittenden Jr., P. J. Coney, and G. H. Davies, eds., Cordilleran metamorphic core complexes: Geological Society of America Memoir 153, p. 445–461.

Tempelman-Kluit, D. J., and D. Parkinson, 1986, Extension across the Eocene Okanagan crustal shear in southern British Columbia: Geology, v. 14, p. 318–321.

Vanderhaeghe, O., J.-P. Burg, and C. Teyssier, 1999, Exhumation of migmatites in two collapsed orogens: Canadian Cordillera and French Variscides. *in* U. Ring, M. T. Brandon, G. S. Lister, and S. D. Willett, eds., Exhumation processes: normal faulting, ductile flow and erosion: Geological Society of London Special Publication 154, p. 181–204.

Wheeler, J. O., and P. McFeely, (compilers), 1991, Tectonic assemblage map of the Canadian Cordillera and adjacent parts of the United States of America: Geological Survey of Canada, Map 1712A, scale 1:2,000,000.

Willett, S., C. Beaumont, and P. Fullsack, 1993, Mechanical model for the tectonics of doubly vergent compressional orogens: Geology, v. 21, p. 371–374.

Cristallini, E. O., A. H. Comínguez, V. A. Ramos, and E. D. Mercerat, 2004, Basement double-wedge thrusting in the northern Sierras Pampeanas of Argentina (27°S) — Constraints from deep seismic reflection, *in* K. R. McClay, ed., Thrust tectonics and hydrocarbon systems: AAPG Memoir 82, p. 65–90.

Basement Double-wedge Thrusting in the Northern Sierras Pampeanas of Argentina (27°S) — Constraints from Deep Seismic Reflection

Ernesto O. Cristallini

Laboratorio de Tectónica Andina, Departamento de Ciencias Geológicas, Facultad de Ciencias Exactas y Naturales, Universidad de Buenos Aires, Buenos Aires, Argentina, CONICET — Consejo Nacional de Investigaciones Científicas y Técnicas, Buenos Aires, Argentina

Alberto H. Comínguez

CONICET — Consejo Nacional de Investigaciones Científicas y Técnicas, Buenos Aires, Argentina, Departamento de Geofísica Aplicada, Facultad de Ciencias Astronómicas y Geofísicas, Universidad Nacional de La Plata, La Plata, Argentina

Victor A. Ramos

Laboratorio de Tectónica Andina, Departamento de Ciencias Geológicas, Facultad de Ciencias Exactas y Naturales, Universidad de Buenos Aires, Buenos Aires, Argentina

Enrique D. Mercerat

Departamento de Geofísica Aplicada, Facultad de Ciencias Astronómicas y Geofísicas, Universidad Nacional de La Plata, La Plata, Argentina

ABSTRACT

Deep seismic reprocessing of industrial lines, combined with surface geologic and structural data, provides the basis of a new tectonic interpretation of the northern Sierras Pampeanas at 27°S latitude. These basement mountain blocks, uplifted during an episode of shallowing of the subduction zone, show an active double-wedge thrusting. Deep seismic data indicate the different vergences of the western and eastern sectors of the Sierra de Aconquija, in the western Sierras Pampeanas at this latitude. Neotectonic evidence reveals that both systems are active, although the western sector has been active since at least middle to late Miocene times and recorded a much greater uplift and horizontal displacement than did the east. The eastern sector, although presently active, recorded only minor uplift and displacement. These facts enable correlation of the events in the northern Sierras Pampeanas with analog and numerical models that predict the behavior of double-wedge thrusting.

A by-product of this analysis is the knowledge that a middle Miocene Atlantic transgression covered the entire region, prior to the uplift of the Sierra de Aconquija. Based on correlation of the foreland basin deposits on both sides of the range, two stages of development are recognized. A single foreland basin covered the study area at an early stage, during the marine transgression (13.5 Ma), and at a late stage, a broken foreland basin developed.

INTRODUCTION

Neotectonic activity is recorded on both sides of the Sierra de Aconquija (5500 m a.s.l. [above sea level]), one of the highest basement uplifts of the Argentine Sierras Pampeanas. This double-wedge thrusting poses an interesting problem for understanding the kinematics of these mountain blocks, which were uplifted during the latest stages of the Andean orogeny. The objective of this chapter is to analyze the deep structure of the Sierra de Aconquija and, in particular, the deep geometry of the wedge-thrust system that bounded this basement mountain block. Analysis of the late Cenozoic synorogenic deposits associated with uplift of this sector of the western Sierras Pampeanas yields important clues toward understanding the general uplift of the Andes at this latitude. As a by-product of this study, it was possible to reconstruct the characteristics of the foreland basin prior to uplift and on this basis to recognize that an important middle Miocene marine transgression covered most of the northern Sierras Pampeanas. The age of these marine deposits supplies additional constraints for uplift of this sector of the Sierras Pampeanas.

The reprocessing of industrial seismic lines obtained by the oil company YPF S.A. was conducted as part of the CAPLI program, an enterprise of YPF S.A. - ICL and the Universities of Buenos Aires and La Plata, performed under the ILP program (Comínguez and Ramos, 1991, 1995; Cristallini et al., 1997).

The study area is located in the northwestern sector of the Sierras Pampeanas of Argentina at 27°S latitude (Figure 1). The Argentine Sierras Pampeanas consist of a series of fault-bounded basement blocks (González Bonorino, 1950a; Jordan and Allmendinger, 1986) that are exposed in the flat-subduction segment of the Andes, between 27 and 33°S latitudes. The area is situated in the northern boundary of this segment, where a transition between a northern normal subduction, with a Wadati-Benioff zone dipping 30°, passes toward the south to a flat subduction segment (Isacks et al., 1982; Jordan et al., 1983).

The Sierra de Aconquija is a Precambrian basement block composed of metamorphic and igneous rocks and bounded on its western side by an active reverse fault. Another basement block is exposed to the southeast of the Sierra de Aconquija, and it is known as the Sierra de Guasayán. This range, bounded in its eastern side by a reverse fault, is only 700 m high, but it also exhibits neotectonic features that indicate active faulting. Both major reverse faults are controlled by previous basement anisotropies such as schistosity, mylonitic zones, and other basement fabrics. For example, in the basement of the Sierra de Aconquija, foliation, schistosity, and the fold axial planes are east dipping, while those in the Sierra de Guasayán have an opposite vergence. Both sets of faults in the western and eastern side of the Sierra de Aconquija defined a double vergent thrust system, as recognized by Mon and Drozdzewski (1999).

Our study is based on the deep seismic reprocessing of several industrial lines located on both sides of the Sierra de Aconquija (Figure 2). The lines on the western side cross the Campo del Arenal a bolsón (a wide intermontane valley with endoreic drainage), which is filled with late Cenozoic synorogenic deposits. The lines also cover the eastern foothills of the Aconquija and adjacent plains. Both sets of lines allow an almost complete transect across the range at 27°S latitude. Reprocessing reveals data down to 13.5 s (two-way traveltime), which is equivalent to a nearly 45-km depth. Surface geologic control was based on the geologic maps produced by González Bonorino (1950b, 1951); Ruiz Huidobro (1960); Battaglia (1982); and Strecker (1987). Some old oil exploration wells, such as Isca Yacu and El Rincón, together with some deep water wells, provided subsurface control (Anonymous, 1949, 1958) for the geologic interpretation of the seismic lines.

GEOLOGIC SETTING

Metamorphic and Igneous Basement

The basement blocks exposed in the northwestern Sierras Pampeanas consist of late Precambrian metamorphic rocks and early Paleozoic granitoids. This basement was uplifted during the Andean orogeny in the Neogene.

The Precambrian basement exhibits some major crustal discontinuities as a result of the accretion of different terranes during latest Proterozoic–early Paleozoic collisions (Ramos, 1988). The old sutures have been partially reactivated during the Tertiary contraction, resulting in a thick-skinned thrust belt (Jordan and Allmendinger, 1986).

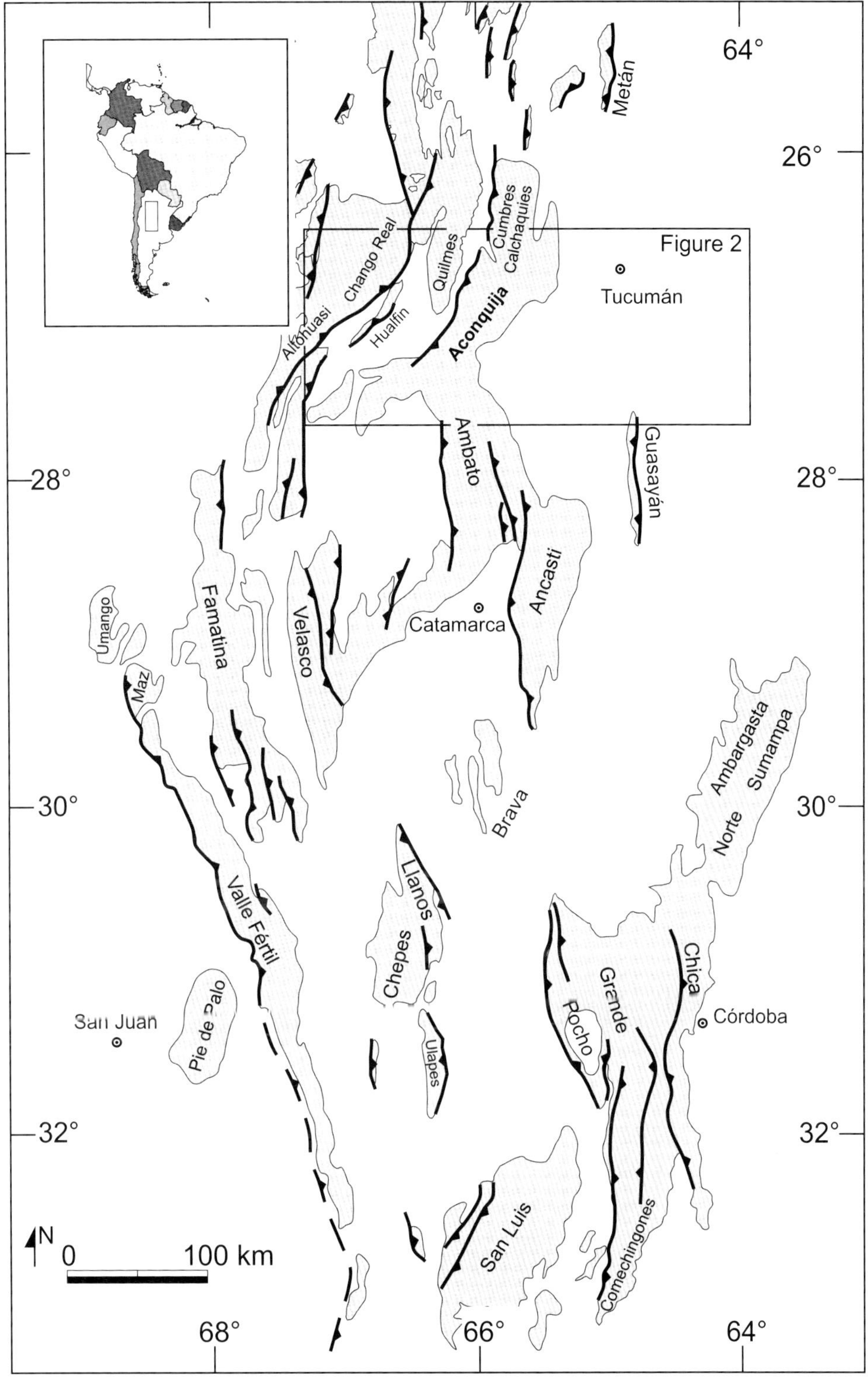

FIGURE 1. Major structural features of the Sierras Pampeanas of Central Argentina, showing location of the study area.

Rock types vary greatly. The Sierra de Guasayán basement is characterized by medium- to high-grade metamorphic rocks intruded by granitoids. The metamorphic rocks are amphibolites, talc schists, metaquartzites, and marbles of the Ancaján Formation. These rocks are associated with quartzphyllites, quartzites, and quartz-chloritoid schists of the Abra de Matienzo Formation. Both units have a penetrative deformation with a north-northeast trend and a dominant west-dipping fabric (Battaglia, 1982).

On the other hand, the Sierra de Aconquija is mainly composed of quartzphyllites, biotite-muscovite schists, and garnet-mica schists. This medium- to low-grade metamorphic basement is also intruded by early Paleozoic granitic apophyses of the Aconquija Batholith (González Bonorino, 1951, Ruiz Huidobro, 1972). The regional trend of schistosity is north-northeast to north-northwest, and it is gently east-dipping in the western sector of the range and steeply east-dipping in the eastern sector. However, there are marked exceptions in the vicinity of the intrusive bodies (González Bonorino, 1972).

At Cumbres Calchaquíes (see location in Figure 2), the basement has similar composition, but in the eastern sector thick marble layers dip 30° east (Ruiz Huidobro, 1972).

The Sierra de Quilmes (or del Cajón) basement (Figure 2) consists of phyllites and mica schists of the Famabalasto Formation, with numerous pegmatite dikes (Turner, 1973). This dominant penetrative fabric has a north-northwest trend and dips 50° to the east (Toselli et al., 1976). This range has no major bounding faults.

The age of the metamorphic basement is constrained by a few K/Ar data, which yield minimum ages for the metamorphism of 580 ± 20 Ma (Linares and González, 1990). The plutonic rocks are arc-related,

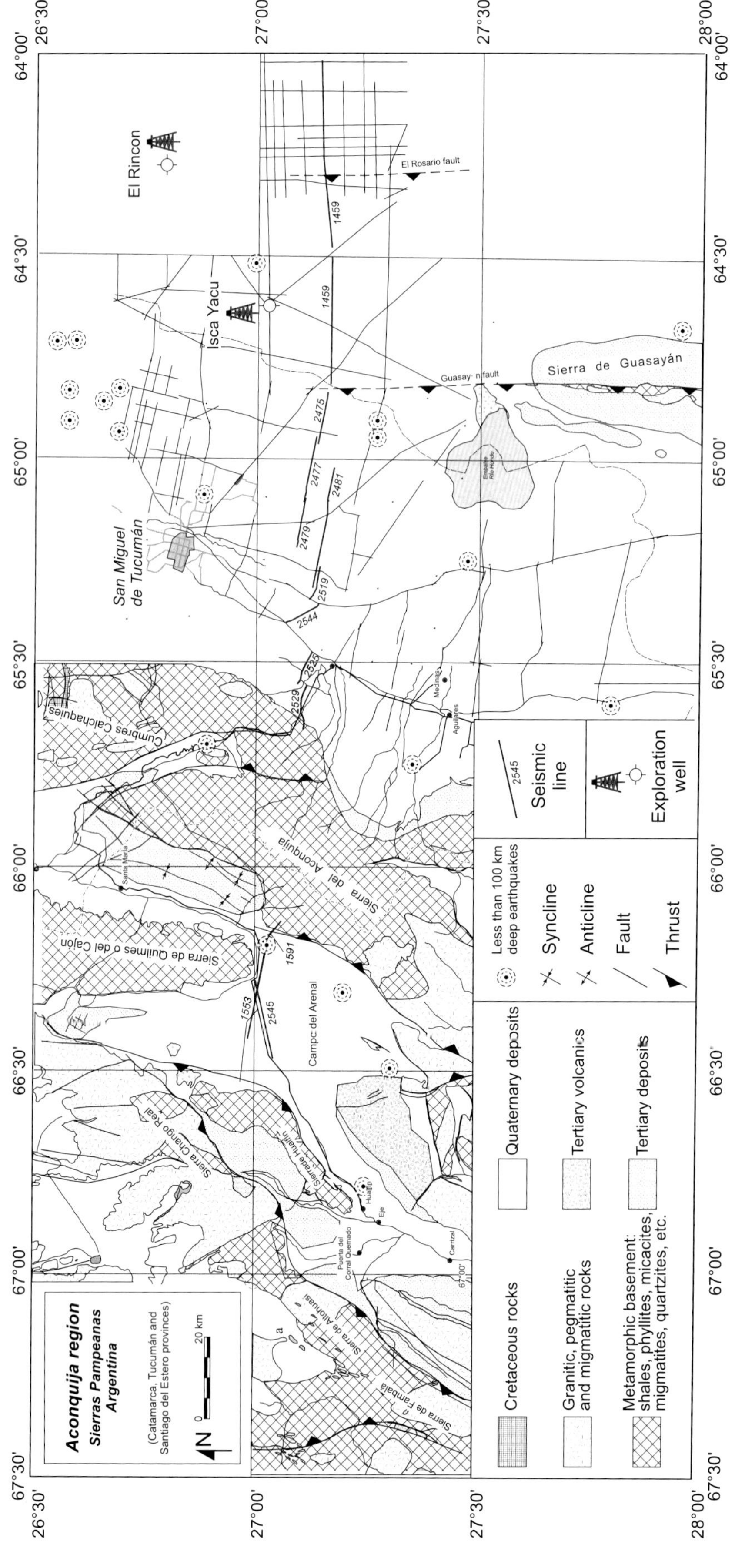

Figure 2. Regional geologic map, based on González Bonorino (1950b, 1951, 1972), Ruiz Huidobro (1966, 1972), and Battaglia (1982). Locations of seismic lines and oil wells are indicated, as well as the epicenters of intraplate earthquakes, based on the IUGS and SALSA program.

syncollisional and postcollisional plutons of early Paleozoic age (Quenardelle and Ramos, 1999).

Paleozoic Deposits

Extensive areas of northwest Argentina expose clastic Silurian-Devonian rocks, which represent a widely developed foreland basin succession as much as 5–6 km thick. These deposits are the result of Late Ordovician Ocloyic compressive deformation (Ramos, 1986). This has been interpreted to be a result of collision of the Arequipa-Antofalla parautochthonous block against the protomargin of Gondwana (Ramos and Basei, 1997). There are different depocenters, such as the subandean foreland Noroeste basin (Vistalli, 1989) and its extension farther south in the Alhuampa Basin. This basin is a strongly asymmetric depocenter, bounded to the west by the Guasayán and El Rosario Faults. It extends to the east about 300 km, where it pinches out against the Río de la Plata craton (Pezzi and Mozetic, 1989). The more than 3500-m-thick Silurian-Devonian deposits have been partially encountered in the El Rincón (location in Figure 2) and Caburé (east of Figure 2) exploration wells. Several sequences of marine shales and sandstones constitute a deltaic progradation system that is dated by trilobites and acritarchs as being Silurian and Early Devonian.

These sequences are unconformably overlain by Carboniferous deposits. A pre-Carboniferous high developed in the western margin of the Alhuampa Basin. The basal units are shales with a prograded sequence of sandstones that overlap the positive areas. Thick sequences of tillites are extensively developed throughout the basin. Based on the microflora, a Carboniferous age was assigned (Russo et al., 1987).

Cretaceous–Early Paleogene Rift Deposits

There are a few outcrops of Cretaceous–early Paleogene deposits in the province of Tucumán (Porto et al., 1982), but there were no exposures along the analyzed seismic lines. The presence of these deposits along the seismic transect is inferred based on the geometry and seismic fabrics, which are similar to that described farther north (Cristallini et al., 1997).

These deposits make up part of the Salta Group, a series of alluvial fan, fluviatile, and lacustrine to shallow-marine sequences associated with a rift system (Bianucci et al., 1981; Salfity et al., 1993). Several isolated troughs are recognized in the early synrift stages, during the Early Cretaceous to Cenomanian, which are progressively linked during the sag phases of sedimentation. The red-bed synrift deposits are associated with alkaline basalts in the Pirgua Subgroup (Galliski and Viramonte, 1988). These rocks are overlain by the Balbuena and Santa Bárbara Subgroups, a set of sandstones, shales, and carbonate deposits that record the marine transgression in Maastrichtian to Paleocene time (Salfity et al., 1993).

Synrift deposits of the Pirgua Subgroup have been observed in the northern sector of the province of Tucumán. The conglomerates and red beds were identified by Keidel (1913) and Bohm et al. (1962), and they were named the El Cadillal Formation by Bossi (1969). Bossi correlated these deposits with the ones exposed in the Sierra de Guasayán (Figure 2). Alkaline basaltic rocks are interlayered in the basal sequence.

Ruiz Huidobro and Porto (1987) described deposits of the Pirgua Subgroup unconformably lying on Precambrian basement rocks in Loma Montosa, 60 km west of Tucumán city. Dinosaurs and bird fossils confirmed a Late Cretaceous age for these deposits (Bonaparte et al., 1977).

In the eastern slope of Cumbres Calchaquíes, other series of red beds assigned to the Pirgua Subgroup (Ruiz Huidobro, 1972) were reinterpreted as part of the Paleogene Balbuena Subgroup (Bossi and Palma, 1982). These beds represent late sag-phase deposits, unconformably underlying the Neogene synorogenic deposits of the Santa María Group, extensively developed in the Cumbres Calchaquíes and in the Santa María Valley.

Other continental sandstones of the Los Cerrillos Formation exposed farther east in the Sierra de Guasayán (Battaglia, 1982) are correlated with the synrift Cretaceous deposits of the Córdoba depocenter (Kay and Ramos, 1996). There is no obvious correlation between these deposits and the synrift facies exposed in Tucumán region.

Based on these isolated outcrops, Porto et al. (1982) presented a Salta Group isopach map of Tucumán, interpolating a thickness of about 400 m along the seismic lines and defining an asymmetric basin that pinches out to the west.

Neogene Synorogenic Deposits

A series of Neogene deposits is poorly exposed along the eastern slope of the Sierra de Aconquija because of dense forest cover. These deposits consist of feldspar-rich fine-grained sandstones, ash-fall tuffs, and a few conglomerates (González Bonorino, 1951). These distal synorogenic deposits unconformably overlie the basement and dip 25° to the east near the mountain front and only 5° away from it. Most proximal facies are coarse conglomerate remnants, with coarse andesitic and granitic clasts that crop out entrenched in some creeks, such as in the Cochuma Valley.

These Neogene deposits are unconformably covered by thick Quaternary alluvium along the latitude of the seismic line. However, they are well exposed to the north, in the vicinities of El Cadillal and in Río Hondo, and to the south in the Sierra de Guasayán, as a result of incipient neotectonic basement uplifts. A series of green clays and gypsum-rich layers is exposed at both sides of the Sierra de Guasayán. These beds, known as the Guasayán Formation, contain middle Miocene forams of the Paraná marine transgression (Battaglia, 1982). The Las Cañas Formation unconformably overlies the previous unit and consists of conglomerates having a local basement provenance, and brown-red mudstones at the top. This unit is gently folded and unconformably overlain by the boulders of the Choya Formation (Beder, 1928), which is composed of locally derived crystalline basement and acidic volcanic clasts up to 30 cm in size. The complete Neogene sequence does not exceed 450 m (Lucero Michaut, 1979).

The foreland basin deposits north of Tucumán city overlie white quartzose sandstones of the Río Loro Formation that have been correlated with the sag-phase deposits of the Salta Group rift. At the Río Salí headwaters, gypsum-rich green clays and siltstones are exposed in beds as much as 1 m thick and interlayered with reddish brown thinly laminated siltstones with low gypsum content. These deposits are known as the Río Salí Formation and their green beds contain *Corbicula*, a fresh-water pelecypod, and osteoderms of *Eosclerocalyptus planus,* found by Peirano (1957), that are similar to the fossils found farther west in Santa María Valley (Bossi, 1969). Keidel (1913) also mentioned remains of fossil fishes in yellowish bituminous marls at the base of this unit. Occasionally conglomeratic gray and light brown sandstones and brown siltstones of the India Muerta Formation transitionally overlie the Río Salí Formation.

The Neogene deposits in the western slope of the Sierra de Aconquija are exposed in Santa María Valley. They are included in the Santa María Group and are represented from base to top by the San José, Las Arcas, Chiquimil, Andalhualá, Corral Quemado, and Yasyamayo Formations (Figure 3) (Ruiz Huidobro, 1972). The fine-grained gray sandstones and green siltstones of the San José Formation have fossils similar to those at the base of Río Salí Formation (Bossi, 1984), as well as middle Miocene forams of the Paraná marine transgression (Bossi and Palma, 1982; Gavriloff and Bossi, 1992). The San Jose overlies the Saladillo Formation, which Strecker (1987) dated at 11 Ma, and it is conformably overlain by red sandstones and conglomerates of the Las Arcas Formation and then by coarse- to medium- to coarse-grained yellowish volcaniclastic sandstones of the Chiquimil Formation. This last unit is subdivided in two members, a fine-grained lower member and a coarser upper member. Magnetostratigraphy dates the boundary between these two members at 6.68 Ma (Strecker, 1987). Based on east-to-west paleocurrents, Strecker interpreted the upper member of Chiquimil Formation to be a response to an incipient uplift of the Sierra de Aconquija-Cumbres Calchaquíes block. Based on paleoclimatic considerations, Strecker (1987) claimed that this block also remained as a low-relief positive block during sedimentation of medium to coarse sandstones and lensoid conglomerates of the Andalhualá Formation. This unit was deposited between 6.02 and 4.8–3.4 Ma and was covered by the coarse conglomerates of the Corral Quemado Formation. These conglomerates mark the final Neogene uplift of the Sierra de Aconquija and Cumbres Calchaquíes (Strecker, 1987). The top of the Corral Quemado Formation has been dated at 2.97 ± 0.6 Ma. Very coarse and well-cemented conglomerates of the Yasyamayo Formation unconformably overlie the Santa María Group and represent the first pedimentation level of Quaternary age on the western slope of the Aconquija (Strecker et al., 1989).

PROCESSING CHARACTERISTICS OF SEISMIC LINES

General Considerations

Seismic data acquired by Yacimientos Petrolíferos Fiscales (YPF S.A.) between 1989 and 1991 were reprocessed. A total of approximately 300 km of seismic lines were analyzed (see location in Figure 2).

Vibrators were used as a unique source in the central sector (southern Tucumán), but the records in Santiago del Estero involved explosives, and in Campo del Arenal (Catamarca) experiments included both vibrators (line 2545) and explosives (lines 1591 and 1553). In general, field parameters changed from one sector to another, and Table A1 in Appendix A presents a synthesis of both Vibroseis and explosive data.

The YPF wells Isca Yacu x-1 and El Rincon x-1 (Figure 4) were used as a control to support the interpretation of some acoustic horizons (see location in Figure 2). However, geologic information from several old deep-water borings beside the lines contributed to the partial understanding of the seismic data.

The different reprocessing techniques applied to the industrial seismic lines are described in Appendix A.

INTERPRETATION OF THE SEISMIC LINES

Basement and Paleozoic Cover

The seismic lines show a clear contact between the basement and the cover (see Figure 5), which mainly consists of unconformably overlying Tertiary synorogenic deposits and synrift sediments of the Late Cretaceous age Pirgua Subgroup. The seismic fabric of the basement is clearly seen in different lines, varying from

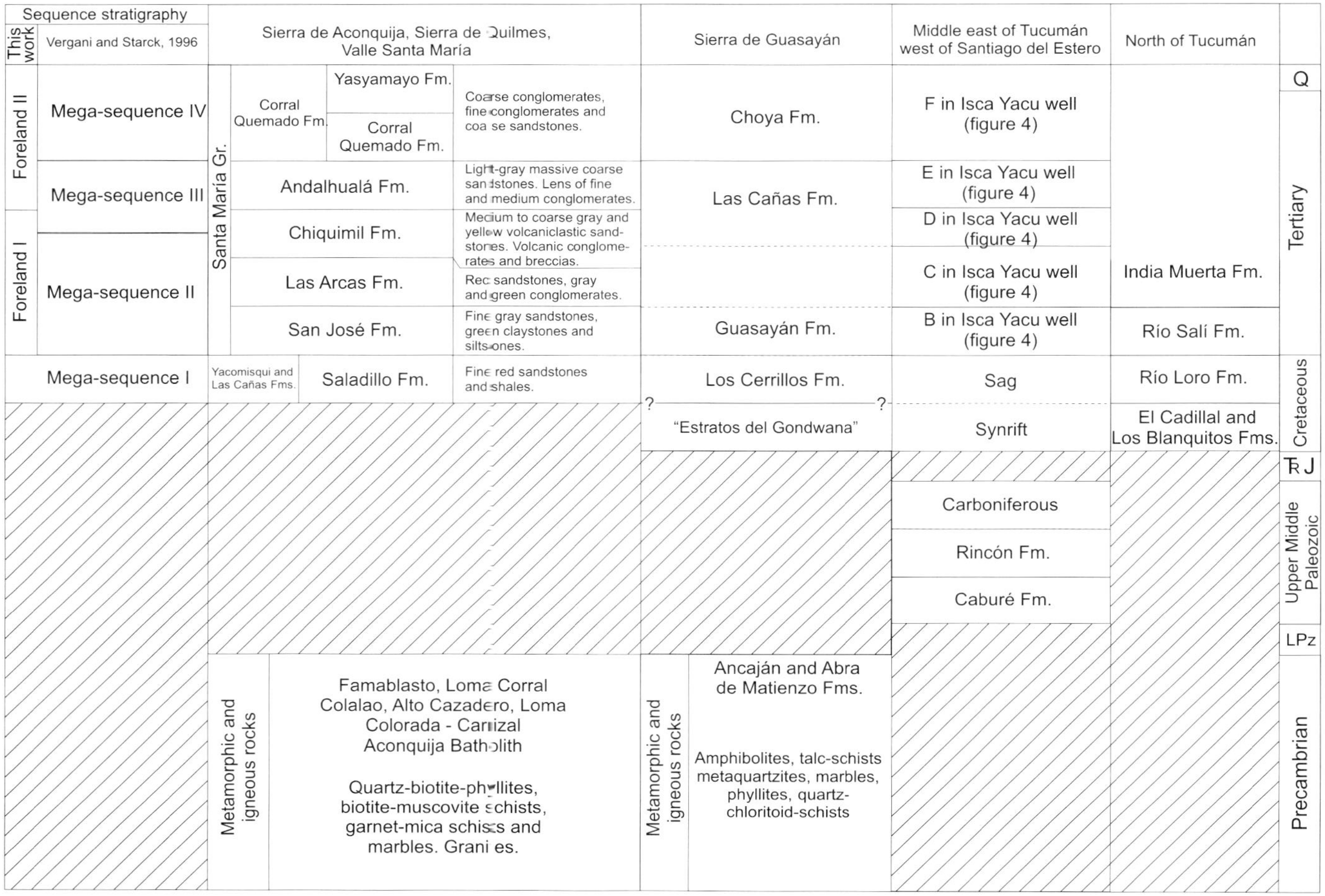

Sequence stratigraphy: This work	Sequence stratigraphy: Vergani and Starck, 1996	Sierra de Aconquija, Sierra de Quilmes, Valle Santa María				Sierra de Guasayán	Middle east of Tucumán west of Santiago del Estero	North of Tucumán	
Foreland II	Mega-sequence IV	Santa María Gr.	Corral Quemado Fm.	Yasyamayo Fm. / Corral Quemado Fm.	Coarse conglomerates, fine conglomerates and coarse sandstones.	Choya Fm.	F in Isca Yacu well (figure 4)		Q / Tertiary
	Mega-sequence III	Santa María Gr.	Andalhualá Fm.		Light-gray massive coarse sandstones. Lens of fine and medium conglomerates.	Las Cañas Fm.	E in Isca Yacu well (figure 4)		Tertiary
Foreland I	Mega-sequence II	Santa María Gr.	Chiquimil Fm.		Medium to coarse gray and yellow volcaniclastic sandstones. Volcanic conglomerates and breccias.	Las Cañas Fm.	D in Isca Yacu well (figure 4)		Tertiary
		Santa María Gr.	Las Arcas Fm.		Red sandstones, gray and green conglomerates.		C in Isca Yacu well (figure 4)	India Muerta Fm.	Tertiary
		Santa María Gr.	San José Fm.		Fine gray sandstones, green claystones and siltstones.	Guasayán Fm.	B in Isca Yacu well (figure 4)	Río Salí Fm.	Tertiary
	Mega-sequence I	Yacomisqui and Las Cañas Fms.	Saladillo Fm.		Fine red sandstones and shales.	Los Cerrillos Fm. ?	Sag	Río Loro Fm.	Cretaceous
						"Estratos del Gondwana" ?	Synrift	El Cadillal and Los Blanquitos Fms.	Cretaceous
									TR J
							Carboniferous		Upper Middle Paleozoic
							Rincón Fm.		Upper Middle Paleozoic
							Caburé Fm.		Upper Middle Paleozoic
									LPz
		Metamorphic and igneous rocks	Famablasto, Loma Corral Colalao, Alto Cazadero, Loma Colorada - Carrizal Aconquija Batholith. Quartz-biotite-phyllites, biotite-muscovite schists, garnet-mica schists and marbles. Granites.			Metamorphic and igneous rocks: Ancaján and Abra de Matienzo Fms. Amphibolites, talc-schists metaquartzites, marbles, phyllites, quartz-chloritoid-schists			Precambrian

FIGURE 3. Stratigraphic correlation chart.

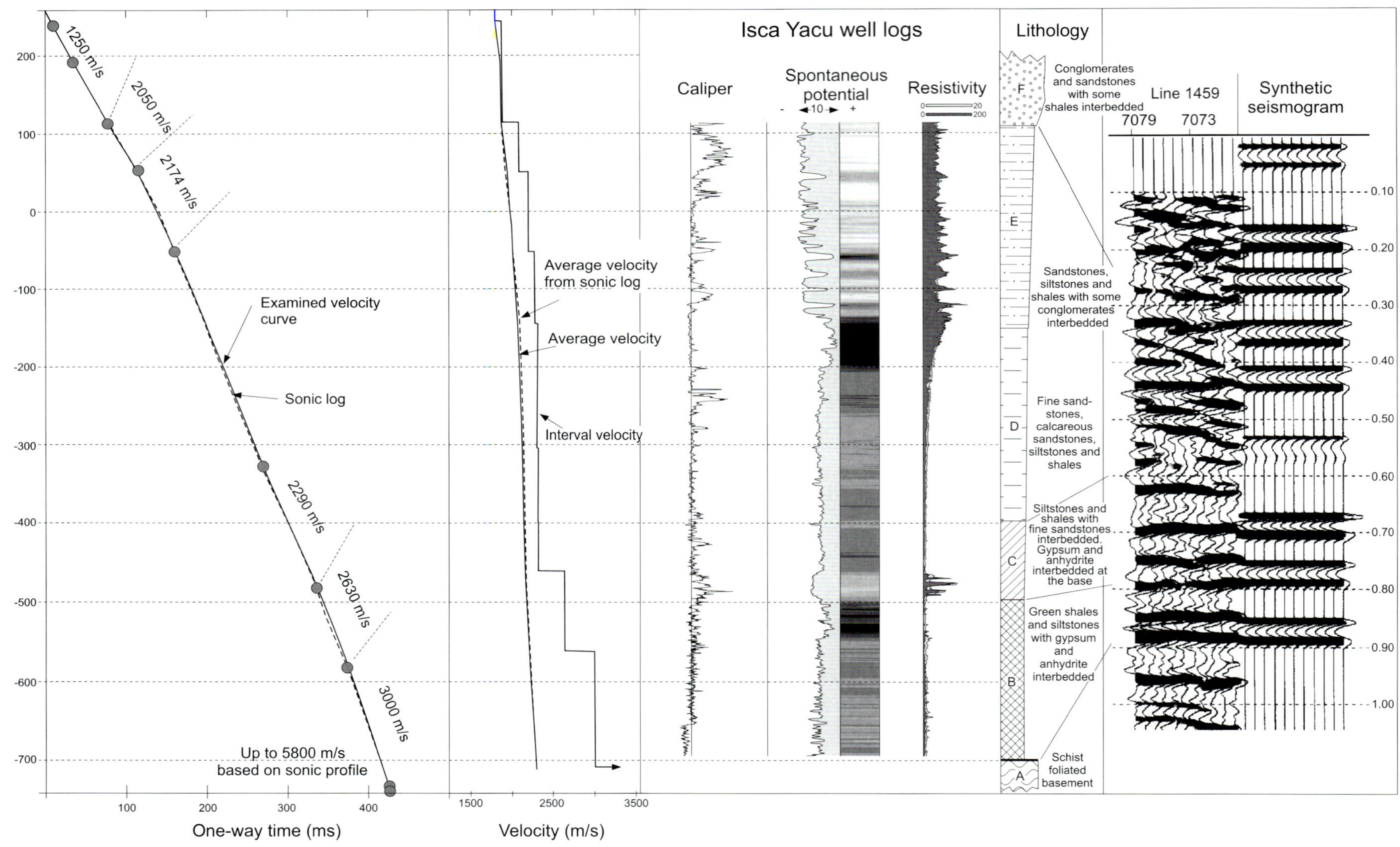

FIGURE 4. Isca Yacu well logs and comparison with the synthetic seismogram and line 1459 record (see discussion in text).

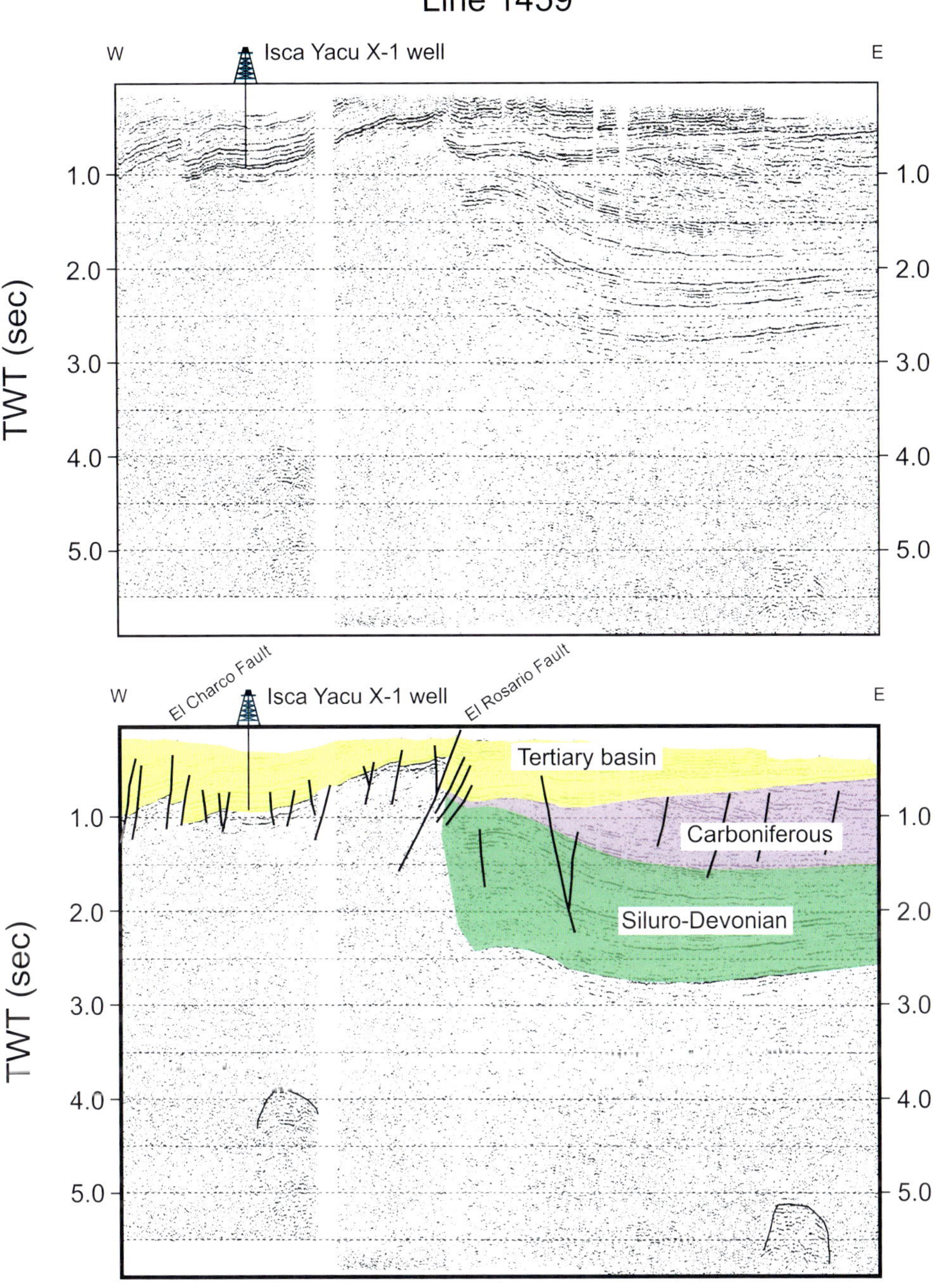

FIGURE 5. Uninterpreted and interpreted versions of seismic line 1459. Note the pinch-out contrast between Carboniferous and Silurian-Devonian deposits.

conspicuous west-dipping reflectors in the eastern side (line 2481, Figure 6) to east-dipping reflectors in the western sector (line 2529, Figure 7) and matching the described main penetrative structural grain of the basement.

The eastern sector of the study area has well-developed Paleozoic cover (Figure 5), but the contact between the basement and the sedimentary sequence is not well defined. Two unconformably overlying sequences are identified in seismic line 1459 (Figure 5). The lower sequence has been correlated with the Silurian-Devonian deposits of the Caburé and El Rincón Formations. They show a typical westward thickening that characterizes the early Paleozoic foreland basin in northwestern Argentina (Vistalli, 1989). The western margin is truncated by the El Rosario Fault, which seems to be a reactivation of the previous middle Paleozoic thrust front that developed during the Ocloyic deformation.

The Carboniferous deposits have a different polarity and pinch out to the west (Figure 5). This feature was noted by Russo et al. (1987) in the Alhuampa Basin. The depocenter of the basin is to the east of the seismic line. One outstanding characteristic of these Carboniferous deposits is the second-order normal faulting, detached at the base of the sequence, that does not affect the Devonian strata.

Westward of the El Rosario Fault, in line 1459, parallel reflectors are present beneath the angular unconformity that separates the Neogene deposits from the basement. These reflectors could be Paleozoic beds, although the bottom of the Isca Yacu well (discussed below) intersected a 3-m interval of steeply dipping, strongly foliated schists. The well was projected into seismic line 1459 (see Figure 2), and therefore the conflict of the subparallel stratified deposits and the foliated schist may be only apparent, or perhaps the schists in this old well are really folded Paleozoic beds.

Cretaceous–Paleogene Sequences

The seismic lines depict a series of stratified deposits with conspicuous thickness variations, which outline a north-south-trending basin at 65°20′ W longitude (Figure 6). These deposits underlie the Neogene foreland basin sequences and are here interpreted to be synrift deposits, based on their half-graben geometry.

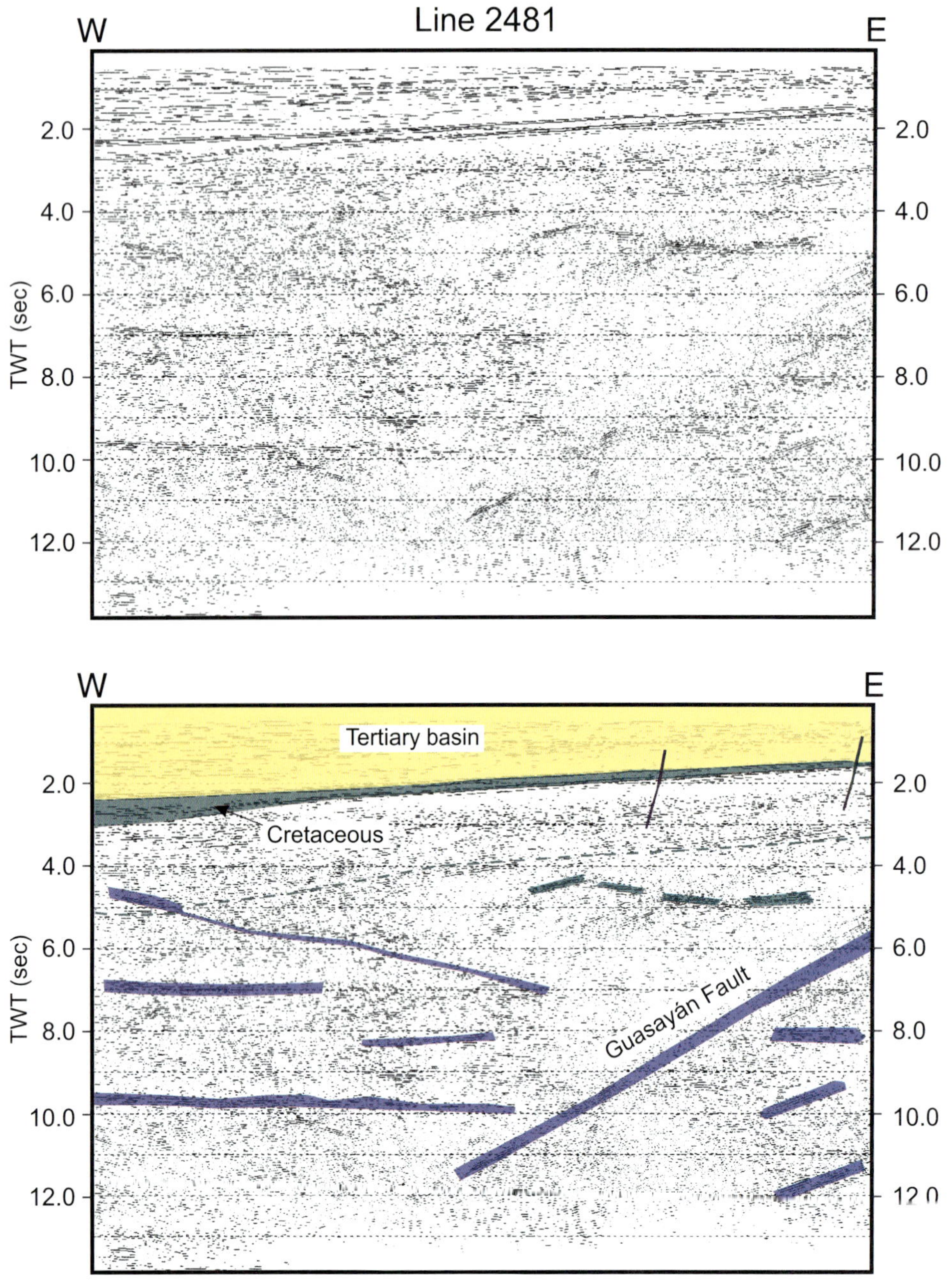

FIGURE 6. Uninterpreted and interpreted versions of seismic line 2481. Note the depth and dip of the Guasayán Fault.

These deposits are the southernmost extension of the Salta Group basin (Porto et al., 1982). The western side of seismic line 2481 (Figure 6) depicts a half-graben fill as much as 400 m thick. This basin is relatively narrow (less than 50 km wide), and the maximum sediment thickness is less than 2000 m.

The Isca Yacu Exploration Well

The Isca Yacu oil well was drilled by Continental Oil Company in 1963. It reached 1000 m in depth, went through 997 m of Tertiary deposits, and crossed 3 m of 70° dipping, strongly foliated schists. It is located 14 km north of seismic line 1459 (at 64°38′W longitude). The sonic well log was used to produce a synthetic seismogram, which is compared with the equivalent section of seismic line 1459, the lithology, resistivity, interval velocity and velocity curves, and other logs (Figure 4).

Based on these data, several sequences have been recognized in the Isca Yacu well. The upper 180 m correspond to conglomerates and sandstones with some interbedded shales (F in Figure 4). These are underlain by 200 m of sandstones, siltstones, and shales, with interbedded conglomerates (E in Figure 4). Approximately at 400 m b.k.b. (below kelly bushing) the proportion of fine material increases (D in Figure 4), replacing the sandstones and conglomerates. This change is clearly seen in the logs, and the shales and siltstones persist down to 800 m b.k.b. Gypsum-rich levels are detected at the base of C (Figure 4). Below this level green shales and siltstones with gypsum-rich beds intercalated with marls characterize the basal sequence A (Figure 4), which was deposited on the strongly foliated schists.

These lower units are correlated with the base of Río Salí Formation in the eastern foreland area and with the San José Formation in the western slope of the Aconquija. These units are typical of the Paraná marine transgression that invaded the region from the Atlantic side

Neogene Synorogenic Sequences

As a result of a late Eocene compression, a series of foreland basins began to develop in northwestern Argentina. Recent studies of the synorogenic deposits identified several tectosedimentary sequences in this region that record the tectonic evolution and uplift of the Andes

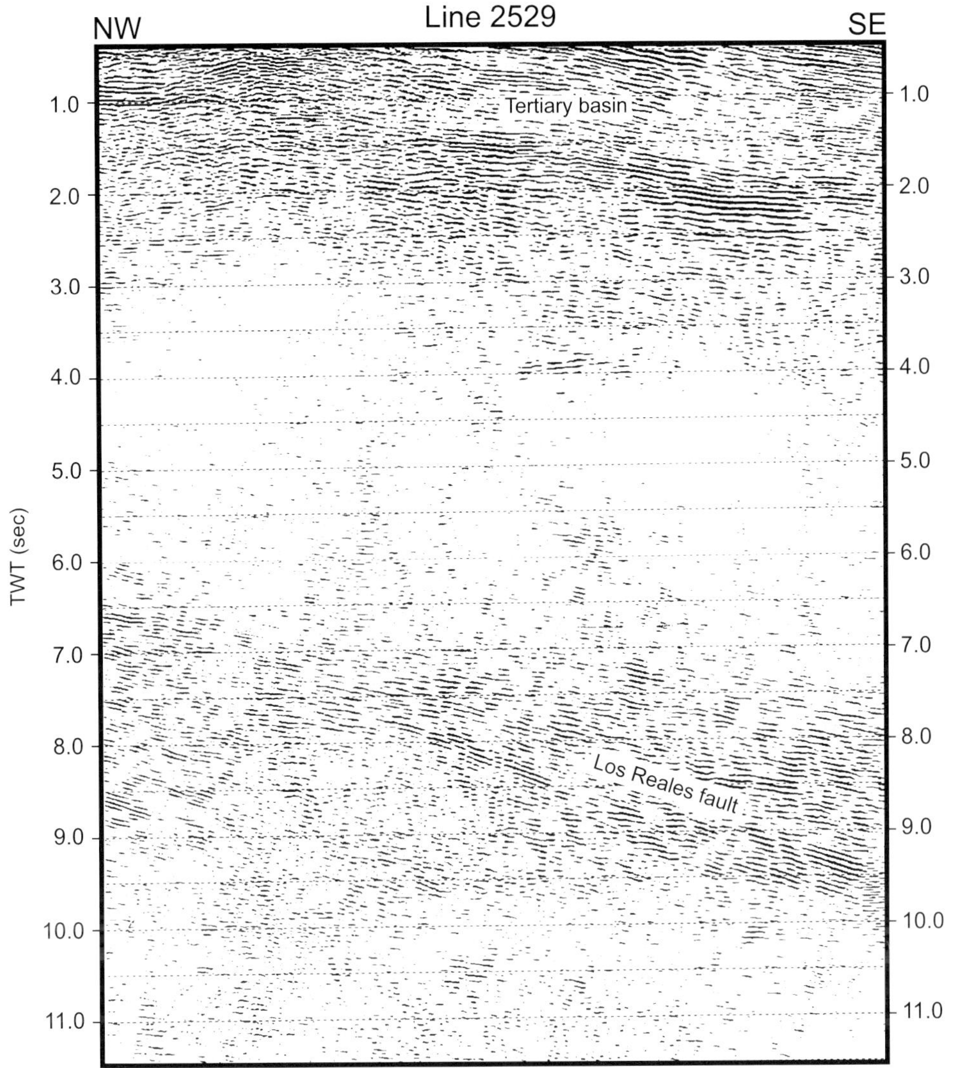

FIGURE 7. Seismic line 2529. Note Los Reales Fault and the regional dip of basement reflectors.

The seismic lines reprocessed in the western side of the Sierra de Aconquija (seismic lines 1553, 1591, and 2545) show three well-defined seismostratigraphic sequences. These sequences are only partially correlated with the proposed megasequences of Starck and Vergani (1996). The lower sequence, a, is deposited on the Precambrian basement, is west-dipping, and gradually pinches out to the east (Figure 8). The parallel seismic fabric, as well as the gradual variations in thickness, indicates a distal west provenance. Although no well intersects these sequences, they are tentatively correlated with the Saladillo, San José, Las Arcas, and the lower member of the Chiquimil Formations of the Santa María Group and therefore are correlated with Megasequences I and II of Starck and Vergani (1996). The lower part of sequence a contains the San José Formation, which records the Paraná marine transgression in the middle Miocene.

The strongly asymmetric overlying sequence, b, thickens and dips to the west (Figure 8). A series of truncations in the lower unit indicates a gentle, angular unconformity between the two sequences. This middle sequence onlaps the lower unit, which may indicate that the middle sequence involves growth strata deposited during the continuing gentle uplift of the source area located to the east of the section. Accordingly, it may correlate with the upper member of the Chiquimil and Andalhualá Formations described by Strecker (1987) farther north and west of the Sierra de Aconquija-Cumbres Calchaquíes, and with Megasequence III of Starck and Vergani (1996).

The upper and third sequence, c, lies unconformably on the middle sequence. Although the seismic fabric does not indicate a definite provenance, the striking parallelism of its seismic reflectors suggests that the source was more remote than that for the middle unit. The upper sequence could be the equivalent of the Corral Quemado and Yasyamayo Formations (Figure 3),

(Starck and Vergani, 1996). At these latitudes, most uplift was controlled by tectonic inversion of Salta Group's normal faults (Grier and Dallmeyer, 1990; Salfity et al., 1993; Cristallini et al., 1997) and by inhomogeneities of the Precambrian–early Paleozoic basement (Hongn and Seggiaro, 1998). As a result, some of these initial depocenters were isolated during most of the Neogene; therefore interbasinal correlation is sometimes quite difficult. The existence of a marine transgression during middle Miocene time provided an excellent level of interbasinal correlation (Ramos and Alonso, 1995), and therefore its identification was a valuable marker for the correlation between western and eastern sequences.

Campo del Arenal Western Sequences

Neogene deposits unconformably overlie the metamorphic basement, the Paleozoic rocks, and the Cretaceous rift sediments.

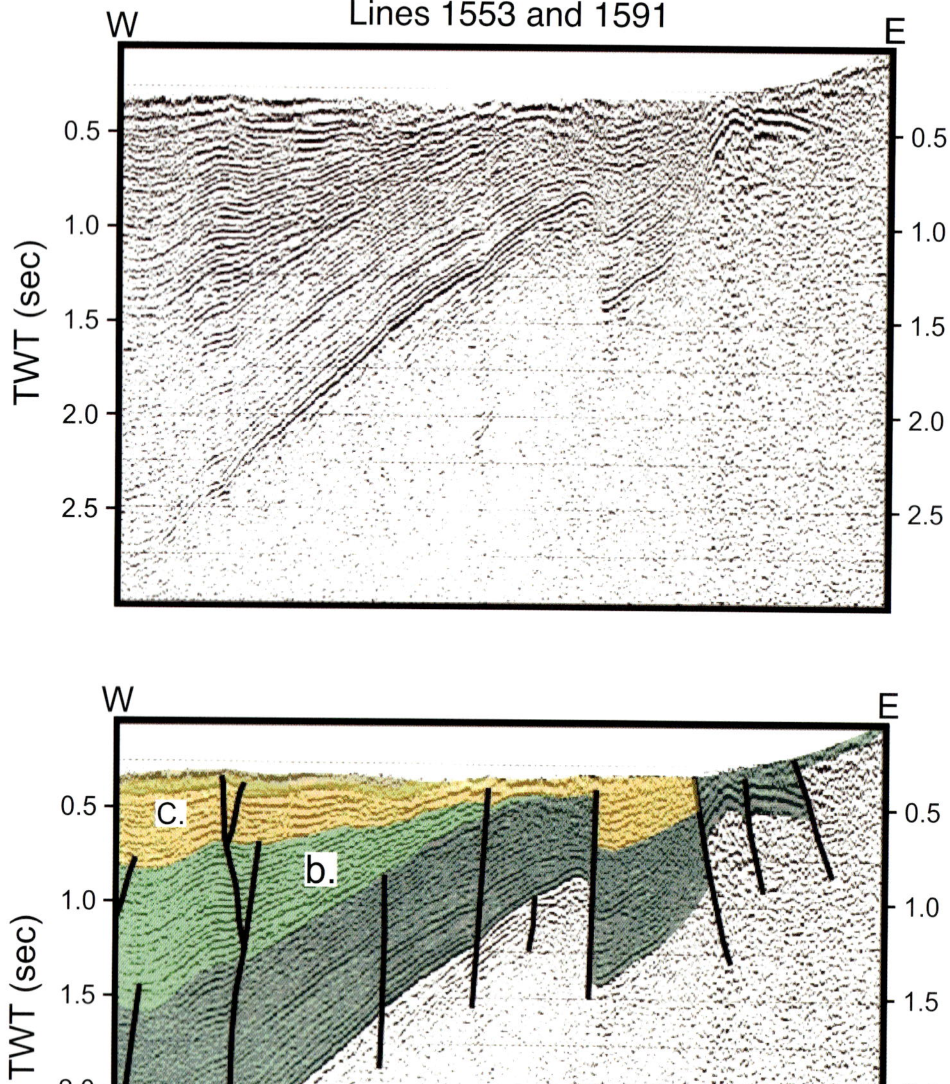

Figure 8. Uninterpreted and interpreted versions of seismic lines 1553 and 1591. Note the change from east vergence to west vergence of the reverse faults. Sequences a, b, and c are discussed in the text.

Foreland basin I sequence, the basal sequence, gently pinches out to the east along the eastern slope of the Sierra de Aconquija and adjacent plains, and it does not overlap the basement in this range. The basal sequences of the Isca Yacu well east of the Aconquija (B and C of Figure 4) can be correlated with the basal sequence a of line 1553 (Figure 8) in the western sector. Thickness and seismic attributes show that both sets of sequences correspond to the development of a common foreland basin, identified here as foreland basin I. These sequences can be correlated with the lower part of Río Salí Formation in the north part of Tucumán, with green clays and gypsum-rich layers of the Guasayán Formation in the Sierra de Guasayán, and with the San José, Las Arcas, and lower member of Chiquimil Formations in the western side of the Aconquija along the Santa María Valley. The basal sequence of these deposits records the Paraná marine transgression.

Foreland basin I was the result of the flexural subsidence related to the basement uplift of the Sierras de Altohuasi and Chango Real (Figure 2) along the eastern border of the Puna high plateau during lower-middle Miocene times. This uplift occurred prior to the marine transgression correlated with the NN5 and dated at 13.5 Ma (Malumián and Nañez, 1996). At that time, the Sierra de Aconquija was not uplifted.

Foreland basin II sequence comprises seismic sequences that onlap transgressively to the east on the western side of the Sierra de Aconquija (b in Figure 8). The equivalent sequence on the eastern side has an opposite transgressive overlap, as shown in Figure 9. It overlaps to the west onto foreland basin I sequences in the headwaters of the Ríos Salí and Colorado. This fact

and may correlate with Megasequence IV of Starck and Vergani (1996).

Eastern Aconquija Sequences

The seismic lines analyzed on the eastern side of the Sierra de Aconquija have two well-defined first-order sequences. Both sequences record the evolution of the foreland basin at this latitude.

indicates that dispersion of the sedimentary fill was antithetic as a consequence of the uplift of the Sierra de Aconquija. However, west of this range foreland basin II is represented by two sequences. The lower one, b, represents a gentle uplift of the Sierra de Aconquija, as observed in Figure 8. The seismic geometry of sequence b indicates that the uplift involved a progressive rotation of the synorogenic deposits to the west. The blind thrust observed in the deepest part of seismic line 2545 (Figure 10) could have been responsible for the initial uplift of the Sierra de Aconquija. Based on the subparallel reflectors of the upper sequence, c, the source is interpreted to have been more remote than in sequence b. If an eastern source is assumed, the mountain front should have retreated, and therefore the fault that uplifted the Aconquija should be younger and to the east of the blind thrust.

As Strecker (1987) established, the lower member of the Chiquimil Formation recorded the initial uplift of the Sierra de Aconquija, but the final uplift was related to the Corral Quemado Formation. The base of this unit was deposited at 3.4 ± 0.5 Ma, and therefore the final uplift of the Aconquija was during the middle to late Pliocene. Evidence for recurrent tectonic movement postdating the main Pliocene deformation is well documented in the pediments of the western slope of the Sierra de Aconquija (Strecker et al., 1989). Large-scale tectonic movements continued to be active until at least 0.6 Ma, as inferred from the large avalanches recorded in the western slope of the Aconquija (Fauqué and Strecker, 1988).

STRUCTURE

Seismic Interpretation

Although several previous studies established the surface structure of the different ranges of the region (González Bonorino, 1950a; Strecker et al., 1989; González and Barreñada, 1993), no data were available on the subsurface structure of the Sierra de Aconquija and surrounding areas.

The eastern seismic lines (1459, 2475, 2477, 2479, 2481, and 2519) show a series of west-dipping reverse faults that intersect and deform the foreland basin deposits (Figures 5, 6, and 11). The most important regional faults are the Rosario and Guasayán Faults (Figure 12). The Rosario Fault is a first-order Neogene structure that is recognized from the central-eastern part of Salta province to the study area, more than 300 km away. It is the present orogenic front of the Andean deformation. It has had important Quaternary activity, as recorded in the morphological expression of recent uplift and in increased erosion detected in the Río Hondo area. Besides this morphological expression, there is important seismic activity in this area. The Rosario Fault has a series of minor splays at the surface (Figure 5) and

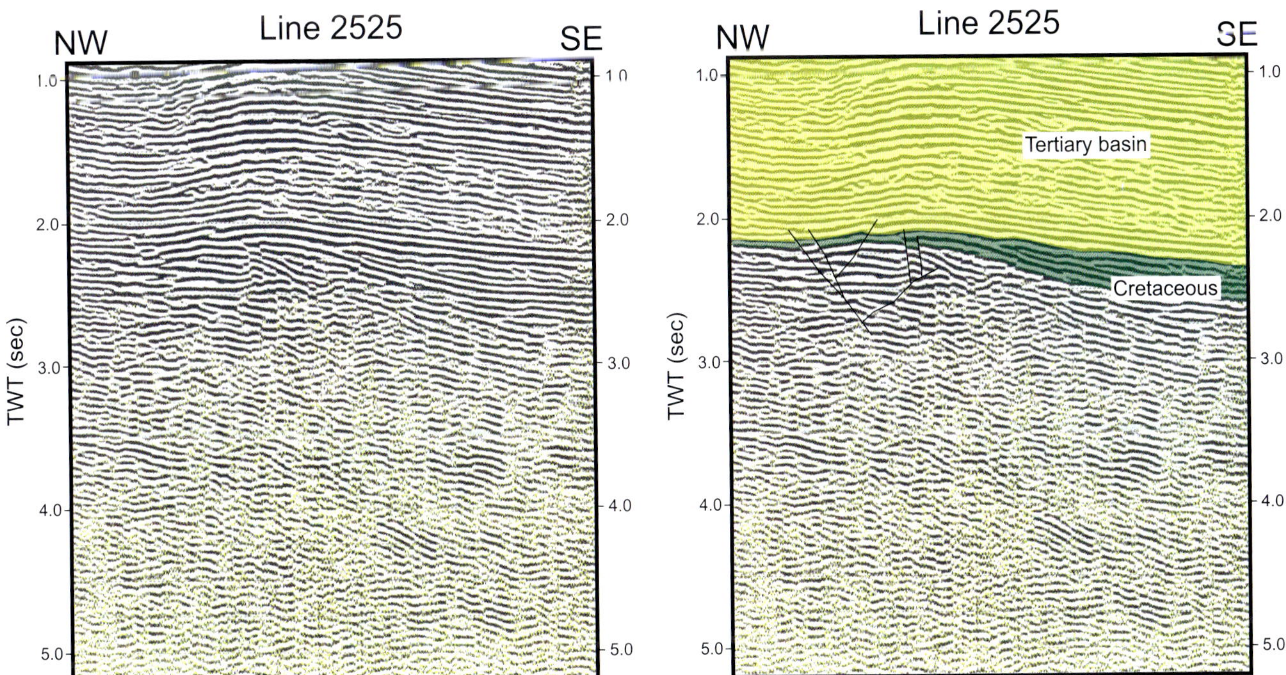

FIGURE 9. Uninterpreted and interpreted versions of line 2525. Note the geometry of the Cretaceous deposits and the western vergence of the structure.

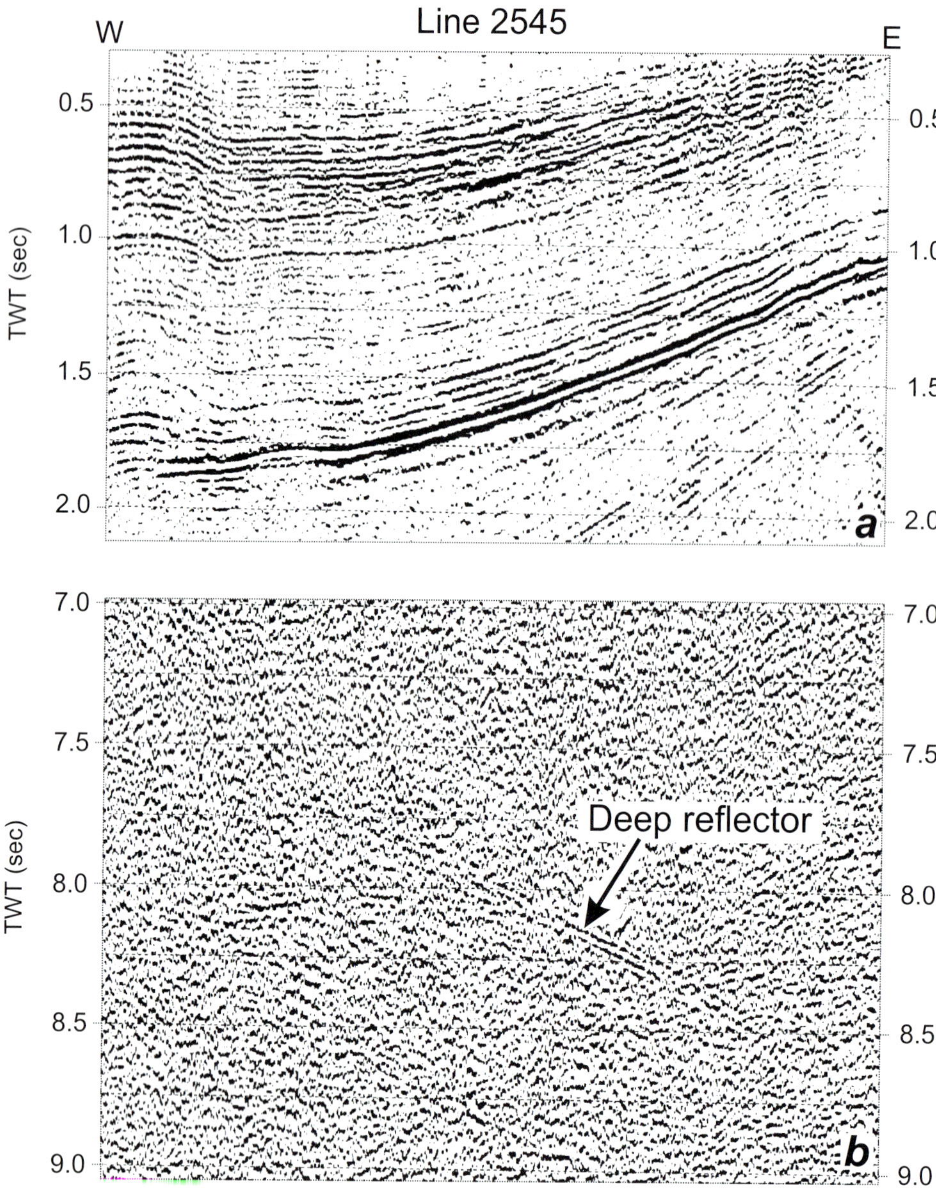

FIGURE 10. Seismic line 2545 (above, upper part, and below, lower part of the line). Note the opposing vergence between the Tertiary strata and the basement oblique reflector.

canas High, which is an important underground water divide, as indicated by several deep water wells. The Guasayán Fault is a north-south-trending structure like most of the basement faults recognized in this study. Oblique faults with important strike-slip components parallel to the Tucumán lineament (Mon, 1976) have not been verified, although their presence cannot be disregarded. Only in the western sector west of the Sierra de Aconquija have some possible minor flower structures been identified (Figure 8) that could indicate some strike-slip displacement along the reverse faults. The Guasayán Fault, as recognized in line 2481 (Figure 6), has a 45° westerly dip down to a depth of 40 km (12 s, TWT). It is important to note that this structure's dip does not decrease at depth, at least as deep as has been detected in the seismic lines (Figure 12). The throw of the Guasayán Fault in the foreland basin strata is not easily observed because the upper part of the fault lies between lines 2475 and 1459 in a seismic gap. However, on geologic grounds its throw is estimated to be approximately 300 m. The other reverse faults, west of the Guasayán Fault, although west-dipping, have small throws of less than 100 m.

records an approximately 500-m offset of the Neogene deposits.

The Silurian-Devonian foreland basin, to the east of the Rosario Fault, indicates that this fault could be a reactivation of the Late Ordovician thrust front that produced this Paleozoic basin. West of the Rosario Fault is a series of east-vergent reverse faults that have offsets of less than 100 m, except the El Charco Fault, which has a displacement of 250 m (Figure 5).

West of these faults is another important regional fault, which is identified in this study as the Guasayán Fault. It is interpreted as being the northern extension of the fault that uplifted the Sierra de Guasayán farther south (Figures 2, 6, and 12). It is responsible for the Ta-

There are no Paleozoic deposits west of the Guasayán Fault. Therefore, the observed half grabens have been interpreted to belong to the Salta Group Cretaceous rift in line 2481 (Figures 6 and 13). The lack of subsurface control prevents the identification of sag-phase deposits, which may have been mapped as part of the Tertiary deposits. A half graben was observed in the western sector of line 2481 (Figure 13), possibly filled with the clastics of Pirgua Subgroup and bounded on the west by an east-dipping normal fault. The remaining east-dipping faults farther to the west have a similar polarity.

The next lines to the west (lines 2455, 2525, and 2529) also contain west-vergent east-dipping reverse faults (Figures 7 and 9). Basement rocks of the eastern slope of the Sierra de Aconquija lie just under the surface along the western sector of line 2529, as inferred by seismic velocities greater than 4000 m/s at the surface. This basement uplift is produced by an east-dipping fault observed in the deep portion of the line that clearly shows a decrease in dip at depth. This fault, which has a seismic expression similar to that of the Guasayán Fault, is correlated with the reverse fault described by González Bonorino (1951) in the Río de Los Reales and Tafí del Valle creeks on the western slope of the Aconquija. This fault uplifts the basement (Figures 2 and 12), and to the northwest this basement overrides Neogene and Quaternary deposits as Strecker et al. (1989) and González and Barreñada (1993) describe.

The deep seismic reprocessing of the lines east of the Sierra de Aconquija has identified subhorizontal reflectors at 19, 25, and 30 km depth (7, 8.5, and 10 s) on line 2481 (Figure 6); at 25, 30, 40, and 52 km depth (8.5, 10, 12.3, and 14.6 s) on line 2455; and at 25 km (8.5 s) on line 2525 (Figure 14). Although these reflectors could be interpreted as brittle-ductile transitions (Comínguez and Ramos, 1991), in many cases the oblique reflectors intersect these lines without changes, as seen in the Guasayán Fault of line 2481 (Figure 6). This could also indicate that these discontinuities were active prior to the present Andean deformation. In other cases, such as with the Los Reales Fault, the dip angle decreases to become subhorizontal and asymptotic with the 10-s horizontal reflector of line 2525 (Figures 7, 12, and 14).

Several lines were reprocessed west of the Sierra de Aconquija in Campo del Arenal. This bolsón, or intermontane valley, is bounded by faults with opposite vergences (Figures 2 and 12). The east side of the bolsón, next to the Sierra de Aconquija, is bounded by a high-angle west-vergent reverse fault, whereas to the west, the Sierra de Hualfín has a reverse fault similar to the Aconquija, but with opposite vergence (González Bonorino, 1951; Strecker et al., 1989, Bossi et al., 1993). The deep seismic data show the interference and coexistence of two fault systems with opposite vergence beneath Campo del Arenal (Figure 10). These systems defined a thick-skinned triangle zone that is similar to the triangle zone identified by Zapata and Allmendinger (1996) between the Precordillera and Sierras Pampeanas.

The Sierra de Quilmes is exposed north of the study lines (2545, 1553, and 1591). Bossi et al. (1993) explain this structure as having been uplifted by a subexposed west-dipping reverse fault that bounds the eastern side of the range. The tip line of this fault is not exposed (Strecker et al., 1989), and the Sierra de Quilmes can be interpreted as a basement anticline uplifted by a blind thrust. The anticlinal geometry can be reconstructed on the basis of the peneplain preserved as a result of the recent uplift of this range (Strecker et al., 1989). The peneplain surface dips to the west on the western slope, but dips more steeply to the east on the eastern slope. Bossi et al. (1993) suggest that this asymmetry is consistent with the east vergence, although east vergence at the surface also can be explained by a high-angle, east-dipping reverse fault located beneath the range. The latter interpretation is more compatible with what is observed farther south, beneath Campo del Arenal, and with the basement fabrics observed in the Sierra de Quilmes. This range is abruptly truncated by a transverse fault in its southern end.

Lines 1553 and 1591 were combined, yielding an almost complete section of Campo del Arenal (Figure 8) where the interference between the two oppositely vergent fault systems is imaged. The western sector is characterized by west-dipping reverse

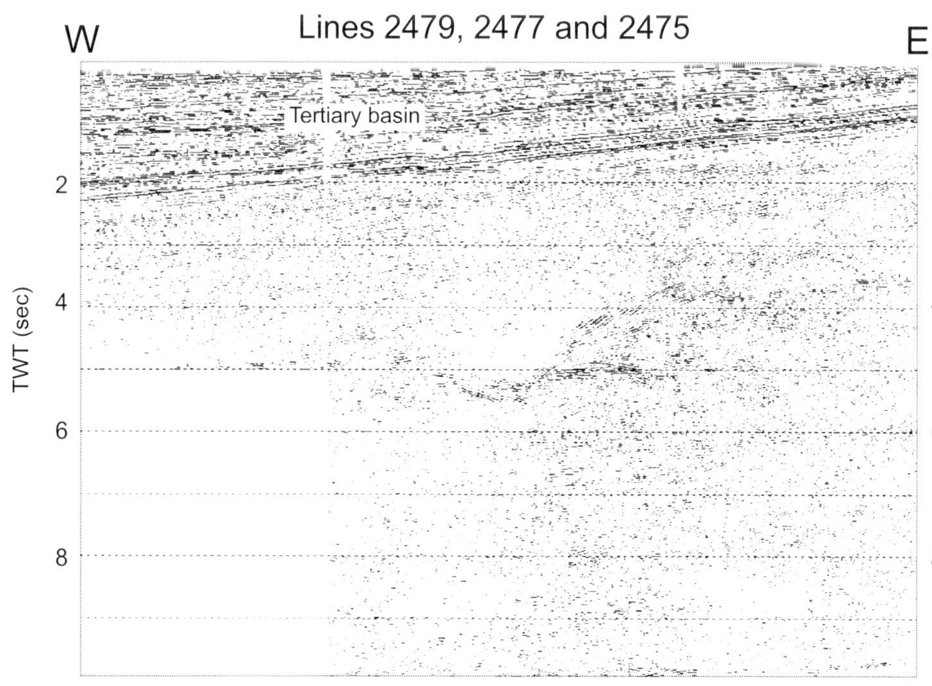

FIGURE 11. Seismic lines 2479, 2477, and 2475. The structures depicted between 3 and 5 seconds are correlated with those farther south identified on line 2481 (Figure 6).

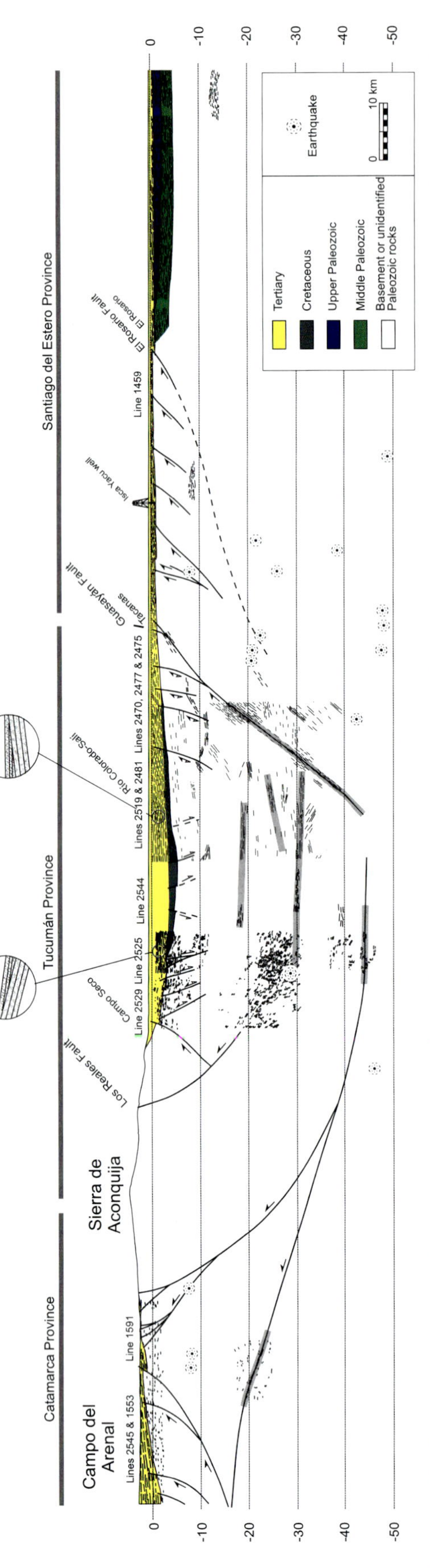

Figure 12. Tracing of main reflectors of the study seismic lines, after depth migration, and interpreted section, with hypocenters projected between 26°30′ and 27°30′S.

faults that intersect most of the foreland basin deposits but do not have surficial expression. On the other hand, the eastern sector shows comparable faults with the opposite vergence but with a clear expression of neotectonic activity (Figure 8). The faults are exposed in the recently uplifted pediments, and there are many other features that indicate active faulting (Strecker et al., 1989).

Line 2545 shows one of the most interesting structures. An east-vergent anticline uplifted by a west-dipping reverse fault is observed in the upper part of the western sector of the line. On the other hand, a small west-vergent imbricate system is present in the eastern portion of the line (Figure 10a). However, in the deeper part of the section (Figure 10b) there is a conspicuous reflector between 20 and 25 km depth that dips 20° to the east. This deep reflector is interpreted to be a reverse fault that propagates and interferes with the east-vergent surficial structure (Figure 12).

Relationships between the Structure and Basement Discontinuities

Hongn and Seggiaro (1998) have demonstrated, in neighboring areas of northwestern Argentina, the importance of crustal discontinuities in the basement serving as controls on the Cretaceous rifting and its later inversion.

The basement fabric, such as schistosity, foliation, or axial-plane cleavage, has strongly controlled the reactivation of faults and the generation of new faults during the Andean orogeny. It has been observed in the study area that the large faults that uplift mountain blocks have the same dip as the dominant basement fabrics. The Sierra de Guasayán has a west-dipping foliation with a north-northeast trend (Battaglia, 1982), and it is uplifted by a west-dipping reverse fault. In the Sierra de Aconquija, the dominant schistosity has a trend that varies from north-northeast to north-northwest and dips gently to the east in the western sector; it dips more steeply in the eastern sector (González Bonorino, 1951). The reverse faults that uplift the range are located in the western slope and dip to the east.

It is also important to mention that the seismic fabric near the Guasayán Fault, as well as in the vicinity of Los Reales Fault (Figures 6 and 7), is parallel to the fault dip and schistosity of metamorphic rocks is exposed at the surface.

The basement fabric has a distinct and opposite vergence in different sectors. The dominant westerly vergence is more compatible with the hypotheses that postulate an east-dipping subduction system during late Proterozoic and early Paleozoic times (Ramos and Vujovich, 1994; Rapela et al., 1998). On the other hand, the dominant basement fabric of the Sierra de Quilmes has a north-northwest trend and dips 50° to the east. The

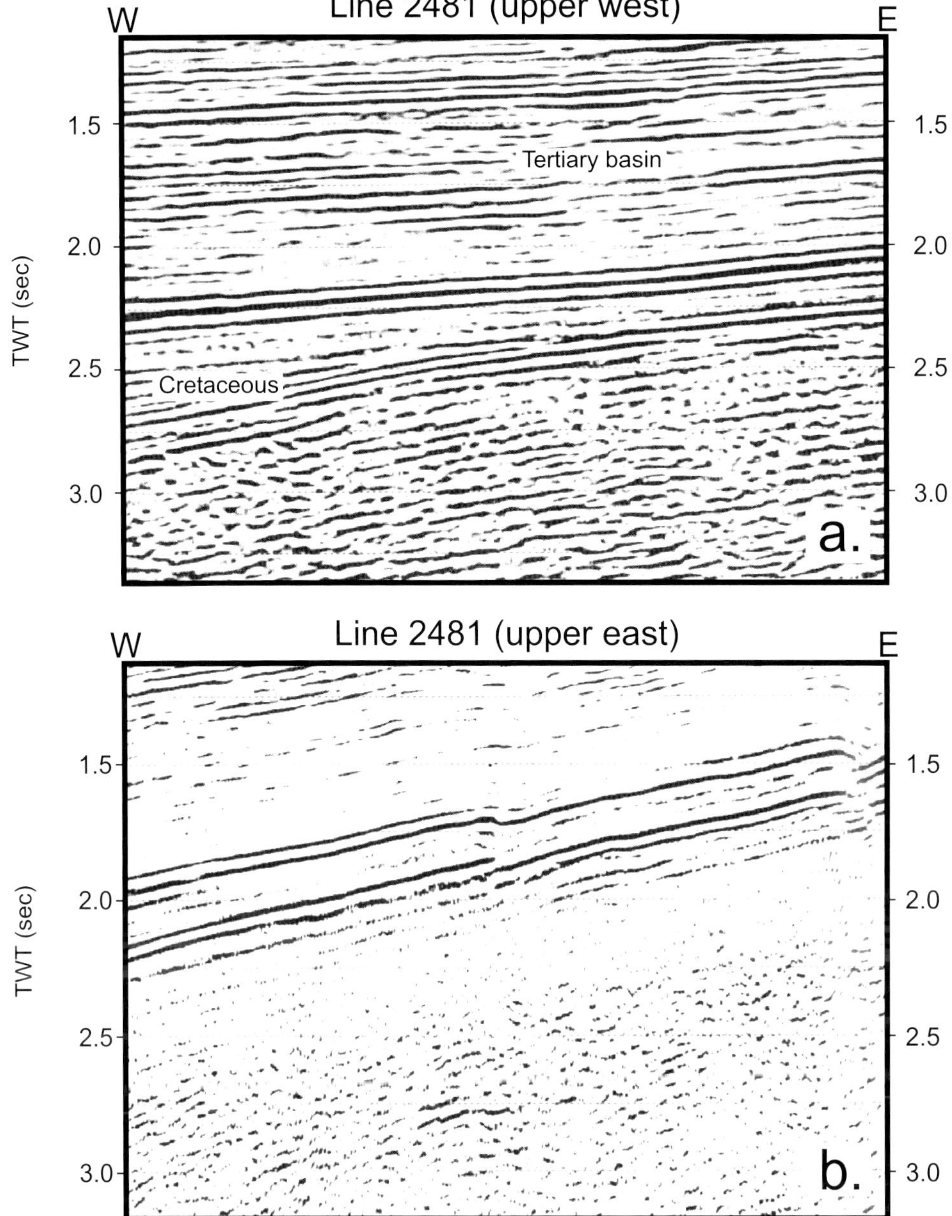

FIGURE 13. Detail of the upper part of seismic line 2481. Above: upper western sector, below: upper eastern sector. Note the interpreted half graben in Cretaceous rocks in the western sector and the high-angle, west-dipping faults that cut the Tertiary sequence.

reverse faults that uplift the range, if the east-vergent hypothesis of Bossi et al. (1993) is accepted, apparently have not controlled the uplift.

DISCUSSION

A series of late Proterozoic–early Paleozoic terrane accretion events has been identified in central and northwestern Argentina (Ramos and Vujovich, 1994; Ramos et al., 2000). These collisions are related to east-dipping subduction and accretion of the Pampia terrane against the Río de La Plata craton to form the protomargin of Gondwana, and the later accretion of the parautochthonous Antofalla-Arequipa terrane to Gondwana. These collisions produced the dominant basement fabrics and the resulting favorable anisotropies that controlled the west-vergent fault system of the western sector of the Sierra de Aconquija.

The seismic data analyzed indicate an important change in vergence in the oblique reflectors between 65°30′ and 65°20′W longitude at these latitudes. The development of the southern end of the half-graben system of the Salta Group rift is located in this area, as established by previous paleogeographic reconstructions (Porto et al., 1982) and the present seismic data. This anisotropy may have controlled the change in vergence.

The steeply dipping, deep, oblique reflectors identified in the reprocessed seismic lines may coincide with major crustal discontinuities that are subparallel to the potential sutures that bounded both sides of the Pampia terrane.

Major Structural Features of the Sierras Pampeanas

Following the early work of González Bonorino (1950a), it was evident that the Sierras Pampeanas are a broad area east of the main Andes that has been uplifted as basement blocks controlled by reverse faults associated with the Late Cenozoic uplift of the Cordillera. Jordan et al. (1983) and Jordan and Allmendinger (1986) proposed that the extension of this foreland deformation was related to the geometry of the Wadati-Benioff zone and confined to the flat-slab segment, and this proposal has been widely accepted. There is a precise coincidence of space and time between the shallowing of the subduction zone and the uplift of the Sierras

Pampeanas. However, the main vergence of the structure that bounded the different ranges was only partially analyzed. González Bonorino (1950a, page 109) assumed that schistosity was the main factor controlling the vergence of the main faults, but with the lack of subsurface data he was unabled to recognize the double vergence of the Sierra de Aconquija.

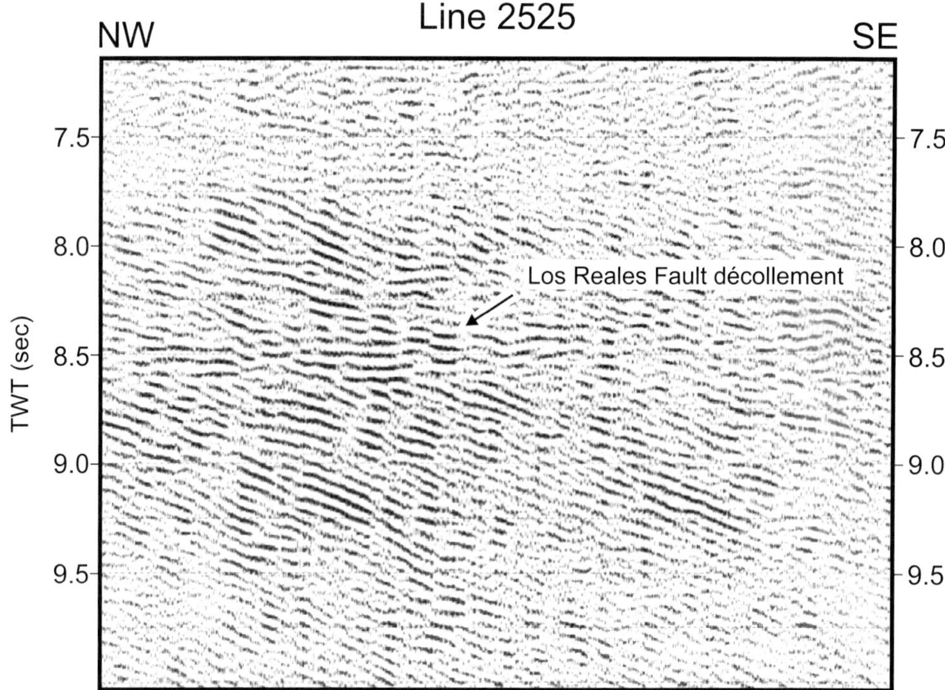

FIGURE 14. Details of deeper part of seismic line 2525. Note the fabric of the oblique basement reflectors. The reflector set at 8.5 s is interpreted as the dêcollment of the Los Reales Fault.

Most of the structures of the Sierras Pampeanas have one dominant vergence (Figure 1) that closely matches the penetrative basement fabric developed during the late Precambrian to early Paleozoic. However, this basement has had a complex structural history during Phanerozoic times. The late Paleozoic–Triassic rifting was concentrated in the hanging wall of the earlier sutures, as seen in the Sierra de Valle Fértil. Thick Triassic basins are preserved in the present upthrown block of the west-bounding Valle Fértil Fault (Ramos, 1994). The downthrown block lacks Triassic deposits, indicating that this fault was the early taphrogenic boundary of the basin and that the Cenozoic fault was inverting the previous normal faulting.

This tectonic inversion is also seen in the Sierras de Chepes, San Luis, and Chica de Córdoba (Figure 1), but it was controlled by the Cretaceous rifting (Schmidt et al., 1995; Ramos, 2000). Most of the Andean faults bounding the Sierras Pampeanas are either inverted previously normal faults or reactivated older penetrative structures. Many of the faults coincide with previous mylonites or shear zones, as described by Schmidt et al. (1995).

Such facts indicate that many, but not all, of the contractional features follow previously normal fault trends, when those trends were favorably orientated for reactivation of east-west contraction. This reactivation, as well as the style of deformation, are similar to the eastern Rocky Mountain blocks of the Laramide region of the United States, as Jordan and Allmendinger (1986) and Schmidt et al. (1995) described. This similarity extends to the regional tectonic setting, because both foreland deformation areas coincide with the precise timing of shallowing of the subduction zone: during the Paleogene in the Laramide region and the Neogene in the Sierras Pampeanas.

Double-wedge Thrusting of the Sierra de Aconquija

As a whole, the structure of the Sierras Pampeanas does not show double-wedge thrusting as a typical feature, with the exception of the Sierra de Aconquija system. This system, as pointed out by Mon and Drozdzewski (1999), is located in the present orogenic front of the Andes at 27°S latitude.

The time constraints indicate that the western faults of the Sierra de Aconquija were the first to be generated, at about 6.7 Ma (Strecker et al., 1989). Recent fission-track data presented by Sobel et al. (1998) have indicated younger ages of 6.3 ± 0.8 Ma. However, based on synorogenic deposits and paleocurrent analyses, Strecker et al. (1989) have identified that the main pulse of uplift occurred during deposition of the Corral Quemado Formation between 3.4 and 2.97 Ma. There is also good evidence of active faulting along the western slope of the Sierra de Aconquija, as indicated by the broken Quaternary pediments described by Strecker et al. (1989) and by large rock avalanche deposits. The avalanches were interpreted to be related to seismic shaking of already preconditioned weak basement rocks in an area dominated by frequent earthquakes (Fauqué and Strecker, 1988).

The structural relief of a few hundred meters and the tectonic offsets on the east-verging faults are subordinate and less important than the west-verging fault

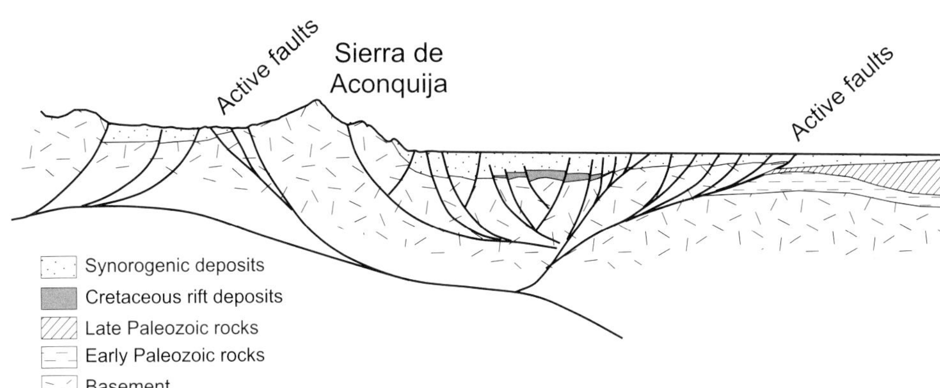

FIGURE 15. Structural scheme that shows the present location of the active faults at both sides of the basement double wedge in the western Sierras Pampeanas (not to scale).

system with its uplift of more than 5000 meters (Figure 12). The east-verging fault system, even if it involved reactivated old Paleozoic faults, seems to be younger. Surficial morphologic expression of neotectonism in the El Rosario Fault, together with the location of thermal springs along the fault, such as the Río Hondo thermal baths, indicates active or incipient faulting.

Taken together, the two fault systems outline active double-wedge thrusting (Figure 15), but this structure has developed asymmetrically because the western component is older. McClay et al. (1999) have reproduced this situation in sand boxes. The experimental models indicate that initially only one system develops, until a critical taper is reached. When a critical taper occurs on the older and steeper flank of the wedge, the oppositely vergent system begins to develop. In our study area, the east-vergent system has a gentler synorogenic topography. Renewal of deformation produces out-of-sequence thrusting activity on both systems. Experiments reproduce step-by-step the double-wedge thrusting observed in the Sierras de Aconquija. Numerical modelling of the well-known Pyrenees ranges also reproduces the different stages of this double-wedge thrusting (Muñoz et al., 1999). In the Pyrenees, it is remarkable that previous anisotropies, such half-graben systems, controlled the location of faulting. The development of the double wedge seems to require a singular point where the deformation begins. In the Aconguija area, a suture within the basement between the Pampia and adjacent terranes (Ramos et al., 2000) could be the singular location that induced double thrusting.

Figure 16 shows the unique location of the double-wedge thrusting of the Aconquija system, in the Sierras Pampeanas, that is developing in the orogenic front of the Andes at these latitudes. The most active part of the system coincides with the blind thrusts that are only partially emergent in the eastern Sierra de Aconquija thrust front. This may indicate that progressive deformation farther south, in the Ambato and Ancasti systems, might produce future double-wedge thrusting. This is partially indicated by the east-vergent faults of the Sierra de Guasayán (see Figure 1 for location).

OIL POTENTIAL

The existence of basement thrusts associated with the Andean uplift in the pericratonic foreland region is considered a plausible process that could contribute to the burial of potential source rocks in the vicinity of the Sierra de Aconquija (Fernández Garrasino et al., 1984). It is well established that the sag facies of the Salta rift system have sufficient organic material and levels of maturation for regional hydrocarbon generation (Uliana et al., 1999). However, in the southern sector of the basin, where the study area is located, the Cretaceous-Paleocene sag facies has experienced insufficient burial by synorogenic deposits to have entered the oil window. Fernández Garrasino et al. (1984) speculated that the tectonic loading of the Sierra de Aconquija thrust could have increased the burial, but the west-vergent thrusts with large displacements coincide with an area lacking Cretaceous rift deposits. The eastern side of the double wedge has east-vergent thrusts with small displacements and relatively high angles that preclude any important burial. Therefore, the sag deposits of this area are not affected by tectonic loading, and the potential source rocks probably are not mature enough to generate hydrocarbons.

CONCLUSIONS

- The deep reprocessing of preexisting industrial seismic lines furnished information down to 50-km depths.
- The deep seismic-reflection data, together with the surface geologic information, permit us to propose a comprehensive structural model that encompasses the upper and lower crust.
- On this basis, an active, doubly vergent-wedge thrust system is proposed for the orogenic front of the Sierras Pampeanas. This asymmetric wedge has

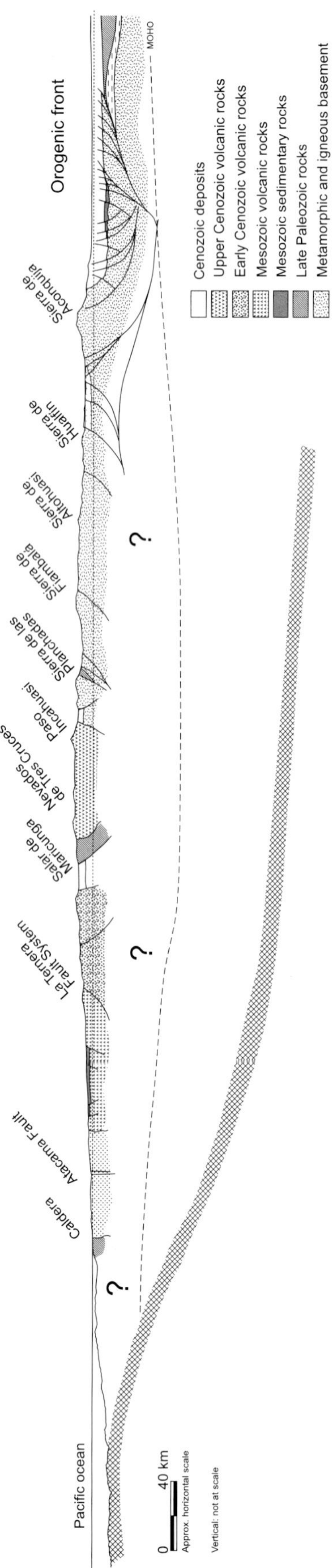

FIGURE 16. Schematic structural cross section of the Andes at 27°S latitude, indicating the location of the Aconquija double-wedge thrust system in the present orogenic thrust front.

an older and steeper western flank that is as high as 5000 m or more, and a gentler and younger eastern flank.

- In spite of the high elevations of this part of the Northern Sierras Pampeanas, shortening is minimal because of the thick-skinned nature of the thrust belt.
- Beneath Campo del Arenal, in the western sector of the study area, a thick-skinned triangle zone is proposed where the western-vergent thrusts are superimposed on the previously east-vergent reverse faults.
- Analysis of the Neogene synorogenic deposits permits identification of two foreland basins: an early, continuous foreland basin that existed between the early and middle Miocene (foreland basin I) and a broken foreland basin after 6.3 Ma (foreland basin II).
- The Paraná marine transgression predated the uplift of the Sierra de Aconquija and most of the uplift of the Sierras Pampeanas, at these latitudes. This constrains the initial uplift and deformation in the double-wedge thrusting as being younger than 13.5 Ma.

ACKNOWLEDGMENTS

Financial support from PIA-CONICET 6202, SECYT-PICT 06729, and logistic support of the Fundación Antorchas are acknowledged. YPF S.A through the CAPLI agreement authorized the use of the industrial seismic lines and gave access to reprocessing facilities. The authors are most grateful to the reviewers Robert Hatcher and Raymond Price for their comments and suggestions.

APPENDIX A

Vibroseis Data

The lines where vibrators were used involved linear upsweeps with dissimilar time length (e.g., 8, 12, or 16 s) but a unique frequency band (12–65 Hz), whereas the field records were characterized by different time lengths (e.g. 13, 14, or 17 s) and a sample rate of 4 ms (Table A1).

The "self-truncating" extended-correlation algorithm (Comínguez and Ramos, 1995) was used to compute cross correlation between the sweep and the records. The original frequency band of 12–65 Hz was preserved for the first 5 s of the seismic section (6 s in the particular case of line 2481), although this band was affected by an upper frequency decreasing from 5 s at a predicted linear rate (Okaya and Jarchow, 1989).

Table A1. Field parameters.

Lines 2475, 2477, 2479, and 2481	
Source:	Vibroseis
Type:	Failing-Y-1100-A
Number:	4
Sweeps:	24
Sweep frequency:	12–65 Hz, linear
Vibrator Time:	8 s
Recording Time:	13/14 s
Recording System:	MSD-10
Number of Channels:	120
Sampling Interval:	4 ms
Station Spacing:	37.5 m
Near Offset:	168.75 m
Far Offset:	2381.25 m
Fold:	60
Spread Configuration:	Symmetric (2381.25 × 168.75-P-168.75 × 2381.25)

Line 2519	
Source:	Vibroseis
Type:	Failing-Y-1100-A
Number:	4
Sweeps:	16
Sweep frequency:	12–65 Hz, linear
Vibrator Time:	12
Recording Time:	17 s
Recording System:	MSD-10
Number of Channels:	120
Sampling Interval:	4 ms
Station Spacing:	37.5 m
Near Offset:	318.75 m
Far Offset:	2531.25 m
Fold:	60
Spread Configuration:	Symmetric (2531.25 × 318.75-P-318.75 × 2531.25)

Lines 2525 and 2529	
Source:	Vibroseis
Type:	Failing-Y-1100-A
Number:	4
Sweeps:	16
Sweep frequency:	12–65 Hz, linear
Vibrator Time:	12
Recording Time:	17 s
Recording System:	MSD-10
Number of Channels:	120
Sampling Interval:	4 ms
Station Spacing:	50 m
Near Offset:	425 m
Far Offset:	3375 m
Fold:	60
Spread Configuration:	Symmetric (3375 × 425-P-425 × 3375)

Line 2544	
Source:	Vibroseis
Type:	Failing-Y-1100-A
Number:	4
Sweeps:	16
Sweep frequency:	12–65 Hz, linear
Vibrator Time:	12
Recording Time:	17 s
Recording System:	MDS-10
Number of Channels:	120
Sampling Interval:	4 ms
Station Spacing:	50 m
Near Offset:	125 m
Far Offset:	3075 m
Fold:	60
Spread Configuration:	Symmetric (3075 × 125-P-125 × 3075)

Line 2545	
Source:	Vibroseis
Type:	Failing-Y-1100-A
Number:	2
Sweeps:	8
Sweep frequency:	12–65 Hz, linear
Vibrator Time:	12
Recording Time:	17 s
Recording System:	MSD-10
Number of Channels:	120
Sampling Interval:	4 ms
Station Spacing:	50 m
Near Offset:	125 m
Far Offset:	3075 m
Fold:	20
Spread Configuration:	Symmetric (3075 × 125-P-125 × 3075)

Line 1459A (lines 1459B, 1459C, and 1459D are characterized by practically the same parameters)	
Source:	Dynamite
Total Charge:	5 kg (multiple holes)
Source Depth:	4–5 m
Charge Interval:	100 m
Recording Time:	6 s
Recording System:	MDS-10
Number of Channels:	96
Sampling Interval:	2 ms
Station Spacing:	50 m
Near Offset:	100 m
Far Offset:	750 m
Fold:	24
Spread Configuration	Quasi-symmetric (2450 × 100-P-400 × 2750)

Table A1. Field parameters (cont.).

Line 1553	
Source:	Dynamite
Total Charge:	5 kg (multiple holes)
Source Depth:	2–3 m
Charge Interval:	100 m
Recording Time:	6 s
Recording System:	MDS-10
Number of Channels:	96
Sampling Interval:	2 ms
Station Spacing:	50 m
Near Offset:	200 m
Far Offset:	2550 m
Fold:	24
Spread Configuration	Symmetric (2550 × 200-P-200 × 2550)
Line 1591	
Source:	Dynamite
Total Charge:	7/10 kg (multiple holes)
Source Depth:	0.50 m
Charge Interval:	50 m
Recording Time:	6 s
Recording System:	MDS-10
Number of Channels:	120
Sampling Interval:	2 ms
Station Spacing:	50 m
Near Offset:	200 m
Far Offset:	3150 m
Fold:	60
Spread Configuration	Symmetric (3150 × 200-P-200 × 3150)

Thus, correlated records of 11–14 s were calculated, depending on the particular set of field parameters characterizing the seismic line.

Prestack Processing

Significant steps in Vibroseis before-stacking processing were:

We used CDP gather equalization, which is a scaling process designed to correct amplitude anomalies of source records. Such anomalies could be produced by variations in the source array, charge amount and depth, etc. In performing the scaling with this process we tried to maintain the relative amplitude of the data, both spatially and temporally.

A balancing process was used to equalize seismic trace amplitudes. With this procedure, areas exhibiting weaker acoustic amplitudes could be strengthened relative to areas of stronger amplitudes. Therefore, after the application of the balancing operation, the variations between shallow and deep areas appeared less severe; however, we tried to conserve the relative amplitude within the boundaries of definite areas (for example: the sedimentary basin, the middle crust, the lower crust, etc.).

Careful monitoring was done to remove noisy zones in each trace of the field gathers. Regardless, because of the low signal-to-noise ratio characterizing this particular trace-sector some difficulties became apparent while we were trying to perceive acoustic horizons in the deeper parts of the crust. Thus the manual trace-mute was done with "Generalized Automatic Trace Editing." A previous statistical analysis made it possible to establish appropriate thresholds for the automatic editing (based, for example, on dominant frequency or on peak to reference amplitude). As a result of combining manual and statistical editing, we could image reliable crust details down to 14 s (42–45 km deep, approximately).

We used a temporary floating datum before normal moveout correction. Application of large static shifts prior to velocity analysis and normal moveout corrections distorts the hyperbolic character of reflected events. This complication is more evident at shallow traveltimes, where the distortion significantly affects velocity measurements. To minimize this distortion, datum statics typically were decomposed into two time components, one that corrected the data to a floating datum (short-wavelength static corrections), and another, involving much larger components, that corrected the data to a final datum (long-wavelength static corrections). The former static was regularly used prior to velocity analysis and normal moveout corrections, while the latter was applied after normal moveout corrections.

The floating datum was established by performing a smoothing operation over a physical datum or surface (for instance, surface elevation or shot-depth surface). Traveltime differences were subsequently computed between the floating datum and the fixed-reference datum. Alternatively, the floating datum was calculated indirectly by averaging static corrections from all the traces belonging to a single common-depth-point gather or group of subsequent gathers. Floating-datum statics (short-wavelength corrections) were finally computed by simple subtraction of the mean CDP-static value from the total-trace static (computed from a fixed datum).

Application of this concept led us to recognize the basement outcrop on both sides of the Aconquija range, when seismic sections as well as interval-velocity profiles were studied.

Prior to channel stacking, a deconvolution operation with unmodified phase (zero-phase deconvolution) was applied, which flattened the amplitude spectrum along the 16–50 Hz frequency band. With this operation, precise seismic-velocity analysis was ensured to as much as 8–9 s of trace length (in turn, a good acoustic response from the crust was found to occur in this sector, e.g., in lines 2545, 2529, 2481 and 2477). Likewise, because phase characteristics of the diffracted signals were kept unmodified, they could be properly focused by migration (Trorey, 1979).

Poststack Processing

We employed the following important aspects in Vibroseis poststack processing.

Calculation and application of residual statics. This procedure included determination of new stacking velocities, new stacking, and a second step of calculating and applying residual statics.

Migration of the section using Claerbout's (1976) finite-difference algorithm.

Seismic section enhancement using FX deconvolution. This process applies a fast Fourier transform to convert a time window for an specified number of traces (that is, a t-x domain) into an f-x domain. For each frequency of the new domain, the complex signal is then enhanced by a predictive filter. An inverse fast Fourier transform restores the signal to the original t-x section producing the attenuation of details that are not predictable from one trace to the next. Time windows of 1000 ms, as well as spatial windows of 200 traces, were used to enhance the continuity of acoustic reflectors. In some cases, the operation was accomplished by the use of a coherency operation—a process designed for signal enhancement by dip scan and semblance scaling.

The use of complex demodulation techniques (Taner and Sheriff, 1977), as an aid to the fault analysis.

Once final migrated and enhanced seismic sections were calculated, the FMED (minimum entropy algorithm with frequency-domain constraints) developed by Sacchi et al. (1996) was used, in some sectors, to improve signal resolution and to match continuous acoustic reflectors with geologic and well information. In particular, FMED was a robust tool to support the seismic-stratigraphic interpretation of the Cenozoic basin.

In addition, as Okaya and Jarchow (1989) illustrated, self-truncating extended correlation (the basic algorithm used in the mathematical vibroseis reprocessing) produces a progressive frequency-band shortening (this fact complements the expected band shortening by frequency absorption in the rock medium). Consequently, the FMED algorithm was also implemented for improving band-deteriorated deep sectors of the seismic sections.

Explosive Source Data

Sources of explosive origin complemented the regional study, specifically in the western sector (line 1459), and in Campo del Arenal (Catamarca Province), where lines 1553 and 1591 considerably reinforced the seismic stratigraphic study of the area.

Pre- and poststacking process steps were analogous to those adopted for lines where vibrators were used. For example, zero-phase deconvolution was applied prior to stacking and this resulted in a flattening of the amplitude spectrum along a 16–46 Hz frequency band. With this operation, precise seismic velocity analysis was ensured to as much as 6 s (maximum time recorded in field experiments). Therefore, the phase characteristics of the diffracted signals were kept unmodified at the early stages of processing. This strategy assured an appropriate focalization of the acoustic targets during migration.

Special procedures, such as finite-difference migration, FX deconvolution, FMED processing, and complex demodulation were applied in a similar way to that of the vibrator lines.

Identification of Shallow Seismic Reflectors

A synthetic seismogram was computed from sonic-log information from YPF well Isca Yacu x-1 (Figure 4). This was a decisive analytical step that helped in the procedure of connecting acoustic reflection events on the seismic sections with the geological interfaces (McCormac et al., 1985). The *P*-wave normal-incidence reflection-coefficient series was computed for the depth interval ranging from 147 to 700 m asl (two-way reflection traveltime from 0.147 to 1.047 s).

A wavelet extracted from field-processed traces was used as the simulated source of the synthetic traces. Figure 4 displays the one-dimensional synthetic seismogram corresponding to the Isca Yacu x-1 well, together with the line-1559 FMED-traces located nearest the well data. Strong peaks on both seismograms identify the top of the main basin sequences, whereas the top of the acoustic basement probably corresponds to the top of the Paleozoic rocks.

REFERENCES CITED

Anonymous, 1949, Diez años de Perforaciones, 1926–1935: Buenos Aires, Ministerio de Industria y Comercio de la Nación, Dirección General de Industria y Minería Publicación, v. 139 (No. 38–S.I.C.), 275 p.

Anonymous, 1958, Perfiles de Perforaciones, período 1904–1915: Buenos Aires, Ministerio de Comercio e Industria de la Nación, Subsecretaría de Minería, Dirección Nacional de Geología y Minería Publicación, v. 146 (No. 121 M.C.I.N.), 236 p.

Battaglia, A., 1982, Descripción Geológica de las Hojas 13f (Río Hondo), 13g (Santiago del Estero), 14g (El Alto), 14h (Villa San Martín) y 15 g (Frías), Santiago del

Estero y Catamarca: Buenos Aires, Servicio Geológico Nacional Boletín, v. 186, p. 1–80.

Beder, R., 1928, La Sierra de Guasayán y sus alrededores. Una contribución a la geología e hidrología de la provincia de Santiago del Estero: Buenos Aires, Dirección General de Minas, Geología e Hidrología Publicación, v. 39, p. 1–171.

Bianucci, H. A., O. M. Acevedo, and J. J. Cerdán, 1981, Evolución tectosedimentaria del Grupo Salta en la subcuenca Lomas de Olmedo (Provincias de Salta y Formosa): VIII° Congreso Geológico Argentino (San Luis) Actas, v. 3, p. 159– 172.

Bohm, K., A. A. Palma, and C. C. Gutiérrez, 1962, Informe geotécnico vinculado con el proyecto de dique de Embalse del Cadillal, provincia de Tucumán: 1° Jornadas Geológicas Argentinas (San Juan) Actas, v. 2, p. 15–24.

Bonaparte, J. F., J. A. Salfity, G. E. Bossi, and J. E. Powell, 1977, Hallazgo de dinosaurios y aves cretácicas en la Formación Lecho de El Brete (Salta) próximo al límite con Tucumán: Acta Geológica Lilloana, v. 14, p. 5–17.

Bossi, G. E., 1969, Geología y estratigrafía del sector sur del valle de Choromoro: Acta Geológica Lilloana, v. 10(2), p. 19–61.

Bossi, G. E., 1984, Terciario, *in* F. G. Aceñolaza, A. Toselli, and G. E. Bossi, eds., Geología de Tucumán: Colegio de Graduados en Ciencias Geológicas de Tucumán, p. 67–80.

Bossi, G. E., and M. Palma, 1982, Reconsideración de la estratigrafía del valle de Santa María, provincia de Catamarca, Argentina: V° Congreso Latinoamericano de Geología (Buenos Aires) Actas, v. 1, p. 155–172.

Bossi, G. E., C. M. Muruaga, J. G. Sanagua, A. Hernando, and A. L. Ahumada, 1993, Geología y estratigrafía de la Cuenca Neógena Santa María–Hualfín (Deptos. Santa María y Belén, Provincia de Catamarca): XII° Congreso Geológico Argentino y II° Congreso de Exploración de Hidrocarburos (Mendoza) Actas, v. 2, p. 156–165.

Claerbout, J. F., 1976, Fundamentals of geophysical data processing: with applications to petroleum prospecting: New York, McGraw Hill, 187 p.

Comínguez, A. H., and V. A. Ramos, 1991, La estructura profunda entre Precordillera y Sierras Pampeanas de la Argentina: evidencia de la sísmica de reflexión profunda: Revista Geológica de Chile, v. 18, p. 3–14.

Comínguez, A. H., and V. A. Ramos, 1995, Geometry and seismic expression of the Cretaceous Salta Rift of Northwestern Argentina, *in* A. J. Tankard, R. Suárez, and H. J. Welsink, eds., Petroleum basins of South America: AAPG Memoir 62, p. 325–340.

Cristallini, E., A. H. Comínguez, and V. A. Ramos, 1997, Deep structure of the Metan – Guachipas region: Tectonic inversion in northwestern Argentina: Journal of South American Earth Sciences, v. 10, p. 403–421.

Fauqué, L., and M. R. Strecker, 1988, Large rock avalanche deposits (Sturzströme, sturzstroms) at Sierra Aconquija, Northern Sierras Pampeanas, Argentina: Eclogae geologicae Helvetiae, v. 82, p. 579–592.

Fernández Garrasino, C., H. Bianucci, and J. Musmarra, 1984, Algunso rasgos geológicos al sur de Salta y este de Tucumán: Boletín de Informaciones Petroleras, Tercera Epoca, Año I, v. 1, p. 62–72.

Galliski, M. A., and J. G. Viramonte, 1988, The Cretaceous paleorift in northwestern Argentina: A petrologic approach: Journal of South American Earth Sciences, v. 1, p. 329–342.

Gavriloff, I. C., and G. E. Bossi, 1992, Las facies lacustres de la Formación San José y Río Salí (Mioceno Medio), NO Argentino y su relación con la ingresión marina Paranaense: VIII° Congreso Latinoamericano de Geología (Salamanca), Simposio, v. 1, p. 78–87.

González Bonorino, F., 1950a, Algunos problemas geológicos de las Sierras Pampeanas: Asociación Geológica Argentina Revista, v. 5, p. 81–110.

González Bonorino, F., 1950b, Geología y petrografía de la Hoja 12d, Capillitas y sus relaciones estructurales (Catamarca): Dirección Nacional de Minería Boletín, v. 70, p. 1–49.

González Bonorino, F., 1951, Descripción geológica de la Hoja 12e Anconquija (Catamarca–Tucumán): Dirección Nacional de Minería Boletín, v. 75, p. 1–49.

González Bonorino, F., 1972, Descripción geológica de la Hoja 13c, Fiambalá, Prov. de Catamarca: Dirección Nacional de Geología y Minería Boletín, v. 127, p. 1–76.

González, O. E., and O. Barreñada, 1993, Geología y estructura de las nacientes del río Amaicha y El Infernillo, Provincia de Tucumán: XII° Congreso Geológico Argentino y II° Congreso de Exploración de Hidrocarburos (Mendoza) Actas, v. 3, p. 72–81.

Grier, M. E., and R. D. Dallmeyer, 1990, Age of the Payogastilla Group: implications for foreland basin development, NW Argentina: Journal of South American Earth Sciences, v. 3, p. 269–278.

Hongn, F. D., and R. E. Seggiaro, 1998, Estructuras del basamento y su relación con el rift cretácico, Valles Calchaquíes, Provincia de Salta: X° Congreso Latinoamericano de Geología y VI° Congreso Nacional de Geología Económica (Buenos Aires) Actas, v. 2, p. 4–9.

Isacks, B., T. Jordan, R. Allmendinger, and V. A. Ramos, 1982, La segmentación tectónica de los Andes Centrales y su relación con la placa de Nazca subductada: V° Congreso Latinoamericano de Geología (Buenos Aires) Actas, v. 3, p. 587–606.

Jordan, T. and R. Allmendinger, 1986, The Sierras Pampeanas of Argentina: a modern analogue of Laramide deformation: American Journal of Science, v. 286, p. 737–764.

Jordan, T. E., B. Isacks, V. A. Ramos, and R. W. Allmendinger, 1983, Mountain building in the Central Andes: Episodes, v. 1983(3), p. 20–26.

Kay, S. M., and V. A. Ramos, 1996, El magmatismo cretácico de las Sierras de Córdoba y sus implicancias tectónicas, XIII° Congreso Geológico Argentino y III° Congreso Exploración de Hidrocarburos (Buenos Aires) Actas, v. 3, p. 453–464.

Keidel, J., 1913, Composición y estructura geológica del Cajón del Cadillal, prov. de Tucumán: Ministerio de Agricultura de la Nación Anales, Sección Geología, Mineralogía y Minería, v. 8, p. 1–45.

Linares, E., and R. R. González, 1990, Catalogo de edades radimétricas de la República Argentina 1957–1987:

Asociación Geológica Argentina Publicaciones Especiales Serie B, Didáctica y Complementaria, v. 19, p. 1–628.

Lucero Michaut, H. N., 1979, Sierras Pampeanas del Norte de Córdoba, Sur de Santiago del Estero, Borde Oriental de Catamarca y Angulo Sudeste de Tucumán, *in* J. C. M. Turner, ed., Segundo Simposio de Geología Regional Argentina: Academia Nacional de Ciencias, v. 1, p. 293–347.

Malumián, N., and C. Nañez, 1996, Microfósiles y nanofósiles calcáreos de la Plataforma Continental, *in* V. A. Ramos, and M. A. Turic, eds., Geología y Recursos Naturales de la Plataforma Continental Argentina: XIII° Congreso Geológico Argentino y III° Congreso de Exploración de Hidrocarburos, Relatorio, p. 73–93.

McClay, K., T. Dooley, and P. Whitehouse, 1999, Analogue modelling of thin and thick-skinned thrust systems: Thrust Tectonics Conference (London) Paper 18, p. 45.

McCormac, M. D., M. G. Justice, and W. W. Sharp, 1985, A stratigraphic interpretation of shear and compresional wave seismic data for the Pennsylvanian Morrow Formation of southeastern of New Mexico, *in* O. R. Berg and D. G. Woolverton, eds., Seismic stratigraphy II – an integrated approach, AAPG Memoir 39, p. 225–239.

Mon, R., 1976, La tectónica del borde oriental de los Andes en las Provincias de Salta, Tucumán y Catamarca, República Argentina: Revista de la Asociación Geológica Argentina, v. 31, p. 65–72.

Mon, R. and G. Drozdzewski, 1999, Cinturones doblevergentes en los Andes del norte argentino— Hipótesis sobre su origen: Revista de la Asociación Geológica Argentina, v. 54, p. 3–8.

Muñoz, J. A., C. Beaumont, F. Storti, and K. McClay, 1999, Thrust partitioning in double-wedge orogens: results from comparison of numerical and analogue models with the tectonic evolution of the Pyrenees: Thrust Tectonics Conference (London) Paper 22, p. 52.

Okaya, D. A., and C. M. Jarchow, 1989, Extraction of deep crustal reflections from shallow Vibroseis data using extended correlation: Geophysics, v. 54, p. 552–562.

Peirano, A., 1957, Observaciones generales sobre la tectónica y los depósitos terciarios del cuadrángulo 26°S-64°30′O – 28°30′S-67°O en el noroeste argentino: Acta Geológica Lilloana, v. 1, p. 61–144.

Pezzi, E. E., and M. E. Mozetic, 1989, Cuencas sedimentarias de la región chacoparanense, *in* G. Chebli and L. Spalletti, eds., Cuencas Sedimentarias Argentinas: Serie Correlación Geológica, v. 6, p. 65–78.

Porto, J., C. Danieli, and O. Ruíz Huidobro, 1982, El grupo Salta en la prov. de Tucumán, Argentina: 5° Congreso Latinoamericano de Geología (Buenos Aires): Actas, v. 4, p. 253–264.

Quenardelle, S., and V. A. Ramos, 1999, The Ordovician western Sierras Pampeanas magmatic belt: record of Precordillera accretion in Argentina, *in* V. A. Ramos and D. Keppie, eds., Laurentia Gondwana Connections before Pangea: Geological Society of America Special Paper 336, p. 63–86.

Ramos, V. A., 1986, El diastrofismo oclóyico: un ejemplo de tectónica de colisión durante el Eopaleozoico en el noroeste Argentino: Revista Instituto Ciencias Geológicas, v. 6, p. 13–28.

Ramos, V. A., 1988, Tectonics of the Late Proterozoic–Early Paleozoic: a collisional history of Southern South America: Episodes, v. 11, p. 168–174.

Ramos, V. A., 1994, Terranes of southern Gondwanaland and their control in the Andean structure (30–33° S lat.), *in* K. J. Reutter, E. Scheuber, and P. J. Wigger, eds., Tectonics of the southern Central Andes, Structure and evolution of an active continental margin: Berlín, Springer Verlag, p. 249–261.

Ramos, V. A., 2000, Las provincias geológicas del territorio argentino, *in* R. Caminos, ed., Geología Argentina: Instituto de Geología y Recursos Minerales, Anales 29(3), p. 41–96.

Ramos V. A., and R. N. Alonso, 1995, El mar paranense en la provincia de Jujuy: Revista Instituto de Geología y Minería, v. 10, p. 73–82.

Ramos, V. A., and M. A. Basei, 1997, Gondwanan, Perigondwanan, and exotic terranes of southern South America: South American Symposium on Isotope Geology (Sao Paulo) Abstracts, p. 250–252.

Ramos, V. A., and G. Vujovich, 1994, The Pampia Craton within western Gondwanland: Proceedings of the First Circum Pacific and Circum Atlantic Terrane Conference, p. 113–116.

Ramos, V. A., M. Escayola, D. Mutti, and G. I. Vujovich, 2000, Proterozoic–early Paleozoic ophiolites in the Andean basement of southern South America, *in* Y. Dilek and E. Moores, eds., Ophiolites and oceanic crust: New insights from field studies and ocean drilling program: Geological Society of America Special Paper 349, p. 331–349.

Rapela, C. W., R. Pankhurst, C. Casquet, E. Baldo, J. Saavedra, and C. Galindo, 1998, Early evolution of the Proto-Andean margin of South America: Geology, v. 26, p. 707–710.

Ruiz Huidobro, O., 1960, El horizonte calcáreo dolomítico en la Provincia de Tucumán: Acta Geológica Lilloana, v. 3, p. 147–171.

Ruiz Huidobro, O., 1966, Contribución a la geología de las Cumbres Calchaquíes y sierra de Aconquija (Tucumán-Catamarca): Acta Geológica Lilloana, v. 8, p. 1–215.

Ruiz Huidobro, O., 1972, Descripción Geológica de la Hoja 11e, Santa María, provincia de Salta: Servicio Nacional Minero Geológico Boletín, v. 134, p. 1–64.

Ruiz Huidobro, O. J., and J. C. Porto, 1987, Geología de las Lomas Montosa–Provincia de Tucumán: X° Congreso Geológico Argentino (Tucumán) Actas, v. 3, p. 201–204.

Russo, A., S. Archangelsky, R. R. Andreis, and A. Cuerda, 1987, Cuenca Chacoparanense, *in* S. Archangelsky, ed., El Sistema Carbonífero de la República Argentina: Academia Nacional de Ciencias, Córdoba, Publicación Especial, p. 197–212.

Sacchi, M. D., D. R. Velis, and A. H. Comínguez, 1996, Minimum entropy deconvolution with frequency-domain constraints, *in* E. A. Robinson and O. M. Osman, eds., Deconvolution II: Tulsa, Oklahoma, Society of Exploration Geophysicists, p. 278–285.

Salfity, J. A., C. R. Monaldi, R. A. Marquillas, and R. E. Gonzáles, 1993, La inversión tectónica del Umbral de los Gallos en la cuenca del Grupo Salta durante la fase incaica: XII° Congreso Geológico Argentino y II° Congreso de Exploración de Hidrocarburos (Mendoza) Actas, v. 3, p. 200–210.

Schmidt, C. J., R. A. Astini, C. H. Costa, C. E. Gardini, P. E. Kraemer, 1995, Cretaceous rifting, alluvial fan sediemntation, and Neogene inversion, southern Sierras Pampeanas, Argentina, *in* A. Tankard, R. Suárez S., and H. J. Welsink, eds., Petroleum basins of South America: AAPG Memoir 62, p. 341–358.

Sobel, E. R., K. Kleinert, and M. R. Streker, 1998, Apatite fission track constraints on exhumation in the Sierras Pampeanas, NW Argentina: American Geophisical Union, 1998 Fall Meeting, Abstract T71B–08.

Starck, D., and G. Vergani, 1996, Desarrollo tectosedimentario del Cenozoico en el sur de la provincia de Salta: XIII° Congreso Geológico Argentino y III° Congreso de Exploración de Hidrocarburos (Buenos Aires) Actas, v. 1, p. 433–452.

Strecker, M. R., 1987, Late Cenozoic Landscape development, the Santa María Valley, Northwestern Argentina: Ithaca, New York,Cornell University, Ph. Dissertation.

Strecker, M. R., P. Cerveny, A. L. Bloom, and D. Malizia, 1989, Late Cenozoic tectonism and landscape development in the foreland of the Andes: Northern Sierras Pampeanas (26°–28°S), Argentina: Tectonics, v. 8, p. 517–534.

Taner, M. T., and R. E. Sheriff, 1977, Application of amplitude, frequency, and other attributes to stratigraphic and hydrocarbon determination, *in* C. E. Payton, ed., Applications to Hydrocarbon exploration: AAPG Memoir 26, p. 301–327.

Toselli, J. N. R., A. J. Toselli, and G. A. Toselli, 1976, Migmatización y metamorfismo en el basamento de la Sierra de Quilmes, al oeste de Colalao del Valle, Provincia de Tucumán, Argentina: Revista de la Asociación Geológica Argentina, v. 31, p. 83–94.

Trorey, A. W., 1979, A simple theory for seismic diffractions: Geophysics, v. 35, p. 762–784.

Turner, J. C. M., 1973, Descripción geológica de la Hoja 11d, Laguna Blanca, provincia de Catamarca: Servicio Nacional de Minería y Geología Boletín, v. 142, p. 1–73.

Uliana, M. A., L. Legarreta, G. A. Laffitte, and H. Villar, 1999, Estratigrafía y geoquímica de las facies generadoras de hidrocarburos en las cuencas petrolíferas de Argentina: IV° Congreso de Exploración y Desarrollo de Hidrocarburos (Mar del Plata) Actas, v. 1, p. 1–62.

Vistalli, M. C., 1989, Cuenca siluro-devónica del Noroeste, *in* G. Chebli and L. Spalletti, eds., Cuencas Sedimentarias Argentinas: Serie Correlación Geológica, v. 6, p. 19–42.

Zapata, T. R., and R. W. Allmendinger, 1996, The thrust front zone of the Precordillera thrust belt, Argentina: a thick-skinned triangle zone: AAPG Bulletin, v. 80, p. 359–381.

6

Mugnier, J.-L., P. Huyghe, P. Leturmy, and F. Jouanne, 2004, Episodicity and rates of thrust-sheet motion in the Himalayas (western Nepal), *in* K. R. McClay, ed., Thrust tectonics and hydrocarbon systems: AAPG Memoir 82, p. 91–114.

Episodicity and Rates of Thrust-sheet Motion in the Himalayas (Western Nepal)

Jean-Louis Mugnier

Laboratoire de Géodynamique des chaînes Alpines et Université Joseph Fourier, Maison des Geosciences, Grenoble Cedex, France

Pascale Huyghe

Laboratoire de Géodynamique des chaînes Alpines et Université Joseph Fourier, Maison des Geosciences, Grenoble Cedex, France

Pascale Leturmy[1]

Laboratoire de Géodynamique des chaînes Alpines et Université Joseph Fourier, Maison des Geosciences, Grenoble Cedex, France

François Jouanne

Laboratoire de Géodynamique des chaînes Alpines et Université de Savoie, Chambéry, France

ABSTRACT

The kinematics and rates of displacement along single faults of the Outer Belt and through the entire thrust belt of the Himalayas of western Nepal have been estimated by structural field work, balanced cross sections, fluviatile terrace-deposit studies, and geodesy. GPS geodetic studies indicate that the shortening is 15 ± 2mm/yr and is perpendicular to the trend of the western Nepal Himalayas and compatible with elastic strain accumulation. This surface shortening would be induced by a 19 mm/yr slip rate, at depth, on the basal detachment beneath the Great Himalayas, whereas the detachment is locked beneath the Lesser Himalayas. In the outer zone, most of the displacement presumably occurs during major seismic events, and the 2.5- to 5-m thrust movements observed locally for imbricate fans branching off the major thrusts of the Sub-Himalayas could reflect surficial rupture during these events. During the Holocene, the Main FrontalThrust (MFT) has been active, but portions of the piggyback thrusts of the Outer Belt also show episodes of activity that vary from a few meters to several tens of meters of displacement. The ratio of MFT shortening to total Himalayan shortening

[1]*Present address*: Université de Cergy-Pontoise, Cergy-Pontoise, France.

varies laterally, from 1 to less than 0.5, and out-of-sequence thrusting occurs in the Sub-Himalayas and possibly in the Lesser Himalayas. This pattern of episodic out-of-sequence reactivation fits with the evolution of a brittle thrust wedge affected by surficial mass transport and/or fluid pressure variation due to fault-valve behavior. The total shortening rate through the Himalayas of western Nepal following the Miocene is 19 mm/yr, with present-day, Holocene, and long-term shortening-rate uncertainties of 2, 6, and 5 mm/yr, respectively.

INTRODUCTION

Numerous field-based studies (see McClay, 1992a for a compilation) have shown that a thrust belt develops by forward nucleation of the frontal ramps that is accompanied by reactivation of inner faults and development of break-back thrusts. In contrast, although earthquake ruptures form the basic increment of tectonic motion (Brune, 1968), very few seismological studies integrate earthquake faulting as a thrust-sequence process. The 1999 Chi-Chi earthquake (Kuo-Fong et al., 1999) in the Taiwan thrust belt, nonetheless, clearly attests to an out-of-sequence surface rupture in an active collision belt. The Himalayas, which have been a collision belt since the Tertiary (Le Fort, 1975; Besse et al., 1984), show out-of-sequence reactivation (Burbank and Beck, 1989; Chalaron et al., 1995; Mugnier et al., 1998) and presently undergo approximately magnitude 8 (M8) earthquakes (Molnar, 1990). Therefore, the Himalayan Belt is a good place to study both earthquake faulting and thrust sequences in the development of a mountain belt. The aim of this chapter is to estimate and compare the displacement rate along single faults and through the entire thrust belt of the Himalayas of western Nepal, at different time scales (during the Miocene-Pliocene, Pliocene-Quaternary, and Holocene) and at the deformation scale determined by seismological and geodetic studies. Uncertainties concerning these displacement estimates are tentatively evaluated, although it is impossible to reconcile the statistical aspect of the uncertainty in geodesy with the scarce field observations used to estimate geologic rates. Comparison of the thrust activity at these different time scales is used to define the respective roles of rheology, external forcing, and large-scale geodynamics in the development of the Himalayan Belt. General controls of thrust-wedge motion are discussed in light of the Himalayan example.

REGIONAL SETTING

The Himalayan Belt of Western Nepal

After the first steps of continental collision between the Indian and Tibetan-Asian plates during the Eocene, the large crustal shortening caused by convergence was asymmetrically taken up, with shortening on the Indian side mainly absorbed by the stacking of the crust into several thrust sheets (Le Fort, 1975).

The main structures of western Nepal consist of several north-dipping thrust faults (Ganser, 1964) which, from south to north, are as follows (Figure 1 and Table 1 for the definition of the abreviations): the Main Frontal Thrust (MFT), the Main BoundaryThrust (MBT), the Mahabarat Thrust (MT), and the Main Central Thrust (MCT). The generally 120° north-trending thrusts are inferred to branch off the major basal detachment (MD on Figure 2) of the Himalayan fold-and-thrust belt (De Celles et al., 1998a). The structural pattern is nonetheless noncylindrical, with several oblique features bounding the nappes (Figure 1b). Transport direction, deduced from stretching lineation data at the base of the MCT (Brunel, 1986; Pecher, 1991) and the MT (Mugnier, 1994, unpublished data), is nearly north–south. Time-space distribution of thrusting generally shows increasingly younger events from north to south, from 24–21 Ma for the MCT to less than 2 Ma for the MFT (see De Celles et al., 1998b for a compilation). Nonetheless, Hodges et al. (1996) and Harrison et al. (1997) demonstrated that faults in the MCT system have undergone late Miocene-Pliocene reactivation, and a major displacement event occurred as recently as 5.5 Ma. Out-of-sequence reactivation is also evident from cross-cutting relationships between ramps and an earlier flat décollement in the Lesser Himalayas of western Nepal décollement (De Celles et al., 1998a; Mugnier et al., 1998). Quaternary faulting in western Nepal has reactivated major faults, such as the MBT or MCT (Nakata, 1989), and/or has been taking place south of the MBT, along the thrust splay of the Sub-Himalayas (Mugnier et al., 1994).

Structural Pattern of the Sub-Himalayan Belt

The Sub-Himalayan Belt is formed by synorogenic sediments of the Siwalik Group (Auden, 1935) and is the outer part of the thin-skinned thrust belt of the Himalayas. The Siwalik Group comprises fluviatile sediments deposited in the foredeep that developed along

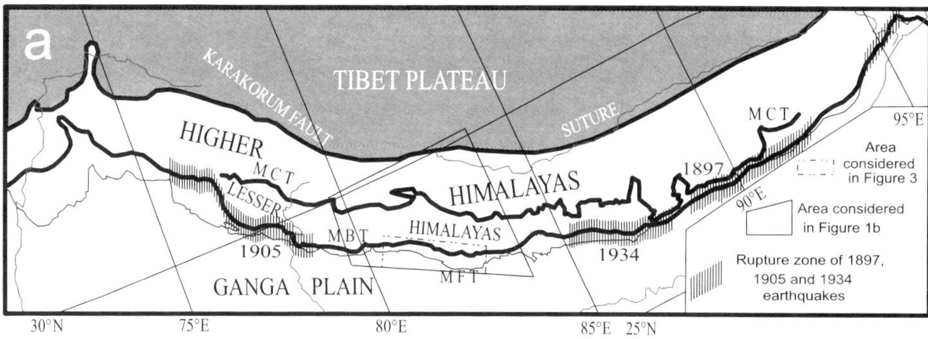

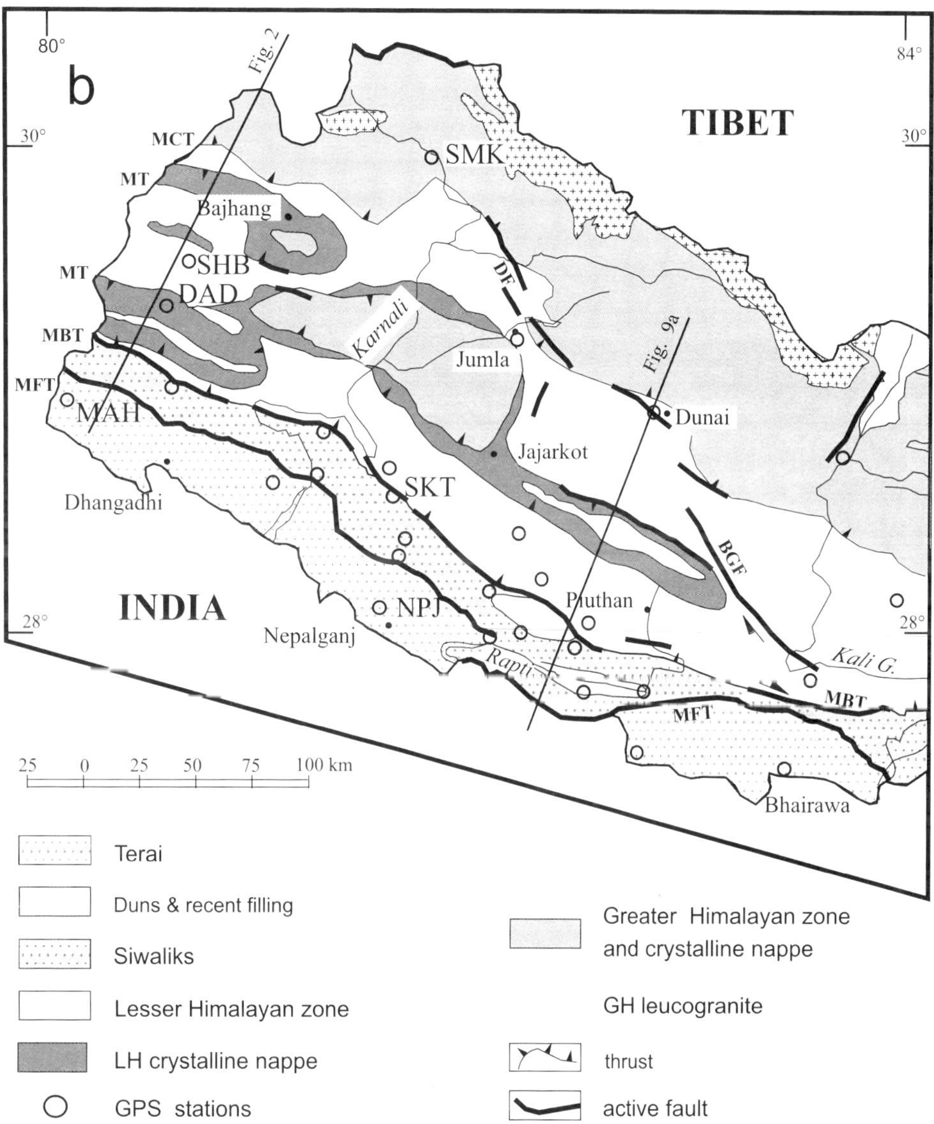

FIGURE 1. (a) Structural sketch of the Himalayas, with location of the area considered and the probable rupture zones of historical earthquakes (modified from Yeats and Lillie, 1991). Structural and active-fault map of western Nepal. Black line refers to the location of Figure 2. Empty dots refer to the location of GPS sites (from Jouanne et al., 1999). SMK and NPJ are sites quoted in the text. MAH, DAD, SHB, and SKT are sites located on Figure 2. MFT, MBT, MT, MCT, DF, and BGF refer to faults (see text and Table 1).

the southern margin of the Himalayan mountain range after the Miocene-Pliocene (Burbank et al., 1996). The Siwalik Group has been dated in several places by paleomagnetic studies (see Gautam and Rosler, 1999, for a compilation of the studies in Nepal). Several north-dipping thrusts delineate tectonic boundaries in the Siwalik Group (Mugnier et al., 1992). The most frontal one is the MFT (Hagen, 1956), whereas the transported thrusts are called the Main Dun Thrusts (MDT) (Herail and Mascle, 1980). Seismic profiles in western India have demonstrated that these faults branch off the outer part of the basal detachment (OD) of the Himalayan fold-and-thrust belt (Powers et al., 1998), and it is assumed that this pattern is also followed in Nepal (Mugnier et al., 1999; Lavé and Avouac, 2000).

The MDT in western Nepal is formed by a succession of five laterally relayed thrusts (from east to west, MDT1 to MDT5 on Figure 3), and the MFT is formed by three segments (MFT1 to MFT3 on Figure 3). These 50- to 100-km-long segments die out laterally into propagating folds (FPF1 and FPF2 on Figure 3) or branch and relay along lateral transfer zones. Along-strike variation in the frontal structural morphology suggests active folds that propagate laterally into the foreland basin (Leturmy et al., 1997; Husson et al., 2004). Because the thrusts are soft-linked in 3-D, the application of Elliott's (1976) rule suggests a maximum displacement for each thrust of the order of 10–20% of the length of the thrust segments, i.e., of the order of 5–15 km.

The superimposition of two décollement levels is indicated in a large portion of the Siwalik Group of

Table 1. Definitions of the abbreviations used in the chapter.

Abbreviation	*Definition*
MFT	Main frontal thrust
MBT	Main boundary thrust
MT	Mahabarat thrust
MCT	Main central thrust
MD	Major basal detachment
MDT	Main Dun thrust
OD	Outer part of the basal detachment
MDT1 to MDT 5	Relayed segments of the main Dun thrust
MFT1 to MFT 3	Relayed segments of the main frontal thrust
FPF1 to FPF2	Relayed fault propagating folds
ID	Intermediate décollement (in Siwalik)
CR	Crustal ramp
2An	Chron of the geomagnetic polarity scale
1r	Chron of the geomagnetic polarity scale
2r	Chron of the geomagnetic polarity scale
BGF	Bari Gad fault
DF	Darma fault
T2b to T4b	Strath surfaces along the Koilabas River
T2t to T2t	Tops of fluviatile deposits (Koilabas River)
SMK	GPS site of Simikhot
NPJ	GPS site of Nepalganj

western Nepal (Mugnier et al., 1999) by the construction of hanging-wall sequence diagrams (as defined by Butler, 1982). The lower décollement (OD) is located at the base of the lower Siwalik member; an intermediate one (ID) is located near the top of the lower Siwalik member in mud-rich levels and forms the base of the sheet at the hanging wall of the MDT1. Duplexes have developed in between (Figure 4).

Motion along MFT and MDT has been deduced from gouge fabrics, R-shear relationships, and slickenside analyses in the fault zones (Huyghe et al., 1998a; Mugnier et al., 1998). The direction of motion is generally 010° to 030°, i.e., nearly perpendicular to the trend of the belt (Figure 3). Strike-slip components of displacement are found locally at the lateral closure of the piggyback Dang and Deukhury Basins. Shortening direction deviates clockwise at the western termination and anticlockwise at the eastern termination of the basins, respectively, suggesting that the strike-slip components only reflect complexities above transfer zones (Huyghe et al., 1998b). These transfer faults delineate at least three distinct portions of the Outer Belt and control MFT segmentation (Figure 3).

SHORTENING ACROSS THE HIMALAYAS

The displacement rate along single faults and through the entire thrust belt of the Himalayas of western Nepal has been estimated at different time scales since the Miocene-Pliocene, during the Pliocene-Quaternary and the Holocene, and at the deformation scale determined by seismological and geodetic studies.

Miocene-Pliocene Shortening Rate through the Lesser Himalayas

In western Nepal, De Celles et al. (1998a) have estimated the shortening through the Lesser Himalayas by use of cross-section balancing of the crustal duplex formed by the Paleozoic to Precambrian rocks. This estimation is nonetheless subject to considerable uncertainty, because the location of the main basal detachment (MD) is not well defined. For instance, the basal shear zone deduced from the aseismic creep zone of the dislocation model proposed by Jouanne et al. (1999) is nearly 5 km deeper than the MD inferred by De Celles et al. (1998a) beneath the Lesser Himalayan duplex (Figure 2). This deeper location coincides with the "Raminatta" formation inferred by De Celles et al. (1998a) at the footwall of the MD. Because the base of the Raminatta is a major décollement level in the Himalayas (Shresta, 1987), this level appears to be a possible candidate for the floor thrust of the Lesser Himalayan duplex. In that case, the stratigraphic thickness of the duplex could be greater than inferred by De Celles et al. (1998a). We are convinced that the present geometry of the Lesser Himalayas is more complex than the simple duplex geometry of Figure 2, and the initial state is not a simple layer-cake stratigraphic pile. Therefore, to estimate shortening in the Lesser Himalayas, we used the total-area balancing method, the results of which are not affected by the complexity of the thrust trajectory. Taking into account a 1000–1400 km^2 total area and a 7–10 km initial thickness of the series, we find that the initial length of the duplex structure would vary from 100 km to 200 km. Because the Lesser Himalayan structures (i.e., structures north of the MBT) above the middle-crust ramp (CR on Figure 2) have a displacement of 50–60 km, the total shortening in the Lesser Himalayas is 150–260 km. Two distinct studies of the origin of the Siwalik Group in western Nepal estimate the Lesser

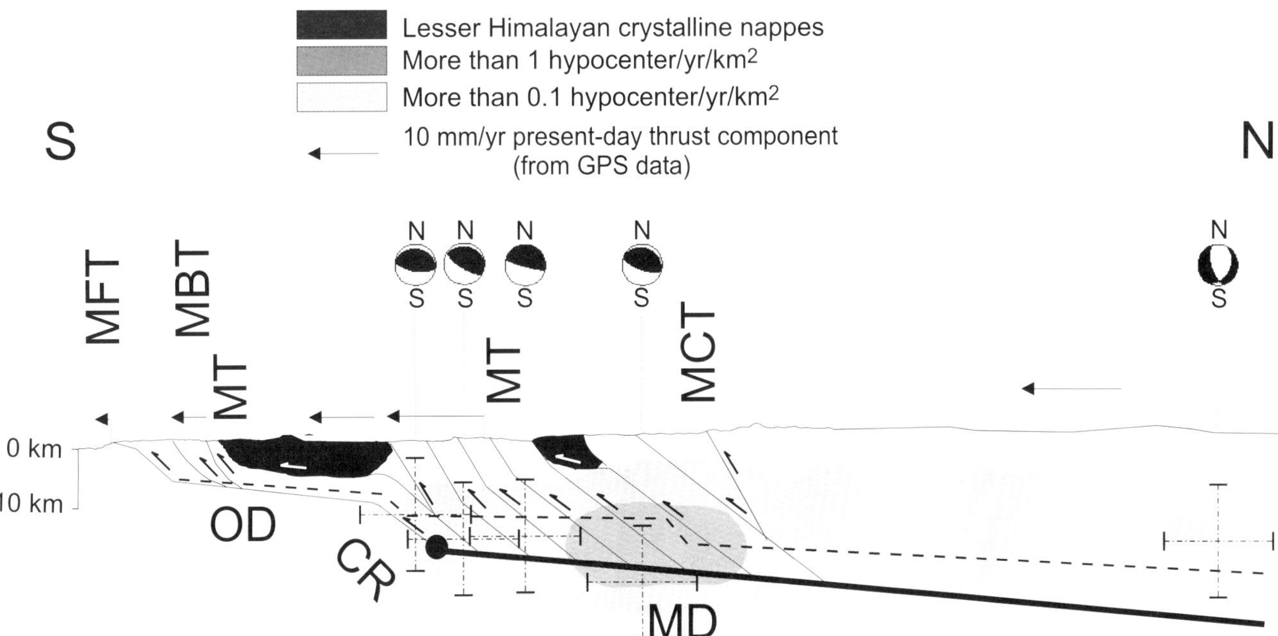

FIGURE 2. Simplified cross section through far-western Nepal. (Location on Figure 1b). Field geology from Shresta (1987) and Upreti and Le Fort, (1999). Microseismicity from Pandey et al. (1999), the density distribution of all events within 30 km from the section are taken into account. Outer décollement is OD; middle-crust ramp is CR; basal detachment (MD) is a thick line from the basal dislocation of Jouanne et al. (1999) or a dashed line from De Celles et al. (1998a). Focal mechanism (in map view) from Harvard CMT (Dziewonski et al., 2000) catalog. Only events between 80° and 82° E longitude are taken into account, and the hypocenters (uncertainty of the order of ± 10 km) are represented by large dashed crosses. Horizontal present-day motion in the plane of the cross section from Jouanne et al. (1999). Only the GPS sites of far-western Nepal (from south to north MAH, SKT, DAD, SHB and SMK on Figure 1) are taken into account and the reference is the GPS site of NPJ.

Himalayan crustal duplex to have been initiated ~11 Ma. The estimates are from K-feldspar sand proportions (De Celles et al., 1998a) and from geochemical isotopic ratios (Huyghe et al., 2001). Therefore, the shortening rate in the Lesser Himalayas ranges from 14 to 24 mm/yr.

Pliocene-Quaternary Tectonics of the Sub-Himalayan Belt

Total Shortening through the Sub-Himalayan Belt of Western Nepal

Along the frontal structure, shortening occurs in a first stage by a fault-propagating fold structure and subsequently evolves toward an emergent ramp stage (Mugnier et al., 1992). Cross sections through fault-propagating fold structures do not show major thickness variations between the nearly vertical forelimb and the gently dipping backlimb (Leturmy, 1997). Furthermore, studies of rock magnetic properties (Gautam and Pant, 1996; Gautam and Rosler, 1999) show that the internal fabric predates folding. Therefore, it is suggested that layer-parallel shortening occurs ahead of the thrust front and does not occur during folding, a deformation pattern identical with that indicated for the South Pyrenean Thrust Belt (Muñoz et al., 1999). Consequently, we believe in our study that the thickness and length of the beds remain constant during folding and thrusting of the Sub-Himalayan belt, which is an approximation that allows line-length balancing of the beds (Hossack, 1979).

Height-balanced cross sections have recently been built through the frontal belt of western Nepal (location on Figure 3 and characteristics in Table 1) (Dhital et al., 1995; Bouvier, 1997; Leturmy, 1997; Mascle et al., 1998). These cross sections are based mainly on field data and sparsely published seismic lines (DMG, 1985; Jackson and Bilham, 1994), and the geometry of the deep structures therefore is not well defined. Nonetheless, line-length balancing of the beds above the duplexes (i.e., at the transition between middle and lower Siwalik members) can be used to estimate shortening independently of the uncertainties that affect the deep geometry and the duplex structures (Figure 4).

Minimum shortening along the eight cross sections varies from 21 to 40 km (Table 2). These variations are interpreted to be mainly induced by underestimation of shortening caused by the erosion that has affected the hanging wall of the Siwalik thrusts and removed the hanging-wall cutoffs. Therefore, it is inferred that the 40-km shortening obtained along the Surai Khola cross section (Figure 4, and A-A′ on Figure 3) is

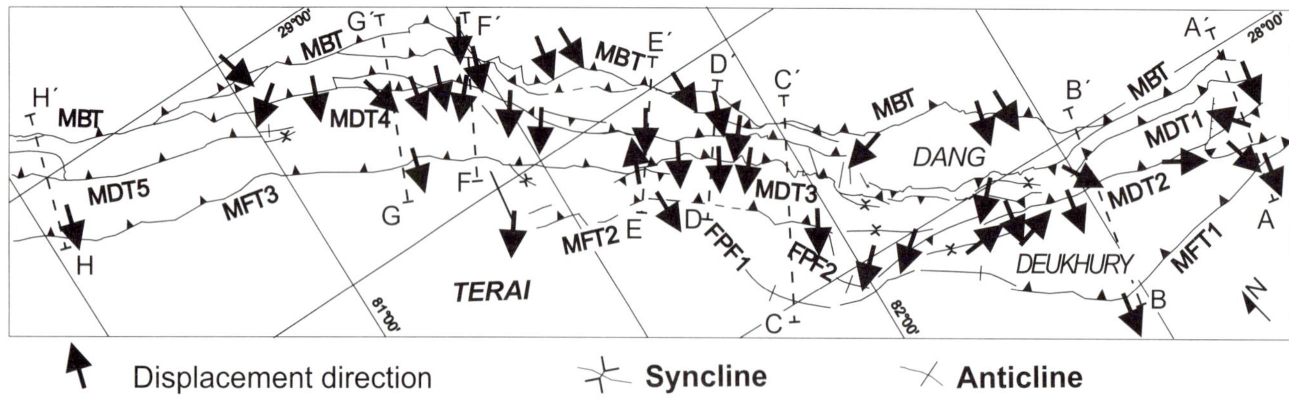

FIGURE 3. Schematic structural map of the Sub-Himalayan Belt of western Nepal. The black bold lines represent the shortening directions (see text), and the dashed lines A-A′ to H-H′ are locations of the cross-sections referred to on Table 2.

the most valid estimate of shortening through the Sub-Himalayan Belt of western Nepal. Furthermore, the three most external sheets of this cross section have been dated by magnetostratigraphy (Appel et al., 1991 for sheet I, and Rosler et al., 1997 for sheets II and III on Figure 4), thereby allowing reliable correlation of the levels used for the line-length-balancing procedure.

Displacement Rate along the Main Frontal Thrust

The combination of paleomagnetic studies and syntectonic sedimentary features helps us estimate the timing of thrusting in the Sub-Himalayan Belt. Evidence is presented here for an unconformity located near the Surai Khola cross section (symbol IV on Figure 4). It has been mapped at approximately 82°40′E and 27°45′N on satellite imagery (Leturmy, 1997). Its lateral extension has been followed eastward up to the Surai Khola cross section. The erosional surface is located in the middle part of the upper Siwalik member and cannot be definitively identified on the outcrops along the Surai Khola. Nonetheless, red soil levels several meters thick and with calcareous nodules, interbedded between conglomeratic levels, could be the expression of this erosional surface. Dip variations of the strata from the bottom to the top of the Surai Khola section (Figure 5) show that the inferred unconformity is located between strata of nearly constant dip (close to 60°) and a stratal unit in which the dip decreases upward. This pattern fits with a model of growth strata above a ramp structure (Zoetemeijer and Sassi; 1992, Burbank and Verges, 1994). The mechanism that induces tilting is still under discussion (Verges et al., 1996; Ford et al., 1997). Progressive rotation during fold amplification is demonstrated for folds located above a detachment (Lickorish and Butler, 1996), whereas Lavé and Avouac (2000) have shown that the fault-bend-fold model (Suppe, 1983) is a robust approximation for the folding of fluviatile deposits located above emergent thrusts whose displacement exceeds the length of the ramp and whose hanging-wall anticlines are highly eroded. Because this latter situation is common in the studied area, the latter model is followed. Tilting is therefore assumed to be linked to a succession of upward-steepening fault segments at the transition between

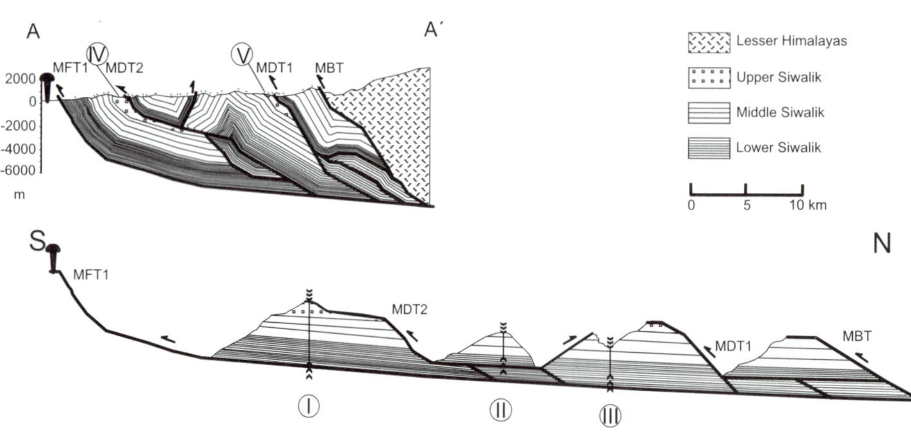

FIGURE 4. Balanced cross section through the Sub-Himalayan Belt of western Nepal (A-A′ on Figure 3). I, II, III on the restored cross section respectively indicate the location of the magnetostratigraphic sections dated by Appel et al. (1991) and Rosler et al. (1997) (III is projected 25 km along strike). IV and V refer to syntectonic sedimentary features (see text).

Table 2. Minimum shortening estimated through the Sub-Himalayan belt of western Nepal. Location of the cross sections on Figure 3.

Cross section	*Initial length (km)*	*Final length (km)*	*Total shortening (km)*	*Adapted from*
A–A′	73	33	40	Dhital et al., 1995
B–B′	68	38	30	Leturmy, 1997
C–C′	64	40	24	Leturmy, 1997
D–D′	44	20	22	Leturmy, 1997
E–E′	44	21	23	Mascle et al., 1998
F–F′	48	21	27	Bouvier, 1997
G–G′	42	25	21	Huyghe et al., 1998a
H–H′	49	23	26	Bouvier, 1997

the décollement and the ramp, and to a displacement greater than the length of the fault segments. The footwall of the unconformity is characterized by a normal polarity related to the 2An chron, as demonstrated by a calibration made from a continuous section of 4000 m below the unconformity (Appel et al., 1991), and by its comparison with a detailed paleontological study (Corvinus, 1993). The hanging wall of the unconformity is characterized by a reverse polarity (Appel et al., 1991) related either to the base of the 1r or to the 2r chron. Initial dating has been modified, taking into account the standard polarity time scale of Cande and Kent (1995) rather than Harland et al. (1989), and is found to be 2.4 Ma for the footwall sediments and 1.8 or 2.4 Ma for the hanging-wall sediments. The beginning of tilting suggests that tectonism has been affecting the MFT for 1.8–2.4 Ma in the Surai Khola area. In this area, a minimum displacement of 10 km is estimated on the MFT from balanced cross sections, and the preservation of the growth structures (IV on Figure 4) implies that displacement is less than 12 km. Therefore, the shortening rate along the MFT has been nearly 4 mm/yr for 2.4 m.y.

Shortening Rate through the Sub-Himalayan Belt

In the inner part of the Sub-Himalayan Belt and near the cross section of Figure 4 (longitude 83°04′E, latitude 27°51′N, symbol V on Figure 4), conglomeratic beds with middle Siwalik boulders larger than 1 m in diameter are located at the footwall of the MDT1 and lie above the lower part of the upper Siwalik, where no sedimentary disturbance has been found. They attest to the erosion of the middle Siwalik strata located on the hanging wall of the moving MDT1. These features postdate the beginning of motion along the MDT1 and precede the thrusting of the lower Siwalik member above the upper Siwalik member. Using the mapping of Dhital et al. (1995) for facies correlation and the magnetostratigraphic studies of Rosler et al. (1997) for dating, we found that the age of the conglomeratic beds ranges from 3.0 to 2.4 Ma. Assuming that the MDT1 is related to the tectonic onset in the Sub-Himalayan Belt, a shortening of 32 to 40 km is found for the last 3 to 2.4 m.y. (i.e., a mean shortening rate between 11 and 17 mm/yr).

Pliocene-Quaternary Out-of-sequence Reactivation

The above estimates show that the shortening rate across the Outer belt is greater than the displacement rate along the MFT, a discrepancy that implies a strong out-of-sequence reactivation component. Kilometer-scale out-of-sequence thrusting has already been demonstrated from tectonic and sedimentary relationships in the Himalayan Thrust Belt of Pakistan (Burbank and Raynolds, 1988) and smaller-scale out-of-sequence thrusting has been demonstrated in western Nepal (Mugnier et al., 1998). An intermediate period of tectonism between the earthquake slip and finite geometry of the thrust system is proposed for these smaller-scale events (Mugnier et al., 1998). This is based on the following evidence: It has been found that a main thrust, generally defined by a gouge zone a few meters thick (Huyghe et al., 1998a), forms locally in the field by hundred-meter-scale imbricate fans (as defined by Dahlstrom, 1970 and McClay, 1992b). Along-strike variations of the imbricate geometry suggest that each splay branches off the major gouge zone of the main thrust. Therefore, the displacement history of the main thrust is recorded by the relationship between splay and sediment during the development of the imbricate fan. For example, the imbricate fan of the Karnali (location on Figure1b) is formed by four splays. Nearly 40 m of reverse displacement occurred along each splay and was followed by a period of quiescence before activity of another splay (Mugnier et al., 1998). The whole development of this imbricate fan suggests a periodic activity for the MDT at longer intervals than the earthquake cycle. Other examples of imbricate fans will be described below.

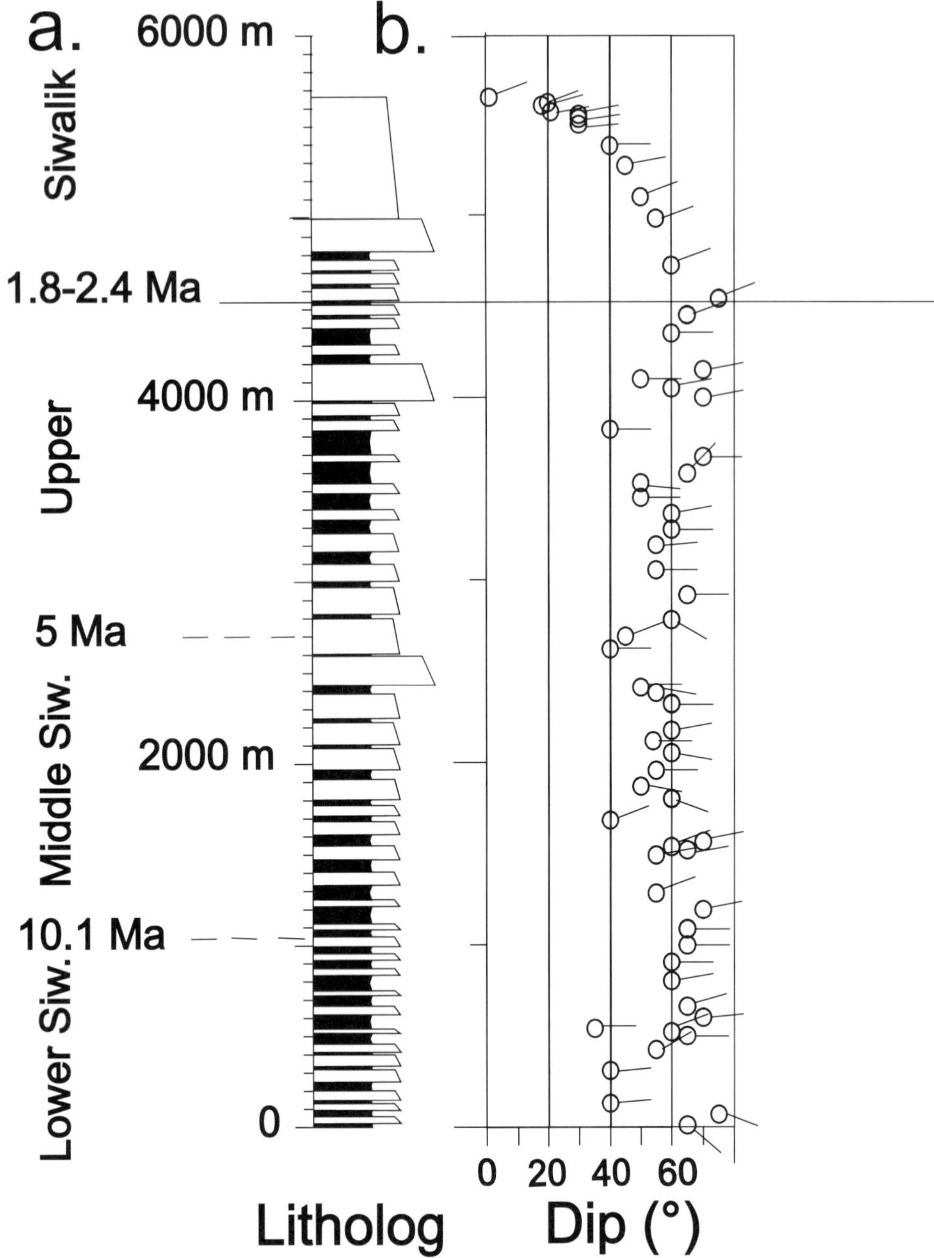

FIGURE 5. The Surai Khola cross section (location on Figure 4). Litholog from Dhital et al. (1995). Dating adapted from Appel et al. (1991) and standard polarity scale of Cande and Kent (1995). Dip and strike variations. Strike is shown by the trend of the lines (north at the top).

Holocene Tectonism of the Sub-Himalayan Belt

The Holocene convergence rate in the Sub-Himalayan belt is deduced from the tectonic uplift of terraces and their geometric relationships to thrusts. Numerous charcoals have been carbon dated in western Nepal (Yamanaka and Yagi, 1984; Leturmy, 1997; Kimura, 1998). Table 3 gives the key relevant dates.

Tectonism of the Frontal Structure

Recent tectonism along the frontal structure is evident in far-western and mid-western Nepal (Nakata, 1989; Mugnier et al., 1999). In the Deukhury area (Figure 3), Holocene tectonism is evidenced by the sediment uplift and tilting. This tectonism affects the five allostratigraphic groups of the piggyback basin, which were regionally mapped by Kimura (1998). These groups have been carbon dated near the Jamnibas cross section (Figure 6a, line B-B′ on Figure 3) and their elevation measured along nearly continuous profiles (Figure 6c). The Deukhury unconsolidated gravels reach their maximum elevation 50 m above the river. In most places, the surface slope of the terraces is very similar to the slope of Jamnibas River. Nonetheless, the tilt of the deposits locally reaches more than 15° (bends A to D on Figure 6c). These local bends of intermediate and older terraces are interpreted as kink folds induced by a change in dip of the basal décollement (Figure 6b). They make possible to estimate a shortening rate of 14 ± 4 mm/yr (Figure 6e; see Appendix A for the calculation and Table 4 for the data). Five terraces are found along the Jamnibas River at the transition between the Siwalik member and the Deukhury Basin. Three of them have been dated along the same vertical at Kakraha (Figure 6d), and two others nearby are dated. The ages of the charcoals are 6945 yr B.P. in the Lamahi Terrace (name from Kimura, 1998), 4740 yr B.P. in the Nimbugaon Terrace, 4172 yr B.P. for the Garwha 1 terrace, 1111 and 877 yr B.P. for the Kakraha terrace, and a modern age (less than 150 yr B.P.) for the present-day terrace. Downstream, the age of the Garwha 2 terrace is 3630 yr B.P. Incision rates of the

Table 3. Age of charcoal samples. Calibration from Stuiver and Reimer (1993).

Location on Figure	*Calibrated age (yr B.P.*)*	*Latitude*	*Longitude*	*Reference*
Figure 6c	3690–3470	27°50.56	82°28.43	Kimura, 1998
Dot 1 on Fig. 6d	<150	27°46.13	82°33.88	Lyon-9523
Dot 2 on Fig. 6d	1177–972	27°46.05	82°33.98	Lyon-9345
Dot 2 on Fig. 6d	925–730	27°46.05	82°33.98	Leturmy, 1997
Dot 3 on Fig. 6d	4417–3927	27°46.05	82°34.01	Lyon-863 (oxa)
Dot 4 on Fig. 6d	4850–4531	27°47.08	82°33.37	Leturmy, 1997
Dot 5 on Fig. 6d	7151–6640	27°35.15	82°16.56	Leturmy, 1997
Star 1 on Fig. 7	7468–7214	27°53.09	82°30.51	Lyon-862 (oxa)
Star 2 on Fig. 7	2340–2130	27°52.72	82°30.49	Leturmy, 1997

**Because 0 yr B.P. is conventionally 1950 years after Jesus Christ, the mean age used in the text, here, and in Tables 4 and 5, is the mean value plus 50 yr.*

Jamnibas River through these terraces are 4–12 mm/yr. To test the validity of the shortening rate, level variations of the Jamnibas River have been calculated at Kakraha (in a frame of reference located in the Indian plate, assuming these level variations to be the difference between incision and calculated uplift; see Appendix A and Lavé and Avouac, 2000). These calculated river level variations are compared (Table 5 and Figure 6e) with the organization of regional sedimentary bodies. A good qualitative correlation has been demonstrated: The Nimbugaon Terrace, which was the widest flood plain in the history of the Dun, aggrades regionally above the alluvial fans of the old Lamahi because of a rise in the trunk stream (the Rapti, location on Figure 1b) and the development of a basinwide flood-plain terrace (Kimura, 1998). The younger Garwha terraces are deposited along river-flood areas carved through the Nimbugaon terrace after a major drop in the river level (Kimura, 1998). Therefore, the development of the piggyback basin fits with river-level fluctuations above an active décollement affected by a 14 ± 4 mm/yr shortening rate.

Main Dun Thrust Tectonics

In far-western Nepal, Holocene fluviatile terraces overlap and seal the MDT4 in the Karnali River area (Figure 1b), whereas between the Deukhury and Dang valleys, Holocene tectonic activity is apparent along the MDT1 (Yamanaka and Yagi, 1984; Leturmy et al., 1997).

Along the Arjun River (location on Figure 6a), the MDT2 comprises at least three splays (Figure 7a). The internal splay has a displacement of a few meters and is synchronous with the deposits of a sandy bed (Figure 7c). Gravel deposits overlap the two northward splays of the MDT and were not deformed during their deposition, thereby suggesting a period of quiescence for the MDT. Reddish deposits overlie a strath surface that affects both lower Siwalik and gravel deposits (Figure 7b). Southward, the reddish beds are highly eroded, whereas gravel deposits lying beneath the strath surface (Figure 7d) tilt 20–40° to the south. This tilt is related to a bend restricted in a narrow area above the more external splay of the MDT, and induces a 23 m uplift of the northern area relative to the southern area. A charcoal sample (star on Figure 7a) in the undisturbed deposits lying above the strath surface was carbon-dated at 7340 yr, whereas sediments remobilized in local depressions have been dated at 2200 yr. This more-than-2200-yr-old terrace dissection suggests that part of the uplift occurred more than 2200 yr ago. This hypothesis is also corroborated by the presence of the aggraded Nimbugaon Terrace south of the bend but not above the Lamahi Terrace (Kimura, 1998). Considering the estimated river-level variations shown in Figure 6e, more than 10 m of uplift would have needed to occur more than 4800 yr ago to prevent the deposition of the Nimbugaon terrace in this area. The mean uplift rate is estimated by dividing the visible vertical offset of the bend (which minimizes the vertical component of motion) by the age of the terrace. A mean uplift rate of 2–3 mm/yr is found. Assuming rigid translation of the thrust's hanging wall (suggested by the fairly undisturbed geometry of the terrace) and considering that the dip of the MDT varies between 40° and 50° in this area, the displacement rate along the fault has been of the order of 2–5 mm/yr over the past 7400 yr (see Equation 2 of Appendix A for the vector components).

Holocene Shortening Rate across the Himalayan Belt

In the Lesser and Great Himalayas, the late Quaternary faulting with the best expression extends from the Bari Gad Fault to the Darma Fault (BGF and DF on Figure 1b) (Nakata, 1989; Yeats and Lillie, 1991), and these faults are located in the prolongation of the

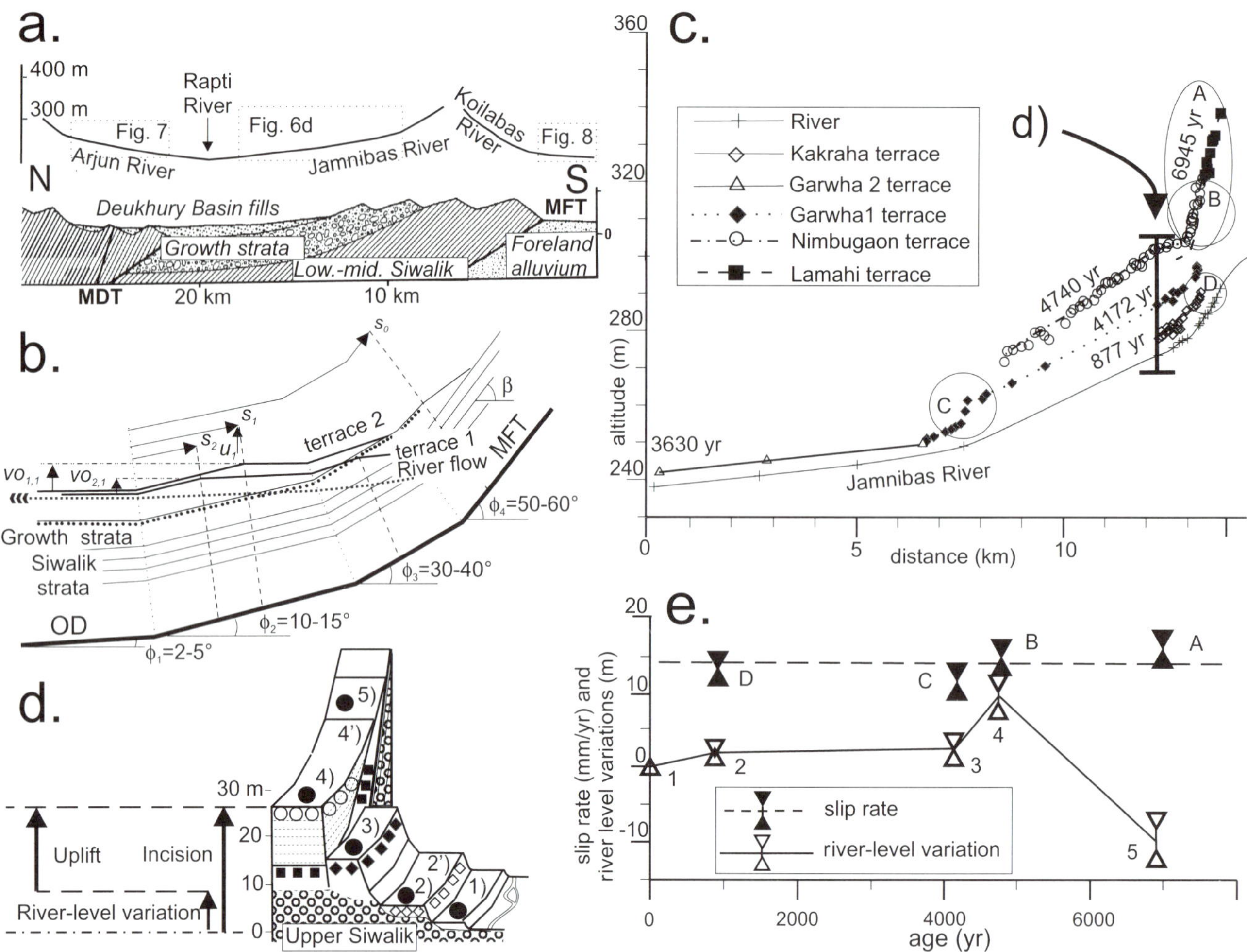

FIGURE 6. Relationships between uplift rates, shortening rates, and incision rates in the Deukhury valley. Simplified geological cross section of the Deukhury Basin (Hatched line B-B′ on Figure 3) adapted from Kimura (1998). Top: Rivers and locations of Figures 6d, 7, and 8. Bottom: Structures of the Sub-Himalayas. Schematic cross section through the frontal structure and the Deukhury Basin, and definition of the parameters used for the calculation of the annex. The relationship between the dip of the Siwalik strata (β), the thrust dip (ϕ), the thrust slip (*s*), and the uplift (*u*) is $s = u / \sin(\phi)$, assuming a fault-bend-fold model. MFT is the Main Frontal Thrust; OD is the Main Outer Detachment. The vertical offset (*vo*) of a terrace folded by a pair of hinge lines (the dotted line is the fixed syncline hinge line, the thin hatched line is the moving anticline hinge line) is $vo_{1,1} = s_1 * (\sin \phi_2 - \sin \phi_1)$ (see Appendix A). Altitude of the terraces along the Jamnibas River. The vertical uncertainty on each measured point (symbol) is of the order of ± 1 m. The lines between dots refer to nearly continuous outcrops. Schematic block diagram at Kakraha through the terraces of the Jamnibas River and relationships between incision, uplift and river level variations (same symbols for the terraces as Figure 6c). The front of the block diagram is parallel to the strike of the syncline structure; Location of Figure 6d is d on Figure 6c. Estimated shortening rates and river-level variations through time at Kakraha (see Appendix A). A to D refer to outcrops of Figure 6c and Table 4. Numbers 1 to 5 refer to outcrops of Figure 6d and Table 5.

right-lateral Karakorum fault zone (Armijo et al., 1989). Late Quaternary motion along the MBT has also been documented from tectonic-sediment relationships in far-western Nepal (Bashial, 1981) and in mid-western Nepal (Nakata, 1989; Mugnier et al., 1994). Geologic movement rates along all these faults of western Nepal are still not well known, but Nakata (1989) tentatively estimates the rates to be less than 1mm/yr by. The slip rate along the MBT and BGF is much smaller than the MDT and MFT slip rates, and the total shortening rate across the entire Himalayas is therefore considered to be equal to the Sub-Himalayan shortening.

The shortening rate through the Sub-Himalayas is approximated by the sum of the slip on the MFT and MDT. Because the OD is assumed to parallel the Siwalik bedding, the slip ratio is equal to 1 between the ramps and the detachment (Suppe, 1983), and no correction is necessary to take into account the dip of the fault. Therefore, the Holocene shortening rate across the Himalayan belt is estimated to be 19 ± 6 mm/yr.

Table 4. **Estimation of the slip rate of the MFT from the bending of the terraces. (Calculation in Appendix A; locations of the outcrops and definitions in Figure 6).**

Outcrop	*Age, t (yr)*	*Initial Slope, α*	*Offset, v_o (m)*	*Dip, β_{j+1} (°)*	*Dip, β_j (°)*	*Calculated slip rate, sr (mm/yr)*
A	6945	~1 %	40	40–30	15–10	14.5–17.5
B	4740	~0.5 %	25	40–30	15–10	13.0–16.0
C	4132	~0.5 %	8	15–10	5–2	10.5–13.5
D	877	~1 %	4.5	40–30	15–10	12.5–16.0

**Because 0 yr B.P. is conventionally 1950 years after Jesus Christ, the mean age used in the text, here, and in Tables 3 and 5, is the mean value plus 50 yr.*

Historic and Instantaneous Tectonism

Present-day and historic tectonism are recorded by geodetic and seismological monitoring, by historical descriptions of earthquake damage, and by active tectonic features.

Seismicity

Major historical damage induced by Himalayan earthquakes has been described in archives since 1255 (Chitrakar and Pandey, 1986). The approximately magnitude 8 (M8) earthquakes of 1897, 1905, 1934, and 1950 seem to result from underthrusting of India under the Himalayas (Molnar, 1987). They ruptured segments of the outer basal detachment (OD on Figure 2), extending for 200 to 300 km along strike and 60 to100 km downdip (Molnar and Pandey, 1989), with an estimated co seismic slip of the order of 3–6 m to 5–10 m (see reviews by Molnar, 1990; and Bilham et al., 1995). A seismic-tectonic scenario for the Himalayas of Nepal suggests that each segment should rupture about every 130 to 350 yr, during approximately M8 earthquakes (Yeats and Thakur, 1998; Pandey et al., 1999). A 500- to 800-km-long seismic gap between the huge 1934 Bihar-Nepal and 1905 Kangra earthquakes (Figure 1a) has not experienced any great earthquake for more than two centuries and possibly not since the huge earthquake of 1255 (Bilham et al., 1995).

Events with magnitude between 5.5 and 7 have been recorded in the Himalayan-Tibet area (Figure 2). Their focal mechanisms (Dziewonski et al., 2000) suggest an east-west extension with a weak strike-slip component in southern Tibet and a shallow (15 ± 10 km from focal depths), gently north-dipping detachment (MD) beneath the Lesser Himalayas (Baranowski et al., 1984). The rupture history of the 1991 ~M7 earthquake (Cotton et al., 1996) illustrates a complex rupture process along the major detachment. It initiated at a depth of 12 ± 3 km beneath the boundary between the Great and Lesser Himalayas, propagated 20 to 25 km toward the outer belt in nearly 12 s, and died out near the bottom of the crustal ramp (CR) that Srivastava and Mitra (1994) have inferred in this area. This detachment may extend northward into the reflectors that are imaged at more than 20 km deep beneath Tibet (Zhao et al., 1993) and that are interpreted to be a shear zone beneath the transitional domain from unstable frictional faulting

Table 5. **Estimation of the Jamnibas River level variations at Kakraha from the incision of the terraces and the results of Table 4. (Calculations in Appendix A; location of the outcrops and definitions on Figure 6).**

Outcrop	*Age, t (yr)*	*Incision, in (m)*	*Incision rate (mm/yr)*	*Dip, β (°)*	*Calculated uplift, u (m)*	*River level variations, rl (m)*
1	0	1,7	–	15–10	0	0
2	877	6	6.8	15 –10	3 –2	1 –3
2′	877	11	12.5	40 –30	8 –7	2 –3
3	4132	15	3.6	15 –10	12 –10	2 –3
4	4740	26	5.5	15 –10	17 –13	7 –11
4′	4740	51	10.8	40 –30	42 –38	7 –11
5	6945	50	7.2	40 –30	67 –61	(–13) – (–8)

**Because 0 yr B.P. is conventionally 1950 years after Jesus Christ, the mean age used in the text, here, and in Tables 3 and 4, is the mean value plus 50 yr.*

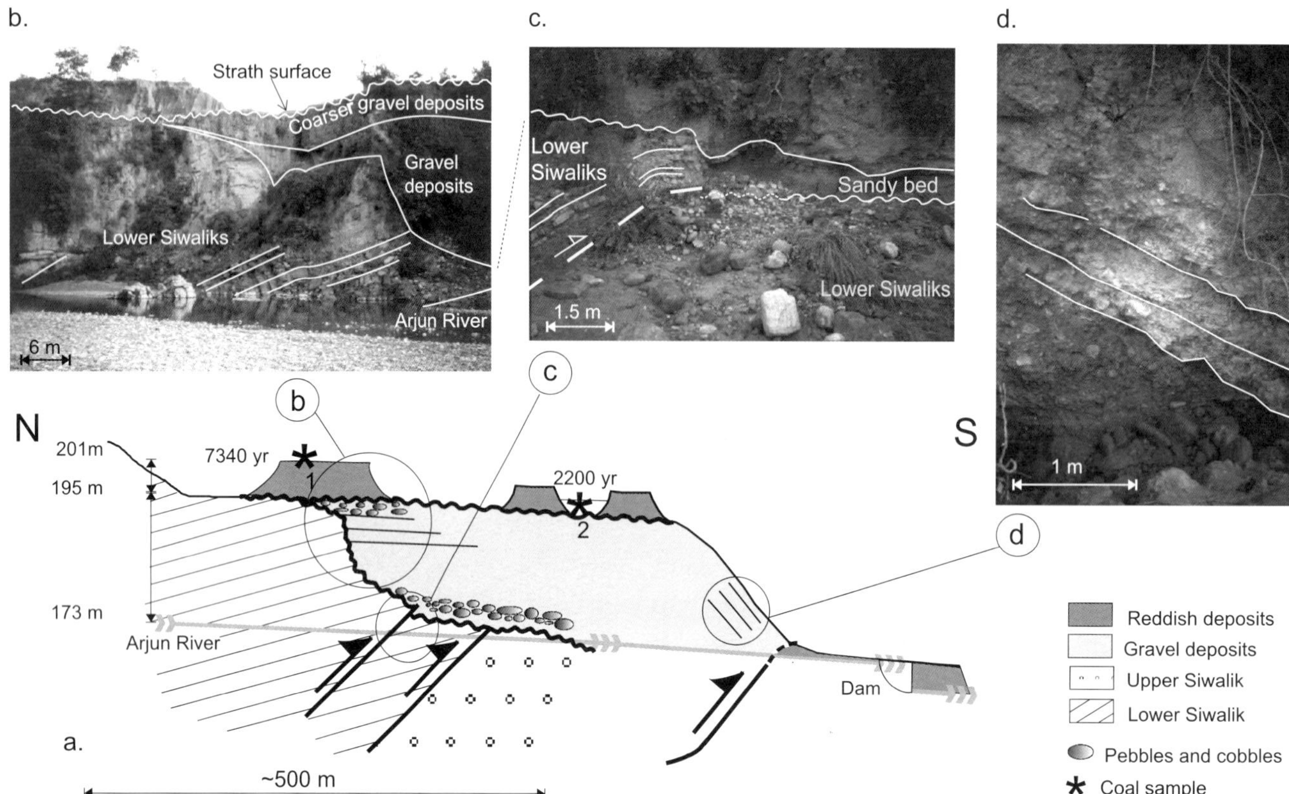

FIGURE 7. Relationships between the strath terraces and the MDT imbricate splay along the Arjun River (location on Figure 6a). Sketch of the relationships between thrusts and deposits; photo of the ~7000-yr strath surface above lower Siwalik and coarse gravels; photo of the MDT splay active during the deposition of a sandy bed; photo of 20–40° tilted beds of gravel deposits.

to localized quasi-plastic shearing flow (Meissner and Strelau, 1982). The seismological network of the Department of Mines and Geology (Kathmandu) shows that intense microseismicity and frequent, medium-size earthquakes (M_L<4) recorded in central Nepal tend to cluster near the crustal ramp in front of the Great Himalayas (Pandey et al., 1995), whereas in western Nepal, microseismicity extends farther beneath the Lesser Himalayas (Figure 2; microseismicity from Pandey et al., 1999). The distribution of the microseisms and the intermediate-magnitude earthquakes beneath the Lesser Himalayas of western Nepal is consistent with the stress built up during the interseismic period that precedes ~M8 earthquakes (Pandey et al., 1999).

Morphological Evidence for Displacement Pulses

Surface ruptures of historic Himalayan earthquakes have not been observed, a fact that is attributed locally to fault-propagation folding (Yeats and Lillie, 1991) absorbing the rupture before it reaches the surface. This lack of surface ruptures is also linked to the difficult field conditions and rapid weathering of outcrops. Trenches recently excavated in eastern Nepal (Nakata et al., 1998) through the MFT suggest that 4–8 m of displacement has ruptured the surface. Slip events on a scale of a few meters have also been described (Delcaillau, 1992) from thrust/Holocene-sediment relationships in central Nepal. The outcrop of Figure 7c could also be related to a natural trench through a single 3-m slip event, considering the 1.5-m vertical offset and the 30° dip of this MDT splay.

Along the Main Frontal Thrust, an imbricate fan has been found along the Koilabas River (location on Figure 6a). The lower Siwalik series thrusts above Quaternary deposits along one splay of the MFT1 that outcrops nicely (Figure 8c). A strath surface (T2b) located 4 m above the river is covered by a 2-m-thick fluviatile level (T2t) that overlaps and seals the MFT splay 1. An older strath terrace (T3) has a 4-m offset induced by a thrust motion along the MFT (Figure 8b). Northward, another splay is evidenced by a gouge zone in the central part of a strongly fractured zone (MFT splay 2 on Figure 8a).

Several encased strath surfaces are found at the hanging wall of splay 2 of the MFT (Figure 8a). Their slope is very similar to the river's slope (nearly 2%), and they have been incised 2 m for T1b, 6 m for T2b, 10 m for T3b, and 11.5 m for T4b by the river. An older terrace (T5) lies 27 m above the river. The abandonment of parallel strath surfaces implies incision at a rate similar

FIGURE 8. The imbricate splay of the MFT at Koilabas (location on Figure 6a). (a) The river (crosses) and terrace profiles (circles for T1, triangles for T2 and T5, diamonds for T3, and square for T4) and their spatial relationships to the splays of the MFT. T2b (empty triangle) and T2t (grey triangle) are for the bottom and the top of Terrace 2, respectively. Rectangles b and c refer to pictures 8b and 8c. (b) Picture of the T3 terrace offset by the MFT. Conversely, the T2 strath surface is not offset. (c) Picture of splay 1 of the MFT (showing a 95° trend and a 55° dip to the north) overlapped by the T2 strath terrace.

to that of uplift (Merrits et al., 1994), and the offset of terrace T3 by the MFT suggests a thrust motion of the same order of magnitude as the incision events. Therefore, the abandonment of the highest strath terraces (T3, T4, and T5) could represent sudden movement (Bull, 1990) along the fault. The T2 strath surface (Figure 8b) located 4 m above the low-flow level corresponds to the river's catastrophic high-flow level (according to inhabitants' experience). Finally, the T1 surface would coincide with the seasonal high flow during the monsoon. Assuming a rigid translation of the hanging wall of the MFT and a fault dip between 45°N (regional dip of the hanging wall beds) and 55°N (dip of the splay observed in the field), the 1.5- to 4-m uplift events would be linked to 2- to 5.5-m slip events along the MFT. The dispersion of these values could relate either to variation of the coseismic slip values or to an artefact resulting from the record of two successive seismic events by one incision event. Nonetheless, they seem to be slightly smaller than the lower bound for coseismic slip along the outer detachment during large earthquakes (Pandey et al., 1999).

Therefore, the rivers are presumably natural trenches through recent surface ruptures of large earthquakes, and as such they indicate that the surface ruptures at the front of the Himalayas either occur at the boundary between Quaternary and Siwaliks sediments (Figure 8b), or lie entirely within either the Quaternary sediments (Nakata et al., 1998) or the Siwaliks (Figure 7a).

Instantaneous Shortening Rate Deduced from GPS Measurements

A GPS network consisting of 29 sites (location on Figure 1) was measured in 1995, 1997, and 1998 (Jouanne et al., 1999), expanding Bilham et al.'s (1997) previous measurements. The 1995/1997 velocities have been obtained by the addition of the normal equations of these campaigns with limits defined to agree with the International Terrestrial Reference Frame defined in 1994 (ITRF94). Uncertainties in velocity have been estimated from the average value of the coordinate repeatabilities for 1995 and 1997 and from the noise of the time series of the Lhassa permanent GPS station, which simulates long-term repeatability. The estimated velocity for the points common with the Colorado University network (Bilham et al., 1997) has been improved using Larson et al.'s (1999) new data. The two sets of velocities (Jouanne et al., 1999 and Larson et al., 1999), both expressed in the ITRF94 reference frame, have been compiled by considering the common point SMK (in the Great Himalayas) and NPJ (in the Terai) as reference. Using these two common points, the geometrical transformation (translation and rotation) between the two velocity sets has been estimated. In a second step, the compiled network has been drawn relative to NPJ (Figure 1b). Vertical motion estimates are not used because of their poor precision, and only the horizontal component is shown. A 15 ± 2 mm/yr velocity vector trending 180° is found for the points of the Great and Lesser Himalayas of far-western Nepal. The points in the Sub-Himalayan Belt have a displacement of generally less than 3 mm/yr after 2 years (i.e., in the range of the noise), although one point suggests 6 ± 2 mm/yr.

The decrease in the horizontal motion from the Great Himalayas to the Sub-Himalayas could be interpreted as an out-of-sequence reactivation of the Lesser Himalayan thrusts, but the mean Himalayan strain between 1991 and 1997 is very low (less than 10 cm/100 km; i.e., less than 10^{-6}) and is more compatible with elastic strain accumulation. Interseismic accumulation of elastic strain has already been suggested in the Himalayas by microseismic studies (Pandey et al., 1995) and geodetic results (Jackson and Bilham, 1994; Bilham et al., 1997; Galahaut and Chander, 1997). It is assumed that the inner and deeper part of the Main Basal Detachment slips smoothly in a pressure-temperature domain where the deformation is mainly ductile, whereas the outer part of the detachment surface only slips episodically during great earthquakes.

DYNAMIC DEVELOPMENTS OF THE HIMALAYAS

Now we will review and discuss the processes that control Himalayan tectonism at different time scales, in light of the above field data.

Elastic Strain Accumulation Released during Major Earthquakes

The assumption of interseismic elastic strain accumulation has been tested in western Nepal (Larson et al., 1999; Jouanne et al., 1999) by using the classical dislocation model (Okada, 1985), whereby an elastic wedge moves above a buried fault surface that is locked at its southern tip and subject to uniform aseismic creep to the north. This dislocation approximates ductile deformation of the lower crust, and its tip is interpreted as a flat-ramp connection (Pandey et al., 1995), a ductile-brittle transition (Bilham et al., 1997), and/or as a zone of fluid pressure variations (Lemonnier et al., 1999). The strike of the dislocation is parallel to the average structural trend in western Nepal; i.e., approximately 120°. A solution that minimizes the differences between simulated and horizontal velocities estimated from the GPS survey (Jouanne et al., 1999) is obtained for a 120° planar dislocation dipping 9° northward, with a tip located at a depth of 17 km below the Lesser Himalayas, and at a distance of 60 km from the MFT (Figure 9a). It creeps by 21 mm/yr in a north-south direction corresponding to a 19 mm/yr thrust component and a 7 mm/yr dextral strike-slip component. The characteristics of this dislocation do not greatly differ from those proposed by Larson et al. (1999) and are based on 15 instead of 8 GPS sites in western Nepal (Figure 9b). Nonetheless, the planar geometry of the ductile shear zone is a basic input assumption of the model, and there is no reason to exclude more complex geometry of the basal ductile zone, such as progressive steepening (Ni and Barazangi, 1984) or ramps (De Celles et al., 1998a; Pandey et al., 1999). Calculated creep along the deep dislocation is an estimation of the shortening rate across the entire Himalayas (Bilham et al., 1997; Jouanne et al., 1999) and fits with the 21 ± 2 mm/yr Holocene displacement rate calculated by Lavé and Avouac (2000) along the MFT in central Nepal. The agreement of this Holocene frontal contraction rate with the interseismic strain accumulation rate at the base of the Great Himalayas suggests that slip on the outer part of the basal detachment (OD) accommodates most of the Himalayan convergence (Lavé and Avouac, 2000).

The coseismic rupture and postseismic motion that follow a major earthquake event would affect the 60-km locked zone and extend a few kilometers into a transition zone between creeping and locked zone (Stuart, 1988; Dragert et al., 1994). The most critical scenario for the seismic gap of western Nepal, therefore, suggests a rupture of 70 km by 500 km and a slip of 15 m, taking into account the slip deficit following the 1255 earthquake. These characteristics would lead to earthquakes as great as M_w 8.5 (Bilham et al., 1995). Nonetheless, several structural transfer zones affect the Himalayas in this area (Huyghe et al., 1998b), and could disturb the propagation of a single rupture (Molnar, 1987). If we consider a more optimistic scenario in which only a 4-m slip deficit has accumulated for two centuries (i.e., since the start of British archiving of natural disasters) and would be released by a succession of magnitude 7 events, more than 50 events would be necessary to release the slip deficit in the 30,000 km^2 seismic gap. That large a cluster of magnitude 7 events is rather improbable.

Creep during the Interseismic Period

The dislocation model is nonetheless a first-order approximation, and the assumption that motion along the outer part of the detachment is locked between

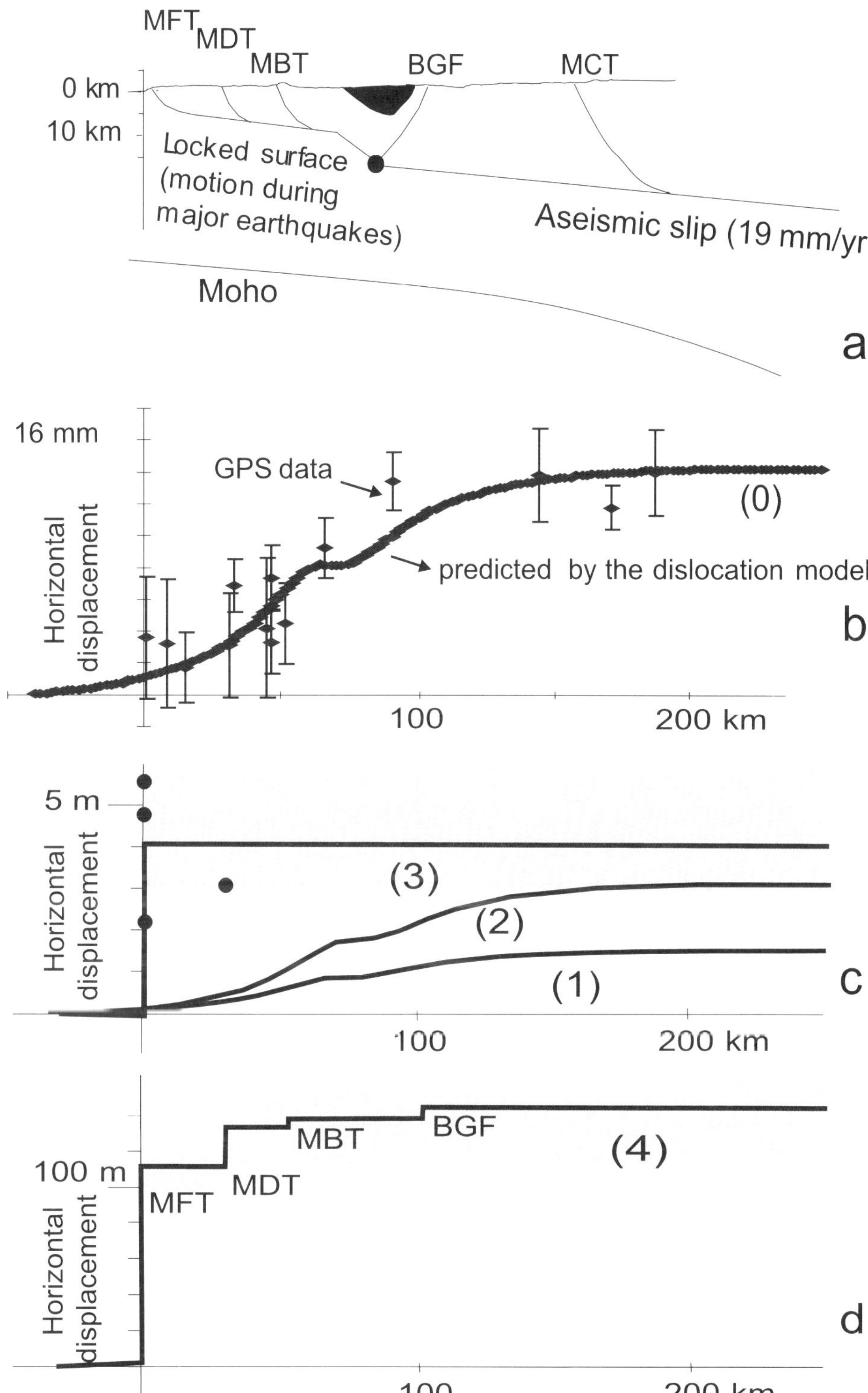

FIGURE 9. Controls on Himalayan tectonism. The curves (0) to (4) refer to the change in horizontal shortening through the Himalayas. (a) Simplified cross section through midwestern Nepal (external part is B-B′ on Figure 3, and corresponds to Figure 6a). (b) Horizontal motion perpendicular to the trend of the belt and deduced from GPS data in western Nepal are compared with those simulated (0) with a dislocation model. This model suggests a detachment that dips 9° northeastward and that creeps by a 19 mm/yr thrust component and a 7 mm/yr dextral slip component (from Jouanne et al., 1999). (c) (1) and (2): sketch of horizontal motion during elastic accumulation between major earthquakes; (3): sketch of the thrust motion along the basal detachment combining elastic strain accumulation, its release, and the slip during major earthquakes. The grey zone refers to the estimated slip of characteristic earthquakes (Pandey et al., 1999) and the black dots to the estimated surface ruptures (from Figure 7c and 8). (d) Sketch of the horizontal thrust motion along the basal detachment for the last ~7000 years, combining motions distributed on the MFT, MDT, MBT, and BGF. The grey zone refers to the maximum and minimum estimates of the motion inferred from the Holocene velocities (see text).

major earthquakes ignores the amounts of deformation accommodated by more frequent, smaller earthquakes and possible aseismic slip. Different data suggest the occurrence of aseismic creep along the detachment beneath the Outer Belt. (1) The vertical uplift detected along and ahead of the MFT use of spirit leveling is interpreted to be the result of aseimic slip along the Outer Detachment (OD) (Jackson and Bilham, 1994). (2) The pressure-solution features and mineralogical transformations observed in the thrust zones of the Sub-Himalayas (Huyghe et al., 1998a) are related to aseismic deformation phenomena. (3) Furthermore, layer-parallel shortening (Gautam and Rosler, 1999) suggests slow propagation of a flat thrust (Geiser, 1988) at the boundary between synorogenic sediments and the Indian basement, which might be related to creep beneath the Sub-Himalayas. Nonetheless, the difference between the GPS geodetic data and the dislocation model (Figure 9b) gives an upper-limit value of 2–6 mm/yr for the deformation part that is not elastically stored and released during the seismic cycle and suggests that the creep component along the Outer Detachment is small.

Thrust Sequence in a Brittle Thrust Wedge Affected by Surficial Mass Transport

Large-scale deformation of a fold-and-thrust belt has been described by the static mechanics of a wedge (Chapple, 1978), and the analytical work of Davis et al. (1983) has popularized the Coulomb-wedge theory. Recent sandbox analogs of the Coulomb wedge (Malavieille et al., 1993; Baby et al., 1995; Storti and McClay, 1995; Mugnier et al., 1997) show that thrust-system kinematics is linked to the sedimentation and erosion conditions and the vertical motions that affect the lithosphere beneath a thin-skinned thrust belt. Numerical modeling of thrust systems that incorporates syntectonic erosion and sedimentation has shown that erosion reduces gravitational and frictional work by locally reducing the overburden above a fault and allows continued activity of the frontal fault or reactivation of the inner ones (Beaumont et al., 1992; Chalaron et al., 1995; Avouac and Burov, 1995). In contrast, sedimentation acts to increase gravitational and frictional work on faults and has the potential to render them inactive. Hardy (1998) has shown that the motion along a ramp is formed by a succession of incremental displacements separated by three types of quiescent periods. The shortest quiescent period is equal to the period of quiescence between two incremental displacements along the basal detachment. The intermediate type is of the order of 10–50 times the period of incremental displacement along the basal detachment and is related to the activity of other thrusts at this time scale. At the scale of this entire period, there is an apparent simultaneity of motion along several thrusts. The third type of quiescent period is several orders of magnitude greater than the period of incremental reactivation. These predicted time-displacement patterns, which are induced by erosion, are comparable to the different thrust sequences reported in the Himalayas by Burbank and Raynolds (1988) for quiescent periods at a scale of more than a million years; by Herail and Mascle (1980) for the apparent simultaneity of motion along several thrusts related to periodic quiescent intervals along a single thrust at a scale of less 0.4 million years (Mugnier et al., 1998); and by Lavé and Avouac (2000) for incremental displacement along a single thrust during the seismic cycle.

Fluid Pressure and Fault Reactivation in a Thrust Wedge

Sets of initially low-dipping thrusts in the wedge may steepen by back-tilting during piggyback transport (Dahlstrom, 1970). This steepening tends to lock up these thrusts and implies high fluid overpressures for continued fault activity (Sibson, 1990). The Coulomb wedge theory involves high fluid pressure to maintain a wedge equilibrium profile of only a few degrees (Davis et al., 1983). Therefore, high-angle active reverse faults in a thrust belt are expected to behave as fluid-pressure-activated valves (Sibson, 1990) where the increase of fluid pressure promotes fault activity and fault activity induces a drop in fluid pressure. This fault-valve behavior may apply to the Himalayan Belt: Numerous piggyback thrusts are very steep in the field, and tension fractures display a very high angle (nearly 70°) to the thrust planes (Huyghe et al., 1998a). Furthermore, the migration of hydrocarbon fluids has been documented along fault contacts of the Sub-Himalayas (Agarwal et al., 1994), and nearly lithostatic fluid pressure has been discovered by drill holes in the Sub-Himalayas (Powers et al., 1998). Therefore, the episodic fault reactivation described in this chapter also could be controlled partly by stress-field variations related to fault-valve behavior.

Convergence between India and Tibet

We have estimated instantaneous, Holocene, Pliocene-Quaternary (0–3 Ma), and long-term (0–11 Ma) shortening rates for north-northeast–south-southwest cross sections through western Nepal (black dots on Figure 10). Their comparison shows that the shortening rate through the Himalayan Belt of western Nepal is permanently in the range of 19–20 mm/yr, with uncertainties that are respectively ± 2 mm/yr, 6 mm/yr, and 5 mm/yr for the instantaneous, Holocene, and Miocene-Pliocene estimates. The Pliocene-Quaternary shortening rate through the Outer Belt is only in the range of 11–17 mm/yr; this lower value may be linked to out-of-sequence reactivation in the Lesser and/or Great Himalayas, as noted by Harrison et al. (1997).

In contrast, the shortening that occurs along the MFT is not constant, and the proportion of the Himalayan shortening that occurs along the MFT can be estimated by the ratio of MFT shortening rate to total Himalayan shortening rate. This ratio varies from one segment of the MFT to another and, for a given segment, from one epoch to another. Only one-fifth of the motion occurred on the MFT during the Pliocene-Quaternary in western Nepal, more than one-half during the last 7000 yr in the same area, and the whole motion occurred along the MFT during the last 9000 yr in central Nepal.

Shortening rates obtained through western Nepal are consistent with those obtained along other cross sections through the Himalayas of India or Nepal (cross on Figure 10, and reference in Table 6), whereas a slightly smaller convergence rate is inferred in the far Western Himalayas of Pakistan. This consistency of results at different

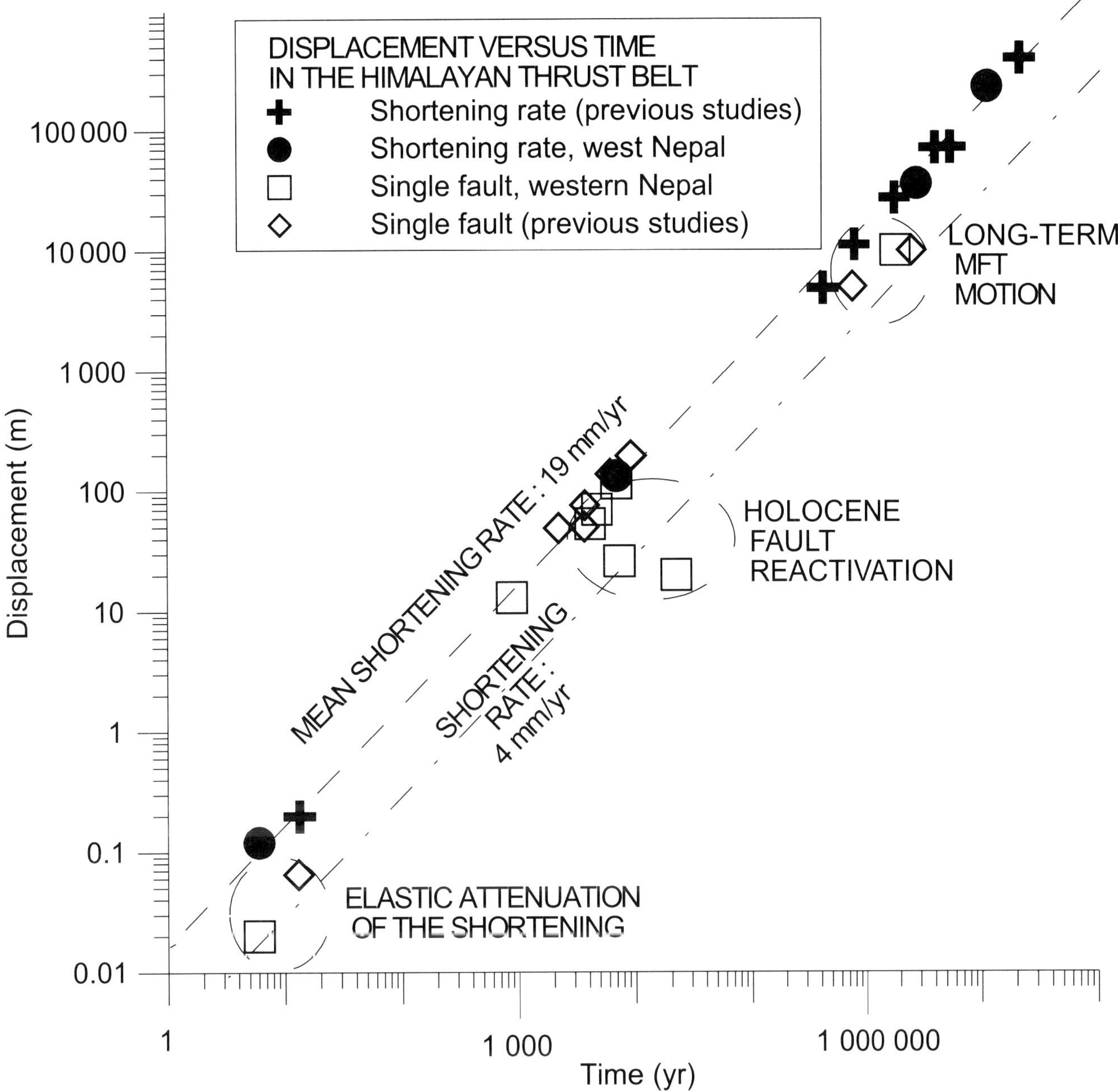

FIGURE 10. Displacement versus time through the Himalayas (log-log scale). Black dots refer to shortening across the Himalayas of western Nepal (this study); crosses refer to other sections of the Himalayas; squares and diamonds refer to motion along a single thrust (squares = this study; diamonds = other studies). Note that the sizes of the symbols are greater than the uncertainties values (see Table 6).

time scales suggests that the plate tectonic motion between India and the Tibetan plateau may occur permanently at a rate of approximately 19 mm/yr for the pure convergence.

An oblique convergence is nonetheless inferred (McCaffrey and Nabelek, 1998), and the strike-slip component seen from the GPS survey in western Nepal is of the order of 5–7 mm/yr (Larson et al., 1999; Jouanne et al., 1999). The convergence presumably is partitioned, with pure thrusting in the Sub-Himalayas and dextral strike-slip along innerward structures. Nakata (1989) described neotectonic features from the Darma to Bari Gad Faults that are in agreement with such a partitioning pattern, but geologic and instantaneous strike-slip rates along the Himalayan faults of western Nepal are unknown.

CONCLUSIONS

From a Dynamic Fault Motion to a Quasi-static Himalayan Wedge

The following scenario is proposed, which takes into account the dynamic fault motion during earthquakes

Table 6. Estimation of displacement through the Himalayas.

Age (yr)	*Total shortening (m)*	*Single fault slip (m)*	*Shortening rate (mm/yr)*	*Location*	*Reference*
~ 20,000,000	~ 390,000		20	Kumaon cross section	Srivastava and Mitra, 1994
~ 11,000,000	~ 231,000		21	Western Nepal cross section	De Celles et al., 1998a
~ 11,000,000	150,000–260,000		19	Western Nepal cross section	This study, Fig. 2
~ 5 100,000	510,00–92,000		14	Potwar cross section	Baker et al., 1988; Jaswal et al., 1997
~ 3,800,000	~ 70,000		18	Eastern Nepal cross section	Schelling and Arita, 1991
2,400,000–3,000,000	32,000–40,000		13	Siwalik western Nepal	This study, Fig. 4
~ 2, 500,000		~10,000	4	Domeli thrust	Johnson et al., 1986
1,800,000–2, 400,000		8,000–10,000	6	MFT1 Surai	Mugnier et al., 1999
1,500,000–1, 900,000	23,000–27,000		16	Siwalik Kangra cross section	Powers et al., 1998
700,000–800,000		~5,000	7	MFT Dehra Dun	Powers et al., 1998
700,000–800,000	~ 11,000		15	Siwalik Dehra Dun cross section	Powers et al., 1998
~ 400,000	~ 4 800		12	Siwalik Potwar cross section	Johnson et al., 1979
~ 22,000		~20	1	MBT western Nepal	Nakata, 1989
8,950–9,450		187–207	21	MFT central Nepal	Lavé and Avouac, 2000
7,214–7,468		15–37	4	MDT2 western Nepal	This study, Fig. 7
6,640–7,151	116–160		19	Siwalik western Nepal	This study, Fig. 6
6,640–7,151		104–126	17	MFT1 western Nepal	This study, Fig. 6
6,000–6 300		117–147	22	MFT central Nepal	Lavé and Avouac, 2000
4,531–4,850		63–79	15	MFT1 western Nepal	This study, Fig. 6
3,927–4,417		46–60	13	MFT1 western Nepal	This study, Fig. 6
3,600–3,800		71–81	21	MFT central Nepal	Lavé and Avouac, 2000
3,448–3,878		40–60	14	MFT Dehra Dun	Wesnousky et al., 1999
2,000–2 400		40–54	22	MFT central Nepal	Lavé and Avouac, 2000
730–925		11.5–14	15	MFT1 western Nepal	This study, Fig. 6
13	0.117–0.234		15	Leveling Higher Himalayas	Jackson and Bilham, 1994
13		0.026–0.091	5	Leveling Lower Himalayas	Jackson and Bilham, 1994
6	0.100–0.125		20	Basal detachment	Jouanne et al., 1999
6		0.012–0.036	3	GPS Siwalik western Nepal	Jouanne et al., 1999
4	0.72–0.88		20	Basal detachment	Bilham et al., 1997

(Figure 9c) and the quasi-static long term evolution (Figure 9d) of the Himalayan thrust wedge.

1) Incremental displacement along the Outer Detachment is induced by each major earthquake rupture that nucleates beneath the Great Himalayas and propagates toward the foreland (Lavé and Avouac, 2000).
2) The reactivation events, ranging in scale from a few meters to several tens of meters (Figure 9d),

that are apparent along the major thrusts of the Sub-Himalayas (Mugnier et al., 1998) may be related to a series of seismic events.

3) If each incremental earthquake displacement (i.e., on the order of a few meters) obeys the law of fracture propagation along the basal décollement, the overall result after a series of such displacements is the same as it would be if the body were inelastic (De Breamaecker, 1987).
4) The choice of ramp that transfers the incremental displacement toward the surface could be controlled by feedback between tectonism and surficial mass transport processes (Beaumont et al., 1992) in the brittle wedge and/or by feedback between fault rupture and fluid pressure (Sibson, 1990).
5) These feedbacks are sensitive to lateral variations in thrust geometry (Chalaron et al., 1995) and promote the diachronous reactivation observed in Nepal from east to west.
6) The convergence between the Indian and Tibetan plateaus has been occurring at a nearly permanent rate (19–20 mm/yr) for the past 10 million yr.

Applicability of the Himalayan Model to Other Mountain Belts

This model of Himalayan evolution could be applied to any mountain belt that is controlled by a dominant aseismic/seismic cycle and by high rates of erosion. The Taiwan thrust belt, characterized by a seismic/aseismic cycle and by high erosion (Suppe, 1981), is probably very similar to the Himalayan Belt.

Subduction thrusts, like the Cascadia Belt (Dragert et al., 1994), are characterized by huge earthquakes but are less affected by erosion, and the frontal accretionary prism is generally dominated by sedimentation. Therefore it is expected that fault-valve behavior could play a more important role than erosion in triggering out-of-sequence reactivation (Sibson, 1990).

For a collision belt like the Alps, which is affected by strong erosion (more than 1 mm/yr) and a weak shortening rate (presently less than 2 mm/yr, according to Jouanne et al., 1994), the crucial question is to assess correctly the seismic hazard by deciphering the creep component versus the seismic component. The creep component could be promoted by the weak rheology of the evaporitic décollement levels, which are more than 100 km wide (Mugnier and Vialon, 1986). In this sense, the Himalayas of Pakistan, which are characterized by a décollement in the Cambrian evaporitic levels (Baker et al., 1988), probably do not follow the above model proposed for the Himalayas of Nepal.

For a collision belt like the Spanish Pyrenees, the shortening rate is small and never exceeds a few millimeters per year (Burbank et al., 1992). Therefore, we might expect that the numerous out-of-sequence events documented by Verges and Muñoz (1990) are linked to erosional controls rather than to earthquake-linked fault-valve behavior.

Nonetheless, a quantitative comparison of these different thrust belts would be necessary to make a clear distinction between the role of erosion and the role of fluids in the out-of-sequence reactivation. Because out-of-sequence reactivation in the frontal thrust belt would significantly increase the thermal gradient (Mancktelow and Grasemann, 1997), and because the seismic activity of the out-of-sequence faults would assist the migration of hydrocarbon fluids (Sibson, 1990; Agarwal et al., 1994), such a comparison would be useful for estimating the petroleum potential of fold-and-thrust belts.

ACKNOWLEDGMENTS

R. Butler and R. Sibson provided insightful reviews. We thank M. R. Pandey (Department of Mines and Geology, Nepal) and P. N. Upreti (Tribhuvan University, Katmandu) for sharing their ideas and experiences in geology of Nepal and for their helps in field work organization. The Institut National des Sciences de l'Univers, CNRS, France is acknowledged for the financial support.

APPENDIX A

The fault-bend-fold model predicts that the slip value is constant for the hanging wall of a thrust sheet parallel to the bedding (Suppe, 1983). Therefore, during the time interval *i*, the slip along a footwall ramp ($s_{i,ramp}$) is equal to the slip along the décollement ($s_{i,dec}$) and does not depend on the dip of the ramp:

$$s_i = s_{i,\,ramp} = s_{i,\,dec} \tag{1}$$

For a thrust sheet that moves above a ramp (with a dip of ϕ_j), the following relation is found between uplift ($u_{i,j}$) and slip (s_i) along the ramp (Mugnier et al., 1992):

$$u_{i,j} = s_j \times \sin(\phi_j) \tag{2}$$

This equation also applies to the uplift of a rigid block translated in a direction parallel to the dip of a ramp.

Assuming a layer-cake stratigraphy for the Siwalik Group, the dip (β_j) of the Siwalik beds in the field gives the estimated dip of the ramp ϕ_j:

$$\phi_j = \beta_j \tag{3}$$

The local bend of a terrace level in a fault-bend-fold model (Zoetemeijer and Sassi, 1992) is related to a pair of hinge lines and is linked to the change in the ramp dip from ϕ_j to ϕ_{j+1}. The vertical offset ($vo_{i,j}$) is the difference in uplift on both sides of the bend that is added to the terrace gradient related to its initial slope (α_ι):

$$vo_{i,j} = s_i \times (\sin\beta_{j+1} - \sin\beta_j) + s_i \times \sin\alpha_\iota \qquad (4)$$

A simple reorganization gives:

$$s_i = vo_{i,j}/(\sin\beta_{j+1} - \sin\beta_j + \sin\alpha_\iota) \qquad (5)$$

Equation (5) is applied to outcrops A to D (Figure 6c), and the slip rate (sr_i) is deduced by dividing the slip value by the age (t_i) of the terrace (Table 4):

$$sr_i = vo_{i,j}/(\sin\beta_{j+1} - \sin\beta_j + \sin\alpha_\iota)/t_i \qquad (6)$$

The tectonic uplift ($u_{i,j}$) is deduced from Equation (2). The variation in the river level (rl_i) (with respect to a frame of reference located at the footwall of the thrust, i.e., the Indian plate) is the difference between the measured incision ($in_{i,j}$) and the calculated uplift (Figure 6c, d) (Lavé and Avouac, 2000). Furthermore, the seasonal variations of the river level are removed, and it is assumed that the incision of the present-day terrace (in_{ac}) represents these variations:

$$rl_i = in_{i,j} - s_i \times \sin\beta_j - in_{ac} \qquad (7)$$

Equation (7) is applied to outcrops 1 to 5. (Figure 6d and Table 5).

REFERENCES CITED

Agarwal, R., D. Prasad, U. Samanta, C. Berry, and J. Sharma, 1994, Hydrocarbon potential of Siwalik basin: Himalayan Geology, v. 15, p. 301–320.

Appel, E., W. Rosler, and G. Corvinus, 1991, Magnetostratigraphy of the Miocene-Pleistocene Surai Khola Siwaliks in West Nepal: Geophysical Journal International, v. 105, p. 191–198.

Armijo, R., P. Tapponnier, and H. Tonglin, 1989, Late Cenozoic right-lateral strike-slip faulting in Southern Tibet: Journal of Geophysical Research, v. 94, p. 2787–2838.

Auden, J. B., 1935, Transverse in the Himalayan: Geological Survey of Indian Record, v. 69, p. 123–167.

Avouac, J. C., and E. Burov, 1995, Erosion as a driving mechanism of intracontinental mountain growth: Journal of Geophysical Research, v. 101, p. 17747–17769.

Baby, P., B. Coletta, and D. Zubieta, 1995, Etude géométrique et expérimentale d'un bassin transporté: exemple du synclinarium de l'Alto Beni (Andes Centrales): Bulletin de la Société Géologique de France, v. 166, p. 797–813.

Baker, D. M., R. J. Lillie, R. S. Yeats, G. D. Johnson, M. Yousuf, and A. Zamin, 1988, Development of the Himalayan frontal thrust zone, Salt Range, Pakistan: Geology, v. 16, p. 3–7.

Baranowski, J., J. Armbruster, L. Seeber, and P. Molnar, 1984, Focal depths and fault plane solutions of earthquakes and active tectonics of the Himalayas: Journal of Geophysical Research, v. 89, p. 6918–6928.

Bashial, R. P., 1981, Geology of Dhangrhi-Dandeldhura road section and its regional significance: Journal of the Nepal Geological Society, v. 1, p. 15–28.

Beaumont, C., P. Fullsack, and J. Hamilton, 1992, Erosional control on active orogens, *in* K. R. McClay, ed., Thrust tectonics: London, Chapman & Hall, p. 419–433.

Besse, J., V. Courtillot, J. P. Pozzi, and Y. Zhou, 1984, Paleomagnetic estimates of Cenozoic convergence in the Himalayan thrusts and Zangbo suture: Nature, v. 311, p. 621–626.

Bilham, R., P. Bodin, and M. Jackson, 1995, Entertaining a great earthquake in western Nepal: Historic inactivity and development of strain: Journal of the Nepal Geological Society, v. 11, p. 73–88.

Bilham, R., K. Larson, J. Freymuller, and Project IDYLIM members, 1997, Indo-Asian collision rates in the Nepal: Nature, v. 386, p. 61–64.

Bouvier, R., 1997, Coupes équilibrées dans le système chevauchant frontal himalayen : Grenoble University, France, Master Memory of Geology, 26 p.

Brune, J. N., 1968, Seismic moment, seismicity and rate of slip along major fault zones: Journal of Geophysical Research, v. 73, p. 777–784.

Brunel, M., 1986, Ductile thrusting in the Himalayas: shear sense criteria and stretching lineations: Tectonics, v. 5, p. 247–265.

Bull, W., 1990, Stream–terrace genesis: implications for soil development: Geomorphology, v. 3, p. 351–367.

Burbank, D., and R. Raynolds, 1988, Stratigraphic keys to the timing of deformation: an example from the northwestern Himalayan foredeep, *in* C. Paola and K. Kleinspehn, eds., New perspective in basin analysis: New York, Springler-Verlag, p. 331–351.

Burbank, D., and R. Beck, 1989, Early Pliocene uplift of the Salt Range, *in* Lawrence, L., J. R. Malinconico, and R. Lillie, eds., Tectonics of the western Himalayas: Special Paper Geological Society of America, v. 232, p. 113–128.

Burbank, D., and J. Verges, 1994, Reconstruction of topography and related depositional systems during active thrusting: Journal of Geophysical Research, v. 99, p. 20281–20297.

Burbank, D., J. Verges, J. Munoz, and P. Bentham, 1992, Coeval hindward- and forward-imbricating thrusting in the central southern Pyrenees, Spain: timing and rates of shortening: Geological Society of America Bulletin, v. 104, p. 3–17.

Burbank, D., R. Beck, and T. Mudler, 1996, The Himalayan foreland basin, *in* Yin, A. and M. Harrison, eds., The tectonic evolution of Asia: Cambridge, U.K., Cambridge University Press, p. 149–188.

Butler, R., 1982, The terminology of structures in thrust belts: Journal of Structural Geology, v. 4, p. 239–245.

Cande, S. C., and D. V. Kent, 1995, Revised calibration of the magnetic polarity time scale for the last Cretaceous and Cenozoic: Journal of Geophysical Research, v. 97, p. 13917–13951.

Chalaron, E., J. L. Mugnier, and G. Mascle, 1995, Control of thrust tectonics in the Himalayan foothills: a view from a numerical model: Tectonophysics, v. 248, p. 139–163.

Chapple, W., 1978, Mechanics of thin-skinned fold and thrust belt: Geological Society of America Bulletin, v. 89, p. 1189–1198.

Chitrakar, G. R., and M. R. Pandey, 1986, Historical earthquakes of Nepal: Bulletin of the Geological Society of Nepal, v. 4, p. 7–8.

Corvinus, G., 1993, The Siwalik Group of sediments at Surai Khola in western Nepal and its palaeontological record: Journal of the Nepal Geological Society, v. 9, p. 21–35.

Cotton, F., M. Campillo, A. Deschamps, and B. Rastogo, 1996, Rupture history and seismotectonics of the 1991 Uttarkashi, Himalayas earthquake: Tectonophysics, v. 258, p. 35–51.

Dahlstrom, C. D. A., 1970, Structural geology in the eastern margin of Canadian Rocky Mountains: Canadian Petroleum Geology Bulletin, v. 18, p. 332–406.

Davis, D., J. Suppe, and F. Dahlen, 1983, Mechanics of fold-and-thrust belts and accretionary wedges: Journal of Geophysical Research, v. 88, p. 1153–1172.

De Breamaecker, J. Cl.,1987, Thrust sheet motion and earthquake mechanisms: Earth and Planetary Science Letters, v. 83, p. 159–166.

De Celles, P. G., G. E. Gehrels, J. Quade, T. P. Ojha, and P. A. Kapp, 1998a, Neogene foreland basin deposits, erosional unroofing and the kinematic history of the Himalayan fold-thrust belt, western Nepal: Geological Society of America Bulletin, v. 110, p. 2–21.

De Celles, P. G., G. E. Gehrels, J. Quade, and T. P. Ojha, 1998b, Eocene-early Miocene foreland basin development and the history of the Himalayan thrusting, Western and Central Nepal: Tectonics, v. 17, p. 741–765.

Delcaillau, B., 1992, Les Siwalik de l'Himalaya du Népal oriental: Paris, Mémoire et Documents de Géographie, éditions du CNRS, 205 p.

Dhital, M. R., A. P. Gajurel, D. Pathak, L. P. Paudel, and K. Kizaki, 1995, Geology and structure of the Siwaliks and Lesser Himalayas in the Surai Khola-Barbanda area, mid-western Nepal: Kathmandu, Nepal, Tribhuvan University, Bulletin of the Department of Geology, v. 4, p. 1–70.

DMG, 1985, Nepal exploration opportunities: Kathmandu, Department of Mines and Geology – Ministry of Industry, His Majesty's Government of Nepal, 52 p.

Dragert, H., R. D. Hyndam, G. C. Rogers, and K. Wang, 1994, Current deformation and the width of the seismogenic zone of the northern Cascadia subduction thrust: Journal of Geophysical Research, v. 99, p. 653–668.

Dziewonski, A. M., G. Ekström, and N. M. Maternovskaya, 2000, Centroid-moment tensor solutions for April-June, 1999: Physics of the Earth and Planetary Interiors, v. 119, p. 161–171.

Elliott, D., 1976, The energy balance and deformation mechanisms of thrust sheets: London, Philosophical Transactions of the Royal Society, v. A283, p. 289–312.

Ford, M., E. Williams, A. Artoni, J. Verges, and S. Hardy, 1997, Progressive evolution of a fault-related fold pair from growth strata geometries: Journal of Structural Geology, v. 19, p. 413–442.

Galahaut, V. K., and R. Chander, 1997, Evidence for an earthquake cycle in the NW outer Himalayas near 78°E longitude, from precision levelling observation: Geophysical Research Letters, v. 24, p. 225–228.

Ganser, A., 1964, Geology of the Himalayas: London, Wiley Interscience Pubs, 289 p.

Gautam, P., and S. R. Pant, 1996, Magnetic fabric of Siwalik Group sediments of Tinau Khola section, west central Nepal: Bulletin of the Department of Geology, Tribhuvan University, Kathmandu, v. 5, p. 21–36.

Gautam, P., and W. Rosler, 1999, Depositional Chronology and fabric of Siwalik group sediments in central Nepal from magnetostratigraphy and magnetic anisotropy: Journal of Asian Earth Sciences, v. 17, p. 659–682.

Geiser, P., 1988, Mechanisms of thrust propagation: some examples and implications for the analysis of overthrust terranes: Journal of Structural Geology, v. 10, p. 829–845.

Hagen, T., 1956, Uber eine uberschiebung der Tertiare Siwaliks über das regente Ganges Alluvium in ost Nepal: Geographica Helvetica, v. 5, p. 217–219.

Hardy, S., 1998, Minimum work, fault activity and the growth of critical wedges in fold and thrust belts: Basin Research, v. 10, p. 365–373.

Harland, W. B., R. L. Armstrong, A. V. Cox, L. E. Craig, A. G. Smith, and D. G. Smith, 1989, the magnetostratigraphic time scale, *in* The Geologic Time Scale: Cambridge, U.K., Cambridge University Press, p. 140–167.

Harrison, T. M., F. J. Ryerson, P. Le Fort, A. Yin, O. M. Lovera, and E. J. Catlos, 1997, a late Miocene-Pliocene origin for the Central Himalayas inverted metamorphism: Earth and Planetary Science Letters, v. 146, E1–E7.

Herail, G., and G. Mascle, 1980, Les Siwalik du Népal central: structures et géomorphologie d'un piémont en cours de déformation: Bulletin Association Géographique Française, v. 431, p. 259–267.

Hodges, K. W., R. Parrish, and M. P. Searle, 1996, Tectonic evolution of the Central Annapurna range, Nepal Himalayas: Tectonics, v. 15, p. 1264–1291.

Hossack, J. R., 1979, The use of balanced cross-sections in the calculation of orogenic contraction: a review: London, Journal of the Geological Society, v. 136, p. 705–711.

Husson, L., J.-L. Mugnier, P. Leturmy, and G. Vidal, 2004, Kinematics and sedimentary balance of the subhimalayan zone, western Nepal, *in* K. R. McClay, ed., Thrust tectonics and hydrocarbon systems: AAPG Memoir 82, p. 115–130.

Huyghe, P., A. Galy, and J. L. Mugnier, 1998a, Microstructures, clay mineralogy and geochemistry of the clay size fraction (<2 μm) of thrusted zones (Karnali area, Siwalik of Western Nepal): Journal of the Nepal Geological Society, v. 18, p. 239–249.

Huyghe, P., J. L. Mugnier, and P. Leturmy, 1998b, Segmentation of the Siwalik thrust belt of Western Nepal and location of piggy-back basins, *in* Peshawar, Proceedings of the 13th Himalayan Karakorum Tibet International Workshop, p. 82–83.

Huyghe, P., A. Galy, C. France-Lanord, and J. L. Mugnier, 2001, Propagation of the thrust system and erosion in the Lesser Himalaya: Geochemical and sedimentological evidence: Geology, v. 29, p. 1007–1010.

Jackson, M. J., and R. Bilham, 1994, Constraints on the Himalayan deformation inferred from vertical velocity fields in Nepal and Tibet: Journal of Geophysical Research, v. 99, p. 13897–13912.

Jaswal, T., R. Lillie, and R. Lawrence, 1997, Structure of the Northern Potwar deformed zone, Pakistan: AAPG Bulletin, v. 81, p. 308–328.

Johnson, G. D., N. M. Johnson, N. D. Opdyke, and R. A. Tahirkheli, 1979, Magnetic reversal stratigraphy and sedimentary tectonic history of the upper Siwalik group, eastern Salt Range and southwestern Kashmir, *in* A. Farah and K. A. DeJong, eds., Geodynamics of Pakistan: Quetta, Geological Survey of Pakistan, p. 149–165.

Johnson, G. D., R. G. Raynolds, and D. W. Burbank, 1986, Late Cenozoic sedimentation in the northwestern Himalayan foredeep: I. Thrust ramping and associated deformation in the Potwar region, *in* P. Allen and P. Homewood, eds., Foreland basins: International Association of Sedimentologists Special Publication, 8, p. 273–291.

Jouanne, F., G. Ménard, and D. Jault, 1994, Present-day deformation of the French northwestern Alps/southern Jura mountains: comparison between historical triangulations: Geophysical Journal International, v. 119, p. 151–165.

Jouanne, F., J. L. Mugnier, M. R. Pandey, J. F. Gamond, P. Le Fort, L. Serrurier, C. Vigny, J. P. Avouac, and Idylhim members, 1999, Oblique convergence in Himalayas of western Nepal deduced from GPS measurements: Geophysical Research Letters, v. 26, p. 1933–1936.

Kimura, K., 1998, Geomorphic development of the Deukhuri Dun, Nepal Sub-Himalayas: The Science Reports of the Toholu University, 7th series (Geography), v. 48, p. 65–83.

Kuo-Fong, M., L. Chyi-Tyi, T. Yi-Ben, S. Tzay-Chyn, and M. Jim, 1999, the Chi-Chi, Taiwan Earthquake: Large surface displacements on an inland thrust fault: Eos, AGU Transactions, v. 80, p. 605–611.

Larson, K., R. Burgmann, R. Bilham, and J. Freymuller, 1999, Kinematics of the India-Eurasia collision zone from GPS measurements: Journal of Geophysical Research, v. 104, p. 1077–1093.

Lavé, J., and J. P. Avouac, 2000, Active folding of fluviatile terraces across the Siwalik hills, Himalayas of central Nepal, implications for Himalayan seismotectonics: Journal of Geophysical Research, v. 105, p. 5735–5770.

Le Fort, P., 1975, Himalayas: the collided range: Present knowledge of the continental arc: American Journal of Science, v. 275, p. 1–44.

Lemonnier, C., G. Marquis, F. Perrier, J. P. Avouac, G. Chitrakar, B. Kafle, S. Sapkota, U. Gautam, D. Tiwari, and M. Bano, 1999, Electrical conductivity around the mid-crustal ramp along the MHT: Geophysical Research Letters, v. 26, p. 3261–3264.

Leturmy, P., 1997, Sédiments et reliefs du front des systèmes chevauchants: modèlisations et exemples du front andin et des Siwalik (Himalayas) à l'Holocène: PhD thesis, Grenoble University, France, 229 p.

Leturmy, P., J. L. Mugnier, and B. Delcaillau, 1997, Development and morphology of fault propagation folds, *in* Proceedings of E.U.G. IX, Strasbourg, 9, 242 p.

Lickorish, W. H., and R. Butler, 1996, Fold amplification and parasequence stacking patterns in syntectonic shoreface carbonates: Geological Society of America Bulletin, v. 108, p. 966–977.

McCaffrey, R., and J. Nabelek, 1998, Role of oblique convergence in the active deformation of the Himalayas and southern Tibet plateau: Geology, v. 26, p. 691–694.

McClay, K. R., 1992a, Thrust tectonics: London, Chapman and Hall, 433 p.

McClay, K. R., 1992b, Glossary of thrust tectonics terms, *in* K. R. McClay, ed., Thrust tectonics: London, Chapman & Hall, p. 419–433.

Malavieille, J., C. Laroque, and S. Calassou, 1993, Modélisation expérimentale des relations tectonique/sédimentation entre bassin avant-arc et prisme d'accrétion: Paris, Comptes Rendus de l'Académie des Sciences, v. 316, p. 1131–1137.

Mancktelow, N., and B. Grasemann, 1997, Time-dependent effects of heat advection and topography on cooling histories during erosion: Tectonophysics, v. 270, p. 167–195.

Mascle, G., E. Chalaron, A. P. Gajurel, P. Leturmy, and J. L. Mugnier, 1998, Paleoseismicity in the Siwaliks: occurrence of major seismic events in the Himalayas of Western Nepal: Journal of the Nepal Geological Society, v. 18, p. 417–430.

Meissner, R., and J. Strelau, 1982, Limits of stress in continental crust and their relation to the depth-frequency relation of shallow earthquakes: Tectonics, v. 1, p. 73–89.

Merrits, D., K. Vincent, and E. Wohl, 1994, Long river profiles, tectonism, and eustasy: A guide to interpret fluviatile terraces: Journal of Geophysical Research, v. 99, p. 14031–14050.

Molnar, P., 1987, The distribution of Intensity associated with the 1905 Kangra earthquake and bounds on the extent of the rupture zone: Journal of the Geological Society of India, v. 29, p. 211–229.

Molnar, P., 1990, A review of the seismicity and the rate of active underthrusting and the deformation of the Himalayas: Journal of the Himalayan Geology, v. 1, p. 131–154.

Molnar, P., and M. R. Pandey, 1989, Rupture zones of Great Earthquakes in the Himalayan region, *in* J. N. Brune, ed., Frontiers of seismology in India: Dehli, Proceedings of Indian Academy of sciences, v. 98, p. 61–70.

Mugnier, J. L., and P. Vialon, 1986, Deformation and displacement of the Jura cover on its basement: Journal of Structural Geology, v. 8, p. 373–387.

Mugnier, J. L., G. Mascle, and T. Faucher, 1992, La structure des Siwalik de l'Ouest Népal: un prisme d'accrétion intracontinental: Bulletin de la Société Géologique de France, v. 163, p. 585–595.

Mugnier, J. L., P. Huyghe, E. Chalaron, and G. Mascle, 1994, Recent movements along the Main Boundary Thrust of the Himalayas: normal faulting in an overcritical thrust wedge?: Tectonophysics, v. 238, p. 199–215.

Mugnier, J. L., P. Baby, B. Coletta, P. Vinour, P. Bale, and P. Leturmy, 1997, Thrust geometry controlled by erosion and sedimentation: a view from analogue models: Geology, v. 25, p. 427–430.

Mugnier, J. L., B. Delcaillau, P. Huyghe, and P. Leturmy, 1998, The break-back thrust splay of the Main Dun Thrust: evidence of an intermediate displacement scale between earthquake slip and finite geometry of thrust systems: Journal of Structural Geology, v. 20, p. 857–864.

Mugnier, J. L., P. Leturmy, G. Mascle, P. Huyghe, E. Chalaron, G. Vidal, L. Husson, and B. Delcaillau, 1999, The Siwaliks of western Nepal: I–Geometry and kinematics: Journal of Asian Earth Sciences, v. 17, p. 629–642.

Muñoz, J. A., J. M. Casas, J. M. Pares, and D. W. Durney, 1999, Internal deformation related to thrust evolution in the Eastern Pyrenees, *in* Proceedings of Thrust Tectonics Conference: Royal Holloway, University of London, London, p. 275.

Nakata, T., 1989, Active faults of the Himalayas of India and Nepal: Geological Society of America Special Paper, v. 232, p. 243–264.

Nakata, T., K. Kumura, and T. Rockwell, 1998, First successful paleoseismic trench study on active faults in the Himalayas, *in* Proceedings of AGU 1998 fall meeting: Eos, AGU Transactions, v. 79, 45 p.

Ni, J., and K. Barazangi, 1984, Seismotectonics of the Himalayan collision zone: Geometry of the underthrusting Indian plate beneath the Himalayas: Journal of Geophysical Research, v. 89, p. 1147–1163.

Okada, Y., 1985, Surface deformation to shear and tensile faults in a half space: Bulletin of the Seismological Society of America, v. 75, p. 1135–1154.

Pandey, M. R., R. P. Tandukar, J. P. Avouac, J. Lavé, and J. P. Massot, 1995, Evidence for recent interseismic strain accumulation on a mid-crustal ramp in the central Himalayas of Nepal: Geophysical Research Letters, v. 22, p. 751–758.

Pandey, M. R., R. P. Tandukar, J. P. Avouac, J. Vergne, and Th. Héritier, 1999, Characteristics of seismicity of Nepal and their seismotectonic implications: Journal of Asian Earth Sciences, v. 17, p. 703–712.

Pecher, A., 1991, The contact between the higher Himalayan crystallines and the Tibetan sedimentary series: Miocene large-scale dextral shearing: Tectonics, v. 10, p. 587–598.

Powers, P. M., R. J. Lillie, and R. S. Yeats, 1998, Structure and shortening of the Kangra and Dehra Dun reentrants, Sub-Himalayas, India: Geological Society of America Bulletin, v. 110, p. 1010–1027.

Rosler, W., W. Metzler, and E. Appel, 1997, Neogene magnetic polarity stratigraphy of some fluviatile Siwalik sections, Nepal: Geophysical Journal International, v. 130, p. 89–111.

Schelling, D., and K. Arita, 1991, Thrust tectonics, crustal shortening, and the structure of the far-eastern Nepal Himalayas: Tectonics, v. 105, p. 851–862.

Shresta, S. B., 1987, Geological map of far Western Nepal, scale 1/250 000: Kathmandu, Department of Mines and Geology – Ministry of Industry, His Majesty's Government of Nepal.

Sibson, R. H., 1990, Conditions for fault-valve behaviour, *in* R. J. Knipe and E. H. Rutter, eds., Deformation Mechanisms, Rheology and Tectonics: Geological Society of London Special Publication 54, p. 15–28.

Srivastava, P., and G. Mitra, 1994, Thrust geometries and deep structure of the outer and lesser Himalayas, Kumaon of the Himalayan fold-and-thrust belt: Tectonics, v. 13, p. 89–109.

Stuiver, M., and P. J. Reimer, 1993, Extended C14 data base and revised CALIB 3.0 C14 age calibration program: Radiocarbon, 35, p. 215–230.

Storti, F., and K. McClay, 1995, Influence of syntectonic sedimentation on thrust wedges in analogue models: Geology, v. 23, p. 999–1002.

Stuart, W. D., 1988, Forecast model for great earthquakes at the Nankai Trough subduction zone: Pure and Applied Geophysics, v. 126, p. 619–641.

Suppe, J., 1981, Mechanics of mountain building and metamorphism in Taiwan: Geological Society China Memory, v. 4, p. 67–89.

Suppe, J., 1983, Geometry and kinematics of fault bend folding: American Journal of Science, v. 283, p. 684–721.

Upreti, B. N., and P. Le Fort, 1999, Lesser Himalayan crystalline nappes of Nepal: problems of their origin, *in* A. Macfarlane, J. Quade, and R. Sorkhabi, eds.: Geological Society of America Special Paper, v. 328, p. 225–238.

Verges, J., and J. Muñoz, 1990, Thrust sequence in the southern central Pyrenees: Bulletin de la Société Geologique de France, v. 8, p. 265–271.

Verges, J., D. W. Burbank, and A. Meigs, 1996, Unfolding: An inverse approach to fold kinematics: Geology, v. 24, p. 175–178.

Wesnousky, S., S. Kumar, R. Mohindra, and V. Thakur, 1999, Uplift and convergence along the Himalayan Frontal Thrust of India: Tectonics, v. 18, p. 967–976.

Yamanaka, H., and H. Yagi, 1984, Geomorphological development of the Dang Dun, Sub-Himalayan zone, southern Nepal: Journal Nepal Geological Society, v. 4, p. 151–159.

Yeats, R. S., and R. Lillie, 1991, Contemporary tectonics of the Himalayan frontal fault system: folds, blind thrusts and the 1905 Kangra earthquake: Journal of Structural Geology, v. 13, p. 215–225.

Yeats, R. S., and V. C. Thakur, 1998, Reassessment of earthquake hazard based on a fault-bend fold model of the Himalayan plate boundary fault: Current Sciences, v. 74, p. 230–233.

Zhao, W., K. D. Nelson, and project INDEPTH Team, 1993, Deep seismic reflection evidence for continental underthrusting beneath southern Tibet: Nature, v. 366, p. 555–559.

Zoetemeijer, R., and W. Sassi, 1992, 2-D Reconstruction of thrust belt evolution using the fault-bend-fold method, *in* K. R. McClay, ed., Thrust tectonics: Chapman & Hall, London, p. 133–140.

Husson, L., J.-L. Mugnier, P. Leturmy, and G. Vidal, 2004, Kinematics and sedimentary balance of the Sub-Himalayan Zone, western Nepal, *in* K. R. McClay, ed., Thrust tectonics and hydrocarbon systems: AAPG Memoir 82, p. 115–130.

Kinematics and Sedimentary Balance of the Sub-Himalayan Zone, Western Nepal

Laurent Husson[1]

Laboratoire de Géodynamique des Chaînes Alpines et Université Joseph Fourier, Maison des Geosciences, Grenoble Cedex, France

Jean-Louis Mugnier

Laboratoire de Géodynamique des Chaînes Alpines et Université Joseph Fourier, Maison des Geosciences, Grenoble Cedex, France

Pascale Leturmy[2]

Laboratoire de Géodynamique des Chaînes Alpines et Université Joseph Fourier, Maison des Geosciences, Grenoble Cedex, France

Gérard Vidal

Ecole Normale Supérieure de Lyon, Laboratoire des Sciences de la Terre, Lyon, France

ABSTRACT

The Sub-Himalayan Zone constitutes a tectonic wedge of synorogenic sediments along the southern edge of the Himalayan Belt. Sediments are incorporated into the prism from the foreland Indo-Gangetic plain, undergo a tectonic cycle within it, and eventually are eroded. The structural sketch map exhibits westward-plunging arcuate structures on the foremost location of the Outer Belt. Investigations from spatial imagery and digital elevation modeling (DEM), together with kinematic data, allow us to calculate velocities for the geomorphologic development. Four velocities rule the general evolution of the wedge. The foremost geomorphic structure (ridge) is the assemblage of elementary structures. The lateral ridge propagation velocity is estimated to be 40 cm/yr, which supports a general cylindrical development of the Outer Belt, in spite of the asymmetrical development of each independent elbow-shaped structure. The sediment's burial history can be quantified from geometric and kinematic data. We emphasize that because of the cylindrical behavior of the prism, extrapolation of the sediment transfer to the entire western Nepal Siwalik is valid. Burial in the foreland basin takes two times longer than the entire tectonic cycle, which only lasts for about 6.5 m.y. Sediments reaching 6×10^{-5} km^3 per year and per linear kilometer accrete along the Siwalik range.

[1]*Present address:* Collège de France - Chaire de géodynamique, Europôle de l'Arbois - Bat Laennec BP 80, 13545 Aix-en-Provence, France.
[2]*Present address:* Université de Cergy-Pontoise, Avenue du Parc, 8, le Campus 95031 Cergy-Pontoise, France.

About 21% of the flowing material within the wedge is captured and withdrawn from it as subducted duplexes. Assuming a steady-state development of the wedge, and according to the Coulomb-wedge theory for the Siwalik, mean erosion rates are estimated to be about 1.9 mm/yr, which is in accord with previous estimates. We emphasize that this consistency supports the overall estimates for the general development of the wedge.

INTRODUCTION

The critical taper formed by orogenic belts can be divided into various parts in their front zone. As is the case for the Sub-Himalayan Zone of western Nepal, the most frontal deformed part often corresponds to an accretionary wedge of synorogenic sediments incorporated from the foreland basin into the mountain belt. This unit constitutes an open system of material, because sediments are added to the wedge from the foreland and removed from it by erosion. According to Dahlen and Barr (1989), mature continental synorogenic accretionary wedges display critical-taper angles and can be in a volumetric steady state. As a consequence, to keep a balanced volume of sediments throughout time within the unit, the processes that withdraw material from the wedge (erosion or duplex subduction) must be strong enough to remove sufficient material from the wedge. This statement is implicitly constrained by the global energy budget for mature prisms, because increasing the volume of the prism implies an increase of its activation energy. Material within the wedge undergoes a burial cycle, from the fresh sediment in the foreland being incorporated into the prism, to the free, eroded sediment at the surface. This pattern is generally associated with thin-skinned deformation styles.

The main morphotectonic units of the Himalayas south of Tibet include the High Himalayas, which are made of crystalline rocks and are bounded southward by the Main Central Thrust (MCT); and the Lesser Himalayas, which are made of metasedimentary rocks and are separated to the south by the Main Boundary Thrust (MBT). The last unit is the Sub-Himalayan Zone Fold-and-thrust Belt, which forms a Neogene accretionary wedge on the southern foothills of the Himalayas that is deformed in a thin-skinned style (Hérail and Mascle, 1980; Hérail et al., 1986; Schelling, 1992; Mugnier et al., 1992, 1999a; Powers et al., 1998).

Characterization of hydrocarbon maturation, migration, and entrapment in fold-and-thrust belts is often limited by poor knowledge of the kinematics of these areas. The burial history and residential time at depth are the key parameters for such problems, which can only be assessed if the global kinematics of the wedge is well understood (Baby et al., 1994; Moretti et al., 1996). Maturation is closely linked to the time span during which kerogen was maintained at depth; migration depends on the fluid paths, which are controlled by the tectonic regime and associated structure; and eventually, entrapment in foreland fold-and-thrust belts is linked to structural buildup. Although the petroleum potential is rather poor in this area (Bashyal, 1998), the Sub-Himalayan Zone is relevant to our general understanding of the deformation schemes of frontal parts of orogenic belts. Most active mountain belts show similar structural sketches, in which tectonic velocities, sedimentation, and erosion rates are in the same range. These parameters are fundamental for the evolution of petroleum systems in fold-and-thrust belts.

The aim of this chapter is to quantify the geometric and kinematic development of the Sub-Himalayan prism in western Nepal. First, we analyze the structural and morphological evolution of the wedge. Second, we estimate the volume of material incorporated from the foreland, the time of residence within the wedge, and the volume of material removed from the prism undergoing erosion, using across-strike geometric and tectonic velocities together with a structural sketch map. The kinematics cannot be restricted to a 2-D forward cycle, because the lateral growth of the structure also controls the sediment-incorporation mechanisms within the wedge. The morphological evolution helps in documenting the global dynamics of the wedge. The overall development of the wedge is best characterized by its frontal and lateral structural and geomorphological velocities.

GEOLOGIC SETTING

The Sub-Himalayan Fold-and-thrust Belt was created by the deformed synorogenic series of the Siwalik Group (Mugnier et al., 1992, 1994; 1999a; Powers et al., 1998), above the northward-subducting Indian plate. The series is located in the foothills of the Himalayas and comprises a succession of east-southeast–west-northwest-trending ridges that are perpendicular to the N10–N30°E compressional axis (Ni and Baraganzi, 1996; Jouanne et al., 1999).

A few ridges are the geomorphological expression of active thrusts. These faults branch off of a gently north-dipping décollement (Raiverman et al., 1994; Galahaut and Chandler, 1992; Biswas, 1994). Therefore, only a few active thrusts beneath south-verging slices accommodate the shortening in the Sub-Himalayan Zone (Figure 1). The fold-and-thrust belt is bounded by the Main Boundary Thrust (MBT) to the north, and the

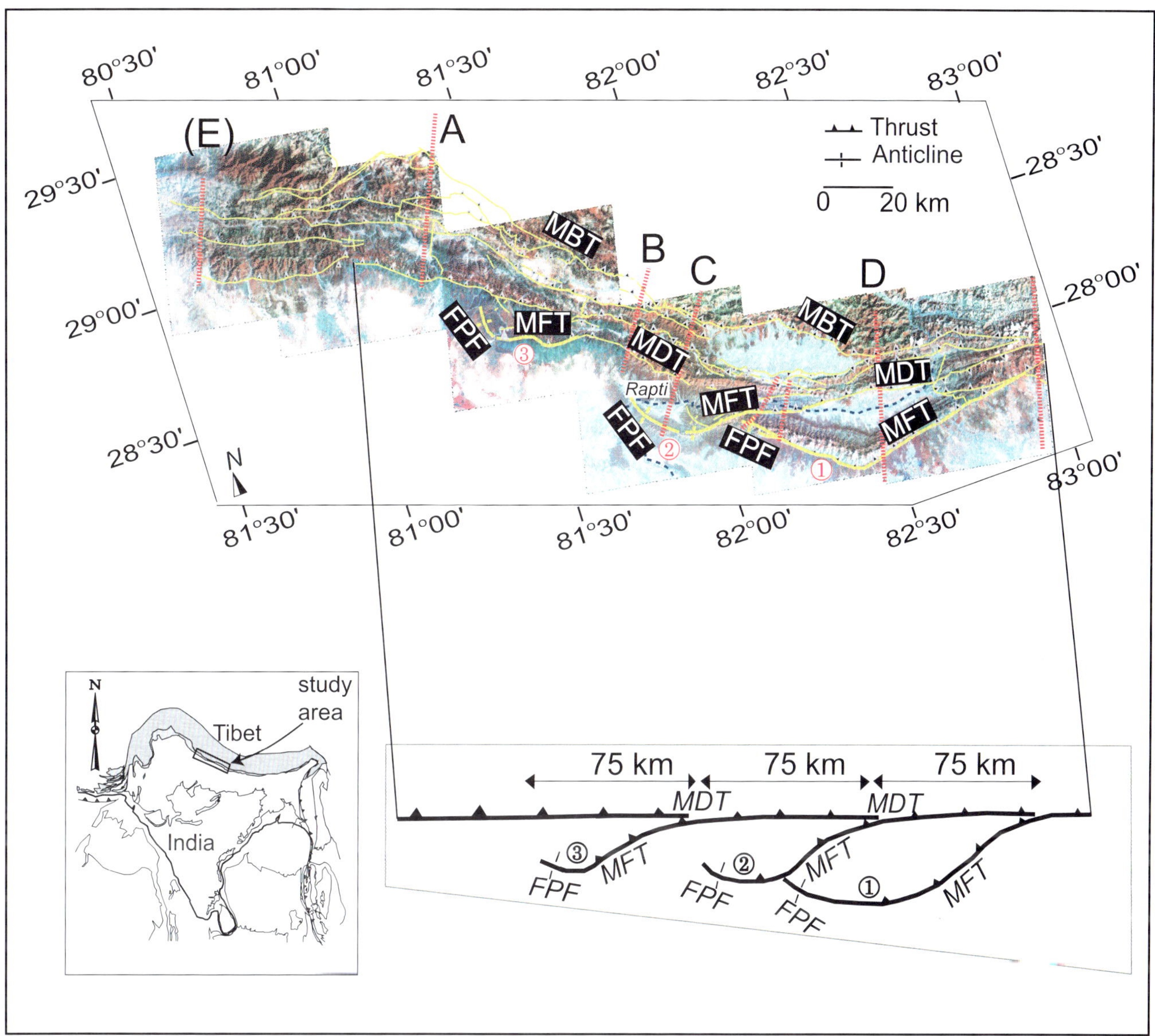

FIGURE 1. Top: Mosaic of SPOT satellite scenes and structural interpretation of the western Nepal Siwalik. Labeled dashed lines A–D are location of balanced cross sections shown in Figure 7. (E) is location of the cross section Figure 6. Unlabelled dashed lines are the locations of other available balanced cross sections. Bottom: Sketch map of west-plunging salients. MDT: Main Dun Thrust; MFT: Main Frontal Thrust; FPF: Fault-related fold (mainly fault-propagation folds).

Main Frontal Thrust (MFT) is the southernmost tectonic feature, characterized by Tertiary sediments overthrusting Quaternary (DeCelles et al., 1998a). Another primary thrust is the Main Dun Thrust (MDT) (Hérail and Mascle, 1980; Delcaillau et al., 1987; Mugnier et al., 1992), which constitutes a succession of laterally relayed thrusts propagating westward as ramp folds. This thrust owes its name to the piggyback basins called duns, which it overthrusts. Previous studies defined the Sub-Himalayan Zone as a fold-and-thrust belt characterized mainly by fault-propagation folds, fault-bend folding, and duplexes (Mascle et al., 1986; Banks and Warburton, 1986; Powers and Lillie, 1996; Powers et al., 1998; Mugnier et al., 1999a).

The prism grows by propagating southward into the foreland Indo-Gangetic plain. The stratigraphic sequence over the Main Décollement (MD) extends from the middle Miocene to the present (Appel and Roesler, 1994; Corvinus, 1994) and consists of a coarsening-upward succession of sandstone and conglomerate as much as 6000 m thick. The Siwalik Group is divided into three informal stratigraphic units. The lower Siwalik unit (middle Miocene) displays paleosols and fluvial channel deposits, the middle Siwalik unit (upper Miocene) is of a flooded-plain to braided-channel type, and the upper Siwalik unit (Pliocene to Pleistocene) is mainly conglomeratic and of a braided-channel type (Hérail et al., 1986; Delcaillau, 1997). This sequence seems to be diachronous in along-strike and across-strike directions (Appel and Roesler, 1994; Corvinus, 1994, DeCelles et al., 1998a, Ohja et al., 2000) and is interpreted to be the Miocene to Holocene sedimentary record of the Himalayan front's southward migration (Hérail et al., 1986; Delcaillau, 1997; Mugnier et al., 1999a).

STRUCTURAL VELOCITIES IN THE HIMALAYAN OUTER BELT

Structural Development

A mosaic of six (60 km × 60 km) SPOT satellite pictures (Figure 1) taken between longitudes 80°30′E and 83°00′E was compiled to complete the previous field structural investigations and create the structural sketch map along a 300-km-long segment of the Himalayan foothills. Analysis of spatial imagery brought out a peculiar recurrent set of en echelon arcuate structures (salients) that are convex toward the foreland, above the MFT (Figure 1). Information from satellite imagery includes the westward transition from the hinterland to the foreland location of the same continuous structures. They change laterally, from north-dipping MDT monoclines on a hinterland setting to ramp folds showing emergent faults (MFT type), and end up in blind-thrust ramp folds. Previous theoretical and field work emphasized that the fault length increases with greater forward displacement over the fault (Elliott, 1976; Walsh and Waterson, 1988; Jackson et al., 1996; Leturmy, 1997; Mueller and Talling, 1997). In the Sub-Himalayan Zone of western Nepal, all the aforesaid structures become progressively less mature toward their western tips (evolution of MDT monoclines to MFT ramp folds). Along the eastern salient, an early Pliocene cartographic unconformity on the backlimb of the fold disappears toward the west (Leturmy, 1997; Mugnier et al., 1999a). In the central salient, a wind gap (Mueller and Talling, 1997) used to be a main water gap for the Rapti River. The current water gap of the Rapti River bypasses the anticline to the west (Leturmy, 1997, Leturmy et al., 1999).

Because the structures are always less evolved toward the west, and considering the field observations, we infer that the salients propagate westward asymmetrically (from the monoclines toward the anticlines); on a map, the development of the frontalmost structures is not cylindrical but is westward-plunging. Fault tips migrate laterally, perpendicularly to the main deformation axis, as the width of the fault increases with its displacement (Elliott, 1976; Walsh and Waterson, 1988; Marchal et al., 1998). If these statements and observations are correct, they imply that greater displacement occurs over the MDT than over the MFT anticlines, and faults propagate westward.

Structural Velocities Over the Fold-and-thrust Belt

The convergence rate of the Indian plate with the Eurasian plate is about 50–55 mm/yr (De Mets et al., 1990). Only a part of it is accommodated in the Sub-Himalayan Zone. The total shortening rate in this area has been estimated by several methods. To constrain the shortening rate in this area, we review the various published results using various methods. Lyon-Caen and Molnar (1985) used the southward migration of the Tertiary basin depocenter to determine a minimum Indo-Himalayan convergence rate of 10–15 mm/yr and a maximum of as much as 20 mm/yr. By subtracting the shortening accommodated on faults in the High Himalayas from the total Indo-Eurasian convergence, Avouac and Tapponier (1993) arrived at a shortening rate of 18 mm/yr. Peltzer and Saucier's (1996) global kinematic model of Asia suggests a convergence rate of 18 mm/yr between the High Himalayas and the Indian plate. Schelling (1992) and Schelling and Arita (1991) determined a rate of 8.4 to 18.6 mm/yr using minimum shortening estimates from cross-section balancing between the High Himalayas and the MFT in eastern Nepal. DeCelles et al. (1998b) estimated a southward forebulge migration rate of 14 to 33 mm/yr during Eocene-Oligocene time. Lavé and Avouac (2000) used terrace folding to estimate a minimum shortening rate of 21 mm/yr for the Holocene period. All these estimates indicate an average shortening rate of 18 mm/yr for the area between the MCT and the MFT, and this value corresponds to the Indo-Himalayan convergence following the initiation of the MCT (20–22 Ma, according to Hodges et al., 1996). After initiation of the MBT, the total shortening has been partitioned between the Lesser Himalayas and the Sub-Himalayan Zone.

From cross-section balancing in western Nepal, Mugnier et al. (1999a) determined minimum shortening of 40 km in the Sub-Himalayan Zone between the MBT and the MFT. Because the onset of shortening in the present Sub-Himalayan Zone postdates the Pliocene upper Siwalik member (DeCelles et al., 1998a; Mugnier et al., 1999a), shortening-rate estimates are at least 17–20 mm/yr for the various cross sections. In northwest India, Powers et al. (1998) synthesized shortening rates that also had been deduced from various cross-section balancing in the Sub-Himalayan Zone between the MBT and the MFT, and they are in the same range (14 ± 4 mm/yr). Such shortening rates in the Sub-Himalayan Zone in turn implies current minor shortening rates in the Lesser Himalayas.

It can be argued that geodetic data from spirit-leveling (Jackson and Bilham, 1994) or GPS measurements (Bilham et al., 1997; Larson et al., 1999) indicate little current shortening of the Sub-Himalayan zone, the maximum convergence being accommodated in the

High Himalayas (about 5 mm/yr, over a total Indo-Himalayan convergence of 19 to 22 mm/yr). However, as several authors have already mentioned (Jackson and Bilham, 1994; Larson et al.,1999; Lavé and Avouac, 2000; and Mugnier et al., 2003), this more likely corresponds to an interseismic period than to the absence of shortening in the Sub-Himalayan Zone. According to Bilham et al. (1997) and Larson et al. (1999), locking below the Lesser Himalayas temporarily precludes foreland shortening, and great earthquakes are expected in order to compensate the lack of rupture south of the MBT. Mugnier et al. (1999c) showed that the mean shortening rate is constant over time (following the onset of deformation in the Sub-Himalayan zone). If correct, these results converge toward a present-day average shortening rate of about 19 mm/yr in the Sub-Himalayan zone, which we consider herein as realistic.

The Sub-Himalayan wedge accommodates shortening by a limited number of thrusts, i.e., the Main Boundary Thrust, the Main Frontal Thrust, and the Main Dun Thrust. Other thrusts currently have minor offsets and do not account for currently important shortening accommodation (see, for example, Mascle et al., 1986; Hérail et al., 1986; Mugnier et al., 1999a).

Scarce short-term velocity estimates are available for the major thrusts. A geomorphological analysis of the ridges overlying the thrusts helps to constrain velocities over the entire wedge. From a 30″ arc DEM (Figure 2a), morphological data are compiled over the ridges overlying the thrusts (Figure 2b). The mean elevation corresponds to the average altitude of each particular thrust-related relief, and the standard deviations of the elevations characterize the distribution of the elevation around this mean value. The morphology of the ridges is directly linked to the tectonic activity (Hurtrez et al., 1999).

The mean elevation increases from the southern MFT periclinal ridge (fault-propagation fold, FPF) to the MFT ridges and to the MDT. The gentle hinterland elevation increase characterizes the average taper geometry of the wedge; it includes the local tectonic activity of the thrusts but also the depth and slope of the Main Dêcollement, MD, which rules the shape of the wedge as a whole (Davis et al., 1983). The standard deviation of the elevation density characterizes the sharpness of the structures: the lower the standard deviation, the sharper the overlying ridge. In other words, if the elevation densities are concentrated around the mean elevation, the flanks of the ridge are abrupt. The lateral termination of the MFT (FPF of Figure 2) is sharp (standard deviation is 100 m), the MFTs have a smoother shape (standard deviations are 145 m and 148 m for the two studied ridges), and the MDTs have an even smoother morphology (standard deviations are 213 m and 228 m). Hurtrez et al. (1999) showed that because of the rapid erosive processes in the Sub-Himalayan Zone, the structures are in a dynamic equilibrium. Erosion compensates for uplift (Figure 3), and the topographic profiles remain constant through time as long as no tangible change occurs in the tectonic velocities or erosion rates. The topographic profile relates to the thrust velocity. For one particular thrust velocity and ramp geometry, there is one, and only one, associated topographic profile for the overlying ridge. This statement applies to mature ridges when the equilibrium profile is reached. Hence, the sharpness of the FPF probably relates to the growing structure (for which an equilibrium profile is not reached, as shown, by its low mean elevation) rather than to its tectonic activity. If the assumption of a dynamic equilibrium law is correct, the shapes of the MFT and MDT reflect their tectonic activity. Because the MFT and MDT ridge morphologies are globally constant, the previous statement indicates that local estimations of thrust velocities for the major thrusts can be extrapolated laterally.

The MBT presently shows major displacement (rocks of the Lesser Himalayas thrust over the Tertiary sediments of the Sub-Himalayan wedge, according to Upreti, 1990; Mugnier et al., 1992). However, Holocene activation of the MBT has locally minor normal sense slip (0.5 mm/yr; Mugnier et al., 1994), thus suggesting that taper is overcritical in this area, which implies that the MBT is not currently accommodating tangible shortening.

As a consequence, the overall shortening is accommodated mainly by the MDT and the MFT. According to the previous discussion on shortening rates, we assume that shortening over the MCT is minor, and that at least 17–20 mm/yr of shortening occurs within the Sub-Himalayan Zone. Sparse velocity data are available for the Sub-Himalayan thrusts: From terrace uplift inversion, Leturmy et al. (1999) suggested velocities in the range of 7–10 mm/yr for both the MFT and MDT in western Nepal, whereas Lavé and Avouac (2000) calculated a velocity of 21 mm/yr for the MFT and found 0 mm/yr for the MDT in eastern Nepal. These differences can be explained by the distribution of shortening versus time over these thrusts. The MDT displays minor out-of-sequence faults that are overlapped by undeformed sediments (Mugnier et al., 1998; Mugnier et al., 1999a), which indicate a nonpermanent tectonic activity. The ratio between faulted sediments and overlapping sediments is 1:3. We assume that it reflects the periods of relative activity to inactivity of the MDT. If shortening in the Sub-Himalayan Zone is partitioned only over the MDT and MFT, the tectonic activity time span reflects the coeval activation of both thrusts (Leturmy, 1997), whereas inactivity indicates periods of fast activation over the MFT alone (Lavé and Avouac, 2000). Coeval activation exists for a third of the time, hence the MDT tectonic velocity $V_{f_{MDT}}$ is one-third of the value given by Leturmy et al. (1999), that is, 2–3 mm/yr. Similar calculations for the MFT indicate a tectonic velocity, $V_{f_{MFT}}$, of about 17 mm/yr.

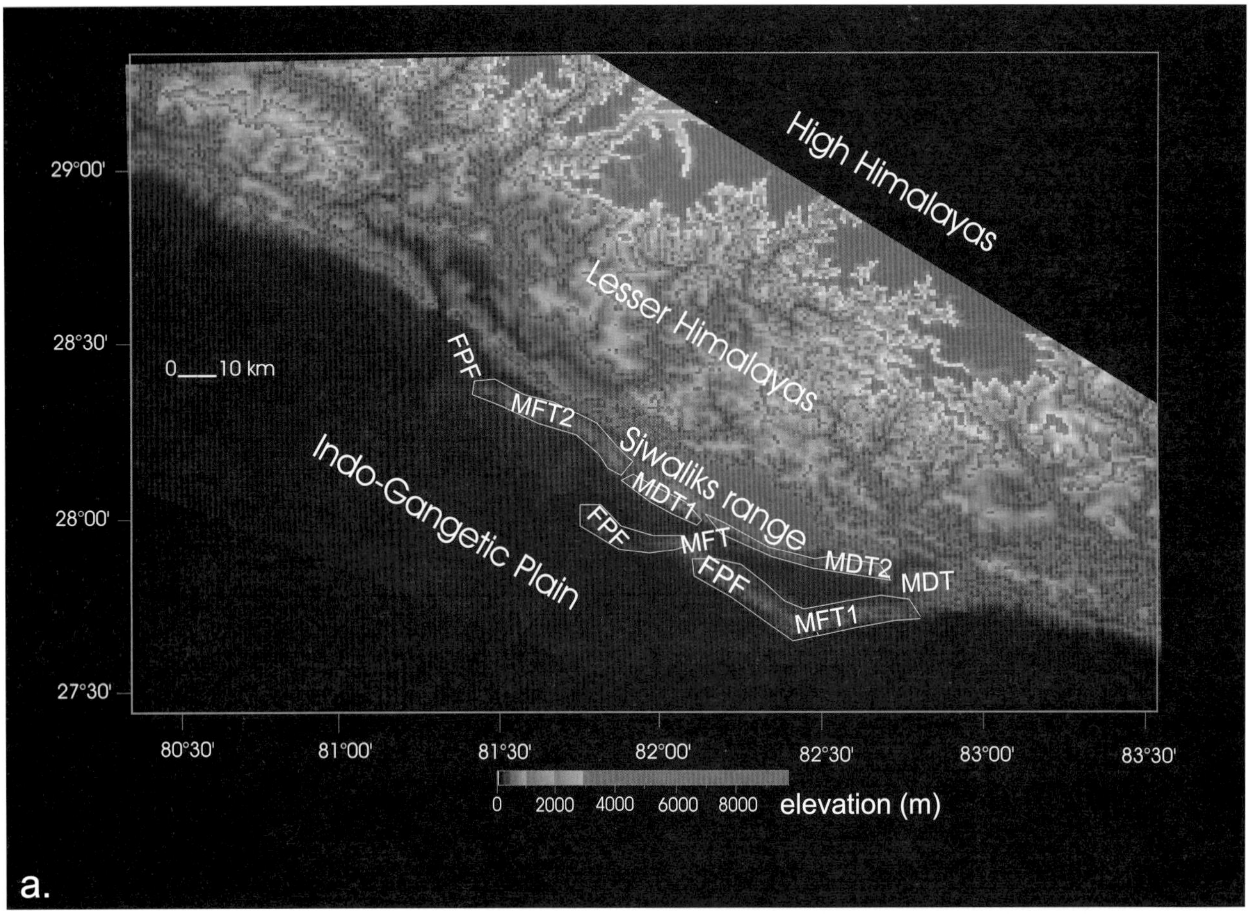

High Himalayas
Lesser Himalayas
Siwaliks range
Indo-Gangetic Plain
FPF
MFT2
MDT1
FPF
MFT
MDT2
MDT
FPF
MFT1
0 10 km
29°00'
28°30'
28°00'
27°30'
80°30' 81°00' 81°30' 82°00' 82°30' 83°00' 83°30'
0 2000 4000 6000 8000 elevation (m)
a.

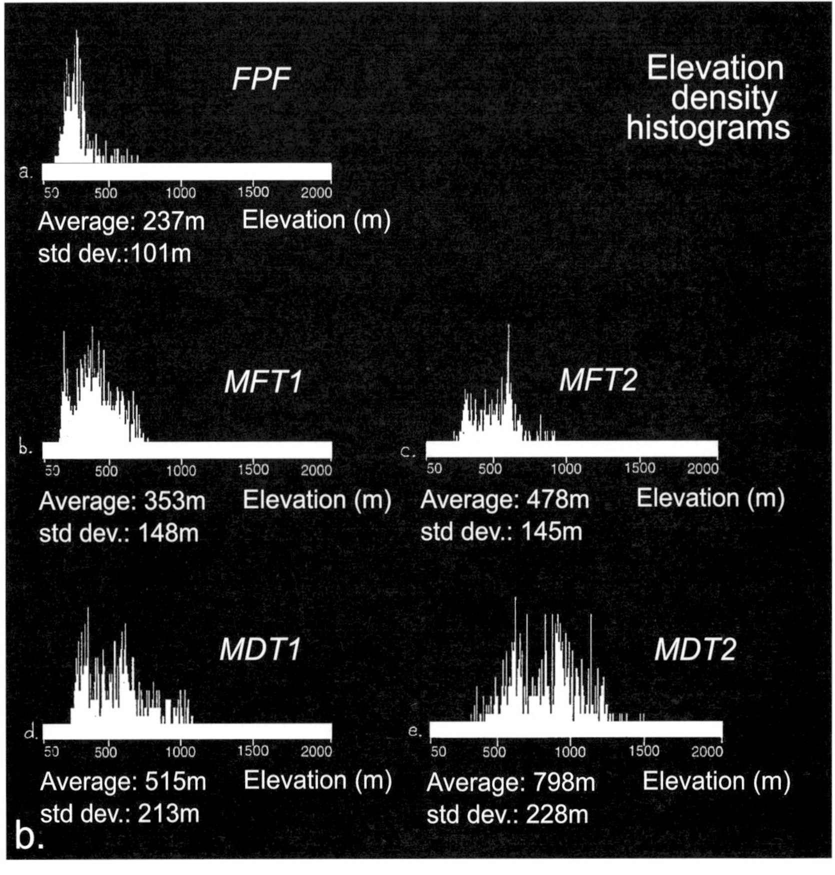

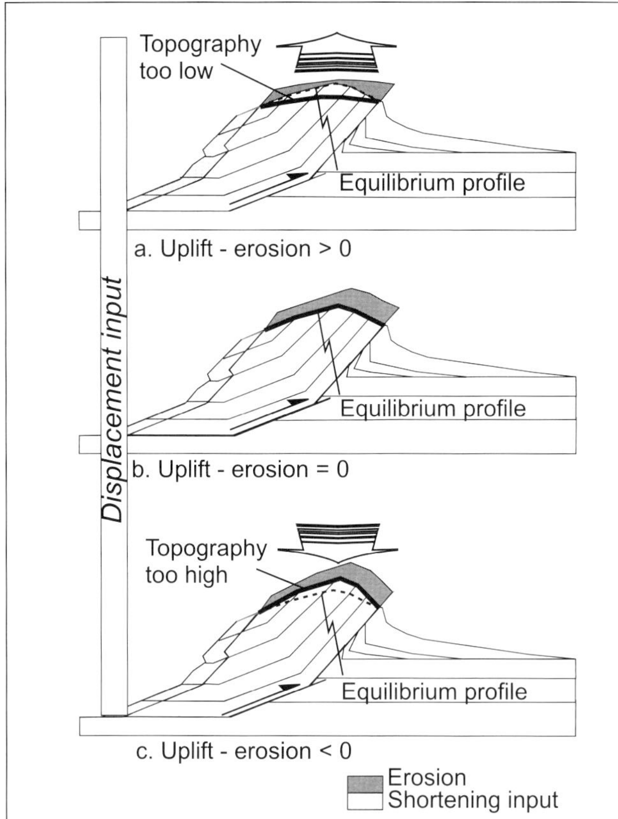

FIGURE 3. Dynamic-equilibrium theory. (a) Topography is too low with respect to the ramp velocity; erosion rates becomes lower than uplift rates. (b) Topography is in a dynamic-equilibrium state, erosion rates balance uplift. (c) Topography is too high, erosion rates become higher than uplift rates. Solid black line is topography, dashed line is equilibrium profile, light gray is the volume input as a result of shortening, dark gray is the erosion response.

Average forward-shortening velocities are available for the main Siwalik thrusts (Figure 4). However, forward velocities do not supply enough information to describe the asymmetrical development of the Sub-Himalayan Zone, and the lateral ramp-fold propagation velocities are needed. Leturmy (1997) calculated lateral-thrust velocities for the structures of the Sub-Himalayan Zone, which are 10 times faster than the forward-thrust velocities. This value is consistent with the observations of average fault displacement versus length given by various authors (Walsh and Waterson, 1988). Therefore, lateral velocity ($V_{l_{MFT}}$) for the MFT is in the range of 170 mm/yr. By analogy, as Elliott (1976) previously suggested, the forward and lateral velocities that control the fault propagation can be compared to the ductile edge and screw dislocations, respectively (Figure 5).

$$V_{f_{MFT}} = 17 \text{ mm/yr}$$

$$V_{l_{MFT}} = 170 \text{ mm/yr}$$

For the sake of simplicity, these velocities will further be named V_1 and V_2, respectively.

These structural velocities rule the westward propagation of the salients, which branch off of the MDT. Such a pattern suggests that the development of the most frontal structures is asymmetrical, because propagation only acts toward the west. The remaining question concerns the cylindrical nature of the overall wedge. How do structures and morphology evolve, over time, in the fold-and-thrust belt of western Nepal, where shortening is perpendicular to the global east-southeast to west-northwest trend of the belt and where structures plunge westward?

HOLOCENE MORPHOLOGICAL VELOCITIES OF THE HIMALAYAN FRONT

Lateral Propagation Velocity of the Morphological Structures

Evidence for an asymmetric westward growth of the salients was described previously. It includes ancient unconformities that are located on the eastern part of the salients but that vanish toward the west; a westward transition from water gaps to wind gaps; the distribution of the drainage pattern; and the systematic maturation of the structures, from the eastern monoclines maturing to fault-related folds showing emergent ramps, to the western ends of fault-related folds evolving with blind ramps.

We emphasize that the en échelon pattern of the thrust belt of western Nepal is linked to the lateral propagation of the imbricate thrusting (see Shaw et al., 1999), perpendicularly to the thrust-sheet motion (Mugnier et al., 1999a), and does not reflect any dextral strike-slip component of the front. MFT salients branch eastward on the MDT and propagate westward on the foremost position. The subcontinuous trend of the MFT corresponds to the most frontal ridge, composed of the en échelon fault-related folds. In other words, the envelope of the ridges overlying the MFT is the southernmost

FIGURE 2. (a) 30″ arc DEM of the studied area. Solid white lines represent sampled areas for statistical analysis of elevation. (b) Elevation density histograms of each sampled structure. FPF: fault-propagation fold ridge, MFT1 and 2 : Main Frontal Thrust ridges, MDT1 and 2 : Main Dun Thrust ridges.

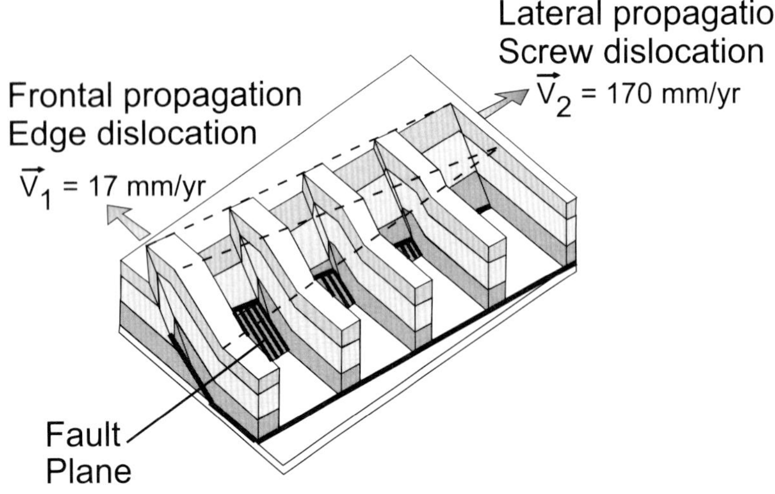

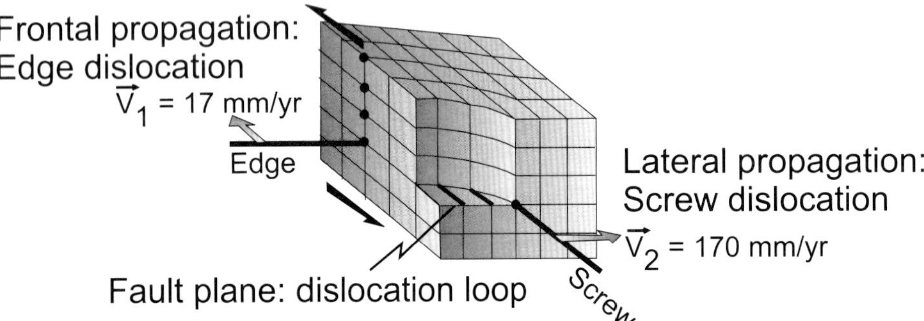

FIGURE 4. Top: Model for fault-propagation fold dynamics (after Suppe and Medwedeff, 1990). V_1 is the MFT forward velocity, V_2 is the MFT lateral propagation velocity. Bottom: analogy with ductile strain dislocations. V_1 is the edge dislocation and V_2 is the screw dislocation.

morphological feature, elongated along a broad N110°E axis. Each salient is a discrete element belonging to the MFT "chain" (Figure 1).

The salients in the studied area are distributed regularly, with a 75-km spacing (Figure 1), which can be defined as a spatial periodicity of the structures. Over the area, the lengths of the bows (from the beginning of the bend to the tip of the structures) show that they are 30 km shorter on each westward step. On the western side of the area (to the west of 81°20′E), no salient is displayed yet, whereas eastward, although some duns forming arcuate structures are displayed, the system is already too evolved. In the studied area, the easternmost salient (salient 1) is already overmature; that is, there is no more available space for lateral propagation to the west of the fold pericline, because it is bounded by the central salient.

Mugnier et al. (1999c) emphasized that shortening rates have remained constant over the Sub-Himalayan Zone through time. This statement is extended to the MDT and MFT tectonic velocities by assuming that a constant strain has been spread over these thrusts, over time. If this is correct, each frontal salient propagated, over time, with forward and lateral MFT-type velocities, and the relative lengths of the salients provide relative ages for these structures. If the 170 mm/yr lateral velocity is globally constant, salients of the en échelon series grow to the west with a 180,000-yr period, which corresponds to the time required to form a 30-km-long structure (the westward decreased length of these structures). These salients can now be regarded as independent, discrete defects that are gradually emerging westward with an average 180,000-yr period. The MFT ridge is created in this area by the assemblage of these defects. Therefore, this morphological structure propagates laterally with its own velocity, depending only on the

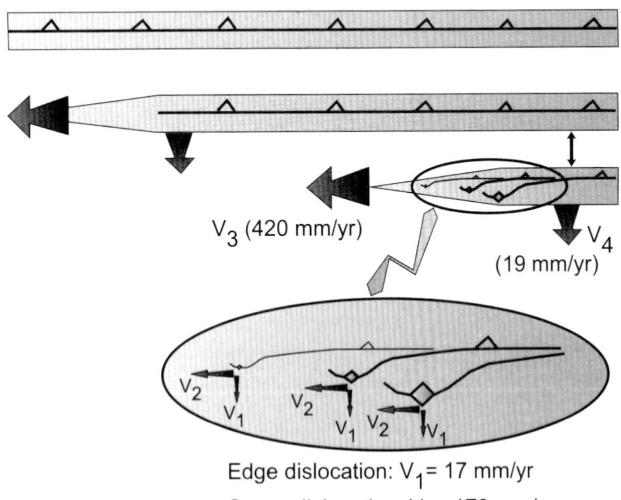

FIGURE 5. Structural and morphological velocities for the Sub-Himalayan wedge's frontal development. V_1 and V_2 are respectively the frontal and lateral propagation velocities for structural growth, and V_4 and V_3 are frontal and lateral velocities for morphological growth.

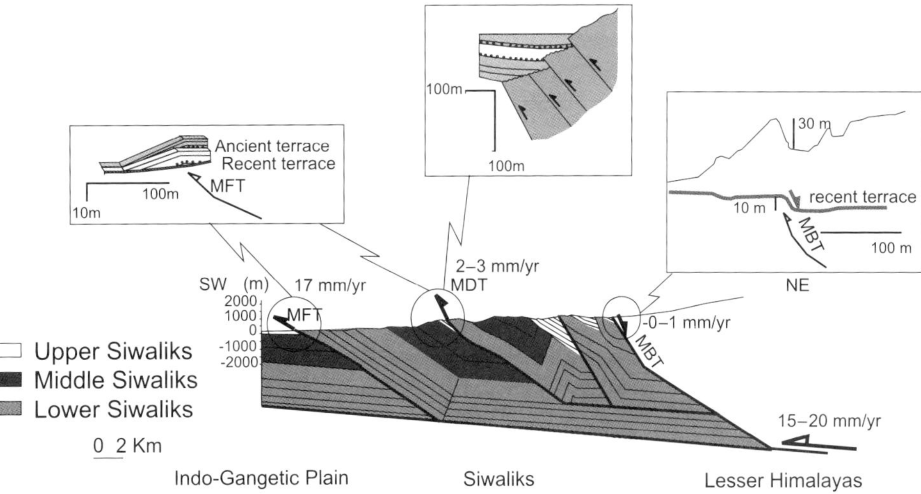

FIGURE 6. Average forward velocities over a typical cross section through the Sub-Himalayan wedge (location in Figure 1). Corresponding values are given by terrace uplift (MFT and MDT, from Leturmy, 1997 and Lavé and Avouac, 2000), sediments seal over out-of-sequence thrusts (MFT and MDT, after Mugnier et al., 1999b), normal-faulted recent terrace (MBT, after Mugnier et al., 1994), and various methods from authors (see text for references).

period of time (180,000 yr) and the spatial offset (75 km) of the independent elements of which it is made. Again, as an analog to the ductile approach, this phenomenon corresponds to dislocation creep. A rough estimate of the lateral morphological propagation velocity (V_3) is now added to the edge and corner structural velocities mentioned above:

$$V_3 = 420 \text{ mm/yr}$$

The morphological ridge, which is an assemblage of the elementary structural defects, propagates laterally more than twenty times faster than the forward structural-propagation velocity and nearly three times faster than the lateral structural-propagation velocity. This, in turn, implies that, even if the structural pattern is asymmetrical (with westward-propagating salients), the overall development of the wedge can be considered to be cylindrical, because the morphological development can be approximated to be instantaneous with respect to structural propagation velocities. However, the proposed value is considered to be a first-order estimate, because we are uncertain about the local structural velocities.

Forward Morphological Propagation Velocity

As for the lateral evolution, the morphological propagation to the south can be defined (Figure 6). Lyon-Caen and Molnar (1985) estimated, from the migration of the flexural Indo-Gangetic plain, that the southward progradation of the deformed area is in the range of 10–15 mm/yr (and is as much as 20 mm/yr). DeCelles et al. (1998a) proposed a southward forebulge migration of 14–33 mm/yr. We assume that the wedge is in a volumetric steady state, and that the taper angle is preserved through time (see Davis et al., 1983 or Dahlen and Barr, 1989). Hence we estimate the southward migration of the morphological front of the foothills to have an average rate of about 19 mm/yr, which is the value we previously debated for the convergence between the Lesser Himalayas and Indian plate. This value is the V_4 forward morphological velocity of the front.

Four velocities thus characterize the development of the Himalayan front. Two structural velocities control the development of individual structures: $V_1 = 17$ mm/yr, and $V_2 = 170$ mm/yr.

Two morphological velocities control the topographical evolution of the fold-and-thrust belt: $V_3 = 420$ mm/yr, and $V_4 = 19$ mm/yr.

The main observation we can infer from these kinematic estimates is that the overall development of the wedge is very fast laterally and is therefore cylindrical on a first-order approximation, in spite of the surface structural pattern displaying west-plunging fault-related folds. As a consequence, an across-strike evaluation of the sediment cycle within the wedge is adapted and is representative of the behavior of the wedge all along its 300-km strike length.

BURIAL CYCLE OF THE SEDIMENTS WITHIN THE WEDGE

Maximum Residence Time

A set of nine balanced cross sections that were based on surface data has been constructed over the Siwalik (Figure 7) (Leturmy, 1997; Mascle et al., 1998; Mugnier et al., 1999a, b). Assuming an "equivalence" hypothesis, lateral along-strike variations observed in the various cross sections represent different stages of the geometric and kinematic history of the analyzed thrust-related fold. This implies that measurement of fold geometries along each cross section can be used to

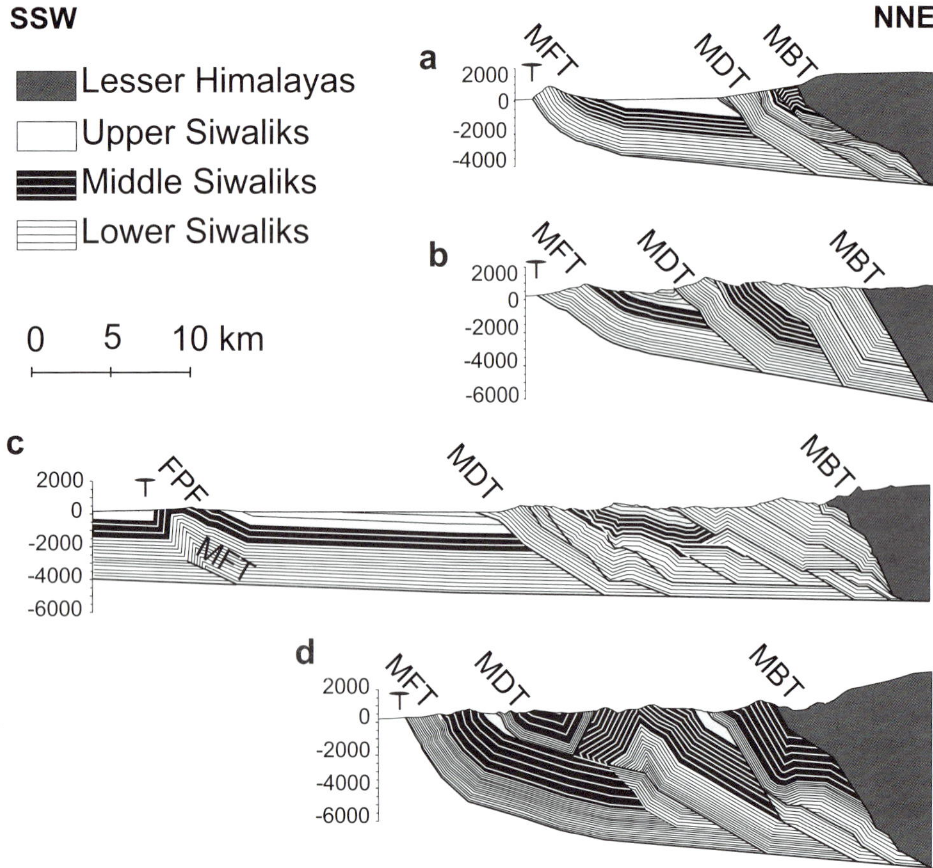

FIGURE 7. Balanced cross sections over the Siwalik (after Leturmy, 1997 and Mugnier et al., 1999a). Other available balanced cross sections are not presented (Mugnier et al., 1999a; Leturmy, 1997, and Husson and Mugnier, 2003). Locations in Figure 1. FPF: fault-propagation fold, MFT: Main Frontal Thrust, MDT: Main Dun Thrust, MBT: Main Boundary Thrust. The pin refers to the point beyond which, it is assumed, no displacement occurred farther south.

determine the kinematics (Poblet et al., 1998). The equivalence hypothesis is extended to all frontal folds, because, it is argued, the kinematics are the same for all the structures in a cylindrical setting. Hence, the whole set of cross sections provides various stages of fault-related fold evolution in the Himalayan fold-and-thrust belt, and together with the tectonic velocities described above, an evolutionary sketch can be constructed (Figure 8). Thrust sheets stack one above the other, with a synchronous development of foreland fault-related folds. New structures develop southward; however, the reactivation of more hinterland thrusts (MDT) implies that hinterland slices gradually tend to be almost entirely eroded (passive remnants are occasionally preserved—see discussion below). Erosion acts as a major control on the evolution of the fold-and-thrust belt. It actually rules the migration of the foremost structures, because it permanently reorganizes the stress field by unloading (Chalaron et al., 1995). Indeed, erosion controls the permanent activation/reactivation regime shifts of the MDT, yielding an average tectonic velocity of 2–3 mm/yr. As soon as a new structure develops toward the south, the previous, juxtaposed hinterland structures enter a reactivation regime. From the geometric analysis, it is estimated that structural growth switches to a foreland fault-propagation fold as soon as the fault begins to crosscut the fold. Suppe and Medwedeff (1990) define this evolution as the breakthrough fault. At this time, both structures are coeval; hinterland structures evolve at low rates (MDT style, at 2–3 mm/yr), whereas the most frontal structures grow at rapid rates (MFT style, at 17 mm/yr). Hence, for a structure, the fast tectonic regime lasts as long as it remains in the foremost position. From the average geometric data obtained from numerous cross sections, we hypothesize that new structures form southward when the ramp is emergent, because two anticlines never coexist along a cross section in the Sub-Himalayan Zone of western Nepal. At least one of the hinges is totally eroded, and only monoclines are preserved.

The burial cycle in such settings can be divided into various stages (Figure 9a), and the associated burial history through time is synthesized in Figure 9b. The residence time presented in the following section is calculated, using average geometries of cross sections and previously described kinematic data, for the tail of a slice implicated in the wedge. Therefore, it corresponds to the maximum residence time within the wedge, because the hinterland part of a thrust sheet is preserved longer.

The first stage of the cycle is the sedimentation in the subsiding foreland basin (stage 1 in Figure 9b), as the Himalayan front migrates toward the south. The Khutia Khola section (Ohja et al., 2000) and the Suraï Khola section (Appel and Roesler, 1994; Corvinus, 1994), on the western edge of the studied area, constitute reference stratigraphic series after the date assignments using magnetostratigraphy (Appel and Roesler, 1994; Ohja et al., 2000) and paleontology (Corvinus, 1994). The sedimentary burial curve used in the present study is derived from the magnetostratigraphic study of Appel

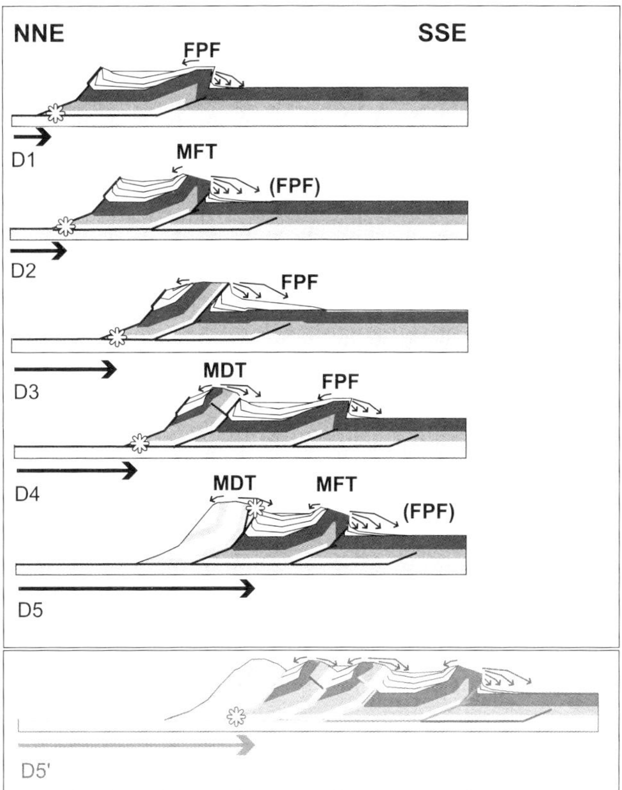

FIGURE 8. Structural evolutionary sketch for the Siwalik of western Nepal, inferred from balanced cross sections. D is the total shortening over the wedge. Stage D5′ (lowest box) is the case where no fault reactivation acts on the wedge (unlike the Sub-Himalayan Zone). Thrust sheets are more numerous, the foremost topographic break is more advanced, the width of the wedge increases. In the Sub-Himalayan Zone, few slices are represented, thus supporting a reactivation hypothesis. White star is the sediment on the tail of the slice. FPF: fault-propagation fold, MFT: Main Frontal Thrust, MDT: Main Dun Thrust.

and Roesler (1994) and Ohja et al. (2000). Although the ages from magnetostratigraphy show strong variations depending on the sampled lithologies (Ohja et al., 2000), it can be estimated that in the Sub-Himalayan Zone, this phenomenon lasts for about 12 m.y. to 13 m.y., and drives sediments to depths as great as 5000 to 6000 m. Next is the tectonic thickening episode at the footwall of the frontal thrust (stage 2, Figure 9b), when this thrust consumes the foreland sedimentary pile. It gradually moves up the hanging-wall of the décollement to the surface, which corresponds to about 5000 m uplift and is partially compensated in this stage by erosion. Beneath the relief of the frontal crest, it subsequently increases the burial depth by as much as 1000–1500 additional meters. Assuming a vertical uplift of the related fold of about 9–12 mm/yr over the ramp (for ramp dips between 30° and 45° and 5000 m total uplift), this phenomenon lasts for 400,000 to 600,000 years. It is considered to be minor with regard to the total cycle. The following stage corresponds to the instant at which the incorporated sediment enters the dynamic part of the cycle, that is, at which the motion of the sediments is governed by faulting and hence undergoes horizontal displacement. In the Sub-Himalayan Belt, most sediments are incorporated below ramp folds and only a very small amount of sediment accumulates on top of growing structures. During this episode, the sediment is transported over the Main Décollement (stage 3, Figure 9b). The total displacement on the Main Décollement is partitioned between the fast MFT (17 mm/yr) and the low MDT (2.5 mm/yr). These velocities are thus given with regard to the thrust-sheet reference. Sliding on this décollement is split into two stages. When the structure is in the foremost position, transportation over the décollement occurs at 17 mm/yr; afterward, displacement undergoes a reactivation regime at 2.5 mm/yr. As we said above, the former episode of transportation lasts from the ramp's initiation to the breakthrough faulting, and the latter lasts until the slice is totally exhumed on an hinterland location. Average lengths are about 10,500 m for the

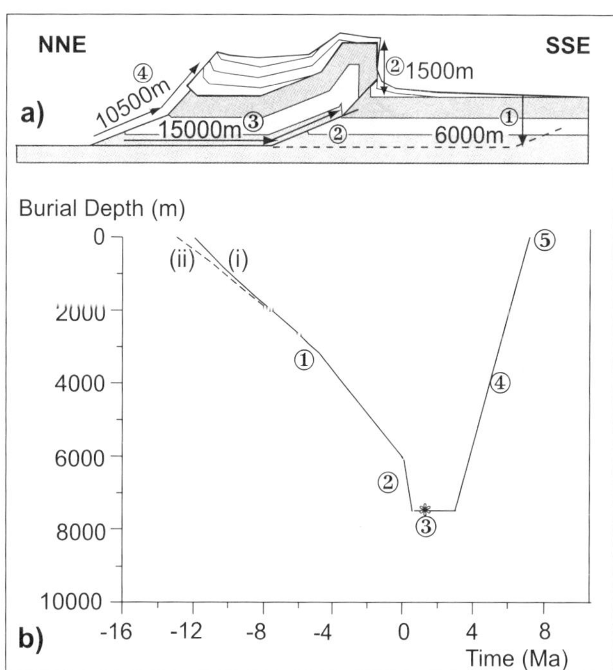

FIGURE 9. Burial evolution of the sediment at the tail of a slice, (a) with regard to the geodetic reference and (b) burial cycle versus time. Stage 1 is sedimentation in the subsiding foreland basin (after (i) Appel and Roesler, 1994 and (ii) Ohja et al., 2000), stage 2 is the tectonic thickening as the sediment is incorporated into the tectonic wedge, stage 3 is slipping on the Main Décollement, stage 4 is exhumation of the sediment in response to erosion. The star represents the instant, during the tectonic cycle, at which velocities change from activation regime (V = 17 mm/yr) to reactivation regime (V = 2.5 mm/yr) for the tail of the slice (not to scale).

ramp and the breakthrough fault, and 15,000 m for the basal décollement. On average, this period initially includes a rapid transportation phase of the tail of the slice at 17 mm/yr over about 10,500 m, and secondly, a slow transportation at 2.5 mm/yr, over 4500 m (15,000 minus 10,500 equals 4500 m). During this episode, the vertical-motion component is very small; the burial depth increases only weakly as a result of piggyback sedimentation, whereas a small uplift component is linked to the slip over the gently dipping basal décollement. Exhumation (stage 4, Figure 9b) of the tail of the slice starts as soon as it begins to climb over the ramp, when sliding over the basal décollement is totally finished and the slice is squeezed between the foremost sheet and the backstop buttress. Average ramp and fault lengths lead to approximately 10,500 m total displacement at that low rate. At that time, new structures have grown forward, and exhumation is realized on both the ramp and the breakthrough fault at low rates of reactivation (2.5 mm/yr). The burial depth subsequently decreases as a function of the ramp and fault angles. An average angle of 45° is used for both the ramp and the fault. The associated vertical displacement gradually leads to the total exhumation and erosion of the thrust sheet. Finally (stage 5 in Figure 9b), surface transportation phenomena are fast enough to be considered instantaneous with respect to the overall cycle.

Therefore, the wedge's development can be described as a steady-state burial cycle, from the incorporation of the foreland sediments to their exhumation and sedimentation in a more distal foreland.

The major burial phase is sedimentation in the foreland, but it lasts fairly long (the average sedimentary rate is only about 500 m/yr). The tectonic thickening, on the other hand, is faster and still increases the burial depth; whereas exhumation is the fastest process of the internal cycle (about 1700 m/m.y. of vertical motion).

The 12-m.y. sedimentation stage in the foreland is about two times longer than the tectonic stage within the wedge. The calculated short time of residence within the prism supports the idea that only a limited number of active thrust sheets control the development of the wedge. Only short-lived slices constitute the actual prism.

Sediment Transfer Balance Within the Wedge

The previously calculated time for the burial/exhumation cycle is an average for the Sub-Himalayan Zone of western Nepal. Using such a sketch of development, we can calculate the material balance within the wedge.

Sediments incorporated from the foreland into the fold-and-thrust belt constitute the main input of material, because piggyback basins represent a weak volume in the Sub-Himalayan Belt. The total input is estimated from kinematic and geometric data, assuming an average velocity of 19 mm/yr of southward migration of the wedge (V_4). In the studied area, the incorporated volume per linear kilometer of the Indo-Gangetic plain (input I), over time, is given by:

$$I = V_4 \times T_s$$

where T_s is the average total thickness of the involved sedimentary pile above the Main Décollement (5000 m).

$$I = 9.5 \times 10^5 \text{ km}^3/\text{yr}$$

Given a steady-state regime for the wedge (Dahlen and Barr, 1989), the input volume, I, equals the output volume, O. The main output is erosion. However, geometric observations from the structural sketch map and the balanced cross sections (Leturmy, 1997; Mugnier et al., 1999a) imply some restrictions for the material balance. A part of the wedge's volume is "captured" either as passive remnants of slices, accreted along the footwall of the MBT, or as duplexes, subducted beneath the MBT. Various settings can be distinguished (Figure 10). The lower Siwalik Formation, at the base of the thrust sedimentary pile, often shows duplexes. If duplex horses are located at the hanging-wall of the Main Internal Décollement ID (Figure 10a), they are unrelentingly eroded as normal stacked slices; on the contrary, if the duplex is located on the footwall of this décollement, it is subducted and consumed by the Lesser Himalayas (Figure 10b) and eventually withdrawn from the prism. Finally, local passive remnants of middle to upper Siwalik sheets are stacked beneath the MBT (Figure 10c) and incorporated into the Lesser Himalayan wedge. Inferences from geometric data (structural sketch map and balanced cross sections) suggest that only subducted duplexes may constitute a significant volume of captured material. On a first-order approximation from geometric data, remnants of slices comprise less than 5% of the total volume of the wedge. On the whole, it can be estimated from the structural sketch map that, within the 300-km length of the studied area, about 55% of the lower Siwalik unit does not reach the surface. In order to keep a balanced structure in the wedge, this unit has to form duplexes, subducted beneath the Lesser Himalayas and subsequently withdrawn from the wedge. Per linear kilometer along the belt, the captured volume (CV) from lower Siwalik duplexes is in the range of:

$$CV = 55\% \times t \times V_4$$

where t is the average stratigraphic thickness of the lower Siwalik unit (2000 m),

$$CV = 2.1 \times 10^5 \text{ km}^3/\text{yr}$$

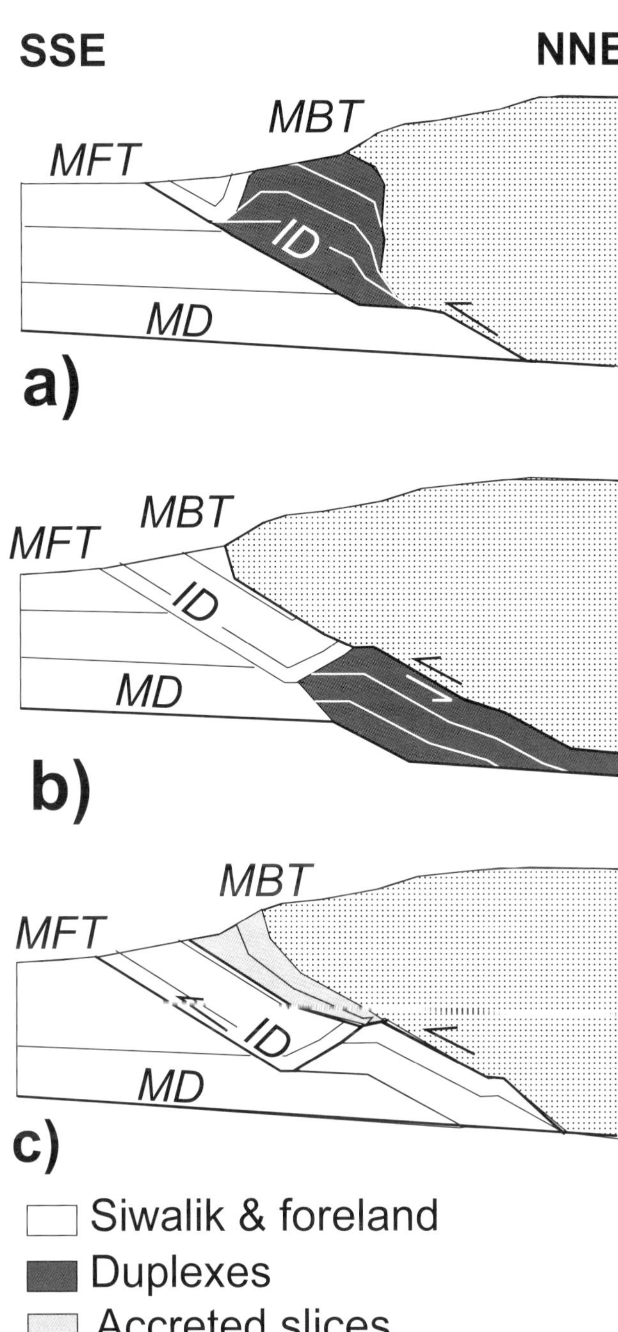

FIGURE 10. Schematic sketch of the material captured within the Siwalik wedge. (a) Duplex is above the main Internal Dêcollement (ID), and is exhumed and eroded as a normal thrust sheet. (b) Duplex is located beneath the ID, and therefore subducted beneath the Lesser Himalayas and withdrawn from the wedge. (c) Remnants of slices are accreted to the Lesser Himalayas, and subsequently incorporated into them. MD: Main Dêcollement, MFT: Main Frontal Thrust, MBT: Main Boundary Thrust.

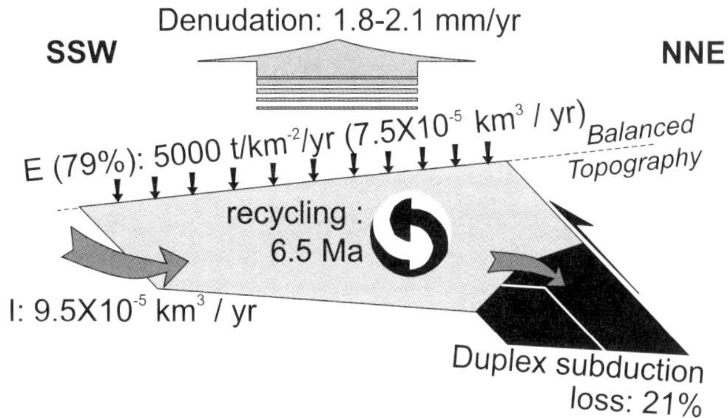

FIGURE 11. Volume, mass, and time transfer cycle for the sediments within the Siwalik wedge. I is the input of material at the MFT; E is erosion and equals the output minus the loss from duplex subduction. The corresponding average denudation is evaluated assuming a steady state of volume and a critical-taper angle for the wedge. Values are given per linear kilometer at the front of the Sub-Himalayan range.

and hence, the current withdrawn material from subduing duplexes, CV, is therefore approximated to be about 21% of the total volume of the Sub-Himalayan Zone of western Nepal. Again, according to a volumetric steady state (see Dahlen and Barr, 1989) and a critical taper angle (Davis et al., 1983; Mugnier et al. 1992; DeCelles and Mitra, 1995) for the prism, the output has to be equivalent to the input I, as topography remains constant versus time. Hence erosion accounts for about 79% of the input I. Sediments totaling 7.5×10^5 km^3 flow through the wedge per year and per linear kilometer along the belt, in western Nepal.

The average width of the Sub-Himalayan Zone in the studied area is 35–40 km. The average erosion rate for the whole wedge is thus in the range of 1.8 to 2.1 mm/yr.

Previous estimates of erosion from the sediment load in rivers are in the range of 1.5 to 2 mm/yr (Delcaillau, 1997), therefore, they support our estimate of erosion. According to an average density of 2.65, the corresponding eroded mass is about 5000 t/km^2/yr. Delcaillau (1997) calculated specific erosion rates for various drainage areas of the Siwalik in the very wide range of 400 to 9500 t/km/yr, that is, 0.1 to 3.5 mm/yr of erosion. This consistency supports the hypothesis of a steady-state regime for the Siwalik wedge and in turn validates the calculated balance (Figure 11).

CONCLUSIONS

The sediment burial cycle was the main focus of our study. However, across-strike estimates for the Sub-Himalayan Zone's sediment cycle are significant if the behavior of the wedge is similar laterally. Structural data inferred from balanced cross sections, a structural sketch map, and spatial imagery reveal a laterally orientated development. Further investigations, including geomorphological observations, allowed us to calculate the velocities of the wedge buildup. We see from this study that the lateral morphological growth is much faster than the lateral propagation of the fault-related folds. Morphological structures are the assemblage of independent structures coalescing with time. These structures propagate laterally in an asymmetric pattern, and the envelope of these elementary "defects" propagates fast enough laterally (in the range of 40 cm/yr) to assume an overall cylindrical development on a wedge scale. Hence, estimates of the sediment burial cycle are justified in such a setting. Accretion is the only input of sediments within the fold-and-thrust belt, and it is about 9.5×10^5 km^3/yr at the front of the Himalayan Belt. An average of 21% of the accreted material is withdrawn in the form of subducted duplexes of the lower Siwalik unit beneath the Lesser Himalayas. The maximum residence time within the wedge is estimated to be 6.5 m.y. for sediments undergoing the total burial/exhumation cycle within the wedge. Assuming a steady-state model for the Sub-Himalayan prism, the subsequent volume of material that completes the cycle led to erosion estimates of about 1.8 to 2.1 mm/yr. This is consistent with previous studies and not only supports the erosion rates presently propounded, but also the overall cycle, and hence, the proposed sketch of the wedge's development.

The evolution of foreland petroleum systems is ruled primarily by these parameters (burial, erosion, and tectonic rates), because they control the thermal regime. According to previous studies (Stüwe et al., 1994; Mancktelow and Grasemann, 1997), sedimentation rates such as those found in the Sub-Himalayan Zone would lower the thermal gradient by about 15% at a 5-km depth, whereas erosion rates in the range of those of the Sub-Himalayan Zone would increase the thermal gradient by about 30%. On the other hand, tectonic velocities are not fast enough to induce perturbations on the thermal field. In such settings, the oil window lies at greater depth in the sedimentary basin than it does in the folded belt, where it is significantly shallower. Given these parameters, one can therefore make general assumptions for kerogen maturation in such settings.

REFERENCES CITED

Appel, E., and W. Roesler, 1994, Magnetic polarity stratigraphy of the Neogene Surai Khola section (Siwalik, SW Nepal): Himalayan Geology, v. 15, p. 63–68.

Avouac, J. P., and P. Tapponier, 1993, Kinematic model of active deformation in central Asia: Geophysical Research Letters, v. 20, p. 895–898.

Baby, P., M. Specht, J.Oller, G. Montemuro, B. Colletta, and J. Letouzey, 1994, The Boomerang-Chapare transfer Zone (recent oil discovery trend in Bolivia): structural interpretation and experimental approach, *in* F. Roure, N. Ellouz, V. S. Shein, and I. Skortsov, eds., Geodynamic evolution of sedimentary basins: International symposium, Moscow, p. 203–218.

Banks, C. J., and J. Warburton, 1986, "Passive-roof" duplex geometry in the frontal structures of the Kirthar and Sulaiman mountain belts, Pakistan: Journal of Structural Geology, v. 8/3–4, p. 229–237.

Bashyal, R. P., 1998, Petroleum exploration in Nepal: Journal of the Nepal Geological Society, v. 18, p. 19–24.

Bilham, R., K. Larson, J. Freymueller, J., and Idylhim members, 1997, GPS measurements of present-day convergence across the Nepal Himalaya: Nature, v. 386/6, p. 61–63.

Biswas, S. K., 1994, Status exploration for hydrocarbons in Siwalik basin of India and future trends: Himalayan Geology, v. 15, p. 283–300.

Chalaron, E., J-L. Mugnier, and G. H. Mascle, 1995, Control of the thrust tectonics in the himalayan foothills; a view from a numerical model: Tectonophysics, v. 248, p. 139–163.

Corvinus, G., 1994, The Surai Khola and Rato Khola fossiliferous sequences in the Siwalik Group, Nepal: Himalayan Geology, v. 15, p. 49–61.

Dahlen, F. A., and T. Barr, 1989, Brittle frictional mountain building, 1. Deformation and mechanical energy budget: Journal of Geophysical Research, v. 94, B4, p. 13906–3922.

Davis, D. A., J. Suppe, and F. A. Dahlen, 1983, Mechanics of fold-and-thrust belts and accretionary wedges: Journal of Geophysical Research, v. 88, p. 1153–1172.

DeCelles, P. G., and G. Mitra, 1995, History of the Sevier orogenic wedge in terms of critical taper models, Northeast Utah and southwestern Wyoming: Geological Society of America Bulletin, v. 107, p. 454–462.

DeCelles, P. G., G. E. Gehrels, J. Quade, T. P. Ojha, and P. A. Kapp, 1998a, Neogene foreland basin deposits, erosional unroofing and the kinematic history of the Himalayan fold-and-thrust belt, Western Nepal: Geological Society of America Bulletin, v. 110/1, p. 2–21.

DeCelles, P. G., G. E. Gehrels, J. Quade, and T. P. Ojha, 1998b, Eocene–early Miocene foreland development and the history of Himalayan thrusting, western and central Nepal: Tectonics, v. 17, p. 741–765.

Delcaillau, B., G. Hérail, and G. H. Mascle, 1987, Evolution geomorphostructurale de chevauchements actifs: le cas des chevauchements intra-Siwalik du Népal Central: Zeitschrift für Geomorphologie, Nue Forsch, Bd. 31, p. 339–360.

Delcaillau, B., 1997, Les fronts de chaîne actives— Genèse des reliefs et relations tectonique-érosion-sédimentation: Thèse d'habilitation à diriger des recherches, Université de Caen, 336 p.

De Mets, C., R. G. Gordon, D. F. Argus, and S. Stein, 1990, Current plate motions: Geophysical Journal International, v. 101, p. 425–478.

Elliott, D., 1976, The energy balance and deformation

mechanisms of thrust sheets: Philosophical Transactions of the Royal Society of London, v. 283, p. 289–312.

Galahaut, V. K., and R. Chandler, 1992, On the active tectonics of the Dehra Dun region from observations of ground elevation changes: Journal of the Geological Society of India, v. 39, p. 61–68.

Hérail, G., and G. Mascle, 1980, Les Siwalik du Népal Central: structures et géomorphologie d'un piedmond en cours de déformation: Bulletin de l'Association Géographique Française, v. 431, p. 259–267.

Hérail, G., G. Mascle, and B. Delcaillau, 1986, Les Siwalik de l'Himalaya du Népal: un exemple de prisme d'accrétion intracontinental, in S.D.L.T.P.: Bordet, v. 47, p. 153–182.

Hodges, K. V., R. R. Parrish, and M. P. Searle, 1996, Tectonic evolution of the Central Annapurna Range, Nepalese Himalayas: Tectonics, v. 15, p. 1264–1291.

Hurtrez, J. E., F. Lucazeau, J. Lavé, and J-P. Avouac, 1999, Investigations of the relationships between basin morphology, tectonic uplift and denudation from the study of an active belt in the Siwalik Hills (Central Nepal): Journal of Geophysical Research, v. 104–B6, p. 12,779–12,796.

Husson, L., and J-L. Mugnier, 2003, Three-dimensional reconstruction from surface data, restoration, and kinematics of the Baisahi passive-roof duplex, W. Nepal: Journal of Structural Geology, v. 25, p. 79–90.

Jackson, M., and R. Bilham, 1994, Constraints on Himalayan deformation inferred from vertical velocity fields in Nepal and Tibet: Journal of Geophysical Research, v. 99–B7, p. 13,897–13,912.

Jackson, J. A., R. Norris, and J. Youngson, 1996, The structural evolution of active fault and fold systems in Central Otago, New Zealand: Journal of Structural Geology, v. 16, p. 1041–1059.

Jouanne, F., J-L. Mugnier, M. R. Pandey, J-F. Gamond, P. Le Fort, L. Serrurier, C. Vigny, J P Avouac, and Idylhim members, 1999, Oblique convergence in the Himalayas of western Nepal deduced from preliminary results of GPS measurements: Geophysical Research Letters, v. 26, p. 1933–1936.

Larson, K. M., R. Bürgmann, R. Bilham, and J. Freymueller, 1999, Kinematics of the Indo-Eurasia collision zone and GPS measurements: Journal of Geophysical Research, v. 104, p. 1077–1093.

Lavé, J., and J. P. Avouac, 2000, Active faulting of fluvial terraces across the Siwaliks Hills, Himalayas of central Nepal: Journal of Geophysical Research, v. 105, p. 5735–5570.

Leturmy, P., 1997, Sédiments et reliefs du front des systèmes chevauchants: Modélisation et exemples du front Andin et des Siwalik à l'Holocène: PhD thesis, UJF-Grenoble I, 235 p.

Leturmy, P., P. Huyghe, J. L. Mugnier, and B. Delcaillau, 1999, An intermediate displacement scale between earthquake slip and finite geometry of thrust systems deduced from a comparison of Quaternary, Holocene and instantaneous rates of shortening in the frontal thrust belt of Himalaya (Western Nepal). Extended abstract in Active Subduction and collision in SE Asia: data and models, Montpellier, May 9–12, Mém. Géosciences— Montpellier, France, v. 14, p. 103–106.

Lyon-Caen, H., and P. Molnar, 1985, Gravity anomalies, flexure of the Indian plate, and the structure, support and evolution of the Himalaya and Ganga basin: Tectonics, v. 4–6, p. 513–538.

Mancktelow, N. S., and B. Grasemann, 1997, Time-dependent effects of heat advection and topography on cooling histories during erosion: Tectonophysics, v. 270, p. 167–195.

Marchal, D., M. Guiraud, T. Rives, and J. Van Den Driessche, 1998, Space and time propagation processes of normal faults, *in* G. Jones, Q. J. Fischer, and R. J. Knipe, eds., Faulting, fault sealing and fluid flow in hydrocarbon reservoirs, Geological Society (London) Special Publication 145, p. 51–70.

Mascle, G., G. Hérail, T. Van Haver, and B. Delcaillau, 1986, Structure et évolution des bassins d'épisuture et de périsuture liés à la chaîne himalayenne: Société Nationale Elf Aquitaine Production, BCREDP 10, Pau, p. 181–203.

Mascle, G., E. Chalaron, A. Gajurel, P. Leturmy, and J-L. Mugnier, 1998, Paleoseismicity in the Siwalik: occurrence of major seismic events in the Himalayas of West Nepal: Bulletin of the Geological Society of Nepal, v. 18, p. 417–430.

Moretti, I., P. Baby, E. Mendez, and D. Zubieta, 1996, Hydrocarbon generation in relation to thrusting in the Sub-Andean Zone from 18 to 22°S, Bolivia: Petroleum Geoscience, v. 2, p. 17–28.

Mueller, K., and P. Talling, 1997, Geomorphic evidence for tear faults accommodating lateral propagation of an active Fault-Bend fold, Wheeler ridge, California: Journal of Structural Geology, v. 19, no. 3–4, p. 397–411.

Mugnier, J. L., P. Huyghe, P. Leturmy, and F. Jouanne, 2004, Episodicity and rates of thrust-sheet motion in Himalayas (western Nepal), *in* K. McClay, ed., Thrust tectonics and hydrocarbon systems: AAPG Memoir 82, this volume, p. 91–114.

Mugnier, J. L., P. Leturmy, G. Mascle, P. Huyghe, E. Chalaron, G. Vidal, L. Husson, and B. Delcaillau, 1999a, The Siwaliks of Western Nepal I: Geometry and kinematics: Journal of Asian Earth Sciences, 17, p. 629–642.

Mugnier, J. L., P. Leturmy, P. Huyghe, and E. Chalaron, 1999b, The Siwaliks of eastern Nepal II: Mechanism of the thrust wedge: Journal of Asian Earth Sciences, 17, p. 643–657.

Mugnier, J-L., P. Leturmy, P. Huyghe, and F. Jouanne, 1999c, A comparison of long-term rate, the Holocene rate and the instantaneous rate of shortening in the frontal thrust belt of Himalaya (Western Nepal): Presented at Thrust Tectonics, Royal Holloway University of London.

Mugnier, J-L., B. Delcaillau, P. Huyghe, and P. Leturmy, 1998, The break-back thrust splay of the Main Dun Thrust: Evidence of an intermediate displacement scale between earthquake slip and finite geometry of thrust systems: Journal of Structural Geology, v. 20, p. 857–864.

Mugnier, J-L., P. Huyghe, E. Chalaron, and G. Mascle, 1994, Recent movements along the Main Boundary Thrust of the Himalayas: normal faulting in an over critical thrust wedge?: Tectonophysics, v. 238, p. 199–215.

Mugnier, J-L, G. Mascle, and T. Faucher, 1992, La structure des Siwalik de l'Ouest Népal: un prisme d'accrétion intracontinental: Bulletin de la Société Géologique de France, v. 163–5, p. 585–595.

Ni, J., and M. Baraganzi, 1996, Seismotectonics of the Himalayan collision zone: geometry of the underthrusting Indian plate beneath the Himalaya: Journal of Geophysical Research, v. 89, p. 1147–1163.

Ohja, T. P., R. F. Butler, J. Quade, P. G. DeCelles, D. Richards, and B. N. Upreti, 2000, Magnetic polarity stratigraphy of the Neogene Siwalik Group at Khutia Khola, far western Nepal: Geological Society of America Bulletin, v. 112, p. 424–434.

Peltzer, G., and F. Saucier, 1996, Present-day kinematics of Asia derived from geologic fault rates: Journal of Geophysical Research, v. 101, B12, p. 27,943–27,956.

Poblet, J., J. A. Munoz, A. Travé, and J. Serra-Kiel, 1998, Quantifying the syntectonic sediments associated with single-layer detachment folds: Journal of Structural Geology, v. 19, p. 369–381.

Powers, P. M., and R. J. Lillie, 1996, Shortening rates within the sub-Himalaya of NW India based on a balanced cross section of the Kangra Dun re-entrant: Himalayan-Karakorum-Tibet Workshop, 11th, Flagstaff, Arizona, p. 120–121.

Powers, P. M., R. J. Lillie, and S. Yeats, 1998, Structure and shortening of the Kangra and Dehra Dun reentrants, Sub-Himalaya, India: Geological Society of America Bulletin, v. 110, p. 1010–1027.

Raiverman, V., A. K. Srivastava, and D. N. Prasad, 1994, Structural style in north-western Himalayan foothills, *in* G. Kumar and N. R. Phadtare, eds.: Siwalik foreland basin of Himalaya, v. 15, p. 263–280.

Schelling, D., 1992, The tectonostratigraphy and structure of the eastern Nepal Himalaya: Tectonics, v. 11–5, p. 925–943.

Schelling, D., and K. Arita, 1991, Thrust tectonics, crustal shortening, and the structure of the far-eastern Nepal Himalaya: Tectonics, v. 10, p. 851–862.

Shaw, J. H., Bilotti, F., and P. A. Brennan, 1999, Patterns of imbricate thrusting: Geological Society of America Bulletin, v. 111, p. 1140–1154.

Stüwe, K., L. White, and R. Brown, 1994, The influence on eroding topography on steady state isotherms. Applications to fission track analysis: Earth and Planetary Science Letters, v. 124, p. 63–74.

Suppe, J., and D. A. Medwedeff, 1990, Geometry an kinematics of fault-propagation folding: Eclogae Geologicae Helveticae, v. 83–3, p. 409–454.

Upreti, B. N., 1990, An outline of the geology of far western Nepal: Journal of Himalayan Geology, v. 1, p. 93–102.

Walsh, J. J., and J. Waterson, 1988, Analysis of the relationship between displacements and dimensions of faults: Journal of Structural Geology, v. 10, p. 239–247.

Marshak, S., 2004, Salients, recesses, arcs, oroclines, and syntaxes—A review of ideas concerning the formation of map-view curves in fold-thrust belts, *in* K. R. McClay, ed., Thrust tectonics and hydrocarbon systems: AAPG Memoir 82, p. 131–156.

Salients, Recesses, Arcs, Oroclines, and Syntaxes—A Review of Ideas Concerning the Formation of Map-view Curves in Fold-thrust Belts

Stephen Marshak

Department of Geology, University of Illinois, Urbana, Illinois, U.S.A.

ABSTRACT

The problem of how map-view curves (variously named salients, recesses, arcs, oroclines, virgations, festoons, bends, oroflexes, and syntaxes) in fold-thrust belts originate has caught the attention of geologists for more than 200 years. This chapter reviews the advances in understanding curves. Early geologists recognized that by understanding curve formation, one might gain insight into the process of orogeny. In recent decades, researchers have proposed several geologically reasonable models to explain curve formation; no single explanation can work for all curves. The majority of curving fold-thrust belts can be called "basin controlled," in that their presence reflects the architecture of the predeformational sedimentary basin from which the curve formed. Factors such as depth to detachment, rock strength, detachment strength, and detachment slope all affect the width of a fold-thrust belt for a given amount of hinterland displacement, as predicted by critical-taper theory. Therefore, along-strike variation in these factors leads to the inception of thrust belts that vary in width along strike, and thus have curved traces. However, not all curved thrust belts are basin controlled. Other causes for curve formation include interaction of a thrust belt with foreland obstacles or promontories, hinterland collision of an indenter, interaction with subsequent strike-slip faults, and warping of the downgoing (underthrust) plate. Not all curve-forming processes lead to "oroclinal" bending of a fold-thrust belt, in that not all curves involve rotation of segments of the thrust belt around a vertical axis. Thus, not all curves are oroclines, where the term "orocline" specifically refers to a mountain belt bent in plan. Basin-controlled curves and curves formed in front of indenters generally initiate with a curved trace, whereas curves formed in response to interactions with foreland obstacles or with strike-slip faults involve oroclinal bending.

INTRODUCTION

In his classic map of the Appalachian Mountains, Bailey Willis (1893) (Figure 1a) artistically emphasized one of the common characteristics of mountain belts—with few exceptions, structural "trend lines" (the map traces of folds hinges, faults, and fabrics) of mountain ranges curve in map view, creating sinuous topographic

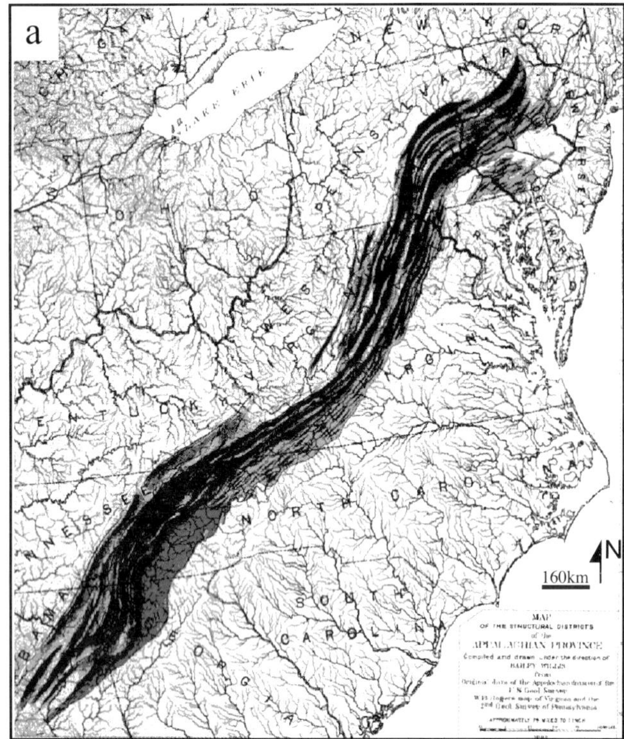

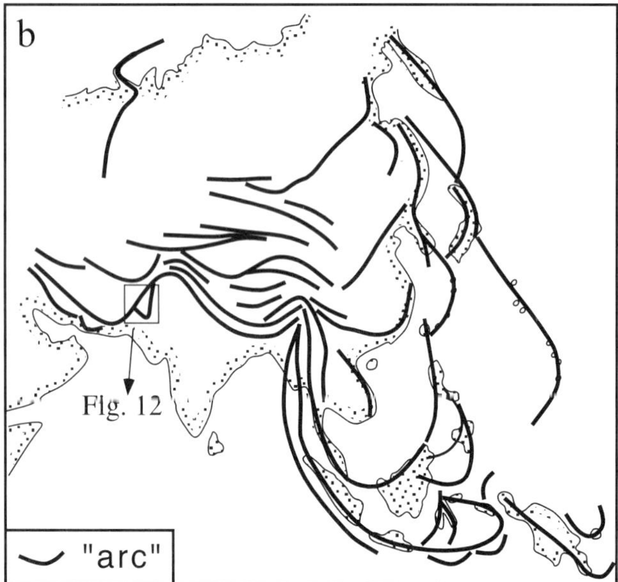

Figure 1. (a) Reproduction of a map of the Appalachian Mountains, eastern U.S.A., by Bailey Willis (1893). Willis used shading to emphasize the morphology of the fold-thrust belt in the days before digital topographic maps became available. The morphology vividly illustrates the salients and recesses of this orogen. (b) Sketch map of Asia, by William Hobbs (1914), illustrating the "arcs" of Asia. Redrafted following the original.

patterns. Numerous geologists have studied mountain-belt curves during the past two centuries, attempting to understand what their presence tells us about the process of orogeny (Figure 1b). Fortunately, paleomagnetic analysis, field structural analysis, laboratory analog modeling, and computer simulations completed after the advent of plate-tectonics theory together have led to a general image of how and why curves form. These studies suggest that not all curves form for the same reason.

The purpose of this chapter is to review the variety of causes for curve formation, to characterize the diversity of structures found in curves, and, where possible, to define the relationship between deformation in a curve and the curve-formation process. I begin by addressing the somewhat confusing terminology used for discussions of curves in the literature, then briefly review the evolution of ideas concerning curves. Finally, I discuss modern interpretations of curves. The paper focuses on curves that involve fold-thrust belts, that is, portions of orogens in which the curving structures are largely "thin-skinned," meaning that their presence does not require bending of the entire crust or lithosphere below. It is clear that various processes can yield curves of similar appearance and that many (possibly most?) fold-thrust belt curves do not involve bending of once-straight orogens around a vertical axis. This chapter draws heavily on work carried out by University of Illinois graduate students.

TERMINOLOGY

Over the years, geologists have used a variety of names for the map-view curves of mountain belts, so to compile a bibliography on curves, one needs to search for many keywords, including: orocline, arc, syntaxis, curve, bend, virgation, salient, festoon, arcuate range, oroflex, and recess. To avoid confusion in this chapter, I first outline the terminology to be used. This terminology represents a modification of that used by Marshak (1988).

A regional progressive change in an orogen's structural trend is a *map-view curve*, regardless of the origin of the change (Figure 2). A *salient* (= an *antitaxial curve*) is a curve that is convex in the direction of transport, while a *recess* (= a *syntaxial curve*, or *syntaxis*) is a curve that is concave in the direction of transport (Figure 2a). In curves defined by structural trends in fold-thrust belts, structures overall verge toward the continental interior. Thus, in orogenic forelands, salients protrude into the continental interior relative to recesses, whereas in accretionary prisms of subduction systems, salients protrude toward the ocean relative to recesses. Salients and recesses are not synonymous with promontories and reentrants—a *promontory* is a protrusion of a continental margin toward the ocean, and a *reentrant* (or, alternatively, an embayment; Thomas, 1983) is an indentation or large bay along a continental margin (Figure 2b).

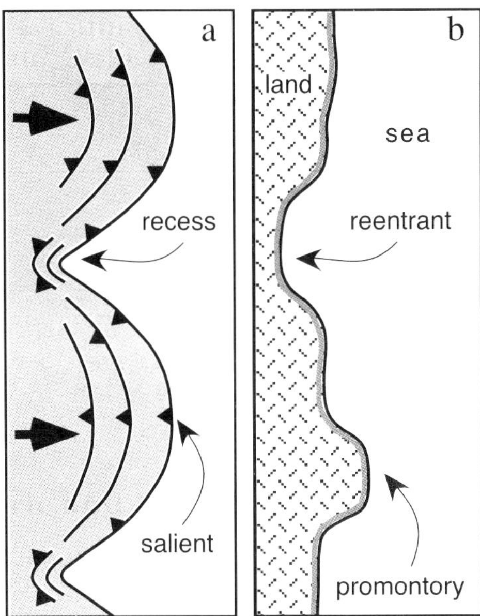

FIGURE 2. Basic terminology for describing map-view curves. (a) The contrast between salients and recesses. Arrows indicate the regional transport direction.

I distinguish salients from recesses solely on the basis of their geometry, therefore these terms can be used without knowledge of the displacement paths leading to curve formation. With cases in which we do have kinematic knowledge of curve formation, we can distinguish between rotational and nonrotational curves on the basis of whether, during curve formation, segments of the belt rotated or did not rotate, respectively, around an imaginary vertical axis (Figure 3). A *rotational curve* (= an *orocline*) forms by progressive bending of once-straight structural trends, so segments of an orogen necessarily rotate around a vertical axis (Figure 3a). A *nonrotational curve* forms without bending of a once-straight orogen. Thus, structural trend lines in a nonrotational curve initiate with a curved trajectory, and their formation does not involve rotation around a vertical axis (Figure 3b) (Marshak, 1988). Note that the term "orocline," introduced by Carey (1955) and used incorrectly by many authors in reference to any curve, should be used only for rotational curves. *Thin-skinned curves* involve rocks above a subhorizontal detachment in the crust, while *thick-skinned curves* involve structures that root in deep basement or involve the entire crust (Gwinn, 1967; Marshak, 1988).

To describe the geometry of a specific curve, Macedo and Marshak (1999) introduced a nongenetic nomenclature. In this chapter, I use only some of these terms (Figure 4). Specifically, the *leading edge* of a curve is the trace of the deformation front, the *apex* of the curve is the point at which the curve has the smallest radius of curvature, the *endpoints* of a curve are two reference points whose positions delimit the boundaries of the curve, the *reference line* (RL) is a chord drawn between two endpoints, and the *amplitude* (A) is the distance between the center of the reference line and the apex of the leading edge.

An examination of fold-thrust belt curves around the world reveals that curve shapes vary greatly. As illustrated by Figures 5 and 6, leading edges of salients, for example, can be parabolic, flat-crested, or irregular; leading edges can be symmetric or asymmetric; and trend line patterns within curves can be parallel, convergent at both endpoints, convergent at one endpoint, divergent, or truncated. Notably, salients are not strictly circular arcs, as many authors assumed following Sollas (1903), who thought of them as portions of

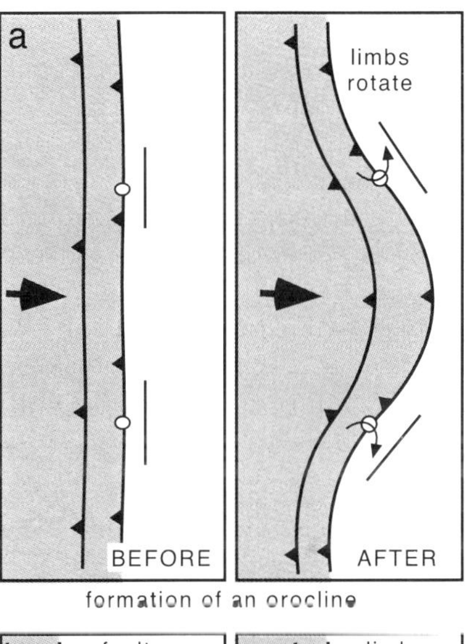

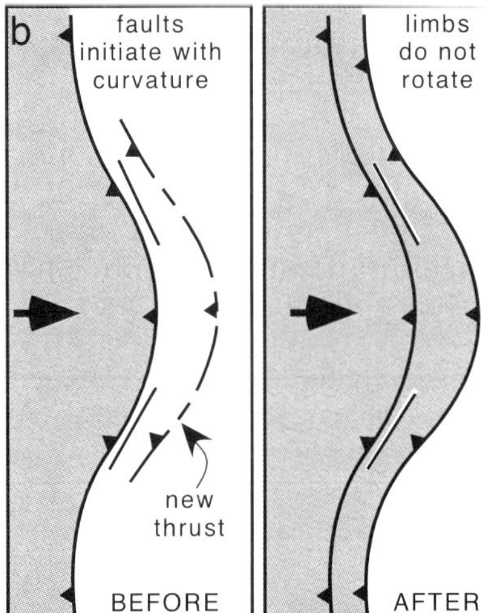

FIGURE 3. The contrast between rotational curves (oroclines) and nonrotational curves. (a) In rotational curves the trace of regional structures changes strike with progressive deformation. The short line segments represent the trends of the thrusts. (b) In nonrotational curves, structures initiate with a curved trace.

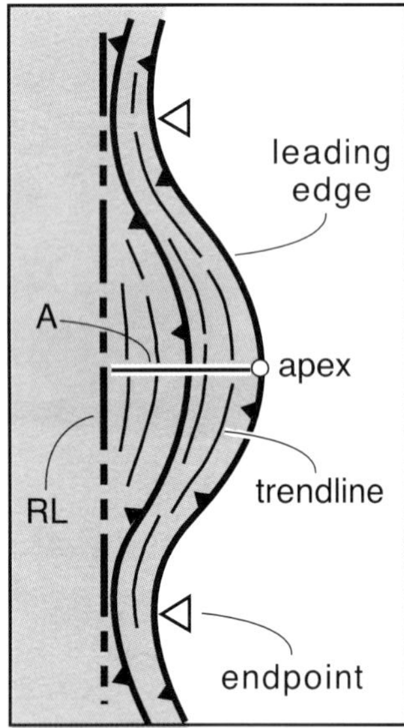

Figure 4. Geometric elements (endpoints, apex, amplitude, leading edge, trend lines) of a single curve. RL = reference line; A = amplitude.

"blisters" on the earth's surface. Recesses tend to be narrower and, in some cases, more angular than salients. Worldwide, amplitudes and reference-line lengths range through one to two orders of magnitude (see Macedo and Marshak, 1999).

HISTORICAL BACKGROUND

Interest in the curving trend lines of mountain belts dates back almost to the appearance of the first maps depicting the geography of mountains. Early geologists believed that trend lines provided a clue to the origin of mountains. For example, the eighteenth century French geologist, Elie de Beaumont, imagined that trend lines represented the geometry of the "fundamental flaws" in the earth's crust that controlled the position and orientation of mountain belts, features that he thought formed in response to vertical movements alone. Presumably, Elie de Beaumont assumed that curves represented the intersections of these flaws. In the nineteenth century, when regional geologic mapping efforts originated, geologists began to speculate that horizontal shortening may be involved in mountain-belt formation. For example, H.D. Rogers (1858) described curves of the Appalachians and made analogies between them and curvilinear ocean waves. Soon after, J.D. Dana (1866) discussed curve formation in his geology textbook. Dana, an adherent to an early version of geosyncline theory, in which mountain belts resulted from horizontal compression at the margin of ocean basins in response to the sinking of ocean basins, suggested that curves formed where mountain belts molded to the nonlinear margin of a preexisting craton. Although Dana did not understand the source of

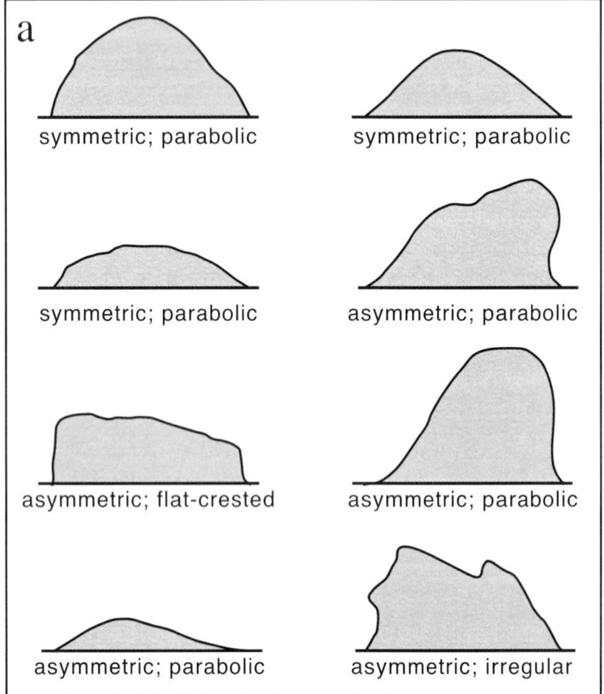

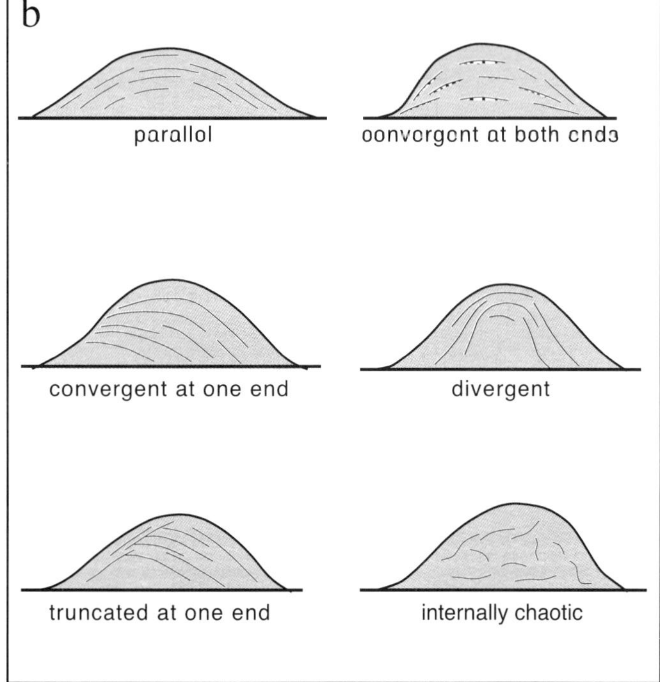

Figure 5. Shapes of salients (from Macedo and Marshak, 1999). (a) Contrasts of leading-edge shapes. (b) Patterns of trend lines within salients.

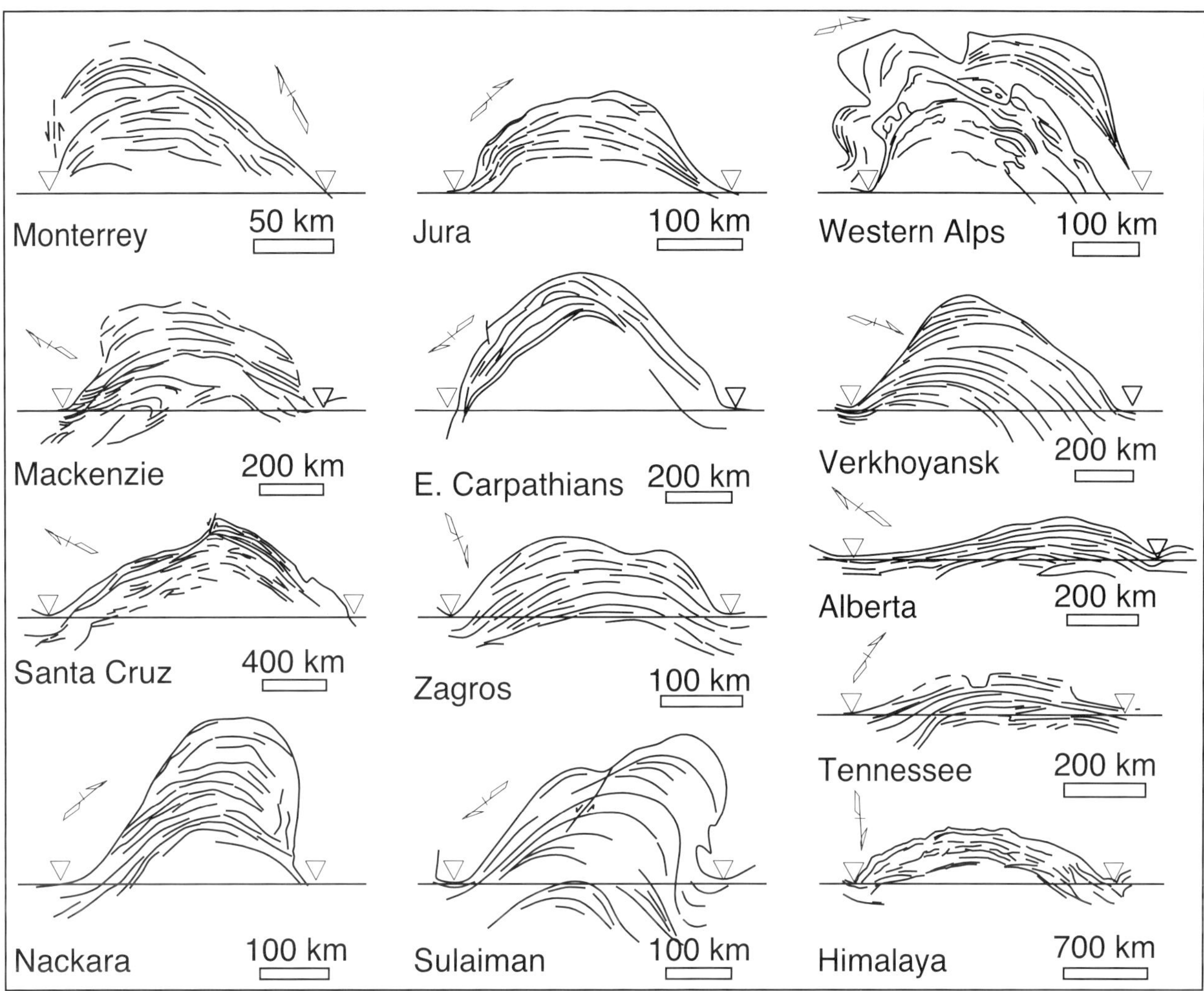

FIGURE 6. Examples of salient shapes (from Macedo and Marshak, 1999). Each image represents a tracing of fault and fold trend lines within the specified salient.

stress in orogeny, his recognition of the importance of promontories and reentrants in controlling the position of mountain-belt curves still holds today.

As the twentieth century dawned, geologists still struggled to understand the basic physical process of mountain building. Did mountains form strictly as a consequence of vertical displacements, or did the process involve significant horizontal displacements as well? Edward Suess and Emile Argand favored the latter idea and used the existence of regional map-view curves (which they called "virgations") in mountain belts as evidence to support their supposition (Suess, 1909; Argand, 1924). They argued that the formation of such curves required glacier-like horizontal flow of the rock comprising mountain belts, because they were greatly impressed by the similarity between the shape of curving regional trend lines in the Alpine orogen and the curving toes of valley glaciers. Argand (1924) went on to distinguish different kinds of mountain-belt curves on the basis of interpreting trend lines as flow lines (Figure 7a). The work of Suess and Argand clearly contains the seeds of understanding for thin-skinned deformation processes. Hobbs (1914; see also Taylor, 1910), however, suggested that instead of viewing the process as lateral spreading of the hanging wall, curves were better viewed in the context of the earth contracting as a consequence of footwall underthrusting. In stark contrast, Bucher (1924) suggested that curves follow the trace of tension cracks formed on an expanding earth.

In his classic book, *Origin of Continents and Oceans,* Alfred Wegener (1929) suggested that regional curves in orogens indicate that continents move. Specifically, he argued that the curves defined by the shape of southern South America, the Scotia Arc, and the Antarctic Peninsula formed because South America and Antarctica drifted west with respect to the Scotia Arc (Figure 7b). Thus, Wegener too used the curvature of orogens as evidence for mobility of the earth's outer layer.

In the succeeding decades, occasional papers focused on explaining the formation of map-view curves.

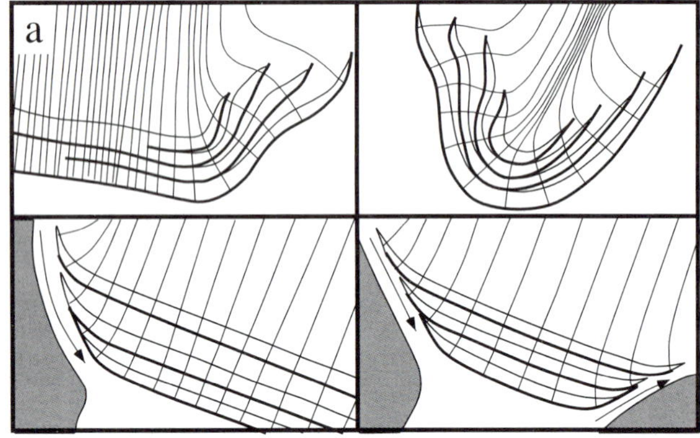

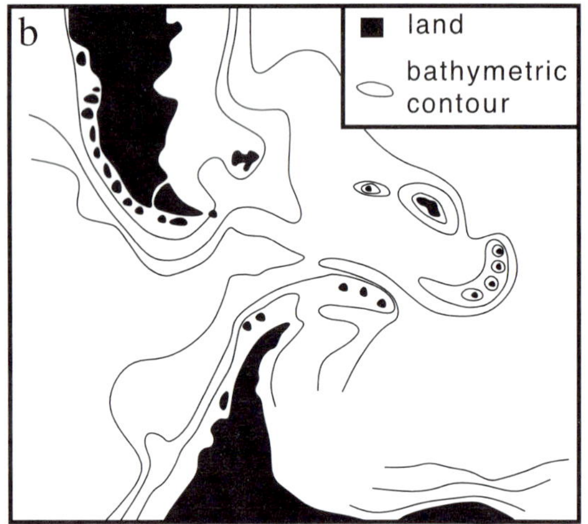

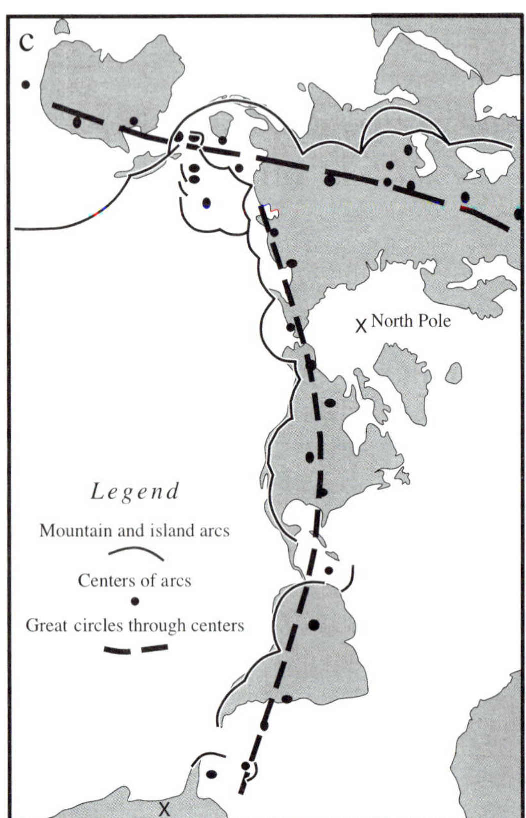

For example, Chamberlain and Shepard (1923) described the results of clay models, in which they simulated curve formation by pushing a rigid indenter into a clay cake. In 1949, J. Tuzo Wilson, later to achieve fame as one of the founders of plate tectonics, revitalized a hypothesis, first proposed by Lake (1931), that curves are small circles formed by the intersection of large thrust-fault planes with the spherical surface of a contracting earth. Wilson viewed curves to be segments of small circles on earth's surface (following Sollas, 1903) and went on to suggest that the centers of all these small circles align along two mutually orthogonal great circles (one running along the west coast of the Americas and one including the Alpine/Himalayan system). According to Wilson (1949, 1950), these great circles represent the principal zones of accommodation for earth's supposed contraction (Figure 7c). The "great circles" that Wilson noted would later be recognized in plate-tectonics theory as major zones of Cenozoic subduction.

In the mid-twentieth century, the curve formation issue once again gained attention, this time because of a paper by S. W. Carey (1955) in which he argued that curves form by the bending of orogens when crustal blocks rotate with respect to one another. Carey, like Wegener, used the presence of curved orogens as evidence for the mobility of the earth's solid surface. He coined the term "orocline" for a map-view curve in a mountain belt formed by physically bending once-straight structural trends.

Carey's work was the last major pre-plate-tectonics consideration of the curve issue. During the early 1960s, some authors addressed the formation of specific curves (e.g., Drake and Woodward, 1963, who incorrectly related curves in the Appalachians to movement on strike-slip faults; and Crosby, 1969, who suggested that

Figure 7. Illustrations from classic papers concerning curve formation. (a) Illustration from Argand (1924) showing his interpretation of curves as if they were formed by glacial-like flow. Each box represents a different curve type, as seen in map view. The heavy lines are faults, the enclosed are folds, and the thin lines are flow lines. The top two boxes represent unconfined flow, while the bottom two boxes represent confined flow, with the fold belt squeezing past or between shaded obstacles. The shaded areas represent foreland obstacles. Redrafted following the original. (b) Illustration from Wegener (1929), illustrating the curves in the Scotia Sea area. He argued that they formed as a result of the westward movement of South America and Antarctica. (c) J. Tuzo Wilson's (1949; 1950) concept of great circle lines of weaknesses, defined by the center points of circular arcs. Wilson, at the time, thought that the earth was contracting by thrusting along these great circles (dashed lines) and that the thrust planes intersected the spherical surface of the earth in circular arcs.

salients are due to radial gravitational-gliding in response to uplift; an idea implied also by Hobbs, 1914). However, only after geologists began to study subduction systems did the curve issue gain global significance once more. Specifically, geologists wondered why many "volcanic arcs" are, in fact, map-view arcs. Are they a consequence of the shape of the downgoing slab, which may pucker when it begins to subduct, like the indentation formed when you push in the surface of a ping-pong ball (Frank, 1968; Yamaoka et al., 1986)? Are they formed by buckling of downgoing slabs (Bayly, 1982)? Or did they form by progressive propagation of the subduction system into the space between subducting chains of seamounts (Voigt, 1973; Hsui and Youngquist, 1985)?

Study of map-view curves in continental orogens took a step forward when new tools became available for studying rock deformation. Strain analysis, kinematic analysis, paleomagnetic analysis, analog (sandbox and clay) modeling, and computer simulation each provide insight into the way rocks move and deform during the formation of a curve.

Paleomagnetic studies, in particular, focus on distinguishing rotational curves from nonrotational curves (Schwartz and Van der Voo, 1983; Bachtadse and Van der Voo, 1986; Eldredge et al., 1985; Klootwijk et al., 1985; Kent, 1988; Bossart et al., 1989; Butler et al., 1995; Gray and Stamatakos, 1997; Van der Voo et al., 1997; Muttoni et al., 1998). Paleomagnetism, in principle, provides a simple test of oroclinal bending. Ideally, oroclinal bending to form a rotational curve should result in the progressive change in orientation of magnetic declination in a given rock unit around the bend (Figure 8a, b). In contrast, the development of nonrotational curves, ideally, would not cause a progressive change in declination. Variation in declination around a curve can be plotted on a declination anomaly graph (Figure 8c) (Eldredge et al., 1985; Lowrie and Hirt, 1986). Unfortunately, paleomagnetic studies are not straightforward. Diagenetic remagnetization resulting from syntectonic brine migration commonly overprints paleomagnetization of rocks in fold-thrust belts (Bethke and Marshak, 1990; Pares et al., 1994), and this overprinting may occur during an intermediate stage of fold-thrust belt development (Stamatakos and Hirt, 1994). Also, as I discuss later, strain within thrust sheets may cause declination to rotate in a way that does not reflect rotation of the fault trace at the leading edge of a thrust sheet (Beck, 1987). Gray and Stamatakos (1997) discuss the complexity of interpreting the application of paleomagnetism to the study of curved orogens.

Structural geologists have focused on defining kinematic and strain patterns in orogenic curves, because data on these patterns potentially define the movement that took place during curve formation. To answer certain questions, kinematic analysis involves defining variations in transport directions as a function of location in a curve. Is there a radial pattern of transport around a curve, or do transport directions trend uniformly in the same direction around a curve? Does the transport direction on a given fault change as a curve evolves, or is transport coaxial on a given fault throughout development of a curve? Ideally, kinematic analysis defines displacement-path trajectories in a curve (Figure 9), which in turn constrain models of curve formation for a given belt (Marshak, 1988; Platt et al., 1989; Marshak and Tabor, 1989; Wise and Werner, 1991; Marshak and Flöttmann, 1996).

Strain analysis involves determining whether the finite strain ellipsoid varies in shape and/or orientation around a curve and determining whether, during curve

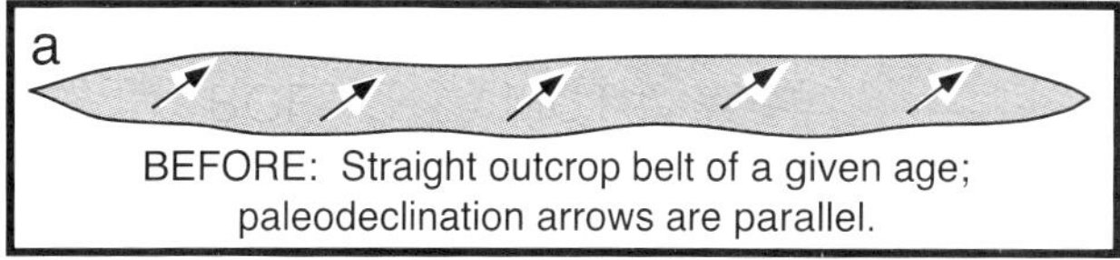

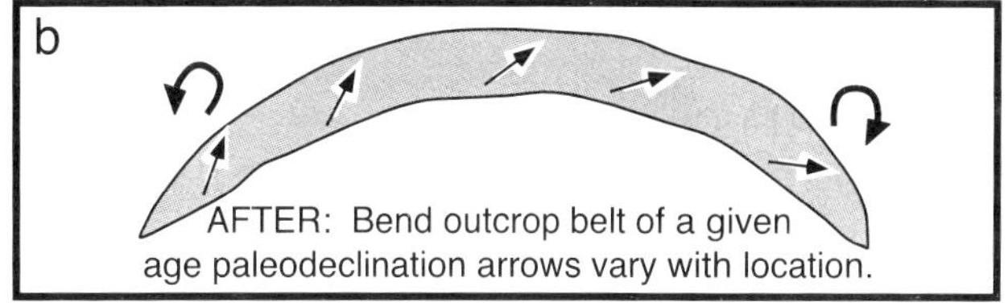

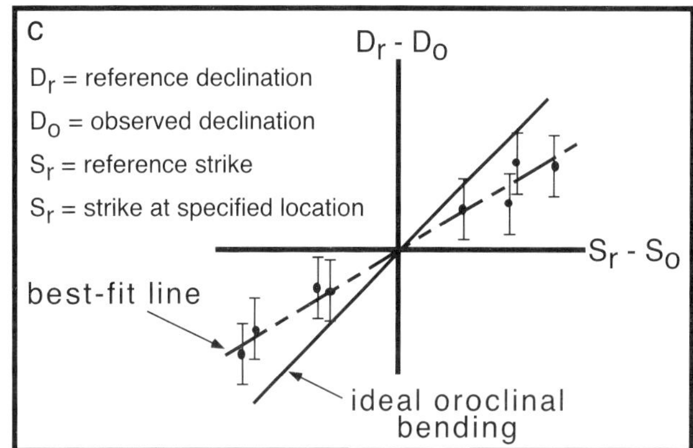

FIGURE 8. Illustration of the consequences of oroclinal bending on the pattern of magnetic declination in a thrust belt. The shaded area represents the outcrop belt of a given unit, in map view, and the arrows represent paleomagnetic declinations. (a) Before bending, all arrows point in the same direction. (b) Bending causes the arrows to change trend along the length of the bend. (c) A graphical comparison of the change in strike vs. the change in declination, by Eldredge et al. (1985). In an "ideal orocline" the amount of bending coincides exactly with the divergence between the strike angle and the reference line of the curve. In most real examples, data define a best-fit line (calculated by linear regression) whose slope is less than that of the line for "ideal oroclinal bending." This difference indicates that rotations of declination are generally less than the angle between strike and the reference line.

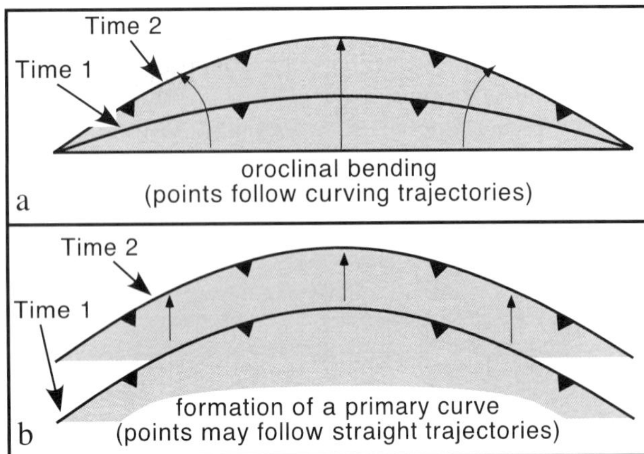

FIGURE 9. Possible displacement-path trajectories in curves shown in map view. The arrows indicate the path of particles during the evolution of the thrust trace from its position at Time 1 to its position at Time 2. Particles must follow curved trajectories in oroclines, but not in nonrotational (primary) curves.

development, curves undergo tangential extension (extension parallel to their trace). As Ries and Shackleton (1976) discussed, the strain distribution in a curve varies depending on whether the curve forms by variations in the amount of foreland transport along strike or by bending of a once-straight belt between two fixed endpoints (Figure 10). Ferrill and Groshong (1993) studied strain distribution to interpret the curves of the Subalpine Chain in France.

Beginning with the work of Chamberlain and Shepard (1923), geologists have used modeling techniques to simulate curve formation and gain intuition into the factors controlling the geometry of curves (Davy and Cobbold, 1988; Marshak et al., 1992; Haq-Saad and Davis, 1997; Lallemand et al., 1992; Zweigel, 1998; Macedo and Marshak, 1999; Lickorish et al., 2002). Most modeling work in the past two decades has used sand, because at the scale of a laboratory model, sand simulates the behavior of Coulomb materials such as the uppermost crust (Coloumb material = a granular material whose bulk strength is provided by friction between grains). Sandbox-model designs generally involve either impressing a backstop into the sand to drive the formation of a thrust wedge, or pulling the mylar substrate that underlies a sand layer beneath a backstop or through a slit in the floor of the sandbox. Some authors (e.g., Marshak et al., 1992) have also used viscous fluids, like chilled glycerine, to study curve formation. Curve formation has also been modeled using computers (Marshak et al., 1992; Macedo, 1997).

In light of the long history of study described above, geologists have defined a great variety of curve-forming processes in fold-thrust belts. I describe these processes below and, where possible, provide an example of the process potentially explaining a real orogenic curve.

BASIN-CONTROLLED CURVES

Several authors have noted that the positions of map-view curves in fold-thrust belts correlate with predeformational characteristics of the sedimentary basin from which the curve formed (Aitken and Long, 1978).

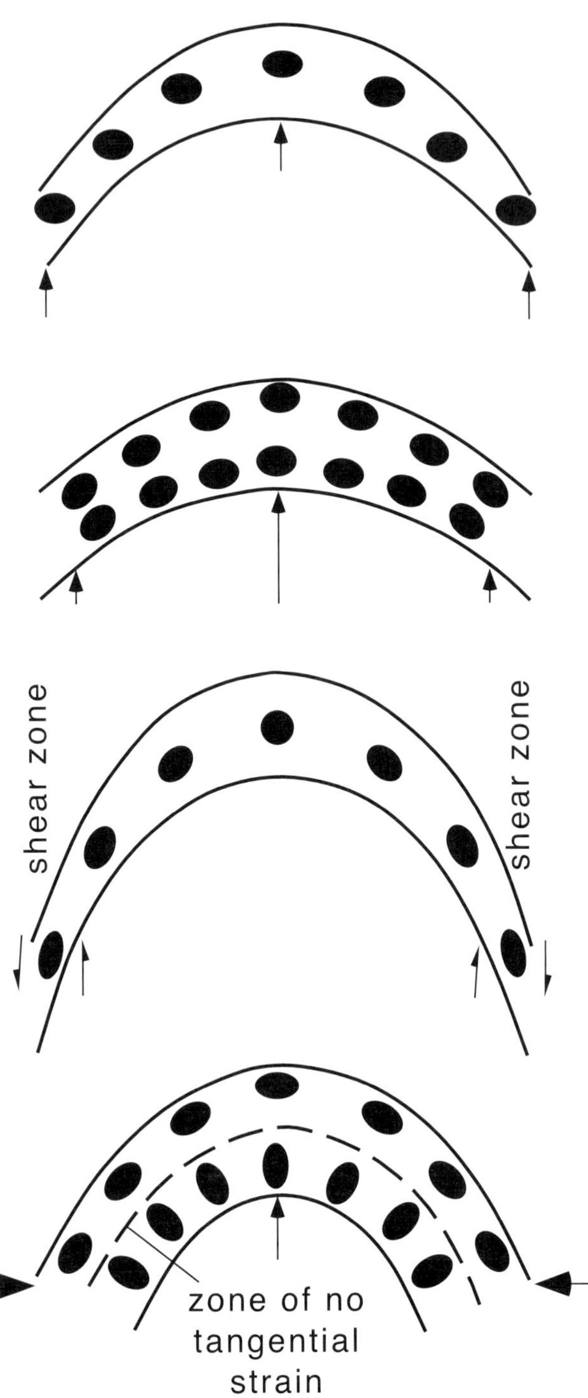

FIGURE 10. Illustration from Ries and Shackelton (1976) illustrating possible strain fields in curves. Redrafted from the original.

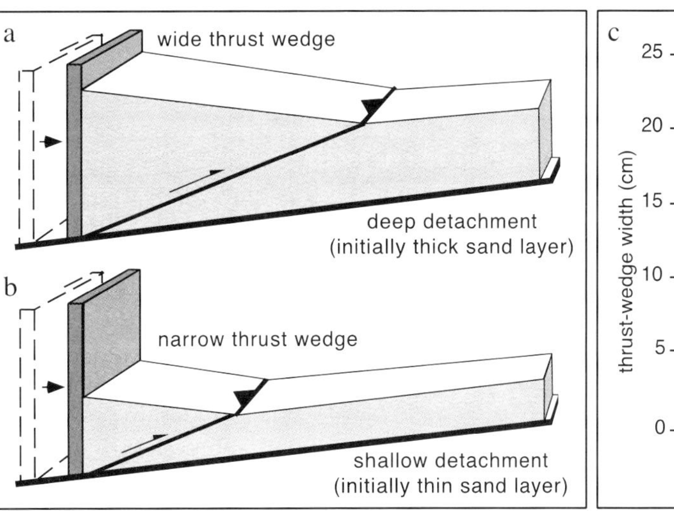

FIGURE 11. The effect of depth-to-detachment (i.e., sediment thickness) on the width of a fold-thrust belt. (a, b) Cross sections of sand models for different original distances between the sand surface and the detachment (modified from Macedo and Marshak, 1999). (c) The observed linear relationship between the distance between the backstop and the trace of the first-formed thrust, and the pre-deformational thickness of sand (from Marshak and Wilkerson, 1992).

This relationship occurs because predeformational characteristics determine the width of a fold-thrust belt for a given amount of upper-crust shortening. This relationship can be understood in the context of critical-taper theory (Davis et al., 1983; Boyer, 1995; Mitra, 1997). Specifically, along-strike changes in the value of the critical-taper angle, the slope of the basal detachment, or the thickness of the sedimentary layer being deformed determine whether a fold-thrust belt is wide or narrow. Wider portions become salients and narrower portions become recesses (Marshak et al., 1992; Marshak and Wilkerson, 1992; Calassou et al., 1993; Boyer, 1995; Mitra, 1997; Macedo and Marshak, 1999). Fold-thrust-belt curves whose position and geometry reflect characteristics of predeformational sedimentary basins are called basin-controlled curves (Macedo and Marshak, 1999).

Along-strike Change in the Depth to Detachment

Macedo and Marshak (1999) examined the shape and geologic setting of many salients around the world (Figure 6) and noted that, in the majority of cases, the apex of the salient coincides with the predeformational depocenter in the sedimentary basin from which the salient formed. This association suggests that curves in fold-thrust belts develop in response to along-strike variations in predeformation sediment thickness—specifically, salients form where sediment thickness is greatest, while recesses form where sediment thickness is less. For example, salients tend to form over deeper passive-margin basins in reentrants and tend to propagate up the axis of rifts trending at a high angle to the fold-thrust belt. Recesses form over promontories. Fold-thrust-belt width depends on depth to detachment simply to maintain volume balance for a given critical taper angle, as simple sandbox models illustrate (Figure 11) (Marshak and Wilkerson, 1992).

The relationship between thicker sedimentary successions and salients can be illustrated in a number of locations around the world (Macedo, 1997). For example, the Pennsylvania salient of the Appalachians formed from a particularly deep portion of the lower Paleozoic eastern North American passive margin (Gray and Stamatakos, 1997; Macedo and Marshak, 1999). The Sulaiman salient of Pakistan formed above a particularly deep portion of a basin (Figure 12) (Davis and Lillie, 1994; Macedo and Marshak, 1999).The Nackara salient of the Adelaide Fold-thrust Belt, South Australia, formed along the axis of a rift trending at a high angle to the fold-thrust belt (Marshak and Flottmann, 1996). The Espinhaço salient of the Araçuaí Belt on the eastern margin of the São Francisco Craton in Brazil coincides with the axis of a rift at a high angle to the belt; and the Helena salient forms where the Sevier Thrust Belt of the western U.S.A. propagates up the axis of a Proterozoic rift trending at a large angle to the trace of the thrust belt. In the case of accretionary prisms along subduction zones, places where the trench has filled with sediment form a wider thrust wedge (prism) than do places where the trench is starved of sediment.

In some locations, the thickness of predeformational strata reflects the structural architecture resulting from an earlier deformational event. For example, in the Quadrilátero Ferrífero of Brazil, Archean and Mesoproterozoic deformation produced a dome-and-keel province in which basement domes are bordered by deep, locally synformal troughs. Neoproterozoic thrusting ($\approx$0.5 Ga) produced salients that telescope strata up the axis of the synformal troughs in places where the troughs lie parallel to the direction of thrusting (Alkmim and Marshak, 1998).

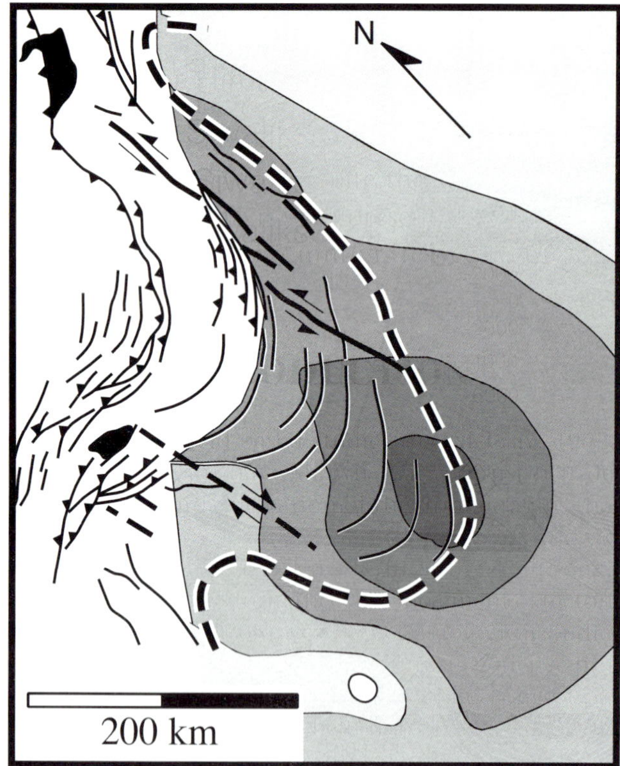

Figure 12. Simplified map of the Sulaiman salient, Pakistan. Location shown in Figure 1b. (adapted from Macedo and Marshak, 1999). The shading represents thickness of predeformational sediment; darker shading is thicker sediment. Note that the apex of the curve coincides with the depocenter. The heavy dashed line is the leading edge of the curve.

Along-strike Changes in Strength of the Detachment

As Davis and Engelder (1985) and Jaumé and Lillie (1988) discussed, the width of a fold-thrust belt depends on the strength of the detachment horizon, because the critical-taper angle decreases as the strength of the detachment horizon decreases. For a given amount of crustal shortening, a portion of a fold-thrust belt that overlies a strong detachment, and that therefore can sustain a larger critical-taper angle, will be narrower than a portion of a fold-thrust belt that overlies a weak detachment (Calassou et al., 1993). Thus, fold-thrust belts protrude farther into the foreland and form salients above a particularly weak glide horizon, such as an evaporite. Because evaporite glide horizons commonly occur in conjunction with thicker sedimentary successions, their contribution may amplify the contribution to a salient's development that is provided by the thickening of a sedimentary section. Places where detachment occurs on midcrustal glide horizons may yield particularly wide orogenic wedges, if partial melting occurs along the glide horizon.

The Sulaiman salient (Figure 12) is an example of the presence of a sedimentary glide horizon localizing a salient (Davis and Lillie, 1994). Other examples include the Zagros Mountains of southern Iran, where the widest portions of the fold-thrust belt coincide with the regions of the belt in which salt diapirs occur (Talbot and Alavi, 1996); the Monterrey salient of the Cordillera Occidental, Mexico (Marrett, 1995; Melnyk and Cameron, 1998), the Salt Range of Pakistan (Jaumé and Lillie, 1988), and the Jura Mountains of Switzerland.

Along-strike Changes in Slope of the Basal Detachment

Changes in the dip of the basal detachment cause the width of the fold-thrust belt to vary. In places where the basal detachment dips more steeply, the thrust belt widens (Boyer, 1995; Mitra, 1997). This relationship occurs because the critical-taper angle of a wedge is the sum of the basal-detachment slope and the surface slope of the wedge. Thus, where the basal-detachment slope is shallow, foreland translation of the backstop causes internal deformation of the thrust wedge before the wedge can become supercritical, and thus before the toe of the wedge can move into the foreland. Thus, the wedge is narrow. But where the detachment slope is steep, the wedge can be supercritical even before the backstop starts to move. Thus, displacement of the backstop immediately causes the wedge to undergo stable sliding into the foreland, and a wide wedge develops (Figure 13). Because an increase in detachment slope angle also causes predeformational basin thickness to increase, the detachment slope angle works, in conjunction with an increasing depth to detachment, to cause a wide wedge.

Boyer (1995) and Mitra (1997) show that along-strike changes in detachment dip may explain the origin of the Wyoming and Provo salients of the Sevier fold-thrust belt in Utah (U.S.A.). They also note that the slope angle may reflect the amount of flexural subsidence of the continental lithosphere that occurs as a result of emplacement of a thrust load, and thus may depend on the flexural strength of the lithosphere beneath the basal detachment, not just on the thickness of predeformational strata.

Along-strike Changes in Strength of the Thrust Wedge

Above, I noted the effect of changing strength of the basal detachment on the width of a thrust wedge, and therefore on the localization of curves. As Boyer (1995) and Mitra (1997) noted, the strength of the rock or sediment comprising the wedge also affects wedge

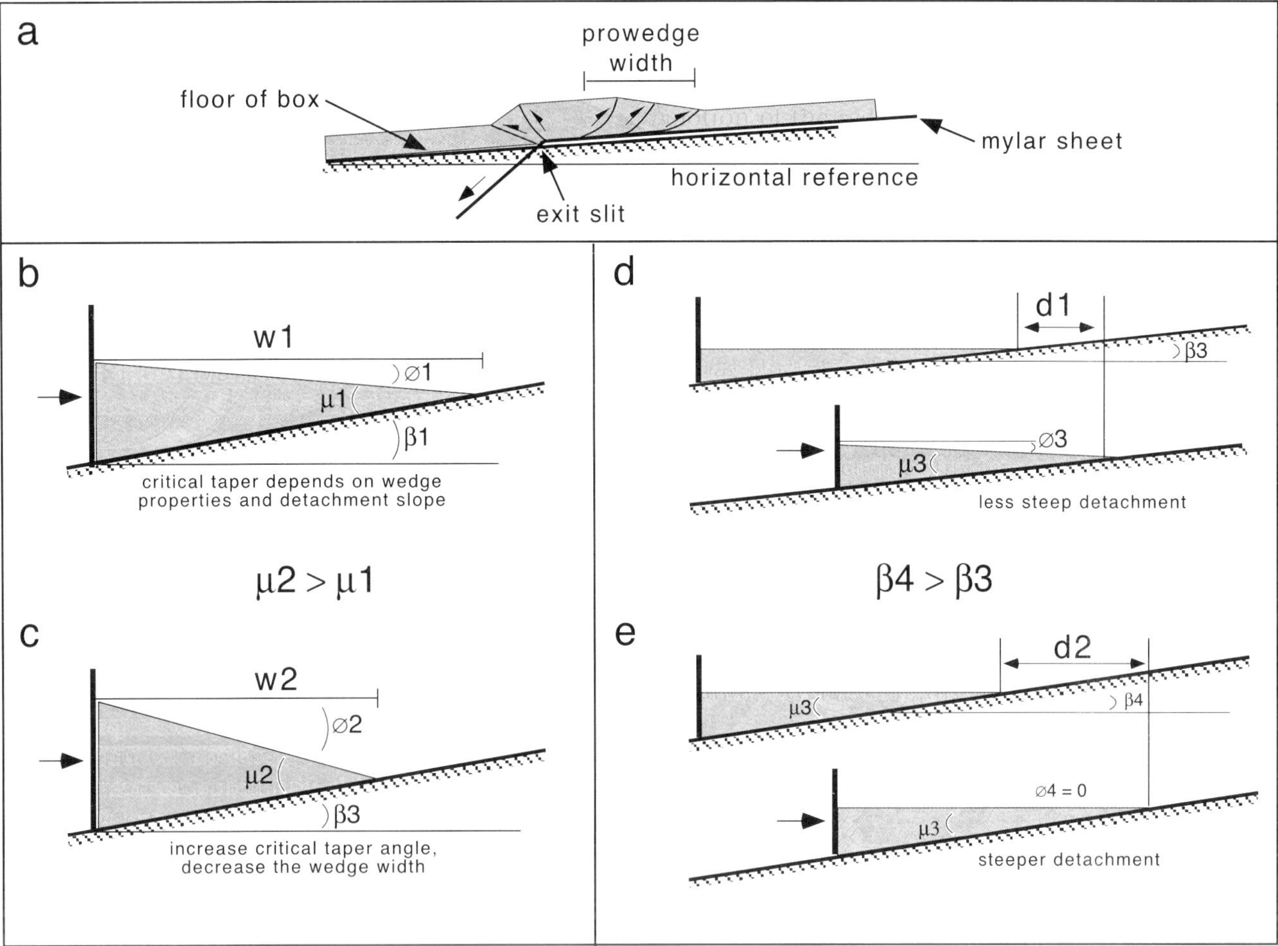

Figure 13. Effects of wedge strength and detachment slope on thrust wedge width. (a) A bivergent thrust wedge formed in sand above a mylar sheet being drawn through a slit in the wood base beneath the mylar. (Marshak et al., 1992; see Willett et al., 1993, for a discussion of bivergent wedges). The following models represent the prowedge of a bivergent sand wedge. (b, c) The width of a wedge depends on the value of the critical-taper angle (μ), which in turn reflects the strength of the wedge or the strength of the basal detachment. A larger critical-taper angle (μ2) leads to the development of a narrower wedge. (d, e) An increase in the detachment slope (β) leads to an increase in the distance that the thrust wedge propagates to the foreland. If the detachment angle is shallow, the thrust wedge thickens internally and thus propagates only a short distance into the foreland for a given displacement of the backstop. If the detachment angle is steep, the wedge does not thicken internally, and thus it propagates a long distance into the foreland for a given displacement of the backstop. w = thrust-wedge width; d = displacement distance ; μ = taper angle; β = detachment slope; ø = surface slope.

width, because strength controls the critical-taper angle. Wedges made of strong rocks have a greater taper angle than wedges made of weak rocks; thus, wedges made of weak rocks propagate relatively farther into the foreland for a given amount of regional shortening (Figure 13). For example, an along-strike change in facies within the predeformational basin that causes an along-strike change from well-indurated sandstone to weak organic shale will lead to the development of curvature, and places where thrusting involves basement tend to result in narrower wedges.

Conceivably, along-strike variations in heat flow that cause along-strike changes in rheology, even in the absence of changes in lithology, can also lead to curve formation. Where rocks are hotter and weaker, thrust wedges have a smaller critical-taper angle. Thus, for a given amount of crustal shortening, the thrust wedge will be wider.

Along-strike Changes in Erosion and Plutonism

Changes in the thickness of the hinterland portion of a thrust wedge can be caused by phenomena that are independent of the predeformational nature of the sedimentary basin. During thrust-wedge growth, the volume of the wedge, or the mass distribution in the

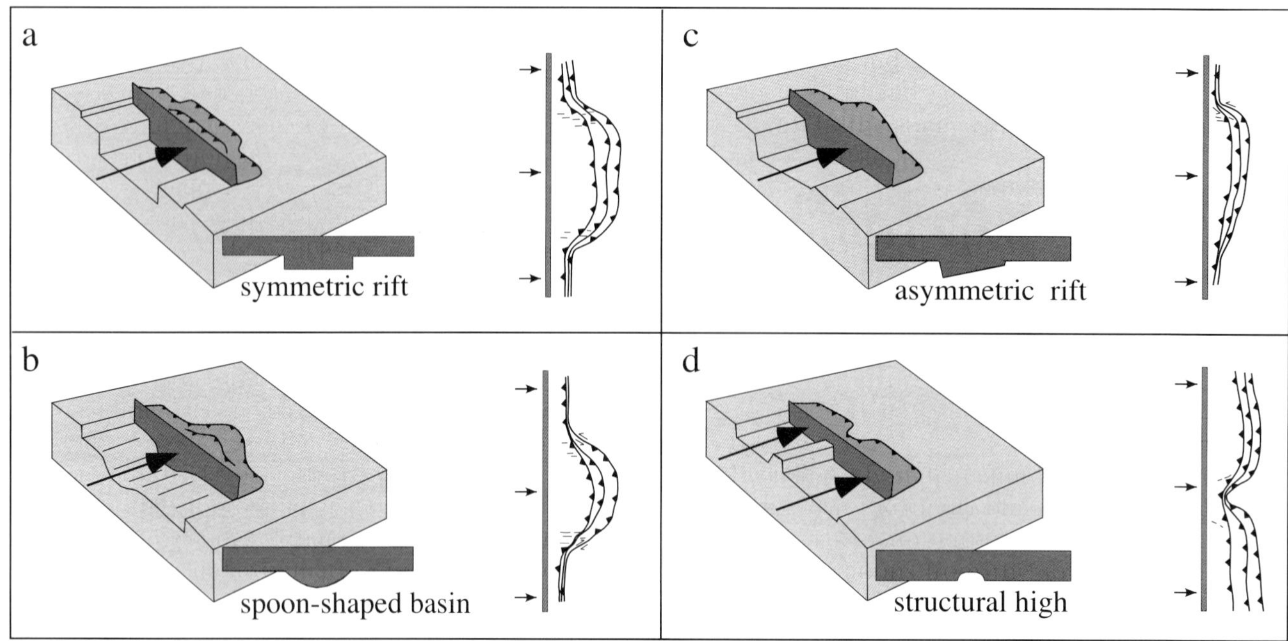

FIGURE 14. Illustration of how the geometry of a thrust-belt curve reflects the geometry of the underlying basin (from Macedo and Marshak, 1999). Sketches to the right of each figure illustrate the traces of faults formed in a sandbox model. (a) A steep-walled aulacogen yields a flat-crested salient. (b) A spoon-shaped basin yields a parabolic salient. (c) An asymmetric aulacogen yields an asymmetric salient. (d) A basement obstacle yields two salients with an intervening recess.

wedge, may change because of erosion. For example, erosion of strata in the hinterland of the wedge and simultaneous accumulation of sediment at the toe of the wedge will decrease the surface slope of the wedge, and therefore will cause the wedge to become subcritical. As a consequence, the wedge will thicken internally, and foreland propagation will stall. Thus, portions of the thrust belt where erosion is slower conceivably may evolve into salients.

Similarly, addition of rock to the hinterland portion of a wedge by intrusion of igneous masses will effectively increase the surface slope and cause the wedge to become supercritical, thereby leading to the foreland advance of the toe of the wedge and the formation of a salient. Intrusion of the Boulder Batholith into the Sevier Fold-thrust Belt may have contributed to the foreland propagation of the Helena salient in Montana.

Shape of Basin-controlled Curves

Sandbox models illustrate that the map-view geometry of a basin-controlled curve reflects the details of basin shape (Figure 14) (Macedo and Marshak, 1999). For example, basins that thin laterally at a gradual rate yield parabolic salients; basins that terminate abruptly at both lateral margins yield flat-crested salients—typically bordered by lateral ramps (which appear as transverse zones in orogens; according to Thomas, 1990; Paulsen and Marshak, 1998a); and basins that terminate abruptly at one end and taper gradually at the other yield asymmetric salients. This relationship can be illustrated by examining the relations on the margin of the Uinta recess (Figure 15) (Paulsen and Marshak, 1999). If the leading edge of a curve encounters local depocenters or patches of a weak glide horizon, the leading edge bulges forelandward locally, creating an irregular leading edge.

Map-view geometry also reflects the direction of convergence during fold-thrust belt development (Macedo and Marshak, 1999). Specifically, convergence perpendicular to the hinterland edge of the basin yields symmetric basins, whereas convergence at an angle to the hinterland edge of the basin yields asymmetric basins (Figure 16). Lateral ramps develop on the margin of salient parallel to the transport direction.

Regardless of the specific aspects of a basin that control differential forelandward propagation of a fold-thrust belt, and therefore localize salients, the trend lines of basin-controlled salients tend to converge toward the endpoints of the salient. In other words, fault traces or fold-hinge traces are more widely spaced at the apex of a basin-controlled salient than at its endpoints. A map of the Sulaiman salient illustrates this point (Figure 12). Note also that thrusts tend to be emergent toward the endpoints and blind toward the apex. The map of the Nackara salient in South Australia shows the same relationship (see Marshak and Flöttmann, 1996).

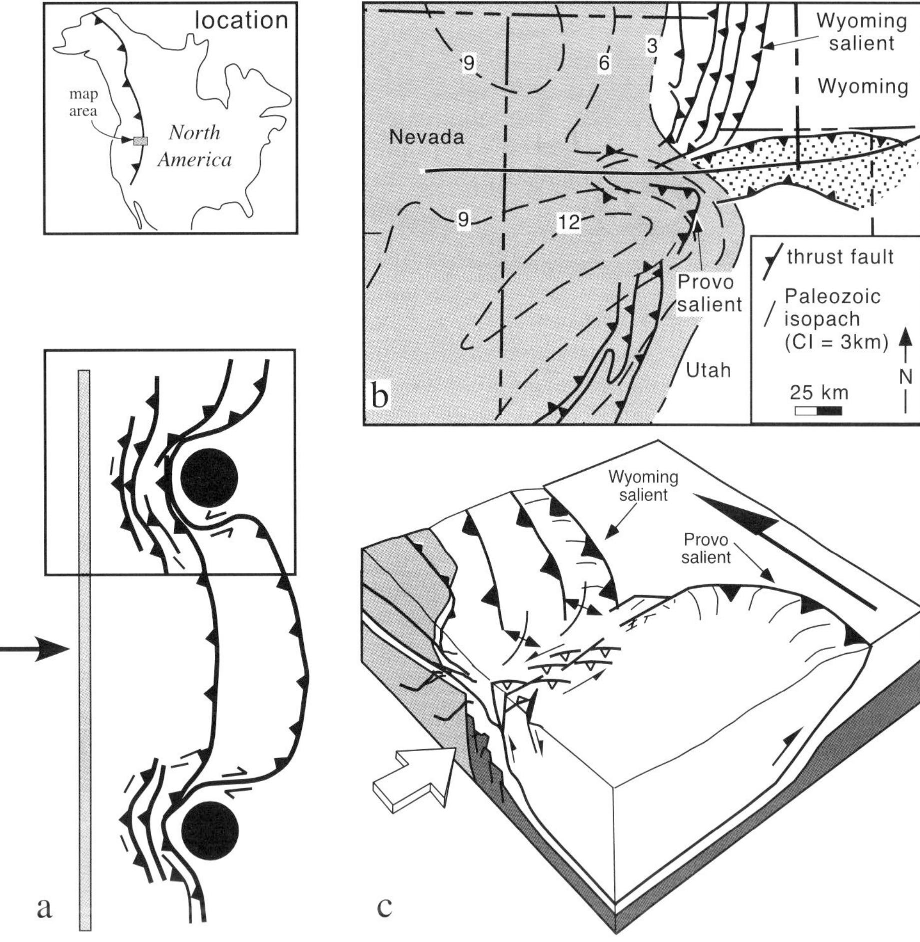

FIGURE 15. A model of recess formation. (a) A map-view sketch of a sandbox model, illustrating a recess forming against obstacles, with a salient forming between the obstacles (from Macedo and Marshak, 1999). The obstacles were metal cylinders buried in the sand. The box represents the area shown in (b). (b) Map of central Utah (inset shows location), illustrating the Uinta recess, formed adjacent to the Uinta Mountains, an uplifted wedge of Precambrian quartzite (from Paulsen and Marshak, 1999). (c) A block showing the contrast between the north side of the Uinta recess and the south side of the Uinta recess, in relation to an asymmetric pre-deformational basement high. Note that the southern limb of the recess, formed over an abruptly deepening basin, evolved into a strike-slip fault while the north limb of the recess, formed over a gently tapering basin, did not but rather formed a gently curving salient (from Paulsen and Marshak, 1999).

a

b

frontal convergence

oblique convergence

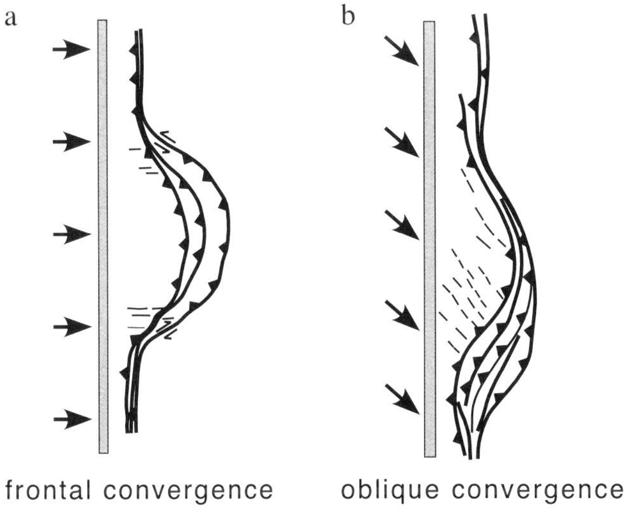

FIGURE 16. (a) Frontal convergence causes a symmetrical salient. (b) Oblique convergence causes an asymmetric salient. One limb of this salient parallels the convergence direction (from Macedo and Marshak, 1999).

CURVES RESULTING FROM IRREGULARITIES ON COLLIDING MARGINS

Indenter-controlled Curves

The impression of a promontory or of an exotic crustal block of limited along-strike length into a sedimentary basin generates a salient to the foreland of the collision, as simple sandbox models illustrate (Figure 17). Impression of an irregular backstop, with promontories and recesses along its length, generates salients and recesses, with salients opposite the promontories and recesses opposite the reentrants on the overriding plate. These indenter-controlled curves develop because of differential shortening along the length of the irregular margin. Shortening begins in advance of a colliding promontory or exotic block before it begins in the surrounding regions. Therefore, the thrust belt has already propagated toward the foreland of the downgoing plate adjacent to the promontory before it begins to propagate in the surrounding regions. Note that, as a consequence, structural trend lines converge at the

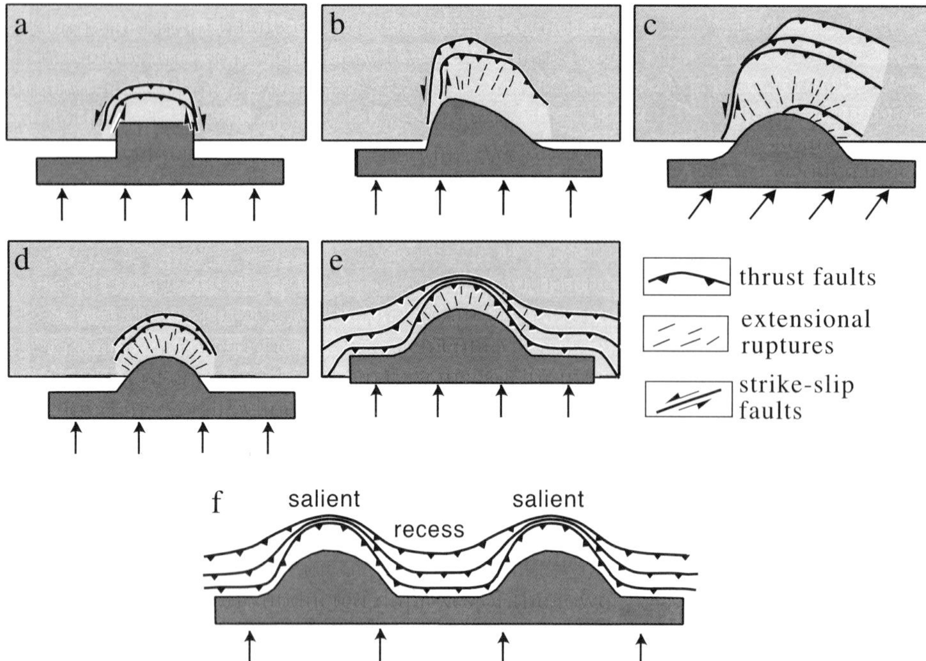

Figure 17. Sketches of sandbox models illustrating the consequences of impressing a hinterland indenter into a sedimentary basin. (a) Frontal convergence of a rectangular indentor. (b) frontal convergence of an asymmetric indentor. (c) Oblique convergence of a smooth indentor. (d) Partial indentation by a smooth indentor. (e) Complete indentation by a smooth indentor—note that structural trend lines converge toward the apex and diverge toward the end points. (f) Linked salients and recesses formed along the length of an indenter with promontories and reentrants (adapted from Macedo and Marshak, 1999).

apex of the curve, and in this way, they contrast with basin-controlled salients (Macedo and Marshak, 1999). As Laubscher (1972) discussed and as civil engineers recognize by studying building foundations, stress trajectories curve in advance of an indenter, and thus curving faults develop (Figure 18).

The shape of the curve resulting from an indenter depends on the indenter's shape and its convergence direction (Figure 17). For example, sandbox models suggest that a rectangular indenter generates a flat-crested salient, whereas a rounded indenter generates a parabolic salient (Macedo and Marshak, 1999). Indenters that collide obliquely generate asymmetric curves, with the side of the salient that parallels the convergence direction evolving into a transpressive fault zone (Dominguez et al., 1999; Lu and Malavieille, 1994). Recent studies have focused on the complexities that form at the corners of indenters and on the effects of rotating the indenter. Zweigel (1998) documented a relationship between the width of the thrust belt, on the frontal and lateral margins of the curve, and the indentation angle (which he defines as the angle between the indentation direction and the normal to the frontal face of the indenter). He also noted that displacement vectors fan around corners, although not by as much as the change in trend of thrust traces. Lickorish et al. (2002) simulate the shape of the western Alps arc by rotating a rectangular indenter around a vertical axis during indentation.

The Indian/Asian collision is, perhaps, the most widely recognized example of indentation (Tapponnier and Molnar, 1976). The rigid craton of India collided with the soft margin of Asia, resulting in the formation of the Himalayas and the Tibet Plateau (Figure 19). Note that major syntaxis formed at each endpoint of the Himalayan arc. This collision can be modeled by a tapered indenting block that underthrusts a soft layer, generating a doubly vergent thrust wedge (Davy and Cobbold, 1988). One side of this wedge forms a salient that arcs into the interior of the soft layer, and one side forms a salient that verges onto the surface of the tapered indenter. The former of these two salients simulates Tibet, which eventually involved a thermally weakened crust and became a broad plateau, and the latter represents the Himalayas. Continued indentation of India into Asia transported both curves toward the interior of Asia, thereby necessitating the formation of transpressional boundaries on either side of the Indian subcontinent. These boundaries terminate in the syntaxes forming at the end of the Himalayan salient.

Margin-controlled Curves

A margin-controlled curve is one whose presence reflects the presence of promontories and recesses to the foreland of the advancing thrust belt. As noted earlier, J.D. Dana (1866) suggested that the sinuousity of the Appalachians developed because mountain belts wrap around the promontories and recesses of the underthrusting North American continental margin—recesses form opposite promontories, and salients form opposite reentrants (Figure 20). This idea was updated by Rankin (1976), Thomas (1977), and Thomas and

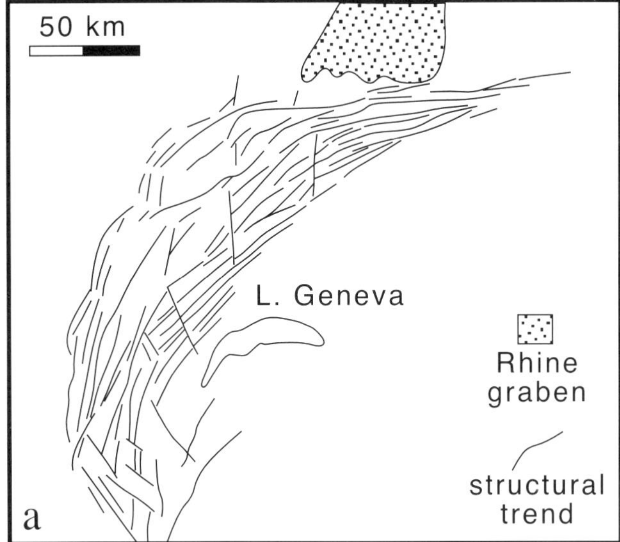

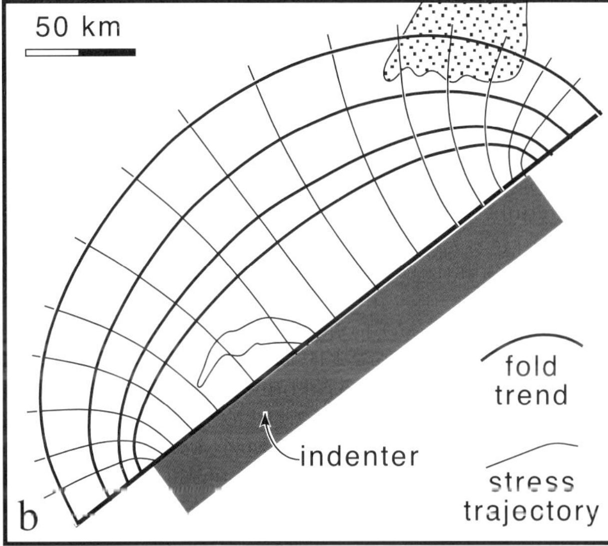

Figure 18. Stress model of the Jura, by Laubscher (1972). (a) Trend lines of the Jura Mountains, Switzerland. (b) Calculated principal-stress trajectories for a rectangular indenter, pushing against an elastic half space. Redrafted from the original.

Whiting (1995). As another example, Royden and Burchfiel (1989) suggested that the pronounced curve of the Carpathians reflects the presence of a deep flysch basin along the southern margin of Europe, whose floor subducted.

In effect, the concept that the predeformational shape of a continental margin controls curve formation is, in part, an alternative way of stating that variations in basin dimensions along a continental margin control curve formation. Recesses on a continent form where the passive-margin basin is wider and deeper, whereas promontories form where the passive-margin basin is narrower and shallower (Thomas, 1977). Therefore, the leading edge of the thrust belt lies farther toward the interior in a reentrant (i.e., forms a salient) than it does at a promontory.

Complex structural assemblages form where fold-thrust belts wrap around the corners of cratonic blocks as the belts advance toward the foreland. The structure of the Fleurieu recess in the Adelaide fold-thrust belt of South Australia illustrates this complexity (Figure 21). At this locality, the westward-propagating fold-thrust belt progressively wrapped around the southeast corner of the Gawler craton, an Archean crustal block. Structural analysis of this curve (Marshak and Flöttmann, 1996) demonstrates that not only does strain magnitude per unit length of cross section increase substantially, relative to the Nackara salient to the north, but the kinematics on faults become more complex. Notably, strike-slip lineations overprint preexisting dip-slip lineations on individual faults in the Fleurieu recess. This relationship suggests that thrust faults that had formed prior to interaction of the Adelaide Fold-thrust Belt with the corner of the Gawler craton were bent in plan-view, and that the bending was accommodated by flexural-slip bending of the fault surfaces around a vertical axis. In addition to the bending of preexisting thrusts, new thrusts formed that cut obliquely across the older ones. A similar history may be illustrated by the evolution of the Carpathians (Ratschbacher et al., 1993).

Obstacle-controlled Curves

Obstacle-controlled curves form where a fold-thrust belt interacts with basement highs in the foreland. A recess forms against the basement high, and a salient forms away from the high (Figure 15). For example, the Uinta recess in Utah formed where the Sevier Fold-thrust Belt interacted with a basement high that existed at the site of the present Uinta-Cottonwood Arch (Paulsen and Marshak, 1998a, b; 1999). Such curves may also develop where an accretionary prism interacts with a subducting seamount chain (Hsui and Youngquist, 1985; Dominguez et al., 1998, 1999). The margin of a recess evolves into a strike-slip fault system, if the edge of the obstacle is steep, or into a gradual curve, if the edge of the obstacle is gentle.

In cases in which the fold-thrust belt forms a salient between two obstacles, it is important to note that faults near the leading edge of the salient *initiate* in rocks to the foreland of the hinterland ends of the obstacles (Figure 15). Thus, the curves develops for the same reason as the curves in a basin-controlled salient. However, pre-interaction faults in the hinterland of the thrust wedge may be forced between the obstacles. If they are, they bend, because shear along the obstacles slows the advance of thrust-sheet portions adjacent to the obstacles relative to portions of the thrust sheet in the region between the obstacles.

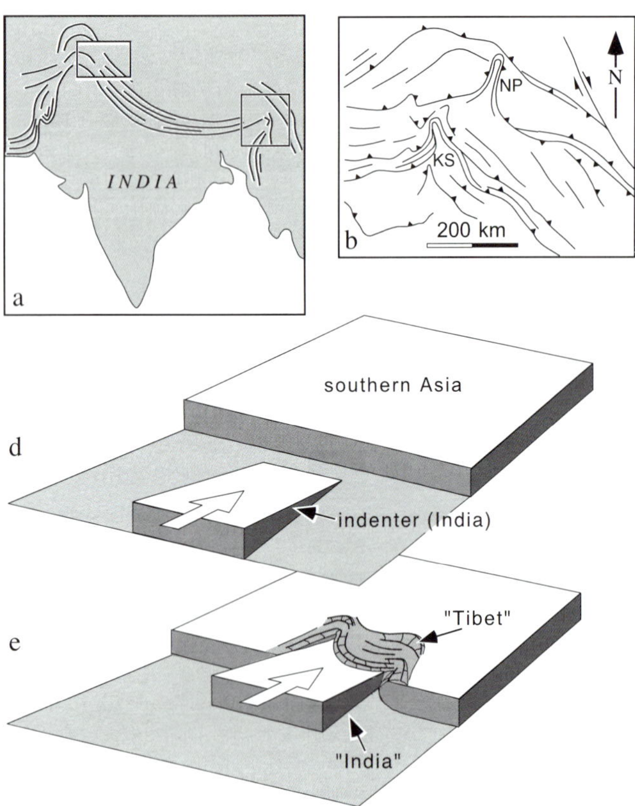

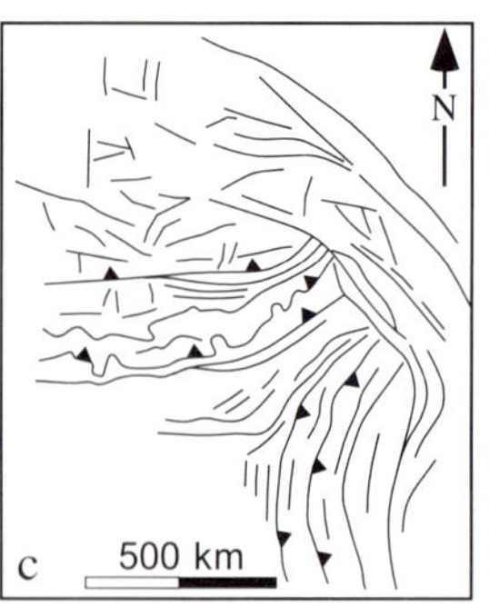

FIGURE 19. (a) Location map of the Himalayas. (b) Structural trend lines of the syntaxis at the northwest end of the Himalayas, illustrating the Kashmir syntaxis and Nanga Parbat (from Searle et al., 1988). (c) Structural trend lines at the east end of the Himalayas (from Sawar and DeJong, 1979). (d, e) Sketch of a model for the development of the Himalayan curves. In this model, India is a wedge-shaped indenter, and the Himalaya/Tibet orogen is a bivergent orogen (see Figure 13a).

Sandbox models show that the shape of a recess depends on the shape of the obstacle (Dominguez et al., 1999; Lickorish et al., 2002). Obstacles with a flat face perpendicular to the convergence direction yield flat-crested recesses, obstacles with a rounded face yield smoothly curving recesses, and pointed obstacles yield angular recesses. The dimensions of an obstacle in the direction of convergence also plays a role in controlling salient shape. Specifically, as Marshak et al. (1992) showed, elongate obstacles confine curved thrusts that formed to the foreland of the obstacle front and thus limit tangential extension, whereas point obstacles do not.

FIGURE 20. Illustration of the concept of a margin-controlled curve. (a) Predeformation condition shows an offshore arc converging with an irregular continental margin. (b) After collision, the orogen molds to the preexisting shape of the margin. Note the distribution of salients and recesses relative to the distribution of predeformational reentrants and promontories.

OTHER POSSIBLE CAUSES OF CURVE FORMATION

Curves Formed at the Intersection of Nonparallel Belts

Different segments of a fold-thrust belt may form at different times, either because convergence or collision occurs at different parts of a margin at different times during a single orogenic event, or because different portions of the fold-thrust belt form in response to entirely separate collisional or convergent events. If the trend lines of different segments are not parallel, or if the different segments are separate salients, a recess forms where the segments overlap. Marshak and Tabor (1989) noted that, in such recesses, younger thrusts carry older faults in their hanging wall and thereby bend the older structures around a vertical axis (Figure 22a, b). In addition, younger cleavages cut across older ones, and older thrusts are reactivated as strike-slip or

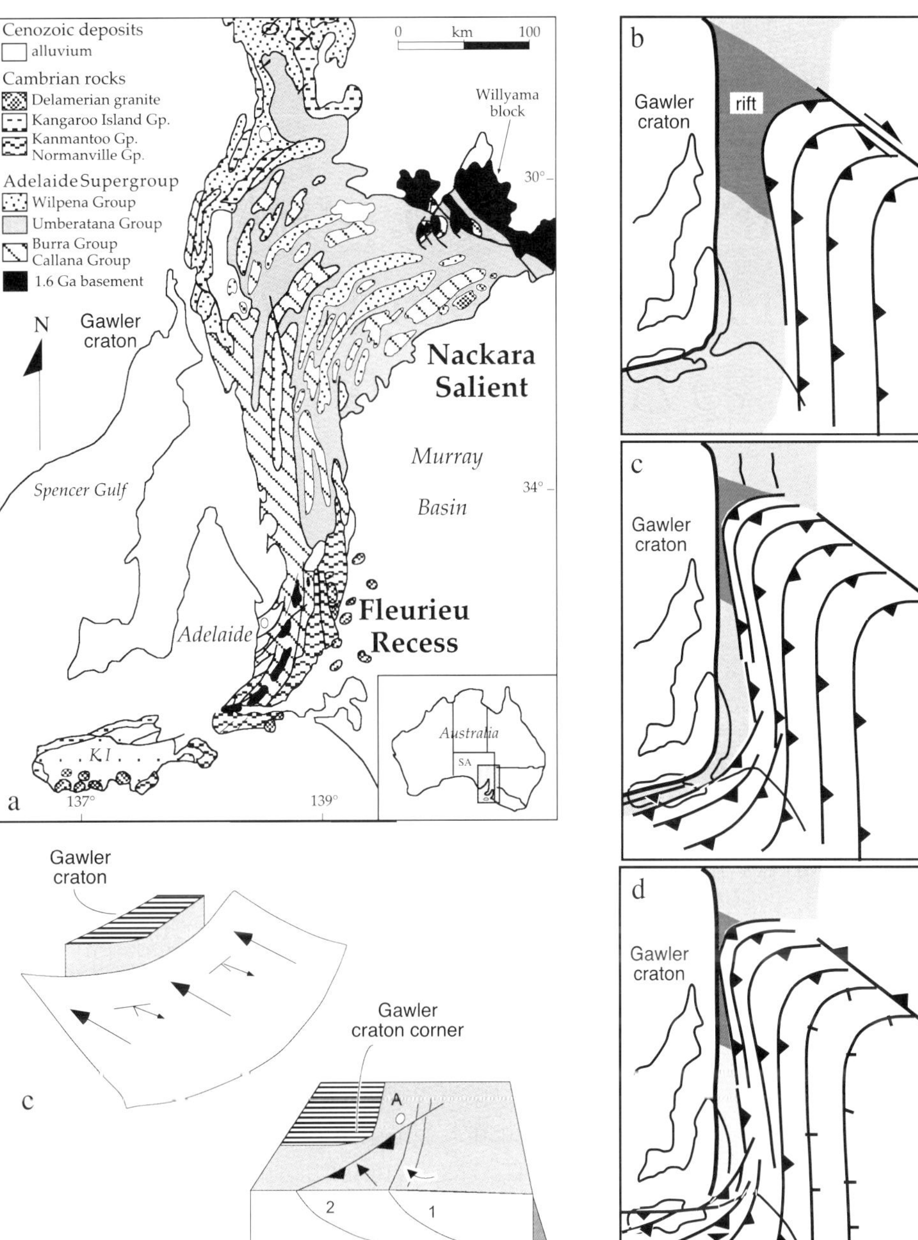

FIGURE 21. Curve formation in the Adelaide Fold-thrust Belt, Australia (from Marshak and Flöttmann, 1996). (a) A structural map of the Adelaide Fold-thrust Belt, illustrating the contrasts between the structure of the Nackara salient and that of the Fleurieu recess. In the recess, the belt is much narrower, strain is greater, and basement has been brought up in thrust slices, while in the salient, emergent thrusting is absent and folding is basement detached. In the recess, younger northeast-trending thrusts cross cut older, more northerly thrusts. (b–d) Sketches illustrating the progressive development of the curves. The Nackara salient forms over a NW-trending rift, containing particularly thick strata (dark shading), while the Fleurieu recess reflects the interaction between the westward propagating fold-thrust belt and the corner of the Gawler craton. Note that cross-cutting faults and strike-slip displacement develop in the Fleurieu recess. (e) Illustration of how lineations become progressively more oblique around the curve of the Fleurieu recess. (f) Illustration showing how movement on a younger thrust rotates an older thrust sheet.

oblique-slip faults. Places where plan-view bending occurs because of overlap of the faults can be called intersection oroclines (Marshak and Tabor, 1989). The Appalachians contain two intersection oroclines—the New York recess, which forms at the juncture of the central and northern Appalachians, and the Roanoke recess, which forms at the juncture of the southern and central Appalachians (Figure 1a).

Araujo and Marshak (1997) studied the Brasilia belt, a Neoproterozoic collisional orogen on the west side of the São Francisco craton in east-central Brazil. They suggest that the Pireneus syntaxis, which is a major regional cusp in the structural trends of the Brasilia belt, represents the intersection between the southern Brasilia belt and the northern Brasilia belt. A similar geometry occurs in the Cape fold belt of southern Africa, where two salients overlap to form a pronounced syntaxis (Figure 22c) (de Beer, 1992).

Curves Formed in Response to Strike-slip Faulting

Movement on a crustal-scale strike-slip fault that cuts across a fold-thrust belt can cause a pronounced curvature to the belt (Figure 23). This process happens either if the thrust belt grows at the same time as the movement on the strike-slip fault or if movement on

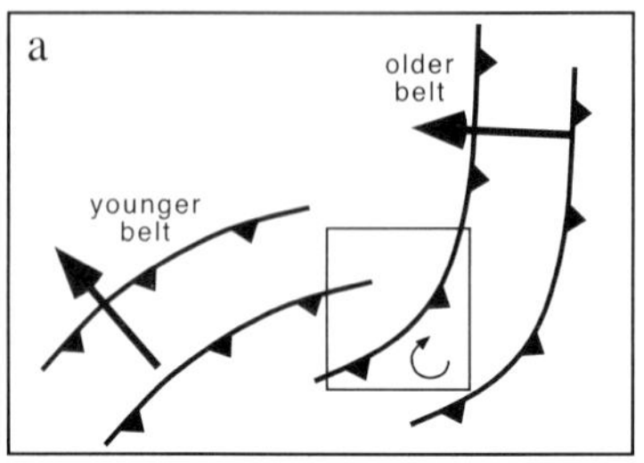

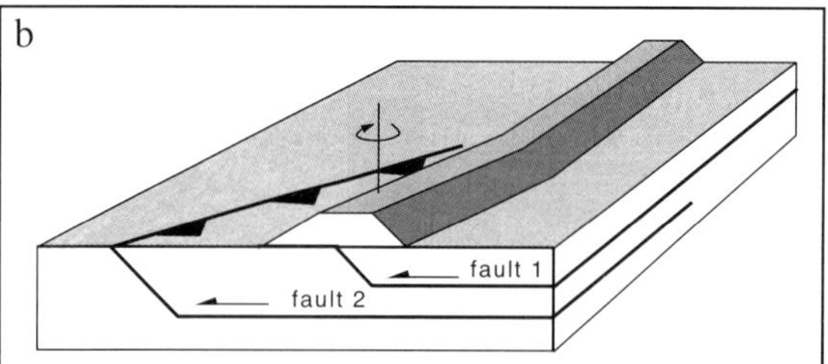

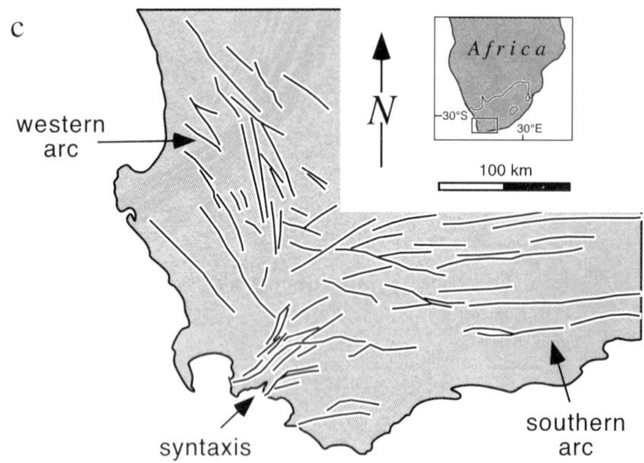

FIGURE 22. (a) An intersection orocline forms where a younger thrust belt overlaps an older one. If thrusts of the younger belt lie to the foreland, movement on them can rotate preexisting thrusts of the older belt. (b) This block diagram illustrates the rotation of an older fold by a younger, noncoaxial thrust (adapted from Marshak and Tabor, 1989). (c) An idealized sketch map of an intersection orocline in southern Africa formed by the overlap of two nonparallel fold-thrust belts (from de Beer, 1992).

the strike-slip fault overprints an already formed thrust belt. Effectively, strike-slip movement subjects the thrust belt to simple shear in plan.

The interaction between the Makran Fold-thrust Belt of southern Pakistan and the Chaman strike-slip system (the transform fault that forms the western boundary of the Indian subcontinent) illustrate the consequences of superimposing strike-slip shear on a thrust belt (Lawrence et al., 1981; Marshak, 1988). In the zone of interaction, thrust traces progressively rotated into parallelism with the Chaman Fault, so that the Makran Belt became an asymmetric salient with one limb asymptotically merging with the Chaman system—this limb undergoes severe thinning, in plan, adjacent to the Chaman Fault.

Several authors have discussed the curvature of southern South America and of the Antarctic Peninsula on opposite sites of the Scotia Sea. Wegener (1929), as noted earlier, argued that these curves formed by bending as the two continents moved westward. Paleomagnetic studies hint that the curves did indeed form as a consequence of bending, and thus that Wegener was correct (Kellogg and Reynolds, 1978; Burns et al., 1980). For example, Cunningham's (1993) structural analysis shows that the pattern of strike-slip faulting in the region is compatible with a model in which the curve of the southern Andes resulted from interplate shearing accommodated by strike-slip faulting. Other candidates for curves related to strike-slip faults include the curve in the Sierra Nevada range where it intersects the Garlock strike-slip fault in southern California (McWilliams and Li, 1985), the curve of structural trends adjacent to the Alpine Fault in New Zealand (Kamp, 1987), and the cuspate folds in the Walker Lane area of Nevada (Albers, 1967; Geissman et al., 1984). Bossart et al. (1988) suggest that the Kashmir syntaxis at the northwest end of the Himalayas formed as a result of superposition of shear on an originally straight thrust belt. Some authors have used displacements on hypothesized buried strike-slip faults to explain curves (e.g., in the Appalachians, Drake and Woodward, 1963; and in the Subalpine Chains of France, Graham, 1978), but in these examples, the curves are better attributed to other causes.

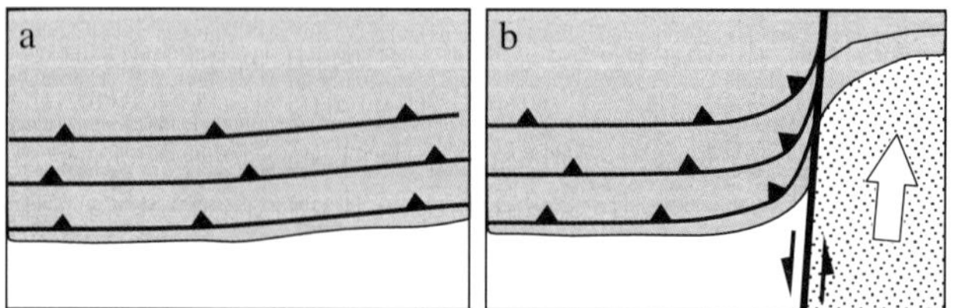

FIGURE 23. An idealized sketch map of the development of a curve in response to the superposition of a strike-slip shear zone on a previously formed fold-thrust belt (modified from Marshak, 1988).

Curves Reflecting the Shape of the Downgoing Slab

Researchers have examined, from a variety of perspectives, the effect of the downgoing plate's shape on the shape of the overlying fold-thrust belt, or accretionary prism. Wilson (1950) suggested that curves form because the downgoing plate is planar, and that the intersection of a plane with the surface of a sphere is a small circle. Frank (1968), Strobach (1973), De Fazio

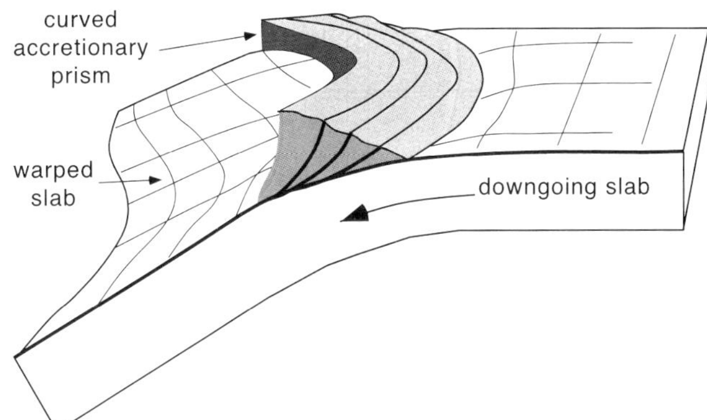

FIGURE 24. A 3-D sketch showing the curve of an accretionary prism formed above a spoon-shaped downgoing slab.

(1974), and Bayly (1982) suggested that the indentation of the surface of a sphere requires buckling of the downgoing plate into a gentle, plunging spoon. Yamaoka et al. (1986) carried this idea farther by suggesting that the subduction-related arcs around the rim of the Pacific plate resemble the crimps formed along the edge of an originally circular foil cap that has been bent to seal the top of a circular cylinder. If the downgoing plate is curved, the thrust wedge must have a curved leading edge in order to maintain contact with the downgoing plate (Figure 24).

Curves as Artifacts of Cross Folding

Some curves may not reflect rotation of structural trends around a vertical axis but may instead be an artifact of the erosion of a regional-scale plunging fold whose axis trends perpendicularly to the thrust front (Figure 25). Such folds may develop as a result of tangential buckling, as occurs at the intersection of oblique ramps with frontal ramps (Apotria, 1995), or in response to a second generation of folding that buckles already formed thrusts. Burg et al. (1997) have suggested that the Nanga Parbat syntaxis of the northwest Himalaya region formed in this way—the suture was folded and then exhumed. Paulsen and Marshak (1998b) suggest that uplift and erosion amplified the curvature of the northern limb of the Uinta recess, because it tilted the originally northeast-trending southern limb of the Wyoming salient so that, after erosion, fault traces trend east–west.

Curves Resulting from Plan-view Buckling

Relatively thin strips of buoyant crust, such as volcanic arcs, conceivably can be buckled in map view, if they are wedged between larger, converging blocks. For example, Silver et al. (1990) suggest that the sinuous trace of the Panamanian Isthmus, between North and South America, formed because the composite island arc was buckled when the two continents converged. They suggest that part of this motion was accommodated by movement on strike-slip faults.

STRESS FIELDS IN CURVES

Several authors have speculated on what stress fields might look like in curved fold-thrust belts and how they might contribute to curve formation. For example, Laubscher (1972) modeled the stress field associated with the Jura salient by considering the consequences of indenting an elastic half-sheet by a rigid indenter (Figure 18). He showed that the stress trajectories would curve and that their orientation would be perpendicular to the presently observed folds. Beutner (1977) modeled the stress field around the Uinta recess, assuming an elastic crust interacting with rigid obstacles in the foreland, and again found curving stress trajectories.

The elastic models described by Laubscher and Beutner do not take the mass of the thrust wedge into account. As a thrust wedge forms, the surface of the wedge becomes elevated relative to the foreland. The resulting gravitational potential energy creates a stress component perpendicular to the leading edge of the wedge. Simplistically, the stress at a point in a salient is

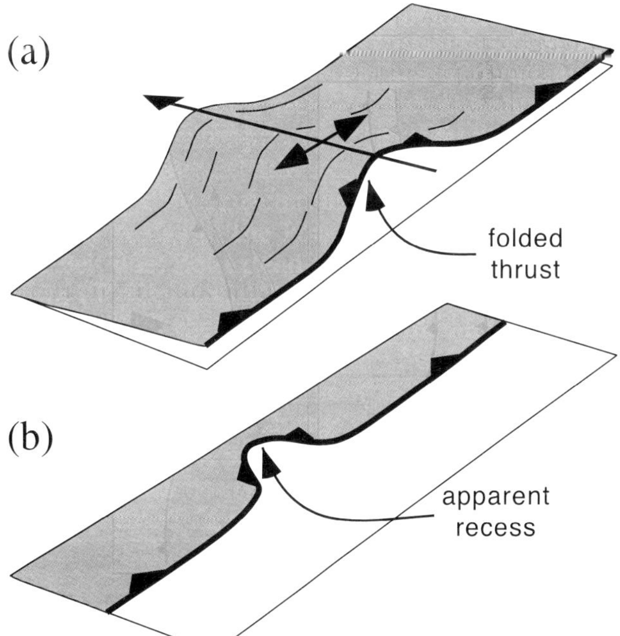

FIGURE 25. A 3-D sketch illustrating a recess forming as a consequence of the erosion of a folded thrust sheet. (a) A warped thrust surface. (b) Outcrop pattern of the thrust surface on a horizontal map plane.

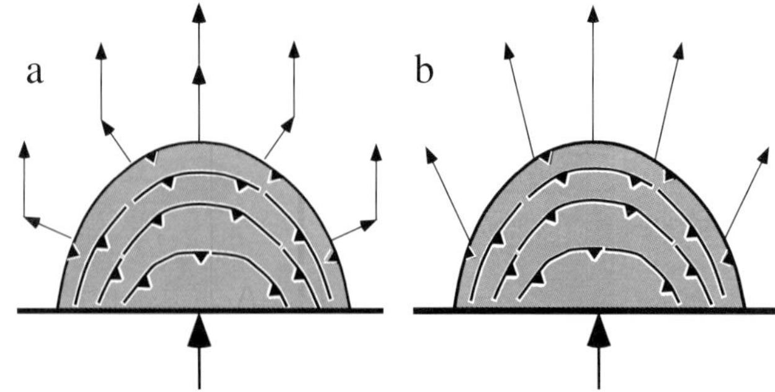

FIGURE 26. Map showing the hypothetical stress field in a thrust wedge. (a) Stress vectors in the map plane. The radial stresses are due to the gravitational potential energy of the thrust wedge, and the parallel vectors represent stresses due to the boundary load. (b) The resultant stress vectors are oblique to the leading edge.

the sum of the foreland direct stress resulting from the boundary load at the hinterland, which may be uniform along the length of the backstop, and the radial stress resulting from the gravitational potential energy of the elevated rock within the wedge (Figure 26) (Platt et al., 1989; Gray and Stamatakos, 1997; Lickorish et al., 2002). The combination of these stresses yields stresses that are oblique to the front of the wedge—stress trajectories can be nearly radial if the tectonic load is small. Notably, kinematic studies of currently forming curves, such as the Himalayan arc, suggest that movement at the leading edge is largely radial (e.g., Seeber et al., 1981).

STRAIN FIELDS IN CURVES

The strain in a curving fold-thrust belt can be complex, because many factors affect the displacement-path trajectories of rock within the curve. Ries and Shackleton (1976) described a number of different kinds of strain patterns possible for curved fold-thrust belts (Figure 10). The strain field depends on whether the curve represents plan-view bending of once-straight trend lines or represents formation of thrusts with originally curved traces; whether tangential extension (stretching parallel to the trace of the belt) occurred during curve formation; whether the reference line between end points of the curve stays constant in length, shortens, or lengthens during curve formation; and whether the curve has had a complex structural evolution, in which rotating indenters or other factors have caused changes in the displacement field over time.

To illustrate the variability in strain fields, consider a few examples. Imagine a curve formed by flattening strata during a thrusting event. Bending the thrust sheet in plan, allowing the end points to move closer together, does not result in tangential extension (Figure 27a), so the long axes of strain ellipses lie parallel to fold trends along the length of the curve. If, however, a strike-slip fault shears part of a once-straight thrust belt, tangential extension accompanies bending, and the aspect ratio of strain ellipses varies along the length of the curve. In the sheared limb, the long axis of the strain ellipse will lie at an angle to trend lines(Figure 27b). If the curve forms by differential foreland displacement along the length of the curve, then strain ellipses parallel structural trends only at the apex of the curve, and the aspect ratio of the ellipses varies around the curve (Figure 20c). Notably, patterns similar to those of Figure 27b can develop even if the thrusts in the belt initiate with a curved trace. For example, in Figure 27d (from Macedo and Marshak, 1999), deformation during initiation of the thrust causes simple shear along the margins of the curve, relative to the apex. Ferrill and Groshong (1993), Gray and Stamatakos (1997), Hindle and Burkhard (1999), and Lickorish et al. (2002) describe strain analyses in curved fold-thrust belts.

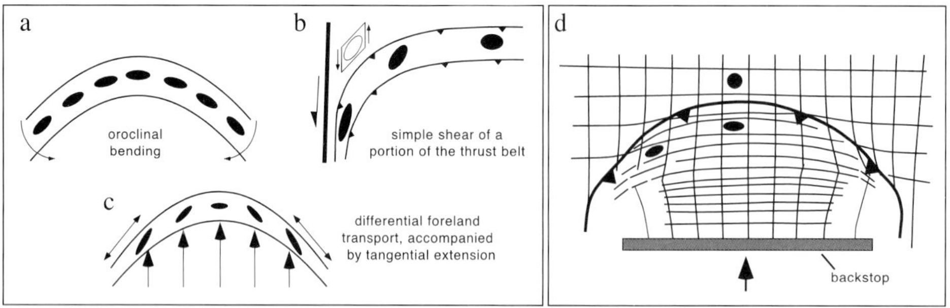

FIGURE 27. Examples of the variability of strain fields in a curving fold-thrust belt, depending on boundary conditions. (a) Strain ellipses remain the same but vary in orientation if a thrust belt oroclinally bends without undergoing tangential extension. (b) Shear at one end of a thrust belt by movement on a later strike-slip fault causes stretching in the limb of the orocline that parallels the younger fault. (c) Differential transport to the foreland causes oroclinal bending and tangential extension. (d) Traces in map view of a sandbox model (from Macedo and Marshak, 1999) showing formation of an indenter-controlled curve. Thin lines are strain markers on the surface of the sand, and ellipses represent strain. Initiation of a curved thrust (nonrotational arc) does not involve bending of the thrust trace, but strain ellipse orientation varies around the trace of the arc. Note that part of this strain develops in advance of the thrust sheet.

Because tangential extension is a likely possibility in curving fold-thrust belts, it is reasonable to ask how such extension can be accommodated in the real world. Marshak (1988) suggested that mesoscopic strike-slip faulting (either on conjugate systems or on parallel arrays) accommodates tangential extension in some localities. Seeber and Pecher (1998) note that in some places, along-strike extension may be accommodated by graben formation perpendicular to the front of the orogen (cf. Powell, 1986). As illustrated in the Makran, along-strike extension is accommodated by slip on strike-slip faults parallel to preexisting thrusts (Lawrence et al., 1981; Marshak, 1988).

CONTRASTS BETWEEN SALIENTS AND RECESSES

A number of examples demonstrate that, in general, recess structural geometries contrast markedly with those of salients. First, recesses are smaller than adjacent salients—to see this contrast, compare the syntaxes at the endpoints of the Himalayan arc to the arc itself (Figure 19) (Bossart et al., 1988). Second, recesses tend to be tighter and/or more angular than salients—to see this contrast, compare the recesses of the Appalachians with the salients of the Appalachians, or compare the Fleurieu recess with the Nackara salient in the Adelaide fold-thrust belt (Figure 21). Third, recesses tend to be structurally more complex than salients—to see this contrast, compare the New York recess with the Hudson Valley fold-thrust belt (Marshak and Tabor, 1989) and the Fleurieu recess to the Nackara salient (Marshak and Flöttmann, 1996). In these examples, the recesses contain transecting cleavages, overprinting of strike-slip movement on preexisting thrust faults, and out-of-sequence faulting. Fourth, the shortening per unit length of cross section is greater in recesses than in salients—the contrast in cross-sectional geometry of the Fleurieu recess and the Nackara salient illustrates this point (Marshak and Flöttmann, 1996). In this example, basement in the recess has been thrust up over sedimentary cover within 20 km of the thrust front in the recess, thrusts are closely spaced and folds are tight, and thrust sheets contain very strong cleavage, whereas in the Nackara salient, folds are gentle, thrusts are blind, the most forelandward basement involvement lies 100 km to the hinterland of the thrust front, and the region contains weak cleavage.

The contrasts between salients and recesses reflect contrasts in the settings in which they form. For example, recesses that form adjacent to obstacles or promontories accommodate more strain in a narrower region (as measured across strike) than do salients that form in broad basins.

THE ISSUE OF OROCLINAL BENDING

An understanding of the variety of geologic settings in which curves form provides a basis for reexamining the question of whether curves are rotational (i.e., formed by oroclinal bending) or nonrotational. Examination of the implications of each of the different kinds of curve formation for rotation of thrust sheets suggests the following.

- *Basin-controlled Salients*: Sandbox models demonstrate that thrust traces of basin-controlled curved fold-thrust belts initiate with curved trend lines. In three dimensions, the thrusts are spoon shaped. The curvature of these belts, therefore, initially reflects the along-strike variation in factors that control thrust-belt width, rather than the bending of the belt.
- *Curves Resulting from Interaction with Indenters*: Sandbox models suggest that thrust traces that define salients formed in front of an indenter initiate with a curved trajectory.
- *Curves Resulting from Interaction with Obstacles*: In the case of curves formed where a thrust belt interacts with paired obstacles in the foreland, thrusts formed to the foreland of the obstacle's face initiate with a curved trajectory, whereas earlier-formed thrusts pushed between the obstacles undergo oroclinal bending. In the case of recesses formed where a fold-thrust belt interacts with a promontory or obstacle in the foreland, field studies suggest that significant oroclinal bending takes place while the fold-thrust belt molds around the obstacle.
- *Curves Resulting from Interaction with Strike-slip Fault*: Curves whose shape reflects shear in response to interaction with a strike-slip fault undergo oroclinal bending during curve formation.
- *Curves Resulting from Thrusting over a Curving Downgoing Plate*: In cases in which the downgoing slab is nonplanar, the thrust wedge that is formed initiates with a curved trajectory.
- *Curves Resulting from Intersection of Two Belts*: If two non-coaxial thrust belts intersect, displacement of thrusts of the younger belt can cause oroclinal bending of thrust sheets in the older belt.

In sum, it appears that oroclinal bending does not occur in the basin-controlled or the indenter-controlled curves, but does occur in curves formed by interactions with strike-slip faults or with obstacles in the foreland. It is important to note, however, that rock rotation may occur on the limbs of basin-controlled or indenter-controlled salients, because of the geometry of strain accumulation during thrust formation. The amount of rotation in the rock presumably is less than that predicted by measuring the angle between the leading edge and the reference line of the belt. Similarly, in true oroclines, bending may be accommodated by shear on an array of parallel strike-slip faults, without rotation of the rocks in between. Considering these caveats, it is no surprise that paleomagnetic results from curved fold-thrust belts have proven difficult to interpret.

THE ISSUE OF ISLAND-ARC CURVATURE

This chapter has focused on the issue of curve formation in fold-thrust belts in the foreland of terrestrial mountain belts. Similar principles apply, however, to explaining the curvature of volcanic arcs and their accretionary prisms, as even the earliest studies of curvature imply (Suess, 1909; Hobbs, 1923). Arc curvature may reflect:

- initial shape of the downgoing slab (Frank, 1968; Voigt, 1973; Tovish and Schubert, 1978; Bayly, 1982; McCabe, 1984; Yamaoka et al., 1986);
- interaction with strike-slip faults (e.g., the Lesser Antilles arc and the Scotia arc are each bordered on the north and south by strike-slip faults; Wegener, 1929; Cunningham, 1993);
- along-strike changes in the thickness of the sedimentary layer to be incorporated in the accretionary prism (Breen, 1989);
- interaction with obstacles, such as seamount chains (e.g., the north and south end points of the Mariana arc coincide with subducting seamount chains, and the western endpoint of the Mariana arc coincides with its intersection with the Emperor seamount chain; Hsui and Youngquist, 1985); and
- buckling of a once straight arc in the vise formed by two converging continents (e.g., the Panamanian isthmus; Silver et al., 1990).

CONCLUSIONS

Fold-thrust-belt curves (salients and recesses) have fascinated geologists since the discipline of geology began, because they provide insight into the process of orogeny. Early studies viewed curves to be a necessary consequence of orogeny. In recent decades, however, it has become clear that curves form in response to local conditions—they reflect basin geometry, the impact of indenters, overprinting by strike-slip faults, buckling of the downgoing slab, overlapping of nonparallel thrust belts, and interaction with foreland obstacles. Basin-controlled curves account for the majority of salients, worldwide. In most examples of such curves, the apex of the salient coincides with a predeformational depocenter or with the subsurface distribution of a weak detachment horizon. The shape of the curve provides constraints on the shape of the predeformational basin. Notably, not all curved fold-thrust belts reflect oroclinal bending. For example, faults in basin-controlled and indenter-controlled curves initiate with curved traces (i.e., they are spoon-shaped in 3-D). But even in nonrotational curves, where the strike of trend lines has not changed since their formation, shear during deformation may cause rock to rotate within thrust sheets on the limb of a curve. Thus, paleomagnetic results from curved fold-thrust belts may be enigmatic. Future kinematic and strain studies of curves, along with paleomagnetic work, will further refine knowledge of curve formation. This work has applications to petroleum exploration, in that the conditions leading to oil-field generation correspond to the conditions leading to basin-controlled salient formation. Thus, major fold-thrust-belt-related oil provinces lie at the apex of basin-controlled salients (Macedo, 1997; Macedo and Marshak, 1999).

ACKNOWLEDGMENTS

This paper incorporates results from the work of my graduate students at the University of Illinois. I draw on publications and dissertations by Juliano Macedo, Tim Paulsen, Scott Wilkerson, and Oswaldo Araujo, and from Master's theses by John Tabor and Dave McEachran, and wish to thank them sincerely for many years of fruitful interactions. I am also very grateful to Thomas Flöttmann and Fernando Alkmim, who have shared their thoughts concerning curved fold-thrust belts. Finally, I wish to thank the authors of numerous published studies (notably those of Steve Boyer and Gautam Mitra) from which I have extracted ideas, and Bob Hatcher, who emphasized the relationship between basin thickness and salients. The title is a modification of one used by Sarwar and DeJong (1979). I am grateful for reviewer's comments by Jacques Malavieille. The contents of this manuscript were presented at the *Thrust Tectonics '99* meeting, generously organized by Ken McClay. The manuscript was submitted for publication in 1999.

REFERENCES CITED

Aitken, J. D., and D. G. F. Long, 1978, Mackenzie tectonic arc—reflection of early basin configuration?: Geology, v. 6, p. 626–629.

Albers, J. P., 1967, Belt of sigmoidal bending and right-lateral faulting in the western Great Basin: Geological Society of America Bulletin, v. 78, p. 143–156.

Alkmim, F. F., and S. Marshak, 1998, Transamazonian orogeny in the southern São Francisco craton region, Minas Gerais, Brazil: evidence for Paleoproterozoic collision and collapse in the Quadrilátero Ferrífero: Precambrian Research, v. 90, p. 29–58.

Apotria, T. G., 1995, Thrust sheet rotation and out-of-plane strains associated with oblique ramps; an example from the Wyoming salient, USA: Journal of Structural Geology, v. 17, p. 647–662.

Argand, E., 1924, La Tectonique de l'Asia (Tectonics of Asia), translated by A. V. Carozzi (1977): New York, Hafner Press, 218 p.

Bachtadse, V., and R. Van der Voo, 1986, Paleomagnetic evidence for crustal and thin-skinned rotations in the European Hercynides: Geophysical Research Letters, v. 13, p. 161–164.

Bayly, B., 1982, Geometry of subducted plates and island arcs viewed as a buckling problem: Geology, v. 10, p. 629–632.

Beck, M. E. Jr., 1987, Tectonic rotations on the leading edge of South America: the Bolivian orocline revisited: Geology, v. 15, p. 806–808.

Bethke, C. M., and S. Marshak, 1990, Brine migrations across North America—the plate tectonics of groundwater: Annual Review of Earth and Planetary Sciences, v. 18, p. 287–316.

Beutner, E. C., 1977, Causes and consequences of curvature in the Sevier orogenic belt, Utah to Montana, *in* Rocky Mountain thrust belt—geology and resources: Guidebook for the Wyoming Geological Association Annual Field Conference 29, p. 153–165.

Bossart, P., R. Ottiger, and F. Heller, 1989, Paleomagnetism in the Hazara-Kashmir syntaxis, northeast Pakistan: Eclogae Geologicae Helvetiae, v. 82, p. 585–601.

Bossart, P., D. Dietrich, A. Greco, R. Ottiger, and J. G. Ramsay, 1988, The tectonic structure of the Hazara-Kashmir syntaxis, southern Himalayas, Pakistan: Tectonics, v. 7, p. 273–297.

Boyer, S. E., 1995, Sedimentary basin taper as a factor controlling the geometry and advance of thrust belts: American Journal of Science, v. 295, p. 1220–1254.

Breen, N. A., 1989, Structural effect of Magdalena fan deposition on the northern Colombia convergent margin: Geology, v. 17, p. 34–37.

Bucher, W. H., 1924, The pattern of the earth's mobile belts: Journal of Geology, v. 32, p. 265–290.

Burg, J. P., P. Davy, P. Nievergelt, F. Oberli, D. Seward, Z. Diao, and M. Meier, 1997, Exhumation during crustal folding in the Namche-Barwa syntaxis: Terra Nova, v. 9, p. 53–56.

Burns, K. L., M. H. Rickard, L. Belbia, and F. Channelawn, 1980, Further paleomagnetic confirmation of the Patagonian orocline: Tectonophysics, v. 63, p. 75–90.

Butler, R. F., D. R. Richards, T. Sempere, and L. G. Marshall, 1995, Paleomagnetic determinations of vertical-axis tectonic rotations from Late Cretceous and Paleocene strata of Bolivia: Geology, v. 23, p. 799–802.

Calassou, S., C. Larroque, and J. Malavieille, 1993, Transfer zones of deformation in thrust wedges: an experimental study: Tectonophysics, v. 221, p. 325–344.

Carey, S. W., 1955, The orocline concept in geotectonics: Proceedings of the Royal Society of Tasmania, v. 89, p. 255–289.

Chamberlain, R. T., and F. B. Shepard, 1923, Some experiments in folding: Journal of Geology, v. 31, p. 490–512.

Crosby, G. W., 1969, Radial movements in the western Wyoming salient of the Cordilleran overthrust belt: Geological Society of America Bulletin, v. 80, p. 1061–1078.

Cunningham, W. D., 1993, Strike-slip faults in the southernmost Andes and the development of the Patagonian orocline: Tectonics, v. 12, p. 169–186.

Dana, J. D., 1866, A texbook of geology: Philadelphia, Theodore Bliss, 354 p.

Davis, D. M., and T. Engelder, 1985, The role of salt in fold-and-thrust belts: Tectonophysics, v. 119, p. 67–88.

Davis, D. M., and R. J. Lillie, 1994, Changing mechanical response during continental collision: active examples from the foreland thrust belts of Pakistan: Journal of Structural Geology, v. 16, p. 21–34.

Davis, D., J. Suppe, and F. A. Dahlen, 1983, Mechanics of fold-and-thrust belts and accretionary wedges: Journal of Geophysical Research, v. 88, p. 1153–1172.

Davy, P., and P. R. Cobbold, 1988, Indentation tectonics in nature and experiment, I: Experiments scaled for gravity: Bulletin of the Geological Institute of the University of Uppsala, N.S., v. 14, 129–141.

de Beer, C. H., 1992, Structural evolution of the Cape fold belt syntaxis and its influence on syntectonic sedimentation in the SW Karroo Basin, *in* J. J. de Wit and I. G. D. Ransome, eds., Inversion tectonics of the Cape fold belt, Karoo and Cretaceous basins of Southern Africa: Rotterdam, Balkema, p. 197–206.

De Fazio, T. L., 1974, Island-arc and underthrust-plate geometry: Tectonophysics, v. 23, p. 149–154.

Dominguez, S., S. E. Lallemand, J. Malavieille, and R. Von Huene, 1998, Upper plate deformation associated with seamount subduction: Tectonophysics, v. 293, p. 207–224.

Dominguez, S., J. Malavieille, and S. E. Lallemand, 1999, Deformation of margins in response to seamount subduction: insights from sandbox experiments, *in* K. R. McClay, ed., Thrust tectonics 99, abstracts: Royal Holloway, University of London, Egham, UK, p. 175–177.

Drake, C. L., and H. P. Woodward, 1963, Appalachian curvature, wrench faulting, and offshore structures: Transactions of the New York Academy of Science, v. 26, p. 48–63.

Eldredge, S., V. Bachtadse, and R. van der Voo, 1985, Paleomagnetism and the orocline hypothesis: Tectonophysics, v. 119, p. 153–179.

Ferrill, D. A., and R. H. Groshong Jr., 1993, Kinematic model for the curvature of the northern subalpine chain, France: Journal of Structural Geology, v. 15, p. 523–541.

Frank, F. C., 1968, Curvature of island arcs: Nature, v. 220, p. 363.

Geissman, J. W., J. T. Callian, J. S. Oldow, and S. E. Humphries, 1984, Paleomagnetic assessment of oroflexural deformation in west-central Nevada and significance for emplacement of allochthonous assemblages: Tectonics, v. 3, p. 179–200.

Graham, R. H., 1978, Wrench faults, arcuate fold patterns, and deformation in the southern French Alps: Proceedings of the Geologists' Association, v. 89, p. 125–142.

Gray, M. B., and J. Stamatakos, 1997, New model for evolution of fold and thrust belt curvature, based on integrated structural and paleomagnetic results from the Pennsylvania salient: Geology, v. 25, p. 1067–1070.

Gwinn, V., 1967, Curvature of marginal folded belts flanking major mountain ranges: accented or caused by lateral translation of epidermal stratified cover?: Geological Society of America Special Paper 115, p. 1–87.

Haq-Saad, S. B., and D. M. Davis, 1997, Oblique convergence and the lobate mountain belts of western Pakistan: Geology, v. 25, n. 1, p. 23–26.

Hindle, D., and M. Burkhard, 1999, Strain, displacement and rotation associated with the formation of curvature in fold belts; the example of the Jura arc: Journal of Structural Geology, v. 21, p. 1089–1101.

Hobbs, W. H., 1914, Mechanics of formation of arcuate mountains: Journal of Geology, v. 22, p. 71–90.

Hobbs, W. H., 1923, The Asiatic arcs: Geological Society of America Bulletin, v. 34, p. 243–252.

Hsui, A. T., and S. Youngquist, 1985, A dynamic model of the curvature of the Mariana Trench: Nature, v. 318, p. 455–457.

Jaumé, S. C., and R. J. Lillie, 1988, Mechanics of the Salt Range-Potwar Plateau, Pakistan: a fold-and-thrust belt underlain by evaporites: Tectonics, v. 7, p. 57–71.

Kamp, P. J. J., 1987, Age and origin of the New Zealand orocline in relation to Alpine fault movement: Geological Society of London Journal, v. 144, p. 641–652.

Kellogg, K. S., and R. L. Reynolds, 1978, Paleomagnetic results from the Lassiter Coast, Antarctica, and a test for oroclinal bending of the southern Antarctic Peninsula: Geological Society of America Bulletin, v. 91, p. 414–420.

Kent, D. V., 1988, Further paleomagnetic evidence for oroclinal rotation in the central folded Appalachians from the Bloomsburg and the Mauch Chunk Formations: Tectonics, v. 7, p. 749–759.

Klootwijk, C. T., P. J. Conaghan, and C. M. Powell, 1985, The Himalayan Arc: large-scale continental subduction, oroclinal bending, and back-arc spreading: Earth and Planetary Science Letters, v. 75, p. 167–183.

Lake, P., 1931, Mountain and island arcs: Geological Magazine, v. 68, p. 34–39.

Lallemand, W. E., J. Malavieille, and S. Calassou, 1992, Effects of oceanic ridge subduction on accretionary wedges: experimental modeling and marine observation: Tectonics, v. 11, p. 1301–1313.

Laubscher, H., 1972, Some overall aspects of Jura dynamics: American Journal of Science, v. 272, p. 293–304.

Lawrence, R. D., S. H. Khan, K. A. DeJong, A. Farah, and R. S. Yeats, 1981, Thrust and strike-slip fault interaction along the Chaman transform zone, Pakistan, *in* K. R. McClay and N. J. Price, eds., Thrust and nappe tectonics: Geological Society of London Special Publication 9, p. 363–370.

Lickorish, W. H., M. Ford, J. Buergisser, and P. R. Cobbold, 2002, Arcuate thrust systems in sandbox experiments; a comparison to the external arcs of the Western Alps: Geological Society of America Bulletin, v. 114, p. 1089–1107.

Lowrie, W., and A. M. Hirt, 1986, Paleomagnetism in arcuate mountain belts, *in* F. C. Wezel, ed., The origin of arcs: Amsterdam, Elsevier, p. 141–158.

Lu, C. Y., and J. Malavieille, 1994, Oblique convergence, indentation and tectonic rotation in Taiwan Mountain belt: Insights from experimental modelling: Earth and Planetary Science Letters, v. 121, p. 477–494.

Macedo, J. M., 1997, Models of continental fold-thrust belt salients: Ph.D. Thesis, University of Illinois, Urbana-Champaign, 295 p.

Macedo, J. M., and S. Marshak, 1999, Controls on the geometry of fold-thrust belt salients: Geological Society of America Bulletin, v. 111, p. 1808–1822.

Marrett, R., 1995, Structure, kinematics, and development of the Sierra Madre Oriental salient, Mexico: Geological Society of America Abstracts with Programs, v. 27, p. 73.

Marshak, S., 1988, Kinematics of orocline and arc formation in thin-skinned orogens: Tectonics, v. 7, p. 73–86.

Marshak, S., and T. Flöttmann, 1996, Structure and origin of the Fleurieu and Nackara arcs in the Adelaide fold-thrust belt, South Australia; salient and recess development in the Delamerian orogen: Journal of Structural Geology, v. 18, p. 891–908.

Marshak, S., and J. Tabor, 1989, Structure of the Kingston orocline in the Appalachian fold-thrust belt, New York: Geological Society of America Bulletin, v. 101, p. 683–701.

Marshak, S., and M. S. Wilkerson, 1992, Effect of overburden thickness on thrust-belt geometry and development: Tectonics, v. 11, p. 560–566.

Marshak, S., M. S. Wilkerson, and A. T. Hsui, 1992, Generation of curved fold-thrust belts: insight from simple physical and analytical models, *in* K. R. McClay, ed., Thrust tectonics: London, Chapman & Hall, p. 83–92.

McCabe, R., 1984, Implications of paleomagnetic data on the collision related bending of island arcs: Tectonics, v. 3, p. 409–428.

McWilliams, M., and Y. Li, 1985, Oroclinal bending of the southern Sierra Nevada Batholith: Science, v. 230, p. 172–174.

Melnyk, M. J., and C. P. Cameron, 1998, Structural geology of a portion of the Monterrey salient, northeastern Mexico; an analog to the Juras: Geological Society of America abstracts with programs, v. 30, n. 7, p. 170–171.

Mitra, G., 1997, Evolution of salients in a fold-and-thrust belt: the effects of sedimentary basin geometry, strain distribution and critical taper, *in* S. Sengupta, ed., Evolution of geological structures in micro- to macro-scales: London, Chapman & Hall, p. 59–90.

Muttoni, G., A. Argnani, D. V. Kent, N. Abrahamsen, and U. Cibin, 1998, Paleomagnetic evidence for Neogene tectonic rotations in the northern Apennines, Italy: Earth and Planetary Science Letters, v. 154, p. 25–40.

Pares, J. M, R. Van der Voo, J. Stamatakos, and A. Perez-Estaun, 1994, Remagnetizations and postfolding oroclinal rotations in the Cantabrian/Asturian Arc, northern Spain: Tectonics, v. 13, p. 1461–1471.

Paulsen, T., and S. Marshak, 1998a, Charleston transverse zone, Wasatch Mountains, Utah; structure of the Provo salient's northern margin, Sevier fold-thrust belt: Geological Society of America Bulletin, v. 110, p. 512–522.

Paulsen, T., and S. Marshak, 1998b, Structure of the Mt. Raymond transverse zone at the southern end of the Wyoming salient, Sevier fold-thrust belt, Utah: Tectonophysics, v. 280, p. 199–211.

Paulsen, T., and S. Marshak, 1999, Origin of the Uinta recess, Sevier fold-thrust belt, Utah: Influence of basin architecture on fold-thrust belt geometry: Tectonophysics, v. 312, p. 203–216.

Platt, J. P., J. H. Behrmann, P. C. Cunningham, J. F. Dewey, M. Helman, M. Parish, M. G. Shepley, S. Wallis, and P. J. Weston, 1989, Kinematics of the Alpine arc and the motion history of Adria: Nature, v. 337, p. 158–161.

Powell, C. M., 1986, Curvature of the Himalayan arc related to Miocene normal faults in southern Tibet: Geology, v. 14, p. 358–359.

Rankin, D. W., 1976, Appalachian salients and recesses: Late Precambrian continental breakup and the opening of the Iapetus Ocean: Journal of Geophysical Research, v. 81, p. 5605–5619.

Ries, A. C., and R. M. Shackleton, 1976, Patterns of strain variation in arcuate fold belts: Philosophical Transactions of the Royal Society of London, pt. A, v. 283, p. 281–288.

Rogers, H. D., 1858, The geology of Pennsylvania, v. 2: Philadelphia, J.B. Lippincott, 1045 p.

Royden, L., and B. C. Burchfiel, 1989, Are systematic variations in thrust belt style related to plate boundary processes? (the western Alps versus the Carpathians): Tectonics, v. 8, p. 51–61.

Sarwar, G., and K. A. DeJong, 1979, Arcs, oroclines, syntaxes: The curvatures of mountain belts in Pakistan, *in* A. Farah and K. A. DeJong, eds., Geodynamics of Pakistan: Quetta, Geological Survey of Pakistan, p. 341–350.

Schwartz, S. Y., and R. Van der Voo, 1983, Paleomagnetic evaluation of the orocline hypothesis in the central and southern Appalachians: Geophysical Research Letters, v. 10, p. 505–508.

Searle, M. P., D. J. W. Cooper, and A. J. Rex, 1988, Collision tectonics of the Ladakh-Zanskar Himalaya: Philosophical Transactions of the Royal Society of London, v. A-326, p. 117–150.

Seeber, L., J. G. Arbruster, and R. C. Quittmeyer, 1981, Seismicity and continental subduction in the Himalayan Arc: American Geophysicical Union Geodynamics Series, v. 3, p. 215–242.

Seeber, L., and A. Pecher, 1998, Strain partitioning along the Himalaya Arc and the Nanga Parbat Antiform: Geology, v. 26, p. 791–794.

Silver, E. A., D. L. Reed, J. E. Tagudin, and D. J. Heil, 1990, Implications of the north and south Panama thrust belts for the origin of the Panama orocline: Tectonics, v. 9, p. 261–281.

Sollas, W. J., 1903, The figure of the Earth: Quarterly Journal of the Geololgical Society of London, v. 59, p. 180–189.

Stamatakos, J., and A. M. Hirt, 1994, Paleomagnetic considerations of the development of the Pennsylvania salient in the central Appalachians: Tectonophysics, v. 231, p. 237–255.

Strobach, K., 1973, Curvature of island arcs and plate tectonics: Aeitschrift für Geophysik, v. 39, p. 819–831.

Suess, E., 1909, The Face of the Earth, translated from German by H. B. C. Sollas and W. J. Sollas: Oxford, Clarendon, 672 p.

Talbot, C. J., and M. Alavi, 1996, The past of a future syntaxis across the Zagros: Geological Society of London Special Publication 100, p. 89–109.

Tapponnier, P., and P. Molnar, 1976, Slip-line field theory and large-scale continental tectonics: Nature, v. 264, p. 319–324.

Taylor, F. B., 1910, Bearing of the Tertiary Mountain belt on the origin of the Earth's plan: Geological Society of America Bulletin, v. 21, p. 179–226.

Thomas, W. A., 1977, Evolution of Appalachian-Ouachita salients and recesses from reentrants and promontories in the continental margin: American Journal of Science, v. 277, p. 1233–1278.

Thomas, W. A., 1990, Controls on location of transverse zones in thrust belts: Eclogae Geologicae Helvetiae, v. 83, p. 727–744.

Thomas, W. A., and B. M. Whiting, 1995, The Alabama Promontory; an example of the evolution of an Appalachian-Ouachita thrust belt recess at a promontory of the rifted continental margin: Geological Association of Canada, Special Paper 41, p. 1–20.

Tovish, A., and G. Schubert, 1978, Island arc curvature, velocity of convergence, and angle of subduction: Geophysical Research Letters, v. 5, p. 329–332.

Van der Voo, R., J. A. Stamatakos, and J. M. Pares, 1997, Kinematic constraints on thrust–belt curvature from syndeformational magnetizations in the Lagos del Valle syncline in the Cantabrian Arc, Spain: Journal of Geophysical Research, v. 102, p. 10,105–10,119.

Voigt, P. R., 1973, Subduction and aseismic ridges: Nature, v. 241, p. 189–191.

Wegener, A., 1929, The origin of continents and oceans, 4th edition: Translated from the 4th revision of the German edition by John Biram: New York, Dover Publications, 246 p.

Willett, S., C. Beaumont, and P. Fullsack, 1993, Mechanical model for the tectonics of doubly vergent compressional orogens: Geology, v. 21, p. 371–374.

Willis, B., 1893, The mechanics of Appalachian structure: U.S. Geological Survey, Annual Report 13, Part 2, p. 211–281.

Wilson, J. T., 1949, An extension of Lake's hypothesis concerning mountain and island arcs: Nature, v. 164, p. 147–148.

Wilson, J. T., 1950, An analysis of the pattern and possible cause of young mountain ranges and island arcs: Proceedings of the Geological Association of Canada, v. 3, p. 141–190.

Wise, D. U., and M. L. Werner, 1991, Pennsylvania salient of the Appalachian piedmont; tectonic transport vectors in relation to curvature: Geological Society of America abstracts with programs, v. 23, no. 1, p. 151.

Yamaoka, K., Y. Fukao, and M. Kumazawa, 1986, Spherical shell tectonics: Effects of sphericity and inextensibility on the geometry of the descending lithosphere: Reviews of Geophysics, v. 24, p. 27–53.

Zweigel, P., 1998, Arcuate accretionary wedge formation at convex plate margin corners: results of sandbox analogue experiments: Journal of Structural Geology, v. 20, p. 1597–1609.

Rowan, M. G., F. J. Peel, and B. C. Vendeville, 2004, Gravity-driven fold belts on passive margins, *in* K. R. McClay, ed., Thrust tectonics and hydrocarbon systems: AAPG Memoir 82, p. 157–182.

Gravity-driven Fold Belts on Passive Margins

Mark G. Rowan
Rowan Consulting, Inc., Boulder, Colorado, U.S.A.

Frank J. Peel
BHP Billiton Petroleum (Americas) Inc., Houston, Texas, U.S.A.

Bruno C. Vendeville[1]
Bureau of Economic Geology, University of Texas at Austin, Austin, Texas, U.S.A.

ABSTRACT

Many passive margins have deep-water, contractional fold belts that formed above salt or shale. Margin failure, accommodated by proximal extension and distal shortening, is caused by some combination of gravity gliding above a basinward-dipping detachment and gravity spreading of a sedimentary wedge with a seaward-dipping bathymetric surface. Gravitational failure is inherently self-limiting, and sedimentation patterns provide fundamental control of deformation. Continued shortening is driven primarily by shelf and upper-slope deposition, which maintains the bathymetric slope and the gravity potential, and by increased basinward tilting. Deformation is retarded or halted by distal thickening of the overburden caused by the folding itself or by lower-slope and abyssal sedimentation. Net shortening amounts and deformation rates are lower than in collisional/accretionary fold belts, because the driving forces are weaker than those induced by lithospheric plate motions.

Structural styles vary but depend largely on the nature of the dêcollement layer, not the driving forces. Fold belts detached on shale typically comprise basinward-vergent thrust imbricates and associated folds because of the relative strength and frictional behavior of the plastic shale. Deformation does not occur until there is sufficient overburden, and it is facilitated by high fluid pressures. In contrast, salt is a viscous material with essentially no strength, which leads to symmetrical detachment folds and early deformation beneath only a thin overburden. Moreover, the surface slope can be reduced by proximal subsidence into salt and distal inflation of salt, and much of the shortening can be accommodated by lateral squeezing of diapirs and salt massifs and by extrusion of salt nappes.

[1]*Present address*: Université de Lille 1, USR des Sciences de la Terre, UNR Processus et Bilans des Domaines Sedimentaires, 59655 Villeneuve d'Ascq Cedex, France.

INTRODUCTION

Passive margins are often characterized by broad zones of deformation in which horizontal translation of the postrift cover is driven by gravitational failure of the margin. The failure is accommodated by a linked system of thin-skinned, proximal (updip) extension and distal (downdip) contraction above one or more detachments. Many dêcollement layers consist of syn- or postrift salt and associated evaporites, whereas others consist of shales that typically are overpressured and undercompacted because of rapid burial.

Distal shortening may be accommodated in a variety of ways (Figure 1), such as by thrusting or folding and the development of deep-water fold belts, lateral squeezing of preexisting diapirs (Vendeville and Nilsen, 1995; Rowan et al., 2000), or by the extrusion, inflation, and shortening of salt nappes (Peel et al., 1995; Trudgill et al., 1999). Other processes, including lateral compaction, pressure solution, or small-scale folding, may also accommodate layer-parallel shortening, although the resulting structures are commonly below the resolution limits of seismic data and their contribution to the overall shortening is likely to be minor.

The focus here is on the deep-water fold belts that are found on many passive margins. Those detached on salt include: the Mississippi Fan (or Atwater) and Perdido Fold Belts of the northern Gulf of Mexico (Wu et al., 1990; Weimer and Buffler, 1992; Peel et al., 1995; Trudgill et al., 1995, 1999; Rowan, 1997; Rowan et al., 2000; Wu and Bally, 2000); fold belts in the Campos, Santos, and Espirito Santo Basins of Brazil (Demercian et al., 1993; Cobbold et al., 1995; Mohriak et al., 1995; Van der Ven et al., 1998; Jackson et al., 1998; Zalán, 1999); fold belts in the Benguela, Kwanza, Congo, Gabon, and Rio Muni Basins of West Africa (Lundin, 1992; Duval et al., 1992; Spathopoulos, 1996; Morley and Guerin, 1996; Marton et al., 2000; Dailly, 2000; Fox and Ashton, 2000; Cramez and Jackson, 2000); and fold belts in the Yemeni Red Sea (Heaton et al., 1995), the Nile deep-sea fan (Sage and Letouzey, 1990; Letouzey et al., 1995; Gaullier et al., 2000a), the Rhône deep-sea fan (Biju-Duval et al., 1979; Le Douaran et al., 1984; Mauffret et al., 1995), the Atlantic margins of Morocco, Mauritania, and Senegal (Tari et al., 2000, 2002; Hall, 2003), and the Scotian margin of eastern Canada (data in Keen and Potter, 1995). Fold belts detached on overpressured shale include: the Mexican Ridges (Buffler et al., 1979; Weimer and Buffler, 1992; B. Trudgill, personal communication, 2000) and Port Isabel (Peel et al., 1995; Peel and Matthews, 1999) fold belts of the western Gulf of Mexico; fold belts in the Sergipe-Alagoas and Pará-Maranhão Basins of Brazil (Guimarães et al., 1989; Zalán, 1999); and fold belts in the Niger Delta (e.g., Evamy et al., 1978; Doust and Omatsola, 1990; Morley and Guerin, 1996; Connors et al., 1998; Wu and Bally, 2000) and the McKenzie Delta in arctic Canada (Dailly, 1976; Damte and Miksha, 2000).

In this chapter, we discuss and illustrate various characteristics of gravity-driven fold belts on passive margins using a combination of seismic examples, cross sections and restorations, experimental models,

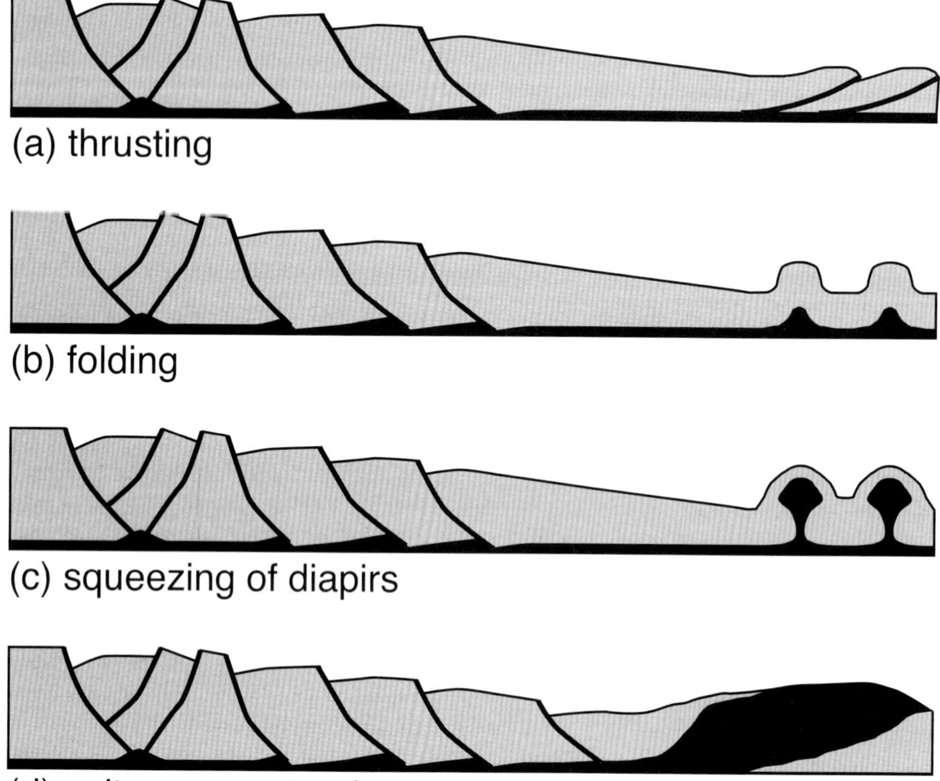

FIGURE 1. Different ways in which gravitational failure of passive margins is expressed at the distal toe of the deforming wedge: (a) thrust-dominated shortening; (b) fold-dominated shortening; (c) lateral squeezing of preexisting diapirs; and (d) inflation and extrusion of an allochthonous salt nappe. Not to scale.

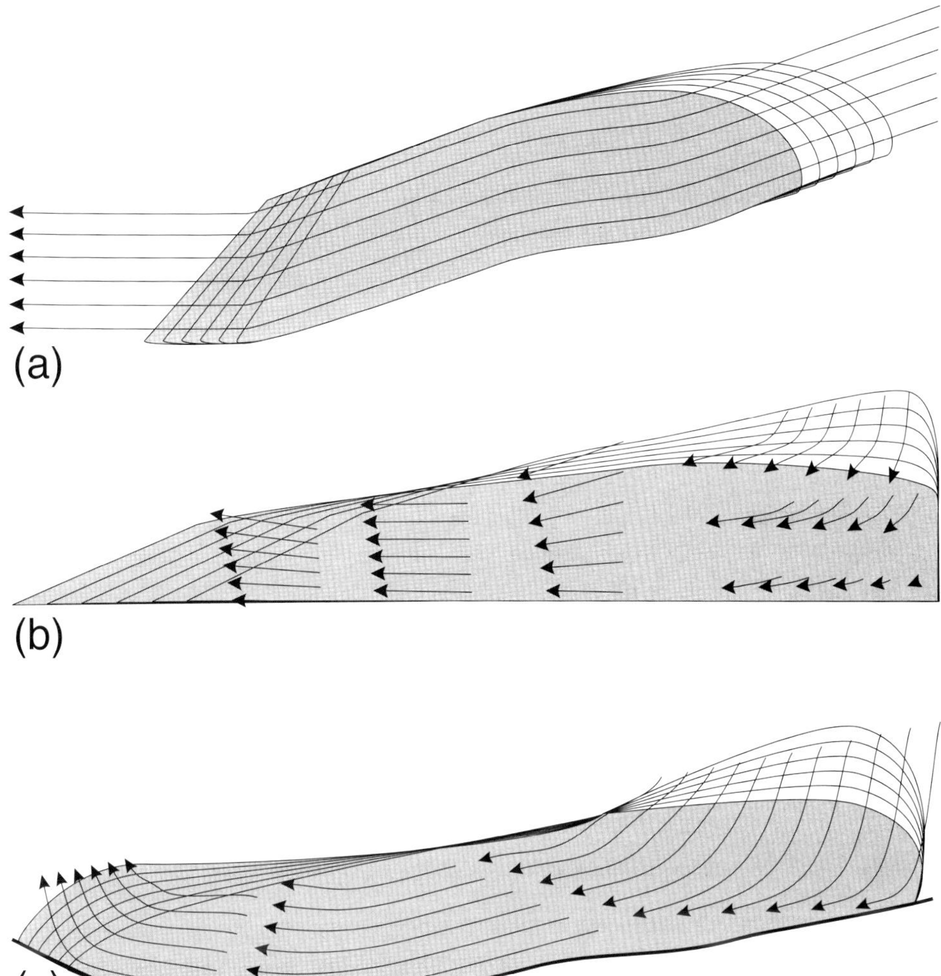

Figure 2. Gravity-driven deformation: (a) gravity gliding, in which a rigid block slides down a detachment; (b) gravity spreading, in which a rock mass distorts under its own weight by vertical collapse and lateral spreading; and (c) mixed-mode deformation. Shaded areas are the final stages and arrows show material movement vectors.

and cartoons. In the following sections, we first address the driving mechanisms of gravity gliding and gravity spreading and use the northern Gulf of Mexico to illustrate how the importance of these processes may vary through time. Second, we focus on salt-detached fold belts and discuss the location, timing, rates, and styles of deformation using examples from the northern Gulf of Mexico and the Angola offshore. Third, we address similar aspects of shale-detached fold belts using examples from the Niger Delta and the northwestern Gulf of Mexico. Finally, we briefly compare and contrast deep-water fold belts to the tectonically driven fold belts of collisional orogens and accretionary prisms. Our ultimate goal is to give readers a better understanding of the processes responsible for deep-water fold belts and the many parameters that influence the observed fold geometries and evolution through time.

DRIVING MECHANISMS

Gravity Gliding and Gravity Spreading

The failure of passive margins is the result of some combination of gravity gliding and gravity spreading. Gravity gliding is strictly defined as the rigid translation of a body down a slope, with displacement vectors parallel to the detachment plane (Figure 2a). Gravity spreading is the vertical collapse and lateral spreading of a rock body under its own weight because of a sloping upper surface (Figure 2b) (DeJong and Scholten, 1973; Ramberg, 1981). In reality, it can be difficult to differentiate between gravity gliding and gravity spreading (Schultz-Ela, 2001). For example, gravity gliding usually involves a surficial slope and thus includes some amount of gravity spreading (Brun and Merle, 1985) (Figure 2c).

The terms gravity gliding and gravity spreading are often used more loosely to characterize the overall deformation of passive margins, such that margins with basinward-dipping detachments are termed gravity-gliding margins (e.g., offshore Angola, Mauduit et al., 1997; and offshore Brazil, Cobbold and Szatmari, 1991, and Demercian et al., 1993); and those having prograding deltas and landward-dipping detachments are considered gravity-spreading margins (e.g., northern Gulf of Mexico; Worrall and Snelson, 1989). Here, we apply the following definitions: the gravity gliding component of the deformation is that component controlled by any basinward slope of the detachment, whereas the gravity spreading component is that controlled by the surficial slope of the seabed (Raillard et al., 1997). Many margins are mixed-mode (Figure 2c), and it is difficult to determine the exact contributions made by gliding

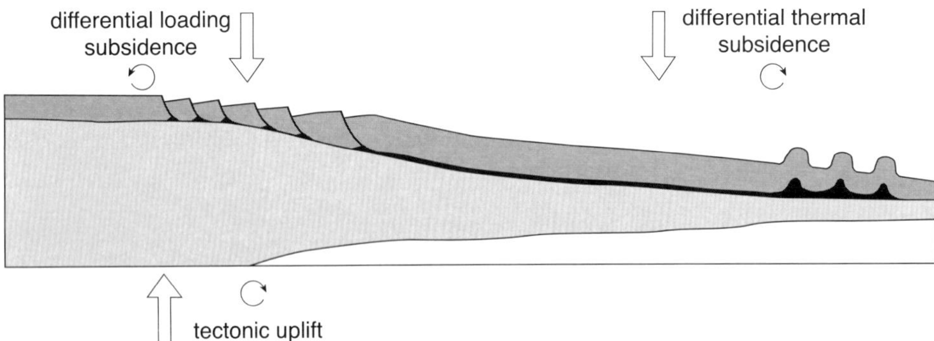

FIGURE 3. Passive margin with basinward-dipping detachment and thus a significant component of gravity gliding. Basinward tilting is enhanced by differential thermal subsidence and cratonic uplift and is reduced by proximal loading subsidence. Not to scale.

and spreading to the total deformation. Nevertheless, we distinguish between the two components for several reasons. First, doing so facilitates analysis of the many factors that influence the deformation. Second, different passive margins have structural styles and histories that were apparently caused by the dominance of one process or the other at various times, as will be discussed below for the northern Gulf of Mexico.

The primary factor in gravity gliding is the dip of the detachment. A steeper dip means that the detachment-parallel component of the weight of the overlying body is greater, and it is therefore easier to overcome the resistance to slip along the basal detachment. The detachment dip may change through time as a result of crustal-scale processes (Figure 3), such as (1) differential thermal subsidence between unthinned craton and oceanic crust, which will cause increased tilt; (2) differential flexural subsidence created by prograding deltaic loads or proximal carbonate buildups, which will decrease tilt; and (3) any episodic tectonic events that cause uplift of the craton and therefore increase tilt (e.g., the Tertiary uplift of west Africa; Duval et al., 1992; Spathopoulos, 1996; Marton et al., 2000; Cramez and Jackson, 2000).

The key parameters in gravity spreading are similar to those in the formation and growth of the critical wedges characteristic of collisional/accretionary fold belts: the dip of the surface slope, the dip of the basal detachment, the coefficient of friction along the detachment, and the internal strength of the wedge (Davis et al., 1983; Dahlen et al., 1984). Updip extension and downdip contraction (Figure 4a) are controlled largely by deposition of sediment. Progradational (proximal) deposition on the outer shelf and upper slope maintains or increases the overall seabed dip, which drives gravity spreading (Figure 4b). In contrast, sediment bypass to the lower slope and abyssal plain decreases the slope and thickens the distal overburden, both of which retard spreading (Figure 4c). Similarly, erosion of the shelf during a major sea-level lowstand will reduce the gravity potential and the tendency of the margin to spread under its own weight.

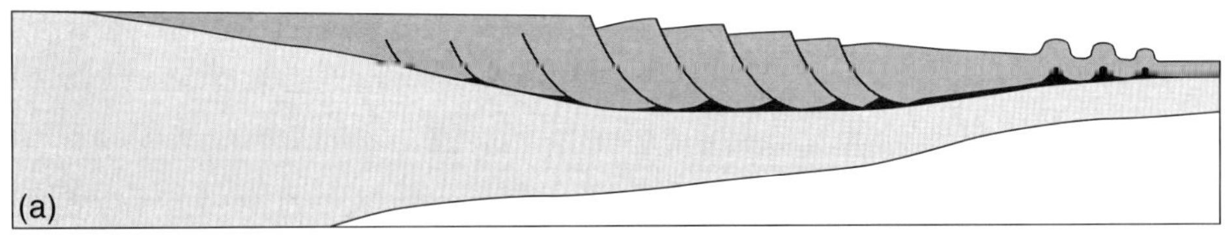

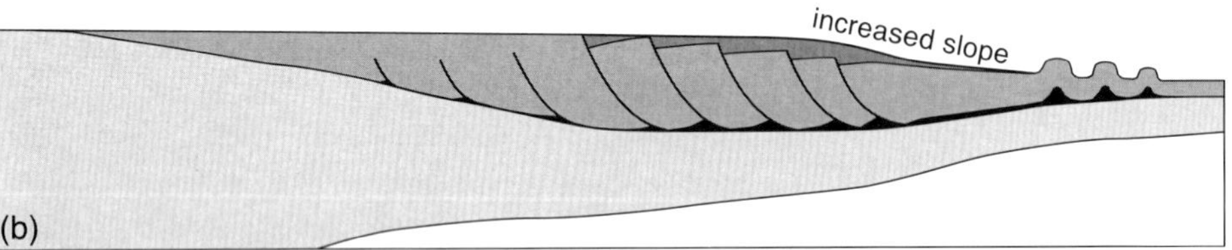

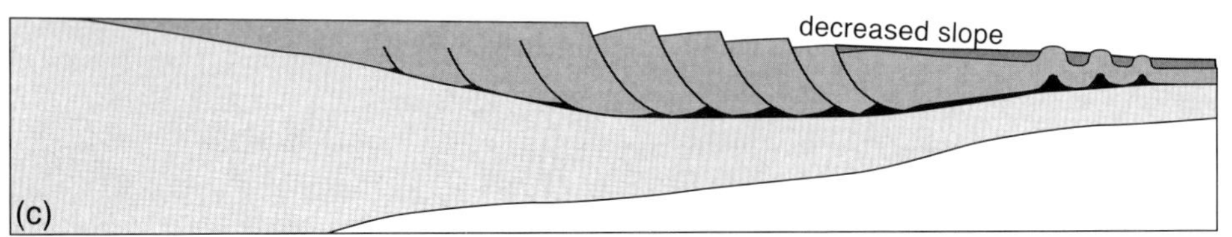

FIGURE 4. Passive margin with failure dominated by gravity spreading. (a) Progradational deposition along the outer shelf and upper slope increases the surficial slope and thus drives further spreading. (b) Upper-slope bypass and distal deposition on the lower slope and abyssal plain reduces the surficial slope and the gravity potential, slowing or stopping spreading. (c) Not to scale.

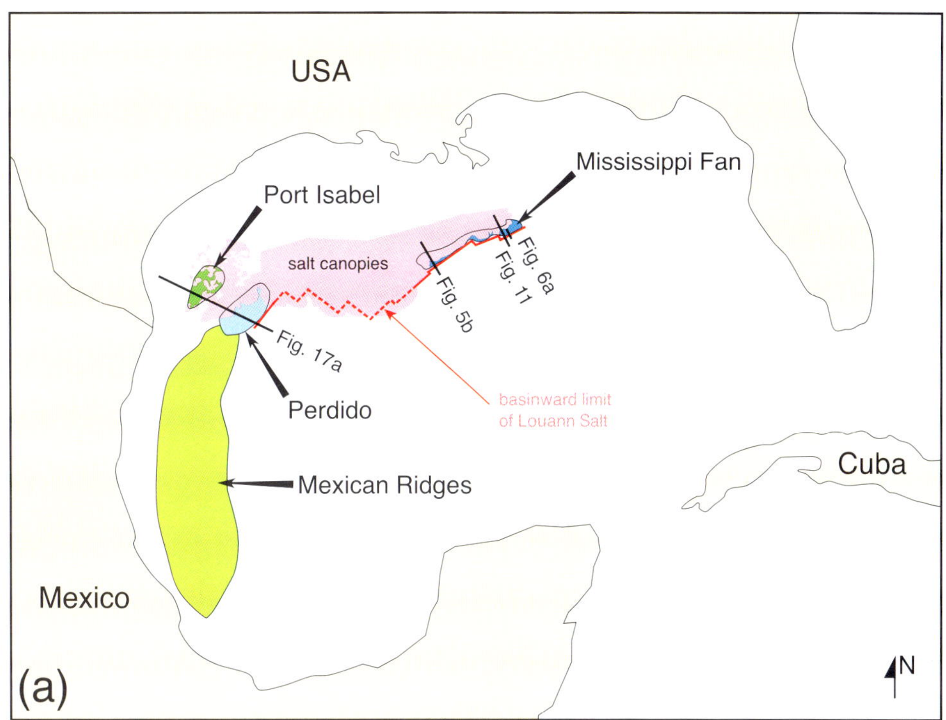

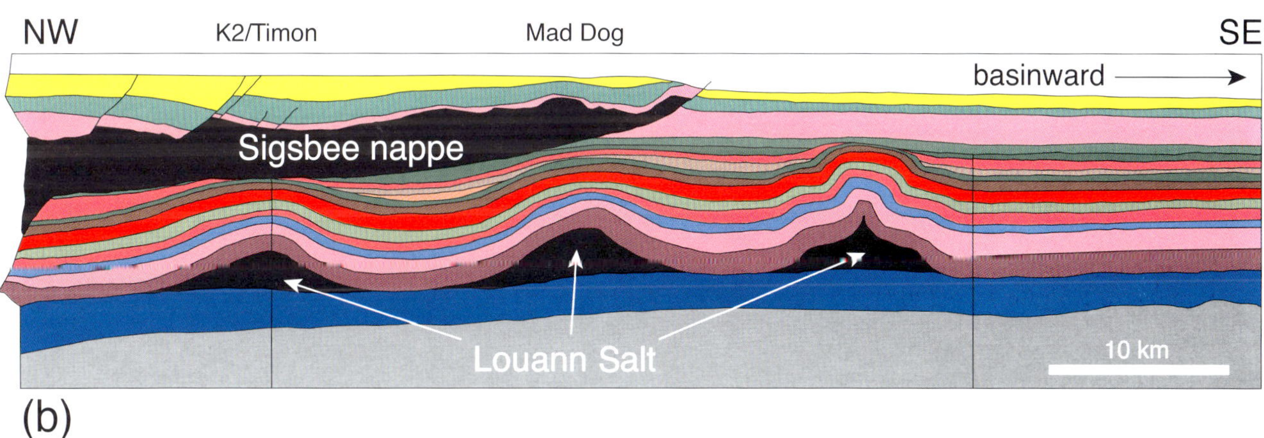

FIGURE 5. (a) Map of the Gulf of Mexico, showing the locations of the salt-based Mississippi Fan and Perdido fold belts (blues) and the shale-based Port Isabel and Mexican Ridges fold belts (greens), the approximate basinward limit of the Louann Salt (red line), the extent of deep-water salt canopies in the northern part of the basin (pink), and the locations of illustrated cross sections. (b) Depth section through the western part of the Mississippi Fan Fold Belt (located on map) showing a series of symmetrical, salt-cored detachment folds between welded synclines. The basinward fold is the same one shown in Figure 12a; the landward two folds, which are announced oil discoveries (Mad Dog and K2/Timon), are overlain by the allochthonous Sigsbee salt nappe, which itself accommodates some shortening at its toe. Salt is in black; other stratigraphic units are not identified because of proprietary interpretation; thin vertical lines are artificial construction lines. No vertical exaggeration; interpretation by F. Peel.

Gravitational Failure in the Northern Gulf of Mexico

The eastern Mississippi Fan Fold Belt provides a nice example of the changing importance of gravity gliding and gravity spreading in driving margin failure. The fold belt is one of several along the northwestern Gulf of Mexico (Figure 5a), and it comprises a regular wave train of detachment folds cored by the Middle Jurassic Louann Salt (Figure 5b).

Structural geometries and thickness patterns observed on seismic data clearly show that the eastern Mississippi Fan Fold Belt had two phases of deformation followed by periods of quiescence (Figure 6a). When

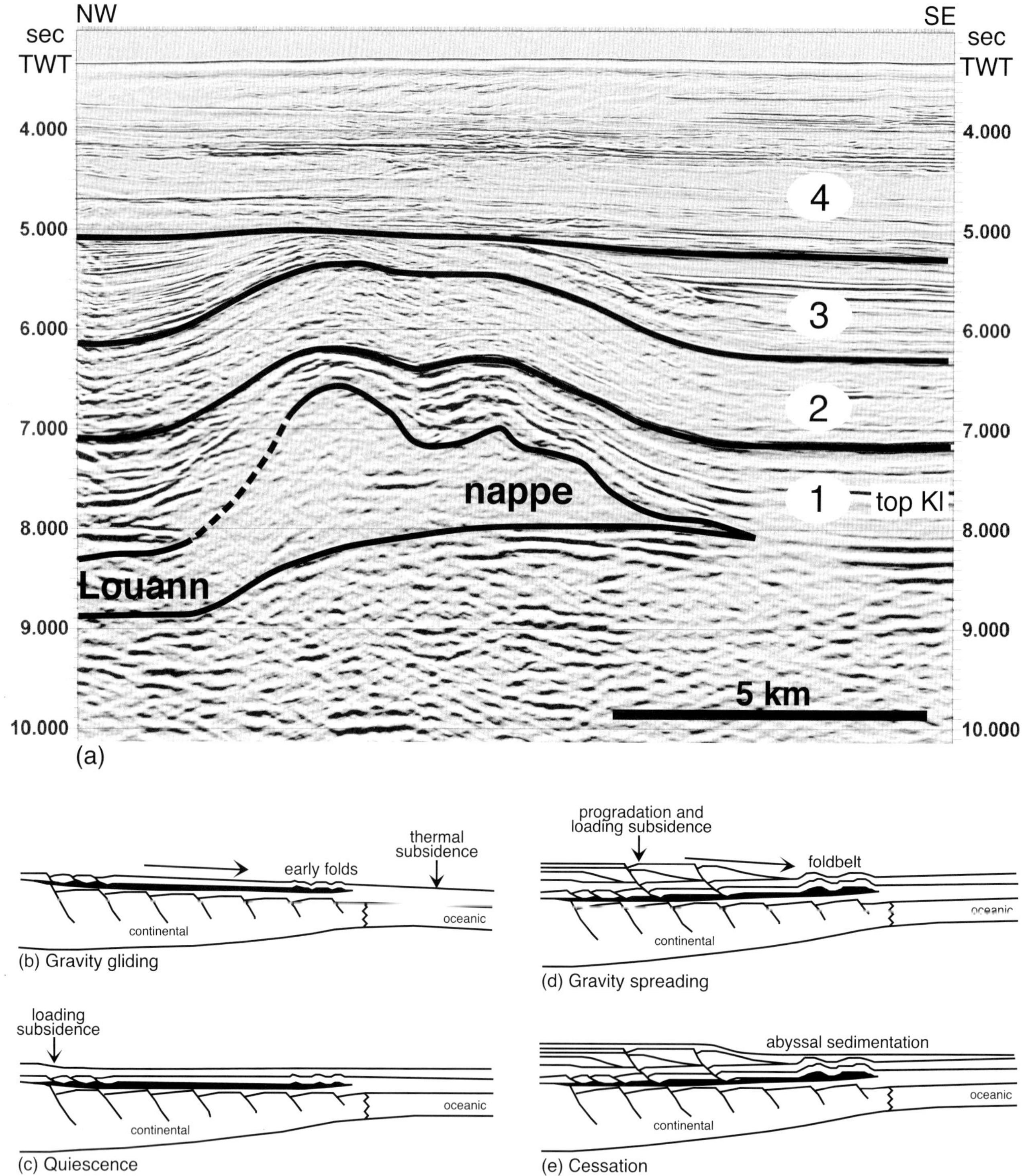

FIGURE 6. Four-stage evolution of the Mississippi Fan Fold Belt. (a) Time-migrated 2-D seismic profile (courtesy of TGS) showing frontal salt-cored fold; thickness patterns show early growth (interval 1), relative quiescence (interval 2), main growth (interval 3), and shut-off by the Mississippi Fan (interval 4). The toe of salt is within the Lower Cretaceous (Kl), showing that the fold occurs above an Upper Jurassic to Lower Cretaceous nappe sourced from the Middle Jurassic Louann Salt. (b–e) Proposed origin (modified from Rowan et al., 2000): (b) early gravity gliding caused by differential thermal subsidence and creating small-wavelength folds; (c) relative quiescence as waning thermal subsidence and proximal loading subsidence decreased the gravity potential; (d) main stage of gravity spreading caused by clastic progradation and resulting in large-wavelength folds; and (e) cessation of deformation as deepwater deposition of the Mississippi Fan reduced the gravity potential. Not to scale.

combined with the history of detachment dip reconstructed by McBride (1998) along a nearby megaregional cross section, the following four-stage history can be interpreted (Rowan et al., 2000). (1) Early, small-wavelength deformation was caused by gravity gliding above a detachment that was tilted basinward during Late Jurassic to Cretaceous cooling of oceanic crust to the south (Figure 6b). (2) There was relative quiescence during the Paleogene to middle Miocene as thermal subsidence waned and the basinward tilt was reversed by proximal loading subsidence (Figure 6c). (3) The quiescence was followed by renewed, large-wavelength shortening caused by gravity spreading as the clastic margin prograded basinward over a landward-dipping detachment during the late Miocene and early Pliocene (Figure 6d). (4) Finally, folding ceased as deep-water, bypass deposition of the Pliocene-Pleistocene Mississippi Fan reduced the bathymetric slope and the gravity potential (Figure 6e; compare to Figure 4c).

Self-limiting Behavior of Gravitational Failure

Both gravity gliding and gravity spreading are inherently self-limiting. Because the gravity potential must be great enough to overcome the internal strength of the sediments and because rocks generally are stronger under compression (Byerlee, 1978; Jaeger and Cook, 1979; Weijermars et al., 1993), it is the strength of the distal overburden that controls whether deformation will occur. Failure shortens and thickens the distal part of the wedge, thereby strengthening it. If the gravity potential remains constant, neither gliding nor spreading continues, and the margin will stabilize.

Gravitational failure is self-limiting in a more fundamental way, because the failure itself reduces the gravity potential. Both proximal extension and distal contraction lessen the dip of the bathymetric slope, thereby decreasing the force that is driving deformation. In the absence of any other process, movement continues only until the stress drops below the failure threshold. For deformation to continue, in the case of gravity spreading, deposition on the shelf and upper-slope is needed to continually maintain or increase the bathymetric slope. In the case of gravity gliding, either proximal extensional basins need to be replenished with sediment, or the basinward tilt of the margin needs to increase.

SALT DÉCOLLEMENT

Mechanics of Salt Flow

Salt deforms as a viscous material having little or no ultimate strength, and it can therefore flow when it is subjected to minimal shear stresses (Urai et al., 1986; Spiers et al., 1990; Weijermars et al., 1993). The overall rheological properties of salt are effectively constant through time, although reduction of both temperature and grain size during deformation may have some minor influence.

A tabular salt layer can deform by Poiseuille or Couette flow, or by a combination of both (Jaeger and Cook, 1979; Weijermars et al., 1993). Poiseuille flow corresponds to vertical thinning and lateral expulsion of salt from beneath sediment depocenters, in a manner similar to flow of a viscous liquid through a narrowing pipe. No lateral translation of the overburden is required. The rate of vertical overburden subsidence (and salt thinning) is proportional to (1) variations in the lithostatic pressure at the base of the overburden (hence, its density and thickness) and (2) the cube of the salt-layer thickness (Vendeville et al., 1993). Therefore, for identical sediment loads, Poiseuille flow is faster where the salt layer is thicker; and for a salt layer of a given thickness, Poiseuille flow is faster where the differential load is greatest.

Couette flow corresponds to layer-parallel simple shear of the salt layer as the overlying sediments are translated seaward. The salt layer neither thins nor thickens. The displacement rate of the overburden during Couette flow is directly proportional to (1) the thickness of the salt layer and (2) the applied shear stress (e.g., the slope-parallel component of the lithostatic pressure at the base of the overburden, in the case of gravity gliding). All other things being equal, deformation by Couette flow is thus faster where the salt layer is thicker.

Timing of Shortening

Because salt is a viscous material with effectively no strength and thus flows under minimal deviatoric stress, deformation can conceivably start immediately after deposition if there is an adequate driving force. This force is provided in most passive margins by early tilting of the margin toward the basin as a result of differential thermal subsidence, so that salt-detached deformation almost always has an early gravity-gliding component. Basinward translation then persists for as long as the gravity potential is maintained by further tilting or sediment progradation. This may take place continuously or episodically, depending on the specific history of a given area.

The weakness of salt also affects the relative timing of different structures in the fold belt. Instead of developing in a forward-propagating sequence, as is typical of shale-detached fold belts, salt-cored folds tend to initiate and grow simultaneously (Davis and Engelder, 1985).

Salt withdrawal and diapir growth typically accompany basinward translation of the overburden, so that the dêcollement layer thins locally as translation progresses. Thus, if the applied shear stress remains constant, Couette flow of the salt and lateral movement of the overburden will slow with time. There is a common misconception that once the salt layer is totally evacuated, thereby forming a weld (Jackson and Cramez, 1989), deformation should cease (e.g., Wu and Bally, 2000). However, the weld is an established detachment plane with listric normal faults and thrust faults soling into it, and it has low shear strength because there are usually patches of remnant salt left after welding. If the gravity potential is large enough to overcome the resistance to shear along the weld, deformation will continue after the salt is evacuated.

Evolution of the Offshore Angolan Fold Belts

The Lower Congo and Kwanza Basins of offshore Angola (Figure 7a) provide instructive examples of some of the timing issues in salt-detached gravitational failure. Both basins have proximal extensional domains and deep-water fold belts above Aptian salt (Duval et al., 1992; Spathopoulos, 1996; Marton et al., 2000; Cramez and Jackson, 2000) (Figures 7b, 7c, 8a).

Thickness variations in the strata just above the salt (Figures 7b, 8a, 8b) indicate that folding started soon, if not immediately, after salt deposition. Moreover, the folds developed simultaneously rather than in a basinward- or landward-propagating sequence. The evolution is seen best in a sequential restoration of one line from the Kwanza Basin, where all folds began growing very early (Figure 8e, f). This early deformation was probably caused by differential thermal subsidence of the margin. Similarly, Cramez and Jackson (2000) illustrate Upper Cretaceous shortening in the middle to lower slope of the Lower Congo Basin and attribute it to early tilt (1–3°) of the detachment layer. The early contraction was linked to synchronous extension in more proximal settings (Duval et al., 1992; Marton et al., 2000; Cramez and Jackson, 2000).

Two lines of evidence prove that basinward translation then continued until the present day. First, the polyharmonic fold geometries and sustained growth show that shortening proceeded as the overburden thickened (Figure 8b–f). Second, progressive migration of growth synclines in the Kwanza Basin mark the gradual translation of the sediment carapace over the Atlantic Hinge Zone, a basinward-dropping step in basement (Peel et al., 1998) (Figure 8b–f).

Although translation (and thus shortening) was ongoing, it did not occur at a constant rate (Duval et al., 1992; Spathopoulos, 1996; Hartman et al., 1998; Marton et al., 2000; Cramez and Jackson, 2000). Early (Upper Cretaceous) translation rates decreased, probably because of the waning of thermal subsidence. However, two pulses of cratonic uplift and associated progradation, in the late Paleogene and late Neogene, increased the gravity potential and led to renewed rapid translation. Note that at least this last phase of movement occurred after portions of the salt layer were welded. Also, detailed analysis of the structural geometries shows that the notion of proximal extension and distal contraction, although largely valid, is simplistic; in reality, extensional and contractional deformation can be superposed in complex patterns (Cramez and Jackson, 2000).

Location of Fold Belts

Deep-water fold belts commonly are found in a zone just landward of the distal pinchout of the original salt layer (Letouzey et al., 1995; Fiduk et al., 1995) (Figure 5a). This pinchout may have an irregular geometry in map view, so that frontal fold trends may be highly variable (Rowan et al., 2000). Shortening may also occur at the toe of the slope, where the compressive stresses are greatest. Thus, there may be a broad zone of deformation, extending from the base of the slope to the salt pinchout, with synchronous fold development (Letouzey et al., 1995). Alternatively, there may be two initial zones of folding, one near the toe of the slope and one near the salt pinchout (Figures 9b and 10a). As the slope progrades through time and the deformation front advances, these two fold belts can eventually coalesce (Vendeville, 2000) (Figures 9a and 10b).

The principal direction of tectonic transport is downslope, which is rarely constant along continental

FIGURE 7. Deepwater contractional deformation in offshore Angola. (a) Map showing the Lower Congo and Kwanza Basins and the locations of seismic profiles and cross sections; thin lines denote lease block boundaries and dashed lines are bathymetric contours. (b) Seismic profile from the Lower Congo Basin (located on map) showing a regular wave train of symmetrical, polyharmonic detachment folds cored by salt (outlined by dashed line). The prominent bathymetric high on the left is underlain by a squeezed diapir that extruded salt basinward. Growth strata show that folding started very early and continued to the present. Adapted from Marton et al. (2000). (c) More basinward seismic profile from the Lower Congo Basin (located on map) showing an inflated salt massif (outlined by dashed line) with a thin, folded overburden comprising the entire suprasalt section. At least some of the salt is allochthonous, having been extruded over the abyssal plain as a nappe, and landward-dipping bands of higher-amplitude reflections (thin dashed lines) within the salt show internal shortening of more competent beds within the salt massif. Profiles from Marton et al. (2000), copyright American Geophysical Union, modified by permission of American Geophysical Union.

margins. Promontories and embayments of the shelf and slope can cause divergent and convergent movement vectors, respectively, resulting in broadly arcuate fold belts (Cobbold and Szatmari, 1991; Cobbold et al., 1995; Spencer et al., 1998). For example, prograding sediment lobes associated with large deltas and deep-sea fans deform by divergent spreading (Vendeville, 2000; Gaullier et al., 2000b). In the case of convergent

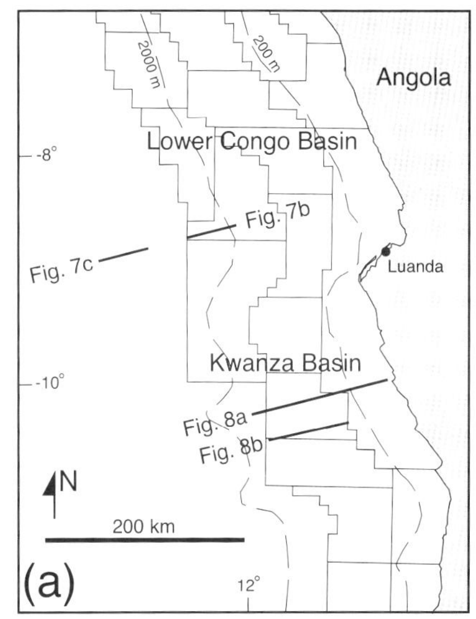

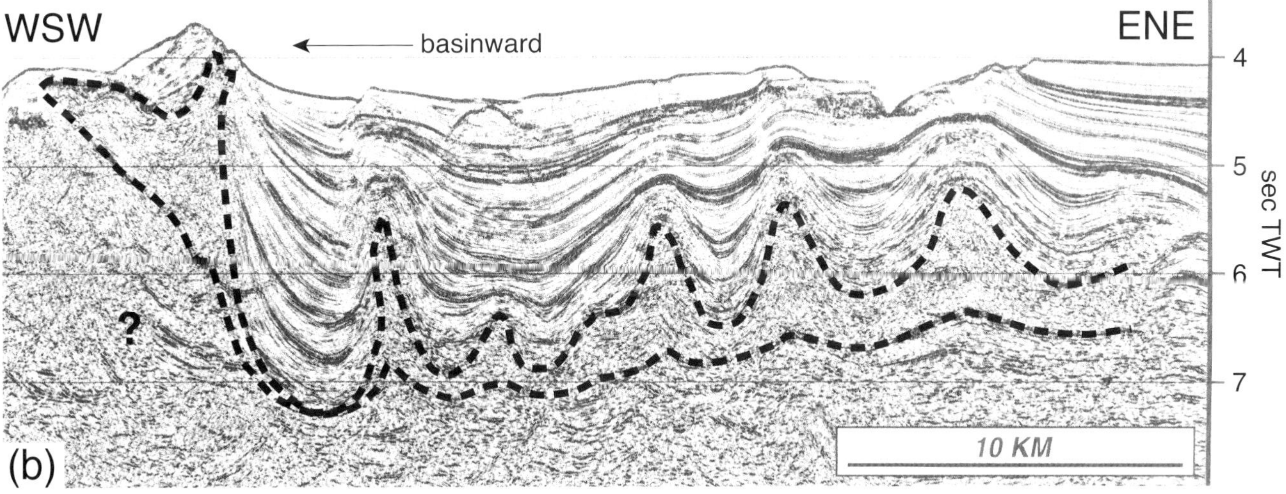

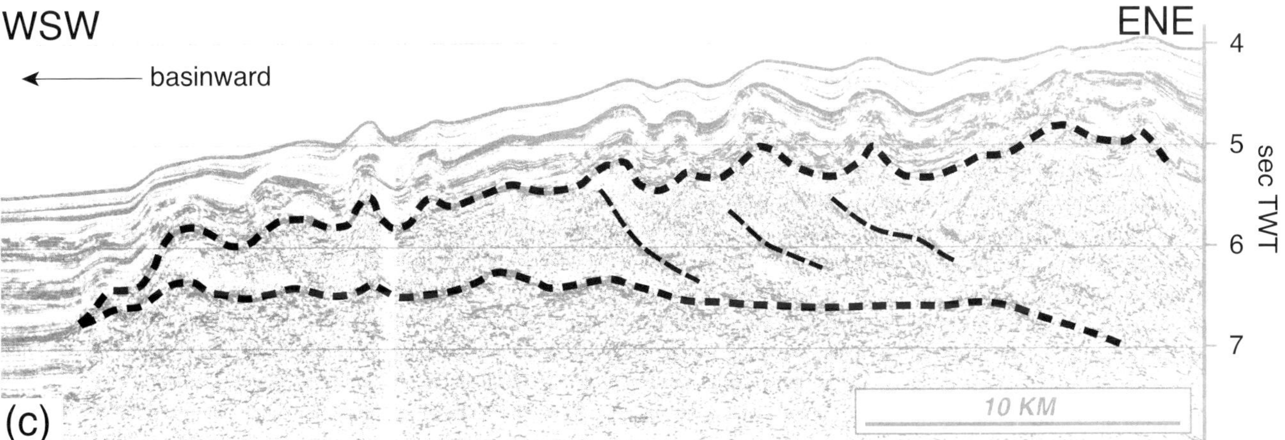

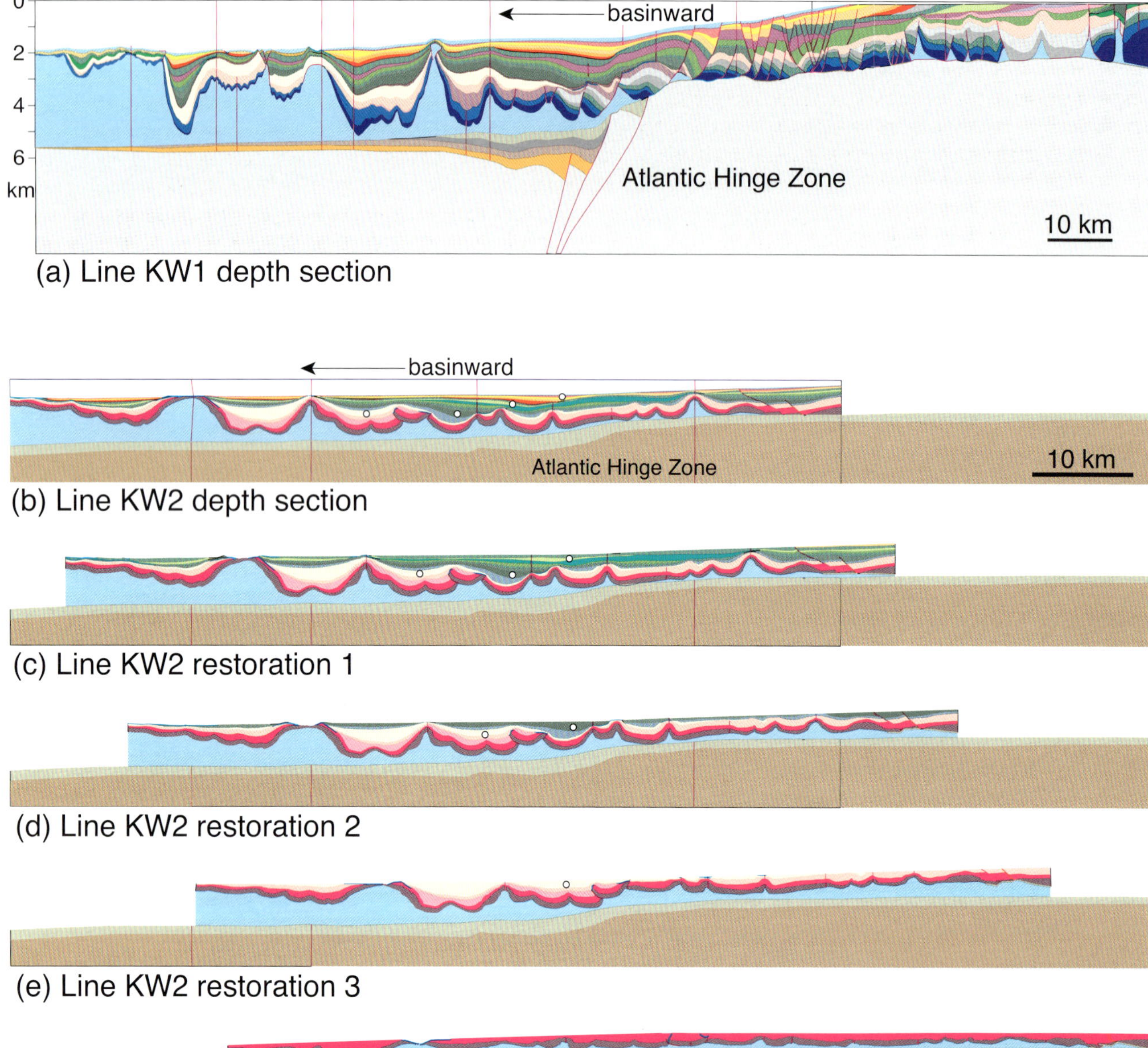

Figure 8. Cross sections from the Kwanza Basin, offshore Angola. (a) Regional depth section (located on Figure 7a) showing linked proximal extension and distal contraction above a basinward-dipping salt layer. The Atlantic Hinge Zone, formed by underlying basement faults, marks the transition between originally thin and thick salt and, at least on this profile, the approximate transition between extension and contraction. Shortening basinward of the hinge zone is accommodated by a combination of detachment folds, partially squeezed diapirs, and a shortened and inflated salt massif with a thin, folded overburden comprising the entire post-salt stratigraphic section. Salt is in light blue; other stratigraphic horizons are not identified because of proprietary interpretation. 4:1 vertical exaggeration; interpretation by F. Peel. (b–f) Depth section and sequential restorations (constructed using Locace) of seismic profile (located on Figure 7a) showing evolution of salt-cored detachment folds and inflated salt (light blue). Early, small-wavelength folds (e–f) gave way to larger-wavelength folds as the overburden thickened (b–d). The amount of lateral transport of the sediment carapace is constrained by the landward migration of growth synclines (small circles) formed over the Atlantic Hinge Zone (Peel et al., 1998). No vertical exaggeration; interpretation and restorations by F. Peel. Thin vertical lines in all panels are artificial construction lines.

translation, fold interference may form dome-and-basin structures.

The location and orientation of deep-water folds also depend on the morphology of the original salt basin. Just as folds form landward of the distal pinchout of salt, they may form wherever the salt changes thickness as a result of the underlying rift architecture. Where the base salt or its equivalent weld ramps up in a basinward direction, the increased resistance to overburden translation will localize folding, landward of the barriers, with possible landward propagation of shortening (Jackson et al., 1998). Similarly, shortening may occur where the basinward dip of the décollement decreases or even reverses because of a crustal-scale hinge (e.g., Figure 8a) (Fox and Ashton, 2000). In either case, steps in the base salt or weld that are oblique to the tectonic transport direction may cause local variations in fold trend.

Salt Evacuation and Inflation

A thick salt layer can have another significant impact on how gravitational failure is accommodated. The gravity potential can be greatly reduced by overburden subsidence into salt in proximal regions and inflation of salt in distal regions (Letouzey et al., 1995), possibly with only minimal lateral translation. This process has been modeled by Ge et al. (1997) and can be seen in the experimental model of Figure 9. Distal salt inflation has been reported from the northern Gulf of Mexico (Hall, 2000) (Figure 11) and offshore Angola (Marton et al., 2000; Cramez and Jackson, 2000) (Figure 8b–f).

Effects of Diapirism and Salt Extrusion

The presence of diapirs, even dormant ones, affects the localization of folds, because diapirs are the weakest parts of the rock volume and locally reduce the overall strength of the overburden. Many folds are associated with diapirs (Figure 12), and Rowan et al. (2000) argue that diapirs in the Mississippi Fan Fold Belt predated the shortening, rather than vice versa. Shortening is preferentially accommodated by the lateral squeezing of preexisting diapirs, driving additional diapirism (Vendeville and Nilsen, 1995; Nilsen et al., 1995; Rowan, 1995; Vendeville, 2000; Rowan et al., 2000; Cramez and Jackson, 2000). In three dimensions, squeezed diapirs act as nucleating points that localize shortening in the adjacent strata, with folds and/or thrusts developing along strike (Figure 12) (Vendeville, 2000).

Where distal salt inflates, forming an early massif, the salt may break through the thin overburden and extrude as a salt nappe that spreads basinward onto the abyssal strata (Ge et al., 1997). Lateral translation into the salt massif and consequent nappe extrusion can accommodate much of the gravity-driven shortening along portions of a margin (Peel et al., 1995; Trudgill et al., 1999). Nappes have been recognized in the Brazilian offshore (Mohriak et al., 1995; Mohriak, 1995) and the Angolan offshore (Marton et al., 2000; Cramez and Jackson, 2000), and there was a paleonappe in the northern Gulf of Mexico (Figure 6a) (Peel 2001; Rowan et al., 2001).

Preexisting diapirs (Figure 13a) can also extrude salt laterally when the rate of upward salt flow is increased by shortening and narrowing of the diapirs.

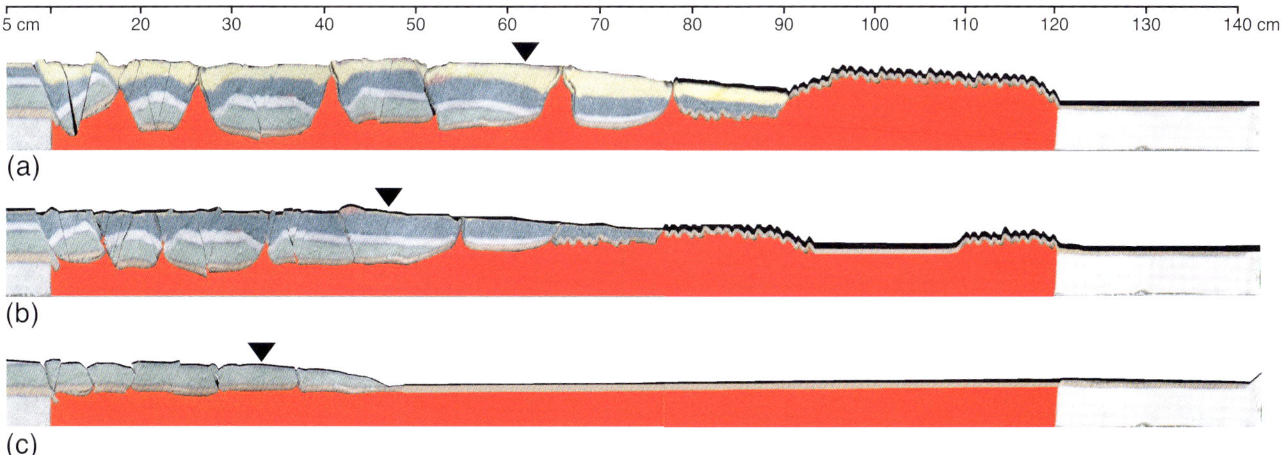

FIGURE 9. Three stages of an experimental model carried out by B. Vendeville. Progradational loading above the thick viscous substrate (red) drives gravity spreading, which is accommodated by proximal subsidence and extension and distal inflation and shortening. Contraction is initially concentrated in two bands, at the toe of the slope and at the distal pinchout of the viscous décollement (b). Progradation of the slope then results in a merged zone of inflation and folding in front of the slope (a). Black triangles denote shelf-slope break; 2:1 vertical exaggeration of original models.

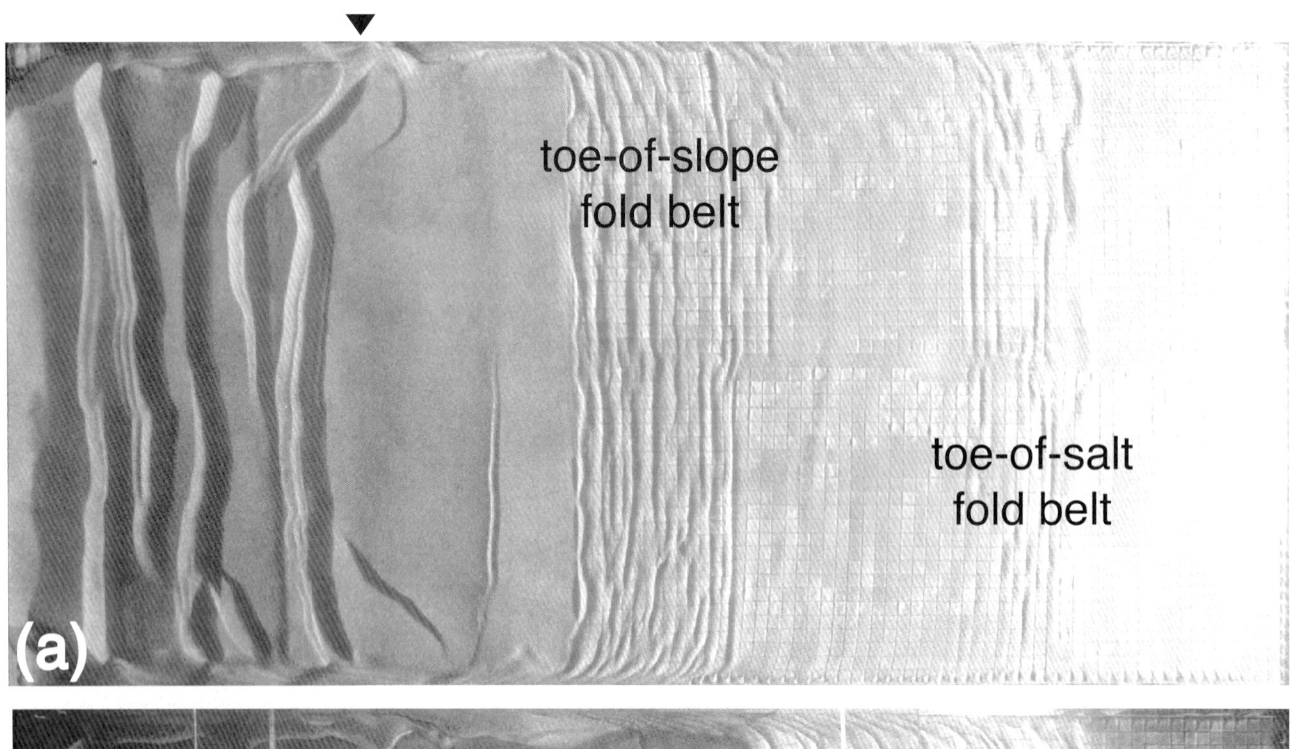

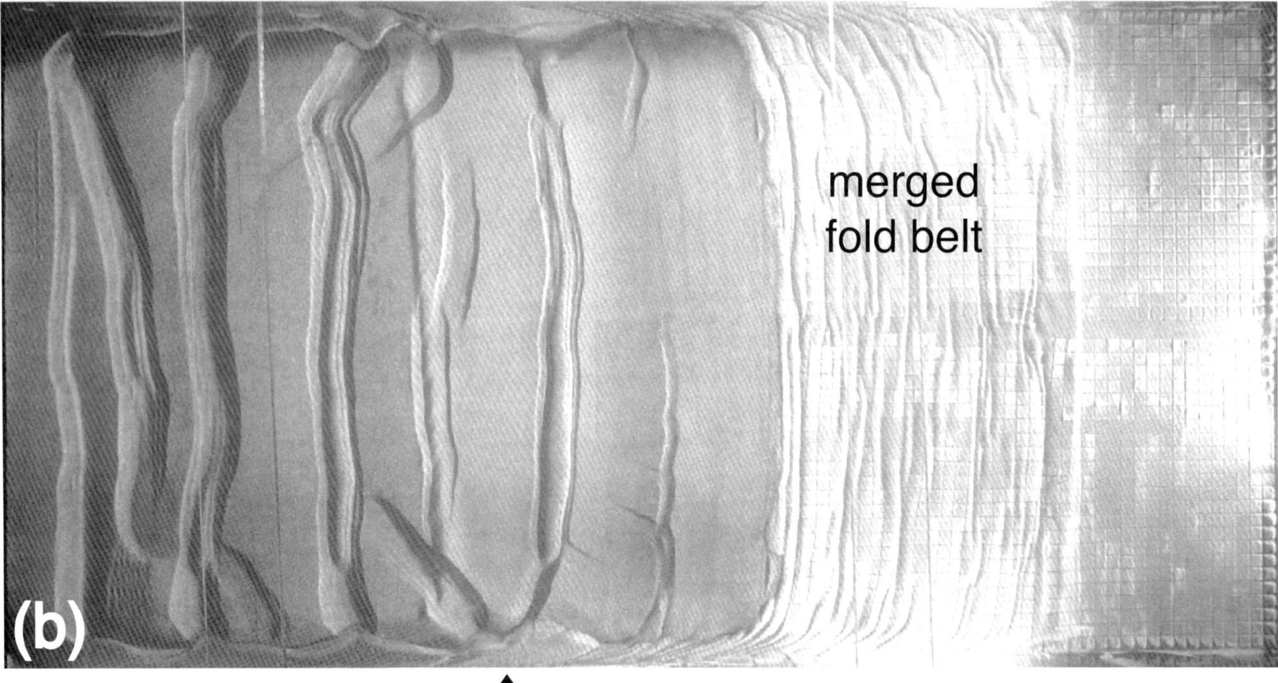

FIGURE 10. Overhead views of two of the stages from the experiment of Figure 9, showing the basinward progression of extensional grabens (left) and contractional folds (right). (a) corresponds to (b) of Figure 9, with two initial bands of folding, and (b) corresponds to (a) of Figure 9, with a merged zone of deformation between the toe-of-slope and the distal pinchout of the viscous layer. Black triangles denote shelf-slope break.

Depending on the salt supply and the diapir spacing, the extruded salt can coalesce to form laterally continuous salt canopies (Figure 13b) (Jackson and Talbot, 1989; Schuster, 1995; Diegel et al., 1995; Rowan, 1995). In turn, a canopy may form a shallow dêcollement level, depending on its base-salt geometry and areal extent. Linked extension and contraction above the allochthonous salt may then accommodate much of the gravitational failure of the seabed slope, thereby reducing the amount of shortening taken up by folding at

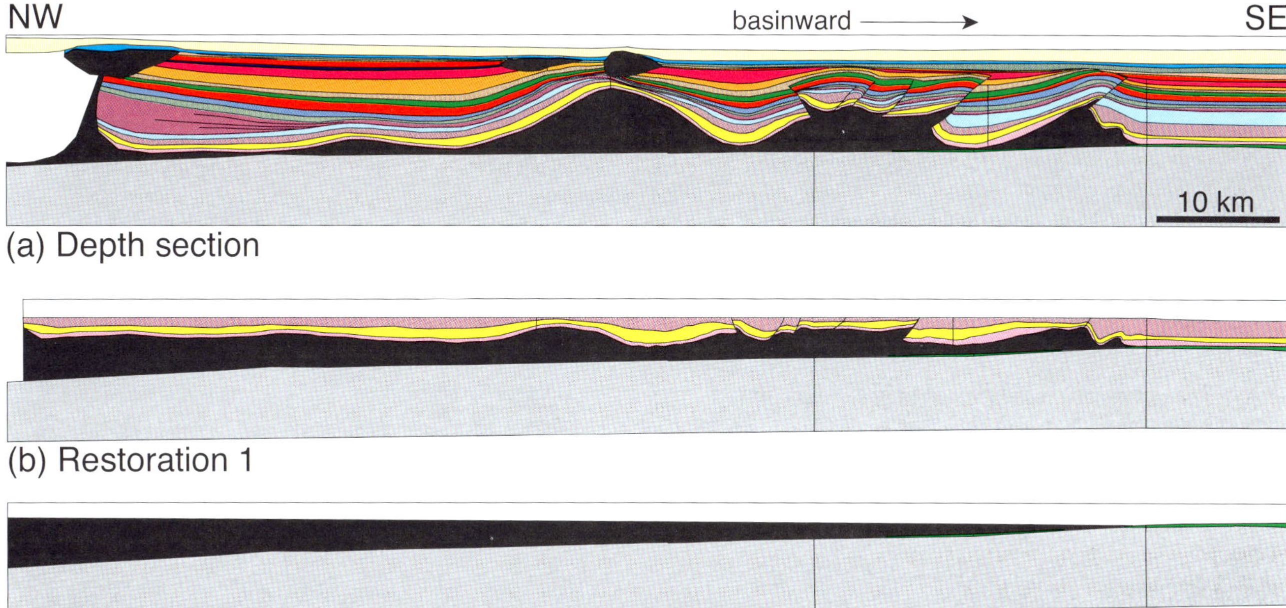

Figure 11. (a) Depth section through the eastern part of the Mississippi Fan Fold Belt (located in Figure 5a) showing both symmetrical, salt-cored detachment folds and faulted detachment folds. The small, shallow blobs of salt over the middle structure are parts of overhanging salt fed from diapirs off-section, similar to the diapir at the left end of the profile. Salt is in black; other stratigraphic units are not identified because of proprietary interpretation. No vertical exaggeration; interpretation by F. Peel. (b–c) Sequential restorations of the early evolution of the depth section in (a) constructed using Locace. Note the early contractional deformation and overall salt inflation (overburden thin compared to abyssal plain) in (b). No vertical exaggeration; restorations by F. Peel. Thin vertical lines in all panels are artificial construction lines.

the deeper, autochthonous level (Figure 13c; see also the shortening at the toe of the allochthonous salt in Figure 5b).

Amounts and Rates of Shortening

Quantitative measurements of the amount of shortening in gravity-driven fold belts are rare and highly variable. They are derived from structural restorations of interpreted seismic profiles. Published results include 10 to 70 km in the Lower Congo Basin (Raillard et al., 1998; Marton et al., 2000); 100 km in the Campos Basin (Demercian et al., 1993); 10 km in the Mississippi Fan Fold Belt (Peel et al., 1995); and greater than 5 km in the Perdido Fold Belt in the northwestern Gulf of Mexico (Peel et al., 1995; Trudgill et al., 1999). Shortening values from the restorations presented here are 6 km for the Mississippi Fan Fold Belt (Figure 11) and 27 to 30 km for the Kwanza Basin (Figure 8b–f).

Shortening accommodated in fold belts does not necessarily represent the total shortening in a linked system of margin failure and thus may not balance the total measured extension in more proximal areas. As we have mentioned, some of the shortening can be accommodated by extrusion of salt nappes, lateral squeezing of diapirs or salt massifs, or shortening above allochthonous salt sheets/canopies. Data from the Lower Congo Basin suggest that shortening of diapirs and a salt massif accounted for 74% of the total shortening (Marton et al., 2000).

Shortening rates are even rarer than net shortening measurements but are perhaps a better measure of gravity-driven deformation. Published values include 0.1–0.2 mm/yr for the northern Gulf of Mexico (Rowan et al., 1993, 2000), and the average rates from the restorations shown here are 0.1–0.5 mm/yr for the Mississippi Fan Fold Belt (Figure 11) and 0.4–0.5 mm/yr for the Kwanza Basin (Figure 8b–f).

Structural Styles of Salt-cored Fold Belts

The structural style of a fold belt is highly dependent on the rheology of the dêcollement layer. A frictional detachment leads to the development of an asymmetric imbricate fan of low-angle thrusts and associated fault-bend and fault-propagation folds (Figure 14a). On the other hand, a viscous basal layer promotes symmetrical detachment folds with only minor, higher-angle faulting (Figure 14b). This is in agreement with critical-taper theory: Strong detachments are characterized by basinward-vergent, low-angle faults, whereas

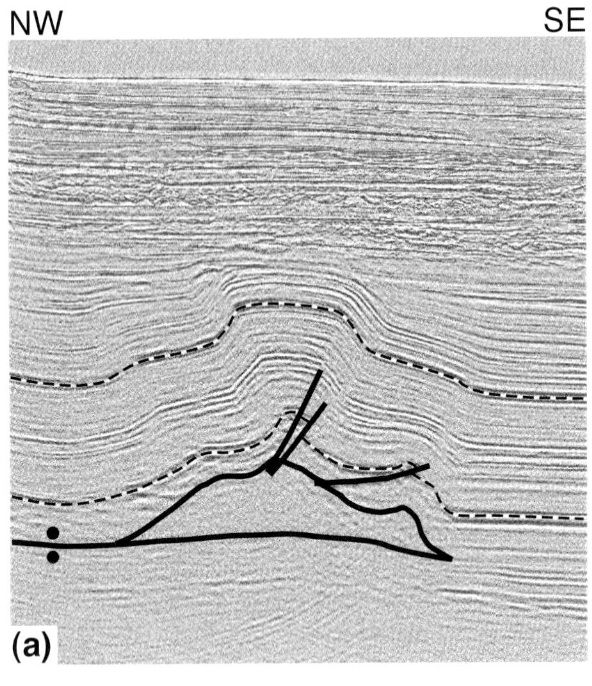

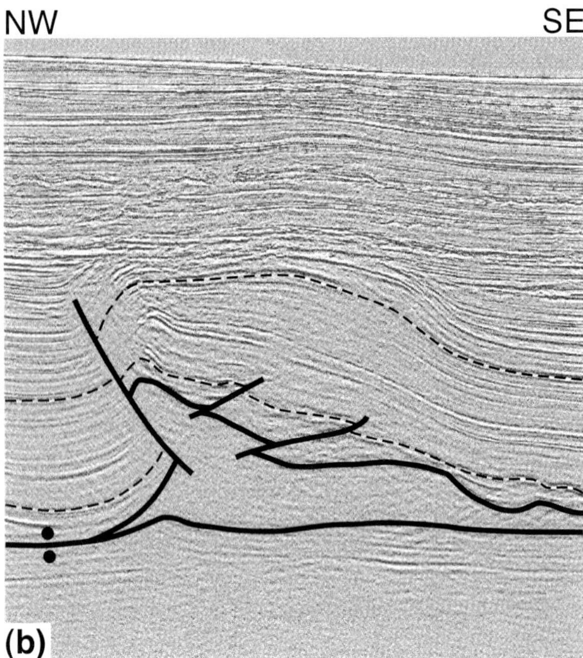

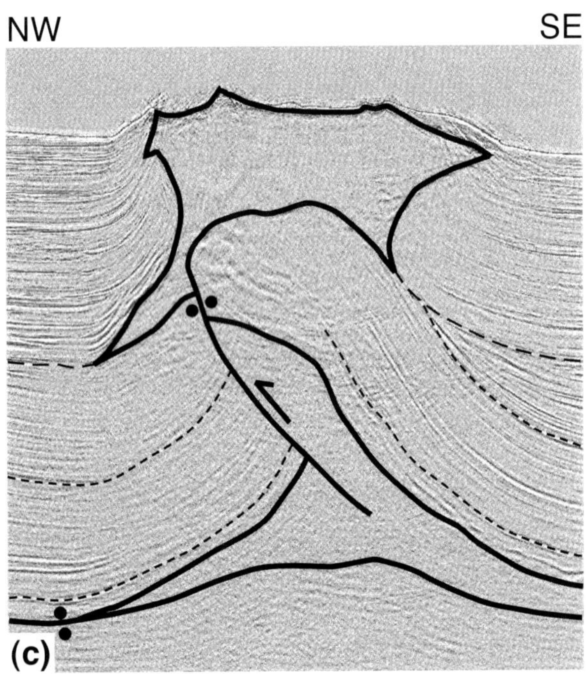

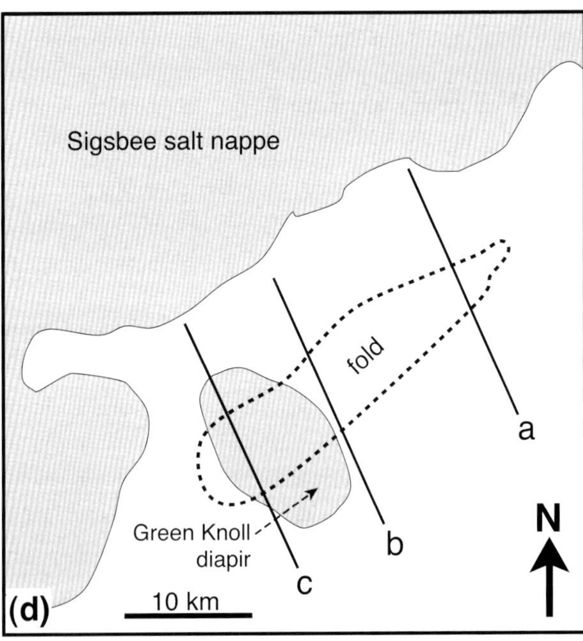

FIGURE 12. Relationship between the Green Knoll diapir and a salt-cored detachment fold in the western Mississippi Fan Fold Belt. (a) Symmetrical geometry with angular fold hinges. (b) Asymmetric geometry with landward-vergent, high-angle reverse fault (the apparent hanging-wall thickening is due to steep out-of-plane dips). (c) Complex relationship between fold geometry and salt diapir, in which the diapir is interpreted to have predated the main contractional phase (as suggested by adjacent growth patterns) and then been squeezed during shortening, thereby driving further salt extrusion. The original diapir is locally welded, with the weld being used as a reverse fault. (d) Map showing the diapir near one end of the fold. For location, the frontal fold of the regional line in Figure 5b is in essentially the same position as profile (a). Seismic data are from a 3-D time volume, courtesy of WesternGeco.

weak detachments result in steeper fault planes having no preferred vergence (Davis and Engelder, 1985; Jaumé and Lillie, 1988).

The structural style also depends on the thickness of the dêcollement layer. A thick salt layer can readily flow from beneath the subsiding synclines toward the

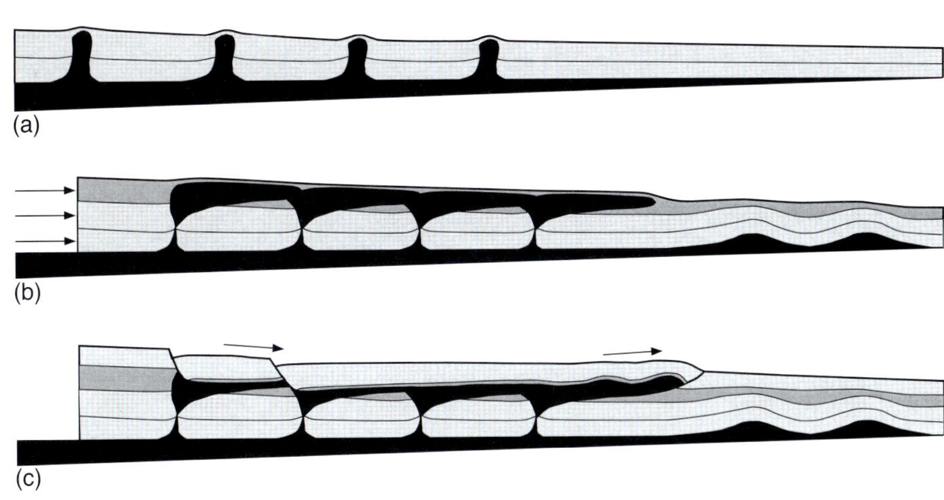

FIGURE 13. Roles of diapirs and allochthonous salt canopies in accommodating gravitational deformation. Diapirs in (a) localize contractional strain, so that net shortening in (b) has been taken up by both squeezing of the diapirs and distal folding. Salt extruded from the narrowing diapirs flows laterally and amalgamates to form an allochthonous canopy. The canopy provides a shallower detachment level for linked extension and contraction, thereby reducing or even halting folding at the deeper level (c).

rising anticlines, to accommodate fold-limb rotation (Homza and Wallace, 1995; Poblet and McClay, 1996). In contrast, a thinner detachment layer is unable to fill growing fold cores, and further shortening is taken up by thrust faulting (Stewart, 1999).

The thickness and rheology of the sediment overburden also impact the structural style. For example, a thin overburden over thin salt can also produce detachment folds, because fold wavelengths and amplitudes are smaller and hence require a smaller amount of mobile material to accommodate limb rotation. The strength of the overburden is important, because the resistance of the overburden to deformation can limit how the deformation is accommodated. Thus, a sedimentary carapace may respond to slow shortening by simple folding and to more rapid shortening by thrust faulting. Other factors include the mechanical stratigraphy within the overburden sequence, including the relative percentages of different lithologies (e.g., carbonates versus fine-grained clastics; Rowan et al., 2000) and the presence of any preexisting structures, such as old normal faults, withdrawal basins, or salt bodies.

In the following sections, we identify three structural assemblages commonly occurring in the distal portions of passive margins that fail above salt dêcollements. We distinguish them based on the thickness of the salt layer.

Thick Salt (Figure 15a)

The characteristic structures where salt is thick enough to fill the cores of growing anticlines are salt-cored detachment folds having a regular wavelength. The folds are generally open because the amount of shortening is modest, and they may be cut by relatively small-displacement, high-angle reverse faults showing

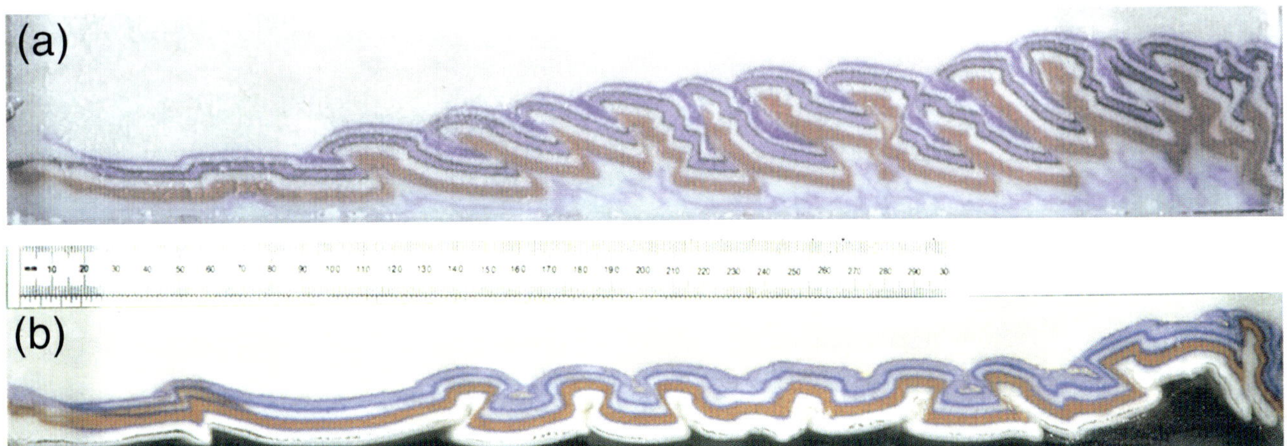

FIGURE 14. Results of two experimental models showing the effect of dêcollement rheology on structural style: (a) a frictional detachment results in asymmetric fault imbricates; and (b) a viscous dêcollement produces symmetrical detachment folds with minor faulting of the limbs. Experiments carried out by B. Vendeville.

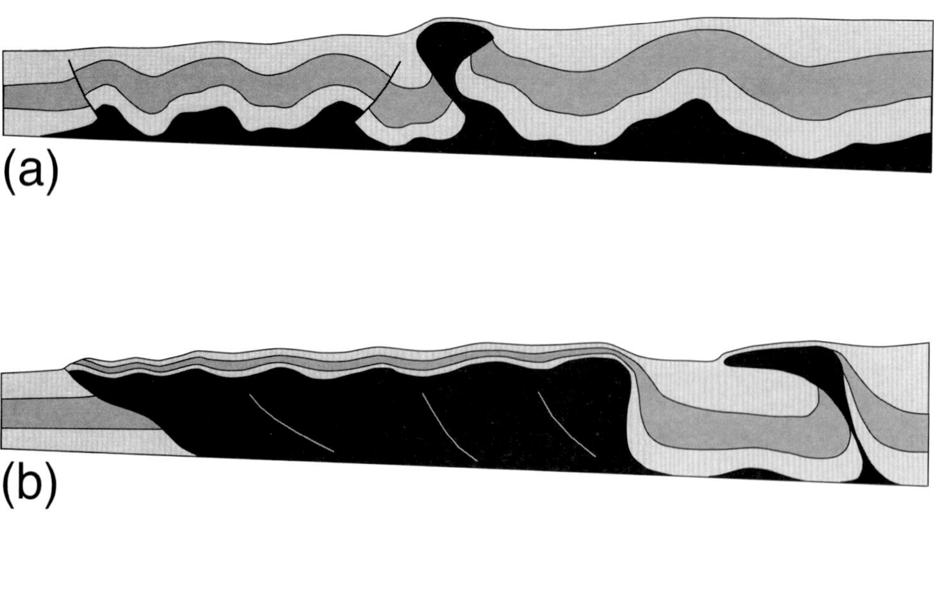

FIGURE 15. Structural styles typical of fold belts developed above salt dêcollements: (a) thick salt with symmetrical detachment folds and shortened diapirs (allochthonous canopies are common but not shown); (b) inflated and thickened salt massifs and nappes; (c) thin salt with both basinward- and landward-vergent thrust faults. See text for discussion.

no preferred vergence. Examples include the Mississippi Fan (Figures 5b, 6a, 11a) and Perdido Fold Belts of the northern Gulf of Mexico and portions of the Angolan Fold Belts (Figures 7b and 8). The folds commonly are polyharmonic, because the wavelength increases as the overburden thickens through time (Figures 7b, 8b–f) (Demercian et al., 1993; Peel and Matthews, 1999; Rowan et al., 2000).

The abundance of salt and the early triggering of deformation beneath a thin overburden usually mean that diapirs are common where salt originally was thick. The diapirs influence the deformation, leading to complex interactions of short fold segments, reverse faults, and diapirs (e.g., Figure 12) (Rowan, 1997; Rowan et al., 2000). Abnormally steep reverse faults typically are associated with diapirs and are interpreted as welds formed by complete squeezing of preexisting diapirs (e.g., Figure 12c). Diapirs squeezed during shortening have been reported from the northern Gulf of Mexico (Rowan, 1995; Rowan et al., 2000; Hall and Smith, 2001), offshore Angola (Duval et al., 1992; Jackson et al., 1998; Marton et al., 2000; Cramez and Jackson, 2000), and the Yemeni Red Sea (Heaton et al., 1995). As stated before, diapir constriction may cause extrusion of allochthonous salt, which may, in turn, amalgamate to form a regional canopy, thereby creating a new detachment level for gravity gliding/spreading (Figure 13).

Not all fold belts formed above thick salt display the early and long-lived history typical of these provinces. In the Perdido Fold Belt, located at the distal pinch out of the Jurassic Louann Salt (Figure 5a), there is a substantial constant-thickness section above the salt, and shortening did not occur until the Paleogene (Trudgill et al., 1999). The lack of early deformation was responsible for the absence of diapirs and the relative simplicity of the folds, compared with the Mississippi Fan Fold Belt (Rowan et al., 2000). The absence of early deformation probably means that initial gravity-driven contraction was accommodated upslope (landward) of the Perdido Fold Belt.

Massive Salt (Figure 15b)

The distal portions of salt-detached margins, located just landward of the salt pinch-out, are sometimes characterized by thickened salt massifs with a folded overburden (e.g., the Lower Congo Basin, Figure 7c). The overburden comprises the complete postsalt section that is thinner than its chronological equivalents both basinward and landward (Marton et al., 2000; Cramez and Jackson, 2000; Hall, 2000). The folds are generally symmetrical because of the thick underlying salt and have shorter wavelengths than elsewhere because the sediment carapace is thinner.

The thickened salt often extends beyond its original depositional pinch-out and ramps up through the abyssal section as a salt nappe, as found in the ultradeep water of offshore Brazil (Mohriak et al., 1995; Mohriak, 1995) and offshore Angola (Marton et al., 2000; Cramez and Jackson, 2000). The massifs commonly contain bands of landward-dipping, high-amplitude reflectors that suggest internal imbrication of the salt (Marton et al., 2000; Cramez and Jackson, 2000). Although the bands could represent parts of the sediment carapace that were incorporated within the salt massif as initially narrower individual salt bodies coalesced during shortening, it is more likely that they are nonevaporitic strata originally embedded within the overall evaporite sequence. There is an apparent lack of balance between the amount of shortening measured in the overburden and that apparent in the thickened salt (Marton et al., 2000; Cramez and Jackson, 2000), but this is explained by the large amount of salt that has moved into the massif from more proximal locations.

Thin Salt (Figure 15c)

As stated above, thin salt has insufficient volume to fill the cores of detachment folds and a significant percentage of the shortening is therefore accommodated by reverse faults (Stewart, 1999). The faults typically are vergent in both directions but may have an asymmetric vergence. Thin salt also means that large, long-lived diapirs are absent, because any early-formed diapirs run out of salt as the adjacent sediment depocenters weld out. In turn, a lack of diapirs means no shallower, allochthonous salt levels that would accommodate some of the gravitational failure.

Gravity-driven fold belts above thin salt appear to be rare. One example is in the middle-slope province of the Lower Congo Basin, where both seaward-vergent and bivergent Cretaceous thrust belts are interpreted (Cramez and Jackson, 2000). The rarity of this style does not mean that most passive margins initially had thick distal salt. Instead, early loading and deformation result in distal salt inflation, so that the resulting structural style is more typical of thick salt.

SHALE DETACHMENT

Mechanics of Shale Décollements

Unlike salt, which is viscous, overpressured shale has a rheological behavior independent of strain rate (Hubbert and Rubey, 1959; Weijermars et al., 1993). Shale is a plastic material, and it deforms only when the deviatoric stress overcomes the strength of the shale. The shear strength of the shale is the product of the shale's coefficient of friction and the effective vertical stress. The effective vertical stress is the lithostatic pressure at the base of the overburden minus the effect of the overpressure. The higher the fluid pressure, the weaker the shale detachment. As the pore-fluid pressure approaches the overburden load (i.e., when the coefficient of pressure nears one), the effective vertical stress tends to zero and the resistance to deformation of the shale becomes minimal. Extreme overpressures may be generated by rapid sedimentation; hydrocarbon generation (Barker, 1990; Morley and Guerin, 1996); tectonic compaction during shortening; and strain softening as clay plates rotate during shear and collapse, thereby removing the lithostatic support of the overburden and increasing the role of the fluid. In highly overpressured shales, the thickness of the décollement layer may affect the resistance to overburden translation, as occurs in the case of salt.

Overpressured Shale in the Niger Delta

The distal portion of the Niger Delta margin has several contractional belts that are detached on highly overpressured shale (Morley and Guerin, 1996; Wu and Bally, 2000). The degree of overpressure is indicated indirectly by the presence of shale diapirs, but more directly by velocity inversion within the décollement layer (Figure 16a). The amount of velocity pushdown beneath thickened masses of shale suggests that fluid pressures approach 98%, or even more, of the overburden pressure.

Closer inspection of the seismic data (Figure 16a) reveals that the décollement layer actually consists of two thick shale intervals separated by a more competent unit that acted as a thin beam during deformation and is complexly imbricated in fold cores (Figure 16b). Farther basinward, the two shale units apparently lose much of their mobility, and the basal detachment ramps up from the base of the lower shale unit to a thin detachment within the upper shale.

Multiple Detachment Levels in the Port Isabel Fold Belt

Because shale layers are common along passive margins, especially in deep-water settings, multiple detachment levels are common. We have just described one example in the Niger Delta; another is the Port Isabel Fold Belt in the northwestern Gulf of Mexico. The basal detachment of the Port Isabel Fold Belt is a lower Eocene shale layer (Peel et al., 1995), but there are also several shallower shale detachments (Figure 17). The result is a complex series of basinward-vergent fault imbricates, fault-bend folds, and fault-propagation folds. The structural style and the apparent lack of any ductile thinning of the shales suggest relatively strong, frictional detachments, but the amount of overpressure at the time of deformation is unknown.

The Port Isabel Fold Belt is anomalous, in that it is a shale-detached fold belt that is locally connected to allochthonous salt detachments. Moreover, it is perched above the deeper Jurassic Louann Salt décollement, which itself cores the Perdido Fold Belt basinward of the Port Isabel Fold Belt (Figure 17a). Eocene to Miocene gravity-driven extensional collapse of the northwestern Gulf of Mexico margin was driven by the progradation of large depositional systems. Listric extensional fault systems soled into several levels, both salt and

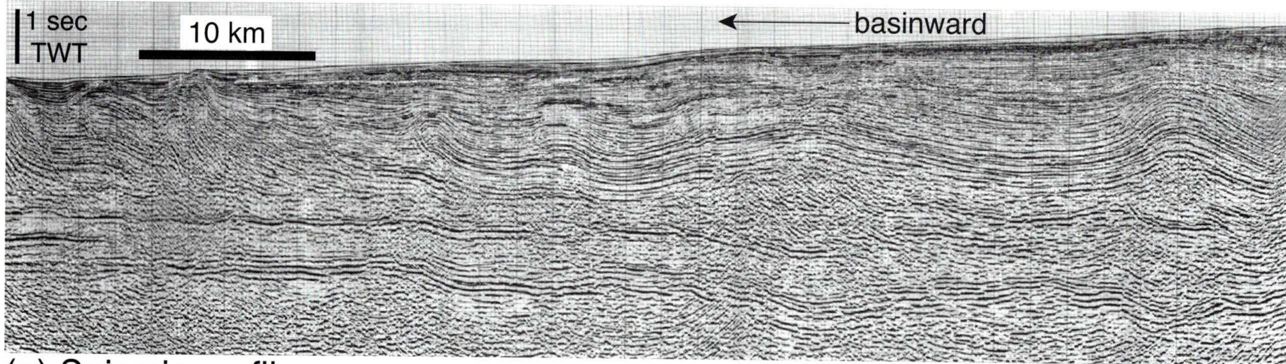

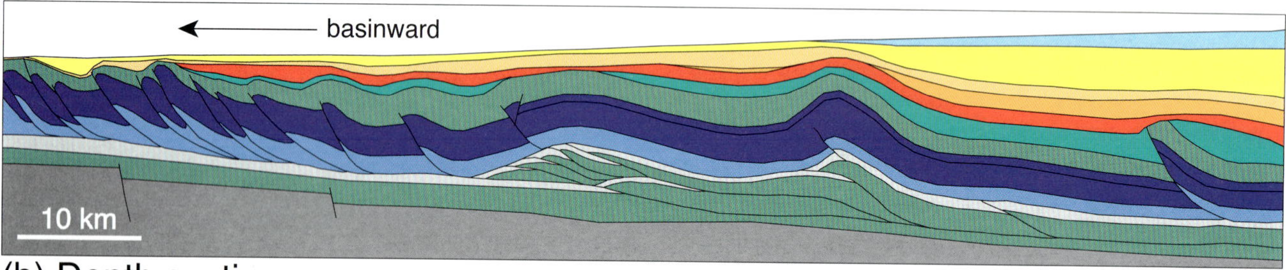

FIGURE 16. Time-migrated 2-D seismic profile (a) and vertically exaggerated (2:1) depth section (b) from the Niger Delta. No map is shown because the location is proprietary. The folds are detached on shale (lower green unit on the depth section) that is locally highly overpressured as indicated by the velocity inversion and associated "push-down" beneath thickened shale on (a). From left to right, the seismic and section show an imbricate fan of basinward-vergent thrusts, a broad fault-bend fold formed where the basal detachment ramps up from the base of the mobile shale layer to a higher level, and a symmetrical detachment fold. Both the fault-bend fold and detachment fold are cored by mobile shale encasing an imbricated competent beam, although the geometry is certainly more complex than shown in (b). Interpretation by F. Peel; data courtesy of Mabon Ltd.

shale, and the resulting distal shortening was accommodated by different processes that overlapped in time (Peel et al., 1995; Trudgill et al., 1999).

Timing and Location of Deformation

In contrast to deformation above salt layers, which usually starts beneath only a thin overburden, shale-detached deformation does not begin until the amount of overpressure can overcome the critical-yield stress of the shale, which happens only after a large thickness of overburden is deposited rapidly enough to raise the fluid pressure sufficiently. Because this usually happens in a delta-front setting and late enough that any basinward tilt of the section has been reversed, shale-detached deformation generally is dominated by gravity spreading rather than gravity gliding.

The relatively late timing of shortening can be seen in seismic data from the Niger Delta (Figure 16a) and the Mexican Ridges Fold Belt of the western Gulf of Mexico (Figure 18a). In both cases, there is a folded and/or thrusted, constant-thickness section between the detachment and the overlying growth strata. This is in contrast to salt-detached fold belts, where thickness changes immediately above the salt record early deformation (e.g., Figures 6a, 7b, 8, 11).

The basinward limit of deformation may occur where, because of a decrease in overpressure, the frictional resistance to sliding increases sufficiently to prevent lateral translation. The boundary between normal and overpressured shales advances basinward through time as the locus of rapid burial progrades into deeper water. Alternatively, the basinward limit of deformation may simply be the front of a propagating wedge of fixed taper angle. In either case, there is a general forward-propagating sequence of fold development, as indicated by the typical landward steepening of fault imbricates (Figure 18b). However, because shales can locally dewater and thus revert to lower pressures, out-of-sequence thrusting and folding are common (Morley and Guerin, 1996; Wu and Bally, 2000).

Shale diapirs are common when the shale is highly overpressured and mobile, as in the Niger Delta (Doust and Omatsola, 1990; Morley and Guerin, 1996; Wu and Bally, 2000). Because the overpressured shale in the diapirs is weaker than the surrounding sediments,

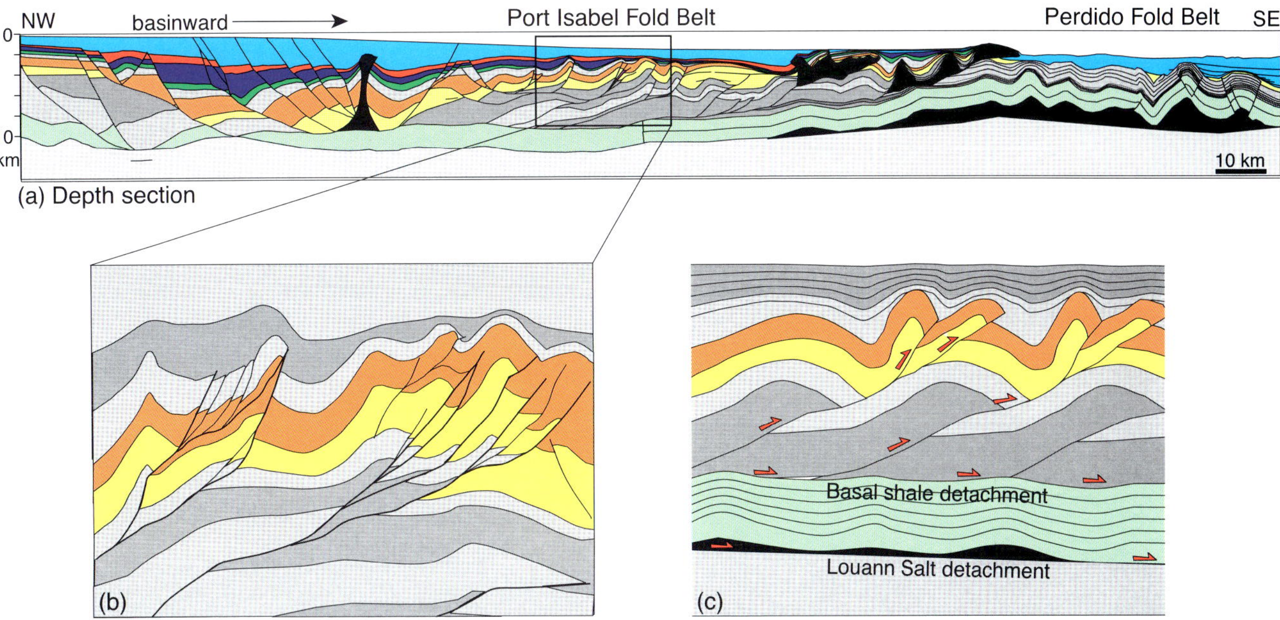

FIGURE 17. Port Isabel Fold Belt, northwestern Gulf of Mexico. (a) Regional depth section showing linked proximal extension and distal contraction. Shortening is accommodated in: the Perdido Fold Belt above the Jurassic Louann Salt; extrusion of, and translation above, allochthonous salt; and the Port Isabel Fold Belt, detached on Eocene shales. 2x1 vertical exaggeration; salt is in black and other stratigraphic units are not identified because of proprietary interpretation. Modified from Peel et al., (1995); location shown in Figure 5a. (b) Tracing of seismic interpretation of the Port Isabel Fold Belt, showing multiple shale detachment levels and associated fault-bend and fault-propagation folds. Interpretation by F. Peel. (c) Schematic cross section (not to scale) showing the relationship of the Port Isabel shale detachments (at the top of the green unit and higher) and the underlying, autochthonous Louann Salt detachment.

shale diapirs probably get squeezed and localize where folds and thrusts develop, just as in the case of salt diapirs. This may be the origin of the "thrusted diapir" province mapped by Morley and Guerin (1996) in the Niger Delta. Alternatively, many shale diapirs may initiate as contractional structures.

Amounts and Rates of Shortening

Measurements of shortening amounts and rates are even rarer for shale-detached fold belts than they are for salt-cored fold belts. Wu and Bally (2000) give no figures but report that shortening in the Niger Delta thrust belt is substantially greater than in the fold belts of the northern Gulf of Mexico. A proprietary restoration of the Niger Delta section presented here (Figure 16) yields about 20 km of net shortening and rates of up to 10 mm/yr.

Structural Styles of Shale-detached Fold Belts

In the following sections, we differentiate three characteristic styles of deep-water fold belts that are detached above shale. We stress that these models are simplistic and that complex combinations of these styles can be found.

Faulted Shale (Figure 19a)

Many fold belts with shale detachments are characterized by basinward-vergent fault imbricates above one or more detachment levels. This marked asymmetry typifies frictional detachments (Figure 14a), and the décollement is inferred to be relatively strong, that is, at most only moderately overpressured. Landward-steepening imbricates suggest a general basinward propagation of deformation.

Where there is a single detachment, fault-propagation folds are the dominant structures. Examples include the Pará-Maranhão Basin of offshore northeast Brazil (Figure 18b) and the frontal portion of the Niger Delta Fold Belt (Figure 16). Where there are multiple detachments, both fault-propagation and fault-bend folds are common, with the latter occurring where deeper detachments ramp up to shallower levels. Examples of fault-bend folds occur in the Port Isabel Fold Belt (e.g., Figure 17c) and the Niger Delta (central structure of Figure 16).

Folded Shale (Figure 19b)

Some shale-detached fold belts are characterized by symmetrical detachment folds rather than by asymmetric fault imbricates. For example, the Mexican Ridges

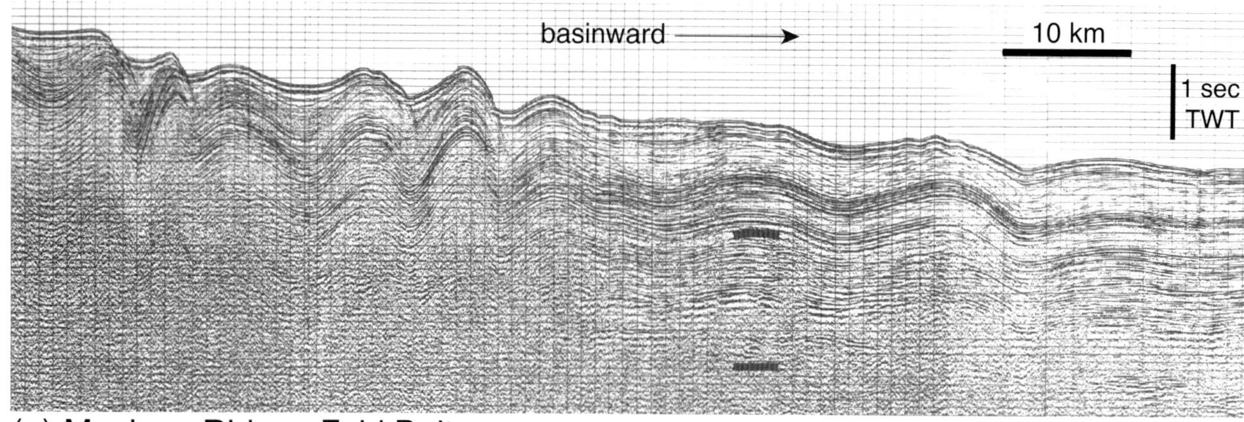

(a) Mexican Ridges Fold Belt

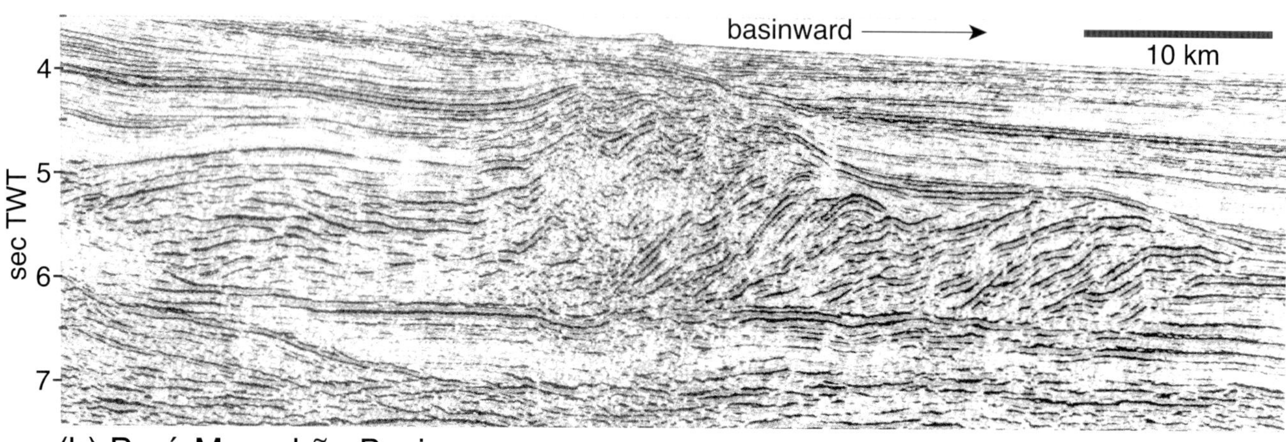

(b) Pará-Maranhão Basin

FIGURE 18. Time-migrated 2-D seismic profiles across shale-detached fold belts showing contrasting geometries that may be related to the thickness and amount of overpressure. (a) Mexican Ridges Fold Belt, western Gulf of Mexico, with symmetric detachment folds. The apparent shorter wavelength of more landward folds is due to tightening and, in some case, faulting of initially broader structures. Adapted from Buffler et al., 1979 (AAPG©1979, reprinted by permission of the AAPG whose permission is required for further use). (b) Pará-Maranhão basin, offshore northeastern Brazil, with asymmetric, stacked thrust imbricates. Adapted from Koyi, "Towards dynamic restoration of geologic profiles: some lessons from analogue modeling," Atlantic Rifts and Continental Margins, AGU Geophysical Monograph 115, p. 115–149, 2000 (originally from Guimarães et al., 1989); copyright (2000) American Geophysical Union, modified by permission of American Geophysical Union.

Fold Belt (Figure 18a) comprises a regular series of anticlines that are cored by mobile shale (Buffler et al., 1979; Weimer and Buffler, 1992; B. Trudgill, personal communication, 2000). The symmetrical structural style suggests a weak décollement layer (Figure 14), that is, a highly overpressured shale, but there is no visible velocity pushdown (Figure 18a). Another possibility is that detachment folding in the Mexican Ridges Fold Belt reflects relatively slow strain rates rather than a weak mobile layer.

Highly Overpressured Shale (Figure 19c)

Fold belts detached on highly overpressured shale contain a variety of structural styles. The variations are probably related to changes in the degree of overpressuring and thus to the weakness and mobility of the décollement layer. The amount of overpressure not only increases landward because of increasing overburden, but also increases through time in any one place (although local dewatering can lead to reversals in these trends).

The basinward limit of high overpressure is marked by a transition from symmetrical detachment folds to asymmetric fault imbricates because of the difference in strength of the décollement layer. Asymmetric, faulted folds may also be found in zones of high overpressure, probably because deformation started while the shale was still relatively strong. Farther landward are belts of shale diapirs that may also accommodate shortening or may even have initiated as contractional structures.

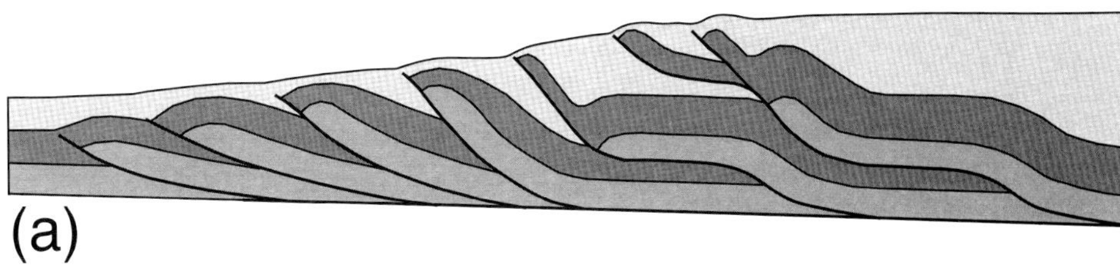

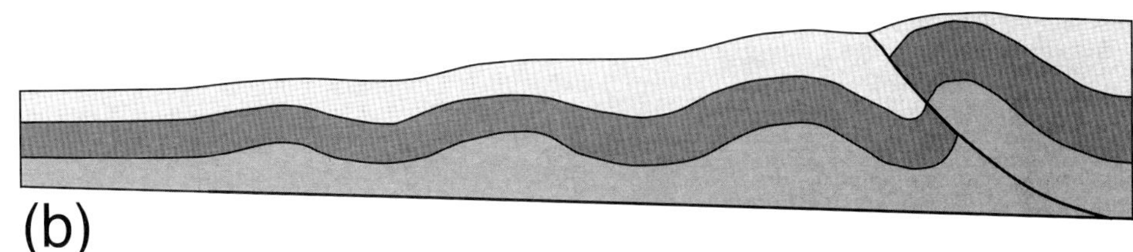

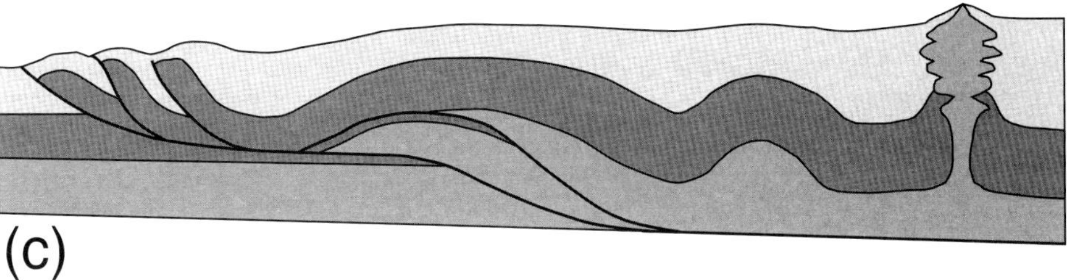

FIGURE 19. Structural styles typical of fold belts developed above shale décollements: (a) thin, relatively strong shale detachments with asymmetric imbricate fans and ramp-flat geometries; (b) thicker shale that forms symmetrical detachment folds, possibly due to slow deformation rates; and (c) thick, highly overpressured shale with a mixture of fold styles and shale diapirs. See text for discussion.

COLLISIONAL/ ACCRETIONARY FOLD BELTS

It is beyond the scope of this chapter to address collisional and accretionary fold belts in any detail, and the reader is referred to other chapters in this volume and previous publications. However, it is worth noting briefly some key similarities and differences that serve to highlight the special characteristics of gravity-driven fold belts on passive margins.

There are many similarities between the two broad categories of fold belts. Both salt and shale (whether overpressured or not) are also the major detachments in mountain belts and accretionary wedges, and the different mechanical behavior of these two décollement types serves to control the timing and style of deformation in the same ways described above. Contraction above a viscous salt layer leads to a broad zone of simultaneously growing symmetrical detachment folds (Davis and Engelder, 1985), whereas frictional shale detachments develop foreland-vergent fault-related folds that typically propagate in a forward sense, albeit with occasional out-of-sequence thrusting (Platt, 1988).

The primary difference is the driving mechanism behind fold belt development. Collisional/accretionary fold belts are controlled by lithospheric plate motions, not by gravitational instability. Although gravity spreading may be involved in the deformation of a tectonic wedge, it is not the cause of shortening but rather a possible result, if the wedge taper is too great (Price, 1981). In other words, crustal-scale shortening creates a wedge that may spread, whereas a passive-margin wedge created by differential sedimentation fails gravitationally and thereby causes distal shortening.

The difference in the driving mechanism has important ramifications. First, the entire crust is involved in collisional/accretionary fold belts, even though the frontal portions are often thin-skinned. Moreover, both the total amount of shortening and the deformation rates are usually substantially higher than in gravity-driven fold belts. Shortening rates reported here are one to two orders of magnitude slower than the tectonic-plate-convergence velocities that drive crustal-scale fold belts. The reason is that the driving forces in passive-margin fold belts are not much higher than the distal sediments' strength that resists deformation. The driving forces are provided by internal instabilities within the deforming wedge, whereas the driving forces in collisional/accretionary fold belts are external and applied to the wedge. Consequently, gravitational failure of passive margins is much more sensitive to slight increases in overburden thickness and strength (i.e., sedimentation patterns) than is the case in fold belts driven by lithospheric plate movements.

CONCLUSIONS

Many passive margins have linked systems of proximal extension and distal contraction. Deformation occurs when the gravity potential is sufficient to overcome both the resistance to slip along the basal detachment and the internal strength of the overburden. The gravity potential is determined in large part by the boundaries of the deforming wedge, namely, the dip of the décollement layer and the dip of the surface slope. Thus, failure is driven by increased basinward tilt and shelf margin progradation and is inhibited by increased landward tilt and distal (bypass) sedimentation.

Gravity gliding is the component of the net deformation that is controlled by a basinward-dipping detachment, and gravity spreading is the component controlled by the seabed slope. Thus, most margins have a significant early component of gravity gliding caused by differential thermal subsidence and the consequent basinward tilting after rifting and oceanic spreading. The gliding component can also be enhanced by cratonic uplift, as occurred during the Tertiary along the Angolan margin. Gravity spreading is always a factor, regardless of the dip of the basal detachment, and is the dominant component in progradational margins. Spreading is also the primary process where there is thick proximal salt that can accommodate significant vertical collapse of the outer shelf and upper slope while more distal salt inflates.

Passive-margin fold belts detach above either salt or shale. Salt is a viscous material and thus flows under minimal deviatoric stresses, so that deformation typically occurs very soon after salt deposition. In contrast, shale is a plastic material and does not yield until the frictional resistance to slip is overcome. Because the coefficient of friction along the décollement is reduced by increased fluid pressure, folding above shale detachments does not commence until there is a sufficient thickness of overburden. A high degree of overpressure greatly weakens the décollement and thus facilitates deformation.

Folds generally form near the toe of the slope and near the basinward limit of the effective detachment layer. In the case of salt, there is typically a broad zone of synchronous folding between the two locations, although there can be some basinward propagation as the margin progrades. In the case of shale, the basinward limit is defined by a decrease in overpressure. The zone of overpressure gradually advances basinward during shelf progradation, so that there is usually a forward-propagating sequence of contractional structures. Other factors that influence the spatial distribution of folding include local barriers to overburden translation, preexisting diapirs or salt walls, and the shifting pattern of major deltaic lobes through time.

Structural styles are largely dependent on the nature of the décollement layer. Salt-cored fold belts are dominated by symmetric, polyharmonic detachment folds, sometimes with high-angle reverse faults cutting one or both limbs. Even when the salt layer is initially thin, distal inflation usually means that detachment folds can continue to grow without running out of salt to fill the fold cores. Margins with salt layers also typically have squeezed diapirs and salt walls, thickened and shortened salt massifs and nappes, and allochthonous salt canopies that accommodate some of the gravitational failure of the slope. Shale-cored fold belts, in contrast, are dominated by asymmetric, basinward-vergent thrust imbricates, multiple detachment levels, and fault-bend and fault-propagation folds. More-symmetric structures cored by mobile shale can develop if deformation is very slow or if the shale is highly overpressured. High overpressures also lead to shale diapirs that probably accommodate some of the shortening.

Many of the same structural styles can be found in the frontal portions of collisional/accretionary fold belts. There is a fundamental difference between the two settings, however. Whereas deep-water fold belts are driven by internal gravitational collapse of passive margins, mountain belts are driven by externally driven collision or subduction of tectonic plates. One result is that the forces responsible for gravity-driven fold belts are much lower than those in plate-tectonic settings and are, in fact, not much higher than the strength of the rocks they deform. Thus, the distribution and rates of active sedimentation along passive margins are of fundamental importance in controlling the location, timing, amounts, and rates of shortening in deep-water settings.

Although their net deformation is relatively minor compared with that of collisional/accretionary fold belts, gravity-driven fold belts are significant tectonic provinces with large structures. Belts of deformation can be more than one hundred kilometers wide and hundreds of kilometers long and can contain anticlines with amplitudes of up to several kilometers. However, gravity-driven fold belts on passive margins are relatively unknown, and it is our hope and expectation that continued deep-water hydrocarbon exploration and associated research will shed more light on these fascinating provinces.

ACKNOWLEDGMENTS

The authors gratefully acknowledge Mabon Ltd., WesternGeco, and TGS for permission to show seismic data, and BHP Billiton Petroleum, Inc., for permission to show the depth sections and structural restorations from the northern Gulf of Mexico and offshore west Africa. Vendeville was funded by the Applied Geodynamics Laboratory consortium at the Bureau of Economic Geology (BEG), which includes Amerada Hess, Anadarko, BHP, BP, Chevron, ENI-Agip, Exxon-Mobil, Marathon, PanCanadian, Petrobras, Phillips, Shell, Texaco, TotalFinaElf, Unocal, and Vastar. We thank Ian Davison and an anonymous reviewer for suggesting significant improvements to the manuscript. The opinions expressed in this paper are those of the authors and do not necessarily represent those of BHP Billiton or the BEG.

REFERENCES CITED

Barker, C., 1990, Calculated volume and pressure changes during the thermal cracking of oil and gas in reservoirs: AAPG Bulletin, v. 74, p. 1254–1261.

Biju-Duval, B. J., J. Letouzey, and L. Montadert, 1979, Variety of margins and deep basins in the Mediterranean, *in* J. S. Watkins, L. Montadert, and P. N. Dickerson, eds., Geologic and geophysical investigations of continental margins: AAPG Memoir 29, p. 293–317.

Brun, J.-P., and O. Merle, 1985, Strain patterns in models of spreading-gliding nappes: Tectonics, v. 4, p. 705–719.

Buffler, R. T., F. J. Shaub, J. S. Watkins, and J. L. Worzel, 1979, Anatomy of the Mexican ridges, southwestern Gulf of Mexico, *in* J. S. Watkins, L. Montadert, and P. W. Dickerson, eds., Geological and geophysical investigations of continental margins: AAPG Memoir 29, p. 319–327.

Byerlee, J. D., 1978, Friction on rocks: Pure and Applied Geophysics, v. 116, p. 615–626.

Cobbold, P. R., and P. Szatmari, 1991, Radial gravitational gliding on passive margins: Tectonophysics, v. 188, p. 249–289.

Cobbold, P. R., P. Szatmari, L. S. Demercian, D. Coelho, and E. A. Rossello, 1995, Seismic and experimental evidence for thin-skinned horizontal shortening by convergent radial gliding on evaporites, deep-water Santos Basin, Brazil, *in* M. P. A. Jackson, D. G. Roberts, and S. Snelson, eds., Salt tectonics: a global perspective: AAPG Memoir 65, p. 305–321.

Connors, C. D., D. B. Denson, G. Kristiansen, and D. M. Angstadt, 1998, Compressive anticlines of the mid-outer slope, central Niger delta: Extended Abstracts Volume, AAPG International Conference and Exhibition, Rio de Janeiro, p. 372.

Cramez, C., and M. P. A. Jackson, 2000, Superposed deformation straddling the continental-oceanic transition in deep-water Angola: Marine and Petroleum Geology, v. 17, p. 1095–1109.

Dahlen, F. A., J. Suppe, and D. Davis, 1984, Mechanics of fold-and-thrust belts and accretionary wedges: cohesive Coulomb theory: Journal of Geophysical Research, v. 89, p. 10,087–10,101.

Dailly, G. C., 1976, A possible mechanism relating progradation, growth faulting, clay diapirism and overthrusting in a regressive sequence of sediments: Canadian Petroleum Geologists Bulletin, v. 24, p. 92–116.

Dailly, P., 2000, Tectonic and stratigraphic development of the Rio Muni basin, Equatorial Guinea: the role of transform zones in Atlantic basin evolution, *in* W. Mohriak and M. Talwani, eds., atlantic Rifts and Continental Margins: American Geophysical Union Geophysical Monograph 115, p. 105–128.

Damte, A., and R. Miksha, 2000, Petroleum fairways associated with stacked toe thrust systems in the Tuktoyaktuk Peninsula, MacKenzie delta area [abs.]: GeoCanada 2000 Conference, Calgary, CD-ROM.

Davis, D. M., and T. Engelder, 1985, The role of salt in fold-and-thrust belts: Tectonophysics, v. 119, p. 67–88.

Davis, D. M., J. Suppe, and F. A. Dahlen, 1983, Mechanics of fold-and-thrust belts and accretionary wedges: Journal of Geophysical Research, v. 88, p. 1153–1172.

DeJong, K. A., and R. Scholten, 1973, Gravity and Tectonics: John Wiley, New York, 502 p.

Demercian, S., P. Szatmari, and P. R. Cobbold, 1993, Style and pattern of salt diapirs due to thin-skinned gravitational gliding, Campos and Santos basins, offshore Brazil: Tectonophysics, v. 228, p. 393–433.

Diegel, F. A., J. F. Karlo, D. C. Schuster, R. C. Shoup, and P. R. Tauvers, 1995, Cenozoic structural evolution and tectono-stratigraphic framework of the northern Gulf Coast continental margin, *in* M. P. A. Jackson, D. G. Roberts, and S. Snelson, eds., Salt Tectonics: a Global Perspective: AAPG Memoir 65, p. 109–151.

Doust, H., and E. M. Omatsola, 1990, Niger delta, *in* J. D. Edwards and P. A. Santogrossi, eds., Divergent/passive margins: AAPG Memoir 48, p. 201–238.

Duval, B., C. Cramez, and M. P. A. Jackson, 1992, Raft tectonics in the Kwanza Basin, Angola: Marine and Petroleum Geology, v. 9, p. 389–404.

Evamy, B. D., J. Haremboure, P. Kamerling, W. A. Knapp, F. A. Molloy, and P. H. Rowlands, 1978, Hydrocarbon habitat of Tertiary Niger delta: AAPG Bulletin, v. 62, p. 1–39.

Fiduk, J. C., B. C. McBride, P. Weimer, M. G. Rowan, and B. D. Trudgill, 1995, Defining the basinward limits of salt deposition, northern Gulf of Mexico, *in* C. J. Travis, H. Harrison, M. R. Hudec, B. C. Vendeville, F. J. Peel, and B. F. Perkins, eds., Salt, sediment and hydrocarbons: Gulf Coast Section Society of Exploration Paleontologists and Mineralogists Foundation 16th Annual Research Conference, p. 53–64.

Fox, J. F., and P. R. Ashton, 2000, Deepwater Gabon – is it an analog for the deepwater Gulf of Mexico, *in* R. Shoup, J. Watkins, J. Karlo, and D. Hall, eds., Integration of geologic models for understanding risk in the Gulf of Mexico, Papers from the Hedberg Conference, 1998 (Galveston): AAPG/Datapages Discovery Series 1 (CD-ROM), AAPG/Datapages, Tulsa.

Gaullier, V., Y. Mart, G. Bellaiche, B. Vendeville, J. Mascle, T. Zitter, and the second leg "PRISMED II" scientific party, 2000a, Salt tectonics in and around the Nile deep-sea fan: insights from the "PRISMED II" cruise, *in* B. C. Vendeville, Y. Mart, and J. L. Vigneresse, eds., Salt, shale, and igneous diapirs in and around Europe: Geological Society (London) Special Publication 174, p. 111–129.

Gaullier, V., B. Vendeville, and J. Mascle, 2000b, 3-D kinematic evolution of delta lobes spreading above salt: inferences from experimental models and natural examples [abs.]: European Geophysical Society, XXV General Assembly, Nice (France), Geophysical Research Abstracts (CD-R), Volume 2, ISSN: 1029–7006.

Ge, H., M. P. A. Jackson, and B. C. Vendeville, 1997, Kinematics and dynamics of salt tectonics driven by progradation: AAPG Bulletin, v. 81, p. 398–423.

Guimarães, P. T. M., E. R. Machado, and S. R. P. Silva, 1989, Interpretação sismo-estratigráfica em águas profundas na Bacia do Pará-Maranhão, *in* Sintex, I Seminário de Interpretação Exploratória, Petrobras Departamento de Exploração, Rio de Janeiro, p. 171–183.

Hall, S. H., 2000, The development of large-scale sub-salt structures in the deep-water western Gulf of Mexico [abs.]: AAPG Annual Convention Official Program, p. A62–A63.

Hall, S. H., 2003, Comparison between salt-related compressional structures in Mauritania and the U.S. and Mexican Gulf of Mexico: AAPG Annual Convention Official Program, v. 12, p. A69.

Hall, S., and T. Smith, 2001, Compressional salt diapirs in the deepwater US Gulf of Mexico [abs.]: AAPG Annual Convention Official Program, v. 10, p. A78.

Hartman, D. A., W. A. Swanson, P. R. Smith, F. J. Goulding, and C. A. Kelly, 1998, Structural development of the continental margin of Congo and northern Angola: Extended Abstracts Volume, AAPG International Conference and Exhibition, Rio de Janeiro, p. 370–371.

Heaton, R. C., M. P. A. Jackson, M. Bamahmoud, and A. S. O. Nani, 1995, Superposed Neogene extension, contraction, and salt canopy emplacement in the Yemeni Red Sea, *in* M. P. A. Jackson, D. G. Roberts, and S. Snelson, eds., Salt tectonics: a global perspective: AAPG Memoir 65, p. 333–351.

Homza, T. X., and W. K. Wallace, 1995, Geometric and kinematic models for detachment folds with fixed and variable detachment depths: Journal of Structural Geology, v. 17, p. 575–588.

Hubbert, M. K., and W. Rubey, 1959, Role of fluid pressure in mechanics of over-thrust faulting, Parts I and II: Geological Society of America Bulletin, v. 70, p. 115–205.

Jackson, M. P. A., and C. Cramez, 1989, Seismic recognition of salt welds in salt tectonic regimes: Gulf Coast Section Society of Exploration Paleontologists and Mineralogists Foundation Tenth Annual Research Conference, Extended Abstracts, p. 66–71.

Jackson, M. P. A., and C. J. Talbot, 1989, Salt canopies: Gulf Coast Section Society of Exploration Paleontologists and Mineralogists Foundation Tenth Annual Research Conference, Extended Abstracts, p. 72–78.

Jackson, M. P. A., C. Cramez, and W. U. Mohriak, 1998, Salt-tectonics provinces across the continental-oceanic boundary in the Lower Congo and Campos basins on the South Atlantic margins [abs.]: Extended Abstracts Volume, AAPG International Conference and Exhibition, Rio de Janeiro, p. 40–41.

Jaeger, J. C., and N. G. W. Cook, 1979, Fundamentals of rock mechanics, 3rd edition: Chapman and Hall, London, 593 p.

Jaumé, S. C., and R. J. Lillie, 1988, Mechanics of the Salt Range-Potwar Plateau, Pakistan: a fold–and-thrust belt underlain by evaporites: Tectonics, v. 7, p. 57–71.

Keen, C. E., and D. P. Potter, 1995, Formation and evolution of the Nova Scotian rifted margin: evidence from deep seismic reflection data: Tectonics, v. 14, p. 918–932.

Koyi, H. A., 2000, Towards dynamic restoration of geologic profiles: some lessons from analogue models, *in* W. Mohriak and M. Talwani, eds., Atlantic rifts and continental margins: American Geophysical Union Geophysical Monograph 115, p. 317–329.

Le Douaran, S., J. Burrus, and F. Avedik, 1984, Deep structure of the northwestern Mediterranean basin: results of a two-ship seismic survey: Marine Geology, v. 55, p. 325–345.

Letouzey, J., B. Colletta, R. Vially, and J. C. Chermette, 1995, Evolution of salt-related structures in compressional settings, *in* M. P. A. Jackson, D. G. Roberts, and S. Snelson, eds., Salt Tectonics: a global perspective: AAPG Memoir 65, p. 41–60.

Lundin, E. R., 1992, Thin-skinned extensional tectonics on a salt detachment, northern Kwanza Basin, Angola: Marine and Petroleum Geology, v. 9, p. 405–411.

Marton, L. G., G. C. Tari, and C. T. Lehmann, 2000, Evolution of the Angolan passive margin, West Africa, with emphasis on post-salt structural styles, *in* W. Mohriak and M. Talwani, eds., Atlantic rifts and continental margins: American Geophysical Union Geophysical Monograph 115, p. 129–149.

Mauduit, T., G. Guerin, J.-P. Brun, and H. Lecanu, 1997, Raft tectonics: the effects of basal slope angle and sedimentation rate on progressive extension: Journal of Structural Geology, v. 19, p. 1219–1230.

Mauffret, A., G. Pascal, A. Maillard, and C. Gorini, 1995, Tectonics and deep structure of the northwestern Mediterranean Basin: Marine and Petroleum Geology, v. 12, p. 645–666.

McBride, B. C., 1998, The evolution of allochthonous salt along a megaregional profile across the northern Gulf of Mexico basin: AAPG Bulletin, v. 82, p. 1037–1054.

Mohriak, W. U., 1995, Salt tectonics structural styles: contrasts and similarities between the South Atlantic and the Gulf of Mexico, *in* C. J. Travis, H. Harrison, M. R. Hudec, B. C. Vendeville, F. J. Peel, and B. F. Perkins, eds., Salt, sediment and hydrocarbons: Gulf Coast Section Society of Exploration Paleontologists and Mineralogists Foundation 16th Annual Research Conference, p. 177–191.

Mohriak, W. U., J. M. Macedo, R. T. Castellani, H. D. Rangel, A. Z. N. Barros, M. A. L. Latgé, J. A. Ricci, A. M. P. Mizusaki, P. Szatmari, L. S. Demercian, J. G. Rizzo, and J. R., Aires, 1995, Salt tectonics and structural styles in the deep-water province of the Cabo Frio region, Rio de Janeiro, Brazil, *in* M. P. A. Jackson, D. G. Roberts, and S. Snelson, eds., Salt tectonics: a global perspective: AAPG Memoir 65, p. 273–304.

Morley, C. K., and G. Guerin, 1996, Comparison of gravity-driven deformation styles and behavior associated with mobile shales and salt: Tectonics, v. 15, p. 1154–1170.

Nilsen, K. T., B. C. Vendeville, and J.-T. Johansen, 1995, Influence of regional tectonics on halokinesis in the Nordkapp Basin, Barents Sea, *in* M. P. A. Jackson,

D. G. Roberts, and S. Snelson, eds., Salt tectonics: a global perspective: AAPG Memoir 65, p. 413–436.

Peel, F. J., 2001, Emplacement, inflation and folding of an extensive allochthonous salt sheet in the late Mesozoic (ultra-deepwater Gulf of Mexico) [abs.]: AAPG Annual Convention Official Program, v. 10, p. A155.

Peel, F. J., and S. Matthews, 1999, Structural styles of traps in deepwater fold/thrust belts of the northern Gulf of Mexico [abs.]: Extended Abstracts Volume, AAPG International Conference and Exhibition, Birmingham, p. 392.

Peel, F. J., C. J. Travis, and J. R. Hossack, 1995, Genetic structural provinces and salt tectonics of the Cenozoic offshore US Gulf of Mexico: a preliminary analysis, *in* M. P. A. Jackson, D. G. Roberts, and S. Snelson, eds., Salt tectonics: a global perspective: AAPG Memoir 65, p. 153–175.

Peel, F. J., M. P. A. Jackson, and D. Ormerod, 1998, Influence of major steps in the base of salt on the structural style of overlying thin–skinned structures in deep water Angola: Extended Abstracts Volume, AAPG International Conference and Exhibition, Rio de Janeiro, p. 366–367.

Platt, J. P., 1988, The mechanics of frontal imbrication: a first-order analysis: Geologische Rundschau, v. 77, p. 577–589.

Poblet, J., and K. McClay, 1996, Geometry and kinematics of single-layer detachment folds: AAPG Bulletin, v. 80, p. 1085–1109.

Price, R. A., 1981, The Cordilleran foreland thrust and fold belt in the southern Canadian Rockies, *in* K. R. McClay and N. J. Price, eds., Thrust and nappe tectonics: Geological Society (London) Special Publication 9, p. 427–448.

Raillard, S., B. C. Vendeville, and G. Guérin, 1997, Causes and structural characteristics of thin-skinned inversion during gravity gliding or spreading above salt or shale [abs.]: AAPG Annual Convention Official Program, p. A95.

Raillard, S., J. J. Biteau, P. Allix, and C. Chevalier, 1998, Lower Congo Tertiary basin – offshore West Africa structural zonation and evolution: Extended Abstracts Volume, AAPG International Conference and Exhibition, Rio de Janeiro, p. 364.

Ramberg, H., 1981, Gravity, deformation and the Earth's crust in theory, experiments and geological application (2nd Edition): London, Academic Press, 452 p.

Rowan, M. G., 1995, Structural styles and evolution of allochthonous salt, central Louisiana outer shelf and upper slope, *in* M. P. A. Jackson, D. G. Roberts, and S. Snelson, eds., Salt Tectonics: a Global Perspective: AAPG Memoir 65, p. 198–228.

Rowan, M. G., 1997, Three-dimensional geometry and evolution of a segmented detachment fold, Mississippi Fan fold belt, Gulf of Mexico: Journal of Structural Geology, v. 19, p. 463–480.

Rowan, M. G., R. Kligfield, and P. Weimer, 1993, Processes and rates of deformation: preliminary results from the Mississippi Fan fold belt, deep Gulf of Mexico, *in* J. M. Armentrout, R. Bloch, H. C. Olson, and B. F. Perkins, eds., Rates of geologic processes: Gulf Coast Section Society of Exploration Paleontologists and Mineralogists Foundation Fourteenth Annual Research Conference, p. 209–218.

Rowan, M. G., B. D. Trudgill, and J. C. Fiduk, 2000, Deepwater, salt-cored fold belts: lessons from the Mississippi Fan and Perdido fold belts, northern Gulf of Mexico, *in* W. Mohriak and M. Talwani, eds., Atlantic rifts and continental margins: American Geophysical Union Geophysical Monograph 115, p. 173–191.

Rowan, M. G., A. Leibold, A. Maccagni, and G. Valenti, 2001, Allochthonous salt, contractional folds, and turtle structures of the Louisiana deepwater province: AAPG Annual Convention Official Program, v. 10, p. A173.

Sage, L., and J. Letouzey, 1990, Convergence of the African and Eurasian plates in the eastern Mediterranean, *in* J. Letouzey, ed., Petroleum and tectonics in mobile belts: Editions Technip, Paris, p. 49–68.

Schultz-Ela, D. D., 2001, Excursus on gravity gliding and gravity spreading: Journal of Structural Geology, v. 23, p. 725–731.

Schuster, D. C., 1995, Deformation of allochthonous salt and evolution of related salt-structural systems, eastern Louisiana Gulf Coast, *in* M. P. A. Jackson, D. G. Roberts, and S. Snelson, eds., Salt tectonics: a global perspective: AAPG Memoir 65, p. 177–198.

Spathopoulos, F., 1996, An insight on salt tectonics in the Angola Basin, South Atlantic, *in* G. I. Alsop, D. J. Blundell, and I. Davison, eds., Salt tectonics: Geological Society (London) Special Publication 100, p. 153–174.

Spencer, J., G. Tari, P. Jeronimo, and B. Hart, 1998, Comparison between offshore Angola and the Gulf of Mexico in terms of salt tectonics: Extended Abstracts Volume, AAPG International Conference and Exhibition, Rio de Janeiro, p. 968.

Spiers, C. J., P. M. T. M. Schutjens, R. H. Brzesowsky, C. J. Peach, J. L. Liezeberg, and H. J. Zwart, 1990, Experimental determination of constitutive parameters governing creep of rocksalt by pressure solution, *in* R. J. Knipe and E. H. Rutter, eds., Deformation mechanisms, rheology and tectonics: Geological Society (London) Special Publication 54, p. 215–227.

Stewart, S. A., 1999, Geometry of thin-skinned tectonic systems in relation to detachment layer thickness in sedimentary basins: Tectonics, v. 18, p. 719–732.

Tari, G. C., P. R. Ashton, K. Coterill, J. S. Molnar, M. Sorgenfrei, P. Thompson, and D. Valasek, 2002, Examples of deep-water salt tectonics from Africa: AAPG Annual Convention Official Program, v. 11, p. A173.

Tari, G., J. Molnar, P. Ashton, and R. Hedley, 2000, Salt tectonics in the Atlantic margin of Morocco: The Leading Edge, v. 19, p. 1074–1078.

Trudgill, B. D., M. G. Rowan, P. Weimer, J. C. Fiduk, P. E. Gale, B. E. Korn, R. L. Phair, W. T. Gafford, J. B. Dischinger, G. R. Roberts, and L. F. Henage, 1995, The structural geometry and evolution of the salt-related Perdido fold belt, Alaminos Canyon, northwestern deep Gulf of Mexico, *in* C. J. Travis, H. Harrison, M. R. Hudec, B. C. Vendeville, F. J. Peel, and B. F. Perkins, eds., Salt, sediment and hydrocarbons: Gulf Coast Section Society of Exploration Paleontologists and Mineralogists Foundation 16th Annual Research Conference, p. 275–284.

Trudgill, B. D., M. G. Rowan, J. C. Fiduk, P. Weimer, P. E. Gale, B. E. Korn, R. L. Phair, W. T. Gafford, G. R. Roberts, and S. W. Dobbs, 1999, The Perdido fold belt, northwestern deep Gulf of Mexico, Part 1: Structural geometry, evolution and regional implications: AAPG Bulletin, v. 83, p. 88–113.

Urai, J. L., C. J. Spiers, H. J. Zwart, and G. S. Lister, 1986, Water weakening effects in rocksalt during long term creep: Nature, v. 324, p. 554–557.

Van der Ven, P. H., C. H. R. Cunha, and A. S. Biassusi, 1998, Structural styles in the Espírito Santo-Mucuri basin, southeastern Brazil: Extended Abstracts Volume, AAPG International Conference and Exhibition, Rio de Janeiro, p. 374–375.

Vendeville, B. C., 2000, Remobilization of salt structures by sediment progradation: experimental models and potential applications to the Gulf of Mexico and other continental margins, *in* R. Shoup, J. Watkins, J. Karlo, and D. Hall, eds., Integration of geologic models for understanding risk in the Gulf of Mexico, Papers from Hedberg Conference, 1998 (Galveston): AAPG/Datapages Discovery Series 1 (CD-ROM), AAPG/Datapages, Tulsa.

Vendeville, B. C., and K. T. Nilsen, 1995, Episodic growth of salt diapirs driven by horizontal shortening, *in* C. J. Travis, H. Harrison, M. R. Hudec, B. C. Vendeville, F. J. Peel, and B. F. Perkins, eds., Salt, sediment and hydrocarbons: Gulf Coast Section Society of Exploration Paleontologists and Mineralogists Foundation 16th Annual Research Conference, p. 285–295.

Vendeville, B. C., M. P. A. Jackson, and R. Weijermars, 1993, Rates of salt flow in passive diapirs and their source layers, *in* J. M. Armentrout, R. Bloch, H. C. Olson, and B. F. Perkins, eds., Rates of geologic processes: Gulf Coast Section Society of Exploration Paleontologists and Mineralogists Foundation 14th Annual Research Conference, p. 269–276.

Weijermars, R., M. P. A. Jackson, and B. C. Vendeville, 1993, Rheological and tectonic modelling of salt provinces: Tectonophysics, v. 217, p. 143–174.

Weimer, P., and R. Buffler, 1992, Structural geology and evolution of the Mississippi Fan Fold Belt, deep Gulf of Mexico: AAPG Bulletin, v. 76, p. 225–251.

Worrall, D. M. and S. Snelson, 1989, Evolution of the northern Gulf of Mexico, with emphasis on Cenozoic growth faulting and the role of salt, *in* A. W. Bally and A. R. Palmer, eds., The Geology of North America: an overview: Geological Society of America, The Geology of North America, v. A, p. 97–138.

Wu, S., and A. W. Bally, 2000, Slope tectonics— comparisons and contrasts of structural styles of salt and shale tectonics of the northern Gulf of Mexico with shale tectonics of offshore Nigeria in Gulf of Guinea, *in* W. Mohriak and M. Talwani, eds., Atlantic rifts and continental margins: American Geophysical Union Geophysical Monograph 115, p. 151– 172.

Wu, S., A. W. Bally, and C. Cramez, 1990, Allochthonous salt, structure and stratigraphy of the northeastern Gulf of Mexico, part II: structure: Marine and Petroleum Geology, v. 7, p. 334–370.

Zalán, P. V., 1999, Seismic expression and internal order of gravitational fold-and-thrust belts in Brazilian deep waters [abs.]: Sixth International Congress of the Brazilian Geophysical Society, Rio de Janeiro [unpaginated].

Analog Modeling of Thrust Systems

10

McClay, K. R., and P. S. Whitehouse, 2004, Analog modeling of doubly vergent thrust wedges, *in* K. R. McClay, ed., Thrust tectonics and hydrocarbon systems: AAPG Memoir 82, p. 184–206.

Analog Modeling of Doubly Vergent Thrust Wedges

K. R. McClay

Fault Dynamics Research Group, Geology Department, Royal Holloway University of London, Egham, Surrey, U.K.

P. S. Whitehouse

Fault Dynamics Research Group, Geology Department, Royal Holloway University of London, Egham, Surrey, U.K.

ABSTRACT

Two-dimensional scaled analog models of doubly vergent thrust systems produced a two-stage evolution for the development of asymmetric thrust wedges. Stage I was the rapid elevation of an axial zone bounded by a major retrovergent thrust system together with closely spaced, low-displacement provergent thrusts in its hanging wall. When a critical elevation of the axial zone was attained, Stage II deformation was characterized by the nucleation and propagation of high-displacement, provergent thrusts to form a critically tapered prowedge. The prowedge taper angle was typically 11–12°, whereas the retrowedge maintained a steep surface taper of 38–42°. During Stage II, the doubly vergent asymmetry increased. This was characterized by the formation of well-developed provergent thrust sheets and slower uplift of the retrowedge system. Overlapping prowedge thrusts showed synchronous displacements.

Addition of synkinematic erosion and/or sedimentation produced dramatic changes to the doubly vergent wedges in the models. There were fewer prowedge thrusts, prolonged prowedge thrust activities, out-of-sequence thrusting, and reactivation of preexisting thrusts. Synkinematic growth strata recorded the progressive evolution of the wedge system and also the activities on individual thrusts. The final geometries of the analog models of doubly vergent thrust wedges closely replicated those of naturally occurring convergent and collisional orogens, such as the Pyrenees, the western Alps, and the Himalayas. The models provide important geometric and kinematic templates for understanding the development of orogenic thrust-and-fold belts. They emphasize the dynamic feedback between surface processes— synkinematic erosion and sedimentation— and thrust-belt evolution. Synchronous thrust activities, out-of-sequence thrusting, and interaction with erosion and sedimentation all need to be taken into account when one evaluates hydrocarbon generation, migration, and entrapment in thrust belts.

INTRODUCTION

Many major hydrocarbon accumulations occur in Phanerozoic fold-and-thrust belts, particularly those that are Cretaceous and younger in age. Classic examples include the Sub-Andean belt of South America—Colombia and Bolivia (Tankard et al., 1995; Dunn et al., 1995); the northern Rocky Mountains of western U.S.A. (cf. Lamerson, 1982); the Canadian Rocky Mountains (Bally et al., 1966; Fermor and Moffat, 1992; Barclay and Smith, 1992); the Zagros Mountains of Iran (Beydoun, 1988; Beydoun et al., 1992); Albania (Velaj et al., 1999); Pakistan (Davis and Engelder, 1985; Jadoon et al., 1994; Jadoon and Frisch, 1997); and Papua New Guinea (Hill et al., 2000; Cole et al., 2000). The structural style of these fold-and-thrust belts is usually described as thin-skinned and lacking in significant involvement of crystalline basement. These thin-skinned foreland fold-and-thrust belts commonly are found at the margins of contractional orogenic belts formed either by oceanic subduction or continent-continent collision. Cross sections through orogenic belts like the Pyrenees (Muñoz, 1992), the Alps (Schmid et al., 1996), the Appalachians (Hatcher et al., 1990), and the Andes (Roeder and Chamberlin, 1995) typically are asymmetric and have marginal fold-thrust belts that verge in opposite directions away from a central, uplifted metamorphic core (Figure 1). These typically are called "doubly vergent orogenic wedges" (after Koons, 1990; Willett et al., 1993). In this chapter, we investigate the 2-D geometric and kinematic evolution of doubly vergent orogenic wedges, using scaled physical models, and we compare the results with natural fold-and-thrust-belt systems. The physical models provide insights into the geometric evolution and kinematics of fold-thrust belts and the sequences and activities of thrust faults within such systems, and, most important, they demonstrate the dynamic coupling between uplift, erosion, and sedimentation in thrust-belt systems (Beaumont et al., 1992; Beaumont et al., 2000a).

Sandbox analog models have proved to be powerful visual tools for simulating complex structures in various terranes. They include extensional fault systems (Faugere and Brun, 1984; Withjack and Jamison, 1986; Serra and Nelson, 1988; Tron and Brun, 1991; McClay and White, 1995; McClay, 1990a, b; McClay et al., 2001), strike-slip faults (Dooley and McClay, 1997; Richard et al., 1995; McClay and Bonora, 2001), and inversion systems (Buchanan and McClay, 1992; McClay, 1995). Previous analog models of thrust systems have included those of Davis et al. (1983), Malavieille (1984), Ballard et al. (1987), Mulugeta (1988), Liu et al. (1992), Calassou et al. (1993), Storti and McClay (1995), Braun and Beaumont (1995), Wang and Davis (1996), Mugnier et al. (1997), Storti et al. (1997), Gutscher et al. (1998), and Storti et al. (2000). In most papers, analog models of thrust wedges have demonstrated the applicability of a simple, critically tapered Coulomb wedge model (Davis et al., 1983; Dahlen, 1990) to the general understanding of thrust-belt geometries and kinematics. Few experimental models, however, have investigated in detail the evolution of doubly vergent Coulomb wedges, and in particular, their dynamic interaction with surface processes—erosion and sedimentation (cf. Mugnier et al., 1997; Storti et al., 2000).

This chapter presents representative results from a comprehensive series of 2-D sandbox models of doubly vergent thrust belts. In particular, models that involve erosion and syntectonic sedimentation are analyzed and compared with natural fold-and-thrust belts, with particular emphasis on the geometries, sequences, and evolutionary patterns of the thrust-fault systems.

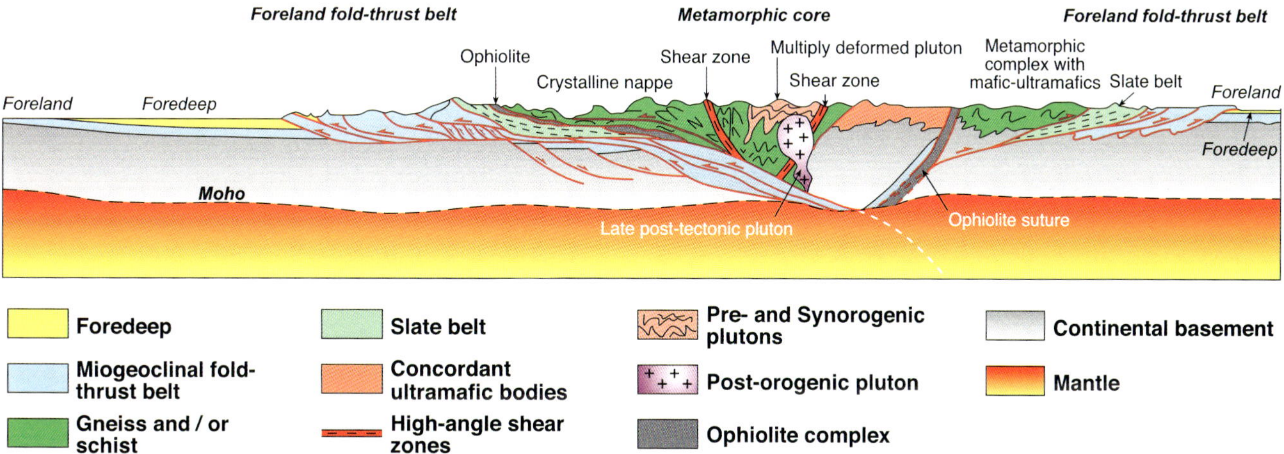

FIGURE 1. Conceptual model of a collisional orogen with doubly vergent fold-and-thrust belts. (Modified after Hatcher and Williams, 1986.)

ANALOG MODELING METHODOLOGY

The modeling methods employed in these experiments follow those developed by Liu et al. (1992), Storti and McClay (1995), and Storti et al. (2000) in the analog modeling laboratories at Royal Holloway, University of London. Sandbox models were constructed in a glass-walled, metal deformation rig with dimensions 150 cm long, 30 cm wide, and 20 cm deep (Figure 2a). Different-colored layers of dry, 190-μm-grain-size quartz sand were mechanically sieved into the deformation apparatus to form an initial sand pack with dimensions 100 × 30 × 2.5 cm (Figure 2a). Dry quartz sand has a coefficient of friction of ~0.55 (McClay, 1990a). It has been widely used to simulate the brittle deformation of sedimentary rocks in the upper 10 km of crust (Figure 2b) (Faugere and Brun, 1984; Vendeville et al., 1987; Liu et al., 1992; McClay, 1990b, 1995; Storti and McClay, 1995; Storti et al., 2000). Deformation was achieved by moving a mylar sheet (coefficient of friction of 0.47), using a system of motor-driven rollers, underneath the sand pack. The sheet was pulled down into the subduction slot in the basement such that the left-hand side of the model was dragged against the stationary right-hand side of the model (Figure 2a). This configuration produces a deformable "backstop," in contrast to the rigid vertical or inclined backstops of many previous models (cf. Davis et al., 1983; Liu et al., 1992). The incremental deformation was photographically recorded every 2.5 mm of contraction. Completed models were gelled and serially sectioned for internal analysis.

Syntectonic sedimentation was introduced to the models during some experiments by layering carefully measured quantities of sand on both the prowedge and retrowedge fault systems. Syntectonic erosion was simulated by removing part of the wedge to give a predetermined 2–5° taper to the prowedge, as well as by reducing the overall height of the wedge. In some models the amount of material removed during the erosion of the prowedge taper was balanced by the equivalent

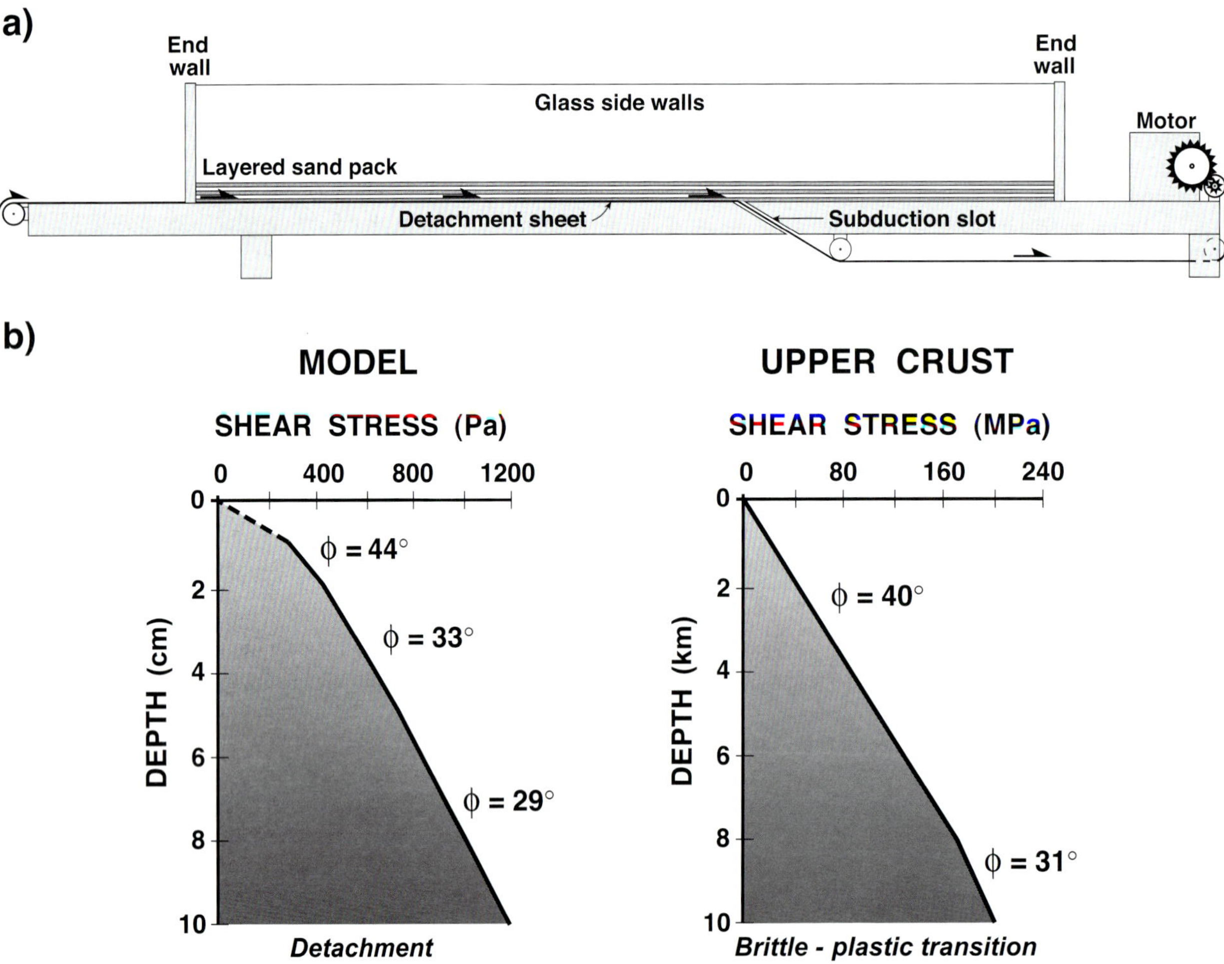

FIGURE 2. (a) Diagram of the experimental apparatus used to simulate doubly vergent thrust wedges. (b) Scaling parameters for analog modeling of upper-crust thrust systems. ϕ = angle of friction.

amount of syntectonic sedimentation at the front of the wedge.

Scaling Parameters

The dry quartz sand has a linear Navier-Coulomb behavior, with an angle of friction of 31° (McClay, 1990a), and it closely simulates the brittle frictional behavior of many rocks in the upper crust. It is essentially cohesionless, but at low stresses it displays a low apparent cohesive strength as a result of surface roughness effects (McClay, 1990a). The linear scaling factor is 10^{-5}, such that 1 cm in the models approximates 1 km of brittle, upper-crust sedimentary rocks (Figure 2b).

Model Parameters

The basic geodynamic framework for doubly vergent orogens, introduced by Koons (1990) and Willett et al. (1993), was used for the description and analysis of the models presented in this chapter (Figure 3a, b). At the first increment of deformation, the mechanical model can be divided into three parts— the moving sand pack to the left of the subduction slot, a stationary sand pack on the right of the subduction slot, and a triangular axial zone bounded by incipient prowedge and retrowedge thrusts in between (Figure 3a). In the final state, after deformation, the mechanical model consists of a prowedge section containing an array of imbricate thrusts, an uplifted axial zone, and a retrowedge that has been overthrust on top of the stationary sand pack (Figure 3b). The prowedge is a Coulomb wedge, as described by Davis et al. (1983), and it is characterized by an upper surface with a specific taper that is a function of the internal friction within the sand pack and the basal friction along the detachment surface (Davis et al., 1983). The retrowedge is characterized by a steeper, upper surface slope that approximates the surface slope failure angle for dry quartz sand (Figure 3b). Figure 3c shows a sketch of a completed model depicting the overall geometry, as shown in Figure 3b. Critical parameters measured for the analysis of the models during and after deformation include the initial sand pack thickness H_0, the overall wedge height H, the prowedge tapers $\alpha1$ and $\alpha2$, the retrowedge taper αR, D_n, the displacement on a particular thrust fault n, LP, the length of the prowedge, and LR, the length of the retrowedge (Figure 3c). In addition, the ramp angles, θ, of individual prowedge thrusts and of the retrowedge thrusts, were also measured. For each model run these parameters were measured at every 1 cm of contraction to quantitatively evaluate the thrust-wedge evolution. The terminology adopted for describing the analog models essentially follows that used by Storti et al. (2000). Table 1 summarizes the main parameters and results for the four series of models presented in this chapter.

RESULTS

The results from four series of experiments are presented and the model parameters for each series are summarized in Table 1. In each series, models were reproduced at least twice to validate reproducibility. Series I experiments were simple, doubly vergent wedges formed from a constant thickness sand pack (Table 1). Series II experiments involved doubly vergent wedges formed from a constant-thickness sand pack with a period of erosion at 33 cm of contraction, such that the prowedge taper was reduced to 5°. Series III experiments involved constant synkinematic deposition on both the prowedge and retrowedge slopes but not on the uplifted axial zone. Series IV experiments involved both synkinematic erosion and sedimentation, such that the material eroded was redeposited on both the pro- and retrowedge slopes.

Series I: Simple Doubly Vergent Thrust Wedge

Initial deformation in the simple doubly vergent wedge model was characterized by the development of a flat-topped anticline bounded by two kink bands— one provergent and dipping at 28° and the other retrovergent and dipping at 35° (Figure 4a–i). With increased contraction, layers in the kink bands became rotated and sheared out along the kink-band limbs forming thrust faults. The retrovergent kink band rapidly evolved into a well-developed, retrovergent thrust fault (Figure 4a–ii). In the early evolution of the model, most of the shortening was taken up by displacement along the single retrovergent thrust together with the formation of closely spaced provergent thrusts. These nucleated successively at the velocity discontinuity at the subduction slot and were carried passively backwards and upward in the hanging-wall of the retrovergent thrust. The retrovergent thrust initiated at a steep ramp angle of 35° but evolved into a lower angled, flatter trajectory once it had emerged on top of the prekinematic sand pack. In this manner, the axial zone of the model was rapidly uplifted until it reached a critical height, H_c, of 6.3 cm at 18 cm of contraction. At this point, the topographic load of the axial zone formed an internal backstop, allowing "forward-breaking" prowedge thrust faults to nucleate. At 20 cm of contraction, the first forward-breaking thrust was clearly visible in the prowedge (Figure 4a–iii). From this point onward, the rate of displacement on the retrovergent thrust diminished, whereas successive provergent thrusts

a) Mechanical Model - First Increment of Deformation

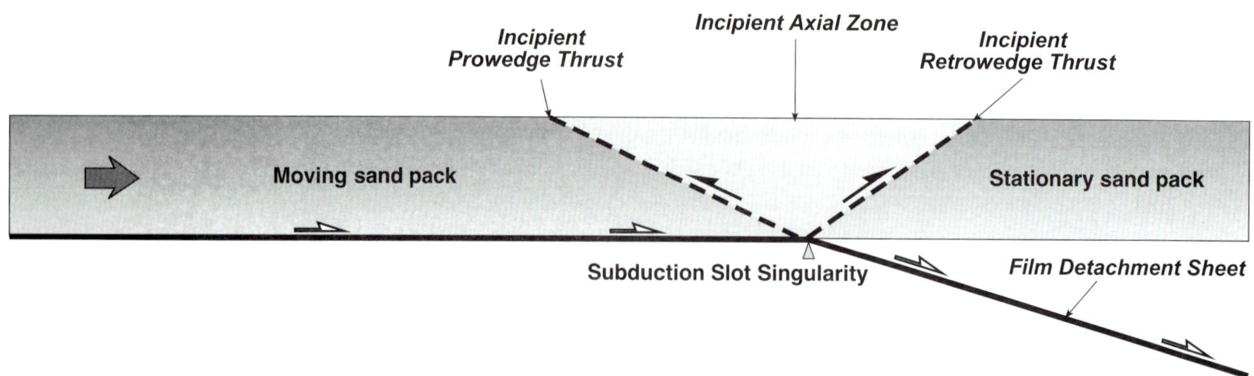

b) Mechanical Model - Final State

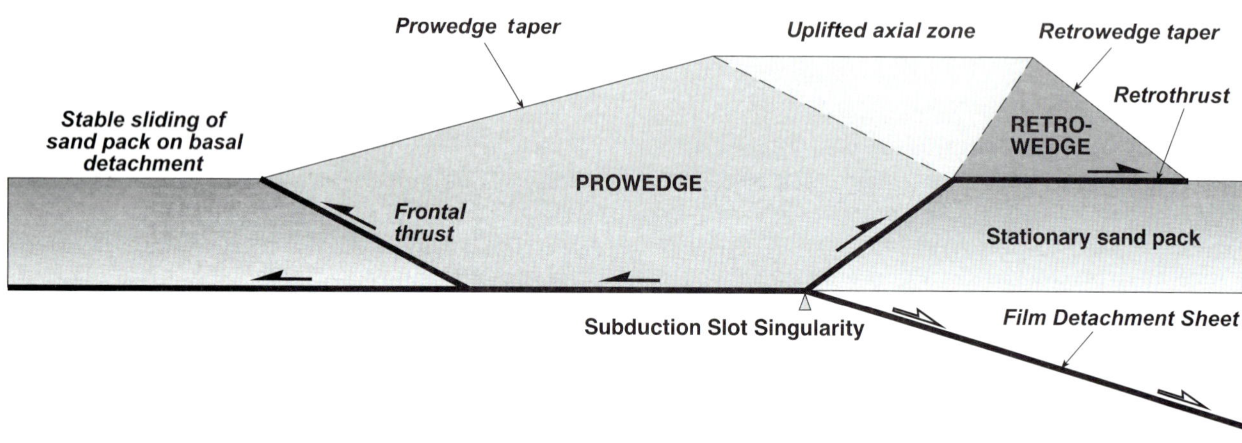

c) Model Parameters

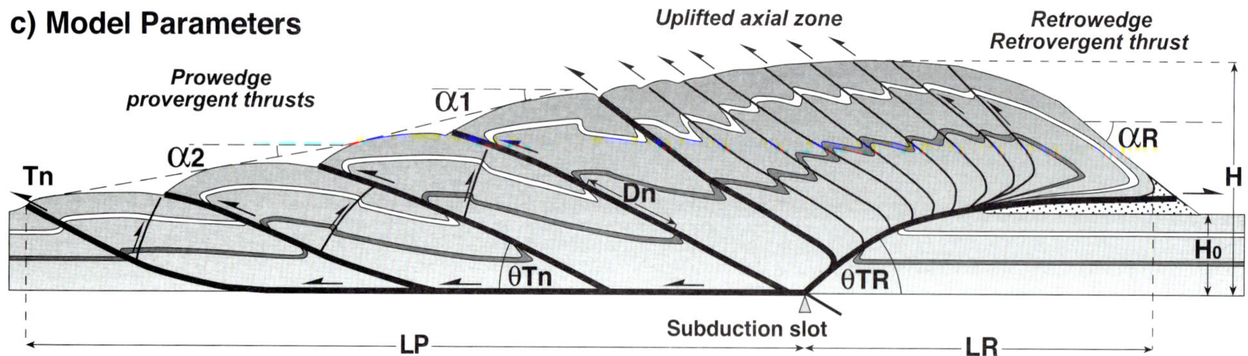

H_0 = Initial height | H = Wedge height | Tn = Major thrust fault

Dn = Fault displacement | LP = Length of Prowedge | LR = Length of retrowedge

α1 = Prowedge taper angle | α2 = Lower Prowedge taper angle | αR = Retrowedge taper angle

θTn = Thrust ramp angle | θTR = Retrothrust ramp angle | Measured marker layer

FIGURE 3. (a) Schematic diagram of the initial state of the experimental model. (b) Schematic diagram of the final state of the experimental model showing the key elements of the doubly vergent wedge system. (c) Interpreted doubly vergent wedge, showing the parameters measured in the analysis of the experimental results.

Experiment Number	Experiment Type	Sandpack Thickness H_o (cm)	Critical Height H_c (cm)	Shortening At Thrust-ramp Initiation (cm), Initial Ramp Angle, & Final Ramp Angle θ							Pre-Erosion/ Sedimentation Height (cm)	Posterosion/ Sedimentation Height (cm)	Pre-Erosion/ Sedimentation Taper Angles			Posterosion/ Sedimentation Taper Angles			Final Wedge Taper Angles			Final Amount of Shortening (cm)	Final Wedge Length (cm)		Final Wedge Height (cm)
				T1	T2	T3	T4	T5	T6	TR			α1	α2	αR	α1	α2	αR	α1	α2	αR		LP	LR	
Series I CC-17	Simple	2.5	6.31	18 24° 37°	20 23° 28°	28 22° 27°	36 23° 23°	42 24° 31°	48 30° 22°	1 35° 35°	-	-	-	-	-	-	-	-	12°	8°	39°	51	29.31	12.98	7.97
Series II CC-18	Erosion At 33 cm Shortening	2.5	6.29	18 25° 28°	22 24° 26°	30 24° 24°	43 21° 29°	50 25° 29°	- - -	1 33° 30°	7.20	5.15	14°	5°	40°	5°	5°	23°	13°	9°	37°	53	24.05	10.81	7.05
Series III CC-06	Continuous Sedimentation	2.5	5.95	12 18° 19°	16 25° 26°	20 39° 29°	23 19° 20°	25 25° 24°	- - -	1 36° 36°	-	-	-	-	-	-	-	-	6°	7°	18°	34	26.63	9.34	7.83
Series IV CC-19	Erosion and Sedimentation At 33 cm Shortening	2.5	5.80	15 22° 27°	19 25° 28°	25 25° 21°	33 25° 24°	46 23° 24°	- - -	1 31° 28°	6.93	4.82	14°	12°	37°	2°	2°	7°	16°	5°	40°	50	23.30	10.31	7.06

Table 1. Summary of experimental parameters and results of experiment series I to IV.

formed in a forward-breaking sequence until the model ceased to deform at 51 cm of contraction (Figure 4a–iv to 4a–vii).

The final geometry of the model in cross section is shown in Figure 4b. The doubly vergent wedge is strongly asymmetric, with a large-displacement, retrovergent thrust, TR, that carries in its hanging wall a closely spaced imbricate fan of provergent thrusts that form the uplifted axial zone between the pro- and retrowedges (Figure 4b). The provergent thrust that bounds the axial uplift is labeled T1. In front of this, there are five more provergent thrusts that form the prowedge. These developed "in sequence"— that is, they formed in a forward-breaking sequence. Each of the prowedge thrusts from T2 to T6 is characterized by a reasonably uniform 22–30° ramp angle, together with a slightly steeper retrovergent "back-thrust" TR. Together these form characteristic boxlike anticlines as each new provergent thrust develops (Figure 4a, b). The final critical taper of the prowedge consists of two parts— a frontal taper, α2, of 8° over the last-formed thrust T6 anticline, and a main prowedge taper, α1, of 12° over the main part of the prowedge (Figure 4b). The retrowedge taper, αR, is much steeper, at 39°.

The progressive evolution of this simple, doubly vergent wedge can be analyzed using the parameters shown in Figure 3c. Figure 5 shows plots of wedge height, thrust-fault displacement and activities, pro- and retrowedge lengths, and wedge taper angles measured during the evolution of the model. In the early stages of experiment CC-17, the height of the wedge's axial zone increased rapidly and nearly linearly until a critical height, H_c, of 6.3 cm was attained at 18 cm of shortening (Figure 5a). Most of this increase in the height of the axial zone was achieved by major displacement on the retrovergent thrust, TR. This first stage in the evolution of the model wedge is termed Stage I. During this stage, deformation was concentrated in uplift of the axial zone until a topographic height developed that was sufficient for the axial zone to act as a backstop and promote development of forward-breaking thrusts to form the prowedge in Stage II (Figures 4 and 5). During Stage II, most of the shortening was focused in the prowedge, with the formation of provergent thrusts T2–T6 (Figure 5b). During this stage, the rate of wedge uplift decreased between 18 and 51 cm of shortening (as indicated by the slope of the wedge height graph), the rate of displacement on the retrovergent thrust TR decreased, and the major provergent thrusts, T1 and T2, nucleated and increased displacement (Figure 5b). From 18 cm of shortening until the end of the deformation at 51 cm of shortening, successive provergent thrusts developed at almost uniform intervals and spacings (Figures 4a and 5b). These thrusts did not develop in a simple piggyback manner; instead, thrust activities overlapped, with as many as three

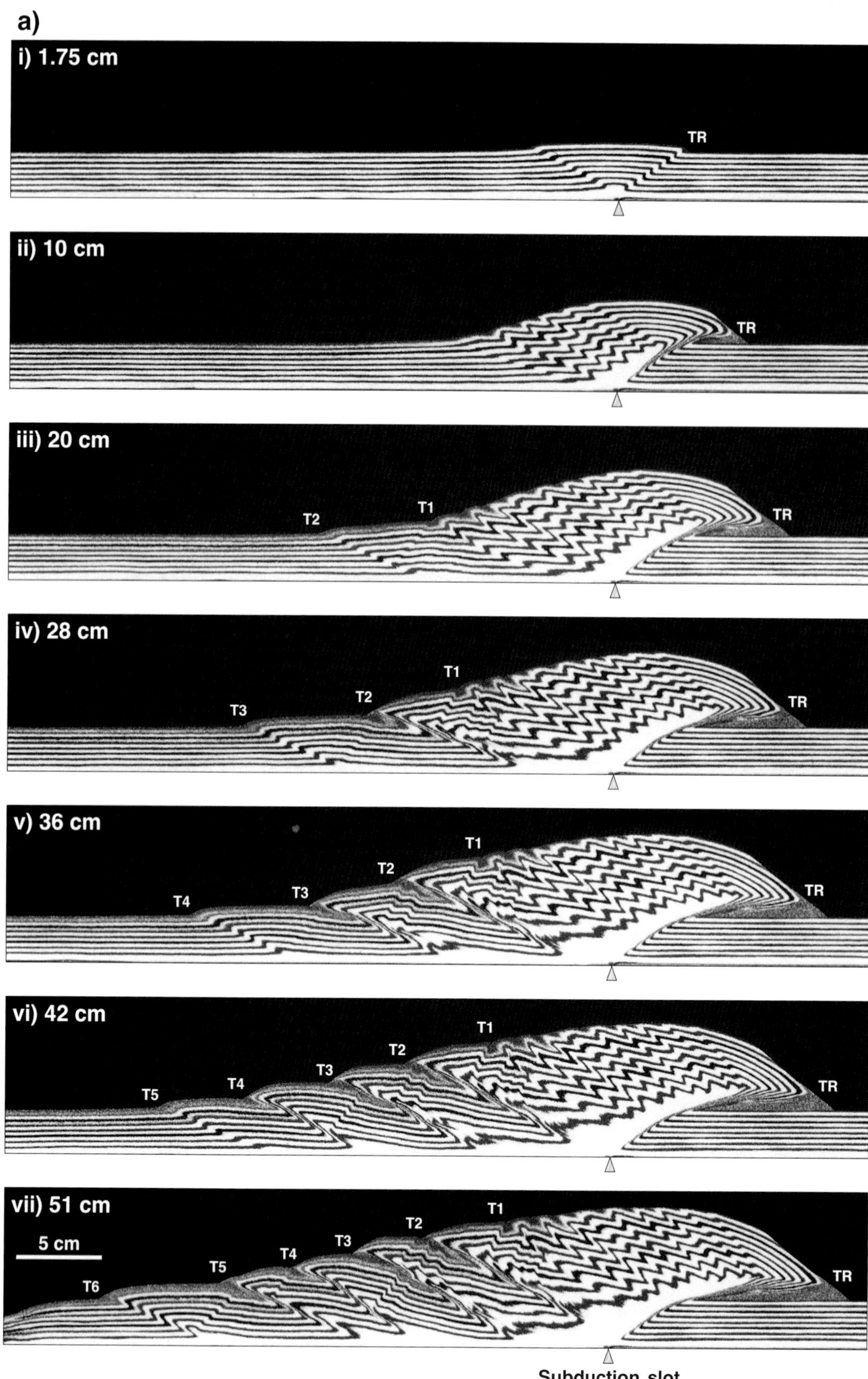

Figure 4. Experiment CC-17. Simple doubly vergent wedge. (a) Sequential photographs showing the progressive evolution of a simple doubly vergent thrust wedge. (b) Vertical cross section and line diagram, showing the final geometry of the model. Prowedge thrust faults are marked T1 to T6, in order of nucleation, and represent a forward-breaking thrust sequence.

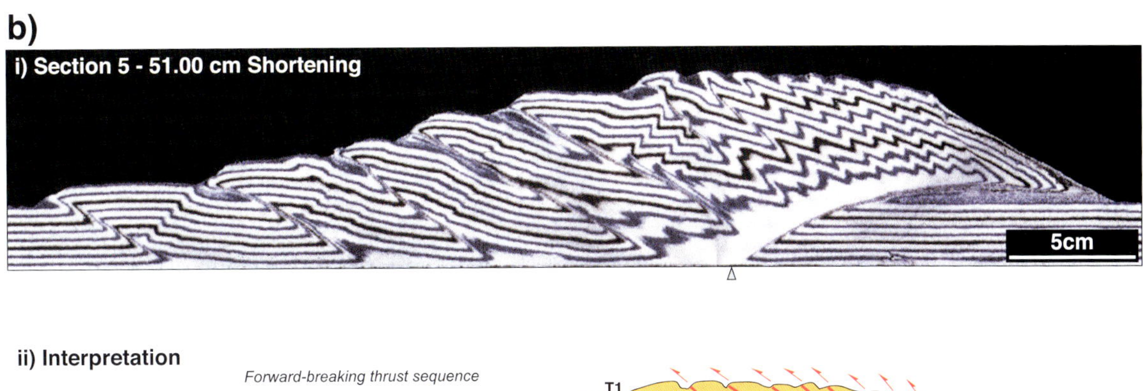

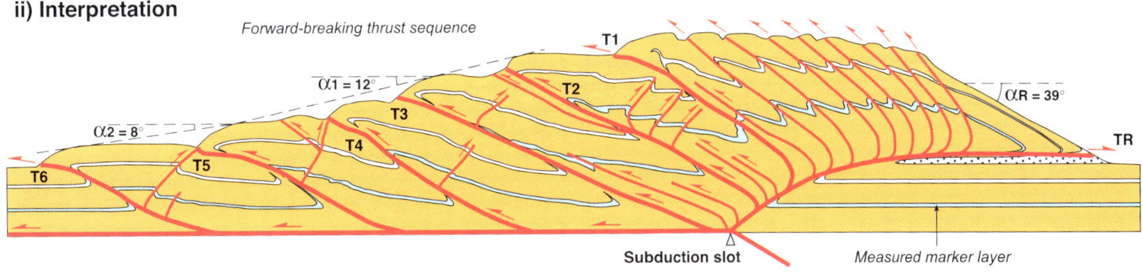

Figure 4. (cont.).

thrusts accommodating displacement at any one time (Figure 5b). During Stage II, there were sharp jumps in the prowedge length that reflect the nucleation of each new prowedge thrust (Figure 5c). The subsequent decrease in LP prior to the development of the next new fault indicates internal shortening within the wedge. The retrovergent wedge continued to increase in length but at a smoothly decreasing rate (Figure 5c). The graphs of wedge taper angles also show the two stages of evolution, with a relatively stable critical taper for the prowedge attained only in Stage II (Figure 5d).

Series II: Doubly Vergent Thrust Wedge with Synkinematic Erosion

The models that were run for Series II experiments were exactly the same as Series I models, in terms of dimensions and design. However, at 33 cm of shortening, once Stage II was well developed, a single, instantaneous period of synkinematic erosion was introduced, thereby reducing the prowedge taper and the overall wedge height (Table 1). Figure 6a shows the progressive evolution of experiment CC-18. Stage I deformation for experiment CC-18 was almost identical to that in experiment CC-17, which was described previously and shown in Figures 4 and 5. Initial deformation produced a flat-topped anticline above a dominant retrovergent thrust, TR, together with a series of closely spaced provergent thrusts (Figure 6a–i). Rapid, near-linear uplift of the axial zone occurred until 18 cm of shortening and the development of the provergent thrust, T1, that bounded the axial uplift. Critical wedge height, H_c, was 6.3 cm at this point. From this point onward, Stage II deformation was characterized by the development of forward-breaking (i.e., "in sequence") provergent thrusts that formed a prowedge with a critical taper of 11–14° (Figure 6a–ii and iii). At 33 cm of shortening, a single erosion event was introduced whereby the height of the axial zone was reduced to 5.15 cm and the prowedge taper reduced to 5° (Figure 6a–iv). Deformation resumed, with prolonged movement along thrust T2 and cessation of movement on thrust T3 (Figure 6a–v). In this manner, rapid axial uplift resumed (i.e., there was a return to Stage I behavior) with displacement evenly distributed on both TR and T2. At 40 cm of shortening, the axial zone had reached an elevation in excess of 6 cm, and Stage II behavior recommenced at 43 cm of shortening with the development of successive, forward-breaking provergent thrusts to continue deformation of the prowedge system (Figure 6a–vi and vii). Internal deformation of the wedge ceased after 53 cm of contraction.

The final model shows five major provergent thrust faults, T1–T5, with T1 bounding the uplifted axial zone (Figure 6b). Four provergent thrusts, T2–T5, form the critically tapered prowedge. Each of these has a smaller, steeper, retrovergent "back thrust" in its hanging wall, and together these define boxlike anticlines at the leading edges of individual thrust sheets (Figure 6b). The overall geometry of the doubly vergent wedge in experiment CC-18 (Figure 6b) was similar to that of the simple doubly vergent wedge (Figure 4b) except that there were fewer thrust faults in CC-18. Detailed analysis of the progressive evolution of experiment CC-18 revealed profiles almost identical to those in

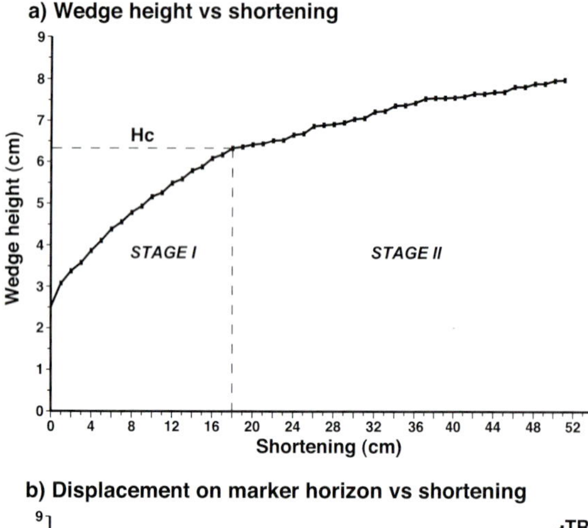

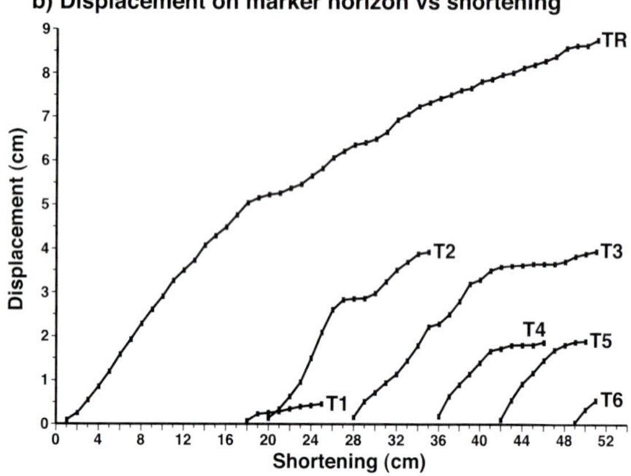

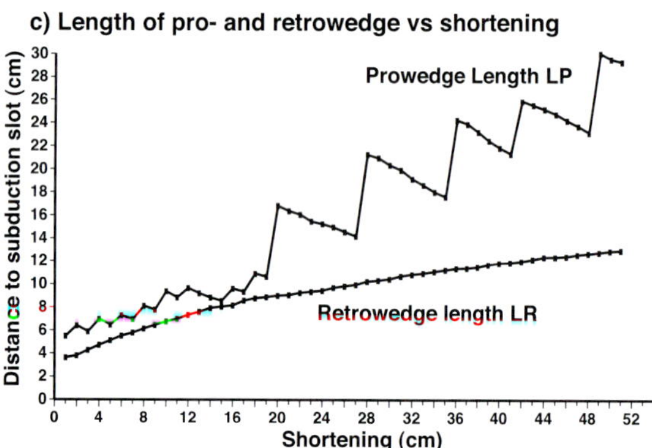

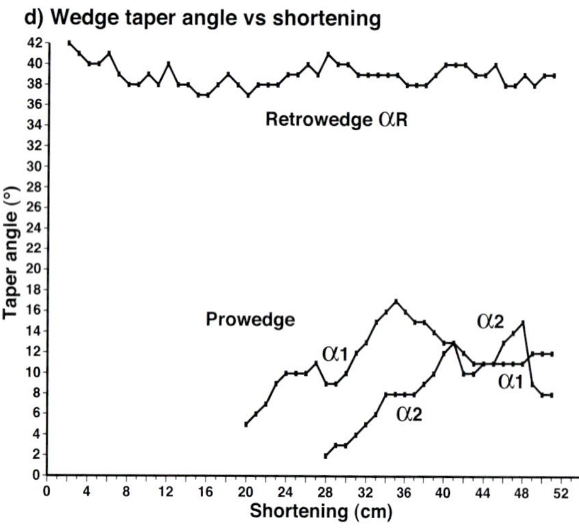

FIGURE 5. Graphs showing the fault activity and wedge geometries during the progressive deformation of experiment CC-17. Refer to Figure 3c for notation. (a) Wedge height versus shortening. (b) Fault displacement measured on a marker horizon, versus shortening. (c) Length of pro- and retrowedge versus shortening. (d) Wedge taper angles versus shortening. Note overlapping thrust fault activities, with up to three faults active synchronously.

experiment CC-17 (Figures 5 and 7) for Stage I and the initial part of Stage II of wedge evolution (up to 33 cm of contraction). However, at 33 cm of contraction in experiment CC-18, the overall wedge height was reduced, as was the critical taper of the prowedge (Figure 7a, b). This initially resulted in thrust T3 switching off, prolonged activity on thrust T2, and an increased thrust displacement rate on the retrothrust, TR, until the critical wedge height H_c was again achieved (Figure 7b). At that point, Stage II activity recommenced, with continued nucleation and propagation on provergent thrusts T4 and T5 (Figure 7b). As in experiment CC-17, the length of the prowedge showed a cyclic evolution reflecting synchronous activity on several overlapping thrust surfaces (Figure 7c). The retrowedge grew smoothly at a decreasing rate, except where it was affected by the 33-cm erosional event (Figure 7c). The critical taper angles for this wedge showed a similar evolution to that in experiment CC-17, except where the overall height and wedge tapers were reduced by erosion at 33 cm of shortening (Figure 7d).

Series III: Doubly Vergent Thrust Wedge with Synkinematic Sedimentation

Series III experiments were similar in dimensions and design to Series I and II experiments, except that during shortening, continuous, synkinematic sedimentation was added to the prowedge and to the retrowedge after every 2 cm of shortening (Table 1). Synkinematic sediments were not deposited on top of the axial zone. The initial deformation of experiment CC-06 was similar to that in the other experiments described above. A flat-topped box fold developed at 1–2 cm of contraction, and this evolved into the uplifted axial zone bounded by a moderately steep retrothrust, TR, which carried many closely spaced provergent thrusts in its hanging wall (Figure 8a–i and ii). In this model, synkinematic sedimentation clearly highlighted the sequential activity of these closely spaced provergent thrusts and also prevented the retrovergent thrust from developing a flat, upper trajectory on top of the prekinematic sand pack (Figure 8a–iii). Throughout the experiment, the retrovergent thrust maintained a steep ramp angle of 35°, propagating through the synkinematic sediments

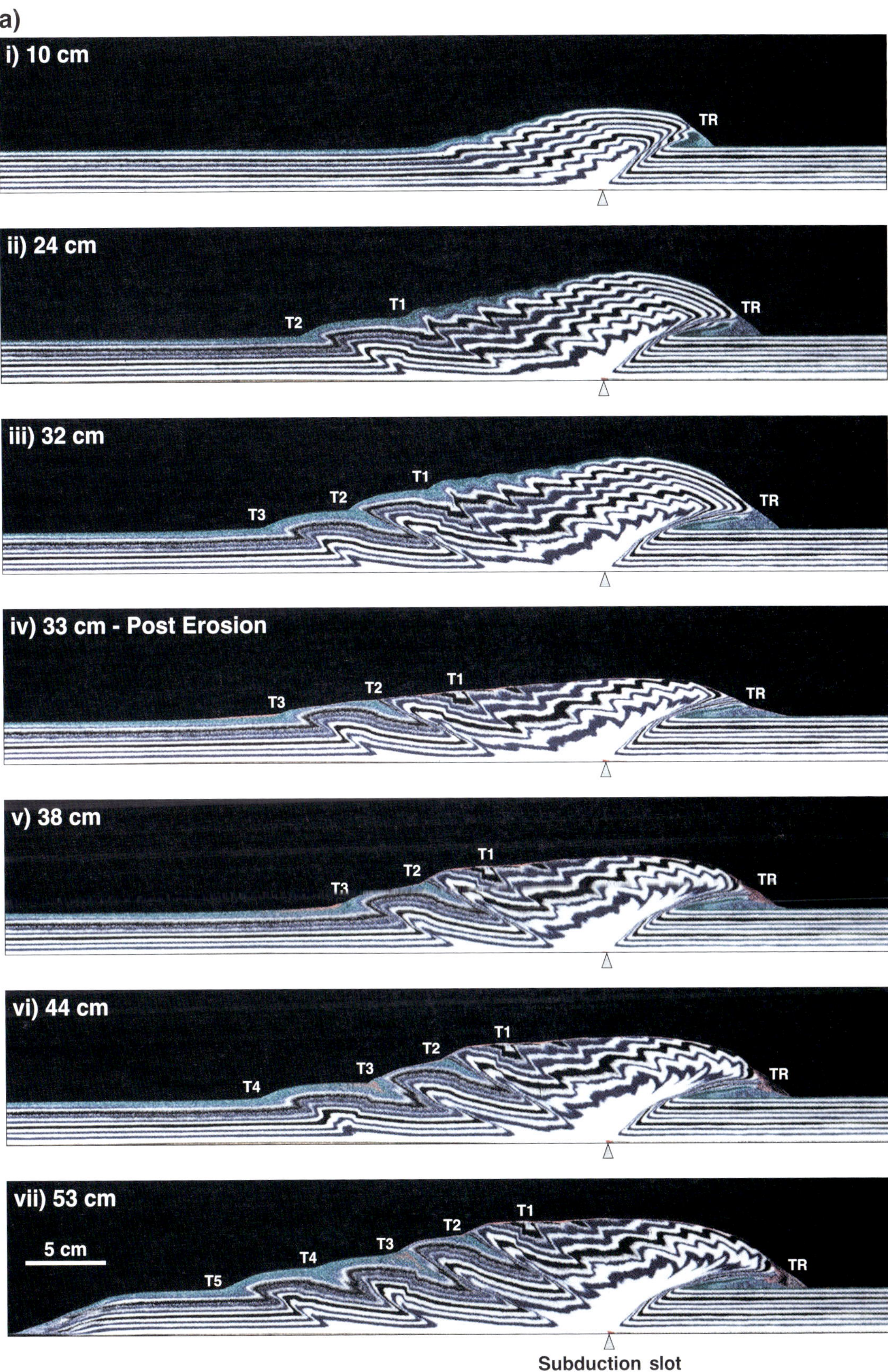

FIGURE 6. Experiment CC-18. Doubly vergent wedge with synkinematic erosion. (a) Sequential photographs, showing the progressive evolution of a simple doubly vergent thrust wedge, with erosion at 33 cm of contraction. (b) Vertical cross section and line diagram, showing the final geometry of the model. Prowedge thrust faults are marked T1 to T5, in order of nucleation.

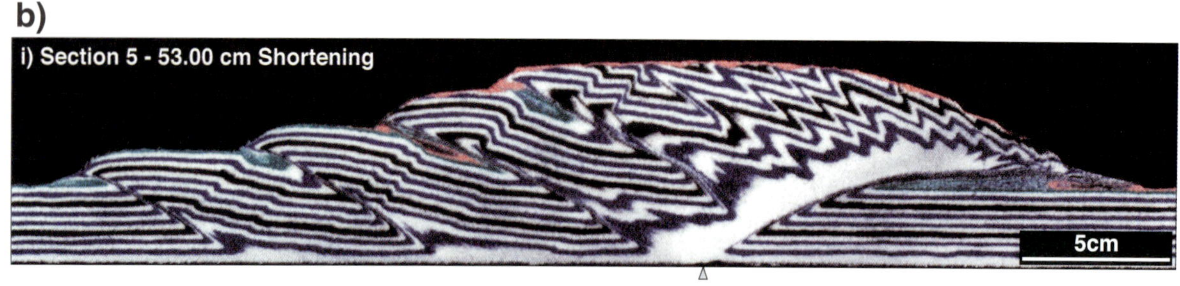

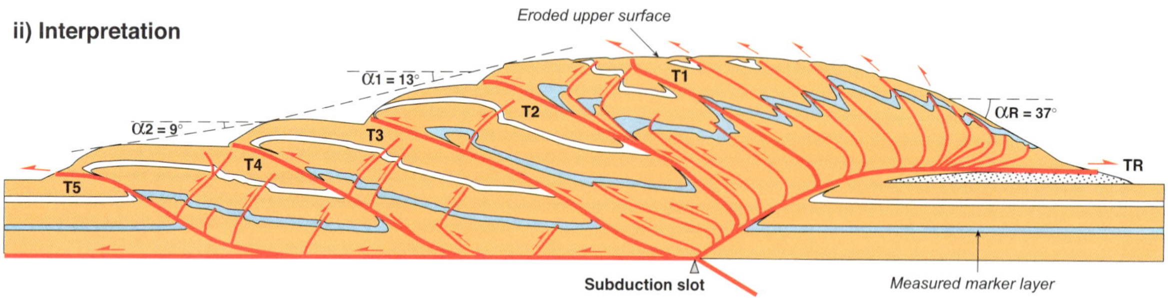

Figure 6. (cont.).

added onto the retrovergent wedge. In addition, a more shallowly dipping shortcut thrust also formed (Figure 8). After 16 cm of shortening, the axial zone had attained a topographic height of 6 cm. At that point a new provergent thrust (T2) nucleated in front of the axial zone and started building a prowedge system (Figure 8a–iv). At the same time, synkinematic sediments were being incrementally added to both the prowedge and retrowedges, thus reducing the overall taper. As a result, a new thrust, T3, nucleated behind thrust T2 (i.e., "out of sequence") (Figure 8a–v), elevating the back section of the prowedge to maintain an overall critical taper of approximately 11°. At 23 cm of shortening, out-of-sequence thrust T4 nucleated. At around 26 cm of contraction, thrust T5 nucleated far in the front of the prowedge (Figure 8a–vi). Deformation continued until 34 cm of contraction (Figure 8a–vii).

The final model shows a triangular doubly vergent wedge. The prowedge taper consists of two components— one produced by the shortening on the provergent thrust faults and the second by synkinematic sedimentation (Figure 8b). The retrovergent thrust maintained a consistently steep, planar trajectory. The provergent thrusts show a pattern of both forward-breaking and break-back nucleation (Figure 8b). Geometric analysis of this model's evolution showed that Stage I deformation, as in the other models described above, occurred until a critical wedge height, H_c, of 6 cm was achieved at 16 cm of shortening, before significant prowedge thrusting developed (Figure 9a). Stage II deformation occurred where prowedge thrusting produced a tapered prowedge when contraction increased beyond 16 cm (Figure 9b). As in the other doubly vergent wedge models, prowedge thrusting was characterized here by simultaneous movement on several overlapping thrusts, but in this experiment some thrusts clearly formed out of sequence (Figure 9b). As in the other wedge models described above, the evolution of the prowedge system was somewhat cyclic, reflecting the activity of key thrust faults (Figure 9c). Both the pro- and retrowedges tapers, however, were progressively reduced by the synkinematic sedimentation. These tapers thus remained subcritical (Figure 9d) because prowedge thrust activities and displacements were not sufficient to attain a theoretical critical wedge taper of 11–12°.

Series IV: Doubly Vergent Thrust Wedge with Synkinematic Sedimentation and Erosion

Series IV experiments were very similar in design to the other series previously described, but Series IV had the addition of synkinematic erosion and sedimentation, as shown by experiment CC-19 (Table 1). From initiation of the experiment to 33 cm of contraction, the evolution of the wedge (Figure 10a–i to iii) was almost identical to that in experiments CC-17 and 18. There was only a minor difference: thrust T3 was slightly less developed compared with that in experiment CC-18 (Figures 6a–iii and 10a–iii). At 33 cm of contraction, the wedge height was reduced to 4.8 cm and the prowedge taper was reduced to 2° (Figure 10a–iv). In contrast to experiment CC-18 (Figure 6a), the volume of material removed by erosion in CC-19 was matched by the volume of synkinematic sedimentation that was added proportionally to both the prowedge and the retrowedge (Figure 10a–iv). Deformation then continued until a maximum shortening of 50 cm. As in experiment CC-18, initial deformation after erosion and sedimentation was focused in the axial zone and at the hinterland of the prowedge, where thrusts T2 and T3 underwent increased displacement to generate the critical wedge height, H_C, necessary for continued prowedge evolution (Figure 10a–iv). Thrust T4 and the retrovergent thrust TR1 remained inactive during this period of axial zone uplift. At around 36 cm of shortening, a new retrovergent thrust, TR2, formed in the hanging wall above thrust TR1 and propagated at a steep ramp angle of 35° through the synkinematic sedimentary layers. Continued, albeit slow, displacement on retrovergent thrusts TR2 and TR3 carried the first-formed main provergent thrust T1 backward, up into the zone of axial uplift (Figure 10a–vi). At 39 cm of shortening, thrust T4 was reactivated, and it deformed the synkinematic sedimentary layers (Figure 10a–v and vi). At 46 cm of contraction, thrust T5 nucleated and increased displacement until the end of deformation at 50 cm of contraction (Figure 10a–vii).

The final model produced an asymmetric, doubly vergent wedge bounded by a steep retrovergent thrust in the retrowedge and by a prowedge with three main thrusts, T3–T5 (Figure 10b). The prowedge was similar in geometry to the prowedge of experiment CC-06 (Figure 8b), but the retrowedge was markedly different in structure. Geometric analysis of the progressive evolution of experiment CC-19 revealed patterns similar to those found in experiments CC-18 and CC-06. The graph of wedge-height evolution for experiment CC-19 (Figure 11a) is almost identical to that for experiment CC-18 (Figure 9a), whereas the plots of fault activities and displacements show distinct differences. Fault displacements and fault-activity plots (Figure 11b) show that in CC-19, thrusts T2 and T3 had prolonged activity as a result of the erosion and sedimentation at 33 cm shortening. Faults T2 to T4 show evidence of out-of-sequence reactivation as a result of the wedge geometry altered by the erosion and sedimentation. Similarly, the wedge lengths (Figure 11c) and wedge taper angles (Figure 11d) show patterns similar to those in experiment CC-18.

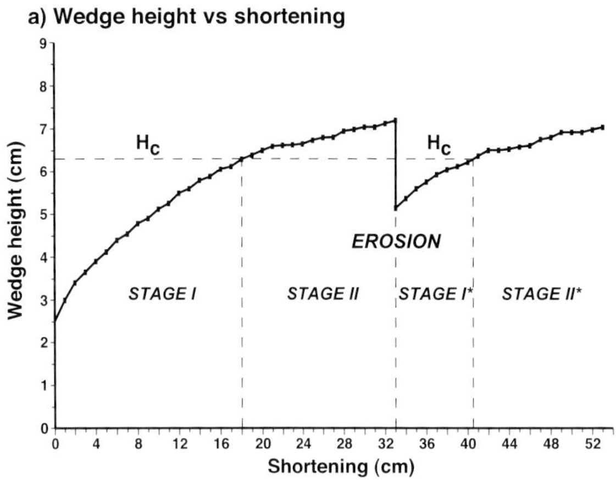

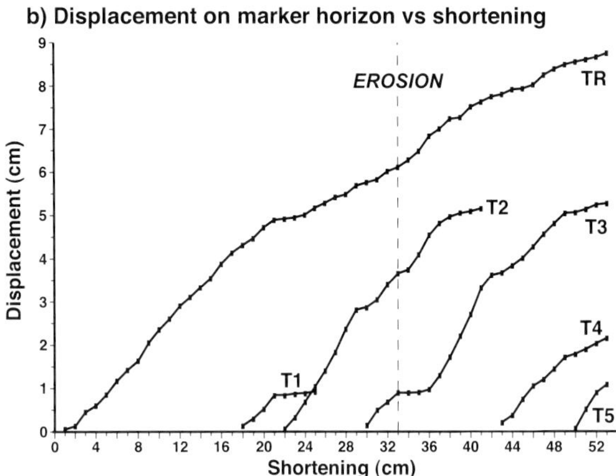

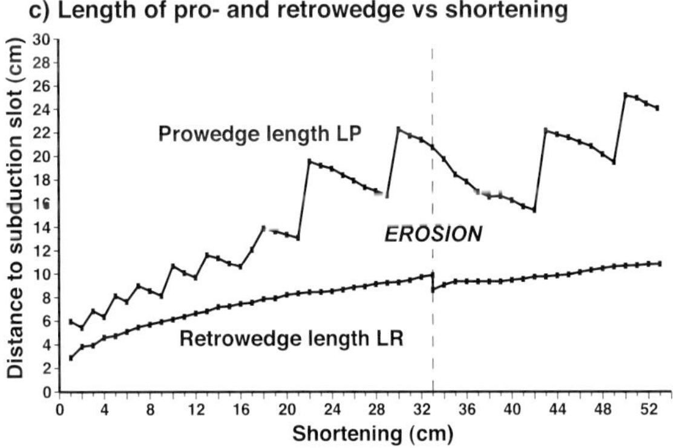

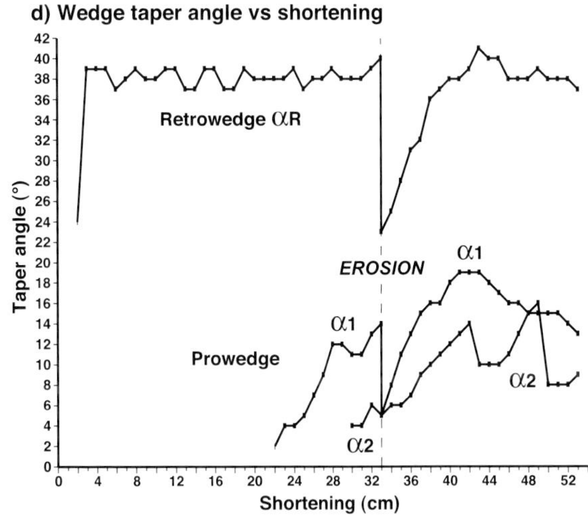

Figure 7. Graphs showing the fault activity and wedge geometries during the progressive deformation of experiment CC-18. Refer to Figure 3c for notation. (a) Wedge height versus shortening. (b) Fault displacement measured on a marker horizon, versus shortening. (c) Length of pro- and retrowedge versus shortening. (d) Wedge taper angles versus shortening. Note prolonged activity on thrust T2, brief cessation of activity on thrust T3, and increased activity on retrothrust TR following erosion.

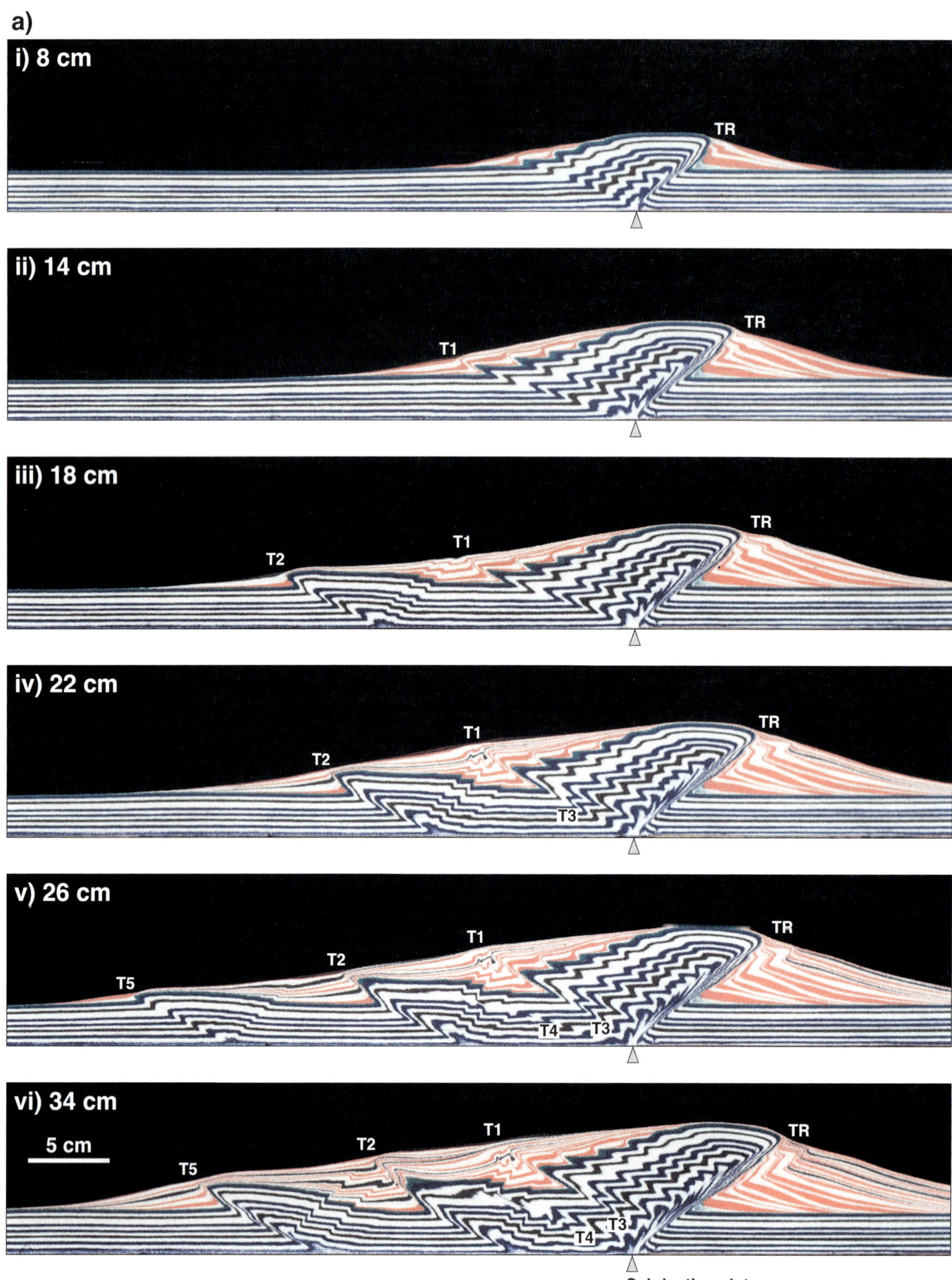

Figure 8. Experiment CC-06. Doubly vergent wedge with synkinematic sedimentation. (a) Sequential photographs showing the progressive evolution of a doubly vergent thrust wedge with synkinematic sedimentation. (b) Vertical cross section and line diagram, showing the final geometry of the model. Prowedge thrust faults are marked T1 to T5, in order of nucleation. Note that thrusts T3 and T4 formed out of sequence.

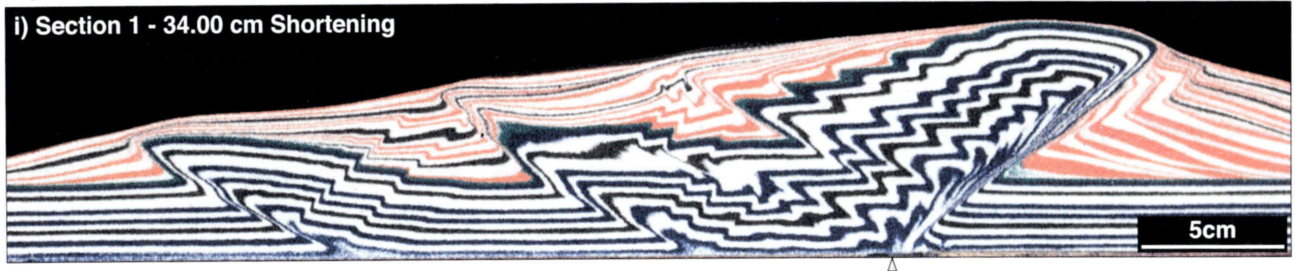

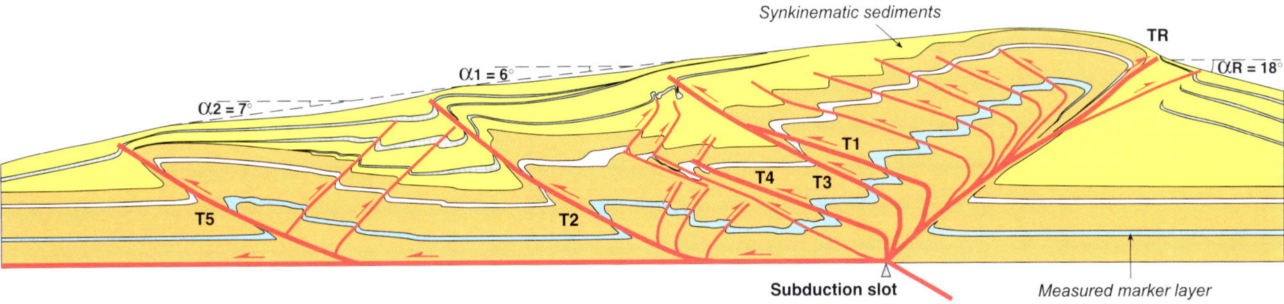

FIGURE 8. (cont.).

DISCUSSION

Analog Models of Doubly Vergent Thrust Belts

The scaled analog models illustrated in this chapter successfully simulated the development of asymmetric, doubly vergent fold-and-thrust belts (Figures 4, 6, 8, and 10). The prowedge region of the models formed by imbrication and underthrusting, whereas the retrowedge was thrust over the foreland region behind the subduction slot. The development of the retrowedge was triggered by the abrupt velocity discontinuity at the subduction slot (Willett et al., 1993).

The kinematic evolution of the analog models of thrust wedges occurred in two stages, I and II, which were similar to those described by Storti et al. (2000). These stages can be seen in all models, including those with synkinematic erosion and sedimentation (Figures 5, 7, 9, and 11). Stage I was characterized by the development of the retrowedge, relatively rapid uplift of the axial zone, and thrusting along the retrovergent thrust. During this stage, closely spaced provergent kink bands and thrust faults nucleated at the subduction slot and were transported backward in the hanging wall of the main active retrovergent thrust. In this manner, the triangular axial zone was uplifted until a critical elevation, H_c, was achieved. At that point, the uplifted axial wedge became the backstop for Stage II deformation, in which a "forward-breaking" thrust sequence generated the critically tapered prowedge. The rate of retrovergent thrusting decreased during this stage, and the uplift rate also decreased and asymptotically approached a value of approximately three times the initial thickness of the layers (i.e., $3 \times H_0$). During Stage II deformation the topography of the models became significantly asymmetric, with a steep retrowedge and a more gently tapered prowedge. The prowedge continued to form by imbricate thrusting during Stage II, until stable sliding over the entire base of the prowedge occurred at shortening values typically around 50–54 cm.

During Stage II deformation, the prowedge rapidly achieved a stable taper of around 11–12°, and this was maintained throughout the life of the experiment (except in models in which it was perturbed by erosion and/or sedimentation). Stage II deformation was dominated by the frontal imbrication of thrust sheets added to the prowedge (e.g., Figures 4, 6, 8, 10). With the nucleation of each new forward-breaking thrust, the prowedge rapidly became wider and the slope, $\alpha 2$, at the very front of the wedge, was less than the critical taper. Then, as the new thrust accumulated displacement and was incorporated into the prowedge, the length of the prowedge decreased, thereby giving rise to the sawtoothed pattern in Figures 5d, 7d, 9d, and 11d.

The doubly vergent wedge models illustrated in this chapter all display a very important pattern of displacement partitioning whereby, in the prowedge, several thrusts were active synchronously, albeit at different displacement rates (e.g., Figures 5b, 7b, 9b, and 11b). Synchronous thrust activity has been discussed

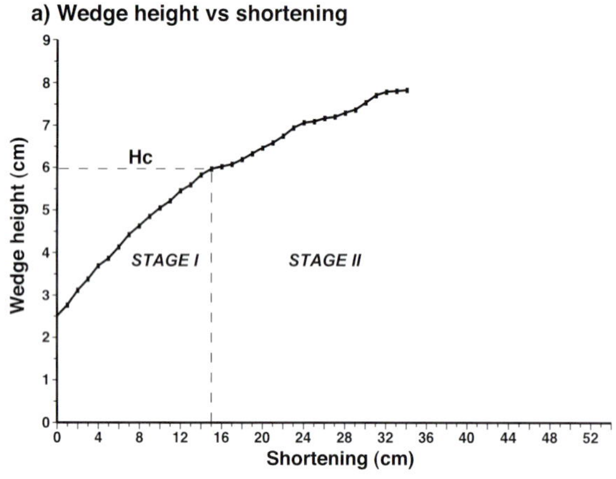

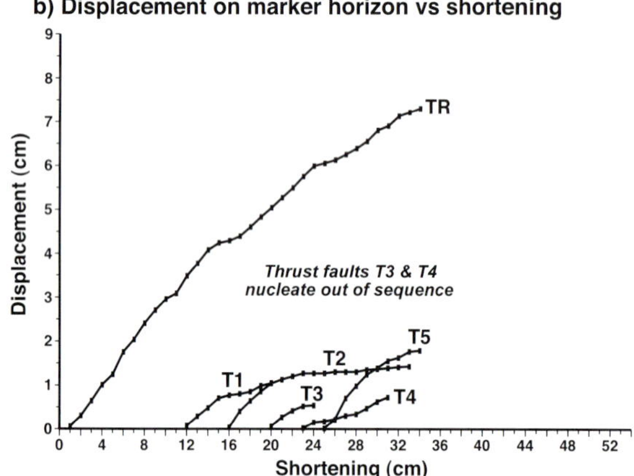

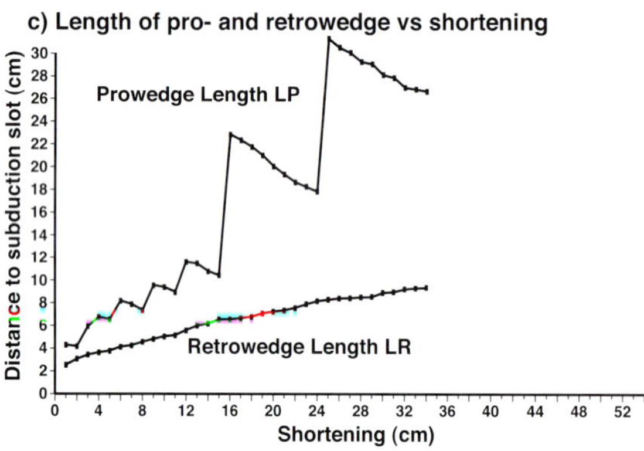

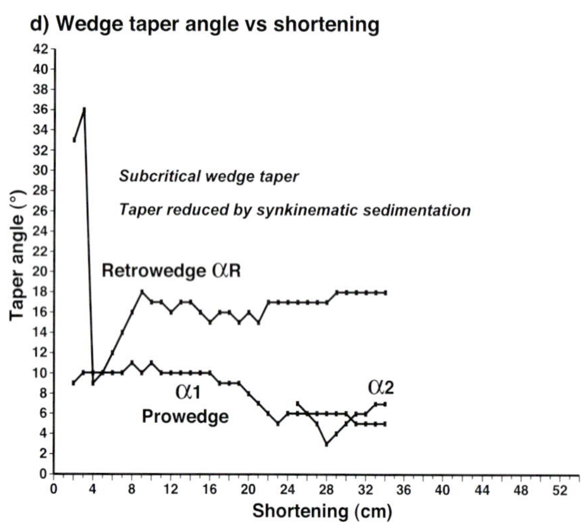

FIGURE 9. Graphs showing the fault activity and wedge geometries during the progressive deformation of experiment CC-06. Refer to Figure 3c for notation. (a) Wedge height versus shortening. (b) Fault displacement measured on a marker horizon, versus shortening. (c) Length of pro- and retrowedge versus shortening. (d) Wedge taper angles versus shortening. Note that thrusts T3 and T4 formed out of sequence.

by Boyer (1992) and may possibly be more common in fold-thrust belts than was hitherto recognized. Evidence of this synchronous thrust activity was recorded by deformation of the synkinematic sediments in experiments CC-06 (Figure 8b) and CC-19 (Figure 11b). Prowedge thrust development in all of the models was characterized by initial kink-band folding, box fold development, and then propagation of through-going discrete thrusts that cut through the limbs of the kink-band folds (e.g., Figure 4a–i to 4a–iiia). Storti et al. (1997) documented similar fold-thrust development in analog thrust models. Storti et al. (2000) presented a detailed analysis of two analog models of doubly vergent wedge systems, together with analytical analyses of the stress systems involved. Our models gave similar results but with the additional component of synkinematic erosion and/or sedimentation.

The effects of erosion and synkinematic evolution were dramatic. In our models, the synkinematic erosion occurred in Stage II during the development of the prowedge (Figures 6 and 10). Erosion reduced both the overall wedge height, to below H_c, and the prowedge critical taper and the number of prowedge thrusts produced. These changes to the shape of the doubly vergent wedge system meant that the wedge returned to Stage I behavior, whereby most of the deformation was concentrated in uplift of the axial region by retrovergent thrusting as well as by provergent thrusting in the hinterland of the prowedge. Dramatic changes were generated in the activities of provergent thrusts, with cessation, reactivation, or prolonged displacement on existing thrusts, and nucleation of out-of-sequence thrusts in the hinterland of the prowedge (Figures 6 and 10). Continued deformation and uplift of the axial region eventually regained the critical height, H_c, and led to a resumption of Stage II activity.

When synkinematic sedimentation was added to the prowedge, it reduced the wedge taper and significantly reduced the number of prowedge thrusts that formed (Figures 8 and 10). In particular, the spacing between the prowedge thrusts increased, and the thrusts then formed large thrust anticlines mantled by synkinematic sediments. In addition, prowedge thrusts ramped directly upward, climbing into the synkinematic sediments and did not form an upper thrust flat (Figure 8). Similarly, the retrowedge thrust propagated upward at

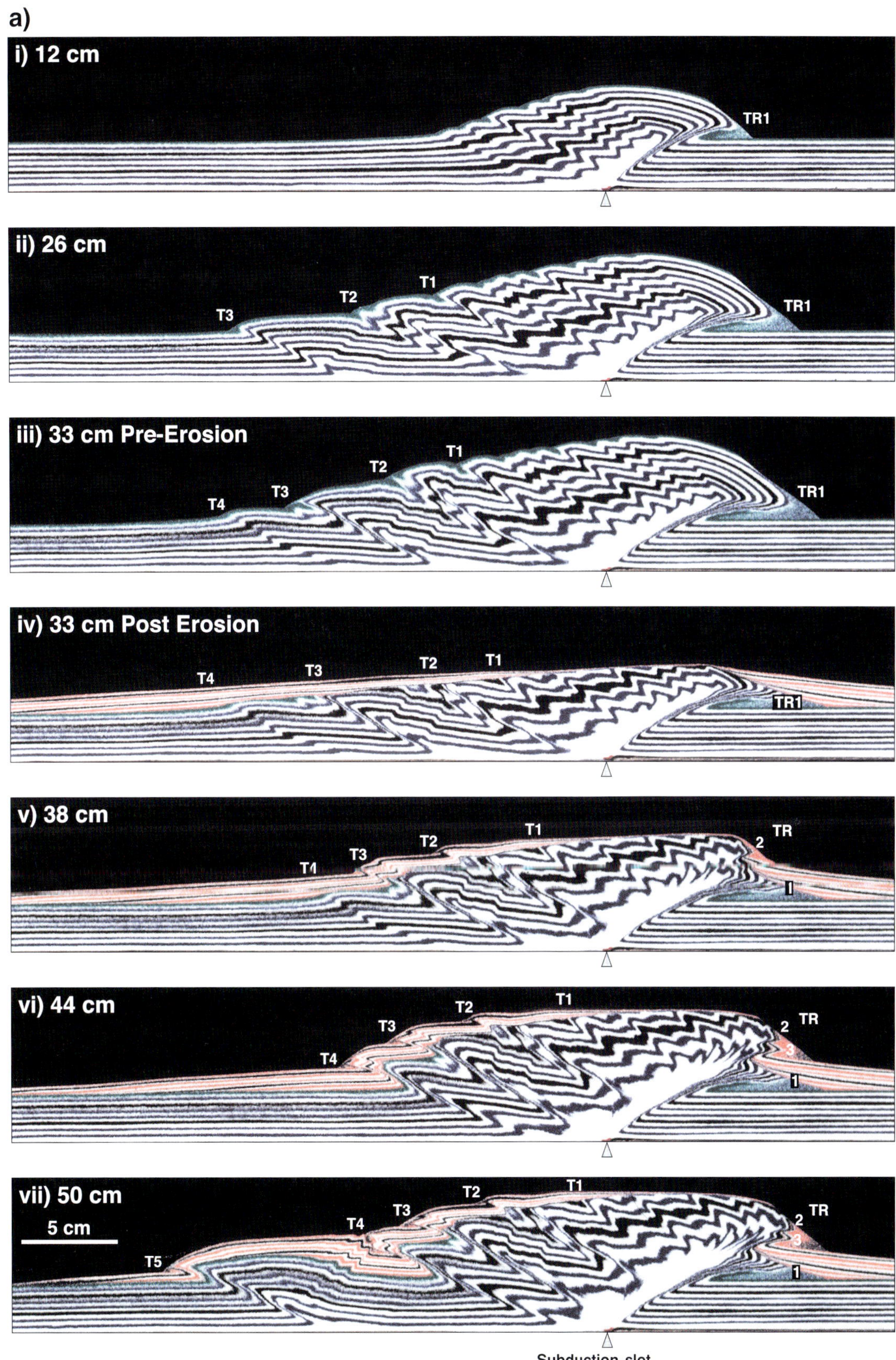

Figure 10. Experiment CC-19. Doubly vergent wedge with synkinematic erosion and sedimentation. (a) Sequential photographs showing the progressive evolution of a doubly vergent thrust wedge, with one phase of synkinematic erosion and sedimentation at 33 cm of contraction. (b) Vertical cross section and line diagram, showing the final geometry of the model. Prowedge thrust faults are marked T1 to T5, in order of nucleation.

b)

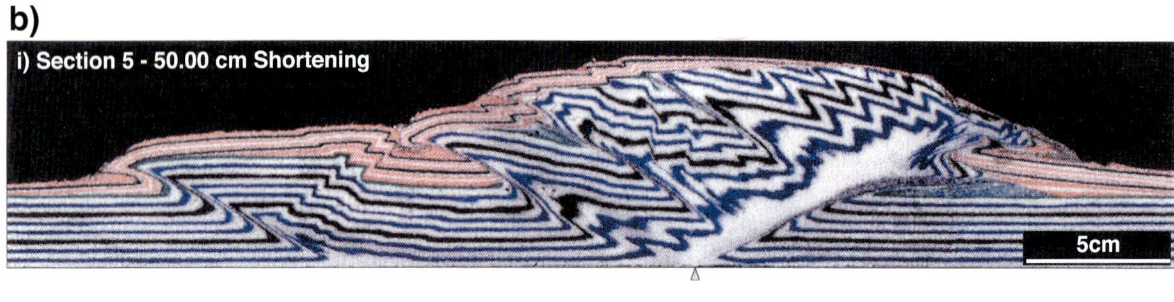

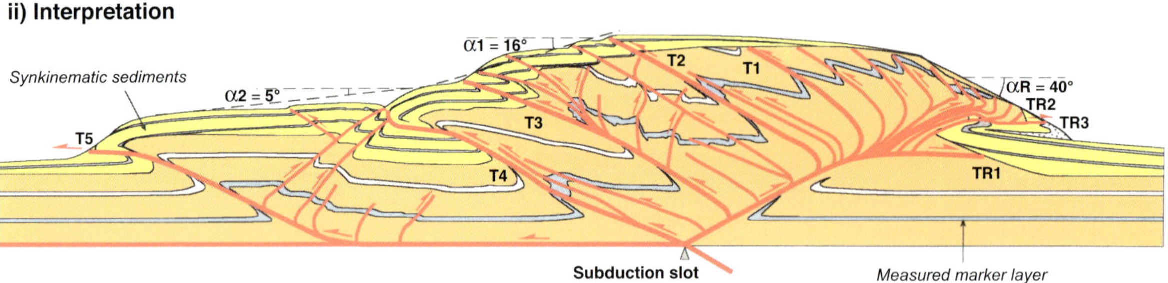

FIGURE 10. (cont.).

angles of around 30–35°, through the synkinematic sediments (Figures 8 and 10). Deformation of these synkinematic growth sequences also recorded the relative timing of the prowedge thrust displacements (Figure 10b). Hardy et al.'s (1998) 2-D numerical models of simple compressional Coulomb wedges also found similar results both for synchronous thrust activities and for the effects of synkinematic sedimentation and erosion.

Limitations of the Analog Models

Whereas the analog models successfully show the progressive development of doubly vergent thrust wedges, there are inherent limitations in the analog modeling techniques that need to be evaluated. The experimental apparatus had a rigid baseplate and thus did not allow the simulation of the flexural and isostatic responses of thrust stack loading and underthrusting of the lithosphere in both the prowedge and retrowedge regions. In addition, pore-fluid pressure effects, thermal effects, and competency contrasts and anisotropies were not simulated in our model experiments. The lack of high pore-fluid pressures probably accounts for the greater wedge taper angles in the models compared with natural thrust wedges (cf. Byrne et al., 1993). Thermal effects and the resultant changes in mechanical behavior with depth cannot be simulated using the current model configuration. Comparisons with Beaumont et al.'s (2000b) finite element models show that our analog-model wedges are very similar in geometry and kinematics to the numerical models that do include thermal effects. Decreases in the thickness of the prekinematic layers produced more closely spaced prowedge thrusts (cf. Liu et al., 1992). The introduction of bed-parallel anisotropies (cf. Storti et al., 2000) or the use of ductile layers may produce more thrusts and folds but inherently does not produce significant changes in the overall doubly vergent wedge geometries.

Comparison with Natural Fold-thrust Belts

Natural doubly vergent thrust systems, such as accretionary prisms and orogenic fold-and-thrust belts, commonly are considerably more complicated than the simple physical models illustrated in this chapter. However, despite the limitations discussed above and the difficulties in applying Coulomb wedge models to thrust systems (e.g., discussions by Price, 1988, and Bombolakis, 1994), the models' kinematic and geometric similarities to natural systems give valuable insights into fold-thrust-belt evolution in a variety of terranes (Storti et al., 2000). In particular, they illustrate the progressive evolution of fold-thrust sheets, synchronous movement on a number of thrust sheets, and dynamic interaction between erosion and sedimentation (Figures 8 and 10).

Many orogens exhibit asymmetric, doubly vergent wedge cross-section profiles, including the Pyrenees (Figure 12a), the Appalachians (Hatcher and Williams, 1986), the Alps (Figure 12b) (Schmid et al., 1996; Pfiffner et al., 2000), and the Himalayas (Figure 12c) (Mattauer, 1986). They all have similar geometric characteristics—a central, uplifted and exhumed metamorphic core (internally highly deformed and intruded by plutons) that is flanked on both sides by fold-and-thrust belts, one of which is usually approximately twice the extent of the other (Figure 13). The wider (i.e., more extensive) fold-thrust belt in these orogens typically has a pro-wedge profile with a critical taper that approximates that of a Coulomb wedge (Davis et al., 1983; Dahlen, 1990). Most of these orogens have undergone considerable uplift and erosion, which has removed the syntectonic growth strata that might have recorded the details of thrust-wedge timing and fault activities. An exception to this is the Spanish Pyrenees. Their unique evolution (Coney et al., 1996) has preserved some of the best examples of thrust-related growth strata (Poblet et al., 1998) that provide critical evidence for the timing of thrust activities within the orogenic wedge (Muñoz, 1992; Meigs et al., 1996).

The Spanish Pyrenees

The Pyrenean orogen in northern Spain and southern France is a Late Cretaceous through Miocene fold-thrust belt that developed as a result of convergence and collision of the African plate with the Iberian plate (Muñoz, 1992). The early stages of contraction in the Pyrenees were dominated by Late Cretaceous inversion of the preexisting Jurassic–Early Cretaceous rift basins (Muñoz, 1992). Thin-skinned thrusting evolved from the inverted fault systems during continued deformation in the Oligocene through the early Miocene. The schematic cross section in Figure 12a shows the strong southward asymmetry of the orogen, with the dominant thrust vergence towards the south. The northern part of the Pyrenees has a steep basal detachment with steeply dipping to overturned thrusts (Figure 12a) and can be considered equivalent to the retrowedge in the analog models (cf. Figures 4 and 6). The axial zone of the Pyrenees is dominated by an uplifted and strongly eroded antiformal stack (the Nogueres Zone, Figure 12a), whereas the southern part has the geometric characteristics of a critically tapered prowedge system. Moreover, detailed analyses of the well-preserved syntectonic growth strata (Poblet et al., 1998) in the Spanish Pyrenees have clearly demonstrated synchronous thrust movements, forward-breaking and reak-back thrust sequences, and out-of-sequence thrusting and reactivation of older thrusts (Vergés and Muñoz, 1990; Muñoz, 1992; Burbank et al., 1992; Vergés et al., 1995; Meigs et al., 1996; Meigs and Burbank, 1997). Beaumont et al.'s (2000b) lithospheric-scale numerical experiments have produced convincing

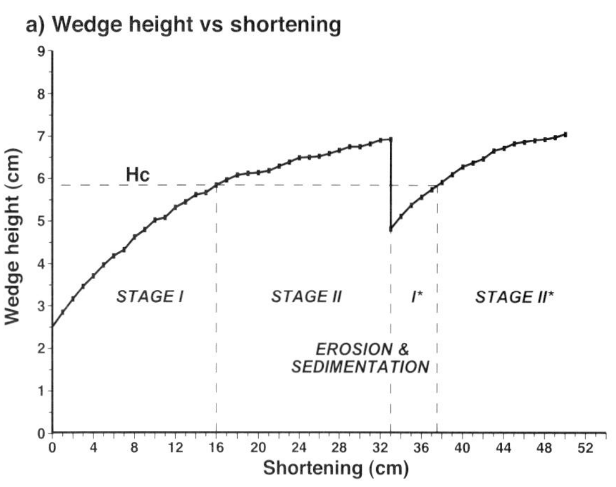

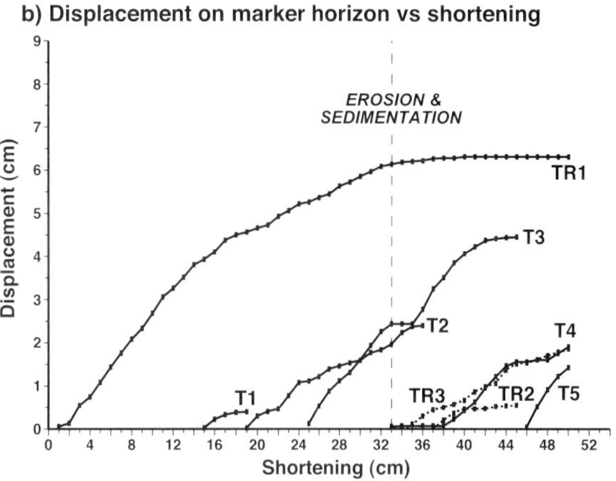

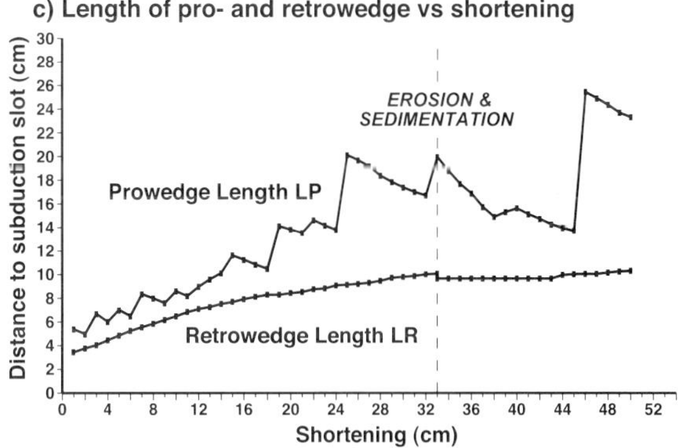

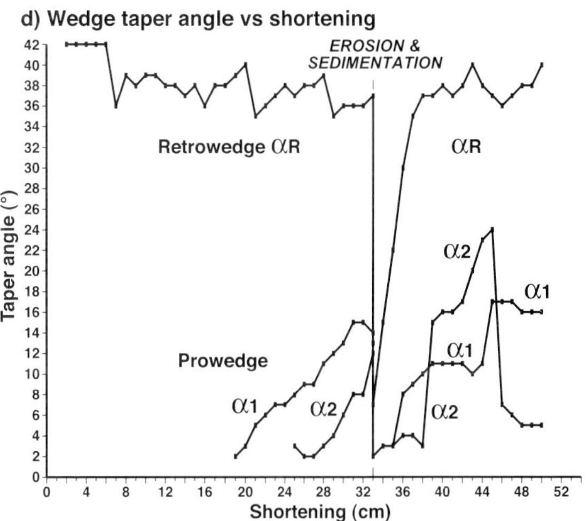

FIGURE 11. Graphs showing the fault activity and wedge geometries during the progressive deformation of experiment CC-19. Refer to Figure 3c for notation. (a) Wedge height versus shortening. (b) Fault displacement measured on a marker horizon, versus shortening. (c) Length of pro- and retrowedge versus shortening. (d) Wedge taper angles versus shortening. Note the increased activity on thrust T2 and the cessation of movement on thrusts T3 and T4 following erosion and sedimentation.

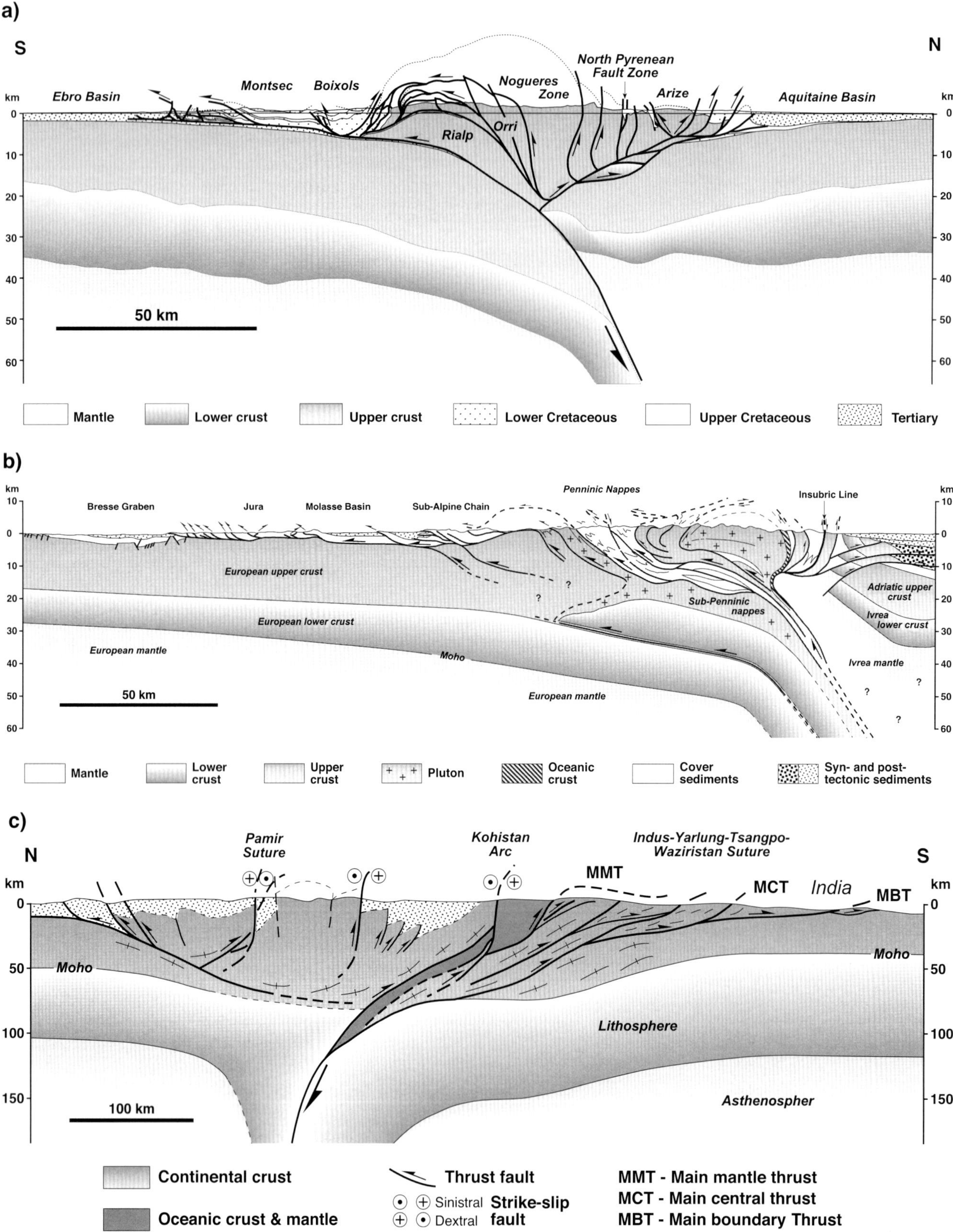

FIGURE 12. Natural examples of doubly vergent wedge systems. (a) Simplified lithospheric-scale cross section through the Pyrenean orogen along the ECORS deep seismic profile section. (Modified after Muñoz, 1992.) (b) Simplified crustal-scale cross section through the western Alps. (Modified from Schmid et al., 1996.) (c) Simplified crustal-scale cross section through the western Himalayas. (Modified after Mattauer, 1986.)

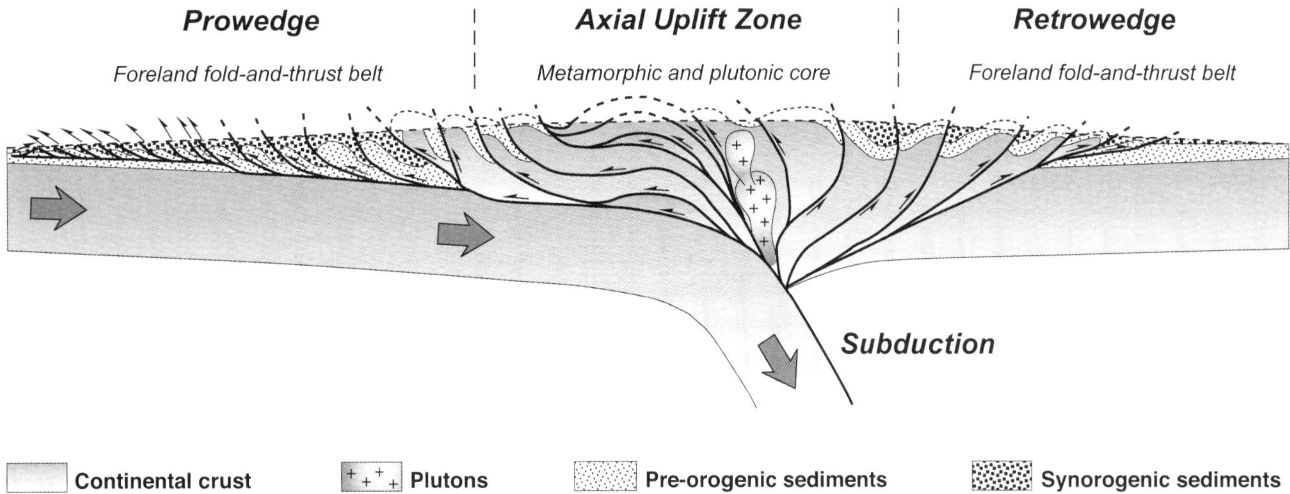

FIGURE 13. Synoptic model of a doubly vergent fold-thrust system in a collisional orogen. An uplifted igneous and metamorphic axial core is flanked by a long prowedge foreland fold-and-thrust belt and a shorter retrowedge foreland fold-and-thrust belt.

evolutionary models and geometries for the Pyrenees. Although our models do not incorporate inversion of a preexisting rift system, they do produce geometries and evolutionary pathways (Figures 4–11) that are similar in style to those derived from field studies (Muñoz, 1992) and from Beaumont et al.'s (2000b) numerical modeling experiments.

Implications for Hydrocarbon Systems

The geometric and kinematic models of thrust-belt evolution, as deduced from the analog models described in this chapter, have major implications for hydrocarbon systems in thrust belts. In particular, the models confirm that, within a fold-thrust belt, several thrusts may be moving at any one time. This has major implications for the timing of thrusting with respect to hydrocarbon generation and migration and for the so-called "piggy-back" rules for construction and restoration of cross sections in thrust belts (Woodward et al., 1989). In addition, the models confirm the dynamic feedback between surface processes and thrust-belt evolution, showing that both synkinematic erosion and sedimentation dramatically affect thrust-belt geometries, thrust timings, and thrust sequences (Hardy et al., 1998). Whereas this study has focused on 2-D thrust systems, the effects of surface processes are expected to be even more important in 3-D and therefore warrant further detailed research.

CONCLUSIONS

The results of the analog modeling of doubly vergent thrust systems has revealed a two-stage evolutionary pathway for the development of thrust belts that result from orthogonal convergence. Stage I is characterized by the rapid elevation of an axial zone that is bounded by a major retrovergent thrust system together with closely spaced provergent thrusts in its hanging wall. When a critical elevation of the axial zone has been achieved, the mechanical balance of the wedge changes and provergent thrusts nucleate and propagate, in front of the axial zone, to form a critically tapered prowedge system. This Stage II deformation is characterized by an increase in the asymmetry of the system, formation of well-developed provergent thrust sheets, and slower uplift and formation of the retrowedge system. Stage II deformation shows synchronous movement on several overlapping prowedge thrusts at any one time.

The addition of synkinematic erosion and/or sedimentation to the system produces dramatic changes to the doubly vergent wedges in models. Both synkinematic erosion and sedimentation produce fewer prowedge thrust faults and, in particular, promote prolonged prowedge thrust activities, out-of-sequence thrust nucleation and propagation, or reactivation of older, preexisting thrusts. Synkinematic growth strata record the progressive evolution of the wedge system and also the activities on individual thrusts. The geometries of the analog models of doubly vergent thrust wedges closely replicate those of convergent and colisional orogens in nature. There are strong similarities between the doubly vergent wedge models that have synkinematic erosion and sedimentation and cross sections through the Pyrenees and the western Alps. Other, less well documented orogens also have similar, asymmetric, doubly vergent profiles. Despite their limitations, the analog models provide important geometric and kinematic templates for understanding the development of fold-and-thrust belts and, in particular, the dynamic feedback between surface processes (synkinematic erosion and sedimentation) and thrust evolution. These factors, together with the concepts of synchronous movement on overlapping thrust surfaces, need to be taken into account when one assesses the hydrocarbon systems in thrust terranes. The models described in this chapter are simple 2-D examples, and further detailed research is needed into 3-D thrust systems with synkinematic erosion and sedimentation.

ACKNOWLEDGMENTS

Research presented in this chapter was supported by the Fault Dynamics Research Group, Geology Department, Royal Holloway University of London. K. McClay also gratefully acknowledges support from BP Exploration. Mike Craker assisted with maintenance and modifications to the deformation apparatus. Fault Dynamics Publication No. 112.

REFERENCES CITED

Ballard, J. F., J. P. Brun, J. Van den Driessche, and P. Allemand, 1987, Propagation des chevauchements au-dessus des zones de décollement: Modèles expérimentaux, Compte Rendu Acadamie des Sciences, v. 305, p. 1249–1253.

Bally, A. W., P. L. Gordy, and G. A. Stewart, 1966, Structure, seismic data and orogenic evolution of the southern Canadian Rockies: Bulletin of Canadian Petroleum Geology, v. 14, p. 337–381.

Barclay, J. E., and D. G. Smith, 1992, Western Canada foreland basin oil and gas plays, *in* R. W. MacQueen and D. A. Leckie, eds., Foreland basins and fold belts: AAPG Memoir 55, p. 81–105.

Beaumont, C., P. Fullsack, and J. H. Hamilton, 1992, Erosional control of active compressional orogens, *in* K. R. McClay ed., Thrust tectonics: London, Chapman & Hall, p. 377–390.

Beaumont, C., H. Kooi, and S. Willett, 2000a, Progress in coupled tectonic-surface process models with applications to rifted margins and collisional orogens, *in* M. A. Summerfield, ed., Geomorphology and global tectonics: New York, John Wiley, p. 29–55.

Beaumont, C., J. A. Muñoz, J. Hamilton, and P. Fullsack, 2000b, Factors controlling the Alpine evolution of the central Pyrenees inferred from a comparison of observations and geodynamical models: Journal of Geophysical Research, v. 105, p. 8121–8145.

Beydoun, Z. R., 1988, The Middle East: regional geology and petroleum resources: Beaconsfield, Buckinghamshire, UK, Scientific Press, 292 p.

Beydoun, Z. R., M. W. Clarke, G. W. Hughes, and R. Stoneley, 1992, Petroleum in the Zagros Basin; a late Tertiary foreland basin overprinted onto the outer edge of a vast hydrocarbon-rich Paleozoic-Mesozoic passive-margin shelf, *in* R. W. MacQueen and D. A. Leckie, eds., Foreland basins and fold belts: AAPG Memoir 55, p. 309–339.

Bombolakis, E. G., 1994, Applicability of critical-wedge theories to foreland belts: Geology, v. 22, p. 535–538.

Boyer, S. E., 1992, Geometric evidence for synchronous thrusting in the southern Alberta and northwest Montana thrust belts, *in* K. R. McClay ed., Thrust tectonics: London, Chapman & Hall, p. 377–390.

Braun, J., and C. Beaumont, 1995, Three-dimensional numerical experiments of strain partitioning at oblique plate boundaries: implications for contrasting tectonic styles in the Southern Coast Ranges, California, and central South Island, New Zealand: Journal of Geophysical Research, v. 100, p. 18059–18074.

Buchanan, P. G., and K. R. McClay, 1992, Experiments on basin inversion above reactivated domino faults: Marine and Petroleum Geology, v. 9, p. 486–500.

Burbank, D. W., J. Vergés, J. A. Muñoz, and P. Bentham, 1992, Coeval hindward- and forward-imbricating thrusting in the south central Pyrenees, Spain: Timing and rates of shortening and deposition: GSA Bulletin, v. 104, p. 3–17.

Byrne, D. E., W. H. Wang, and D. M. Davis, 1993, Mechanical role of backstops in the growth of forearcs: Tectonics, 12, p. 123–144.

Calassou, S., C. Larroque, and J. Malavieille, 1993, Transfer zones of deformation in thrust wedges: an experimental study: Tectonophysics, v. 221, p. 325–344.

Cole, J. P., M. Parish, and D. Schmidt, 2000, Sub-thrust plays in the Papuan fold belt: the next generation of exploration targets, *in* P. G. Buchanan, A. M. Grainge, and R. C. N. Thornton, eds., Papua New Guinea's Petroleum industry in the 21st century: Proceedings of the Fourth PNG Petroleum Conference, Port Moresby, 29th to 31st May, 2000, p. 87–100.

Coney, P. J., J. A. Muñoz, K. R. McClay, and C. A. Evenchick, 1996, Syntectonic burial and post-tectonic exhumation of the southern Pyrenees foreland fold-thrust belt: Journal of the Geological Society of London, v. 153, p. 9–16.

Dahlen, F. A., 1990, Critical taper model of fold-and-thrust belts and accretionary wedges: Annual Review of Earth and Planetary Sciences, v. 18, p. 55–99.

Davis, D., and T. Engelder, 1985, The role of salt in fold/thrust belts: Tectonophysics, v. 119, p. 67–88.

Davis, D., J. Suppe, and F. A. Dahlen, 1983, Mechanics of fold/thrust belts and accretionary wedges: Journal of Geophysical Research, v. 88, p. 1153–1172.

Dooley, T., and K. R. McClay, 1997, Analog modeling of strike-slip pull-apart basins: AAPG Bulletin, v. 81, p. 804–826.

Dunn, J. F., K. G. Hartshorn, and P. W. Hartshorn, 1995, structural styles and hydrocarbon potential of the sub-Andean thrust belt of southern Bolivia, *in* A. J. Tankard, R. Suarez Soruco, and H. J. Welsink, eds., Petroleum basins of South America: AAPG Memoir 62, p. 523–543.

Faugere, E., and J. P. Brun, 1984, Modelisation experimentale de la distention continentale: Compte Rendu Acadamie des Sciences, v. 229, Ser. II, p. 365–370.

Fermor, P. R., and I. W. Moffat, 1992, The Cordilleran collage and the foreland fold-and-thrust belt, *in* R. W. MacQueen and D. A. Leckie, eds., Foreland basins and fold belts: AAPG Memoir 55, p. 81–105.

Gutscher, M. A., N. Kukowski, J. Malavieille, and S. Lallemand, 1998, Material transfer in accretionary wedges from analysis of a systematic series of analog experiments: Journal of Structural Geology, v. 20, p. 407–416.

Hardy, S., C. Duncan, J. Masek, and D. Brown, 1998, Minimum work, fault activity and the growth of critical wedges in fold-and-thrust belts: Basin Research, v. 10, p. 365–373.

Hatcher, R., and R. T. Williams, 1986, Mechanical model for single thrust sheets: Geological Society of America Bulletin, v. 97, p. 975–985.

Hatcher, R. D., P. H. Osberg, A. A. Drake, P. Robinson, and W. A. Thomas, 1990, Tectonic map of the U.S Appalachians: Geological Society of America, The geology of North America, v. F-2, Plate 1, scale 1/2,500,000.

Hill, K. C., M. S. Norvick, J. T. Keetley, and A. Adams, 2000, Structural and stratigraphic shelf-edge hydrocarbon plays in the Papuan fold belt, *in* P. G. Buchanan, A. M. Grainge, and R. C. N. Thornton, eds., Papua New Guinea's petroleum industry in the 21st century: Proceedings of the Fourth PNG Petroleum Conference, Port Moresby, 29th to 31st May, 2000, p. 67–85.

Jadoon, I. A. K., and W. Frisch, 1997, Hinterland-vergent tectonic wedge below the Riwat Thrust, Himalayan foreland, Pakistan: Implications for hydrocarbon exploration: AAPG Bulletin, v. 81, p. 438–448.

Jadoon, I. A. K., R. D. Lawrence, and R. J. Lillie, 1994, Seismic data, geometry, evolution and shortening in the active Sulaiman fold-and thrust belt of Pakistan: AAPG Bulletin, v. 78, 758–774.

Koons, P. O., 1990, Two sided orogens: collision and erosion from sandbox to the Southern Alps, New Zealand: Geology, v. 18, p. 679–682.

Lamerson, P. R., 1982, The Fossil Basin area and its relationship to the Absaroka thrust fault system, *in* R. B. Powers, ed., Geologic studies of the Cordilleran thrust belt: Rocky Mountain Association of Geologists, p. 279–340.

Liu, H., K. R. McClay, and D. Powell, 1992, Physical models of thrust wedges. *in* K. R. McClay ed., Thrust tectonics: London, Chapman & Hall, p. 71–81.

Malavieille, J., 1984, Modelisation experimentale des chevauchements imbriques: Application aux chaines de montagnes: Bulletin de la Société Géologique de France, v. 7, p. 129–138.

Mattauer, M., 1986, Intracontinental subduction, crust-mantle décollement and crustal-stacking wedge in the Himalayas and other collision belts, *in* M. P. Coward and A. C. Ries, eds., Collision tectonics: Geological Society of London Special Publication 19, p. 37–50.

McClay, K. R., 1990a, Deformation mechanics in analogue models of extensional fault systems, *in* R. J. Knipe and E. H. Rutter, eds., Deformation mechanisms, rheology and tectonics: Geological Society of London Special Publication 54, p. 445–453.

McClay, K. R., 1990b, Extensional fault systems in sedimentary basins: a review of analogue model studies: Marine and Petroleum Geology, v. 7, p. 206–233.

McClay, K. R., 1995, The geometries and kinematics of inverted fault systems: A review of analogue model studies, *in* J. G. Buchanan and P. G. Buchanan, eds., Inversion tectonics: Geological Society of London Special Publication 88, p. 97–118.

McClay, K. R., and M. White, 1995, Analogue models of orthogonal and oblique rifting: Marine and Petroleum Geology, v. 12, p. 137–151.

McClay, K. R., and M. S. Bonora, 2001, Analogue models of restraining stepovers in strike-slip fault systems: AAPG Bulletin, v. 85, no. 2, p. 233–260.

McClay, K. R., T. Dooley, R. Gloaguen, P. Whitehouse, and S. Khalil, 2001, Analogue modelling of extensional fault architectures: Comparisons with natural rift fault systems, *in* K. C. Hill and T. Bernecker, eds., Eastern Australian basins symposium 2001, A refocused energy perspective for the future: Petroleum Exploration Society of Australia Special Publication, p. 573–584.

Meigs, A. J., and D. W. Burbank, 1997, Growth of the South Pyrenean orogenic wedge: Tectonics, v. 16, p. 239–258.

Meigs, A. J., J. Verges, and D. W. Burbank, 1996, Ten-million-year history of a thrust sheet: Geological Society of America Bulletin, v. 108, p. 1608–1625.

Mugnier, J. L., P. Baby, B. Colletta, P. Vinour, P. Bale, and P. Leturmy, 1997, Thrust geometry controlled by erosion and sedimentation: A view from analog models: Geology, v. 25, p. 427–430.

Mulugeta, G., 1988, Modeling the geometry of Coulomb thrust wedges: Journal of Structural Geology, v. 10, p. 847–859.

Muñoz, J. A., 1992, Evolution of a continental collision belt: ECORS-Pyrenees crustal balanced cross section, *in* K. R. McClay, ed., Thrust tectonics: London, Chapman & Hall, p. 235–246.

Pfiffner, O. A., S. Ellis, and C. Beaumont, 2000, Collision tectonics in the Swiss Alps: insight from geodynamic modeling: Tectonics, v. 19, p. 1065–1094.

Poblet, J., J. A. Muñoz, A. Travé, and J. Serra-Keil, 1998, Quantifying the kinematics of detachment folds using 3D geometry: Application to the Mediano anticline (Pyrenees, Spain): GSA Bulletin, v. 110, p. 111–125.

Price, R. A., 1988, The mechanical paradox of large overthrusts: Geological Society of America Bulletin, v. 100, p. 1898–1908.

Richard, P. D., M. A. Naylor, and A. Koopman, 1995, Experimental models of strike-slip tectonics: Petroleum Geoscience, v. 1, p. 71–80.

Roeder, D., and R. L. Chamberlin, 1995, Structural geology of sub-Andean fold-and-thrust belt in northwestern Bolivia, *in* A. J. Tankard, R. Suarez Soruco, and H. J. Welsink, eds., Petroleum basins of South America: AAPG Memoir 62, p. 459–479.

Schmid, S. M., O. A. Pfiffner, N. Froitzheim, G. Schönborn, and E. Kissling, 1996, Geophysical-geological transect and tectonic evolution of the Swiss-Italian Alps: Tectonics, v. 15, p. 1036–1064.

Serra, S., and R. A. Nelson, 1988, Clay modeling of rift asymmetry and associated structures: Tectonophysics, v. 153, p. 307–312.

Storti, F., and K. R. McClay, 1995, Influence of syntectonic sedimentation on thrust wedges in analog models: Geology, v. 23, p. 999–1002.

Storti, F., F. Salvini, and K. R. McClay, 1997, Fault-related folding in analogue models of thrust wedges: Journal of Structural Geology, v. 19, p. 583–602.

Storti, F., F. Salvini, and K. R. McClay, 2000, Velocity-partitioned synchronous thrusting and thrust polarity reversal in experimental doubly vergent thrust wedges accreted at different syntectonic sedimentation rates: implications for natural orogens: Tectonics, v. 19, p. 378–396.

Tankard, A. J., M. A. Uliana, H. J. Welsink, V. A. Ramos, M. Turic, A. B. França, E. J. Milani, B. B. de Brito Neves, N. Eyles, J. Skarmeta, H. Santa Ana, F. Weins, M. Cirbian, O. Lopez Paulsen, G. J. B. Germs, M. J. De Wit, T. Machacha, and R. McG. Miller, 1995, Structural and tectonic controls of basin evolution in southwestern Gondwana during the Phanerozoic, *in* A. J. Tankard, R. Suarez Soruco, and H. J. Welsink, eds., Petroleum basins of South America: AAPG Memoir 62, p. 5–52.

Tron, V., and J. P. Brun, 1991, Experiments on oblique rifting in brittle-ductile systems: Tectonophysics, v. 188, p. 71–84.

Velaj, T., I. Davison, A. Serjani, and I. Alsop, 1999, Thrust tectonics and the role of evaporites in the Ionian zone of the Albanides: AAPG Bulletin, v. 83, p. 1408–1425.

Vendeville, B., P. R. Cobbold, P. Davy, J. P. Brun, and P. Choukroune, 1987, Physical models of extensional tectonics at various scales, *in* M. P. Coward, J. F. Dewey and P. L. Hancock, eds., Continental extensional tectonics: Geological Society of London Special Publication 128, p. 95–107.

Vergés, J., and J. A. Muñoz, 1990, Thrust sequences in the southern central Pyrenees: Bulletin de la Société Géologique de France, v. 8, p. 265–271.

Vergés, J., H. Millán, E. Roca, J. A. Muñoz, M. Marzo, J. Cirés, T. Den Bezemer, R. Zoetemeijer, and S. Cloetingh, 1995, Eastern Pyrenees and related foreland basins: Pre-, syn- and post-collisional crustal-scale cross sections: Marine and Petroleum Geology, v. 12, p. 893–915.

Wang, W. H., and D. M. Davis, 1996, Sandbox model simulation of forearc evolution and noncritical wedges: Journal of Geophysical Research, v. 101, p. 11329–11339.

Willett, S., C. Beaumont, and P. Fullsack, 1993, Mechanical model for the tectonics of doubly vergent compressional orogens: Geology, v. 21, p. 371–374.

Withjack, M. O., and W. R. Jamison, 1986, Deformation produced by oblique rifting: Tectonophysics, v. 126, p. 99–124.

Woodward, N. B., S. E. Boyer, and J. Suppe, 1989, Balanced geological cross sections: Short Course in Geology: v. 6, Washington, D.C.,USA, American Geophysical Union, 132 p.

11

Koyi, H. A., M. Sans, A. Teixell, J. Cotton, and H. Zeyen, 2004, The significance of penetrative strain in the restoration of shortened layers—Insights from sand models and the Spanish Pyrenees, *in* K. R. McClay, ed., Thrust tectonics and hydrocarbon systems: AAPG Memoir 82, p. 207–222.

The Significance of Penetrative Strain in the Restoration of Shortened Layers—Insights from Sand Models and the Spanish Pyrenees

Hemin A. Koyi

Hans Ramberg Tectonic Laboratory, Department of Earth Sciences, Uppsala University, Uppsala, Sweden

Maura Sans

Departament de Geologia Dinamica, Geofisica i Paleontologia, Universitat de Barcelona, Barcelona, Spain

Antonio Teixell

Departament de Geologia (Geotectonica), Universitat Autonoma de Barcelona, Bellaterra, Spain

James Cotton[1]

BP Exploration, Chertsey Road, Sunbury-on-Thames, Middlesex, U.K.

Hermann Zeyen

Dép. des Sciences de la Terre, Université de Paris-Sud, Orsay cedex, France

ABSTRACT

Dynamic restoration is achieved when one accounts for the changes that occur in area or volume during deformation. In contractional areas, layer-parallel shortening (LPS) cannot always be easily estimated or measured, although it is a significant component of deformation, as is gravitational compaction. Five model analogs with known initial dimensions and boundary conditions were shortened from one end. Profiles of these models were used to (1) estimate the amount of layer-parallel compaction (LPC), the main modality of layer-parallel shortening in granular analog materials; (2) outline variation of LPC with depth, lateral location, and percentage shortening; and (3) estimate the effect of lithology on LPC.

During progressive deformation, a modeled accretionary wedge, which formed during the shortening of the models, did not undergo homogeneous compaction; instead, loss of area varied in both space (with depth and laterally) and time. Balancing the area of sequential sections of one of the sand models, which was shortened above a high-friction

[1]*Present address*: BP Trinidad and Tobago LLC, Queens Park Plaza, Port of Spain, Trinidad and Tobago, W.I.

basal décollement, shows that the layers experienced tectonic compaction during deformation and lost as much as 17% of their cross-sectional area during 50% bulk shortening.

Restoration of two model profiles shows that LPC is three times greater in the model with high basal friction than in the model with low basal friction. In models where a sand layer was embedded within a viscous layer (a Newtonian material simulating rock salt), the layer accommodated all the shortening by folding and underwent no significant LPC.

Examples from the Spanish Pyrenees are used to illustrate the significance of LPS in restoring profiles of contractional areas. In the eastern Spanish Pyrenees, on the basis of deformed raindrop marks and burrows, from 16 to 23% of total shortening is estimated to be by LPS, whereas only 6 to 10% of the total shortening is accommodated by folding.

Model results illustrate the lateral and temporal variations of penetrative strain within shortened layers. Outlining this heterogeneous distribution of penetrative strain and any associated volume loss is important in distinguishing areas of reduced porosity, which are significant for hydrocarbon exploration.

INTRODUCTION

Bed length measurement and area balancing are two effective tools for interpreting the deformation history and determining the amount of shortening in cross sections of contractional areas (Dahlstrom, 1969; Hossack, 1979; Dixon, 1982; Cooper et al., 1983; Woodward et al., 1985, 1986, 1989; Suppe, 1985; Ramsay and Huber, 1987; Baker et al., 1988; Marshak and Woodward, 1988; Moretti et al., 1990; Mitra, 1978b, 1994; Jaswal et al., 1997; Meigs et al., 1996; Mukul and Mitra, 1998). In most applications, either the length or the area of the shortened layers is kept constant during reconstruction of the balanced cross sections.

Deformation in fold-and-thrust belts is accommodated by three main components: layer-parallel shortening, folding, and thrusting. If bed lengths are kept constant between the deformed and restored sections, one only accounts for deformation by folding and thrusting, so that layer-parallel shortening is not considered when the amount of bulk shortening is quantified. If strain markers are available, penetrative strain can be included in the balanced cross sections to restore the amount of shortening (Hossack, 1978; Mitra, 1990; Protzman and Mitra, 1990; Howard, 1993; Homza and Wallace, 1997).

Using examples from the Norwegian Caledonides, Hossack (1979) emphasized the necessity of areal restorations and pointed out possible errors that may be introduced if area decreased during shortening. Fischer and Coward (1982) quantified strain distribution in the Scottish Caledonides and concluded that prethrusting strain was not homogeneously distributed in the wedge. From deformed burrows, these authors measured a maximum local shortening of as much as 33%. In an excellent paper, Cooper et al. (1983) discussed the significance of layer-parallel shortening in balancing geologic cross sections and concluded that both bed length and area must be analyzed to recognize layer-parallel shortening. On the basis of stratal thickness calculations in a small-scale duplex in a limestone quarry, and assuming that area was conserved, Cooper et al. (1983) concluded that, in a total shortening of 49%, 27% was accommodated by LPS and the rest by imbrication.

Woodward et al. (1986) documented how strain varied in different thrust sheets in the southern Appalachians. In incompetent formations, they measured axial ratios of strain ellipse that increased, from external to internal thrust sheets, by R = 1.2 to R = 2.8.

In the shallow thrust system of the Oslo region, Morley (1986) recorded a maximum shortening by pressure solution of 15%, variably distributed within the belt. However, he suggested an average value of 5% for limestones across the entire section.

Mitra (1994) also emphasized the effects of LPS in thrust-belt restoration. In a regional cross section of the Sevier belt, he used different kinds of strain markers to show that LPS varies between 10 and 30% in the belt. McNaught and Mitra (1996) used finite strain data to document an LPS component of about 15% in the Meade thrust sheet of the Sevier Thrust Belt.

Because the initial stages of sand models are well documented, they are easy to compare with their later stages to quantify the components of strain. Many workers have used sand models to study different aspects of accretionary wedges and/or fold-and-thrust belts (Davis et al., 1983; Dahlen et al., 1984; Malavieille, 1984; Karig 1986; Zhao et al., 1986; Mulugeta and Koyi, 1987, 1992; Colletta et al., 1991; Liu et al., 1991; Koyi, 1995; Storti and McClay, 1995; Gutscher et al., 1996; Storti et al., 1997; Koyi et al., 2000; Lohrmann et al., 2003). In this study, layer-parallel compaction (LPC) is quantified in five sand models, which have been shortened from one end, to comment on the distribution of layer-parallel shortening (LPS) and to illustrate the significance of area loss within accretionary prisms and fold-thrust belts. LPC is quantified by measuring the cross-sectional area of deformed layers and comparing them to their undeformed initial areas. In some models, bed length is restored to estimate amount of penetrative strain in comparison with the other components of shortening (imbrication and folding).

MODELS AND THEIR LIMITATIONS

We have quantified layer-parallel compaction (LPC) in five sand models with different initial configurations (Table 1). All models consisted of passively layered, loose sand and were shortened from one end. After the models were shortened, they were sectioned and photographed for analysis.

Dry, loose sand is a suitable material to simulate the brittle Coulomb behavior of shallow crustal rocks (Hubbert, 1937, 1951; Horsfield, 1977; McClay and Ellis, 1987; Ellis and McClay, 1988; Mandl, 1988; Mulugeta, 1988; Cobbold et al., 1989; Weijermars et al., 1993). Dry sand has a Navier-Coulomb rheology and an angle of friction similar to that of sedimentary rocks (cf. McClay, 1990). The rheology of loose sand may be approximated by the Coulomb equation (Hubbert, 1937; Cobbold et al., 1989; Weijermars et al., 1993):

$$\tau = \tau_0 + \sigma \tan \phi \qquad (1)$$

where τ is the shear strength, τ_0 is the cohesive strength, σ is the normal stress, and ϕ is the angle of internal friction. For the material used in the models, the cohesive strength, and the angle of internal friction, see Table 1.

This study is a two-dimensional approach that assumes no movement along strike, because movement perpendicular to transport direction is significantly small relative to movement parallel to it. In this experimental approach, we have omitted the additional complicating factors of erosion, deposition, material anisotropy resulting from facies changes, and time variations in pore-fluid pressure ratios across the wedge. All the models were deformed on a rigid horizontal planar substrate and therefore do not account for slopes or irregularities in natural dêcollements. Model 4 was shortened above a ductile substrate of a Newtonian silicone polymer (SGM36; Weijermars et al., 1993) that simulated rock salt or overpressured shale. In the description of the models, these variations are distinguished. Because gravity compaction is minimal in sand models, lateral compaction in full-scale prisms can be overestimated. In general, penetrative strain in a rock unit is strongly dependent on lithology, temperature, and fluid content. In our models, we have quantified LPC as a representative of penetrative strain. The effects of temperature and fluid content are not taken into account in the models. However, in one of the models (model 3), materials with different mechanical properties are used to study the influence of lithology on the amount and distribution of LPC.

To estimate the amount of layer-parallel compaction in the models, the shortening that results from thrusting and folding of each layer was restored. The resulting bed-length was compared with the initial, known length of the layers. The difference between the two lengths gave the amount of layer shortening (bed-length change) without taking the thickening of the layer into account. However, during restoration, most of the shortening is not accompanied by thickening of the layers, but is instead accompanied by an area loss. Hence, it can be described as layer-parallel compaction (LPC). Area loss was calculated by comparing the area in a profile of the deformed model with the area in a profile in the undeformed, initial stage.

LAYER-PARALLEL COMPACTION IN THE MODELS' THRUST WEDGES

During compression, a layer can accommodate the shortening in three ways: by layer-parallel shortening

Table 1. Mechanical properties of materials used, and dimensions of, the five models.

Model	*Thickness (cm)*	*Nature of detachment*	*Coefficient of internal friction /viscosity*	*% Bulk shortening*
1	0.7	frictional	loose sand (0.57)	47
2	1.5	frictional	loose sand (0.57)	20
3	2.5	frictional	glass beads (0.37)	27
4	2	frictional/viscous	loose sand (0.57)/ SGM36 (5×10^4 Pa s)	35
5	3.5	frictional	loose sand (0.57)	21

(LPS), folding, or thrusting. However, the intensity and amount of shortening accommodated in each way depend on the mechanical properties of the layer and its surrounding host, and on the boundary conditions. In many natural cases, three internal deformation mechanisms are common to the emplacement of imbricate sheets. An initial phase of layer-parallel shortening precedes or is simultaneous with propagation of the sole thrust (Cooper et al., 1983; Williams and Chapman, 1983; Marshak and Engelder, 1985; Nickelsen, 1986; Geiser, 1988a, b; Evans and Dunne, 1991). Next, a second phase of non-layer-parallel shortening and bending strain occurs (Wiltschko, 1981; Sanderson, 1982; Suppe, 1983; Kilsdonk and Wiltschko, 1988). Finally, there is a later phase of pure-shear shortening and simple shear (parallel to the sole thrust) that accompany the thrust-sheet transport (Elliot, 1976; Mitra and Elliott, 1980; Coward and Kim, 1981; Sanderson, 1982). In most natural examples, it is relatively easy to determine the amount of shortening that occurs by folding and thrusting. In the presence of appropriate strain markers, it is also easy to determine penetrative strain and LPS. However, on seismic profiles, unlike the folding and thrusting components, layer-parallel shortening is more difficult to estimate. Layer-parallel shortening can be accommodated by plastic deformation of mineral grains and formation of fabric, reduction of porosity (secondary or tectonic compaction), and by dissolution of minerals (for example, formation of stylolites). To address the penetrative strain in the Appalachian orogenic belt, Mitra (1978a) described regional variations in deformation mechanisms (e.g., pressure solution, dislocation creep, and grain boundary sliding) that affect sandstones and quartzites in that area. Penetrative strain, as a general phenomenon during the evolution of fold-and-thrust belts, has been described and quantified by many workers (Helmstedt and Greggs, 1980; Mitra et al., 1984; Geiser, 1988a, b; Mitra, 1988; Protzman and Mitra, 1990; Evans and Dunne, 1991; Gary and Mitra, 1993; Howard, 1993; Thorbjornsen and Dunne, 1997; and others).

Unlike in nature, in a model it is easy to estimate layer-parallel penetrative strain by comparing the initial and final stages of that model. Each of the five models described in this chapter studies one element that influences the amount or distribution of LPC within a model fold-and-thrust belt. We show here that, at any given stage of model deformation, area loss is not constant. It varies laterally and with depth, amount of bulk shortening, and material properties.

Variation of LPC with Depth

Model 1, which was used to study the variation of LPC with depth, contained passively colored homogeneous sand with a constant total thickness of 5 mm (Mulugeta and Koyi, 1992; Koyi, 2000). Sequential sections were eroded using a vacuum cleaner, and then photographed, at every 1.5% increment of bulk shortening. This technique exposed the three-dimensional geometry of the imbricate sheets, so that fine-grained sand and thin (0.2-mm-thick) individual sand layers could be used to quantify area loss and penetrative strain. The model was shortened to a total of 47% bulk shortening (Figure 1). For more details of deformation and sectioning of this model, see Koyi (1995).

In a profile of the final stage of the deformed model, bed-length restoration was conducted for three layers located at different stratigraphic horizons (Figure 2). Bed-length restoration was used to partition the amount of deformation accommodated by each of the three strain components: layer-parallel compaction, folding, and thrusting. The effects of imbrication and folding were removed by measuring the segment length of each deformed layer and adding them together. By comparing this "restored" length with the deformed length, the amount of shortening by imbrication and folding was calculated. This restored length was then compared with the initial (undeformed) length of the model to calculate the amount of shortening accommodated by layer-parallel compaction. In all the cases, the restored length of the beds was shorter than their initial length (Figure 2).

To quantify the change in the mode of deformation with depth, longitudinal strain was partitioned into three layers located at different stratigraphic levels (top, middle, and bottom) (Figure 2). Earlier, Koyi (1995, 2000) had documented that layer-parallel compaction dominates in deeper levels of model accretionary

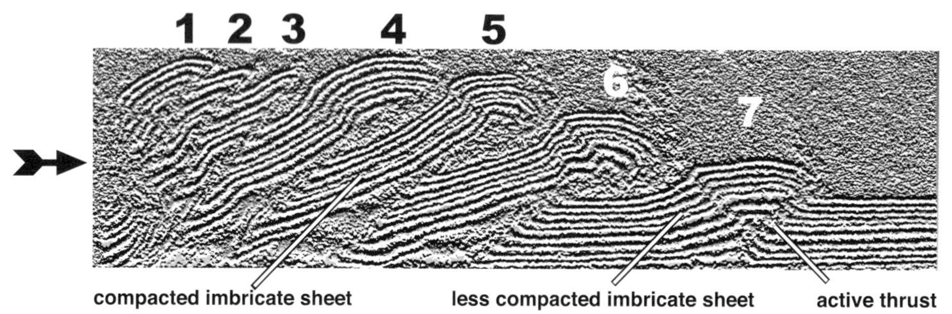

Figure 1. A relief image of a profile of model 1 after 40% shortening, showing seven imbricate sheets. The numbers indicate the sequence in which the imbricates have formed. Arrow shows direction of shortening.

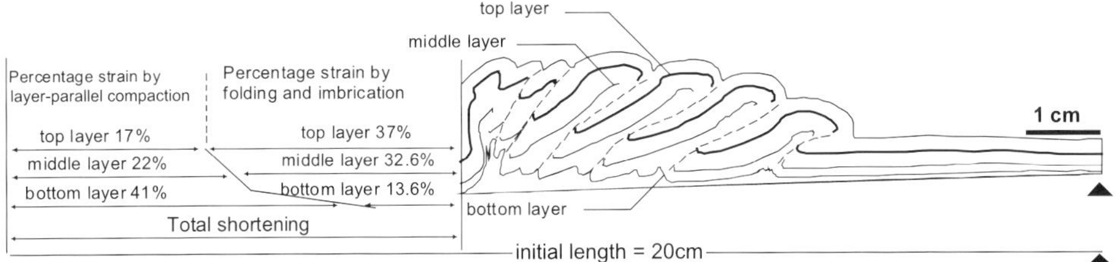

FIGURE 2. A line drawing of a profile of model 1 at 44% bulk shortening. Three layers (top, middle, and bottom) are restored to quantify the three components of shortening (layer-parallel shortening, folding, and imbrication). Note that the top layer deforms mainly by folding and imbrication, in contrast to the bottom layer, which deforms mainly by layer-parallel thickening.

wedges, whereas imbrication prevails at shallower levels. The current analysis demonstrates that the shallow (top) layer could be restored back to 83% of its initial length (Figure 2). The remaining 17% of the initial length, which was not restorable, was consumed by layer-parallel compaction during deformation of the model. This latter strain component was recoverable in the model because its initial dimensions were known. The deep layer, on the other hand, could only be restored to 59% of its initial length (Figure 2). In this layer, the entire remaining (unrestored) 41% of the layer length was consumed by layer-parallel compaction. This comparison illustrates that the mode of deformation within a modeled imbricate stack changes with depth. It also points out that balancing the bed lengths of different layers in natural profiles of shortened areas may yield different percentages of shortening. Because, in the absence of a good estimate of LPC, deeper layers are restored to shorter initial lengths than are shallower layers; they show a higher percentage of bulk shortening.

Lateral Variation of LPC

Model 1 was also used to quantify the lateral change in the amount of penetrative shortening. In this model, the area of an imbricate (in this case, imbricate number 4) was measured at the onset of its formation (and throughout its deformation) until a new imbricate (number 5) formed in front of it. During the same time interval, the area of the model wedge was measured to compare loss of area within the wedge and the imbricate during the same period of deformation (Figure 3).

Restoration of the model wedge between two stages of imbricate formation shows that 76% of the total area lost within the wedge was accommodated within the youngest of the imbricates at the toe (Figure 3). During the same period of deformation, the entire wedge lost only 3.9% of its area, of which 3% was accommodated by the youngest of the imbricates. This imbricate sheet lost 10.5% of its initial area during the same period of deformation (Figure 3). These results suggest that the newly accreted materials at the toe of the wedge underwent secondary tectonic compaction, whereas the rear part of the wedge, which contained an older stack of imbricate sheets and was already strongly compacted, could not accommodate much more tectonic compaction with deformation. In general, cumulative compaction was highest at the rear of the wedge, whereas incremental compaction was highest at the toe area. In other words, as the wedge grows, incremental compaction localizes within the newly accreted material at the toe of the model. The rear portion of the wedge is transported almost as a rigid block, without much strain, while the wedge accretes material at the toe. Naturally, synkinematic erosion or any change in the basal slope or friction may relocate the deformation and thereby alter this scenario.

FIGURE 3. Plots of (a) wedge area (triangles) and (b) imbricate area (circles) versus bulk shortening of model 1. At 38% shortening, the wedge loses approximately 4% of its initial area. During the same period, the youngest imbricate loses 10.5% of its initial area, which is equal to approximately 3% of the total area of the wedge. This suggests that much of the area lost within the wedge is accommodated within the youngest imbricates at the toe.

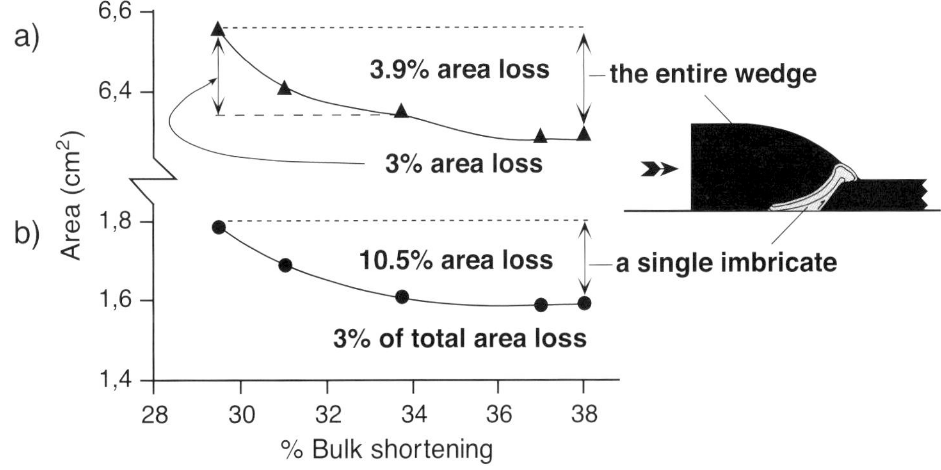

Change of LPC with Percentage of Bulk Shortening and with Material Properties

Model 2 was used to illustrate how the penetrative strain varies with the amount of bulk shortening. In this model, LPC was measured for two layers located at different stratigraphic levels in two sections shortened to different percentages of bulk shortening (Figure 4). As expected, LPC increased with increasing percentage of bulk shortening. However, there was a significant difference in the change in LPC between the two layers (Figure 4). At 5% bulk shortening, the shallow layer accommodated only 1% of the bulk shortening by LPC, whereas at 20% bulk shortening, LPC in the same layer was seven times greater (7%). The difference in LPC in the deeper layer between two stages of bulk shortening was only fourfold. Comparing the area of the initial stage of the model with its area at the final stage of deformation shows that there is an area loss of 1% at 5% bulk shortening and a loss of 4% at 20% bulk shortening (Figure 5).

To quantify the change in LPC with lithology, model 3 was prepared using two materials with different mechanical properties: loose sand and glass beads. This model was shortened to 27% bulk shortening. To simulate a rock that does not undergo much compaction during shortening, we used a layer of well-sorted glass beads, which as a medium is mechanically weaker [has a lower coefficient of internal friction (0.37)] than loose sand used in our models. This model consisted of a 4-mm-thick layer of beads overlain by three layers of loose sand, each of which were 5 mm thick. The model was shortened from one end to a total shortening of 27%. A section of this model was area-balanced to measure the amount of area lost in each layer (Figure 5). The analysis showed that the glass-bead layer preserved its cross-sectional area and underwent no significant amount of LPC, although it was located at deeper levels, where penetrative strain is usually dominant in other models such as models 1 and 2 (Figure 5). Instead, the glass-beads layer accommodated the bulk shortening by layer-parallel thickening. The sand layers, on the other hand, underwent different amounts of LPC. The deeper layer lost 5% of its area, whereas the shallow layer showed an area loss of only 1.5%. This demonstrates that the area lost in different lithologies is accommodated differently depending on how each lithology compacts.

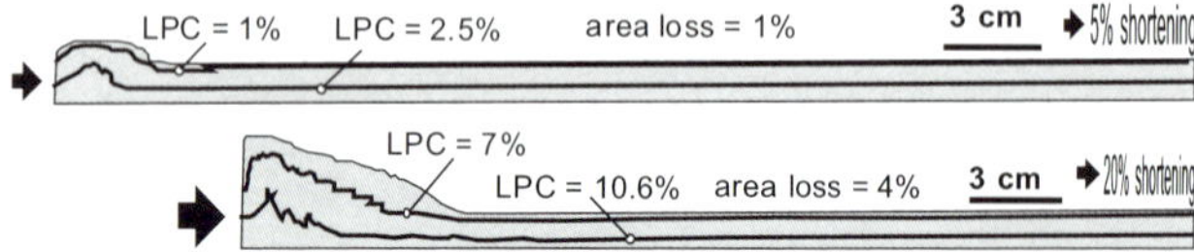

FIGURE 4. Line drawings of two profiles of model 2 at two different percentages of bulk shortening. As expected, LPC increases with the increase of the amount of bulk shortening and with depth.

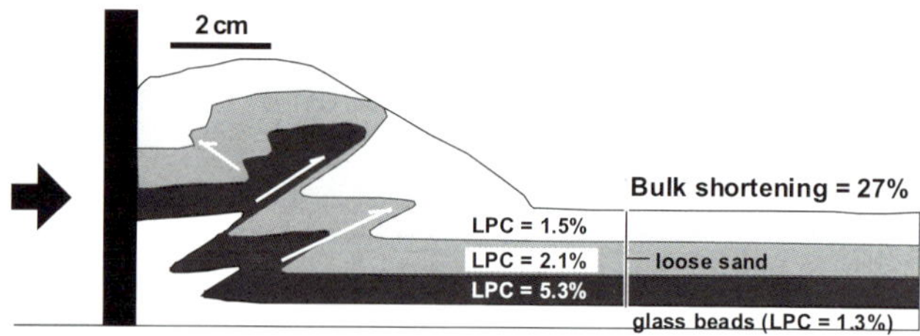

FIGURE 5. Line drawing of a profile of model 3, consisting of layers of loose sand overlying a layer of mechanically weaker glass beads. The layer of beads undergoes significantly less LPC than do the sand layers, but thickens instead.

Oblique Sections

For us to achieve a correct restoration, we need for the section to be parallel or nearly parallel to the transport direction. This is because a profile that is oblique to the shortening direction is naturally longer than profiles parallel to the shortening direction; hence, it is expected to have a longer initial bed length. Many times, only oblique profiles (profiles not parallel to the shortening direction) are available, and some workers apply the restoration technique to available seismic profiles even though they are not parallel to the shortening direction. Recently, Jaswal et al. (1997) restored a composite seismic image across the north Potwar deformation zone (NPDZ) and Dhurnal structure of the Himalayan foreland of Pakistan. The seismic image consisted of three segments: a northern north-south segment parallel to the shortening direction, a middle segment making an angle of 30° with the northern segment (but at a high angle to the structures), and a southern segment making an angle of 40° with the northern segment. By balancing this composite section, Jaswal et al. (1997) suggested a minimum amount of shortening across the NPDZ to be more than 55 km, at a rate of 18 mm/yr. Wissinger et al. (1998) conducted a similar exercise on a reflection seismic line that was oblique by 55 to 60° to the strike of the thrusts in the central Brooks Range in Arctic Alaska. Wissinger et al. (1998) restored the deeper structures and estimated a minimum shortening of 500 to 600 km in the Brooks Range. Price (1981) stated that the amount of tectonic shortening can be estimated from sections that make an angle of less than 30° with the actual direction of net tectonic displacement. According to Cooper (1983), a 20° obliquity between the balanced section and the transport direction will not greatly affect bulk-strain calculations on the balanced section of the structures.

Below, we restore a model profile that is oblique to the direction of shortening. We compare it with a

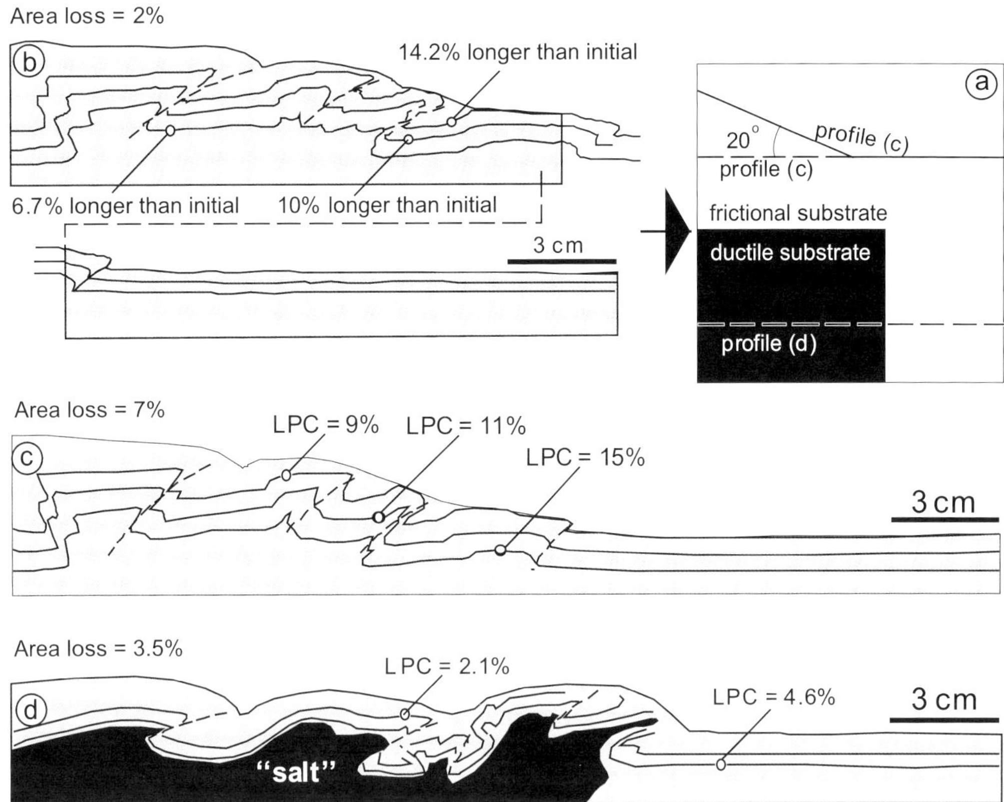

FIGURE 6. (a) A schematic diagram of model 4 in plan view, showing the locations of three profiles (b, c, and d) cut in the two domains (frictional and viscous substrates) for restoration after 35% bulk shortening of the model. (b) Line drawing of a profile, cut in the frictional-substrate domain, consisting of two segments (a rear segment oblique to shortening direction and a frontal segment parallel to the shortening direction). Note that after restoration, the layers in this profile are longer than the initial length of layers in a profile parallel to transport direction (see c). (c) Line drawing of a profile in the frictional-substrate domain cut parallel to the shortening direction. The layers in this profile have undergone LPC and therefore cannot be restored to their initial length. Note that the deeper layers undergo more LPC than the shallow layers do. (d) Line drawing of a profile cut in the viscous-substrate domain. LPC shortening in these layers is significantly less than their equivalents shortened above a frictional substrate in (c).

restored profile cut parallel to the shortening direction and illustrate the discrepancy that can result from restoring oblique sections.

In model 4, we studied the role of frictional and viscous (salt) dêcollements in governing deformation style. In this model, a package of layers of loose sand was deformed: one half was shortened on another sand layer and the second half was shortened above a viscous layer (SGM36, a Newtonian viscous material) (Figure 6a). This model simulated the deformation history of the Potwar and Salt Range of Pakistan (Cotton and Koyi, 2000). Here, we use the model to illustrate (1) the effect of restoration of oblique sections (those not parallel to the transport direction; Figure 6b), and (2) the effect of a ductile substrate on the mode of deformation and amount of penetrative strain within the overlying units (Figure 6d). As with the other models, this model was shortened from one end. After 35% shortening, three profiles were prepared for restoration (Figure 6). Two profiles were cut parallel to the shortening direction: one in the viscous substrate domain and the other in the frictional-substrate domain (Figure 6c and d). The third profile was a composite section consisting of two segments: a segment making an angle of 30° with the shortening direction, and a second segment parallel to the shortening direction (Figure 6b). All three profiles were restored and compared with the initial stage of the model to calculate the penetrative strain accommodated by layer-parallel compaction (Figure 6). Results for the parallel profile cut in the frictional domain were similar to those for other models, in which layer-parallel compaction increased with depth (Figure 6c). In this profile, a deep layer accommodated 15% of the shortening by penetrative strain, compared with 9% for a shallow layer (Figure 6c). The composite section, cut in the frictional-substrate domain, showed totally different results. After restoration, the layers in this section, as expected, were longer than the initial length of the layers in the section that was parallel to the shortening direction. The shallow layer was 14% longer, whereas the deep layer was 6.7% longer than the initial layer (Figure 6b). The amount of shortening in this section differs from that in the section cut parallel to the shortening direction by approximately 25%. This difference is because the composite section, which consists of a segment oblique to the shortening direction, is initially longer than any section parallel to the shortening direction. Hence, its restored version does not display the correct amount of shortening.

The area of each section is compared with the area of an undeformed section. The parallel profile showed an area loss of 7% (Figure 6c). The oblique section, however, showed a very small amount of area loss (2%, Figure 6b). This discrepancy is because this section is oblique to the transport direction and hence shows a larger initial area. These results demonstrate quantitatively that neither bed length nor area balancing of oblique sections gives the correct amount of bulk shortening.

Change of LPC and LPS with Basal Friction

In model 4, a section that had been cut in the viscous substrate domain was restored and compared with its initial stage to estimate the amount of LPC with depth (Figure 6d). Here, the amount of deformation accommodated by LPC, even at the deeper levels, was significantly smaller (Figure 6d). In the deeper layer, only 5.6% of the deformation was accommodated by penetrative strain, compared with 2% for the shallow layers. Compared with the section cut in the frictional substrate domain (where LPC was 15% for the deeper layer and 9% for the shallow layer; Figure 6c), these figures suggest that layer-parallel shortening in the overlying sediments decreases significantly in the presence of a viscous substrate. This is because the ductile layer decreases the friction along the basal dêcollement and eases forward propagation of the deformation front, instead of contributing to a wedge buildup.

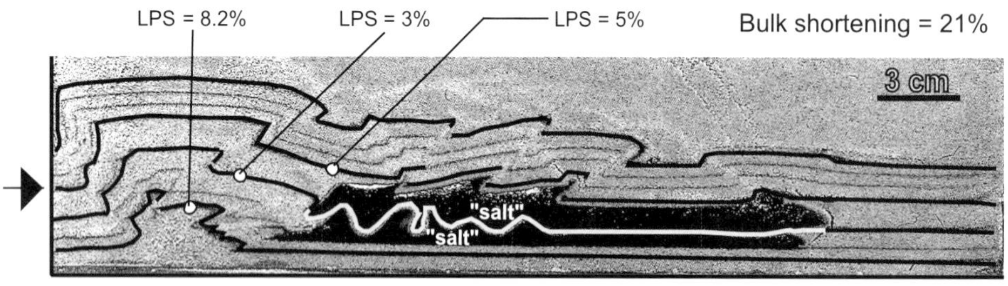

FIGURE 7. Photograph of a profile of model 5, consisting of two viscous layers simulating salt embedded between layers of loose sand. This model was shortened by as much as 21%. The three layers outlined on the profile have undergone different amounts of layer-parallel compaction (LPC), depending on their stratigraphic location and their relation to the viscous layers. The middle layer, which is embedded between the two viscous layers, has undergone the least LPC, because it accommodated most of the deformation by folding. The deepest layer has undergone the maximum amount of LPC.

In another model (model 5), which consisted of alternating sand and ductile layers, LPC within the layers varied significantly depending on the location of the sand layers relative to the ductile layers (Figure 7). In this model, alternating layers of loose sand and a viscous material (SGM36) were shortened, from one end, to 21% bulk shortening (Figure 7). During shortening, the sand layer, which was embedded between two viscous layers, underwent the least LPC. It is feasible that even that amount of LPC in this layer is accommodated mainly by the rear portion of the layer, which was not embedded between the viscous layers (Figure 7). Had the entire layer been embedded between the two viscous layers, it would have accommodated the shortening only by folding. In this model, a deeper layer underlying the two viscous layers underwent more LPC than the other layers (Figure 7). This model reemphasizes how significant the presence of a viscous layer is in partitioning strain within nonevaporitic frictional sediments. Because a viscous substrate flows easier than the rest of the model, it provides a low basal friction, which allows the sole thrust to propagate farther without the necessity of building a steep wedge. The low tapered wedge therefore undergoes less internal compaction and hence less penetrative strain and LPC.

NATURAL EXAMPLES

Two examples from the Pyrenean orogenic belt are discussed here to illustrate the significance of LPS: the Pyrenean hinterland and the south-central Pyrenean Fold-and-thrust Belt.

Pyrenean Hinterland

The structures of the Pyrenean hinterland consist of an Alpine antiformal stack involving mainly Paleozoic rocks and a Mesozoic-Cenozoic cover (Teixell, 1996; Teixell and Koyi, 2003). These rocks were deformed during the Hercynian orogeny. In the western Axial Zone, the Mesozoic-Cenozoic sediments, which from bottom to top consist of Upper Cretaceous carbonates, calcareous shale, Marbore Sandstone, Paleocene limestone, and Eocene turbidites, form an imbricate stack verging to the south. In this area, with its good exposure, it is relatively easy to estimate the amount of shortening accommodated by thrusting and folding

(a)

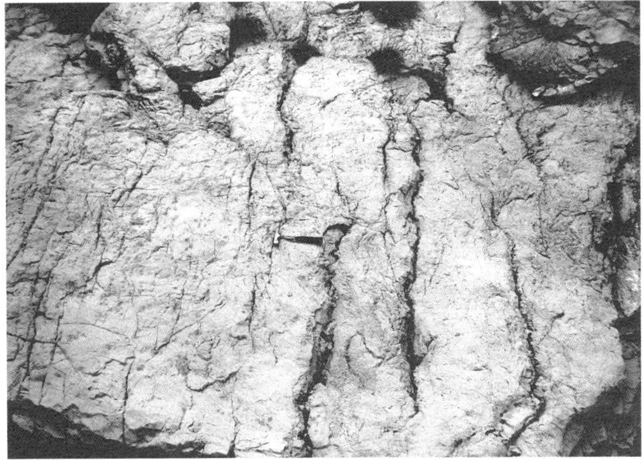

(b)

FIGURE 8. (a) An oblique photograph of the bedding plane of a limestone layer of the Pyrenean hinterland (Paleocene cover of the western Axial Zone), showing stylolites normal to the bedding. (b) Photograph of a profile of a limestone layer showing stylolite peaks oblique to bedding (horizontal). Note the inclined peaks of the stylolites, which make an angle of 60° with the bedding.

(Teixell and Koyi, 2003). Evidence for penetrative strain is visible in all the units and positions, in addition to the fold forelimb regions, which show a strong heterogeneous strain (Alonso and Teixell, 1992). The turbidites and the sandstone units show a northerly dipping tectonic fabric that is especially well developed in the turbidites. The carbonates, on the other hand, have not developed any visible cleavage, but show two kinds of stylolite joints: a set perpendicular to bedding and another set parallel to bedding (Figures 8a, 8b, and 9). The bedding-perpendicular stylolite joints are interpreted to have formed during early stages of layer-parallel shortening, when the carbonates were deformed by penetrative strain before folding and imbrication (Figure 9). The teeth in these stylolite joints are typically smaller (<1 cm) (Figure 8a). Evans and Dunne (1991) recorded meso-scale bed-perpendicular stylolites in the North Mountain thrust sheet of the central Appalachians and also interpreted them as evidence for an early layer-parallel shortening event. The bedding-parallel stylolite joints, on the other hand, are accompanied by vertical calcite veins and are interpreted to have formed as a result of tectonic loading by older overlying imbricates (Figure 9). The teeth of these joints are significantly larger (up to 10 cm) and are dipping in the direction of transport (Figures 8b and 9). Only the normal set formed by layer-parallel lateral shortening (Figures 8a and 9). These structures indicate that all units in the Pyrenean alpine hinterland have accommodated part of their deformation by penetrative strain. In the absence of systematic quantitative indicators, not accounting for penetrative strain will result in only a partial restoration of the deformation.

The South-central Pyrenean External Fold-and-thrust Belt

Calculating the amount of layer-parallel shortening in the external areas of a fold-and-thrust belt usually is not an easy exercise. In the south Pyrenees, different techniques, such as analysis of the strain markers, fissility ratios, and anisotropy of the magnetic susceptibility have been used to estimate layer-parallel shortening in several traverses (Casas et al., 1996; Sans, 1999). Here, we will discuss the results of the analysis of the sedimentary strain markers along the Cardener River traverse to quantify layer-parallel shortening in the deformed south Pyrenean foreland.

The south Pyrenean front developed a triangle zone that coincides with the presence of three partially superposed evaporitic levels at depth (Vergés, 1993; Sans et al., 1996). This triangle zone widened as successive thrusts that developed at the pinch-out of each evaporitic formation front (Beuda, Cardona, and Barbastro) were abandoned and the sole detachment climbed from

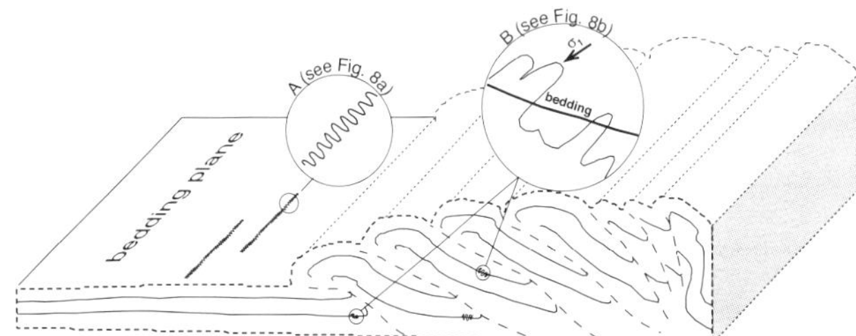

FIGURE 9. Schematic line drawing showing the location of the two sets of stylolites observed in the Pyrenean carbonates. Set (A), which is normal to bedding, is interpreted to have formed as a result of layer-parallel shortening before folding and imbrication, whereas set (B) formed as a result of tectonic loading of older imbricates.

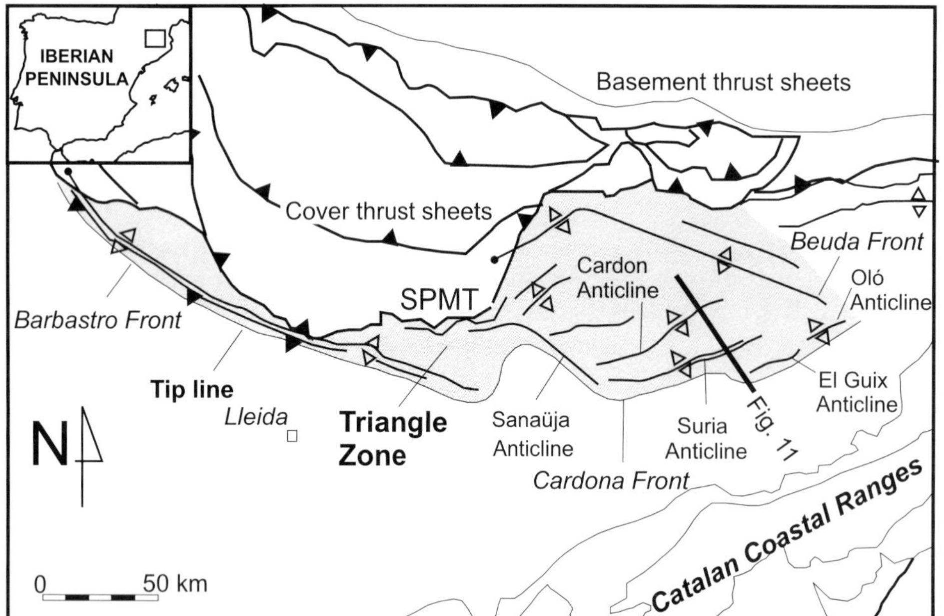

FIGURE 10. Map of the south Pyrenean triangle zone. Inset shows the Iberian Peninsula and location of the study area in the Pyrenees. The black bar shows the sampled section across the Cardona front, from under the detachment horizon to the Cardona Anticline. SPMT stands for South Pyrenean Main Thrust.

lower evaporitic formations in the north to higher ones in the south (Figure 10). In order to quantify LPS, we have analyzed samples from different parts of this area. Samples were collected in a north-south section that crosses the Cardona thrust front (Figures 10 and 11). The southernmost samples were collected from below the detachment level (Cardona salt formation), whereas to the north, three different anticlines above this detachment horizon were sampled. The three structures sampled are, from south to north, the El Guix Anticline, the Suria Anticline, and the Pinós-Cardona Anticline (Figure 11). The El Guix and Suria Anticlines are the frontal structures in the southeastern part of the thrust front (Figure 10). Both of these anticlines have a thrust-wedge geometry characterized by a north-verging structure in the south and a south-verging structure in the north separated by a narrow syncline (Figure 11). The Cardona Anticline (Figure 11) is a south-verging detachment anticline developed where the Cardona salt is thickest (center of the evaporitic basin).

The preserved sedimentary pile is less than 500 m thick in the frontal thrust-wedge (El Guix Anticline) and more than 2 km thick in the northernmost syncline (north of the Cardona Anticline). The scarce vitrinite analysis in the basin indicates, however, that there was a sedimentary thickness of approximately 2.5 km over the Cardona salt at El Guix Anticline (Vergés et al., 1998). These data are consistent with the preserved sedimentary thickness in the northern syncline and the small (<0.5°) sedimentary slope of the deformed sediments (Sáez, 1987). The sampled sediments are fine-grained red and gray sandstones that contain burrows

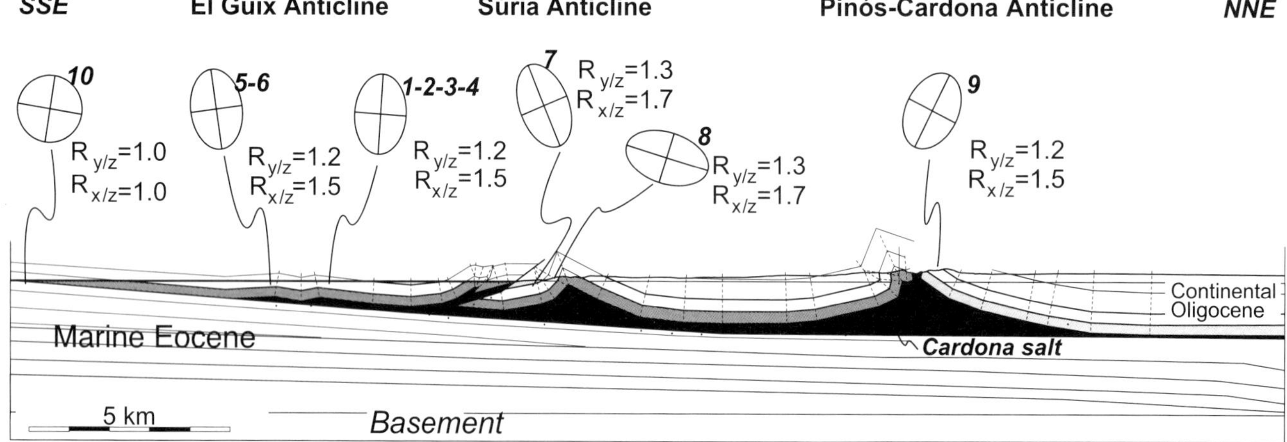

FIGURE 11. Regional section of the Cardona thrust front. Ellipses show the Z/X section of the strain ellipsoid. Numerical values show the ellipticity (R = y/z) of the y/z section measured on the bedding plane, and the ellipticity in the x/z plane (R = x/z) calculated for the vertical section. Ellipses of locations 1, 2, 3, 4, and 5 are projected along strike from the eastern section. Ellipses from locations 6, 7, 8, 9, and 10 lie in the plane of the section.

Table 2. Ellipticity (R) and orientation (φ) values of the strain ellipses in the Cardona thrust front.

Location	*No. of samples*	*Marker type*	*R*	*φ*	*No. of measurements*
		El Guix- Sallent			
1	1	Burrows	1.2	73	12
2	8	Burrows Oxidation marks Rain drops	1.2	71	186
3	3	Burrows	1.2	78	76
4	1	Burrows	1.2	72	14
5	2	Burrows	1.2		104
		El Guix- Callús			
6	1	Burrows	1.2		27
		Súria			
7	1	Burrows	1.3	47	10
8	2	Burrows	1.3	47	10
		Cardona			
9	4	Burrows	1.2	62	102
		Manresa			
10	3	Burrows	1.0		56

and other strain markers. These sediments belong to the alluvial/fluvial and lacustrine sequences that filled the south-Pyrenean foreland basin. Although the sampled levels are not the same through the whole structure, the samples have been collected from an interval of 500 to 800 m from the top of the salt, and the estimated age for these sediments is early Oligocene.

Sampling is irregularly distributed through the structures, because the presence of strain markers is highly dependent on lithology. The frontal thrust wedge (El Guix Anticline) was profusely sampled, especially in the north (1 sample every 25 m), whereas no strain markers have been found in the central syncline (Figure 11). The El Guix Anticline was also sampled 10 km to the west (Figure 10), in younger sediments (1000 m above the top of the salt). The Súria Anticline was sampled in the footwall of the main back thrust, and the Cardona Anticline was sampled in the northern limb of the anticline (Figure 11).

The oriented samples were photographed and drawn for analysis by the IMAT program developed by the Serveis Cientifico-Tecnics of the Universitat de Barcelona (Durney, 1995). This program analyzes the image and determines the long and short axes, the area and the perimeter of the strain marker, and orientation of the long axis with respect to a reference line. These data are then processed by the GRFRFP program, developed at Barcelona University (Durney, 1995), which calculates the mean elongation value (Rf) and orientation (φ) by interactive retrodeformation of the set of ellipses to an isotropic distribution (where ln (Rf)/2φ vector mean is equal to zero).

The results presented here will be grouped in structural localities, which are separated by main thrusts or structural elements (Table 2 and Figure 11). To quantify penetrative strain, the Z axis is assumed to be contained in the bedding planes, because incipient cleavage is normal to bedding (Sans and Vergés, 1995). Also, a plane strain value is assumed for the area ($Y = 1$). This assumption is based on (1) lack of extensional structures perpendicular to the folds along strike; (2) absence of folds in the salt layer along strike that would indicate flow of the salt in a direction different from the overburden transport direction; and (3) minimum deformation of the halite grain shape (strain markers) in a section perpendicular to the shortening direction (Miralles, 1999).

The calculated ellipticity of the strain ellipsoid in the YZ section (bedding plane, Table 2) changes from 1.0 below the detachment to 1.2/1.3 above the detachment. This indicates the effective decoupling achieved by the Cardona salt formation. In contrast, there is no significant increase in ellipticity across the frontal thrust wedge, where the ellipticity is R = 1.2 in all the sampled localities—even in those measured along strike in a stratigraphically different position. There is also no significant increase between the frontal thrust wedge and the northernmost anticline located 30 km to the north. There is a slight increase in the Súria Anticline, where the samples were collected very close to the main back thrust. The deformation of these strain ellipses was achieved before folding and thrusting, because after unfolding, the strain ellipses in neighboring localities have the same orientation and values in beds that are horizontal as they have in vertical or inclined beds. The orientation of the strain ellipsoid has a good gathering in the three sampled anticlines (El Guix, Cardona, and Suria). The azimuth of the *Y* axis varies from N74 to N47 and N62, respectively. These orientations are parallel to the fold axes at the sampled localities. The change in orientation between the three different folds can also be seen in map view (Figure 10).

Layer-parallel shortening calculated from the elongation parameters suggests 16% penetrative strain in the frontal thrust wedge (El Guix Anticline) and the northern anticline (Cardona Anticline), and 23% in the Súria Anticline. In contrast, in the same section, shortening accommodated by folding and thrusting is 5–6% in the El Guix and Cardona Anticlines and 10% in the Suria Anticline.

DISCUSSION

Penetrative strain is a significant element of shortening in many fold-thrust belts (Mitra, 1988, 1990). In this chapter, we apply the results of sandbox models to quantify the relative amount and distribution of layer-parallel strain within a fold-thrust belt or an accretionary prism. We assume that the layer-parallel compaction (LPC) observed in the sandbox models (which usually show little bed thickening) corresponds in style to that part of layer-parallel shortening (LPS) which is not accommodated by thickening of sedimentary layers in nature. It is worthwhile to underline, however, that the amount of LPC observed in the models does not equate to the amount of LPS in a natural case. However, by monitoring LPC in the models, we assume that model results can illustrate the distribution and style of penetrative strain within areas that have undergone compressional tectonism.

In layers of loose sand, the amount of LPC depends on shape and size of the sand grains and the degree of sorting. The variation in LPC between different models in which different types of loose sand are used is therefore attributed to variations in these parameters. However, the variation in LPC within the same model (e.g., LPC variations with depth in models 1, 2, and 4) in which the same type of sand is used, cannot be attributed to these features, but must be the result of heterogeneous distribution of strain within the sedimentary pile of the model during its shortening.

In the sand layers of the models, penetrative strain is accommodated by compaction through volume loss and porosity reduction. In nature, as documented from many thrust belts, this secondary (tectonic) compaction is likely to be accommodated for by volume loss through pressure solution and porosity reduction, but much of the layer-parallel penetrative shortening involves layer thickening. Mitra's (1988) theoretical derivation of the relationship between finite strain and original and final porosity indicated that little porosity is preserved in rocks whose strain (R) exceeds 1.5.

Comparison of the restored length with the initial length of the shortened model layers overestimates the amount of shortening that results from LPC, because layer thickening is not accounted for. However, comparing the cross-sectional area between the deformed and the undeformed profiles accounts for any thickening of the layer during shortening. Then the correct amount of LPC can be estimated, because the decrease in the cross-sectional area in the deformed profile is attributed to layer-parallel compaction.

Layer-parallel compaction in our models, which were shortened above a frictional dêcollement, is concentrated in front of and at the leading edge of the deformation front, where relatively less-compacted sediments are accreted (Figure 3). This is similar to the deformation style observed in Nankai accretionary prism, where shortening of the soft sediments in the prism toe is pervasive and is not just accommodated by displacement along imbricate thrusts and shear bands (Karig and Lundberg, 1990). Similarly, the active toe area in the wedge of model 1 accommodated most of the penetrative strain that the wedge experienced (Figure 3).

Within the same model (model 4), bed-length restoration of a section shortened above a viscous substrate showed only 3.5% area loss, compared with 7% area loss in the section shortened above a viscous substrate (Figure 6c and d). This suggests that there may be relatively less LPS in layers shortened above a viscous substrate than in similar layers shortened above a frictional substrate. Therefore, LPS in lithologies deformed above, for example, a layer of salt, is expected to be relatively less than it would be if these units were deformed above a frictional substrate. This is supported by the fact that estimated LPS in the external areas of the Spanish Pyrenees, which are not shortened above a layer of salt, is greater (30%) compared with LPS (20%) in the areas shortened above a layer of salt.

However, the example of the Spanish Pyrenees shows that even in the presence of a salt layer, a considerable amount (about 20%), of penetrative strain may be recorded in the layers shortened above the salt. This relatively high amount of LPS above a ductile layer is attributed to the thickness of the viscous layer and its distribution. A thin ductile layer, whose viscous flow is retarded by viscous drag, promotes more penetrative strain in the overlying units than a thicker viscous layer, which flows more easily. Before overcoming the basal friction, early shortening is partly consumed by penetrative strain within the layers. Hence, penetrative strain would be more effective.

The variation of strain with different modeling materials shown in model 3 could be applied to natural cases where incompetent lithologies undergo more penetrative strain than do more competent units. A good example of this is given in the North Mountain thrust sheet of the central Appalachians, where the incompetent Ordovician Martinsburg Formation shows significantly greater strain (primarily plane strain) than do the more competent carbonates of the same thrust sheet (Woodward et al., 1986).

The effects of LPS must also be taken into account when one is interpreting porosity or sonic-velocity profiles in terms of apparent exhumation. In many inverted sedimentary basins (Bulat and Stoker, 1987; Menpes and Hillis, 1995; Tonghban, 2000) exceptionally high sonic velocities have been interpreted as remnants of earlier burial that has been removed by erosion. It is supposed that sedimentary rocks compact irreversibly when they are submitted to the pressure of overburden. This compaction decreases porosity and therefore increases sonic velocities. A comparison of velocity profiles in undisturbed and inverted basins allows us to model a minimum overpressure that the rock has suffered. This overpressure (P_o) can then be converted into minimum eroded overburden (h_o) using the formula $h_o = P_o/(g\rho)$, where g is the gravitational acceleration and ρ the mean density of the overburden.

In view of our results regarding LPS in shortened areas, this relationship should, however, be used with caution. Exhumation is certainly related to uplift. The uplift may be caused by deep processes, acting mainly vertically (e.g., heating of the lithosphere, or underplating), which will generally not increase the pressure in the overlying rocks by an important amount. However, in areas with significant horizontal shortening, inverted faults and thrusting would inevitably increase the pressure in the whole overthrust column. The horizontal force per unit length necessary to overcome gravitational and frictional forces along a thrust are given by the following formula (Turcotte and Schubert, 1982, p. 354):

$$F_x = \frac{f\rho g(1-\lambda)H^2}{\sin(2\beta) - f(1\ \cos(2\beta))} \quad (2)$$

where f is the frictional coefficient (usually 0.7–0.85), ρ the average density of the hanging-wall rocks, g the gravitational acceleration, λ the pore pressure coefficient, H the thickness of the thrust sheet, and β the angle of the thrust fault with the horizontal.

The minimum force is necessary for an angle $\beta = 0.5 * \arctan(1/f)$, giving

$$\mathrm{F_x} = \frac{\mathrm{f}\rho \mathrm{g}(1-\lambda)\mathrm{H}^2}{\sqrt{1+\mathrm{f}^2} - \mathrm{f}} \quad (3)$$

The average tectonic stress amounts to $\sigma_{xx} = \mathrm{F_x/H}$ and can be written as a function of the lithostatic pressure $\mathrm{P} = \rho \mathrm{gH}$ at the base of the thrust sheet:

$$\sigma_{xx} = \frac{\mathrm{f}}{\sqrt{1+\mathrm{f}^2} - \mathrm{f}} * (1-\lambda) * \mathrm{P} \quad (4)$$

In situations of reasonable f and hydrostatic pore pressure, the average tectonic horizontal stress is between 50% and 100% of the lithostatic pressure at the base of the thrust sheet. Because this additional stress may increase the pressure by the same amount or more than would a possible eroded overburden, it is highly probable that exhumation estimates based on sonic velocity interpretations often are too large. In other words, in shortened or inverted areas, estimates of burial depth that are based on the compaction of units may be too high when the effect of horizontal compaction resulting from lateral shortening is not accounted for.

CONCLUSIONS

Analyses of sequential sections of sand models demonstrate the following.

1) Layer-parallel shortening, achieved mainly by layer-parallel compaction, is a significant component of deformation in sand models.
2) LPC dominates at deep levels, imbrication prevails at shallow levels.
3) Layers shortened above a weak, viscous substrate undergo less LPC relative to those shortened above a frictional dêcollement.
4) Penetrative strain varies in space (vertically and laterally), in time, with the type of modeling granular material, and with bulk shortening. At any given time, LPC is accommodated mainly within the youngest of the imbricate sheets.

In the sand layers of the models, LPC is accommodated by volume loss through porosity reduction. In nature, as documented from many thrust belts, this secondary (tectonic) compaction is likely to be accommodated by volume loss through pressure solution and porosity reduction, although a significant component of layer-parallel shortening may be balanced by layer thickening. This compactional deformation mechanism plays a significant role in determining the porosity and permeability of reservoir rocks in fold-thrust belts.

When applied to nature, our model results indicate that penetrative strain is heterogeneous in space and time in fold-thrust belts. Analyses of strain markers in the Spanish Pyrenees show that as much as 20% of the bulk shortening may be accommodated by LPS in the external areas.

ACKNOWLEDGMENTS

Thanks are due to Professors C.J. Talbot and Dr. A. Skelton for reading and commenting on this manuscript. This manuscript benefited from the critical and thorough reviews and useful suggestions of Peter Cobbold and Fabrizio Storti. The Université Paris-Sud financed a visit by HAK to Orsay. The Swedish Research Council (VR) has funded HAK.

REFERENCES CITED

Alonso, J. L., and A. Teixell, 1992, Forelimb deformation in natural examples of fault-propagation folds. , *in* K. R. McClay, ed., Thrust tectonics: London, Chapman & Hall, p. 175–180.

Baker, D. M., R. J. Lillie, S. R. Yeats, G. D. Johnson, M. Yousuf, and A. S. H. Zamin, 1988, Development of the Himalayan frontal thrust zone: Salt Range, Pakistan: Geology, v. 16, p. 3–7.

Bulat, J., and S. J. Stoker, 1987, Uplift determination from interval velocity studies, *in* J. Brooks and K. Glennie, eds., Petroleum geology of North West Europe, Graham and Trotman, London, p. 293–305.

Casas, J. M., D. Durney, J. Ferret, J., and J. A. Muñoz, 1996, Determinación de la deformación finita en la vertiente sur del Pirineo oriental a lo largo de la transversal del río Ter: Geogaceta, v. 20 (4), p. 803–805.

Cobbold, P., E. Rossello, and B. Vendeville, 1989, Some experiments on interacting sedimentation and deformation above salt horizons: Bulletin Societ Geologique France, v. V, p. 453–460.

Colletta, B., J. Letouzey, R. Pinedo, J. F. Ballard, and P. Balé, 1991, Computerized X-ray tomography analysis of sandbox models: Examples of thin-skinned thrust systems: Geology, v. 19, p. 1063–1067.

Cooper, M. A., 1983, The calculation of bulk strain in oblique and inclined balanced sections: Journal of Structural Geology, v. 5, p. 161–165.

Cooper, M. A., M. R. Garton, and J. R. Hossack, 1983, The origin of the Basse Normandie duplex, Boulonnais, France: Journal of Structural Geology, v. 5, p. 139–152.

Cotton, J., and H. A. Koyi, 2000, Modeling thrust fronts above ductile and frictional detachments: Application to structures in the Salt Range and Potwar Plateau, Pakistan: Geological Society of America Bulletin, v. 112, p. 351–363.

Coward, M. P. and J. H.,Kim, 1981, Strain within thrust sheets, *in* K. R. McClay and N. J. Price, eds., Thrust and nappe tectonics: an international conference: Geological Society of London Special Publication 9, p. 275–292.

Dahlen, F. A., J. Suppe, and D. M. Davis, 1984, Mechanics of fold and-thrust belts and accretionary wedges (continued): Cohesive Coulomb Theory: Journal of Geophysical Research, v. 89, p. 10087–10101.

Dahlstrom, C. D. A., 1969, Balanced cross sections: Canadian Journal of Earth Sciences, v. 6, p. 743–747.

Davis, D., J. Suppe, and F. A. Dahlen, 1983, Mechanics of fold-and-thrust belts and accretionary wedges: Journal of Geophysical Research, v. 88, p. 1153–1172.

Dixon, J., 1982, Regional structural synthesis, Wyoming salient of western overthrust belt: AAPG Bulletin, v. 66, p. 1560–1580.

Durney, D. W., 1995, GRFRFP: a program for statistical analysis and plotting of elliptical shape data, version 5.4, December 1995. Barcelona University, Barcelona, Spain (unpublished).

Elliott, D., 1976, The energy balance and deformation mechanisms of thrust sheets: Royal Society of London Philosophical Transactions, series A, v. 283, p. 289–312.

Ellis, P. G., and K. R. McClay, 1988, Listric extensional fault systems—results of analogue model experiments: Basin Research, v. 1, p. 55–70.

Evans, M. A., and W. M. Dunne, 1991, Strain factorization and partitioning in the North Mountain thrust sheet, central Appalachia, U.S.A.: Journal of Structural Geology, v. 13, p. 21–35.

Fischer, M. W., and M. P. Coward, 1982, Strains and folds within thrust sheets: an analysis of the Heilam sheet, northwest Scotland, *in* G. D. Williams, ed., Strain within thrust belts: Tectonophysics, v. 88, p. 291–312.

Gary, M. B., and G. Mitra, 1993, Migration of deformation fronts during progressive deformation: evidence from detailed structural studies in the Pennsylvania Anthracite region, U.S.A.: Journal of Structural Geology, v. 15, no. 3–5, p. 435–449.

Geiser, P. A., 1988a, Mechanisms of thrust propagation: some examples and implications for the analysis of overthrust terranes: Journal of Structural Geology, v. 10, p. 829–845.

Geiser, P. A., 1988b, The role of kinematics in the construction and analysis of geological cross sections in deformed terranes, *in* G. Mitra and S. Wojtal, eds., Geometries and mechanisms of thrusting: Geological Society of America Special Paper 222, p. 47–76.

Gutscher, M. A., N. Kukowski, J. Malavieille, and S. Lallemand, 1996, Cyclical behaviour of thrust wedges: Insight from basal friction sandbox experiments: Geology, v. 24, p. 35–138.

Helmstedt, H., and R. G. Greggs, 1980, Stylolitic cleavage and cleavage refraction in lower Paleozoic carbonate rocks of the great valley, Maryland: Tectonophysics, v. 66, p. 99–114.

Homza, T. X., and W. K. Wallace, 1997, Detachment folds with fixed hinges and variable detachment depth, northeastern Brooks Range, Alaska: Journal of Structural Geology, v. 19, p. 337–354.

Horsfield, W. T., 1977, An experimental approach to basement controlled faulting: Geologie en Mijnbouw, v. 56, p. 363–370.

Hossack, J. R., 1978, The correction of stratigraphic sections for tectonic finite strain in the Bygdin area, Norway: Journal of the Geological Society of London, v. 135, p. 229–241.

Hossack, J. R., 1979, The use of balanced cross section in the calculation of orogenic contraction: Journal of Geological Society of London, v. 136, p. 705–711.

Howard, J. H., 1993, Restoration of cross sections through unfaulted, variably strained strata: Journal of Structural Geology, v. 15, p. 1331–1342.

Hubbert, K., 1937, Theory of scaled models as applied to geologic structures: Geological Society of America Bulletin, v. 48, p. 1459–1519.

Hubbert, K., 1951, Mechanical basis for certain familiar geological structures: Geological Society of America Bulletin, v. 62, p. 355–372.

Jaswal, T. J., R. J. Lillie, and R. D. Lawrence, 1997, Structure and evolution of the northern Potwar deformed zone, Pakistan: AAPG Bulletin, v. 81, p. 308–328.

Karig, D. E., 1986, Physical properties and mechanical state of accreted sediments in the Nankai Trough, Southwest Japan Arc: Geological Society of America Memoir 166, p. 117–133.

Karig, D.E., and N. Lundberg, 1990, Deformation bands from the toe of the Nankai accretionary prism: Journal of Geophysical Research, v. 95 (B6), p. 9099–9109.

Kilsdonk, B., and D. V. Wiltschko, 1988, Deformation mechanisms in the southeastern ramp region of the Pine Mountain Block, Tennessee: Geological Society of America Bulletin, v. 100, p. 653–664.

Koyi, H., 1995, Mode of internal deformation in sand wedges: Journal of Structural Geology, v. 17, p. 293–300.

Koyi, H. A., 2000, Towards dynamic restoration of geologic profiles: some lessons from analogue modeling, *in* M. Webster and M. Talwani, eds., Atlantic Rifting and Continental Margins, Geophysical Monograph Series, v. 115, p. 334.

Koyi, H. A., K. Hessami, and A. Teixell, 2000, Epicenter distribution and magnitude of earthquakes in fold-thrust belts: insights from sandbox model, Geophysical Research letters, v. 27, p. 273–276.

Liu, H., K. R. McClay, and D. Powell, 1991, Physical models of thrust wedges, *in* K. R. McClay, ed., Thrust tectonics: London, Chapman & Hall, p. 71–81.

Lohrmann, J., N. Kukowski, J. Adam, and O. Onken, 2003, The impact of analogue material properties on the geometry, kinematics, and dynamics of convergent sand wedges: Journal of Structural Geology, v. 25, p. 1691–1711.

Malavieille, J., 1984, Modélisation experimentale des chevauchements imbriqués: Application aux chaines de montagnes: Geological Society of France Bulletin, v. 7, p. 129–138.

Mandl, G., 1988, Mechanics of tectonic faulting: Elsevier, Amsterdam, 407 p.

Marshak, S., and T. Engelder, 1985, Development of cleavage in a fold-thrust belt in eastern New York: Journal of Structural Geology, v. 7, p. 345–359.

Marshak, S., and N. B. Woodward, 1988, Introduction to cross section balancing, in basic methods, *in* S. Marshak and G. Mitra, eds., Structural geology: Engelwood Cliffs, NJ, Prentice Hall, p. 303–332.

McClay, K. R., 1990, Deformation mechanics in analogue modelsof extensional fault systems, *in* R. J. Knipe and E. H. Rutter, eds., Deformation mechanisms, rheology and tectonics: Geological Society of London Special Publication 54, p. 445–453.

McClay, K. R., and P. G. Ellis, 1987, Geometries of extensional fault systems developed in model experiments: Geology, v. 15, p. 341–344.

McNaught, M. A., and G. Mitra, 1996, The use of finite strain data in constructing a retrodeformable cross-section of the Meade thrust sheet, southeastern Idaho, U.S.A.: Journal of Structural Geology, v. (18)5, pp. 573–583.

Meigs, A., J. Verges, and D. W. Burbank, 1996, Ten-million-year history of a thrust sheet: Geological Society of America Bulletin, v. 108, p. 1608–1625.

Menpes, R. J., and R. R. Hillis, 1995, Qualification of Tertiary exhumation from sonic velocity data, Celtic Sea/south-Western Approaches, *in* J. G. Buchanan and P. G. Buchanan, eds., Basin inversion: Geological Society of London Special Publication 88, p. 191–207.

Miralles, L., 1999, Estudi de la fàbrica de roques d' halita: aplicació a la conca Potàssica Catalana: Unpublished Ph.D. Thesis, Universitat de Barcelona, 213 p.

Mitra, G., 1978a, Ductile deformation zones and mylonites: the mechanical processes involved in the deformation of crystalline basement rocks: American Journal of Science, v. 278, p 1057–1084.

Mitra, S., 1978b, Microscopic deformation mechanisms and flow laws in quartzites within the South Mountain anticline: Journal of Geology, v. 86, p. 129–152.

Mitra, S., 1988, Effects of deformation mechanisms on reservoir potential in central Appalachian overthrust belt: AAPG Bulletin, v. 72, p. 536–554.

Mitra, S., 1990, Fault propagation folds: Geometry, kinematic evolution and hydrocarbon traps: AAPG Bulletin, v. 74, p. 921–945.

Mitra, G., 1994, Strain variation in thrust sheets across the Sevier fold-and-thrust belt (Idaho-Utah-Wyoming): implications for section restoration and wedge taper evolution: Journal of Structural Geology, v. 6, p. 51–61.

Mitra, G., and D. Elliott, 1980, Deformation of basement in the Blue Ridge and the development of the South Mountain cleavage, *in* D. R. Wones, ed., Proceedings of "TheCaledonides in the USA" Memoir, v. 2: Virginia Polytechnic Institute, Department of Geological Sciences, p. 307–311.

Mitra, G., W. A. Yonkee, and D. J. Gentry, 1984, Solution cleavage and its relationship to major structures in the Idaho-Wyoming-Utah thrust belt: Geology, v. 12, p. 354–358.

Moretti, I., S. Wu, and A. W. Bally, 1990, Computerised balanced cross section LOCACE to reconstruct an allochthonous salt sheet, offshore Louisiana: Marine and Petroleum Geology, p. 371–377.

Morley, C. K., 1986, A classification of thrust fronts: AAPG Bulletin, v. 70, p. 12–25.

Mukul, M., and G. Mitra, 1998, Finite strain and strain variation analysis in the Sheeprock thrust sheet; an internal thrust sheet in the Provo salient of the Sevier fold-and-thrust belt, central Utah: Journal of Structural Geology, v. 20, p. 385–405.

Mulugeta, G., 1988, Modeling the geometry of Coulomb thrust wedges: Journal of Structural Geology, v. 10, p. 847–859.

Mulugeta, G., and H. Koyi, 1987, Three-dimensional

geometry and kinematics of experimental piggyback thrusting: Geology, v. 15, p. 1052–1056.

Mulugeta, G., and H. Koyi, 1992, Episodic accretion and strain partitioning in a model sand wedge: Tectonophysics, v. 202, p. 319–333.

Nickelsen, R. P., 1986, Cleavage duplexes in the Marcellus shale of the Appalachian foreland: Journal of Structural Geology, v. 8, p. 361–371.

Price, R. A., 1981, The Cordilleran foreland thrust and fold belt in the southern Canadian Rocky Mountains, *in* K. McClay and R. A. Price, eds., Thrust and nappe tectonics: Geological Society of London Special Publication 9, p. 427–448.

Protzman, G. M., and G. Mitra, 1990, Strain fabric associated with the Meade thrust sheet: implications for cross-section balancing: Journal of Structural Geology, v. 12, p. 403–417.

Ramsay, J. G., and M. I. Huber, 1987, The techniques of modern structural geology, volume 2, Folds and fractures: Oxford, Academic Press, 700 p.

Sanderson, D. J., 1982, Models of strain variation in nappes and thrust sheets: A review: Tectonophysics, v. 88, p. 201–233.

Sáez, A., 1987, Estratigrafía y sedimentología de las formaciones lacustres del tránsito Eoceno Oligoceno del NE de la Cuenca del Ebro. PhD thesis. Universitat de Barcelona, 352 p.

Sans, M., 1999, From thrust tectonics to diapirism. The role of evaporites in the kinematic evolution of the Eastern South-Pyrenean front: Unpublished PhD Thesis Universitat de Barcelona, 197 p.

Sans, M., and J. Vergés, 1995, Fold development related to contractional salt tectonics southeastern Pyrenean thrust front, Spain, in M. P. A. Jackson, D. G. Roberts, and S. Snelson, eds., Salt tectonics: a global perspective: AAPG memoir 65, p. 369–378.

Sans, M., J. A. Muñoz, and J. Verges, 1996, Triangle zone and thrust wedge geometries related to evaporitic horizons (southern Pyrenees): Bulletin of Canadian Petroleum Geology, v. 44 (2), p. 375–384.

Storti, F., and K. McClay, 1995, The influence of sedimentation on the growth of thrust wedges in analogue models: Geology, v. 23, p. 999–1002.

Storti, F., F. Salvini, and K. McClay, 1997, Fault related folds in sandbox analogue models of thrust wedges: Journal of Structural Geology, v. 19, p. 583–602.

Suppe, J., 1983, Geometry and kinematics of fault-bend folding: American Journal of Science, v. 283, p. 684–721.

Suppe, J., 1985, Principles of structural geology: Englewood Cliffs, New Jersey, Prentice Hall, 537 pp.

Teixell, A., 1996, The Ansó transect of the southern Pyrenees: basement and cover thrust geometries: Journal of Geological Society of London, v. 153, p. 301–310.

Teixell, A., and H. A. Koyi, 2003, Experimental and Field study of the effects of lithological contrasts on thrust-related deformation: Tectonics, v. 22, p. 1054–1073.

Thorbjornsen, K. L., and W. M. Dunne, 1997, Origin of a thrust-related fold: geometric vs, kinematic tests: Journal of Structural Geology, v. 19, p. 303–319.

Tonghban, H., 2000, Estude du soulevement au neogene de la bordure continentale norvegienne: approche stratigraphique et quantification, Ph.D. thesis, Université Paris-Sud, 320 p.

Turcotte, D. L. and G. Schubert, 1982, Geodynamics: New York, J. Wiley & Sons, 450 p.

Vergés, J. 1993, Estudi geologic del vessant sud del Pirineu oriental i central. Evolució cinemàtica en 3·D. Unpublised Ph.D. thesis, Universitat de Barcelona, 203 p.

Vergés, J., M. Marzo, T. Santaeulària, J. Serra-Kiel, D. W. Burbank, J. A. Muñoz, and J. Giménez-Montsant, 1998, Quantified vertical motions and tectonic evolution of the SE Pyrenean foreland basin., *in* A. Mascle, C. Puigdefàbregas, H. P. Luterbacher, and M. Fernàndez, eds., Cenozoic foreland basins of Western Europe: Geological Society of London Special Publication 134, p. 107–134.

Weijermars, R., 1986, Flow behavior and physical chemistry of bouncing putties and related polymers in view of tectonic laboratory applications: Tectonophysics, v. 124, p. 325–358.

Weijermars, R., M. P. A. Jackson, and B. Vendeville, 1993, Rheological and tectonic modeling of salt provinces: Tectonophysics, v. 217, p. 143–174.

Williams, G. D., and T. J. Chapman, 1983, Strain developed in the hanging walls of thrusts due to slip/propagation rate: a dislocation model: Journal of Structural Geology, v. 5, p. 563–571.

Wiltschko, D. V., 1981, Thrust sheet deformation at a ramp: summary and extension of an earlier model, *in* K. R. McClay and N. J. Price, eds., Thrust and nappe tectonics, Geological Society of London Special Publication 9, p. 55–64.

Wissinger, E. S., A. R. Levander, J. S. Oldow, G. S. Fuis, and W. J. Lutter, 1998, Seismic profiling constraints on the evolution of central Brooks Range, Arctic Alaska, *in* J. S. Oldow and H. G. Lallemant, eds., Architecture of the Central Brooks Range fold and thrust belt, Arctic Alaska, Geological Society of America Special Paper, 324, p. 269–291.

Woodward, N. B., S. E. Boyer, and J. Suppe, 1985, An outline of balanced cross sections, 2nd edition: University of Tennessee Department of Geological Sciences Studies in Geology, 11, 170 p.

Woodward, N. B., D. R. Grey, and D. B. Spears, 1986, Including strain data in balanced cross-sections: Journal of Structural Geology, v. 8, p. 313–324.

Woodward, N. B., S. E. Boyer, and J. Suppe, 1989, Structural lithic units in external orogenic zones: Tectonophysics, v. 158, p. 247–267.

Zhao, W. L., D. M. Davis, F. A. Dahlen, and J. Suppe, 1986, Origin of convex accretionary wedges: evidence from Barbados: Journal of Geophysical Research, v. 91, p. 10246–10258.

12

Dixon, J. M., 2004, Physical (centrifuge) modeling of fold-thrust shortening across carbonate bank margins — timing, vergence, and style of deformation, *in* K. R. McClay, ed., Thrust tectonics and hydrocarbon systems: AAPG Memoir 82, p. 223–238.

Physical (Centrifuge) Modeling of Fold-thrust Shortening Across Carbonate Bank Margins—Timing, Vergence, and Style of Deformation

John M. Dixon

Experimental Tectonics Laboratory, Dept. of Geological Sciences and Geological Engineering, Queen's University, Kingston, Ontario, Canada

ABSTRACT

Physical models constructed of plasticine and silicone putty are deformed in a centrifuge at 4000 *g* in an investigation of the influence of lateral facies boundaries (simulating the margins of carbonate platforms) on the evolution of superimposed fold-thrust structures. Initial configurations represent vertically aggraded, prograded, or retrograded platform margins, and the fold-thrust deformation is directed either from the basin toward the platform or from the platform toward the basin. The effects of mechanical layering in the platform facies, the strength of the basal décollement, and more complex facies distributions are also investigated.

With the basin on the hinterland side of the platform, shortening first propagates across the basin facies, but an anomalous anticlinal structure tends to form along the facies boundary at an early stage of shortening. After the fold-thrust shortening has propagated across the facies boundary and into the platform facies, the anticline at the margin evolves into a large, foreland-verging structure that carries the basin facies over the platform margin. With the basin on the foreland side of the platform, the first compressional structure is an anticline situated in the basin facies along the facies boundary, even though this is distant from the hinterland end of the system. Shortening by folding propagates from this structure across the basin facies toward the foreland. Subsequently, the platform facies also undergoes foreland-propagating shortening. The dip of the boundary controls whether the platform facies is thrust over, under, or into the basin facies. The strength of the basal décollement within "prereef" strata affects the style and symmetry of structures in both the basin and platform facies and also affects the timing and size of the margin-related structure.

A structure that develops early in the shortening process may be a prospective exploration target, because it has the potential to retain an early hydrocarbon charge. The geometry (facing and dip) of the facies boundary controls whether the localized structure is a fold or major thrust and whether it is foreland- or hinterland-vergent. Therefore, knowing the geometry of a facies boundary may aid explorationists in predicting the geometry of target structures associated with it. Conversely, the structural configuration may aid us in interpreting the original distribution of lithofacies.

INTRODUCTION

Fold-thrust belts such as the Canadian Rocky Mountains result from lateral compression of horizontally layered successions of sedimentary rock that were deposited as passive-margin sequences following crustal extension, rifting, and subsidence (Bally et al., 1966; Price, 1981). They contain significant petroleum resources in structural traps. Understanding the geometry and kinematics of fold-thrust belts and the mechanics of the fold-thrust process provides important constraints on predicting subsurface structure and locating potential hydrocarbon reservoirs. Construction of accurate balanced and restored cross sections also depends on knowledge of the relative ages of the structures.

Compressional structures reflect the mechanical architecture of the passive margin: The form and wavelength of folds and the spacing of thrust ramps are largely controlled by the thicknesses and relative strengths of the strata. In laterally homogeneous strata, folds and thrusts typically develop in a regular hinterland-to-foreland sequence: Contractional structures at the trailing edge of the deformed belt are the oldest, and those at the leading edge are the youngest (Dahlstrom, 1970). This foreland-directed sequence of nucleation is consistent with the critically tapered Coulomb wedge analysis of fold-thrust belts, in which the wedge is treated as a homogeneous continuum (Davis et al., 1983). It is also observed in physical models (for sandbox experiments, see Davis et al., 1983; Liu et al., 1991; and for centrifuge experiments, see Liu and Dixon, 1990; Dixon and Liu, 1991) with laterally uniform strata. Despite the overall "break-forward" nucleation sequence, it should be noted that structures in the hinterland portion of the belt may continue to accumulate horizontal shortening concurrently with those closer to the foreland (Dixon and Liu, 1991; Storti et al., 2000; Price, 2001). Such "out-of-sequence" faulting (Morley, 1988) is necessary to accomplish thickening of the accretionary wedge as its toe propagates.

Mechanical heterogeneities, such as lateral lithofacies changes in the stratified cover sequence can disrupt the normal foreland propagation sequence by influencing the location, timing, and kinematic evolution of compressional structures. Facies boundaries can have a variety of orientations relative to the vergence of the fold-thrust belt. The platform margin can face toward the hinterland or toward the foreland. The facies boundary can dip vertically—a vertically aggraded boundary; toward the basin—a "retrograde" boundary; or toward the platform—a "prograde" boundary. A well-known example of a steep facies boundary that faces toward the hinterland is the "Kicking Horse Rim" facies change between the Middle Cambrian Chancellor Formation (slate/phyllite) and the Cathedral to Waterfowl Formations (carbonates) that marks the stratigraphic and structural transition between the Main Ranges and the Western Main Ranges of the central Canadian Rocky Mountains (Cook, 1975). Other, smaller-scale examples are the buildups of the Upper Devonian Peechee Member (Southesk Formation, Fairholme Group) in the southern Canadian Rocky Mountains. The Peechee buildups are of particular interest because of the variety of margin geometries and facings relative to the vergence of Rocky Mountain structures (see Price, 1964; Mountjoy, 1978; and McLean and Mountjoy, 1993 for detailed reviews).

With fold-thrust shortening verging from basin toward platform, one might anticipate that deformation should, at least initially, be concentrated in the weaker basin facies, with the platform margin acting as a buttress. This is what is observed at the Kicking Horse Rim in the central Canadian Rockies (Cook, 1975), where pelites of the Chancellor Formation were strongly folded and pervasively recrystallized into slates and phyllites in the Porcupine Creek Anticlinorium on the hinterland side of a vertically aggraded carbonate bank margin. One might also anticipate that a retrograde margin (dipping toward the basin) would enhance transport of the basin facies onto the platform by nucleating a foreland-verging thrust or fold. A key question is whether a prograde margin (dipping toward the platform) could nucleate a hinterland-verging structure (fold or thrust) that transports the platform facies over the basin facies.

It is less obvious where deformation should commence in a system in which the stronger platform facies is situated on the hinterland side of the weaker basin facies. A mechanically competent platform and the presence of a subreef décollement horizon might facilitate transmission of compression across the intact platform and cause shortening in the basin, whereas a weaker platform above more competent subreef strata might allow shortening to initiate in the platform despite the presence of weaker basin facies on the foreland side.

Given that hydrocarbon traps commonly are formed by antiformal fold closures and fault seals, both of which can be associated with thrust ramps, the explorationist needs to be able to predict the locations of these structures in the subsurface. Therefore, any insight into mechanisms that localize thrust ramps, and any constraints on the subsurface positions of such structures, are potentially useful. As well, the timing of trap formation relative to the generation-migration sequence of hydrocarbons has a bearing on the prospectivity of those traps.

This chapter reports a systematic investigation of the effects of two principal variables, to determine their effects on the timing of nucleation, size, and vergence of structures that may form along the margin as a result of fold-thrust shortening. These variables are the facing of the platform margin relative to the vergence of the fold-thrust shortening (i.e., toward the hinterland or toward the foreland), and the dip of the facies boundary (i.e., vertical, retrograde, or prograde). The effects of the mechanical character of the platform facies (massive or laminated), the strength of the basal décollement, and the plan geometry of the platform margin (perpendicular or oblique to the shortening direction, or more complex) are also surveyed.

The literature documenting field-based studies of deformation associated with platform margins is very limited. The focus of investigations of such features has been primarily stratigraphic; thus, it is not surprising that undeformed exposures have received the most attention. Because there is a lack of structural data from natural prototypes, the models described here are intended to be generic rather than detailed analogs for specific prototype examples. The focus is on the large-scale effects of the gross mechanical contrast across platform margins with various geometries. No attempt is made to examine the role of details, such as the interfingering of tongues of reef material in the basin facies.

CONFIGURATIONS OF THE MODELS AND EXPERIMENTAL TECHNIQUE

The centrifuge modeling technique was suggested by Hubbert (1937) and was first applied to the study of tectonic processes by Ramberg (1967). Dixon and Summers (1985) described the 20,000-*g* centrifuge at Queen's University. The reader is referred to McClay (1976) and Dixon and Summers (1985, 1986) for discussion of the rheological properties of the model materials plasticine and silicone putty. Dixon and Summers (1985), Dixon and Tirrul (1991), and Dixon and Liu (1991) discussed the scaling relationships between prototype and model foreland fold-thrust-belt systems applicable to centrifuge models of the type I will describe here. I will then briefly summarize the model configuration and the experimental procedure. Table 1 lists the model scaling ratios applicable to this study. Given the uncertainties in the rheological scaling, the linear scale ratio of 10^{-6} should be taken as order-of-magnitude only.

Figure 1 shows the dimensions and initial configurations of the physical models. They were constructed of Harbutt's Gold Medal™ Plasticine and Dow-Corning silicone putty (dilatant compound 3179), the plasticine simulating competent rocks such as limestone and dolostone and the silicone putty simulating incompetent rocks such as shale (Dixon and Summers, 1985). The models included three stratigraphic units. The lowest unit had medium bulk competence and consisted of interlayered plasticine and silicone putty (in a ratio of 3:1), simulating a "prereef" or subreef facies of interlayered carbonate and shale. The middle unit contained a lateral facies boundary. One side, with low bulk competence, consisted of interlayered plasticine and silicone putty (in a ratio of 2:1), simulating an "off-reef" or basin shaly facies. The other side, with high bulk competence, consisted either of plasticine with colored layers to serve as passive markers with no mechanical significance, simulating a reef or platform facies of massive carbonate, or interlayered plasticine and silicone putty in a ratio of 5:1, simulating a platform facies with thin shaly interbeds. The dip and facing of the facies boundary relative to the shortening direction, and the stratigraphic character of the platform facies, were the variables in the present study (Table 2). The uppermost stratigraphic unit was a stratified plasticine/silicone putty unit identical to the lowest one. It simulated a "postreef" facies of interlayered carbonate and shale. In all experiments, the model's stratigraphic succession

Table 1. Applicable model scaling ratios.

Quantity	*Ratio, model/prototype*	*Equivalence (model ≡ prototype)*
Length	$l_r \approx 1.0 \times 10^{-6}$	1 mm ≡ 1 km (order-of-magnitude only)
Specific gravity (mass)	$\rho_r = 0.6$	1.60 ≡ 2.67 (bulk value for whole stratigraphic column)
Time	$t_r \approx 1.0 \times 10^{-11}$	1 hr ≡ 12 m.y.
Strain rate	$\varepsilon_r \approx 1.0 \times 10^{11}$	$10^{-3}\ s^{-1} \equiv 10^{-14}\ s^{-1}$ (for example)
Effective viscosity	$\mu_r \approx 2.4 \times 10^{-14}$	2.4×10^2 Pa s ≡ 10^{16} Pa s (for example)
Acceleration	$a_r = \mu_r/(\rho_r\ l_r\ t_r) = 4.0 \times 10^3$	4000 ***g*** ≡ 1 ***g***
Stress	$\sigma_r = \rho_r\ l_r\ a_r \approx 2.4 \times 10^{-3}$	(calculated from other ratios)

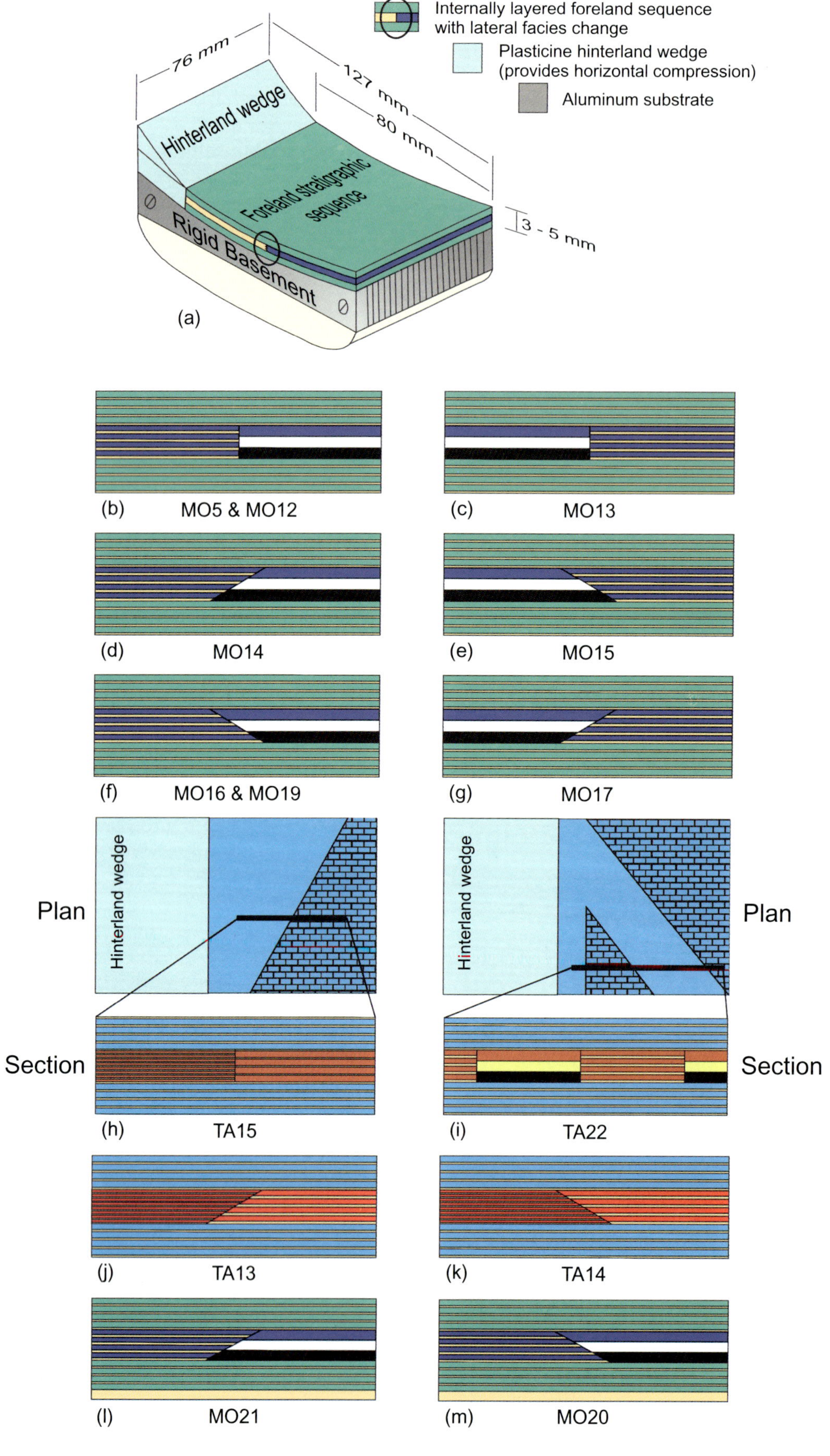

Internally layered foreland sequence with lateral facies change
Plasticine hinterland wedge (provides horizontal compression)
Aluminum substrate
76 mm
127 mm
80 mm
3 - 5 mm
Hinterland wedge
Foreland stratigraphic sequence
Rigid Basement
(a)
(b) MO5 & MO12
(c) MO13
(d) MO14
(e) MO15
(f) MO16 & MO19
(g) MO17
Plan
Hinterland wedge
Section
(h) TA15
(i) TA22
(j) TA13
(k) TA14
(l) MO21
(m) MO20

Table 2. Dip and facing of the facies boundary in selected models of the MO and TA series.

Dip of facies boundary (reflecting relationship between sedimentation and sea-level change)	*Facing of facies boundary relative to fold-thrust vergence*	
	Basin on the hinterland side (i.e., shortening directed toward the platform)	*Basin on the foreland side (i.e., shortening directed toward the basin)*
Vertical (vertical aggradation)	MO5, MO12 TA15 (oblique margin) TA22 (complex plan geometry)	MO13 TA22 (complex plan geometry)
Dipping at 30° towards the basin (retrogradation)	MO14 TA13 (laminated platform) MO21 (0.5 mm silicone putty layer at base)	MO15
Dipping at 30° towards the platform (progradation)	MO16 TA14 (laminated platform) MO19 (part of décollement coated with vaseline) MO20 (0.5 mm silicone putty layer at base)	MO17

Note: All models have the platform margin trending perpendicular to the shortening direction and a thin basal silicone putty lamina unless otherwise noted.

rested on a rigid base (the aluminum base plate), and the shortening occurred above a basal décollement. The base either was not lubricated or was coated with vaseline or a thicker layer of silicone putty to simulate a low-strength basal décollement.

The models were deformed at between 2500 and 4000 *g* in the centrifuge of the Experimental Tectonics Laboratory at Queen's University. The models were subjected to as many as six stages (numbered I to VI) of horizontal compression from one end by the collapse and lateral spreading of a "hinterland wedge" of plasticine (see Figure 1a). Each experimental stage involved about 180 s during which the centripetal acceleration rose from 0 to 4000 ***g***, then 300 s at 4000 ***g***, followed by deceleration. The hinterland wedge began to collapse and spread laterally at about 2500 ***g*** and reached a stable profile after about 200 s at 4000 ***g***. After each stage, a few transverse sections were cut and photographed to document the progressive evolution of structures. The sections were reassembled and the hinterland wedge reconstructed to its initial slope (to restore its gravitational potential) prior to each successive stage of shortening. After the final stage, each model was cut into serial sections spaced at an interval of 0.125 in. (3.0 mm for models MO20 and MO21).

Because centrifuged models are deformed in a rotating reference frame, it is necessary to evaluate whether Coriolis acceleration influences the experiments. In the present study, models being deformed at 4000 ***g*** were rotating at 3300 rpm at a radius of 0.325 m (the base of the model). The maximum particle velocity in the radial direction, associated with amplification of a fold (for example), was on the order of 3×10^{-6} m s^{-1}. This corresponds to a ratio of Coriolis acceleration to centrifugal acceleration on the order of 5×10^{-8}; consequently, the Coriolis effect can be ignored (see also Ramberg, 1967, p. 46–48). For this reason, and because the models were consistently placed in the centrifuge in the same orientation relative to the rotor's rotation direction, contrasts between structures with foreland and hinterland vergence can be interpreted as being structurally significant.

In the centrifuge technique, the models are laterally confined on four sides by the rigid walls of the sample chamber in the centrifuge rotor. During advanced stages

Figure 1. Dimensions and initial configurations of the physical models described in this chapter. Part (a) shows the general geometry and dimensions. Circled area encloses the facies boundary. Lower and upper stratigraphic units comprise alternating laminae of plasticine and silicone putty, with a thickness ratio of 3:1. In the middle unit, the facies boundary separates a competent platform facies (three thick plasticine laminae or interlayered plasticine and silicone putty in a ratio of 5:1) and an incompetent basin facies (interlayered plasticine and silicone putty [2:1]). The collapsing hinterland wedge spreads horizontally from left to right. Parts (b) to (m) show enlarged schematic sections of the facies boundaries: (b) and (c) vertically aggraded facies boundary; (d) and (e) retrograde boundary; (f) and (g) prograde boundary; (h) oblique boundary; (i) complex boundary with oblique channel; (j) and (k) retrograde and prograde boundaries with layered platform facies; (l) and (m) retrograde and prograde boundaries with massive platform facies and weak décollement horizon (0.5-mm silicone-putty layer (pink) at the base).

of shortening, the models tend to be squeezed against the foreland end wall. This tends to produce structures that are relatively upright and highly compressed, in contrast to those that form in natural fold-thrust belts lacking a rigid boundary at the foreland end. Nevertheless, the models allow comparison of the effects of different internal stratigraphic configurations. Further, the influence of the end wall only becomes significant at late stages of shortening.

EXPERIMENTAL RESULTS

Platform-verging Deformation

Models of the MO series had massive platform units, and representative examples will be discussed in detail. Models MO5 and MO12 (Figure 2) have a vertical platform margin (reflecting vertical aggradation) and similar stratigraphic sequences, the variations being the total thickness of the foreland succession (3 mm and 5 mm) and the thickness and number of interbedded plasticine and silicone putty laminae in the prereef, basin, and postreef units. These stratigraphic differences primarily influence the wavelength of fold and thrust structures and have little effect on structural style and strain partitioning. Models MO14 and MO16 (Figure 3) have stratigraphic successions like that of MO12, but the facies boundary dips at 30° toward the basin (i.e., in a retrograde geometry) in the case of MO14 and at 30° toward the platform (i.e., in a prograde geometry) in MO16. Models TA13 and TA14 (Figure 4), with anisotropic (laminated) platform units, will be discussed briefly for comparison. The facies boundary dips at 30° toward the basin (a prograde margin) in TA13 at 30° toward the platform (a retrograde margin) in TA14.

Massive Platform Unit

The general pattern of deformation is similar in all models with massive platform units and deformation verging from basin to platform, regardless of the dip of the platform margin (Figures 2 and 3). Shortening begins as buckle folding in the laminated basin facies adjacent to the hinterland wedge. Early in the shortening history, a conspicuous anticline nucleates at the site of the facies boundary, well ahead of the foreland-propagating fold structures (Figures 2a, b, and 3b, c, i, j). This structure evolves into an overturned anticline with foreland vergence that carries basin facies over the platform margin (Figures 2c, d, and 3d, e, l, m). The foreland-propagating shortening eventually propagates across the facies boundary and is accommodated by foreland-verging thrusts in the platform facies. Subsequently, in all cases, the overturned anticline that formed at the basin margin is transported over the platform facies on a prominent thrust localized by the facies boundary, while thrust-shortening of the platform continues (Figures 2e, f, and 3f, g, n).

The structural style of each stratigraphic unit is controlled by its mechanical character (Figures 2 and 3). Shortening is accommodated by buckling in laminated units (including the prereef, basin facies, and postreef strata), whereas the massive platform unit deforms first by layer-parallel shortening and then by foreland-vergent thrust faulting. Back thrusts (not shown in the figures) also develop in some models, mainly at the foreland end of the platform unit. The dominance of foreland-vergent structures (folds and thrusts) is because of the relatively high strength of the basal contact against the base plate (see below).

Whereas the overall pattern is similar, there are significant local variations in the vicinity of the facies boundary. The vertical boundary (Figure 2) initially localizes a single upright anticline and acts as a buttress against which the basin facies is thickened until it is able to spill over onto the platform when the major thrust develops. A small slice of basin facies is left in the footwall of the thrust, at the base of the facies boundary (see, e.g., model MO12 at stage V in Figure 2f). Such a strain shadow on the hinterland side of the platform margin probably contributed to the exquisite preservation of fossils of the Burgess Shale fauna in the Canadian Rocky Mountains near Field, British Columbia (Powell et al., 1998).

The retrograde platform margin (Figure 3a–g) also localizes a prominent fold at the earliest stage, far in front of any other shortening structures. The fold structure forms precisely above the most outboard portion of the platform unit, which in this case is the base of the unit. At later stages of horizontal shortening, significant displacement accumulates on the thrust that follows the inclined facies boundary. Despite the fact that the facies boundary is synthetic to the sense of foreland-vergent thrusting, the major boundary thrust in this model develops by linkage of adjacent folds in the basin strata and still leaves a small slice of basin facies strata in its footwall adjacent to the facies boundary (MO14, stage V, Figure 3g).

It is not unexpected that the retrograde margin enhances foreland-vergent thrusting, but the obvious question is whether prograde geometry can nucleate a hinterland-verging thrust (a back thrust) carrying the platform facies over the basin facies. At the earliest stage, MO16 (Figure 3i, j) develops a prominent anticline above the facies boundary, far in front of any other contractional structures. As in MO14, this structure is localized precisely above the most outboard portion of the platform unit, which in this case is the top of the unit. By stage III (Figure 3l) the shortening has been transferred upward into the postreef strata. In stage IV (Figure 3m), this structure is a major through-going foreland-verging

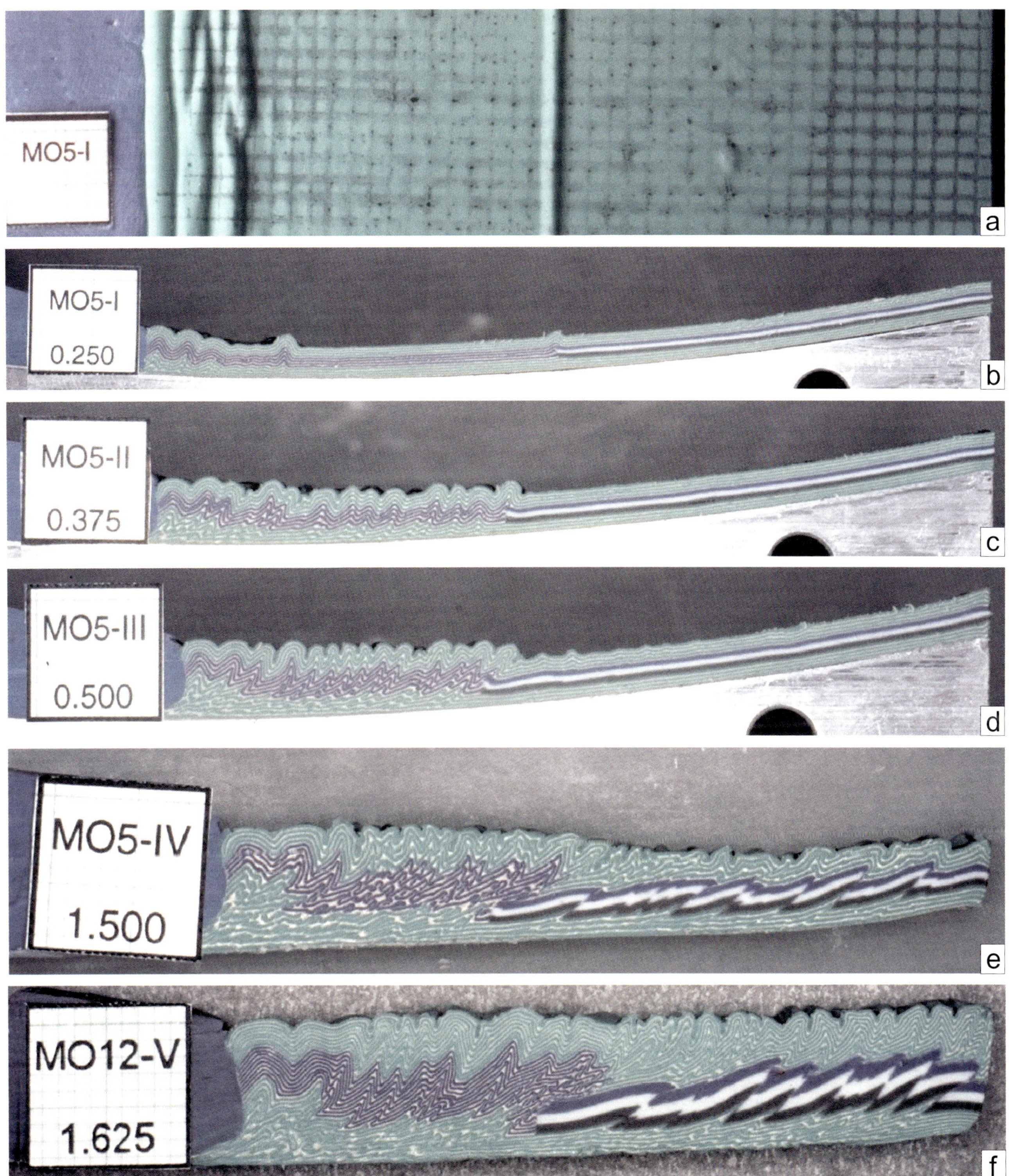

Figure 2. Progressive evolution of models MO5 and MO12, which had vertical facies boundaries and deformation verging toward the platform. (a) Part of the surface of MO5 after stage I, illuminated from the left. (b) MO5, Stage I, section 0.250. (c) MO5, Stage II, section 0.375. (d) MO5, Stage III, section 0.500. (e) MO5, Stage IV, section 1.500. (f) MO12, Stage V, section 1.625. Labels are 10 mm wide.

thrust cutting over the outboard end of the platform. The ramp is localized by the facies boundary but does not follow it (because of its antithetic dip). Finally, in stage V (Figure 3n), the overriding thrust becomes well established and the outboard edge of the platform is folded back and overturned. The facies-boundary thrust leaves

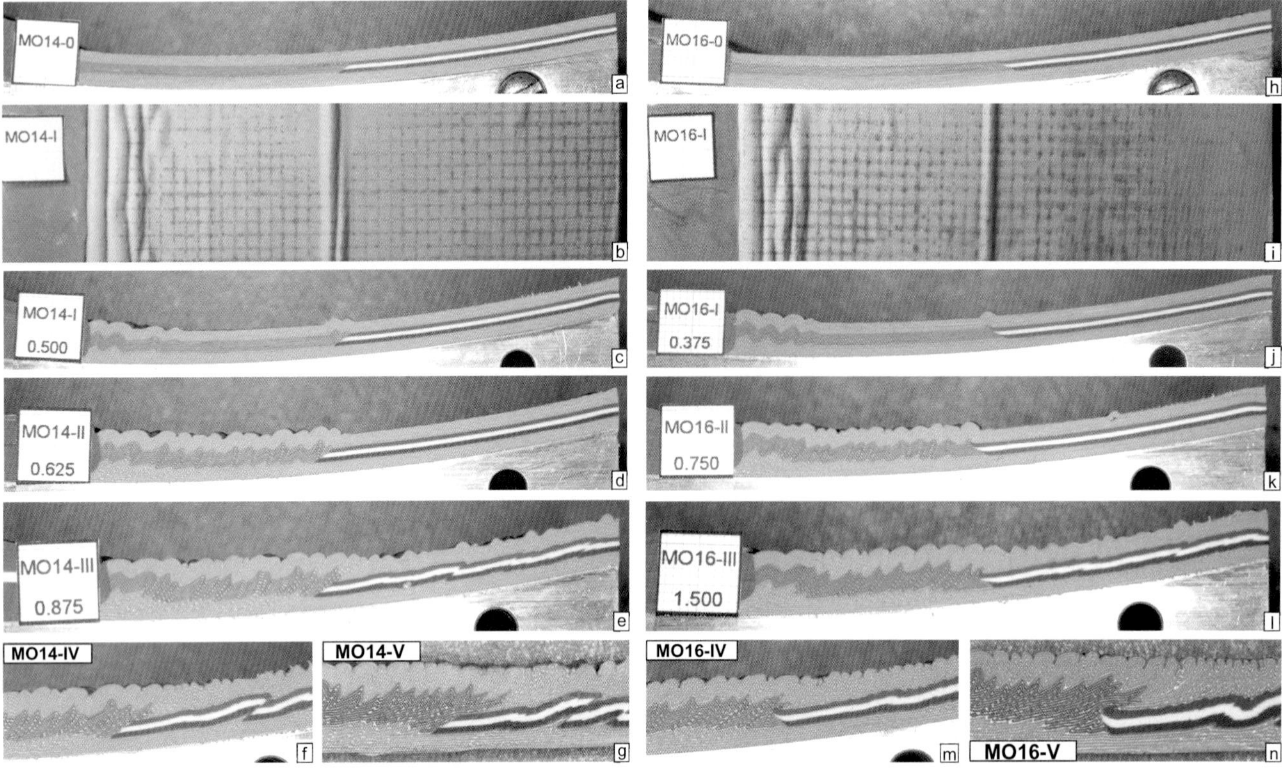

FIGURE 3. Progressive deformation of models with massive platform units and deformation verging toward the platform. (a)–(g) Model MO14 (retrograde facies boundary). (a) Section at stage 0. (b) Part of the surface after stage I, illuminated from the left. (c) Stage I, section 0.500. (d) Stage II, section 0.625. (e) Stage III, section 0.875. (f) Stage IV, section 0.750 (detail). (g) Stage V, section 1.500 (detail). (h)–(n) Model MO16 (prograde facies boundary). (h) Section at stage 0. (i) Part of the surface after stage I, illuminated from the left. (j) Stage I, section 0.375. (k) Stage II, section 0.750. (l) Stage III, section 1.500. (m) Stage IV, section 1.125 (detail). (n) Stage V, section 1.000 (detail). Labels are 10 mm wide.

a relatively large slice of basin facies strata in its footwall, sheltered beneath the overhanging facies boundary (model MO16, stage V, Figure 3n). The prograde margin exhibits no tendency toward back-thrusting of the platform unit over the basin facies. If hinterland-verging displacement is not possible on a surface that dips at 30° toward the foreland, this suggests that the basal décollement is relatively strong (Tirrul, 1983). The effects of reducing the shear strength of the basal décollement horizon are discussed below.

The different margin configurations are reflected in contrasting strain partitioning. The total shortening across the whole foreland and the shortening across the platform are fairly consistent, from one stage to the next. However, as the geometry of the margin changes from prograde to vertical to retrograde, the amount of thrust transport of basin strata over the platform margin increases and the amount of vertical thickening in the basin facies adjacent to the margin decreases.

Layered Platform Unit

In contrast to the MO-series models, which have strong, massive platform units, models TA13 and TA14 have weaker, layered platform units (Figure 4). The facies boundaries initially dip at 30° toward the basin (a retrograde margin) in TA13 and at 30° toward the platform (prograde margin) in TA14.

The general pattern of deformation is similar, with folding restricted to the basin area initially and then

FIGURE 4. Models TA13 and TA14, with laminated platform unit and retrograde and prograde margin, respectively. Deformation verges toward the platform. (a) Section of TA13 at stage 0. (b) Part of the surface of TA13 at stage I, illuminated from the left. (c) TA13, stage I, section 0.625. (d) Part of the surface of TA14 at stage I, illuminated from the left. (e) TA14, stage I, section 1.000. (f) TA13, stage III, section 0.875 (detail). (g) TA14, stage III, section 0.875 (detail). Labels are 10 mm wide.

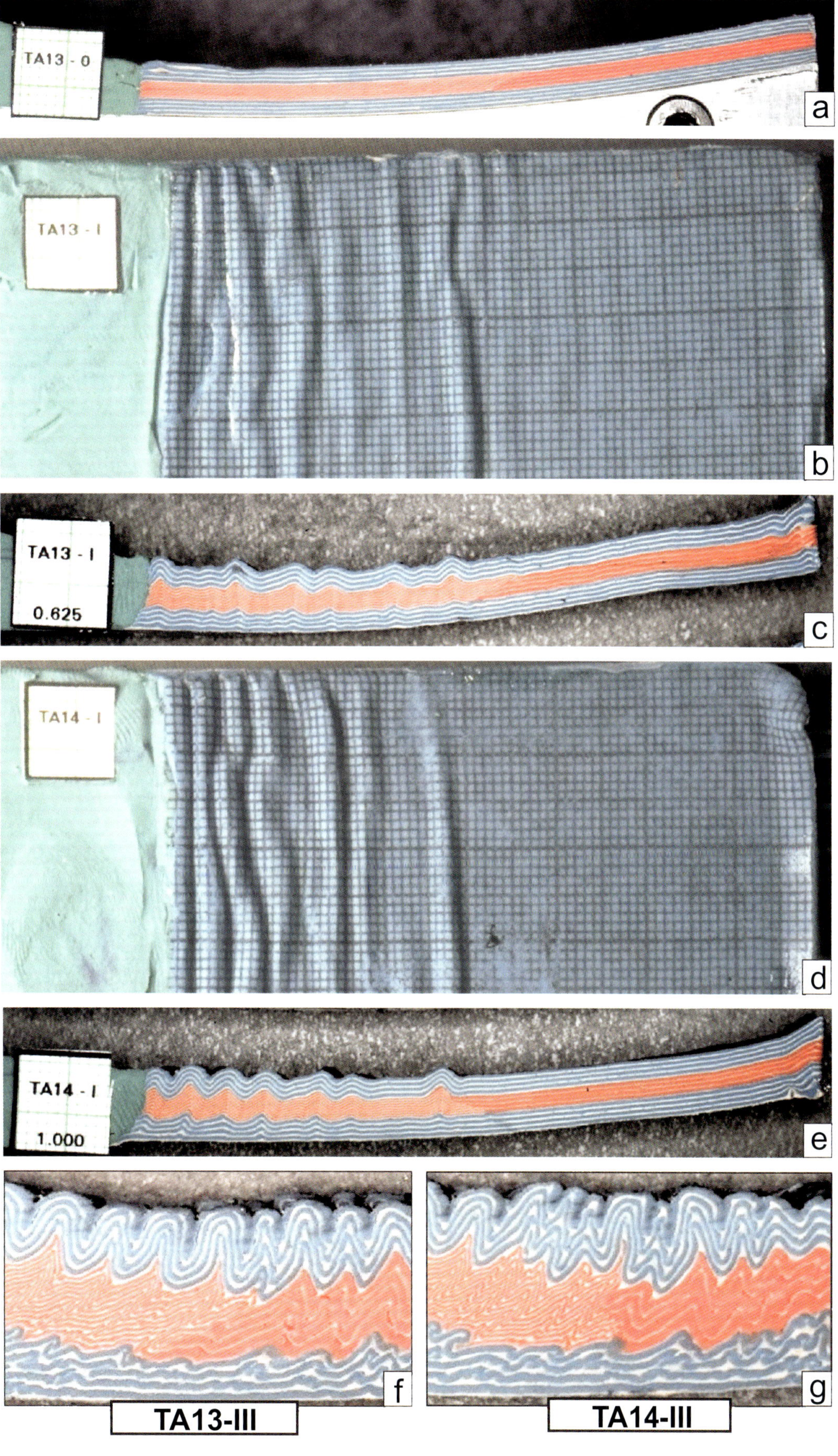

TA13 - 0
a
TA13 - I
b
TA13 - I
0.625
c
TA14 - I
d
TA14 - I
1.000
e
f
g
TA13-III
TA14-III

propagating across the facies boundary into the platform (Figure 4). The principal difference between the massive and layered platforms is the deformation style. Whereas the massive platforms shorten by thrusting, the layered ones shorten by flexural-slip folding. With a layered platform, the facies boundary localizes an early fold (Figure 4b–e), but this structure is not developed as far ahead of the foreland-propagating deformation front as in the models with massive platforms. However, as in the case of the massive platform, the precise location of this fold is controlled by the geometry of the margin. This fold develops outboard of the most distal portion of the reef, regardless of whether this horizon is at the base or the top of the reef unit. There is no prominent dislocation associated with the platform margin, even at the final stage (Figure 4f, g). Rather, the margin is, by and large, overprinted by folds.

Basin-verging Deformation

Models MO13, MO15, and MO17 (Figure 5) are stratigraphic equivalents of MO12, MO14, and MO16, respectively, but with the relative positions of the platform and basin facies reversed so that the vergence is toward the basin. In this configuration, shortening by buckling develops very early at the facies boundary (Figure 5a, b, i, and n), even though this is situated far from the advancing front within the platform. In other words,

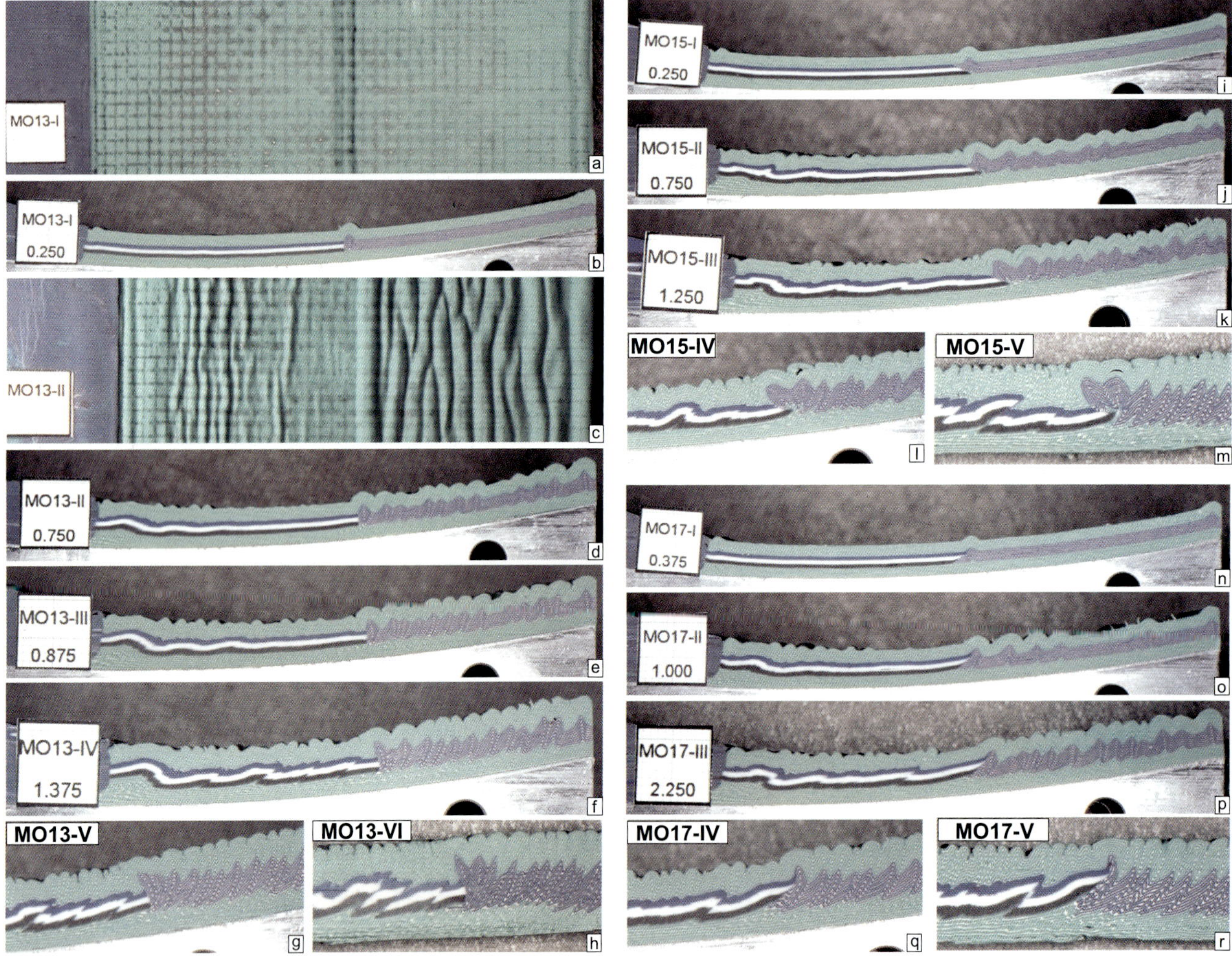

FIGURE 5. Progressive deformation of models with massive platform units, deformation verging toward the basin. (a)–(h) Model MO13 (vertical facies boundary). (a) Part of the surface after stage I, illuminated from above. (b) Stage I, section 0.250. (c) Part of the surface after stage II, illuminated from the left. (d) Stage II, section 0.750. (d) Stage III, section 0.875. (f) Stage IV, section 1.375. (g) Stage V, section 1.500. (h) Stage VI, section 2.250. (i)–(m) Model MO15 (retrograde facies boundary). (i) Stage I, section 0.250. (j) Stage II, section 0.750. (k) Stage III, section 1.125. (l) Stage IV, section 1.500 (detail). (m) Stage V, section 1.875 (detail). (n)–(r) Model MO17 (prograde facies boundary). (n) Stage I, section 0.375. (o) Stage II, section 1.000. (p) Stage III, section 2.250. (q) Stage IV, section 2.000 (detail). (r) Stage V, section 2.125 (detail). Labels are 10 mm wide.

the horizontal displacement is transferred through the platform facies, which is virtually undeformed at the early stages. Its stiffness must inhibit buckling deformation in underlying prereef strata, which therefore deform mainly by simple shear. The first-formed structure is a prominent anticline situated adjacent to the platform margin, localized above the distal edge of the platform, regardless of whether this horizon is at the base or the top of the reef unit. In all three cases, the buckle-shortening propagates across the basin toward the foreland end of the model in the early stages of deformation (Figure 5a–d, i, j, and n, o). Shortening of the platform by foreland-vergent thrusting begins slightly later (stages II and III), but it also propagates from hinterland toward foreland while shortening and thickening of the basin are still in progress. At stage II, the two deformation fronts are propagating simultaneously toward the foreland in these models (see Figure 5d, j, and o).

With shortening from platform to basin, the geometry of the platform margin proves to have a stronger effect on the local structural evolution. In the case of vertical aggradation (MO13), the conspicuous anticline in the basin strata adjacent to the platform margin develops a pronounced divergent fan-fold configuration. The backlimb verges over the platform margin toward the hinterland because the platform unit is forced into the basin strata as a tectonic wedge (Figure 5f–h). The fold at the retrograde platform margin (model MO15; Figure 5i–m) evolves into a more pronounced divergent fan-fold structure. This structure subsequently is transported over the margin of the platform on a prominent hinterland-verging thrust that is localized by the mechanical boundary between the basin and platform units. In other words, the platform is tectonically wedged beneath the basin facies. The facies boundary at the prograding platform's margin (model MO17; Figure 5n–r) evolves into a prominent foreland-verging thrust and the margin of the platform is carried up over the basin strata. In none of these cases is there a strain shadow in the basin facies adjacent to the platform margin.

Facies Boundaries with More Complex Plan Geometry

The models described above had linear facies boundaries trending perpendicular to the direction of propagation of the fold-thrust belt. For comparison, two other models were constructed with more complex plan geometries.

Model TA15 (Figure 6a–c) has a layered reef facies (like models TA13 and TA14) and a vertical facies boundary striking 30° obliquely to the axis of the fold-thrust belt. It was deformed with platform-verging shortening. During stage I (Figure 6a, b), folds develop in the basin facies adjacent to and parallel to the front of the collapsing and spreading hinterland wedge. However, there is also a prominent fold, with an oblique trend, situated above the platform margin some distance from the front of the advancing hinterland wedge. The facies boundary clearly constitutes a heterogeneity that is sufficient to nucleate folding far ahead of the folds that are nucleating and propagating serially in front of the advancing hinterland wedge boundary. Subsequently, the shortening propagates throughout the model, involving both the basin and platform facies. The wavelength of the folds is larger in the more competent reef facies than in the less competent basin facies. By stage III (Figure 6c) the whole model is saturated with folds, most of which are parallel to the wedge front, but the oblique fold that nucleated above the bank margin retains its identity through the full deformational history. There is interference between this early-formed oblique fold and the later-formed folds, and this leads to the possibility of structural culminations along the reef front.

Model TA22 (Figure 6d–f) has vertical facies boundaries and a complex plan distribution of basin and platform units comprising an elongate channel trending into the platform facies obliquely to the direction of overthrusting. This model approximates the geometry (but not the details of the stratigraphy) of the Cline Channel in the Upper Devonian Fairholme Group in the central Canadian Rocky Mountains (Mountjoy, 1978). Because of the obliquity in model TA22, the overthrusting is directed from basin to platform in some areas and from platform to basin in others. The schematic diagram in Figure 1(i) shows the initial configuration (stage 0) in plan and section. The deformation begins first in the basin areas and propagates into the postreef unit. The plan view at stage I (Figure 6d) shows how the weaker basin facies absorbs a significant amount of shortening. The competent platform unit remains little deformed until late in the deformation history, and it also shelters the prereef strata from deformation (Figure 6f). The massive platform unit is shortened by thrusting rather than by folding. Prominent thrust dislocations root at the outboard edges of the carbonate bank and ramp upwards across the postreef strata on the hinterland-facing margins, but there is a back fold of basin and postreef strata above the foreland-facing margin. Even when the total shortening is extreme (Figure 6e, f), contrasts in size and style of structures above the platform and basin facies are evident.

Influence of the Properties of the Basal Décollement

All of the models described above have a basal lamina of silicone putty in contact with the base plate and were shortened at similar strain rates. Thus they have similar values of basal cohesion and resistance to sliding.

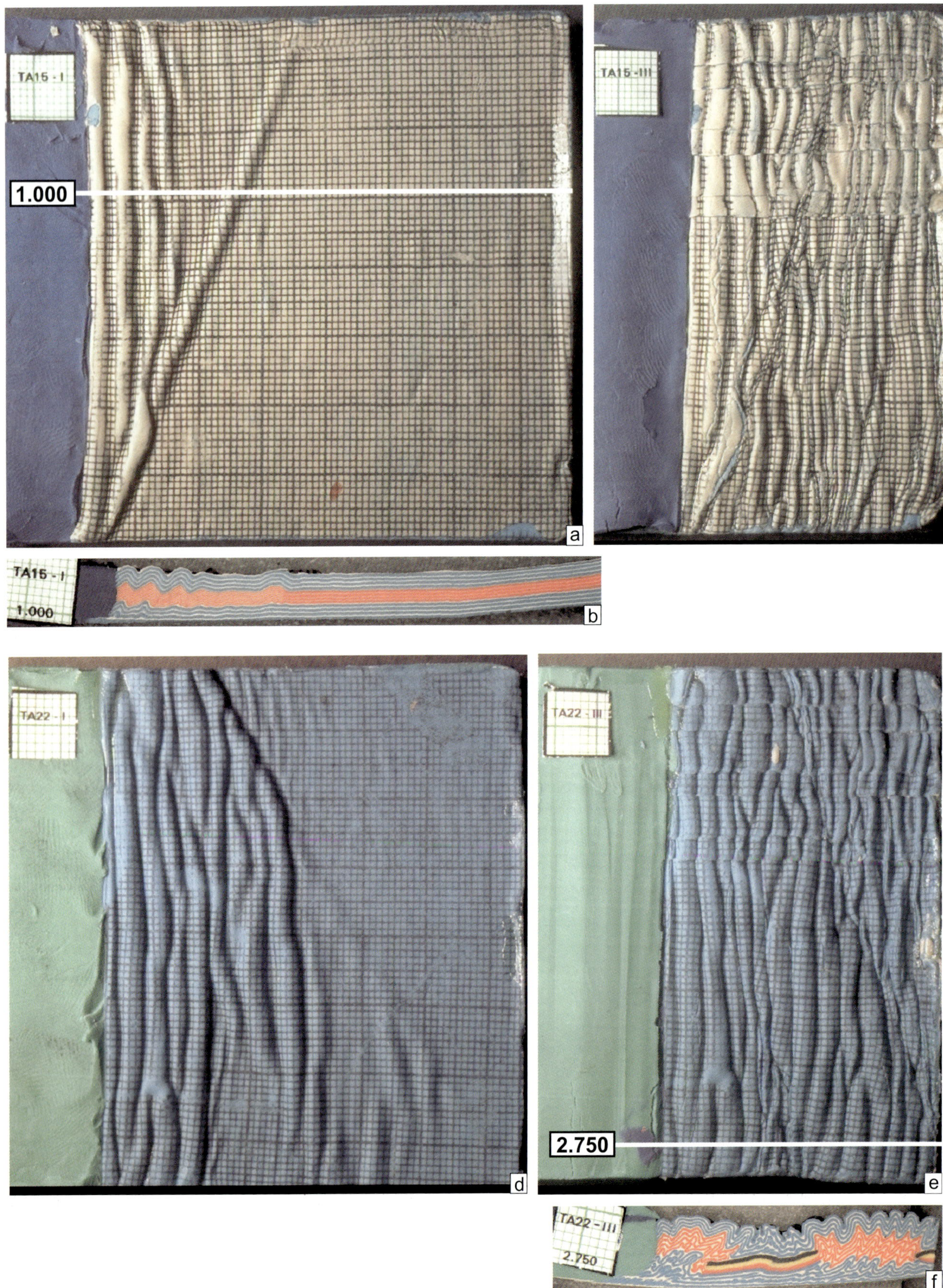
TA15 - I
1.000
a
TA15 - III
c
TA15 - I
1.000
b
TA22 - I
d
TA22 - III
2.750
e
TA22 - III
2.750
f

The influence of these variables on deformational style can be investigated by modifying the nature of the basal contact. This has also been investigated, for laterally uniform strata, by varying the basal cohesion and basal pore pressure in analytical models (Chapple, 1978; Davis et al., 1983) and by varying the character of the basal sheet (mylar, sand-paper) or the basal layer of the stratigraphic column (sand, mica flakes, silicone putty) in sandbox experiments (Davis et al., 1983; Liu et al., 1991).

Model MO19 (Figure 7a–g) has the same stratigraphic configuration as model MO16 (Figure 3h–n). However, half of its basal contact (the region between sections 0.000 and 1.500, i.e., the lower half of the image in Figure 7a) is coated with vaseline petroleum jelly. The lubricated and unlubricated portions develop contrasting structural styles. At early stages, the deformation front propagates farther toward the foreland above the weaker décollement, accommodated by a lateral gradient in the amplitude and tightness of folds in the basin facies (Figure 7a). The unlubricated portion evolves much like MO16 (with foreland-verging folds in the basin portion and foreland-verging thrusts in the platform; see Figure 7b–d). In contrast, the lubricated portion develops structures that are more symmetrical (upright folds in the basin facies, and upright folds and fore- and back-thrusts in the platform; see Figure 7e–g). This behavior is consistent with the expected effect of a low-strength basal décollement (see Tirrul, 1983), which reduces the resolved shear stress on the base and rotates the principal compressive stress axis toward the horizontal. However, there is still no back-thrusting at the facies boundary: the platform unit is driven into the basin facies, and its leading edge is overturned in the footwall of a prominent forethrust carrying basin strata over the platform (Figure 7g).

In models MO20 and MO21 (Figure 7h–m), the basal décollement lies at a stratigraphic horizon that is not only weak but also mobile (for example, a relatively thick unit of weak shale or salt). These models have stratigraphic configurations and facies boundaries with prograde (Figure 7h–j) and retrograde (Figure 7k–m) geometry similar to models MO14 and MO16, respectively, but they differ in that the basal décollements consist of a 0.5-mm layer of silicone putty. Overthrusting is directed from the basin toward the platform. The purpose of these experiments is to investigate whether a weak, ductile décollement horizon promotes increased and localized activation of the facies boundary as a thrust, either foreland- or hinterland-vergent. In both models, the weak substrate promotes formation of upright structures (akin to those above the vaseline-coated décollement in model MO19). The mobile décollement unit enhances the amplitude of fold structures because it is able to flow laterally into anticlinal cores. It also reduces the taper of the accretionary wedge (consistent with theory; see above). The low-strength décollement beneath the platform also allows an increase in the amount of shortening within the platform and a concomitant reduction in the amount of overthrusting of the basin facies over the platform. Lowering the effective strength of the platform reduces the tendency for shortening to ramp upward above the platform, thus rendering the mechanical heterogeneity of the facies change less significant.

No analogous tests have been run with shortening directed toward the basin. However, comparison between the two sets MO12-14-16 and MO13-15-17 leads to the expectation that a low-strength basal décollement would facilitate foreland-directed translation of the undeformed platform and exaggerate the early deformation in the basin facies distant from the hinterland end of the system. Thus, in that situation, the weak décollement should enhance the mechanical significance of the facies boundary.

The models demonstrate that the rheological character of the stratigraphic unit that is the locus of décollement beneath the level of the facies boundary also influences the timing, intensity, and style of deformation at the boundary and in the adjacent basin and platform domains.

SUMMARY

The models show that, in general, the basin facies is much more readily deformed than the platform facies, regardless of the facing of the boundary relative to the vergence of the shortening. The laminated, incompetent basin facies shortens by buckling, as does a laminated, competent platform, while a massive, competent platform unit is prone to thrusting.

The mechanical contrast between the basin and platform facies interferes with the normal pattern of

FIGURE 6. Complex boundary geometries. (a)–(c) Model TA15 (laminated platform, vertically aggraded boundary trending obliquely [at 60°] to the compression direction). (a) Top surface at stage I, illuminated from the left. (b) Stage I, section 1.000. (c) Top view at stage III, illuminated from the left. Note persistence of obliquely trending structures associated with the facies boundary. (d)–(f) Model TA22 (massive platform, vertically aggraded boundaries in a complex plan geometry including a narrow channel filled with basin facies trending obliquely to the shortening direction). (d) Top view at stage I, illuminated from the left. (e) Top view at stage III, illuminated from the left. (f) Section 2.750 at stage III. White lines in (a) and (e) mark positions of sections (b) and (f). Labels are 10 mm wide.

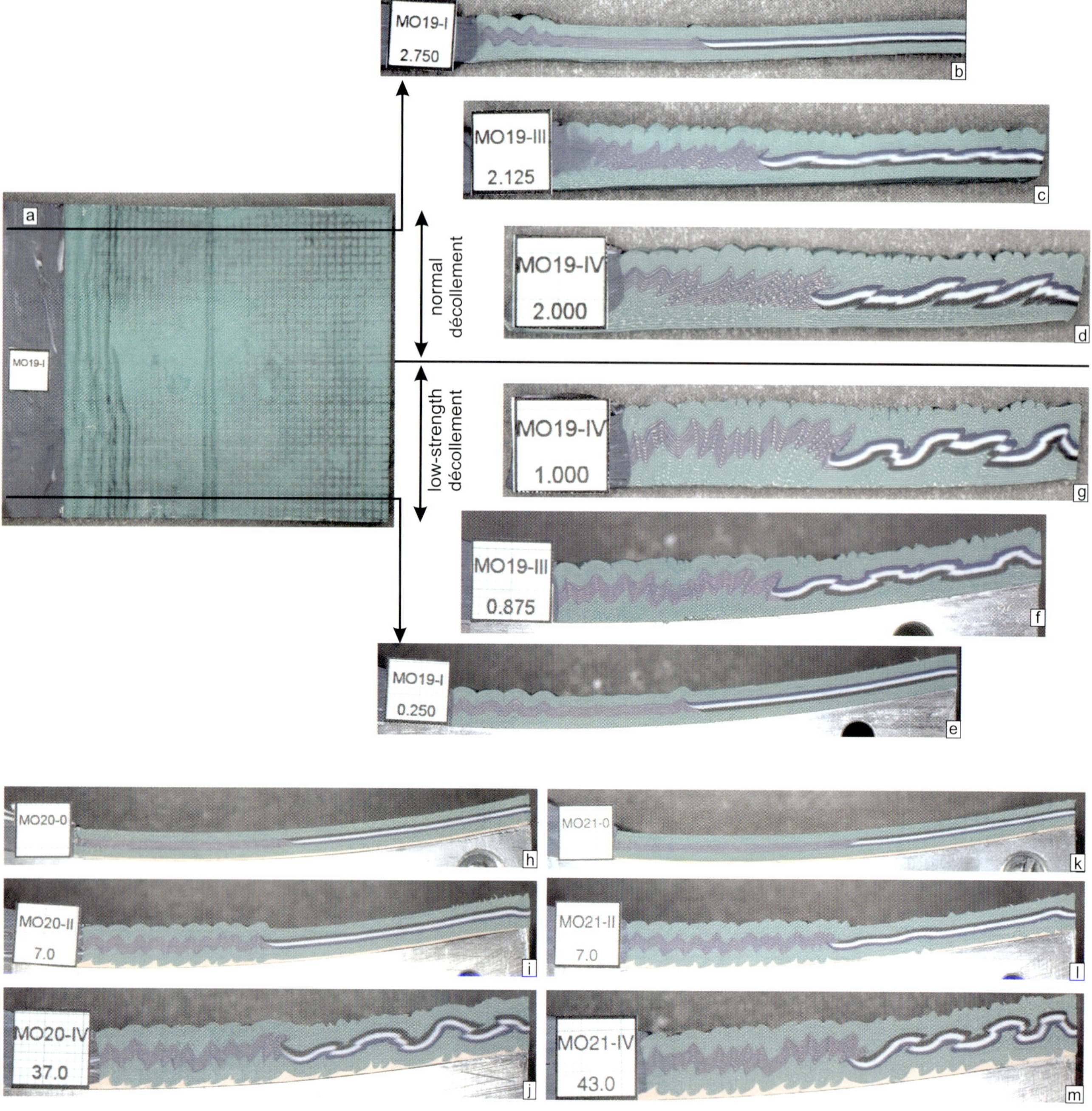

FIGURE 7. Effects of the strength of the décollement horizon. (a)–(g) Progressive evolution of model MO19 (prograde facies boundary, deformation verging toward the platform). One half of the basal contact (between sections 0.000 and 1.500, i.e., the lower half of the image in [a]) was coated with vaseline petroleum jelly to reduce its shear strength. (a) Surface view at stage I, illuminated from above. (b) Stage I, section 2.750 (high-strength base). (c) Stage III, section 2.125 (high-strength base). (d) Stage IV, section 2.000 (high-strength base). (e) Stage I, section 0.250 (low-strength base). (f) Stage III, section 0.875 (low-strength base). (g) Stage IV, section 1.000 (low-strength base). (h)–(m) Progressive evolution of models MO20 and MO21 (prograde and retrograde facies boundaries, respectively; deformation verging toward the platform). The lowest stratigraphic unit was a 0.5-mm thick lamina of silicone putty (pink), simulating an incompetent basal unit such as shale or salt. MO20, section at stage 0 (h); section 7.0 at stage II (i), section 37.0 at stage IV (j). MO21, section at stage 0 (k); section 7.0 at stage II (l); section 43.0 at stage IV (m). Labels are 10 mm wide. For MO20 and 21, section positions are given in millimeters relative to one edge of the model.

foreland-verging deformation that is observed in a laterally homogeneous stratigraphic sequence. With the platform on the foreland side, deformation propagates across the basin, and the platform is not deformed until late in the compressional process. With the platform on the hinterland side, the competent platform tends to be displaced rigidly, and basin deformation propagates from the facies boundary toward the foreland. Foreland-verging deformation in the platform commences later, despite the fact that the platform is situated at the hinterland side of the system. In this configuration, two foreland-propagating deformation fronts can be active concurrently.

The facies boundary is a significant structural heterogeneity. Whether the deformation verges across the basin toward the platform or across the platform toward the basin, a fold usually nucleates within the basin facies adjacent to the facies boundary at an early stage. This structure generally becomes overturned and can evolve into a major thrust. The fold and related thrust commonly grow to an anomalously large size relative to those that form in either the basin or the platform.

The orientation of the facies boundary strongly affects the style of deformation that is localized by this structural heterogeneity. When the direction of overthrusting is from platform to basin, a progradationally dipping facies boundary acts as a strong mechanical discontinuity that evolves into a thrust dislocation of the same orientation. However, a retrograde-dipping or vertical boundary tends to produce tectonic wedging of the platform into the basin strata. In contrast, where the direction of overthrusting is from the basin toward the platform, foreland-vergent thrusting of the basin strata over the platform occurs regardless of the dip of the mechanical boundary, and a local area of low strain can be preserved in the basin facies on the hinterland side of the platform margin.

The strength of the basal décollement surface influences the style of deformation and the rate of shortening. A moderate-strength basal contact consisting of a thin lamina of silicone putty in contact with the base plate produces foreland-verging folding in the basin facies and foreland-verging thrusting in the competent platform. A weaker basal contact (the normal basal unit coated with vaseline or resting on a thick lamina of silicone putty) promotes upright folding in the basin facies and a combination of fore- and back-thrusting and upright buckling of the platform. The rate of shortening is inversely related to the strength of the basal contact.

The structure localized by a linear facies boundary striking perpendicular to the compression direction is parallel to the trend of the fold-thrust belt and is distinguishable from the superimposed structures mainly by its anomalous size, style, and/or vergence. A linear boundary striking obliquely to the shortening direction can nucleate a localized structure that trends obliquely to the superimposed fold-thrust structures, and the existence of the facies boundary can be discerned from the interference between the two structural trends. More complex facies distributions, such as basin reentrants into the platform, also influence the distribution of strain and structural style, with the less-competent basin areas tending to be more strongly deformed than the platform areas.

IMPLICATIONS FOR HYDROCARBON EXPLORATION

The experiments suggest that the structural heterogeneity represented by a facies boundary between a shale basin and a carbonate platform can have a strong impact on superimposed fold-thrust structures. The boundary can nucleate structures (folds or thrusts) relatively early in the shortening process, in some cases prior to the formation of structures by the normal process of foreland propagation. Such structures may be especially prospective exploration targets: Their early development in the fold-thrust process increases the likelihood that they were formed in time to trap an early hydrocarbon charge, and they can have anomalously large volumes. The present results demonstrate that the geometry (facing and dip) of the facies boundary also influences the local deformation and controls, for example, whether the localized structures are folds or major thrusts and whether they are foreland- or hinterland-vergent. Interference between the structure localized by an oblique boundary and the superimposed fold-thrust structures can result in local structural culminations with excellent closure.

An understanding of the geometry of primary facies boundaries may assist us in predicting the geometry of prospective facies-boundary-related structures in the subsurface. Conversely, recognition of boundary-related structures on the basis of their anomalous size, style, and vergence may assist us in delineating the distribution of lithofacies in the subsurface.

ACKNOWLEDGMENTS

This research was conducted as part of the Fold-Fault Research Project (FRP), a collaboration between D.C. Lawton and D.A. Spratt of the University of Calgary and the author of Queen's University. FRP is funded by a consortium of petroleum industry firms and the Natural Sciences and Engineering Research Council of Canada (NSERC). NSERC provided a capital equipment grant (to JMD) for installation of the centrifuge and also provides ongoing support for operation of the Experimental Tectonics Laboratory at Queen's University, through grant RGPIN-9146. It has been a pleasure to work with summer research assistants Julia Blackburn (1996–1998), Maggie Oliphant (1997–1998), and Tom Anthony (1995), and their contributions to the present work are greatly appreciated. Ray Price reviewed an early version of the manuscript and suggested ways to improve its clarity. Peter Cobbold, Fabrizio Storti, and Francesco Salvini reviewed the manuscript for publication, and I acknowledge their helpful comments with thanks.

REFERENCES CITED

Bally, A. W., P. L. Gordey, and G. A. Stewart, 1966, Structure, seismic data and orogenic evolution of the southern Canadian Rocky Mountains: Bulletin of Canadian Petroleum Geology, v. 14, p. 337–381.

Chapple, W. M., 1978, Mechanics of thin-skinned fold-and-thrust belts: Geological Society of America Bulletin, v. 89, p. 1189–1198.

Cook, D. G., 1975, Structural style influenced by lithofacies, Rocky Mountain main ranges, Alberta-British Columbia: Geological Survey of Canada Bulletin, v. 273, 73 p.

Dahlstrom, C. D. A., 1970, Structural geology in the eastern margin of the Canadian Rocky Mountains: Bulletin of Canadian Petroleum Geology, v. 18, p. 332–406.

Davis, D., J. Suppe, and F. A. Dahlen, 1983, Mechanics of fold-and-thrust belts and accretionary wedges: Journal of Geophysical Research, v. 88, p. 1153–1172.

Dixon, J. M., and S. Liu, 1991, Centrifuge modelling of the propagation of thrust faults, *in* K. R. McClay, ed., Thrust tectonics: London: Chapman & Hall, p. 53–70.

Dixon, J. M., and J. M. Summers, 1985, Recent developments in centrifuge modelling of tectonic processes: equipment, model construction techniques and rheology of model materials: Journal of Structural Geology, v. 7, p. 83–102.

Dixon, J. M., and J. M. Summers, 1986, Another word on the rheology of silicone putty: Bingham: Journal of Structural Geology, v. 8, p. 593–595.

Dixon, J. M., and R. Tirrul, 1991, Centrifuge modelling of fold-thrust structures in a tripartite stratigraphic succession: Journal of Structural Geology, v. 13, p. 3–20.

Hubbert, M. K., 1937, Theory of scale models as applied to the study of geological structures: Geological Society of America Bulletin, v. 48, p. 1459–1520.

Liu, H., K. R. McClay, and D. Powell, 1991, Physical models of thrust wedges, *in* K. R. McClay, ed., Thrust tectonics: London, Chapman & Hall, p. 71–81.

Liu, S., and J. M. Dixon, 1990, Centrifuge modelling of thrust faulting: strain partitioning and sequence of thrusting in duplex structures, *in* R. J. Knipe, ed., Deformation mechanisms, rheology and tectonics: Geological Society of London Special Publication 54, p. 431–444.

McClay, K. R., 1976, The rheology of Plasticine: Tectonophysics, v. 33, p. T7–T15.

McLean, D. J., and E. W. Mountjoy, 1993, Upper Devonian buildup-margin and slope development in the southern Canadian Rocky Mountains: Geological Society of America Bulletin, v. 105, p. 1263–1283.

Morley, C. K., 1988, Out-of-sequence thrusts: Tectonics, v. 7, p. 539–561.

Mountjoy, E. W., 1978, Upper Devonian reef trends and configurations in the western portion of the Alberta basin, *in* I. A. McIlreath and P. C. Jackson, eds., The Fairholme carbonate complex at Hummingbird and Cripple Creek: Canadian Society of Petroleum Geologists, p. 1–30.

Powell, W. G., C. M. Henderson, and D. A. Glatiotis, 1998, Recrystallization and metamorphism of the Stephen Formation and its implications for the preservation of the Burgess Shale fossils (extended abstract): GeoTriad'98 (joint convention of Canadian Society of Petroleum Geologists, Canadian Society of Exploration Geophysicists and Canadian Well-Logging Society), Abstracts volume, p. 441–442.

Price, R. A., 1964, The Devonian Fairholme-Sassenach succession and evolution of reef-front geometry in the Flathead–Crowsnest Pass area, Alberta and British Columbia: Bulletin of Canadian Petroleum Geology, v. 12, p. 427–451.

Price, R. A., 1981, The Cordilleran foreland thrust and fold belt in the southern Canadian Rocky Mountains, *in* K. R. McClay and N. J. Price, eds., Thrust and nappe tectonics: Geological Society of London Special Publication 9, p. 427–448.

Price, R. A., 2001, An evaluation of models for the kinematic evolution of thrust and fold belts: structural analysis of a transverse fault zone in the Front Ranges of the Canadian Rockies north of Banff, Alberta: Journal of Structural Geology, v. 23, p. 1079–1088.

Ramberg, H., 1967, Gravity, deformation and the earth's crust: London: Academic Press, 214 p.

Storti, F., F. Salvini, and K. McClay, 2000, Synchronous and velocity-partitioned thrusting and thrust polarity reversal in experimentally produced, doubly-vergent thrust wedges: Implications for natural orogens: Journal of Structural Geology, v. 19, p. 378–396.

Tirrul, R., 1983, Structure cross-sections across Asiak foreland thrust and fold belt, Wopmay orogen, District of Mackenzie: Geological Survey of Canada Paper 83-1B, p. 253–260.

13

Dixon, J. M., and D. A. Spratt, 2004, Deformation at lateral ramps and tear faults — centrifuge models and examples from the Canadian Rocky Mountain Foothills, *in* K. R. McClay, ed., Thrust tectonics and hydrocarbon systems: AAPG Memoir 82, p. 239–258.

Deformation at Lateral Ramps and Tear Faults—Centrifuge Models and Examples from the Canadian Rocky Mountain Foothills

John M. Dixon

Dept. of Geological Sciences and Geological Engineering, Queen's University, Kingston, Ontario, Canada

Deborah A. Spratt

Dept. of Geology and Geophysics, The University of Calgary, Calgary, Alberta, Canada

ABSTRACT

Scaled physical models demonstrate how a thrust sheet accommodates itself to displacement over ramp systems consisting of two frontal-ramp segments linked by a vertical tear fault or a gently dipping lateral ramp. Four stratigraphic units of alternating competence, composed of thin layers of plasticine and silicone putty, rest on a rigid base. Ramp systems are cut in the basal (competent) stratigraphic unit during assembly. The models are built at a linear scale ratio of $\approx 10^{-6}$ (1 mm $\approx$ 1 km) and deformed in a centrifuge at 4000 g. Matched models are serially sectioned transversely or longitudinally, to constrain the structure in 3-D.

Structures in the hanging wall and footwall are diagnostic of the geometry of the ramp system. The thrust sheet develops prominent structural culminations associated with the frontal ramps. The culminations terminate abruptly above the buried transverse structure, and associated folds plunge in opposite directions off the ends of the culminations. The leading edge of the competent lowest unit in the thrust sheet tends to become overturned toward the foreland. In plan view, it mimics the geometry of the ramp system. However, the leading-edge segments become curved, reflecting a decrease in thrust displacement on each frontal-ramp segment toward the lateral structure.

The transverse structures appear to be rotated in response to 3-D strain comprising longitudinal gravitational spreading and stretching of the ramp-anticline culminations, as well as local longitudinal spreading of the footwall of the trailing frontal ramp adjacent to the transverse structure. Both frontal ramps propagate laterally past the transverse structure. The leading frontal ramp propagates into the footwall beneath the main thrust surface, and the trailing ramp propagates into the thrust sheet itself. Lateral propagation of the ramps generates local fault-propagation fold structures that are potential hydrocarbon traps.

The models suggest that a transverse link between two frontal ramps is unstable. Once it has rotated out of the transport direction, the linking fault is no longer favorably oriented to transfer slip between the frontal-ramp segments, so it is abandoned and the two frontal ramps propagate laterally to form a lap-joint displacement-transfer zone. This begs the question: Can a tear fault or lateral ramp develop as a link between two frontal ramps in laterally uniform strata, or must such a structure be localized by a transverse heterogeneity such as a preexisting fault or facies boundary?

Some features of the models are also seen in the Limestone Mountain area, southern Alberta Foothills, where the Brazeau Thrust fault and its splays ramp upward laterally to the southeast through lower Paleozoic carbonates. The Limestone Mountain and Marble Mountain culminations terminate abruptly to the southeast, at these blind lateral ramps. In strike section, the lateral structure appears to change its orientation along strike, from southwest to northeast, from a shallow northwest-dipping lateral ramp to a steep and even overturned (southeast-dipping) tear fault. With its variable dip, the Brazeau lateral ramp resembles a combination of the lateral-ramp and tear-fault physical models, but the variable dip is inferred to be a primary feature rather than a result of out-of-plane rotation.

Serial transverse and longitudinal sections of the models provide constraints on 3-D geometry that will aid geoscientists in interpreting wells and seismic sections through these complex structures and other similar situations, in the Canadian Rockies and other fold-thrust belts.

INTRODUCTION

To a first approximation, structures in fold-thrust belts tend to have cylindrical symmetry and reflect two-dimensional strain. In detail, however, thrusts and folds vary in profile and displacement, and also terminate along-strike. Displacement is transferred from one structure to another, for example by slip on lateral and oblique ramps (Gwinn, 1964; Dahlstrom, 1969, 1970). As the quality and quantity of seismic and well data from fold-thrust belts such as the Canadian Rocky Mountains increase, it is becoming recognized that lateral and oblique ramps are more common than previously thought (Bégin et al., 1996; Fermor, 1999).

Thrust faults generally transect a stratified sedimentary sequence along a staircase or ramp-flat trajectory (Rich, 1934; Douglas, 1950; Dahlstrom, 1970). Modern terminology applied to thrust ramps is summarized by McClay (1991) (Figure 1a). A frontal ramp is oriented such that its dip is parallel to the transport direction of the thrust sheet. A thrust can also change stratigraphic level along the strike of the thrust belt. A lateral (or "transverse" or "sidewall") ramp strikes parallel to the thrust transport direction, and an oblique ramp strikes obliquely to that direction. A transverse structure with a high cutoff angle is commonly called a tear fault.

Ramp (or cutoff) angles (between the bedding and the fault plane) vary through a wide range. Frontal-ramp cutoff angles for large- and small-scale thrusts in a variety of lithologic units in the Canadian Rockies have been studied systematically by Norris (1958, 1964), Price (1967), Bielenstein (1969), and Spratt and Lawton (1996), among others. Consistent with these studies, Fermor (1999) notes that frontal and oblique ramps tend to intersect the bedding at between 10° and 30°, with maximum frequency near 20° (although strain and folding of the thrust sheet can cause hanging-wall cutoff angles to be substantially increased). Lateral ramps can have any cutoff angle up to 90°, but the frequency distribution is not well known.

A thrust sheet that is transported on a fault comprising frontal ramp segments with transverse and oblique linkages conforms to the footwall by undergoing internal deformation that reflects the footwall's geometry. This deformation is manifest at the large scale in plunge reversals and structural culminations or depressions marked by abrupt changes of stratigraphic level in the hanging wall. For example, a tear fault should give rise to a tear in the lower part of the thrust sheet, where it ascends the frontal ramps, and to local culminations, terminated by areas of opposing plunge, higher in the sheet (Dahlstrom, 1970; Fermor, 1999) (Figure 1b). One should also expect local three-dimensional straining of the rock mass by complex folding and/or fracturing (e.g., Butler, 1982).

Some well-known structural features in the Rocky Mountains (see Figure 2) have been attributed to thrust

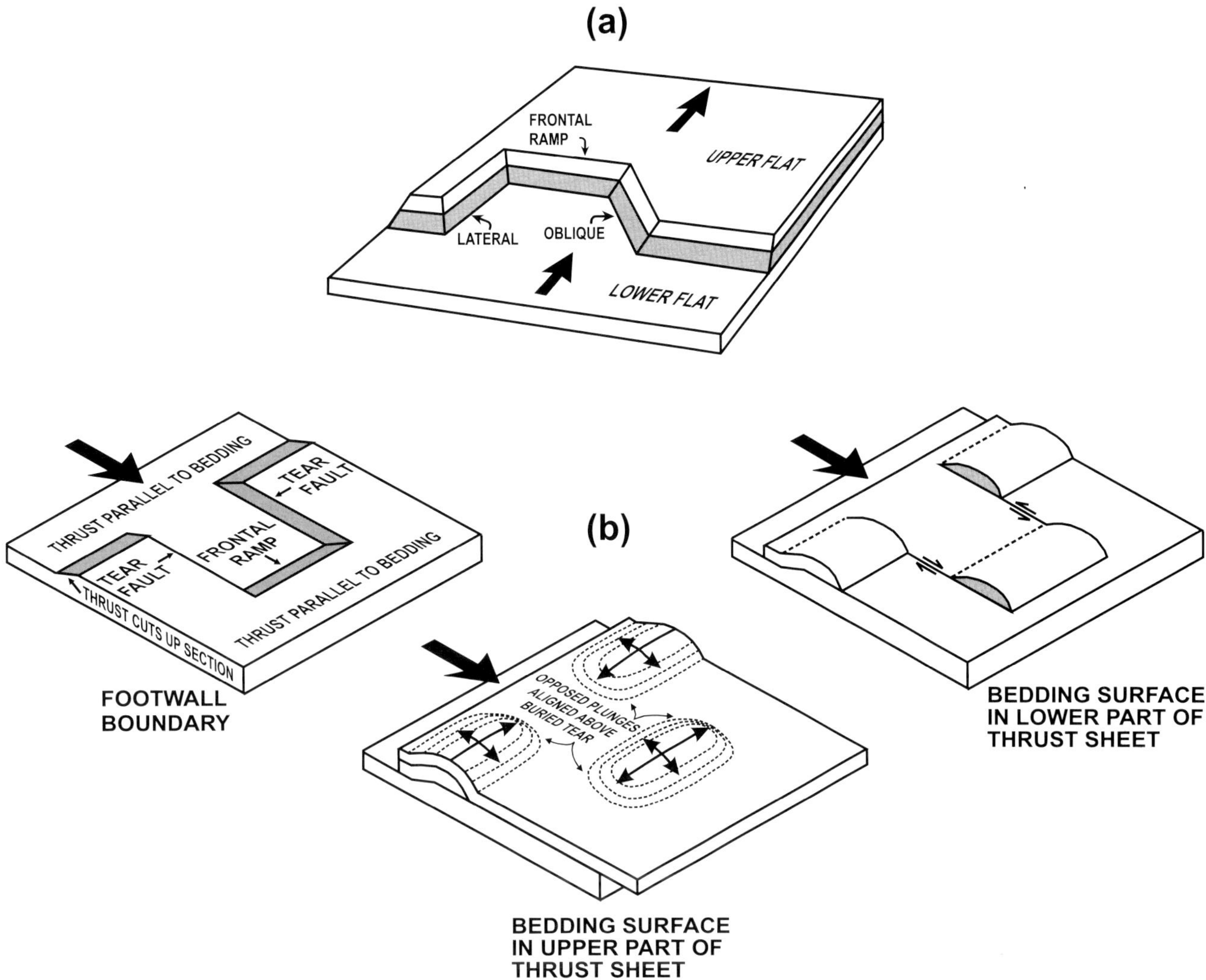

FIGURE 1. (a) Frontal, lateral, and oblique ramps on the footwall of a thrust fault (after McClay, 1991). If the lateral ramp had a vertical dip, it would be called a tear fault. Heavy arrows indicate transport direction. (b) Structures in a thrust sheet due to tear faults in the underlying thrust fault (modeled on the Brazeau Thrust in the Canadian Rocky Mountain Foothills; after Dahlstrom, 1970). In all cases the heavy arrow indicates the transport direction of the thrust sheet.

displacement over tear faults or lateral ramps. For example, the southward plunge at the south end of the Brazeau Range, the adjacent Panther Culmination with its northward plunge, and the eastward salient of the overlying McConnell Thrust sheet have all been interpreted to reflect a vertical tear fault in the underlying Brazeau Thrust fault (Dahlstrom, 1970; Figure 2). The large eastward salient of the central Lewis Thrust sheet is bounded to the north and south by lateral ramps. At North Kootenay Pass, structures in the Lewis sheet plunge steeply northward where the Lewis Thrust cuts up-section to the north on an oblique ramp (Price, 1964). At the southern boundary of the Lewis salient (at Marias Pass, not shown on Figure 2), the fault cuts up-section southward on a lateral ramp (Dahlstrom, 1970).

It is more difficult but nonetheless important to recognize smaller-scale and sub-surface lateral ramps and tear faults. For example, the Turner Valley Thrust cuts up-section to the south on a lateral ramp through Paleozoic carbonates at the south end of the Turner Valley structure (Figure 2). Fermor (1999) discussed some of these and other examples in some detail.

As is apparent from these examples, attention is often focused on hanging-wall structures associated with lateral ramps and tear faults. Perhaps because footwall exposure tends to be poorer, corresponding footwall structures and deformation are less well known. However, despite the fact that rocks in the footwall of a thrust fault tend to be less deformed than those in the hanging wall, transport of a thrust sheet over a footwall of complex geometry can produce local stress concentrations that could cause localized footwall deformation.

It can be argued that a thrust fault should adopt a configuration of thrust flats linked by frontal/lateral/oblique ramps and/or tear faults that minimizes the

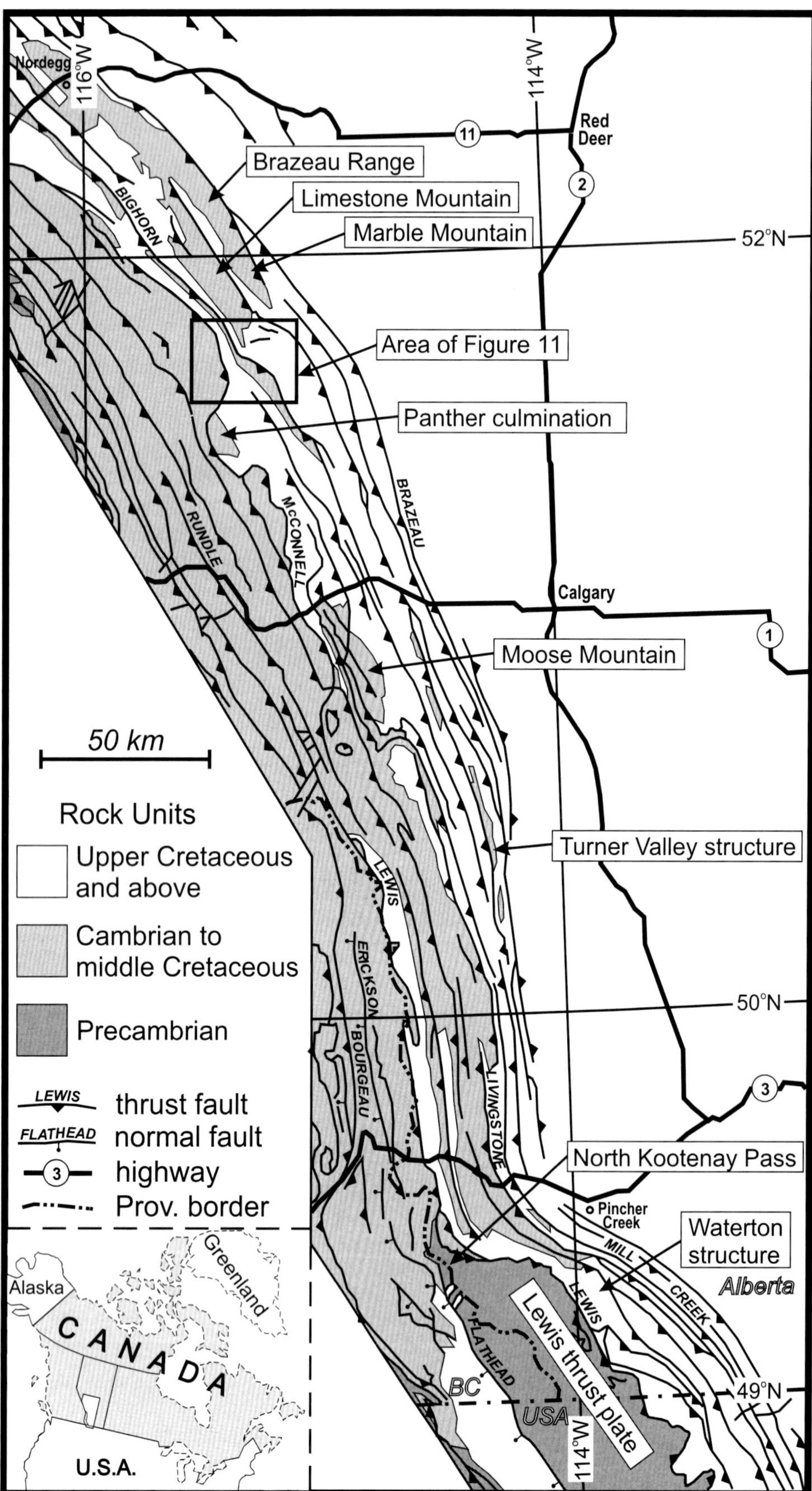

FIGURE 2. Simplified geologic map of part of the Canadian Rocky Mountains fold-thrust belt in southwestern Alberta, Canada, and adjacent areas of British Columbia and the U.S.A. (based on Wheeler and McFeely, 1991), showing structural culminations and other features related to transverse linkages between major thrusts, mainly at the level of the Paleozoic carbonate units. Inset map shows location within North America.

total strain energy associated with displacement of the thrust sheet and deformation of the hanging wall and footwall rocks (Elliott, 1976; Fermor, 1999). This does not imply that there should not be locally anomalous strain in the hanging wall and footwall (especially because the system's configuration evolves as thrust displacement increases). Also, as lateral and oblique ramps and tear faults represent deviations from the ideal cylindrical nature of fold-thrust structures, they have the potential to induce deviations from the ideal plane strain that characterizes the fold-thrust belt as a whole. This has implications for section balancing.

Lateral and oblique ramps also generate a more complex pattern of juxtaposition of units in hanging wall and footwall. The evolution of these structural and stratigraphic relationships in turn affects migration fluid pathways and trapping potential. Thus, it is important that the hydrocarbon explorationist have a good understanding of the details of kinematics associated with thrusting over nonlinear ramp systems.

Physical modeling offers a means to determine the geometry and kinematics of structures: The models have known initial configurations and are deformed under controlled boundary conditions; their evolution can be monitored; and the final geometry can be documented in detail from serial sections cut in different orientations. Here we report a series of models designed to document how motion of a thrust sheet over lateral ramps and tear faults is accommodated by deformation in the sheet and in the underlying footwall. By characterizing the expected suite of associated structures, we hope to make it easier to identify and interpret these features in nature.

In nature, lateral ramps and tear faults likely develop for a variety of reasons. They may be stratigraphically or lithologically controlled or influenced by preexisting structures, such as early normal faults. Alternatively, they may nucleate in otherwise laterally continuous strata as the thrust system evolves. The aim of this study is to document the effects of these transverse structures, rather than to investigate the factors that cause them. Nevertheless, the models have implications regarding the initiation of these structures.

MODELING PROCEDURE

The physical models were constructed of Harbutt's Gold Medal™ Plasticine and Dow Corning silicone putty (Dilatant Compound 3179) to represent part of a foreland stratigraphic succession at a scale of about 1:1,000,000. They were deformed at 2500–4000 ***g*** in a large centrifuge, to achieve dynamic scaling. The principles of scale modeling, the centrifuge-modeling technique, and the rheology of the model materials have been described elsewhere (Ramberg, 1981; Dixon and Summers, 1985, 1986; Dixon and Tirrul, 1991). The model ratios applicable to this work are listed in Table 1.

The overall dimensions and configuration of the model assembly are shown in Figure 3. Horizontal shortening of the layered foreland sequence is driven by gravitational collapse and spreading of the "hinterland wedge" of plasticine. Because the models are deformed in stages, the structural evolution can be monitored. Each stage corresponds to the amount of shortening that is achieved by flattening of the wedge at 4000 ***g***. The wedge must be reconstructed to restore its gravitational potential for a subsequent stage of shortening. Because the wedge spreads in the general direction of its topographic slope, and because the side-wall contacts between the foreland block and the walls of the model chamber are lubricated with petroleum jelly to reduce friction, the deformation of the model's foreland has overall plane-strain geometry.

The models have a foreland stratigraphic sequence consisting of four internally laminated units of alternating competence, resting on a rigid base. The stratigraphic column is illustrated in Figure 4a. The competent units (Units I and III) are composed of plasticine modeling clay, which represents competent rock units such as massive carbonates or coarse clastics. The incompetent units (II and IV) are made of interlayered laminae of silicone putty and plasticine, which simulate incompetent, thinly bedded lithologic units such as shale interbedded with limestone or sandstone.

This sequence approximates the overall mechanical characteristics of the stratigraphic succession in the eastern Front Ranges and Foothills of the southern Canadian Rocky Mountains. Unit I can be taken to represent the massive lower Paleozoic (Middle Cambrian–Mississippian) platform carbonates (Cathedral Formation to Rundle Group). Unit II represents the overlying, well-bedded and less competent Pennsylvanian to lower Upper Cretaceous terrigenous clastics (Rocky Mountain Supergroup to Wapiabi Formation). Unit III represents the coarse clastics of the Upper Cretaceous Brazeau Group, and unit IV the Tertiary Paskapoo Formation. Note that the model units correspond to gross mechanical units rather than formal stratigraphic units or sequences. At a

Table 1. Applicable model scaling ratios.

Quantity	*Ratio, model/prototype*	*Equivalence (model ≡ prototype)*
Length	$l_r \approx 1.0 \times 10^{-6}$	1 mm ≡ 1 km (order-of-magnitude only)
Specific gravity (mass)	$\rho_r = 0.6$	1.60 ≡ 2.67 (bulk value for whole stratigraphic column)
Time	$t_r \approx 1.0 \times 10^{-11}$	1 hr ≡_12 m.y.
Strain rate	$\varepsilon_r \approx 1.0 \times 10^{11}$	$10^{-3}\ s^{-1} \equiv 10^{-14}\ s^{-1}$ (for example)
Effective viscosity	$\mu_r \approx 2.4 \times 10^{-14}$	2.4×10^{2} Pa s ≡ 10^{16} Pa s (for example)
Acceleration	$a_r = \mu_r/(\_{r}\ l_r\ t_r) = 4.0 \times 10^{3}$	4000 ***g*** ≡ 1 ***g***
Stress	$\sigma_r = \rho_r\ l_r\ a_r \approx 2.4 \times 10^{-3}$	(calculated from other ratios)

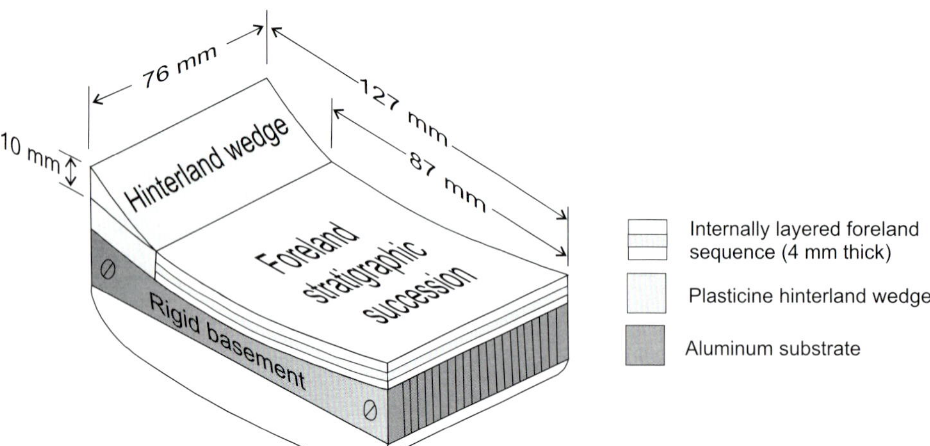

FIGURE 3. Schematic diagram of model assembly. The basement block is composed of aluminum plates, each 3 mm thick, which can be separated so that the model can be cut into transverse serial sections at that spacing. Details of the foreland stratigraphy and structure are shown in Figure 4.

model length ratio of approximately 1:1,000,000, the 4-mm-thick stratigraphic column in the models represents roughly 4 km in nature. Given other uncertainties, the scaling is sufficiently flexible that the models can reasonably be applied to prototype systems that are larger or smaller by a factor of two or so.

Our aim was to see the structural effects of ramp systems of specific geometries. In similar models with planar, laterally continuous layering (e.g., Dixon and Liu, 1991), folds tend to be linear, and thrust systems comprise flats in incompetent units and fairly linear foreland- and hinterland-verging frontal thrust ramps that transect the competent units. Although displacement transfer does occur at lateral tip lines (Dixon and Liu, 1991; Aydemir, 1998), lateral and oblique ramps typically do not develop spontaneously in the centrifuge models. Therefore we needed to force this geometry for the purpose of our study.

The desired ramp system was initiated by including a precut discontinuity during model assembly. Unit I was prepared in two separate pieces of contrasting colors (so that the hanging-wall and footwall portions would be readily distinguishable). Before the two pieces were assembled, the matching hanging-wall and footwall cutoffs were carved with a knife to the desired shape. The ramp surfaces and the lower flat (between unit I and the base plate) were coated with petroleum jelly to ensure that the precut discontinuity had low cohesion.

We describe four models (designated MO28 to MO31; Figure 4b) in this paper. The lowest stratigraphic unit contained the precut, foreland-verging thrust-ramp system, consisting of two frontal segments that dipped at 22° connected by a transverse segment. The frontal segment closer to the foreland side of the model will be referred to as the leading frontal ramp, and that closer to the hinterland side will be called the trailing ramp. Two models were constructed in each of two configurations (Table 2 and Figure 4b): (1) a transverse tear-fault with vertical dip (MO28 and MO29); and (2) a lateral ramp dipping at 22° (MO30 and MO31). Similar models with oblique ramps will be described elsewhere.

Each model underwent three stages of shortening that were designated stage I, II, and III. After each stage, the top surface was photographed using oblique lighting to enhance surface relief, and several transverse sections were cut and photographed. After the final stage, the models of each ramp configuration were cut into either transverse serial sections or combined longitudinal and transverse serial sections (Table 2 and Figure 4b), so that the deformation associated with the transverse structure could be well constrained. We tried to achieve equal total shortening in each pair of models that had the same initial ramp geometry (Table 2), so that the same final structural configuration would be sampled by the transverse and longitudinal sections.

We will examine the two ramp configurations individually, focusing on hanging-wall structures, as seen on the top surface, and on hanging-wall and footwall structures, as seen in transverse and longitudinal sections. In the following description, we adopt the convention that the overthrusting verged from west to east. The locations of transverse sections are designated in millimeters (7.0, 10.0, 13.0, . . . , 67.0) relative to the south edge of the model, and transverse sections are viewed looking north (i.e., with the hinterland on the left). The positions of longitudinal sections are designated in millimeters (3, 6, 9, . . . , 54) relative to the front edge of the hinterland wedge, and longitudinal sections are viewed looking west (i.e., toward the hinterland).

TRANSVERSE TEAR FAULT

Models MO28 and MO29 contained two frontal-ramp segments linked by a transverse tear fault, in the initial configuration shown in Figure 3b. This configuration may be akin to the south end of the Brazeau

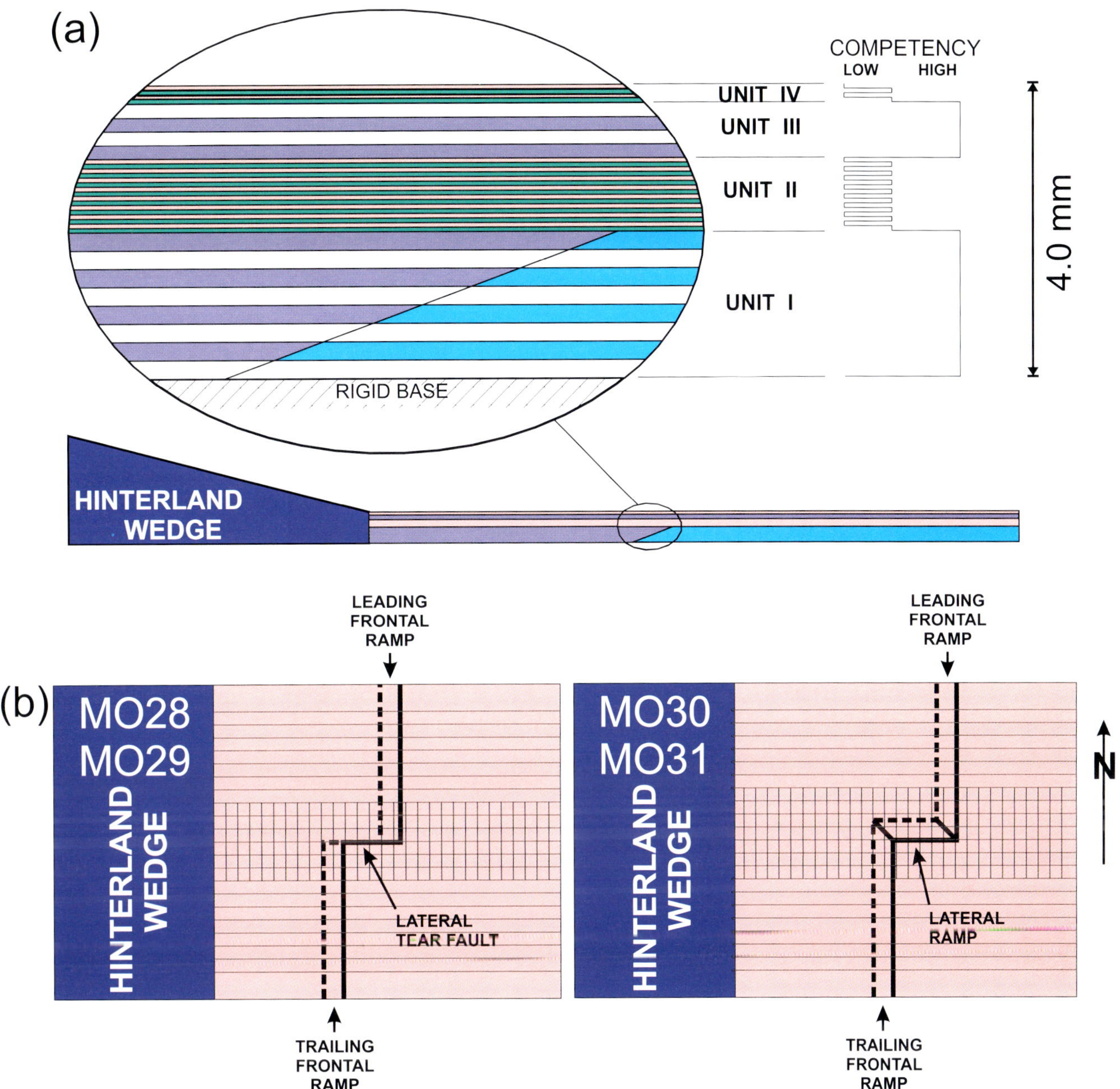

FIGURE 4. (a) Schematic transverse section of models MO28–MO31. Enlargement shows internal stratification and position of the trailing frontal ramp (a precut discontinuity in unit I). Relative competence of units is shown by stratigraphic column at right. (b) Schematic plan views of the two model configurations, showing positions of ramp systems in unit I (top of ramp = heavy solid line; base of ramp = heavy dashed line). Light lines show transverse (west-east) and longitudinal (north-south) sections to be cut at 3-mm spacing after deformation. Note that longitudinal sections are shown schematically, because the models become foreshortened during deformation.

Range near Marble Mountain, as interpreted by Dahlstrom (1970) and Fermor (1999); see Figure 1b.

Surface Structures

The evolution of the surface structures associated with the transverse tear fault in model MO28, through stages I to III, is shown in Figure 5a, c, and e. The initial shortening (stage I) produces two prominent linear ridges (Figure 5a), which are surface expressions of ramp anticlines forming above the frontal ramps. The ridges have abrupt rectangular terminations at the tear fault, where the surface layers of the model dip to the north and south, respectively.

As shortening increases (stage II, Figure 5c), the ridges increase in width as a result of growth of the frontal-ramp anticlines. The model's surface is also marked by parallel ridges and troughs of shorter wavelength

Table 2. Index of models: configuration of model ramp systems and section layout, final foreland width (initial width 86 mm in all cases), and total foreland shortening after three stages of shortening.

Model #	*Ramp type, relative strike*	*Dip of transverse ramp*	*Final width of foreland*	*Total shortening*	*Section array*
MO28	lateral tear fault, parallel to transport direction	vertical	61 mm	25 mm	transverse
MO29			65 mm	21 mm	longitudinal and transverse
MO30	lateral ramp, parallel to transport direction	22°	58 mm	28 mm	longitudinal and transverse
MO31			58 mm	28 mm	transverse

that reflect buckling of the upper competent unit III. Most of these folds have linear, horizontal hinges, but those near the ends of the frontal-ramp anticlines plunge north or south away from the ridges and display a subtle fanning of their hinge trends. This pattern is analogous to the southward plunge at the south end of the Brazeau Range and the northward plunge of the Panther culmination (Figure 2), which are attributed by Dahlstrom (1970) to a tear in the underlying Brazeau fault.

These patterns are more evident by the final stage (stage III, Figure 5e [Model MO28] and 6a [Model MO29]). In addition, the leading edges of the frontal-ramp anticlines are marked by a persistent short-wavelength trough-ridge pair that is the surface expression of a fault-propagation fold above the tip of the main thrust (as will become evident in transverse serial sections). Another notable feature of the surface structure is that, although the precut frontal ramps were straight, the ridges that formed above the frontal ramps become curved in plan, and are convex toward the foreland (Figures 5a, c, and e, and 6a). As will be shown below, this curvature reflects nonideal behavior in the vicinity of the tear fault.

Transverse Serial Sections Remote from the Tear Fault

We will look first at the general pattern of structural evolution remote from the tear fault, and then will examine how the structures in its immediate vicinity differ.

Transverse sections show that material in the hanging wall of the main ramping thrust undergoes a combination of thrust transport, folding, and layer-parallel shortening. During stage I (Figure 5b), the thrust sheet is translated rigidly on the floor thrust and undergoes fault-bend folding as it moves up the pre-cut frontal ramps. The leading tip of the fault begins to propagate upwards through unit II, and the leading edge of the thrust sheet develops into a fault-propagation fold.

During stage II (Figure 5d), rigid translation increases, but the thrust sheet is also internally deformed by layer-parallel shortening and minor buckling in unit I and minor buckling in the upper units. On the foreland side of the main ramp, some shortening is accommodated by buckling of units III and IV above a décollement horizon in unit II. The main thrust branches into two ductile shear-zone splays: a minor one with steep dip that dies out upward into the fault-propagation fold, and a major flat décollement zone within unit II that is overlain by a train of detachment folds in units III and IV.

These patterns are amplified in stage III. Figure 5f shows a series of nine serial sections through the central portion of model MO28. Sections 28.0 to 36.0 transect the trailing frontal ramp, and sections 40.0 to 46.0 transect the leading frontal ramp. Figure 6b shows four equivalent transverse sections through model MO29, with sections 22.0 and 25.0 through the trailing ramp and 46.0 and 49.0 through the leading ramp. Apart from the lateral position of the ramp, all of these sections are very similar. The fault-propagation fold in units III and IV (the trough-ridge pair on the foreland side of the main ramp anticline referred to above) develops strong foreland vergence and an overturned forelimb as the upper (ductile) splay of the main thrust develops.

Several new features develop during stage III. First, the leading edge of unit I in the hanging wall becomes an overturned fold. This structure resembles the overturned leading edge of Paleozoic strata in some Rocky Mountains thrust sheets such as the Brazeau sheet at Marble Mountain in the Canadian Rocky Mountain Foothills (Figure 2). Second, the portion of unit I that overlies the lower flat (i.e., on the hinterland side of the frontal ramps) shortens internally by conjugate thrusting on steeply dipping thrusts of small displacement. Finally, on the foreland side of the main frontal ramps,

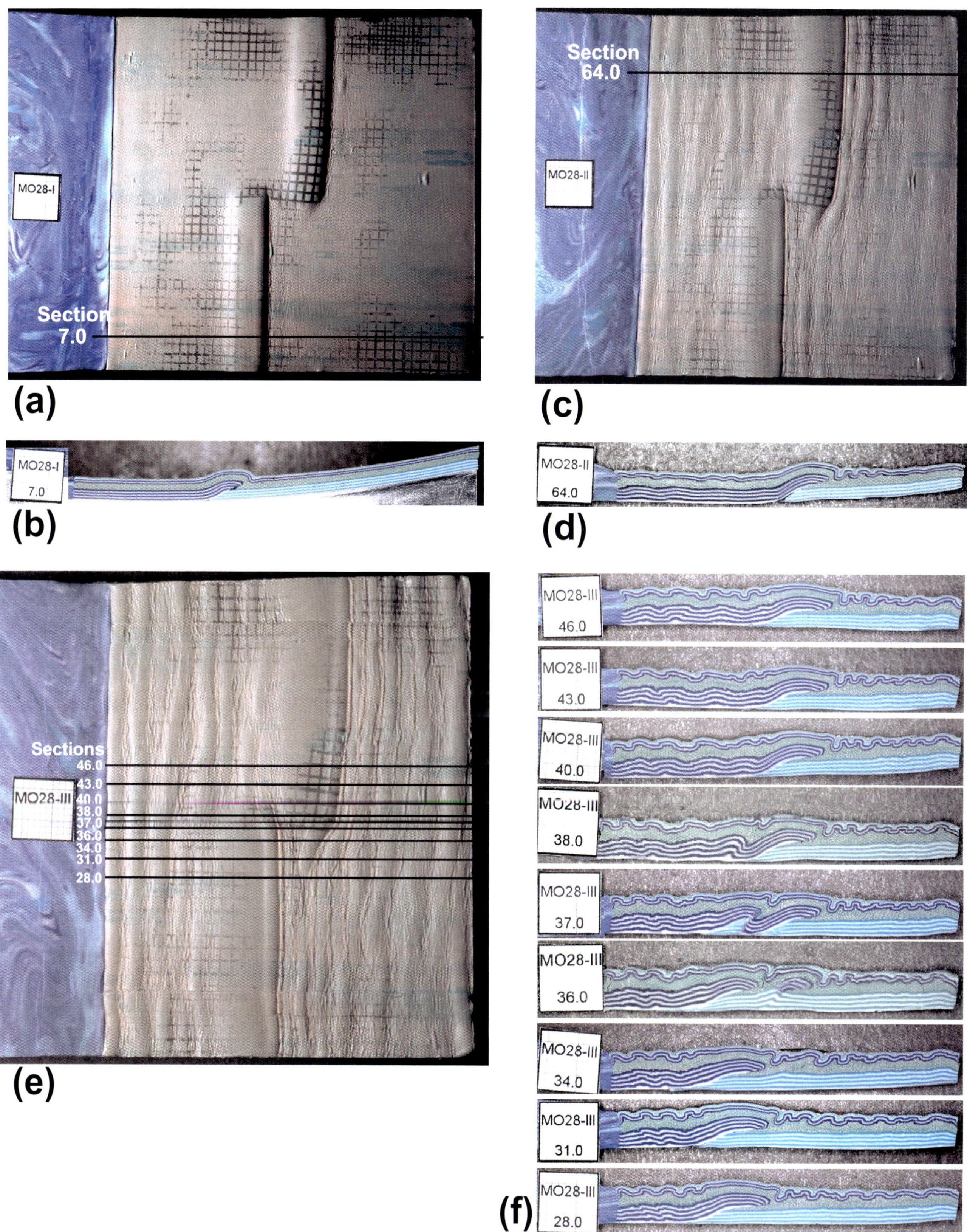

FIGURE 5. Model MO28, with a transverse tear fault linking two frontal ramps. (a) Top surface view (illuminated from the left to enhance surface relief) at stage I. (b) Transverse section 7.0 at stage I. (c) Top surface view at stage II. (d) Transverse section 64.0 at stage II. (e) Top surface view at stage III. (f) Serial transverse sections 28.0 to 46.0 at stage III. Labels are marked in millimeter squares for scale. Horizontal lines in (a), (c), and (e) indicate the positions of transverse sections shown in (b), (d), and (f).

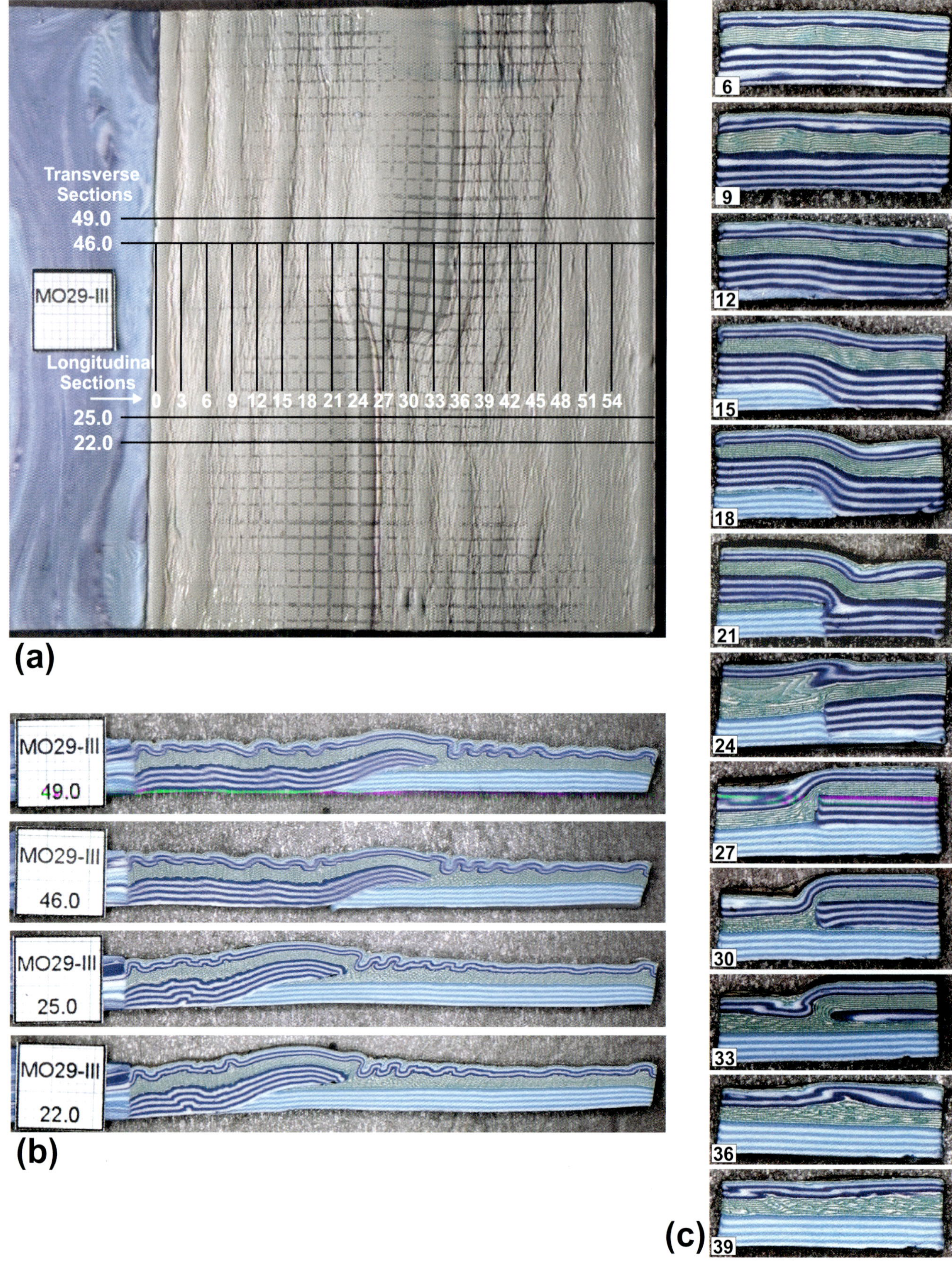

Transverse Sections
49.0
46.0
MO29-III
Longitudinal Sections
0 3 6 9 12 15 18 21 24 27 30 33 36 39 42 45 48 51 54
25.0
22.0
(a)
MO29-III 49.0
MO29-III 46.0
MO29-III 25.0
MO29-III 22.0
(b)
6
9
12
15
18
21
24
27
30
33
36
39
(c)

unit I in the footwall undergoes some layer-parallel shortening but is otherwise essentially unstructured.

Transverse Serial Sections near the Tear Fault

We will now focus on the area near the tear fault (which was initially situated in the position of transverse section 37.0). Sections 34.0, 36.0, 37.0, and 38.0 of Model MO28 (Figure 5f) have features that are directly associated with the tear fault.

The most notable feature of section 37.0 (Figure 5f) is that it intersects both of the frontal footwall ramps and the tear fault. Likewise, section 36.0 intersects both of the frontal hanging wall ramps. These observations are unexpected, because we know that the tear fault was initially oriented parallel to the orientation of these transverse sections. For both the trailing and leading frontal ramps and the tear fault to be intersected by the same transverse section, the tear fault must have been rotated clockwise about a vertical axis. This rotation was a result of northward extension of unit I in the footwall of the trailing frontal ramp and southward spreading of unit I in the hanging wall of the leading frontal ramp during thrust displacement, possibly as a result of gravitational spreading of the two ramp-anticlinal culminations. Evidence of flattening and northward extension of the block in the footwall of the trailing ramp is also found in the longitudinal sections of model MO29 (see below). The rotation of the tear fault and the longitudinal spreading in footwall and hanging wall require that, in the vicinity of the tear fault, the strain and displacement deviated from the ideal 2-D pattern.

In an ideal system of frontal ramps linked by a transverse tear fault, one would expect the displacement on the two frontal segments to be equal and constant, because there is one upper plate and one lower plate, and the two frontal segments do not overlap across strike. In other words, displacement should be completely transferred from one frontal ramp to the other, along the tear fault. In the model, however, along-strike stretching of the hanging-wall panel of the leading ramp and the footwall panel of the trailing ramp causes the two frontal-ramp segments to evolve to an overlap situation. Also, displacement is transferred gradually because of along-strike displacement gradients on the frontal ramps. The overall plan curvature (which is convex to the foreland) of the two ramp-anticline segments (see above and Figure 5e) reflects these gradients. Contraction is also accommodated by buckling and layer-parallel strain, both of which can vary independently both along and across strike. The strain pattern is clearly more complex in detail than would be expected in an idealized system.

There are two other anomalous features associated with the tear fault. Not only have the leading ramp's hanging wall and the trailing ramp's footwall been extended along strike, but the two frontal ramps have also propagated along strike past their initial terminations against the tear fault. Figure 5f shows that the leading frontal ramp has propagated southward along strike through unit I in section 36.0 (where it is clearly visible as a thrust that cuts the full thickness of unit I) and into the upper part of unit I in section 34.0 (where it is present as an upwarp in the upper few beds of unit I). Likewise, in section 38.0, we see a prominent anticline in unit I to the west (i.e., on the hinterland side) of the leading frontal ramp. This fold appears to be a fault-propagation fold: It has an overturned forelimb through which a foreland-verging thrust fault is propagating. This structure is the northward-propagating tip (north of the tear fault) of the trailing frontal footwall ramp. The along-strike displacement gradients and the lateral propagation of the two frontal ramps are shown schematically in Figure 7.

Lateral propagation of the two frontal ramps across the tear fault — one northward into the thrust sheet and the other southward into its footwall — simplifies the geometry of the thrust system. The tear fault is being abandoned in favor of two imbricate listric thrusts that root into a common décollement. The reorganization may be particularly favored when the transverse tear fault is rotated out of the transport direction as a result of gravitational spreading of the structural culminations above the frontal ramps, as discussed above. Once the tear is abandoned, the new configuration has the geometry of a lap-joint displacement-transfer zone (Dahlstrom, 1970).

Longitudinal Sections across the Tear Fault

Models MO28 and MO29 had the same initial configuration and similar shortening histories, and the three-dimensional structure in the vicinity of the tear

FIGURE 6. Model MO29 at stage III. (a) Top surface. Black lines indicate positions of transverse sections 22.0 to 49.0 (shown in b) and longitudinal sections 0 to 54 (see part c). (b) Transverse sections 22.0 to 49.0. Labels are marked in mm squares for scale. (c) Longitudinal sections 6 to 39 (hinterland to foreland) through the central portion of the model. The observer is facing toward the hinterland. The sections are 18 mm in length and spaced at 3-mm intervals. See text for discussion.

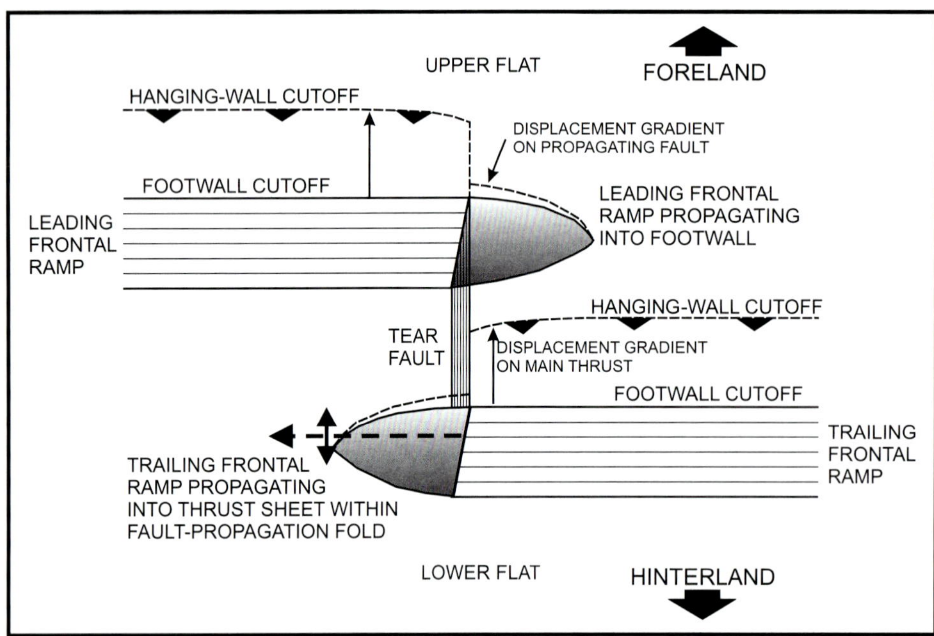

FIGURE 7. Schematic diagram (vertical view) showing fault configuration and displacement gradients in a system consisting of two frontal ramps linked by a transverse tear fault (based on model MO29, Figure 5). The ramp system is marked by ruling that simulates structure contours; the tear fault is drawn with a steep rather than vertical dip to aid visualization. Solid lines with barbs represent hanging-wall cutoffs of the bedding plane that initially would have corresponded with the top of the ramp system. Note displacement gradients adjacent to the tear fault. The shaded areas represent the laterally propagating frontal ramps (also with displacement gradients). No attempt is made to show rotation of the tear fault or local deformation of the footwall adjacent to the intersection of the trailing frontal ramp and the tear fault, both of which are features of the physical model.

fault can be further constrained by viewing longitudinal sections of MO29 together with transverse sections 28.0 to 46.0 of MO28. However, the models are not identical (partly because of slightly different total shortening), so the different views cannot be combined rigorously. After stage III, longitudinal sections were cut at 3-mm intervals through the central portion of model MO29, between transverse sections 28.0 and 46.0, in the positions shown in Figure 6a. Sections 6 through 39 are illustrated in Figure 6c (viewed toward the hinterland).

Sections 6 and 9 (Figure 6c) show the full stratigraphic column in the thrust sheet, on the hinterland side of the ramp system, and give distorted views of small-displacement thrusts in unit I and folds and minor thrusts in units II–IV that are better seen in the transverse sections (Figure 6b). The main thrust sheet climbs the trailing frontal ramp through unit I and into unit II in sections 12 to 18. It is deformed into a progressively steepening monocline that faces to the north toward the lower flat behind the leading frontal ramp. Beneath the monocline, there is significant deformation of the footwall adjacent to the precut tear fault. In section 15, the footwall material has been compressed vertically and extruded northward beneath the monocline.

Section 21 catches the full height of the tear fault in unit I. To the south, the hanging wall includes most of the stratigraphic column (the lowest bed of unit I is missing) in thrust contact with the lower half of unit II and all of unit I in the footwall. To the north, the thrust sheet contains the full stratigraphic column in the hanging wall on the lower flat behind the leading frontal ramp. At the transition, the thrust sheet has a steep, north-facing monocline in its upper units. However, the lower part of unit I is offset by a south-dipping reverse fault that must have formed by upward propagation of the tear fault in a scissoring motion as the south part of the sheet climbed the trailing ramp (e.g., Dahlstrom, 1970) (Figure 1b).

The south part of section 24 (Figure 6c) is an oblique longitudinal section through the overturned fault-propagation fold that wraps around the leading edge of unit I above the trailing frontal ramp. The canoe-shaped closure in unit II arises because this arcuate, overturned anticline is decapitated by the plane of section. At a higher level, the Z-shaped closure in unit III indicates that here the fold is oblique to the plane of section. The disrupted area in unit II beneath the canoe-shaped closure marks the position of the upward-propagating thrust fault (compare with transverse sections 31.0 and 34.0 in Figure 5f).

Sections 24 to 33 (Figure 6c) catch the thrust sheet as it climbs the leading frontal footwall ramp through unit I and into unit II. The monocline in the upper units of the thrust sheet dips to the south in this region, and Figure 6a shows that surface folds plunge to the south. In sections 24 to 30, unit II has been vertically offset (north side up) by a fault that has propagated upward above the precut tear fault as unit I in the hanging wall was lifted during displacement up the leading frontal ramp. However, units III and IV are simply draped across the tear fault as a south-facing monocline. Section 27 is located at the top of the leading frontal ramp.

Here, the thrust carries the full stratigraphic column in its hanging wall above the full thickness of unit I in the footwall. This section is in a location equivalent to the position of the upwarp of footwall unit I seen in sections 36.0 and 34.0 of MO28 (Figure 5f; see above). Footwall unit I in MO29 is about 10% thicker to the south of the tear fault in section 27 than in section 30 (Figure 6c). This probably reflects southward propagation of the leading frontal ramp across the tear fault, although perhaps to a lesser extent here than in model MO28 (see above).

From sections 30 to 33, the thickness of unit I in the hanging wall decreases as the thrust cuts higher on the hanging-wall ramp. There is a longitudinal gradient as well as a transverse one: Note that in section 30, the fault carries all eight beds of unit I at the north end of the section, but to the south (adjacent to the tear fault) it carries only the six upper beds. In section 33, there is a southward-facing closure of the uppermost three beds of unit I. This is a highly oblique section through the overturned fold at the leading edge of unit I (compare with transverse section 40.0 in Figure 5f). The obliquity is a consequence of the arcuate shape of this folded leading edge in plan (Figure 6a).

In sections 36 and 39 (Figure 6c), undeformed strata of unit I constitute the footwall beneath a zone of décollement in unit II, which is overlain by folded and thrusted units III and IV (compare with transverse sections 46.0 and 49.0 in Figure 6b). The folds and thrusts are seen in these longitudinal sections because they are arcuate in plan and therefore trend slightly obliquely to these section planes (see Figure 6a).

LATERAL RAMP

Models MO30 and MO31 were constructed with two frontal-ramp segments linked by a lateral ramp with a dip of 22° (Figure 4). Figures 8 and 9 show top-surface views and transverse and longitudinal sections of these models. The following discussion will focus on comparisons with the tear-fault models.

Surface Structures

The surface structures that developed in model MO31 through three stages of deformation (Figure 8a–c) and

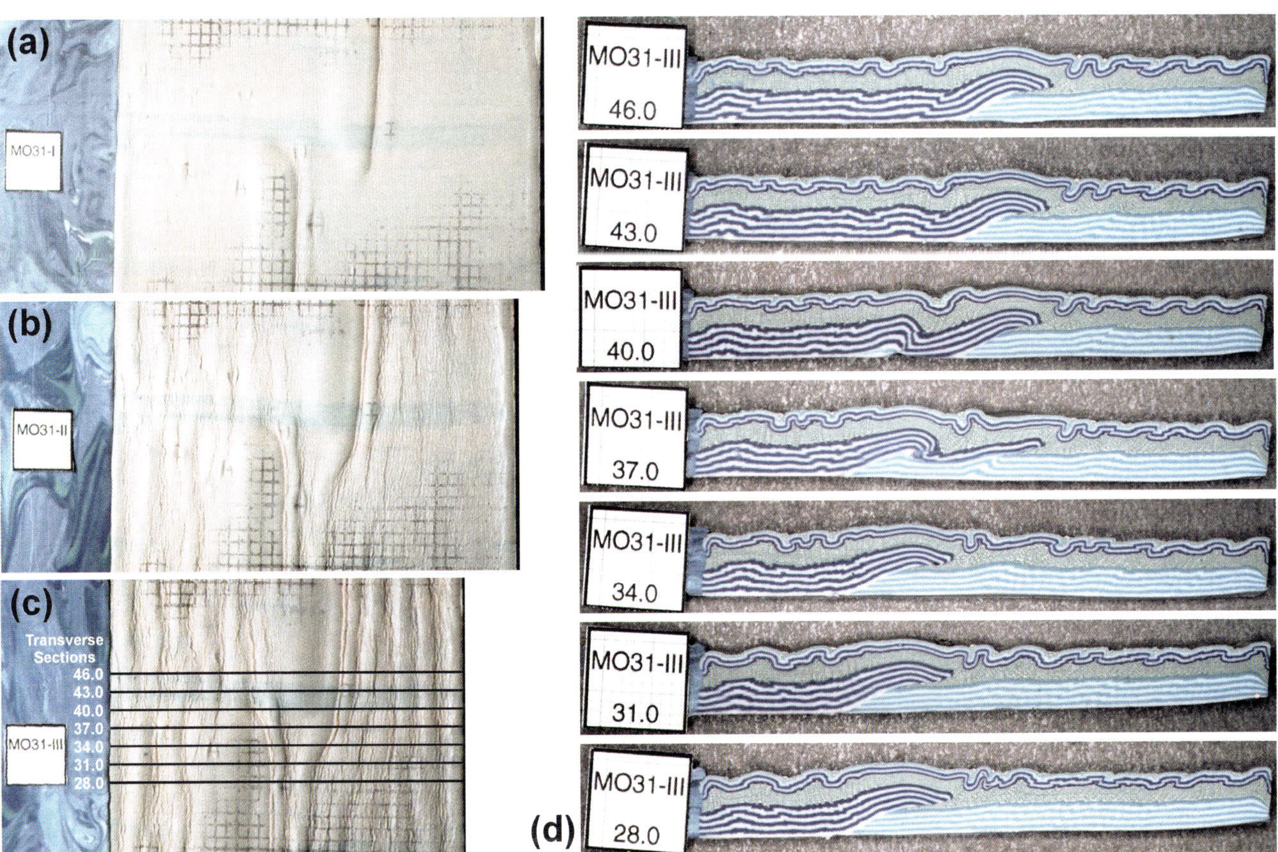

FIGURE 8. Model MO31, with a lateral ramp linking two frontal ramps. (a)–(c) Surface views (illuminated from the left to enhance topographic relief) of the central portion of the model after stages I to III. Black lines in (c) show positions of transverse sections. (d) Transverse serial sections through model MO31 at stage III. The labels are marked with 1-mm squares for scale. See text for discussion.

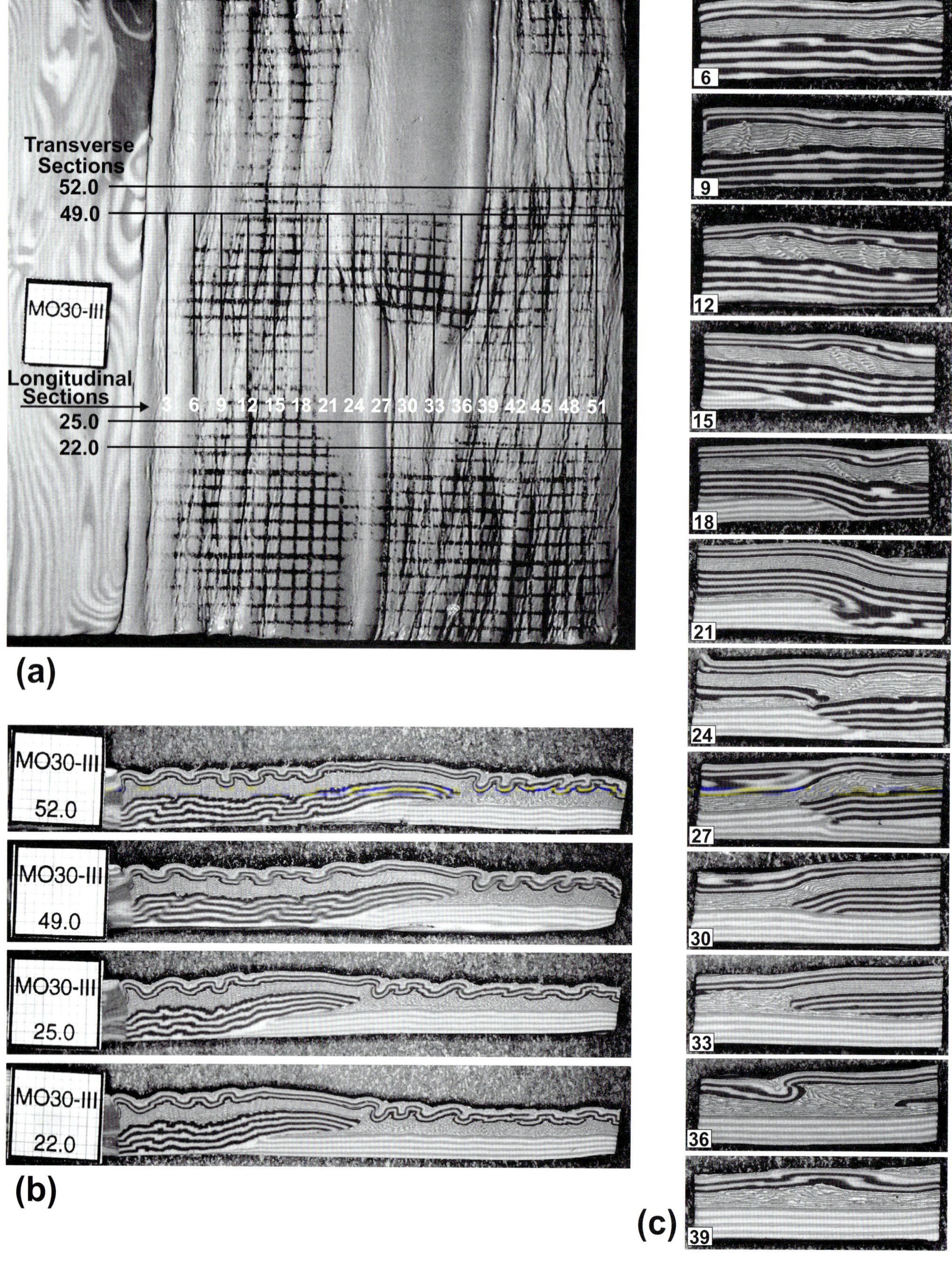
Transverse Sections
52.0
49.0
MO30-III
Longitudinal Sections
25.0
22.0
3 6 9 12 15 18 21 24 27 30 33 36 39 42 45 48 51
(a)
MO30-III 52.0
MO30-III 49.0
MO30-III 25.0
MO30-III 22.0
(b)
6
9
12
15
18
21
24
27
30
33
36
39
(c)

MO30 at stage III (Figure 9a) are very similar to those of the tear-fault models (Figures 5 and 6). One subtle difference is visible in the early stages of thrust displacement. The two major culminations above the frontal ramps have wedge-shaped terminations adjacent to the lateral ramp (Figure 8a and b), in contrast to the square terminations adjacent to the tear fault (Figure 5a and c). This arises because the respective lines of intersection between the fault plane on the lateral ramp and the fault planes on the frontal ramps trend obliquely to the transport direction, so the hinges of the associated fault-bend folds in the thrust sheet also trend obliquely.

Transverse Serial Sections Remote from the Lateral Ramp

Transverse sections 28.0, 31.0, 43.0, and 46.0 of model MO31 (Figure 8d) and 22.0, 25.0, 49.0, and 52.0 of model MO30 (Figure 9b) intersect the frontal ramps some distance along strike from the lateral ramp. The structures resemble those in equivalent positions in the tear-fault models discussed above (Figures 5f and 6b).

Transverse Sections near the Lateral Ramp

Sections 34.0 and 40.0 of model MO31 are situated on either side of the lateral ramp, and section 37.0 intersects it at about three-quarters of its height (Figure 8d). Apart from the fact that the lateral ramp dips gently northward, whereas the tear fault in MO28 is vertical, the associated structures are remarkably similar. For example, section 40.0 is located north of the base of the lateral ramp and should only intersect the leading frontal ramp. However, in that section there is a small piece of the trailing ramp in the footwall of the thrust. This requires that the lateral ramp has been deflected clockwise from its original orientation (striking parallel to the transport direction) and that the footwall of the trailing ramp was extended to the north. There is also evidence of lateral propagation of the two frontal ramps, the leading one to the south into the footwall (in section 34.0) and the trailing one to the north as a fault-propagation fold with an overturned forelimb in the hanging wall (in section 40.0). Thus, it appears that the lateral ramp (like the tear fault) is an unstable configuration that tends to be replaced by overlapping imbricate thrusts rooted on the same décollement. The principal difference between the tear-fault and lateral-ramp models is that the along-strike gradients are steeper in the former: Structural changes that occur between sections 36.0 and 38.0 in MO28 are equivalent to those between sections 34.0 and 40.0 in MO31.

Longitudinal Sections across the Lateral Ramp

The central portion of Model MO30 was sectioned longitudinally at 3-mm spacing (Figure 9a and c). The overall structural pattern is similar to that of the tear-fault case (Figure 6c), except that the along-strike gradients are gentler because of the low (22°) dip of the lateral ramp.

Sections 6 and 9 intersect the thrust sheet on the hinterland side of the trailing ramp. Sections 12 to 21 intersect the trailing frontal ramp as the thrust climbs progressively through unit I and into the upper flat in unit II. Sections 18 and 21 show the full height of the lateral ramp. The gentle northward drape of the thrust sheet contrasts with the steep, north-facing monocline in the tear-fault model.

Sections 24 to 36 intersect the leading frontal ramp as the thrust climbs progressively through unit I and into the upper flat in unit II. Hanging-wall strata at the lateral-ramp cutoff of unit I have become a gently south-dipping monocline in sections 30 and 33. This lateral fold closure, coupled with the overturned fold at the frontal-ramp hanging-wall cutoff, gives the thrust sheet excellent three-dimensional structural closure in this area.

The longitudinal sections of model MO30 provide useful constraints on the 3-D geometry of several structures seen in transverse sections of the central portion of MO31. The fault in the overturned forelimb of the fault-propagation fold in sections 37.0 and 40.0 (Figure 8d) was interpreted to represent northward-propagation of the trailing frontal ramp. Longitudinal section 21 (Figure 9c) transects an equivalent structure in model MO30: It appears as a flat fault in the lower part of hanging-wall unit I, north of the lateral ramp. The trace of this fault drops to a lower elevation to the north (left). This would be consistent with the propagating fault having a slightly oblique strike (to the north-northeast)

Figure 9. Model MO30 (lateral ramp linking two frontal ramps) after stage III. (a) Top surface (illuminated from the left to enhance surface relief). Black lines mark transverse and longitudinal sections. (b) Transverse sections 22.0 to 52.0, remote from the lateral ramp. Labels are marked with 1-mm squares for scale. (c) Longitudinal sections 6 to 39 (hinterland to foreland) through the central portion of model MO30 after stage III, showing structures associated with the lateral ramp that links two frontal ramps. The observer is facing toward the hinterland. Each section is 21 mm in length. See text for discussion.

relative to the north-striking frontal ramp. The leading frontal ramp was inferred to be propagating laterally to the south as a foreland-verging thrust in footwall unit I in sections 37.0 and 34.0 of MO31 (Figure 8d). In longitudinal section 27 of model MO30 (Figure 9c), the equivalent structure may be represented by disrupted layering in the middle of footwall unit I beside the lateral ramp.

The southern hanging-wall panel of unit I overlaps the northern one above the lateral ramp in section 24 of MO30 (Figure 9c). This suggests that the structural culmination above the trailing frontal ramp has undergone some northward stretching. This probably resulted from northward gravitational spreading of the southern culmination.

Finally, it was pointed out above that the structural culmination above the leading frontal ramp is gently curved in plan (Figure 9a), and is convex toward the foreland. The leading edge of unit I occurs in a different longitudinal position in sections 33 and 36 (Figure 9c), reflecting its oblique trend relative to the section planes in that area.

AN EXAMPLE FROM THE CANADIAN ROCKY MOUNTAINS: THE BRAZEAU THRUST

Most of the major oil and gas fields that Dahlstrom (1970) identified in the fold-and-thrust belt occur in structural culminations or duplexes that terminate along strike (Figure 2). However, detailed subsurface studies to determine whether a particular structure is bounded by blind tear faults or lateral ramps are rare. The southern edge of Paleozoic strata carried in the Brazeau Thrust sheet, between the Limestone Mountain culmination and the Williams Creek syncline, was cited by Dahlstrom (1970) (Figures 1b, 2) as an example of a tear fault. This feature has come to be known as the Marble-Mountain "swing-back" of the Brazeau Thrust.

Bégin and Spratt (2002) examined a portion of the southern edge of the Brazeau sheet in detail, using 2-D seismic data, well logs, and surface mapping (Figure 10). In transverse (dip) section (e.g., dip line JR-101, Figure 11a), the thrust sheet above the main frontal ramp of the Brazeau Thrust is cut by fore and back thrusts, and it carries a fault-propagation fold associated with the overlying James Thrust. These fold and fault structures within the sheet and above its major frontal (footwall) ramp are very similar to those observed in transverse section 43.0 of model MO31 (Figure 8d).

Bégin and Spratt (2002) focused on the geometry of the Marble Mountain swing-back structure, as seen in two longitudinal (strike) sections interpreted from 2-D seismic lines JR-110 and JR-111 (Figure 11b and 11c). At its eastern end, the structure is interpreted to have a steep southeasterly dip. At the level of the Paleozoic strata, this interpretation resembles a section drawn in the same location by F.R. Frey of Shell Canada Ltd in 1979 (but first published in a paper by Fermor [1999]). However, in Bégin and Spratt's interpretation, the fault is inferred to die out upward and link into a roof thrust, such that overlying strata are draped over it from north to south. On the other hand, Frey showed the steep fault cutting higher-level thrust faults as well as young strata in the Brazeau Thrust sheet. Section JR-110 (Figure 11b) resembles sections 27 and 30 of model MO29 (Figure 6c). In particular, it is notable that the model fault in section 27 developed an overturned (southward) dip. In both sections 27 and 30 the fault dies out upward and does not cut the high-level strata, which merely drape southward in a lateral culmination wall or south-facing monocline.

Along the southwestern portion of the Limestone Mountain culmination (intersected by dip line JR-111, Figure 11c), Bégin and Spratt interpret the Brazeau Thrust as having a lateral flat-ramp-flat geometry. The Limestone Mountain structure is interpreted to have been elevated on a lateral ramp that cut up-section to the southeast, with the Brazeau Thrust merging with the James Thrust (roof thrust). Section JR-111 generally resembles section 27 of model MO30 (Figure 9c), although the physical model is simpler in that it lacks multiple detachment horizons.

The contrast between sections JR-110 and JR-111 (Figure 11b and 11c) suggests that the dip of the northeast-striking transverse structure steepens and even becomes overturned, along strike, from southwest to northeast, as Figure 12 illustrates schematically. At first glance, the change in dip resembles the dip change observed along the strike of the tear-fault in model MO29, which was ascribed to longitudinal spreading of the culmination developed above the leading frontal ramp. However, the dip change along the Brazeau lateral ramp is much greater, and, more significantly, the sense of the rotation is opposite to that in the model. Therefore, the change in dip of the Brazeau lateral ramp is more likely a primary feature of the geometry of the transverse structure. The Brazeau lateral ramp thus probably had a primary geometry that resembles a combination of the initial configurations of the lateral-ramp and tear-fault models in the present study.

The hypothesis that the Brazeau lateral ramp had a complex, twisted initial geometry can be investigated by constructing models in which the precut transverse fault has a variable dip. A change in dip of the primary fault structure should give rise to diagnostic accommodation structures in the lateral edge of the thrust sheet while it is transported toward the foreland over a footwall of varying geometry. Factors that would lead to such a complex initial geometry are open to speculation.

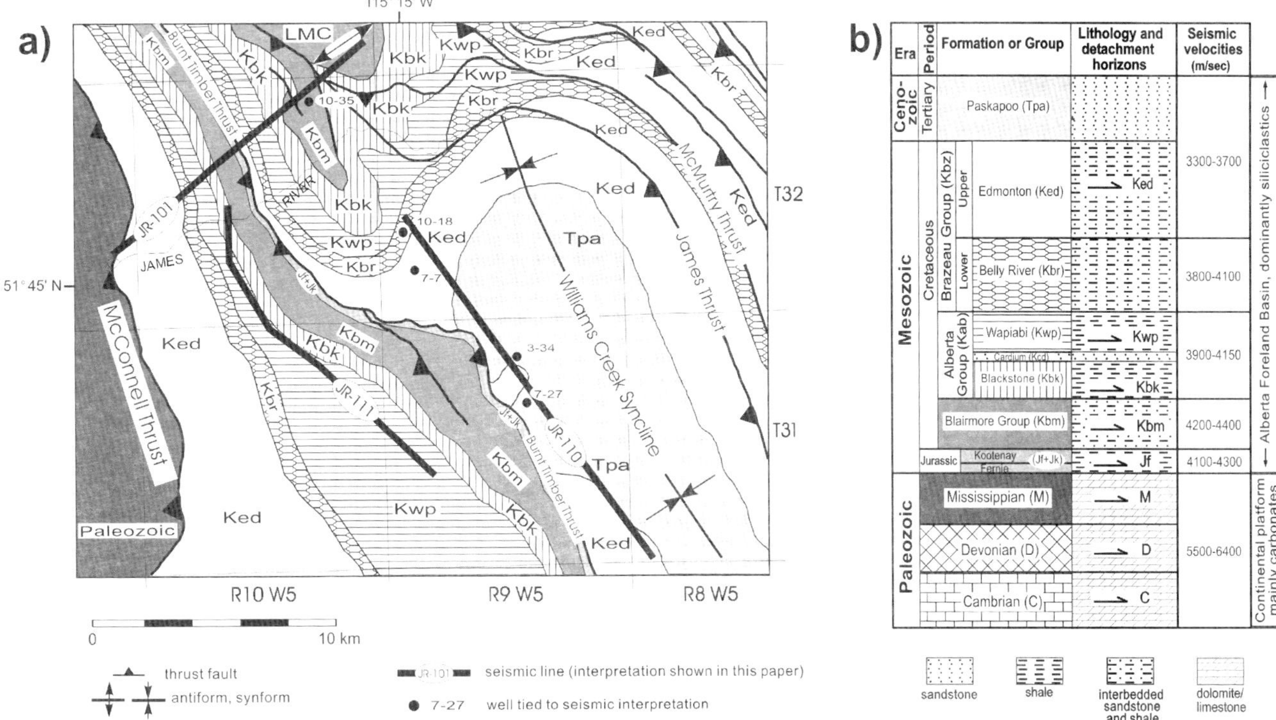

FIGURE 10. (a) Geological map of part of the Alberta Foothills in the vicinity of James River (location shown in Figure 2). Thick straight lines labeled JR-101, JR-110, and JR-111 show positions of 2-D seismic lines, interpretations of which are shown in Figure 11. Bottom-hole locations of deviated wells tied to sections JR-101 and JR-110 are shown as black dots. (b) Generalized stratigraphic column showing formation names and abbreviations, generalized rock types, positions of the main detachment horizons, and seismic velocities. After Bégin and Spratt (2002).

SUMMARY AND DISCUSSION

Here we will focus on contrasts between deformation patterns associated with the transverse linkages between frontal ramps, and we will draw attention to some geometrical and mechanical implications.

The main thrust sheets in both tear-fault and lateral-ramp models developed two prominent structural culminations associated with the frontal ramps. The culminations terminate laterally in steep monoclines that have opposing facing, and folds in the thrust sheet exhibit plunge reversal. The lateral terminations of the culminations are steeper above the vertical tear fault than above the shallow-dipping lateral ramp.

The leading edges of the competent lowest unit in the thrust sheet develop overturned folds with foreland vergence. The plan form of the leading edge reflects the geometry of the ramp system. The frontal ramps initially were straight, and the total shortening is essentially constant across the width of each model. The leading edges in the hanging wall develop arcuate forms that are convex toward the foreland and that reflect a decrease in the displacement on each frontal-ramp segment toward the lateral structure. This arises because deformation in the vicinity of the lateral structure causes the strike of the lateral structure (tear fault or lateral ramp) to be rotated out of the thrust transport direction. Once the two frontal ramps overlap, their individual displacements must decrease along strike in the overlap region, if the total displacement remains uniform along strike.

Several factors may contribute to the rotation of the transverse structures. Material in the footwall of the trailing frontal ramp, adjacent to the transverse structure, undergoes vertical flattening and along-strike extension because of stress localized beneath the overriding thrust sheet as it accommodated itself to the sharp deflection in its footwall. In addition, the thrust-sheet culminations may undergo gravitational spreading (vertical flattening and longitudinal stretching) to alleviate their excess topography.

The models demonstrate that there can be local 3-D strain of footwall and hanging-wall blocks adjacent to the transverse structure, and that the transverse structure can be rotated about a vertical axis. This has implications for section balancing: It would be very difficult (if not impossible) to draw a balanced section along section 37.0 of model MO28 (Figure 5f).

A transverse linkage between two frontal thrust ramps—either a tear fault or lateral ramp—appears to be unstable. The models show that in both cases the two frontal ramps propagate along strike across the position of the transverse structure (Figure 6). The leading ramp

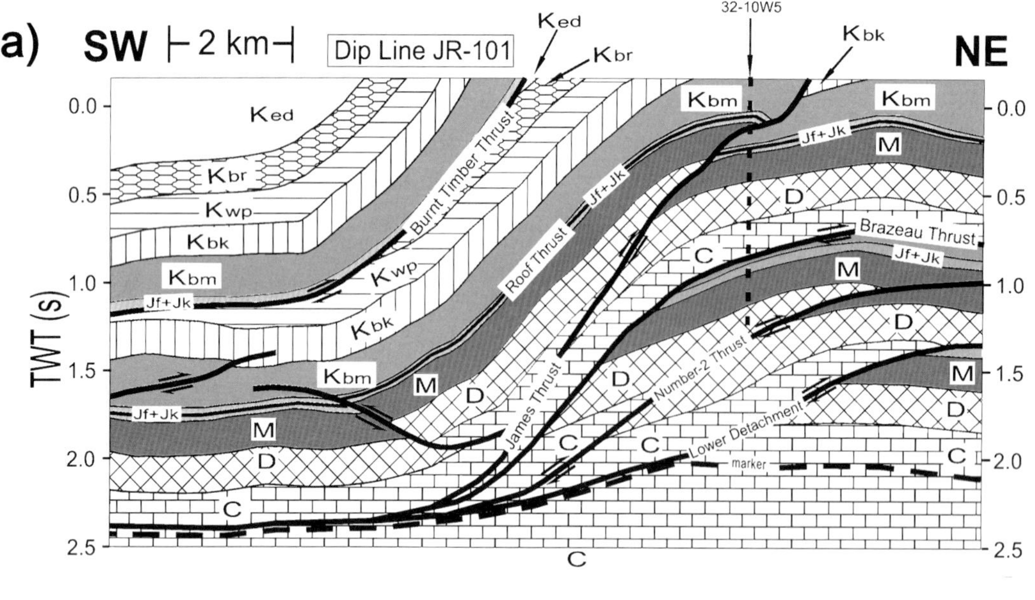
a)
SW
2 km
Dip Line JR-101
Ked
Kbr
10-35-
32-10W5
Kbk
NE
Kbm
Jf+Jk
M
D
Brazeau Thrust
Burnt Timber Thrust
Roof Thrust
James Thrust
Number-2 Thrust
Lower Detachment
marker
Kwp
Kbk
C
TWT (s)
0.0
0.5
1.0
1.5
2.0
2.5

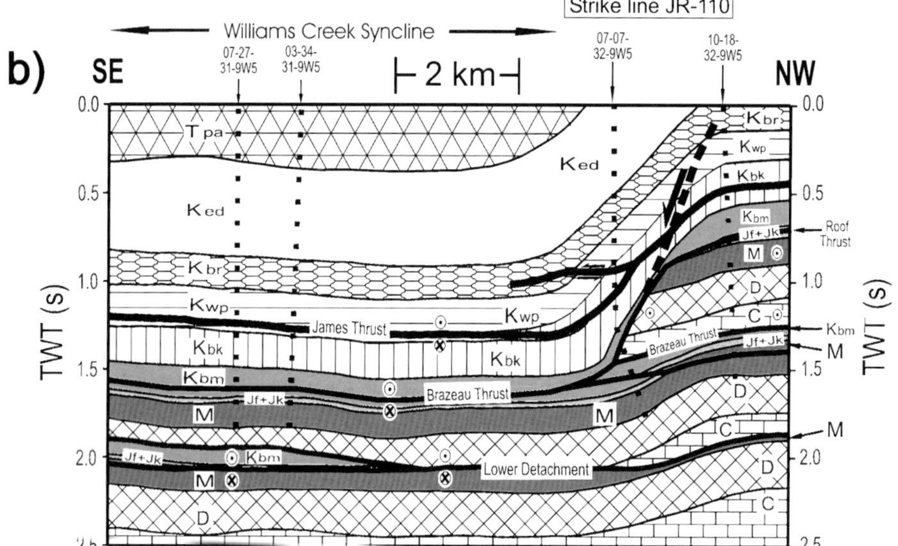
Strike line JR-110
Williams Creek Syncline
b)
SE
07-27-
31-9W5
03-34-
31-9W5
2 km
07-07-
32-9W5
10-18-
32-9W5
NW
Tpa
Ked
Kbr
Kwp
Kbk
Kbm
Jf+Jk
M
D
C
James Thrust
Brazeau Thrust
Lower Detachment
Roof Thrust
TWT (s)
0.0
0.5
1.0
1.5
2.0
2.5

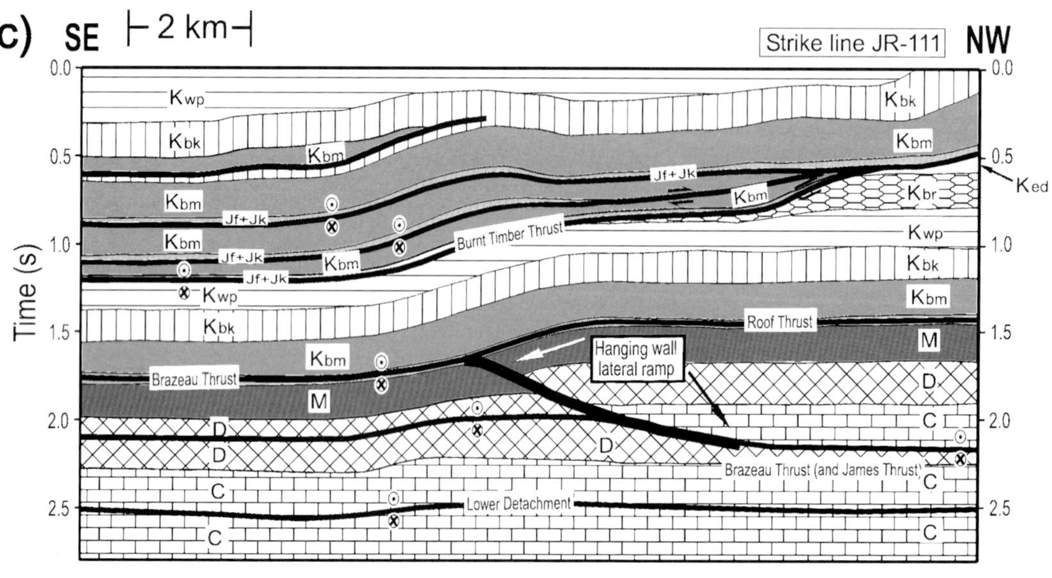
c)
SE
2 km
Strike line JR-111
NW
Kwp
Kbk
Kbm
Jf+Jk
Burnt Timber Thrust
Kbr
Ked
Roof Thrust
Hanging wall
lateral ramp
Brazeau Thrust
M
D
C
Brazeau Thrust (and James Thrust)
Lower Detachment
Time (s)
0.0
0.5
1.0
1.5
2.0
2.5

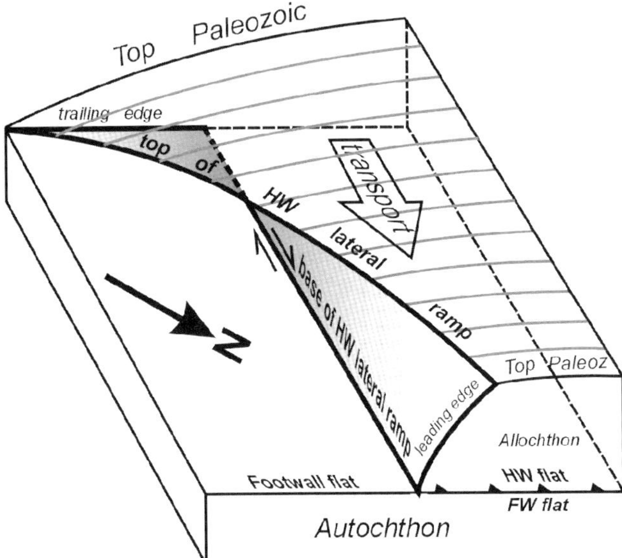

FIGURE 12. Schematic representation of the along-strike change in geometry of the transverse structure along the southeast margin of the Brazeau Thrust sheet at the level of the Paleozoic strata, based on dip sections JR-111 and JR-110 (Figure 11b and 11c). The structure is interpreted to change from a northwest-dipping lateral ramp to a vertical and then an overturned, steeply southeast-dipping tear fault. See text for discussion.

propagates laterally into the footwall, whereas the trailing ramp propagates laterally into the thrust sheet. As the transverse structure is rotated out of the transport direction, it is no longer favorably oriented to transfer slip from one frontal segment to the other, so it is progressively abandoned and replaced by the two frontal thrusts in a lap-joint displacement-transfer zone. Because the two overlapping frontal ramps root into a common basal detachment and merge into a common roof décollement in unit II, they constitute a duplex structure (admittedly of limited strike length). This duplex forms by lateral propagation of two simultaneously active thrusts, rather than by the conventional break-forward or break-back mechanism whereby one thrust replaces the other. However, as the two thrusts propagate laterally, one of them might eventually be abandoned in favor of the other, and the final outcome would be the same.

The lateral propagation of the two frontal-ramp segments gives rise to structures in the hanging wall and footwall of the main thrust that may have trapping potential. In the models, the hanging-wall structure (a large fault-propagation fold) that results from lateral propagation of the trailing frontal ramp is prominent, and one would expect that an equivalent structure in nature would be easy to map or image seismically. On the other hand, in the models, the footwall imbricate formed by lateral propagation of the leading frontal ramp is more obscure. Nevertheless, in the right circumstances such a structure might provide both structural closure and seal.

In models with tear-fault and lateral-ramp configurations, the frontal-ramp segments propagate along strike in a pattern of deformation that suggests that these configurations are inherently unstable. This raises the question: Can a transverse tear fault or lateral ramp arise on its own in homogeneous rock in the base of a thrust sheet, or must such a feature always be nucleated by a primary heterogeneity such as a lithologic boundary (e.g., a change in facies or cementation) or a primary structure (e.g., an early normal fault)?

CONCLUSIONS

The scaled physical models presented here demonstrate mechanisms by which a thrust sheet accommodates itself to displacement over ramp systems that comprise two frontal-ramp segments linked by either a vertical tear fault or a gently dipping lateral ramp. Structures in the hanging wall and footwall are diagnostic of the geometry of the ramp system. The models may serve as benchmarks against which natural examples can be compared, and that will assist geoscientists in interpreting those examples.

Most of the geometric relationships in the models are consistent with commonly held conceptions of the structural effects of thrust displacement over tear faults and lateral ramps. Perhaps the most interesting relationship observed in the models is that ramp systems in which frontal-ramp segments are linked by tear faults and lateral ramps appear to be unstable, the transverse link tending to be abandoned through a process of along-strike propagation of the frontal-ramp segments in a lap-joint configuration. It is also noteworthy that

FIGURE 11. Geological cross sections based on interpretations of 2-D seismic lines in the James River region (locations shown in Figure 10a; after Bégin and Spratt, 2002). (a) Dip section JR-101 shows the main frontal ramp of the Brazeau Thrust sheet. The sheet is cut by forethrusts and carries a fault-propagation fold in the James thrust. Compare with transverse section 43.0 of Model MO31 (Figure 8d). (b) Strike section JR-110 shows the lateral termination of the Brazeau sheet as a steeply southeast-dipping (overturned) tear fault. Compare with sections 27 and 30 of model MO29 (Figure 6c). (c) Strike section JR-111 shows the lateral termination of the Brazeau sheet as a gently northwest-dipping lateral ramp. Compare with section 27 of model MO30 (Figure 9c). See text for discussion.

lateral ramps and tear faults generate localized culminations in the thrust sheet. The lateral-ramp geometry appears to have better potential to produce a culmination with large structural closure and good trapping potential. It is thus very important to interpret natural transverse structures correctly.

ACKNOWLEDGMENTS

This work was conducted as part of the Fold-Fault Research Project (FRP), an industry-sponsored collaboration between the University of Calgary and Queen's University. We are grateful for financial support from the Industry sponsors of FRP and the Natural Sciences and Engineering Research Council of Canada (NSERC). Maggie Oliphant, Julia Blackburn, Simone Garneau and Marie Wardman participated in this project as research assistants to JMD in the Experimental Tectonics Lab at Queen's University and their contributions are greatly appreciated. Peter Fermor provided a preprint of his paper on three-dimensional aspects of structure in the Alberta Foothills and Front Ranges. Reviews of the manuscript by Peter Cobbold, Fabrizio Storti and Francesco Salvini are greatly appreciated.

REFERENCES CITED

Aydemir, E. O., 1998, Investigation of strain related to displacement transfer and along-strike variation using 3-d seismic interpretation, physical modelling and computer-graphics visualization: M.Sc. Thesis, Queen's University, 174 p.

Bégin, N. J., D. C. Lawton, and D. A. Spratt, 1996, Seismic interpretation of the Rocky Mountain thrust front near the Crowsnest deflection, southern Alberta: Bulletin of Canadian Petroleum Geology, v. 44, p. 1–13.

Bégin, N. J., and D. A. Spratt, 2002, Role of transverse faulting in along-strike termination of Limestone Mountain Culmination, Rocky Mountain thrust and fold belt, Alberta: Journal of Structural Geology, v. 24, p. 689–707.

Bielenstein, H. U., 1969, The Rundle thrust sheet, Banff, Alberta—a structural analysis: Unpublished Ph.D. Thesis, Queen's University, Kingston, Ontario, 149 p.

Butler, R. W. H., 1982, Hangingwall strain: a function of duplex shape and footwall topography: Tectonophysics, v. 88, p. 235–246.

Dahlstrom, C. D. A., 1969, Balanced cross sections: Canadian Journal of Earth Sciences, v. 6, p. 743–757.

Dahlstrom, C. D. A., 1970, Structural geology in the eastern margin of the Canadian Rocky Mountains: Bulletin of Canadian Petroleum Geology, v. 18, p. 332–406.

Dixon, J. M., and S. Liu, 1991, Centrifuge modelling of the propagation of thrust faults, *in* K. R. McClay, ed., Thrust tectonics: London: Chapman & Hall, p. 53–70.

Dixon, J. M., and J. M. Summers, 1985, Recent developments in centrifuge modelling of tectonic processes: equipment, model construction techniques and rheology of model materials: Journal of Structural Geology, v. 7, p. 83–102.

Dixon, J. M., and J. M. Summers, 1986, Another word on the rheology of silicone putty: Bingham: Journal of Structural Geology, v. 8, p. 593–595.

Dixon, J. M., and R. Tirrul, 1991, Centrifuge modelling of fold-thrust structures in a tripartite stratigraphic succession: Journal of Structural Geology, v. 13, p. 3–20.

Douglas, R. J. W., 1950, Callum Creek, Langford Creek, and Gap map areas, Alberta: Geological Survey of Canada, Memoir 255, 124 p.

Elliott, D., 1976, The energy balance and deformation mechanisms of thrust sheets: Philosophical Transactions of the Royal Society of London, A, v. 283, p. 289–312.

Fermor, P., 1999, Aspects of the three-dimensional structure of the Alberta Foothills and Front Ranges: Geological Society of America Bulletin, v. 111, p. 317–346.

Gwinn, V. E., 1964, Thin skinned tectonics in the Plateau and northwestern Valley and Ridge Provinces of the central Appalachians: Geological Society of America Bulletin, v. 75, p. 863–900.

McClay, K. R., 1991, Glossary of thrust tectonics terms, *in* K. R. McClay, ed., Thrust tectonics: London, Chapman & Hall, p. 419–433.

Norris, D. K., 1958, Structural conditions in Canadian coalmines: Geological Survey of Canada, Bulletin 44, 54 p.

Norris, D. K., 1964, Microtectonics of the Kootenay Formation near Fernie, British Columbia: Bulletin of Canadian Petroleum Geology, v. 12, p. 383–398.

Price, R. A., 1964, Flexural-slip folds in the Rocky Mountains, southern Alberta and British Columbia: Geological Survey of Canada, Reprint 78, 16 p.

Price, R. A., 1967, The tectonic significance of mesoscopic subfabrics in the southern Canadian Rocky Mountains of Alberta and British Columbia: Canadian Journal of Earth Sciences, v. 4, p. 39–70.

Ramberg, H., 1981, Gravity, deformation and the earth's crust in theory, experiments and geological applications, 2nd ed.: London, Academic Press, 452 p.

Rich, J. L., 1934, Mechanics of low-angle overthrust faulting illustrated by Cumberland thrust block, Virginia, Kentucky and Tennessee: AAPG Bulletin, v. 18, p. 1584–1596.

Spratt, D. A., and D. C. Lawton, 1996, Variations in detachment levels, ramp angles and wedge geometries along the Alberta thrust front: Bulletin of Canadian Petroleum Geology, v. 44, p. 313–323.

Wheeler, J. O., and P. McFeely, comps., 1991, Tectonic assemblage map of the Canadian Cordillera and adjacent parts of the United States of America: Geological Survey of Canada, Map 1712A, scale 1:2,000,000.

Spratt, D. A., J. M. Dixon, and E. T. Beattie, 2004, Changes in structural style controlled by lithofacies contrast across transverse carbonate bank margins — Canadian Rocky Mountains and scaled physical models, *in* K. R. McClay, ed., Thrust tectonics and hydrocarbon systems: AAPG Memoir 82, p. 259–275.

Changes in Structural Style Controlled by Lithofacies Contrast Across Transverse Carbonate Bank Margins—Canadian Rocky Mountains and Scaled Physical Models

Deborah A. Spratt

Department of Geology and Geophysics, University of Calgary, Calgary, Alberta, Canada

John M. Dixon

Department of Geological Sciences, Queen's University, Kingston, Ontario, Canada

Edward T. Beattie

ABSTRACT

Profound variations in fold and fault geometries are mapped across a transverse lithofacies boundary in the Upper Devonian (Frasnian) Fairholme Group between hydrocarbon-bearing dolostones of the Southesk and Cairn Formations (Leduc and Nisku reefs) and off-reef shales of the Mount Hawk, Perdrix, and Flume Formations (Cline Channel) in the North Saskatchewan River area. These lithofacies constitute only 10% of the thickness of the stratigraphic package, but, during compression subparallel to the bank margin, they affected the deformational style of the entire supercrustal wedge, including units below them as well as above them. This area is ideal for studying lateral changes in structural style because the transition occurs over a short distance (<1.5 km) within individual thrust sheets, where the overlying lithologies, percent shortening, burial depth and thermal history remain constant along strike. The excellent exposures are at the same stratigraphic and structural level, and the Palliser Formation contains peloidal strain markers, making it possible to quantify the changes in structural geometry in cross sections and also to determine the mechanisms of deformation in outcrops and in thin sections. Upper Devonian (Famennian) and Mississippian platform carbonates overlying the reef facies are concentrically folded, and thrust faulting is the primary mechanism of shortening. Fault-bend folding broadly warps strata and the rare folds with wavelengths shorter than 500 m are complexly faulted in their cores. Where the platform carbonates overlie the off-reef shale facies, they are tightly folded above detachment faults. Pressure-solution cleavage is strongly developed and fractures are filled with calcite, thereby significantly reducing the porosity and permeability of reservoir rocks that overlie the incompetent shale facies. Scaled physical models deformed by the centrifuge technique reproduce the variations in structural style seen across transverse lithofacies boundaries and illustrate how a fault-propagation fold can convert into a detachment fold along strike. The position and geometry of the underlying reef margin are evident in plan view at early stages of deformation, but structure contour mapping and detailed structural analysis are required to locate the underlying reef margin at later stages of deformation in the models and real-world examples.

INTRODUCTION

Prediction of the size, geometry, and lateral continuity of potential structural traps in fold-and-thrust belts requires knowledge of the mechanical stratigraphy and three-dimensional distribution of the weak and strong layers involved. Variations in fold and fault geometries related to changes in lithology have been documented in many mountain belts (Mountjoy, 1960, 1992; Currie et al., 1962; DeSitter, 1964; Bally et al., 1966; Ramsay, 1967; Dahlstrom, 1970; Royse et al., 1975; Cook, 1975; Harris and Milici, 1977; Price, 1981; Woodward and Rutherford, 1989). However, many of the studies are regional and involve rocks from different structural levels; thus the changes in style cannot be attributed to stratigraphy alone. Thompson (1981) and Pfiffner (1993) examine differences in stratigraphy and structural style along strike, but because in both studies the transects are tens of kilometers apart, we cannot see how the geometrical changes occur between their cross sections, and the lithologic variations are not restricted to one unit. Studies that can demonstrate a clear correspondence between the presence of a weak unit and a predictable fold style typically involve a salt horizon (Butler, 1992; Vergés et al., 1992; Velaj et al., 1999), and the limited areal distribution of the evaporite basins limits the lateral extent of folds in these belts. Changes in structural style over carbonate platform margins have been noted by Mountjoy (1960), Schönborn (1992, 1999), and Ghisetti and Vezzani (1997), among others. All of these studies indicate that thrusts and harmonic folds predominate in areas with higher proportions of competent strata, whereas disharmonic folds and bedding-parallel detachments accommodate additional shortening as the proportion of weaker lithofacies increases. And, where a particularly weak unit is involved, significant changes in the trends of folds and faults occur near the unit's lateral limit.

The goal of this chapter is to examine how fault-related fold geometries can change where the variation in mechanical stratigraphy occurs within individual structures or exploration areas. Obviously, the structural style of the unit undergoing the lithologic transition is expected to change, but we are also interested in determining what effect a transverse lithofacies boundary may have on the fold style, scale, closure, and reservoir quality of overlying and underlying units. We investigate changes in fault-related fold style across carbonate lithofacies boundaries oriented at low angles to the thrust transport direction in well-exposed areas of the Front Ranges of the southern Canadian Rockies, approximately 200 km northwest of Calgary, Alberta, Canada (Figure 1), and in physical analog models deformed in the centrifuge.

Geologic Setting

In the Front Ranges of southern Alberta, a southwestward-thickening wedge of supracrustal rocks, comprising strata that range in age from Cambrian to Tertiary, rests on the gently southwest-dipping Hudsonian basement surface of the North American craton. This supracrustal wedge is subdivided into two stratigraphic packages that represent two fundamentally different tectonostratigraphic settings. The lower package consists of Cambrian to Lower Jurassic sediments that were deposited along the western passive continental margin of ancient North America. The upper tectonostratigraphic package consists of Upper Jurassic to Tertiary sediments (primarily siliciclastics) deposited during the Columbian and Laramide orogenies. The western portion of the supracrustal wedge was progressively deformed in a compressional regime during accretion of allochthonous terranes to the North American continent. Deformation began in Middle Jurassic time and continued into the Paleocene. The Rocky Mountain fold-and-thrust belt comprises sedimentary-cover rocks of North American origin that were deformed under very low-grade metamorphic conditions. They were transported eastward onto the craton on a décollement above the crystalline basement. The deformation style is typical of thin-skinned fold-and-thrust belts, with low-angle thrusts and a foreland-directed progression of thrusting associated with detachment, fault-propagation, fault-bend, and break-thrust folding. For a thorough description and

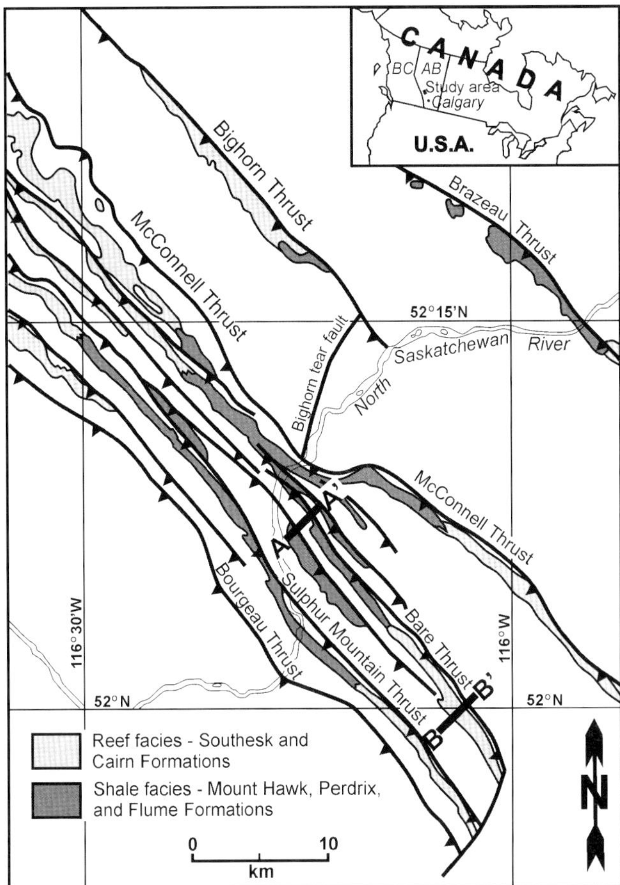

FIGURE 1. Geologic map, modified from Woodhead and Wright-Broughton (1967), showing thrust faults and shale and reef facies of the Devonian Fairholme Group in the vicinity of the Cline Channel and North Saskatchewan River. A–A′ marks the location of the cross section through the shale facies at Mount Michener (Figure 4a); B–B′ marks the location of the cross section through the reef facies at Rough Mountain (Figure 4b). Inset: location map.

discussion of the sedimentology and tectonics of the Western Canada Sedimentary Basin, see Mossop and Shetsen (1994); the chapter by Price (1994) provides an excellent summary of the tectonic framework and history of the basin.

Deformation of the lower package, primarily carbonates, is the focus of this paper. In the classic transect along the Bow Valley west of Calgary, it is common to see all or most of the package carried intact in major thrust sheets, and the thickness of the carbonate beam constrains the height of thrust ramps and dominant wavelength of folds (Currie et al., 1962; Bally et al., 1966; Dahlstrom, 1970). This is also the case in the southeastern part of the North Saskatchewan River study area in central Alberta, where Devonian Fairholme through Mississippian Rundle Group strata are folded and faulted together (Figures 2, 3b, and 4b). There are two shale-dominated units, the Devonian Mount Hawk and Mississippian lower Banff Formations (Figure 2, southeast portion), that act as only minor detachment zones within the multilayer. Figure 3b shows how the Mississippian Banff Formation (Mbf) accommodates disharmony between the top of the Palliser Formation and the base of the Rundle Group, whereas the underlying Devonian Fairholme (Df) and Cambrian units and the overlying Mississippian Rundle Group (Mru) strata remain as parallel dip panels. In contrast, the increase in number and thickness of shale units northwest of the facies boundary in the Devonian Fairholme Group gives rise to pervasive, tight folds (Figures 2 [northwest portion], 3a, and 4a). Mountjoy (1960, 1992) documented similar features in the Jasper area Front Ranges, 150 km to the northwest.

Reservoir Units

The Fairholme carbonate complex is not the only Paleozoic exploration target in western Canada (Mossop and Shetson, 1994). The overlying Devonian and Mississippian strata contain several hydrocarbon reservoir units that extend into the Interior Plains of southern Alberta, where exploration is more straightforward than in the fold-and-thrust belt. Subsurface equivalents of the Cairn and Southesk Formations of the Upper Devonian (Frasnian) Fairholme Group are the oil- and gas-bearing Leduc and Nisku reefs of the Woodbend and Winterburn Groups. The Wabamun Group in the Interior Plains is the subsurface equivalent of the Upper Devonian (Famennian) Palliser Formation. Production from the Mississippian Rundle Group is primarily from the Debolt-Elkton and Shunda-Pekisko intervals in the subsurface. The breached folds in the Front Ranges serve as analogs for buried structural targets in the Foothills Belt and contain the same reservoir units that have been exploited in the Plains.

Douglas (1956) described the abrupt Fairholme facies boundary in the hanging wall of the Bighorn Thrust on the north side of the Cline Channel, which is the zone of Fairholme shale facies rocks that are localized in the vicinity of the present-day North Saskatchewan River valley, 200 km northwest of Calgary, in the central Alberta Rocky Mountains and Foothills (Figures 1 and 2). The Cline Channel is similar in scale to the intermediate basin in the Friuli platform of the Dolomites, described by Schönborn (1999), but it has a less complex tectonic history. Douglas (1956) noted that the Bighorn Thrust ramps from Upper Cambrian into Devonian strata just south of the facies boundary. Fox (1969) recognized that, where the Perdrix (Dpx) shale is present, the competency of the entire Paleozoic section is reduced to such an extent that remarkably tight folds involving the Devonian Palliser and Mississippian Rundle carbonates occur. Where the Perdrix shale is absent, the

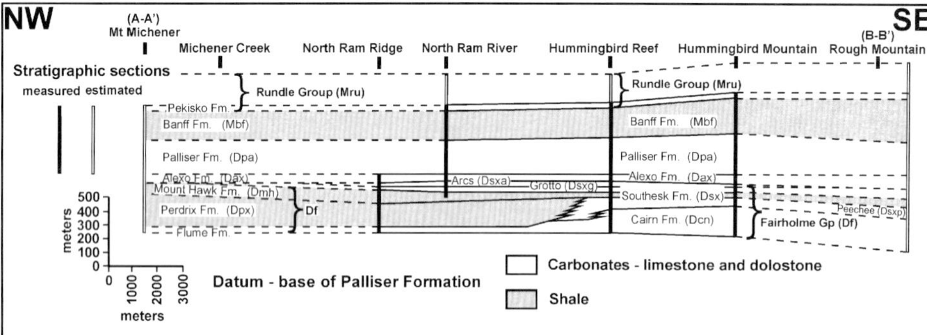

FIGURE 2. Longitudinal stratigraphic section in the hanging wall of the Bare Thrust between Mount Michener (shale facies) at A–A′ and Rough Mountain (reef facies) at B–B′. See Figure 1 for location. Vertical exaggeration = 5×.

Paleozoic structures are much simpler, with the Palliser-Rundle succession acting as a strong beam. Mountjoy (1960) noted similar relationships and also documented changes in the structural style of underlying Cambrian units across competent to less competent lithofacies boundaries. Verrall (1968) and Jones and Workum (1978) provide structural field guides for the area.

Field Mapping

The Cline Channel is an ideal area for studying the effects that facies variations at carbonate bank margins have on the development of different structural styles during compression. The bank margin orientation was roughly parallel to the compression direction, so we can examine and compare lateral changes in structural style within individual thrust sheets where the percentage of shortening, the structural burial depth, and the thermal history remain constant along strike (Figures 1 and 2). Folds, faults, and transverse and oblique bank margins are well exposed in the McConnell, Bare, and Sulphur Mountain Thrust sheets (Figures 1 and 3). The exposures are at the same stratigraphic and structural level, and the Palliser Formation contains peloidal strain markers, making it possible to quantify the changes in structural geometry in cross sections and to determine the mechanisms of deformation in outcrops and in thin sections.

To acquire the control necessary for accurate documentation of fold and facies geometries, the area was mapped at several scales and stratigraphic sections were measured at four localities along strike in the Bare Thrust sheet (Figure 2; see also McLaren, 1955; Hargreaves, 1959, 1968; Workum, 1978; Beattie, 1984). The Upper Devonian (Frasnian) Fairholme Group shows a profound facies change, from dolostones of the Cairn and Southesk Formations in the south, to thick shales of the Perdrix and Mount Hawk Formations in the Cline Channel to the north. Although the overlying (Famennian) Palliser Formation and Mississippian Banff Formation and Rundle Group maintain uniform thicknesses with no obvious lithologic variation across the area, they also exhibit a dramatic change in structural style over the facies boundary (Figures 2 and 3). The 1:50,000-scale maps of Mountjoy and Price (1974a, b) and Price and Mountjoy (1977a, b) are excellent for locating large structures, but they do not contain enough dip information to determine detailed fold geometries. Beattie (1984) remapped the Paleozoic rocks in the hanging wall of the Bare Thrust along the entire length of the Ram Range at a 1:50,000 scale, the structures near Rough Mountain and Mt. Michener at a 1:10,000 scale, and complex structures at 1:1,000 and 1:2,000 scales. Down-plunge projection of data points was based on analysis of stereoplots

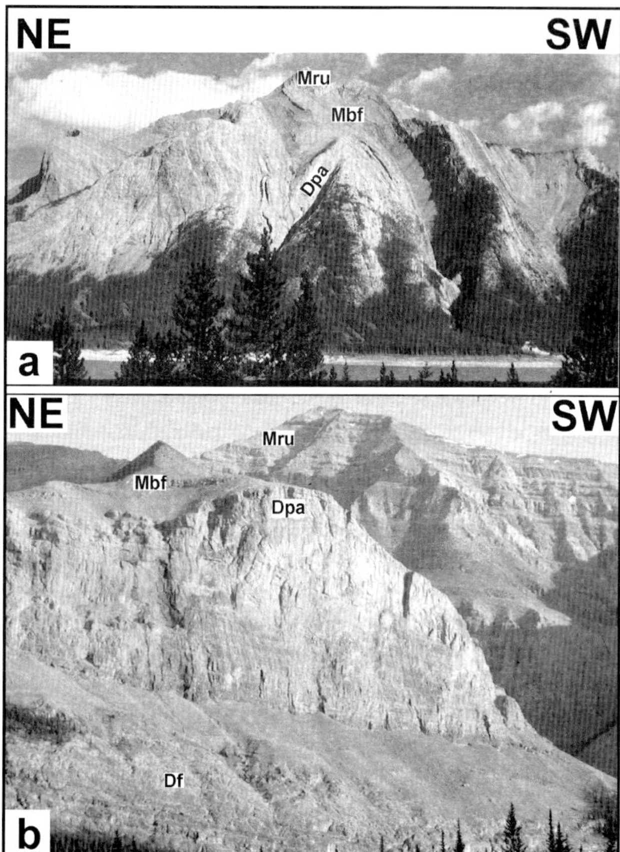

FIGURE 3. Photographs of structures developed above the (a) shale facies at Mount Michener and (b) reef facies at Gable Mountain, in the McConnell Thrust sheet southeast of the Cline Channel. Symbols as in Figure 2.

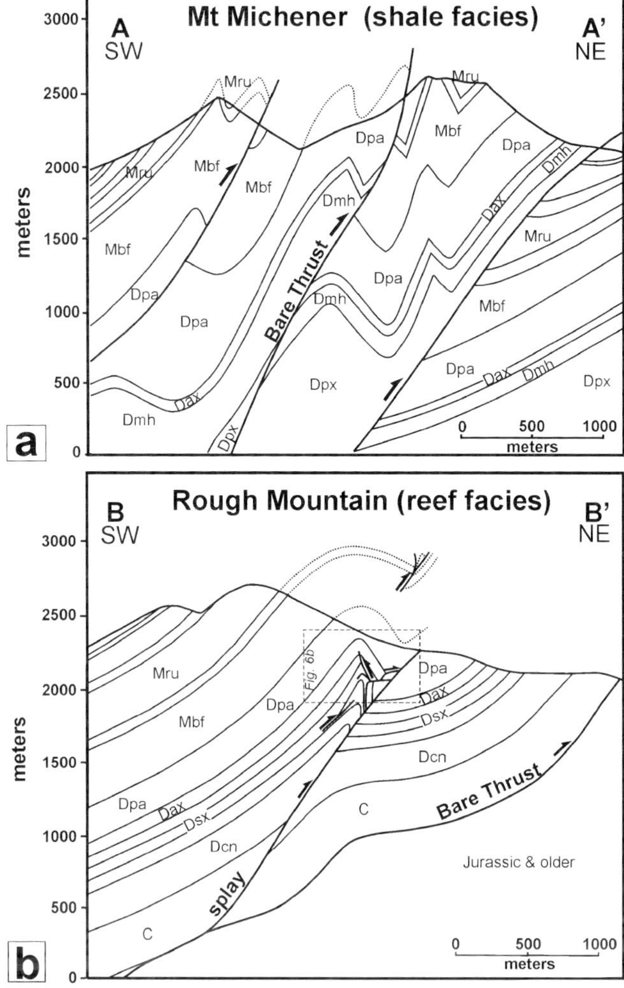

FIGURE 4. Cross sections through the Ram Range, showing the change in structural style of Devonian and Mississippian rocks overlying the facies boundary in the Fairholme Group. Both sections are at the same scale and V = H. See Figure 1 for locations and Figure 2 for stratigraphic abbreviations. (a) Cross section A–A′ at Mt. Michener, showing tight similar folds above the shale facies. (b) Cross section B–B′ through Rough Mountain, exhibiting open parallel folds and brittle faulting above the reef facies.

of poles to bedding to determine a projection axis for each structural domain (Figure 5). Orientation data are remarkably consistent along strike, with a mean fold and fault trend of 317° across all domains in both areas. The mean dip of bedding panels in thrust sheets with no folding is 44°W, and where beds are folded, plunges average 7°NW above the reef facies and steepen to 25°NW over the shale facies. Down-plunge projections of folds exposed on steep slopes were constructed in the field at scales of 1:1,000 to 1:2,000, using a plane table and alidade, by modifying the trigonometry to produce sections perpendicular to the fold axes (Figure 6) instead of maps. Constructing the sections in the field makes it possible to see exactly where more control points are needed to constrain the fold geometry and to quantify the variations in bed thickness in different units around the folds.

FOLD-FAULT GEOMETRIES

Right sections through areas of the reef and shale facies clearly illustrate the significant differences in structural style (Figures 4 and 6).

Above the reef facies in the southeastern part of the study area, concentric (Class 1B of Ramsay, 1967) folds with nearly constant bed thicknesses have developed in the Palliser Formation and thrust faulting is the primary mechanism of shortening (Figures 3b, 4b, 5b, and 7). Fault-bend folding broadly warps strata in the area and the rare folds with wavelengths shorter than 500 m are complexly faulted in their cores (Figures 4b and 6b). At first glance, the structure in Figure 4b appears to be a fault-propagation fold with its core modified by brittle accommodation or bourrage (stuffing with excess material). Displacement does die out upward into a fold, but this well-constrained structure (its ramp angle, fold interlimb angle, shortening, and strain distribution are all known) is not consistent with Jamison's (1987) diagrams for fault-propagation folding or transported fault-propagation folding when partially restored (untilted relative to horizontal footwall). Like other natural folds described by Alonso and Teixell (1992), Mountjoy (1992), Groshong and Epard (1994), Erslev and Mayborn (1997), and Fischer et al. (1992), this structure is a hybrid and displays flattening and elements of detachment, break-thrust, and fault-propagation folding. The proportion of each folding mechanism also changes along strike, even in individual folds, and especially over the lithofacies transition, as we will show below in discussions of the field and model data.

To the northwest, above the shale facies, structures in the Palliser Formation have a similar fold geometry (Class 2 of Ramsay, 1967) and faults serve mainly as detachment surfaces for folding (Figures 3a, 4a, 6a, and 7). These natural folds plot in the "no solution" zone of Jamison's (1987) diagrams for detachment folding because of their significant flattening, which is manifested by thinning of backlimbs as well as forelimbs. Figure 7a shows nearly equal thinning of the limbs, with some layers slightly thinner in the backlimb. The overall deformation is more ductile than in areas above the reef facies; pressure-solution cleavage is pervasive in the Palliser limestones and Banff shales (Figure 8), and calcite fills extension fractures oriented perpendicular to and parallel to the fold axes. Axial ratios of deformed Palliser peloids range from 2:1 in the limbs to 3.7:1 in the hinge zones over the shale facies and from 1:1 in the

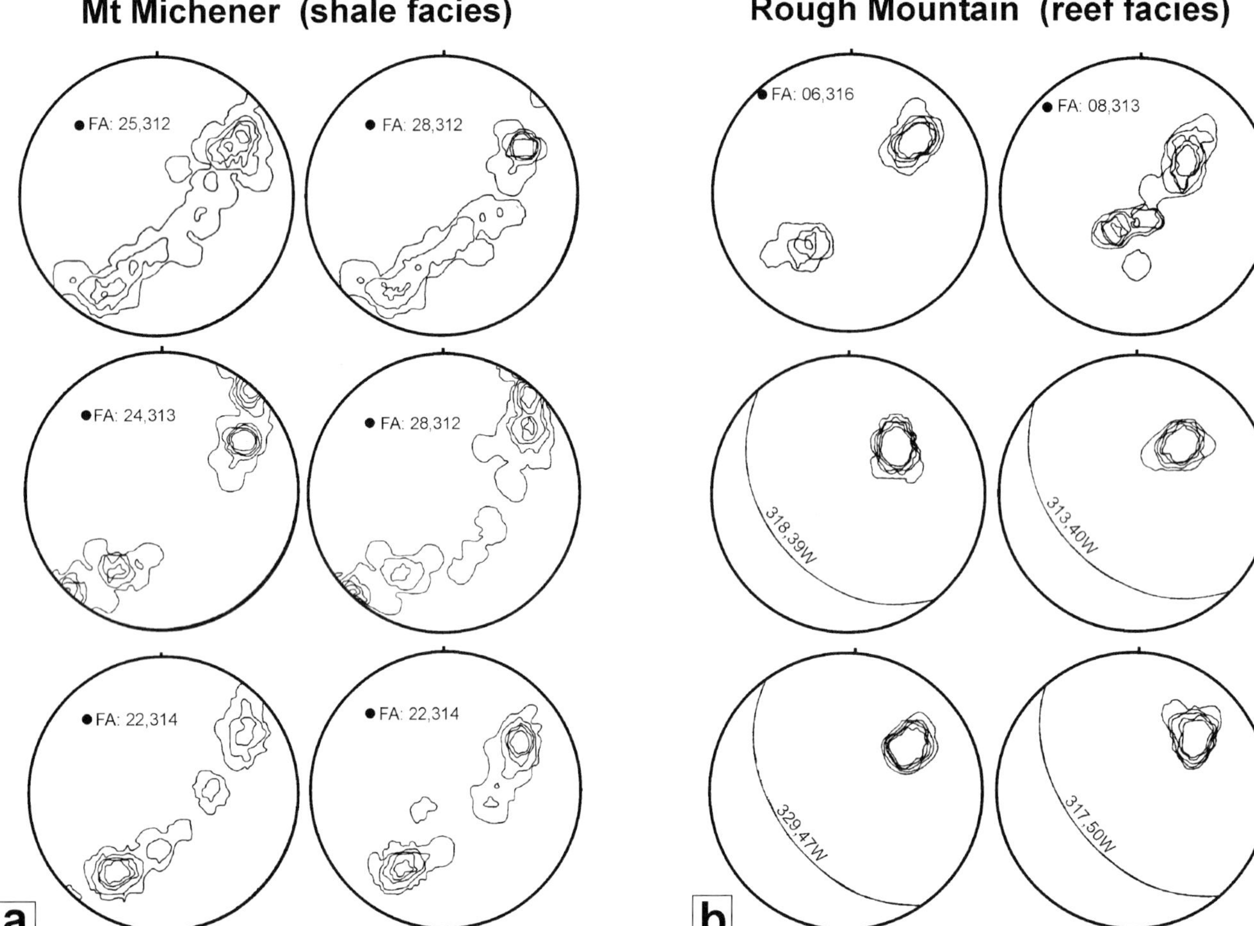

FIGURE 5. Stereoplots of poles to bedding in folds and thrust sheets, showing (a) cylindrical folding of units above the shale facies in the Mount Michener area, and (b) localized folding and large dip panels above the reef facies in the Rough Mountain area, with mean plane and fold axis orientations labeled.

limbs to 2.4:1 in the hinges of folds over the reef facies (Figure 6).

Wavelength, Amplitude, and Tightness of Folds

The wavelengths of folds in the area vary considerably, but the average values show that, for folds in the Palliser Formation, wavelength differs significantly between folds above the shale facies and folds above the reef facies of the Fairholme Group. The mean wavelength above the shale facies in the Michener Creek area is 285 m (±80 m), whereas the mean wavelength for folds above the reef facies in the Rough Mountain area and in Banff National Park is 610 m (±210 m). The Rough Mountain folds have amplitudes of less than 500 m, whereas the Mt. Michener folds vary from 200 m to nearly 1000 m in amplitude. Obviously if two folds have similar amplitudes but different wavelengths, the shorter-wavelength fold will be much tighter. Tightness is quantified by measuring the interlimb angle. The Rough Mountain anticline has an interlimb angle of 80°, whereas interlimb angles of folds in the Mt. Michener area range from 10° to 72° and average 45°. The interlimb angle increases across the Mt. Michener area, with folds being progressively tighter to the west. Folds in the Mississippian strata on the west side of Mt. Michener have interlimb angles of 10° to 20°.

From an exploration perspective, knowledge of lithologic variations in the strata below a reservoir unit is important for estimating the potential size of structural targets because folds developed over a weak layer may be less than half the size of folds developed over a strong layer, even if the thickness and lithology of the overlying reservoir unit remain constant.

Thickness Variations Around Folds

Internal deformation can enhance reservoir quality (in the case of open fractures) or reduce it (in the case of layer-parallel shortening, pressure solution, and vein-filling), in all, or portions of, a folded layer. Thus it is important to determine which is more likely to have

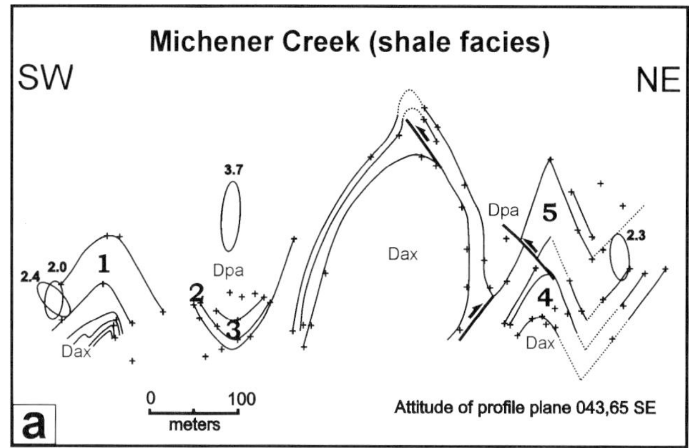

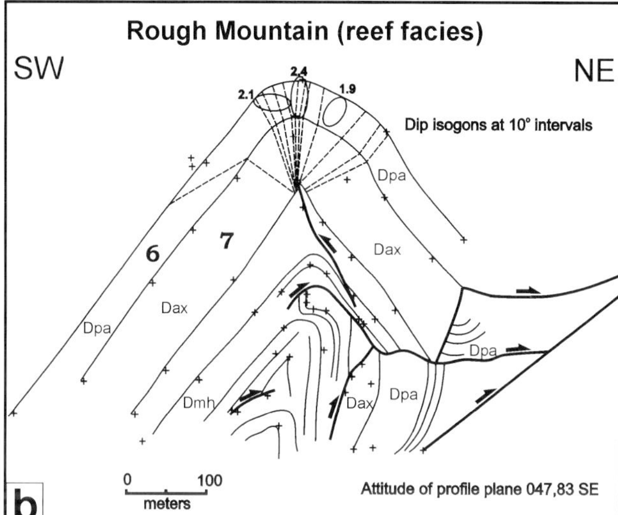

FIGURE 6. Down-plunge projection cross section of plane table data, showing the details of fold geometries and structural styles at (a) Michener Creek (above shale facies), and (b) Rough Mountain (above reef facies). Abbreviations as in Figure 2. Pluses (+) are surveyed control points. Strain ellipses for deformed peloids shown with their axial ratios. Both sections are at the same scale and V = H. Beds #1–7 are analyzed in Figure 7.

occurred. Jamison (1987) provides graphs that simplify estimation of the amount of limb thinning that occurs in fault-related folds and that require only basic geometrical information (ramp angle, limb dips, and interlimb angle). However, none of these natural folds are ideal fault-bend, fault-propagation, or detachment folds, and they plot in the no-solution zones of these graphs. However, Ramsay (1967) devised a method, based on geometry alone, that evaluates limb strain in natural folds and can be applied to the analysis of fold-style changes related to mechanical stratigraphy in exploration belts (Thompson, 1989).

Beds in the Rough Mountain anticline maintain a nearly constant thickness around the fold, whereas the beds in the cores of the Michener Creek folds thicken considerably (Figure 6). Bed thickness ordinarily is

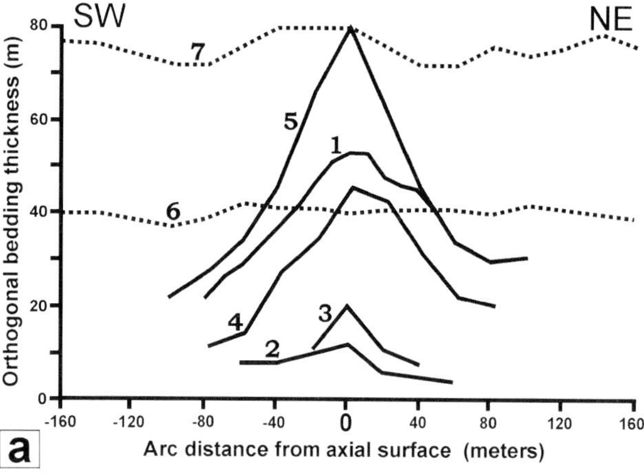

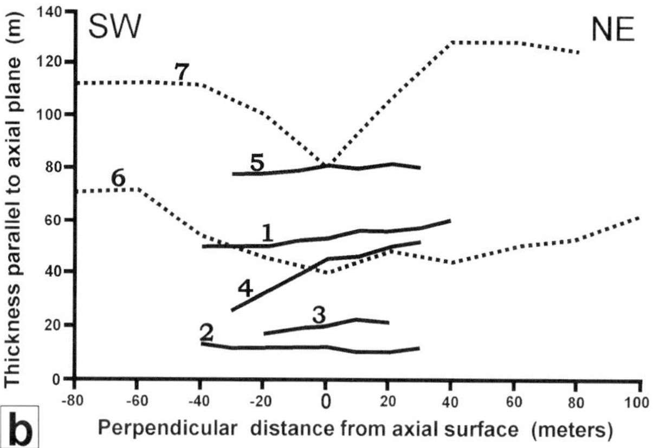

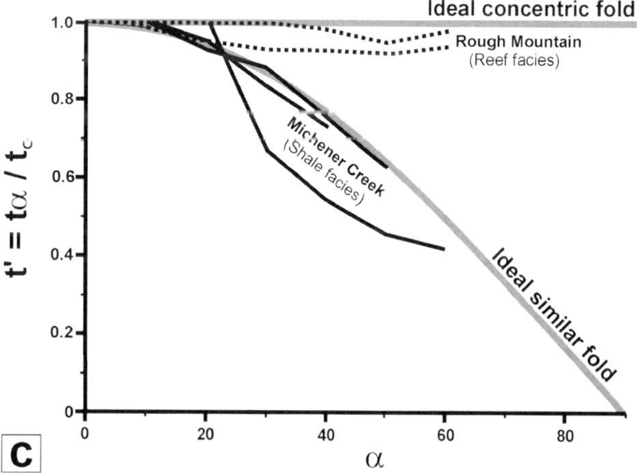

FIGURE 7. Graphs of thickness measurements around folded beds #1–7 of Figure 6. (a) Orthogonal thickness versus arc distance from the axial surface; concentric folds plot as horizontal lines. (b) Thickness parallel to the axial plane versus perpendicular distance from the axial surface; similar folds plot as horizontal lines. (c) $t' = t_\alpha/t_o$ versus dip angle (α), where t′ is the ratio between the bed thickness measured perpendicular to dip, for any dip angle α, and the bed thickness measured at the fold hinge (t_o). Data from surveyed continuous beds above the shale facies (Figure 6a, beds #1–5) are shown as solid lines; data from folded layers above the reef facies (Figure 6b, beds #6–7) are shown as dashed lines.

FIGURE 8. Pressure-solution cleavage in Lower Banff shales on Mount Michener. (a) Outcrop photograph; scale bar = 1 m. (b) Close-up of cleavage planes and folded layering; scale bar = 5 cm.

measured between, and perpendicular to, two parallel bedding planes. This value is known as the orthogonal thickness, t, and the variation in bed thickness between the two areas is clear in the graphs of orthogonal thickness versus arc distance from the axial surface (Figure 7a). In structural traps with steep to overturned forelimbs, bed thicknesses commonly vary, but if the rock is deformed in a brittle manner, the porosity and permeability of the reservoir unit may be enhanced. However, bed thickness variations are also associated with ductile deformation, pressure solution, and filled fractures that decrease the porosity and permeability of the reservoir. Structures with damaged reservoir rocks can be singled out, on the basis of fold geometry alone, by plotting the thickness parallel to the axial plane against the perpendicular distance from the axial surface. One can easily recognize rocks that have undergone ductile deformation, such as by pressure solution, because their fold geometry yields horizontal lines on the graph (Figure 7b).

Ramsay (1967) developed other graphical methods for determining fold geometries that are especially useful for comparing beds of vastly different thickness. He used the value t', which is the ratio of the thickness measured at any dip angle, t_α, to the thickness measured at the fold hinge, t_o, such that $t' = t_\alpha/t_o$. On Figure 7c, where t' is plotted against the dip angle, the ideal concentric fold will plot with a constant t' value of 1, whereas the ideal similar fold will plot along a line such that $t' = \cos \alpha$. The data for the two folds developed in the Palliser Formation above the Fairholme reef facies at Rough Mountain both plot as nearly concentric (parallel) folds. Data from three of the four folds at Michener Creek, above the shale facies, plot as ideal similar folds. The fourth fold, which is the large anticline near the center of Figure 6b, plots below the ideal similar fold curve, but the fold is nearly isoclinal and pressure solution in the limbs and faulting in the hinge have modified its shape. Intense pressure-solution cleavage is well-developed in Banff shales and limestone layers where they overlie Palliser strata above the shale facies (Figure 8), but macroscopic evidence of cleavage is not observed along strike where these same shale and limestone layers overlie the Fairholme reef facies (Beattie, 1984).

Even in areas with excellent exposure, data gaps resulting from erosion, cover, or lack of subsurface data commonly occur in critical parts of the structures found over transverse and oblique facies boundaries. Physical analog modeling using the centrifuge technique allows us to visualize the complex geometries of these structures more clearly and to investigate how they developed.

CONFIGURATIONS OF THE MODELS AND EXPERIMENTAL TECHNIQUE

The centrifuge modeling technique was suggested by Hubbert (1937) and first applied to the study of tectonic processes by Ramberg (1967). The 20,000-g centrifuge at Queen's University was described by Dixon and Summers (1985). The reader is referred to McClay

Table 1. Applicable model scaling ratios.

Quantity	*Ratio model/prototype*	*Equivalence (model ≡ prototype)*
length	$l_r \approx 2.0 \times 10^{-6}$	1 mm ≡ 0.5 km
specific gravity (mass)	$\rho_r = 0.6$	1.60 ≡ 2.67 (bulk value for whole stratigraphic column)
time	$t_r \approx 1.0 \times 10^{-11}$	1 hr ≡ 12 m.y.
strain rate	$\dot{\varepsilon}_r \approx 1.0 \times 10^{11}$	$10^{-3}\ s^{-1} \equiv 10^{-14}\ s^{-1}$ (for example)
viscosity	$\mu_r \approx 4.8 \times 10^{-14}$	4.8×10^2 Pa s ≡ 10^{16} Pa s (for example)
acceleration	$a_r = \mu_r/(\rho_r\ l_r\ t_r) = 4.0 \times 10^3$	4000 ***g*** ≡ 1 ***g***
stress	$\sigma_r = \rho_r\ l_r\ a_r \approx 4.8 \times 10^{-3}$	(calculated from other ratios)

(1976) and Dixon and Summers (1985, 1986) for discussion of the rheological properties of the model materials plasticine and silicone putty, respectively. Dixon and Summers (1985), Dixon and Tirrul (1991), and Dixon and Liu (1992) discussed the scaling relationships between prototype and model foreland fold-thrust-belt systems applicable to centrifuge models of the type described here. A brief summary of the model configuration and the experimental procedure follows. Table 1 lists the model scaling ratios applicable to this study. The linear scale ratio is set at 2×10^{-6} for this study, on the basis of the thickness of the stratigraphic column of interest.

Figure 9 shows the dimensions and initial configurations of the physical models. They are constructed of Harbutt's Gold Medal™ Plasticine simulating competent rocks such as limestone and dolostone, and Dow-Corning silicone putty (dilatant compound 3179) simulating incompetent rocks such as shale (Dixon and Summers, 1985). The models include eight stratigraphic units that represent the complete section from the level of the Middle Cambrian (which overlies the basal décollement) to the Jurassic Fernie Formation (Table 2 and Figure 9a). Competent units (Dcn, Dpa, Mru, P-T) are represented by massive plasticine, and less-competent units (C, Dpx, Dmh, Mbf, JLK) are represented by interlayered plasticine and silicone putty, with the ratio of the two constituents' laminae thicknesses ranging from 3:1 to 1:1 to approximate the bulk competency of each stratigraphic unit. Following Dixon's (1997) technique, the unit above the model Cambrian is constructed in two parts with different competencies to simulate the transverse facies boundary between the Cairn/Southesk carbonates and the Perdrix shale of the Cline Channel to the north. The dip of the facies boundary initially was approximately vertical, and its trend was parallel to the direction of overthrusting imposed on the models. The model stratigraphic succession rests on a rigid base (the aluminum base plate) and the shortening occurs above a basal décollement (Figure 9b).

The models were deformed at between 2500 and 4000 ***g*** in the centrifuge of the Experimental Tectonics Laboratory at Queen's University. They were subjected to three stages (numbered I to III) of horizontal compression from one end by the collapse and lateral spreading of a "hinterland wedge" of plasticine (see Figure 10).

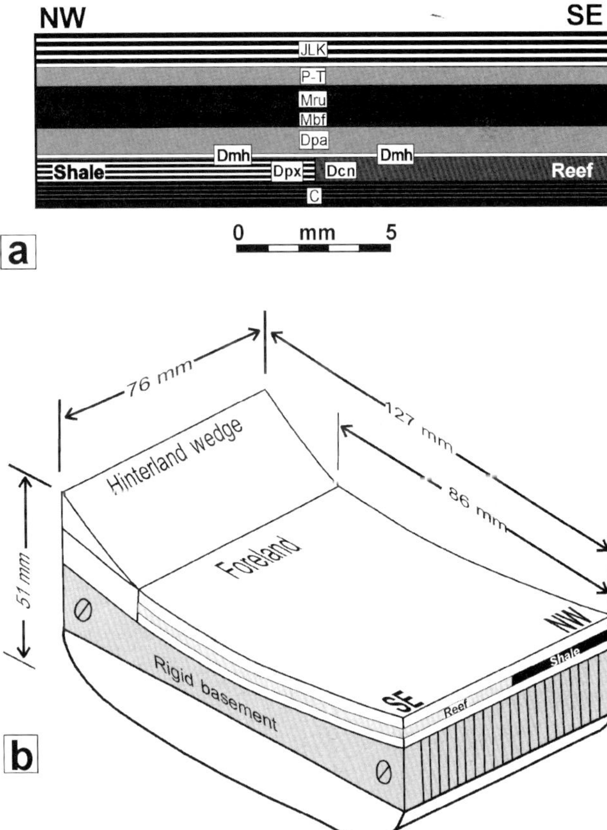

FIGURE 9. Model configurations. (a) Longitudinal section across the facies boundary. Solid black and gray units are plasticine (strong), striped units are interbedded plasticine and white silicone putty (weak). C = Cambrian, Dpx = Perdrix, Dcn = Cairn/Southesk, Dmh = Mt. Hawk, Dpa = Palliser, Mbf = Banff, Mru = Rundle, P–T = Permian and Triassic, JLK = Jurassic and Lower Cretaceous. (b) Schematic 3-D diagram showing model dimensions and configuration with vertical facies boundary perpendicular to the leading edge of the driving wedge. Shale facies shown in black, reef facies in gray. Layering is not to scale; refer to Table 2 and Figure 9a for unit thicknesses and compositions.

Table 2. Stratigraphic column for the physical models.

Unit	*Ratio (P:sp)*	*Number of laminations*	*Final thickness (mm)*	*Nature (m)*
Jurassic/Lower Cretaceous (JLK)	1:1	8	0.84	400
Permian-Triassic (P-T)	1:0	1	0.52	250
Rundle (Mru)	1:0	1	0.83	400
Banff (Mbf)	3:1	8	0.50	250
Palliser (Dpa)	1:0	1	0.85	400
Mt Hawk (Dmh)	0:1	1	0.11	50
Cairn/Sx (Dcn)	*1:0*	*1*	*0.62**	*250*
or Perdrix (Dpx)	*1:1*	*8*	*0.62**	*250*
Cambrian (C)	3:1**	8	0.79	350
TOTAL:			**~5.00 mm**	**2350 m**

Length ratio ~2×10^{-6} (1 mm = 0.5 km).
P = competent Plasticine, sp = incompetent silicone putty.
**These are nominal values predicted on basis of rolling sequence. In fact, the Perdrix unit ended up thicker than the Cairn/Southesk because of differences in the relative strengths during rolling.*
***Silicone putty lamina on lower side, in contact with the aluminum base. No other lubrication.*

Each experimental stage involved about 180 s, during which the centripetal acceleration rose from 0 to 4000 ***g***, and then 300 s at 4000 ***g***, followed by deceleration. The hinterland wedge began to collapse and spread laterally at about 2500 ***g*** and reached a stable profile after about 200 s at 4000 ***g***. After each stage, a few transverse sections were cut and photographed to document the progressive evolution of structures. The sections were reassembled and the hinterland wedge was reconstructed to its initial slope (to restore its gravitational potential) prior to each successive stage of shortening. After the final stage, each model was cut into serial sections spaced at an interval of 3.0 mm.

MODELING RESULTS

Stages of Deformation

The initial geometry, progressive development, and 3-D geometry of structures that develop over a transverse facies boundary are shown in Figure 10. The position of the boundary is indicated by the dashed white lines in the plan views (Figure 10a, b, and c) and in the strike section as it appears prior to deformation (Figure 10d). Plan views of the model at deformational stages I, II, and III show how the deformation front propagates toward the foreland, from left to right (Figure 10a, b, and c). A large anticline develops over the shale facies, near the hinterland, early in the deformation, and it grows as more folds develop in front of it (Figure 10a, b, c, e, and f, slices 43.0–55.0). At stage I (Figure 10e), shortening occurs by bedding-parallel detachment in the weak layers and buckling of the stronger layers. The deeper layers have longer fold wavelengths than the uppermost unit (model Jurassic-Lower Cretaceous), which has short wavelengths but widely spaced folds (Figure 10a, CC5-I). Folds near the surface appear to initiate over the very broad synclines best seen in the black layer (model Mississippian Rundle Group) in Figure 10e. As deformation progresses to stage II, the wavelengths of near-surface folds remain nearly constant, but more folds develop between earlier folds near the hinterland and across the model (Figure 10b, CC5-II). This indicates that, in the models at least, there is not a simple hinterland-to-foreland progression of deformation; simultaneous folding and out-of-sequence faulting occur as well. At stage III, the number of folds does not increase significantly, but tightness increases as wavelength decreases.

Model Fold Styles

Over the reef facies, the fold amplitudes and wavelengths are more consistent across the model and the thrust sheets and most of the folds are foreland verging (Figure 10f, slices 19.0–31.0; Figure 11b), just like those in the Gable Mountain and Rough Mountain areas (Figures 3b, 4b, and 6b). Splay faults branch from the basal décollement above the base-plate and lose displacement upward into folds (Figures 10f and 11b). Minimal faulting occurs in the model Palliser unit and it conforms to the shape of the underlying fault-propagation folds.

In comparison, folds that develop over the shale facies are more upright, and some are slightly inclined toward the foreland and others toward the hinterland. All have developed over a bedding-parallel detachment

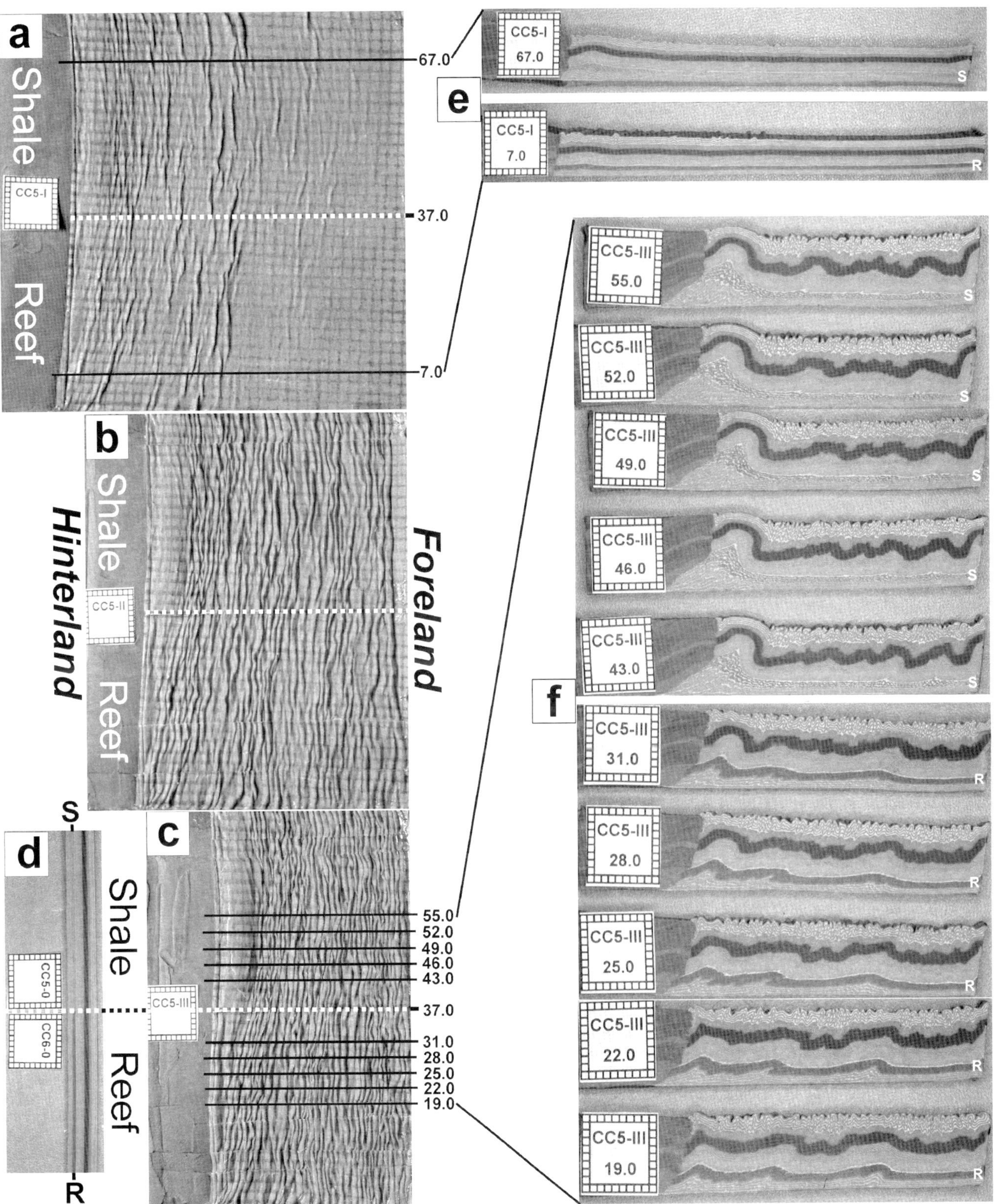

Figure 10. Plan views of (a) stage I, (b) stage II, and (c) stage III deformation. (d) Longitudinal section of undeformed stratigraphy. (e) Transverse sections of structures above the shale (S) and reef (R) facies at stage I deformation. (f) Transverse serial sections above the shale (S) and reef (R) facies at stage III deformation. Model labels are 1 cm square with mm divisions.

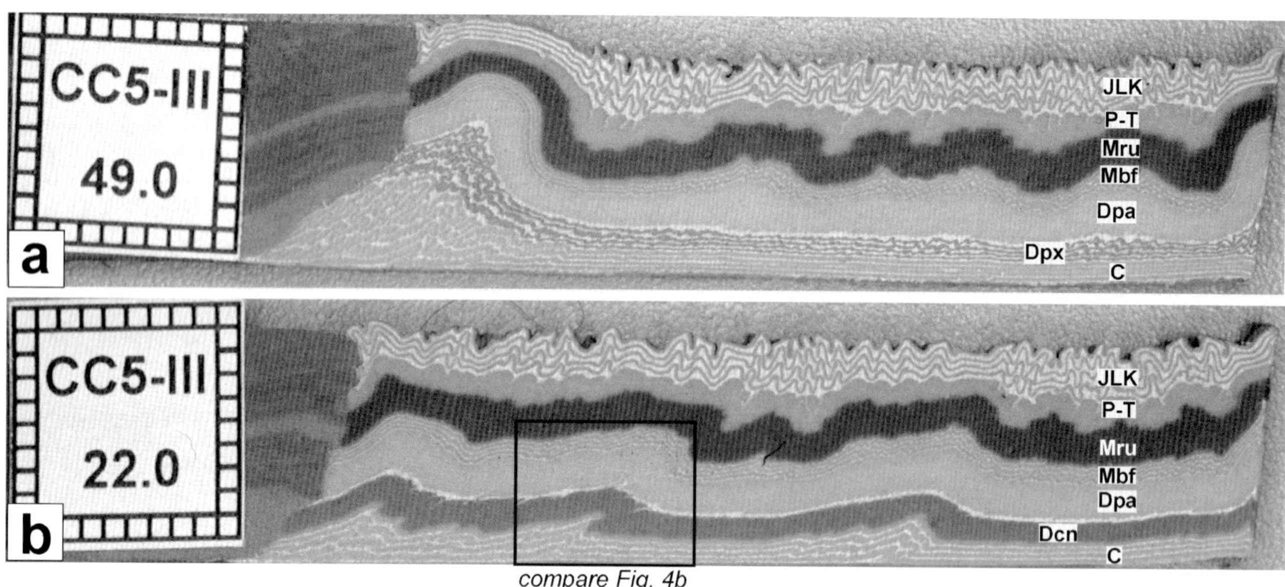

FIGURE 11. Close-ups of representative slices through the (a) shale facies (section 49.0) and (b) reef facies (section 22.0). Dark gray plasticine layer (Dcn) in 11b represents the Fairholme reef facies above the basal Cambrian unit; it is replaced by weaker Perdrix (Dpx; interlayered medium gray plasticine and white silicone putty) in 11a. All other layers are the same composition and thickness across the model; see Table 2 and Figure 9a for details and abbreviations. Model labels are 1 cm square with mm divisions.

within the model Fairholme shale facies, and display ductile hinge thickening that is especially obvious in the model Palliser unit, just as it is in the Michener Creek area (Figures 3a, 4a, and 6a).

Model folds involving the reef facies (Figure 11b) compare very well with those in the field area (Figures 3b, 4b, 6b) in type, geometry, and scale. Both are foreland-verging fault-propagation folds with minor complications in their cores. Using the scaling ratio of Table 2, the frame of Figure 4b is superimposed on Figure 11b to illustrate equivalent scales and geometries.

Modeling of structures over the shale facies is not quite as comparable. The model displays the same upright, similar, detachment fold geometry seen in the field (Figures 3a, 4a, 6a), and the wavelengths of the folds in the middle of Figure 11a are approximately half as large as in Figure 11b, but the details of the structures do not correspond well. This is because of limitations of the scale of the models and the materials used. Although this is the most complex stratigraphy modeled to date, the thinnest lamination represents 50 m of strata, which is an order of magnitude larger than the thinnest unit mapped in the field. Thus the small-wavelength structures of Figure 6a could not be modeled. It also appears that the rheology we chose for the Perdrix shale (Dpx) analog is a bit too weak relative to the overlying plasticine Palliser (Dpa), because the Palliser is detached from the Perdrix in the model yet folds with the Palliser as a multilayer in nature (Figure 4a). The natural examples also contain more faults, but many may have been late features because they cut tight folds (Figures 4a, 6a).

Examination of the serial sections, in conjunction with the plot of orthogonal thickness of the Palliser versus arc distance from the hinterland edge of the model, for representative slices through the shale and reef facies, reveals several key points (Figures 10f, 11, and 12). The folds above the shale facies have more hinge thickening and there are more of them. Discounting the edge effects at the right-hand side of the model, there are three main anticlines in the Dpa/Mbf above the reef facies (Figures 10f, 11b) and four or five above the shale facies (Figures 10f, 11a). The folds can be traced from section to section, with the fault-propagation folds involving the reef facies changing along strike into detachment folds above the shale facies. Individual folds are translated farther toward the foreland above the shale facies.

Blind Lateral Ramps

The basal detachment below the reef facies is at a deeper level and therefore must ramp up-section to the northwest from near the base of the model Middle Cambrian unit in slice 31.0 to near the top of the Devonian Perdrix shale in slice 43.0 (Figures 10f and 11). This would be analogous to a 600-m-high lateral ramp in nature (Table 2) and is of the same order as the lateral ramp farther east in the Bighorn Thrust, documented by Douglas (1956), which ramps from Upper

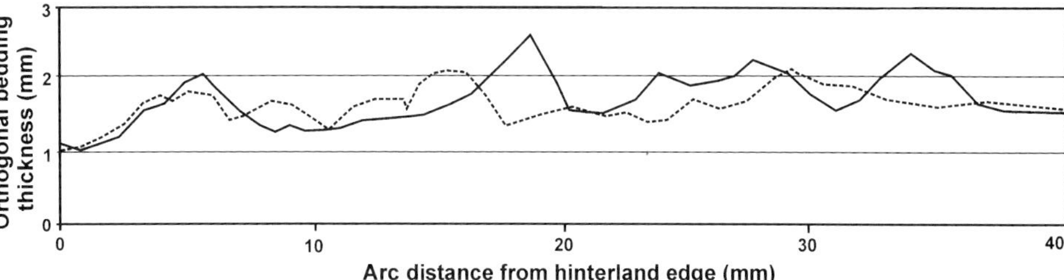

FIGURE 12. Plot of orthogonal thickness of the model Palliser unit versus arc distance from the hinterland edge of the stage III model. Solid line represents Palliser above the shale facies in section 49.0 (Figure 11a); dashed line represents Palliser above the reef facies in section 22.0 (Figure 11b).

Cambrian into Devonian strata. The Bare Thrust also follows a detachment horizon within the Upper Cambrian in the study area (Figure 4b) (Beattie, 1984) but cuts up-section to a flat in the Devonian Perdrix shales and ramps through overlying Paleozoic strata at Mt. Michener (Figure 4a) (Mountjoy and Price, 1974b; Beattie, 1984). Note that these lateral ramps are not accompanied by any obvious deviation in the map traces of the Bare and Bighorn Thrusts (Figure 1).

Exploration Significance of the Models

Geometries of deep layers in the model have important implications for exploration even though strata beneath the Devonian Fairholme Group are not a target in the Alberta Foothills. The model Cambrian unit was created with a uniform thickness and moderately strong bulk rheology (made up of interlayered strong plasticine and weak silicone putty) across the entire model to avoid introducing into the study more variables of the effect of the Devonian facies boundary. However, the model Cambrian is folded and thrusted along with the overlying stronger reef facies, yet remains undeformed beneath the weaker shale facies. Thus the facies transition in the Devonian Fairholme Group affects the structural style of the entire multilayer, including units below it as well as those above it, even though the Fairholme unit makes up only 10% of the model's thickness. When exploring for deep structures, it is important to better constrain the lithologies of all units above the basement, because one or more weak layers within the stratigraphic package can affect the structural style of the entire supracrustal wedge (Thompson, 1981; Price, 1981, 1994; Spratt and Lawton, 1996). The models also show that the surface geology may be of little help in identifying the positions of deep, blind facies transitions or lateral ramps beneath the Foothills, except in the earliest stages of deformation (Figure 10a). Folds in the top two layers at stages II and III of deformation remain remarkably consistent in style and orientation along strike. Over the shale facies, a slight bowing of axial traces toward the foreland is evident when the entire width of the Foothills model is viewed (Figures 10b and c) but would not be apparent when studying smaller areas or plays. Tight folds such as these are found in Cretaceous clastics at the surface in the Foothills (Bally et al., 1966; Dahlstrom, 1970), but most of the Triassic and younger strata have been eroded from the North Saskatchewan River study area. Examination of the serial sections suggests that quantifying the 3-D geometry of a deeper layer, such as the top of the Palliser, would help to better constrain the position of the underlying facies boundary. The Devonian Palliser Formation is also particularly well exposed in the study area and is a prolific oil and gas producer in parts of the subsurface of Alberta where it is known as the Wabamun Group.

STRUCTURE-CONTOUR MAPS

Structure-contour maps of the top of the Palliser have been constructed over both the model (Figure 13a) and the field data (Figure 13b) to illustrate the orientations, continuity, and closure of structures. They reveal broadly sinuous fold and fault traces over the facies transition, with structures above the shale facies translated farther toward the foreland. The sinuosity is more pronounced than that observed on the surface of the models and at higher structural levels on the ground. Although a few of the folds and thrusts have propagated or coalesced across the facies transition, several structures terminate above or near the position of the underlying bank margin. Many of these structures have strike lengths that are 10–50% of the length of the through-going structures, thereby limiting the size of potential structural traps.

COMPARISON WITH OTHER BELTS

Mitra (1990) states that variations in mechanical properties can cause fault-propagation folds to change shape and flatten into detachments within incompetent strata. Serial sections of our models (Figure 10f) show how this transition occurs along strike. Mitra (1990) shows Appalachian examples of fault-propagation folds

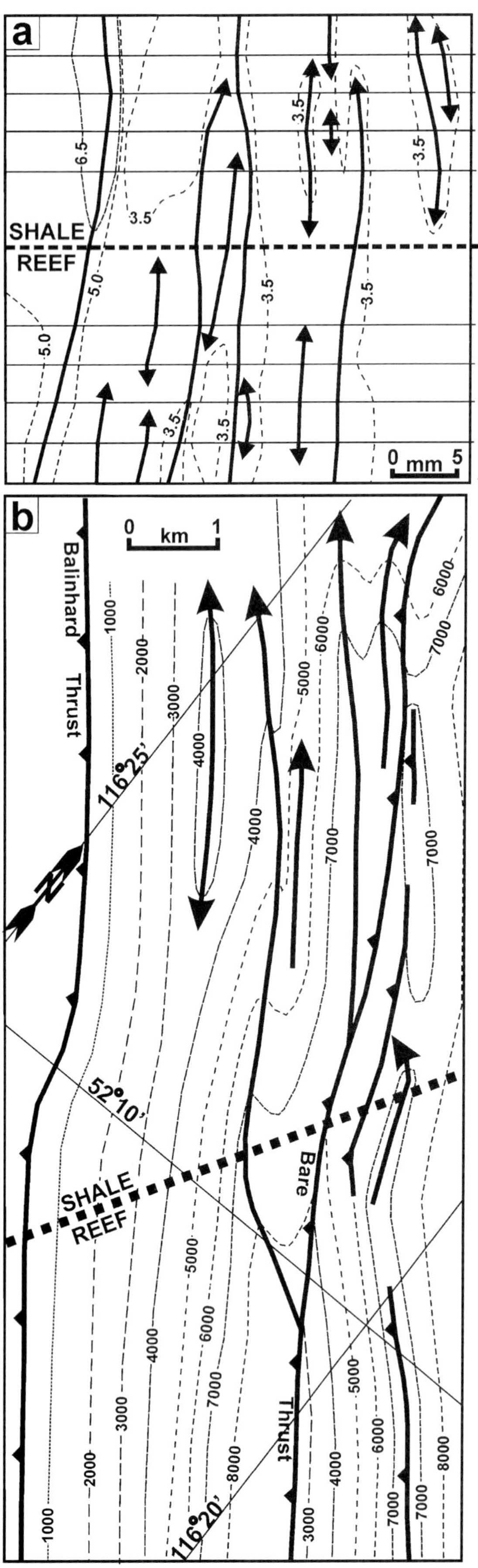

modified by thrusting (and back thrusting) that look like small-scale versions of the moderately shortened folds involving competent lithofacies in the Rough Mountain area (Beattie, 1984; Price and Mountjoy, 1977a, b) and the Idaho-Wyoming thrust belt (Woodward and Rutherford, 1989). Brittle bourrage in the cores of folds (e.g., Figure 6b) resembles structures of similar scales in limestones of the French subalpine chains, for which Butler (1992) proposes a structural sequence of detachment folding, followed by fault-propagation folding, followed by break thrusting. Our models and Tavarnelli's (1997) work in the Appenines support this relative timing of map-scale structures, but also include a stage of amplification and tightening of the detachment folds prior to break thrusting. However, one should not attempt to deduce the relative timing of a structure solely on the basis of its geometry, because the fault-propagation folds and detachment folds developed simultaneously, along strike from each other, in our models.

On a more regional scale, the change from fault-dominated structures involving the competent facies to detachment folds above the incompetent facies is also observed across platform-basin transitions in the Dolomites (Schönborn, 1992, 1999), Alps (Pfiffner, 1993; Velaj et al., 1999), Appenines (Ghisetti and Vezzani, 1997), Pyrenees (Alonso and Teixell, 1992; Vergés et al., 1992), and other parts of the Rockies (Cook, 1975; Royse et al., 1975; Mountjoy, 1960, 1992; Woodward and Rutherford, 1989). One thing we do not see in surface maps of the field area (Beattie, 1984; Price and Mountjoy, 1977a, 1977b) is the pronounced swing in thrust and fold traces observed near facies boundaries in the external Alps (Butler, 1992; Velaj et al., 1999) and Pyrenees (Vergés et al., 1992). This is partly because salt in these other areas provides a greater competency contrast than shale does, partly because of the later deformational stage and larger amount of shortening preserved in the Canadian Front Ranges, and partly because the deeper structures in our study are obscured by overlying disharmonic structures. Our models show that folds developed over the shale facies occur farther to the foreland than do those above the reef facies at early stages of deformation (Figure 10a), but the difference

Figure 13. Structure-contour maps. (a) Contoured top of the Palliser unit in stage III of the model, constructed from blowups of serial sections 19.0–55.0 (Figure 10f). Contour interval is 1.5 mm. (b) Contoured top of the Palliser Formation in the Mt. Michener area, based on maps and sections of Beattie (1984) and Mountjoy and Price (1974b). Contours are shown as dashed lines, anticlinal traces with solid lines, plunging fold axes with arrows, and thrust traces with triangular barbs on the upthrown side. Contour interval is 1000 ft (305 m) (base maps have Imperial topographic contours).

becomes less pronounced as the folds tighten and propagate laterally (Figure 10b and c). Stripping off the overlying, disharmonically folded strata does reveal a small but distinct swing in axial traces near the transverse lithofacies boundary at the level of the Palliser, as seen in structure contour maps of our model and the field data (Figure 13). It is difficult to directly compare the structural details developed over the Cline Channel with those above the intermediate basin of the Friuli platform, because that portion of the Dolomites has been further modified by transtension (Schönborn, 1999).

CONCLUSIONS

Strong, competent rocks of uniform thickness and lithology, like the Palliser Formation and Rundle Group, are generally thought of as controlling the geometry of fold-and-thrust belt structures and having predictable geometries and wavelengths (Currie et al., 1962). However, their fold geometry also depends on the lithology of the underlying rocks and varies drastically along strike over transverse facies boundaries. When a layer of weak rock is introduced that comprises as little as 10% of the stratigraphic package, it can significantly change the structural style of the entire package, including units below it as well as those above it. Where the stronger lithofacies adds to the thickness of the competent beam, it is involved in the thrusting and large-scale folding, whereas the same beds below the weaker lithofacies remain undeformed. This requires the spontaneous development of a lateral ramp; the basal décollement ramps up-section along strike, from a detachment in Cambrian strata above the crystalline basement to a detachment in the Devonian shale facies. Lateral ramps as high as 600 m, 15–25% of the height of the supercrustal wedge, have developed in the field area and scale models, yet they are not easily delineated by surface mapping. Structure-contour maps of the deeper layers are more useful for locating buried facies boundaries than are representations of the ground/model surface. Sinuous fault and fold traces develop over the facies transition, with structures over the shale facies developed farther toward the foreland; the sinuosity is more pronounced closer to the facies boundary. Few folds and thrusts propagate across the facies transition; several terminate above or near it, and their strike lengths are 10–50% of the length of through-going structures, which significantly reduces the size of potential structural traps.

In the modeled structures, there is not a simple hinterland-to-foreland progression of deformation; simultaneous folding and out-of-sequence faulting occur as well. Initially, in the models, broad buckles form from hinterland to foreland and advance farther toward the foreland above the shale facies. With increased shortening, detachment folds over the shale facies are simultaneously amplified while additional out-of-sequence folds and thrusts develop over the reef facies. Detachment folds that formed early over the shale facies propagate laterally and develop into, and coalesce with, fault-propagation folds initiated beneath the reef facies. At 50% shortening, most of the natural and model structures are hybrids that are flattened in addition to displaying elements of detachment, fault-propagation, and break-thrust folding in cross sections and along strike. Thus it would be futile to use only one of the ideal fault-related fold models to characterize and construct sections through these real folds.

The structures at Rough Mountain and their analog models, which are underlain by the carbonate reef facies of the Fairholme Group, approximate ideal concentric folds accompanied by brittle bourrage that formed by fault-propagation, fault-bend, and break-thrust folding. They are representative of the typical fold-and-thrust styles observed in the Canadian Rockies south and north of the study area. The folds at Mt. Michener and their physical analogs, which are underlain by the shale facies of the Fairholme Group, geometrically approximate ideal similar folds with interlimb angles of 10°–72°. In the tight folds, pressure solution and ductile strain reduce the porosity of the reservoir rocks, and calcite filling of fractures decreases their permeability. These differences in fabric and reservoir quality cannot be attributed to changes in burial temperatures and pressures, because they are found only 1.5–10 km apart, along strike, in the same rock units, at the same structural level, and in the same thrust sheet. Both areas are shortened by 50%, so the brittle-ductile transition must result from a change in strain rate, with episodic faulting occurring in strata over the reef facies and more-continuous folding occurring along strike where the strata are detached from the underlying shale facies.

The average wavelength of folds above the reef facies is twice that of folds above the shale facies, both in the models and natural examples, so folds detached above the shale facies are generally tighter and smaller than structural targets developed over the reef facies. This is to be expected because of the decreased thickness of the competent beam, but one must also watch for tectonically induced changes in reservoir quality associated with the change in wavelength. In the field and in the models, many structures on both sides of the lithofacies boundary terminate near the boundary, displaying closure of folds in the overlying Mississippian reservoir units with plunges of as much as 28°. A few of the larger folds continue across the boundary for several kilometers and would also be prospective targets if encountered in the subsurface. However, if such brittle, fault-cored folds are tracked along strike to where they become detachment folds, tectonically reduced porosity and permeability could limit their exploitation potential.

ACKNOWLEDGMENTS

Financial support for this work was provided by industry sponsors of the Fold-Fault Research Project (FRP) and the Natural Sciences and Engineering Research Council of Canada. We thank Eric Mountjoy and Mark Cooper for their helpful reviews of the manuscript.

REFERENCES CITED

Alonso, J. L., and A. Teixell, 1992, Forelimb deformation in some natural examples of fault-propagation folds, *in* K. R. McClay, ed., Thrust tectonics: London, Chapman & Hall, p. 175–180.

Bally, A. W., P. L. Gordy, and G. A. Stewart, 1966, Structure, seismic data and orogenic evolution of southern Canadian Rocky Mountains: Bulletin of Canadian Petroleum Geology, v. 14, p. 337–381.

Beattie, E. T., 1984, Structural style affected by Devonian facies changes, Ram Range, Alberta: M.Sc. Thesis, University of Calgary, 146 p.

Butler, R. W. H., 1992, Structural evolution of the western Chartreuse fold and thrust system, NW French subalpine chains, *in* K. R. McClay, ed., Thrust tectonics: London, Chapman & Hall, p. 287–298.

Cook, D. G., 1975, Structural style influenced by lithofacies, Rocky Mountain Main Ranges, Alberta–British Columbia: Geological Survey of Canada, Bulletin 233, 73 p.

Currie, J. B., H. W. Patnode, and R. P. Trump, 1962, Development of folds in sedimentary strata: Geological Society of America Bulletin, v. 73, p. 655–674.

Dahlstrom, C. D. A., 1970, Structural geology of the eastern margin of the Canadian Rocky Mountains: Bulletin of Canadian Petroleum Geology, v. 18, p. 331–406.

DeSitter, L. U., 1964, Variations in tectonic style: Bulletin of Canadian Petroleum Geology, v. 12, p. 263–278.

Dixon, J. M., and S. Liu, 1992, Centrifuge modelling of the propagation of thrust faults, *in* K. R. McClay, ed., Thrust tectonics: London, Chapman & Hall, p. 53–70.

Dixon, J. M., and J. M. Summers, 1985, Recent developments in centrifuge modelling of tectonic processes: equipment, model construction techniques and rheology of model materials: Journal of Structural Geology, v. 7, p. 83–102.

Dixon, J. M., and J. M. Summers, 1986, Another word on the rheology of silicone putty: Bingham: Journal of Structural Geology, v. 8, p. 593–595.

Dixon, J. M., and R. Tirrul, 1991, Centrifuge modelling of fold–thrust structures in a tripartite stratigraphic succession: Journal of Structural Geology, v. 13, p. 3–20.

Dixon, J. M., 1997, Physical (centrifuge) modelling of the influence of lateral facies boundaries on folding and thrusting, and implications for hydrocarbon trapping: American Association of Petroleum Geologists Annual Meeting, Official Program, 6, p. A28.

Douglas, R. J. W., 1956, Nordegg, Alberta: Geological Survey of Canada Paper 55-34, 31 p.

Erslev, E. A., and K. R. Mayborn, 1997, Multiple geometries and modes of fault-propagation folding in the Canadian thrust belt: Journal of Structural Geology, v. 19, p. 321–335.

Fischer, M. P., N. B. Woodward, and M. M. Mitchell, 1992, The kinematics of break-thrust folds: Journal of Structural Geology, v. 14, p. 451–460.

Fox, F. G., 1969, Some principles governing interpretation of structure in the Rocky Mountain orogenic belt, *in* P. E., Kent, G. E. Satterthwaite, and A. M. Spencer, eds., Time and place in orogeny: London, Geological Society of London, 311 p.

Ghisetti, F., and L. Vezzani, 1997, Geometrie deformative ed evoluzione cinematica dell'Appennino centrale: Studi Geologici Camerti, v. 14, p. 127–154.

Groshong, R. H., and J.-L. Epard, 1994, The role of strain in area-constant detachment folding: Journal of Structural Geology, v. 16, p. 613–618.

Hargreaves, G. E., 1959, Devonian section at Hummingbird Reef, *in* H. Hornford, ed., Canadian Rockies, Bow River to North Saskatchewan River, 16th Annual Field Conference Guidebook: Calgary, Alberta Society of Petroleum Geologists, 168 p.

Hargreaves, G. E., 1968, Nisku lithofacies of Rocky Mountains, Alberta, *in* G. H. Austin, ed., Moose Mountain-Drumheller, 9th Annual Field Conference Guidebook: Calgary, Alberta Society of Petroleum Geologists, 196 p.

Harris, L. D., and R. C. Milici, 1977, Characteristics of thin-skinned style of deformation in the southern Appalachians and potential hydrocarbon traps: USGS Professional Paper 1018, p. 1–40.

Hubbert, M. K., 1937, Theory of scale models as applied to the study of geological structures: Bulletin of the Geological Society of America, v. 48, p. 1459–1520.

Jamison, W. R., 1987, Geometric analysis of fold development in overthrust terrains: Journal of Structural Geology, v. 9, p. 207–219.

Jones, P. B., and R. H. Workum, 1978, Geological guide to the central Foothills and Rocky Mountains of Alberta: Calgary, Canadian Society of Petroleum Geologists, 61 p.

McClay, K. R., 1976, The rheology of Plasticine: Tectonophysics, v. 33, p. T7–T15.

McLaren, D. J., 1955, Devonian formations in the Alberta Rocky Mountains between Bow and Athabasca Rivers: Geological Survey of Canada Bulletin, v. 35, 59 p.

Mitra, S., 1990, Fault-propagation folds: geometry, kinematic evolution, and hydrocarbon traps: AAPG Bulletin, v. 74, p. 921–945.

Mossop, G. D., and I. Shetsen, compilers, 1994, Geological atlas of the Western Canada Sedimentary Basin: Calgary, Canadian Society of Petroleum Geologists and Alberta Research Council. Online version: http://www.ags.gov.ab.ca/AGS_PUB/ATLAS_WWW/ATLAS.HTM (accessed March 7, 2003).

Mountjoy, E. W., 1960, Structure and stratigraphy of the Miette and adjacent areas, eastern Jasper National Park, Alberta: Ph.D. thesis, University of Toronto, 249 p.

Mountjoy, E. W., 1992, Significance of rotated folds and thrust faults, Alberta Rocky Mountains, *in* S. Mitra, ed., Structural geology of fold-and-thrust belts: Baltimore, The Johns Hopkins University Press, p. 207–223.

Mountjoy, E. W., and R. A. Price, 1974a, Whiterabbit Creek map 1388A (east half): Ottawa, Geological Survey of Canada, 1:50,000 scale, 2 sheets.

Mountjoy, E. W., and R. A. Price, 1974b, Whiterabbit Creek map 1389A (west half): Ottawa, Geological Survey of Canada, 1:50,000 scale, 2 sheets.

Pfiffner, O. A., 1993, The structure of the Helvetic nappes and its relation to the mechanical stratigraphy: Journal of Structural Geology, v. 15, p. 511–521.

Price, R. A., 1981, The Cordilleran foreland thrust and fold belt in the southern Canadian Rocky Mountains, *in* K. R. McClay and N. J. Price, eds., Thrust and nappe tectonics: Geological Society of London Special Publication 9, p. 427–448.

Price, R. A., 1994, Cordilleran tectonics and the evolution of the Western Canada Sedimentary Basin, chapter 2 *in* G. D. Mossop and I. Shetsen, compilers, Geological atlas of the Western Canada Sedimentary Basin: Calgary, Canadian Society of Petroleum Geologists and Alberta Research Council. Online version: http://www.ags.gov.ab.ca/AGS_PUB/ATLAS_WWW/A_CH02/CH_02.HTM (accessed March 7, 2003).

Price, R. A., and E. W. Mountjoy, 1977a, Siffleur River map 1465A (east half): Ottawa, Geological Survey of Canada, 1:50,000 scale, 2 sheets.

Price, R. A., and E. W. Mountjoy, 1977b, Siffleur River map 1466A (west half): Ottawa, Geological Survey of Canada, 1:50,000 scale, 2 sheets.

Ramberg, H., 1967, Gravity, deformation and the earth's crust as studied by centrifuge models: London, Academic Press, 214 p.

Ramsay, J. G., 1967, Folding and fracturing of rocks: New York, McGraw-Hill, 568 p.

Royse, F. Jr., M. A. Warner, and D. L. Reese, 1975, Thrust belt structural geometry and related stratigraphic problems—Wyoming, Idaho, northern Utah: Rocky Mountain Association of Geologists, Symposium on Deep Drilling Frontiers in Central Rocky Mountains, p. 41–54.

Spratt, D. A., and D. C. Lawton, 1996, Variations in detachment levels, ramp angles and wedge geometries along the Alberta thrust front: Bulletin of Canadian Petroleum Geology, v. 44, p. 313–323.

Schönborn, G., 1992, Kinematics of a transverse zone in the southern Alps, Italy, *in* K. R. McClay, ed., Thrust tectonics: London, Chapman & Hall, p. 299–310.

Schönborn, G., 1999, Balancing cross sections with kinematic constraints; the Dolomites (northern Italy): Tectonics, v. 18, p. 527–545.

Tavarnelli, E., 1997, Structural evolution of a foreland fold-and-thrust belt: the Umbria-Marche Appenines, Italy: Journal of Structural Geology, v. 19, p. 523–534.

Thompson, R. I., 1981, The nature and significance of large "blind" thrusts within the northern Rocky Mountains of Canada, *in* K. R. McClay and N. J. Price, eds., Thrust and nappe tectonics: Geological Society (London) Special Publication 9, p. 449–462.

Thompson, R. I., 1989, Stratigraphy, tectonic evolution and structural analysis of the Halfway River map area (94B), northern Rocky Mountains, British Columbia: Ottawa, Geological Survey of Canada, Memoir 425, 119 p.

Velaj, T., I. Davison, A. Serjani, and I. Alsop, 1999, Thrust tectonics and the role of evaporites in the Ionian Zone of the Albanides: AAPG Bulletin, v. 83, p. 1408–1425.

Vergés, J., J. A. Muñoz, and A. Martínez, 1992, South Pyrenean fold-and-thrust belt: Role of foreland evaporitic levels in thrust geometry, *in* K. R. McClay, ed., Thrust tectonics: London, Chapman & Hall, p. 255–264.

Verrall, P., 1968, Geological compilation, Bow River to North Saskatchewan River, Alberta, *in* H. Hornford, ed., Canadian Rockies, Bow River to North Saskatchewan River, 16th Annual Field Conference Guidebook: Calgary, Alberta Society of Petroleum Geologists, 168 p.

Woodhead, M. P., and C. Wright-Broughton, 1967, A map showing subsurface distribution of the Devonian in part of the Canadian Rocky Mountains with a reconstruction of the reef facies of the disturbed belt: Calgary, Alberta Society of Petroleum Geologists, 1:506,880 scale, 1 sheet.

Woodward, N. B., and E. Rutherford, 1989, Structural lithic units in external orogenic zones: Tectonophysics, v. 158, p. 247–267.

Workum, R. H., 1978, Hummingbird, *in* I. A. McIlreath and P. C. Jackson, eds., The Fairholme carbonate complex at Hummingbird and Cripple Creek: Calgary, Canadian Society of Petroleum Geologists, 87 p.

15

Yamada, Y., and K. McClay, 2004, 3-D Analog modeling of inversion thrust structures, *in* K. R. McClay, ed., Thrust tectonics and hydrocarbon systems: AAPG Memoir 82, p. 276–301.

3-D Analog Modeling of Inversion Thrust Structures

Yasuhiro Yamada[1]

Fault Dynamics Research Group, Geology Department, Royal Holloway University of London, Egham, Surrey, United Kingdom

Ken McClay

Fault Dynamics Research Group, Geology Department, Royal Holloway University of London, Egham, Surrey, United Kingdom

ABSTRACT

The geometries and kinematic evolution of 3-D inversion thrust structures have been modeled using 3-D sandbox analogs of hanging-wall deformation above footwall blocks with both concave-up and convex-up listric geometries. Extension over 3-D concave-up listric detachments produced characteristic rollover anticlines and crestal-collapse graben systems that parallel the along-strike, sinusoidal or cuspate plan geometries of the detachment breakaways. Three-dimensional inversion by horizontal contraction produced asymmetric, thrust-fault-bounded, hanging-wall inversion anticlines with curved axial traces that also follow the plan-view shape of the extensional breakaways. The main detachments were reactivated during the inversion, and new, steep thrust segments propagated upward from the detachment breakaways. Shallow to moderately dipping hanging-wall back thrusts also propagated outward from the crestal-collapse graben systems. The periclinal inversion anticlines exhibited two plunge culminations above the most concave sections of the main detachment surface. Extension above a 3-D convex-up, listric detachment surface produced a hanging-wall syncline together with a narrow, complex crestal-collapse graben system. In plan view, the axes of the hanging-wall syncline and the crestal-collapse graben formed parallel to the sinusoidal detachment breakaway. Inversion of this system produced a broad, thrust-fault-bounded anticline that shows along-strike plunge culminations. The main detachment surface was reactivated, and a moderately dipping thrust propagated upward through the syninversion strata. Segmented, hanging-wall back thrusts formed subparallel to the main detachment breakaway. Vertical and horizontal sections through the completed models were used to construct 3-D synoptic models for these inversion systems. The results of the analog experiments compare well with published examples of 3-D inversion structures from petroleum basins in the North Sea, Indonesia, and Argentina.

[1]*Present address:* Dept. of Civil and Earth Resources Engineering, Kyoto University, Kyoto, Japan.

INTRODUCTION

Inversion structures develop where there is a change in the stress field, from extension to contraction or vice versa (Williams et al., 1989). This transformation of the stress field usually causes reactivation of preexisting faults in the opposite mode as well as the development of new faults. Inversion-related structures are formed in both the hanging walls and footwalls of the reactivated faults. In most sedimentary basins, the common mode of inversion is "positive inversion" (Williams et al., 1989), in which preexisting extensional faults are reactivated as contractional faults (Figure 1). In this chapter, the term inversion is used for positive inversion— that is, extension followed by contraction. Inversion structures have been recognized in many sedimentary basins in a variety of tectonic environments (Lowell, 1985; Cooper and Williams, 1989; Buchanan and Buchanan, 1995). In particular, inversion has been invoked to explain basement involvement in some fold-thrust belts, such as in Papua New Guinea, (Buchanan and Warburton, 1996), in the sub-Andean Neuquen Basin, Argentina (Manceda and Figueroa, 1995), and in the Syrian Arc Fold Belt, Egypt (Ayyad, 1997). Inversion produces thrust structures and related hanging-wall anticlines, which are important trap-forming systems in many hydrocarbon basins, including the southern North Sea (Ziegler, 1987), Indonesia (Ginger et al., 1993; Phillips et al., 1997), and in the Neuquen and Cuyo Basins of Argentina (Manceda and Figueroa, 1995; Uliana et al., 1995).

The aims of the research presented in this chapter were to develop geometric and kinematic models of 3-D inversion structures using scaled physical models. In particular, this study focused on the inversion of 3-D listric fault systems to produce structural templates for seismic interpretation of inversion systems. Previous experimental studies of inverted extensional fault systems have been highly successful in developing templates for the kinematic evolution of 2-D inversion structures (Koopman et al., 1987; McClay, 1989; Buchanan, 1991; McClay and Buchanan, 1992; Buchanan and McClay, 1991, 1992; Mitra, 1993; Eisenstadt and Withjack, 1995; McClay, 1995). In contrast, however, the nature of 3-D inversion structures is poorly understood. The research presented here developed from an earlier investigation by Buchanan (1991), who conducted a comprehensive series of 2-D inversion experiments.

EXPERIMENTAL PROCEDURE

A comprehensive suite of sandbox models has been used to simulate the development of inversion above both concave-up and convex-up 3-D listric fault systems. Each experiment was run at least three times to check reproducibility and to permit vertical sectioning in two directions as well as horizontal sectioning. The deformation rig (Figure 2) was designed to model the inversion structures of various 3-D listric fault geometries. These were defined by rigid footwall blocks where the fault surfaces were sinusoidal or cuspate in plan view and either concave-up or convex-up in cross section (Figure 3). Two synchronized electric motors were used to move the rigid footwall blocks at a constant displacement rate of 4.16×10^{-3} cm^{-1}. For extension, the footwall block was moved out from underneath the sand pack, and the motion was reversed for the inversion phase of the experiment (Figure 2). The detachment surface was composed of a series of rayon strips parallel to the movement direction. In this manner, space problems and local complexities resulting from the 3-D geometries of the footwall blocks were minimized. The deformation rig had initial dimensions of $100 \times 60 \times 20$ cm.

Dry, cohesionless quartz sand (average grain size 190 μm) was used as the modeling material. Quartz sand has been successfully used to model brittle, upper-crust deformation in a wide variety of tectonic environments. These include extension (Horsfield, 1977, 1980; McClay and Ellis, 1987a, b; Ellis and McClay, 1988; McClay, 1990a), strike-slip (Naylor et al., 1986; Richard et al., 1995), thrust faulting (Colletta et al., 1991; Huiqi et al., 1992; Lallemand et al., 1992), salt tectonics (Vendeville and Jackson, 1992a, b; Jackson and Vendeville, 1994; Alsop, 1996), and inversion tectonics (Koopman et al., 1987; McClay, 1989; Buchanan and McClay, 1991, 1992; McClay and Buchanan, 1992; McClay, 1995). The models were constructed from alternating layers of white and colored sand that were mechanically sieved into the apparatus. Prerift strata consisted of blue, white, and black sand layers, whereas synextensional strata were red, black, and white sand layers. At the end of 10 cm of extension in the models, a uniform, 9-mm-thick, green-colored postrift layer was added. During the contraction phase, synkinematic layers were added after each 1 cm of shortening, such that the inversion anticline was just covered by syninversion strata. The scaling ratio of the models to natural structures is approximately 10^{-5}, thus, 1 cm in the model scales to ~1 km in nature (McClay, 1990b).

Three different 3-D listric fault geometries were analyzed in this study. Series I experiments used a fault block that was concave-up in cross section and had a sinusoidal fault trace in plan view (Figure 3a). Series II experiments had a footwall fault block that was concave-up in cross section but had a plan view shape consisting of two distinct, spoon-shaped concave elements (Figure 3b). Series III experiments had a convex-up listric fault in cross section and a sinusoidal trace in plan view (Figure 3c). During the experiments, the free top surface of the models was photographed

POSITIVE INVERSION

a. Extension

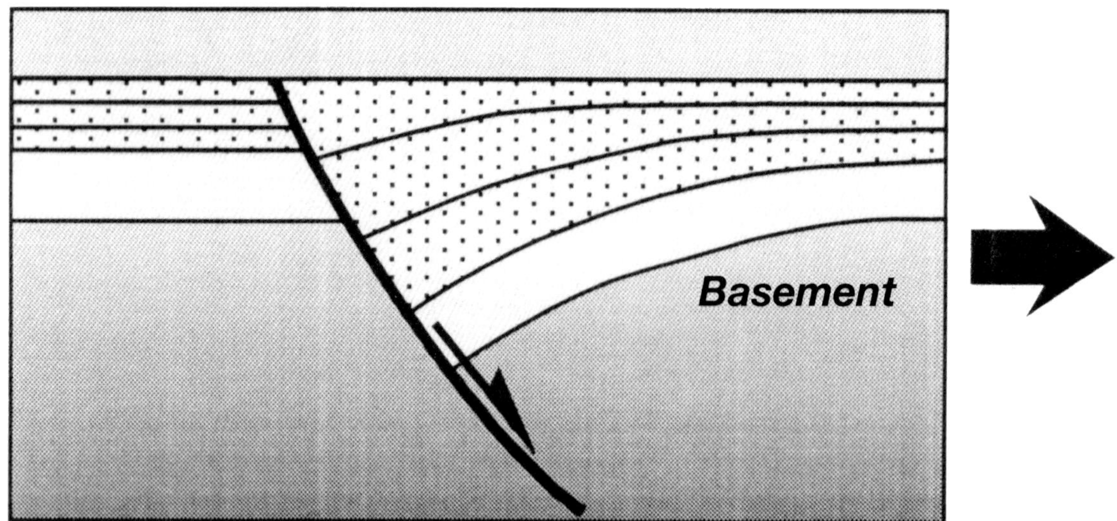

b. Extension

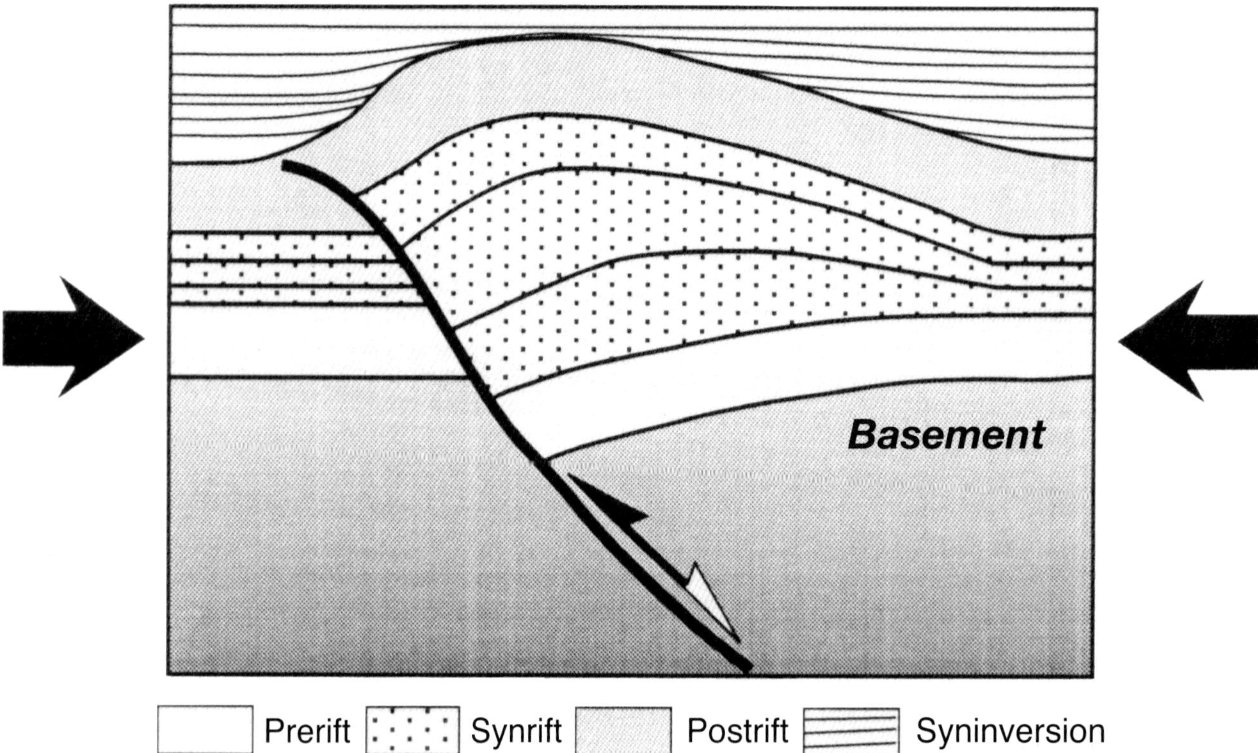

Figure 1. Schematic diagrams of (a) extension, and (b) positive inversion structure, showing synextension and syninversion growth wedges.

after every 10 mm of displacement of the footwall block, in both the extension and inversion phases. Synkinematic sediments were added incrementally during both the extension and inversion phases. The completed models were serially sectioned in three orientations: two vertical and one horizontal.

The scaled physical models described in this chapter have important limitations that must be borne in mind

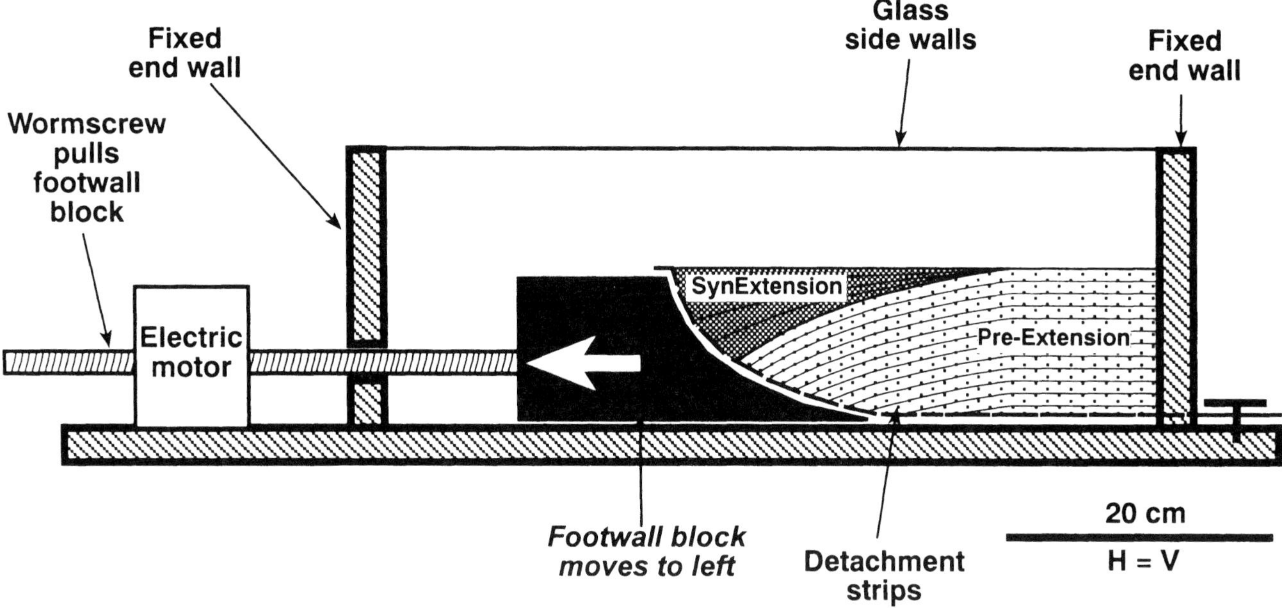

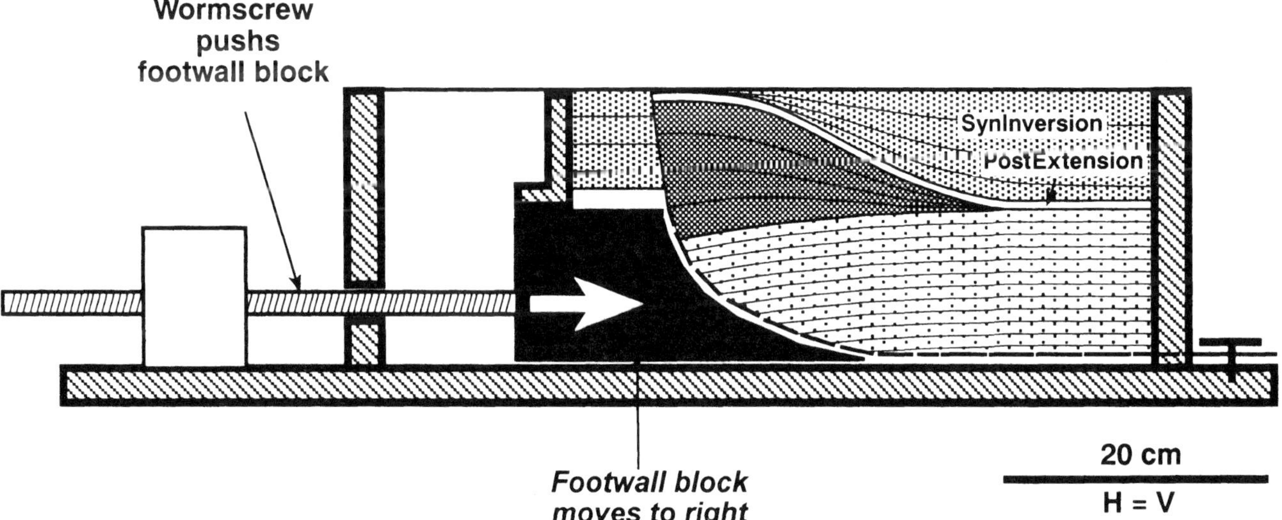

Figure 2. Experimental apparatus. Deformation rig configurations for extension and inversion by horizontal contraction.

when applying the results to natural inversion structures. The main limitations are that pore-fluid pressures (which are important during inversion tectonics, e.g. Sibson, 1985, 1995), isotatic and flexural responses, and thermal effects, cannot be simulated in the physical models (cf. Buchanan, 1991; McClay, 1996). Despite these limitations, the scaled analog models produce geometries and kinematic pathways that can be directly compared with natural inversion systems and, as such, can provide valuable insights into inversion tectonics and the formation of basement-involved thrust systems.

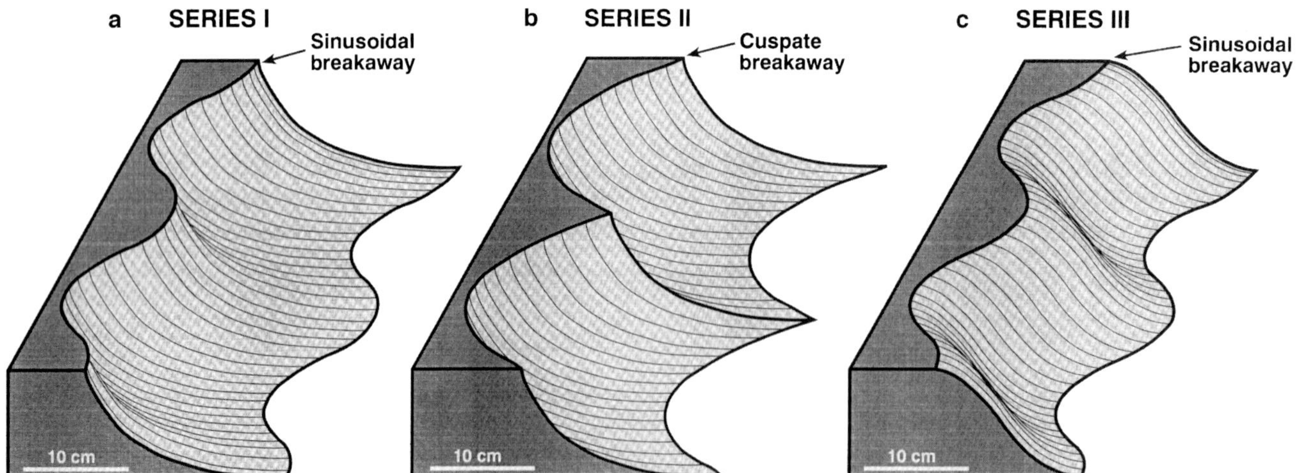

FIGURE 3. Schematic diagrams of the three 3-D footwall block geometries used in this study. (a) Series I— sinusoidal breakaway geometry with a listric, concave-up detachment surface; (b) Series II— cuspate breakaway geometry with a listric, concave-up detachment surface; and (c) Series III— sinusoidal breakaway geometry with a listric, convex-up detachment surface.

EXPERIMENTAL RESULTS

Series I Experiments

Extension

Series I experiments involved 10 cm of extension above a sinusoidal, concave-up listric detachment system followed by 10 cm of horizontal contraction. To illustrate the characteristic geometries of the hanging-wall deformation above a concave-up listric detachment, the results of an extension-only experiment were analyzed for Figure 4. During extension above a concave-up listric detachment fault, the hanging wall formed a classic rollover anticline together with a crestal-collapse graben system (Figure 4). In 3-D, the rollover anticline showed little variation along strike, with a well-developed wedge of synextension sediments forming a large half-graben structure together with planar antithetic and convex-up synthetic faults in the crestal-collapse graben (Figure 4). Greatest growth of the synextensional layers occurred where the main detachment surface was strongly concave in plan view, whereas least growth occurred above the salient in the main detachment surface (Figure 4). After 10 cm of extension, the upper surface of the model showed strongly segmented crestal-collapse graben faults in a sinusoidal pattern that was subparallel to the main detachment breakaway (Figure 5a). The dominant trends of the extensional faults in the crestal-collapse graben were at a high angle to the extension direction, except for local small, extensional faults around the concave embayment in the main detachment surface (Figure 5a). Small, low-displacement reverse faults also formed above the salient in the main detachment surface (Figures 4 and 5a). Broadly similar rollover anticlines and crestal-collapse graben systems have been well described from 2-D models of concave-up listric detachment systems (McClay and Ellis, 1987a; Ellis and McClay, 1988; and McClay 1989, 1990a).

Inversion

After 10 cm of extension, the Series I models were inverted by horizontal contraction (Figure 2). The main detachment fault was reactivated as a thrust fault and propagated upward, initially through the postextension layer and then through the syninversion layers as they were progressively added to the model (Figure 5b–f). As was revealed in serial cross sections through the inverted model, this fault formed a new, high-angle thrust that cut through the postextension and syninversion layers and propagated at 60°, which is the initial cutoff angle of the detachment breakaway (Figure 6). The main characteristics of the inversion structure were the steep reverse fault of the reactivated detachment surface, an asymmetric anticline in the hanging wall above this thrust, and segmented, hanging-wall-vergent back thrusts that propagated upward out of the original crestal-collapse graben system. The inversion anticline was strongly asymmetric, with a gently dipping backlimb and a steeply dipping forelimb (Figure 6).

Progressive evolution of the inversion structures, as was revealed by successive plan views of the upper surface of the model, showed the rapid reactivation and upward propagation of the main detachment surface together with the formation of the asymmetric hanging-wall anticline (Figure 5b, c). Segmented, hanging-wall-vergent back thrusts formed early in the inversion history and became more linked as the amount of contraction

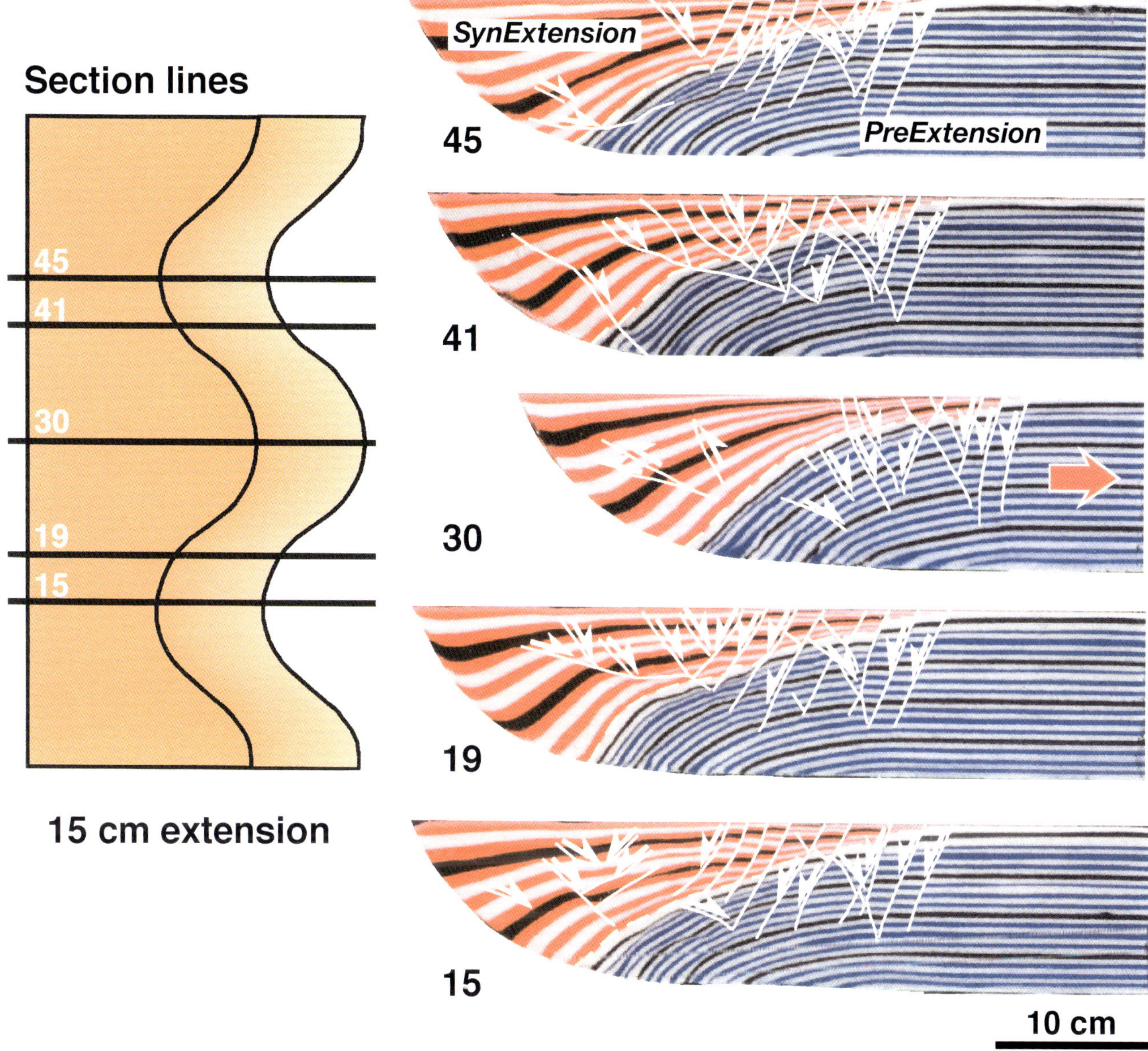

FIGURE 4. Series I, Experiment L68 – Extension. Serial vertical sections through a completed, concave-up, listric, extension-only model, showing a characteristic rollover anticline and crestal-collapse graben system. Note that reverse faults formed during extension (Section 30). The section numbers correspond to the distance in centimeters from a model's margin.

increased (Figure 5 d–f). Small-displacement extensional faults formed on the crest of the inversion anticline and also on the steep, front limb of the anticline.

In vertical cross sections, the inversion structures were remarkably similar along strike (Figure 6). The inversion anticline was slightly more asymmetric over the more concave elements of the main detachment surface and slightly tighter above the salient in the detachment surface. In all sections, back thrusts were seen to nucleate upward from the crestal-collapse graben system, and minor footwall shortcut thrusts developed in front of the main reactivated detachment surface (Figure 6). The geometry of the main inversion anticline was an "arrowhead" or "harpoon structure," as has been classically described for typical inverted extensional half grabens (Badley et al., 1989; McClay, 1995). The inversion anticline characteristically had thin syninversion strata on the crest and thicker syninversion strata on the flanks (Figure 6). At the end of the inversion phase of deformation, the listric detachment fault remained just in net extension (<1 cm) where it was measured at the top of the pre-extension sequence (Figure 6).

Horizontal sections through the completed models showed that the uplift was not uniform during inversion, with the main hanging-wall anticline showing a

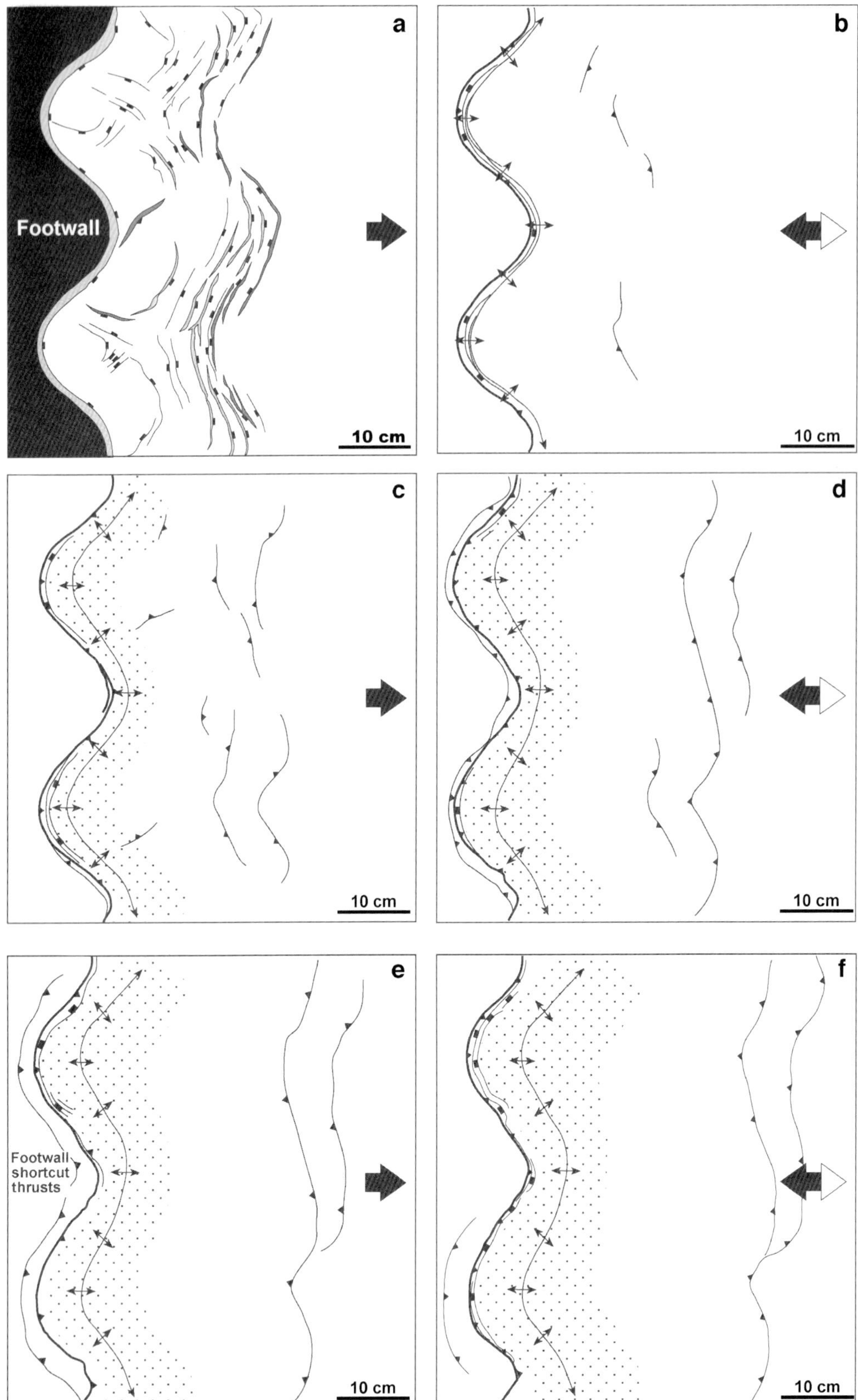

FIGURE 5. Series I, Experiment I304— Inversion. Line diagrams of the top surface of the model: (a) at the end of 10 cm of extension (preexisting faults before inversion); (b) after 2 cm of horizontal contraction (initial stage of fault reactivation); (c) after 4 cm of horizontal contraction; (d) after 6 cm of horizontal contraction; (e) after 8 cm of horizontal contraction; and (f) end of the experiment after 10 cm of horizontal contraction. The dotted area represents the characteristic flat surface in the hanging wall, just next to the main thrust.

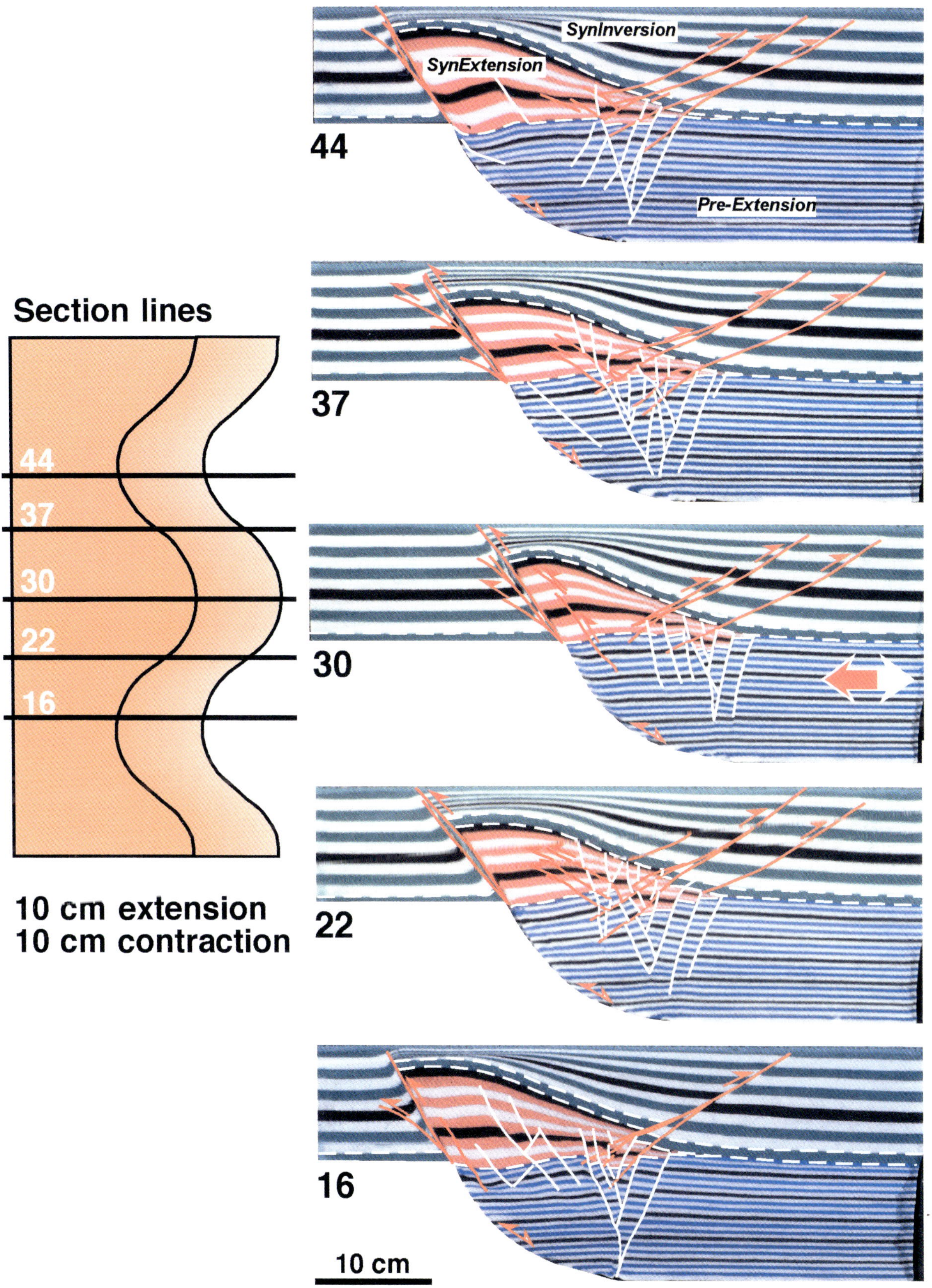

FIGURE 6. Series I, Experiment I309. Serial vertical sections through the central sectors of a completed concave-up, listric inversion experiment, showing the characteristic harpoon structure, reactivated detachment fault and thrust fault, as well as the back-thrust system. Note that the amount of uplifting during inversion varies. The section numbers correspond to the distance in centimeters from a model's margin.

strongly sinusoidal axial trace and two marked plunge culminations (Figure 7). Cross section I304-5 was cut at the top of the footwall block and showed the complexity of the extensional faulting associated with both the crestal-collapse graben and with the strongly concave elements of the main detachment fault surface (Figure 7a and b). Contractional faults (except for the main detachment surface) were difficult to distinguish at this level (Figure 7b). Section I304-7, taken higher in the model, had a simpler fault pattern, with the main detachment thrust forming a clearly defined thrust footwall to the asymmetric inversion anticline. Segmented and linked back thrusts occurred on the right-hand side of the model, whereas convex-up, footwall shortcut thrusts formed beneath the main detachment thrust surface (Figure 7c and d).

Series II Experiments

Extension

Series II experiments involved 10 cm of extension above a doubly cuspate concave-up listric detachment system, followed by 10 cm of horizontal contraction. As in Series I experiments, extension produced classic listric-fault-bounded half grabens with classic rollover anticlines and crestal-collapse geometries. In cross-sectional view, they were almost identical to the vertical sections shown in Figure 4. In plan view, the extensional architecture was characterized by two well-developed crestal-collapse grabens associated with each concave element of the main detachment system (Figure 8a). A less well developed crestal-collapse graben formed above the sharp salient in the main detachment surface. The extensional faults defining these crestal-collapse systems were strongly segmented but overall were subparallel to the plan view trace of the main fault surface (Figure 8a). Small, low-displacement reverse faults also formed above the sharp salient in the main detachment surface (Figure 8a).

Inversion

After 10 cm of extension, the Series II models were inverted by 10 cm of horizontal contraction. At the early stages of inversion, the main detachment fault propagated upward at a constant angle of 60° (the same angle as the extensional breakaway) (Figure 8b). An asymmetric hanging-wall inversion anticline developed with a steep forelimb and a gently dipping backlimb. The axial trace of the anticline was parallel to the plan view of the main detachment surface. At 4 cm of contraction, shallowly dipping back thrusts developed on the upper surface of the model (Figure 8c). These increased in strike length with increased shortening and had traces that were relatively straight compared with the main detachment thrust fault (Figure 8d and e). Small-displacement shortcut thrusts developed in the footwall to the main detachment fault, after 6 cm of shortening (Figure 8d). At the end of 10 cm of contraction, the inversion was characterized by a major steep thrust (the reactivated detachment surface) with an asymmetric hanging-wall anticline and shallow to moderately dipping, hanging-wall-vergent back thrusts on the gently dipping anticline limb (Figure 8f). Small-displacement, slightly convex-up footwall shortcut faults developed in front of the main thrust fault.

In cross section, the inversion profiles of Series II experiments (Figure 9) were almost identical to those of Series I experiments (Figure 6). In vertical cross sections, the inversion structures were remarkably similar along strike (Figure 9). The inversion anticline was slightly more asymmetric over the more concave elements of the main detachment surface and less well developed over the sharp salient in the detachment surface. As occurred in Series I experiments, back thrusts were seen to have nucleated upward from the crestal-collapse graben system. Minor, convex-up footwall shortcut thrusts developed in front of the main reactivated detachment fault (Figure 9). The geometry of the main inversion anticline was an arrowhead or harpoon structure, with thin syninversion strata on the crest and thicker syninversion strata on the flanks. At the end of the inversion phase of deformation, the listric detachment fault remained just in net extension (<1 cm), as was the case in Series I experiments (Figure 9).

The two horizontal sections in Figure 10 show that the maximum inversion uplift was focused in the immediate hanging wall of the main detachment system's maximum concave-up portions. Section I314-5, taken at the level of the top of the footwall block, showed mainly the synextensional fault arrays, with complex patterns of crestal-collapse faulting above the main concave parts of the main detachment system (Figure 10a and b). Section I314-7, taken through the inversion anticline, showed two doubly plunging, anticlinal culminations focused on the most concave parts of the reactivated main detachment surface (Figure 10c and d). Small-displacement extensional faults developed on the crest of the anticlines. As is revealed in the vertical cross sections, these inversion anticlines were strongly asymmetric, with maximum uplift and maximum inversion focused in the immediate hanging wall of the main footwall vergent thrust surface (Figure 9).

Series III Experiments

Extension

Series III experiments involved 10 cm of extension above a sinusoidal convex-up listric detachment system, followed by 10 cm of horizontal contraction. The extensional geometries produced by hanging-wall

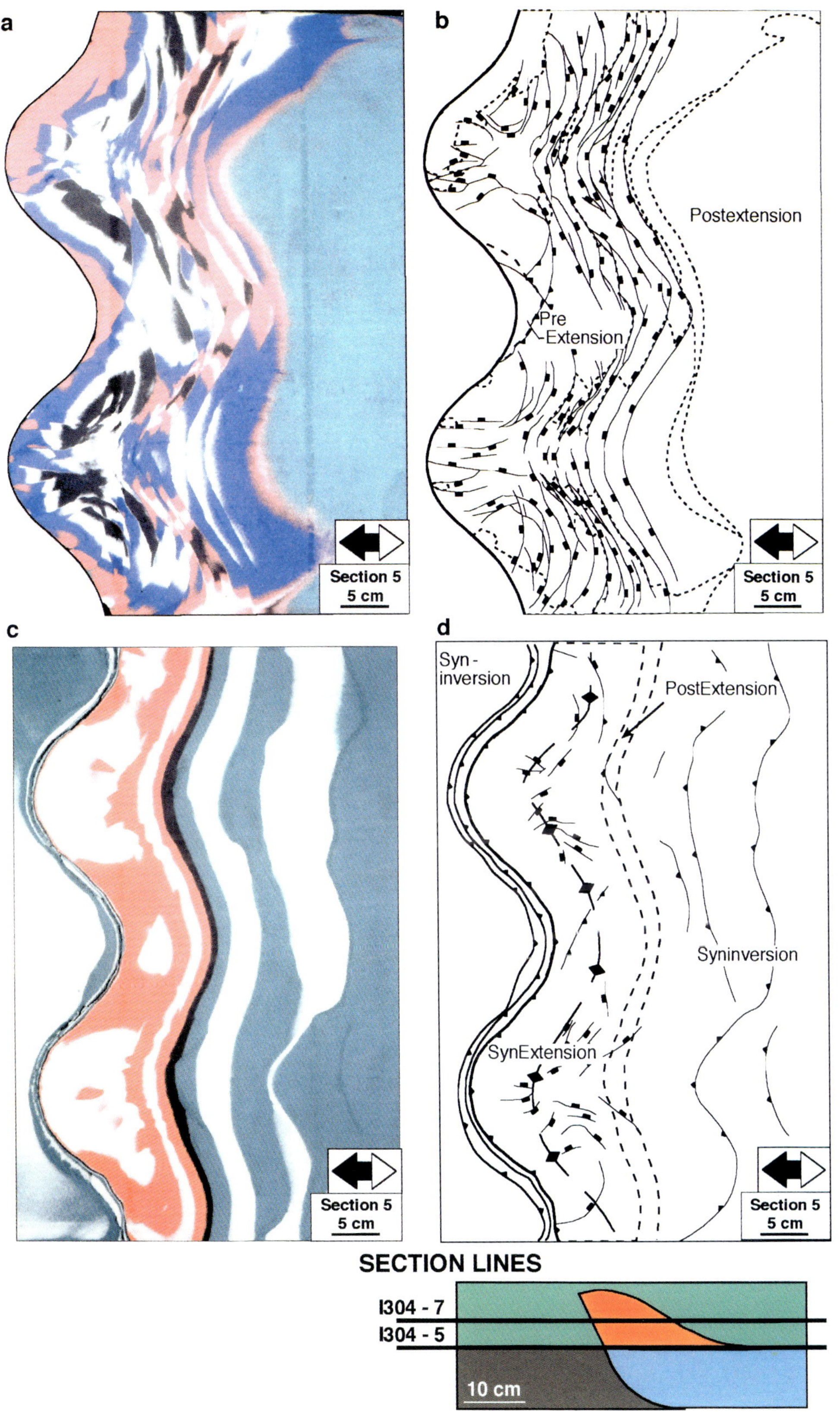

FIGURE 7. Series I, Experiment I304. Horizontal sections and interpretations: (a) Photograph of section 5 cut at 10 cm above base of the model— at the position of the top of the footwall block; (b) line diagram interpretation of (a); (c) photograph of section 7 cut at 14 cm above the base of the model; and (d) line diagram interpretation of (c). Section 5 shows that most preexisting faults before inversion were preserved (cf. Figure 5a). The inversion anticline has two marked culminations (Section 7).

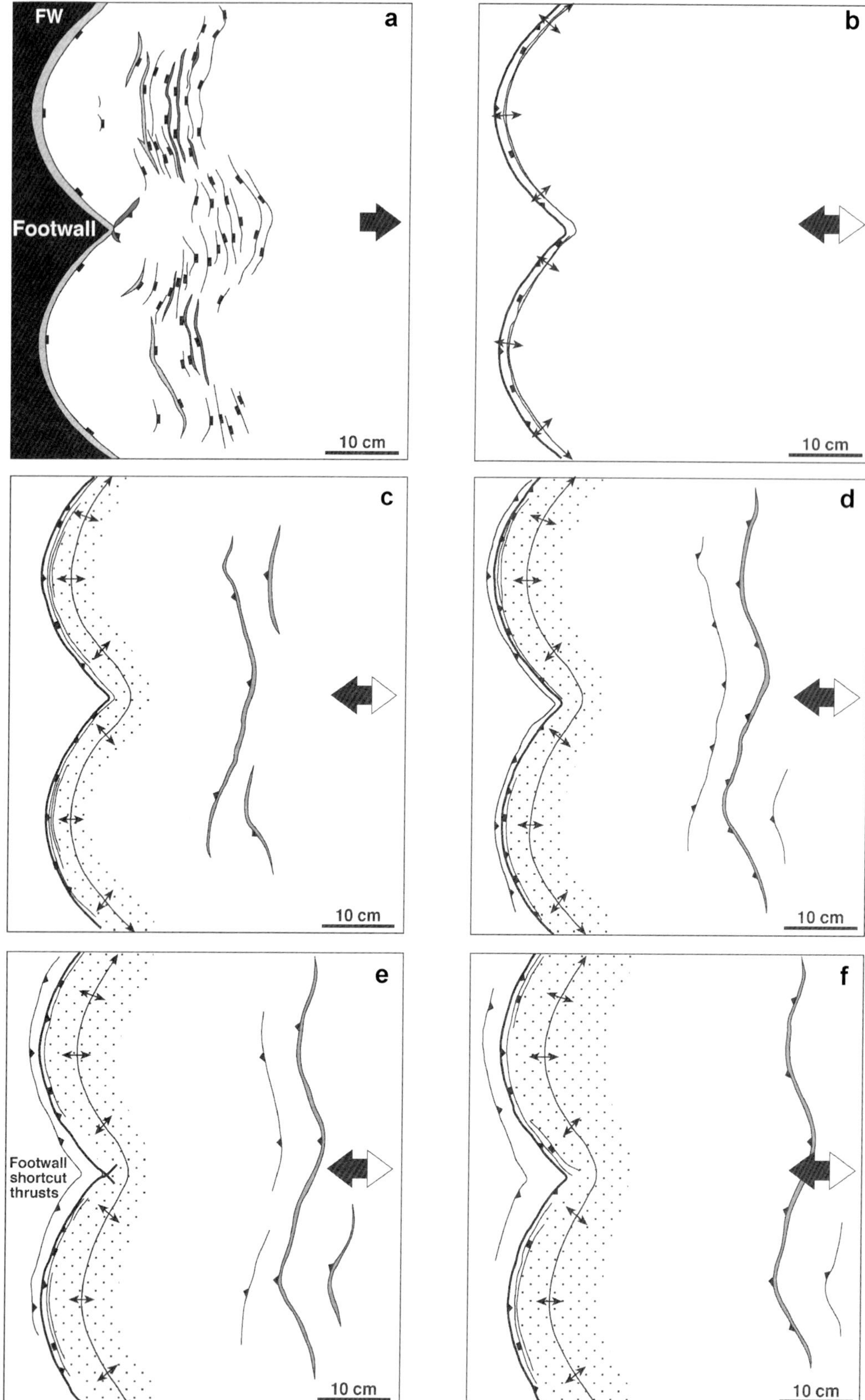

FIGURE 8. Series II, Experiment I313— Inversion. Line diagrams of the top surface of the model: (a) At the end of 10 cm of extension (preexisting faults before inversion); (b) after 2 cm of horizontal contraction; (c) after 4 cm of horizontal contraction; (d) After 6 cm of horizontal contraction; (e) after 8 cm of horizontal contraction; and (f) end of the experiment, after 10 cm of horizontal contraction. The dotted area represents the characteristic flat surface in the hanging wall, just next to the main thrust.

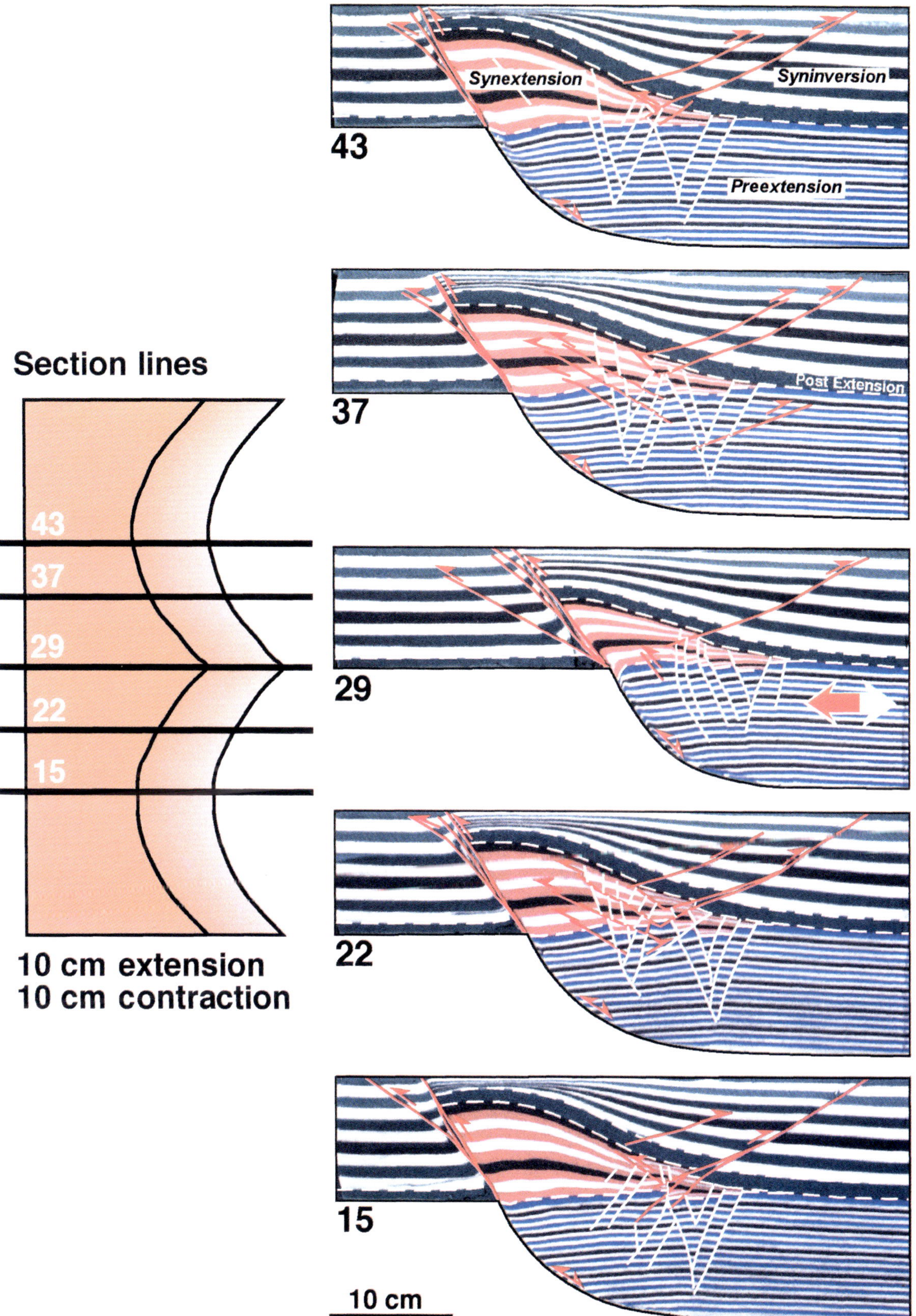

Figure 9. Series II, Experiment I312. Serial vertical sections through the central sectors of a completed concave-up, listric inversion experiment, showing the characteristic harpoon structure, reactivated detachment fault and thrust fault, as well as the back-thrust system. Note the variation in hanging-wall uplift during inversion. The section numbers correspond to the distance in centimeters from a model's margin.

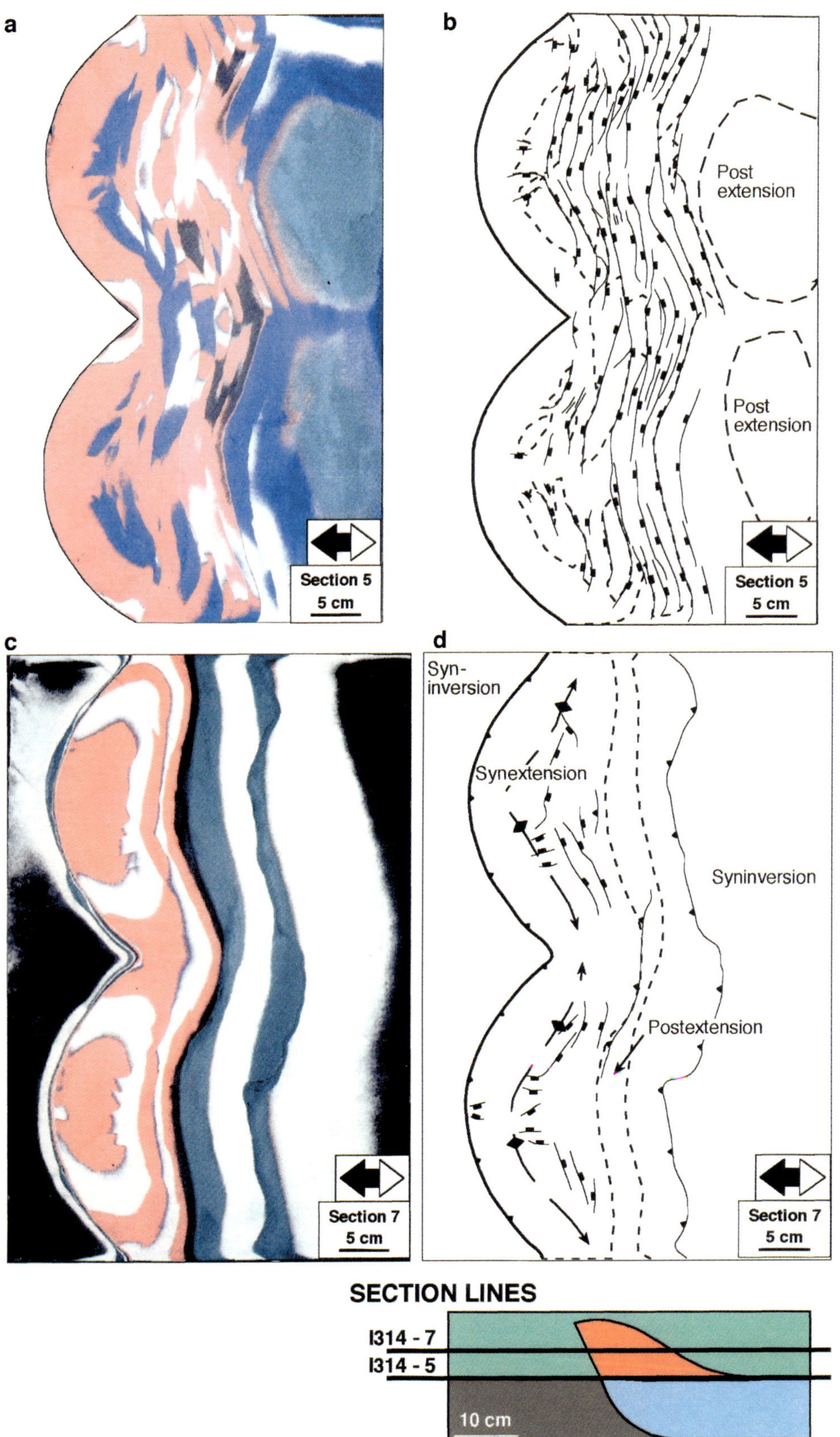

FIGURE 10. Series II, Experiment I314. Horizontal sections and interpretations: (a) Photograph of section cut at 10 cm above the base of the model— at the position of the top of the footwall block; (b) line diagram interpretation of (a); (c) photograph of section cut at 14 cm above the base of the model; and (d) line diagram interpretation of (c). Section 5 shows that most preexisting faults before inversion were preserved (cf. Figure 8a). The inversion anticline has two marked culminations (Section 7).

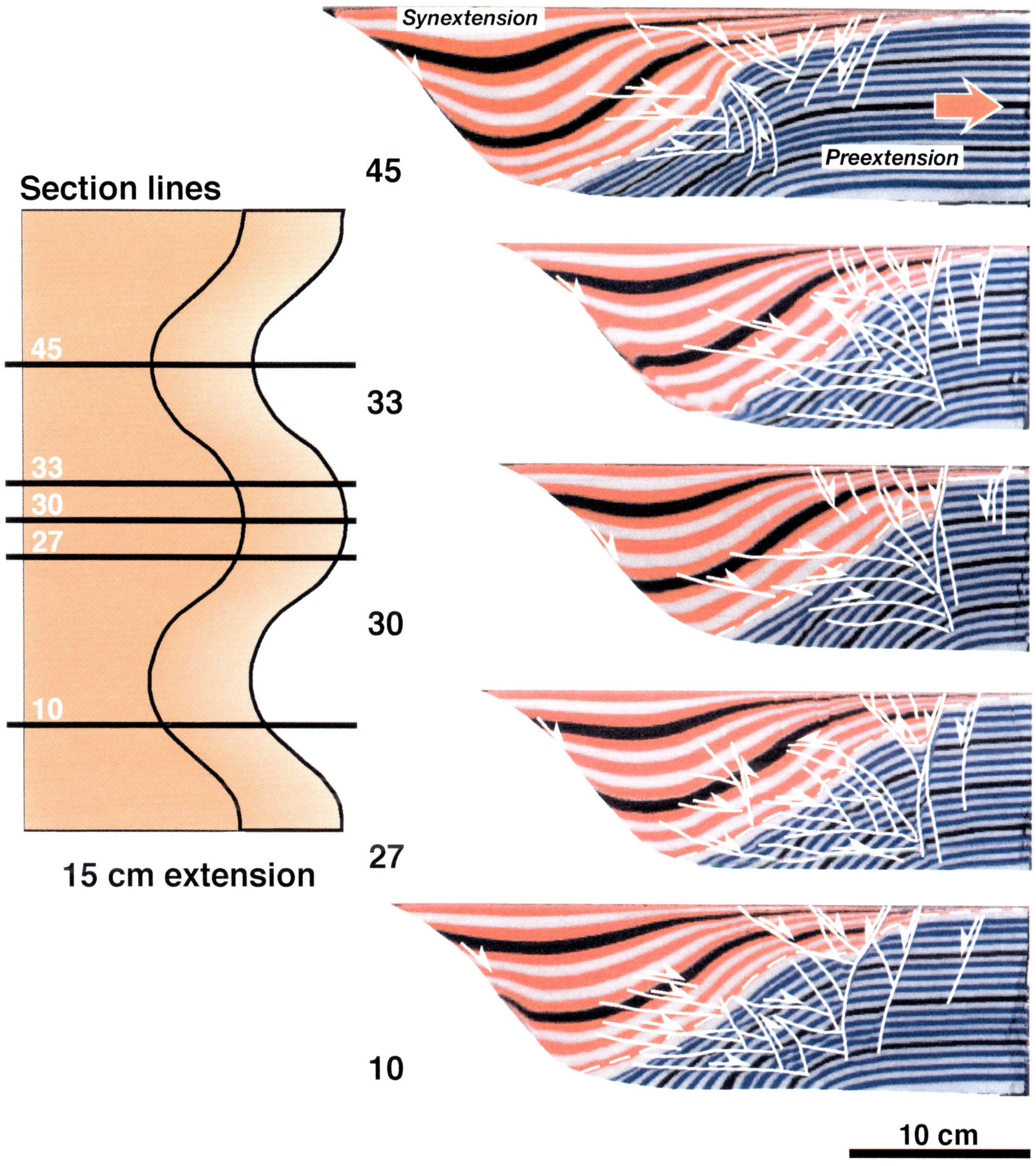

FIGURE 11. Series III, Experiment L72— Extension. Serial vertical sections through a completed, convex-up, listric, extension-only model, showing a characteristic hanging-wall syncline and a narrow, complex crestal-collapse graben system.

deformation above a convex-up listric detachment were markedly different from those for a concave-up listric detachment. The experiment shown in Figure 11 was a model vertically sectioned, at the end of extension, to analyze the 3-D hanging-wall geometries. Above the convex-up listric detachment, a broad hanging-wall syncline of synextensional strata developed, together with a complex crestal-collapse graben system (Figure 11). Steep, planar antithetic faults defined the right-hand margin of the crestal-collapse structure, whereas shallow-dipping, almost flat-lying synthetic faults define the left-hand margin of the crestal-collapse structure. Early-formed crestal-collapse extensional faults have been strongly rotated, with some antithetic faults overturned to give an apparent reverse offset within the prekinematic layers (Figure 11). The hanging-wall syncline and crestal-collapse graben system showed little change in structural style along the strike of the model (Figure 11). In plan view, at the end of extension, the axial trace of the hanging-wall syncline was parallel to the footwall breakaway fault system (Figure 12a). The crestal-collapse graben faults were highly segmented and curved subparallel to the main detachment breakaway.

Inversion

After 10 cm of extension, the Series III models were inverted by 10 cm of horizontal contraction. Initial contraction of the model reactivated the main detachment surface as a thrust fault that propagated upward at a moderate angle through the postextension and syninversion layers (Figure 12b). A broad, flat-topped hanging-wall anticline developed, initially with an axial trace parallel to the main detachment breakaway (Figure 12b). With increased shortening, the main footwall-vergent thrust continued to propagate outward and upward, but the axial trace of the hanging-wall anticline became progressively straighter (Figure 12c–e). After about 6 cm of shortening, a hanging-wall-vergent thrust developed on the upper surface of the model. This had only a slightly curved trace (Figure 12d). At the end of inversion, the model consisted of a broad, nearly symmetric, hanging-wall anticlinal uplift bounded by moderately dipping, footwall-vergent and hanging-wall-vergent thrusts (Figure 12f).

Serial, vertical cross sections through the completed model showed almost identical inversion geometries along strike (Figure 13). The inversion structure consisted of a broad, thrust-bound symmetric anticline with thin syninversion strata on the crest and thicker syninversion strata on the flanks (Figure 13). The main detachment surface was reactivated and propagated upward as a new footwall-vergent thrust, at a constant angle of 30°, through both the postextension and the syninversion strata (Figure 13). Moderately dipping hinterland-vergent back thrusts propagated outward from the crestal-collapse graben in the prekinematic layers. Maximum hanging-wall uplift occurred in the regions above the strongly concave parts of the main detachment surface. Minor bypass thrusts developed in the hanging wall next to the main thrust fault, but, because the footwall-vergent thrust ramp angle was only 30°, no footwall shortcut thrusts formed (Figure 13).

Horizontal sections through the completed model revealed structural styles that were significantly different from those in the Series I and II experiments. Section I316-5, taken at the level of the top of the footwall block, primarily showed the synextensional crestal-collapse-graben fault systems and the main hanging-wall-vergent back-thrust system (Figure 14a, b). Horizontal section I316-7, which cut through the main inversion structure, showed two major plunge culminations within the broad inversion anticline. It is noteworthy that this inversion anticline had an axial trace that was subparallel to the main sinusoidal detachment surface but was more symmetric (Figure 14c, d) than the concave-up inversion systems (Figure 10).

Inversion Growth Sequences

Syninversion growth strata were mechanically sieved into the models after each centimeter of contraction. This simulated uniform background sedimentation, whereby the upper surface of each growth increment was horizontal. In the models, there was no syninversion erosion of the crests of the anticlines. Growth-strata patterns showed that there was differential uplift while inversion progressed, and recorded the relative rotational patterns of the limbs of the inversion folds. Serial vertical sections through the completed models revealed growth-strata geometries for each experimental series (Figures 6, 9, and 13).

In vertical cross sections, Series I–III experiments produced very similar growth-strata patterns (Figures 6, 9, and 13). Thinnest growth strata were found on the crests of the hanging-wall anticlines, whereas the thickest growth strata formed on the undeformed footwall block and on the very right-hand part of the hanging wall where there was no internal deformation. In the Series I and II experiments, the thickest accumulations of growth strata were found in the region of the hanging-wall salients— where there was the least vertical uplift during inversion (Figures 6 and 9). Progressive limb rotation during the growth of the inversion anticlines was revealed by outwardly fanning growth wedges on all of the backlimbs of the inversion folds (Figures 6, 9, and 13). For the concave-up detachment geometries, front-limb growth structures were not well defined, but, where they were visible, these also showed fanning growth wedges indicating progressive limb rotation. The best frontlimb growth structures were found on the convex-up experiments where fanning growth wedges were well developed (Figure 13).

Figure 15 shows selected thickness maps of syninversion strata for all three series of experiments. These vertical isopach maps show that, for all three series of experiments, the greatest syninversion uplift occurred in the hanging walls immediate to the concave embayments in the main detachment surface, regardless of whether the main fault surface was concave or convex-up in cross section (Figure 15). The contours also show that, for the concave-up experiments (Series I and II), the crest and the axial trace of the inversion anticline

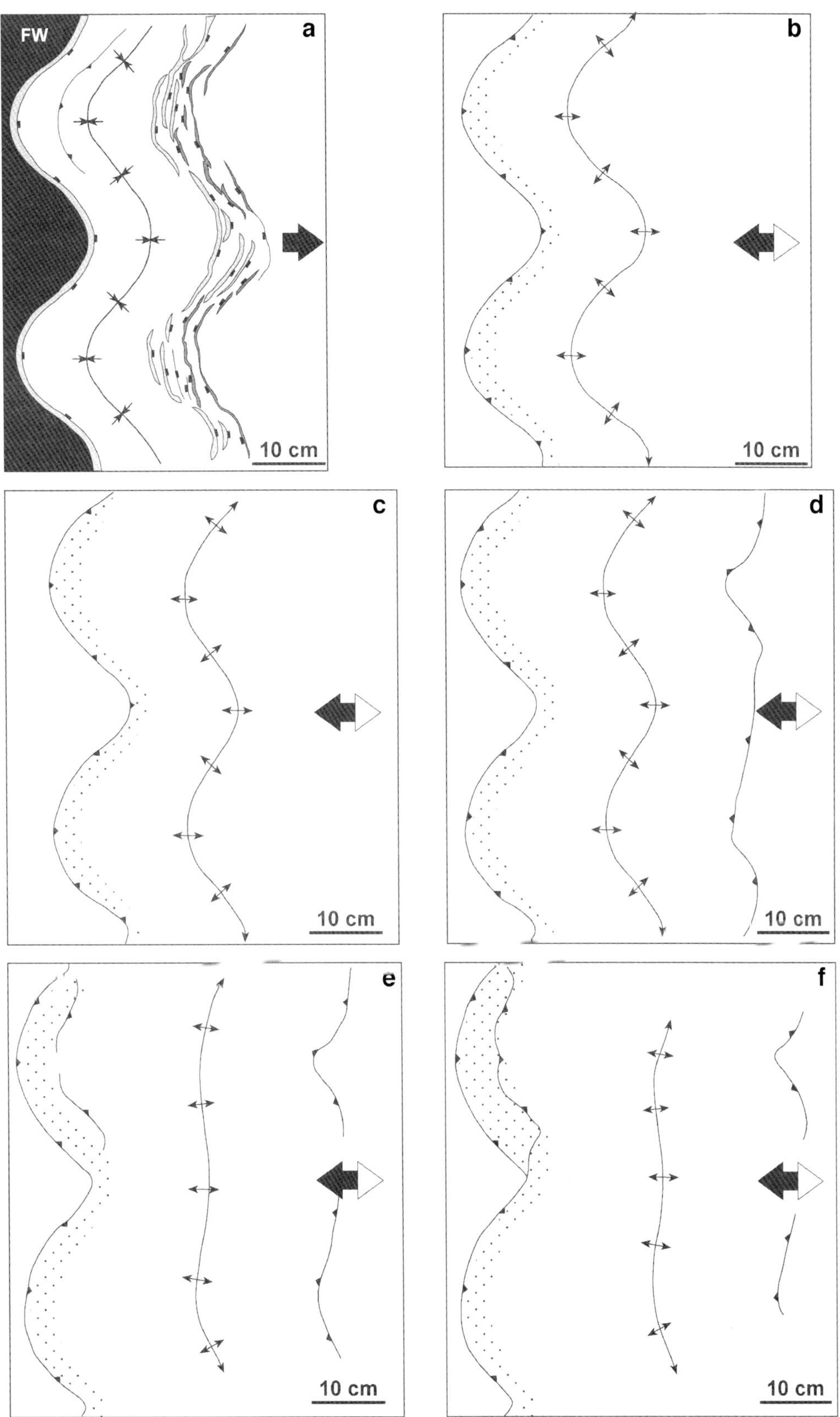

Figure 12. Series III, Experiment I315— Inversion. Line diagrams of the top surface of the model: (a) At the end of 10 cm of extension (preexisting faults before inversion); (b) after 2 cm of horizontal contraction; (c) after 4 cm of horizontal contraction; (d) after 6 cm of horizontal contraction; (e) after 8 cm of horizontal contraction; and (f) end of the experiment, after 10 cm of horizontal contraction. The dotted area represents the characteristic flat surface in the hanging wall, just next to the main thrust.

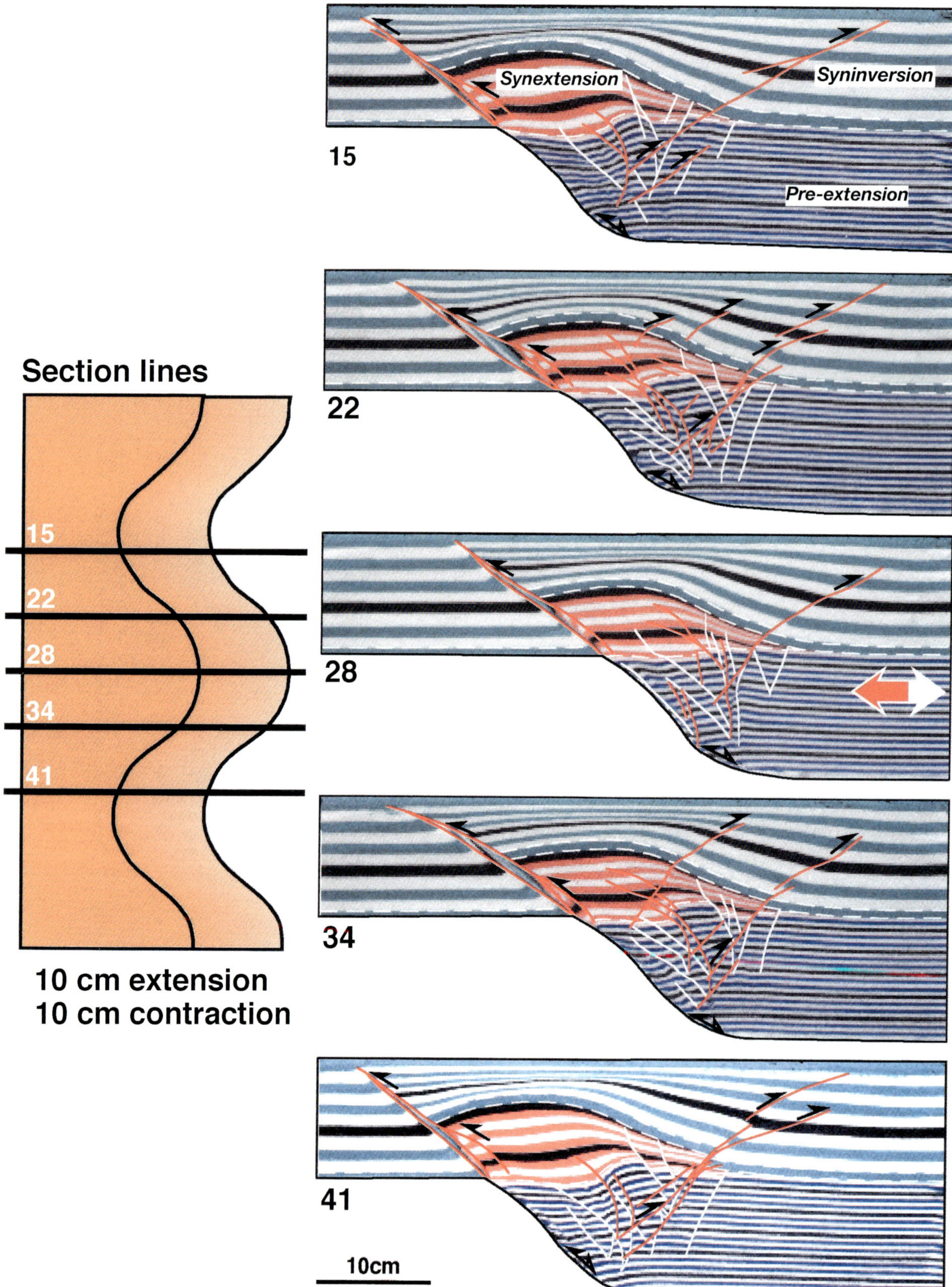

FIGURE 13. Series III, Experiment I311. Serial vertical sections through the central sectors of a completed convex-up, listric inversion experiment, showing the characteristic broad harpoon structure, reactivated detachment fault and thrust fault, as well as the back-thrust system. Note the variation in hanging-wall uplift during inversion. The section numbers correspond to the distance in centimeters from a model's margin.

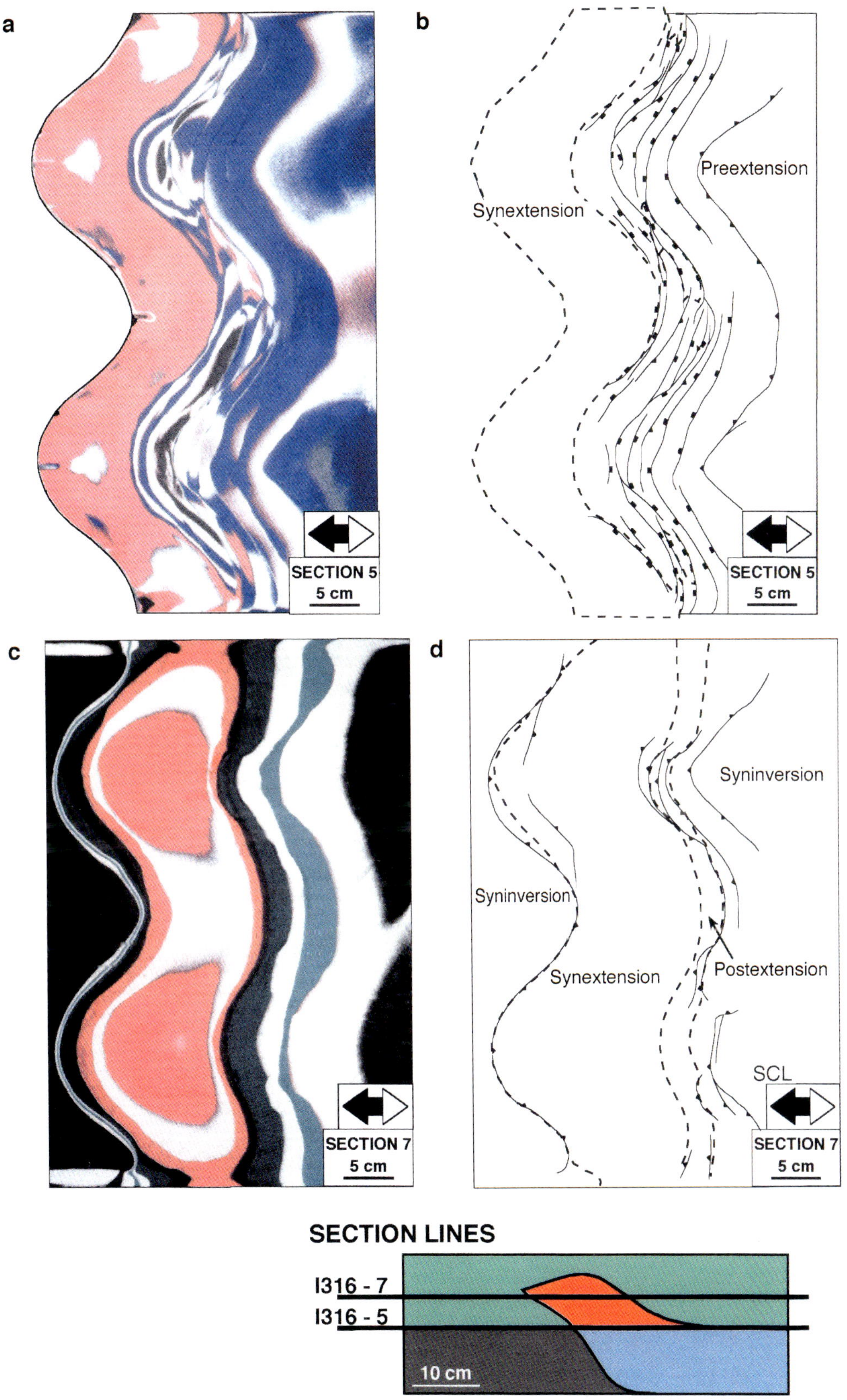

FIGURE 14. Series III, Experiment I316. Horizontal sections and interpretations: (a) Photograph of a section cut at 10 cm above base of the model— at the position of the top of the footwall block; (b) line diagram interpretation of (a); (c) photograph of a section cut at 14 cm above the base of the model; and (d) line diagram interpretation of (c). Section 5 shows that most preexisting faults before inversion were preserved (cf. Figure 12a). The inversion anticline has two marked culminations (Section 7).

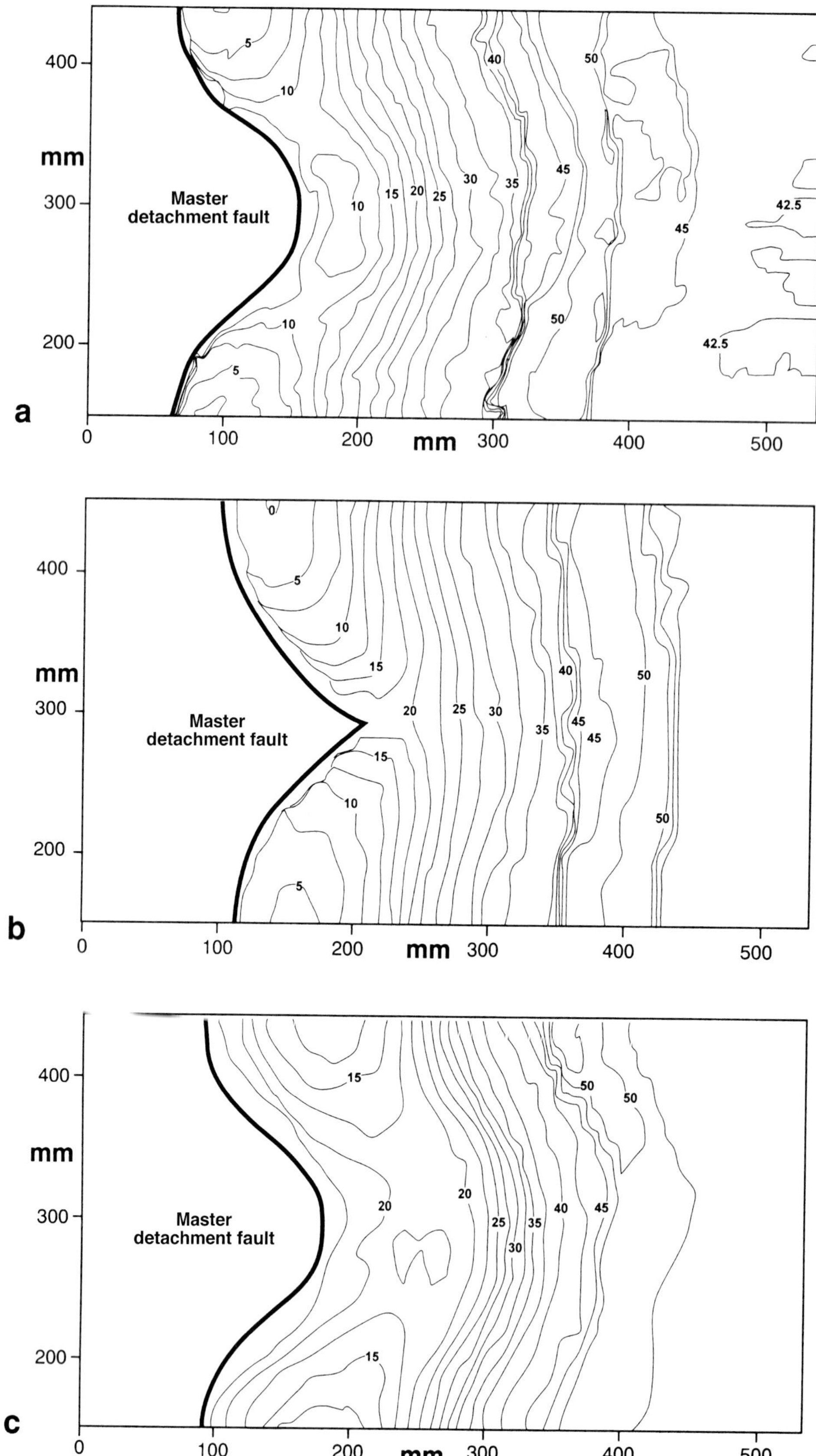

Figure 15. Maps of syninversion stratigraphy, showing vertical thickness variations around the thrust-bound anticlines. Contours are of thickness in millimeters for the lowermost five layers of the syninversion sequence. (a) Series I, Experiment I309. (b) Series II, Experiment I312. (c) Series III, Experiment I311.

were very close to the main thrust fault, whereas these features were significantly further into the hanging wall for the convex-up experiments (Figure 15). The doubly plunging nature of the hanging-wall inversion anticlines was also revealed by the thickness plots, in which major plunge culminations were developed above the concave embayments in the main detachment surface. Small, second-order culminations (elevations < 5 mm) also developed opposite the convex salients in the sigmoidally shaped main detachment surfaces (Figure 15a and c).

DISCUSSION

In this section, the results of the analog experiments are analyzed and discussed. Three-dimensional conceptual models for inversion thrust systems are developed, and the implications for hydrocarbon systems associated with inversion thrust structures are reviewed.

3-D Extension Geometries

3-D extension above Series I concave-up listric detachment faults with a sigmoidal plan view geometry produced a classic rollover anticline and crestal-collapse graben geometry (Figure 4). Vertical serial sections showed very little structural variation along strike. The vertical cross sections from the 3-D models are very similar to cross sections from 2-D dip-slip extensional models published by Ellis and McClay (1988). Therefore, we can expect to have problems if we attempt to identify complex 3-D extension structures solely from single, or widely spaced, 2-D seismic lines.

Plan view maps of the experimental crestal-collapse graben systems show strongly segmented extensional faults that are subparallel to the geometry of the main extensional breakaway (Figure 5a). Similar rollover architectures and extensional fault patterns are found for the Series II experiments (Figure 8a). In both Series I and II experiments, complex 3-D, obliquely striking reverse faults developed around the along-strike convex elements of the detachment surface to accommodate the oblique-slip extension in these regions (e.g., Figures 5a and 8a). In contrast, Series III experiments produced radically different extensional structures. Hanging-wall deformation above a convex-up listric detachment fault produced a hanging-wall syncline together with a complex crestal-collapse graben system (Figure 11). As was the case in the Series I and II experiments, there was little along-strike change in the hanging-wall geometry. In plan view, the strongly segmented, crestal-collapse extensional faults mimic the geometry of the detachment breakaway (Figure 12a). The results of the extensional phases of deformation show that the profile section of the main detachment surface has the most dramatic effects on the fault and rollover architectures. However, plan-view fault geometries are largely controlled by the along-strike shape of the main detachment surface. Local oblique extensional faults form around concave elements in the main detachment surface.

3-D Inversion Geometries

All three series of experiments produced similar 3-D inversion geometries. In each model, the dominant feature was a thrust-bounded harpoon anticlinal structure that displayed along-strike plunge variations as a result of variable uplift during inversion (Figures 5, 8, and 12). In each experiment, the main extensional detachment fault was reactivated, in contraction, such that a new thrust propagated upward at the same angle as that of the upper part of the extensional fault. The thrust maintained a constant angle through the postextension and synextension strata. Where the thrust dipped at ~60°, shortcut thrusts developed in the footwall (Series I and II experiments; Figures 5 and 9). In the Series II experiments, however, the thrust propagated upward at 30° and no footwall shortcuts developed (Figure 13). Low-angle to moderately dipping, hanging-wall vergent back thrusts developed in all models (Figures 6, 9, and 13). Typically, these nucleate from the tips of antithetic extensional faults that bound the crestal-collapse grabens in the synextensional sequence.

The hanging-wall inversion anticlines are consistently asymmetric, with inclined axial surfaces and a dominant vergence toward the footwall (Figures 6, 9, and 13). They are fault-propagation folds that formed as a new thrust fault propagated upward through the postextension and syninversion strata. Concave-up listric detachments (Series I and II) produce inversion anticlines that have maximum uplift above the edge of the footwall block, narrow forelimbs, and curved, gently dipping backlimbs (Figures 6 and 9). In contrast, convex-up listric detachments (Series III) produce inversion anticlines that have maximum uplift above the most convex part of the footwall detachment surface, wide forelimbs, and gently dipping backlimbs (Figure 13). In Series III experiments, the inversion anticlines are less asymmetric, broader, and have their crests further into the hanging wall, away from the detachment breakaway. In all experiments, syninversion growth strata thin over the anticlinal crests and thicken down both the backlimb and forelimbs of the structures. The growth-strata wedges show outward-fanning geometries, which indicate progressive limb rotation during development of the inversion anticlines (Figures 6, 9, and 13).

The horizontal sections through the completed models revealed the geometries of the hanging-wall inversion anticlines in detail (Figures 7, 10, and 14). In all experiments, hanging-wall uplift varied along the strike of the detachment surface and consistently was greatest above the most concave parts of the detachment surface. The stratal patterns on these horizontal sections graphically illustrate the along-strike plunge culminations generated by differential inversion uplift. Anticlinal axial traces follow the geometry of the detachment-surface footwall breakaway (Figures 7, 10, and 14).

The analog model's results demonstrate that the 3-D inversion's structural architecture is controlled by the geometry of the main listric detachment surface. Maximum inversion uplift occurs where there previously was maximum extension-generated subsidence. Progressive inversion uplift is accomplished by translation of the hanging wall back up the reactivated extensional fault and the new thrust fault that propagates from the extensional breakaway. During inversion, progressive back-rotation of the hanging wall produces fanning, syninversion growth-strata patterns.

Vertical sections through all three series of experiments show that, for each model, the cross-section structures are remarkably similar along strike (Figures 6, 9, and 13). Interpretations of complex 3-D inversion structures from single or widely spaced 2-D seismic lines would be extremely difficult, and therefore 3-D seismic data would be needed.

3-D Synoptic Models of Inversion Structures

The results of the analog experiments presented in this chapter have been used to construct 3-D synoptic models for the hanging-wall extension and inversion geometries above each type of detachment surface (Figure 16). In each of the models, the main inversion features are controlled by the geometry of the main detachment surface. Maximum inversion and uplift characteristically occur above the strongly concave elements in the detachment surface, thus producing doubly plunging, anticlinal culminations in the hanging wall (Figure 16). These synoptic models may be used as templates for structural interpretation of 3-D inversion systems where good 3-D seismic data are poor or absent.

Series I Synoptic Models

Figure 16a summarizes the 3-D geometries of extension and inversion above a sinusoidal, concave-up, listric detachment fault system. The extension phase shows the main rollover anticline and the sigmoidally shaped, crestal-collapse graben system. Note also the oblique reverse faults adjacent to the convex salient in the main detachment surface. The inversion structure is a 3-D harpoon structure dominated by a steeply dipping main thrust fault and an asymmetric, doubly plunging anticline in its hanging wall (Figure 16a). Slightly sinusoidal back thrusts occur in the hanging wall of the inversion structure. Small-displacement shortcut thrusts occur in the footwall to the main thrust fault.

Series II Synoptic Models

Figure 16b summarizes the 3-D geometries of extension and inversion above a doubly cuspate, concave-up, listric detachment fault system. The basic elements are similar to the Series I synoptic models (Figure 16a). The extension phase shows the main rollover anticline and the segmented, cusp-shaped, crestal-collapse graben system. Note also the minor thrust system formed at the cusp in the 3-D master-fault geometry. The inversion structure is a 3-D harpoon dominated by a steeply dipping main thrust fault and an asymmetric, doubly plunging anticline in its immediate hanging wall (Figure 16b). Segmented and slightly sinusoidal back thrusts occur in the hanging wall of the inversion structure. Small-displacement shortcut thrusts also occur in the footwall to the main thrust fault.

Series III Synoptic Models

Figure 16c summarizes the 3-D geometries of extension and inversion above a sinusoidal, convex-up, listric detachment fault system. The extension phase shows the main, hanging-wall syncline and the segmented, sigmoidal, crestal-collapse graben system. Note the well-developed relay ramps between overlapping extensional fault segments. Complex fault systems, including reverse faults, form between the detachment surface and the crestal-collapse graben system. The inversion structure is a 3-D harpoon dominated by a moderately dipping main thrust fault and a doubly plunging, broad anticline in the hanging wall (Figure 16c). Segmented and slightly sinusoidal back thrusts occur in the hanging wall of the inversion structure. Footwall shortcut thrusts do not occur in this model.

Analog Models Compared with Natural Inversion Structures: Implications for Hydrocarbon Systems

2-D analog models of inversion structures have been successfully compared with a wide variety of natural structures (Buchanan, 1991; Buchanan and McClay; 1991; Buchanan and McClay, 1992; McClay, 1995, 1996). In particular, these studies emphasized the complexities of 2-D inversion geometries produced by simple dip-slip extension followed by dip-slip inversion. Natural examples of 3-D inversion systems, however, have not been well described in the literature, despite their importance in generating structural traps for hydrocarbon systems in basins such as the southern North

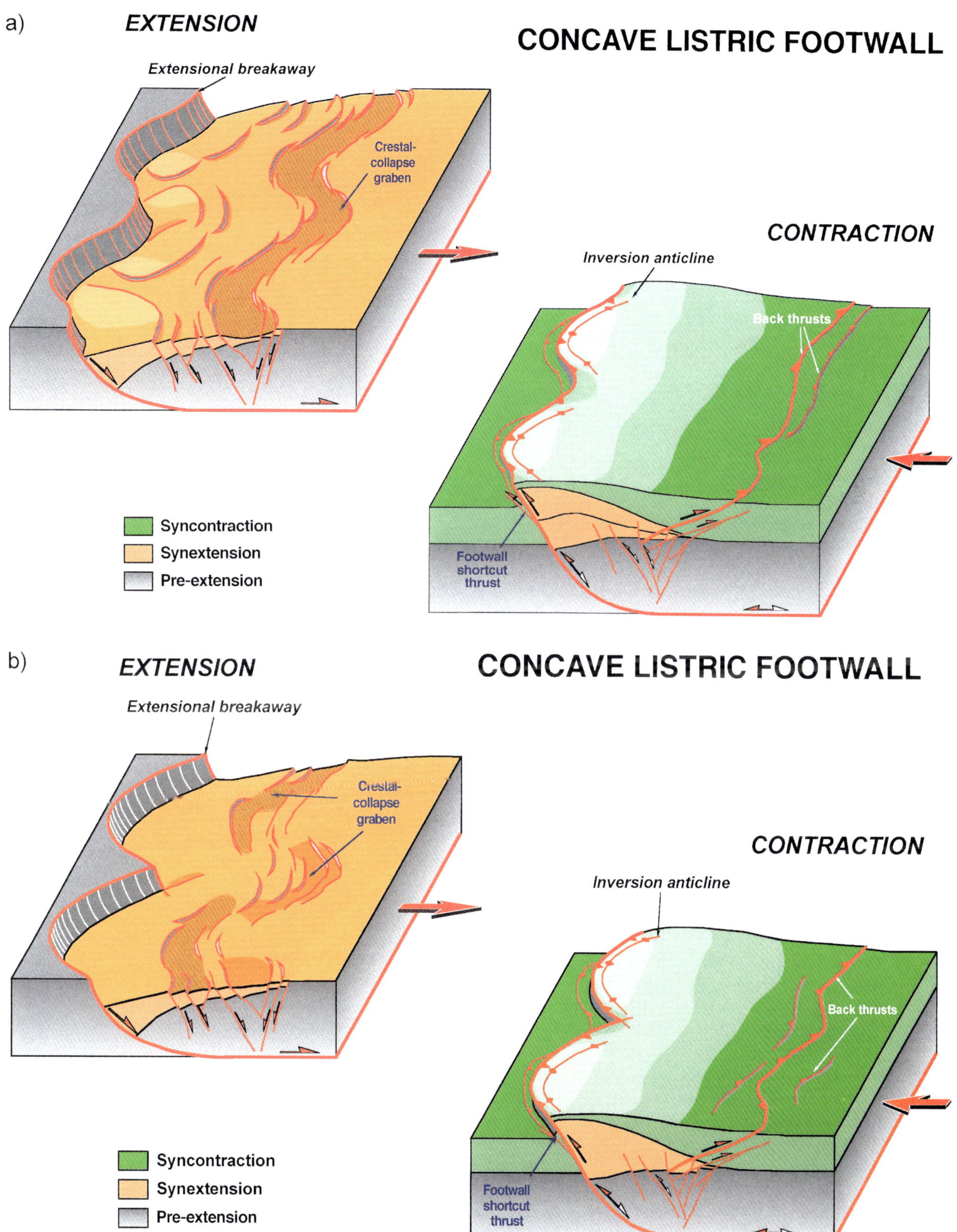

FIGURE 16. 3-D synoptic models derived from the experimental results, showing hanging-wall extension and inversion. (a) Series I experiments. (b) Series II experiments. (c) Series III experiments.

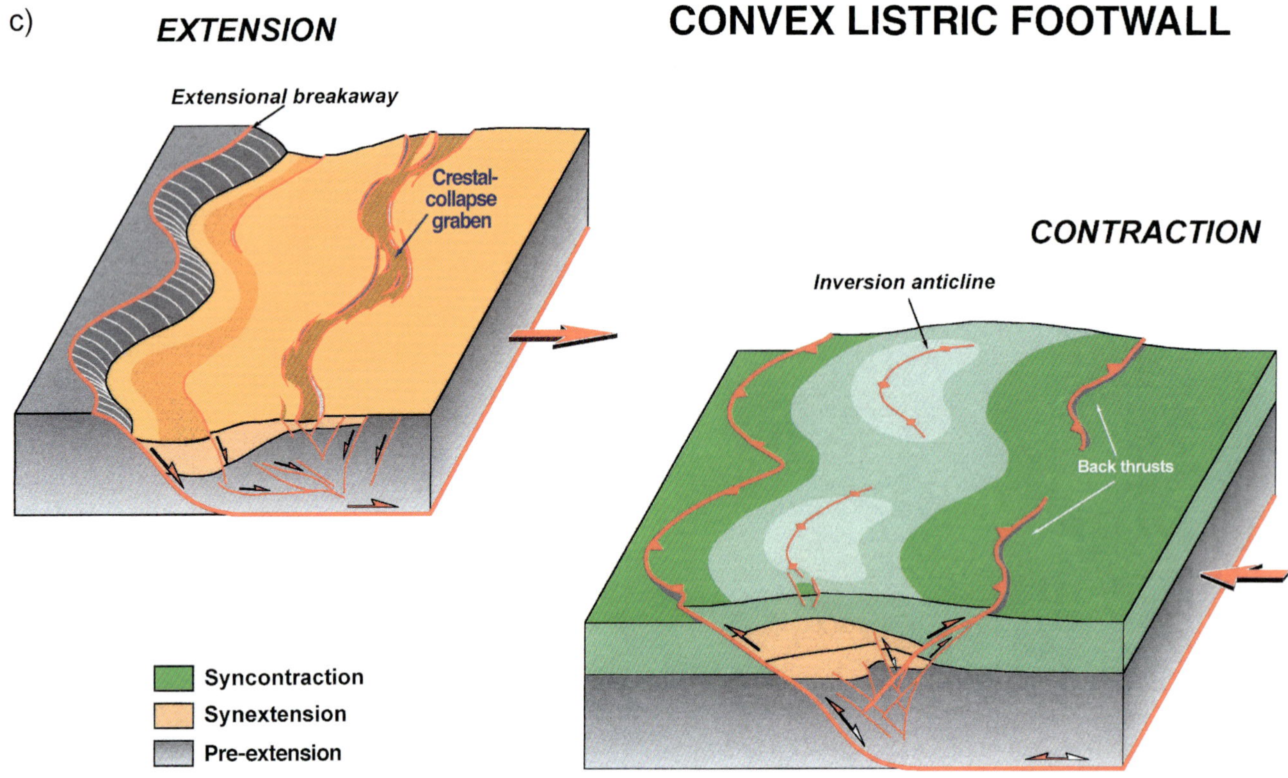

FIGURE 16. (cont.).

Sea (Ziegler, 1987; Badley et al., 1989) in the West Natuna Sea, Indonesia (Dickerman , 1993, Ginger et al., 1993; Phillips et al., 1997), the Malay Basin (Tjia, 1999), and in the Neuquen and Cuyo Basins, Argentina (Vergani et al., 1995; Uliana et al., 1995).

Yamada (1999) described a 3-D example of inversion structures from the Niigata-Akita back-arc basin in the Sea of Japan and concluded that the published seismic sections of Okamura et al. (1995) were similar in structural style to the 3-D analog models of concave-up listric faults (e.g., Figures 6–10). In particular, the bounding thrust structures appeared to have propagated upward at the same angle as that of the main extensional fault system (Yamada, 1999). Similar, steep thrust-fault systems have been observed in a number of inversion structures in the West Natuna Sea, Indonesia, where important structural traps are formed in the anticlinal hanging walls above the steep inversion thrust system (Ginger et al., 1993; Phillips et al., 1997). Published examples of 3-D time slices from the Sembilang field in the West Natuna Sea (Dickerman, 1993) also show a dome structure that has developed above a convex salient in the main fault system. This structure is very similar to that found adjacent to the convex salient in the main detachment surface the Series I experiments (Figure 15a).

Maps of inversion structures in the Cuyo Basin, Mendoza province, Argentina (Uliana et al., 1995) show well-developed periclinal (doubly plunging) anticlines in the hanging walls of inverted half-graben bounding faults. These are generally asymmetric anticlines bounded by a steep thrust fault. They form important structural traps for hydrocarbon accumulations (Uliana et al., 1995). Similarly, in the San Jorge Basin, Santa Cruz and Chubut provinces, Argentina, the north–south-trending San Bernardo Fold Belt is characterized by numerous doubly plunging inversion anticlines (Homovc et al., 1995; Figari et al., 1999). In particular, the Perales field is a doubly plunging anticline with maximum uplift in the hanging wall above the most concave part of the inverted extensional fault-thrust system (Homovc et al., 1995), and therefore is very similar in style to the Series I experiments described in this paper.

In the southern part of the Norwegian sector of the North Sea, the Valhall and Hod fields are developed in doubly plunging anticlines that occur in the hanging wall of the steeply dipping Skrubbe fault zone (Farmer and Barkved, 1999). This is an inverted Lower Jurassic extensional fault where variable uplift and outer-arc extension during inversion has produced a complex hanging-wall anticline. The field is developed in the syninversion Cretaceous Chalk reservoir, whose deposition is controlled by the crestal extensional graben systems (Farmer and Barkved, 1999). As in the Series I models, maximum uplift in the inversion structures appears to be focused in the hanging-wall sectors above the more concave elements of the steep thrust system.

The discussion above highlights the marked lack of detailed analyses of natural 3-D inversion thrust-bounded anticlines, despite their importance in generating structural traps for hydrocarbons in many basins. To validate and further improve the experiments, additional detailed 3-D seismic studies of inverted structures are needed for comparison with the analog models presented in this chapter.

CONCLUSIONS

Extension and inversion above concave-up and convex-up listric detachment surfaces were modeled using scaled sandbox experiments. Extension over 3-D concave-up listric detachments produced characteristic rollover anticlines and crestal-collapse graben systems. In 3-D, the crestal-collapse-graben fault systems are strongly segmented and parallel the along-strike, sinusoidal or cuspate plan geometries of the detachment breakaways. Three-dimensional inversion by horizontal contraction produced asymmetric, thrust-fault-bounded, hanging-wall inversion anticlines. The anticlinal axial traces mimic the plan-view shape of the extensional breakaways. The main extensional detachment faults were reactivated during the inversion. New, steep thrust segments propagated upward at a constant angle from the detachment breakaways. Shallow to moderately dipping, hanging-wall vergent back thrusts also propagated outward from the crestal-collapse graben system. The model inversion anticlines were periclinal in form, with two plunge culminations that correspond to positions of maximum 3-D concavity in the geometry of the footwall blocks.

In contrast, extension above a 3-D, convex-up, listric detachment surface produced a characteristic hanging-wall syncline together with a narrow, complex crestal-collapse graben system. In plan view, the axial trace of the hanging-wall syncline and the faults of the crestal-collapse graben system parallel the sinusoidal shape of the detachment breakaway. Inversion of this system by horizontal contraction produced a broad, thrust fault-bounded anticline. The main detachment surface was reactivated, and a moderately dipping thrust propagated upward through the syninversion strata. The hanging-wall anticline was less asymmetric compared with the concave-up systems, but it also showed similar along-strike plunge culminations. Segmented, hanging-wall back thrusts also formed with plan view geometries subparallel to the main detachment breakaway.

In all models, syninversion growth strata thin across the crest of the anticlines and are thickest either in the footwall to the main thrust fault or at the extremities of the backlimb of the inversion anticline. Vertical and horizontal sections through the models were used to construct 3-D synoptic models for these inversion systems. These may be used as templates for seismic interpretation where there is poor resolution or insufficient coverage to permit a full 3-D analysis.

The analog-model results demonstrate that the 3-D inversion thrust architecture is controlled by the geometry of the main listric detachment surface— that is, by basement control. Maximum inversion uplift occurs where there previously was maximum extension-generated subsidence. Progressive inversion uplift is accomplished by translation of the hanging wall back up the reactivated extensional fault and along the new thrust fault that propagates from the extensional breakaway. During inversion, progressive back rotation of the hanging wall produces fanning, syninversion growth-strata patterns. The results of the analog-model experiments compare well with published examples of 3-D inversion thrust structures from petroleum basins in Indonesia, Japan, Argentina, and the North Sea.

ACKNOWLEDGMENTS

This work is based on Y. Yamada's Ph.D. research, which was supported by the Fault Dynamics Project sponsored by ARCO British Limited, PETROBRAS U.K. Ltd., BP Exploration, Conoco (U.K.) Limited, Mobil North Sea Limited, and Sun Oil Britain. Y. Yamada also acknowledges funding from JNOC and JAPEX. K. McClay gratefully acknowledges support from BP Exploration. Brian Adams and Howard Moore are thanked for constructing and maintaining the deformation apparatus. T. Dooley and P. Whitehouse are thanked for constructive reviews. Fault Dynamics Publication No. 113.

REFERENCES CITED

Alsop, G. I., 1996, Physical modelling of fold and fracture geometries associated with salt diapirism, *in* G. I. Alsop, D. J. Brundell, and I. Davison, eds., Salt tectonics: Geological Society of London Special Publication 100, p. 227–241.

Ayyad, M. H., 1997, Inverted basins in northeast Egypt; Geology and hydrocarbon prospectivities: Ph.D. Thesis, Univ. of Cairo, 299 p.

Badley, M. E., J. D. Price, and L. C. Backshall, 1989, Inversion, reactivated faults and related structures: seismic examples from the southern North Sea, *in* M. A. Cooper and G. D. Williams, eds., Inversion tectonics: Geological Society of London Special Publication 44, p. 201–219.

Buchanan, P. G., 1991, Geometries and kinematic analysis of inversion tectonics from analogue model studies: Ph.D. Thesis, University of London, 416 p.

Buchanan, J. G., and P. G. Buchanan, 1995, Basin inversion: Geological Society of London Special Publication 88, 596 p.

Buchanan, P. G., and K. R. McClay, 1991, Sand box experiments of inverted listric and planar fault systems: Tectonophysics, v. 188, p. 97–115.

Buchanan, P. G., and K. R. McClay, 1992, Experiments on basin inversion above reactivated domino faults: Marine and Petroleum Geology, v. 9, p. 486–500.

Buchanan, P. G., and J. Warburton, 1996, The influence of pre-existing basin architecture in the development of the Papuan fold and thrust belt: implications for petroleum prospectivity, *in* P. G. Buchanan, ed., Petroleum exploration, development and production in Papua New Guinea: Proceedings of the Third PNG Petroleum Convention, Port Moresby, 9th–11th September, 1996, p. 89–109.

Colletta, B., J. Letouzey, R. Pinedo, J. F. Ballard, and P. Bale, 1991, Computerized X-ray tomography analysis of sandbox models: Examples of thin-skinned thrust systems: Geology, v. 19, p. 1063–1067.

Cooper, M. A., and G. D. Williams, 1989, Inversion tectonics: Geological Society of London Special Publication 44, 376 p.

Dickerman, K. M., 1993, The utilization of 3-D seismic for small fields in the South Natuna Sea Block B: Proceedings of the Indonesian Petroleum Association 93, v. 1, p. 660–678.

Eisenstadt, G., and M. O. Withjack, 1995, Estimating inversion: results from clay models, *in* J. G. Buchanan and P. G. Buchanan, eds., Basin inversion: Geological Society of London Special Publication 88, p. 119–136.

Ellis, P. G., and K. R. McClay, 1988, Listric extensional fault systems— results of analogue model experiments: Basin Research, v. 1, p. 55–70.

Farmer, C. L., and O. I. Barkved, 1999, Influence of syndepositional faulting on thickness variations in Chalk reservoirs— Valhall and Hod field, *in* A. J. Fleet and S. A. R. Bold, eds., Petroleum geology of Northwest Europe: Proceedings of the 5th Conference, Geological Society of London, p. 949–957.

Figari, E., et al., 1999, Los sistemas petrolereos de la cuenca del Golfo San Jorge: sintesis estructural, estratigrafica y geoquimica, en Actas de IV Congresso de Exploracion y Desarrollo de Hidrocarburos, InstitutoArgentino del Petroleo y del Gas, v. 1, p. 197–237.

Ginger, D. C., W. O. Ardjakusumah, R. J. Hedley, and J. Pothecary, 1993, Inversion history of the West Natuna Basin: examples from the Cumi-Cumi PSC: Proceedings of the Indonesian Petroleum Association 93, v. 1, p. 635–658.

Horsfield, W. T., 1977, An experimental approach to basement controlled faulting: Geologie en Mijnbouw, v. 56, p. 363–370.

Horsfield, W. T., 1980, Contemporaneous movement along crossing conjugate normal faults: Journal of Structural Geology, v. 2, p. 305–310.

Homovc , J. F., P. A. Conforto, P. A. Lafourcade, and L. A. Chelotti, 1995, Fold belt in the San Jorge Basin, Argentina: an example of tectonic inversion, *in* J. G. Buchanan and P. G. Buchanan, eds., Basin inversion, Geological Society of London Special Publication 88, p. 235–248.

Huiqi, L., K. R. McClay, and D. Powell, 1992, Physical models of thrust wedges, *in* K. R. McClay, ed., Thrust tectonics: Chapman & Hall, London, p. 71–81.

Jackson, M. P. A., and B. C. Vendeville, 1994, Regional extension as a geologic trigger for diapirism: Bulletin of the Geological Society America, v. 106, p. 57–73.

Koopman, A., A. Specksnijder, and W. T. Horsfield, 1987, Sand box studies of inversion tectonics: Tectonophysics, v. 137, p. 379–388.

Lallemand, S. E., J. Malavieille, and S. Calassou, 1992, Effects of oceanic ridge subduction on accretionary wedges: Experimental modelling and marine observations: Tectonics, v. 11, p. 1301–1313.

Lowell, J. D., 1985, Structural styles in petroleum exploration: Tulsa, O.G.C.I., 460 p.

Manceda, R., and D. Figueroa, 1995, Inversion of the Mesozoic Neuquen rift in the Malargue fold and thrust belt, Mendoza, Argentina, *in* A. J. Tankard, R. Suarez Soruco, and H. Welsink, eds., Petroleum basins of South America, AAPG Memoir 62, p. 36 –382.

McClay, K. R., 1989, Analogue models of inversion tectonics, *in* M. A. Cooper and G. D. Williams, eds., Inversion tectonics: Geological Society of London Special Publication 44, p. 41–59.

McClay, K. R., 1990a, Extensional fault systems in sedimentary basins: a review of analogue model studies: Marine and Petroleum Geology, v. 7, p. 206–233.

McClay, K. R., 1990b, Deformation mechanics in analogue models of extensional fault systems, *in*, E. H. Rutter and R. J. Knipe, eds., Deformation mechanisms, rheology and tectonics: Geological Society of London Special Publication 54, p. 445–454.

McClay, K. R., 1995, The geometries and kinematics of inverted fault systems: a review of analogue model studies, *in* J. G. Buchanan and P. G. Buchanan, eds., Basin inversion: Geological Society of London Special Publication 88, p. 97–118.

McClay, K. R., 1996, Recent advances in analogue modelling: uses in section interpretation and validation, *in* P. G. Buchanan and D. A. Nieuwland, eds., Modern developments in structural interpretation, validation and modelling: Geological Society of London Special Publication 99, p. 201–225.

McClay, K. R., and P. G. Buchanan, 1992, Thrust faults in inverted extensional basins, *in* K. R. McClay, ed., Thrust tectonics: London, Chapman & Hall, p. 93–121.

McClay, K. R., and P. G. Ellis, 1987a, Analogue models of extensional fault geometries, *in* M. P. Coward, J. F. Dewey, and P. L. Hancock, eds., Continental extensional tectonics: Geological Society of London Special Publication 28, p. 109–125.

McClay, K. R., and P. G. Ellis, 1987b, Geometries of extensional fault systems developed in model experiments: Geology, v. 15, p. 341–344.

Mitra, S., 1993, Geometry and kinematic evolution of inversion structures: AAPG Bulletin, v. 77, p. 1159–1191.

Naylor, M. A., G. Mandl, and C. H. K. Sijpesteijn, 1986, Fault geometries in basement-induced wrench faulting under different initial stress states: Journal of Structural Geology, v. 7, p. 737–752.

Okamura, Y., M. Watanabe, R. Morijiri, and M. Satoh, 1995, Rifting and basin inversion in the eastern margin of the Japan Sea: The Island Arc, v. 4, p. 166–181.

Phillips, S., L. Little, E. Michael, and V. Odell, 1997, Sequence stratigraphy of Tertiary petroleum systems in the West Natuna Basin, Indonesia, *in* J. V. C. Howes and R. A. Noble, eds., Petroleum systems of SE Asia and Australasia: Indonesian Petroleum Association, p. 381–390.

Richard, P. D., M. A. Naylor, and A. Koopman, 1995, Experimental models of strike-slip tectonics: Petroleum Geoscience, v. 1, p. 71–80.

Sibson, R. H., 1985, A note on fault reactivation: Journal of Structural Geology, v. 7, p. 751–754.

Sibson, R. H., 1995, Selective fault reactivation during basin inversion: potential for fluid redistribution through fault-valve action, *in* J. G. Buchanan and P. G. Buchanan, eds., Basin inversion: Geological Society of London Special Publication 88, p. 3–19.

Tjia, H. D., 1999, Geological setting of Peninsula Malaysia, *in* The petroleum geology and resources of Malaysia: Petronas, p. 139–170.

Uliana, M. A., M. E. Arteaga, L. Legaretta, J. J. Cerdan, and G. O. Peroni, 1995, Inversion structures and hydrocarbon occurrence in Argentina, *in* J. G. Buchanan and P. G. Buchanan, eds., Basin inversion: Geological Society of London Special Publication No. 88, p. 211–233.

Vendeville, B. C., and M. P. A. Jackson, 1992a, The rise of diapirs during thin-skinned extension: Marine and Petroleum Geology, v. 9, p. 331–353.

Vendeville, B. C., and M. P. A. Jackson, 1992b, The fall of diapirs during thin-skinned extension: Marine and Petroleum Geology, v. 9, p. 354–371.

Vergani, G. D., A. J. Tankard, H. J. Belotti, and H. Welsink, 1995, Tectonic evolution and palaeogeography of the Neuquen basin, Argentina, *in* A. J. Tankard, R. Suarez Soruco, and H. Welsink, eds., Petroleum basins of South America: AAPG Memoir 62, p. 383–402.

Williams, G. D., C. M. Powell, and M. A. Cooper, 1989, Geometry and kinematics of inversion tectonics, *in* M. A. Cooper and G. D. Williams, eds., Inversion tectonics: Geological Society of London Special Publication 44, p. 3–15.

Yamada, Y., 1999, 3-D analogue modelling of inversion structures: Ph.D. Thesis, Royal Holloway, University of London, 743 p.

Ziegler, P. A., 1987, Compressional intra-plate deformations in the Alpine foreland— an introduction: Tectonophysics, v. 137, p. 1–5.

Fault-related Folds in Thrust Systems

Suppe, J., C. D. Connors, and Y. Zhang, 2004, Shear fault-bend folding, *in* K. R. McClay, ed., Thrust tectonics and hydrocarbon systems: AAPG Memoir 82, p. 303–323.

Shear Fault-bend Folding

John Suppe
Department of Geosciences, Princeton University, Princeton, New Jersey, U.S.A.

Christopher D. Connors
Department of Geology, Washington and Lee University, Lexington, Virginia, U.S.A.

Yikun Zhang[1]
Department of Geosciences, Princeton University, Princeton, New Jersey, U.S.A.

ABSTRACT

Shear fault-bend folding produces ramp anticlines with very distinctive shapes. They are characterized by long, gentle backlimbs that dip less than the fault ramp, in contrast with classical fault-bend folding. Backlimb dips and limb lengths increase progressively with fault slip, by a combination of limb rotation and kink-band migration. We summarize and apply two simple end-member theories of shear fault-bend folding involving a weak décollement layer of finite thickness at the base of a ramp: (1) simple-shear fault-bend folding, in which the layer undergoes an externally imposed bedding-parallel simple shear with no basal fault, and (2) pure-shear fault-bend folding, in which this basal layer slides above a basal fault and shortens and thickens above the ramp, with no externally applied bed-parallel simple shear. In the limit of large displacement, the fold geometry in pregrowth strata approaches the geometry of classical fault-bend folding, with a backlimb dip that approaches the ramp dip. However, even in these cases, growth strata may record the history of limb rotation that is characteristic of a shear fault-bend fold heritage. We demonstrate that these theories are in agreement with well-imaged seismic examples from the Nankai Trough and Cascadia accretionary wedges, which show substantial shears (40–65°) over stratigraphic intervals of a few hundred meters.

INTRODUCTION

There has been a long-standing intuition that thrust sheets might undergo substantial internal deformation even while they are displaced over their footwalls. This intuition is exemplified by Figure 1, a diagram from David Elliott (1976) showing a ramp anticline— what might now be called a fault-bend fold— with significant layer-parallel simple shear but negligible layer-parallel shortening and thickening (pure shear) within the hanging wall. Indeed, internal deformation in this drawing dominates over basal fault slip in the total displacement of the hanging wall, leading to an output fault slip that exceeds the input fault slip, in spite of the fact that the folding itself consumes slip.

Without discussing the mechanical motivation and merits of this intuition of important internal deformation, suffice it to say that if such internal deformation

[1]*Present address*: Department of Earth Sciences, Nanjing University, Nanjing, China.

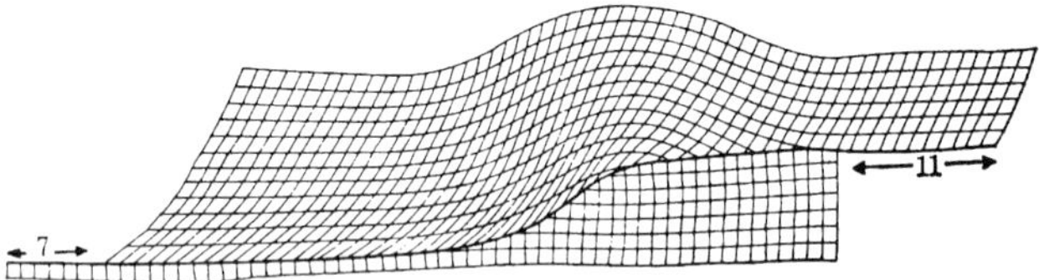

FIGURE 1. There has been a long-standing intuition that thrust sheets might undergo substantial internal deformation even as they are displaced over their footwalls. This schematic ramp anticline from Elliott (1976) exemplifies this intuition by showing the possibility of substantial internal deformation of the hanging wall, in addition to that directly associated with fault displacement. The contributions of layer-parallel simple shear to the displacement of the top of the model (9 units) dominate over the input fault slip (7 units), which leads to an output fault slip (11 units) that is greater than the input, in spite of the fact that folding consumes fault slip.

exists, then the resulting shapes of ramp anticlines should be sensitive to the proportions of fault slip, simple shear, and pure shear. Because of this potential importance, the original balanced theories of fault-bend and fault-propagation folding explicitly included the possibility of layer-parallel simple shear (Suppe, 1983; Suppe and Medwedeff, 1990). Shear in fault-related folds has been considered theoretically by many authors, including Jamison (1987), Mitra and Namson (1989), Mitra (1990, 1992), Mosar and Suppe (1991), Poblet and McClay (1996), Storti and Poblet (1997), and Tamagawa et al. (1998).

In contrast with this substantial theoretical effort, it is remarkable that most successful, rigorous applications of fault-related folding theory to well-documented structures have involved no shear. Most cases in our experience match the no-shear theory rather closely when the fault shape and fold shape are both well constrained by data (one possible but not wholly convincing exception is given in Suppe, 1984). For example, Mitra (1992, 1988) points out that a well-constrained cross section of the Pine Mountain thrust sheet in the southern Appalachians deviates from classical no-shear fault-bend folding by about 1° (with a ramp angle of 17°, the no-shear theory predicts a front-limb dip of 18.8°, whereas the observed frontlimb dip is 18°). This discrepancy from a perfect no-shear solution could result from a forward shear in the frontlimb of about 5° (Figure 2), which would be difficult to prove unequivocally. Mitra did not assign this discrepancy to shear; he suggested that the small deviation might be an effect of dilation as a result of fracturing in the hanging wall.

Mitra's Pine Mountain example is typical of many well-constrained structures in fold-and-thrust belts in which externally imposed layer-parallel shear has been found to be second-order or absent, presumably reflecting the fact that thrust faults are commonly much weaker than their hanging walls (cf. Davis et al., 1983). Rather than further document such examples of negligible shear, it is the purpose of this chapter to show that there are some widespread thrust-belt environments in which hanging-wall shear is a very important process. In particular, we summarize simple end-member theories of pure-shear and simple-shear fault-bend folding following Suppe (2004). We then show that several well-imaged seismic examples from the deep-water Nankai Trough and Cascadia accretionary wedges agree closely with shear fault-bend fold theory, with shears of 40–65° over stratigraphic intervals of a few hundred meters.

CLASSICAL FAULT-BEND FOLDING WITH SIMPLE SHEAR

Fault-related fold theories involving layer-parallel simple shear are most easily developed using the concept of effective-cutoff and fault-bend angles. Effective-cutoff and fault-bend angles are the angles that would exist if we could apply the shear before folding. If we use effective angles rather than the actual angles, then classical no-shear fault-bend folding theory applies immediately without modification. It's that simple.

The concept of effective-cutoff angles is shown in Figure 3. The figure shows a classical fault-bend fold with an externally imposed simple shear, α_e. If, in a thought experiment, we apply this shear before folding, thereby causing a displacement d of the top of the hanging wall, then the fault-bend angle ϕ will be modified to an effective fault-bend angle, ϕ_{ef}, and the initial cutoff angle θ

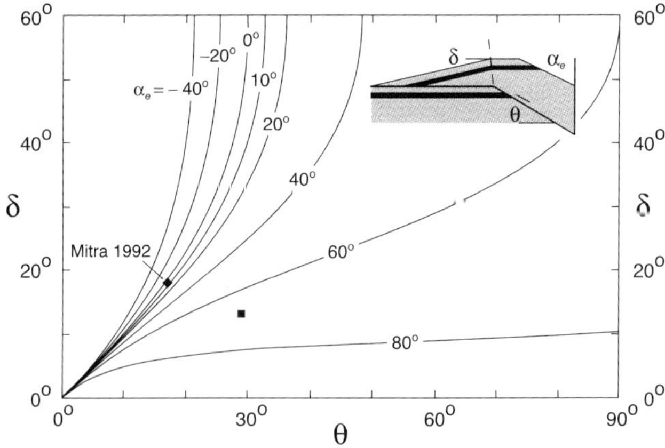

FIGURE 2. Anticlinal fault-bend folding at a ramp-flat transition, including possible externally imposed layer-parallel simple shear α_e. The parameters for the inset schematic fold are shown as the black square. Data for the Pine Mountain thrust sheet from Mitra (1992) deviate slightly from the no-shear theory (with a ramp angle of 17°, the no-shear theory predicts a frontlimb dip of 18.8°, whereas the observed frontlimb dip is 18°). This deviation could result from a forward shear of about 5°, but it would be difficult to prove unequivocally. This example is typical of many structures that show negligible shear.

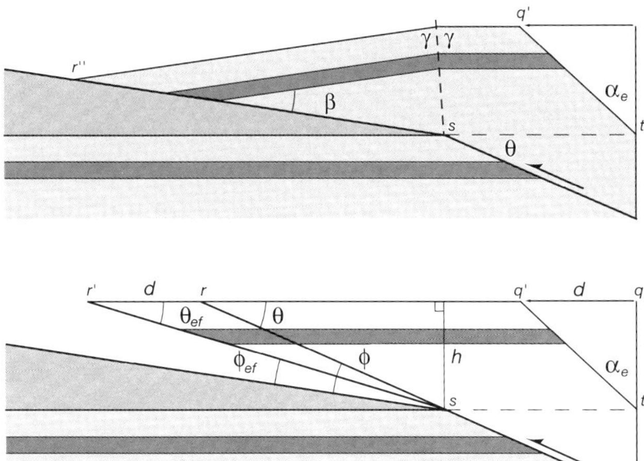

FIGURE 3. Anticlinal fault-bend fold with an externally applied simple shear α_e. The correct balanced fold shape, given the fault shape and shear, can be determined immediately from classical no-shear fault bend folding, if we use *effective-cutoff* θ_{ef} and *fault-bend angles* ϕ_{ef}, rather than the ordinary angles θ and ϕ. The effective angles are those that would exist if we could apply the shear before folding, as shown in the lower figure. Figure 2 was computed using effective angles.

will be modified to an effective-cutoff angle, θ_{ef}. These effective angles may be inserted directly into the no-shear fault-bend fold equations (Suppe, 1983) to compute the fold shape (β, γ). Alternatively, the no-shear fault-bend-folding graph (Figure 7 of Suppe, 1983) can be employed directly to solve subsurface structural problems in cross section, using the effective angles. Variations on this concept of effective angles are used to develop the simple-shear and pure-shear fault-bend-fold theories summarized in this chapter. Figure 2 was computed using effective angles.

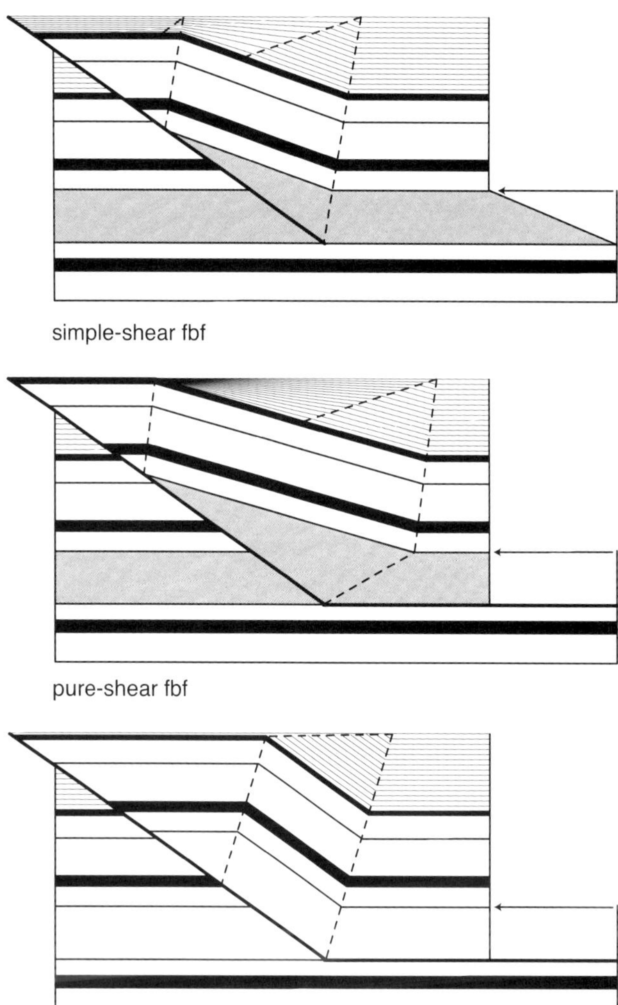

FIGURE 4. Simple-shear and pure-shear fault-bend folds characteristically show backlimb dips that are less than the ramp dip, in contrast with classical no-shear fault-bend folds. The two shear end members involve distributed deformation of a weak décollement layer of finite thickness at the base of a fault ramp, shown in gray. In *simple-shear fault-bend folding,* the décollement layer undergoes an externally imposed bedding-parallel simple shear with no basal fault slip. In *pure-shear fault-bend folding,* this basal layer slides above a basal fault and shortens and thickens above the ramp, with no externally applied simple shear. The growth strata show that shear fault-bend folds undergo a combination of progressive limb rotation and limb lengthening by kink-band migration, whereas classical fault-bend folds grow solely by kink-band migration.

SHEAR WITHIN A BASAL DÉCOLLEMENT OF FINITE THICKNESS

The two end-member types of fault-bend folding considered here— pure-shear and simple-shear fault-bend folding— involve distributed deformation of a weak décollement layer of finite thickness at the base of a fault ramp (Figure 4). In simple-shear fault-bend folding, the décollement layer undergoes an externally imposed bedding-parallel simple shear with no basal fault slip. Thus the simple-shear end member has no bedding-parallel fault— just a ramp with slip going to zero at its base. In pure-shear fault-bend folding, this basal layer slides above a basal fault and shortens parallel to bedding and thickens perpendicular to bedding above the ramp, with no externally imposed simple shear. In both cases, the weak layer is overlain by normal strata that conserve layer thickness and bed length and undergo no externally imposed shear.

These two end members correspond to the Type 2 and Type 3 fault-bend folds of Jordan and Noack (1992), who presented some theory of the end members and their

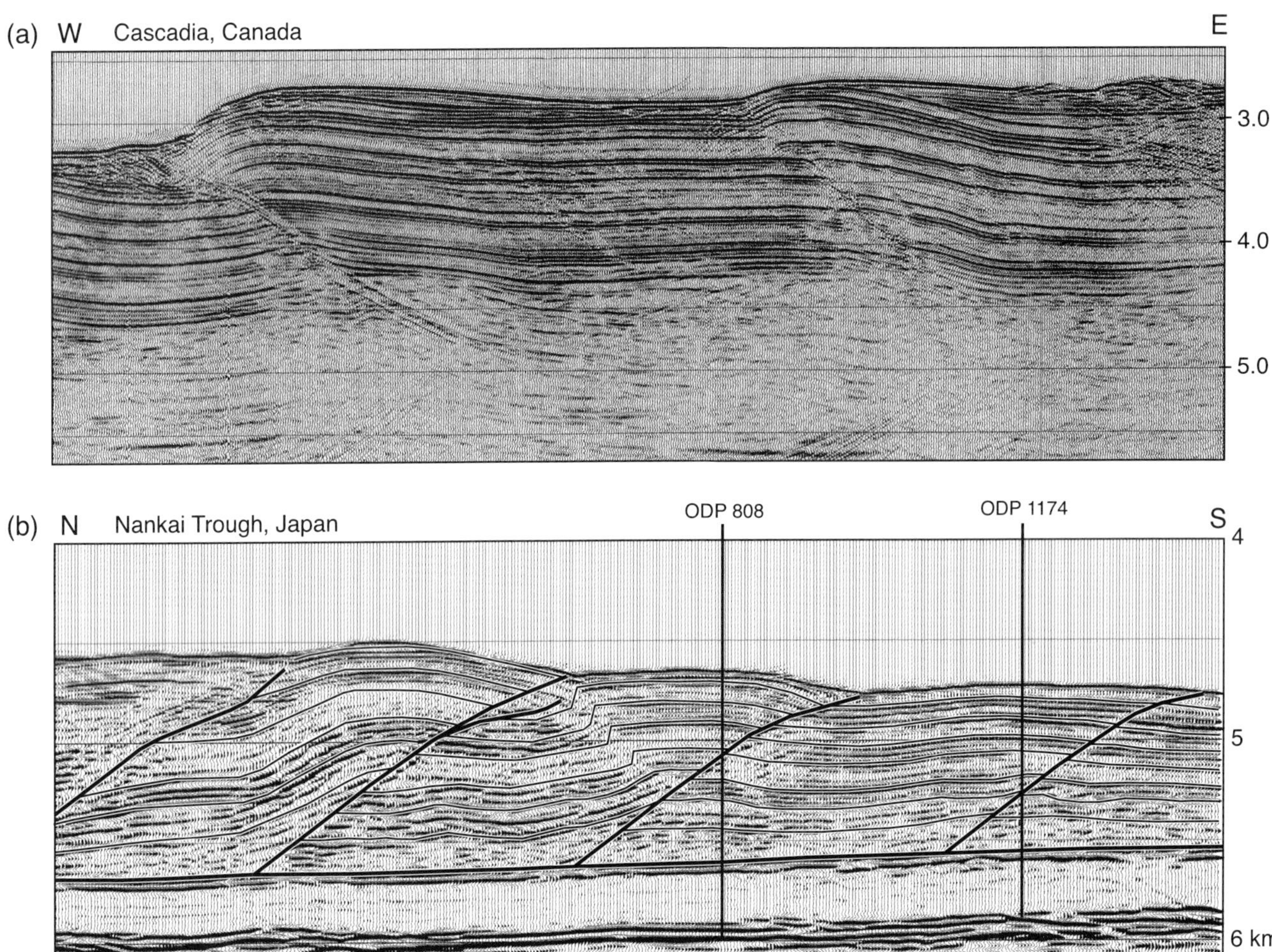

Figure 5. These folds display the qualitative characteristics of shear fault-bend folds, including backlimbs that dip much less than the fault dip and a steeply dipping narrow frontlimb. (a) Above is a time section of two ramp anticlines from the Cascadia accretionary wedge, offshore western Canada (line 89-04 of Hyndman et al., 1994). (b) Below is a depth-migrated section (h = v) of line NT62-8 from the Nankai Trough accretionary wedge, offshore Shikoku, Japan (Moore et al. 1990, 1991). This line passes through Ocean Drilling Project sites ODP-808 and ODP-1174 (Moore et al., 1991, 2001; Shipboard Scientific Party, 2001).

more complex intermediates, together with applications involving the basal ductile décollement of the Jura. The idea that simple-shear fault-bend folding involves a basal ductile décollement (Type 2) was, according to Jordan and Noack (1992), first introduced by Malavieille and Ritz (1989) and Taboada et al. (1990), who were primarily concerned with the strain paths within the ductile layer. A simple-shear fault-bend folding theory also was developed by Wayne Narr (personal communication, 1989). Early articulations of pure-shear fault-bend folding include those of Serra (1977) and Suter (1981). A continuous gradation is theoretically possible between a classical fault-bend folding end member and the pure-shear and simple-shear end members, as discussed by Jordan and Noack (1992). They also discussed heterogeneous shear. To our knowledge, the concepts of pure-shear and simple-shear fault-bend folding have not been widely applied. Therefore, this chapter summarizes key elements of a more complete and accessible theory after Suppe (2004) and shows that several well-imaged structures agree well with the theory.

Pure-shear and simple-shear fault-bend folding both produce a fold geometry for ramp anticlines that is normally quite different from classical fault-bend folding (Figure 4). They display backlimb dips that are shallower— often much shallower— than the fault-ramp dips, yet they may have a steep, narrow frontlimb at the top of the ramp. In contrast, a classical no-shear fault-bend fold stepping up from a décollement will have backlimb dips that are equal to the ramp dip. With excellent seismic data, these contrasting structural styles are easily differentiated. For example, seismic images of fault-related folds of the Cascadia subduction zone of western Canada and the Nankai Trough of Japan (Figure 5) show long, gentle backlimbs, steep fault ramps, and narrow, steep frontlimbs that are quite unlike classical fault-bend folds.

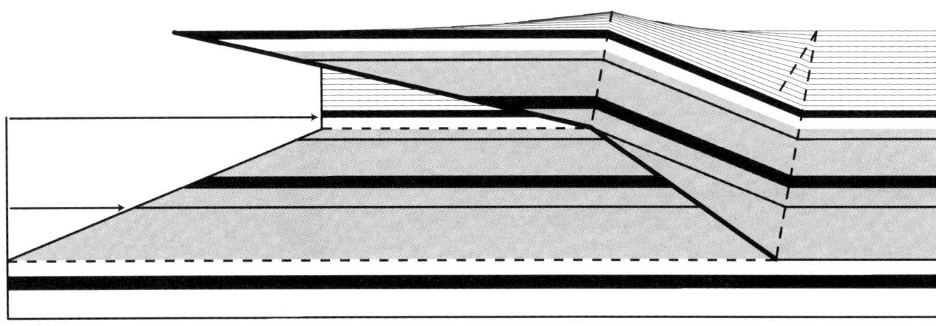

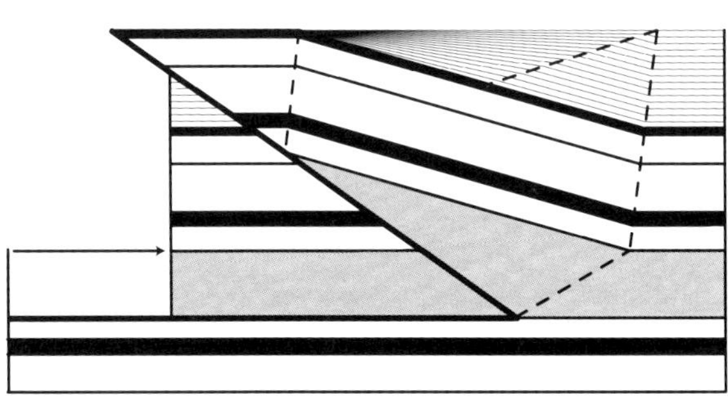

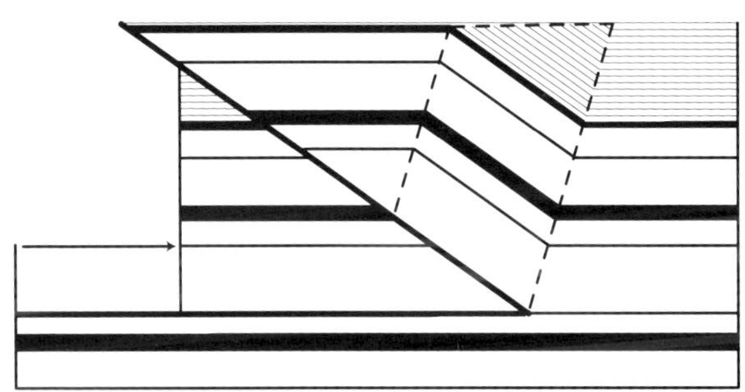

FIGURE 6. Wedge-thrust fault-bend folds show identical fold-limb geometries and kinematics to their forward-thrust equivalents (compare Figures 4 and 6). The simple-shear wedge case is more complex because the footwall fault-ramp is folded progressively by a footwall shear that drives strong migration of the anticlinal axial surface and limb rotation, which accounts for the different geometry of growth strata.

Shear fault-bend folding is also capable of producing wedge structures, as shown in Figure 6. Identical theories describe both the forward-thrust and wedge-thrust structures, except that, in the simple-shear end member, footwall shear deforms the fault ramp and requires a thicker shear zone, as described in Appendix A. Otherwise, the geometry and kinematics of the fold limbs are identical between the forward-thrust and wedge-thrust theories (compare Figures 4 and 6). Examples of wedge-thrust shear fault-bend folds are known from the Niger Delta (Connors et al., 1998) and from southern Taiwan (J. Suppe, 2004, unpublished works), but are not considered in this chapter.

SHEAR FAULT-BEND FOLDING THEORY

Simple-shear Fault-bend Folding

The concept of simple-shear fault-bend folding is motivated by the idea that a weak décollement layer at the base of a ramp— for example, an evaporite layer of substantial thickness— may behave more like a shear zone than like a discrete fault surface. The assumptions of the simplest possible simple-shear fault-bend-folding theory are identical to classical fault-bend folding (conserving layer thickness and bed length, with angular fault bends and fold hinges), except that a décollement layer of finite thickness undergoes an externally imposed homogeneous simple shear, α_e. Under these conditions, the backlimb dip, δ_b, is directly related to the ramp dip, θ, and the shear, α_e, by

$$\cot \alpha_e = \frac{\sin \delta_b}{2C} \left| \left[\frac{1}{\sin \delta_b \cot \theta + 1 - \cos \delta_b} \right]^2 - \left[\frac{1}{\sin \delta_b \cot \theta + 1 - \cos \delta_b} \right] \right| \quad (1)$$

where $C = 0.5$ is a parameter discussed below. A brief derivation of this equation, which is also valid for pure-shear fault-bend folding but with a different value of C, is given in Appendix A, and is from J. Suppe (2004, unpublished works). A graph of this relationship is shown in Figure 7, and a graph of the shape of a ramp anticline as a function of shear is shown in Figure 8. Note in these figures that, at very large shear— which, for example, could correspond to having a very thin décollement— the fold shape asymptotically approaches that of classical fault-bend-folding theory, with the back dip δ_b equal to the fault dip θ. The Pine Mountain thrust studied by Mitra (1988, 1992), reviewed above, lies near this asymptotic classical solution.

Pure-shear Fault-bend Folding

The concept of pure-shear fault-bend folding (Figure 4) is motivated by the idea that the deformation of

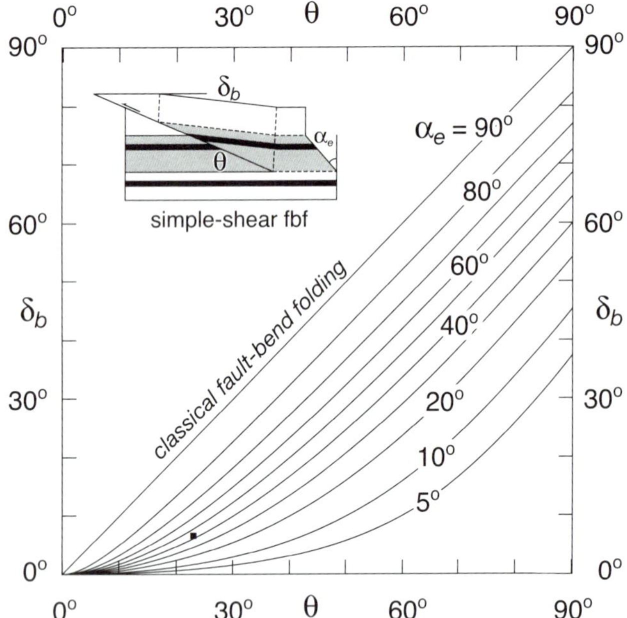

FIGURE 7. Relationship between ramp dip θ back-dip δ_b, and shear α_e for simple-shear fault-bend folding. The inset drawing of a fault ramp corresponds to the angles shown by the black square ($\theta = 23°$, $\delta_b = 6.5°$ and $\alpha_e = 42°$).

intermediate combinations of pure shear and simple shear, is

$$A_o = Cdh \tag{2}$$

where $C = 1$ for pure shear and $C = 0.5$ for homogeneous simple shear. Heterogeneous simple shear may have values of C greater or less than 0.5. Furthermore, using the appropriate numerical values of this parameter C, equation (1) more generally describes the relationship between backlimb dip δ_b, ramp dip θ, and a dimensionless input displacement or shear, $\alpha = \tan^{-1} d/h$, for both pure shear and homogeneous simple shear and intermediate combinations, as outlined in Appendix A. Figure 9 gives a graph of this relationship between limb dip δ_b and ramp dip θ as a function of input displacement, for the pure-shear case.

The pure-shear shortening and thickening of the décollement layer within the backlimb requires that the back-synclinal axial surface not bisect the syncline within the basal layer, whereas the axial surface does bisect the syncline within the overlying strata (Figure 4). The dip, ψ, (Figure 9), of the synclinal axial surface within a weak décollement layer might be confined locally to the rock volume in the immediate vicinity of a fault ramp where stresses are high, in contrast with the simple-shear end member in which shear enters the structure from the hinterland, deforming an arbitrarily large volume of rock. The theory is also motivated by observations of structures in the field (Serra, 1977) and by the results of analog and numerical mechanical models of thrust ramps with a weak décollement layer (Liu and Dixon, 1992; Liu et al., 1992; Strayer and Hudleston, 1997).

The assumptions of the simplest possible pure-shear fault-bend-folding theory are identical to those for classical fault-bend folding (conserving layer thickness and bed length, with angular fault bends and fold hinges), except that a décollement layer of finite thickness h undergoes a pure-shear shortening parallel to bedding above the ramp, with corresponding thickening perpendicular to bedding. There is an input rigid-body fault slip d with no externally imposed simple shear α_e, in contrast with simple-shear fault-bend folding (Figure 9). Thus the area of shortening of the décollement layer in the pure-shear theory is dh, which is twice the area of shortening in the homogeneous simple-shear case and which causes pure-shear folds to grow more rapidly than simple-shear folds (compare Figure 9 with Figure 7).

More generally, the area of shortening for both the pure-shear and simple-shear end members, as well as

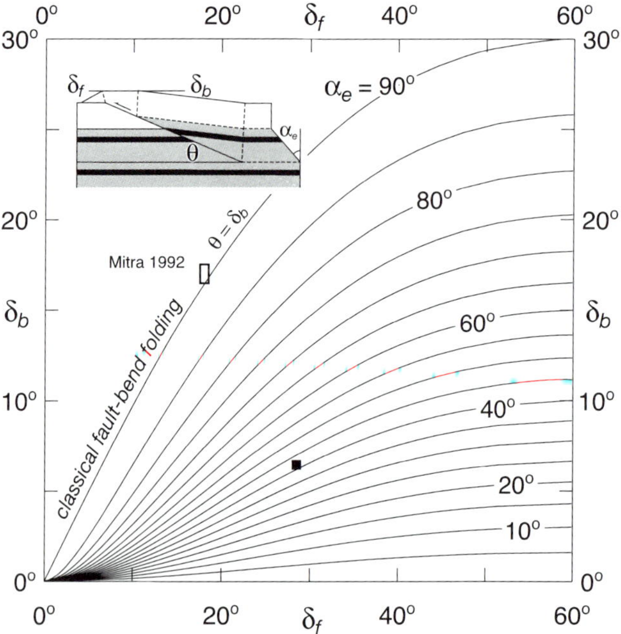

FIGURE 8. Relationship between back dip δ_b, front dip δ_f, and shear α_e for simple-shear fault-bend folding. The inset ramp anticline corresponds to the angles shown by the black square ($\theta = 23°$, $\delta_f = 28.4°$, $\delta_b = 6.5°$ and $\alpha_e = 42°$). Data from a cross section of the Pine Mountain thrust (Mitra, 1992) lie near the classical fault-bend folding theory ($\theta = 17°$, $\delta_f = 18°$, and $\delta_b = 17°$), which would correspond to a décollement layer of zero thickness, and therefore $\alpha_e = 90°$. The ramp dip θ is given by the curve $\alpha_e = 90°$; in this case, $\theta = \delta_b$.

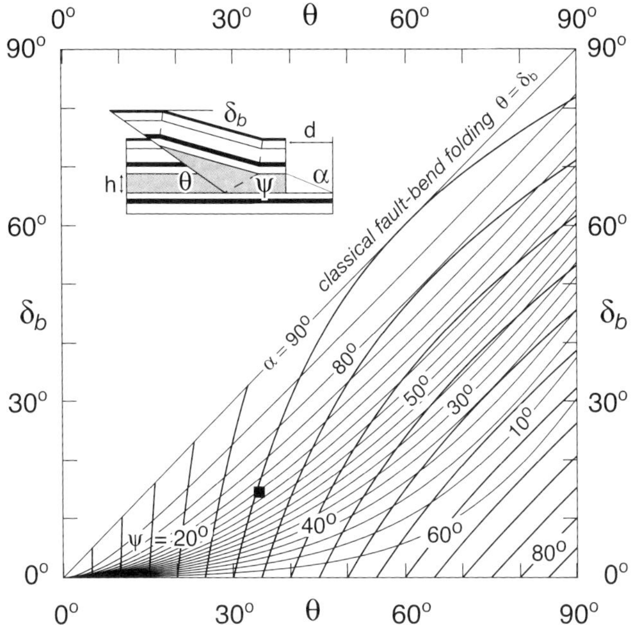

FIGURE 9. Relationship between ramp dip, θ, back-dip, δ_b, and dip of the back syncline, ψ, within the weak décollement layer for pure-shear fault-bend folding. The inset drawing of a fault ramp corresponds to the angles shown by the black square ($\theta = 34^\circ$, $\delta_b = 15.5^\circ$, $\alpha = 68^\circ$ and $\psi = 30^\circ$).

the décollement layer is given by

$$\cot\psi = 2C\left[\cot\theta + \frac{1}{\sin\delta_b} - \cot\delta_b\right] - \cot\theta \qquad (3)$$

where $C = 1$ for pure shear (see Appendix A). A graph of this equation for axial-surface dip, ψ, is given in Figure 9.

EVOLUTION OF SHEAR FAULT-BEND FOLDS

In contrast with classical fault-bend folds, shear fault-bend folds display a progressive increase in backlimb dip with increasing fault slip. This evolution is shown in Figures 10 and 11, which are graphs of backlimb dip δ_b as a function of increasing displacement d of the top of a décollement layer of finite thickness h. For

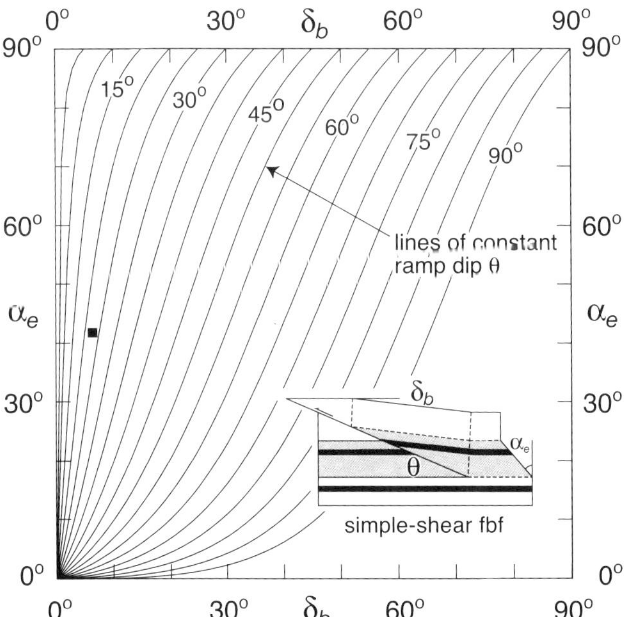

FIGURE 10. The normal evolution of simple-shear fault-bend folds is represented by lines of constant fault (ramp) dip θ (which remains constant during fold growth, unless there is footwall deformation). The backlimb dip δ_b progressively increases with increasing shear α_e until, at infinite shear $\alpha_e = 90^\circ$, the back dip δ_b asymptotically approaches the ramp dip θ. The simple-shear fault-bend fold shown as an inset has a ramp dip of 23°, therefore, it has tracked along the $\theta = 23^\circ$ curve, reaching its present back dip of $\delta_b = 6.5^\circ$ at an angular displacement of $\alpha_e = 42^\circ$.

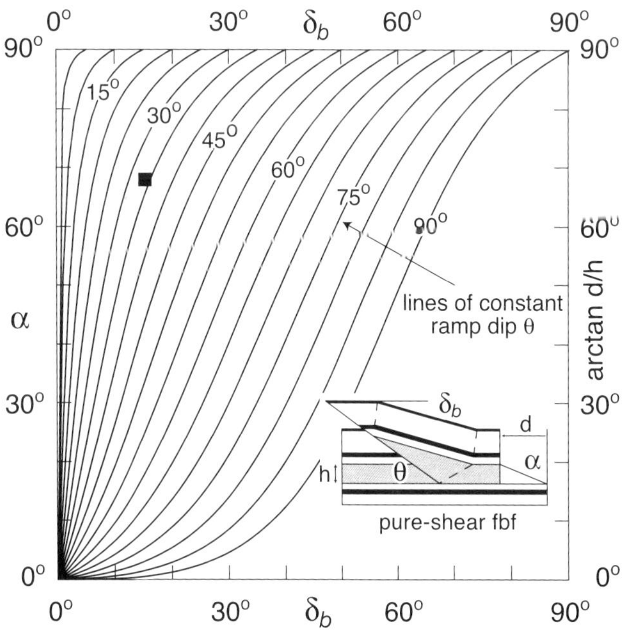

FIGURE 11. The normal evolution of pure-shear fault-bend folds is represented by lines of constant fault (ramp) dip θ (which remains constant during fold growth, unless there is footwall deformation). The backlimb dip δ_b progressively increases with increasing angular displacement (shear) $\alpha = \tan^{-1} d/h$ until, at infinite displacement d/h ($\alpha = 90^\circ$), the back dip δ_b asymptotically approaches the ramp dip θ. The pure-shear fault-bend fold shown as an inset has a ramp dip of 34°, therefore, it has tracked along the $\theta = 34^\circ$ curve, reaching its present back dip of δ_b of 15.5° degrees at an angular displacement α of 68°.

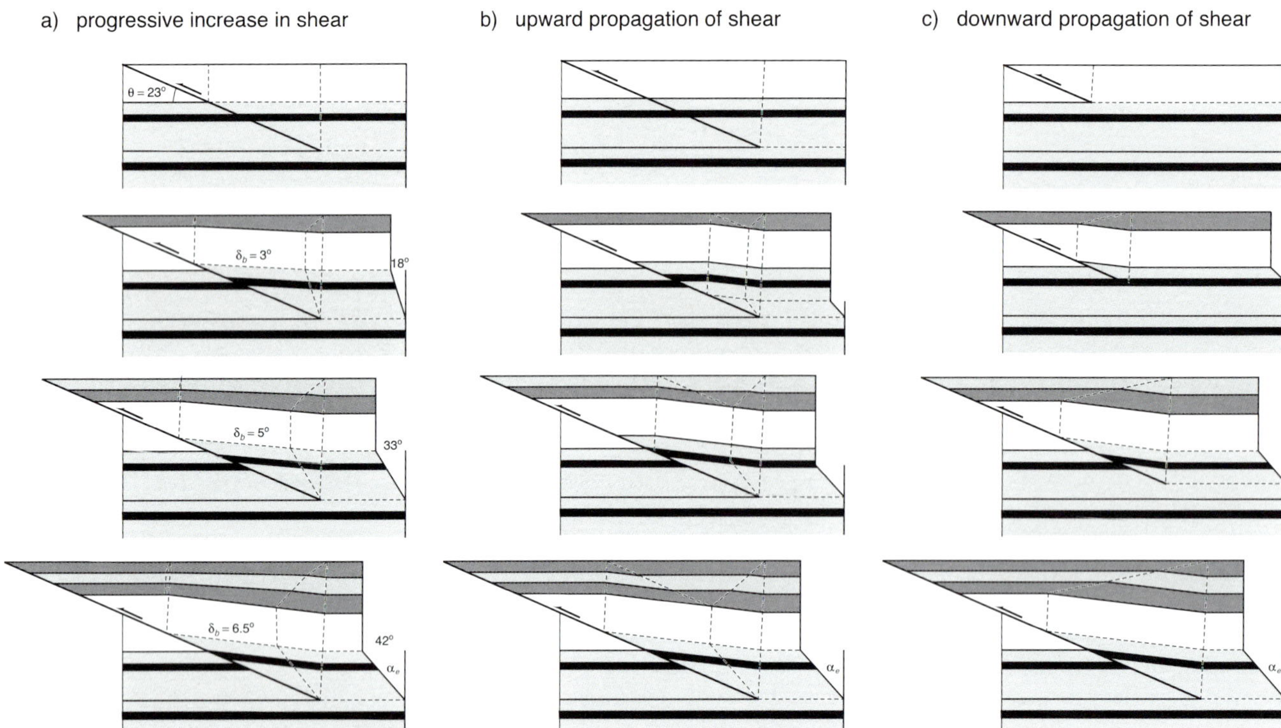

FIGURE 12. Growth models of simple-shear fault-bend folds with contrasting kinematic histories of the shear zones. (a) The left-hand drawings show a progressive increase in shear within a shear zone of constant thickness, leading to limb rotation and kink-band migration. It is the progressive change in shear that leads to limb rotation. (b) The drawings in the center show a progressive upward propagation of shear with constant angular shear, leading to kink-band migration, but no limb rotation. (c) The right-hand drawings show a progressive downward propagation of shear with constant angular shear, leading to kink-band migration of the syncline but no limb rotation.

the pure-shear fault-bend folds, as well as mixtures of pure and simple shear, this externally imposed displacement is conveniently represented in dimensionless form as an angular displacement, $\alpha = \tan^{-1} d/h$, which is immediately comparable to the externally imposed simple shear α_e in the simple-shear end member.

The evolution of any specific shear fault-bend fold normally would follow along a line of constant ramp dip, θ, as shown in these graphs (Figures 10 and 11), unless of course the ramp geometry changes with time because of footwall deformation. For example, in Figure 10 the simple-shear fault-bend fold shown as an inset in the graph has a ramp dip of 23°, therefore it has tracked along the $\theta = 23°$ curve, reaching its present back dip of $\delta_b = 6.5°$ at an angular displacement of $\alpha_e = 42°$. Note that at a very large displacement (approaching infinite shear $\alpha_e = 90°$), the backlimb dip asymptotically approaches the back dip of classical no-shear fault-bend folding, which is the ramp dip θ.

Similarly, the backlimb dips δ_b of pure-shear fault-bend folds approach the ramp dip θ at large displacement (Figure 11). For example, the pure-shear fault-bend fold shown as an inset in the graph of Figure 11 has a ramp dip of 34°, therefore it has tracked along the $\theta = 34°$ curve, reaching its present back dip δ_b of 15.5° at an angular displacement α of 68°.

GROWTH STRATA IN SHEAR FAULT-BEND FOLDS

Because of the progressive increase in both back dip and limb length, models of sedimentation over shear fault-bend folds commonly show a combination of limb rotation and kink-band migration (Figures 4 and 6). However, models of the simple-shear end member have the potential to show more variability than the pure-shear end member because of the variety of possible shear-zone kinematics. Several simple-shear models are shown as examples in Figure 12, involving (1) fixed shear-zone width with progressive increase in shear and (2) upward and (3) downward propagation of the shear zone. The progressive increase in shear causes the limb rotation. Pure-shear models display a pattern of growth strata that is similar to the fixed-thickness shear-zone models with progressive increase in shear (compare Figures 4 and 12). A seismic image from central California

FIGURE 13. Seismic image of a fold limb, showing a combination of limb rotation and kink-band migration, similar to shear fault-bend fold models (compare Figures 4, 6, and 12), from Antelope Hills, San Joaquin Valley, California (Medwedeff, 1988). A similar seismic image showing a combination of kink-band migration and limb rotation is published by Zapata and Allmendinger (1996, *their* Figure 13) from the Bermejo anticline in Argentina.

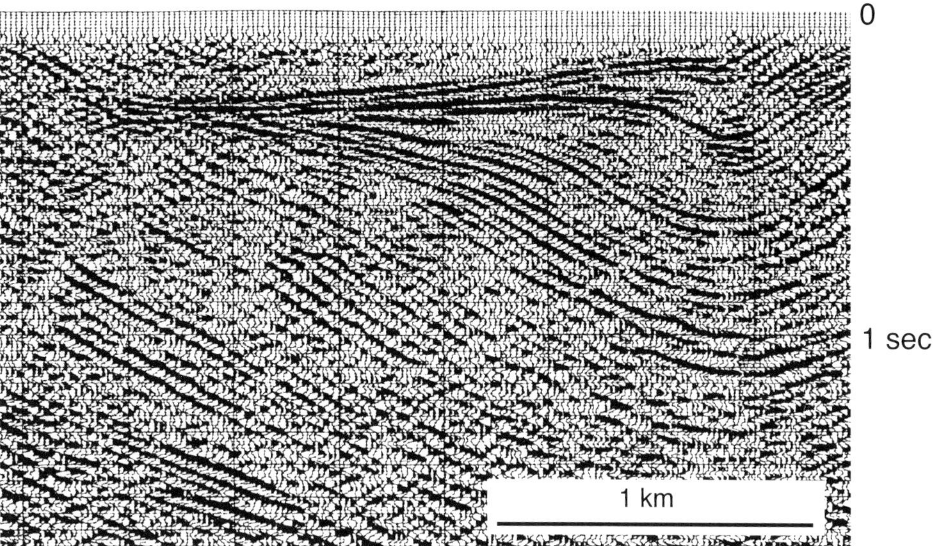

that resembles these models, showing a combination of limb rotation and kink-band migration, is shown in Figure 13 and another from the Bermejo anticline in Argentina is published by Zapata and Allmendinger (1996, their Figure 13).

In the case of large displacement, the fold geometry in pregrowth strata approaches the geometry of classical fault-bend folding, with a backlimb dip that approaches the ramp dip. However, even in these cases growth strata may record the history of limb rotation that is characteristic of a shear fault-bend fold heritage, if the sedimentation rate is rapid enough (Figure 14).

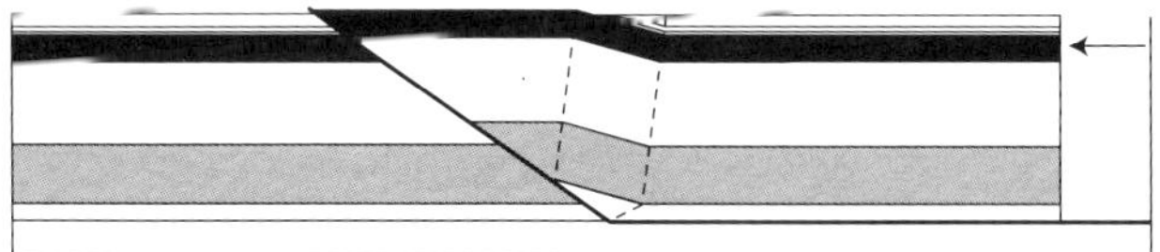

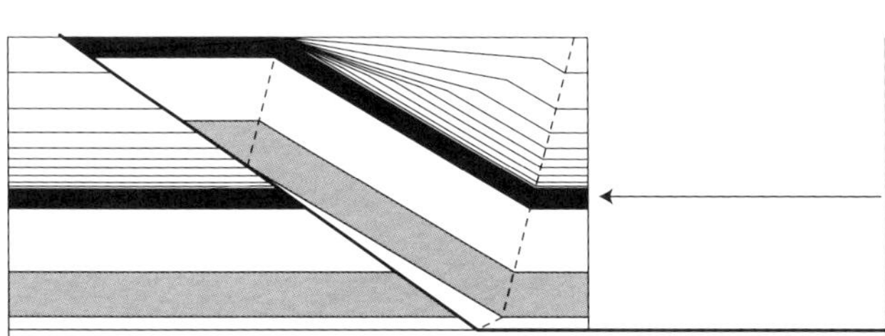

FIGURE 14. Growth model of a pure-shear fault-bend fold with large displacement such that the backlimb dip approaches the ramp dip. The geometry in pregrowth strata is similar to a classical no-shear fault-bend fold, but the limb rotation recorded in the growth strata indicates that it is a shear fault-bend fold. The sedimentation rate increases upward relative to the slip rate.

EFFECTS OF FAULT CURVATURE AND VARIABLE SHEAR

All the models shown so far have straight fault ramps of constant dip and homogeneous shear. We present several models illustrating the effects of curved fault ramps and heterogeneous simple shear to show how such complexities might be identified in data. For comparison, notice that Elliott's (1976) model (Figure 1) shows a combination of a curved fault, heterogeneous simple shear, and input fault displacement.

Figure 15 displays a set of models with a curved fault ramp. Figure 15a is a classical no-shear fault-bend fold, which requires the existence of a progressively widening curved band a–a' at the base of the ramp in which bedding in the hanging wall is parallel to the fault. In contrast, the simple-shear end member shows no such band but shows a fault slip that asymptotically reaches zero at the base of the ramp (Figure 15b and c). Therefore, the effects of fault curvature and shear are distinguishable in suitable seismic images and in models. For example, notice that such a curved band a–a' is seen in Elliott's (1976) model (Figure 1).

The effects of shear-zone thickness are also shown in Figure 15b and c. At constant displacement, the effect of increased shear-zone thickness is to reduce the limb dips above the hanging-wall cutoff of the shear zone and to reduce the area of the anticline's structural relief. Notice that the input area of shortening decreases from (a) to (c) in Figure 15.

The effects of heterogeneous simple shear are shown in Figure 16. The shear within a stratigraphic interval determines the dip of beds overlying the hanging-wall cutoff of that interval. Notice that increasing shear

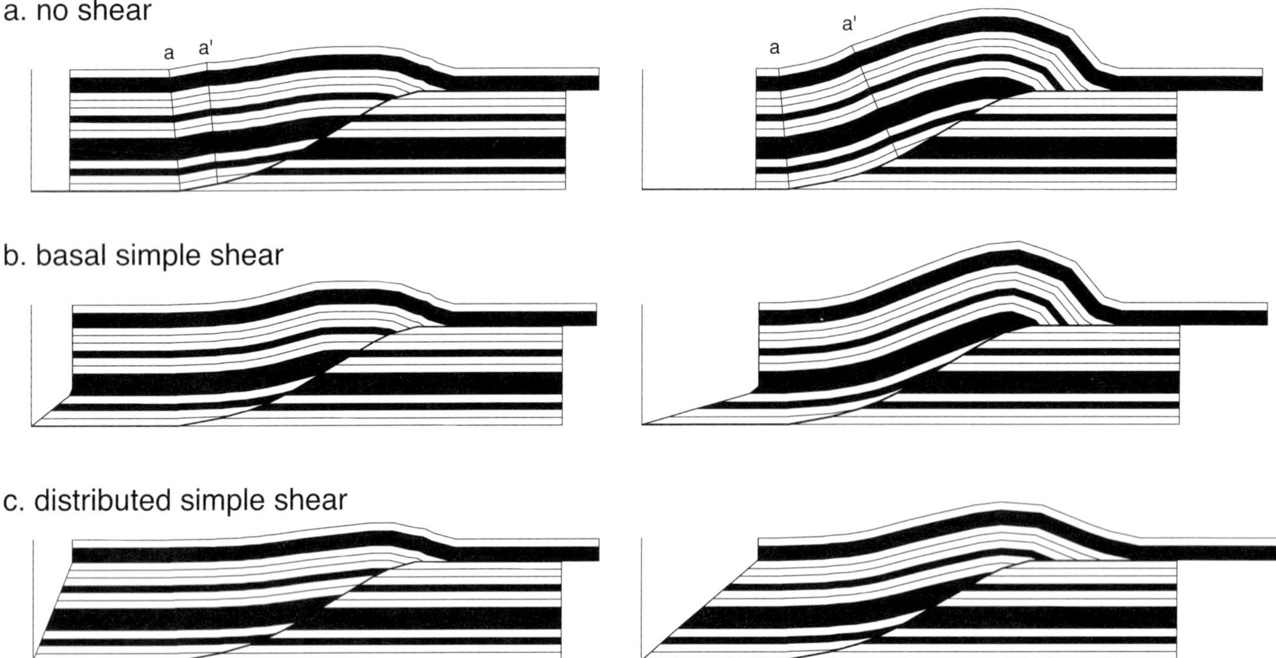

FIGURE 15. Models of curved-ramp fault-bend folding, with and without simple shear. The case of no shear (a) requires the progressive growth of a curved band a–a' at the base of the ramp, with bedding parallel to the fault, which is absent in end-member shear fault-bend folds such as (b) and (c). Also, slip goes to zero at the base of the ramp in the shear cases (b) and (c) with no bedding-parallel fault, whereas a basal fault of slip a–a' exists in the nonshear case (a). Therefore the effects of fault curvature and shear are distinguishable. Figures (b) and (c) also illustrate the effect of increasing shear zone thickness at constant displacement, which is to reduce the limb dips above the hanging-wall cutoff of the shear zone and to reduce the area of structural relief of the anticline. Notice that the input area of shortening decreases from (a) to (c).

increases the dip in the backlimb until it reaches the ramp dip at infinite shear. In contrast, increasing shear reduces the dip in the frontlimb. More generally, forward hanging-wall shear reduces the hanging-wall cutoff angle until it reaches zero at infinite shear.

TRANSMISSION OF SHEAR TO THE HINTERLAND

Models of ramp anticlines like that of Elliott (1976) (Figure 1) or the simple-shear end member (Figures 4, 15, and 16) are characterized by shear that is externally imposed— in a geometric sense— from the hinterland. Therefore, we briefly consider the implications of this hinterland transmission for the shapes of adjacent fault ramps and ramp anticlines to help us understand how this phenomenon of shear transmission could be identified in data.

If a series of foreland-propagating simple-shear fault-bend folds form on the same décollement layer, then their shear is in series and progressively adds toward the hinterland. As shown in Figure 17, the shear from a younger, more forward fold deforms both the footwall and hanging wall of an adjacent ramp in the hinterland. This reduces the backlimb dip of the hinterland structure, because both the ramp dip within the décollement layer and the effective fault-bend and cutoff angles are reduced. Thus, an additive hinterland transfer of shear should produce a distinctive pattern of hinterland reduction of ramp dips within the décollement layer, and in many cases, it should produce a reduction in backlimb dips.

However, this additive hinterland transfer of shear is not required. Just as a simple-shear fault-bend fold transforms fault slip on the ramp into shear within the décollement layer, any fault ramp in the hinterland

FIGURE 16. Model of simple-shear fault-bend fold with heterogeneous shear. The shear within a stratigraphic interval determines the dips of beds overlying the hanging-wall cutoff of that interval. Notice that increasing shear increases dip in the backlimb but decreases the dip in the frontlimb.

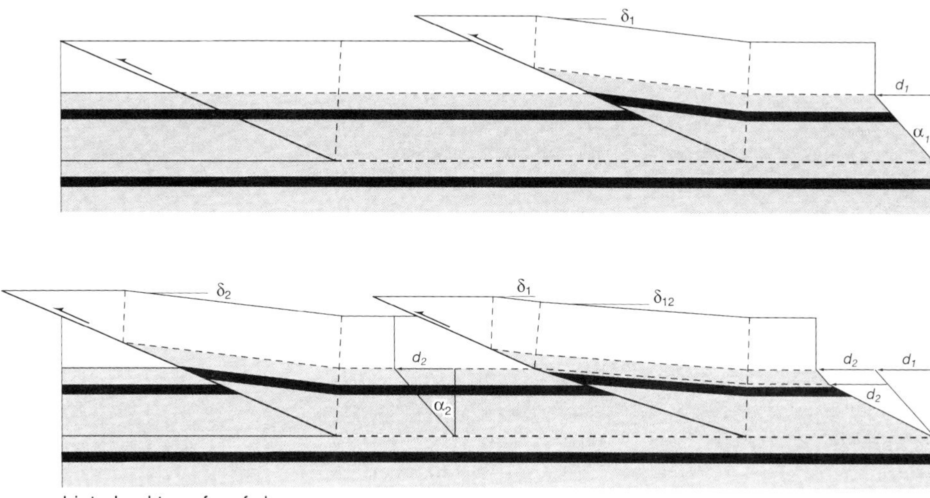

FIGURE 17. Hinterland transfer of shear. Shear from a younger simple-shear fault-bend fold toward the foreland may deform the footwall, fault ramp, and hanging wall of an adjacent fold toward the hinterland, thereby reducing its effective cutoff and ramp angles and back dip. In this model, the younger shear is transferred through the adjacent structure to produce a progressive increase in shear toward the hinterland. In contrast, Figure 18 shows a hinterland termination of shear.

can absorb that décollement shear quantitatively, as shown in Figure 18.

QUALITATIVE IDENTIFICATION OF PURE-SHEAR AND SIMPLE-SHEAR FAULT-BEND FOLDS

We can interpret candidate shear fault-bend folds by comparing them both qualitatively and quantitatively with the simple models, bearing in mind that nature might produce more-complex structures that are still fundamentally shear fault-bend folds. For example, Jordan and Noack (1992) discuss the possibility of complex heterogeneous mixtures of the end-member types involving different shear behavior of each stratigraphic level, thereby leading to more complex fold geometries. Here we point out several key qualitative features of shear fault-bend folds that will be used in our seismic interpretations below.

The most straightforward qualitative method of distinguishing the pure-shear and simple-shear end members is to observe the shape of the synclinal axial surface within the décollement layer (Figure 4). In simple-shear models the axial surface bisects the syncline because layer thickness is everywhere conserved. In contrast, the syncline's axial surface does not bisect in pure-shear models because the décollement layer thickens above the ramp. This causes the back syncline, at stratigraphic levels above the décollement layer, to be displaced toward the hinterland of the fault bend (Figure 4).

Another distinctive aspect of both end-member models is that the anticlinal axial surface marking the

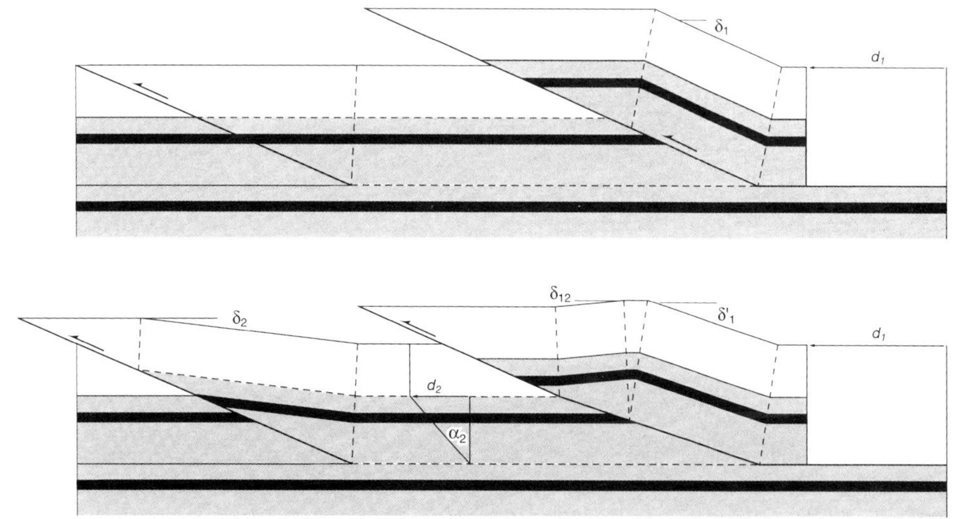

FIGURE 18. Hinterland termination of shear. In contrast with Figure 17, this model shows that shear from a younger simple-shear fault bend in the foreland can terminate at a fault in the hinterland, which might happen if the décollement layer is stronger in the hinterland or if a discrete basal fault is weak relative to the layer. Only the hanging-wall geometries differ between Figures 17 and 18; the footwall geometries are identical.

top of the back fold limb terminates at the fault at the top horizon of the décollement layer. Therefore, the inflection in dip of the syncline's axial surface should be at this same top horizon in pure-shear fault-bend folds (Figure 4). We will see this property displayed in a Nankai Trough example, below.

CASCADIA ACCRETIONARY WEDGE

Low-taper, sediment-rich accretionary wedges commonly have ramp anticlines that qualitatively resemble shear fault-bend folds. Examples are the Cascadia wedge offshore western Canada (Hyndman et al., 1994) and the Nankai wedge offshore Japan (Moore et al., 1990, 1991), both of which are shown in Figure 5. We begin by analyzing one structure from the Cascadia accretionary wedge that was well imaged as part of the Canadian Lithoprobe Project (Hyndman et al., 1994). Observe Figure 19, which is a depth-converted version of part of line 89-05. An along-strike image of the same structure in time is shown on the right-hand side of Figure 5 (line 89-04). The time section of Figure 19 was depth converted by use of a heterogeneous depth stretch based on a smoothed velocity model that incorporates velocity analysis from Yuan et al. (1994) and stacking velocities. Both seismic lines appear to be oriented within 20° of true dip direction of this structure, according to maps of Hyndman et al. (1994) and axial surface mapping (Shaw et al., 1994). In contrast, the left-hand Cascadia structure in Figure 5 is much more oblique (~40°) and hence is elongated.

The structure in Figure 19 shows a backlimb geometry that qualitatively agrees with simple-shear models (compare with Figure 4). The backlimb dip, δ_b, of 5–13° is substantially less than the ramp dip of θ = 35–40°, indicating a shear fault-bend fold with a back syncline that approximately bisects the syncline and emanates from the base of the fault ramp, which eliminates pure-shear models. The fault picks that constrain the fault geometry are shown and rule out strongly listric fault interpretations. Furthermore, fault slip goes to zero or nearly zero at the base of the ramp, indicating that no significant basal fault exists.

This structure is more complex than the simple models of Figure 4, because the fault ramp is not straight but is instead composed of two segments that dip 35° and 40°. Furthermore, the backlimb has two kink bands, *ab* and *bc*, of different dips. Let us begin by treating each segment separately. Applying the homogeneous simple-shear theory (Figure 10), we find that a backlimb dip δ_b of 11–12° within the lower kink band *ab* and a lower ramp dip θ of 35° predict an external simple shear α_e of 31–32°. The upper backlimb dip δ_b of 5° within the upper kink band *bc* together with an upper ramp dip θ of 40° predicts an external simple shear α_e of about 8°.

We test these predictions by line-length restoration of many beds in Figure 19, which shows that shear is heterogeneous on several scales. The lowest 340 m, which terminates in kink-band *ab*, has a somewhat heterogeneous shear that is overall about 31° and agrees with the 31–32° obtained from dips via theory above. The next 450 m, which terminates at the fault in kink-band *bc,* has a shear of about 8° and agrees with 8° obtained via theory. Thus the measurements from line-length restoration of many reflectors are in very good agreement with application of the homogeneous simple-shear theory to limb and fault dips.

About 120 m of growth strata has accumulated over the base of the backlimb during deformation and record limb rotation and kink-band migration (see also Figure 5). Furthermore, the reflector geometry indicates that the next shallow hinterland thrust to the right (east) deactivated just before the initiation of our structure (Figure 19).

NANKAI TROUGH ACCRETIONARY WEDGE

The frontal ramp anticline of the Nankai Trough (Figure 20) is a scientifically well-documented structure

FIGURE 19. (a) Depth-converted seismic image of part of the toe of the Cascadia accretionary wedge offshore western Canada, produced as part of the Lithoprobe Project (Hyndman et al. 1994) (line 89-05; h = v). (b) Small left-hand inset shows key fault picks that constrain the ramp geometry and indicate that it is not listric. (c) Small right-hand inset shows unfolding of the hanging wall based on conservation of bed length. The hanging wall has been unfolded to the regional dip, which allows measurement of the shear based on the deformed hanging-wall cutoff. The shear profile shows the deformation of a line originally perpendicular to bedding before deformation. An overall simple shear α_e of 31° is observed in the lowest 340 m that terminates at the fault in kink band *ab*, which agrees well with the value of 31–32° predicted via theory (Figure 10) from the backlimb dip δ_b of 11–12° for kink-band *ab* and a lower fault dip θ of 31°. A simple shear α_e of 8° is observed in the next 450 m that terminates at the fault in kink band *bc,* which agrees well with a shear of 8° predicted via theory from the backlimb dip δ_b of 5° for kink-band *bc* and an upper fault dip θ of 40°. (d) Interpreted section. Notice that fault slip goes to zero at the base of the ramp indicating that no significant bedding parallel fault exists. Shallow growth strata show limb rotation. Shallow reflector geometry indicates that growth began immediately upon termination of slip on the shallow thrust to the right (east).

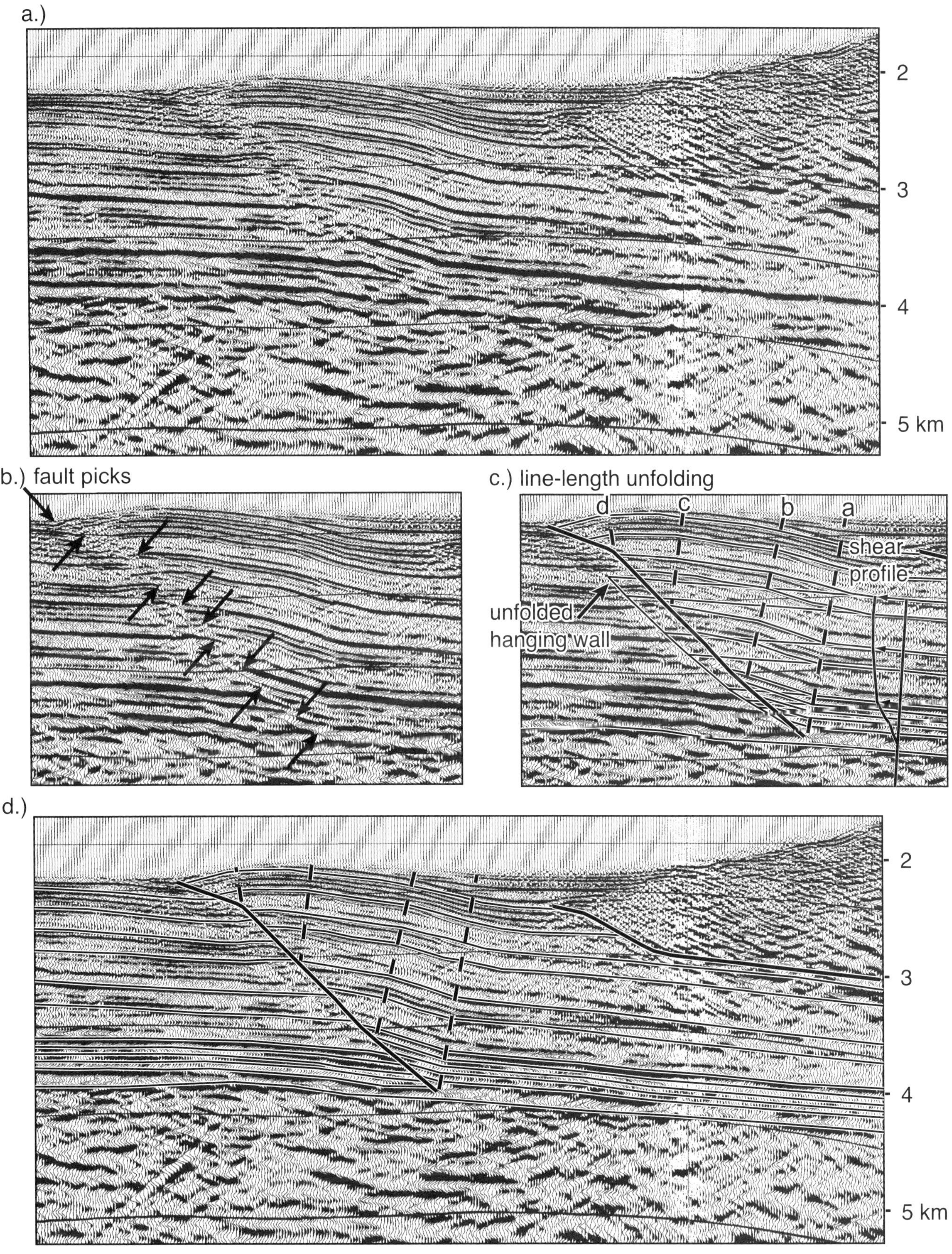
a.)
2
3
4
5 km
b.) fault picks
c.) line-length unfolding
d
c
b
a
shear
profile
unfolded
hanging wall
d.)
2
3
4
5 km

because it is the site of Ocean Drilling Project hole ODP-808/808I and extensive scientific studies in ODP Legs 131, 190, and 196. The borehole, which was extensively cored and logged, lies along the seismic line, and both the thrust ramp fault zone and the underlying décollement stratigraphic interval were cored. The depth seismic line is within 5–10° of the dip direction, as shown by Sea Beam bathymetry (Moore et al., 1991, 2001).

It should be noted that there is significant deformation prior to, and in addition to, the development of well-established dominant fault ramp structures in the front of the Nankai wedge (Figure 5). This is shown by the significant deformation of the footwall of the ODP-808 thrust and of the next seaward anticline drilled by ODP hole 1174, for which the fault ramp has yet to accumulate much displacement. Elsewhere in the Nankai Trough, Morgan et al. (1995) and Morgan and Karig (1995) have documented similar distributed deformation ahead of the frontal thrust ramp. Distributed deformation and minor wedge faulting are proportionately less significant to the total hanging-wall structure of ODP-808; nevertheless, significant wedging can be seen in the lower backlimb, similar to that in the footwall (Figure 20).

We interpreted the ODP-808 fault-ramp location based on reflector geometry and borehole data that constrain the ramp location and reflector correlations (see fault picks in Figure 20). The fault is a 60-m-wide zone with overturned beds (Tiara et al., 1991a, p. 139). Such seismic data would be interpreted traditionally with a curved, sled-runner fault geometry. However, careful examination of reflector geometry in comparison with curved fault models (Figures 1 and 15) shows that the fault cannot be substantially curved near the base of the ramp. Furthermore, recent, not-yet-published, 3-D seismic data show nearly straight faults abruptly taking off of the detachment (Gulick et al., 2000, also see Bangs et al., 1999, Moore et al., 1999, 2001). Therefore we proceed with a nearly straight-ramp interpretation (Figure 20).

The depth-migrated seismic image of the frontal thrust ramp (Figure 20) is qualitatively similar to shear fault-bend-folding models. In particular, the backlimb dip, $\delta_b \approx 13^\circ$, is substantially less than the ramp dip, $\theta = 35^\circ$. Furthermore, the back syncline in the highly reflective interval of the hanging wall is displaced substantially to the hinterland of the base of the ramp, in a way that is qualitatively similar to pure-shear fault-bend fold models, as discussed in the preceding section (compare Figure 4). Furthermore, the anticlinal axial surface at the top of the backlimb terminates at the fault at the horizon of the bend in the back syncline, in agreement with pure-shear models.

On the basis of the observed backlimb dip $\delta_b \approx 13^\circ$ and the ramp dip $\theta = 35^\circ$, we compute the dip of the back synclinal axial surface in the weak décollement layer to be $\psi = 31^\circ$. This axial surface intersects the back syncline in the overlying lid near the base of the more highly reflective interval (see pure-shear model, Figure 20). This point of intersection, according to the pure-shear model, should mark the top of the décollement interval. If we trace the horizon of intersection updip to the fault ramp, we see that it terminates at the fault at the termination of the back anticlinal axial surface, in agreement with theory. Therefore the first-order shape of the structure agrees quantitatively with the pure-shear end-member model. The back-dip and ramp angles predict a pure shear dimensionless fault slip $d/h \approx 1.7$ ($\alpha = \tan^{-1} d/h \approx 59^\circ$). The location of the top of the décollement layer yields a décollement thickness h of about 230 m, which indicates an input fault slip d of about 390 m.

Core from ODP-808 shows an intensely deformed basal detachment zone that is 19 m thick (Moore et al., 1991; Tiara et al., 1991b). Overlying strata that would correspond to the stratigraphic interval of the décollement layer of the pure-shear interpretation is composed of Shikoku Basin muds, and the overlying, more highly reflective interval is composed of Shikoku Basin turbidites. Analysis of faulting in ODP-808 core and

FIGURE 20. (a) Depth-migrated seismic image of the toe of the Nankai Trough accretionary wedge at Ocean Drilling Project site ODP-808, offshore Japan (Moore et al., 1991) (line NT62-8; h = v). (b) The small left-hand figure shows constraints on fault-ramp location based on reflector geometry at locations shown by arrows and by the location in ODP-808 core. (c) Small right-hand figure shows best-fitting pure-shear model discussed in text. (d) Annotated seismic image showing key observations indicating a pure-shear fault-bend fold origin. Note that the shallow structure is somewhat detached. The deep backlimb geometry qualitatively agrees with pure-shear models (compare Figure 4) with a backlimb dip $\delta_b = 11–13^\circ$ that is less than the ramp dip $\theta = 35^\circ$ and a back syncline that is displaced substantially to the hinterland of the fault bend, which eliminates simple-shear models. Simple-shear models are less favored by the observation of a 19-m-thick detachment zone observed in the ODP-808 core. The backlimb dip $\delta_b \approx 13^\circ$ and ramp dip $\theta = 35^\circ$ yield a pure-shear fault-bend fold prediction of the back synclinal axial surface in the décollement layer of $\psi = 31^\circ$, which quantitatively agrees with the seismic image as shown in the right-hand figure. The location of the top of the décollement is indicated by the inflection in the back syncline, which agrees with the location indicated by the fault cutoff of the back anticline, as predicted by pure-shear models (compare Figure 4). The back-dip and ramp angles predict a pure shear dimensionless fault slip $d/h = 1.7$ ($\alpha = \tan^{-1} d/h \approx 59^\circ$). The location of the top of the décollement layer yields a décollement thickness h of about 230 m, which indicates an input fault slip d of about 390 m.

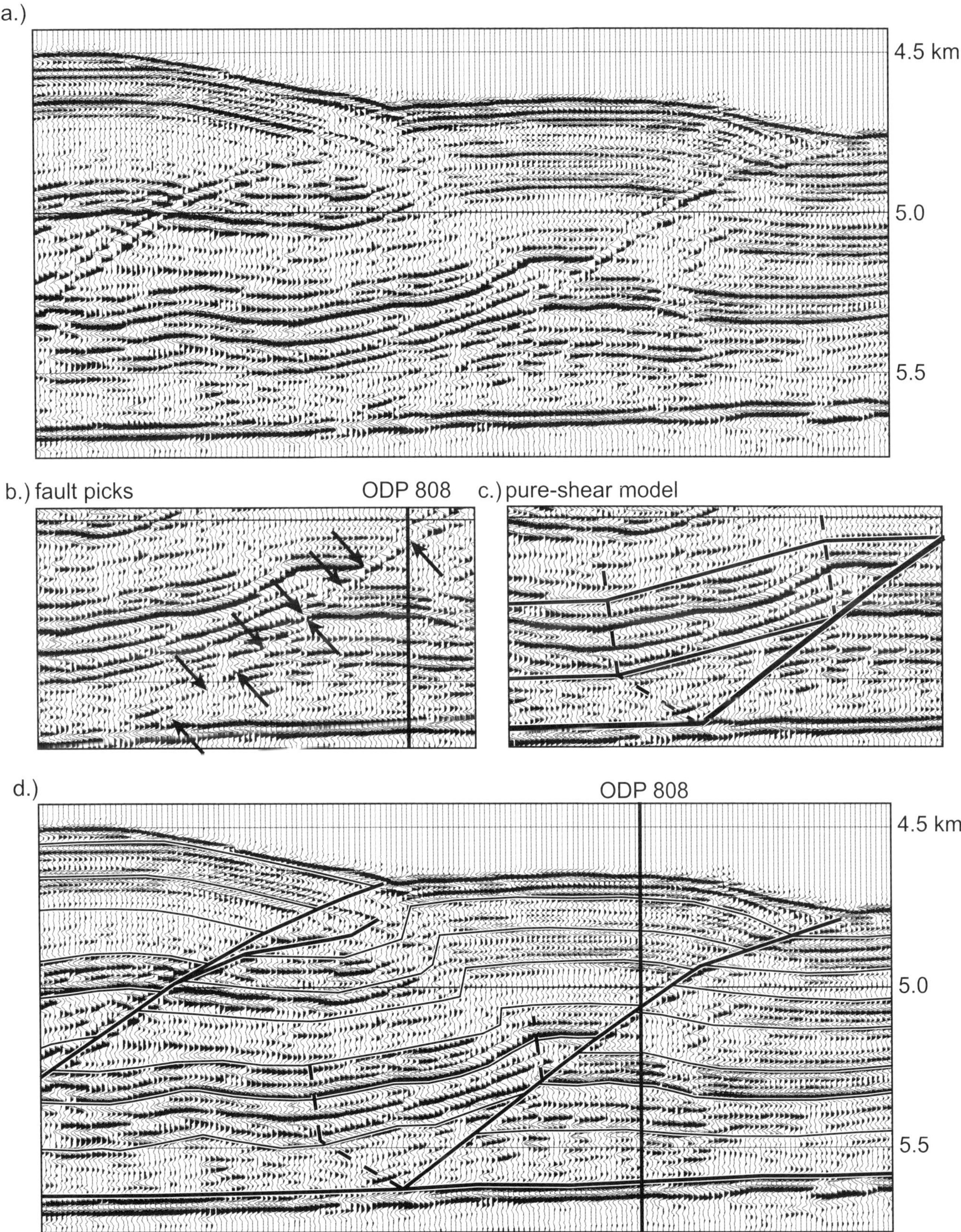
a.)
4.5 km
5.0
5.5
b.) fault picks
ODP 808
c.) pure-shear model
d.)
ODP 808
4.5 km
5.0
5.5

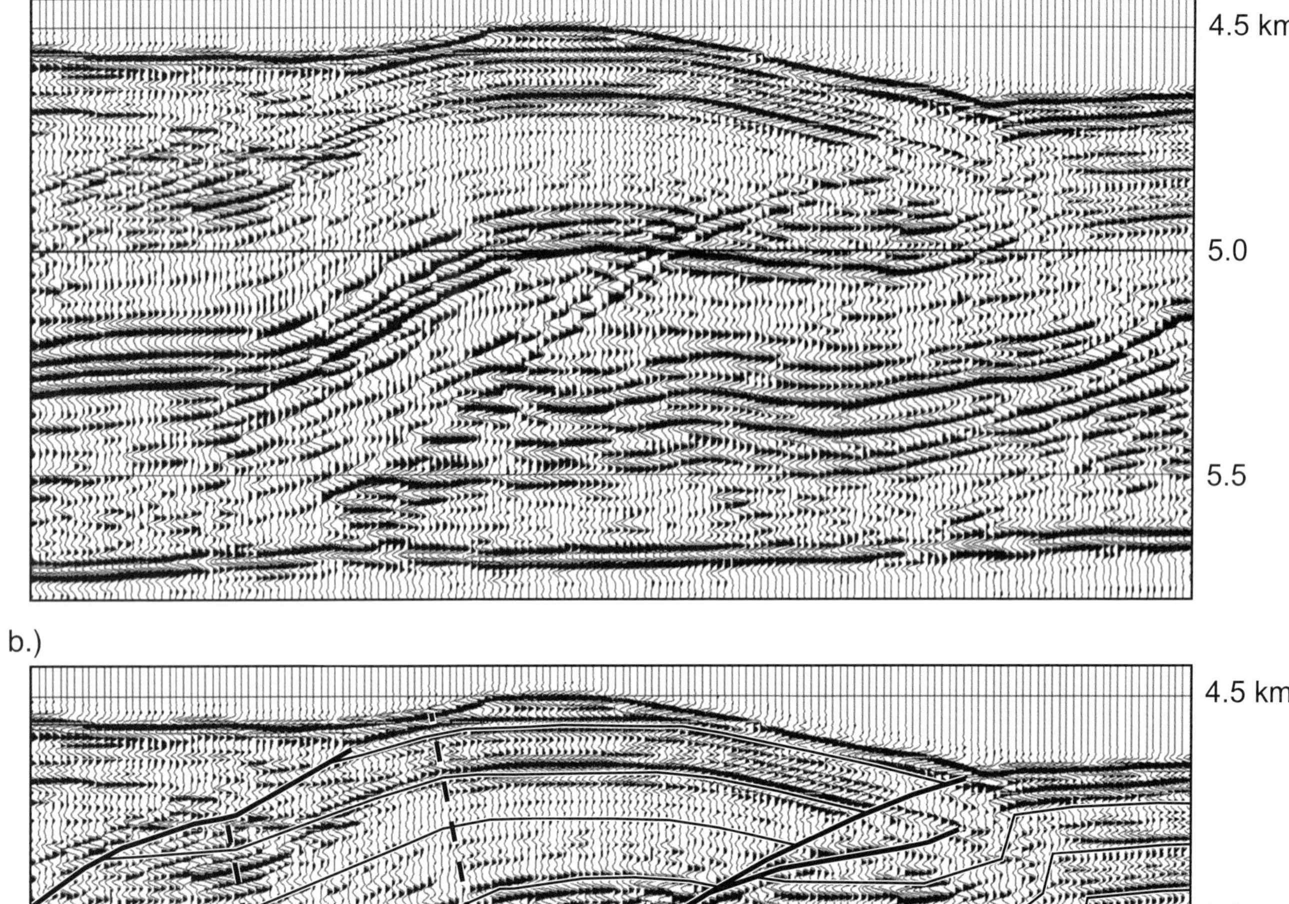

FIGURE 21. (a) Depth-migrated seismic image of structure just interior to that of Figure 20 (Moore et al. 1991) (line NT62-8; h = v). (b) Annotated seismic showing a backlimb geometry that qualitatively agrees with simple-shear models (compare Figure 4). The backlimb dip δ_b = 23–25° is less than the ramp dip θ = 39°, and the back syncline approximately bisects the syncline and emanates from the fault bend, which eliminates pure-shear models. The backlimb dip δ_b = 23–25° and ramp dip θ = 39° predict a simple shear of α_e = 60–65°. The stratigraphic level of the top of the shear zone is indicated by the fault cutoff of the back anticline, as shown by simple-shear models (compare Figure 4), which yields a décollement thickness h of about 190 m. The displacement of the top of the shear zone is therefore $d = h \tan \alpha_e$, or about 330–410 m.

image logging indicates a higher density of faulting below the turbidites (Tiara et al., 1991a, Figure 155; Shipboard Scientific Party 2001).

The next thrust ramp to the north of ODP-808 (Figure 21) shows a similar ramp dip but a substantially different hanging-wall geometry that qualitatively agrees with simple-shear models (compare Figure 4). The backlimb dip δ_b = 23–25° is less than the ramp dip θ = 39°, indicating a shear fault-bend fold, but the fact that the back syncline approximately bisects the syncline and emanates from the fault bend eliminates pure-shear models. Furthermore, the steepness of the backlimb qualitatively indicates very substantial shear, indeed, more than in the Cascadia example, which had a similar ramp dip of 35–40° but limb dips of only 5° and 12° at two stratigraphic levels, thereby indicating shears α_e of 8° and 31° (Figure 19).

Applying the simple-shear theory (Figure 10), we find that a backlimb dip δ_b of 23–25° together with a ramp dip θ of 40° predicts a simple shear α_e of about 63–67°. Furthermore, the stratigraphic level of the top of the shear zone is indicated by the fault cutoff of the back anticlinal axial surface (as shown by simple-shear models— compare Figure 4), which yields a décollement thickness h of about 190 m and is therefore similar to that of the previous structure. The displacement d of the top of the shear zone is $h \tan \alpha_e$, or about 330–410 m. This very substantial shear, of $\alpha_e = 63–67°$, of course requires substantial transfer of shear toward the hinterland. Unfortunately, the structures to the north are not well imaged and we cannot test this implication at present.

DISCUSSION

Shear fault-bend folding produces very distinctive geometries for ramp anticlines. They are characterized by long, gentle backlimbs that dip less than the fault ramp does, in contrast with classical fault-bend folding. Backlimb dips and limb lengths progressively increase with fault slip, by a combination of limb rotation and kink-band migration. Application of simple end-member shear fault-bend fold models to depth-seismic images from the Nankai Trough and Cascadia accretionary wedges shows agreement between theory and data, with substantial shear ($\alpha = 40–65°$) over stratigraphic intervals a few hundred meters thick. Furthermore, the structural styles of many ramp anticlines in the compressive toe of the deep-water Niger Delta are qualitatively very similar to shear fault-bend fold models (Connors et al., 1998). Therefore, we suggest that shear fault-bend fold concepts may have wide applicability, in spite of their limited application to date. Shear fault-bend folding may be particularly relevant at the toes of passive margins, which are important areas of present-day oil exploration.

ACKNOWLEDGMENTS

We are especially grateful to Roy D. Hyndman of the Pacific Geoscience Centre for providing us with copies of the Cascadia seismic lines, and to Greg F. Moore of the University of Hawaii for providing the Nankai Trough line. We are also extremely grateful for the very careful and thoughtful reviews of the manuscript by Josep Poblet of the University of Oviedo and Dave Waltham of Royal Holloway. The final manuscript was substantially improved because of their stimulating reviews. Finally, we take pleasure in thanking Ken R. McClay for having the vision and energy to host three stimulating Thrust Tectonics conferences and to see the volumes through to publication.

APPENDIX A

We present here a brief derivation of the simple-shear and pure-shear fault-bend folding theories, as a single combined theory following J. Suppe (2004, unpublished works), using the geometric elements shown in Figures 22 and 23. The theory is conceptually similar to that of Jordan and Noack (1992) but differs substantially in detail. A thrust steps up from a bedding-parallel detachment at an angle θ. There is a basal deformable layer of original thickness h that is allowed either to (1) undergo homogeneous simple shear parallel to bedding or (2) to shorten and thicken to h' within the backlimb by pure shear parallel to bedding. In contrast, the overlying strata undergo no externally applied simple shear and instead conserve bed length and layer thickness.

We now apply the constraints of conservation of bed length and area, and of continuity, to obtain the basic equations of simple-shear and pure-shear fault-bend folding. The two theories are end members differing only by the numerical value of a parameter C that describes the area of shortening of the basal deformable layer after a displacement d of its top from q to q'

$$A_o = Cdh \tag{4}$$

where $C = 1/2$ for homogeneous simple shear and $C = 1$ for pure shear. Other values of C are possible because of heterogeneity or because of mixtures of end members, but here we only consider the two special cases.

For purposes of the derivation, we partition the deformation into an imaginary horizontal displacement followed by folding. The geometry after a horizontal simple shear $\alpha_e = \tan^{-1} d/h$ is shown in the bottom of Figure 22. In the pure-shear case, the geometry after the same horizontal displacement d is shown in the bottom of Figure 23, for which the same function $\alpha = \tan^{-1} d/h$ can be defined. The hanging-wall cutoff of the top of the basal deformable layer is displaced a distance d, from r to r, producing an area of shortening of $A_o = dh/2$ in the homogeneous simple-shear case and $A_o = dh$ in the pure-shear case, which is the area of overlap between the hanging wall and footwall in our imaginary state before folding.

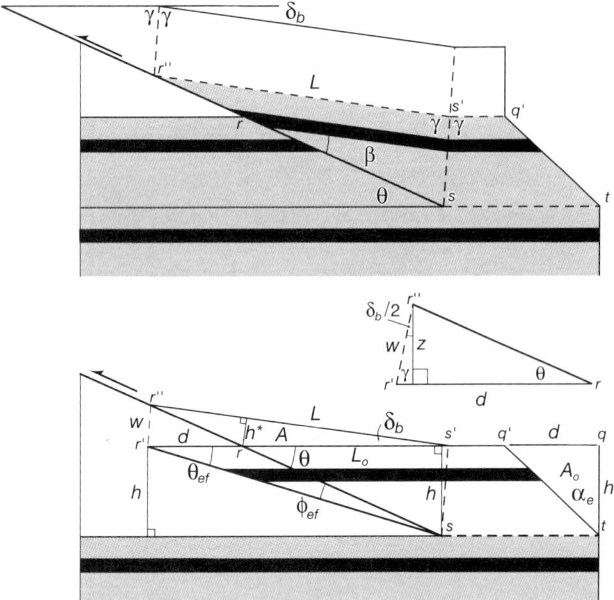

FIGURE 22. Geometric elements used in the derivation of the simple-shear fault-bend folding theory (after J. Suppe, 2004, unpublished works). The present deformed shape is shown in the top drawing, with a ramp dip θ, backlimb dip δ_b, and limb length L at the top of a basal layer of thickness h. The derivation factors the deformation into two nonphysical steps shown in the bottom drawing. (1) An external shear, α_e, is applied before folding of the hanging-wall block, to modify the initial cutoff angle θ to an effective cutoff angle θ_{ef} and to create an effective fault bend ϕ_{ef}. (2) The hanging wall is then folded by flexural slip; in particular, the bed segment L_o at the top of the basal layer is rotated to become the backlimb of length L. The folded hanging wall conforms to the fault, with no voids or overlap, subject to several balancing constraints. The area of shortening, A_o, is constrained to be equal to the area of structural relief A and to the area of overlap before folding (area of triangle $\Delta rr's$). There is conservation of layer thickness across axial surfaces ($\gamma = 90^\circ - \delta_b/2$), and bed length is conserved ($L_o = L$). The middle drawing is an enlargement of the details of triangle $\Delta rr'r''$ in the bottom drawing.

After folding, there is no overlap. The hanging-wall cutoff of the top of the deformable basal layer is displaced a distance w, from r' to r'', such that the bed segment $r's' = L_o$ is rotated to become $r''s' = L$ with a limb dip δ_b, as is shown in the bottom of Figures 22 and 23.

By applying conservation of bed length to triangle $\Delta r'r''s'$ we get

$$w = 2L\sin(\delta_b/2) \tag{5}$$

and by applying the law of sines to triangle $\Delta rr'r''$ we get

$$\frac{\cos((\delta_b/2) - \theta)}{d} = \frac{\sin\theta}{2L\sin(\delta_b/2)} = \frac{\sin\gamma}{rr''}. \tag{6}$$

Simplifying, we obtain the ratio of displacement d to deformed limb length L

$$d/L = \sin\delta_b \cot\theta + 1 - \cos\delta_b. \tag{7}$$

By constraining the area of structural relief A to be equal to the area of shortening, $A_o = Cdh$, we have

$$A = Cdh = L(L-d)\frac{\sin\delta_b}{2} \tag{8}$$

and

$$\cot\alpha = \frac{h}{d} = \frac{\sin\delta_b}{2C}\left[\left[\frac{L}{d}\right]^2 - \left[\frac{L}{d}\right]\right]. \tag{9}$$

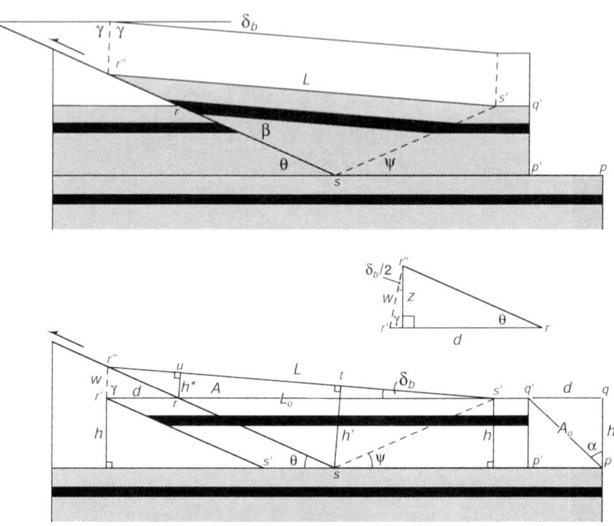

FIGURE 23. Geometric elements used in the derivation of pure-shear fault-bend folding theory (after J. Suppe, 2004, unpublished works). The present deformed shape is shown in the top drawing, with a ramp dip θ, backlimb dip δ_b, and limb length L at the top of a basal deformable layer of initial thickness h. The derivation factors the deformation into two nonphysical steps shown in the bottom drawing. (1) A dimensionless fault slip, $\alpha = \tan^{-1}d/h$, is applied before folding of the hanging-wall block, to produce an overlap of the footwall of area $A_o = dh$ (area of parallelogram $rr's's$). (2) The hanging wall is then folded by shortening and thickening of the basal layer in bedding-parallel pure shear above the ramp, reaching a thickness h'. In particular, the bed segment L_o at the top of the basal layer is rotated to become the back limb of length L. The folded hanging wall conforms to the fault, with no voids or overlap, subject to several balancing constraints. The area of shortening A_o is constrained to be equal to the area of structural relief A and to the area of overlap before folding. There is continuity of layers across the back syncline in the deformable basal layer, such that $h'/h = \sin(\psi + \delta_b)/\sin\psi$. Above this basal layer, there is conservation of layer thickness across axial surfaces ($\gamma = 90^\circ - \delta_b/2$), and bed length is conserved ($L_o = L$). The middle drawing is an enlargement of the details of triangle $\Delta rr'r''$ in the bottom drawing.

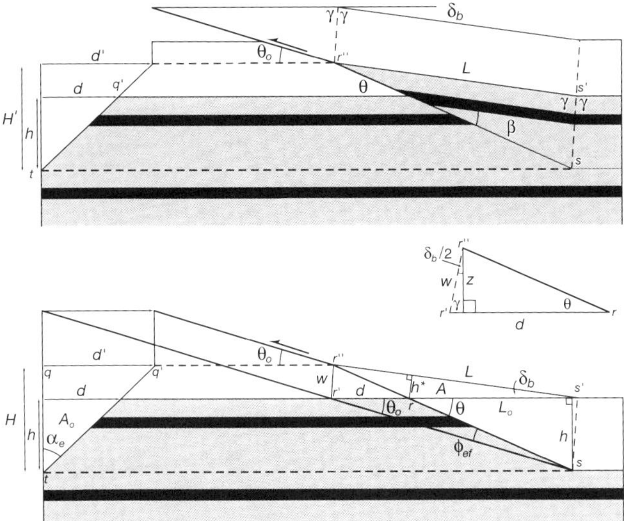

FIGURE 24. Geometric elements used in the derivation of the wedge simple-shear fault-bend fold theory (after J. Suppe, 2004, unpublished works). The present deformed shape is shown in the top drawing, with a deformed ramp dip θ, undeformed ramp dip θ_o, backlimb dip $\delta_{b'}$, and limb length L at the top of a basal layer of initial thickness h. The derivation factors the deformation into two nonphysical steps shown in the bottom drawing. (1) An external shear α_e, is applied to the footwall before folding of the hanging-wall block, to modify the initial cutoff angle θ_o to the deformed cutoff angle θ and create an effective fault bend ϕ_{ef}. After shear, the fault has a change in dip at point r''. The footwall shear zone has a thickness H' and a displacement d'. (2) The hanging wall is then folded by flexural slip. In particular, the bed segment L_o at the top of the basal layer is rotated to become the backlimb of length L. The geometry and kinematics of the fold limb in the hanging-wall are substantially different from the case of hanging-wall shear (Figure 22). This is because there is progressive deformation of the fault, which drives rapid migration of the fault bend (point r''), thereby producing distinctive growth stratal geometries (Figure 6). The middle drawing is an enlargement of the details of triangle $\Delta rr'r''$ in the bottom drawing.

Combining with (7)

$$\cot\alpha = \frac{\sin\delta_b}{2C}\left[\left[\frac{1}{\sin\delta_b\cot\theta + 1 - \cos\delta_b}\right]^2 - \left[\frac{1}{\sin\delta_b\cot\theta + 1 - \cos\delta_b}\right]\right] \qquad (10)$$

which is the same as (1), the basic equation for dimensionless displacement α as a function of ramp dip θ and limb dip δ_b. The parameter $C = A_o/dh$ is 1 for pure-shear and 1/2 for homogeneous simple-shear.

In the simple-shear case, the back syncline within the basal deformable layer bisects the fold hinge at an angle $\gamma = 90^\circ - \delta_b/2$, reflecting the fact that layer thickness is conserved. In contrast, in the pure-shear case, the basal deformable layer of original thickness h is allowed to shorten and thicken to h' within the backlimb, by pure shear parallel to bedding, whereas the overlying strata conserve bed length and layer thickness.

By continuity of bedding across the back synclinal axial surface of orientation ψ,

$$\frac{h'}{h} = \frac{\sin(\psi + \delta_b)}{\sin\psi}. \qquad (11)$$

By conservation of area in the deformed backlimb,

$$\frac{1}{2}h'L = \frac{1}{2}h(L - d) + Cdh \qquad (12)$$

and by combining with (7) and (11) we have

$$\cot\psi = 2C\left[\cot\theta + \frac{1}{\sin\delta_b} - \cot\delta_b\right] - \cot\theta \qquad (13)$$

which is the same as (3), the basic equation for the orientation of the back synclinal axial surface ψ as a function of ramp dip θ and limb dip δ_b. In the pure-shear case ($C = 1$), equation (13) reduces to

$$\cot\psi = \cot\theta + 2\tan\delta/2 = \cot\theta + 2\cot\gamma \qquad (14)$$

and in the homogeneous simple-shear case ($C = 1/2$), it reduces to

$$\cot\psi = \tan\delta/2 = \cot\gamma$$

and

$$\psi = 90^\circ - \delta/2 = \gamma. \qquad (15)$$

The equations outlined above also hold for footwall wedges, producing identical fold-limb geometry and, in the pure-shear case, identical kinematics (compare Figures 4 and 6). However, in the simple-shear wedge case, the fault is deformed and the shear zone is thicker, H, than in the forward simple-shear case, h, because of footwall shear (see Figures 6 and 24). The undeformed fault dip θ_o of the simple-shear wedge is

$$\cot\theta_o = \tan\alpha_e + \cot\theta. \qquad (16)$$

The fault slip rr'' is from equation (6)

$$rr'' = \frac{d \sin \gamma}{\cos((\delta_b/2) - \theta)} = \frac{h \tan \alpha_e \cos(\delta_b/2)}{\cos((\delta_b/2) - \theta)}$$

$$= \frac{h \tan \alpha_e}{\cos \theta + \tan(\delta_b/2) \sin \theta} \quad (17)$$

and the total thickness of the footwall shear zone is

$$H = z + h = rr'' \sin \theta + h = h \left[\frac{\tan \alpha_e}{\cot \theta + \tan(\delta_b/2)} + 1 \right]. \quad (18)$$

REFERENCES CITED

Bangs, N. L., A. Taira, S. Kuramoto, T. H. Shipley, G. F. Moore, K. Mochizuki, S. S. Gulick, Z. Zhao, Y. Nakamura, J. O. Park, et al., 1999, U.S.-Japan collaborative 3-D seismic investigation of the Nankai Trough plate-boundary interface and shallowmost seismogenic zone: EOS, Transactions of the American Geophysical Union, v. 80, p. F569.

Connors, C. D., D. B. Denson, G. Kristiansen, and D. M. Angstadt, 1998, Compressive anticlines of the mid-outer slope, central Niger Delta: AAPG Bulletin, v. 82, p. 1903.

Davis, D., J. Suppe, and F. A. Dahlen, 1983, Mechanics of fold-and-thrust belts and accretionary wedges: Journal of Geophysical Research, v. 88, p. 1153–1172.

Elliott, D., 1976, The motion of thrust sheets: Journal of Geophysical Research, v. 81, p. 949–963.

Gulick, S. P., N. L. Bangs, T. H. Shipley, Z. Zhao, Y. Nakamura, G. F Moore, S. Kuramoto, A. Taira, D. J. Hills, and J. Park, 2000, 3-D structural geometry and fluid flow indicators from the 1st out-of-sequence thrust to the deformation front of the Nankai accretionary prism: preliminary results from the Nankai Trough 3-D Seismic Experiment: EOS, Transactions of the American Geophysical Union, v. 81, p. F1067.

Hyndman, R. D., G. D. Spence, T. Yuan, and E. E. Davis, 1994, Regional geophysics and structural framework of the Vancouver Island margin accretionary prism: Proceedings of the Ocean Drilling Program, Part A: Initial Reports, v. 146, p. 399–419.

Jamison, W. R., 1987, Geometric analysis of fold development in overthrust terranes: Journal of Structural Geology, v. 9, p. 207–219.

Jordan, P., and T. Noack, 1992, Hangingwall geometry of overthrusts emanating from ductile décollements, *in* K. R. McClay, ed., Thrust tectonics: London, Chapman & Hall, p. 311–318.

Liu S., and J. Dixon, 1992, Centrifuge modelling of the propagation of thrust faults, *in* K. R McClay, ed., Thrust tectonics: London, Chapman & Hall, p. 53–69.

Liu H., K. R McClay, and D. Powell, 1992, Physical models of thrust wedges, *in* K. R McClay, ed., Thrust tectonics: London, Chapman & Hall, p. 71–81.

Malavieille, J., and J. F. Ritz, 1989, Mylonitic deformation of evaporites in décollements: examples from the Southern Alps, France: Journal of Structural Geology, v. 11, p. 583–590.

Medwedeff, D. A., 1988, Structural analysis and tectonic significance of late-Tertiary and Quaternary compressive growth folding, San Joaquin Valley, California: Ph.D. Dissertation, Princeton University, 184 p.

Mitra, S., 1988, Three-dimensional geometry and kinematic evolution of the Pine Mountain thrust system, Southern Appalachians: Geological Society of America Bulletin, v. 100, no. 1, p. 72–95.

Mitra, S., 1990, Fault-propagation folds: geometry, kinematic evolution, and hydrocarbon traps: AAPG Bulletin, v. 74, no. 6, p. 921–945.

Mitra, S., 1992, Balanced structural interpretations in fold and thrust belts, *in* S. Mitra and G. W. Fischer, eds., Structural geology of fold and thrust belts: Baltimore, The Johns Hopkins University Press, p. 53–77.

Mitra, S., and J. Namson, 1989, Equal-area balancing: American Journal of Science, v. 289, p. 563–599.

Moore, G. F., T. H. Shipley, P. L. Stoffa, D. E. Karig, A. Taira, S. Kuramoto, H. Tokuyama, and K. Suyehiro, 1990, Structure of the Nankai Trough accretionary zone from multichannel seismic reflection data: Journal of Geophysical Research, v. 95, p. 8753–8765.

Moore, G. F., D. E. Karig, T. H. Shipley, A. Taira, P. L. Stoffa, and W. T. Wood, 1991, Structural framework of the ODP Leg 131 area, Nankai Trough: Proceedings of the Ocean Drilling Program, Part A: Initial Reports, v. 131, p. 15–20.

Moore, G. F., A. Taira, S. Kuramoto, T. H. Shipley, and N. L. Bangs, 1999, Structural setting of the 1999 U.S.-Japan Nankai Trough 3-D seismic reflection survey: EOS, Transactions of the American Geophysical Union, v. 80, p. F569.

Moore, G. F., A. Taira, N. L. Bangs, S. Kuramoto, T. H. Shipley, C. M. Alex, S. S. Gulick, D. J. Hills, T. Ike, S. Ito, et al., 2001, Data report: structural setting of the Leg 190 Muroto transect, *in* Moore, G. F., Taira, A., Klaus, A., et al., Proceedings of the Ocean Drilling Program, Initial Reports v. 190. ⟨http://www-odp.tamu.edu/publications/190_IR/chap_02/chap_02.htm⟩, (accessed February 2003).

Morgan, J. K., D. E. Karig, and A. Maniatty, 1995, The estimation of diffuse strains in the toe of the western Nankai accretionary prism; a kinematic solution: Journal of Geophysical Research, v. B99, p. 7019–7032.

Morgan, J. K., and D. E. Karig, 1995, Kinematics and a balanced and restored cross-section across the toe of the eastern Nankai accretionary prism: Journal of Structural Geology, v. 17, p. 31–45.

Mosar, J., and J. Suppe, 1991, Role of shear in fault-propagation folding, *in* K. R. McClay, ed., Thrust tectonics: London, Chapman & Hall, p. 123–132.

Poblet, J., and K. McClay, 1996, Geometry and kinematics of single-layer detachment folds: AAPG Bulletin, v. 80, p. 1085–1109.

Serra, S., 1977, Styles of deformation in the ramp regions of overthrust faults: Wyoming Geological Association Guidebook, 29th Annual Field Conference, p. 487–498.

Shaw, J. H., S. C. Hook, and J. Suppe, 1994, Structural trend analysis by axial surface mapping: AAPG Bulletin, v. 78, p. 700–721.

Shipboard Scientific Party, 2001, Leg 196 Preliminary Report: Deformation and fluid flow processes in the Nankai Trough accretionary prism: logging while drilling and advanced CORKs. Ocean Drilling Project Preliminary Report, v. 96 [Online]. ⟨http://www-odp.tamu.edu/publications/prelim/196_rel/196PREL.PDF⟩.

Storti, F., and J. Poblet, 1997, Growth stratal architectures associated to décollement folds and fault-propagation folds; inferences on fold kinematics: Tectonophysics, v. 282, p. 353–373.

Strayer, L., and P. Hudleston, 1997, Numerical modeling of fold initiation at thrust ramps: Journal of Structural Geology, v. 19, p. 551–566.

Suppe, J.,1983, Geometry and kinematics of fault-bend folding: American Journal of Science, v. 283, p. 684–721.

Suppe, J., 1984, Seismic interpretation of compressively reactivated normal fault near Hsinchu, western Taiwan: Petroleum Geology of Taiwan, no. 20, p. 85–96.

Suppe, J., and D. A. Medwedeff, 1990, Geometry and kinematics of fault-propagation folding: Eclogae Geologicae Helvetiae, v. 83 (Laubscher vol.), p. 409–454.

Suter, M., 1981, Strukturelles Querprofil durch den nordwestlichen Faltenjura, Mt-Terri-Randüberschiebung-FreibergeL: Eclogae Geologicae Helvetiae, v. 74, p. 255–275.

Taboada, A., J. F. Ritz, and J. Malavieillie, 1990, Effect of ramp geometry on deformation in a ductile décollement level: Journal of Structural Geology, v. 12, p. 297–302.

Tamagawa, T., T. Matsuoka, and Y. Tamura, 1998, Geometrical shape of fault-bend folding with simple shear deformation in the thrust sheet (in Japanese): Geoinformatics (Joho Chishitsu, Osaka), v. 9, p. 3–11.

Tiara, A., I. Hill, J. V. Firth, et al., 1991a, Site 808: Proceedings of the Ocean Drilling Program, Part A: Initial Reports, v. 131, p. 71–269.

Tiara, A., I. Hill, J. V. Firth, et al., 1991b, Sediment deformation and hydrogeology at the Nankai accretionary prism: synthesis of ODP Leg 131 shipboard results: Proceedings of the Ocean Drilling Program, Part A: Initial Reports, v. 131, p. 273–285.

Yuan, T., G. D. Spence, and R. D. Hyndman, 1994, Seismic velocities and inferred porosities in the accretionary wedge sediments at the Cascadia margin: Journal of Geophysical Research, v. 99, p. 4413–4427.

Zapata, T., and R. W. Allmendinger, 1996, Growth stratal records of instantaneous and progressive limb rotation in the Precordillera thrust belt and Bermejo basin, Argentina: Tectonics, v. 15, p. 1065–1083.

17

Wallace, W. K., and T. X. Homza, 2004, Detachment folds versus fault-propagation folds, and their truncation by thrust faults, *in* K. R. McClay, ed., Thrust tectonics and hydrocarbon systems: AAPG Memoir 82, p. 324–355.

Detachment Folds versus Fault-propagation Folds, and Their Truncation by Thrust Faults

Wesley K. Wallace

Geophysical Institute and Department of Geology and Geophysics, University of Alaska, Fairbanks, Alaska, U.S.A.

Thomas X. Homza

EnCana Oil & Gas (U.S.A.) Inc., Anchorage, Alaska, U.S.A.

ABSTRACT

Asymmetric anticlines with steep to overturned forelimbs are a common element of fold-and-thrust belts, but typically their origin must be interpreted on the basis of incomplete knowledge of their geometry and kinematics. Many such folds are interpreted to be fault-propagation folds, but their known characteristics fit as well or better with interpretation as a detachment fold or a thrust-truncated example of either fold type. A fault-propagation fold forms by propagation of a ramp tip, so a ramp on which displacement decreases upward to a tip is consistent with this fold type. Fault-propagation fold models generally assume that hinges migrate with respect to the rock, especially in synclines, and that limbs, especially backlimbs, do not rotate with fold growth. A detachment fold forms above a décollement that may have a fixed tip or a propagating tip or that may extend beyond the limits of the fold. Nonparallel thickening in the anticlinal core and lack of a ramp are characteristic of a detachment fold. Detachment-fold models assume either fixed or migrating hinges and either fixed or rotating limbs, although rotating limbs and at least a fixed anticlinal hinge seem best supported by natural examples.

Truncation and displacement of a preexisting fold by a thrust fault modifies fold geometry and makes it more difficult to determine a fold's origin. A ramp results from truncation of the forelimb of an existing anticline, so a ramp does not, in itself, rule out a detachment-fold origin. An anticlinal forelimb in the hanging wall may be steepened either by displacement over a convex-upward bend in the underlying thrust or by the thrust being folded into an antiform. Thrust truncation of an existing anticlinal forelimb may result in a footwall syncline, but most fault-propagation fold models require either an abandoned ramp tip or significant strain within the forelimb to account for a footwall syncline. Origin as a detachment fold is possible if a footwall syncline is present, especially if an abandoned ramp tip is absent, or if the anticlinal core is internally thickened.

The structural style and mechanical stratigraphy of a region may provide additional useful information for determining fold origin. Knowledge about a specific fold may be insufficient to determine its origin, but other, less ambiguous examples in an area may indicate the most likely possibilities to consider. Detachment folds are likely where a competent unit overlies a much less competent unit, and fault-propagation folds may be more likely in evenly layered rocks that have relatively high competency but weak layer interfaces.

The origin of asymmetric map-scale folds in the northeastern Brooks Range of Alaska is difficult to determine on the basis of their geometry alone, especially because most of them have been truncated by thrust faults. The presence of thickened anticlinal cores, the absence of ramp tips, the presence of remnant uncut detachment folds, a mechanical stratigraphy characterized by a competent unit over an incompetent unit, and a transition from unbroken detachment folds to thrust-truncated asymmetric folds, together suggest that the thrust-truncated folds originated mainly as detachment folds rather than as fault-propagation folds.

INTRODUCTION

The major types of thrust-related folds, including fault-propagation and detachment folds, differ fundamentally in how they form. Interpretation of a particular fold by type has real implications both for its final geometry and for the progressive evolution of that geometry, or its kinematics (Figure 1). Correct identification of fold type is important for reconstructing the geometry of unknown parts of folds, either at the surface or in the subsurface, and for interpreting their kinematics. Distinguishing among fold types is essential for practical applications, including petroleum exploration and production and the assessment of seismic hazards.

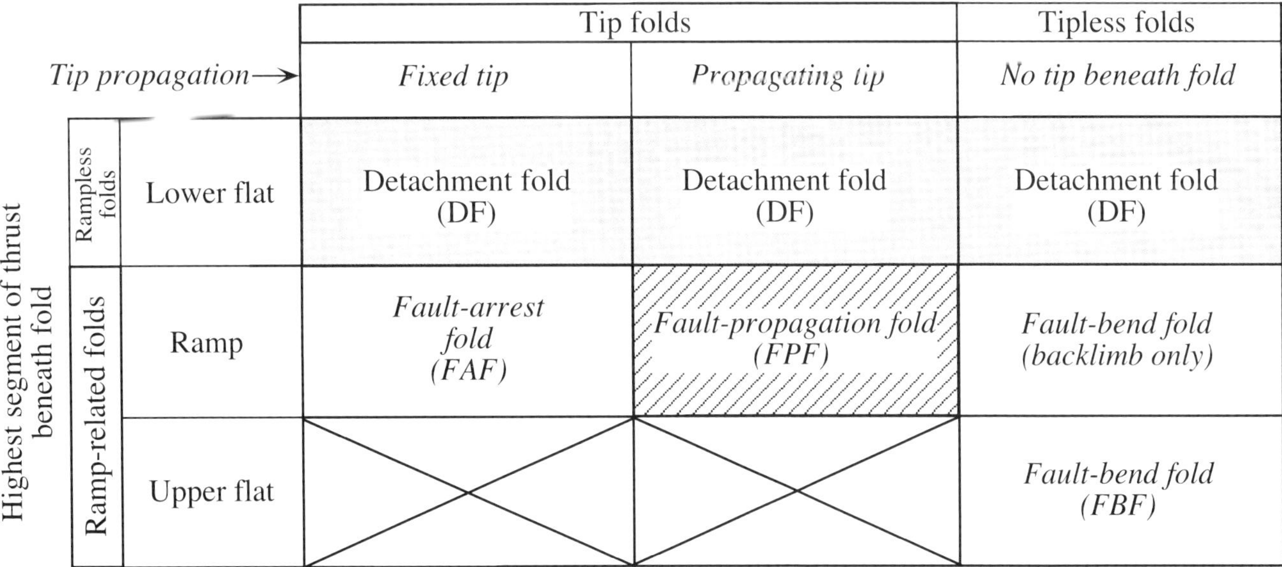

FIGURE 1. Thrust-related folds can be distinguished on the basis of two essential characteristics: The geometry of the thrust that underlies the fold and the relation between folding and the kinematics of the thrust tip. The folds are distinguished geometrically depending on whether they form above a thrust with (1) no ramp (décollement), (2) a lower flat and ramp, or (3) a lower flat, ramp, and upper flat. The folds are distinguished kinematically depending on whether they form above (1) a fixed tip, (2) a propagating tip, or (3) a thrust that continues beyond the fold. The term "fault-arrest fold" is after Thorbjornsen and Dunne (1997). The other fold types are discussed in the text. Terms that include a significant kinematic component in their definition are shown in italics.

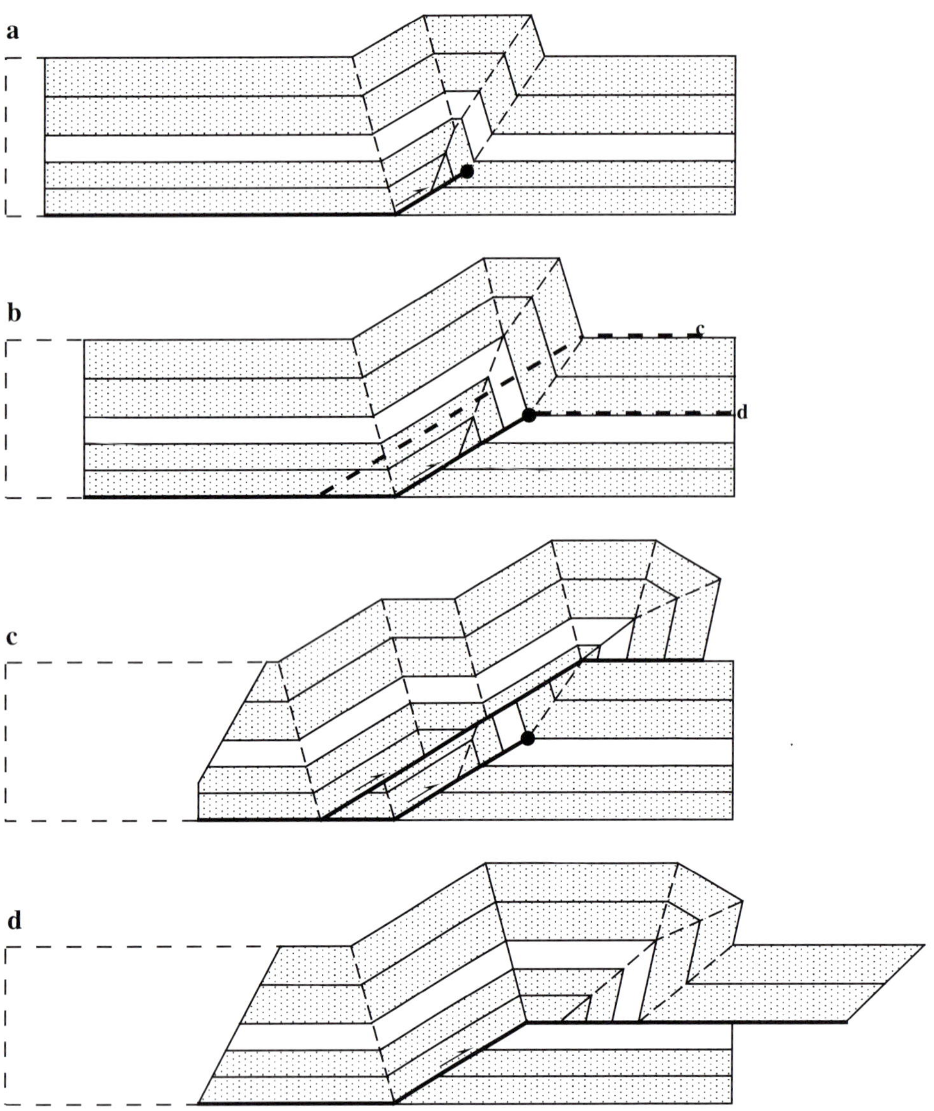

FIGURE 2. Kinematic evolution of a thrust-truncated fault-propagation fold. White layer is a marker that has no mechanical significance. (a) and (b) Evolution of a fold formed by propagation of a ramp tip according to the model of Suppe and Medwedeff (1990). Note that a unique depth to the lower flat is determined by the fold geometry. (c) and (d) Two different modifications of the fold in (b) by truncation and displacement along the trajectories shown by dashed lines (identified by letter labels). (c) Breakthrough above and hindward of the ramp tip in (b) is shown to illustrate formation of a footwall syncline from a fault-propagation fold and to show that identical geometries in the upper part of the fold can be attained by breakthrough of either fault-propagation or detachment folds (Figure 3c). An extra flat panel forms in the backlimb of (c) because the breakthrough cuts to a higher stratigraphic level in the backlimb of (b). Simple rotation of the hanging wall ramp over the footwall ramp-to-upper-flat transition results in the layer-parallel shear gradient shown. (d) Breakthrough is from the ramp tip to an upper flat, thereby destroying the ramp tip. Simple rotation of the hanging-wall ramp is accommodated by a gradient in layer-parallel shear as in (c).

Similarly, fold type must be correctly identified in order to understand better the deformational processes in fold-and-thrust belts by assessing the origin, characteristics, and evolution of thrust-related folds. Interpretation of fold type typically must be based on incomplete knowledge of the geometry and kinematics of a natural fold, and commonly it is guided by comparison with geometric-kinematic models.

Numerous geometric-kinematic models have been published for fault-propagation folds (Figure 2, Table 1) (Jamison, 1987; Chester and Chester, 1990; Mitra, 1990; Suppe and Medwedeff, 1990; Erslev, 1991), and the concept has become very widely recognized and used. Many examples of asymmetric thrust-related anticlines with steep forelimbs have been interpreted to have originated as fault-propagation folds in these and many other papers (Huftile, 1991; Alonso and Teixell, 1992; Baby et al., 1992; McConnell, 1994; Zapata and Allmendinger, 1996; Erslev and Mayborn, 1997; Spang and McConnell, 1997). However, these interpretations typically are supported primarily by consistency between a model and an incomplete knowledge of a natural fold. Two other important possibilities that may be consistent with asymmetric thrust-related anticlines with steep forelimbs are (1) detachment folding (Figure 3 and Table 1) (Jamison, 1987; Mitra and Namson, 1989; Dahlstrom, 1990; Epard

Table 1. Geometric and kinematic characteristics of some representative fault-propagation and detachment-fold models. Comparison of these characteristics against those of natural folds can serve as a basis for discriminating between fault-propagation and detachment folds, and for assessing the applicability of specific models. To facilitate comparison, the same format is used for both types. (A) Fault-propagation folds.

A. Fault-propagation folds

	Constant thickness (Suppe and Medwedeff, 1990; Jamison, 1987)	*Variable thickness (Mitra, 1990)*	*Variable forelimb thickness (Jamison, 1987; Mitra, 1990; Chester and Chester, 1990)*	*Trishear (Erslev, 1991; McConnell, 1994)*
Mechanical stratigraphy	All effectively competent	Uniform competency	Uniform competency	Uniform competency
Bed thickness	Constant	Uniform thickening	Forelimb thinning	Thinned HW, thickened FW in forelimb, forelimb dip decreases upward from ramp tip
Footwall ramp	Yes	Yes	Yes	Yes
Footwall syncline	No	No	No	Yes
Ramp tip	Yes	Yes	Yes	Yes
Ramp cutoff angles	HW high, FW low	HW high, FW low	HW high, FW low	HW & FW same
Detachment depth	Constant	Variable	Constant	Fixed*
Fixed vs. migrating hinges	Synclines migrate, anticline fixed below upward migrating branch	Synclines migrate, anticline fixed below upward migrating branch	Synclines migrate, anticline fixed below upward migrating branch	Fixed forelimb hinges
Crest evolution	Narrows	Narrows	Narrows	Constant*
Forelimb rotation	No	Yes	Yes	Yes
Backlimb rotation	No	No	No	No*
Arc length	Increases	Increases	Increases	Increases*
Wavelength	Increases	Increases	Increases	Increases*
Height	Increases	Increases	Increases	Increases

**Characteristics not addressed by these models, but inferred assuming backlimb forms over ramp as in other models*

Table 1. (cont.) (B) Detachment folds.

	B. Detachment folds			
	Constant dip, variable length (model 1 of Poblet and McClay, 1996; Mitra and Namson, 1989)	***Variable dip, constant length (Model 2 of Poblet and McClay, 1996; De Sitter, 1956; Homza and Wallace, 1995, 1997)***	***Variable dip, variable length (Model 3 of Poblet and McClay, 1996; Dahlstrom, 1990)***	***Variable thickness (Epard and Groshong, 1995)***
Mechanical stratigraphy	Competent over incompetent	Competent over incompetent	Competent over incompetent	All effectively incompetent
Bed thickness	Constant	Constant	Constant	Variable
Footwall ramp	No	No	No	No
Footwall syncline	No	No	No	No
Ramp tip	No	No	No	No
Ramp cutoff angles	None	None	None	None
Detachment depth	Constant*	Variable	Constant*	Constant*
Fixed vs. migrating hinges	At least 2 migrate	All fixed	At least 2 migrate	Either
Crest evolution	Widens	Widens	Widens	No crest
Forelimb rotation	No	Yes	Yes	Yes
Backlimb rotation	No	Yes	Yes	Yes
Arc length	Increases	Constant	Increases	Decreases or increases#
Wavelength	Increases	Decreases	Decreases	Decreases
Height	Increases	Increases then decreases†	Increases then decreases†	Increases

**Assumed but not required*
†Decreases when forelimb overturns
#Depends on shortening and distance above detachment

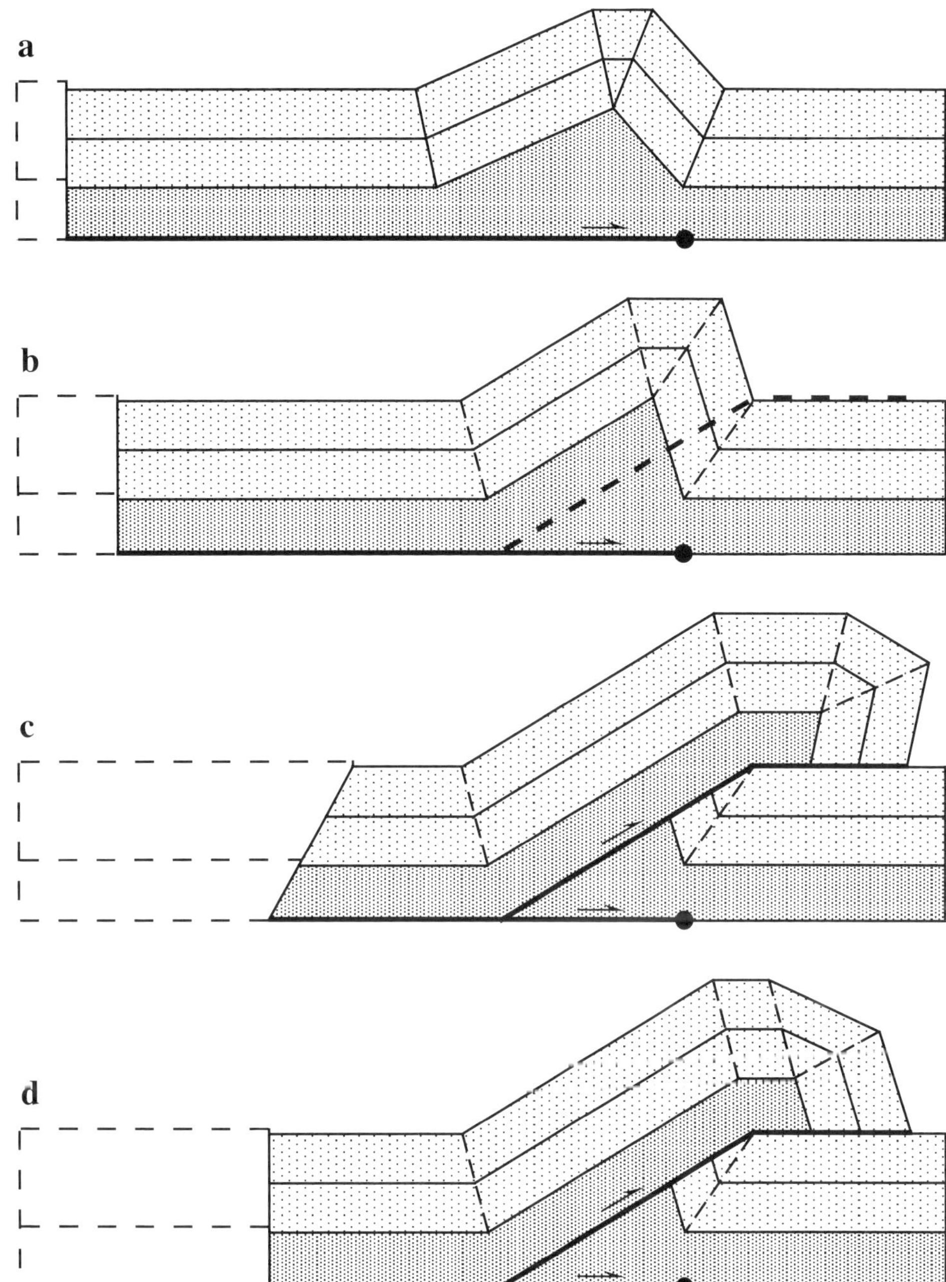

FIGURE 3. Kinematic evolution of a thrust-truncated detachment fold. Lower dark gray layer is incompetent, upper light gray layer is competent. (a) and (b) Evolution of a detachment fold with fixed hinges and rotating limbs (Homza and Wallace, 1997 and model 2 of Poblet et al., 1997). The level of the décollement was arbitrarily chosen to differ from the fixed detachment depth of the fault-propagation fold in Figure 2. This is to emphasize that, although the geometry of the upper parts of the folds in Figures 2b and 3b are identical, the depth to the décollement of a detachment fold is not uniquely determined by fold geometry, if detachment depth varies during fold growth by structural thinning or thickening of the incompetent layer. (c) and (d) Two different modifications of the fold in (b) by truncation and displacement along the trajectory shown by the dashed line. In (c), simple rotation of the hanging-wall ramp over the footwall ramp-to-upper-flat transition results in the layer-parallel shear gradient shown. In (d), collapse and extension of part of the forelimb and flat crest of the fold allow rotation of the hanging-wall ramp without a layer-parallel shear gradient. The trajectory of the breakthrough in the backlimb of (b) is at the same stratigraphic level as the lower flat, so that a planar backlimb forms in (c) and (d).

and Groshong, 1995; Homza and Wallace, 1995, 1997; Poblet and McClay, 1996) and (2) the breakthrough and displacement of an existing fold by a thrust fault (Figures 2c and d and 3c and d) (Willis, 1893; Jamison, 1987; Suppe and Medwedeff, 1990; Mitra, 1990; Fischer et al., 1992; Morley, 1994). Interpretation of a fold as a simple fault-propagation fold without explicitly considering these other possibilities has become common, despite the fact that detachment folds and thrust truncation have both been long recognized and have been addressed in numerous recent papers on thrust-related folds. More effort is needed in considering the full range of possibilities that could reasonably apply, given what is known about a particular fold. Correct interpretations are more likely with a clearer and wider recognition of the fundamental distinctions between fault-propagation folds and detachment folds, and of the fact that breakthrough of these fold types can represent a significant change in the process that can substantially modify fold geometry and kinematics.

The main purpose of this chapter is to review the essential geometric and kinematic characteristics that are inherent in the concepts of fault-propagation folds, detachment folds, and their truncation by thrust faults.

These characteristics serve as a basis for comparing and contrasting the fold types and for making distinctions among them. The focus of this chapter is not on presenting either extensive new data or elaborate quantitative models, but rather on illustrating some of the uncertainties of interpretation with examples of natural folds. Our objectives here are to (1) draw attention to a broader range of possibilities that should be considered when one is interpreting the origin of natural thrust-related folds, (2) assess the basic distinctions among the models, (3) review the criteria that could be used for discriminating among the possibilities, and (4) provide a more systematic framework for developing and testing models for the origin of natural thrust-related folds.

The chapter includes four main parts. First, it reviews the geometric and kinematic characteristics of fault-propagation and detachment folds and how these characteristics can be used to distinguish between the two fold types. Second, it discusses modification of fold geometry by thrust faults that break through the folds, with emphasis on breakthrough of fault-propagation and detachment folds. Third, it presents natural examples, from the northeastern Brooks Range of Alaska, of asymmetric map-scale folds, many of which are truncated by thrust faults, to illustrate some of the uncertainties inherent in interpreting fold type. Finally, with an emphasis on petroleum exploration, it briefly discusses the practical significance of distinguishing between fault-propagation and detachment folds and of recognizing thrust truncation of folds.

COMPARISON OF FAULT-PROPAGATION AND DETACHMENT FOLDS

Fault-propagation Folds

As its very name indicates, a fault-propagation fold by definition must originate as a consequence of propagation of a fault tip (Figures 1, 2; Table 1) (Jamison, 1987; Chester and Chester, 1990; Mitra, 1990; Suppe and Medwedeff, 1990; Erslev, 1991). Furthermore, the definition is generally taken to apply more narrowly to folds formed by propagation of the tip of a ramp. Fold and fault geometry alone is not sufficient to demonstrate that these essential kinematic conditions have been met. However, it is very difficult in practice to demonstrate that a fold has grown synchronously with and as a consequence of propagation of a ramp tip, and it is impossible in many natural folds even to determine whether a ramp tip is or was present. Consequently, many of the natural folds that have been identified as fault-propagation folds have not even been shown to possess ramp tips. Indeed, very few have been demonstrated to have grown as a consequence of propagation of a ramp tip.

The requirement that fold growth be a consequence of propagation of a ramp tip places some important conditions on the geometry and kinematics of fault-propagation folds (Table 1). Most important, the fold must overlie a ramp that terminates upward at a tip (Figures 1 and 2). The cutoff lengths of the layers cut by the fault must differ across the fault in order to accommodate the decrease in displacement up the ramp (Williams and Chapman, 1983; Chapman and Williams, 1984). Suppe and Medwedeff's (1990) constant-thickness model accomplishes this, while maintaining parallel fold geometry, by having the fault cut layers in the hanging wall at a higher angle than in the footwall, a geometry that is attained by migration of the forward synclinal hinge with the ramp tip (Figure 2). Numerous other models (Jamison, 1987; fixed-axis model of Suppe and Medwedeff, 1990; Mitra, 1990; Chester and Chester, 1990; Erslev, 1991) incorporate this basic geometry plus strain that results in changes in bed thickness distributed in a variety of ways, but most commonly in the fold forelimb. In all of these models, the dip of the backlimb is fixed with respect to the footwall ramp during fold growth, although it is not necessarily parallel to the ramp. Forelimb rotation can be accommodated in some of the models by layer-thickness changes that result from strain (Mitra, 1990; Erslev, 1991).

The origin of fault-propagation folds also has important implications for the fold hinges. First, transport of the hanging wall from the lower flat up the ramp will result in the rocks of the hanging wall rolling through a synclinal hinge that is fixed with respect to the footwall, as in the hindward part of a fault-bend fold (Figure 2). The growth of the fold in response to upward propagation of the ramp tip is geometrically most easily accommodated by forward migration of the forward synclinal hinge with the ramp tip, and this is a feature of most models for fault-propagation folds (Suppe and Medwedeff, 1990; Mitra, 1990; Chester and Chester, 1990). Propagation of the ramp tip by parallel folding is possible by a combination of different cutoff angles in the hanging wall and footwall and a forward synclinal hinge that migrates with the ramp tip (Suppe and Medwedeff, 1990). Other solutions require strain to shorten the hanging wall and/or extend the footwall relative to the fault to attain the necessary upward decrease in displacement across the ramp. The anticlinal hinge may be fixed above the point at which the thrust in the hanging wall changes from a lower flat to a ramp. However, most models assume that this hinge branches upward to form a flat crest and that this branch point resides at the stratigraphic level of the ramp tip and thus migrates up section as the ramp propagates (Figure 2) (Suppe and Medwedeff, 1990; Mitra, 1990; Chester and Chester, 1990).

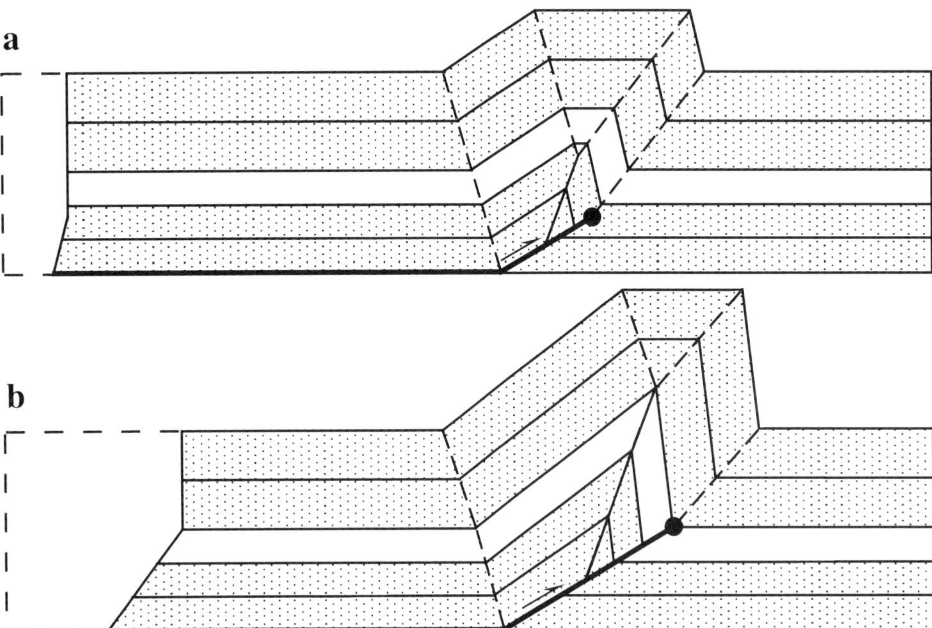

Figure 4. Conceptual model of a rotating-limb fault-propagation fold. White layer is a marker that has no mechanical significance. The Suppe and Medwedeff model (1990) of constant-thickness fault-propagation folding has been modified by allowing thickening of limbs toward the anticlinal hinge in the part of the section that is cut by the ramp. This thickening is a response to a layer-parallel shear gradient. A conceptual model for a fixed-hinge, rotating-limb fault-propagation fold is also possible, but would be more complex geometrically.

Although origin of a fold by ramp propagation is perhaps most easily explained by models with migrating synclinal hinges and a nonrotating backlimb, an important question that remains is whether it is possible for a fault-propagation fold to form with multiple fixed hinges and rotation of both limbs. Erslev and Mayborn (1997) and Woodward (1997) have proposed the possibility that folds with these properties form during propagation of a ramp. Thickening of both limbs toward the anticlinal hinge and, in the Erslev and Mayborn (1997) model, layer-parallel shear, are the mechanisms that allow fixed hinges and limb rotation. Geometries compatible with these models have been demonstrated in natural folds that display ramp tips and have been interpreted to have formed by tip propagation (Williams and Chapman, 1983; McConnell et al., 1997; Tavarnelli, 1997). Although they formed above ramps, these folds display nonparallel thickening in the core of the fold. This is an important geometric characteristic of detachment folds and could represent a true hybrid between fault-propagation folds and detachment folds. However, the question remains: Do the thickness changes in the anticline accommodate ramp propagation, or do the thickness changes only represent an additional increment of fold shortening superimposed on a fault-propagation fold formed by one of the more "conventional" models (Figure 4).

Unfortunately, the geometric and kinematic characteristics predicted by the various models for fault-propagation folds remain to be thoroughly tested against natural examples. It is difficult to find examples of map-scale folds where the displacement gradient across the ramp can be determined or a ramp tip can be documented (Williams and Chapman, 1983; Chapman and Williams, 1984; Ellis and Dunlap, 1988; Tavarnelli, 1997). The ramp and ramp tip are not documented in many of the natural examples cited as fault-propagation folds, so it is not possible to demonstrate that displacement decreases upward to a ramp tip. Thickness changes and kinematics have been documented in relatively few natural examples that have been interpreted to be fault-propagation folds (Alonso and Teixell, 1992; Erslev and Mayborn, 1997). Even these examples do not provide evidence that the folds originated by ramp-tip propagation, and evidence for fixed-hinge fold growth (Erslev and Mayborn, 1997) is not consistent with most models for fault-propagation folds.

Detachment Folds

Detachment (or décollement) folds form above a décollement as a result of structural thickening by means other than parallel folding or displacement over a coherent ramp (Figures 1, 3; Table 1) (Epard and Groshong, 1995; Homza and Wallace, 1995; Poblet and McClay, 1996). This thickening may include penetrative strain or folding and faulting at a scale significantly smaller than that of the fold. Although they are not mechanical models, most geometric-kinematic models for detachment folds either imply or explicitly assume a mechanical stratigraphy with a strong competency contrast that is directly reflected by a fold geometry with a parallel-folded competent layer above a nonparallel-folded incompetent layer (Figure 3, Table 1) (Jamison, 1987; Mitra and Namson, 1989; Dahlstrom, 1990; Homza and Wallace, 1995, 1997; Poblet and McClay, 1996).

However, at least one model broadens the concept to include folds in which nonparallel fold geometry characterizes the full vertical extent of the fold and competency contrast plays no role at that scale (Epard and Groshong, 1995). These models define a range of geometries for the upper part of the folds that includes the range of geometries modeled for fault-propagation folds, although a much wider range of geometries is possible with detachment folds. It is important to note that, although kinematics plays an important role in the definition of detachment folds, the geometry alone is sufficient to document "incompetent" thickening and lack of a ramp, provided that the geometry of the detachment and the overlying incompetent unit are known.

The models and documented natural examples of detachment folds span a wide range of kinematic possibilities (Table 1). Two key aspects of kinematics are whether hinges remain fixed or migrate with respect to the rock and whether limbs rotate or remain constant in dip (Epard and Groshong, 1995; Homza and Wallace, 1995; Poblet and McClay, 1996). The interplay between fold geometry and cross-sectional area as shortening increases depends largely on these kinematic characteristics. Poblet and McClay (1996) and Poblet et al. (1997) summarize a range of possibilities using three models for folds with angular hinges and straight limbs (Table 1). These possibilities are (1) constant limb dip and at least two migrating hinges, a model in which detachment depth normally varies with increasing shortening (Homza and Wallace, 1995) but in which, under certain conditions, detachment depth can remain constant with increasing shortening (Figure 5a) (e.g., Mitra and Namson, 1989); (2) fixed hinges, which necessitate rotating limbs and variable detachment depth as fold cross-sectional area initially increases and then decreases (Figure 5b) (e.g., Homza and Wallace, 1995); or (3) an intermediate model with at least two migrating hinges and rotating limbs, and in which detachment depth can remain constant with increasing shortening (Dahlstrom, 1990). Interpretation of which of these kinematic models applies to examples of natural detachment folds has been based mainly on small-scale kinematic indicators or growth stratigraphy. Evidence for fixed hinges and rotating limbs has been shown for a number of folds, although the synclinal hinges in some of these may have migrated while the anticlinal hinge remained fixed (Rowan and Kligfield, 1992; Poblet and Hardy, 1995; Royse, 1996; Anastasio et al., 1997; Homza and Wallace, 1997; Poblet et al., 1997). Changes in depth to detachment as a result of thinning or thickening of the incompetent unit, a consequence expected with fixed-hinge fold growth, have also been demonstrated for some folds (Wiltschko and Chapple, 1977; Homza and Wallace, 1997). We are not aware of examples of detachment folds whose growth with constant limb dip by hinge migration has been well documented, although such folds likely exist. Each of the models may apply to some natural detachment folds, depending on controls that are yet to be determined.

Another important kinematic variable in the origin of natural detachment folds is the relationship between fault propagation and fold growth. Relatively little attention has been paid to this aspect of detachment fold growth, probably in large part because it is very difficult to assess the history of fault propagation along a décollement. Detachment folds are commonly assumed to form at a décollement tip and to grow as a consequence of tip propagation (Figure 5a). However, formation of a detachment fold is kinematically possible above a décollement tip that is either fixed or propagating, or even above a décollement that extends beyond the limits of the fold (Figures 1, 5). Growth of a detachment fold requires only that a gradient in displacement on the décollement reflect the conversion of fault slip into fold shortening. The leading edge of this gradient could be represented either by a fault tip or by a point on the décollement at which excess slip passes beyond the fold, to be accommodated by other structures. Furthermore, the leading edge of this gradient (tip or otherwise) need not propagate during fold growth, as would be the case with any fold in which the forward synclinal hinge is fixed with respect to the rock. While tip propagation could be expected with some migrating-hinge folds, it would not be expected with fixed-hinge folds.

Distinction between Fault-propagation and Detachment Folds

To demonstrate that a fold originated as a fault-propagation fold necessitates showing that the fold grew as a consequence of ramp propagation, which requires evidence about fold kinematics that is very difficult to obtain in practice (Figures 1, 2). On the other hand, a detachment fold may form whether or not a décollement tip propagates or even is present below the fold (Figures 1, 5). The presence of a ramp, with an upward decrease in displacement to a ramp tip, are geometric conditions that must be met by a fault-propagation fold (Figure 2). However, these conditions alone do not uniquely establish the origin of a fold as a fault-propagation fold, because they could also be met by a fold that grew above a fixed ramp tip (a "fault-arrest fold," Figure 1) (Wickham, 1995; Thorbjornsen and Dunne, 1997) or by a preexisting fold that has been modified by propagation of a ramp tip. In contrast, a fold can be shown to be a detachment fold on geometric grounds alone if a coherent ramp is absent and a nonparallel fold geometry above a décollement characterizes at least the lower part of the fold (Figure 3). However, the geometric conditions that are necessary

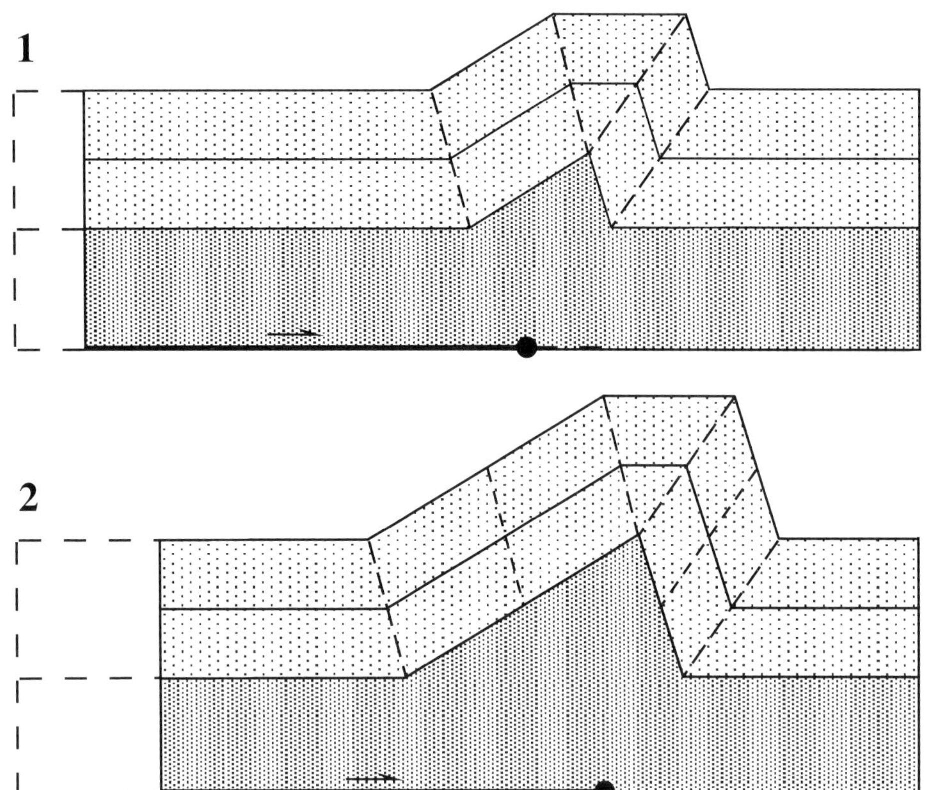

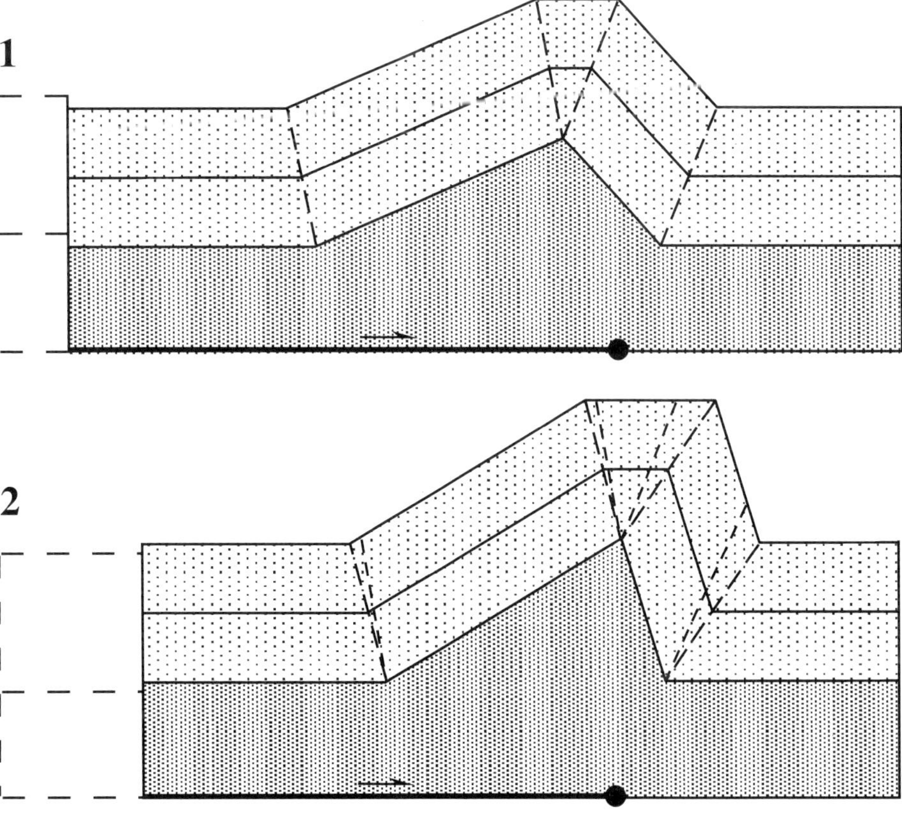

Figure 5. Different kinematic models for detachment folds. Lower, dark gray layer is incompetent, upper, light gray layer is competent. (a) Kinematics of a migrating hinge, nonrotating limb detachment fold (as in model 1 of Poblet et al., 1997). Constant limb dip and constant detachment depth are maintained by outward migration of the synclinal hinges. Short-dashed lines in (2) indicate the position of the synclinal hinges in (1). The forward synclinal hinge migrates with a propagating décollement tip. (b) Kinematics of a fixed-hinge, rotating-limb detachment fold (as in model 2 of Poblet et al., 1997). Fixed hinges require steepening of limbs and a change of structural thickness of the incompetent layer to accommodate change in the area of the fold core. Hinges are fixed to points at the incompetent-competent layer contact, but must rotate to allow parallel folding of the competent layer. Short-dashed lines in (2) indicate the position of the synclinal hinges in (1). The décollement tip does not propagate with fold growth.

for a fault-propagation fold and are diagnostic for a detachment fold can be documented only if the geometry of the core of the fold is well known. Unfortunately, this is the very part of a natural fold that most commonly is unknown and must be reconstructed. The incompetent unit in detachment folds generally is recessive, so that the décollement and fold core only rarely are well exposed. Ramps more commonly are exposed, but exposures sufficient to document both hanging-wall and footwall cutoffs are unusual, and exposures of definitive ramp tips are very rare. Steep forelimbs and complex deformation in the cores of both detachment and fault-propagation folds result in poorly defined reflectors that generally make interpretation of geometry in seismic reflection data ambiguous. Similarly, well data lack sufficient resolution, in most cases, to resolve geometry in the fold core uniquely.

What other criteria might be used to distinguish whether a fold originated as a fault-propagation or detachment fold, if the geometry of the fold core is unknown (Mitchell and Woodward, 1988; Storti and Poblet, 1997)? According to existing geometric-kinematic models, fault-propagation folds are likely to form with a nonrotating backlimb and at least a migrating hindward synclinal hinge (Table 1). The models also do not account for a rotating forelimb, a fixed forward synclinal hinge, and a fixed anticlinal hinge without a flat crest in fault-propagation folds, if the forelimb has maintained constant thickness. On the other hand, observational evidence indicates that at least some detachment folds form with rotating limbs, at least a fixed anticlinal hinge, and perhaps fixed synclinal hinges, characteristics that are consistent with a number of the existing geometric-kinematic models for detachment folds (Table 1). These characteristics may be useful guides toward a preferred intepretation. However, they certainly should not be regarded as diagnostic, because they are based largely on models and have only limited support from observations. Furthermore, folds that are not pure end-member fault-propagation or detachment folds likely exist, such as hybrids with incompetent thickening in the anticlinal core above a propagating ramp (e.g., Figure 4) or thickening of the anticlinal core above a décollement by multiple small-displacement propagating ramps. A wide range of overlapping kinematic possibilities can be envisioned for both fold types, and more observational evidence is required to ascertain which of these possibilities apply to natural examples of each fold type and under what conditions they apply.

The local structural style and the mechanical stratigraphy are perhaps the most readily applied criteria for assessing whether a fold is likely to have originated as a fault-propagation or detachment fold, yet these criteria are not necessarily routinely considered. A specific fold may not be known well enough to determine its origin, but other, better known and less ambiguous examples in an area may indicate the most likely alternatives to be considered. Detachment folds clearly should be considered likely where a competent unit overlies a much less competent unit, because abundant observational evidence as well as extensive research on the mechanics of buckle folds indicates that this mechanical stratigraphy favors formation of folds with the characteristics of detachment folds (Ramsay and Huber, 1987; Price and Cosgrove, 1990; Hudleston and Lan, 1993). Despite the fact that this mechanical stratigraphy applies to many folds interpreted as fault-propagation folds, the alternative possibility, that they are in fact detachment folds, often is not addressed. On the other hand, fault-propagation folding may be favored by a mechanical stratigraphy that is homogeneous but strongly anisotropic, such as one that consists of relatively thin, strong layers of even thickness and with weak interfaces (Chester et al., 1991). Fault-propagation folding may be further favored by the preexistence of a ramp beneath unbroken layers of this mechanical stratigraphy. These mechanical stratigraphic characteristics may be useful starting points, but more work is required to define better the relationship between mechanical stratigraphy and fault-propagation or detachment folding. In particular, to what extent does fault-propagation folding occur in a mechanical stratigraphy that is uniformly weak, and does the degree of anisotropy influence whether fault-propagation folding occurs in such rocks? Fault-propagation folding in weak rocks could occur by essentially the same processes as in detachment folds and could have results not addressed by existing models, such as backlimb rotation.

BREAKTHROUGH OF FAULT-PROPAGATION AND DETACHMENT FOLDS

Breakthrough of anticlinal forelimbs by thrust faults has long been recognized as a common feature in fold-and-thrust belts, a well-known example being the "break-thrust" of Willis (1893) (Figure 6a). More recently, numerous authors have recognized such breakthrough as a means by which fault-propagation or detachment folds may be modified as they evolve. These authors have proposed a wide variety of terms, geometries, and processes by which such breakthrough may occur (e.g., "transported folds," Jamison, 1987; "breakthrough" of folds, Suppe and Medwedeff, 1990; "translated folds," Mitra, 1990; "break-thrust folds," Fischer et al., 1992; McNaught and Mitra, 1993; Morley, 1994; Liu and Dixon, 1995; Thorbjornsen and Dunne, 1997; Erslev and Mayborn, 1997; Storti et al., 1997). Although some authors consider products of thrust breakthrough to be a separate type of thrust-related fold (e.g., the "break-thrust folds" of Fischer et al., 1992), breakthrough will be considered here to be a modification of a preexisting fold by thrust truncation that represents a change in geometry and/or process from that of the original fold.

Despite the widespread recognition that existing folds may be truncated by thrust faults, folds commonly are interpreted to be one of the primary types of

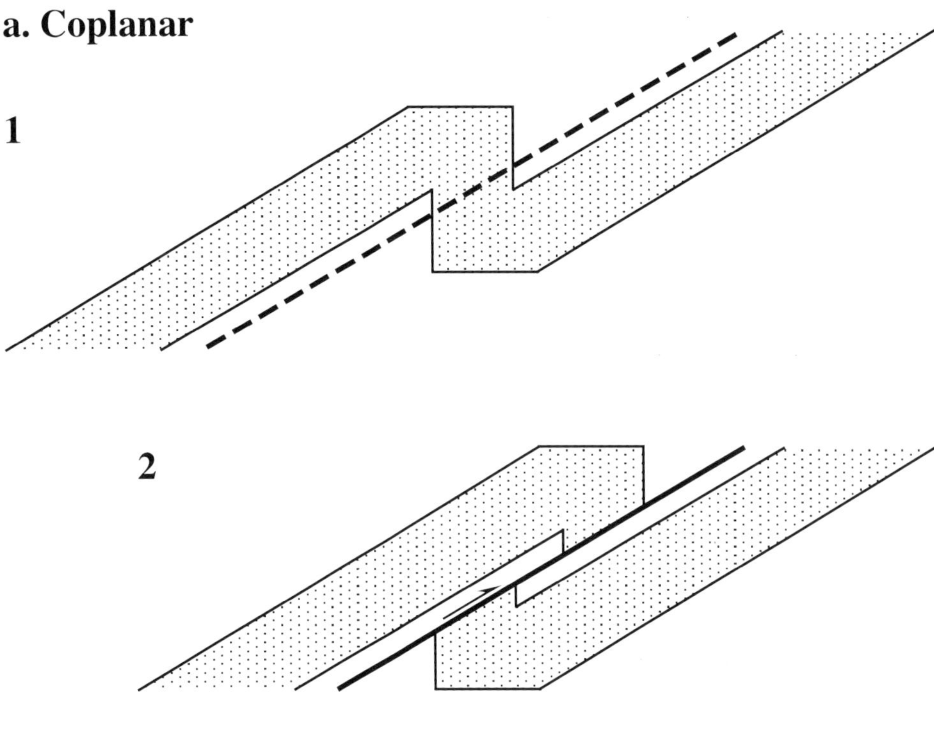

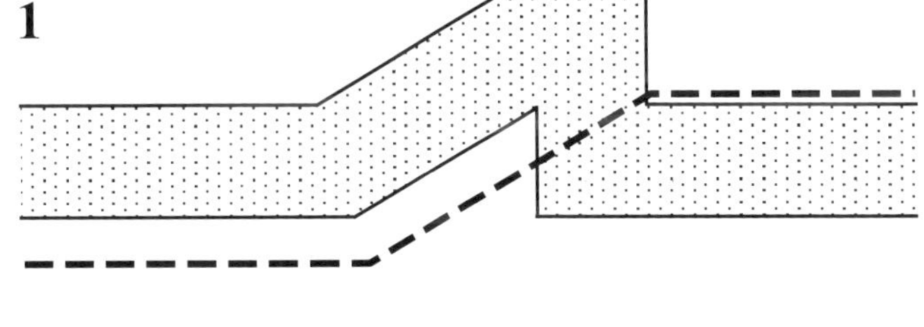

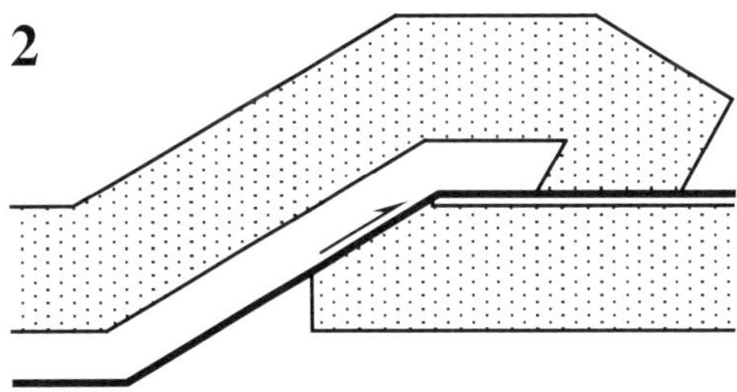

FIGURE 6. Two end-member possibilities for breakthrough of folds. In each, the fault trajectory with respect to bedding is from lower flat to ramp to upper flat. (a) Coplanar breakthrough. This is the geometry of a "break-thrust," as illustrated by Willis (1893). (b) Noncoplanar breakthrough. A fault-bend fold is superposed on the truncated fold in the hanging wall as a result of its displacement over bends in the footwall.

thrust-related fold based on consistency with a model. Such an interpretation fails to consider the possibility of thrust breakthrough and the significant modification of fold geometry and kinematics that may have resulted. In particular, a steep to overturned forelimb and the presence of a thrust beneath the fold commonly are used as a basis for interpreting folds as fault-propagation folds. However, fault-propagation folds may also result from thrust breakthrough of a fold that may or may not have originated by ramp-tip propagation.

Many different geometries of forelimb breakthrough can be envisioned (Suppe and Medwedeff, 1990), but only two important end-member cases will be considered here (Figures 6, 7). The first case is a forelimb breakthrough that is coplanar both with the lower hanging wall flat in the backlimb and with the upper footwall flat (Figures 6a, 8), and is the geometry illustrated by Willis (1893) for a "break-thrust." The second case is a noncoplanar forelimb breakthrough from a lower flat to an upper flat via a ramp (Figures 6b, 9). Geometries similar to both of these types of breakthrough are common in fold-and-thrust belts throughout the world, and numerous examples have been interpreted to be the result of thrust truncation of existing folds (Fischer and Coward, 1982; Chapman and Williams, 1984; Kulander and Dean, 1986; Alonso and Teixell, 1992; Butler, 1992; Pfiffner, 1993; Morley, 1994; Tavarnelli, 1997; Thorbjornsen and Dunne, 1997; Wallace et al., 1997).

A breakthrough that is coplanar does not geometrically require any modification of fold geometry beyond truncation and displacement of the fold (Mitra, 1990), although modifications as a result of fault propagation, drag, or other processes can be envisioned. In the absence of such modification, displacement will

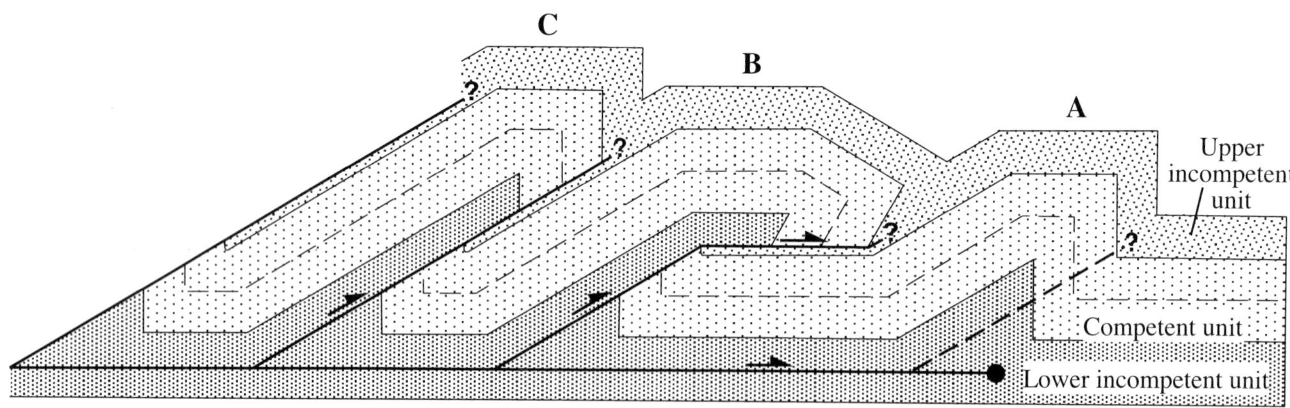

FIGURE 7. Coplanar and noncoplanar breakthrough of detachment folds. Truncated anticlines B and C may originate from a detachment fold with the geometry of anticline A by displacement on thrust faults with trajectories as shown by the dashed line. In anticline B, breakthrough and displacement of the anticline onto an upper flat that is not coplanar with the ramp has resulted in formation of a fault-bend fold with an anomalously steep forelimb. In anticline C, breakthrough and displacement of the anticline onto an upper flat that is coplanar with the ramp have resulted in truncation but no other modification of fold geometry. Tilting of the upper flat may have occurred either before or after breakthrough, as a normal consequence of thrust imbrication. Figure after Wallace et al. (1997).

remain constant along the length of the breakthrough (Hedlund, 1997; McConnell et al., 1997). If the lower-hanging-wall flat in the backlimb is coplanar with the upper-footwall flat, a special situation results in which the fault is planar but a ramp exists with respect to layering, because the fault has cut across the forelimb of an existing fold (Figures 6a, 8). This geometry of breakthrough can arise in various ways. Formation or later modification of the fold (for example, by an underlying fold or fault) prior to breakthrough can result in structural elevation of the future lower hanging-wall flat to the level of the upper footwall flat. This situation is common where breakthrough occurs in a fold that is parasitic to a larger fold, as in many outcrop-scale folds (e.g., Figure 8 and some of the examples in McConnell et al., 1997). In the case of map-scale thrust-related folds, this situation may arise if breakthrough occurs to an upper flat that is tilted to the same dip as in the backlimb (Figure 7). Tilting of the upper flat can occur either before or after breakthrough, as a result of folding ahead of the point of fold breakthrough. Upper-flat tilting is a normal consequence of the formation of fold trains or imbricate thrusts, irrespective of the sequence in which the individual structures form (i.e., piggyback, break-back, or synchronous). If tilting post-dates breakthrough, a noncoplanar breakthrough may be modified to a coplanar breakthrough.

In the case of noncoplanar breakthrough considered here (Figures 6b, 9), the footwall ramp is defined both by the trajectory of the thrust with respect to layering and by bends in the thrust. Breakthrough to an upper flat that is not coplanar with the ramp might be expected where that flat is a mechanically weak horizon (Figure 9) (Chester et al., 1991). This situation is common in nature, as is reflected by the applicability of the fault-bend fold model (Suppe, 1983) to many natural folds. The ramp-to-upper-flat bend in the fault is the same as in a fault-bend fold, but the resulting fold geometry differs from that of a fault-bend fold formed in flat-lying strata in that a preexisting fold must be modified by displacement over a fault bend. Translation of the forelimb of a fold over a footwall ramp-to-upper-flat bend will result in bending of already folded layering, and the geometric consequences are difficult to predict. A variety of conceptual possibilities have been suggested (Jamison, 1987; Suppe and Medwedeff, 1990; Mitra, 1990; Homza, 1992) or can be envisioned, but they have yet to be thoroughly compared with and tested against natural examples. Nonetheless, some general geometric characteristics of noncoplanar fold breakthrough can be predicted (Figures 2c and d, 3c and d). The forelimb will remain at least as steep and likely will steepen, depending on how bending is accommodated. For many combinations of forelimb dip and ramp angle, accommodation of bending will require some combination of collapse or extension of the forelimb or fold crest (Figure 3d), or an increase up-section in layer-parallel shear (Figures 2c and d, 3c). As in a conventional fault-bend fold, the backlimb will maintain a constant dip during displacement and may be parallel to the footwall ramp, and the length of the backlimb and the flat crest will be related to fault displacement. Thus, breakthrough of this type will result in superposition of a fault-bend fold geometry on an existing fold (Figure 9), thereby resulting in a forelimb that is steeper than would be predicted for a normal mode I fault-bend fold (Suppe, 1983; Jamison, 1987). Folding of the thrust underlying a displaced fold, as commonly occurs in fold-and-thrust belts, can also result in a very similar geometry in the hanging wall.

Figure 8. Example of a "break thrust" from Cape Liptrap, Victoria, Australia. The forelimb of a parasitic fold has broken through along bedding planes in the long limb of a larger fold. The breakthrough is nearly coplanar and displays constant displacement across the ramp and its bounding axial surfaces.

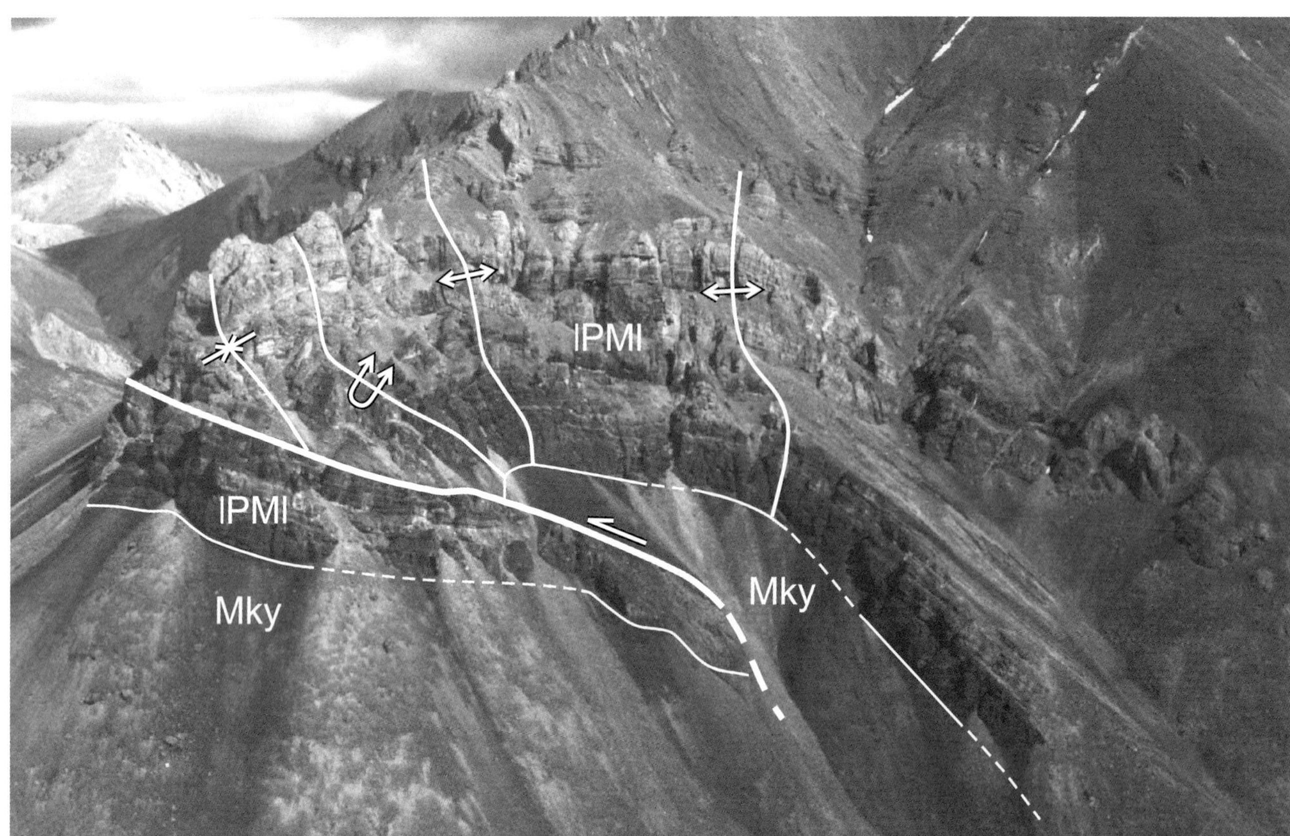

FIGURE 9. View east of a truncated and displaced fold southwest of Bathtub Ridge in the northeastern Brooks Range of Alaska. Breakthrough of an existing fold has resulted in the geometry of a fault-bend fold with an anomalously steep forelimb. Breakthrough has occurred from the incompetent shale unit (Mky: Kayak Shale) to a décollement within the overlying competent carbonate unit (PMl: Lisburne Limestone). This is inferred to be a truncated detachment fold, based on the mechanical stratigraphy and regional structural style. However, it is atypical of the area in the absence of a footwall syncline, probably reflecting breakthrough near the synclinal hinge.

Breakthrough of an existing fold results in the presence of a hanging wall anticline and, in some cases, a footwall syncline, but these folds do not owe their origin to the breakthrough itself. If the geometry of a truncated fold resembles the geometry predicted by one of the models for a type of thrust-related fold, it is easy to assume that the fold formed according to that model. However, breakthrough, if it has occurred, must also be recognized and incorporated for a geometric and kinematic interpretation to be correct. For example, superposition of a fault-bend fold geometry on an existing fold may result from noncoplanar breakthrough, but the fault-bend fold model does not fully account for the final geometry of that fold. The formation of the original fold and its modification by truncation and displacement must be incorporated for the interpretation to be complete. The need to recognize breakthrough and see through its consequences further complicates the already difficult task of correctly interpreting fold origin. Fold geometry, determination of relative timing of folding versus faulting, and kinematic evidence all may help reveal that a fold does not fit one of the thrust-related fold models, but has instead been modified by breakthrough. The same types of evidence, coupled with restoration of the prebreakthrough geometry, may allow interpretation of the origin of the prebreakthrough fold.

Breakthrough of Fault-propagation Folds

The conceptually simplest form of breakthrough of a fault-propagation fold is from the ramp tip to an upper flat (Figure 2d) (i.e., the "décollement breakthrough" of Suppe and Medwedeff, 1990; also represented in models by Jamison, 1987; Mitra, 1990; and Mercier et al., 1997). In this type of breakthrough, the ramp tip is destroyed when the hanging wall ramp is displaced over the upper footwall flat. A decrease in displacement up section, as measured between the hanging wall and footwall ramps, may record ramp propagation. However, bending is likely to result from displacement over the ramp-to-upper-flat bend formed by breakthrough. This will modify the original displacement gradient and make it difficult to assess the extent to which the observed displacement gradient resulted from ramp propagation. Correct interpretation of the geometry of this type of breakthrough may be difficult, if it is not recognized that a hanging-wall ramp has been displaced onto an upper-footwall flat— an oversight that can occur easily because the geometry of the core of so many natural folds is unknown. In this case, the fold could be interpreted to be an unbroken fault-propagation fold, among other possible incorrect interpretations. The footwall ramp remains undeformed in most fault-propagation fold models, so a footwall syncline would not be expected to result from ramp-tip breakthrough of folds formed according to these models. The notable exception is folds formed according to "trishear" models of fault-propagation folding (Erslev, 1991; McConnell, 1994; Spang and McConnell, 1997), but the distinctive strain patterns required by these models might facilitate recognition of folds to which those models apply.

A footwall syncline could be produced in folds formed according to any of the fault-propagation fold models, if the breakthrough occurred above and hindward of the original ramp and ramp tip (Figure 2c), although this probably is less likely than a breakthrough from ramp tip to upper flat. In this case, a footwall syncline would be present where the breakthrough cut through the forelimb of the fold above the ramp tip. However, the ramp, ramp tip, and the lower part of the hanging-wall anticline of the original fold would be stranded in the footwall of the breakthrough and would be distinctive indicators of this type of breakthrough.

Breakthrough of Detachment Folds

Breakthrough of detachment folds has been observed or suggested in many recent papers (Jamison, 1987; Mitra, 1990; Homza, 1992; Wallace, 1993; Rowan, 1997; Wallace et al., 1997). Other examples have been attributed to "break-thrusting" (Fischer et al., 1992; Morley, 1994; Woodward, 1997), although the term "break-thrust" indicates only breakthrough of an existing fold and does not identify how that fold originated. The descriptions of structures in these papers suggest that they could be more explicitly interpreted to represent breakthrough of detachment folds. A number of papers have suggested a kinematic progression from initial detachment folding, to breakthrough by fault-propagation folding, to post-breakthrough fault-bend folding (McNaught and Mitra, 1993; Liu and Dixon, 1995; Storti et al., 1997; Woodward, 1997). Modification of an existing fold by fault-propagation folding is conceptually possible but would be difficult either to model or to demonstrate in natural folds. To focus on the geometry of the breakthrough thrust and not the more difficult question of the process of breakthrough, we will limit our discussion here to the geometrically and kinematically simpler end-member case in which propagation of the breakthrough thrust does not in itself result in folding (implying a very high ratio of fault propagation rate to slip rate).

Numerous authors have suggested that rotating-limb folds will tend to lock as their interlimb angle decreases to within the range of ~110 to 60°, as a result of some combination of increase in interlayer angular shear and decrease in fold cross-sectional area with decreasing interlimb angle (De Sitter, 1956; Ramsay, 1974; Rowan and Kligfield, 1992; Homza and Wallace, 1997; Woodward, 1997). Increasing resistance to folding, in turn, may cause thrusting to become the favored means for accommodating additional shortening. Fold asymmetry may both favor breakthrough and result in preferential breakthrough of the short, steep limb. This suggests that thrust breakthrough could be a likely consequence late in the evolution of rotating-limb detachment folds, particularly if they are asymmetric.

The most significant geometric consequence of breakthrough of a detachment fold is formation of a ramp where none existed previously (Figures 3, 9)— a ramp that was not present during origin of the truncated fold, unless some folding accompanied ramp propagation. If the ramp has played no role in formation of the original fold, how it cuts across the fold may be unrelated to the geometry of the fold. In other words, the breakthrough thrust need not parallel the backlimb or any of the existing fold hinges, although one of these may provide a path of least resistance. Breakthrough occurs most commonly through the anticlinal forelimb, which means that the bedding cutoff angle is not simply the ramp dip, but a potentially much higher angle determined by the combination of ramp dip and forelimb dip. Although the cutoff geometry might superficially resemble that in one of the thrust-related fold models, the relations between cutoff angle and ramp dip based on that model would not be applicable in the case of breakthrough.

Another very important geometric consequence of breakthrough of a detachment fold is that it provides a geometrically and kinematically simple mechanism for the formation of footwall synclines. A syncline will be stranded in the footwall if the breakthrough thrust cuts through any part of the anticlinal forelimb. Footwall synclines are a common feature of fold-and-thrust belts worldwide (Fischer and Coward, 1982; Chapman and Williams, 1984; Kulander and Dean, 1986; Butler, 1992; Pfiffner, 1993; Morley, 1994; Tavarnelli, 1997; Wallace et al., 1997), yet the existing models for thrust-related folds do not account for them well. Although footwall synclines commonly are observed or inferred in folds interpreted to be fault-propagation folds, only the trishear models account for their existence (Erslev, 1991; McConnell, 1994; Spang and McConnell, 1997). However, these models require highly nonparallel folding, and thus do not fit well with the common assumption that footwall synclines have simple parallel fold geometry. Although McNaught and Mitra (1993) have proposed a model for formation of footwall synclines by a progression from detachment folding to fault-propagation folding, a simpler explanation may be that many footwall synclines have formed simply by breakthrough of detachment folds, without a requirement for breakthrough by fault-propagation folding.

Distinction between Thrust-truncated Fault-propagation and Detachment Folds

Whether a thrust-truncated fold originated as a detachment fold or a fault-propagation fold may be very difficult to determine. An answer may be possible if prebreakthrough geometry can be restored fully and accurately and kinematic evidence is available from small-scale structures or growth stratigraphy. However, as a practical matter, less-direct evidence probably will have to be used in most cases to reach the most likely interpretation. Such evidence may be easier to find in the case of thrust-truncated detachment folds, so we will discuss them first.

Mechanical stratigraphy and evidence for incompetent behavior are probably the most obvious indicators that one should consider a detachment-fold interpretation. A known strong competency contrast within the stratigraphy suggests the potential for detachment folding. The presence of detachment folds elsewhere in the region may provide direct evidence that detachment folding has occurred in the appropriate part of the section, and remnant unbroken detachment folds may be preserved locally in the area of the truncated fold being assessed. Flats of thrust faults commonly are located in the incompetent units that form the lower part of detachment folds. Also, some thrust sheets may carry enough of the incompetent unit to preserve fold cores that display the incompetent behavior and structural thickening characteristic of detachment folds. The presence of a footwall syncline suggests that a detachment-fold origin should be considered, particularly if such synclines are a common feature of the structural style of the region. Absence of evidence for fault-propagation folds may also suggest a detachment-fold origin. This would include the absence of ramp tips in the area, the lack of a displacement gradient in the case of a coplanar breakthrough, and the truncation of fold axial surfaces with a geometry unlikely to arise from breakthrough from the ramp of a fault-propagation fold.

Direct evidence supporting a fault-propagation-fold origin would include the presence of ramp tips in the area and the existence of a displacement gradient in the case of a coplanar breakthrough. If a footwall syncline is present, distinctive strain patterns or abandoned ramps and ramp tips would be expected if the fold originated as a fault-propagation fold. Fault-propagation folds, either preserved in the local area or characteristic of the regional structural style, would support a fault-propagation-fold interpretation. A homogeneous but strongly anisotropic mechanical stratigraphy would be consistent with formation of fault-propagation folds, but too little is known about the mechanical stratigraphic controls on fault-propagation folds to rule out their formation in other types of mechanical stratigraphy. The relatively limited possibilities for the geometry of truncation of axial surfaces as a result of breakthrough from the ramp of a fault-propagation fold (Suppe and Medwedeff, 1990) may be a useful discriminator.

ORIGIN OF ASYMMETRIC THRUST-RELATED FOLDS IN THE BROOKS RANGE OF ALASKA: DETACHMENT FOLDS OR FAULT-PROPAGATION FOLDS?

The Brooks Range of northern Alaska is the northern continuation of the Rocky Mountain Fold-and-thrust Belt. The northeastern part of this range provides a useful example of the difficulties that can arise in interpreting the origin of asymmetric thrust-related folds with steep to overturned forelimbs, particularly where most of these folds have been truncated by thrust faults. The region also illustrates how regional structural style and mechanical stratigraphy may aid us in reaching a preferred interpretation for the origin of such folds within an area.

The "Continental Divide thrust front" (Figure 10) marks the structural boundary between the main axis of the Brooks Range and a major salient in the northeastern Brooks Range (Wallace and Hanks, 1990). Symmetric detachment folds that are not cut by thrust faults dominate within a part of the stratigraphic section that is exposed throughout the northeastern Brooks Range, whereas asymmetric folds of uncertain origin dominate the same part of the stratigraphic section throughout the northern part of the main axis of the Brooks Range. The origin of the asymmetric folds to the south is the main question here, but an understanding of the detachment folds to the north is important in attempting to answer that question.

The detachment folds of the northeastern Brooks Range are summarized here but have been described elsewhere in more detail (Wallace and Hanks, 1990; Wallace, 1993; Homza and Wallace, 1997; Wallace and Homza, 1998). The detachment folds formed in the

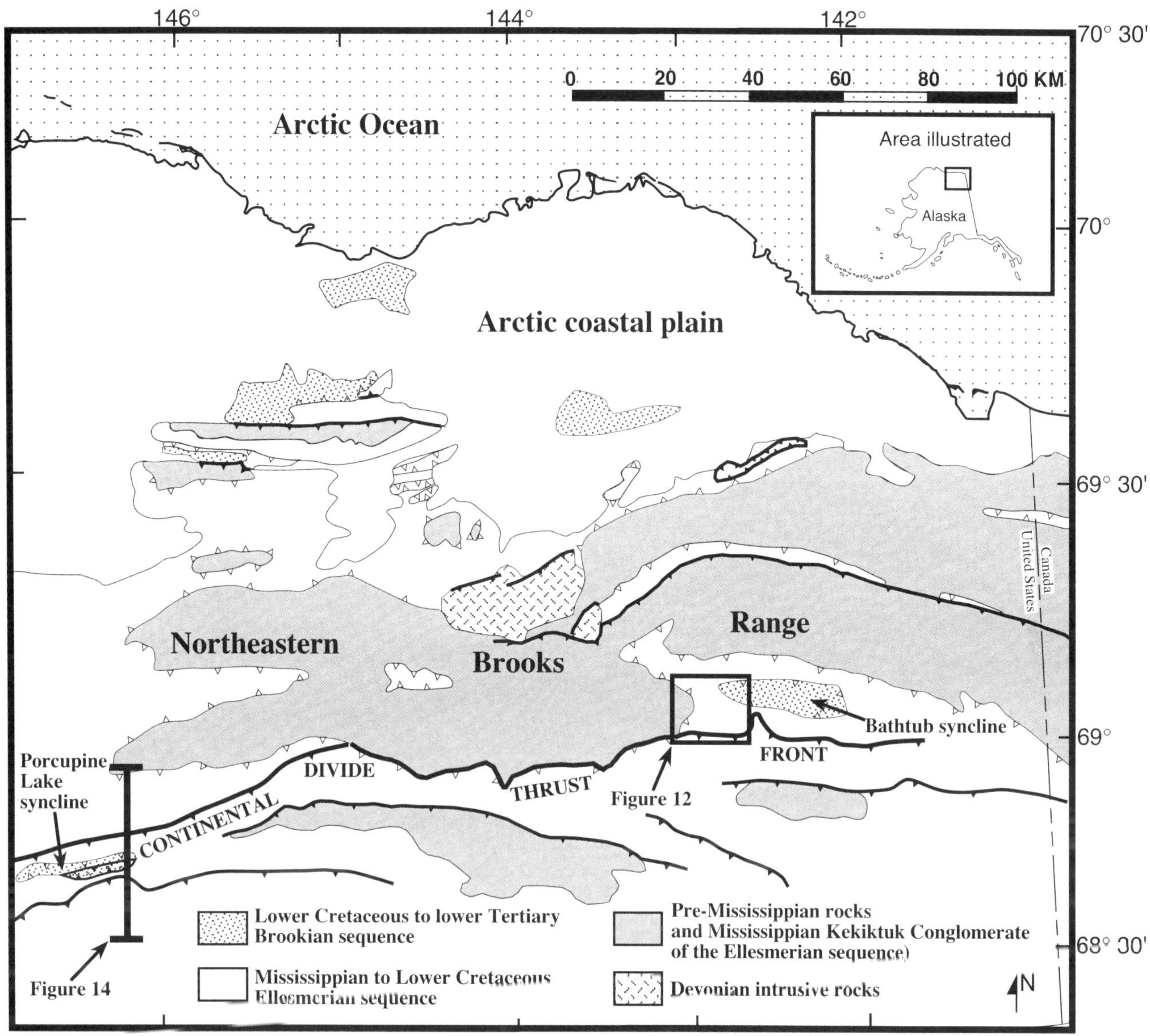

FIGURE 10. Generalized tectonic map of the northeastern Brooks Range (modified from Wallace and Hanks, 1990). The units that form detachment folds and thrust-truncated detachment folds are near the base of the Ellesmerian sequence. These rocks are characterized by symmetric detachment folds north of the "Continental Divide thrust front" and by thrust faults that have truncated asymmetric folds to the south. Solid teeth represent older-over-younger thrust faults, whereas open teeth represent younger-over-older décollements.

incompetent Mississippian Kayak Shale (~110–300 m thick) and the overlying competent Mississippian to Pennsylvanian Lisburne Limestone (~1400–1700 m thick). These units are detached from an underlying competent unit that consists of the Mississippian Kekiktuk Conglomerate and the unconformably underlying pre-Middle Devonian structural basement. A regional-scale duplex in the lower competent unit probably formed synchronously with detachment folding and has resulted in folding of the basal detachment of the detachment folds at a much larger wavelength than that of the detachment folds.

The detachment folds display ranges in wavelength of ~1–3 km and in interlimb angle of ~130–0°; thus they represent a broad range in shortening. The folds range in geometry from simple, straight-limbed folds (Figure 11) to more-complex folds with curved hinges and parasitic folds. These variations in geometry do not appear to be related to the amount of shortening, except that box-fold geometries are common where shortening is least. Most of the folds are upright and symmetric and do not display a strong or consistent sense of vergence throughout the region. The folds are interpreted to have formed with fixed hinges and rotating limbs, based on meso- to microstructures (Homza and Wallace, 1997). Very few thrust faults are present, despite the very great shortening locally. The few thrust faults that are present typically

FIGURE 11. View east of Straight Creek Anticline, an unbroken detachment fold in the northeastern Brooks Range. The basal detachment of the fold lies above the exhumed surface of the Kekiktuk Conglomerate (Mkt) in the right foreground. The incompetent Kayak Shale (Mky) is thickened in the core of the anticline, as reflected by smaller-scale folding of a competent sandstone layer within the shale. The competent Lisburne Limestone (PMl) forms an angular anticline and rounded synclines above the shale. This fold is described in more detail by Homza and Wallace (1997).

are associated with local asymmetry and commonly are out-of-syncline thrusts.

The Kayak Shale and Lisburne Group display a very different structural style, south and southwest of the northeastern Brooks Range, despite an overall similarity in their mechanical stratigraphy. Throughout the northern part of the main axis of the Brooks Range, the structural style of these units is dominated by thrust faults (Mull et al., 1987; Oldow et al., 1987; Moore et al., 1994; Kelley and Brosgé, 1995; Wallace et al., 1997). Thrust sheets generally dip hindward (to the south) and are characterized by steep to overturned, leading hanging-wall anticlines and trailing footwall synclines. These thrust sheets are bounded below by a flat in the Kayak Shale and above by a flat in Permian to Triassic shale-rich units that overlie the Lisburne Limestone.

The "Continental Divide thrust front" (Wallace and Hanks, 1990) forms an abrupt boundary between fold-dominated and thrust-dominated structure within the Kayak Shale and Lisburne Limestone (Figure 10). Geometries characteristic of fault-bend folds, except for excessively steep forelimbs, are common just south of this boundary and reflect bending of thrust sheets over a nonplanar thrust, either because of ramp-to-upper-flat transitions in the footwall or folding of the thrust. Farther south, within the thrust-dominated region, adjacent thrust sheets commonly are separated by essentially planar thrusts and thus have about the same dip. Figure 7 schematically illustrates the changes in structural style from unbroken detachment folds, to fault-bend folded truncated folds, to folds truncated by planar imbricate thrusts.

The examples presented below are from two areas that span the Continental Divide thrust front (Figure 10) and that range from unbroken detachment folds to thrust-truncated asymmetric folds. These examples illustrate both the characteristics of the folds and the evidence that can be used to reach a preferred interpretation for the origin of the thrust-truncated asymmetric folds.

Bathtub Ridge Area

Homza (1992) studied the transition from unbroken detachment folds to thrust-truncated asymmetric folds in detail in the area southwest of Bathtub Ridge

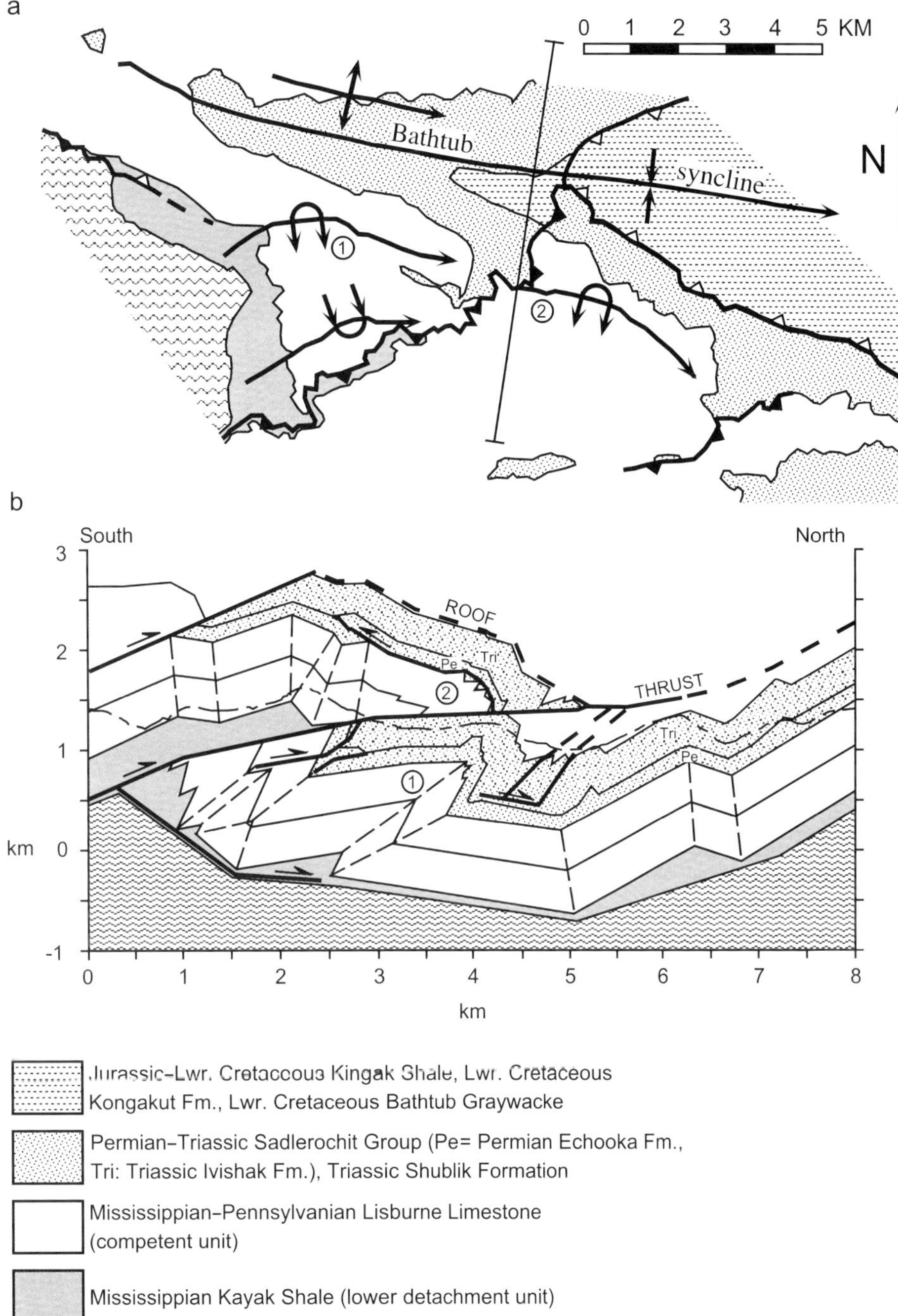

Figure 12. Geologic map (a) and cross section (b) of the southwest Bathtub Ridge area, modified from Homza (1992). Location of map area is shown in Figure 10. Line of section for (b) is shown in (a). (a) An unbroken detachment fold (1) and thrust-truncated detachment fold (2) are exposed in down-plunge projection above and east of basement in the southwestern part of the map. (b) A truncated asymmetric detachment fold (2) has been displaced and rotated above a broadly curved ramp-to-upper-flat transition. A truncated syncline and an unbroken detachment fold (1) lie in the footwall to the north.

(Figure 10). His study revealed that the transition is marked by an unbroken detachment fold (Fold 1, Figure 12) that is bounded to the south by a thrust fault. The detachment fold is strongly asymmetric, with a steep to overturned forelimb that displays some parasitic folding (Figures 12b, 13a). No ramp is evident in the Kayak Shale that is exposed in the core of the fold, but the underlying décollement is not exposed. A thrust fault cuts up from basement to the Kayak Shale along strike with the forelimb of the fold to the northwest (Figure 12a), which suggests that the fold could be a fault-propagation fold. However, this thrust appears to flatten up-section in the Kayak, which argues that the overlying fold is not a fault-propagation fold but instead may have nucleated by thickening of Kayak over the change in structural relief on the basement.

A down-plunge exposure provides an unusually complete cross-sectional view of the structure of the thrust that lies to the south of the unbroken detachment fold (Figure 12). In the footwall, the thrust cuts from a lower flat near the base of the Kayak, across a large overturned footwall syncline in the Lisburne, to an upper flat above the Lisburne. The Kayak displays significant structural thickening in the steep limb of the syncline. The same thrust trajectory in the hanging wall is marked by a large overturned hanging-wall anticline (Fold 2, Figure 12). This fold has a complex geometry in detail, but its wide crest and long, gentle backlimb reflect fault-bend folding over a curved ramp-to-upper-flat transition. Unlike the fold to the north, this thrust does not overlie a fault that cuts upward

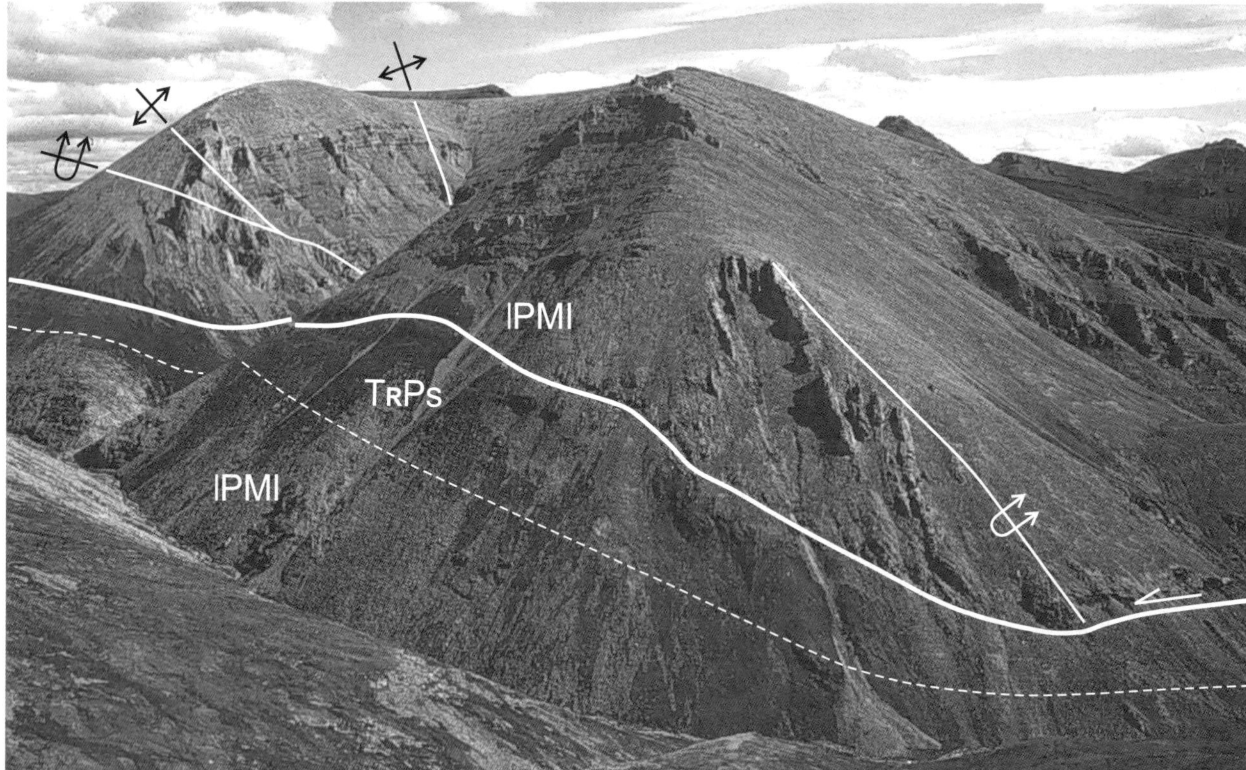

FIGURE 13. Folds south of Bathtub Ridge. (a) View east of the unbroken detachment fold shown in Figure 12. The incompetent Kayak Shale (Mky) in the core of the fold is not cut by a ramp. A long, gentle backlimb is bounded to the north by a steep to overturned forelimb with a parasitic anticline-syncline pair in the overlyinging competent Lisburne Limestone (PMl). (b) Oblique view to the southeast of a simple thrust-truncated fold south of Bathtub syncline. The hanging-wall anticline is in the competent Lisburne Limestone (PMl) and displays a relatively simple geometry that is similar to that shown in Figure 3c. The truncated fold rests on an upper footwall flat in incompetent beds of the Sadlerochit Group (TrPs) (cont.).

FIGURE 13. (cont.) (c) View east of a complex thrust-truncated fold south of Bathtub syncline. The hanging-wall anticline is in the competent Lisburne Limestone (PMl) and displays complex parasitic folds similar to those in unbroken detachment folds in the northeastern Brooks Range. Some of these are truncated by a minor thrust near the top of the structure. Truncated Lisburne at the base of the structure rests on an upper footwall flat in incompetent beds of the Sadlerochit Group (TrPs). (d) Oblique view to the northeast of a footwall syncline south of Bathtub Syncline. The competent Lisburne Limestone (PMl) is overlain by a small remnant of the Sadlerochit Group (TrPs) in the core of the syncline. The syncline is defined by a long, gentle limb to the north, a flat bottom, and a short, overturned limb to the south. The axial surfaces converge upward to define an inverted box-fold geometry. Gently south-dipping beds in the hanging wall are visible in the distance, to the right.

from basement. Instead, it lies over the long backlimb of a basement anticlinorium.

The exceptional down-plunge view of the structure in this area suggests truncation of an asymmetric anticline that had a long, gentle backlimb, a thickened core of Kayak Shale, and an overturned forelimb with parasitic folds. Truncation and displacement on a convex-upward thrust led to superposition of a fault-bend-fold geometry and further steepening of the forelimb of the hanging-wall anticline. The similar, but smaller, detachment anticline that is exposed to the north in the footwall did not break through.

Numerous other examples of thrust-truncated folds are exposed along strike to the east, south of Bathtub Ridge, although the basal detachment is not exposed in this area. Hanging-wall anticlines display a range of geometries, from simple to complex, and commonly display superposed fault-bend-fold geometries (Figure 13b, c). Footwall synclines typically display strongly overturned backlimbs and flat bottoms bounded by two synclinal hinges that merge upward to define an inverted box-fold geometry (Figure 13d).

Porcupine Lake Area

Exceptional exposures along the Marsh Fork of the Canning River near Porcupine Lake document the transition from unbroken detachment folds to thrust-truncated folds particularly well (Figures 10, 14). A long train of unbroken, symmetric detachment folds is exposed to the north of the Continental Divide thrust front (Figure 14). The average structural relief in this area is relatively constant, which suggests that the basal detachment is relatively flat. To the south, the thrust front is marked by the abrupt appearance of strongly asymmetric hanging-wall anticlines, which commonly form antiformal stacks near the boundary (Figure 14). Hanging-wall anticlines commonly are preserved at the boundary, and some of these display a superposed fault-bend fold geometry (Figure 15a). These anticlines typically display steep to overturned forelimbs that consist of at least two dip panels ahead of the flat crest of the structure (Figures 3c, 15a) and that commonly have an additional narrow, strongly overturned panel near the basal thrust. Some forelimbs are cut at high angles by multiple thrusts. Some are cut by calcite veins as much as 5 m thick that are parallel to, and commonly localized near, anticlinal hinges and that may reflect extension of the hanging wall as it bent over an underlying nonplanar thrust.

Two asymmetric anticline-syncline pairs south of the thrust front (Figures 14, 15b) suggest two different origins. The southern fold pair has a chevron geometry, with disharmonic folding in the fold core and asymmetric parasitic folding in the long backlimb. No thrust faults are present in the exposed part of this fold pair. Its characteristics are consistent with origin as a detachment fold, although one or more ramp tips could conceivably be present at depth. The core of the anticline in the northern fold pair is cut by at least four small-displacement thrust splays, including one in the backlimb, one near the anticlinal hinge, one with a tip in the synclinal hinge, and one that flattens into bedding beneath the syncline. The characteristics of this fold pair are consistent with origin as a very complex fault-propagation fold, but so far no other folds have been found in the region that show similar characteristics in their cores.

Origin of the Asymmetric Folds

Little has been published about interpretation of the asymmetric folds in the thrust sheets of the northern Brooks Range according to models for thrust-related folds, but their characteristics make it easy to assume that they originated as fault-propagation folds (e.g., Figure 1 of Suppe and Medwedeff, 1990). Homza (1992), Wallace (1993), and Wallace et al. (1997) described structures from parts of this region and suggested instead that they are thrust-truncated detachment folds (although an earlier fault-propagation fold interpretation is illustrated by Wallace (1993) in his Figure 5). Several lines of evidence support origin of these folds as detachment folds and are exemplified by the Bathtub Ridge and Porcupine Lake areas. The mechanical stratigraphy is favorable for detachment folding, and the Kayak and Lisburne form detachment folds throughout the northeastern Brooks Range. Unfaulted detachment folds are present locally within the region dominated by thrust faults, and incompetent thickening of Kayak Shale is common in hanging-wall anticlines. Ramp tips are rarely observed. Footwall synclines are common and generally display fold geometry that appears approximately parallel. Probably the strongest evidence that these folds formed by breakthrough of detachment folds is the transition that is observed from unfaulted detachment folds in the northeastern Brooks Range to thrust-truncated folds to the south, in the eastern part of the main axis of the Brooks Range (Homza, 1992; Wallace, 1993; and Wallace et al., 1997). Although they are rare, thrust-truncated detachment folds observed north of the Continental Divide thrust front display geometries very similar to those observed to the south.

The evidence presented here suggests that most of the asymmetric folds south of the Continental Divide thrust front formed as detachment folds, although at least a few could have originated as fault-propagation folds. However, this interpretation is not immediately obvious simply from observations of individual folds in the field, despite exceptionally good exposures, nor can it be demonstrated definitively to be the correct interpretation, either for individual examples of folds

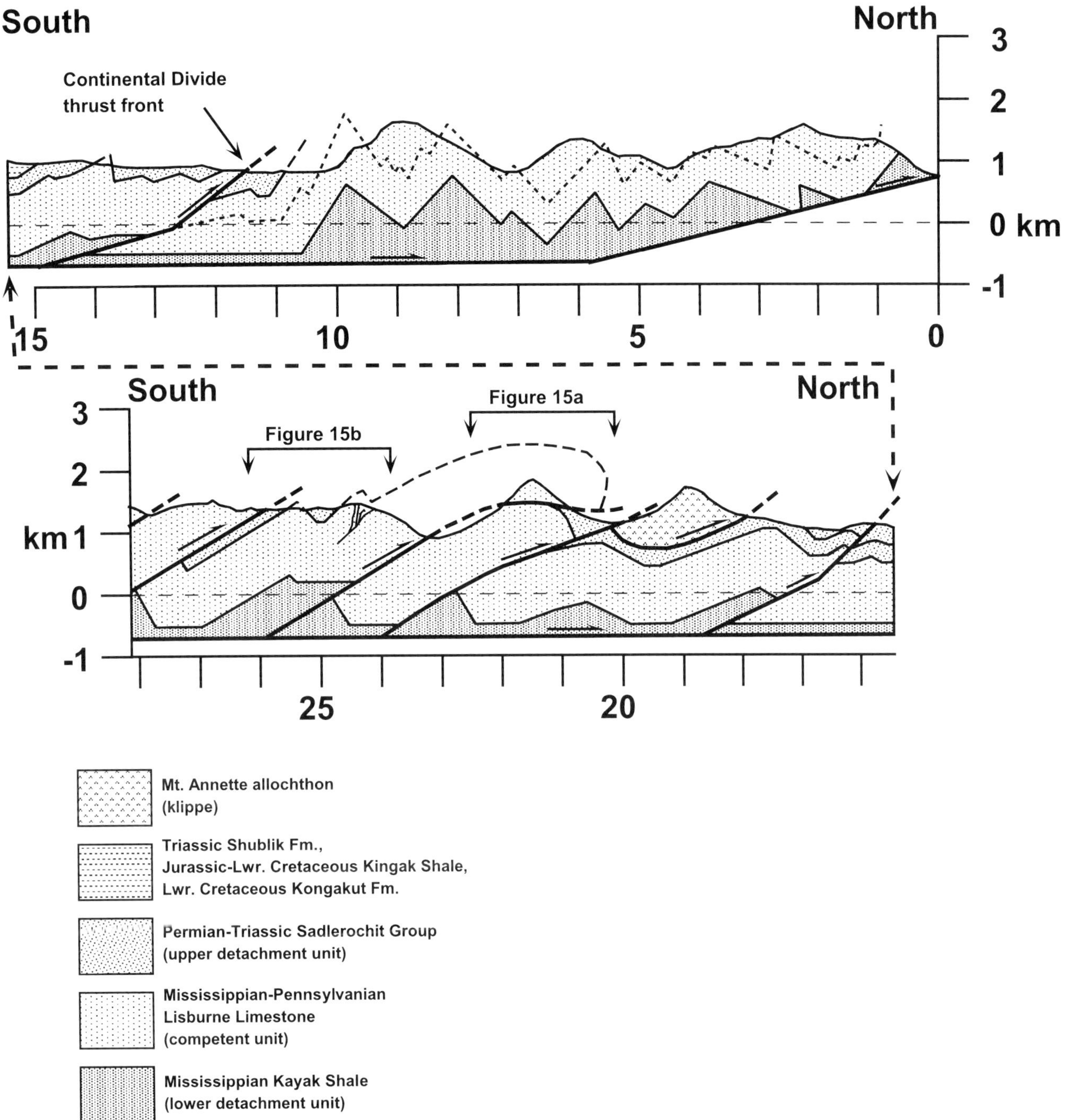

FIGURE 14. Cross section of Porcupine Lake area. Line of section is shown in Figure 10. The upper (northern) part shows unbroken upright and symmetric detachment folds typical of the northeastern Brooks Range, north of the Continental Divide thrust front. The lower (southern) part shows thrust-truncated asymmetric folds typical of the main axis of the Brooks Range, south of the Continental Divide thrust front. The largely eroded thrust-truncated anticline is shown in Figure 15a, where it is preserved to the east. Two folds within this thrust sheet, including an unbroken fold and a fold with small-displacement truncation, are shown in Figure 15b.

or for the folds as a group. These uncertainties are the result of practical limitations that are to be expected in assessing the origin of comparable folds anywhere. Geometrically critical parts of fold cores are only rarely exposed, original fold geometry has been modified by thrust truncation and displacement, and important information on kinematics is unavailable because the topography and exposures prevent access, appropriate structures or markers are absent, and/or insufficient time is available to obtain meaningful results.

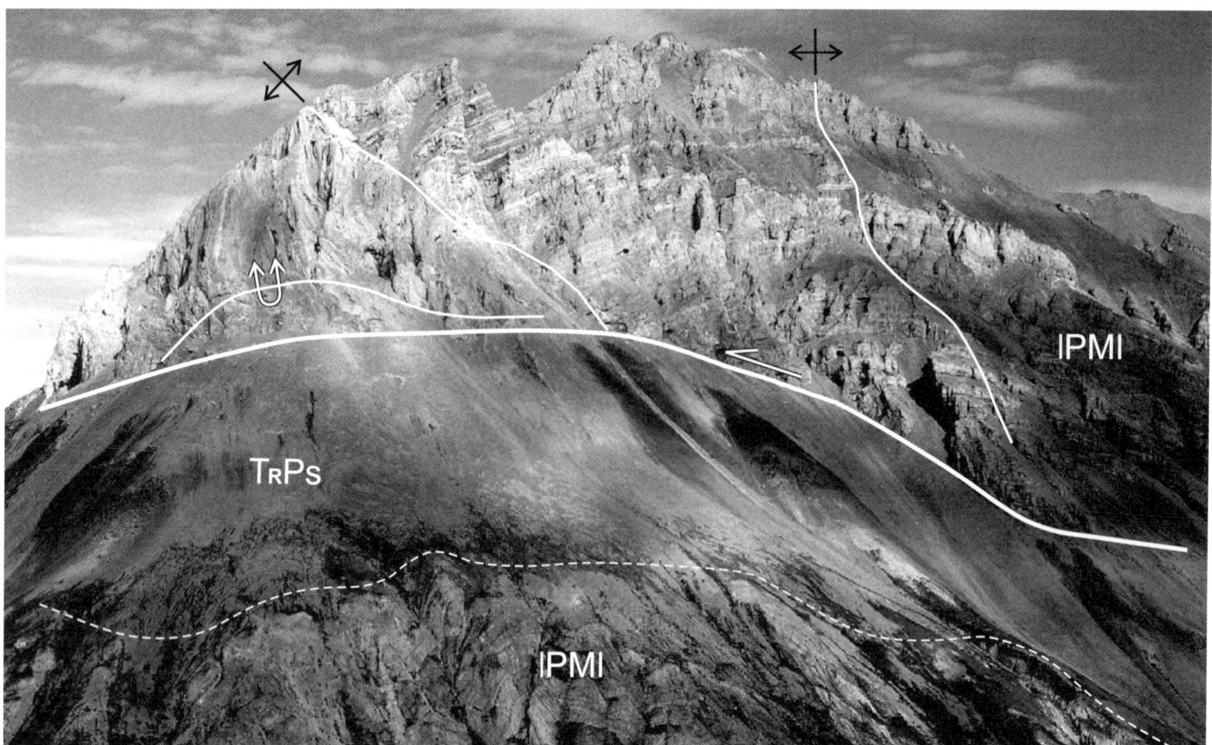

FIGURE 15. Folds south of the Porcupine Lake syncline (location shown in Figure 14). (a) Oblique view to the southeast of a thrust-truncated anticline. The hanging-wall anticline is in the competent Lisburne Limestone (PMl) and displays a geometry typical for noncoplanar breakthrough of asymmetric folds in the region. Discordant white layers, especially near the middle anticlinal hinge, are large calcite veins. The truncated fold rests on an upper footwall flat in incompetent beds of the Sadlerochit Group (TrPs). (b) View to the west of two fold pairs within a thrust sheet. Asymmetric chevron folds in the competent Lisburne Limestone (PMl) are locally overlain by the Sadlerochit Group (TrPs). The southern anticline is not cut by thrust faults and displays disharmonic folding in its core and parasitic folds in its backlimb. The northern anticline is cut by several small-displacement thrust faults, including one with a ramp tip at the base of the forelimb.

PRACTICAL SIGNIFICANCE, ESPECIALLY FOR PETROLEUM EXPLORATION AND PRODUCTION

Why does it matter if we correctly interpret a particular natural fold as a fault-propagation fold or detachment fold, or if we recognize whether that fold has been truncated by a thrust fault? The different types of thrust-related folds, including fault-propagation and detachment folds, differ fundamentally in how they form. Consequently, assignment of a name to a fold has specific implications for its origin, as well as for its geometry and kinematics. Fold names commonly are assigned using only partial knowledge of fold geometry, and the interpretation may be based primarily on consistency of that partially constrained geometry with a particular geometic-kinematic model for a fold type. However, geometry alone is not sufficient for interpreting fold kinematics or mechanics, and it rarely is sufficient for determining uniquely either the fold type or the applicable geometic-kinematic model. Incorrect identification of fold type or use of an inappropriate geometic-kinematic model can lead to incorrect reconstruction of unknown parts of the fold geometry and incorrect assumptions about fold evolution. Truncation of an existing fold by a thrust fault raises similar problems: Truncation may result in significant modification of original fold geometry, but the available information commonly is not sufficient to lead to a unique interpretation of the geometry and kinematics of a truncated fold.

A schematic example illustrates a wide range of interpretations that are possible, given a relatively simple fold geometry (Figure 16). Surface exposures are very limited in vertical extent, and fold cores commonly are not exposed (Figure 16a). Subsurface data, such as seismic-reflection profiles, may provide information with a greater vertical extent, but critical gaps in data may exist at depth and in areas of steep dip or other structural complexity (Figure 16b). Four different interpretations of the data are shown, including a detachment fold (Figure 16c), a fault-propagation fold (Figure 16d), a thrust-truncated detachment fold (Figure 16e), and a thrust-truncated fault-propagation fold (Figure 16f). Although the upper parts of these interpretations are identical, they differ at depth in terms of fold and thrust geometry, detachment depth, and mechanical stratigraphy. In addition, the thrust-truncated folds are very different from the unmodified folds. Many other interpretations that are not shown here are consistent with the data and would yield further variations in these characteristics.

This example illustrates the importance of assessing multiple interpretations of a given structure: Showing that the available knowledge of a natural structure is consistent with a specific fold type or a particular geometic-kinematic model is not sufficient to demonstrate that the interpretation is correct. Widespread attention to fault-propagation folding has led to widespread use of the concept for interpretation. However, other concepts, such as detachment folding and thrust truncation of an existing fold, may also be consistent with the available data. Comparing different interpretations may provide a basis to identify a preferred interpretation or to determine a means of testing the different interpretations. Alternatively, it simply may not be feasible to discriminate among multiple interpretations.

The petroleum-related implications of an interpretation of fold type or geometric-kinematic model depend largely on the specific details of that interpretation, but a few general comments can be made. The importance of correctly interpreting the geometry of a hydrocarbon trap is quite obvious. Assuming the wrong fold type or geometric-kinematic model, or failing to recognize that a fold has been truncated and displaced by a thrust fault, can result in incorrect interpretation of trap geometry, including location, shape, size, depth, and vertical extent. How the different stratigraphic units are involved in the structure relates strongly to the interpretation and has implications not only for the trap, but also for reservoir, seal, source, and migration. Detachment folds can be expected to involve incompetent stratigraphic units in at least their lower parts. These units commonly are evaporite, shale, or argillaceous limestone that may serve as poor reservoir and good seal, and they may provide a hydrocarbon source low in the structure if they consist of organic-rich shale or carbonate. Fault-propagation folds may not involve incompetent stratigraphic units, although the mechanical stratigraphy to be expected with this fold type is less clear.

Different interpretations will have different implications for the presence and geometry of faults and for whether those faults enhance or impede the migration of fluids or result in compartmentalization of reservoirs. For example, faults in a thrust-truncated detachment fold are likely to involve incompetent rock and thus are more likely to be barriers to fluid migration than are faults in a fault-propagation fold or thrust-truncated fault-propagation fold. Interpretation of fold kinematics also has important implications for enhancement or degradation of reservoir porosity and permeability. In particular, whether hinges have migrated or remained fixed with respect to the rock will strongly influence the orientation and distribution of fold-related fractures and penetrative strain.

Structures that either have been or could be interpreted as fault-propagation folds, detachment folds, or thrust-truncated variants of either are found in fold-and-thrust belts worldwide, many of them in important

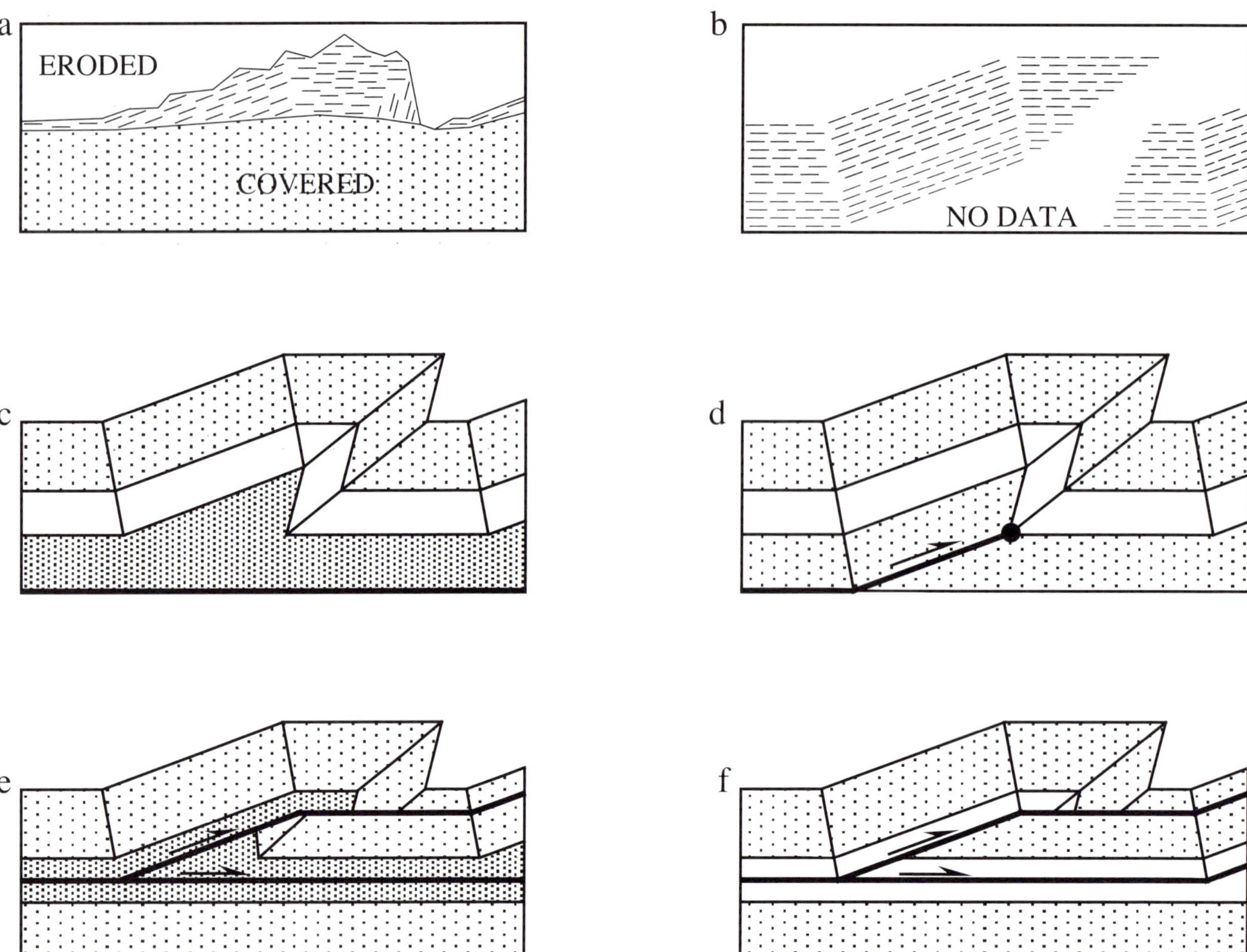

FIGURE 16. Schematic diagram of multiple intepretations possible from incomplete data. Dark gray layer is incompetent, light gray and white layers are competent. White layer is a marker with no mechanical distinction from the light gray layer. (a) Typical data that might be available from good surface exposure of a fold. Fold geometry is well-defined in exposures of a competent unit high in the fold, but lower parts of the structure are not exposed, including the anticlinal core, the synclinal hinge, and lower stratigraphic units. (b) Typical data that might be available from a seismic-reflection profile. Fold geometry is well defined in gently dipping panels in the upper part of the fold, and some stratigraphic distinctions are defined by differences in the character of reflectors. Data are absent or ambiguous at depth and in areas of steep dip or complex structure. (c) Detachment fold. (d) Fault-propagation fold. (E) Thrust-truncated detachment fold. (F) Thrust-truncated fault-propagation fold.

petroleum provinces. Many such structures have been described in the Rocky Mountain Fold-and-thrust Belt in the United States and Canada. Structures in the Idaho-Wyoming-Utah Thrust Belt display geometries that suggest they are either detachment or fault-propagation folds, including thrust-truncated examples of each (Lamerson, 1982; West and Lewis, 1982; Royse, 1993). Both fault-propagation and detachment fold origins have been proposed for structures here and in the nearby Montana thrust belt (Boyer, 1986; Mitchell and Woodward, 1988; Mitra, 1990; Suppe and Medwedeff, 1990; Woodward, 1992). In the foothills of the Canadian Rockies, the Turner Valley Anticline is cited as an example of a fold that overlies a ramp on which displacement decreases upward (Williams and Chapman, 1983; Ellis and Dunlap, 1988), although an alternative interpretation lacks such a displacement gradient (Tippet et al., 1985). The displacement gradient has been used as evidence to interpret the structure as a fault-propagation fold (Jamison, 1987; Mitra, 1990; Suppe and Medwedeff, 1990), although the cross-sectional representation of the structure as a thrust-truncated fold with a footwall syncline and thickening in both limbs could allow other interpretions. Both fault-propagation-fold and detachment-fold interpretations have been proposed for other structures in the Canadian Rockies and its foothills (Jamison, 1987; Poblet and McClay, 1996; Erslev and Mayborn, 1997; Rait and Dixon, 1999), and mechanical stratigraphy has been interpreted as a strong control on fold type in this region (Thompson, 1981; McMechan and Thompson, 1989; Spratt and Dixon, 1999).

In the western Transverse Ranges of southern California, the Ventura Avenue Anticline has been interpreted both as a detachment fold (Yeats, 1983) and, more recently, as a fault-propagation fold (Yeats et al., 1988; Huftile and Yeats, 1995). The nearby Sulphur Mountain Anticlinorium has been interpreted as a thrust-truncated fault-propagation fold that formed above a décollement at the same stratigraphic level as in the Ventura Avenue Anticline (Huftile, 1991). Although both the Ventura Avenue Anticline and Sulphur Mountain Anticlinorium have been interpreted as fault-propagation folds, their geometry and mechanical stratigraphy appear to be more consistent with existing models for detachment folds. The complex structure in this region provides many other examples for which multiple interpretations of thrust-related folds have been proposed or could be considered. Identifying the best interpretation is relevant not only to the extensive petroleum production in the region, but also to the assessment of seismic hazard (Yeats et al., 1997).

Two developing petroleum provinces also illustrate the importance of discriminating between fault-propagation and detachment folds and of recognizing thrust truncation. Multiple model-based interpretations have already played a big role in exploration in Papua-New Guinea (Hill et al., 1996), where surface and seismic-reflection data are both difficult to obtain. Structures developed above basement in this area have been interpreted to have resulted from breakthrough of either fault-propagation or detachment folds (Hill et al., 1996; Buchanan et al., 1999; Keetley et al., 1999). Detachment folds and fault-propagation folds, commonly truncated by thrust faults, have been interpreted in various areas and stratigraphic levels of the fold-and-thrust belt along the eastern side of the Andes (Baby et al., 1992; Tankard et al., 1995; Zapata and Allmendinger, 1996; Colletta et al., 1999; Rojas et al., 1999).

These represent just a few examples of areas where multiple interpretations of fault-propagation folding, detachment folding, and thrust truncation have been proposed or can be envisioned based on the available information. The applicability of multiple interpretations to the folds in these areas probably reflects some combination of data that are sufficiently ambiguous to allow multiple interpretations and also the presence of different types of folds within the same region, depending on local variations in stratigraphic or other controlling factors. Critical assessment of interpretations and work aimed at defining discriminating characteristics and controlling factors will be needed to develop a clearer picture of which fold types actually occur where.

SUMMARY AND CONCLUSIONS

Distinguishing between fault-propagation and detachment folds may be difficult, particularly if the geometry and kinematics of the critical core area of the fold are not known. The essential characteristic of a fault-propagation fold is origin by propagation of a ramp tip, but this is difficult to demonstrate in practice. The presence of a ramp and a ramp tip and an upward decrease in displacement are consistent with, but not diagnostic of, a fault-propagation fold (Figure 2). Absence of a ramp and, instead, nonparallel fold geometry above a décollement in at least the lower part of the fold are diagnostic of a detachment fold (Figure 3).

A fault-propagation fold is more likely to form with a nonrotating backlimb and migrating forward and hindward synclinal hinges. A detachment fold is more likely to form with rotating limbs and at least an anticlinal hinge that is fixed. However, both types of folds may form according to more than one model of hinge and limb kinematics. These characteristics should only be used as general guides, until more is learned about the kinematics of natural fault-propagation and detachment folds.

Mechanical stratigraphy and regional structural style may be useful practical supplements to geometry and kinematics in assessing the origin of a natural fold. Incompetent units, especially where overlain by a competent unit, are prone to detachment folding. Fault-propagation folds may be favored by a homogeneous but strongly anisotropic mechanical stratigraphy, particularly if underlain by an existing ramp. Although the origin of a particular fold may be unclear, less-ambiguous examples elsewhere in the region may provide a useful guide to the most likely interpretation.

Breakthrough of existing folds probably is very common in nature and should be recognized as a modification of one of the basic types of thrust-related folds, such as fault-propagation or detachment folds. A misinterpretation may result if one of these models is assumed where breakthrough is not recognized because the geometry of the fold core is not known. Forelimb breakthrough in which the lower and upper flats are coplanar can arise in a variety of ways and may result in little modification of fold geometry other than truncation and displacement of the forelimb (Figures 6a, 7). Noncoplanar breakthrough to an upper flat, on the other hand, can result in significant modification of fold geometry (Figures 6b, 7). Displacement on such a breakthrough may form what appears to be a fault-bend fold with an anomalously steep forelimb (Figures 2c and d, 3c and d). This is the result of superposition of a fault-bend-fold geometry on an existing fold, which may include steepening of the forelimb by rotation, as well as a variety of other possible strain patterns.

Thrust truncation renders the distinction between origin as a fault-propagation or detachment fold even more difficult than in unbroken folds, although the same criteria apply. In particular, introduction of a ramp where none existed previously may make a thrust-truncated detachment fold even more likely to be misinterpreted as a fault-propagation fold. Restoration of the geometry and determination of the kinematics prior to breakthrough is the most direct approach to determining fold origin, but may not be practical. Unbroken fault-propagation or detachment folds elsewhere in the region are a useful guide to the most likely interpretation. Partial remnants, such as ramp tips or incompetent thickening in fold cores, may be helpful clues. Truncation of a detachment fold is more likely to yield a footwall syncline than is truncation of a fault-propagation fold. Mechanical stratigraphy remains a potentially useful indicator of fold origin, and flats of thrust faults may help identify incompetent intervals above which detachment folds may have formed.

Much of what is stated here should be obvious. However, many of these points frequently seem to be overlooked when natural folds are interpreted according to thrust-related fold type and geometric-kinematic models. In particular, folds are commonly interpreted to be fault-propagation folds on the basis of superficial consistency with a model, when other interpretations may account for what is known about the fold as well or better. Definitive distinction between fault-propagation and detachment folds clearly can be difficult, and breakthrough of folds renders discrimination even more difficult. It is not sufficient simply to assume a particular origin of a fold on the basis of ambiguous geometric criteria. It is critically important that the distinctions between fold types be recognized, that the full range of reasonable possibilities be considered, including detachment fold, fault-propagation fold, breakthrough, or whatever other models might be applicable, and that evidence be explicitly sought and presented to support any interpretation of fold origin.

ACKNOWLEDGMENTS

The Brooks Range field work that inspired this paper was supported over the years by many different sources, including the petroleum industry sponsors of the Tectonics and Sedimentation group at the University of Alaska, the U.S. Department of Energy, the National Science Foundation, the Petroleum Research Fund, the U.S. Geological Survey Trans-Alaska Crustal Transect (TACT) project, and the Alaska Division of Geological and Geophysical Surveys. We're grateful to Scott Broadwell for scanning the photographs and to Paul Atkinson for drafting the interpretations on the photographs. We thank Rick Groshong and Geoff Rait for their helpful reviews, as well as an anonymous reviewer whose comments led us to clarify the objectives of the paper. Ken McClay deserves special thanks for organizing the Thrust Tectonics 99 conference and editing this volume.

REFERENCES CITED

Alonso, J. L., and A. Teixell, 1992, Forelimb deformation in some natural examples of fault-propagation folds, *in* K. R. McClay, ed., Thrust tectonics: London, Chapman & Hall, p. 175–180.

Anastasio, D. J., D. M. Fisher, T. A. Messina, and J. E. Holl, 1997, Kinematics of décollement folding in the Lost River Range, Idaho: Journal of Structural Geology, v. 19, p. 355–368.

Baby, P., G. Hérail, R. Salinas, and T. Sempere, 1992, Geometry and kinematic evolution of passive roof duplexes deduced from cross section balancing: Example from the foreland thrust system of the southern Bolivian subandean zone: Tectonics, v. 11, p. 523–536.

Boyer, S. E., 1986, Styles of folding within thrust sheets: examples from the Appalachian and Rocky Mountains of the USA and Canada: Journal of Structural Geology, v. 8, p. 325–339.

Buchanan, P., J. Cole, and M. Parrish, 1999, Structural models applied in the interpretation of the Papuan fold and thrust belt, PNG: Implications for petroleum prospectivity, *in* K. R. McClay, ed., Thrust Tectonics 99 Programme: Royal Holloway University of London, April 26–29, 1999, p. 74.

Butler, R. W. H., 1992, Structural evolution of the western Chartreuse fold and thrust system, NW French Subalpine chains, *in* K. R. McClay, ed., Thrust tectonics: London, Chapman & Hall, p. 287–298.

Chapman, T. J., and G. D. Williams, 1984, Displacement-distance methods in the analysis of fold-thrust structures and linked-fault systems: Journal of the Geological Society of London, v. 141, p. 121–128.

Chester, J., and F. Chester, 1990, Fault propagation folds above thrusts with constant dip: Journal of Structural Geology, v. 12, p. 903–910.

Chester, J., J. M. Logan, and J. H. Spang, 1991, Influence of layering and boundary conditions on fault–bend and fault–propagation folding: Geological Society of America Bulletin, v. 103, p. 1059–1072.

Colletta, B., J. Letouzey, J. Soares, and M. Specht, 1999, Detachment versus fault propagation folding: Insights from the sub-Andean Ranges of southern Bolivia, *in* K. R. McClay, ed., Thrust Tectonics 99 Programme, Royal Holloway University of London, April 26–29, 1999, p. 106–107.

Dahlstrom, C. D. A., 1990, Geometric constraints derived from the law of conservation of volume and applied to evolutionary models for detachment folding: AAPG Bulletin, v. 74, p. 336–344.

De Sitter, L. U., 1956, Structural geology: New York, McGraw-Hill, 552 p.

Ellis, M. A., and W. J. Dunlap, 1988, Displacement variation along thrust faults: implications for the development of large faults: Journal of Structural Geology, v. 10, p. 183–192.

Epard, J.-L., and R. H. Groshong Jr., 1995, Kinematic model of detachment folding including limb rotation, fixed hinges and layer–parallel strain: Tectonophysics, v. 247, p. 85–103.

Erslev, E. A., 1991, Trishear fault-propagation folding: Geology, v. 19, p. 617–620

Erslev, E. A., and K. R. Mayborn, 1997, Multiple geometries and modes of fault–propagation folding in the Canadian thrust belt: Journal of Structural Geology, v. 19, p. 321–335.

Fischer, M. P., N. B. Woodward, and M. M. Mitchell, 1992, The kinematics of break–thrust folds: Journal of Structural Geology, v. 14, p. 451–460.

Fischer, M. W., and M. P. Coward, 1982, Strains and folds within thrust sheets: An analysis of the Heilam sheet, northwest Scotland: Tectonophysics, v. 88, p. 291–312.

Hedlund, C. A., 1997, Fault–propagation, ductile strain, and displacement–distance relationships: Journal of Structural Geology, v. 19, p. 249–256.

Hill, K. C., R. J. Simpson, R. D. Kendrick, P. V. Crowhurst, P. B. O'Sullivan, and I. Saefudin, 1996, Hydrocarbons in New Guinea, controlled by basement fabric, Mesozoic extension and Tertiary convergent margin tectonics, *in* P. G. Buchanan, ed., Petroleum exploration, development and production in Papua New Guinea: Proceedings of the third PNG Petroleum Convention, Port Moresby, September, 1996, p. 63–76.

Homza, T. X., 1992. A detachment fold–truncation duplex southwest of Bathtub Ridge–northeastern Brooks Range, Alaska. Unpublished M.S. thesis, University of Alaska, Fairbanks.

Homza, T. X., and W. K. Wallace, 1995, Geometric and kinematic models for detachment folds with fixed and variable detachment depths: Journal of Structural Geology, v. 17, p. 475–588.

Homza, T. X., and W. K. Wallace, 1997, Detachment folds with fixed hinges and variable detachment depth, northeastern Brooks Range, Alaska: Journal of Structural Geology, v. 19, p. 337–354.

Hudleston, P. J., and L. Lan, 1993, Information from fold shapes: Journal of Structural Geology, v. 15, p. 253–264.

Huftile, G. J., 1991, Thin–skinned tectonics of the upper Ojai Valley and Sulphur Mountain area, Ventura basin, California: AAPG Bulletin, v. 75, p. 1353–1373.

Huftile, G. J., and R. S. Yeats, 1995, Convergence rates across a displacement transfer zone in the western Transverse Ranges, Ventura basin, California: Journal of Geophysical Research, v. 100, no. B2, p. 2043–2067.

Jamison, W. R., 1987, Geometric analysis of fold development in overthrust terranes: Journal of Structural Geology, v. 9, p. 207–219.

Keetley, J., K. Hill, C. Nguyen, and T. Murray, 1999, 3d structural modelling and restoration of detachment folds and forelimb thrusts from Cape Liptrap, Australia, and the PNG fold belt, in K. R. McClay, editor, Thrust Tectonics 99 Programme, Royal Holloway University of London, April 26–29, 1999, p. 233–236.

Kelley, J. S., and W. P. Brosgé, 1995, Geologic framework of a transect of the central Brooks Range: Regional relations and an alternative to the Endicott Mountains allochthon: AAPG Bulletin, v. 79, p. 1087–1116.

Kulander, B. R., and S. L. Dean, 1986, Structure and tectonics of central and southern Appalachian Valley and Ridge and Plateau provinces, West Virginia and Virginia: AAPG Bulletin, v. 70, p. 1674–1684.

Lamerson, P. R., 1982, The Fossil basin and its relationship to the Absaroka thrust system, Wyoming and Utah, *in* R. B. Powers, ed., Geologic studies of the Cordilleran thrust belt: Rocky Mountain Association of Geologists, p. 279–340.

Liu, S., and J. M. Dixon, 1995, Localization of duplex thrust–ramps by buckling: Analog and numerical modelling: Journal of Structural Geology, v. 17, p. 875–886.

McConnell, D. A., 1994, Fixed–hinge, basement–involved fault–propagation folds, Wyoming: Geological Society of America Bulletin, v. 106, p. 1583–1593.

McConnell, D. A., S. A. Kattenhorn, and L. M. Benner, 1997, Distribution of fault slip in outcrop–scale fault–related folds, Appalachian Mountains: Journal of Structural Geology, v. 19, p. 257–267.

McMechan, M. E., and R. I. Thompson, 1989, Structural style and history of the Rocky Mountain fold and thrust belt, *in* B. D. Ricketts, ed., Western Canada sedimentary basin: Canadian Society of Petroleum Geologists, p. 47–72.

McNaught, M. A., and G. Mitra, 1993, A kinematic model for the origin of footwall synclines: Journal of Structural Geology, v. 15, p. 805–808.

Mercier, E., F. Outtani, and D. F. De Lamotte, 1997, Late-stage evolution of fault-propagation folds: principles and example: Journal of Structural Geology, v. 19, p. 185–193.

Mitchell, M. M., and N. B. Woodward, 1988, Kink detachment fold in the southwest Montana fold and thrust belt: Geology, v. 16, p. 162–165.

Mitra, S., 1990, Fault–propagation folds: geometry, kinematic evolution, and hydrocarbon traps: AAPG Bulletin, v. 74, p. 921–945.

Mitra, S., and J. S. Namson, 1989, Equal–area balancing: American Journal of Science, v. 289, p. 563–599.

Moore, T. E., W. K. Wallace, K. J. Bird, S. M. Karl, C. G. Mull, and J. T. Dillon, 1994, Geology of northern Alaska, Chapter 3, *in* G. Plafker and H. C. Berg, eds., The geology of Alaska: The Geology of North America: Geological Society of America, v. G1, p. 49–140.

Morley, C. K., 1994, Fold-generated imbricates: examples from the Caledonides of southern Norway: Journal of Structural Geology, v. 16, p. 619–631.

Mull, C. G., D. H. Roeder, I. L. Tailleur, G. H. Pessel, A. Grantz, and S. D. May, 1987, Geologic sections and maps across Brooks Range and Arctic Slope to Beaufort Sea: Geological Society of America Map and Chart Series MC–28S, scale 1:500,000.

Oldow, J. S., C. M. Seidensticker, J. C. Phelps, F. E. Julian,

R. R. Gottschalk, K. W. Boler, J. W. Handschy, and H. G. Ave Lallemant, 1987, Balanced cross sections through the central Brooks Range and North Slope, Arctic Alaska: Tulsa, Oklahoma, AAPG, 19 p., scale 1:200,000.

Pfiffner, O. A., 1993, The structure of the Helvetic nappes and its relation to the mechanical stratigraphy: Journal of Structural Geology, v. 15, nos. 3–5, p. 511–521.

Poblet, J., and S. Hardy, 1995, Reverse modelling of detachment folds; application to the Pico del Aguila anticline in the South Central Pyrenees (Spain): Journal of Structural Geology, v. 17, p. 1707–1724.

Poblet, J., and K. McClay, 1996, Geometry and kinematics of single–layer detachment folds: AAPG Bulletin, v. 80, p. 1085–1109.

Poblet, J., K. McClay, F. Storti, and J. A. Munoz, 1997, Geometries of syntectonic sediments associated with single–layer detachment folds: Journal of Structural Geology, v. 19, p. 369–381.

Price, H. J., and J. W. Cosgrove, 1990, Analysis of geological structures: Cambridge. Cambridge University Press, 502 p.

Rait, G., and J. M. Dixon, 1999, Fold/fault kinematics in the foothills and front ranges, Canadian Rocky Mountains, *in* K. R. McClay, ed., Thrust Tectonics 99 Programme, Royal Holloway University of London, April 26–29, 1999, p. 300–301.

Ramsay, J. G., 1974, Development of chevron folds: Geological Society of America Bulletin, v. 85, p. 1741–1754.

Ramsay, J. G., and M. I. Huber, 1987, The techniques of modern structural geology, volume 2: folds and fractures: New York, Academic Press, 392 p.

Rojas, L., N. Muñoz, J. P. Radic, and K. McClay, 1999, The stratigraphical controls in the transference of displacement from basement thrusts to the sedimentary cover in the Malargue fold–thrust belt, Neuquen basin, Argentina: The Puesto Rojas oilfield example, *in* K. R. McClay, ed., Thrust Tectonics 99 Programme, Royal Holloway University of London, April 26–29, 1999, p. 119–121.

Rowan, M. G., 1997, Three–dimensional geometry and evolution of a segmented detachment fold, Mississippi fan foldbelt, Gulf of Mexico, Journal of Structural Geology, v. 19, p. 463–480.

Rowan, M. G., and R. Kligfield, 1992, Kinematics of large-scale asymmetric buckle folds in overthrust shear: an example from the Helvetic nappes, *in* K. R. McClay, ed., Thrust tectonics: London, Chapman & Hall, p. 165–174.

Royse, F. Jr., 1993, An overview of the geologic structure of the thrust belt in Wyoming, northern Utah, and eastern Idaho, *in* A. W. Snoke, J. R. Steidtmann, and S. M. Roberts, eds., Geology of Wyoming: Geological Survey of Wyoming Memoir no. 5, p. 272–311.

Royse, F. Jr., 1996, Detachment fold train, Reed Wash area, west flank San Rafael Swell, Utah: An example of a limb-lengthening, roll-through folding process on the eastern margin of the Sevier thrust belt: The Mountain Geologist, v. 33, p. 45–64.

Spang, J. H., and D. A. McConnell, 1997, Effect of initial fault geometry on the development of fixed-hinge, fault-propagation folds: Journal of Structural Geology, v. 19, p. 1537–1541.

Spratt, D. A., and J. M. Dixon, 1999, Changes in structural style controlled by lithofacies contrast across oblique carbonate bank margins: Canadian Rocky Mountains and scaled physical models, *in* K. R. McClay, ed., Thrust Tectonics 99 Programme, Royal Holloway University of London, April 26–29, 1999, p. 102–103.

Storti, F., and J. Poblet, 1997, Growth stratal architectures associated to décollement folds and fault–propagation folds. Inferences on fold kinematics: Tectonophysics, v. 282, p. 353–373.

Storti, F., F. Salvini, and K. McClay, 1997, Fault-related folding in sandbox analogue models of thrust wedges: Journal of Structural Geology, v. 19, p. 583–602.

Suppe, J., 1983, Geometry and kinematics of fault–bend folding: American Journal of Science, v. 283, p. 684–721.

Suppe, J., and D. A. Medwedeff, 1990, Geometry and kinematics of fault-propagation folding: Eclogae Geologicae Helvetiae, v. 83, p. 409–454.

Tankard, A. J., R. Suarez Soruco, and H. J. Welsink, eds., 1995, Petroleum basins of South America: AAPG Memoir 62, 792 p.

Tavarnelli, E., 1997, Structural evolution of a foreland fold-and-thrust belt: the Umbria–Marche Appennines, Italy: Journal of Structural Geology, v. 19, p. 523–534.

Thompson, R. I., 1981, The nature and significance of large "blind thrusts" within the northern Rocky Mountains of Canada, *in* K. R. McClay and N. J. Price, eds., Thrust and nappe tectonics: Geological Society of London Special Publication 9, p. 449–462.

Thorbjornsen, K. L., and W. M. Dunne, 1997, Origin of a thrust–related fold: Geometric vs. kinematic tests: Journal of Structural Geology, v. 19, p. 303–319.

Tippet, C. R., P. B. Jones, and F. R. Frey, 1985, Strains developed in the hanging wall of thrusts due to their slip/propagation rate: a dislocation model: discussion: Journal of Structural Geology, v. 7, p. 755–758.

Wallace, W. K., 1993, Detachment folds and a passive-roof duplex: Examples from the northeastern Brooks Range, Alaska, *in* D. N. Solie and F. Tannian, eds., Short notes on Alaskan geology 1993: Alaska Division of Geological and Geophysical Surveys Geologic Report 113, p. 81–99.

Wallace, W. K., and C. L. Hanks, 1990, Structural provinces of the northeastern Brooks Range, Arctic National Wildlife Refuge, Alaska: AAPG Bulletin, v. 74, p. 1100–1118.

Wallace, W. K., and T. X. Homza, 1998, Detachment folds with fixed hinges and variable detachment depth, northeastern Brooks Range, Alaska: Reply: Journal of Structural Geology, v. 20, p. 1591–1595.

Wallace, W. K., T. E. Moore, and G. Plafker, 1997, Multistory duplexes with forward dipping roofs, north central Brooks Range, Alaska: Journal of Geophysical Research, v. 102, p. 20, 773–20, 796.

West, J., and H. Lewis, 1982, Structure and palinspastic

reconstruction of the Absaroka thrust, Anschutz Ranch area, Utah and Wyoming, *in* R. B. Powers, ed., Geologic studies of the Cordilleran thrust belt: Rocky Mountain Association of Geologists, p. 633–639.

Wickham, J., 1995, Fault displacement-gradient folds and the structure at Lost Hills, California (U.S.A.): Journal of Structural Geology, v. 17, p. 1293–1302.

Wiltschko, D. V., and W. M. Chapple, 1977, Flow of weak rocks in the Appalachian Plateau folds: AAPG Bulletin, v. 61, p. 653–670.

Williams, G. D., and T. J. Chapman, 1983, Strains developed in the hanging wall of thrusts due to their slip/propagation rate; a dislocation model: Journal of Structural Geology, v. 5, p. 563–571.

Willis, B., 1893, The mechanics of Appalachian structure: U.S. Geological Survey, Annual Report 13 (1891–1892), Part 2, p. 211–281.

Woodward, N. B., 1992, Deformation styles and geometric evolution of some Idaho-Wyoming thrust belt structures, *in* S. Mitra and G. W. Fisher, eds., Structural geology of fold and thrust belts: Baltimore, The Johns Hopkins University Press, p. 191–206.

Woodward, N. B., 1997, Low-amplitude evolution of break-thrust folding: Journal of Structural Geology, v. 19, p. 293–301.

Yeats, R. S., 1983, Large-scale Quaternary detachments in Ventura basin, southern California: Journal of Geophysical Research, v. 88, p. 569–583.

Yeats, R. S., G. J. Huftile, and F. B. Grigsby, 1988, Oak Ridge fault, Ventura fold belt, and the Sisar décollement, Ventura basin, California: Geology, v. 16, p. 1112–1116.

Yeats, R. S., K. Sieh, and C. R. Allen, 1997, The geology of earthquakes: New York, Oxford University Press, 568 p.

Zapata, T. R., and R. W. Allmendinger, 1996, Thrust-front zone of the Precordillera, Argentina: A thick-skinned triangle zone: AAPG Bulletin, v. 80, p. 359–381.

18

Allmendinger, R. W., T. Zapata, R. Manceda, and F. Dzelalija, 2004, Trishear kinematic modeling of structures, with examples from the Neuquén Basin, Argentina, *in* K. R. McClay, ed., Thrust tectonics and hydrocarbon systems: AAPG Memoir 82, p. 356–371.

Trishear Kinematic Modeling of Structures, with Examples from the Neuquén Basin, Argentina

Richard W. Allmendinger
Department of Earth and Atmospheric Sciences, Cornell University, Ithaca, New York, U.S.A.

Tomás Zapata[1]
YPF S.A., Departamento Exploración Regional Oeste, Neuquén, Argentina

René Manceda[2]
Maxus Energy, Dallas, Texas, U.S.A.

Francisco Dzelalija[1]
YPF S.A., Dto. Exploración Regional Oeste, Neuquén, Argentina

ABSTRACT

We have expanded previous trishear fault-propagation-fold forward models by allowing additional structural complexity in the form of multiple ramps and flats, variable propagation-to-slip ratio (P/S) and trishear angle, as well as multiple faults in a single section. The resulting forward models simulate characteristics of real structures very well. Our six-parameter grid search overcomes a long-standing obstacle to the application of trishear by providing a scientifically objective way of choosing the correct parameters to apply to real structures. By grid searching forward models with known initial parameters, we investigate the sensitivity of trishear to various parameters. These experiments highlight the importance of P/S in determining fold shape: A change in P/S of just 0.3 produces a change in fold shape that is equivalent to that produced by a 15–20° change in trishear angle. Application of the trishear model to contractional and extensional structures of the Neuquén Basin highlights its utility for predicting (1) strain and strain path, (2) fracture orientation and distribution, (3) fault-slip magnitude, and (4) fault nucleation point/décollement depth. Our study of large basement-cored producing anticlines such as Filo Morado–Pampa Tril emphasizes an important point: Trishear and parallel kink-fold geometries can be compatible when applied at different scales. Trishear provides a bulk description of the

[1]*Present address*: Repsol-YPF, Buenos Aires, Argentina.
[2]*Present address*: APEX Petroleum Inc., Englewood, Colorado, U.S.A.

deforming zone on the thickened, triangular eastern flank of the fold but makes no explicit prediction about how the strain is accommodated. In these structures, the strain is variably accommodated by tight folding, duplexing, and flow of evaporites and depends significantly on the thickness of the "competent" units in the section.

INTRODUCTION

Fault-propagation folding, wherein rocks are folded in response to the diminishing slip at, and propagation of, a fault tip line, has been recognized by structural geologists and petroleum explorationists for at least 50 years. Such structures were formally described in the Foothills of the Canadian Rocky Mountains, where they constitute an important member of a suite of hydrocarbon-producing structures (Gallup, 1951). More recently, in the wake of the 1993 Northridge earthquake in southern California, earthquake seismologists have come to appreciate the significance of blind thrusts and fault-propagation folds. Fault-propagation folds are inherently more complicated kinematically than are fault-bend folds, because a mechanism must be found that accommodates the rapidly diminishing slip on the fault plane. If parallel-fold, kink-band migration kinematics are assumed, the range of possible geometries associated with a given fault is quite restricted (Suppe and Medwedeff, 1990). Yet, several features commonly associated with field examples of fault-propagation folds — footwall synclines, thickness and dip variations on forelimbs, and the like — indicate that, in detail, the parallel-kink-fold model is, in some cases, too restrictive.

An alternative model of fault-propagation folding, called "trishear," has been proposed by Erslev (1991; Erslev and Rogers, 1993). In this model, the diminishing slip is accommodated by deformation of the anticlinal forelimb in a triangular zone of distributed shear. Trishear can produce many of the features that are difficult to explain with parallel kink folding (Allmendinger, 1998a; Hardy and Ford, 1997; Erslev, 1991). In addition, trishear allows great flexibility in the choice of a key parameter, the propagation-to-slip ratio (P/S). In parallel-kink fault-propagation folding, P/S is completely fixed by the geometry; for a given fault ramp angle, there is only one possible P/S. In trishear, one can specify any P/S for any fault geometry. Furthermore, strain in parallel-kink folding is entirely homogeneous (within a dip domain), whereas trishear has much greater potential for predicting heterogeneous strain distribution. This means that, where strain is manifest as fracturing, trishear may potentially be more useful in predicting the nature and location of fractured reservoirs.

This paper follows up on our earlier work on trishear deformation (Allmendinger, 1998a) and builds on the insight of Erslev (Erslev, 1991; Erslev and Rogers, 1993) and Hardy and Ford (1997). All analyses are carried out using the computer program "TrishearPPC" (Allmendinger, 1998b). We provide tests of our inverse-grid search model and demonstrate application of various aspects of trishear analysis, using examples from the Neuquén Basin of Argentina. Of all the trishear parameters, we demonstrate that the most important is P/S. This parameter not only exerts the primary control on fold shape but can tell us where a fault nucleated at depth in the crust. This nucleation point has significant implications for depth to dêcollement, fault reactivation, and perhaps to earthquake seismicity. Furthermore, we show that trishear and parallel-fold models are not necessarily incompatible when applied at different scales.

REVIEW OF TRISHEAR KINEMATICS

Trishear is strictly a kinematic model that, like other kinematic models such as kink fault-propagation folding, lacks a complete mechanical foundation. Trishearlike structures were originally called "forced folds" (Stearns, 1978), and several experimental studies using rock or clay as analog materials have been conducted (Withjack et al., 1990; Friedman et al., 1980). Mechanical analyses of forced folds have been carried out by Reches and Johnson (1978), Rodgers and Rizer (1981), Patton and Fletcher (1995), and Johnson and Johnson (2002), among others.

The specifics of trishear kinematics can be found in several key publications (Zehnder and Allmendinger, 2000; Allmendinger, 1998a; Hardy and Ford, 1997; Erslev and Rogers, 1993; Erslev, 1991). In this paper, we use the velocity description of trishear as described by Hardy and Ford (1997) and clarified by Zehnder and Allmendinger (2000). All the material points in the hanging wall of the fault move with constant velocity, and all points in the footwall are fixed. To conserve area, the triangular shear zone at the tip of the fault is symmetric with respect to the fault (Erslev and Rogers, 1993; Erslev, 1991). Within this zone, all points on a ray that goes through the tip line experience the same velocity (Figure 1). Along a tie line across the trishear zone perpendicular to the fault projection, the velocity diminishes linearly to zero, with an angular variation equal to half that of the trishear zone itself (Hardy and Ford, 1997). Zehnder and Allmendinger (2000) showed that, in addition to the simple linear-velocity field, an unlimited number of other velocity fields also satisfy the trishear boundary conditions, including center-concentrated trishear, which may be a continuous formulation of Erslev's (1991) "heterogeneous" trishear. Furthermore, they showed that both symmetric and asymmetric trishear are possible.

In Erslev's (1991) original description, the trishear zone was attached either to the hanging wall or to the footwall. Hardy and Ford (1997) showed that these were two special cases in a spectrum of possible ratios of propagation to slip (P/S). Footwall-fixed trishear zones have P/S = 0, and hanging wall fixed zones have P/S = 1. However, the range of possible P/S is virtually infinite. Although the concept of P/S has been in the literature since the work of Elliott (1976) and Williams and Chapman (1983), it remains little appreciated by the structural geology community. We explore the importance of this parameter for trishear, below.

Six parameters control the geometry of a linear-velocity field trishear fold: fault ramp angle, the fault slip, the *X* and *Y* positions of the tip line, the trishear apical angle, and P/S (Figures 1 and 2); a seventh parameter describes the degree of center concentration. Although several of these parameters are relatively easy to determine if the final geometry of the structure is well known, the physical characterization of the trishear apical angle and the P/S ratio have proved elusive. The difficulty of defining these two parameters has been one of the greatest barriers to widespread acceptance of trishear. Because simple trishear deformation is reversible, Allmendinger (1998a) proposed an inverse method to extract all six parameters from the final geometry of the fold-fault couple. A brute-force grid search is applied, systematically varying some, or all, of the six parameters. The goodness of fit of each unique combination of parameters is evaluated by simple least squares linear regression, using χ^2 as a statistic. The combination that minimizes χ^2 is that which most nearly returns a key bed within the structure to a linear (but not necessarily horizontal) initial geometry (in 2-D).

INVERSE MODELING OF FORWARD MODELS

With any grid search procedure, numerous questions arise. Is the solution unique? Are there local minima in the solution space that could be mistaken for the global minimum? How sensitive is the solution to different parameters? The answers to these questions not only clarify the limitations on the inverse method but also illuminate the nature of trishear kinematics itself. Here, we highlight an initial test of uniqueness and sensitivity of the inverse method. The primary vehicle for doing so is to carry out the inversion on forward models generated with precisely known trishear parameters. The relative importances of trishear angle and propagation-to-slip ratio are emphasized, precisely because they are so poorly understood.

First, we produce a simple trishear forward model allowing no variations, during faulting, of any of the parameters (Figure 2). This model is then read back into our computer program as a deformed section. The ramp angle and tip line position are known, so we search over a reasonably range of P/S, trishear angle, and displacement. Table 1 lists the input parameters and the grid search matrix; in all, more than 185,000 unique combinations were tested. This yields a three-dimensional matrix of χ^2 values,

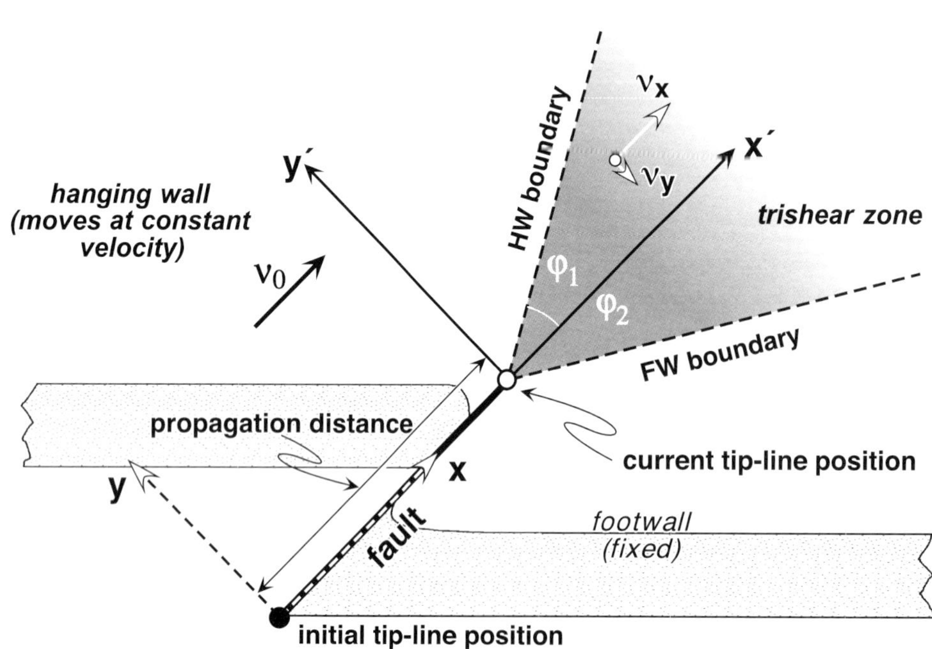

FIGURE 1. The velocity-based description of trishear kinematics first presented by Hardy and Ford (1997) and used in this paper (also Allmendinger, 1998a; Allmendinger, 1998b).

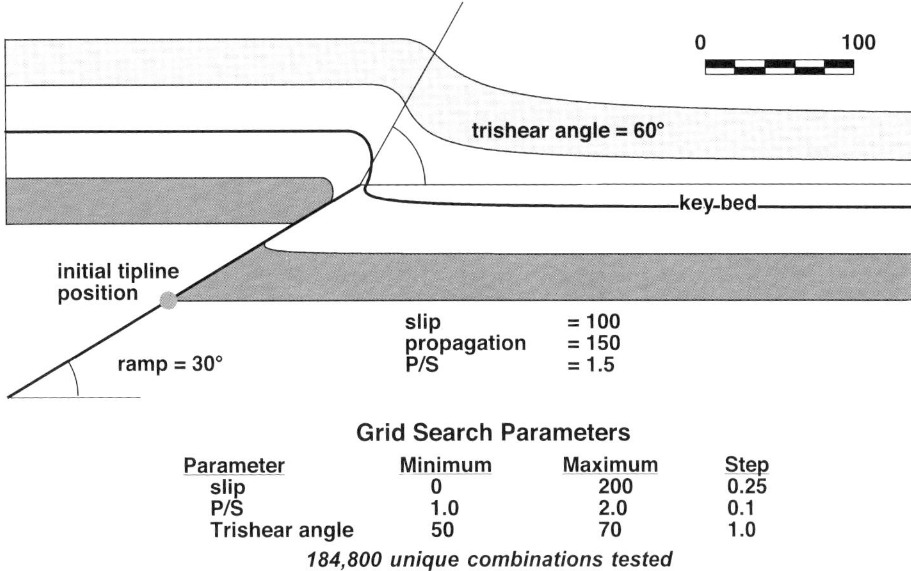

Grid Search Parameters

Parameter	Minimum	Maximum	Step
slip	0	200	0.25
P/S	1.0	2.0	0.1
Trishear angle	50	70	1.0

184,800 unique combinations tested

FIGURE 2. Example of a simple trishear forward model, showing the six key parameters that define the geometry of any trishear fold: (1) slip, (2) ramp angle, (3 and 4) *X* and *Y* coordinates of the tip line, (5) P/S, and (6) trishear angle. Because no changes in fault dip are permitted, the fold produced is a simple monocline. This section is also the starting forward model that was fed into our grid-search inverse procedure to test of the accuracy of that technique and the sensitivity of the various trishear parameters. The table below the section shows the grid of values searched.

with axes of displacement, trishear angle, and P/S (Figure 3).

Test of Grid Search

The most basic objective of this exercise is to see if the best model parameters identified by the inverse procedure are the same as the input parameters to the forward model. In general, the inverse procedure works well— commonly, the parameters found are the same as, or only slightly different from, those of the starting forward model. But the inversion also highlights a subtle but important point: Trishear is a continuous deformation, but the computer models it as a series of discrete displacement steps. This slip increment can be varied, producing only subtle changes in forward-model fold geometry, even when it is increased by a factor of ten (Figure 4). The inversion, however, is very sensitive to this parameter. With relatively large slip increments (equal to 1 or 2% of the total slip, in our case), the best-fitting inverse model commonly renders one parameter, usually the trishear angle, incorrectly. The error is small; the best-fit trishear angle underestimates the input value by just a degree or two. Matching all the parameters correctly in our experiments commonly required a slip increment equal to between 0.5 and 0.25% of the total slip.

When a small enough slip increment is used, the inversion always finds the correct input parameters for the forward model. In all tests using a simple starting model with horizontal beds, we have never observed local minima.

Sensitivity to Various Parameters

Inverse modeling of forward models also allows us to examine the relative importance of various parameters. As seen in the two-dimensional slices through the matrix of χ^2 values (Figure 3b, c), the solution for the best displacement is very robust. A model can have a value of P/S or a trishear angle that is very far off, and the correct displacement can still be determined. This results because the majority of points on the digitized bed lie outside of the folded region, whereas P/S and trishear angle affect only the folded area. In these simple forward models, determining the displacement requires nothing more than measuring the correct throw.

The plots also show that P/S and trishear angle are inversely correlated (Figure 3a). Comparable folds can be produced with a low P/S and high trishear angle or vice versa. Both low trishear angles and small P/S will produce more deformation on the forelimb. So, if one parameter is reduced, the other has to be increased to maintain equivalent geometries. However, the propagation-to-slip ratio is much more tightly constrained by the inversion than is the trishear angle (Figure 3b, c). A small change in P/S has a more profound change on fold shape than does a small change in trishear angle (although this comparison is admittedly akin to comparing apples to oranges). As shown in Figure 3, a 0.3 variation in P/S is approximately equivalent to a 15–20° change in trishear angle.

This result is physically and intuitively satisfying. It is easy to conceive of the physical and mechanical constraints on P/S. On the other hand, the trishear apical angle's exact nature and its physical controls remain enigmatic. It would be much more worrisome if the method depended more highly on a parameter that we still don't understand!

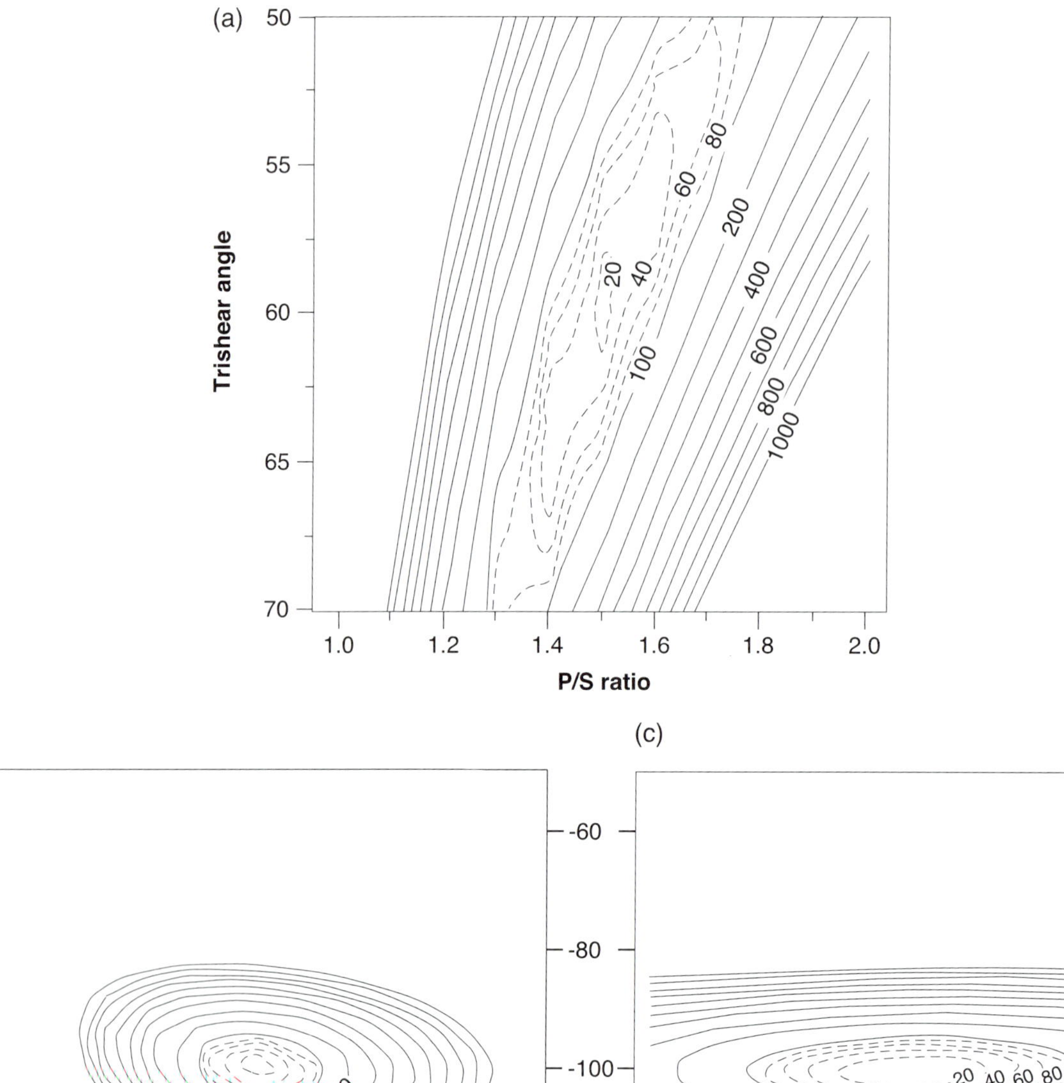

FIGURE 3. Two-dimensional slices through the 3-D matrix of statistical values produced by grid searching the forward model over the range of values shown in Figure 2. Slices are constructed (a) at best displacement (100 units), (b) at best trishear angle (60°), and (c) at best P/S (1.5). In each diagram, contours are in intervals of 100 χ^2 values; dashed contours below $\chi^2 = 100$ are in intervals of 20. The banding in the gray shading benenath the contours reflects the actual size of the grid. The best model has a $\chi^2 = 0.23$. The best-fit parameters found by the grid search are exactly those used to construct the original forward model; no significant local minima are seen. Note the inverse relationship between trishear angle and P/S in (a), the robustness of the solution for displacement in (b) and (c), and the significantly more tightly constrained solution for P/S than for trishear angle, highlighted by comparing (b) and (c).

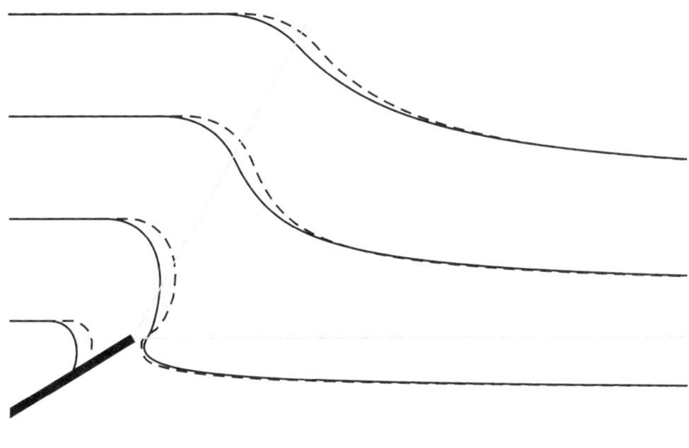

FIGURE 4. Magnified detail of the fold in Figure 2, showing how an tenfold increase in slip increment used to generate the model produces only a subtle change in fold geometry. The solid bedding lines were produced with a slip increment equal to 1% of the total slip; the dashed bedding lines result when a slip increment equal to 10% of the total slip is used. For the grid searching of the forward model (Figures 2 and 3), a slip increment of 0.5% was used.

THE IMPORTANCE OF P/S

As shown in the previous section, P/S exerts the most important control on the amount of strain in a trishear structure: the lower the P/S, the longer a point will reside in the trishear zone. In more general terms, the lower the P/S, the more opportunity the ductile bead will have to deform (Williams and Chapman, 1983). Even if the triangular shear zone description of deformation at a tip line is not accurate, it is likely that P/S will continue to be an important constraint on the geometry and strain in fault-propagation folds.

Different Fold Types in a Continuum of P/S Variation

The notion of P/S is potentially a unifying concept in the analysis of fault-related folds. Structural geologists commonly use discrete and largely unrelated kinematic models to analyze detachment, fault-propagation, and fault-bend folds— the three types of folds found in thrust belts (Suppe, 1983; Dahlstrom, 1990; Suppe and Medwedeff, 1990; Poblet and McClay, 1996). Yet, if we assume that deformation over a ramp occurs as a fault-bend fold, all three types of folds can be produced simply by varying P/S, as shown in Figure 5. That is, these three types of folds are, in reality, points along a continuum of possible P/S values. The fault-bend fold occurs because the tip line propagates rapidly enough that virtually no strain accrues within the trishear zone before the tip line passes by the point. P/S ratios as low as 15 result in practically perfect parallel fault-bend folds (Figure 5c). In contrast, the detachment fold forms because the tip line does not move at all, even as the slip accumulates on the fault. It is equivalent to nailing one end of a carpet to the floor and pushing on the other, producing a bulge in the rug.

Propagation Distance and Fault Nucleation Point

Contained within P/S is another important parameter: the distance that the tip line of the fault has propagated from its initial crack. There are at least three possible interpretations of this fault nucleation point.

First, the fault may nucleate on a décollement horizon, usually the tip of a propagating décollement (Figure 6a) but perhaps also from some flaw along it. The resulting fold has a fault-bend backlimb and a trishear fault-propagation-fold forelimb. By modeling the shape of the fold and determining the appropriate P/S, one can determine the depth to the décollement horizon. In our studies of Neuquén basement anticlines, described below, we use this aspect of trishear restorations to infer where in the crust the master faults must have originated.

Second, the fault may reactivate a preexisting fault (Figure 6b). In this case, the initial tip line position of the new fault should coincide with that of the old fault; for instance, at the base of postrift strata in the case of a reactivated normal fault. With a tip line in this position, the structure should display little if any folding of the interface itself, because that layer would be located completely outside the trishear zone at the start of the second deformation (Figure 7). Reactivation versus formation of a new fault is a commonly debated issue. Trishear inverse modeling has the potential to resolve this question: Fold shape differs significantly, if the fault nucleates at some depth in the crust (a new fracture), from the shape it has if the initial tip line propagates from a position much closer to the strata of interest (compare Figure 6b, c). This approach is not without pitfalls, however. It is possible that a new fault would form at depth and then experience very rapid tip line propagation to the base of the postrift strata but with very little slip (large P/S). Subsequently, the tip might propagate slowly relative to the slip, producing a fold geometry that would appear to result from reactivation.

Finally, it is possible to have propagation of the fault tip lines, both up-section and down-section, from an initial crack or flaw in the crust (Figure 6c; see also Figure 4 of Williams and Chapman, 1983). In a perfectly symmetric case, fault slip would die out toward

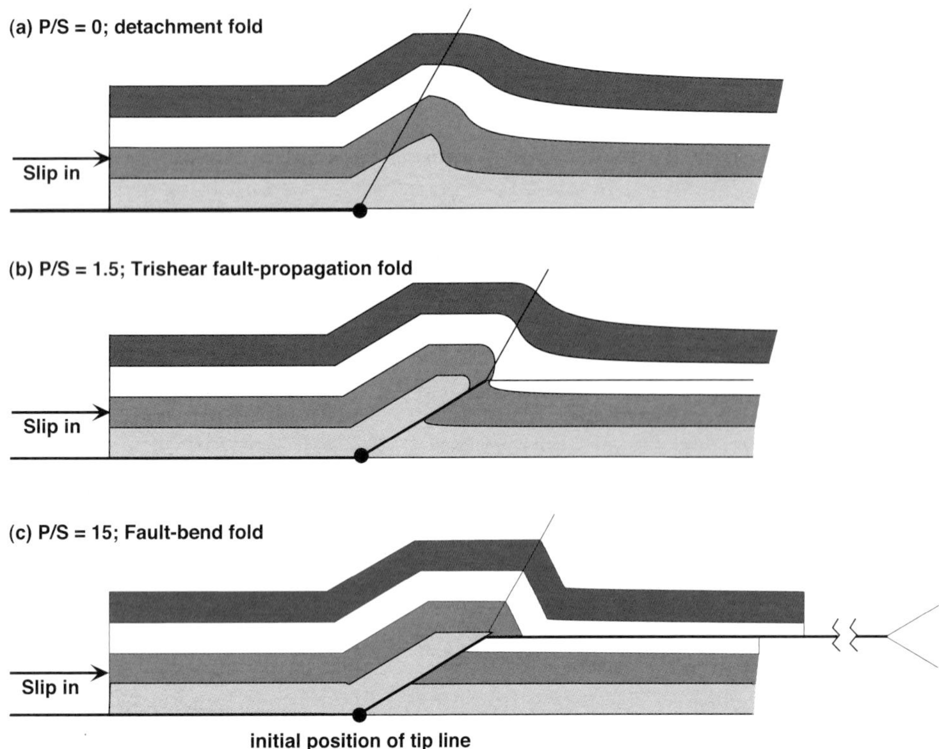

FIGURE 5. Three cross sections illustrating the importance of P/S. All parameters in each of the three sections are identical except for P/S, which is allowed to vary as shown. Thus, simply by varyiing P/S, one can produce (a) detachment folds (P/S ≈ 0), (b) trishear fault-propagation folds at low values of P/S, and (c) virtually perfect fault-bend folds at moderately high values of P/S (> ~10). Negative values of P/S, combined with negative slip, produce normal faults.

both tip lines (Figure 6c). However, in this case, the hanging wall outside of the trishear zones is uniformly uplifted for an infinite distance to the left of the nascent fault (holding the footwall fixed). Although constant velocity displacement of the hanging wall is a characteristic of all kinematic models, it is physically unrealistic, particularly for small displacements where the hanging wall deforms elastically, as in the case of fold growth during a single earthquake. The important point is that a return to regional on the backlimb of a structure does not require the presence of a dêcollement if fault slip dies out downward as well as upward— a point also made by Williams and Chapman (1983).

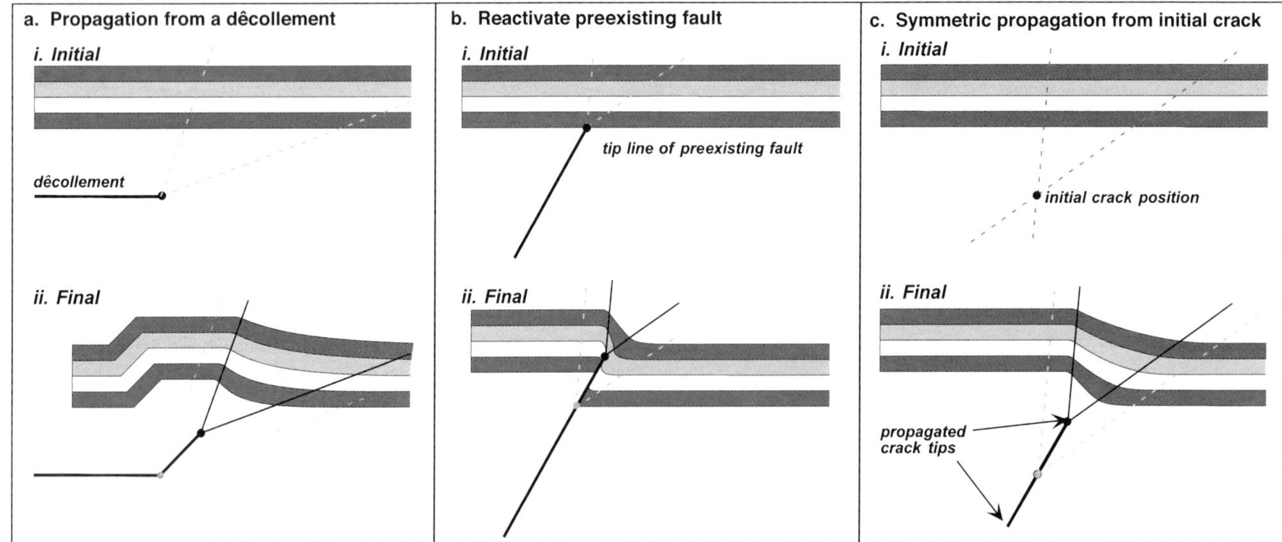

FIGURE 6. Different modes of fault nucleation and propagation. Modeling the fold shape yields slip and propagation, allowing one to determine objectively which mode is responsible. (a) Propagation of the tip line from a dêcollement; (b) propagation from the tip of the old fracture for the case of reactivation; (c) symmetric propagation from an initial flaw within the crust, with displacement diminishing downward as well as upward in mirror-symmetry trishear zones. Case (c) could also occur as asymmetric propagation, particularly where the nucleation point is near the level of maximum strength in the middle crust. See text for discussion.

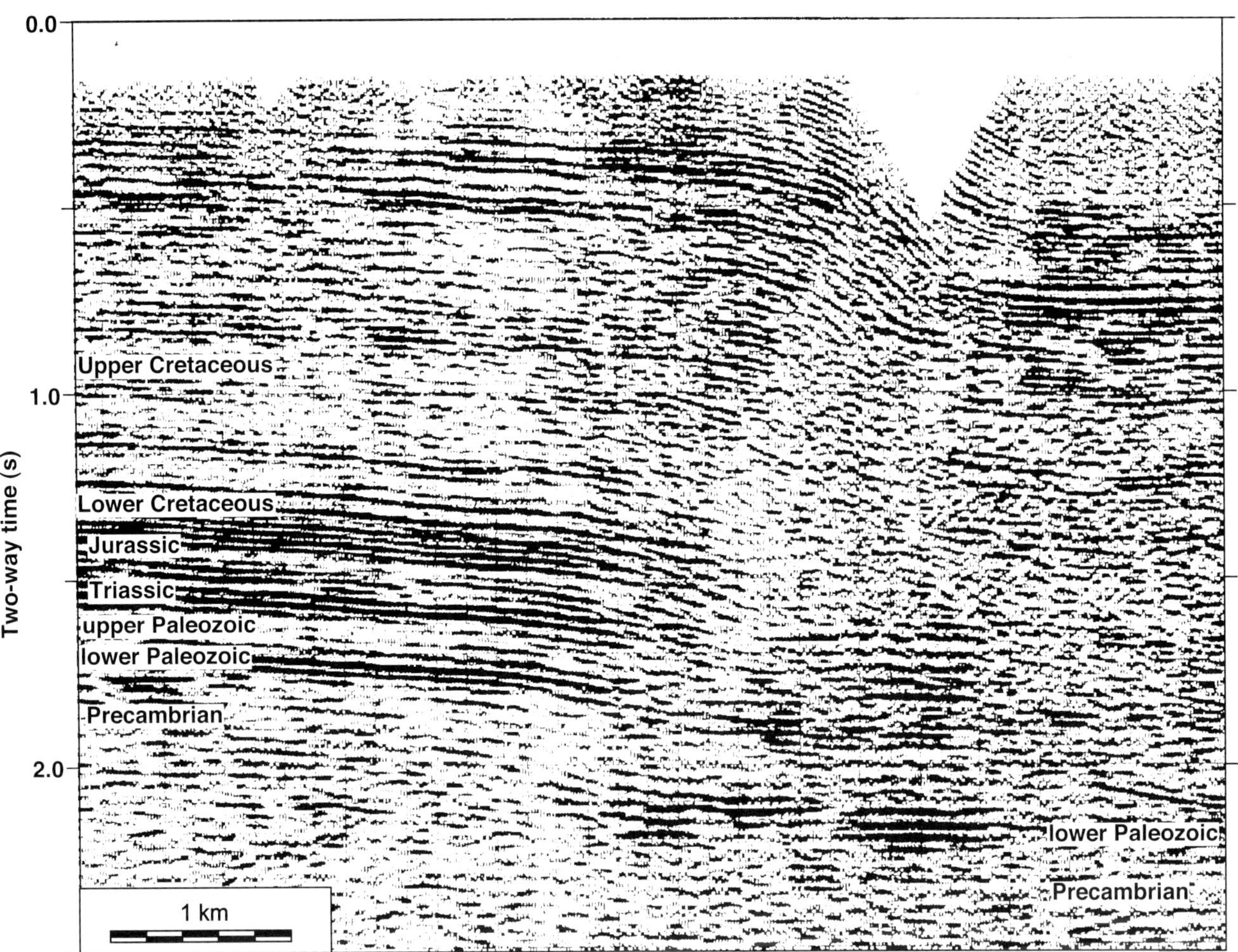

FIGURE 7. Seismic-reflection section across a monocline in the Sand Wash Basin, Colorado (modified after Livesey, 1985). Note that the lower Paleozoic/Precambrian interface is not significantly folded, whereas younger strata are folded, suggesting that the structure was formed by a reactivated fault in Precambrian basement and did not form by propagation from middle-crust depths.

COMPLEX TRISHEAR FORWARD MODELS

In its simplest state, a trishear fold resembles a monocline (Figure 2). This familiar incarnation of trishear, although of interest for modeling structures of, for example, the Colorado Plateau, belies the true power of the method. Allmendinger (1998a) has already demonstrated the geometric complexities that result when any of the six parameters are allowed to vary during slip on the fault. Here, we permit still greater complexity by allowing multiple faults in a section that can dip in either direction. In general, the forelimbs of structures are modeled with trishear, and the backlimb geometries are treated as fault-bend folds (Suppe, 1983). For simplicity's sake in modeling, shear throughout the hanging wall is always parallel to the active ramp. Where there are multiple, in-sequence imbrications, however, this means that shear will be oblique to bedding on all but the lowest imbrication, producing thinned bedding. Nonetheless the sections are still area balanced.

Two examples of complex forward models (Figure 8) illustrate the broad array of geometries that can be reproduced. In the first, two imbrications face each other across a type of triangle zone. In the second, we show a décollement lift off the crest of an underlying fault-bend fold. Although the geometries are far more complex than is possible with any other forward-modeling scheme, the models are fully area balanced and contain a rich record of strain, deformation path, and indirectly, fracture history.

STRAIN AND FRACTURE PREDICTIONS OF TRISHEAR MODELS

As a kinematic model, trishear can only predict particle paths. From those, we can calculate two types of information that are particularly pertinent to petroleum exploration: strain magnitude and lines of no finite elongation. In parallel-kink models of folding, the strain is homogeneous and completely dependent on bedding dip. Cross sections constructed with such a model in mind thus provide almost no predictive capabilities for fracture distribution and orientation. In contrast, because the strain in trishear is quite heterogeneous and variable in magnitude, the method potentially can provide a much higher degree of prediction. The strain, of course, can be accommodated by any deformation mechanism. In the following discussion, however, we will assume that the strain in competent units is produced by fracturing; we are motivated by the fractured reservoirs of such interest to the petroleum industry.

To date, we have carried out only limited tests of the correlation between strain predicted in a trishear model and fracture concentration in a real structure. One example is a small structure in the Neuquén Basin (Figure 9). This small structure is modeled as a trishear fault-propagation fold with a subsequent synclinal breakthrough. The geometric match between model and observation is good, as shown by the overlay of the model on the seismic and by the correspondence with the dipmeter survey (Figure 9a). The displacements are modest, but the strains are significant. The structure has been drilled and logged using FMI. By overlaying the borehole information on the trishear model cross section with contours of maximum strain intensity, we can test the correlation between predicted strain magnitude and maximum fracture density. Within the interval logged (only certain formations were), the maximum fracture density coincides with the maximum strain magnitude encountered along the trajectory of the borehole. The model further suggests that even higher fracture concentrations should be encountered in the footwall of the structure. This is a general result of the trishear model: The highest strains are found in the footwall of thrust faults and the hanging wall of normal faults. If the other conditions for reservoir formation — seal, migration paths, thermal history, and the like— are also present, then the best fractured reservoirs should be found in the footwall, or at the tip lines, of thrust faults.

As Allmendinger (1998a) describes, lines of no finite elongation (LNFE) can be excellent proxies for shear-plane orientation and shear sense. He was able to match the fracture orientation developed in Withjack et al.'s (1990) clay models of extensional forced folds extremely well. One observation in Withjack et al.'s (1990) model, which is also predicted by trishear numerical models (Allmendinger, 1998a) and Patton and Fletcher's (1995) forced-fold mechanical model, is that high-angle reverse faults should be present in the hanging walls of steep normal faults. Excellent examples of these types of structures can be seen in the seismic reflection data from

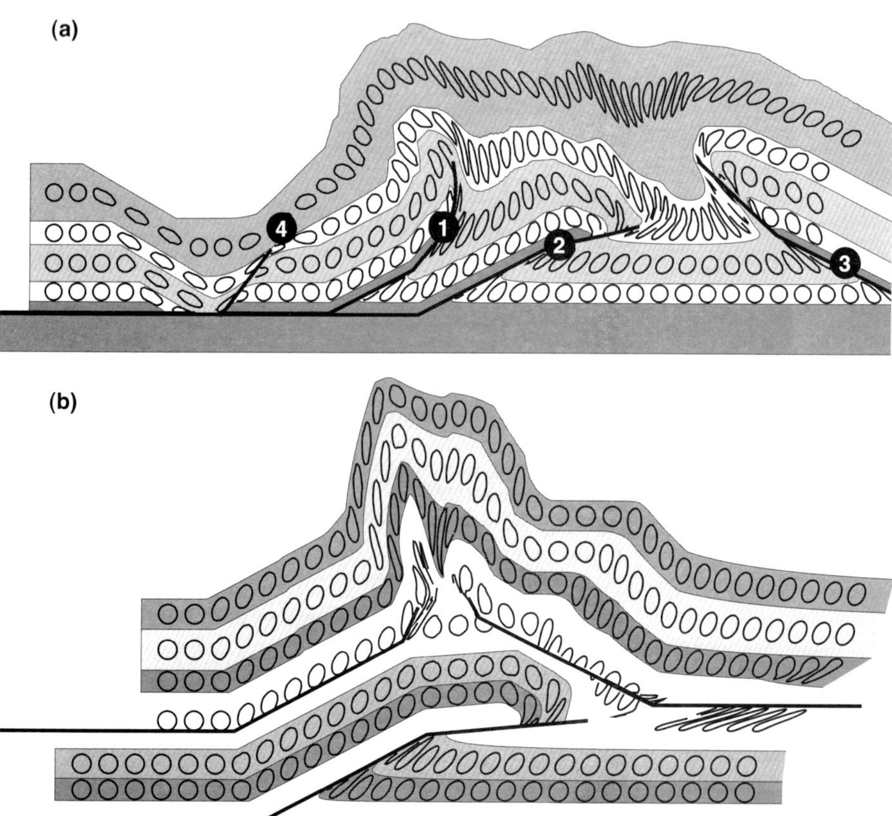

FIGURE 8. Examples of the complex forward models that can be produced when trishear fault-propagation folding is combined with multiple ramps and multiple faults of varying orientations. (a) Three thrust faults and one normal fault, all using the same décollement horizon. Numbers indicate the order for formation of the faults; number 4 is a normal fault. All faults had constant P/S (= 1.5) and trishear angle (50°). Ellipses show the complex strain variations. (b) Complex forward model of a lift-off detachment fold formed on the crest of a fault-bend fold with a trishear forelimb. White layer in the middle of the model is the weak layer. The faults in this layer have a P/S = 0.3. Structures such as this may be similar to the vertical-sided anticlines formed in the Sub-Andean belt of southern Bolivia (e.g., Belotti et al., 1995).

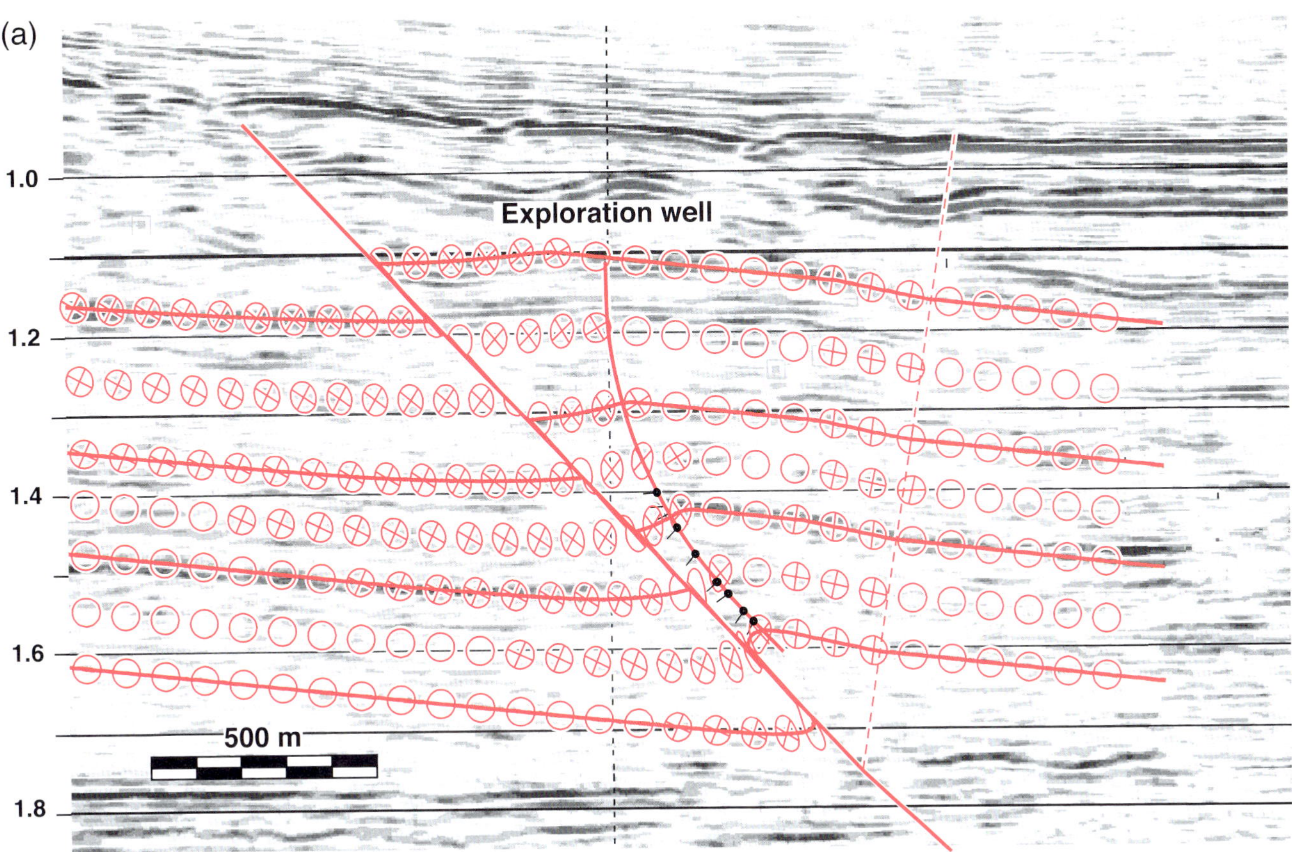

Figure 9. Small thrust structure from the Neuquén Basin and drilled by YPF, S.A. (a) Seismic-reflection section is in time, but is displayed at a spatial scale of 1:1 for the interval modeled. The superimposed trishear model was constructed as a trishear fault-propagation fold with a minor ramp and a subsequent synclinal breakthrough. (b) Trishear model with strain intensity, in terms of maximum principal stretch, contoured. The position of the major concentration of fractures in the well, as determined by FMI analysis, is shown as a large black dot along the trajectory of the well. Note, however, that the entire hole was not logged.

the Neuquén Basin (Figure 10). Commonly, such reverse faults are interpreted to be evidence of incipient thrust inversion of the normal faults. However, viewed from a trishear perspective, this is probably a misinterpretation. We show an additional example of the use of LNFE in the following section.

APPLICATION OF PARALLEL AND TRISHEAR FOLD MODELS AT DIFFERENT SCALES

A common misconception holds that trishear and parallel folding are two mutually exclusive kinematic fold models. Although this is true when applied at the same scale, when the two models are applied at different scales, they can be very complementary. One can interpret the geometric complexities of a deformed zone with parallel folding, duplexing, and imbricate faulting, while at the same time modeling the broader scale geometries of the first-order stratigraphic packages with trishear. Here, we demonstrate this two-pronged approach with our study of the Filo Morado structure of the western Neuquén Basin.

The Filo Morado field occurs on the forelimb of Pampa Tril, a large basement-cored anticline located in the western part of the Neuquén Basin, at the eastern margin of Andean deformation. It is the southernmost of a series of basement-cored domes that define the Andean deformation front for nearly 200 km northward into southern Mendoza province (Manceda and Figueroa, 1995). The stratigraphic section of this part of the Neuquén Basin (Vergani et al., 1995) is dominated by several thick, incompetent units, including the Auquilco, Huitrín, and lower Rayoso evaporite units, and the thick shale sequences of the Agrio and Vaca Muerta Formations. The latter unit is the primary source rock for the basin and is known to be a prominent dêcollement horizon. Within this sequence are three

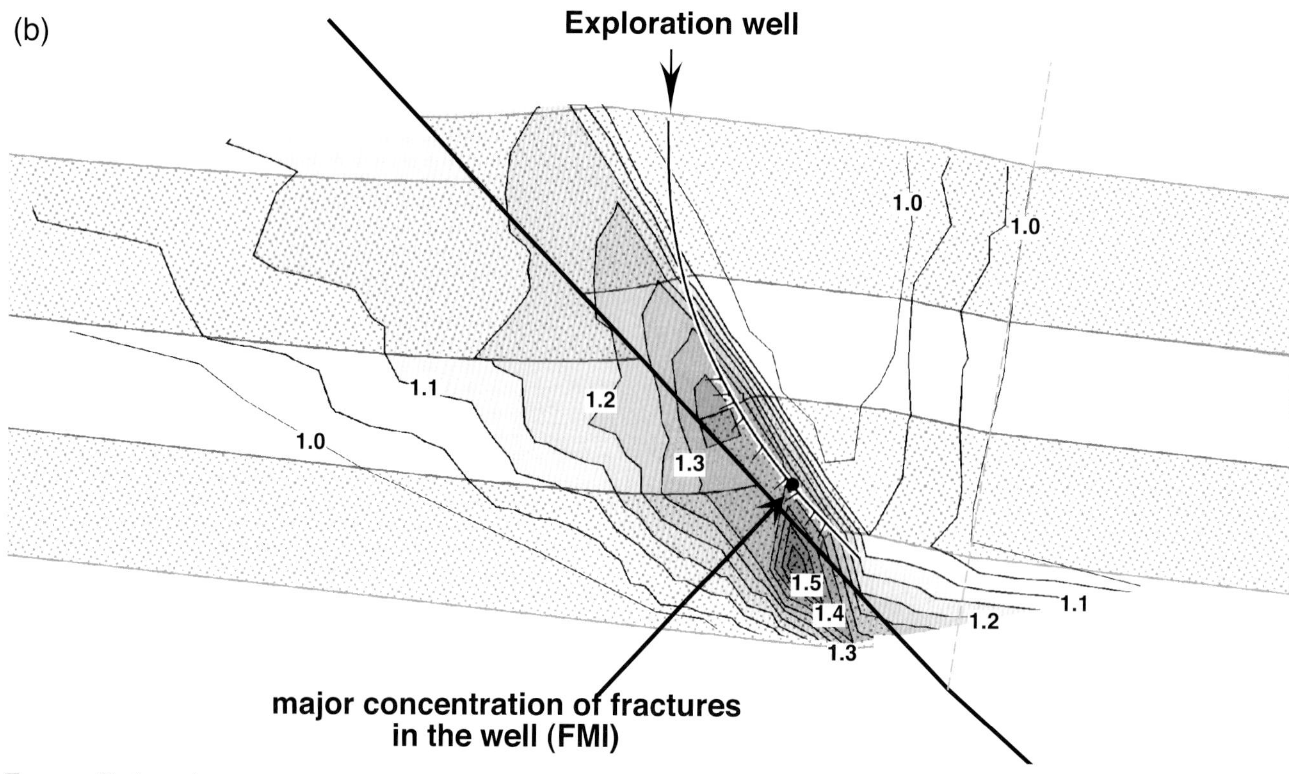

FIGURE 9. (cont.).

thin competent sandstone members, any one of which can act as a reservoir: the Mulichinco Formation, the Avilé member of the Agrio Formation, and the Troncoso member of the Huitrín. Capping the thick incompetent sequence are the Cretaceous Neuquén Group's continental clastic strata, locally 3000 m thick, which are well known for their prolific dinosaur fauna. This stratigraphic architecture clearly influences the style of deformation within the basin.

At the surface, the dips on the forelimb of the Pampa Tril–Filo Morado anticline dip gently (~30–40°) eastward, belying the complexity at depth (Figure 11a). Seismic and well data, however, show that the strata steepen in the subsurface and are complexly deformed; the top of basement (here the Choiyoi igneous rocks) dips vertically. The structure of Filo Morado has been described as a "triangle zone" (Viñes, 1990), but with its basement involvement and lack of clear back thrusting, it is quite different from Alberta-style triangle zones (Jones, 1982). The first-order geometric characteristics of Filo Morado–Pampa Tril are similar to trishear fault-propagation folds: forelimb-thickening and downward-steepening bedding dips.

To model Filo Morado as a trishear fold, we started by applying our inverse method to model the curvature of the top of basement and the top of Vaca Muerta. Because the basement fault is not imaged beneath Pampa Tril–Filo Morado (Figure 11b), we had to search for the tip line position and ramp dip as well. The full six-parameter grid search required about 24 hours of computer time to complete, but yielded basic information unavailable by any other means: the fault geometry and magnitude of displacement necessary to produce the curvature at the top of basement. The procedure also yielded a first-order estimate of the total slip, P/S, and trishear angle. After obtaining these basic parameters, we made adjustments to the forward model to fine tune our ultimate solution. These adjustments were made on the basis of intuition. Although it is certainly technically possible to write a computer program to try a new ramp angle, a different P/S, or a different trishear angle at every tip line position, such a procedure would take tens of thousands of years with currently available computing power! In our experience, few structures can be analyzed just by applying the grid search. More commonly, the procedure is as in this case: The parameters are narrowed down by use of the grid search and then fine tuned in the forward model to yield the ultimate fit.

The resulting trishear solution fits the basic geometry of the structure quite well (Figure 12). The curvatures of the lower surfaces, as well as the top of the Neuquén Group, are matched quite well. The only significant deviation is at the base of the Neuquén-Rayoso clastic sequence. The trishear model predicts that this surface should be more folded than the borehole data document. We suspect that this local mismatch is due to lateral, nonplane strain flow of the thick Huitrín-Rayoso evaporitic sequence. Of course, the trishear model does not begin to duplicate the geometric complexity in the Vaca Muerta–Agrio–Huitrín stratigraphic interval. This level of detail is well modeled by parallel folding, duplexing, and minor back thrusting (Figures 11, 12). Even so, the trishear model matches the strain magnitude and orientation quite well: The minimum stretch in the Vaca Muerta–Mulichinco duplex (0.39) is very similar to the local minimum stretch of 0.36 in the trishear model, and the orientations are similar (Figure 12). The predicted shear-fracture orientations from lines of no finite elongation in the trishear model also fit the observed cross-section-scale faults remarkably well. Note, in particular, the closely spaced suite of small displacement back thrusts located between the forelimb duplex and the top of basement. Their curving trajectories are predicted almost perfectly by the lines of no finite elongation.

Thus parallel trishear fold kinematics have yielded complementary information about this structure. From the parallel folding, we get a far more detailed view of the geometric complexities of the structure. This view can be as detailed as well data and seismic data permit. From trishear, we get strain magnitude, strain history, deformation path, and shear fracture orientation. Although trishear does not give us detailed geometry, it has allowed us to make a reasonably well-constrained interpretation of the basement-fault geometry and orientation. Because this fault is not resolved in seismic data (Figure 11b) and is not pierced in the borehole, the parallel-fold approach gives us no constraint on its geometry.

The trishear model gives a slip of 8.7 km and a P/S = 1.9, indicating that the tip line propagated nearly twice as far (16.4 km) as the fault slipped. When this value is combined with the fault dip, we can predict that the basement fault responsible for the structure nucleated at about 9.4 km below sea level. This figure is significant for two reasons. Previously, there was no objective

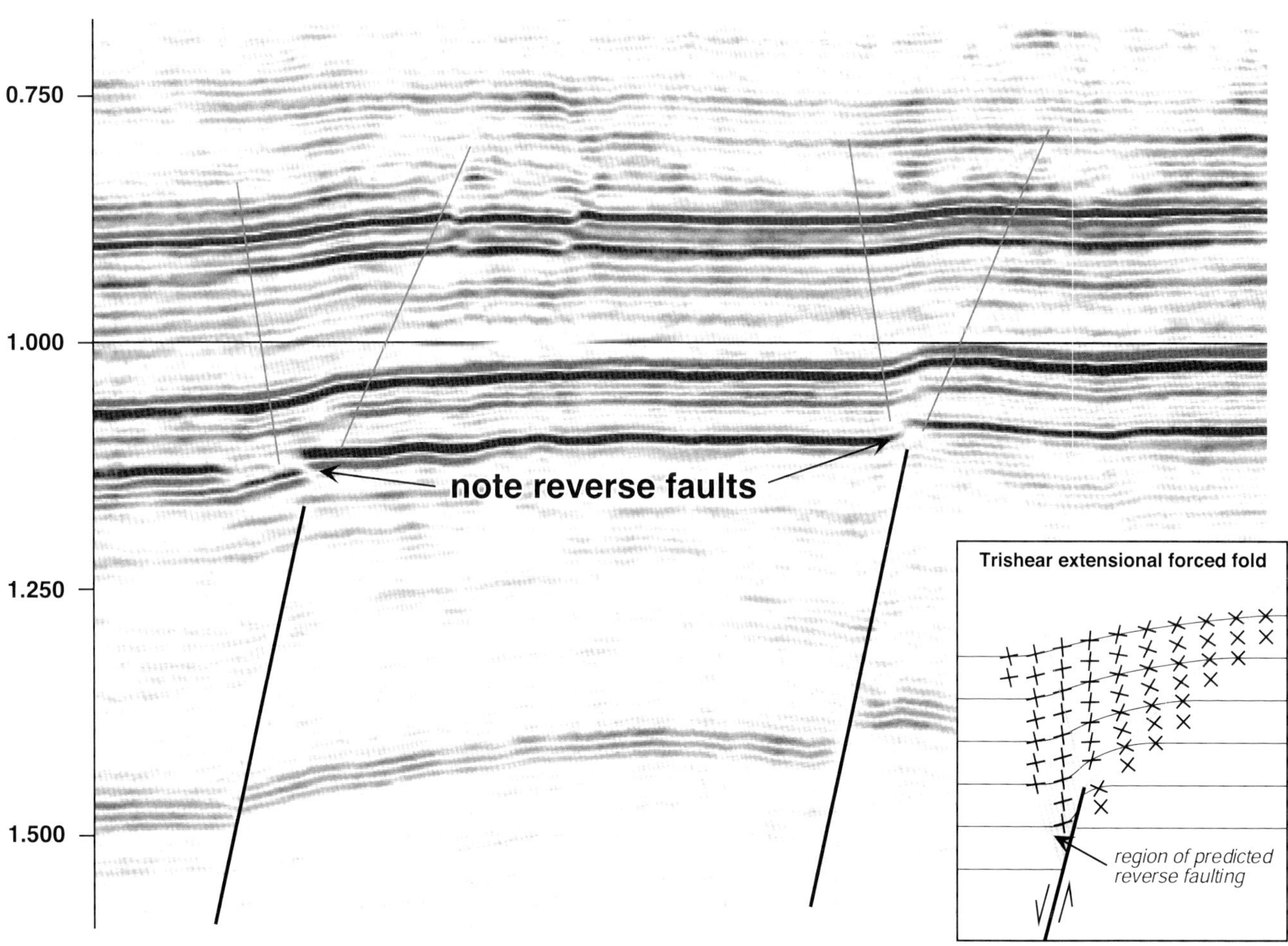

FIGURE 10. Seismic-reflection section from the Neuquén Basin, showing a set of trishear normal faults. Of particular significance, note the high-angle reverse faults near the tip line of the normal faults. This relationship is predicted by the orientations of lines of no finite elongation in deformation associated with trishear normal faults (see inset figure in lower-right corner) and has been observed in analog clay models of extensional forced folds (Withjack et al., 1990).

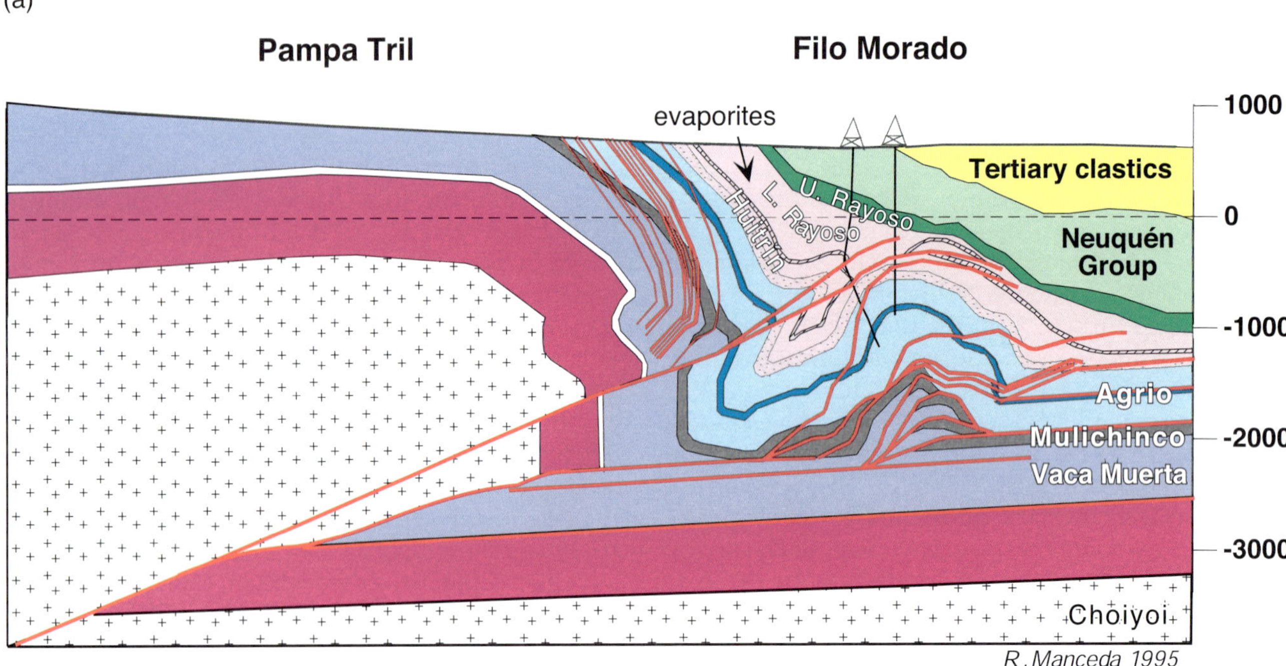

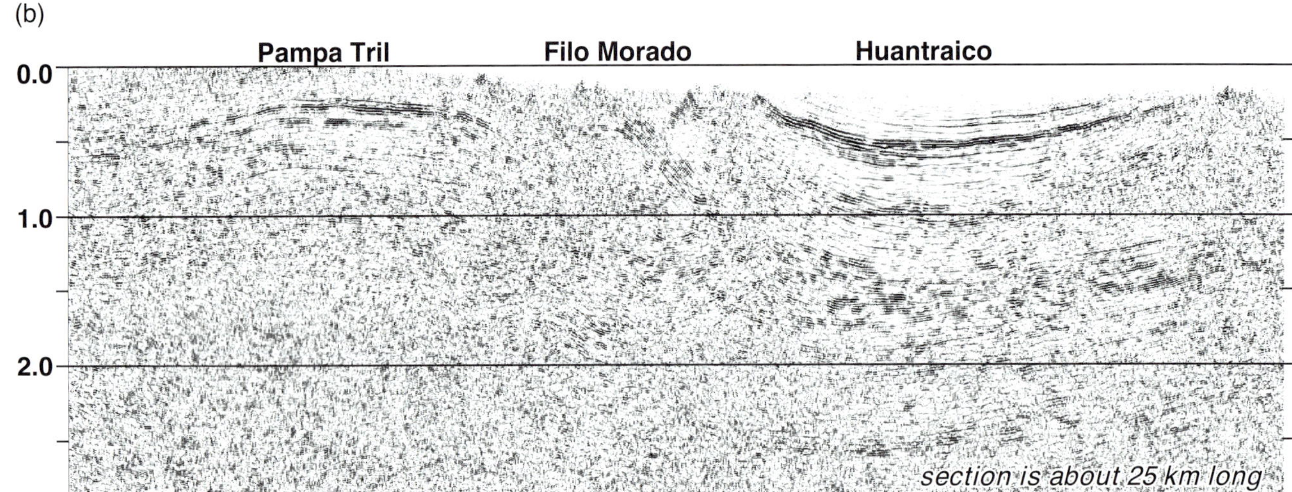

FIGURE 11. (a) Cross section of the Filo Morado structure in the western part of the Neuquén Basin. Section is controlled by tens of exploration and production wells, only a few of which are shown. Note the downward-steepening dips from Tertiary down to the basement surface (pluses pattern), the anticlinal stack in Vaca Muerta and Agrio Formations, the extremely variable thickness of the evaporites (Huitrín and lower Rayoso Formations), and the imbricate fan of closely spaced curving back thrusts on the forelimb of Pampa Tril Anticline. (b) Regional seismic line showing the loss of resolution beneath the structure, a hint of the downward-steepening dips, and the total lack of information on the position of the tip line and orientation of the fault.

estimate of the depth to the décollement for these basement anticlines that line the western margin of the Neuquén Basin. Although the trishear model does not require a décollement, there must be some reason why the fault nucleated at middle-crust depths. Secondly, this depth of well within the basement strongly indicates that these are newly formed faults and not reactivated Mesozoic rift structures. Were these reactivated normal faults, they would have propagated very little from their initial, precontraction position in the older Mesozoic units. We cannot rule out the possibility that this fault nucleated at the position of an old rift décollement, but no independent evidence exists for a rift décollement and its location at depth in the crust.

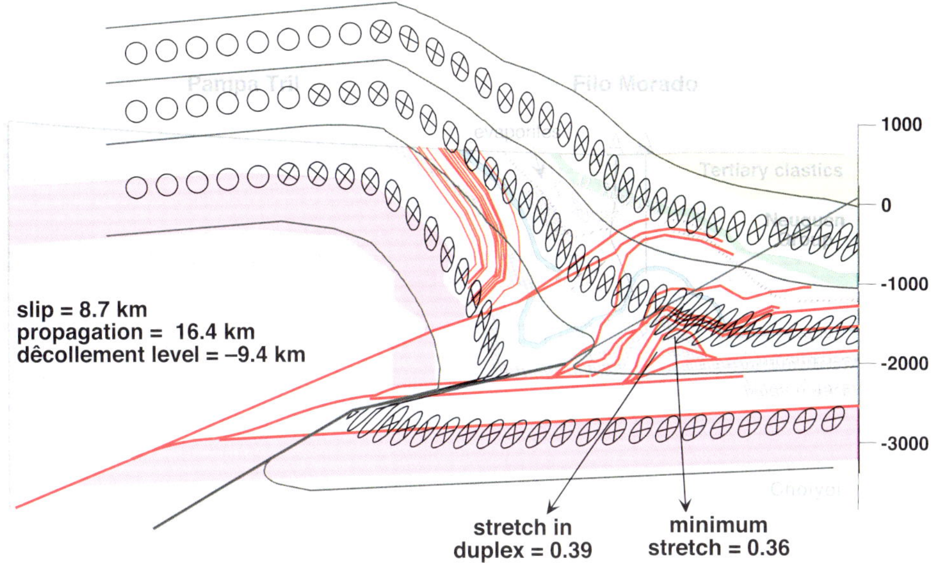

FIGURE 12. Trishear model of the Filo Morado structure. The small "Xs" in the strain ellipses are lines of no finite elongation. Note how the LNFEs on the forelimb of Pampa Tril Anticline curve around toward the west, mapping out extremely closely the curving back thrusts in Figure 11a.

sufficient to invert for anything more complex than simple monoclines with constant trishear parameters during slip. In our experience, few real structures evolved with constant parameters. Thus, our approach is to carry out a grid search of individual, digitized folds in a structure as just the first step in narrowing down the range of possible trishear parameters (particularly, P/S and trishear angle, but also tip line position and ramp angle, if necessary). Then, the geometry is "fine tuned" in a forward model by varying the parameters, adding ramps and flats, and so forth. Commonly, this fine tuning is considerably more time-consuming than the initial grid search, because it must be done iteratively, by hand.

Preliminary tests to relate fracture concentration to strain magnitude show great promise. Trishear is one of the few fold kinematic models that predict heterogeneous strain distribution; parallel-kink-fold models predict only homogeneous strain as a function of limb dip. Furthermore, shear-fracture orientation appears to be very well predicted by lines of no finite elongation in trishear models of extensional and contractional folds. Trishear successfully predicts small reverse faults in the hanging walls of normal faults (as does a mechanical model for extensional forced folds [Patton and Fletcher, 1995]) and small normal faults in the footwalls of thrusts. Conceivably, one could also predict opening mode fractures (i.e., joints) by assuming that they should be oriented perpendicularly to the long axis of the finite-strain ellipse. However, we have not carried out that particular test. Of course, fracturing in reservoir units may be completely unrelated to fold kinematics. In those cases, neither trishear nor any other fold kinematic model, nor any mechanical fold model, would provide any insight.

Finally, trishear and parallel folding can be complementary when applied at different scales. First-order stratigraphic units may display thickening and thinning on the forelimb and leading syncline, but in detail, those thickness changes may be produced by parallel folding, duplexing, or imbrication.

Trishear can be applied to thick- or thin-skinned structures and to normal or reverse faults. As field study and imaging of structures in mountain belts become

CONCLUSIONS

Erslev's (1991) trishear forward model was long treated by structural geologists as a mere curiosity, but it has considerable power and broad applicability. Hardy and Ford (1997) provided a glimpse of that power when they first explicitly considered the propagation-to-slip ratio in the context of trishear deformation. We have shown here that P/S is the primary constraint on fold shape. Furthermore, in the past we have treated the three common fault-related fold types— dêcollement folds, fault-propagation folds, and fault-bend folds— using completely separate geometric and kinematic analyses. But, when viewed in the context of P/S, these three fold types are simple, discrete points along a continuum. Finally, P/S combined with slip defines the propagation distance of the tip line. This parameter can be used to evaluate where a fault nucleated and whether it is a reactivation of an older fault.

Simple trishear structures are nothing more than monoclines. However, when trishear parameters can vary during fault slip, when faults are permitted to have ramps and flats, and when multiple faults can be plotted on a single section, the range of possible structural geometries is exceptionally broad. These structural geometries are, at least superficially, very realistic in appearance. It would be nice if the grid-search approach to extracting the best-fit trishear parameters from real structures could be applied to complex structures such as these. Unfortunately, computing power remains in-more refined, it is clear that our simple structural models are not adequate to explain many of the details of those structures. Trishear, which can mimic those complexities, is likely to see increasing use in the future.

ACKNOWLEDGMENTS

We are greatly indebted to YPF S.A. for permission to publish the seismic data displayed herein. Numerous discussions with Stuart Hardy, Ernesto Cristallini, Dan Litzenberg, Ben Brooks, John Suppe, and others have substantially increased our understanding of trishear. Reviews by Eric Erslev and two anonymous reviewers helped to improve the manuscript. We thank Ken McClay for organizing Thrust Tectonics 99 and inviting our participation. We are grateful to the U.S. National Science Foundation for partial support of this research, through grants EAR-9614759 and EAR-9814348 to R. W. Allmendinger. Acknowledgment is also made to the donors of the Petroleum Research Fund, administered by the ACS, for partial support of this research.

REFERENCES CITED

Allmendinger, R. W., 1998a, Inverse and Forward numerical modeling of trishear fault-propagation folds: Tectonics, v. 17, p. 640–656.

Allmendinger, R. W., 1998b, TrishearPPC, 2.9.7: Ithaca, New York, R. W. Allmendinger.

Belotti, H. J., L. L. Saccavino, and G. A. Schachner, 1995, Structural styles and petroleum occurrence in the Sub-Andean fold and thrust belt of northern Argentina, *in* A. J. Tankard, R. Suárez S., and H. J. Welsink, eds., Petroleum basins of South America: AAPG Memoir 62, p. 545–555.

Dahlstrom, C. D. A., 1990, Geometric constraints derived from the law of conservation of volume and applied to evolutionary models for detachment folding: AAPG Bulletin, v. 74, p. 336–344.

Elliott, D., 1976, The energy balance and deformation mechanisms of thrust sheets: Philosophical Transactions of the Royal Astronomical Society, v. A–283, p. 289–312.

Erslev, E. A., 1991, Trishear fault-propagation folding: Geology, v. 19, p. 617–620.

Erslev, E. A., and J. L. Rogers, 1993, Basement-cover geometry of Laramide fault-propagation folds, *in* C. J. Schmidt, R. B. Chase, and E. A. Erslev, eds., Laramide basement deformation in the Rocky Mountain foreland of the Western United States: Geological Society of America, Special Paper 280, p. 125–146.

Friedman, M., R. H. H. I. Hugman, and J. Handin, 1980, Experimental folding of rocks under confining pressure, part VIII— forced folding of unconsolidated sand and lubricated layers of limestone and sandstone: Geological Society of America Bulletin, v. 91, p. 307–312.

Gallup, W. B., 1951, Geology of Turner Valley oil and gas field, Alberta, Canada: AAPG Bulletin, v. 34, p. 797–821.

Hardy, S., and M. Ford, 1997, Numerical modelling of trishear fault-propagation folding and associated growth strata: Tectonics, v. 16, p. 841–854.

Johnson, K. M., and A. M. Johnson, 2002, Mechanical models of trishear-like folds: Journal of Structural Geology, v. 24, p. 277–287.

Jones, P. B., 1982, Oil and gas beneath east-dipping underthrust faults in the Alberta Foothills, Canada, *in* R. B. Powers, ed., Geologic studies of the Cordilleran thrust belt, Denver, Rocky Mountain Association of Geologists, p. 61–74.

Livesey, G. B., 1985, Laramide structures of southeastern Sand Wash Basin, *in* R. R. Gries and R. C. Dyer, eds., Seismic exploration of the Rocky Mountain Region, Denver, Rocky Mountain Association of Geologists, p. 87–94.

Manceda, R., and D. Figueroa, 1995, Inversion of the Mesozoic Neuquén rift in the Malargüe fold and thrust belt, Mendoza, Argentina, *in* A. J. Tankard, R. Suárez S., and H. J. Welsink, eds., Petroleum basins of South America: AAPG Memoir 62, p. 369–382.

Patton, T. L., and R. C. Fletcher, 1995, Mathematical block-motion model for deformation of a layer above a buried fault of arbitrary dip and sense of slip: Journal of Structural Geology, v. 17, p. 1455–1472.

Poblet, J., and K. McClay, 1996, Geometry and kinematics of single-layer detachment folds: AAPG Bulletin, v. 80, p. 1085–1109.

Reches, Z., and A. M. Johnson, 1978, Development of monoclines: Part I: Structure of the Palisades Creek branch of the East kaibab monocline, Grand Canyon, *in* V. Matthews, III, ed., Laramide folding associated with basement block faulting in the western United States: Geological Society of America Memoir 151, p. 273–311.

Rodgers, D. A., and W. D. Rizer, 1981, Deformation and secondary faulting near the leading edge of a thrust fault, *in* K. R. McClay and N. J. Price, eds., Thrust and nappe tectonics: Geological Society of London Special Publication 9, p. 65–77.

Stearns, D. W., 1978, Faulting and forced folding in the Rocky Mountains foreland, *in* V. Matthews, III, ed., Laramide folding associated with basement block faulting in the western United States: Geological Society of America Memoir 151, p. 1–37.

Suppe, J., 1983, Geometry and kinematics of fault-bend folding: American Journal of Science, v. 283, p. 684–721.

Suppe, J., and D. Medwedeff, 1990, Geometry and kinematics of fault-propagation folding: Eclogae Geologicae Helvetiae, v. 83, p. 409–454.

Vergani, G., A. J. Tankard, H. J. Belotti, and H. J. Welsink, 1995, Tectonic evolution and paleogeography of the Neuquén Basin, Argentina, *in* A. J. Tankard, R. Suárez S., and H. J. Welsink, eds., Petroleum basins of South America: AAPG Memoir 62, p. 383–402.

Viñes, R. F., 1990, Productive duplex imbrication at the Neuquina basin thust belt front, Argentina, *in* J. Letouzey, ed., Petroleum and tectonics in mobile belts, Editions Technip., p. 69–80.

Williams, G., and T. Chapman, 1983, Strains developed in the hangingwalls of thrusts due to their slip/propagation rate; a dislocation model: Journal of Structural Geology, v. 5, p. 563–571.

Withjack, M. O., J. Olson, and E. Peterson, 1990, Experimental models of extensional forced folds: AAPG Bulletin, v. 74, p. 1038–1054.

Zehnder, A. T., and R. W. Allmendinger, 2000, Velocity field for the trishear model: Journal of Structural Geology, v. 22, p. 1009–1014.

19

Poblet, J., M. Bulnes, K. McClay, and S. Hardy, 2004, Plots of crestal structural relief and fold area versus shortening — A graphical technique to unravel the kinematics of thrust-related folds, *in* K. R. McClay, ed., Thrust tectonics and hydrocarbon systems: AAPG Memoir 82, p. 372–399.

Plots of Crestal Structural Relief and Fold Area versus Shortening—A Graphical Technique to Unravel the Kinematics of Thrust-related Folds

J. Poblet
Departamento de Geología, Universidad de Oviedo, Oviedo, Spain

M. Bulnes
Departamento de Geología, Universidad de Oviedo, Oviedo, Spain

K. McClay
Fault Dynamics Research Group, Department of Geology, Royal Holloway, University of London, Egham, Surrey, U.K.

S. Hardy
Department of Earth Sciences, The University of Manchester, Manchester, U.K.

ABSTRACT

A new method is presented for unraveling some aspects of the kinematic evolution of thrust-related folds. This technique consists of measuring crestal structural relief, shortening, and the area of a fold for different amplification stages, and then plotting the crestal structural relief and the fold area versus the shortening. One of the main advantages of this technique is that the data can be obtained from different sources: a section across a fold with associated syntectonic sediments, different sections across a fold that underwent a lateral shortening gradient, or different sections across a fold at different amplification stages. This technique has been applied to theoretical, natural, and experimental thrust-related folds. The analyses carried out show that each fold had a different kinematic evolution. They show also that, except for the theoretical examples, the kinematic evolution may be very complex, because different folding mechanisms may operate during fold amplification and increases/decreases of fold area and thickening/thinning of beds may occur. Therefore, to model or sequentially restore natural/experimental thrust-related folds, we recommend the application of techniques such as the one we propose here, when possible. This would avoid the automatic application of forward models based on kinematic assumptions and geometric resemblance between the model and the actual example.

INTRODUCTION

Structural geologists attempt to decipher the kinematics of thrust-related folds for economic, social, and scientific reasons. Thrust-related anticlines are economically important because they may store major hydrocarbon and mineral accumulations. Therefore, oil/mining explorationists need to establish fold kinematics to test the position of the source rocks and the validity of migration routes, to analyze the formation of the structural traps, and so on (Buchanan, 1996). Within tectonically active zones undergoing shortening, folds form as the surface expression of coseismic deformation during thrust-related earthquakes in the upper crust. Therefore, in regions where folds form incrementally during each earthquake, our understanding of fold kinematics plays an essential role when we assess potential seismic hazards (Lin and Stein, 1989). From the academic point of view, information on the kinematics of folds within a fold-and-thrust belt enables us to gain insight into the evolution of orogenic wedges. An understanding of the kinematics of thrust-related folds supplies data on the progressive deformation the rocks have undergone. In addition, analyses of the kinematics of natural and experimental folds are needed for testing and refining the theoretical models that have been proposed to account for the evolution of thrust-related folds.

Although the importance of determining the kinematics of thrust-related folds is widely accepted, often we are still unable to determine unambiguously the history of fold amplification. The challenge is to develop methods that provide sufficient data and resolution for us to unravel the kinematics of thrust-related folds. In this paper, we describe a new method that provides qualitative and quantitative information on some aspects of fold evolution. (1) It consists of determining the magnitude of crestal structural relief, shortening, and fold area during different stages of fold evolution by using previous and new techniques described here; and (2) once they have been determined, both the crestal structural relief and the fold area must be plotted versus shortening to obtain the fold uplift and variation of fold area that occurred with increasing shortening. The curves obtained can be analyzed in detail and compared with curves obtained for theoretical models or for other natural/experimental examples.

This approach has several advantages over other techniques for unraveling fold kinematics. (1) The data needed to construct these graphs can be obtained from different sources, such as growth folds, folds involving a lateral shortening gradient, and folds "frozen" at different stages of fold amplification, in the case of theoretical/physical models. (2) Only accurate sections across a fold are needed to apply this technique. (3) Unlike the measurements of limb/crest length and limb dip, which require the folds to be approximated by kinklike structures, the technique we propose here does not require any previous treatment and can be applied both to kinklike and rounded-hinge folds. (4) Our technique can be used in the case of natural, experimental, and theoretical folds. (5) It allows us to decipher the kinematics of any type of thrust-related fold (fault-bend folds, fault-propagation folds, detachment folds, etc.). (6) The results obtained depend neither on the fold amplification rates (i.e., uplift, shortening, etc.) nor on the syntectonic sedimentation rates and ages of the syntectonic sediments.

The method we present here is applied to two theoretical thrust-related folds, six natural thrust-related folds, and three physical experiments of thrust-related folds.

MEASURING CRESTAL STRUCTURAL RELIEF, SHORTENING, AND FOLD AREA

The regional datum must be defined before we deal with the crestal structural relief, shortening, and fold area. The regional datum is the elevation of a particular stratigraphic horizon where it is not involved in the tectonic structure (McClay, 1992). (1) The crestal structural relief of a fold is the elevation of a particular stratigraphic horizon in the anticline crest, with respect to the regional datum defined by the same stratigraphic horizon (Figure 1). (2) Assuming constant bed length

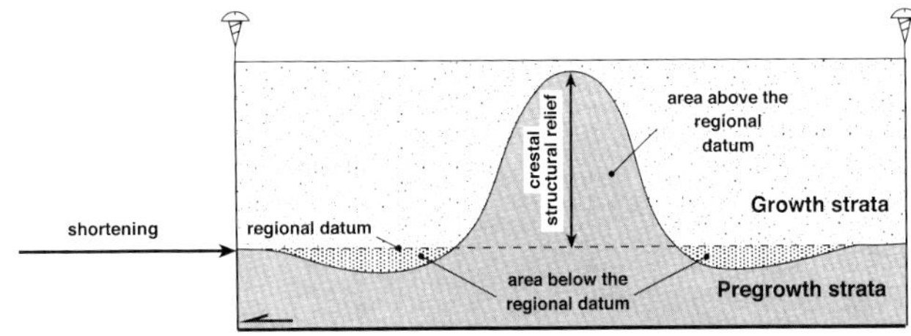

FIGURE 1. Crestal structural relief, shortening, and fold area (above and below the regional datum) for thrust-related anticlines.

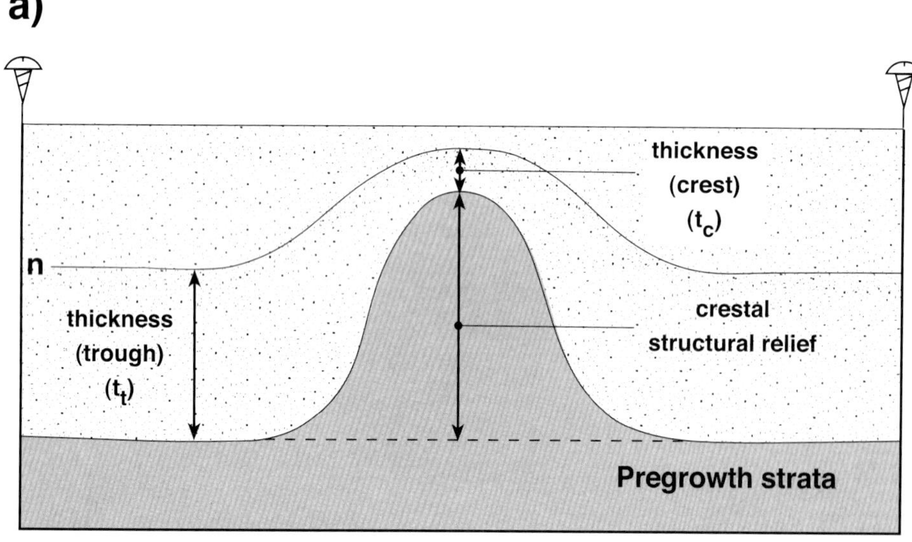

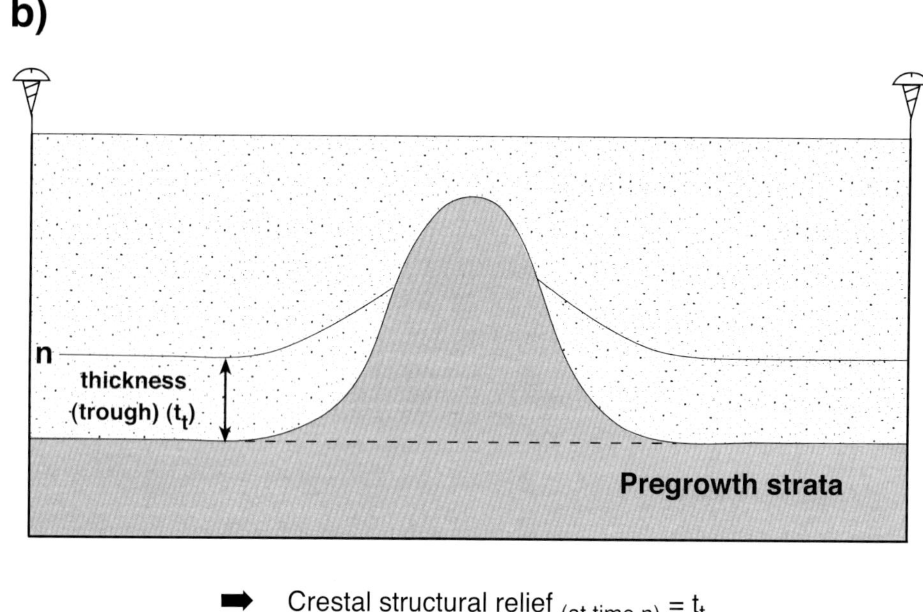

FIGURE 2. Procedure to estimate crestal structural relief using overlapping syntectonic sediments (a), and onlapping syntectonic sediments (b) associated with thrust-related anticlines. To estimate the crestal structural relief when growth beds onlap the fold, we have assumed that the crest of the anticline did not produce any topographic or bathymetric positive relief, that is, the beds "filled to the top" the space available for deposition (see Masaferro et al., 1999, 2002 for a discussion of this topic when estimating crestal structural relief).

during deformation, the shortening is the total length of a section, once the folded beds have been unfolded and the movement along the faults has been restored, minus the present-day length of the section (Figure 1). (3) The area-conservation principle predicts that the area that results from multiplying the shortening by the detachment depth (displaced area) must equal the area beneath a particular stratigraphic horizon uplifted above the regional datum minus the area above the same horizon dropped below the regional datum. Therefore, to estimate the correct displaced area, we must measure the uplifted area above the regional datum minus the depressed area below the regional datum (Figure 1).

One can measure the crestal structural relief, shortening, and fold area at different stages of fold amplification, using three different strategies that we describe below.

Growth Folds

To quantify the crestal structural relief of the growth/pregrowth strata boundary at different stages of fold evolution, we employ a technique proposed by Masaferro et al. (1999; their Figure 12) that has been partially used by Poblet and Hardy (1995), Schneider et al. (1996), Poblet et al. (1998), and Masaferro et al. (2002). This technique assumes that the growth horizons were deposited horizontally, that their thicknesses in the basin adjacent to the anticline and in the crestal area remained constant during fold amplification, and that the intersection lines between the onlapping growth horizons and the pregrowth beds did not move along the fold limbs—that is, no shear occurred on the growth/pregrowth surface. This technique (Figure 2) enables us to calculate the crestal structural relief of a fold during deposition of the growth beds. By subtracting the thickness of all the growth

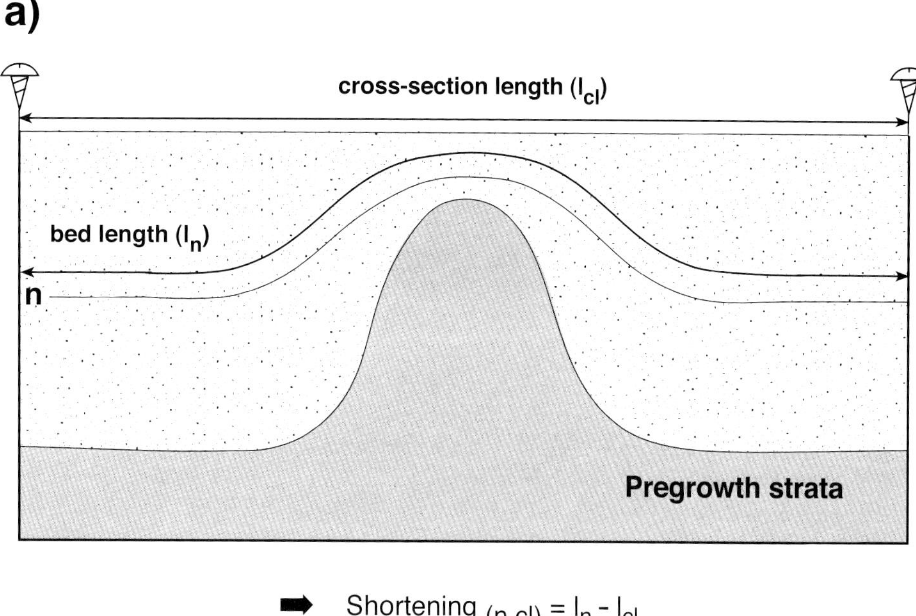

FIGURE 3. Procedure to estimate shortening using overlapping syntectonic sediments (a), and onlapping syntectonic sediments (b) associated with fault-propagation and detachment folds. To estimate the shortening when growth beds onlap the fold, we have assumed that the crest of the anticline did not produce any topographic or bathymetric positive relief, that is, the beds "filled to the top" the space available for deposition (see Masaferro et al., 1999, 2002 for a discussion of this topic when estimating shortening). Shortening $_{(n,cl)}$ = shortening undergone from deposition of growth bed *n* to present day.

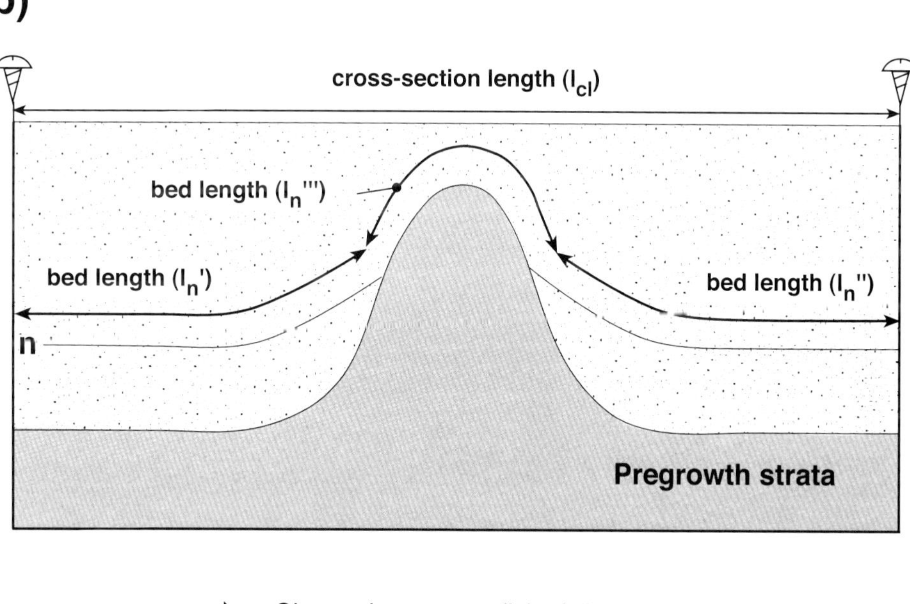

beds deposited on the anticline's crest, prior to horizon *n,* from the thickness of the same growth beds deposited in the basin adjacent to the anticline, we obtain the crestal structural relief of a fold at time *n*. In the case of horizons deposited at time *n* that do not extend over the fold crest because they onlap pregrowth strata, the crestal structural relief estimated using the procedure above assumes that the crest of the anticline did not cause any topographic or bathymetric positive relief, that is, beds "filled to the top" the space available for deposition.

To estimate the shortening undergone by the fold during deposition of the growth beds, we employ two techniques that are based on the following assumptions: the growth horizons were deposited horizontally, they maintained their cross-sectional lengths during fold amplification, and no shear occurred on the growth/pregrowth surface.

The first technique for estimating the shortening was proposed by Masaferro et al. (1999; their Figure 14) and is similar to the procedures used by Rowan et al. (1993), Schneider et al. (1996), and Butler and Lickorish (1997). It is valid for detachment folds and thrust-propagation folds, as long as the growth beds are not offset by the thrust. This technique (Figure 3) enables calculation of the shortening undergone by a fold from time *n* to the present day by subtracting the length of the present-day cross section from the folded length of the growth horizon deposited at time *n*. In the case of growth beds that onlap pregrowth strata, the folded length of the growth horizon deposited at time *n* should include the length of the growth/pregrowth strata boundary measured from the intersection point between the pregrowth strata and the onlapping growth bed in one fold limb to the equivalent point in the other fold limb. This

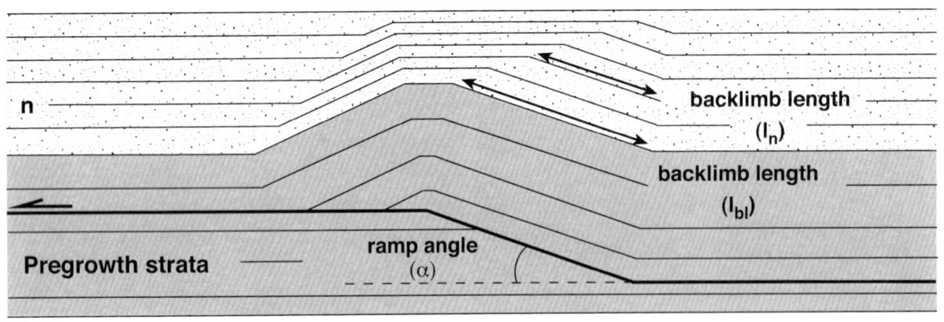

FIGURE 4. Procedure to estimate shortening using syntectonic sediments associated with fault-bend folds that have not reached the crestal broadening stage. Shortening $_{(n,bl)}$ = shortening undergone from deposition of growth bed n to present day.

type of measurement assumes that the crest of the anticline did not cause any topographic or bathymetric positive relief, that is, the beds "filled to the top" the space available for deposition.

The second technique for estimating the shortening is illustrated in Figure 4 and is used to determine shortening in the case of fault-bend folds that have not reached the crestal-broadening stage. This technique enables one to calculate the shortening undergone by a fold, from time n to present day, by subtracting the backlimb length of the growth horizon deposited at time n from the backlimb length of the pregrowth/growth strata boundary.

To estimate the fold area delimited by the growth/pregrowth strata boundary during deposition of the growth beds, that is, at different stages of fold amplification, we present a new procedure displayed in Figures 5 and 6. This technique assumes that the growth horizons were deposited horizontally, that their area remained constant with time, and that no shear occurred on the growth/pregrowth surface. The fold area beneath the growth/pregrowth strata boundary at a specific time, n, can be obtained by subtracting the area occupied by the growth sediments prior to horizon n from the total area beneath growth horizon n. The total area beneath a growth horizon deposited at time n results from multiplying the folded length of the growth horizon n by the thickness of the growth beds deposited in the basin adjacent to the anticline prior to horizon n. For horizons deposited at time n that do not extend over the fold crest because they onlap pregrowth strata, the total area beneath horizon n must be computed using the folded length of the growth horizon n on both sides of the anticline, plus the length of the growth/pregrowth strata boundary measured from the intersection point between the pregrowth strata and the onlapping growth bed in one fold limb, to the equivalent point in the other fold limb. This type of measurement assumes that the crest of the anticline did not cause any topographic or bathymetric positive relief, that is, the beds "filled to the top" the space available for deposition.

The strategies described above can be applied to natural, experimental, and theoretical growth anticlines, but there is the inconvenience that they only compute the fold evolution recorded by the syntectonic sediments. The results obtained using these strategies include a number of uncertainties, because the calculations do not take into account the occurrence of paleodepositional slopes; variations of thickness, bed length, and area resulting from tectonic processes and/or compaction; or shear along surfaces that have a high viscosity contrast (growth/pregrowth surface). These processes may occur in nature and physical experiments to a certain extent. However, modeling them would require data that in many cases are hardly available, and would require more-complex interactive algorithms, which are beyond the scope of this study. The work presented here is a first approach to obtain qualitative/quantitative information on some aspects of the kinematics of thrust-related folds, using solely the growth-strata geometry.

Folds with a Lateral Gradient of Shortening

The second strategy for unraveling fold kinematics assumes that observed variations in fold geometry along strike are equivalent to geometric variations in time (see Means, 1976; Elliot, 1976). This implies that, to determine fold kinematics, we can use measurements of crestal structural relief, shortening, and fold area for each of a series of cross sections in regions where a lateral gradient of shortening occurs. Because stratigraphy strongly influences the evolution of structures, this technique should be applied only in folds where the lithostratigraphic features of the folded rocks do not vary significantly along strike. This assumption is controversial, and although in some cases it has yielded viable kinematic models of thrust-related folds (Wojtal and Mitra, 1988; Shaw and Suppe, 1994; Shaw et al., 1994; Pashin et al., 1995; Poblet et al., 1998), it cannot explain the geometry of other natural examples (Fischer and Woodward, 1992). Some of the results we present later in this chapter depend critically on the validity of this

FIGURE 5. Procedure to estimate fold area using overlapping syntectonic sediments (a), and onlapping syntectonic sediments (b) associated with thrust-related anticlines. To estimate the fold area when growth beds onlap the fold, we have assumed that the crest of the anticline did not produce any topographic or bathymetric positive relief, that is, the beds "filled to the top" the space available for deposition.

a)

bed length (l_n)

n

thickness (trough) (t_t)

area (growth) (a_g)

Pregrowth strata

➡ Core area $_{(\text{at time n})} = (l_n \times t_t) - a_g$

b)

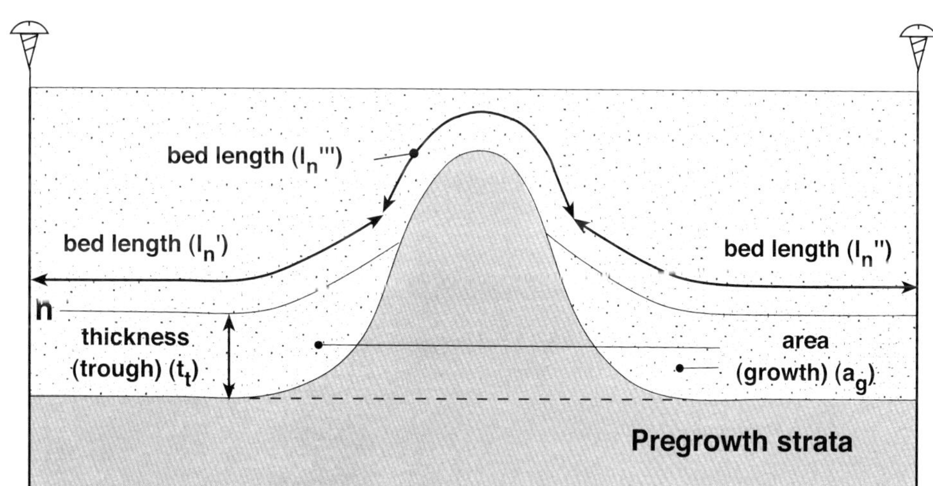

➡ Core area $_{(\text{at time n})} = ([l_n' + l_n'' + l_n'''] \times t_t) - a_g$

assumption. This method can be used for natural, experimental, and theoretical anticlines, and assumes that the bed-length, thickness, and area of the pregrowth horizons remained constant during fold amplification.

Folds "Frozen" at Successive Amplification Stages

The third strategy for unraveling fold kinematics consists of measuring the crestal structural relief, shortening, and fold area for each of a series of cross sections obtained at different stages of fold amplification. This method can only be applied to experimental and theoretical anticlines, and assumes that the bed length, thickness, and area of the pregrowth horizons remained constant during the entire fold amplification.

APPLICATION TO THRUST-RELATED FOLDS

Before we analyzed examples of thrust-related folds using the strategies described above, we plotted the crestal structural relief, shortening, and fold area of a number of theoretical models to compare them with the actual examples. The following kinematic forward models of single parallel folds related to thrusts are plotted in Figure 7:

1) a simple-step, fault-bend fold in the crestal uplift stage, formed by hinge migration according to the model proposed by Suppe (1983);
2) a simple-step, fault-bend fold in the crestal broadening stage, formed by hinge migration according to the model proposed by Suppe (1983);

a)

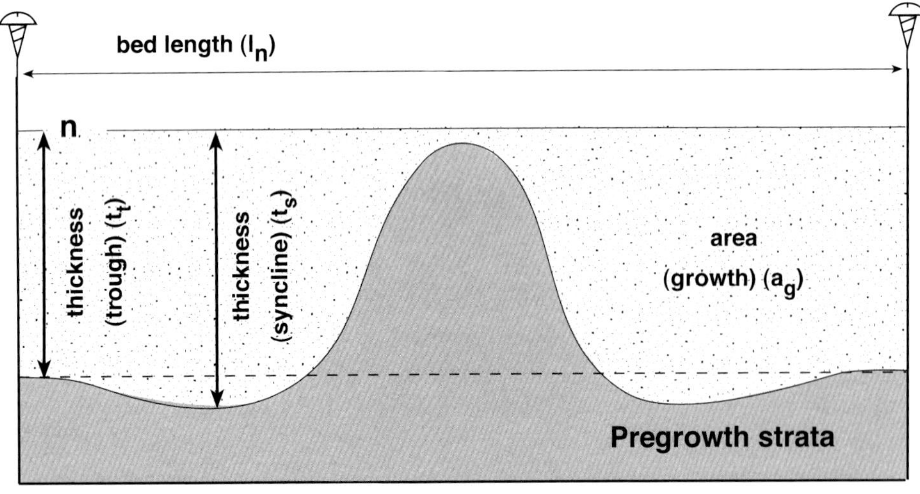

$$\text{Area}_{\text{(at time n)}} = ([l_n \times t_s] - a_g) - (l_n \times [t_s - t_t]) = (l_n \times t_t) - a_g$$

FIGURE 6. Procedure to estimate fold area using overlapping syntectonic sediments (a), and onlapping syntectonic sediments (b) associated with thrust-related anticlines. In this case the adjacent synclinal hinges shift vertically during fold amplification. Unlike Figure 5, in which the cross sections show the final state of fold amplification, in this case the cross sections show the fold geometry during deposition of growth horizon *n*. To estimate the fold area when growth beds onlap the fold, we have assumed that the crest of the anticline did not produce any topographic or bathymetric positive relief, that is, the beds "filled to the top" the space available for deposition.

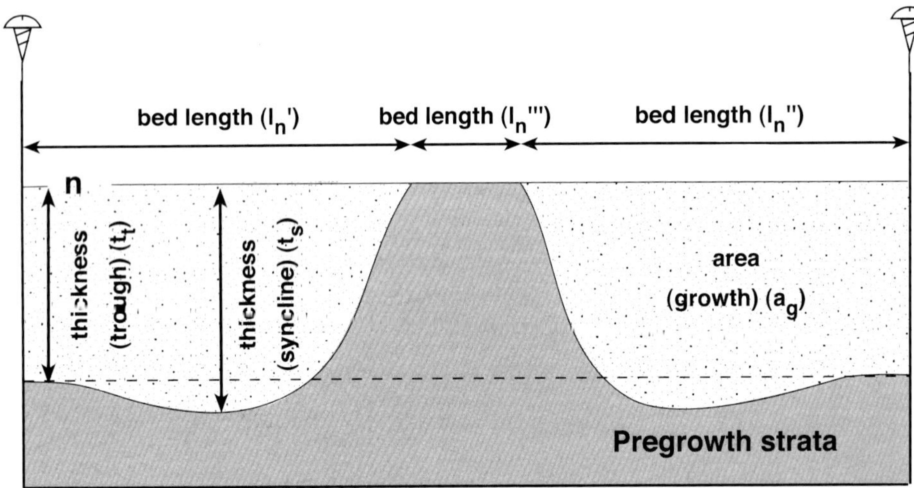

$$\text{Area}_{\text{(at time n)}} = \{[(l_n' + l_n'' + l_n''') \times t_s] - a_g\} - ([l_n' + l_n'' + l_n'''] \times [t_s - t_t]) = ((l_n' + l_n'' + l_n''') \times t_t) - a_g$$

3) a simple-step, fault-propagation fold, formed by hinge migration according to the model proposed by Suppe and Medwedeff (1990);
4) a detachment fold developed by hinge migration, formed according to the models proposed by Homza and Wallace (1995), and Poblet and McClay (1996);
5) a detachment fold developed by limb rotation (De Sitter, 1956), formed according to the models proposed by Hardy and Poblet (1994), Homza and Wallace (1995), Poblet and Hardy (1995), and Poblet and McClay (1996); and
6) a detachment fold developed by a combination of hinge migration and limb rotation (Beutner and Diegel, 1985; Dahlstrom's mechanism [1990]), formed according to the models proposed by Homza and Wallace (1995), Poblet and Hardy (1995), and Poblet and McClay (1996).

To normalize the graphs illustrated in Figure 7, the value of each geometric parameter at a particular stage of fold amplification has been divided by the value attained at the final stage and the result multiplied by 100. The fold-area-versus-shortening graph (Figure 7b) shows the area beneath the fold core located above the regional datum. With fault-bend folds and

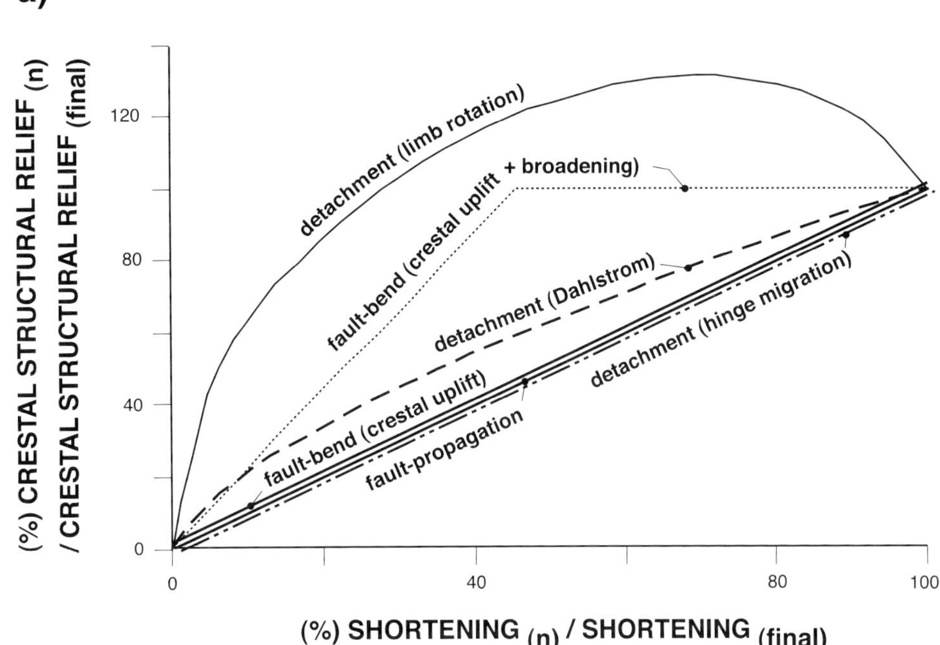

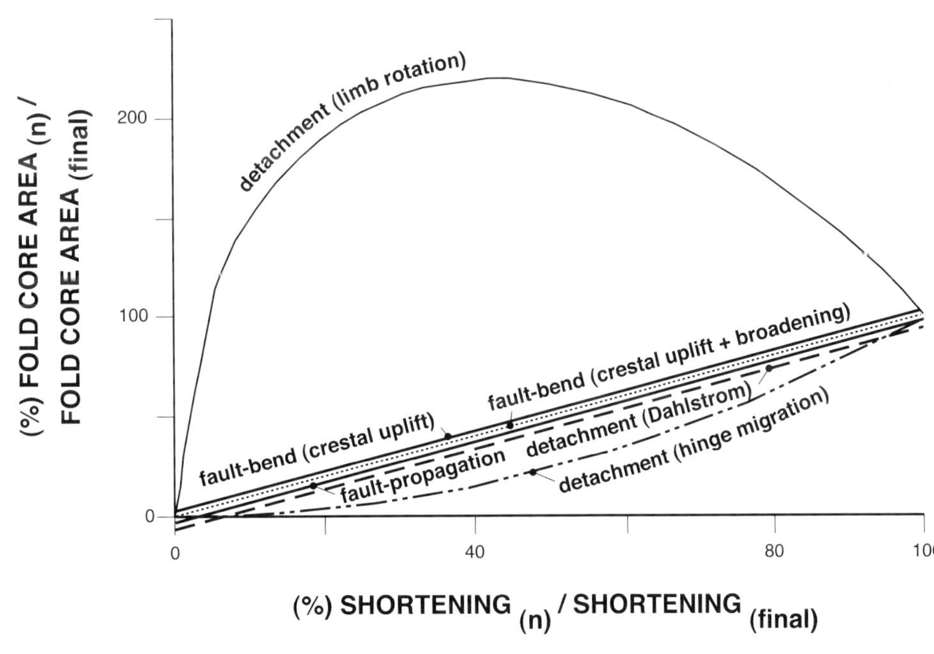

FIGURE 7. Plots of crestal structural relief versus shortening (modified after Bulnes and Poblet, 1999) (a), and fold area versus shortening (b) for six types of theoretical kink-like folds related to thrusts.

fault-propagation folds, and with detachment folds formed by the combination of hinge migration and limb rotation, the displaced area equals the fold area during all the amplification stages; that is, the fold area versus shortening is a linear function. In the case of single-detachment anticlines formed purely by limb rotation or by hinge migration, the fold-area-versus-shortening graph is not linear because of the fold amplification mechanisms used. Thus, in many amplification stages, the uplifted area does not equal the displaced area (Wiltschko and Chapple, 1977; Jones, 1987; Mitra and Namson, 1989; Dahlstrom, 1990; Butler, 1992; Groshong and Epard, 1994; Homza and Wallace, 1995; Poblet and Hardy, 1995; Bulnes and Poblet, 1999). However, in nature and in physical experiments of detachment anticlines with a ductile unit in their core, two phenomena may occur. (1) The ductile unit may flow along the cross section's plane (inside/outside/within the area enclosed within the bounding lines) and/or along planes other than that of the cross section (Norris, 1971). This may cause thickening or thinning of the ductile unit beneath the adjacent synclinal hinges, thereby leading to formation of synclines above or below the regional datum (Wiltschko and Chapple, 1977). (2) The cross section's area of the ductile unit located in the anticline's core may increase or decrease as a result of compaction and/or internal deformation. These two phenomena may provoke extremely variable relationships between the area uplifted or depressed above or below the regional datum and the shortening for these types of folds.

Theoretical Thrust-related Folds

Using growth strata to check the validity of our estimates of crestal structural relief, shortening, and fold area, we analyzed a fault-bend fold constructed using

Suppe's (1983) equations, and a fault-propagation fold constructed using Suppe and Medwedeff's (1990) equations (Figures 8 and 9). Both folds exhibit growth strata associated with them, which we added according to Suppe et al.'s (1992) method. The data have been collected from Suppe et al. (1992; their Figures 12b and 17c). The plots in Figures 8b, 8c, 9b, and 9c show that both the fold uplift and the variation of fold area have a linear relationship with shortening that is similar to the relationship in the models of thrust-bend folds and the thrust-propagation fold plotted in Figure 7. This proves the validity of the strategy presented above using growth strata. The slight dispersion of the measured points in Figures 8 and 9 is the result of measurement and drawing inaccuracies.

Natural Thrust-related Folds

Six natural examples will be used to show the applicability of the strategies described above.

1) The Pico del Aguila Anticline, mapped by Puigdefàbregas (1975), Anastasio (1987), Dobson (1990), Barnolas et al. (1991), Millán et al. (1994), and Poblet and Hardy (1995), is a field example of a detachment growth anticline located in the External Sierras that corresponds to the southern border of the Jaca Basin (southern Pyrenean Fold-and-Thrust Belt). The most active petroleum exploration in the Jaca Basin took place during the 1970s and 1980s. However, commercial well and seismic data are sparse, except from some specific areas. One gas field, which is still productive, was found in the northern part of the Jaca Basin.
2) The Mediano Anticline, mapped by Garrido (1973), Martínez-Peña (1991), Teixell and Barnolas (1995), and Poblet et al. (1998), is a field example of a detachment growth anticline located in the Ainsa Basin (southern Pyrenean Fold-and-Thrust Belt). Oil exploration in the Ainsa Basin was carried out from the 1950s to the 1980s, but seismic profiles and wells are very limited. Further geologic research in the Ainsa Basin, partially funded by some oil companies, was conducted during the 1990s and focused mainly on acquiring data from this field analog for reservoir information.
3) The Santaren Anticline, illustrated in Ball et al. (1985), Echevarría-Rodríguez et al. (1991), and Masaferro et al. (1999, 2002) is a seismic example of a detachment growth anticline located in the Bahamas foreland. It corresponds to the frontal fold of the north edge of the Cuban Fold-and-Thrust Belt. Exploration for hydrocarbons in Cuba began at the end of the nineteenth century and continues today. Since then, several on- and offshore fields have been discovered that are still being exploited. A few scattered seismic profiles across the Santaren Anticline are available, acquired by academic institutions and commercial companies during the 1970s and 1980s.
4) The frontal fold of the Mississippi Fan Fold Belt in the northern Gulf of Mexico is a seismic example of a detachment growth anticline that was illustrated in Weimer and Buffler (1992), Rowan et al.

a)

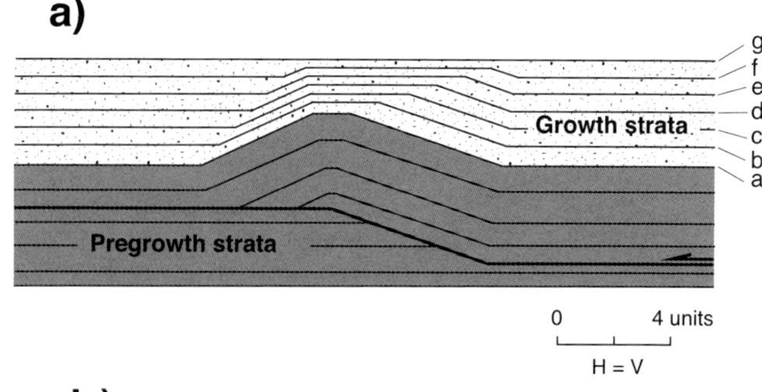

b)

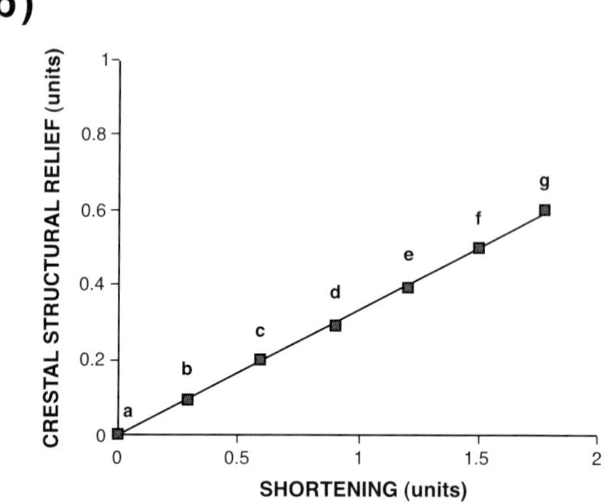

c)

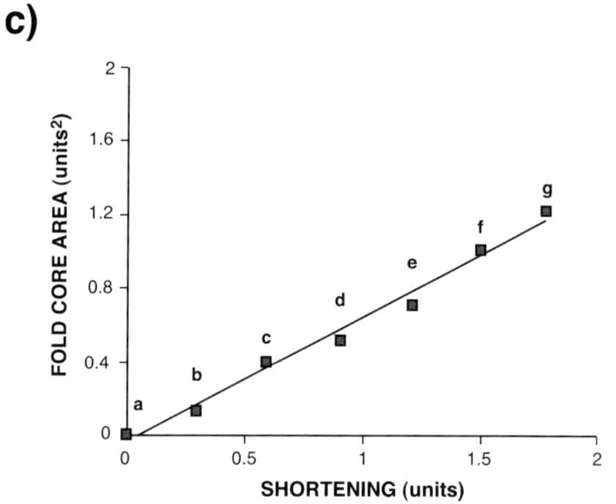

FIGURE 8. (a) Theoretical fault-bend fold (modified after Suppe et al., 1992), (b) plot of crestal structural relief versus shortening, and (c) plot of fold area versus shortening for the theoretical fault-bend fold illustrated in (a). Growth horizons have been used to take the measurements. Best-fit lines are displayed.

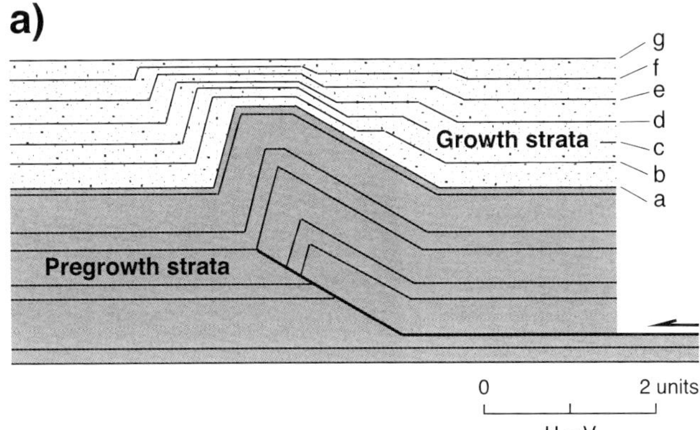

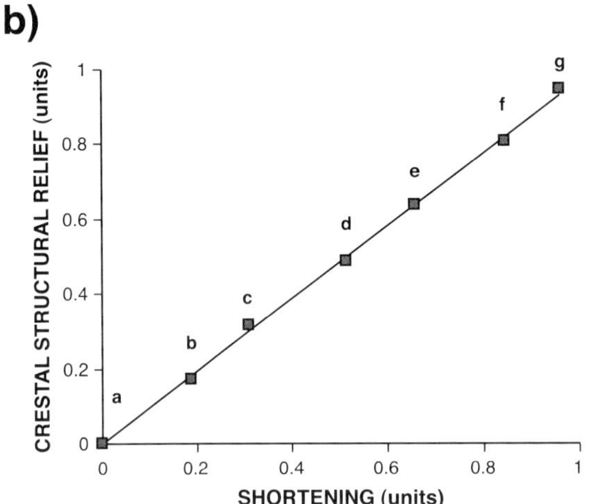

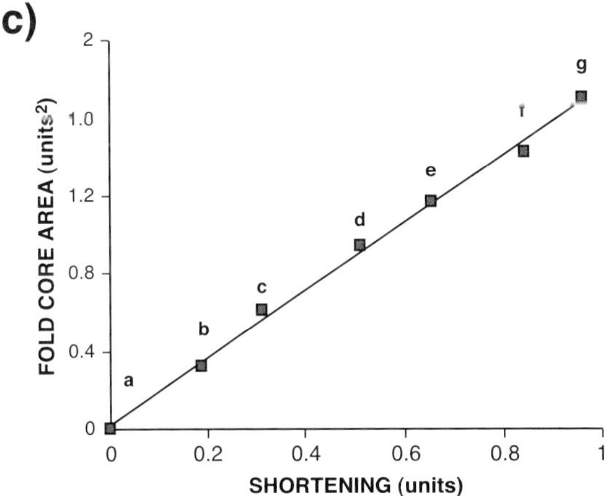

FIGURE 9. (a) Theoretical fault-propagation fold (modified after Suppe et al., 1992), (b) plot of crestal structural relief versus shortening, and (c) plot of fold area versus shortening for the theoretical fault-propagation fold illustrated in (a). Growth horizons have been used to take the measurements. Best-fit lines are displayed.

(1993), and Rowan (1997). The Mississippi Fan Fold Belt is of particular interest to the oil industry, and its petroleum potential has been recognized since the 1980s. A well plus two closely spaced seismic grids, collected by geophysical companies during the 1980s, cover the anticline analyzed here.

5) The Wheeler Ridge Anticline is a field example of a growth fault-bend fold that was illustrated in Namson and Davis (1988), Medwedeff (1992), Mueller and Suppe (1997), and Mueller and Talling (1997). It is the frontal fold of the north side of the Transverse Ranges Fold-and-Thrust Belt formed in relation with the San Andreas Fault, south-central California. The economic interest of this anticline as a potential trap for hydrocarbons is unquestionable, because it is penetrated by a high density of logged oil wells.
6) One of the frontal folds of the Apennine Fold Belt in the South Adriatic Basin is a seismic example of a growth fault–propagation fold that was illustrated in Schwander (1989). This area has been a locus of oil exploration, because some of the sediments that filled in the Apennine foredeep contain important biogenic gas accumulations. A seismic profile acquired by an oil company during the 1980s runs across the anticline.

Pico del Aguila Anticline

The west limb of the Pico del Aguila Anticline is analyzed here because the growth strata associated with this structure are well exposed in this limb (Figure 10), and their geometry allows us to estimate accurately the crestal structural relief, shortening, and fold area during deposition of different growth beds.

Paleomagnetic data document a regional clockwise rotation of the External Sierras during formation of the Pico del Aguila Anticline (Hogan, 1991; Pueyo et al., 1994, 1996; Larrasoaña et al., 1996). Séguret (1972) and other authors proposed a structural hypothesis that explains this rotation. As the South Central Pyrenean Unit moved to the south, it pushed and detached the footwall, thereby giving rise to a fold train that extends from the Ainsa Basin to the External Sierras (Figure 10a). The displacement of this thrust sheet was greatest in the central part and decreased toward the west. This caused progressive clockwise rotation of the western margin of the thrust sheet, the tectonic transport vector in this area, and the folds formed. As a result, the Pico del Aguila Anticline acquired its present north-south orientation. This structural evolution may complicate its kinematic interpretation based on the plots described above, because the estimations of crestal structural relief, shortening, and fold area require accurate cross sections constructed perpendicular to the fold axis and parallel to the tectonic transport vector. In areas involving rotation around vertical axes coeval to fold amplification, there are two situations in which it is extremely complicated or impossible to construct sections across folds that are valid for the structural analysis

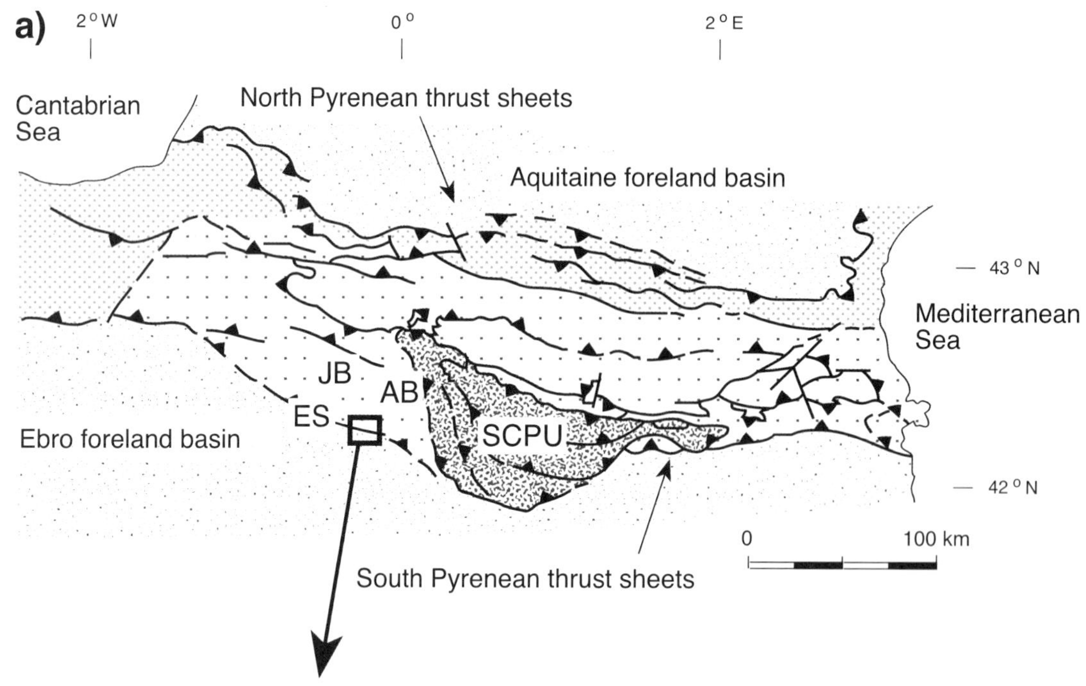

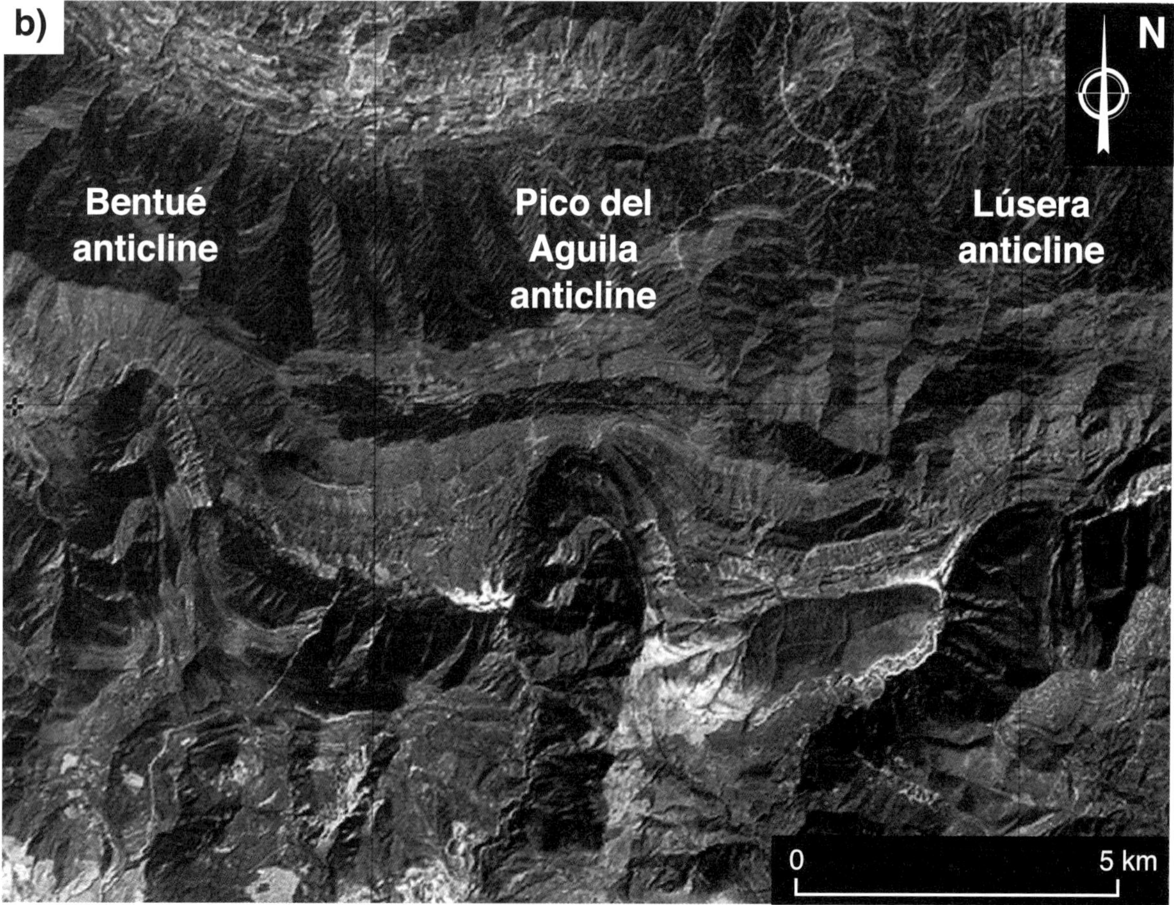

FIGURE 10. (a) Tectonic sketch of the Pyrenean Ranges. SCPU = South Central Pyrenean Unit, AB = Ainsa Basin, JB = Jaca Basin, and ES = External Sierras. (b) Satellite image of part of the External Sierras, southern Pyrenees, Spain. The scale is not accurate.

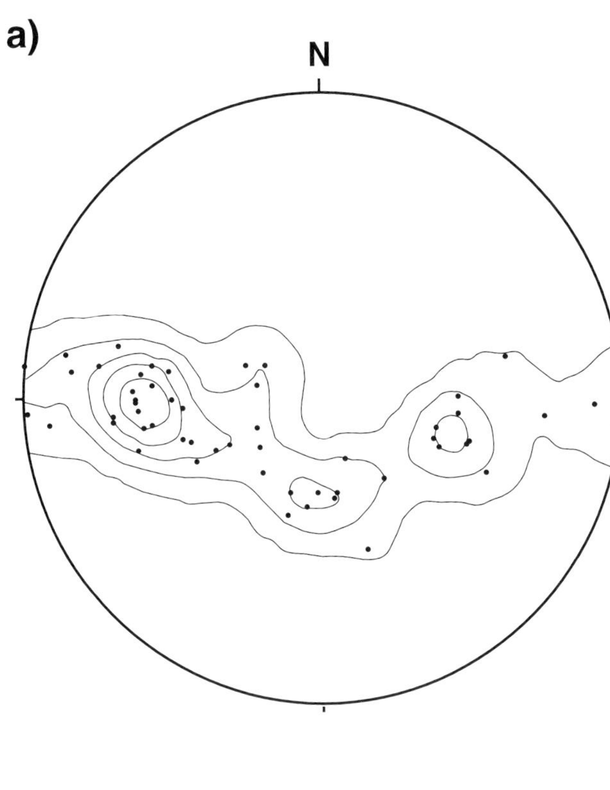

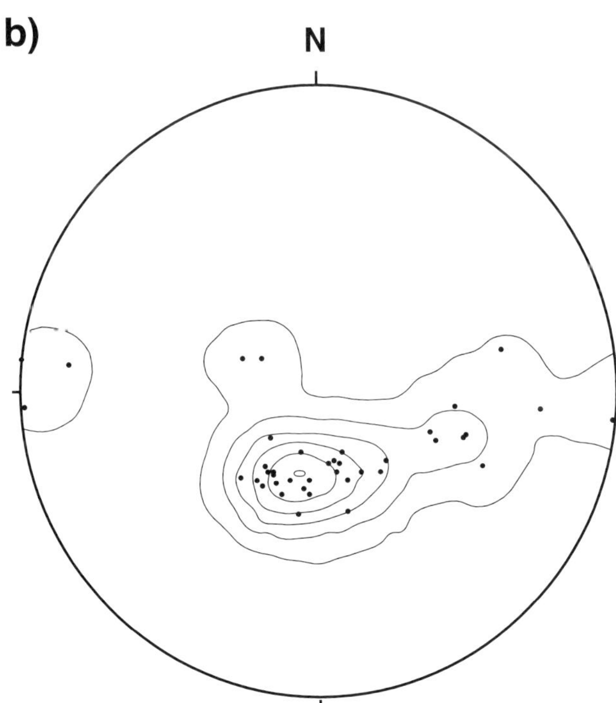

FIGURE 11. Stereoplots of the poles to bedding measured in the pregrowth strata located in the anticlinal hinge and limbs of the Pico del Aguila Anticline (54 data, contours at 1, 3, 5, 7, and 9%) (a), and in the growth strata located around the west synclinal hinge of the Pico del Aguila Anticline (40 data, contours at 1, 4, 7, 10, and 13%) (b). Dips and strikes taken from Anastasio (1987), Dobson (1990), and Poblet and Hardy (1995).

described above: (1) unequal rotation of the limbs of the folds around vertical axes resulting in conical folds, and (2) migration/rotation around vertical axes of the fold hinges leading to a progressive reorientation of the crests and troughs through time. We will show that neither of these two situations occurred during the amplification of the Pico del Aguila Anticline.

Measurements of the pregrowth beds show that the Pico del Aguila Anticline is a slightly conical fold with a small semi-apical angle, because the best-fit circle does not coincide exactly with a great circle (Figure 11). In addition, both the pregrowth beds and each synsedimentary bed show approximately coincident north-south fold axes that plunge from 26° to 30° to the north. This disagrees with the Pico del Aguila Anticline having a strongly conical shape produced during syntectonic sedimentation, which would result in a lower plunge of the fold axis in progressively younger synsedimentary beds and a progressive thickening of the growth strata toward the north. Moreover, if we assume that the growth strata were deposited horizontally, it follows that the fold was formed with an approximately flat-lying axis, because the growth strata in the core of the adjacent syncline presently dip about 30° to the north. Unfortunately, no data are available on the amount of clockwise rotation the limbs of the Pico del Aguila Anticline have undergone. However, paleomagnetic data collected in the anticline located immediately west of the Pico del Aguila Anticline (the Bentué Anticline) (Figure 10b) show that both limbs underwent approximately the same amount of clockwise rotation (Pueyo et al., 1996). If we assume that the history of the Bentué Anticline is comparable to that of the Pico del Aguila Anticline, this is in accordance with the slightly conical shape of the Pico del Aguila Anticline.

Significant hinge migration/rotation around a vertical axis during the clockwise rotation of the Pico del Aguila Anticline would have resulted in anomalous positions of the thickest and thinnest portions of some syntectonic beds as a consequence of the progressive reorientation of the depocenters. Thus, the maximum thickness of some syntectonic beds should be located in the anticline's limbs or in the anticline's crest, and the minimum thickness should be located in the anticline's limbs or in the basins adjacent to the anticline. Such "irregularities" have not been identified in the Pico del Aguila Anticline, thus ruling out the hypothesis of significant hinge migration/rotation around a vertical axis during the clockwise rotation of the External Sierras.

The observations presented above suggest that the Pico del Aguila Anticline and adjacent basins behaved as an almost rigid block during the clockwise rotation that affected the External Sierras. Significant, unequal migration/rotation is unlikely around a vertical axis of the anticline's limbs and the hinges that bound the anticline. Because the semi-apical angle is small, the Pico

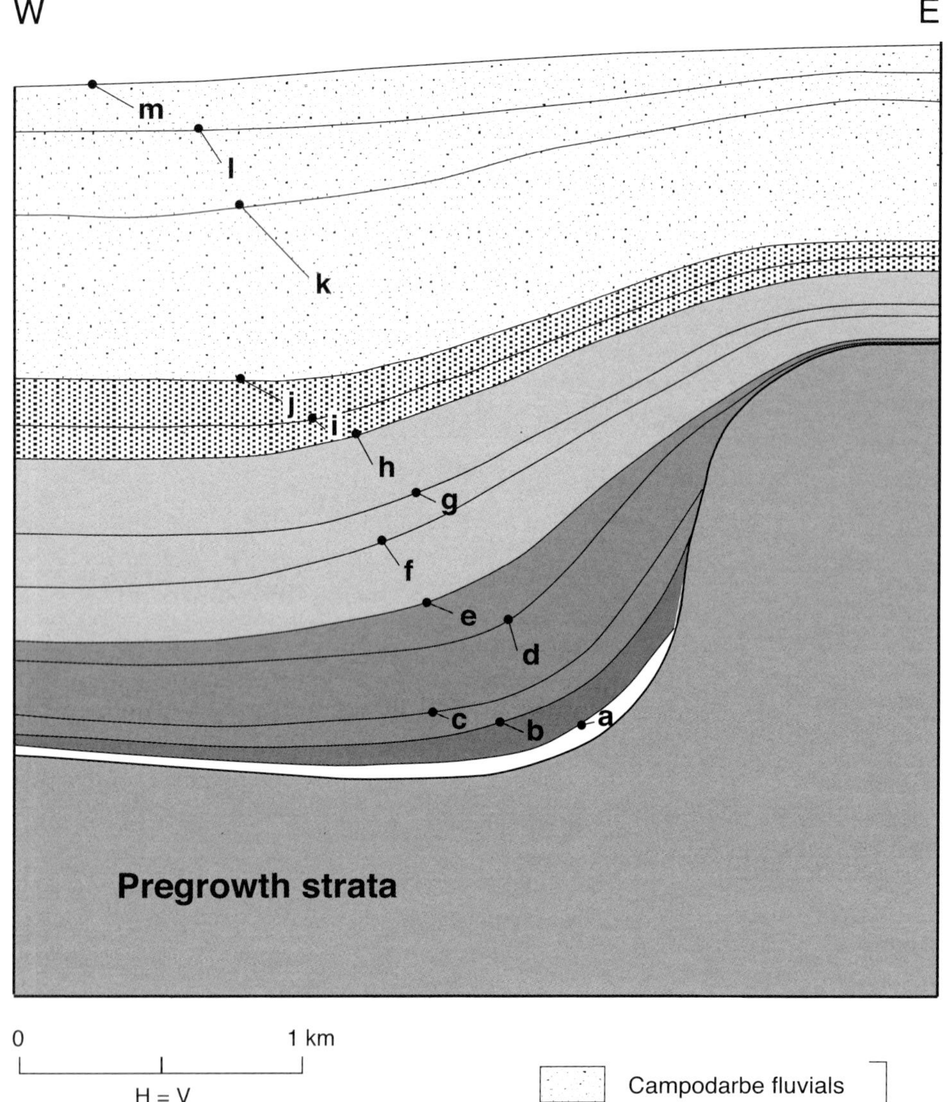

FIGURE 12. Down-plunge view of the Pico del Aguila Anticline, constructed using Poblet and Hardy (1995) geologic data.

del Aguila Anticline can be considered to be an approximately cylindrical fold, for the purposes of our analysis. Because its fold axis plunges to the north, a profile perpendicular to the fold axis has been constructed using Ramsay and Huber's (1987) method and Poblet and Hardy's (1995) field data (Figure 12). Taking into account the above considerations, we believe that the down-plunge profile of the Pico del Aguila Anticline shows the nearly correct cross-sectional shape of the fold. Therefore, the effect of its slightly conical shape is negligible in the measurements of layer thicknesses, bed lengths, and areas needed to estimate the crestal structural relief, shortening, and fold area.

Before we apply the strategies described above, we will present a number of features to unravel some aspects of the kinematics of the Pico del Aguila Anticline. This structure is an almost isoclinal fold (Millán et al., 1994; Poblet and Hardy, 1995) formed by a ductile unit that underlies a thick, competent package (Puigdefàbregas, 1975; Anastasio, 1987; Dobson, 1990; Millán et al., 1994; Poblet and Hardy, 1995). The geometry of the growth strata (Figures 10b and 12) resembles the "progressive unconformities" defined by Riba (1976), and it also resembles the geometry obtained in some forward models of detachment folds (Hardy and Poblet, 1994; Torrente and Kligfield, 1995; Hardy et al., 1996; Poblet et al., 1997; Storti and Poblet, 1997). These features allow us to interpret the Pico del Aguila Anticline as a detachment fold that involved an important amount of limb rotation. A remarkable feature of this anticline is that the maximum thickness of some growth beds occurs in the west synclinal hinge (Figure 12). This indicates that this synclinal hinge sank more than the rest of the adjacent west basin during fold amplification. This has been observed in natural examples (Prucha, 1968; Frey, 1973; Wiltschko and Chapple, 1977) and physical experiments (Blay et al., 1977; Abbassi and Mancktelow, 1992) of folds involving limb rotation. In these folds, the fold core area increases dramatically during the initial amplification stages, so that ductile material migrates to fill in the anticline's core. The resulting withdrawal of ductile material from below the adjacent synclinal hinges causes the synclinal hinges to sink. Figure 13a illustrates the position of the boundary

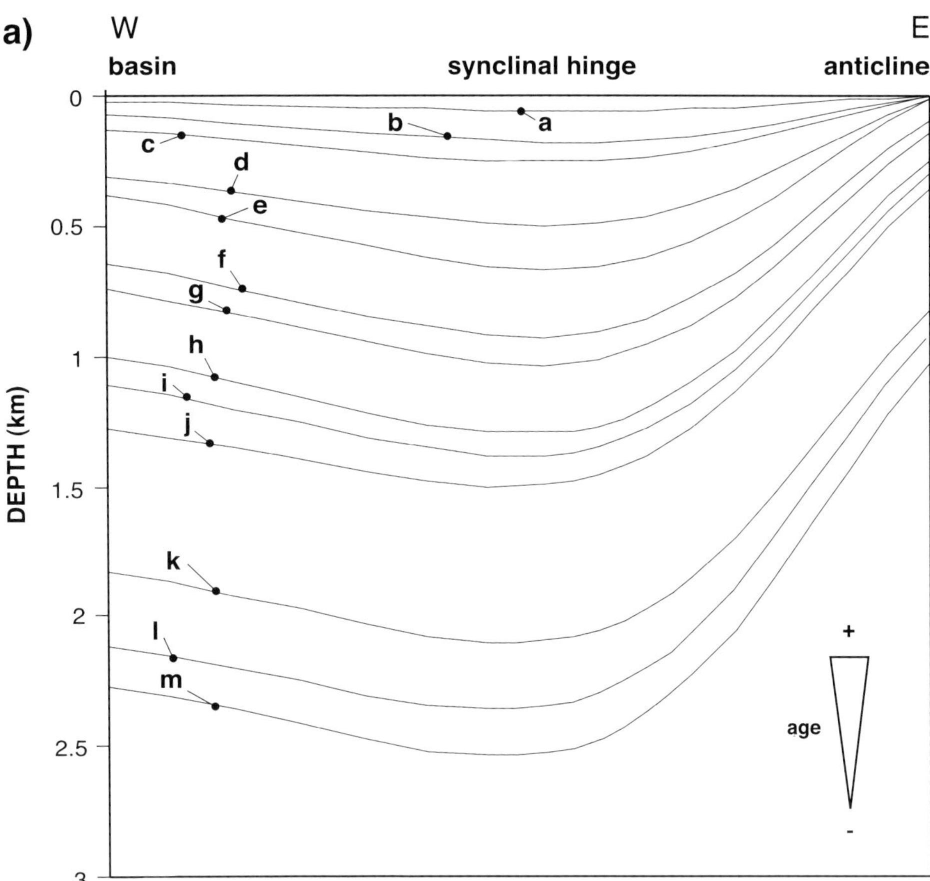

FIGURE 13. (a) Plot that illustrates the depth of the top of the pregrowth strata during deposition of different growth beds associated with the Pico del Aguila Anticline. The *x*-axis is not to scale and this is the reason why the curves do not illustrate the proper geometry of the top of the pregrowth strata. (b) Plot of thickness of individual growth beds in the west basin minus their thickness in the west synclinal hinge of the Pico del Aguila Anticline. Both graphs have been constructed using the thickness measured in the down-plunge view in Figure 12.

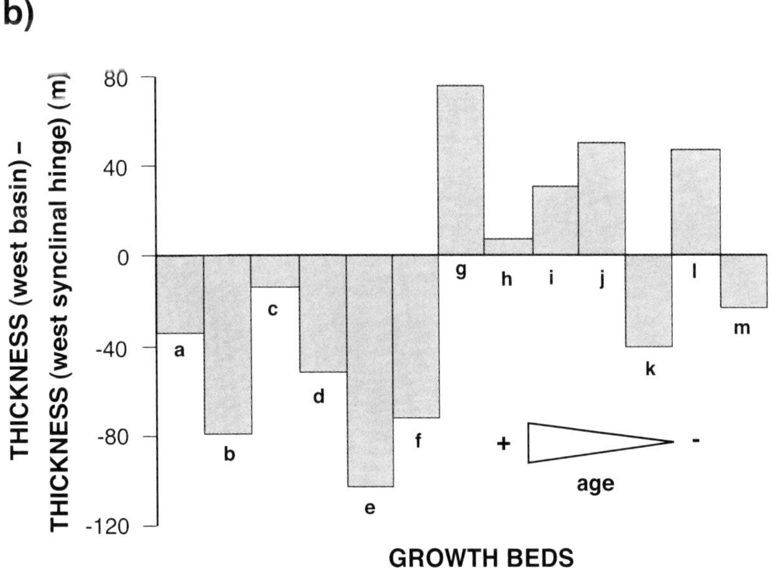

between the pregrowth and growth beds during different stages of fold amplification, and Figure 13b shows the difference in thickness between the basin and the west synclinal hinge for the growth beds analyzed. The graph in Figure 13a assumes that the boundary between the pregrowth and growth strata was horizontal at the beginning of syntectonic sedimentation. Both plots show that the west synclinal hinge underwent significant sinking, especially during the initial stages of fold amplification.

According to Poblet and Hardy's (1995) conclusions and the observations listed above, the Pico del Aguila Anticline appears to be a detachment fold that formed mainly by limb rotation. However, the trend of the crestal-structural-relief-versus-shortening curve obtained for the rounded-hinge Pico del Aguila Anticline (Figure 14a) does not match the plots obtained for the theoretical, sharp-hinge detachment folds (kink, chevron, and box folds) that involve limb rotation (Figure 7a). During the initial stages, the Pico del Aguila Anticline uplifted slowly until two-thirds of the total shortening (approximately 650 m) was reached. At this specific value of shortening, the process accelerated substantially, beginning a sudden explosive fold uplift. Pure kink, chevron, and box folds may uplift differently from other types of rounded-hinge folds, even if the formation mechanism is similar (mainly limb rotation,

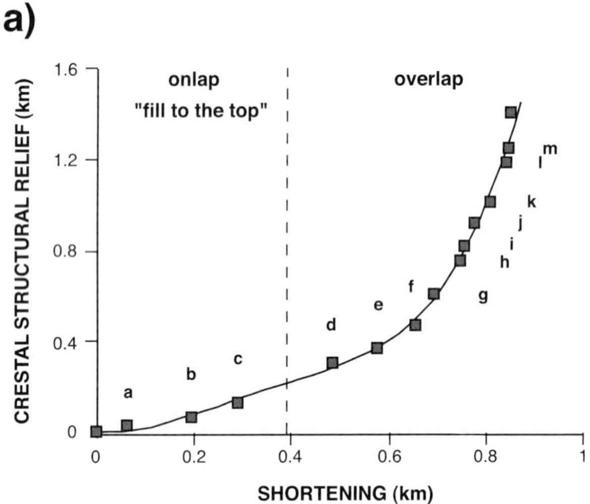

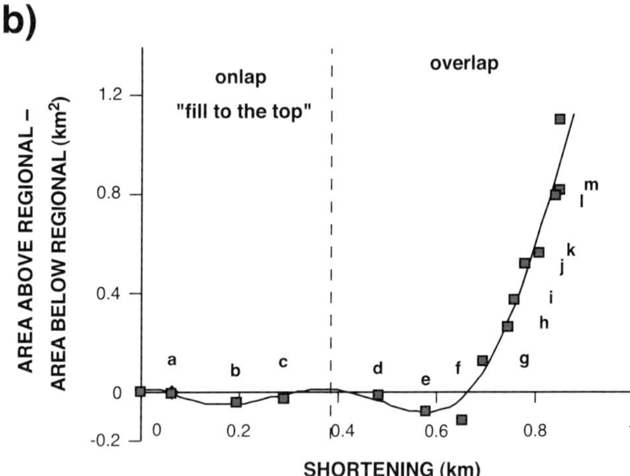

FIGURE 14. Plots of crestal structural relief versus shortening (a), and fold area versus shortening (b) for the west limb of the Pico del Aguila Anticline. Growth horizons have been used to take the measurements. The point with the greatest crestal structural relief and fold area corresponds to the present-day fold geometry. Best-fit lines are displayed. To estimate the crestal structural relief, shortening, and fold area during the initial stages (when growth beds onlap the fold), we have assumed that the crest of the anticline did not produce any topographic or bathymetric positive relief, that is, the beds "filled to the top" the space available for deposition.

in this particular case), thereby causing the discrepancy we observed.

The fold-area-versus-shortening curve obtained for the Pico del Aguila Anticline (Figure 14b) does not resemble that of detachment folds involving limb rotation (Figure 7b), either. During the initial amplification stages, the area uplifted above the regional datum was smaller than the area depressed below the regional datum. This agrees with the plot in Figure 13b, which indicates that during deposition of the six oldest growth beds, their thickness was substantially greater in the west synclinal hinge than in the adjacent basin. The fact that the crestal structural relief of the anticline was small and increased slowly during these initial stages (Figure 14a) might have contributed to making the depressed area larger than the uplifted area. A situation with depressed areas that are larger than uplifted areas suggests that the ductile unit flowed outside the area enclosed within the bounding lines and/or to planes other than those of the cross section. This phenomenon was a likely occurrence, because the Pico del Aguila Anticline amplified synchronously with other anticlines in the External Sierras (Millán et al., 1994), and ductile material may have migrated to fill in the cores of adjacent folds. During the last stages of its development, the area of the Pico del Aguila Anticline increased in approximately linear fashion (Figure 14b). This coincided with the acceleration in the fold uplift (Figure 14a).

Mediano Anticline

The second natural example we present is the Mediano Anticline, because four subparallel sections across the anticline show different geometries (Figure 15a) and are available in Poblet et al. (1998; their Figure 5). To derive the crestal structural relief, shortening, and fold area, we will assume that these different cross sections illustrate the anticline's kinematic evolution. Poblet et al.'s (1998) analysis of the Mediano Anticline, the geometry of the pregrowth and growth strata, and the mechanical stratigraphy all indicate that the Mediano Anticline is a detachment fold cored by a ductile unit that formed by limb rotation and minor hinge migration. The plots illustrated in Figure 15b and c corroborate this interpretation. Figure 15 presents the Mediano Anticline data together with curves obtained for theoretical detachment folds with the same final geometry as that of the Mediano Anticline. The curves obtained for the Mediano Anticline fall within the range defined by detachment folds formed by pure limb rotation and detachment folds formed by a combination of hinge migration and limb rotation, according to Dahlstrom's (1990) mechanism. However, they are more similar to folds formed by limb rotation than to folds formed by hinge migration, which points out that the Mediano Anticline involved a major component of limb rotation.

Santaren Anticline

The third natural example we document is the north limb of the Santaren Anticline, because it is a well-imaged growth fold (Figure 16a). The data were collected from a good quality seismic profile illustrated

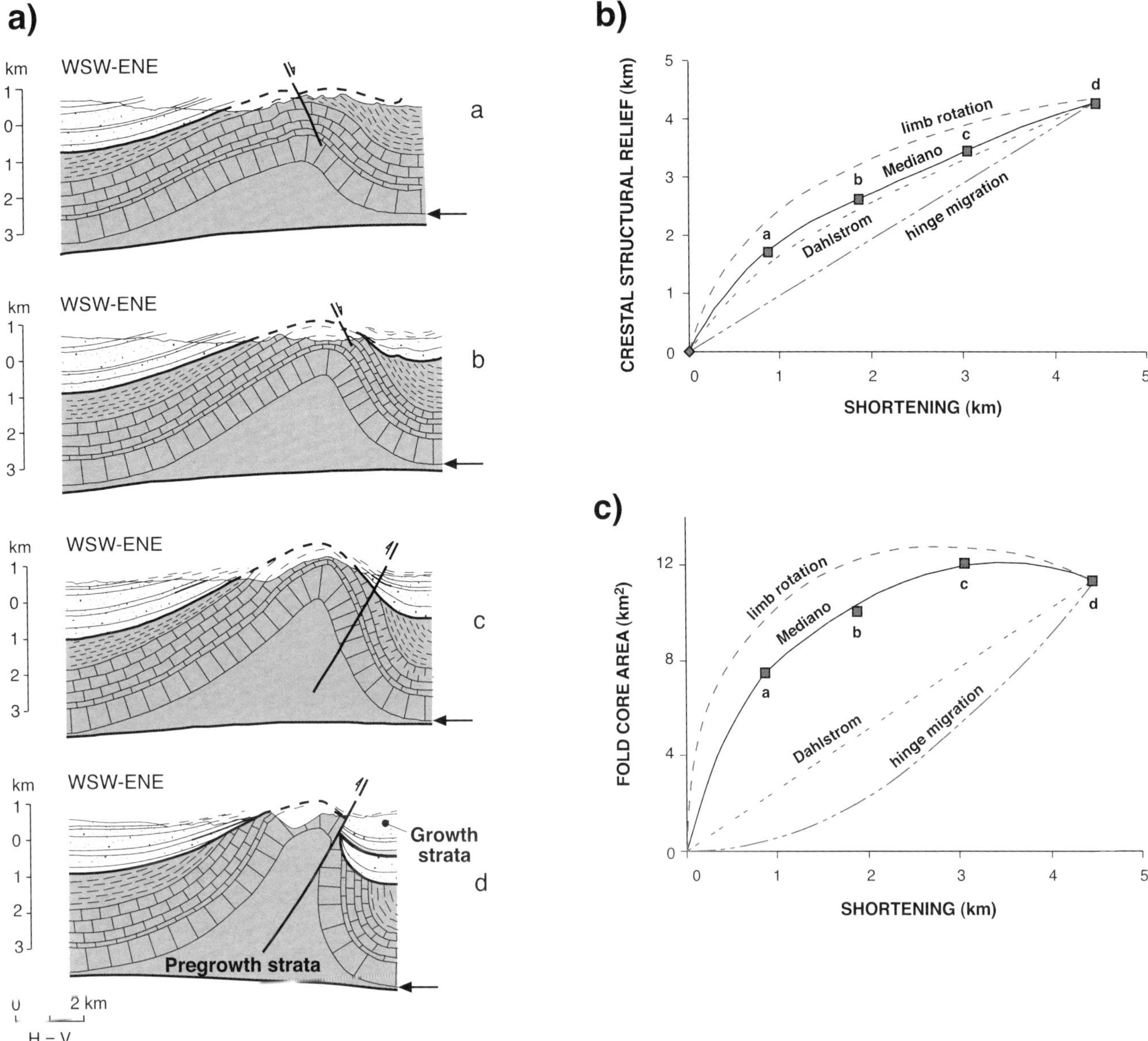

FIGURE 15. (a) Four sections from north (section *a*) to south (section *d*) across the Mediano Anticline (after Poblet et al., 1998 modified), (b) plot of crestal structural relief versus shortening, and (c) plot of fold area versus shortening for the Mediano Anticline. The arrows indicate the horizon used to take the measurements. Best-fit lines are displayed. Three theoretical detachment anticlines have also been plotted and have the same final geometry formed by hinge migration, limb rotation, and a combination of hinge migration and limb rotation according to Dahlstrom's (1990) mechanism.

in Masaferro et al. (1999; their Figure 8). The analysis we present here uses the syntectonic sediments associated with the anticline and shows the fold evolution during deposition of the growth beds. The geometry of the pregrowth and growth strata allowed Masaferro et al. (1999) to conclude that the Santaren Anticline is a detachment fold that involved some limb rotation. The plots of crestal structural relief versus shortening and of fold area versus shortening (Figure 16b, c) are similar and indicate that this anticline exhibits a two-step behavior. The first period (beds mainly onlapping the anticline) corresponds to the fold evolution until it reached 95% of its total shortening. This initial period is characterized by curves that have a steeper slope at the beginning than at the end. This kinematic path coincides with the trends obtained for theoretical detachment folds that involve a certain amount of limb rotation (Figure 7). The second period (beds overlapping the anticline) is characterized by very steep curves that are almost linear or whose slope increases with increasing shortening. This path is similar to that observed in the case of detachment folds formed mainly by hinge migration (Figure 7). Although not perfectly imaged, this two-step evolution may be in accordance with the geometry of the growth strata associated with the Santaren Anticline (Masaferro et al., 1999; their

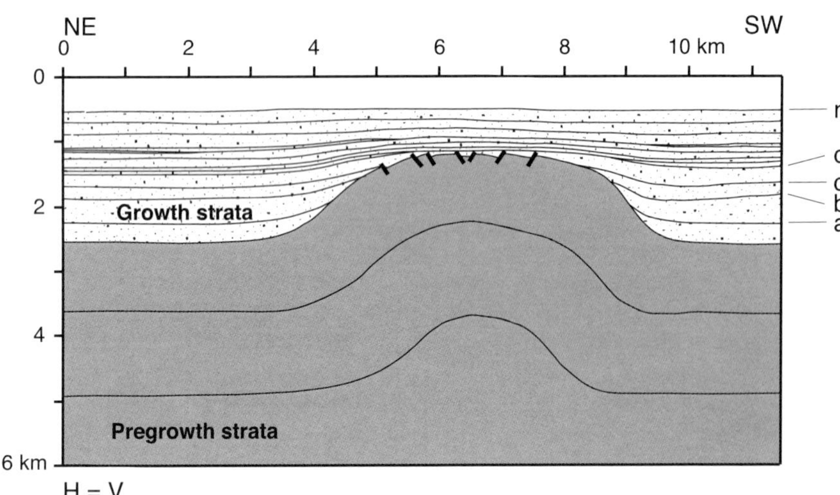

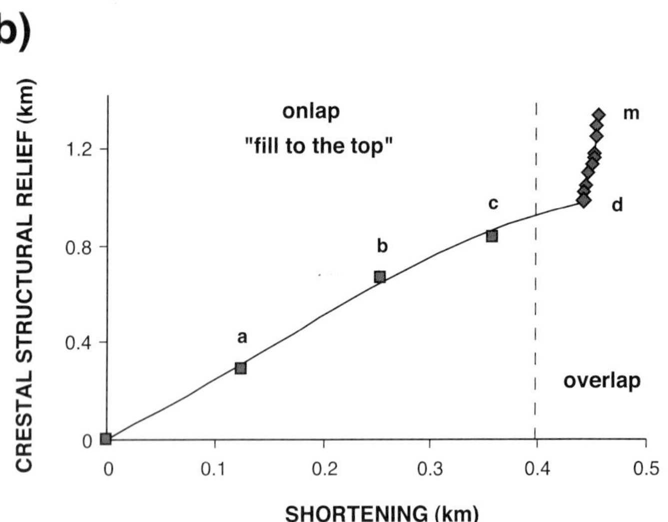

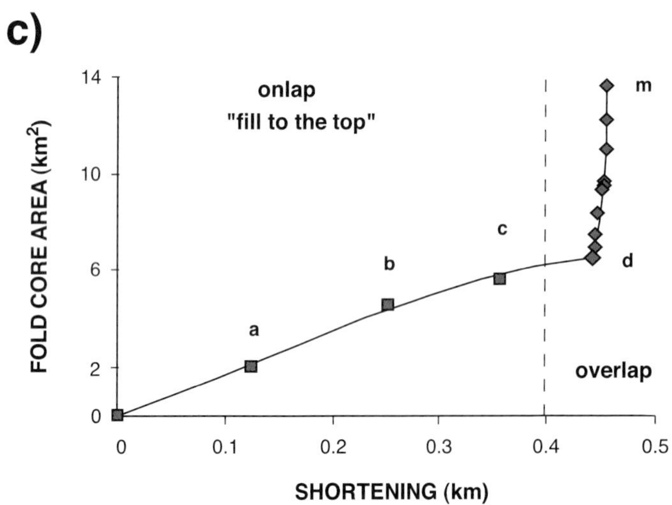

FIGURE 16. (a) Section across the Santaren Anticline (modified after Masaferro et al., 1999), (b) plot of crestal structural relief versus shortening, and (c) plot of fold area versus shortening for the north limb of the Santaren Anticline. Growth horizons have been used to take the measurements. Best-fit lines are displayed. To estimate the crestal structural relief, shortening, and fold area during the initial stages (when growth beds onlap the fold), we have assumed that the crest of the anticline did not produce any topographic or bathymetric positive relief, that is, the beds "filled to the top" the space available for deposition.

Figure 8). The onlapping growth beds tend to thin gradually toward the fold limbs and decrease their dip progressively upward, thereby indicating limb rotation. The overlapping growth strata consist mainly of panels of rocks with approximately constant thickness. The thickness decreases slightly above the anticline's crest, indicating a major component of hinge migration (Poblet et al., 1997). Masaferro et al. (1999) suggested that the Santaren Anticline is still active. If this is correct, the functions illustrated in Figure 16 do not represent the complete evolution of the Santaren Anticline.

Frontal Anticline of the Mississippi Fan Fold Belt

Six parallel sections across the frontal anticline of the Mississippi Fan Fold Belt, derived from good quality seismic profiles, are illustrated in Rowan (1997; his Figure 10). The geometry of the pregrowth and growth strata allowed Rowan (1997) to conclude that this anticline is a salt-cored detachment fold formed by a combination of hinge migration and limb rotation that was cut by some reverse faults. The crestal structural relief, shortening, and fold area were obtained from (1) two sections across the anticline that show syntectonic sediments

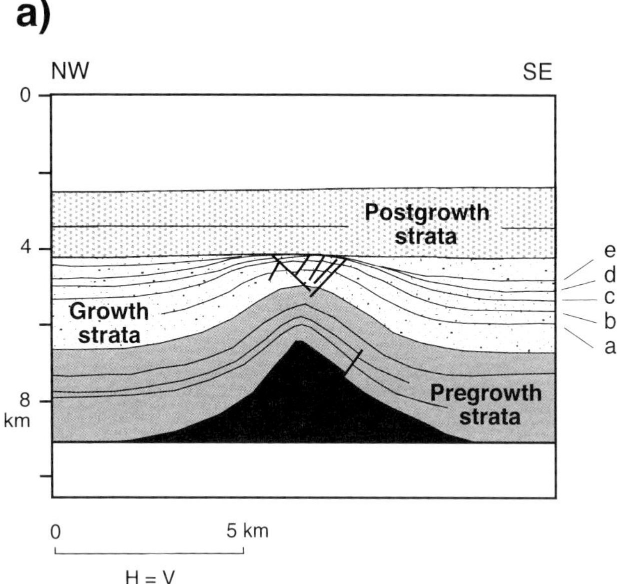

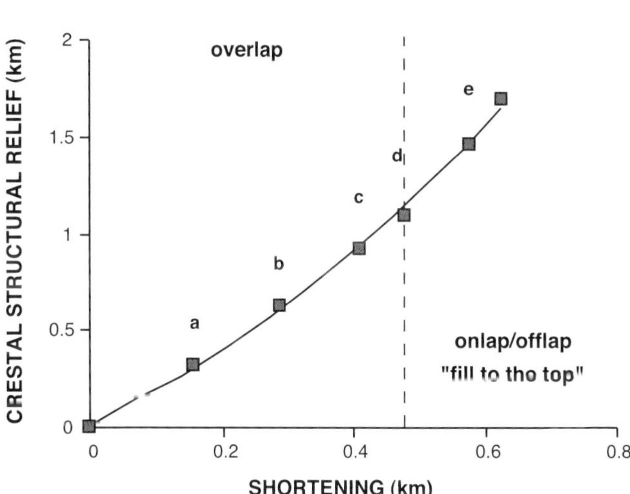

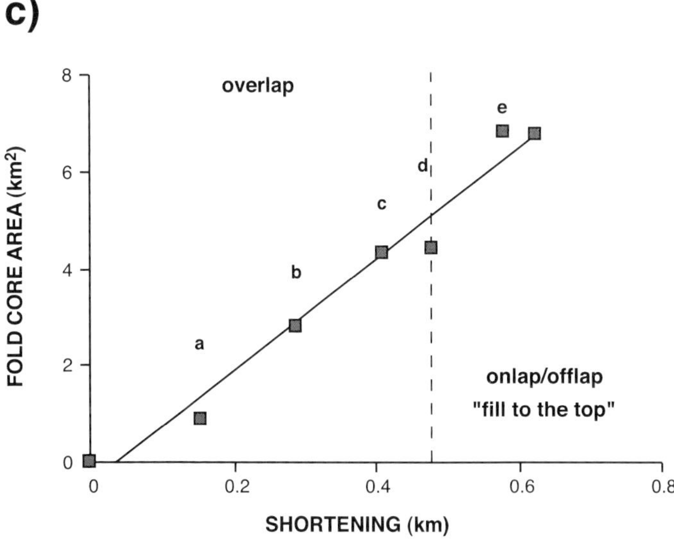

FIGURE 17. (a) Section across the frontal anticline of the Mississippi Fan Fold Belt (modified after Rowan, 1997), (b) plot of crestal structural relief versus shortening, and (c) plot of fold area versus shortening for the frontal anticline. Growth horizons have been used to take the measurements. The point with the greatest crestal structural relief and fold area corresponds to the present-day fold geometry. Best-fit lines are displayed. To estimate the crestal structural relief, shortening, and fold area during the final stages (when growth beds offlap the fold), we have assumed that the crest of the anticline did not produce any topographic or bathymetric positive relief, that is, the beds "filled to the top" the space available for deposition. This cross section corresponds to cross section *b* in Figure 19a.

associated with the fold (Figures 17 and 18), and (2) five cross sections that we assume reflect the kinematic evolution of the anticline (Figure 19). The results obtained in the first case (analysis of the syntectonic sediments) show the evolution of the anticline during deposition of the growth beds. The fold-core-area-versus-shortening functions derived from the analysis of the syntectonic sediments are linear (Figures 17c and 18c). Therefore, they support Rowan's (1997) hypothesis, because they coincide with those of theoretical detachment folds that formed by a combination of hinge migration and limb rotation according to Dahlstrom's (1990) mechanism (Figure 7b). This implies that the displaced area equaled the fold core area during all the amplification stages, and, therefore, that the ductile material that cores the anticline may have flowed only along the transport direction (cross-section planes) with no oblique/transversal flow. The crestal-structural-relief-versus-shortening plots derived from the syntectonic sediments show that the rate of crestal uplift increases with increasing shortening (Figures 17b and 18b). This trend obtained for the rounded-hinge anticline analyzed here does not match the plots obtained for theoretical, sharp-hinge detachment folds (kink, chevron, and box folds) that involve hinge migration and limb rotation (Figure 7a). As we theorized with the Pico del Aguila Anticline, pure kink, chevron, and box folds may uplift in a manner different from that of other types of folds with rounded hinges, even if the formation mechanism is similar (hinge migration plus limb rotation, in this particular case).

The cross sections across the frontal anticline of the Mississippi Fan Fold Belt exhibit a progressive increase in shortening from section *a* to section *e* in Figure 19a. However, two observations render unreliable the fold-kinematics characterization in which we assume that these different

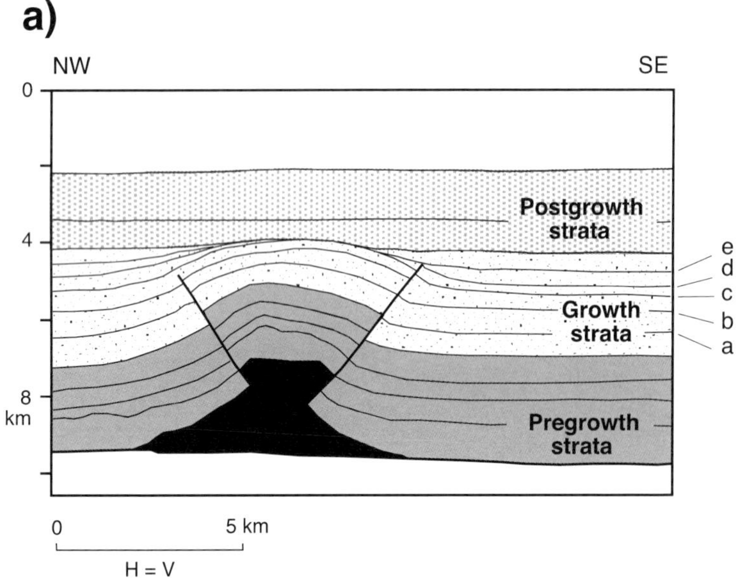

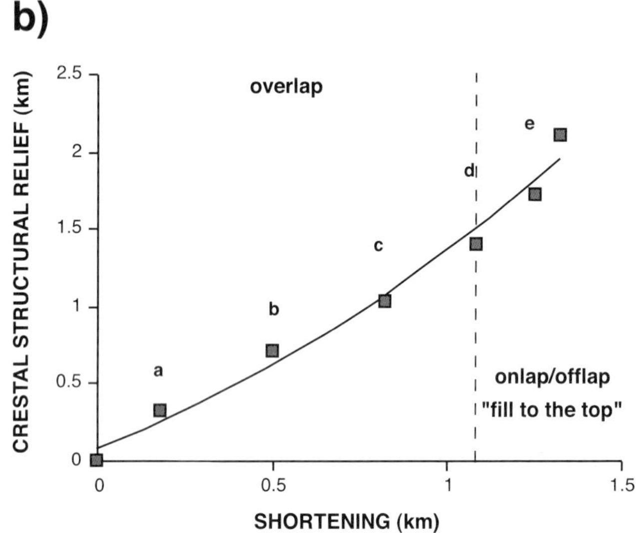

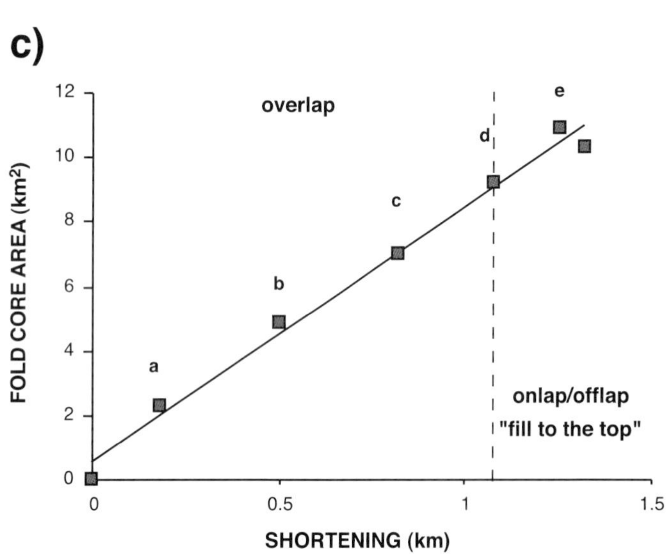

FIGURE 18. (a) Section across the frontal anticline of the Mississippi Fan Fold Belt (modified after Rowan, 1997), (b) plot of crestal structural relief versus shortening, and (c) plot of fold area versus shortening for the frontal anticline. Growth horizons have been used to take the measurements. The point with the greatest crestal structural relief and fold area corresponds to the present-day fold geometry. Best-fit lines are displayed. To estimate the crestal structural relief, shortening, and fold area during the final stages (when growth beds offlap the fold), we have assumed that the crest of the anticline did not produce any topographic or bathymetric positive relief, that is, the beds "filled to the top" the space available for deposition. This cross section corresponds to cross section *d* in Figure 19a.

cross sections along strike reflect the fold evolution. (1) Contour maps presented by Rowan (1997) show that the three-dimensional geometry of the fold is segmented in an en échelon pattern. These maps illustrate separate culminations related to the major faults that cut the anticline (see cross sections in Figure 19a). (2) The crestal uplift curves in Figures 17b and 18b show that the cross sections in Figures 17a and 18a are not different stages of the same kinematic evolution. Thus, both uplift functions show the same trend, but the specific values do not coincide (Figures 17b and 18b). For instance, the crestal uplift reaches a value of 1 km when the shortening undergone by the cross section in Figure 17a is slightly greater than 0.4 km (Figure 17b). On the contrary, the crestal uplift reaches a value of 1 km when the shortening undergone by the cross section in Figure 18a reaches the range 0.8–0.9 km (Figure 18b). However, to check the fold analysis assuming kinematic equivalence along strike, we have measured the crestal structural relief, shortening, and fold core area for a particular horizon in five different cross sections. According to this analysis, the rate of variation of crestal structural relief decreases with shortening (Figure 19b), whereas the fold core area increases linearly with respect to the shortening (Figure 19c). The curves obtained from the analysis of the syntectonic sediments and the curves obtained assuming kinematic equivalence along strike may not record the same periods of fold evolution. However, we compare them below to visualize the differences in the results obtained. The path and

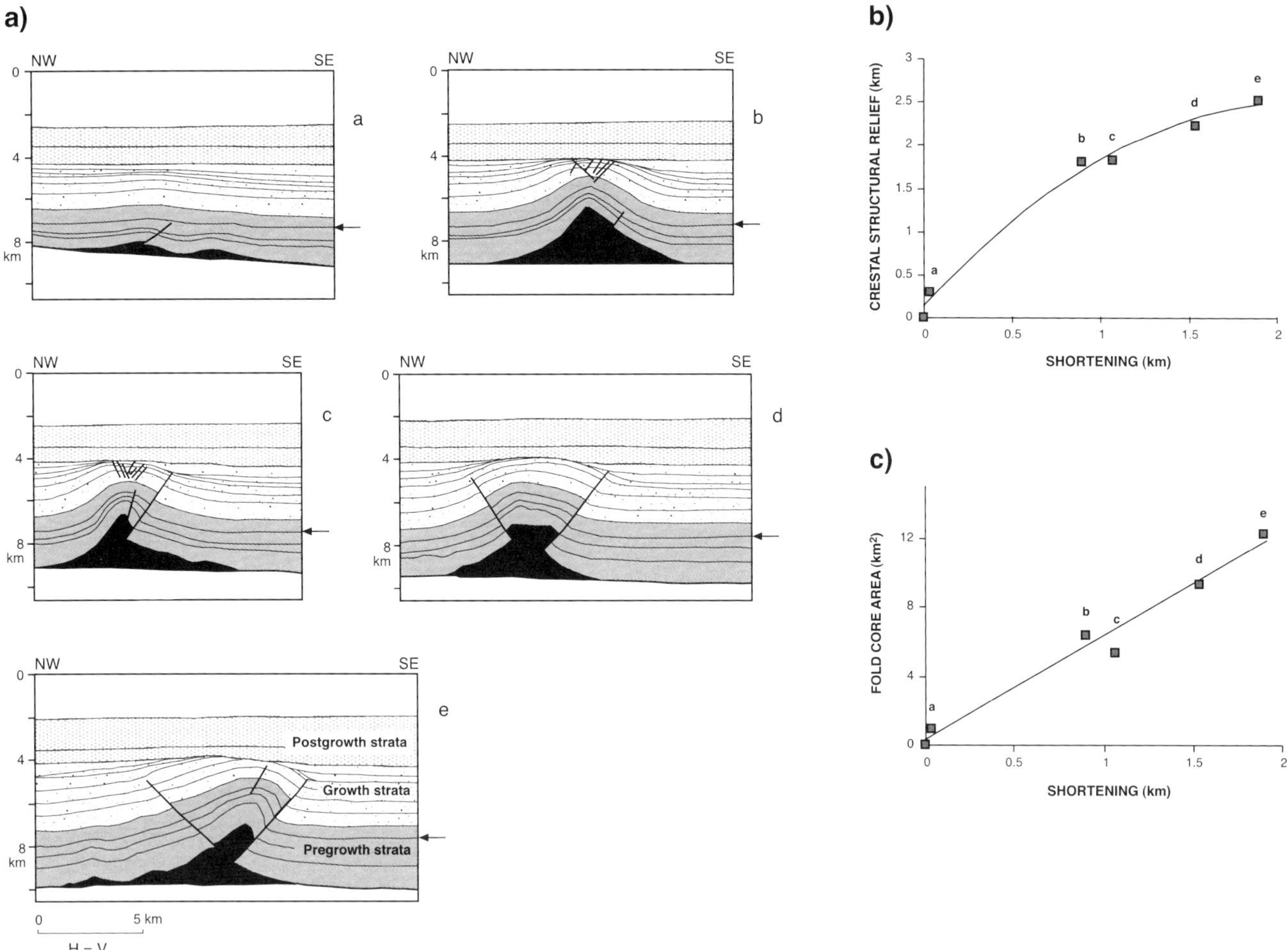

FIGURE 19. (a) Five sections from east (section *a*) to west (section *e*) across the frontal anticline of the Mississippi Fan Fold Belt (modified after Rowan, 1997), (b) plot of crestal structural relief versus shortening, and (c) plot of fold area versus shortening for the frontal anticline. The arrows indicate the horizon used to take the measurements. Best-fit lines are displayed. The cross sections illustrated in Figures 17a and 18a correspond to the cross sections *b* and *d*, respectively.

values of the fold-core-area-versus-shortening curves derived from the analysis of the syntectonic sediments (Figures 17c and 18c) coincide with the curve obtained from the analysis of several cross sections (Figure 19c). However, the path of the crestal-structural-relief-versus-shortening curve is different between the two types of analysis (syntectonic sediments and several cross sections along strike) (Figures 17b, 18b, and 19b). This means that the variation of crestal structural relief derived from the analysis of different cross sections along strike may be wrong.

Wheeler Ridge Anticline

We analyze and present here the eastern termination of the Wheeler Ridge Anticline, because four subparallel sections (Figure 20a), well constrained by surface exposures, topographic data, and subsurface information, are available in Medwedeff (1992; his Figures 1.7a, 1.9a, 1.10a, and 1.11a). The geometry of the anticline led Namson and Davis (1988) and Medwedeff (1992) to conclude that the Wheeler Ridge Anticline is a fault-bend fold. The geometry of the pregrowth beds, together with the geometry and position of the growth sediments (terraces and alluvial fan ridges) associated with the anticline, suggest that this fold amplified by hinge migration (Mueller and Suppe, 1997; Mueller and Talling, 1997). Geomorphic features and soil ages indicate that the anticline grew by lateral propagation of the fold tip from west to east (Medwedeff, 1992; Mueller and Talling, 1997; Keller et al., 1998, 1999). In theory, this makes the estimates of crestal structural relief, shortening, and fold area viable, assuming that different north-south sections across the anticline from east (section *a*) to west (section *d*) illustrate its kinematic evolution. The graphs in Figure 20b and c do not exhibit a simple kinematic evolution. During the initial period (cross sections *a* and *b* in Figure 20a), the crestal structural relief increases linearly with shortening (Figure 20b), whereas the rate of variation of fold core area

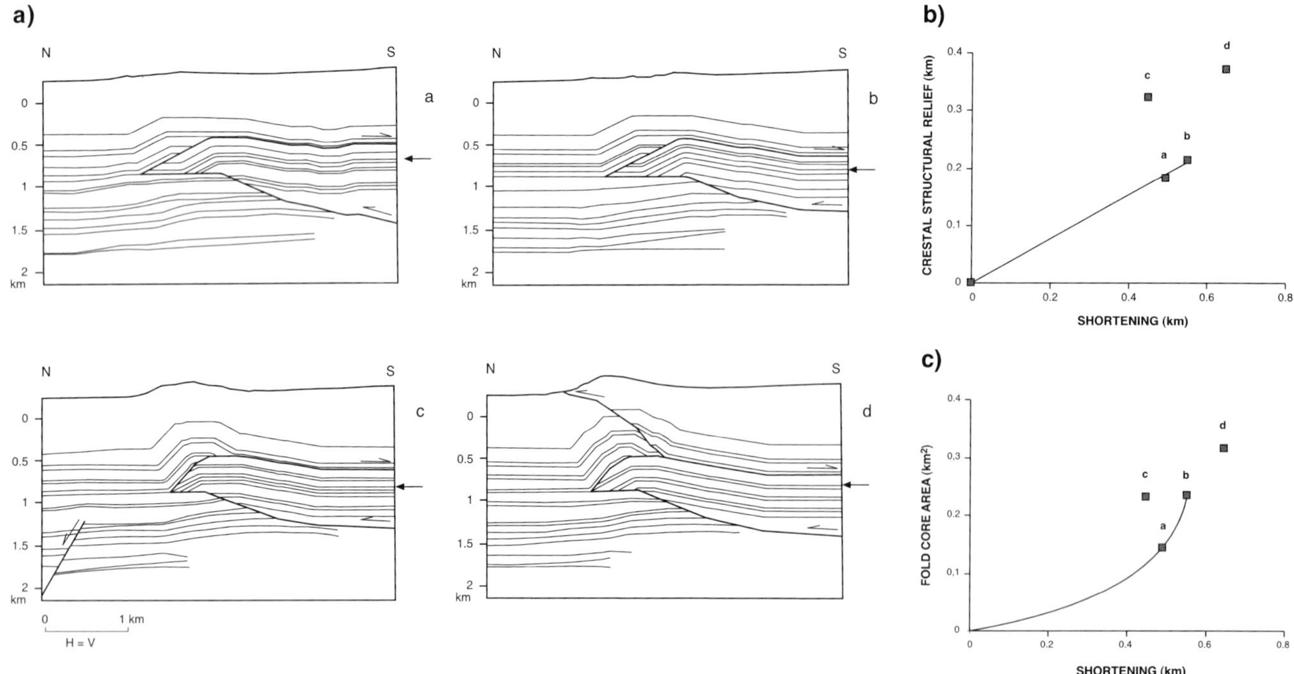

FIGURE 20. (a) Four sections from east (section *a*) to west (section *d*) across the Wheeler Ridge Anticline (modified after Medwedeff, 1992), (b) plot of crestal structural relief versus shortening, and (c) plot of fold area versus shortening for the Wheeler Ridge Anticline. The arrows indicate the horizon used to take the measurements. Best-fit lines are displayed.

becomes steeper with increasing shortening (Figure 20c). During this initial period, the crestal-structural-relief-versus-shortening curve displays a kinematic evolution similar to that of simple fault-bend folds formed by hinge migration in the crestal uplift stage (Figure 7a). However, the variation of fold area does not coincide with that followed by simple fault-bend folds (Figure 7b), possibly because the Wheeler Ridge Anticline corresponds to a wedge thrust (Medwedeff, 1992). From cross section *b* to cross section *c* in Figure 20a, a retreat in the amount of shortening occurs in both the structural-relief and fold-area graphs. This indicates that there is not a progressive lateral gradient of shortening from cross section *a* to *d*, and therefore, the different sections across the fold, from east to west, do not illustrate its kinematic evolution. This can also be inferred from the geometry of the faults related to the fold. In the cross sections in Figure 20a, the uppermost footwall flat associated with the north-directed thrust is shorter in cross section *c* than in cross section *b*. In addition, Figure 20a shows that the footwall flat associated with the south-directed thrust is located at higher stratigraphic horizons in cross section *c* than in cross section *d*.

The purpose of analyzing this particular example and assuming that different cross sections reflect its kinematic evolution is to show that:

1) This technique allows us to corroborate or discard a progressive increase/decrease of shortening in one direction along strike, because it requires us to measure the shortening of different sections across a fold, along strike.
2) When we apply this technique to thrust-related folds in an area where a lateral gradient of shortening occurs, it is crucial to know the fault geometry. Even if the shortening obtained from different cross sections along strike indicates that a lateral shortening gradient occurs, these cross sections may not illustrate the kinematic evolution of the fold.

Frontal Anticline of the Apennine Fold Belt

A section across one of the frontal folds of the Apennine fold belt (Figure 21a), derived from a good quality seismic profile, is available in Schwander (1989; his Figure 6). Syntectonic sediments associated with this anticline have been used to unravel its kinematic evolution during deposition of the growth beds. The geometry of the anticline illustrated in the cross section indicates that it may correspond to a fault-propagation fold. Because the syntectonic sediments are wedge-shaped (Figure 21a), different crestal-structural-relief-versus-shortening curves have been obtained. In outline, the rate of variation of crestal structural relief measured with respect to the bottom of the southwest basin (gray dots in Figure 21b) increases with shortening. On the

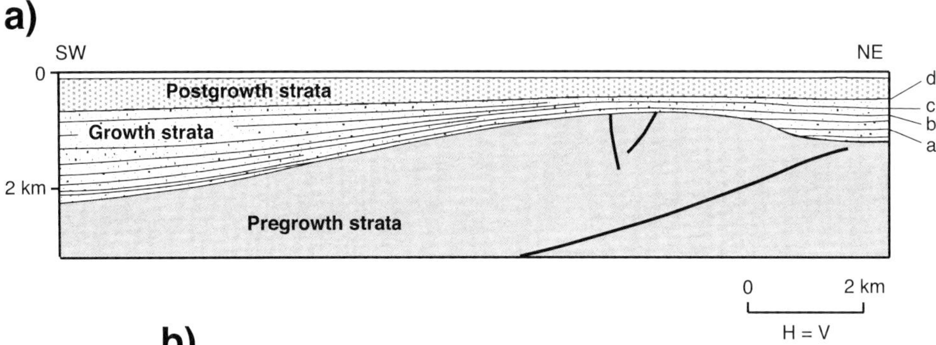

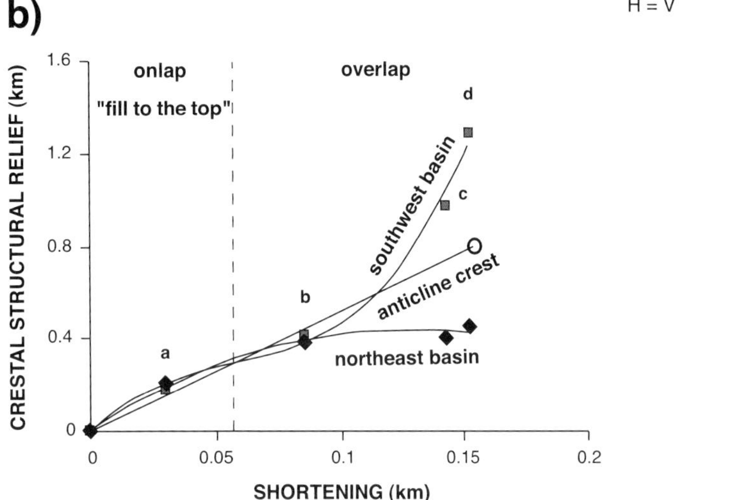

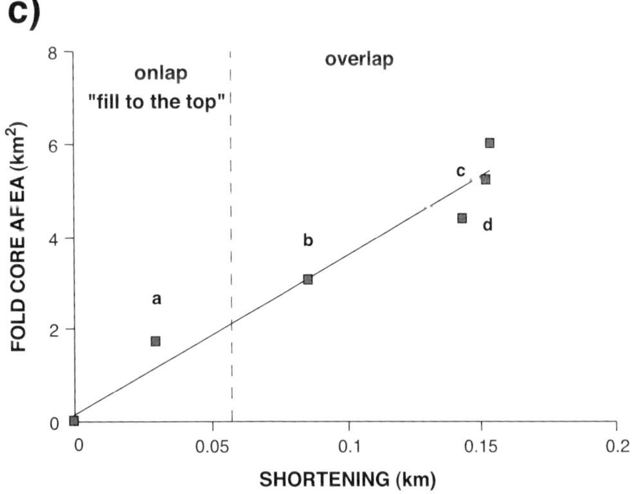

Figure 21. (a) Section across one of the frontal anticlines of the Apennine fold belt (modified after Schwander, 1989), (b) plot of crestal structural relief versus shortening, and (c) plot of fold area versus shortening for the frontal anticline. Growth horizons have been used to take the measurements. The point with the greatest crestal structural relief (empty circle) and fold area corresponds to the present-day fold geometry. Shown are the crestal structural reliefs, measured with respect to the southwest basin, with respect to the northeast basin, and below the anticline crest. Best-fit lines are displayed. To estimate the crestal structural relief, shortening, and fold area during the initial stages (when growth beds onlap the fold), we have assumed that the crest of the anticline did not produce any topographic or bathymetric positive relief, that is, the beds "filled to the top" the space available for deposition.

contrary, the rate of variation of crestal structural relief measured with respect to the bottom of the northeast basin (black dots in Figure 21b) decreases with shortening. The rate of variation of crestal structural relief measured below the anticline's crest, perpendicular to the regional datum, has been constructed using the common points derived from the adjacent basins plus the present-day crestal structural relief (empty circle in Figure 21b). This function is approximately linear (Figure 21b). During the initial stages of fold amplification (onlapping growth beds and oldest overlapping growth bed), the amount of crestal structural relief with respect to both basins was similar. However, during the last stages of fold amplification (youngest overlapping growth beds), the southwest basin subsided faster than the northeast basin did, with respect to the anticline's crest. The variation of fold core area with shortening is approximately linear (Figure 21c). The paths of the crestal structural relief and fold core area functions are in accordance with those of theoretical fault-propagation folds formed by hinge migration (Figure 7).

Experimental Thrust-related Folds

Three thrust-related folds that were developed in physical experiments show the applicability of the strategies described above. The first example is a

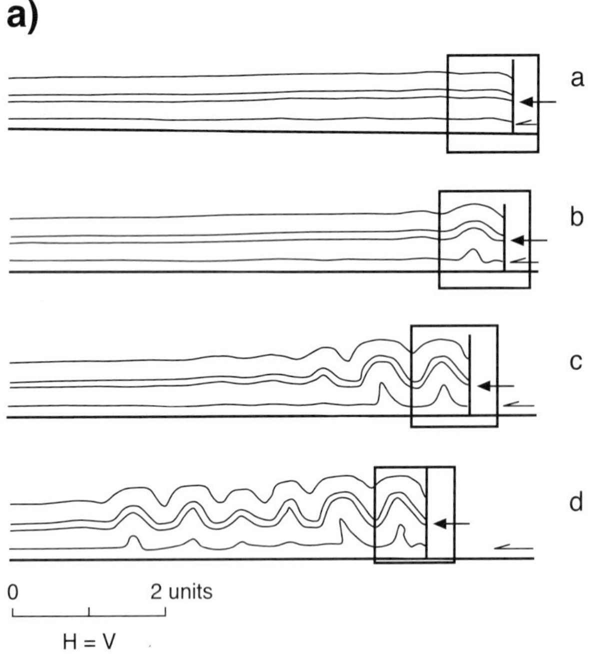

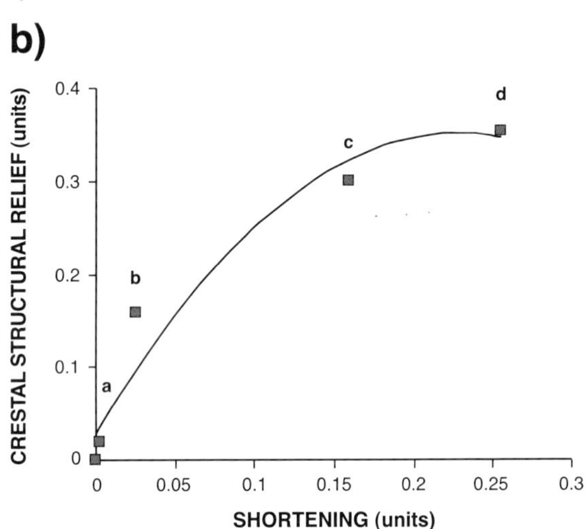

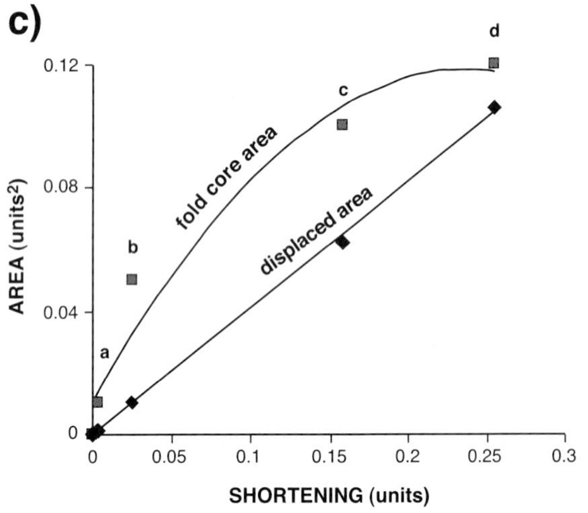

detachment anticline developed in a lubricated gelatin multilayer. The data to analyze this structure have been collected from Price and Cosgrove (1990; their Figure 10.41) using different sections across the anticline at different stages of fold amplification (Figure 22a). The second example is a single-step fault-bend fold developed in a multilayer system composed of limestones, lead, and sandstones. In this case, the data have been collected from Chester et al.'s (1991; their Figure 7) measurements of different stages of fold amplification (Figure 23a). The third example is a single-step fault-propagation fold that was developed in a multilayer system composed of limestones, mica, and sandstones. The data have been collected from Chester et al.'s (1991; their Figure 8) measurements of different stages of fold amplification (Figure 24a).

The graphs in Figures 22, 23, and 24 plot the results obtained from our application of the techniques described above to these three experimental folds. The crestal-structural-relief-versus-shortening curves for the physical experiments analyzed show distinctive trends. Thus, the detachment fold uplifts rapidly during the initial amplification stages, but the uplift rate decreases with increasing shortening (Figure 22b). The path of this curve resembles the path of the curve for theoretical detachment folds involving limb rotation (Figure 7a). In the case of the fault-bend fold, the crestal structural relief increases linearly with increasing shortening (Figure 23b). The crestal structural relief of the fault-propagation fold also increases linearly with increasing shortening (Figure 24b). These paths are consistent with the trend of the curves obtained for theoretical thrust-ramp folds (Figure 7a).

The fold-area-versus-shortening curves (Figures 22c, 23c, and 24c) have trends relatively similar to the crestal-structural-relief-versus-shortening curves (Figures 22b, 23b, and 24b). The area of the detachment anticline increases substantially during the initial amplification stages, but the rate decreases during the last stages (Figure 22c). Again, the path of this curve resembles the path of the curve obtained for theoretical detachment folds involving a certain amount of limb rotation (Figure 7b). The curves that represent the area of the fault-bend and fault-propagation folds versus shortening are approximately linear (Figures 23c and

FIGURE 22. (a) Four line drawings of an experimental detachment fold train (modified after Price and Cosgrove, 1990), (b) plot of crestal structural relief versus shortening, and (c) plot of area versus shortening for the experimental detachment anticline enclosed in a box in (a). Both the fold area and displaced areas are shown. The arrows indicate the horizon used to take the measurements. Best-fit lines are displayed. The shortening values used to construct the displaced area curve correspond to the shortening estimated for the anticline analyzed.

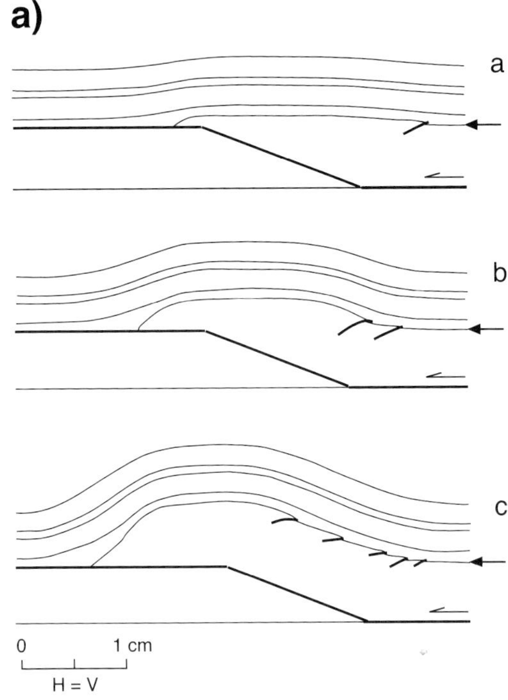

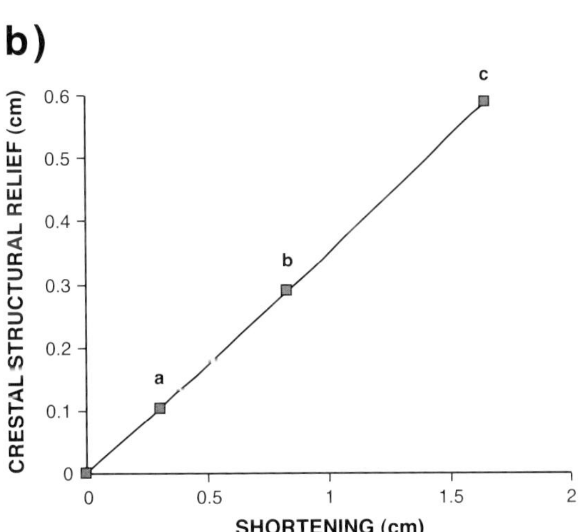

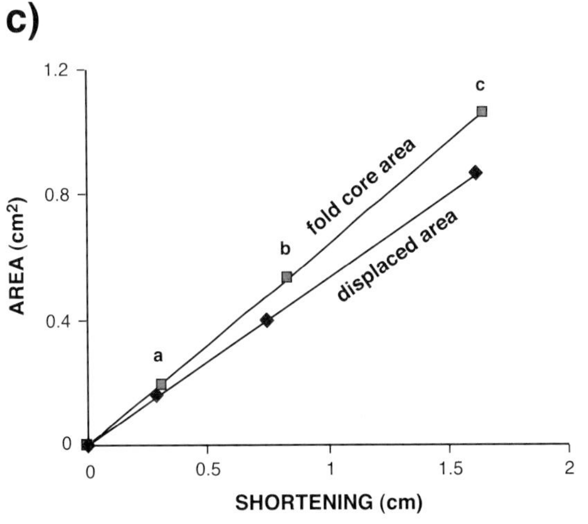

24c), like the trend obtained for theoretical thrust-ramp folds (Figure 7b).

In the three experimental folds studied, the distance from the regional datum to the lower detachment remains approximately constant during the whole experiment, that is, the detachment depth does not vary during fold amplification. This implies that the displaced area (detachment depth multiplied by shortening) should equal the area uplifted above the regional datum in the core of the anticline (Chamberlin, 1910, 1919). In the fold-area-versus-shortening plots, the displaced areas for each fold have been shown (Figures 22c, 23c, and 24c). The area of the experimental detachment fold remains above the displaced area during the entire fold evolution (Figure 22c). To check whether the trend of the curve is influenced by errors in the shortening measurements resulting from bed-length changes during deformation, we measured the total length of the beds from one edge to the other in the four stages of the experiment (Figure 22a). Bed lengths increased/decreased by a value < 3.4 %, which means that the assumption of constant bed length during deformation is approximately valid. This indicates that increases occurred in the cross-sectional area as a result of deformation and/or flow of the ductile unit inside the area enclosed within the lines that bounded the fold analyzed. The area of the experimental fault-bend fold is slightly larger than the displaced area, whereas the area of the experimental fault-propagation fold is slightly smaller than the displaced area. These differences appear to enlarge with increasing shortening. Because no ductile material is involved in these experimental thrust-ramp folds, longitudinal and/or transversal flow of material is unlikely. The fault-bend fold exhibits tensional faults in the outer arc (Chester et al., 1991; their Figure 7), which may have contributed to an overestimate of the area. Moreover, slight bed thickening in the core of the fault-bend fold is not unlikely. This would have caused slight overestimates of the bed length, and it would explain why the shortening measured using bed lengths (gray dots in Figure 23c) is slightly greater than the shortening recorded by the back-end wall of the experiment (black dots in Figure 23c). On the contrary, the fault-propagation fold involved slight bed thinning in the

FIGURE 23. (a) Three line drawings of an experimental fault-bend fold (modified after Chester et al., 1991), (b) plot of crestal structural relief versus shortening, and (c) plot of area versus shortening for the experimental fault-bend fold illustrated in (a). Both the fold area and displaced areas are shown. The arrows indicate the horizon used to take the measurements. Best-fit lines are displayed. The shortening values used to construct the displaced area curve correspond to the amount of movement of the back end wall in the experiment.

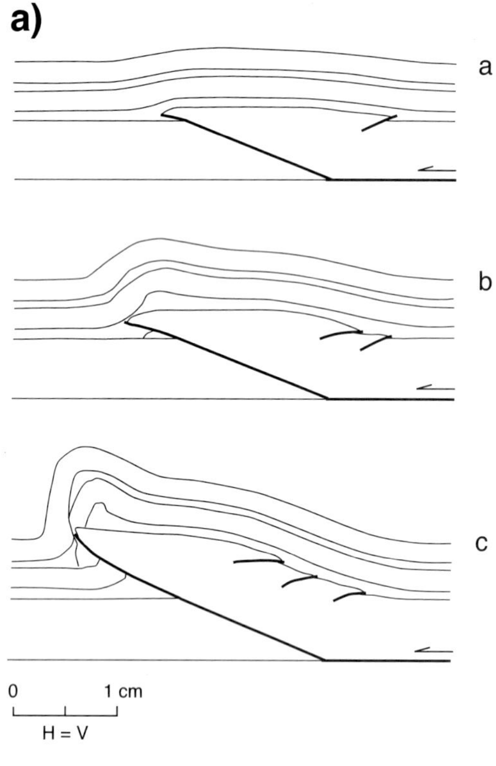

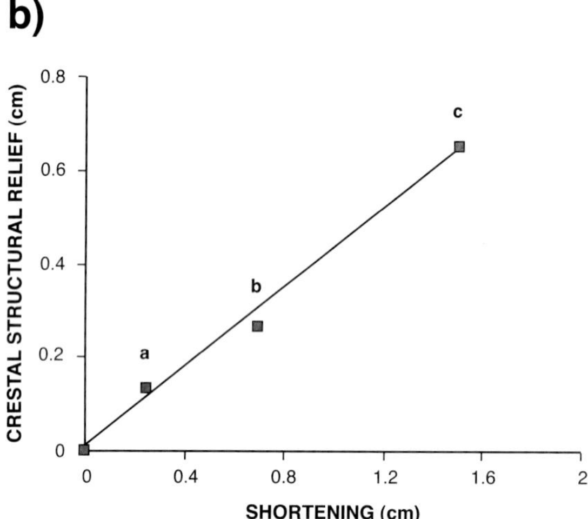

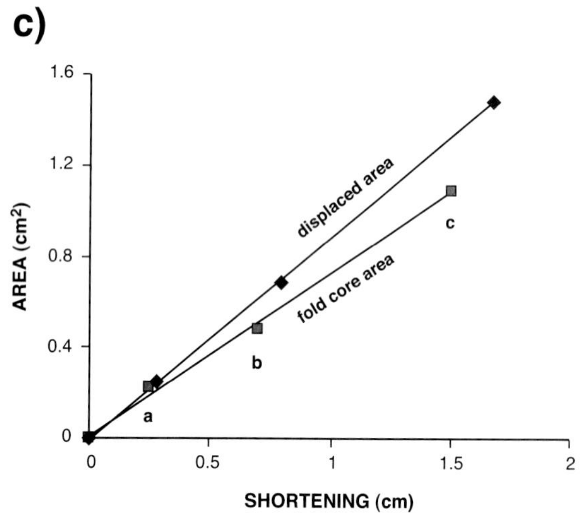

anticline's core and/or the area was underestimated because of the occurrence of minor reverse faults (Chester et al., 1991; their Figure 8). This would have caused slight underestimates of the bed length, and it is the reason why the shortening measured using bed lengths (gray dots in Figure 24c) is slightly smaller than the shortening recorded by the back end wall of the experiment (black dots in Figure 24c).

CONCLUSIONS

The thrust-related folds analyzed here show variable fold uplift and fold area trends, which implies that each fold had a different kinematic evolution. This suggests that no theoretical models should be taken for granted when trying to decipher the kinematics of natural or experimental thrust-related folds, even if there is a close geometric resemblance between the theoretical and the actual example. We believe that the kinematic evolution of actual examples must be analyzed using techniques such as the ones described in this chapter and other techniques previously published, and avoiding kinematic assumptions when possible.

The results presented in this study reflect that some of the forward, kinematic models of sharp-hinge thrust-related folds used to date may be too simplistic to simulate actual structures. Different deformation mechanisms may operate during formation of a single fold; cross-sectional area may not be maintained because of compaction, deformation, and/or migration of ductile material in longitudinal/transverse direction(s); thickening/thinning may occur during fold amplification; and so on. More-complex algorithms to simulate the kinematic evolution of different types of rounded-hinge folds are required to predict the evolution of some natural examples.

The analyses carried out in this chapter have implications on the sequential restoration of individual thrust-related folds. In particular, balancing and sequentially restoring some individual thrust-related folds in 2-D reveal these as imprecise procedures because of inequality between the displaced area and the area uplifted/depressed by deformation. In nature, we observe only

Figure 24. (a) Three line drawings of an experimental fault-propagation fold (modified after Chester et al., 1991), (b) plot of crestal structural relief versus shortening, and (c) plot of area versus shortening for the experimental fault-propagation fold illustrated in (a). Both the fold area and displaced areas are shown. The arrows indicate the horizon used to take the measurements. Best-fit lines are displayed. The shortening values used to construct the displaced area curve correspond to the amount of movement of the back end wall in the experiment.

the final deformation state, and many classical restoration methods do not incorporate a precise estimation of the amount of thickening and/or thinning during different stages of fold amplification. The best-fit lines obtained in the plots of crestal structural relief and fold area versus shortening allow us to estimate two of these parameters, if the other one is known, for any stage during fold evolution. These data are a useful tool for constructing sequential restorations of thrust-related folds.

The results we obtained by applying the techniques presented in this paper to theoretical, natural, and experimental examples of thrust-related folds must be taken with caution because of the assumptions inherent in the methods presented. Nevertheless, in our opinion, it has been proved that these techniques are relatively easy to apply and allow us to gain insight into some aspects of the kinematics of thrust-related folds.

ACKNOWLEDGMENTS

Suggestions by R. Groshong and an anonymous referee substantially improved the original manuscript. F. Bastida, J. Aller, J.L. Alonso, and A. Marcos are also thanked for their comments. We acknowledge financial support by Projects AMB98-1012-CO2-O2 (Actividad sismotectónica, estructura litosférica y modelos de deformación varisca y alpina en el NO de la Península Ibérica) and PB98-1557 (Mecanismos de plegamiento: teoría y aplicaciones en geología económica y regional) funded by the Spanish Ministry for Education and Culture, and support by the Acción Integrada Hispano-Británica HB1999-0038 (Cinemática de pliegues y estructuras menores asociadas a cabalgamientos a partir del estudio de materiales sintectónicos y de su modelización) funded by the Spanish Ministry for Education and Culture and the British Council.

REFERENCES CITED

Abbassi, M. R., and N. S. Mancktelow, 1992, Single layer buckle folding in non-linear materials—I, experimental study of development from an isolated initial perturbation: Journal of Structural Geology, v. 14, p. 85–104.

Anastasio, D., 1987, Thrusting, halotectonics and sedimentation in the External Sierras, Southern Pyrenees, Spain: Ph.D. thesis, John Hopkins University, U.S.A., 181 p.

Ball, M. M., R. G. Martin, W. D. Bock, R. E. Sylwester, R. M. Bowles, D. E. Taylor, L. Coward, J. E. Dodd, and L. Gilbert, 1985, Seismic structure and stratigraphy of northern edge of Bahaman-Cuban collision zone: AAPG Bulletin, v. 69, p. 1275–1294.

Barnolas, A., J. M. Samsó, A. Teixell, J. Tosquella, and M. Zamorano, 1991, Evolución sedimentaria entre la cuenca de Graus-Tremp y la cuenca de Jaca-Pamplona: I Congreso del Grupo Español del Terciario, Vic, Libro-Guía Excursión 1, p. 74–114.

Beutner, E. C., and F. A. Diegel, 1985, Determination of fold kinematics from syntectonic fibers in pressure shadows, Martinsburg slate, New Jersey: American Journal of Science, v. 285, p. 16–50.

Blay, P., J. W. Cosgrove, and J. M. Summers, 1977, An experimental investigation of the development of structures in multilayers under the influence of gravity: Journal of the Geological Society of London, v. 133, p. 329–342.

Buchanan, J. G., 1996, The application of cross section construction and validation within exploration and production: a discussion, *in* P. G. Buchanan and D. A. Nieuwland, eds., Modern developments in structural interpretation, validation and modelling: Geological Society of London Special Publication 99, p. 41–50.

Bulnes, M., and J. Poblet, 1999, Estimating the detachment depth in cross sections involving detachment folds: Geological Magazine, v. 136, p. 395–412.

Butler, R. W. H., 1992, Evolution of Alpine fold-thrust complexes: a linked kinematic approach, *in* S. Mitra and W. Fisher, eds., Structural geology of folds and thrust belts: Baltimore, The John Hopkins University Press, p. 29–44.

Butler, R. W. H., and W. H. Lickorish, 1997, Using high-resolution stratigraphy to date fold and thrust activity: examples from the Neogene of south-central Sicily: Journal of the Geological Society of London, v. 154, p. 633–643.

Chester, J. S., J. M. Logan, and J. H. Spang, 1991, Influence of layering and boundary conditions on fault-bend and fault-propagation folding: GSA Bulletin, v. 103, p. 1059–1072.

Chamberlin, R. T., 1910, The Appalachian folds of central Pennsylvania: Journal of Geology, v. 18, p. 228–251.

Chamberlin, R. T., 1919, The building of the Colorado Rockies: Journal of Geology, v. 27, p. 225–251.

Dahlstrom, C. D. A., 1990, Geometric constraints derived from the law of conservation of volume and applied to evolutionary models for detachment folding: AAPG Bulletin, v. 74, p. 336–344.

De Sitter, L. V., 1956, Structural geology: New York, McGraw-Hill, 552 p.

Dobson, J., 1990, The dynamics of foreland fold and thrust belts, Southern Canadian Rocky Mountains, SW Alberta, Canada: Ph.D. Thesis, University of London, U.K., 300 p.

Echevarría-Rodríguez, G., G. Hernández-Pérez, J. O. López-Quintero, J. G. López-River, R. Rodríguez-Hernández, J. R. Sánchez-Arango, R. Socorro-Trujillo, R. Tenreyro-Pérez, and J. L. Yparraguirre-Pena, 1991, Oil and gas exploration in Cuba: Journal of Petroleum Geology, v. 14, p. 259–274.

Elliot, D., 1976, The energy balance and deformation mechanisms of thrust sheets: Philosophical Transactions of the Royal Society of London, v. A, 283, p. 289–312.

Fischer, M. P., and N. B. Woodward, 1992, The geometric evolution of foreland thrust systems, *in* K. McClay, ed., Thrust tectonics: London, Chapman & Hall, p. 181–189.

Frey, M. G., 1973, Influence of Salina Salt on structure in New York-Pennsylvania part of the Appalachian Plateau: AAPG Bulletin, v. 57, p. 1027–1037.

Garrido, A., 1973, Estudio geológico y relación entre tectónica y sedimentación del secundario y terciario de la vertiente meridional pirenaica en su zona central: Ph.D. Thesis, Universidad de Granada, Spain, 395 p.

Groshong, R. H. Jr., and J.-L. Epard, 1994, The role of strain in area-constant detachment folding: Journal of Structural Geology, v. 16, p. 613–618.

Hardy, S., and J. Poblet, 1994, Geometric and numerical model of progressive limb rotation in detachment folds: Geology, v. 22, p. 371–374.

Hardy, S., J. Poblet, K. McClay, and D. Waltham, 1996, Mathematical modelling of growth fault-related folds, *in* P. G. Buchanan and D. A. Nieuwland, eds., Modern developments in structural interpretation, validation and modelling: Geological Society of London Special Publication 99, p. 265–282.

Hogan, P. J., 1991, Geochronologic, tectonic, and stratigraphic evolution of the Southwest Pyrenean Foreland Basin, Northern Spain: Ph D. thesis, University of Southern California, U.S.A., 220 p.

Homza, T. X., and W. K. Wallace, 1995, Geometric and kinematic models for detachment folds with fixed and variable detachment depths: Journal of Structural Geology, v. 17, p. 575–588.

Jones, P. B., 1987, Quantitative geometry of thrust and fold belt structures: AAPG Methods in Exploration, no. 6, 26 p.

Keller, E. A., R. L. Zepeda, T. K. Rockwell, T. L. Wu, and W. S. Dinklage, 1998, Active tectonics at Wheeler Ridge, southern San Joaquin Valley, California: GSA Bulletin, v. 110, p. 298–310.

Keller, E. A., L. Gurrola, and T. E. Tierney, 1999, Geomorphic criteria to determine direction of lateral propagation of reverse faulting and folding: Geology, v. 27, p. 515–518.

Larrasoaña, J. C., E. L. Pueyo, J. del Valle, H. Millán, J. M. Parés, A. Pocoví, and J. Dinarés, 1996, Datos magnetotectónicos del Eoceno de la cuenca de Jaca-Pamplona: resultados iniciales: Geogaceta, v. 20, p. 1058–1061.

Lin, J., and R. S. Stein, 1989, Coseismic folding, earthquake recurrence, and the 1987 source mechanism at Whittier Narrows, Los Angeles Basin, California: Journal of Geophysical Research, v. 94, p. 9614–9632.

Martínez-Peña, B., 1991, La estructura del límite occidental de la unidad surpirenaica central: Ph.D. Thesis, Universidad de Zaragoza, Spain, 380 p.

Masaferro, J. L., J. Poblet, M. Bulnes, G. P. Eberli, T. H. Dixon, and K. McClay, 1999, Palaeogene-Neogene/present day (?) growth folding in the Bahamian foreland of the Cuban fold and thrust belt: Journal of the Geological Society of London, v. 156, p. 617–631.

Masaferro, J. L., M. Bulnes, J. Poblet, and G. P. Eberli, 2002, Episodic folding inferred from syntectonic carbonate sedimentation: the Santaren anticline, Bahamas foreland: Sedimentary Geology, v. 146, p. 11–24.

McClay, K., 1992, Glossary of thrust tectonic terms, *in* K. McClay, ed., Thrust tectonics: London, Chapman & Hall, p. 419–433.

Means, W. D., 1976, Stress and strain: New York, Springer-Verlag, 339 p.

Medwedeff, D. A., 1992, Geometry and kinematics of an active, laterally propagating wedge thrust, Wheeler Ridge, California, *in* S. Mitra and G. W. Fisher, eds., Structural geology of fold and thrust belts: Baltimore, The John Hopkins University Press, p. 3–28.

Millán, H., M. Aurell, and A. Meléndez, 1994, Synchronous detachment folds and coeval sedimentation in the Prepyrenean External Sierras (Spain): a case study for a tectonic origin of sequences and systems tracts: Sedimentology, v. 41, p. 1001–1024.

Mitra, S., and J. S. Namson, 1989, Equal-area balancing: American Journal of Science, v. 289, p. 563–599.

Mueller, K., and J. Suppe, 1997, Growth of Wheeler Ridge anticline, California. Geomorphic evidence for fault-bend folding behavior during earthquakes: Journal of Structural Geology, v. 19, p. 383–396.

Mueller, K., and P. Talling, 1997, Geomorphic evidence for tear faults accommodating lateral propagation of an active fault-bend fold, Wheeler Ridge, California: Journal of Structural Geology, v. 19, p. 397–411.

Namson, J. S., and T. L. Davis, 1988, Seismically active fold and thrust belt in the San Joaquin Valley, central California: GSA Bulletin, v. 100, p. 257–273.

Norris, D. K., 1971, Comparative study of the Castle River and other folds in the eastern cordillera of Canada: Bulletin of the Geological Survey of Canada, v. 205, 58 p.

Pashin, J. C., R. H. Groshong Jr., and S. Wang, 1995, Thin-skinned structures influence gas production in Alabama coalbed methane fields: Intergas '95, v. 9508, p. 39–52.

Poblet, J., and S. Hardy, 1995, Reverse modelling of detachment folds; application to the Pico del Aguila anticline in the South Central Pyrenees (Spain): Journal of Structural Geology, v. 17, p. 1707–1724.

Poblet, J., and K. McClay, 1996, Geometry and kinematics of single-layer detachment folds: AAPG Bulletin, v. 80, p. 1085–1109.

Poblet, J., K. McClay, F. Storti, and J. A. Muñoz, 1997, Geometries of syntectonic sediments associated with single-layer detachment folds: Journal of Structural Geology, v. 19, p. 369–381.

Poblet, J., J. A. Muñoz, A. Travé, and J. Serra-Kiel, 1998, Quantifying the kinematics of detachment folds using three-dimensional geometry: application to the Mediano anticline: GSA Bulletin, v. 110, p. 111–125.

Price, N. J., and J. W. Cosgrove, 1990, Analysis of geological structures: New York, Cambridge University Press, 502 p.

Prucha, J. J., 1968, Salt deformation and décollement in the Fir Tree Point anticline of central New York: Tectonophysics, v. 6, p. 273–299.

Pueyo, E. L., J. M. Parés, H. Millán, and A. Pocoví, 1994, Evidencia magnetotectónica de la rotación de las Sierras Exteriores Altoaragonesas: II Congreso del Grupo Español del Terciario, Jaca, Comunicaciones, p. 185–189.

Pueyo, E. L., H. Millán, A. Pocoví, and J. M. Parés, 1996, Correcciones geométricas en magnetotectónica: filtrado de rotaciones aparentes debidas a pliegues: Geogaceta, v. 20, p. 1054–1057.

Puigdefàbregas, C., 1975, La sedimentación molásica en la cuenca de Jaca: Pirineos, v. 104, p. 1–88.

Ramsay, J. G., and M. I. Huber, 1987, The techniques of modern structural geology, volume 2: folds and fractures: London, Academic Press Limited, 700 p.

Riba, O., 1976, Syntectonic unconformities in the Alto Cardener, Spanish Pyrenees: a genetic interpretation: Sedimentary Geology, v. 15, p. 213–233.

Rowan, M. G., 1997, Three-dimensional geometry and evolution of a segmented detachment fold, Mississippi Fan foldbelt, Gulf of Mexico: Journal of Structural Geology, v. 19, p. 463–480.

Rowan, M. G., R. Kligfield, and P. Weimer, 1993, Processes and rates of deformation: preliminary results from the Mississippi Fan Fold belt, deep Gulf of Mexico: GCSSEPM Foundation 14th Annual Research Conference, Rates of Geologic Processes, p. 209–218.

Schneider, C. L., D. Hummon, R. S. Yeats, and G. L. Huftile, 1996, Structural evolution of the northern Los Angeles basin, California, based on growth strata: Tectonics, v. 15, p. 341–355.

Schwander, M. M., 1989, The southern Adriatic Basin, offshore Italy, *in* A. W. Bally, ed., Atlas of seismic stratigraphy, volume 3: AAPG Studies in Geology no. 27, p. 112–115.

Séguret, M., 1972, Etude tectonique des nappes et séries décollées de la partie centrale du versant sud des Pyrenées: Montpellier, Publications Ustela, Série geologie structurale 2, 155 p.

Shaw, J. H., and J. Suppe, 1994, Active faulting and growth folding in the eastern Santa Barbara Channel, California: GSA Bulletin, v. 106, p. 607–626.

Shaw, J. H., S. C. Hook, and J. Suppe, 1994, Structural trend analysis by axial surface mapping: AAPG Bulletin, v. 78, p. 700–721.

Storti, F., and J. Poblet, 1997, Growth stratal architectures associated with décollement folds and fault-propagation folds. Inferences on fold kinematics: Tectonophysics, v. 282, p. 353–373.

Suppe, J., 1983, Geometry and kinematics of fault-bend folding: American Journal of Science, v. 283, p. 684–721.

Suppe, J., and D. A. Medwedeff, 1990, Geometry and kinematics of fault-propagation folding: Eclogae Geologicae Elvetiae, v. 83, p. 409–454.

Suppe, J., G. T. Chou, and S. C. Hook, 1992, Rates of folding and faulting determined from growth strata, *in* K. McClay, ed., Thrust tectonics: London, Chapman & Hall, p. 105–121.

Teixell, A., and A. Barnolas, 1995, Significado de la discordancia de Mediano en relación con las estructuras adyacentes (Eoceno, Pirineo central): Geogaceta, v. 18, p. 34–37.

Torrente, M. M., and R. Kligfield, R., 1995, Modellizzazione predittiva di pieghe sinsedimentarie: Bolletino de la Societá Geologica Italiana, v. 114, p. 293–309.

Weimer, P., and R. T. Buffler, 1992, Structural geology and evolution of the Mississippi Fan fold belt, deep Gulf of Mexico: AAPG Bulletin, v. 76, p. 225–251.

Wiltschko, D. V., and W. M. Chapple, 1977, Flow of weak rocks in Appalachian Plateau folds: AAPG Bulletin, v. 61, p. 653–670.

Wojtal, S., and G. Mitra, 1988, Nature of deformation in some fault rocks from Appalachian thrusts, *in* G. Mitra and S. Wojtal, eds., Geometries and mechanics of thrusting, with special reference to the Appalachians: GSA Special Paper 222, p. 17–33.

Shaw, J. H., E. Novoa, and C. D. Connors, 2004, Structural controls on growth stratigraphy in contractional fault-related folds, *in* K. R. McClay, ed., Thrust tectonics and hydrocarbon systems: AAPG Memoir 82, p. 400–412.

Structural Controls on Growth Stratigraphy in Contractional Fault-related Folds

John H. Shaw
Dept. of Earth & Planetary Sciences, Harvard University, Cambridge, Massachusetts, U.S.A.

Enrique Novoa
Departamento de Ciencia de La Tierra, Gerencia de Produccion y Exploracion, PDVSA-INTEVEP, Los Teques, Estado Miranda, Venezuela

Christopher D. Connors[1]
Texaco Exploration, Bellaire, Texas, U.S.A.

ABSTRACT

We describe local structural controls on deposition above contractional fault-related folds that yield patterns of stratigraphic onlaps, pinch-outs, and facies transitions that are diagnostic of folding mechanism. In folds that grow by kink-band migration, stratigraphic onlaps and pinch-outs that formed at emergent fold scarps are incorporated into fold limbs and aligned along growth axial surfaces. In contrast, the positions of these same features in structures that grow primarily by limb rotation are more variable and are controlled directly by sedimentation-to-uplift ratio. We present kinematic models and natural examples that integrate seismic reflection data and well control to describe these structural influences on growth stratigraphy. An understanding of this interplay between local deformation and deposition helps us infer the positions of subtle pinch-outs that may provide hydrocarbon traps and can yield a detailed history of structural development.

INTRODUCTION

Patterns of deformed growth strata record the timing and kinematics of deformation, much as magnetic anomalies record the process of sea-floor spreading (Suppe et al., 1992). Thus, growth strata are commonly used to define the ages and rates of deformation as well as to infer folding mechanisms (Medwedeff, 1989; Shaw and Suppe, 1994, 1996; Hardy et al., 1996; Vergés et al., 1996; Schneider et al., 1996; Ford et al., 1997; Suppe et al., 1997; Novoa et al., 2000).

Deformation also can influence the internal stratigraphy of growth strata. In cases where uplift rate locally exceeds sedimentation rate, folding and faulting can produce surface scarps that control sediment accommodation space and influence local depositional systems

[1]*Present address*: Dept. of Geology, Washington and Lee University, Lexington, Virginia, U.S.A.

(Burbank et al., 1996). For example, fan deposits may be ponded behind scarps, whereas the path of channels may be diverted to run parallel to the strike of the structures (Figure 1). These processes yield deposits restricted to one side of a fold or fault scarp. These deposits can be expressed as seismic sequences that onlap fold limbs, and in certain depositional environments may correspond to abrupt changes in sedimentary facies. In deep marine environments, for example, sand-rich fan and channel deposits may be restricted to the structurally low side of the scarps, whereas deposition on the fold crests may be dominated by hemipelagic muds. This would produce abrupt facies transitions across fold limbs. These stratigraphic patterns, consisting of onlapping growth sequences and sand pinch-outs on the flanks of structural highs, are commonplace. Yet, they are often modified by subsequent folding, which obscures the original depositional geometries. Two mechanisms—kink-band migration and limb rotation— are commonly employed to model the kinematics of this folding. Both mechanisms produce distinctly different patterns of deformed pinch-outs, onlaps, and facies boundaries in growth strata. We present kinematic models and natural examples to illustrate how these stratigraphic patterns form and are obscured by deformation via kink-band migration and limb rotation. Our intent is to show that both the internal stratigraphy and geometry of growth strata can be diagnostic of folding mechanisms. Conversely, we demonstrate that fold kinematics may influence the local distribution of sedimentary facies in a growth structure.

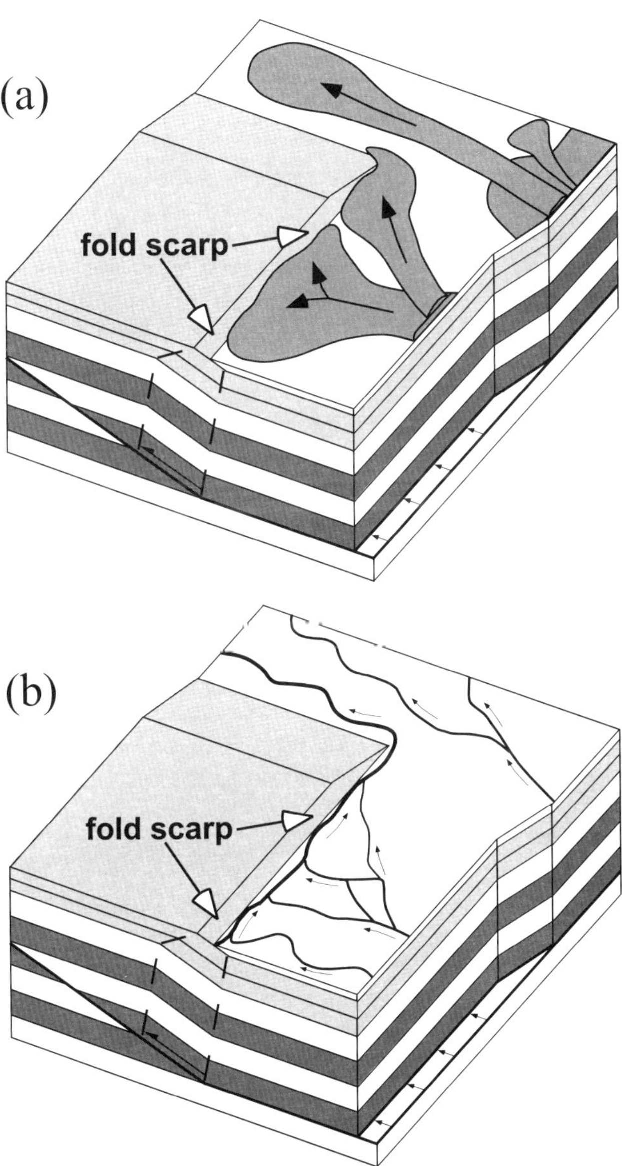

Figure 1. Perspective views of three-dimensional fault-bend fold models, showing the influence of emergent fold scarps on the distribution of growth sediments. (a) Ponding of fan deposits behind a fold scarp. (b) Deflection of channels to run along strike of the fold scarp.

MODELED STRATIGRAPHIC PATTERNS IN GROWTH STRUCTURES

In fault-related folds that develop purely by kink-band migration, fold limbs widen through time while maintaining a fixed dip (Suppe et al., 1992). Material is incorporated into the fold limb by passing through an active axial surface, which at depth is generally pinned to a bend or tip of a fault (Suppe, 1983; Suppe and Medwedeff, 1990). In cases where fold uplift rate exceeds syntectonic sedimentation rate, each increment of folding produces a discrete fold scarp located where the active axes project to the surface. Subsequent deposits then onlap the fold scarp, producing stratigraphic pinch-outs above the fold limb. In cases where anticlinal axial surfaces are active, growth strata remain undeformed and successive pinch-outs migrate up dip toward the anticlinal axis (Figure 2). In contrast, when synclinal axial surfaces are active, fold scarps and stratigraphic pinch-outs are displaced laterally and folded as they are incorporated into widening limbs (Suppe et al., 1992). Growth horizons between the positions of the onlaps and the active axial surface are folded concordantly with the underlying strata. New scarps form at the surface above the active, synclinal axial surfaces. Filling of each new accommodation space, followed by repeated folding and deposition, yields a series of stratigraphic pinch-outs at "paleoscarps" extending downward into the growth structure. These pinch-outs align along the fold's growth axial surface (Figure 2), which marks the particles originally deposited along the active fold hinge and subsequently incorporated into the fold limb (Suppe et al., 1992). Thus growth axial surfaces, which often

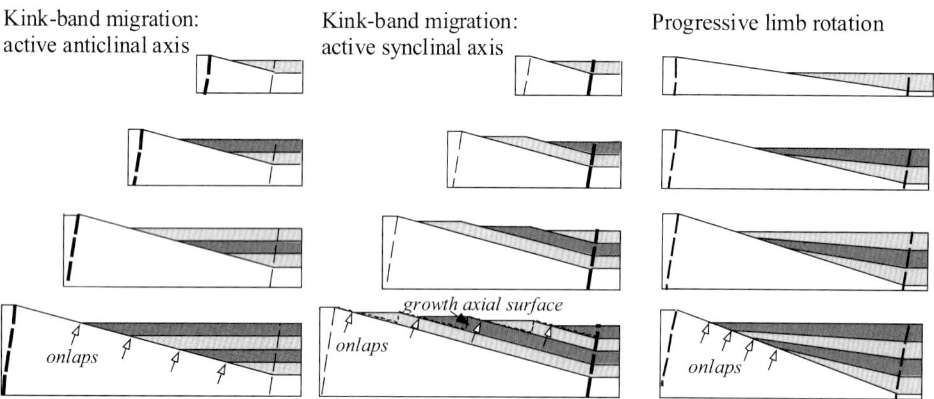

FIGURE 2. Kinematic models illustrating the onlap patterns in growth folds that develop by kink-band migration and limb rotation. In both of the kink-band-migration models, the sedimentation-to-uplift ratio is 0.75. In the limb-rotation model, the sedimentation rate is held constant while the uplift rate decreases as a function of the limb rotation (Hardy and Poblet, 1994).

correspond to a marked change in bed dip, can delineate these stratigraphic pinch-outs and associated facies boundaries in folds developed by kink-band migration.

Different stratigraphic patterns are expected in folds that grow primarily by limb rotation with fixed hinges (Figure 2), including certain types of detachment (Dahlstrom, 1990; Hardy and Poblet, 1994; Poblet et al., 1997; Storti and Poblet, 1997) and trishear (Erslev, 1991; Hardy and Ford, 1997; Allmendinger, 1998) folds. In these structures, rotation is distributed across the entire fold limb, with each increment of deformation. Thus fold scarps are diffuse, rather than being localized along active axial surfaces, as is the case in kink-band migration folds. Lacking discrete fold scarps, the positions of stratigraphic pinch-outs along rotational fold limbs are controlled primarily by the sedimentation-to-uplift ratio (Hardy et al., 1996; Storti and Poblet, 1997) and not by the positions of active axial surfaces, as is the case in kink-band migration folds. Hardy and Poblet (1994) demonstrated that in fixed-axis, limb-rotation models with sedimentation-to-uplift ratios nearly equal to one, pinch-outs form near anticlinal axial surfaces and corresponding growth horizons exhibit a fanning of limb dips. When sedimentation-to-uplift ratios are less than one, onlaps or pinch-outs form on the fold limb between the bounding axial surfaces (Figure 2). Uplift rates in these folds generally decrease through time, at constant shortening rates (Hardy and Poblet, 1994). Thus, when sedimentation rates are constant, onlapping growth horizons generally migrate up dip toward the anticlinal axial surface. In cases of increasing sedimentation rate relative to shortening rate, the onlaps will also tend to migrate upward toward the anticlinal axis. Onlaps will migrate toward the synclinal axis of a fold limb only when sedimentation rate relative to shortening rate is very low or decreases substantially through time.

The simple, end-member models shown in Figure 2 demonstrate that onlap patterns vary substantially between kink-band migration and limb-rotation folds, at constant rates of sedimentation and shortening. In natural cases, however, sedimentation and tectonic rates may vary over the life of folds. These variations can complicate onlap patterns. Thus, to distinguish fold kinematics, onlap relations should be considered in combination with the overall geometry of growth folds. For example, a kink-band migration fold with an active synclinal axis will preferentially form onlaps at fold scarps that develop where the synclinal axis meets the land surface. These onlaps are subsequently translated away from the synclinal axis by fold growth. As a result, onlaps generally migrate upward in section, toward the synclinal axis. Moreover, these onlaps are localized along dip changes in their corresponding stratigraphic horizons, where beds change from having primary sedimentary dip to being parallel to the underlying fold limb. This dip change defines the fold's growth axial surface and is not expected to occur in folds developed solely by limb rotation. Thus, patterns of onlaps migrating toward synclinal axes, combined with marked bed-dip changes and narrowing-upward, parallel fold geometries in growth strata, may be considered diagnostic of folding by kink-band migration with active synclinal axes. In contrast, a limb-rotation fold with constant sedimentation rate relative to shortening rate will typically have onlaps that migrate upward toward the anticlinal axis. This onlap pattern is similar to that of a kink-band migration fold with an active anticlinal axis. However, in growth, the kink-band migration structure does not exhibit a fanning of limb dips like that which is present in a limb-rotation structure (Figure 2). Thus, patterns of onlaps migrating toward anticlinal axes, combined with a shallowing-upward geometry of limb dips in growth strata, may together be considered diagnostic of folding by limb rotation. In the following section, we use these contrasting patterns of onlaps and fold geometry to interpret folding mechanisms in natural examples of both kink-band migration and limb-rotation structures. In the final example, we also compare these macroscopic fold patterns with sedimentary facies transitions observed in wells.

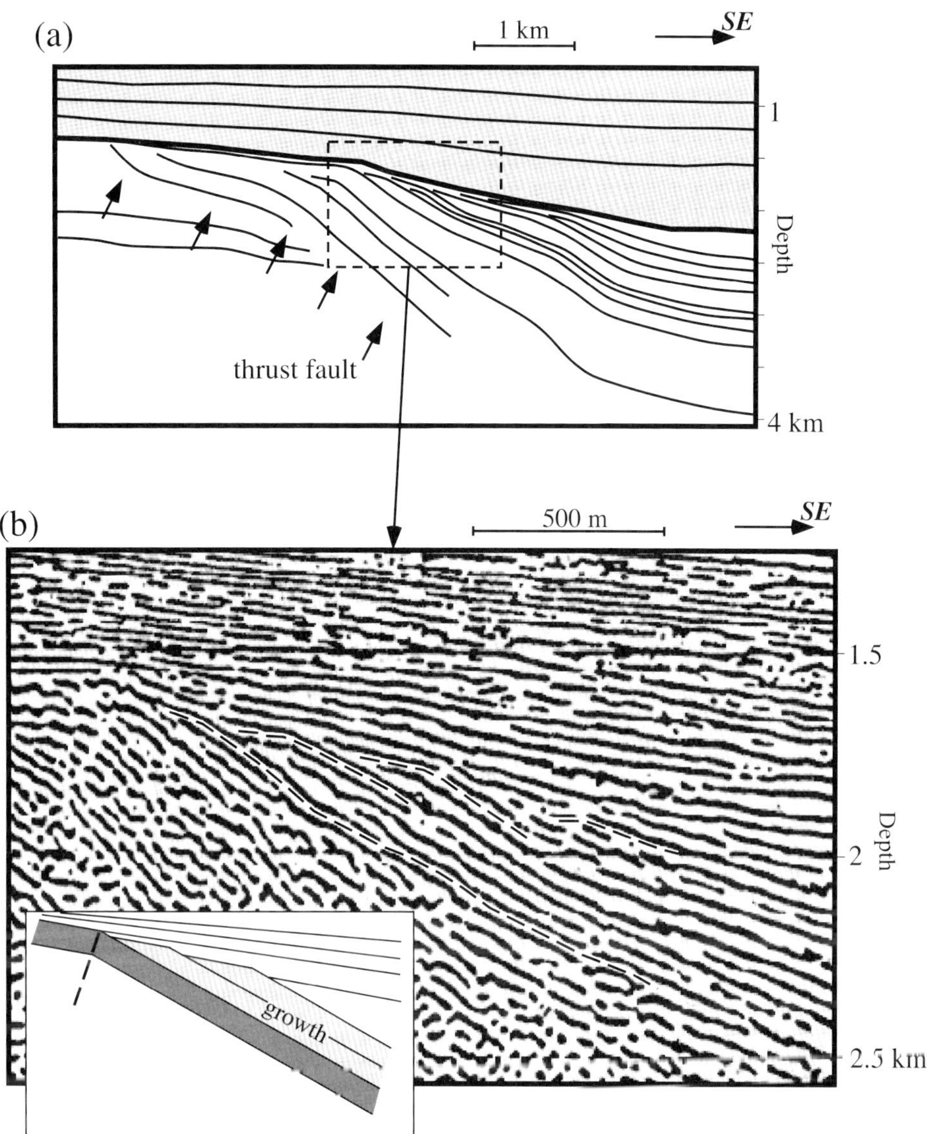

Figure 3. (a) Folded growth strata on the limb of an anticline interpreted on a seismic image from the western Jungar Basin, China. (b) 3-D prestack, depth-migrated seismic-reflection data (Zhang et al., 1996) imaging onlaps within the growth section. Highlighted growth sequences onlap the fold limb. To the east, these growth sequences change dip, becoming parallel with the underlying limb. This growth pattern is analogous to that modeled for folding by kink-band migration with an active synclinal axis (Figure 2 and inset). Younger strata above the onlapping growth units may record post-tectonic drape or a later phase of structural growth involving a component of limb rotation.

OBSERVED STRATIGRAPHIC PATTERNS IN GROWTH STRUCTURES

Jungar Basin, China

Stratigraphic sequences that onlap a fold limb are resolved in a prestack-migration, three-dimensional seismic survey (Figure 3) from the Jungar Basin, Xingjiang Province, northwestern China (Zhang et al., 1996). The stratigraphic section consists of alternating coarse- and fine-grained Mesozoic continental deposits that onlap an east-dipping fold limb at the western margin of the basin. The basin margin is defined by a southeast-vergent fold-and-thrust belt of Mesozoic age that holds large oil reserves in faulted and folded Jurassic and Triassic sandstones (Fan, 1991).

In the imaged structure, the onlaps migrated upward and laterally toward the synclinal axis of the fold (Figure 3). These growth strata dip gently as they onlap the fold limb, yet downdip, each growth unit is folded parallel to the underlying limb. We discount that these features might be clinoforms or related sedimentary features (Mitchum et al., 1977), because the beds dip more than 25°. Rather, we interpret the pattern of onlaps and the fold geometry of each of these growth horizons to reflect folding by kink-band migration with an active synclinal axis, as shown in Figure 2.

Each onlapping sequence in the Jungar example is about 100 m thick, and thus the fold scarps must have been at least this high. Moreover, each onlapping sequence must have been deposited during a period of little or no folding to yield the observed growth pattern. Without precise age control, we cannot determine the time it took to form such scarps or to fill the local accommodation space with sediment. Nevertheless, we can constrain these time periods by considering reasonable rates of uplift. Assuming very high rates of uplift (10 mm/yr), each scarp must have taken 10,000 years to form. Perhaps more reasonable uplift rates of 1 mm/yr

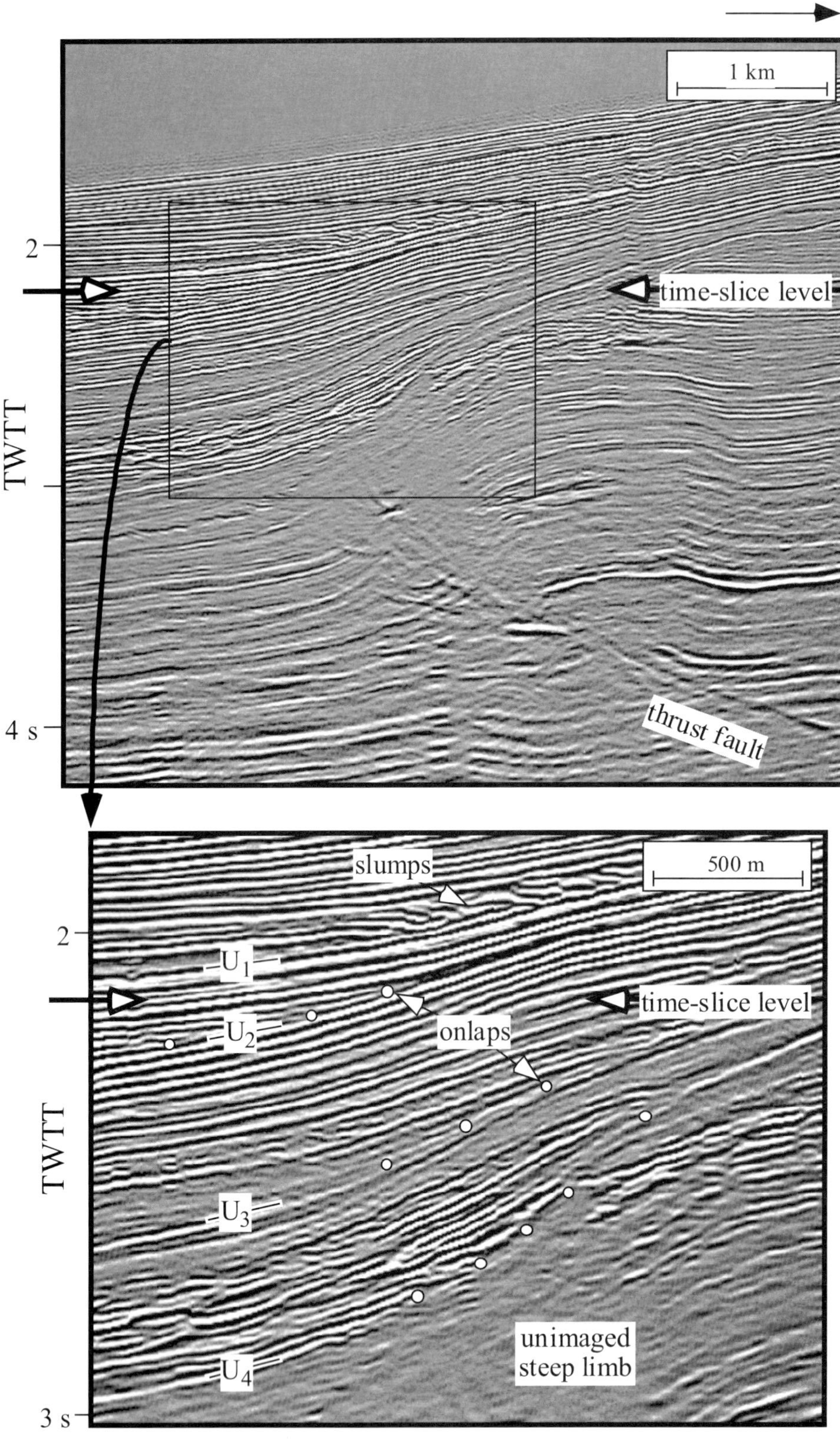

FIGURE 4. Migrated seismic-reflection profile imaging growth strata on the forelimb of a fault-related anticline, Niger Delta. Growth sequences are bounded by unconformities U1 through U4, and exhibit internally concordant bedding. Upward shallowing of bed dips reflects periods of fold amplification by limb rotation between the periods of deposition for each of these growth sequences. Onlaps at three of the major unconformities migrate upward toward the anticlinal axis of the fold, comparable with the limb rotation model of Figure 2. The time-slice level indicates the position of the map-view image shown in Figure 7.

or less would require 100,000 years or more for scarp formation. This implies that long (≥ 10,000 yr) periods of tectonic activity with little or no deposition alternated with periods of rapid deposition and/or tectonic quiescence during the formation of this fold.

Niger Delta

A poststack, three-dimensional seismic image resolves the forelimb of a contractional fault-related fold developed above a toe thrust in the deep-water Niger Delta (Figure 4). The thrust ramps upward from a basal detachment in the Acata Formation, a hemipelagic marine shale that overlies oceanic crust. Thrusting and related contractional folding consume slip on the detachment that was produced by gravity-driven extension and slumping on the continental shelf (Figure 5) (Lehner and de Ruiter, 1977; Doust and Omatsola, 1990). Growth strata in the contractional folds typically consist of alternating muds and sands in channel and basin-floor fan deposits.

In contrast to the previous example, growth strata on the forelimb of this fold show evidence of progressive limb rotation. The growth strata are composed of

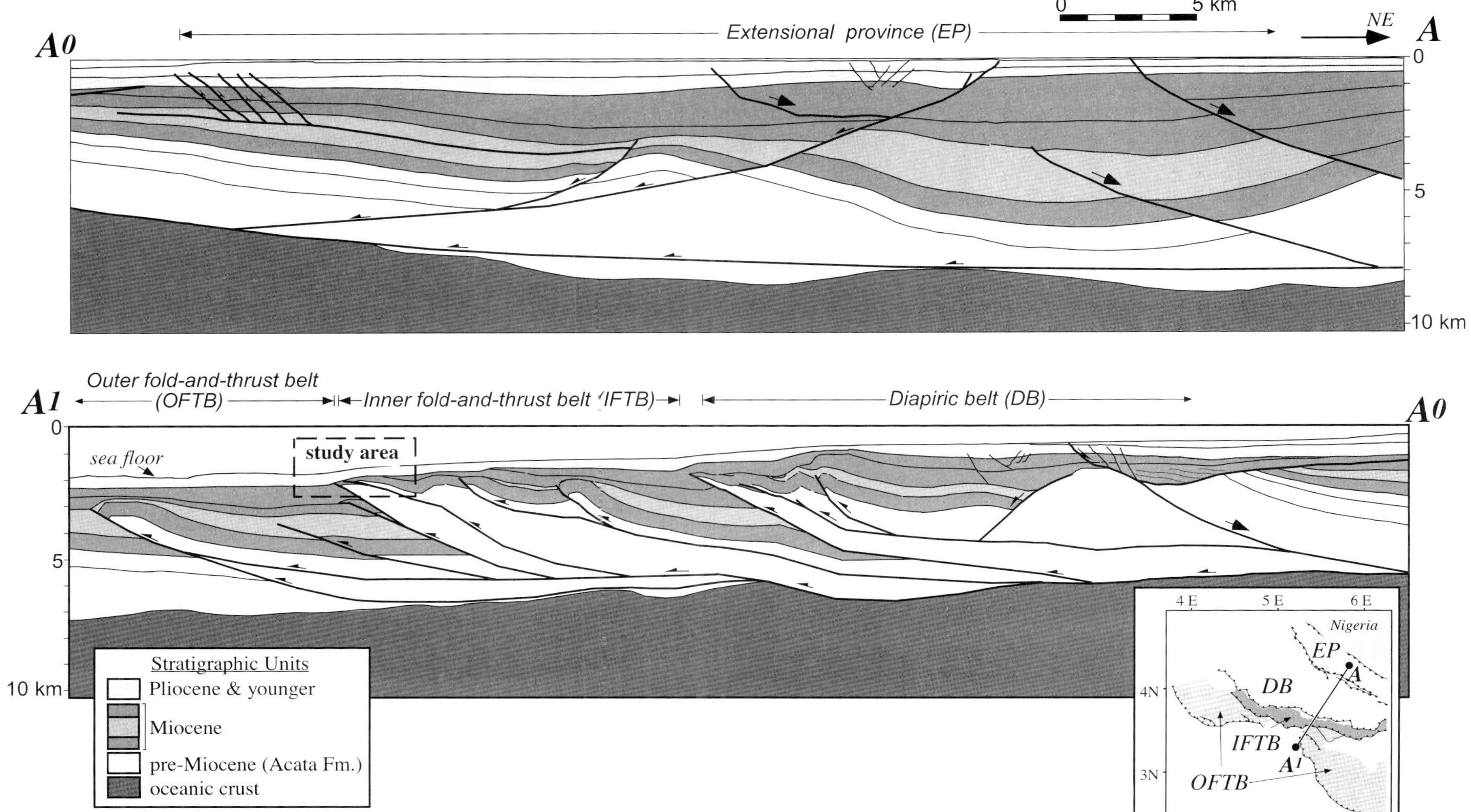

Figure 5. Regional geologic cross section across the Niger Delta, showing a detachment above oceanic crust that links extensional structures on the shelf to toe-thrust systems in deeper water. The structure shown in Figure 4 lies in the inner fold-and-thrust belt (IFTB).

several sequences, bounded by unconformities, that onlap the underlying fold limb. The dips of the sequences shallow upward, suggesting that the fold amplified between the deposition of each of these units. Pinch-outs, which are denoted by onlapping reflections, occur at the base of each major growth sequence. Pinch-outs migrate toward the anticlinal crest along each major unconformity, which is consistent with the pattern of onlaps expected for limb-rotation folds with constant or increasing rates of sedimentation relative to shortening (Figure 2). Notably, pinch-outs are not associated with prominent dip changes in their corresponding seismic horizons or in the underlying fold limb, as was the case in kink-band migration models (Figure 3). Thus, we interpret that this fold developed with a component of limb rotation.

The presence of several major unconformities with onlapping sequences reflects an episodic relationship between sedimentation and folding in this forelimb. Although the dips of sequences shallow upward, consistent with limb rotation, the bed dips within each sequence are roughly parallel. This pattern indicates that the fold did not simply amplify during the deposition of each sequence, but rather that it grew largely between the deposition of each sequence. Thus, the growth structure exhibits periods of very low sedimentation-to-uplift ratio alternating with periods of deposition in the absence of substantial deformation.

This growth history can be explained as a function of the forelimb's orientation relative to the direction of sediment transport (Burbank et al., 1996). The forelimb dips in the direction of regional sediment transport, which is southwest from the coast toward the abyssal plain (Doust and Omatsola, 1990). Thus, instead of sediments continuously ponding against the fold limb, sediment may, at times, have been restricted to the local accommodation space on the opposite side of the fold (i.e., above the backlimb). In effect, the forelimb occupied a depositional "shadow zone" behind the emergent anticline (Figure 6). Periods of rapid uplift of the fold crest would favor restriction of the sediment to the backlimb. Thus, periods of rapid fold amplification would correlate with periods of little or no deposition on the forelimb. In contrast, when uplift stopped or decreased substantially, the backlimb accommodation space would fill. Subsequently, sediments could breach the crest of the structure and be deposited on the forelimb (Figure 6). As is resolved in a three-dimensional seismic time slice, the onlapping sequences in this example are fan deposits fed by a canyon that breached the anticlinal crest (Figure 7). In this depositional scenario, forelimb deposits should correlate with periods of little or no uplift. This is consistent with the observation that the growth sequences exhibit no internal fanning of dips, as would be expected if they were deposited during limb rotation. In periods of rapid fold growth, sediments were generally not deposited on the forelimb. As a result, each depositional sequence has a distinctly shallower dip than that of the underlying sequence.

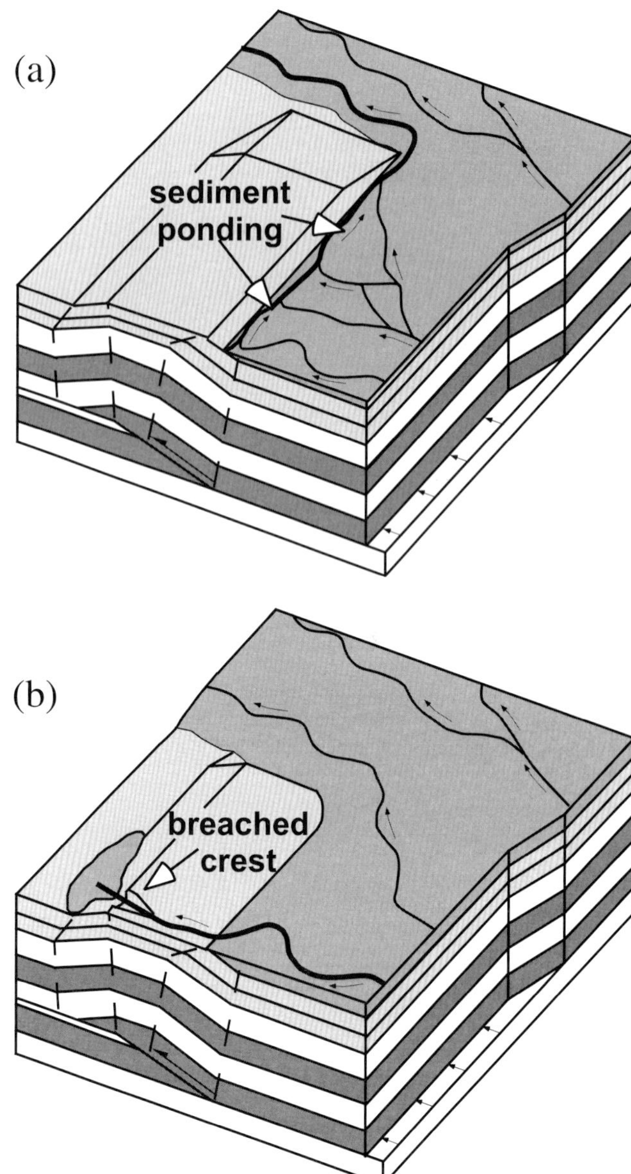

FIGURE 6. Perspective views of three-dimensional sequential models describing the depositional history proposed for the Niger Delta example shown in Figure 4. (a) Ponding of deposits behind a backlimb fold scarp restricts sediments to the backlimb accommodation space. No sediments are deposited above the forelimb. (b) After the backlimb accommodation space is filled, sediments are transported over the fold crest and deposited above the forelimb. Although a simple fault-bend fold structure formed by kink-band migration is shown, these general growth patterns also may occur in folds developed solely, or in part, by limb rotation.

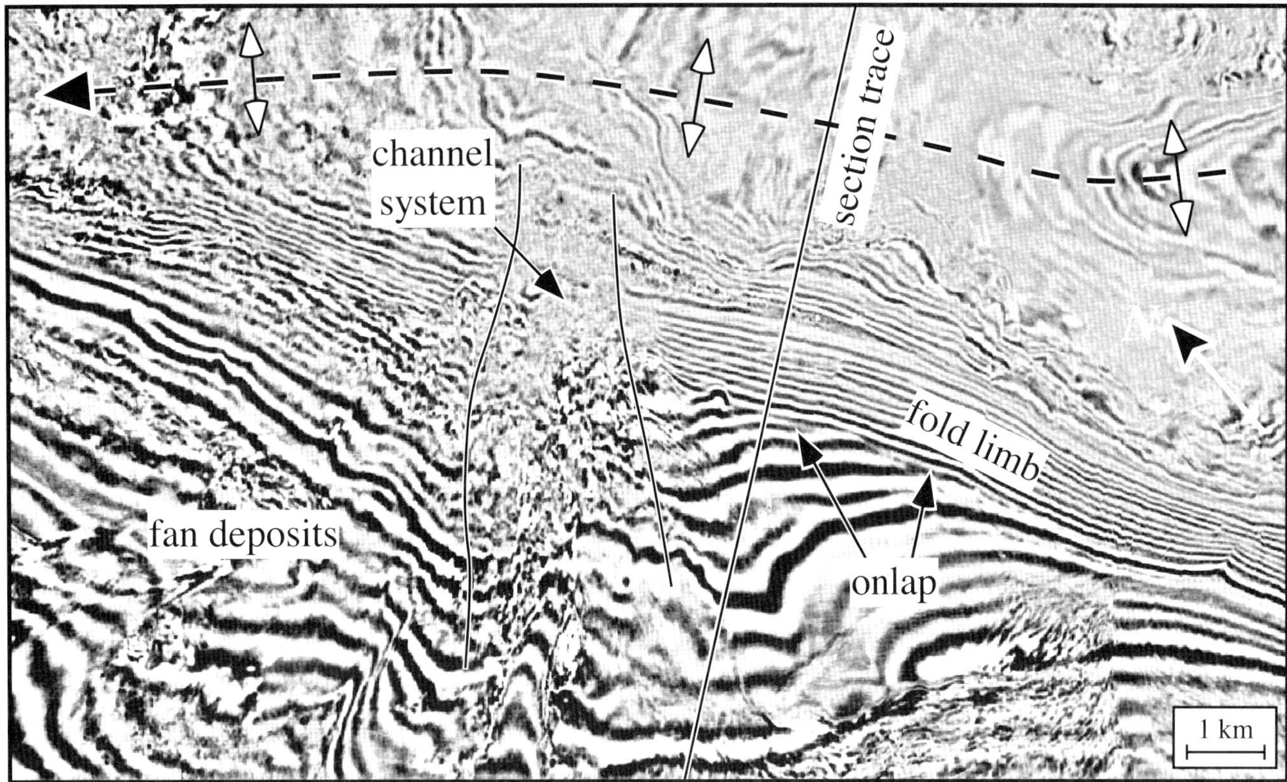

FIGURE 7. Three-dimensional seismic-reflection time slice across the growth anticline imaged in Figure 4. A southwest-flowing channel system breaches the anticlinal crest, producing fan deposits that onlap the forelimb. A series of these onlapping sequences imaged in Figure 4 records several distinct periods when the fold crest was breached by channel systems.

Transverse Ranges, California

Our final example is a Tertiary growth fold along the Offshore Oak Ridge trend in the Santa Barbara Basin, California (Figure 8). The basin is a southwestern extension of the Transverse Ranges Fold-and-thrust Belt of south-central California, which developed during Pliocene through Quaternary contraction (Yeats, 1983; Namson and Davis, 1988). Growth strata in the Offshore Oak Ridge structure contain thick sandstone sections interbedded with pelagic shales and turbidites (Sangree and Widmier, 1977). The fold exhibits an upward-narrowing limb that is interpreted to reflect folding by kink-band migration (Shaw and Suppe, 1994; Novoa et al., 1995, Novoa et al., 2000). No major stratigraphic onlaps on the fold limb are observed, however, at the resolution of the seismic image. Thus, we investigate the internal stratigraphy of the growth strata using well control to determine whether finer stratigraphy reflects the folding mechanism. We define the structural geometry and distribution of sedimentary facies in this fold using seismic-reflection images and well control, including dipmeter, sonic, spontaneous potential, and resistivity logs. The seismic profile in Figure 9 was depth converted using the sonic logs and stacking velocities. The seismic image closely correlates with posted dipmeter data in the wells. Thus, we suggest that the velocities and depth conversion are sound and provide an accurate picture of the structural geometry and precise ties of stratigraphy in the wells to the seismic image.

The distribution of sand and shale facies in the Pliocene-Quaternary Pico Formation of the Offshore Oak Ridge trend is defined by electric logs (Figure 9). The lower part of the middle Pico Formation contains mostly sandstones with thin shale partings, whereas the uppermost middle Pico and upper Pico sections consist almost entirely of shale. The intervening middle Pico section, however, contains a lateral change from sandstone- to shale-dominated facies across the Offshore Oak Ridge fold. Prominent sandstone intervals in well 1, between horizons S and X, are also present in the north-dipping fold limb (wells 2 and 4). However, these sandstones are absent on the fold crest (well 6) and to the south of the Oak Ridge trend. This facies change occurs abruptly across the growth axial surface (A′) of the Oak Ridge fold (Figure 9, inset), which is defined by the upward change from north-dipping to horizontal beds recorded in the seismic image and by dipmeter. The position of these pinch-outs along the growth axial surface, combined with the parallel dips of growth strata in the fold limb, are consistent with patterns modeled

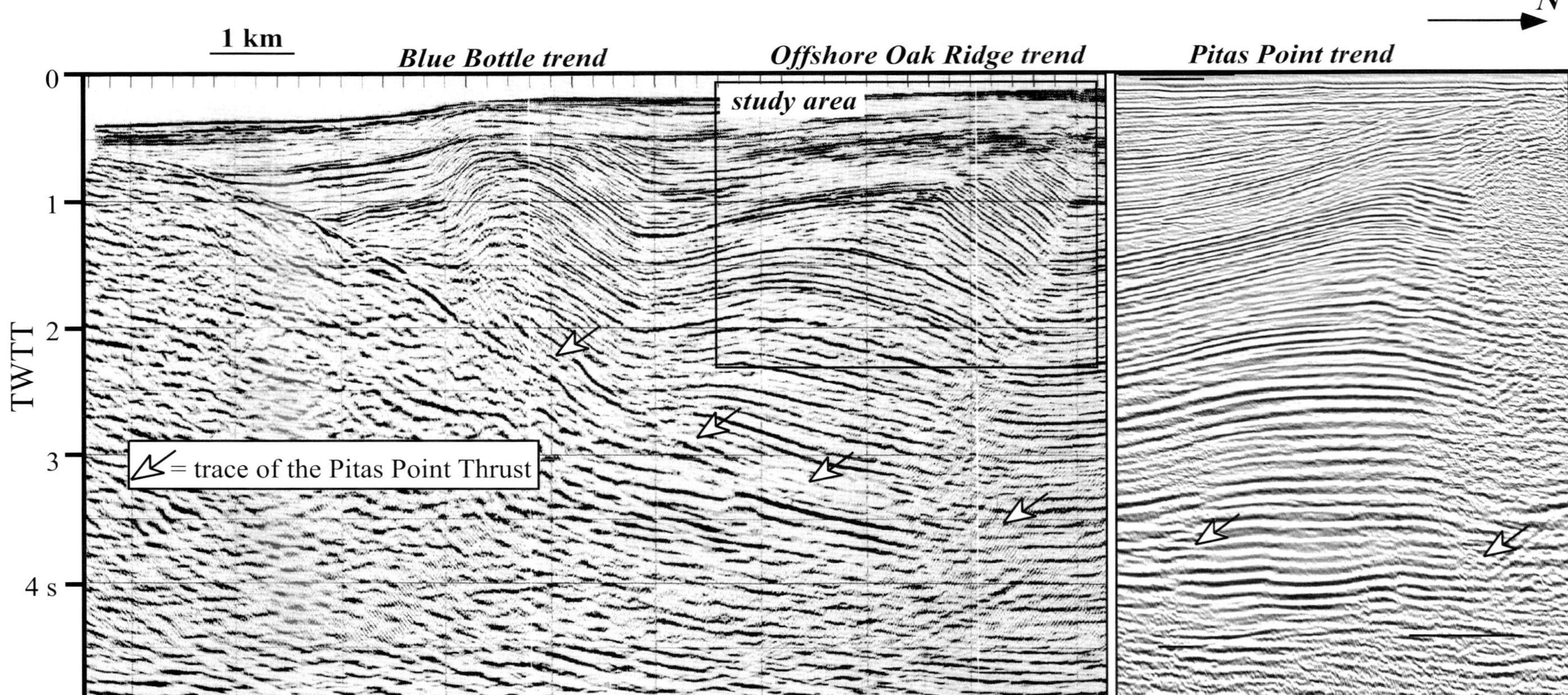

Figure 8. Three-dimensional, migrated seismic profiles across the eastern Santa Barbara Channel imaging a series of contractional growth folds, including the Offshore Oak Ridge structure (Shaw and Suppe, 1994; Shaw et al., 1996). The Offshore Oak Ridge structure developed above a series of vertically stacked blind-thrust faults, including the Pitas Point thrust. The position of the Pitas Point thrust is defined by reflector truncations and downward-terminating dip domains.

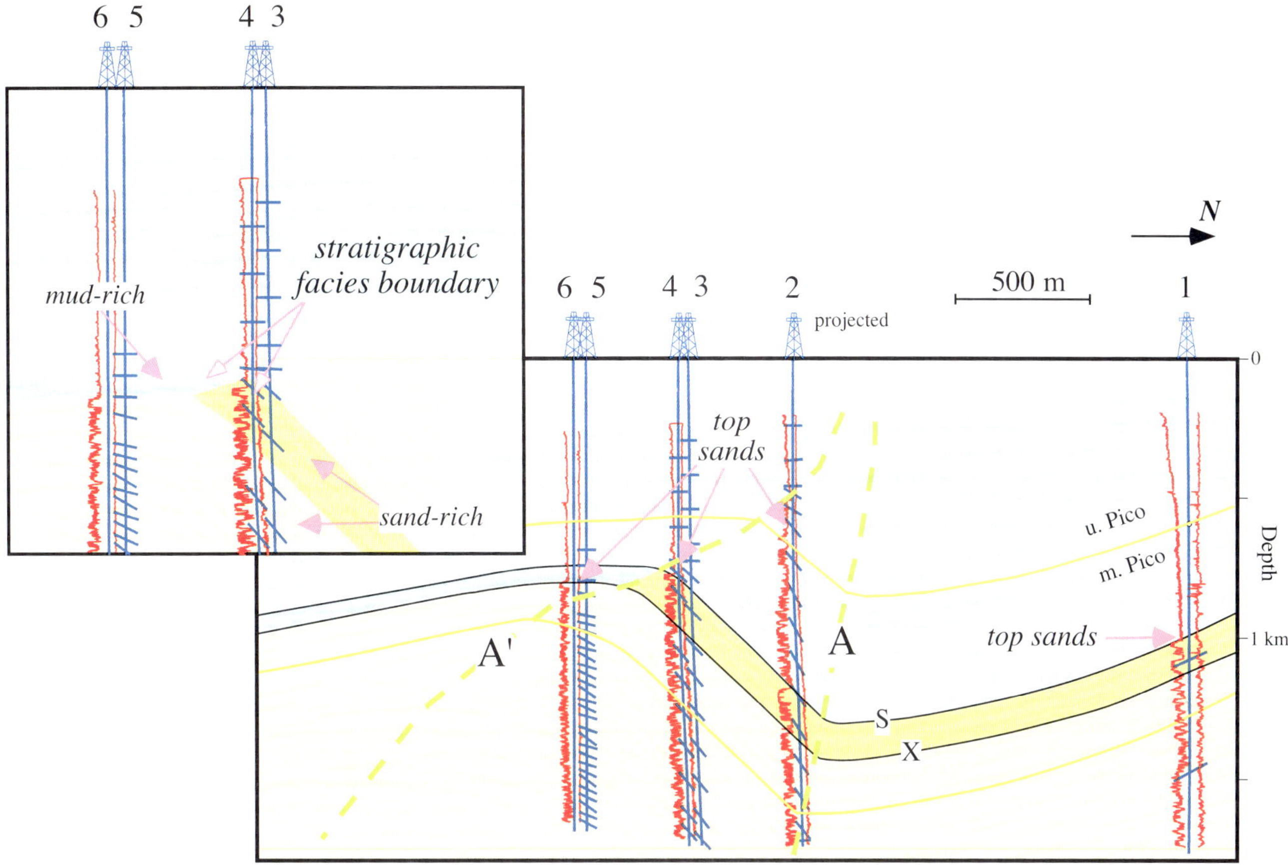

FIGURE 9. Structurally controlled facies change and sandstone pinch-outs on the Offshore Oak Ridge trend, Santa Barbara Channel, California, U.S.A. The migrated seismic-reflection profile is displayed in depth with a vertical exaggeration of 2:1. Spontaneous potential (SP) and resistivity logs are shown on the left and right sides, respectively, of selected wells. Shifts to the left in the SP logs indicate sands. Note how, in well 6, the uppermost sandstones in wells 1, 2, and 4 have changed to shale across the growth axial surface (A′). Enlarged section at left shows facies change across the growth axial surface. Wells: 1 = Texaco 234-7; 2 = Conoco SBC 21; 3 = Conoco SBC 34; 4 = Exxon SBC 39B; 5 = Arco SBC 39B; 6 = Phillips SBC 2. Data courtesy of the former Texaco, Chevron, and Arco.

by kink-band migration with an active synclinal axis (Figure 2). Thus, we suggest that the facies change was localized by ponding of turbidite sands behind the fold scarp (Figure 10). Subsequent folding by kink-band migration has translated these pinch-outs, which formed above the active axial surface A, to their present position near the anticlinal crest along the growth axial surface (A′).

Sandstone pinch-outs between horizons S and X occur in the uppermost part of the Pliocene turbidite sequence. This stratigraphic position correlates with a period of lower sediment influx, when deposits were unable to fill the local accommodation space produced by the fold scarp, as shown in Figure 1a. In contrast, older turbidite deposits overflowed the local accommodation space and thus are present on both sides of the fold (Figure 9). The apparent decrease in deposition rate, and/or increase in fold growth rate, is recorded in Figure 10 by the shape of the growth axial surface (A′), which shallows upward into the middle Pico section (Suppe et al., 1992; Shaw and Suppe, 1996).

Above horizon S, sandstones occur in well 2 that are not present to the north or south of the Offshore Oak Ridge trend (Figure 9). These discontinuous sands have a log character consistent with a more channelized facies than do the deeper ponded sands, and may be derived from submarine channel systems that were deflected to run along the base of the seafloor fold scarps (Figure 1b). The transition from turbidite- to channel-dominated deposits probably reflects a decrease in water depth caused by infilling of the basin, a trend that is well documented in other deep-water environments (Mitchum, 1985).

SUMMARY AND CONCLUSIONS

We have described how the internal stratigraphy and geometry of growth structures are influenced by the kinematics of fold growth. When sedimentation-to-uplift ratios are less than or equal to one, stratigraphic

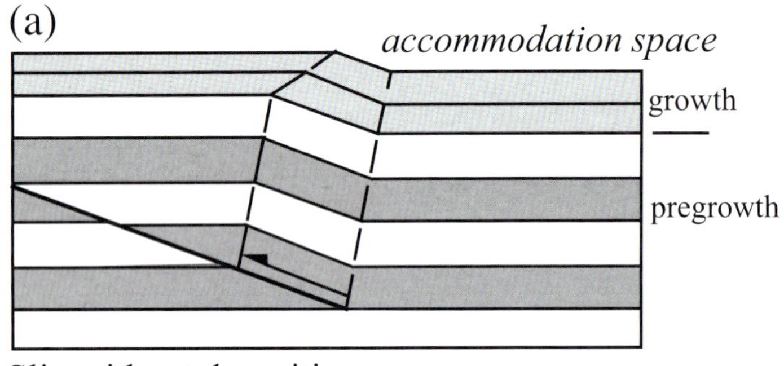

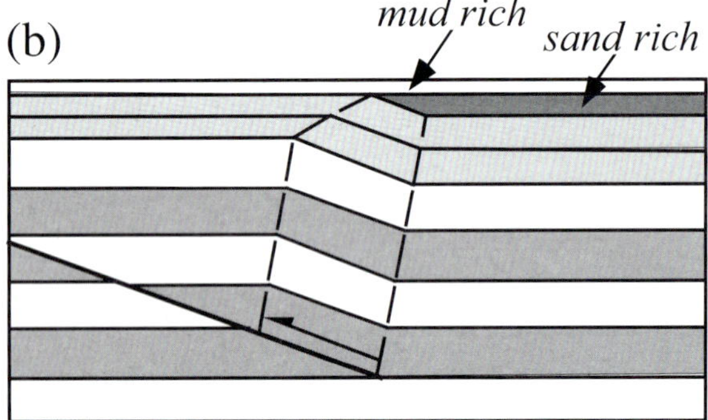

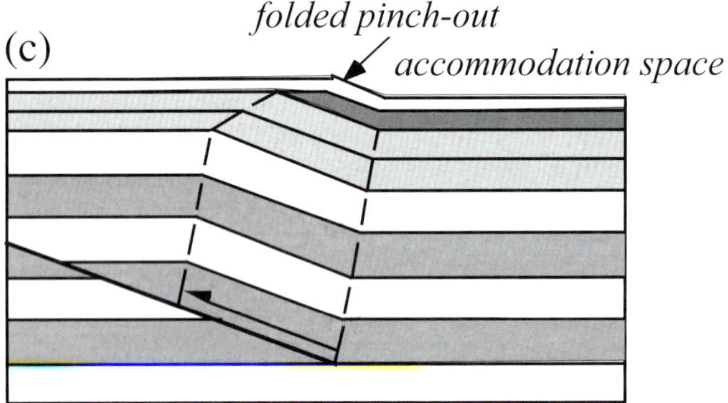

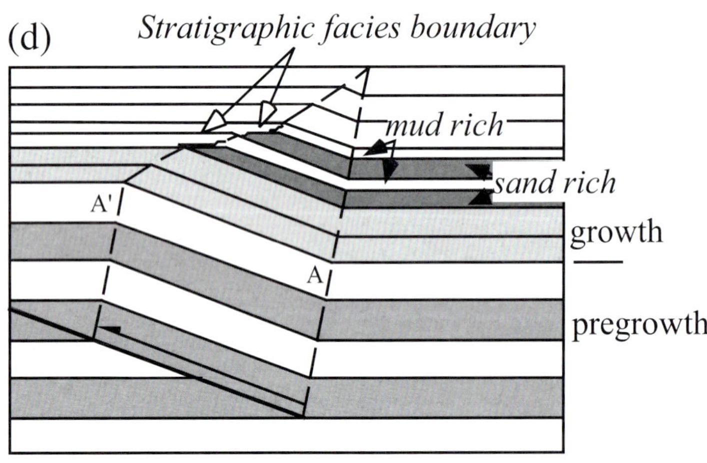

FIGURE 10. Sequential models illustrating the development of the Offshore Oak Ridge structure by kink-band migration and fault-bend folding with deposition. (a) Folding generates a surface scarp and local accommodation space. (b) Sand-rich deposits pond behind the fold scarp and fill the accommodation space, as described in Figure 1. After the accommodation space is filled, mud-rich sediments blanket the structure. (c) Another episode of folding generates a surface scarp and local accommodation space, which is subsequently filled with sand-rich deposits. The sandy units pinch out onto the fold limb. (d) Deformation folds and translates the sand pinch-outs, which are aligned along the axial surface A′. Muds are deposited concurrent with this last phase of deformation.

pinch-outs and associated facies transitions may develop on fold limbs. These simple stratigraphic patterns are modified by subsequent folding. In folds that develop by kink-band migration, onlaps and corresponding stratigraphic pinch-outs develop at discrete fold scarps associated with active axial surfaces. Progressive deformation folds and translates these onlaps, so that they are aligned with prominent dip changes that occur along the fold's growth axial surface (Suppe et al., 1992). In contrast, folds that develop purely by limb rotation have onlap patterns controlled primarily by the relative rates of sedimentation and deformation. In general, onlaps migrate toward the anticlinal hinge of the fold, in limb-rotation structures. Moreover, these onlaps are not associated with prominent dip changes in their corresponding stratigraphic horizons, as was the case for kink-bend migration structures. Thus, we can expect differences in the patterns of onlapping stratigraphic sequences, as well as in macroscopic fold geometry, between folds that develop by kink-band migration and those that develop by limb rotation. We used these insights to interpret that two folds, in southern California and China, developed by kink-band migration. In contrast, an example from Nigeria showed a growth fold that has amplified with a component of limb rotation.

In addition to distinguishing fold kinematics, insights into the internal stratigraphy of growth structures can help to resolve a detailed history of structural development. The large size (≈100 m) of fold scarps inferred in the example from China, and the punctuated limb rotations observed in the Nigeria example, suggest that relative rates of sedimentation and deformation varied substantially over the growth of these structures. These changes may reflect episodic sedimentation. However, they may also result from long periods of heightened tectonic activity followed by quiescence. In contrast, small scarps and pinch-outs (<2 m) in the Offshore Oak Ridge trend may reflect individual episodes of fold growth, perhaps driven by earthquake rupture on the underlying faults. Alternatively, the patterns could be generated by steady-state folding with episodic turbidite deposition. Detailed absolute-age control for the stratigraphic sections, combined with the observed fold patterns, would be needed to resolve between these sedimentological and tectonic influences. Both factors, however, clearly influence the geometry and internal stratigraphy of growth strata. All three of the examples that we document exhibit variable rates of tectonism and shortening, which may be common in contractional growth structures.

Stratigraphic facies changes and pinch-outs in growth structure also can provide hydrocarbon traps. Although these features can be observed directly in some seismic-reflection profiles (Figure 3), we demonstrate that they can exist beyond the imaging resolution of typical seismic-reflection data and thus must be corroborated by well control (Figure 9). Nevertheless, knowledge of the folding mechanism can be used to predict the local distribution of these traps in certain types of structures. For example, subtle pinch-outs can align along growth axial surfaces in folds that are formed by kink-band migration (e.g., fault-bend and fault-propagation folds), as illustrated by our example from California. These growth axial surfaces are readily observed on seismic images as marked changes in the dips of seismic reflections. Thus, macroscopic fold patterns can be used to identify areas that may offer stratigraphic traps, which could then be evaluated by subsequent drilling. These subtle traps can occur downdip on the flanks and limbs of structures. They are seldom tested by wells on the fold crest, and thus may provide exploration and development opportunities even in mature petroleum provinces.

ACKNOWLEDGMENTS

This research was supported by Texaco, Inc., and PDVSA, with data generously provided by Arco, Chevron, BP, Statoil, and Texaco. The authors thank John Suppe and Stephen Hook for helpful insights into growth folding and Kevin Bishop for assistance with processing of the seismic profile from California. Ken McClay and two anonymous reviewers provided helpful suggestions for improving the manuscript.

REFERENCES CITED

Allmendinger, R. W., 1998, Inverse and forward numerical modeling of trishear fault-propagation folds: Tectonics, v. 17, no. 4, p. 640–656.

Burbank, D., A. Meigs, and N. Brozovic, 1996, Interaction of growing folds and coeval depositional systems: Basin Research, 8, p. 199–223.

Dahlstrom, C. D. A., 1990, Geometric constraints derived from the law of conservation volume and applied to evolutionary models for detachment folding: AAPG Bulletin, v. 74, p. 336–344.

Doust, H., and E. Omatsola, 1990, Niger Delta, *in* J. D. Edwards and P. A. Santogrossi, eds., Divergent/passive margins: AAPG Memoir 48, p. 201–238.

Erslev, E. A., 1991, Trishear fault-propagation folding: Geology, v. 19, no. 6, p. 617–620.

Fan, C. L., 1991, Reservoir models in the Jungar Basin, Xin-jiang, north-west China: Marine and Petroleum Geology, v. 8, p. 107–114.

Ford, M., E. A. Williams, A. Artoni, J. Vergés, and S. Hardy, 1997, Progressive evolution of a fault-related fold pair from growth strata geometries, Sant Llorenç de Morunys, SE Pyrenees: Journal of Structural Geology, v. 19, no. 3–4, p. 413–441.

Hardy, S., and M. Ford, 1997, Numerical modeling of trishear fault-propagation folding and associated growth strata: Tectonics, v. 16, no. 5, p. 841–854.

Hardy, S., and J. Poblet, 1994, Geometric and numerical model of progressive limb rotation in detachment folds: Geology, v. 22, p. 371–374.

Hardy, S., J. Poblet, K. McClay, and D. Waltham, 1996, Mathematical modeling of growth strata associated with fault-related fold structures, *in* P. G. Buchanan and D. A. Nieuwland, eds., Modern developments in structural interpretation, Validation and modeling: Geological Society of London Special Publication 99, p. 265–282.

Lehner, P., and P. A. C. de Ruiter, 1977, Structural history of Atlantic margin of Africa: AAPG Bulletin, v. 61, p. 963–981.

Medwedeff, D. W., 1989, Growth fault-bend folding at southeast Lost Hills, San Joaquin Valley, California: AAPG Bulletin, v. 73, p. 54–67.

Mitchum, R. M., 1985, Seismic stratigraphic expression of submarine fans, *in* O. R. Berg and D. G. Woolverton, eds., Seismic stratigraphy II: An integrated approach to hydrocarbon exploration: AAPG Memoir 39, p. 274.

Mitchum, R. M., P. R. Vail, and S. Thompson III, 1977, Seismic stratigraphy and global changes of sea level, Part 2: The depositional sequence as a basic unit for stratigraphic analysis, *in* C. E. Payton, ed., Seismic stratigraphy— applications to hydrocarbon exploration: AAPG Memoir 26, 516 p.

Namson, J., and T. L. Davis, 1988, A Structural transect of the western Transverse Ranges, California: Implications for lithospheric kinematics and seismic risk evaluation: Geology, v. 16, p. 675–679.

Novoa, E., J. Suppe, K. Mueller, and J. Shaw, 1995, Axial surface mapping of active folds in the Santa Barbara Channel, *in* 1995 Thrust ramps and detachment faults in the Western Transverse Ranges: SCEC workshop Abs., p. 19.

Novoa, E., J. Suppe, and J. H. Shaw, 2000, Inclined-shear restoration of growth folds: AAPG Bulletin, v. 84, p. 787–804.

Poblet, J., K. McClay, F. Storti, and J. A. Muñoz, 1997, Geometries of syntectonic sediments associated with single layer detachment folds: Journal of Structural Geology, v. 19, p. 369–381.

Sangree, J. B., and J. M. Widmier, 1977, Seismic interpretation of clastic depositional facies, *in* C. E. Payton, ed., Seismic stratigraphy – applications to hydrocarbon exploration: AAPG Memoir 26, 516 p.

Schneider, C. L., C. Hummon, R. S. Yeats, and G. L. Huftile, 1996, Structural evolution of the northern Los Angeles basin, California, based on growth strata: Tectonics, v. 15, p. 341–355.

Shaw, J., and J. Suppe, 1994, Active faulting and growth folding in the eastern Santa Barbara Channel, California: Bulletin of the Geological Society of America, v. 106, p. 607–626.

Shaw, J., and J. Suppe, 1996, Earthquake hazards of active blind-thrust faults under the central Los Angeles basin, California: Journal of Geophysical Research, v. 101, p. 8623–8642.

Shaw, J. H., S. C. Hook, and J. Suppe, 1996, Structural trend analysis by axial surface mapping: Reply to Discussion of Shaw, J. H., S. Hook, and J. Suppe, 1994, Structural trend analysis by axial surface mapping in AAPG Bulletin, v. 78, p. 700–721: AAPG Bulletin, v. 80, p. 780–787.

Storti, F., and J. Poblet, 1997, Growth stratal architectures associated with decollement folds and fault-propagation folds: Inferences on fold kinematics: Tectonophysics, v. 282, p. 353–373.

Suppe, J., 1983, Geometry and kinematics of fault-bend folding: American Journal of Science, v. 283, p. 684–721.

Suppe, J., and D. A. Medwedeff, 1990, Geometry and kinematics of fault-propagation folding: Eclogae Geologicae Helvetiae , v. 83, p. 409–454.

Suppe, J., G. T. Chou, and S. C. Hook, 1992, Rates of folding and faulting determined from growth strata, *in* K. R. McClay, ed., Thrust tectonics: London, Chapman & Hall, p. 105–121.

Suppe, J., F. Sàbat, J. A. Muñoz, J. Poblet, E. Roca, and J. Vergés, 1997, Bed-by-bed growth by kink-band migration: Sant Llorenç de Morunys, Eastern Pyrenees: Journal of Structural Geology, v. 19, p. 443–461.

Vergés, J., D. W. Burbank, and A. Meigs, 1996, Unfolding: An inverse approach to fold kinematics: Geology, v. 24, no. 2, p. 175–178.

Yeats, R. S., 1983, Large-scale Quaternary detachments in Ventura basin, southern California: Journal of Geophysical Research, v. 88, no. B1, p. 569–583.

Zhang, G., L. Liu, and Z. Wu, 1996, The preliminary application of GeoDepth prestack depth migration in data processing in Xinjiang area: Proceedings of the 1996 International Geophysical Conference, Society of Exploration Geophysicists/Chinese Society of Petroleum Geophysicists, Urumqi, P.R.C, p. 51–56.

21

Strayer, L. M., S. G. Erickson, and J. Suppe, 2004, Influence of growth strata on the evolution of fault-related folds — Distinct-element models, *in* K. R. McClay, ed., Thrust tectonics and hydrocarbon systems: AAPG Memoir 82, p. 413–437.

Influence of Growth Strata on the Evolution of Fault-related Folds— Distinct-element Models

Luther M. Strayer[1]
Department of Geosciences, Princeton University, Princeton, New Jersey, U.S.A.

S. Gregg Erickson[2]
Department of Geosciences, Princeton University, Princeton, New Jersey, U.S.A.

John Suppe
Department of Geosciences, Princeton University, Princeton, New Jersey, U.S.A.

ABSTRACT

We construct a series of distinct-element models that consist of bonded assemblies of elastic particles that simulate the brittle deformation associated with large-scale, fault-related folding over a rigid footwall. The initial rock mass is simulated by a series of discrete, circular, elastic, frictional particles, bonded in shear and tension and capable of progressive fracture during loading. This produces discontinuities within the simulated rock mass: Regions of intact material are separated by discrete faults or fault zones. The simulations are fully dynamic. Material properties are assigned as microproperties of and between particles; elastic stiffness, friction, and bond strength (shear and tensional) describe particle interactions. Because the macroscopic elastic, failure, and flow properties are not directly determined by the microproperties, uniaxial and biaxial compression tests must be conducted on a representative specimen to determine macroscopic parameters, such as unconfined compressive strength, Young's modulus, friction angle, and cohesion.

To determine the effect of growth strata on the evolution of the underlying fold, growth strata with thicknesses equivalent to 0%, 25%, 50%, 75%, and 100% of the elevation of the associated anticline above the upper footwall flat are added to the upper surface of the models. The presence or absence of growth strata has a significant effect on the final geometry of the associated fault-related fold. The growth strata provide resistance to forelandward translation, causing folds to tighten from open, gentle ramp anticlines with no or thin growth strata to strongly overturned fault-propagation-type folds with thick growth sequences. The nature and intensity of fracturing is strongly

[1]*Present address*: Department of Geological Sciences, California State University, Hayward, Hayward, California, U.S.A.
[2]*Present address*: Science and Math, Sullivan County Community College, Loch Sheldrake, New York, U.S.A.

affected by growth-strata thickness. All models have effectively identical early stages, as a result of thin or no growth, and initial deformation at the thrust ramp is always by back thrusting. This changes with increased burial. (1) With no growth strata, back thrusting continues, although fault dips become shallower with time. (2) With intermediate thicknesses of growth strata, back thrusting gives way to, or alternates with, bedding-parallel shear. (3) With thick growth strata and thus high confining pressure, brittle deformation is inhibited and replaced by elastic deformation of the particles. These models produce realistic-looking packages of growth strata. Well-developed growth triangles are common in backlimbs, and are cut by the faults that originate in the pregrowth strata. The growth strata commonly show angular unconformities and stratigraphic pinch-outs. Overturned forelimb growth strata are significantly deformed, recording brittle deformation associated with progressive growth of the underlying fold.

INTRODUCTION

Much of the work concerning the relationship between growth strata and associated underlying folds has focused on the geometry of the growth-strata packages and has been based on kinematics and rates of fold growth (Suppe et al., 1992, 1997; Hardy and Poblet, 1994; Storti and Poblet, 1997), or on growth-strata architecture as a constraint on the kinematic evolution of the associated fold by sequential back-stripping and unfolding of the growth strata (DeCelles et al., 1991; Verges et al., 1996; Ford et al., 1997; Erickson et al., 1998). Numerical modeling of growth strata has, to date, focused on applying forward kinematic models to predict growth-strata architecture (Hardy and Poblet, 1994; Hardy et al., 1996; Storti and Poblet, 1997), or on using restoration by an area-conserving block-packing technique (Rouby et al., 1993), which treats the growth strata as an assemblage of rigid blocks separated by discontinuities, to determine the displacement field (Erickson et al., 1998). In all of these approaches, it is tacitly assumed that the growth strata play a passive role in the deformation, recording fold growth rates and kinematics but not affecting the evolution of the fold. However, thick growth-strata packages in some thrust belts, such as the Ebro Basin of northeastern Spain (Williams et al., 1998), might significantly affect the kinematic evolution of developing folds by supplying resistance to forelandward translation (Barrier, 1998) or by altering the buckling wavelength by increasing sedimentation and therefore increasing the layer thickness (Lawton et al., 1999).

Unscaled analog rock models (Chester et al., 1991) and continuum numerical models (Strayer and Hudleston, 1997a) show that resistance to foreland translation supplied by near-field boundary conditions has a significant effect on the final geometry of fault-related folds. Unlimited forelandward slip produces an open fault-bend-style fold with an emergent thrust surface. If foreland translation is restricted, which effectively buttresses the emergent fault-related fold, the forelimb of the resultant fold is vertical to overturned, thereby producing a fault-propagation-style fold with a nonemergent thrust. In analog sand and silicone-putty models of fault-related folds under various sedimentation rates, Barrier (1998) showed that high sedimentation rates form fault-propagation-style folds, while low sedimentation rates favor fault-bend-style folds.

The goal of this study is to produce a set of numerical models that investigate the influence of growth strata as a foreland buttress during the evolution of fault-first fault-related folds. For this modeling we use the two-dimensional finite-difference distinct-element code PFC2D (Particle Flow Code in 2 Dimensions) (Itasca Consulting Group, 1999a), which is based on, and originated from, the numerical code TRUBAL (Cundall, 1988). The code treats geologic materials as an assembly of bonded circular particles that can progressively fracture, fault, and flow, depending on the material properties and applied boundary conditions. PFC2D and modified forms of TRUBAL have been used to simulate faulting processes (Cundall, 1989), large-scale folding and faulting of fold-thrust belts (Strayer and Hudleston, 1997b), deformation in accretionary prisms (Morgan, 1997), fault-gouge formation (Morgan and Boettcher, 1999), and normal faulting (Saltzer and Pollard, 1992).

CODE DESCRIPTION

PFC2D is based on the distinct-element method (DEM) of Cundall and Strack (1979) and initially was developed to study constitutive behavior and flow of granular material. The code simulates a rock mass in two dimensions by treating it as an assembly of frictional, elastic, circular disk-shaped particles that may be bonded in shear and tension. This bonded assembly represents the initially continuous rock mass, capable of fracturing and faulting by progressive bond breakage during loading. It allows us to monitor the evolution of both continuous and discontinuous deformation in shallow-level tectonic regimes. The discrete, particle-based nature of this approach also allows us also to model the deposition and erosion of sediments.

The DEM uses a time-stepping, explicit-finite-difference scheme. At each time step, for each particle, Newton's second law:

$$F_i = m(a_i - g_i) \tag{1}$$

is integrated to yield new velocities and positions based on the forces, F_i, acting upon a particle, where m is mass, and g_i and a_i (i = 1, 2) are gravitational and particle acceleration, respectively. Contact forces result from a linear spring-force displacement law, yielding the new positions of a given particle pair. The normal force is given by:

$$F_i^n = K^n \; U^n n_i \tag{2}$$

where K^n is the normal particle stiffness, U^n is the normal displacement, and shear force is given by:

$$F_i^s = K^s U^s n_i \tag{3}$$

The code solves the full dynamic equations of motion and thus easily treats physical instability, such as localization/faulting.

The particle physics is described by (1) linear spring behavior of particle contacts (shear and bulk modulus), (2) Coulomb sliding between particles (coefficient of friction), and (3) optional bonding between particles (tensile and shear strength) (Figure 1). The particles are bounded by elastic, frictional walls that can be fixed or velocity controlled. Because this is a 2-D formulation, the models can represent either (1) a series of 3-D spheres whose centers are aligned on a common plane or (2) a series of disks with constant or unit thickness. For these models, we will use the unit-thickness disk mode.

Models are constructed by filling a region that is bounded by walls with a set of elastic, frictional particles. In order to prevent problems caused by crystallization of hexagonal close-packed systems, we vary the radii of the particles over a normal distribution, ranging from 100% to 75% of the maximum particle radius. The initial stress state is achieved by a process of either (1) radius expansion to create an initial isotropic stress state, used here for creating biaxial test specimens; or (2) gravity settling, which results in linearly increasing pressure with depth, used here for large-scale fault-related fold models.

Because of the discrete, particle-based nature of the model, specification of material properties and boundary conditions is more difficult than with available continuum methods (Itasca Consulting Group, 1999a, b, 2000). The user may specify microproperties that control particle-particle interaction (e.g., particle shear and bulk modulus, coefficient of friction, and bond strength), but the user has no way to directly prescribe the macroproperties of the model, such as Young's modulus (E), unconfined compressive strength (UCS), cohesion (C_0), coefficient of friction (μ), porosity, and the initial stress state. Creating the initial bonded assembly, therefore, is a painstaking iterative, trial-and-error process.

The resultant equilibrium assemblies are then bonded in shear and tension. Because the particle sizes vary, it is necessary to treat bond strength in terms of stress instead of force. The contact area between two particles is calculated from the averaged diameter of the two contacting particles and their unit thicknesses. The force that defines bond strength is divided by the contact area, so that the size of the particles in a model assembly

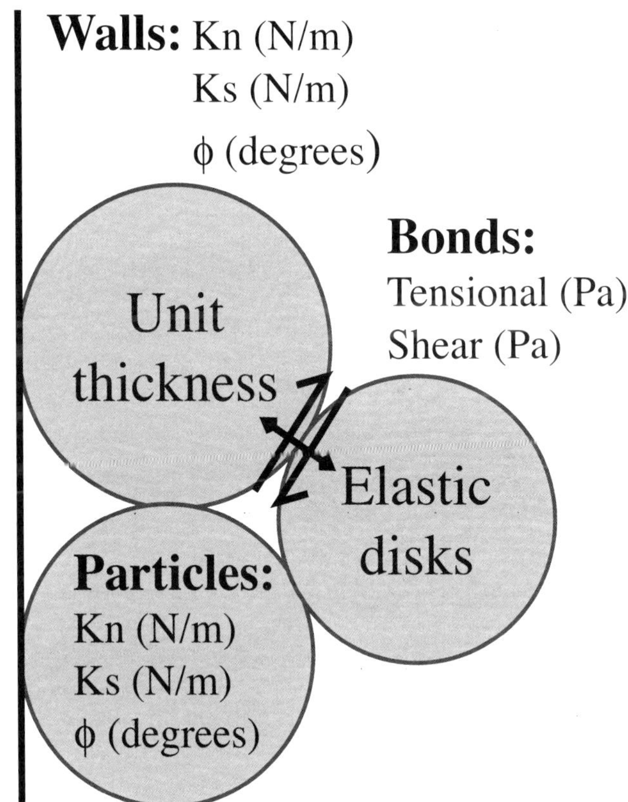

Figure 1. Microscopic elements of the model. Model components include frictional, linear-elastic, disks that act as springs in compression; frictional, elastic walls that are either fixed or velocity controlled; and contact bonds that resist tensional separation and shear displacement. Bonds are applied after the particles achieve static gravitational equilibrium, creating an initial solid that is capable of progressive fracture during loading. Kn and Ks are the particle and wall bulk and shear moduli, respectively. Particle and wall friction angle ϕ is simply a convenient representation of the critical ratio of shear to normal force required for a contact to slide.

does not affect the stress state significantly (Potyondy and Cundall, 1999), given enough particles for flow and localization. The contact bonding that we use simply connects particles along a line between their centers, thus, bonds cannot support a moment. Bonds are broken when either the shear or tensile bonding force is exceeded. For these simulations, broken bonds remain broken, although it would be possible to include recementation.

The porous material that results from this initial assembly generation process is defined solely by a simple set of microproperties whose rock mass properties are not directly derived from the values of the microproperties. To determine the mechanical properties of the subsequent assemblies, we must analyze them in a manner analogous to the way material properties are determined for true rock specimens, that is, through strength testing in a biaxial test rig.

Continuum data, such as stress and strain rate, cannot exist at every point in a discontinuous porous material. As a result, for these and other data, such as porosity, fraction of sliding contacts, and coordination number (number of contacts/particle), a measurement-circle logic uses averaging procedures to convert discrete microdata to continuum data. Measurement circles can be placed anywhere within the material and simply need to contain enough particles to make the averaging accurate. Histories of these measurement-circle data can be used to analyze model behavior and also to control model runs by feedback and servo-control. We use measurement circles at five locations (Figure 2d) to monitor the evolution of the full stress tensor σ_{ij}, porosity, and the fraction of particle-particle contacts that are currently sliding.

LARGE-SCALE MATERIAL PROPERTIES AND TESTING

During the early evolution of the modeling program, we settled on a simple set of criteria to guide the development of an initial, stable, no-growth-strata model that is the basic foundation of this study. These criteria are that

1) there is an overall fault-bend-fold style geometry;
2) there is a hanging wall that conforms smoothly to the footwall shape — no gaps are allowed to form adjacent to fault bends;
3) no crevasses or cliffs are allowed in the material — these can form as the result of improper scaling of bond strength;
4) the hanging wall slides up the ramp instead of failing at the rear — this assumes that the model deforms as a kilometer-scale, end-loaded structural beam, rather than by critical-taper processes;
5) there is a realistic failure mechanism at the lower ramp hinge: either back thrusting or flexural slip; and
6) the material is not stronger than laboratory specimens of sedimentary rocks.

The models must, at a minimum, appear realistic and deform by geologic mechanisms.

The fact that the model assembly is mechanically defined just by microproperties requires that this sort of modeling is necessarily iterative. Typically, there is an initial stage of wide-ranging parameter studies that focuses on the results of varying particle elastic moduli and bond strength, with the goal of simply finding combinations that look geologically reasonable. During this time, we simultaneously run uniaxial and biaxial compressive-strength tests on specimens that have size distributions, elastic moduli, friction, and bond strength identical to those of the large-scale model. These strength experiments yield the Young's modulus, E, and the unconfined compressive strength, UCS, of our large-scale models. Initial efforts to use strength and elastic stiffness values that reasonably match published values for limestone and sandstone rock masses (Jaeger and Cook, 1969) produced models that were clearly too strong. By iterating between the full-scale thrust and compressive-strength models, we adjusted the particle elastic stiffness and the shear and normal bonding until we created models with the geometries and many familiar structural features of fault-related folds. Through this process, we settled on a well-behaved model with bulk elastic moduli and bonding strength values that are about two orders of magnitude lower than published laboratory values. Macroscopic strength parameters E and UCS are effectively controlled by particle spring constants, which are generally of the same order of magnitude as Young's modulus.

Unlike many continuum numerical codes, in which macroscopic material properties are the primary input, distinct-element models are defined by microproperties. Similarly, the mechanical behavior of the particle assembly is not predefined by specifying a constitutive rule; instead, the character of the particle assembly itself defines its own mechanical behavior. Thus, the peak strength and the form of the stress-strain curve must also be determined by testing. Using a biaxial testing environment (Potyondy and Cundall, 1999), we determine Young's modulus, E; unconfined compressive strength, UCS; angle of internal friction, ϕ; and cohesion, C_o.

The fact that laboratory values prove to be too strong for large-scale structures should come as no surprise. The size of laboratory testing rigs requires that specimens be chosen to avoid discontinuities, and as a result, they require small dimensions. In principle, as the rock mass becomes larger, it becomes weaker,

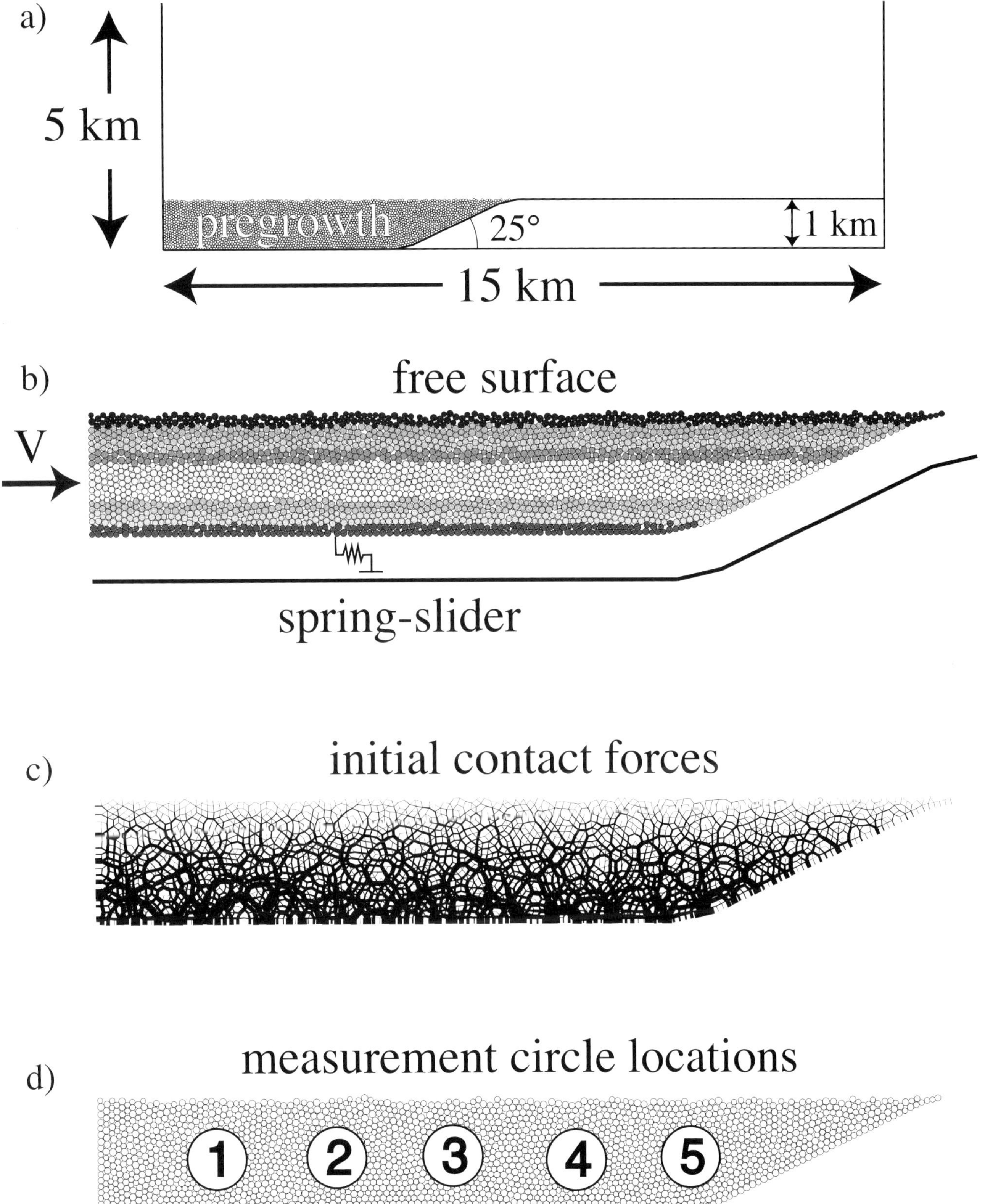

FIGURE 2. Model configuration and boundary conditions. (a) Overall model dimensions are 15 km in length and 5 km in height. Pregrowth model assembly is 1 km thick and consists of 2367 particles surrounded and controlled by wall segments. The fault is composed of wall segments with a ramp dip of 25°. (b) Boundary conditions are: left side is velocity controlled, top is stress-free, basal fault is frictional, and gravity is applied. Colored layering is passive and has no mechanical significance. (c) Contact forces after gravitational settling show a linear increase in pressure with depth. (d) Location of measurement circles in initial pregrowth assembly. Measurement circles are 0.5 km in diameter.

because it contains more flaws. In the extreme, elastic moduli values as low as 1% of published laboratory values may be appropriate for large-scale rock-mass modeling problems (C. Fairhurst, personal communication, 1992). By taking into account the degree and nature of fracturing, engineers and geologists (Rubin, 1992; Schultz, 1996) working with large fractured rock masses (i.e., mine sites and tunnels) can estimate the reduced, in-situ, large-scale elastic moduli and unconfined compressive strength of a rock mass by use of a variety of theoretical and empirically based methods (Barton et al., 1974). The most common rock-quality rating system is the Rock Mass Rating (RMR) system (Serafim and Pereira, 1983; Hoek and Brown, 1988; Bieniawski, 1989), which ranges from RMR = 100 for "laboratory specimen free of discontinuities," to RMR = 85 for "tightly interlocked undisturbed rock with unweathered joints spaced at 1–3 m," to RMR = 3 for "numerous heavily weathered joints spaced at less than 50 mm" (Bieniawski, 1989). Using the RMR value, the large-scale modified Young's modulus is determined from the equation:

$$E_m = 10^{(RMR-10)/40} \quad \text{(Serafim and Pereira, 1983)}, \quad (4)$$

and the unconfined compressive strength (σ_c = UCS) for large rock masses can be calculated using:

$$\sigma_1 = \sigma_3 + \sigma_c(m_b\sigma_3 + s\,\sigma_c)^{1/2} \quad \text{(Hoek and Brown, 1988)}, \quad (5)$$

where m_b and s are empirical constants that depend on the composition, structure, and surface conditions of the rock mass.

For a heavily fractured and slickensided outcrop, equations 4 and 5 predict values of E and UCS that are at least two orders of magnitude smaller than those of a similar laboratory sample (Bieniawski, 1989, p. 64 and 181). As an engineering tool, the RMR system is designed to characterize rock masses that range from tens to hundreds of meters in scale. Because our idealized field-prototype fault-related fold is an order of magnitude larger in scale than a large tunnel or mine site, and because rock-mass strength scales inversely with size, it is not difficult to extend the concept of the RMR to rocks in thrust belts. The two-orders-of-magnitude decrease in strength for regions of rock represented by 40-m-diameter particles, in our models, is quite reasonable. The use of reduced-strength parameters has implications for our particle size, in that larger particles will contain more flaws per area than smaller ones. Ideally, if the model's particles were the same size as the intact, continuous blocks that make up the actual rock mass, then the laboratory strength values would be appropriate. The continuous (elastic) deformation of the reduced-strength particles represents, *in the field*, actual small-scale deformation on natural joints and faults within particle-diameter-size portions of rock (40 m and 20 m for pregrowth strata and growth strata, respectively). This deformation will not be accounted for by discrete behavior in the model, but instead is accounted for by the increased elastic deformation of individual particles.

Our biaxial test assembly consists of about 1300 particles and is 1.5 times taller than it is wide (Figure 3). The tests are run in the absence of gravity, and thus the actual dimensions of the specimen are unimportant as long as they are packed identically to (have the same distribution of particle sizes as) the full-scale model. Because bond strength is specified in units of stress, the strength of the whole assembly is effectively insensitive to the diameter of the particles (Itasca Consulting Group, 1999a); we simply need to use enough particles to simulate granular behavior. The postfailure behavior is, however, sensitive to the particle size, given fixed specimen dimensions: Assemblies with fewer, larger-sized particles will tend to form fractures that are more irregular relative to the fractures formed by a similar specimen with more and smaller particles. The shape of the stress-strain curve at failure is strongly influenced by the standard deviation of the contact bond strength: A small or zero standard deviation results in a sharp failure peak, and large standard deviations produce broad peaks.

The biaxial test's boundary conditions consist of rigid, velocity-controlled walls on the top and bottom that move toward each other, and side walls that apply confining pressure to the sides of the assembly. To simulate soft confinement, the side walls have their normal stiffness set at 1% and 10% of the particle stiffness for the unconfined and confined tests, respectively. The confining stress is maintained by a numerical servo-control mechanism operating on the side walls.

Figure 3 shows the typical evolution of one of these biaxial tests (with a confining pressure of 0.5-km depth). Microcracks first form prior to peak load at about 12% axial shortening, during the linear, elastic part of the stress-strain curve (Figure 3a). At about 16% axial shortening, the specimen fails (Figure 3b) and loses cohesion. Continued shortening produces significant plastic softening, and a well-developed, through-going fault zone cuts diagonally across the specimen (Figure 3c).

Young's modulus, E, and the unconfined compressive strength, UCS, are determined from a single biaxial test run, in which the confining pressure is set to a very small value equivalent to a few meters of overburden. Young's modulus, E, is taken as the slope of the elastic, linear part of the axial stress–axial strain curve (Figure 4), giving a value for E = 0.3 GPa. The unconfined compressive strength, UCS, is taken as the peak value of the axial stress at failure, where UCS = 2.7 MPa. The macroscopic angle of internal friction for the particle assembly is derived from a series of biaxial compression tests performed at different confining pressures, equivalent to depths of 0.0 km, 0.25 km, 0.50 km, and 1.0 km (Figure 5a). The values of σ_1 and σ_2 at failure are plotted as the mean stress, $(\sigma_1 + \sigma_2)/2$, over the differential stress, $(\sigma_1 - \sigma_2)/2$, (Figure 5b), and the angles of internal friction and cohesion are determined from the following relationships:

$$\phi = \sin^{-1}(m) \tag{6}$$

$$C_o = b/\cos(\phi) \tag{7}$$

where m is the slope of the best-fit curve through the data points and b is the y-intercept (Craig, 1992). For

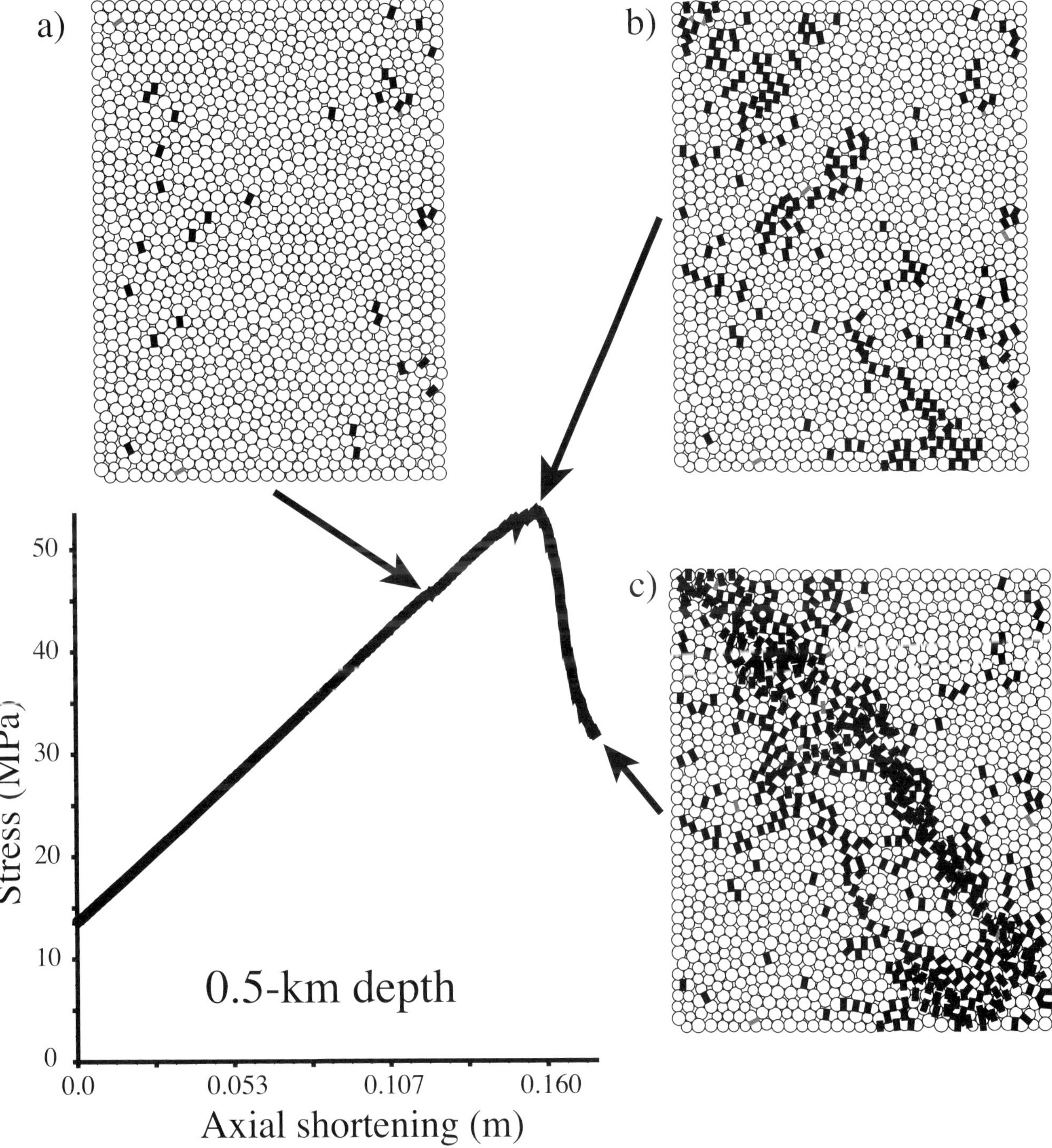

FIGURE 3. Example of a typical biaxial strength test performed with confining pressure equivalent to 0.5 km overburden, showing the stress-axial shortening curve. Short black lines represent microcracks or broken contact bonds. (a) Initial microcracks occur during linear elastic loading. (b) Test assembly at peak stress. c) Test assembly with a through-going fault zone that runs diagonally through the specimen after the loss of cohesion and softening behavior.

this assembly, the angle of internal friction $\phi = 32.6°$ and the cohesion $C_o = 4.3$ MPa.

The values of UCS and E resulting from these biaxial tests are about two orders of magnitude smaller than published values of E and UCS for typical sedimentary rocks (UCS = 140 MPa for limestone, and UCS = 62 MPa for sandstone [Jaeger and Cook, 1969, p. 146] and Young's moduli of E = 53 GPa for limestone and E = 96 GPa for sandstone [Jaeger and Cook, 1969, p. 139]). The values of ϕ and C_o, on the other hand, are in line with published values for internal friction and cohesion and do not reflect our attempts to weaken the rock mass. While E and UCS are effectively insensitive to particle size, the angle of internal friction is not. Coarser assemblies will produce faults or fault zones with more asperities than do similar-size assemblies with fine particles. The particle packing and the distribution of particle size also affects the geometry of faults or fault zones in a similar way, because smaller particles that are nested adjacent to larger ones will smooth the fault surfaces. It is thus the nature of particle assembly packing that controls the macroscopic angle of internal friction, and macroscopic cohesion is a function of both particle bonding and macroscopic internal-friction angle.

The mode of initial failure (tensile or shear) between individual particles in these models can be controlled by the relative magnitudes of contact bond strength. It has been well demonstrated that natural rocks fail by a process of initial random tensile microcracking, followed by coalescence of the microcracks to form through-going shear fractures or faults (Horii and Nemat-Nasser, 1985; Reches and Lockner, 1994). For this reason, we

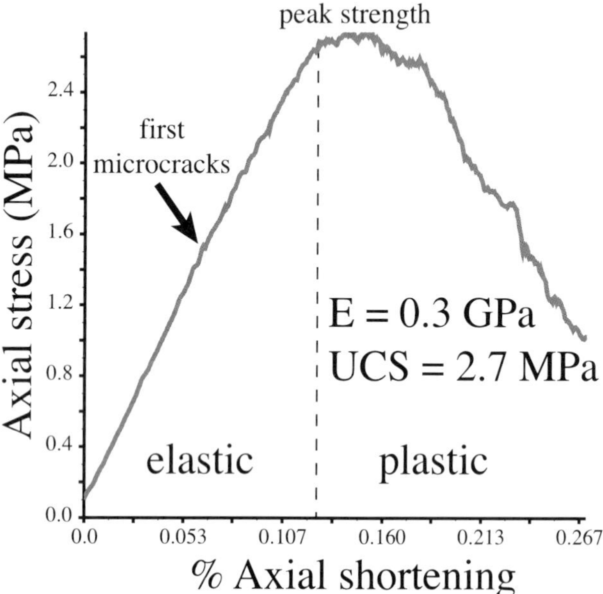

FIGURE 4. Stress-strain curve for the uniaxial compression test, which yields a Young's modulus, E = 0.3 Gpa, and an unconfined compressive strength, UCS = 2.7 MPa.

have adjusted the bonding such that failure in these models is strongly dominated by tensile processes. Nearly all (>90%) of the broken contact bonds, which represent microcracks and microfaults (shown as short black lines in some figures) in the biaxial tests and fault models, represent Mode I or tensile fractures, and all faults and fault zones (Mode II) are formed by their interaction and coalescence.

INITIAL STATE, BOUNDARY CONDITIONS, AND SEDIMENTATION

The initial configuration of the models presented here consists of a coarse-particle (with a 40-m maximum diameter, which is limited by computational time) pregrowth hanging wall that moves over an effectively rigid frictional surface composed of frictional wall segments. The thickness of the pregrowth strata and the elevation of the upper footwall flat is 1 km (Figure 2a). The fault ramp intersects the lower flat 5 km from the left wall. The footwall is constructed by a series of wall segments, creating a compound ramp that dips 25° along the main segment and 12.5° on short segments at the ramp hinges, to round the ramp-flat transitions (Figure 2b).

The initial model assembly is created by loosely filling the space above the lower footwall flat and footwall ramp with about 5000 particles. The particles are allowed to settle under the force of gravity until they come to an equilibrium that is defined to be when the maximum unbalanced force in any part of the model drops to a negligible value. Particles above 1 km elevation are then deleted. The model responds to this unloading by rebounding vertically, and again it is cycled to equilibrium. This deletion/rebound sequence is repeated until the upper surface lies at 1 km elevation and the model is at equilibrium. The resultant lithostatically loaded assembly has contact forces that increase linearly with depth (Figure 2c) and are distributed in force chains that branch downward. Passive, colored 100-m-thick "stratigraphy" is used simply to visually track deformation and has no mechanical significance. At this stage, the contact bonds are applied.

Boundary conditions for these models are quite simple. The footwall flats and ramp are elastic and frictional ($\phi = 16°$); the left wall moves toward the right with a velocity of 5 m/time step; the upper surface is free; and gravity is applied to the entire model (Figure 2b).

For this study, we ran a series of models that have different syntectonic sedimentation rates. One model (M1), which acts as the control simulation, has no sediment added during the run. The four other models have differing amounts of sediment added during the runs.

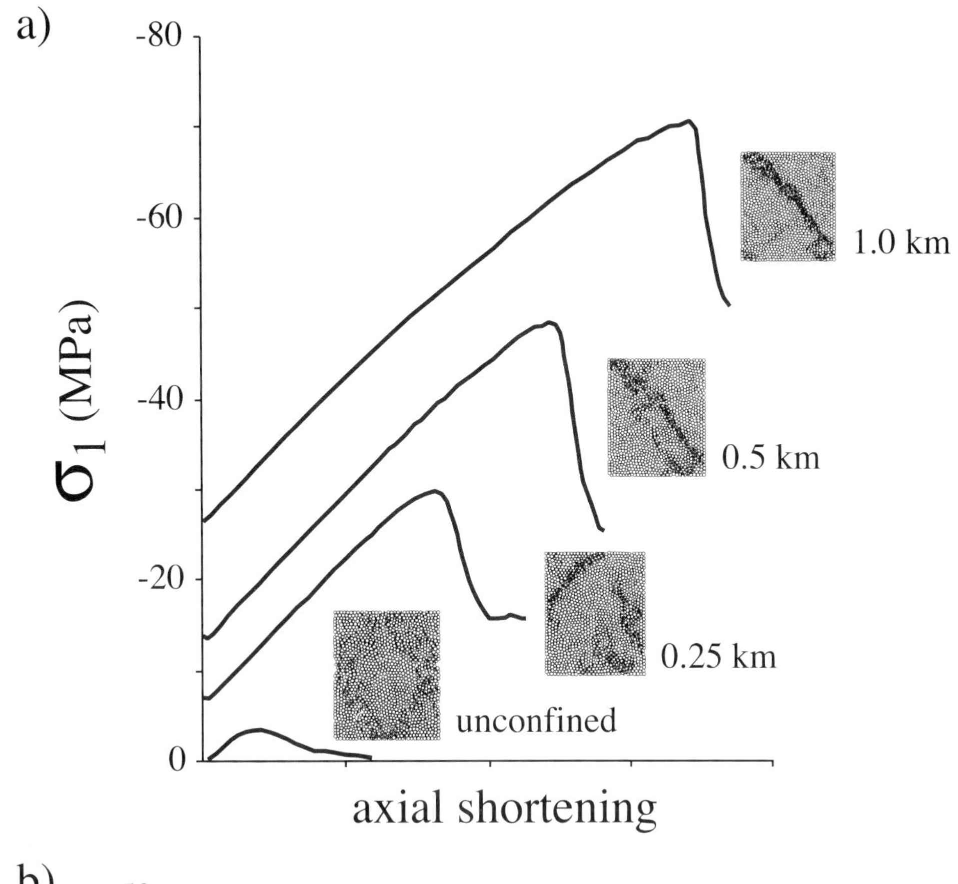

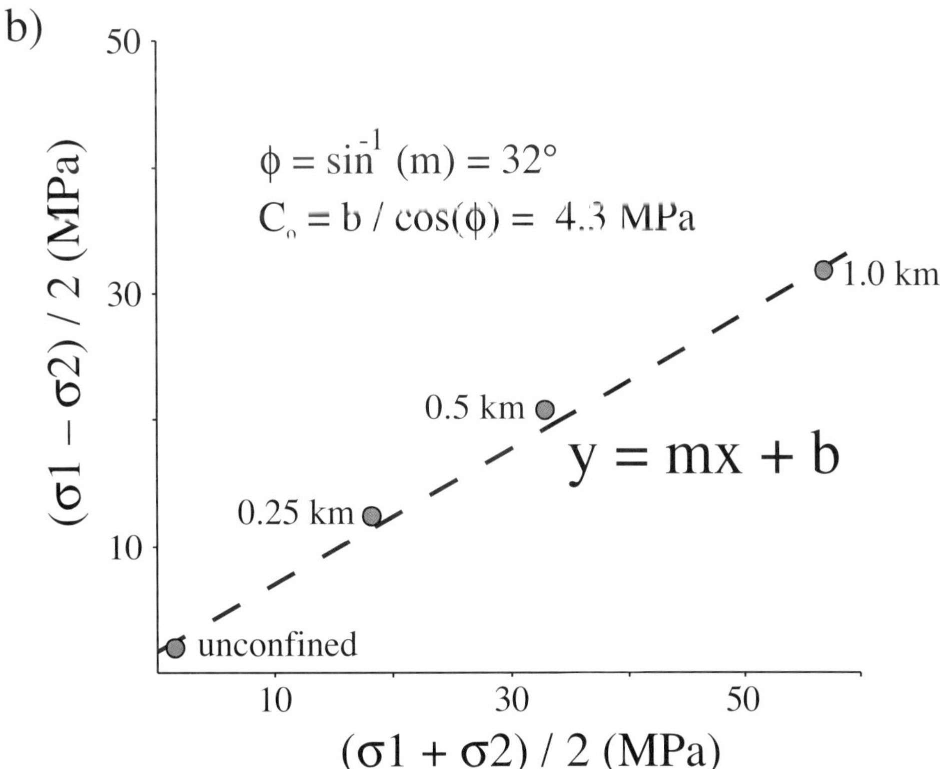

Figure 5. (a) Stress-strain curves for the four strength tests and resulting specimens, showing the distribution of fractures within the specimens. With increasing confining pressure, failure becomes more localized: unconfined material experiences granular flow, while at 1 km, failure is by discrete faulting. (b) The angle of internal friction, ϕ, and cohesion, C_o, are determined by plotting the mean stress against the differential stress from four biaxial strength tests performed at confining pressures equivalent to 0.0 km, 0.25 km, 0.5 km, and 1.0 km depth.

The added sediment ranges in thickness from 25%, to 50%, to 75%, and to 100% (models M2 through M5) of the altitude of the crest of the growing fault-related fold above the upper footwall flat. For example, the fault-related fold in the 100% model (M5) is completely buried to the level of the top of the pregrowth strata, and the 50% model (M3) has the upper half of the pregrowth-strata fold exposed.

Sediment is added to the upper surface of the models after every 250 m of displacement (5%), but this does not begin until the first 750 m, which yields eight growth layers. During sedimentation, the shortening stops and the underlying pregrowth and any growth-strata particles present are fixed in x and y. The newly added sediments are allowed to come to equilibrium, after which they are eroded off to the appropriate level, and again the model is cycled to equilibrium to accommodate rebound. At this point the underlying particles are freed in x, y, and spin, and they are allowed to come to equilibrium, sometimes causing minor faulting in response to the newly applied sediment load. Displacement is then resumed for another 250 m, and this cycle is continued for a total displacement of 2.5 km (50% shortening). Newly added growth strata are not bonded until after the next layer is deposited, in an attempt to simulate lithification with burial.

MODEL RESULTS

Geometry

The most obvious result of this modeling study is the effect of different thicknesses of growth strata on the final shape of the resultant fault-related fold (Figure 6). There is a smooth progression from fault-bend style (M1) to fault-propagation style (M5) geometry, as growth-strata thickness increases. Fault slip goes from being almost completely transferred up to the upper footwall flat in the fault-bend fold where the fault is emergent, to being lost upward into the overturned limb of the fault-propagation fold and adjacent growth strata. The growth strata clearly act to buttress the deformation, resisting forelandward translation of the emergent fault-related fold.

In the earliest stages of all models, the nascent fault-related fold has a kink-style geometry, with effectively straight limb segments separated by somewhat discrete hinges and zones of back thrusting. The early kinklike geometry quickly gives way to folds with rounded shapes in models M1 through M3, which have thin or no growth packages (Figures 7 and 8). The models with thicker growth sequences tend to maintain more kinklike geometries through their evolution (Figure 9). In the lower sedimentation rate models, near-surface regions tend to become unbonded with continuing deformation and uplift, subsequently eroding and redepositing under the force of gravity. The increased pressure applied to the pregrowth strata by burial under the thicker growth strata suppresses surface unbonding and faulting, and maintains the integrity of the pregrowth structural beam. The initial, flat-topped anticline that forms early in model M5 is nearly rigidly back-rotated about 13°(CCW). The rotation occurs by a combination slip on the main thrust and minor normal movement on the first-formed back thrust, and this is clearly reflected in the overlying wedge-shaped growth strata.

Faulting

Because of the incremental nature of sedimentation in these models, all of the models have similar early evolution. Even in the higher-sedimentation-rate models (models M4 and M5), the early stages are marked by only a small increase in overburden resulting from the applied growth strata. As a result, a number of features are common to all models—most importantly, early-stage back thrusting. Figures 6, 7, 8, and 9 show back thrusting in all models. In all cases, the first back thrusts form almost immediately upon movement of the hanging wall up the thrust ramp. These faults form with an initial dip of about 50° to 55°. In model M1, and to a lesser extent model M2, the early-formed back thrusts are shallowed by later, large-scale downslope movement of unbonded particles, as evidenced by significant bed thinning, extensional faulting, and surface-slope reduction of the emergent anticline (Figures 6 and 7). Back thrusts cause noticeable thickening in the backlimbs of all models, resulting in bedding that does not remain parallel to, but instead dips at a shallower angle than, that of the underlying fault. At the base of the pregrowth packages are persistent bedding-parallel zones of fracture that originate at the lower ramp hinge and are transported up the ramp. These appear to be analogous to damage zones seen in the hanging walls of Morse's (1977, 1978) analog models and Erickson et al.'s (2001, 2004) numerical-continuum models.

Figure 6. Final results of models M1 through M5, with 0%, 25%, 50%, 75% and 100% growth strata. Shown are the pregrowth material (large particles) and growth strata material (small particles), and small black lines representing broken contact bonds or fractures. There is a clear progression from open, fault-bend-style folding with no or little growth strata, to overturned, fault-propagation-style folding with thicker growth sequences. Faults and damage zones are visible as discrete back thrusts, bedding-parallel faults, or diffuse zones of broken bonds.

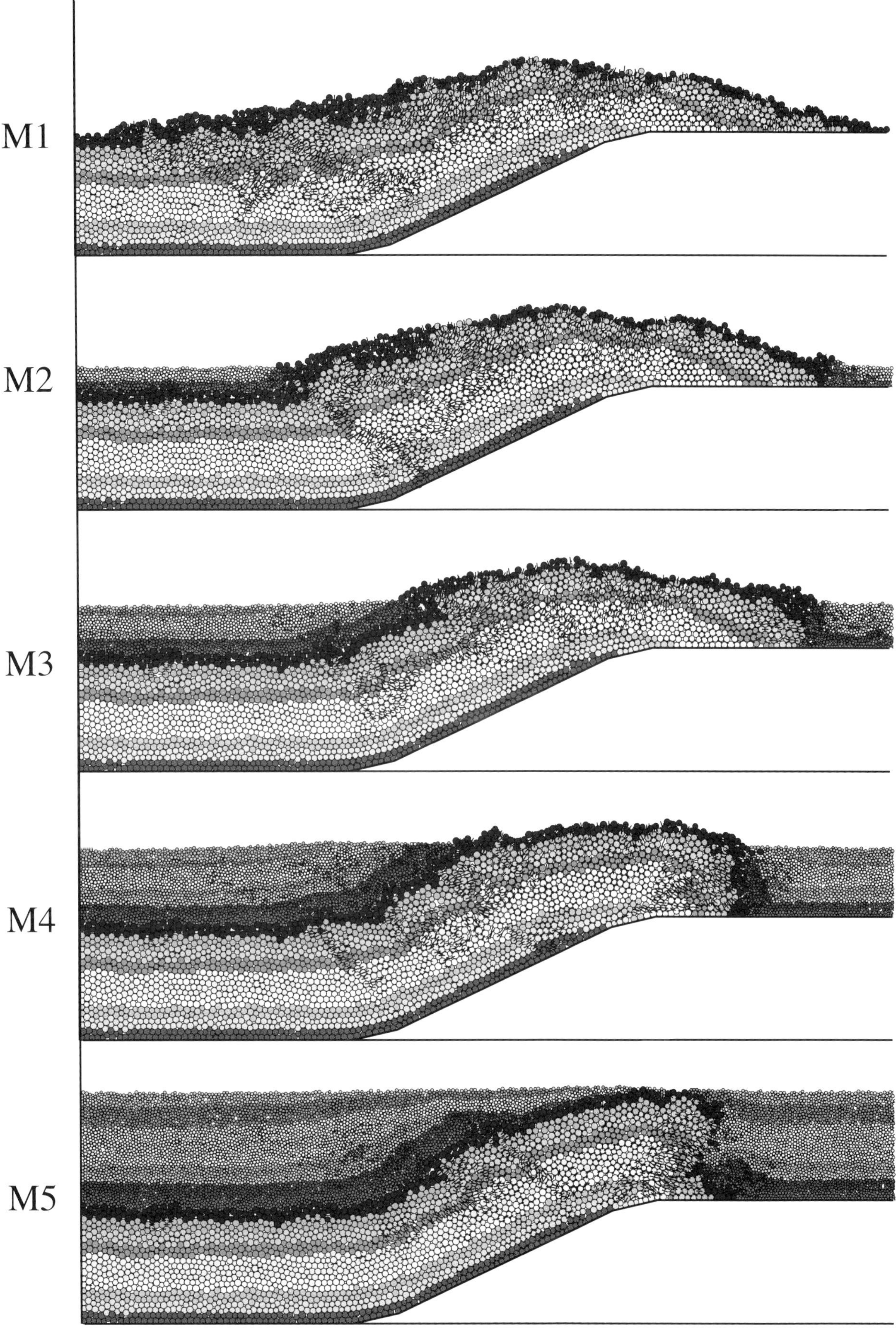
M1
M2
M3
M4
M5

In the models with higher sedimentation rates (M4 and M5), nonemergent or blind thrust faults are oriented subparallel to the fault ramp. These upward-widening fault zones lose displacement in the overturned pregrowth anticline and overlying forelimb growth strata, suggesting trishear (Erslev, 1991).

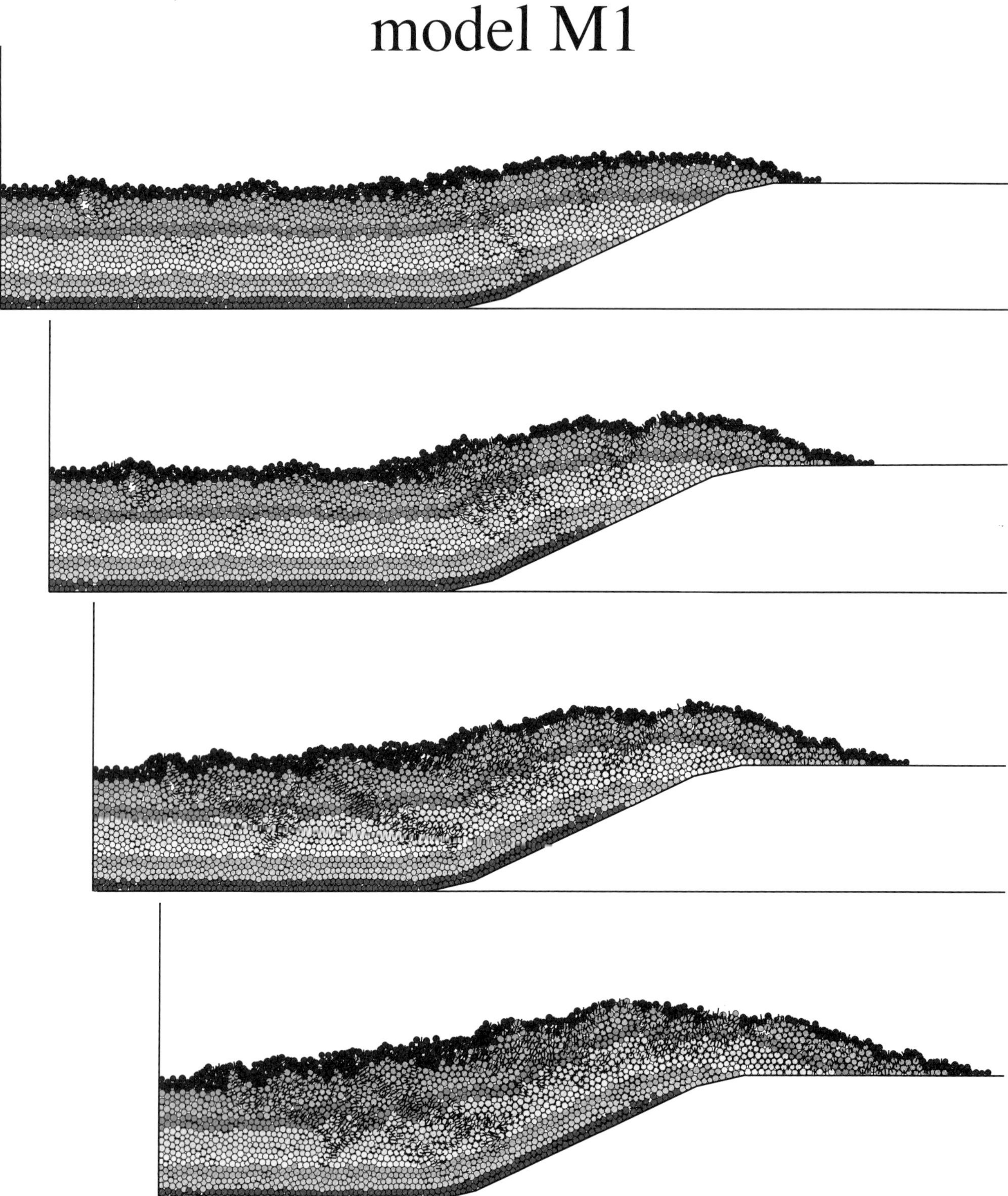

Figure 7. Evolution of the base model M1 with no growth strata. Back thrusting occurs immediately upon hanging-wall movement up the fault ramp. The earliest anticline is somewhat box-shaped, but becomes rounded with continued thrusting and near-surface unbonding of particles. An outer-arc graben begins to form early and remains active throughout the simulation. Initially steep back thrusting gives way to lower-angle faults as the simulation progresses.

There is a distinct alternation between the deformation mechanisms of back thrusting and bedding-parallel (flexural) slip in the pregrowth material as it moves up the fault ramp. This is clearly visible in Figures 8 and 9, where, in nearly all cases, early back thrusting is followed by flexural slip and by further deformation

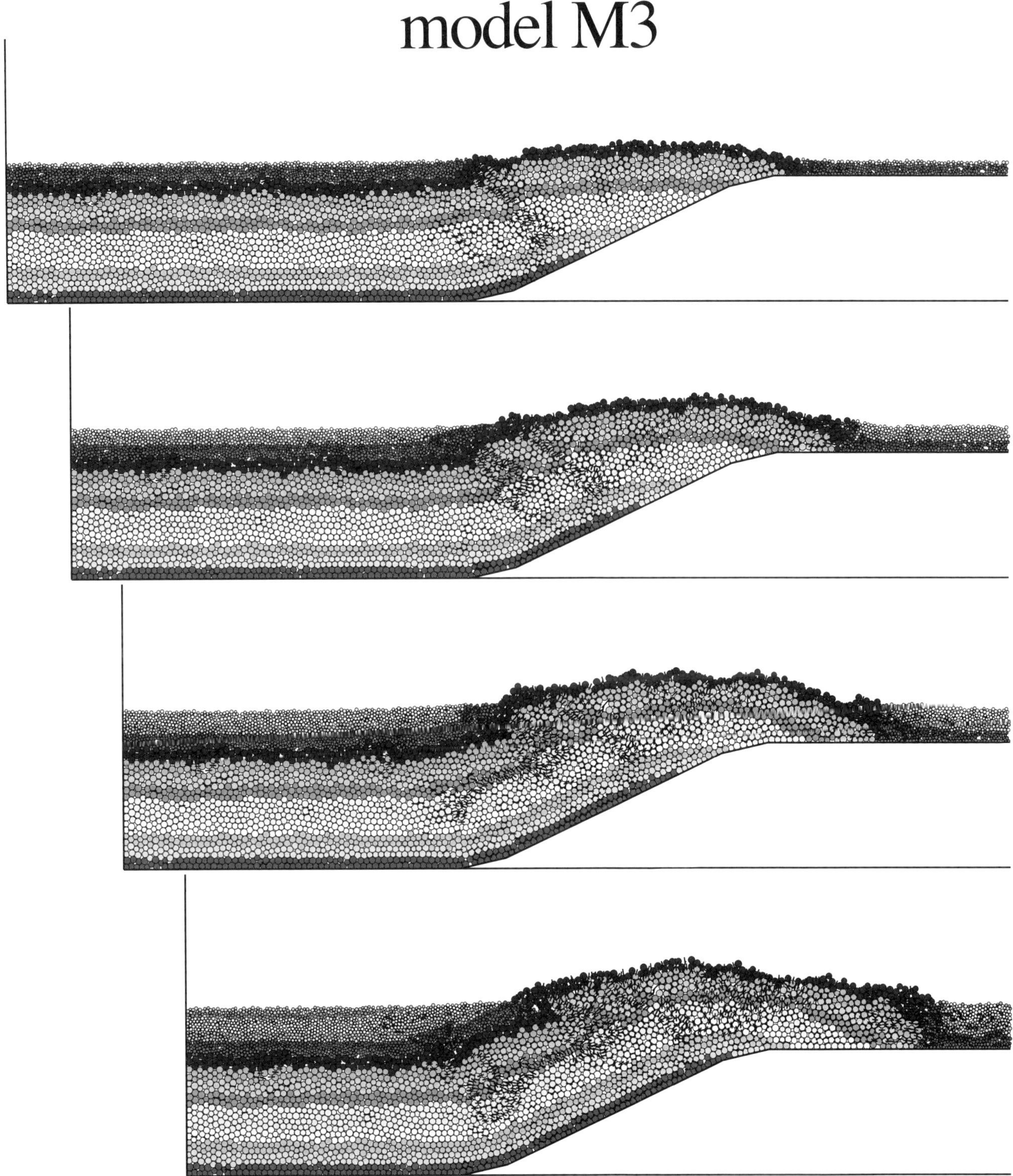

FIGURE 8. Evolution of the 50%-growth-strata model M3. This model begins with pregrowth deformation that is nearly identical to model M1, but later differs as a result of sediment accumulation. There is some minor overturning of the tip of the forelimb as a result of buttressing against growth strata. Early, spaced back thrusting gives way to bedding parallel faulting with increased shortening and burial. Outer-arc graben formation in the emergent anticline is visible.

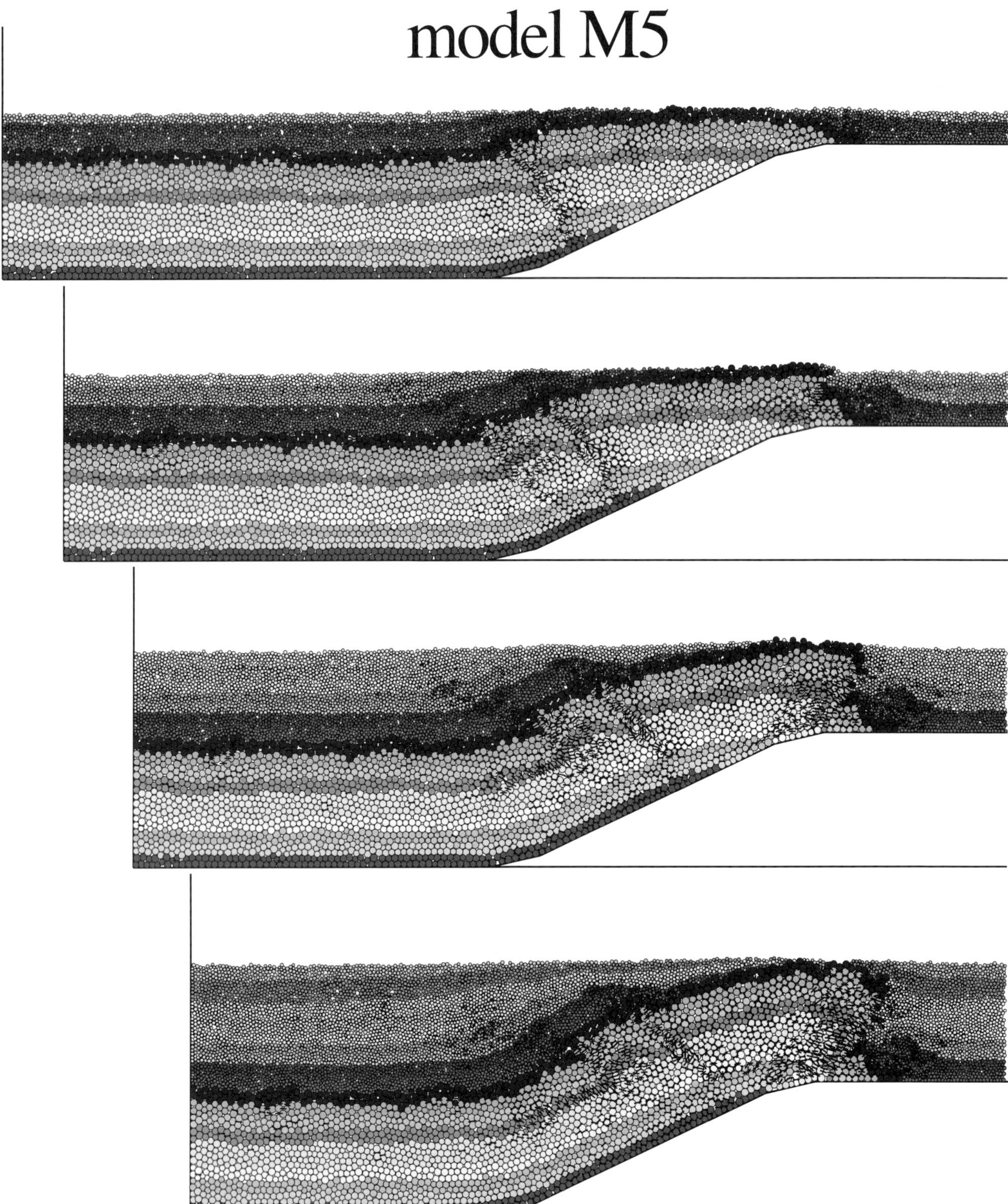

Figure 9. Evolution of the 100%-growth-strata model M5. Again, this model has an early stage that is similar to the other models (M1–M4), but with perhaps a slightly more flat-topped, open anticline. In the second stage, the hanging-wall tip is buried by growth strata and is thrust over by the main bulk of the anticline. Back thrusting extends into the overlying growth strata, causing significant deformation in the backlimb growth triangle. In the third stage, the once flat-topped anticline is back-rotated by thrusting over the forelimb growth strata, and minor normal motion on the first-formed back thrust, and this is clearly recorded in the overlying growth strata. In the final stage, back thrusting and, to a lesser extent, bedding-parallel faulting appear to be suppressed by the overburden supplied by the thick growth sequence.

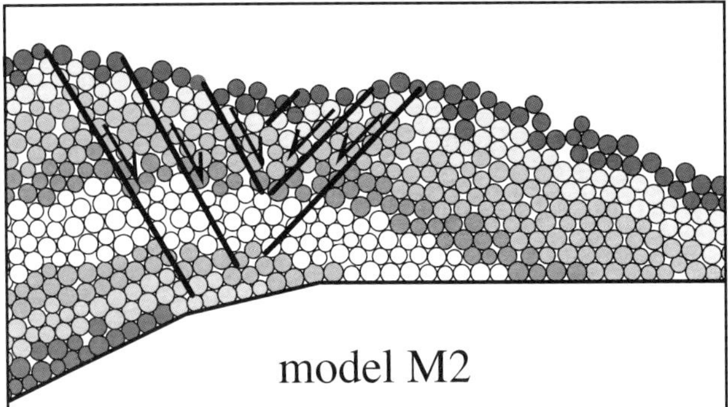

Figure 10. Detail of outer-arc graben in model M2. Note significant thinning of highest stratigraphic levels.

alternating between the two modes. Because there is no preexisting anisotropy in the pregrowth material, flexural-slip deformation occurs spontaneously, likely as a result of neutral surface bending, where the highest shear stress occurs at the center of the flexed beam. This sort of behavior has also been seen in the clay-cake models of Kuenen and de Sitter (1938) and in the continuum-numerical models of Strayer and Hudleston (1997a) and Erickson et al. (2001, 2004), even when a homogeneous and isotropic continuum material is used to study fault-related folding.

During the evolution of all models, there is some extension in the outer arcs of the evolving anticlines. This is quite clear in models M1 through M3, but is especially clear in model M2, where there is a pronounced graben on the anticlinal crest (Figure 10). There is a small amount of extension evident in models M4 and M5 (see Figures 6 and 9) that occurs during overturning of the forelimb.

Measurement Circles and Parameter Histories

We placed 0.5-km-diameter measurement circles at five locations within the pregrowth material to monitor the stress tensor (σ_{ij}), thus yielding the normal stresses in the *x* and *y* directions (σ_x, σ_y) and associated shear stress (τ_{xy}), principal stresses (σ_1, σ_2), mean stress ($\sigma_m = (\sigma_1 + \sigma_2)/2$), differential stress ($\sigma_d = \sigma_1 - \sigma_2$), maximum shear stress (τ_{max}), the angle between σ_1 and σ_x (θ), porosity, and the contact sliding fraction (a measure of the amount of faulting) during the model evolution (Figure 2d). These circles allow us to compare the responses of the different regions of the pregrowth material. The accuracy of these data depend on the size and location of the measurement circles. The current set of measurement circles provides insight only into the general stress state of five specific regions ($\sim$0.2 km^2 area) of the model. Here, we emphasize the insight that can be gained into the kinematic and mechanical evolution of fold structures by use of only a small suite of parameters derived from a few measurement circles and requiring minimal postprocessing.

We compare model M1, the basic, fault-bend fold with no growth strata, with model M5, the completely buried fault-propagation fold, because the two evolved so differently. Within those two models, we focus on results from measurement circles 1 (MC1) and 5 (MC5) (Figure 2d), because they too have quite different histories. When dealing with model M5, which has growth strata, we trim-out the data or time steps from when particle motions are fixed during sedimentation. This only removes flat segments when stresses are constant, not those time steps during which the model is coming to equilibrium. Other responses to the growth sedimentation are preserved, as evidenced by the clear periodic oscillations of the stress and porosity signals of model M5 (Figure 11). For nomenclature, we will use a code alluded to above to specify the model number and measurement-circle number referred to in the form of M#MC#. For example, M1MC5 refers to the fifth measurement circle of model M1.

Figure 11 shows the mean stress, σ_m; shear stress, τ_{xy}; porosity; and the sliding fraction for models M1 and M5 at all five measurement circles. Although the curves for both models M1 and M5 have similar shapes, there are differences, most of which are attributable to the elevated pressure caused by increased overburden of the growth strata. The effects of the sedimentation are clear for M5 where there is a nearly linear increase in lithostatic load and overall a higher mean stress than in M1. There is an early general increase in mean stress and shear stress in both the control model M1, and in M5, that is matched by an early decrease in porosity and a low sliding fraction. This represents a period of elastic loading, which is followed by a sudden decrease in mean and shear stress with brittle failure, and an increase in porosity of at least 5% in M1MC5 compared with a 2% porosity increase in M5MC5, and a significant increase in the amount of faulting, as indicated by the sliding fraction. The pressure increase from the added growth strata inhibited failure by increasing the mean stress (shifts the Mohr circle to the right, away from the failure envelope), suppressing dilatant brittle deformation. Both models M1 and M5 have similar shear stresses, as reflected in the magnitudes of the shear stresses within each measurement circle. The presence of growth strata, in general, suppresses failure.

Figure 12 presents the evolution of principal stress in MC1 and MC5 for both models. Plots of principal stresses during the evolution of M1 and M5 show similar general form, but distinctly different behavior between the lower hanging-wall flat (MC1) and the core of the emergent anticline (MC5). In MC1, σ_1 increases as a function of loading in the *x*-direction for both models, but addition of growth strata in M5 causes a

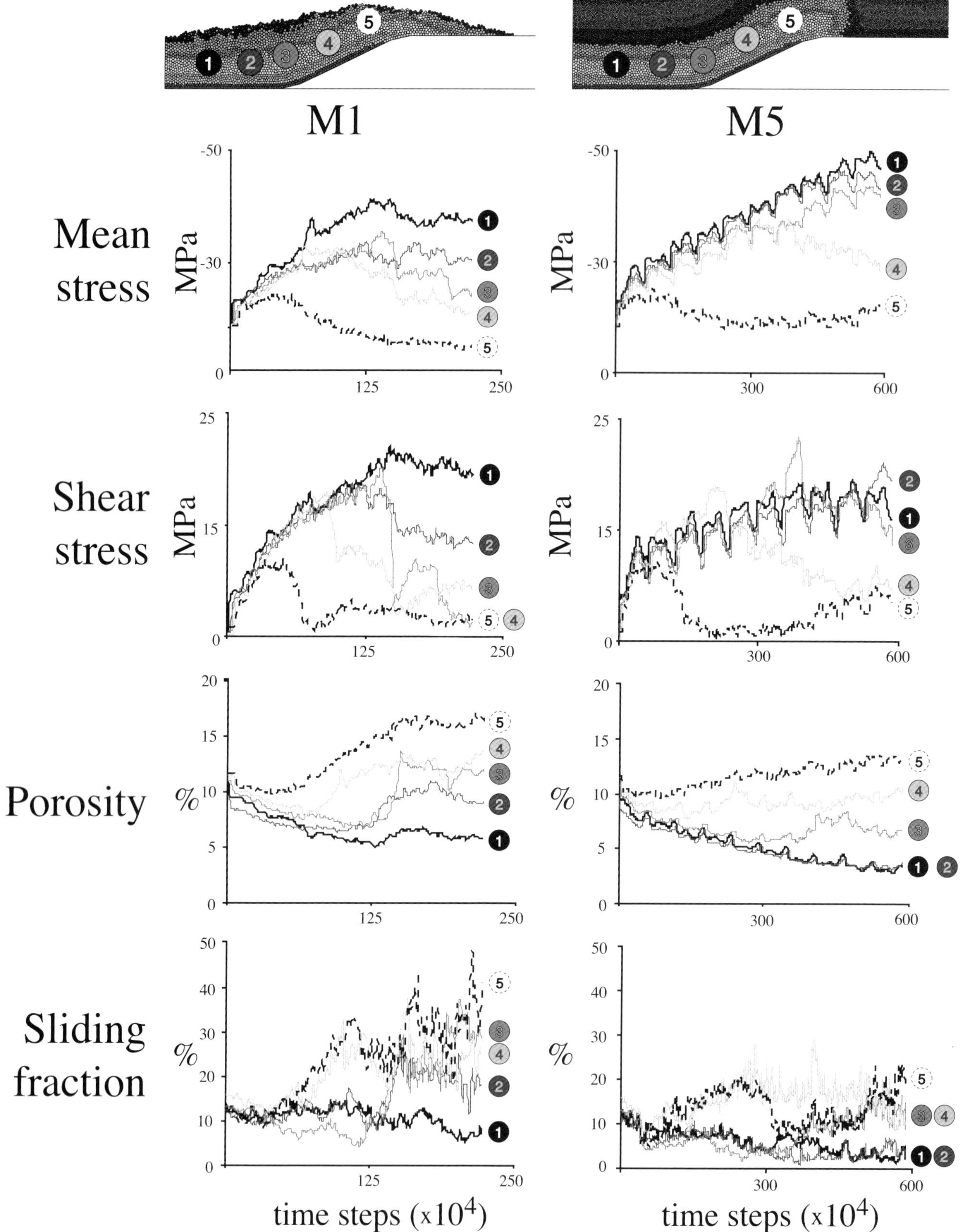

FIGURE 11. Measurement-circle results from the end-member models M1 (no growth strata) and M5 (100% growth strata), showing mean stress ($(\sigma_1 + \sigma_2) / 2$), shear stress ($\tau_{xy}$), porosity, and the fraction of sliding contacts. In general, the response of both models is quite similar, but mean stress increases and porosity and sliding fraction decrease in model M5 relative to M1, as a result of increased pressure applied by the thick growth sequence.

nearly linear increase in the magnitude of the principal stresses, whereas in the model with no growth strata (M1), the stress magnitudes flatten out. During this stage of both models, the magnitude of the principal stresses are lower in general than in other measurement circles because of uplift along the ramp, and because the shapes of the curves of M1 and M5 are similar.

These stress graphs (Figure 12) can be treated as a group of serial Mohr diagrams through time, tracking only the position through time of σ_x, σ_y, σ_1, and σ_2. In this way, we can get a feeling for the evolution of the Mohr circle parameters: differential stress (diameter of the Mohr circle, difference between σ_1 and σ_2 on graphs) and mean stress or pressure (Mohr circle center, average of σ_1 and σ_2 on graphs) through the model run. By also tracking the angle between σ_1 and the horizontal direction, θ, we have a complete picture of the two-dimensional stress state in each measurement circle.

There are portions of both of the MC5 principal-stress curves where σ_1 and σ_2 overlap (isotropic stress state). For both MC5 curves, these periods of low shear stresses often coincide with periods when the vertical normal stress, σ_y, exceeds the horizontal normal stress, σ_x. Figure 12 shows that, for M1MC5 after the σ_1, σ_2 curves merge, σ_x and σ_y cross. After this, this part of the hanging wall is dominated by vertical rather than horizontal stresses, where the angle θ becomes generally subvertically oriented ($\theta > 45°$). The emergent anticline changes from being dominated by subhorizontal compression to subhorizontal extension. Although the material represented by the measurement circles never experiences any tensile stresses during or after these reversals, the stress state predicts extensional failure modes, which in fact are manifest by outer-arc extension of the hanging walls (see Figure 10). Compare this with M5MC5 where, although σ_1 and σ_2 merge and σ_x and σ_y cross several times, in general σ_x remains greater than σ_y and extension in the anticline surrounded by growth strata is not significant. During this time, the orientation of σ_1 does switch from being initially subhorizontal ($\theta < 45°$) to being subvertical ($\theta > 45°$), but remains near 45°.

Distinctive bulbs form by the σ_1 and σ_2 curves in the early stages of the simulations. The MC5 curves for M1 and M5 indicate that both hanging walls begin with similar stresses that mark an initial period of loading (Figure 12), wherein $\sigma_x > \sigma_y$, and σ_1 increases and is oriented subhorizontally ($\theta < 45°$). At the same time, σ_2 decreases. The bulbs represent the initial period of elastic loading, followed by, in both models, an almost complete loss of cohesion that is marked by (1) a decrease in τ_{max}; (2) an increase in porosity with a simultaneous increase in the sliding fraction; (3) a trend toward isotropic stress states, evidenced by the merging of σ_1 and σ_2, and (4) inability to sustain large differential stresses—loss of strength. With about 25–30% (M1) and 15% (M5) of possible sliding contacts being active and a majority of contacts bonds broken, these hanging-wall regions are unable to resist deformation in a state of plastic failure and thus have low strength. There is evidence of softening (in M1MC5) and of late hardening as a result of continued burial (in M5MC5) by addition of growth strata.

One of the features that stands out from these measurement-circle data is the bumpy nature of the curves, seen clearly in data from model M1 and, excluding oscillations from the sedimentation process, model M5. This irregularity represents stick-slip behavior, or earthquakes, in the models. Because the code formulation is fully dynamic, and because the fault surfaces contain particle-size asperities, these earthquakes are a natural result of failure and slip. Although we make no attempt here to quantify this phenomenon, it is possible to do so (Hazard et al., 1998). Future work will investigate this phenomenon.

In summary, analysis of measurement-circle data allows us to compare the stress states of the two extreme models (M1 and M5) and of the two extreme measurement circles (MC1 and MC5). The presence of growth strata has a strong effect on stress magnitudes and the intensity and distribution of faulting, both in the emergent anticline and above the lower hanging-wall flat. Without growth strata to supply confining pressure, the emergent anticline changes from being dominated by subhorizontal stresses to being dominated by subvertical stresses (M1MC5), which is consistent with extension seen in the emergent anticline. The lateral pressure supplied by growth strata allows the stress state to remain constant and nearly isotropic during a large part of the simulation (see M5MC5). The nature of porosity and the amount of faulting varies considerably with the presence of growth strata, compared with the no-growth-strata model (M1). Secondary porosity as a result of material failure and subsequent fault slip is clearly inhibited by the confining pressure caused by the thick growth strata. Measurement circles provide a powerful means of monitoring the "pulse" of these simulations, even with simple parameters.

Growth Strata

Because this modeling focuses on how growth strata affect the kinematic and mechanical evolution of the pregrowth strata, rather than on the growth strata themselves, there are no measurement circles in the growth layers. We are thus unable to monitor the stresses there, but we still can observe the geometry and faulting patterns. Here, we examine model M4 (which has a 75% sedimentation rate), because the pregrowth strata are slightly emergent, because it has the most interesting forelimb growth package, and because the architecture

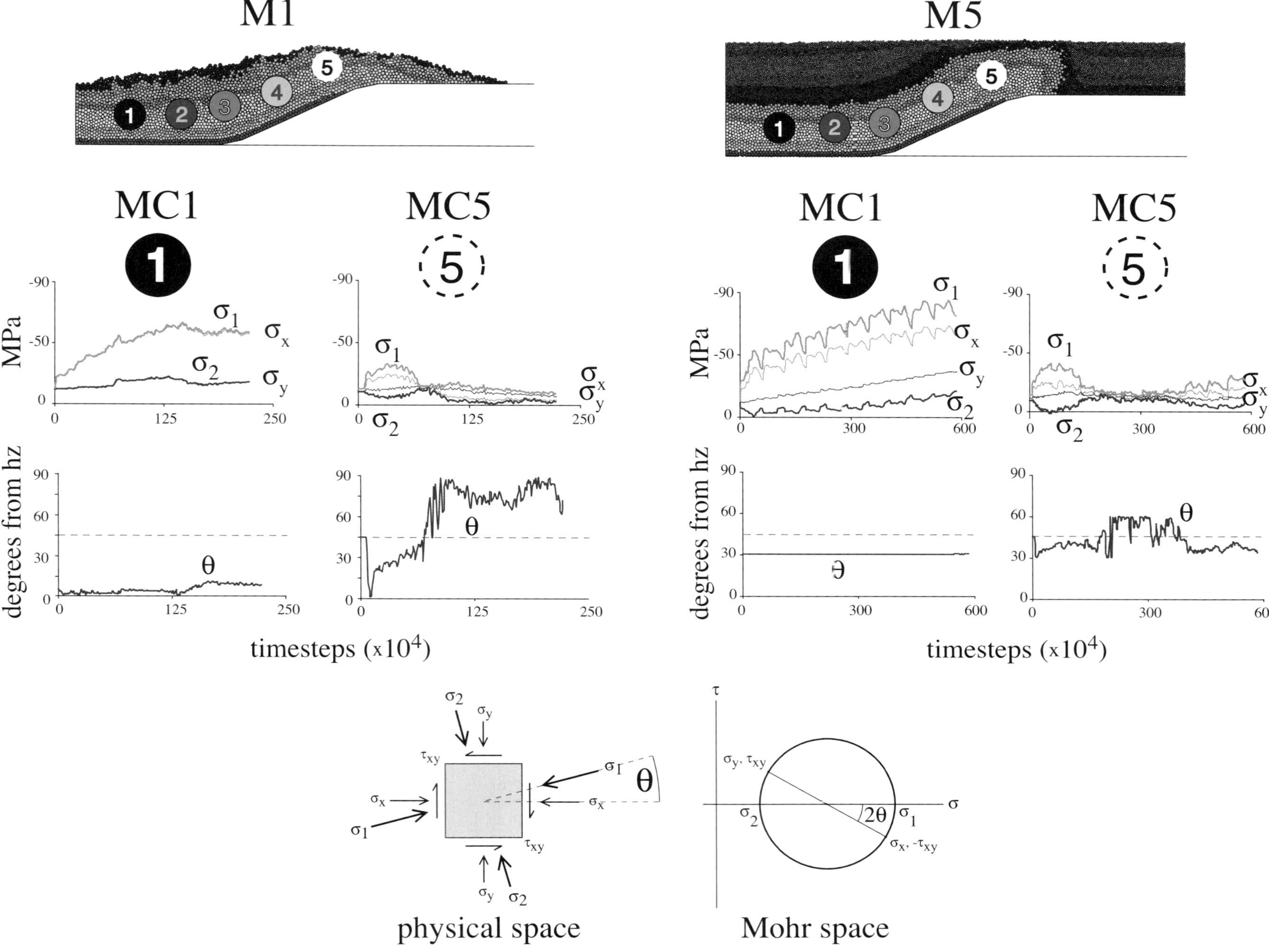
M1
M5
MC1
MC5
MPa
degrees from hz
timesteps (x10^4)
physical space
Mohr space

of the growth strata is quite detailed and nicely represented by the ~100-m-thick beds (Figure 13). Perhaps the most striking feature of the growth-strata packages is how realistic the geometry of the growth strata appear given the model's simple initial configuration and boundary conditions. Even given the coarse discretization represented by 20-m-average-diameter growth particles, there are clear examples of unconformities, stratigraphic pinch-outs, onlaps, and offlaps, which record the complex folding of the underlying pregrowth strata. Distinct growth triangles are present in both back- and forelimbs, and some axial planes are of a compound style — that is, noncontinuous and oriented differently through the fold (Ford et al., 1997). Not surprisingly, deformation is concentrated in the growth triangles of both limbs of the fold, where in some places the "rock mass" is reduced to individual particles, suggesting very closely spaced fault planes and slip zones. This faulting may also occur as bedding-parallel slip on nonbonded growth bedding planes. The patterns of faulting, deformation, and displacement (not shown) in the overturned limbs and adjacent growth strata of models M4 and M5 are strongly suggestive of trishear (Hardy and Ford, 1997).

In the backlimb of model M4 (Figure 13), there appears to be some evidence of early hinge migration (Suppe et al., 1997), marked by early, constant-thickness bedding and by later limb rotation (Ford et al., 1997), as indicated by thickness changes in the growth strata. Nearly all of the local topography on the main pregrowth-growth unconformity results from back thrusts originating in the pregrowth strata, which ultimately extend up into the growth strata and form minor folds. Although M4 starts with a fault-bend fold geometry, the forelimb of this fold is overturned, and the earliest, now-overturned forelimb growth strata record the progressive rotation of the forelimb. The forelimb growth triangle zone is also strongly faulted and has a second-order growth anticline. This general pattern of deformation is consistent with results from Ford et al. (1997), who observe conjugate reverse faults in the overturned limb and synclinal core of a fault-related fold pair that has growth strata of similar scale and form, in the Spanish Pyrenees. Because of the discrete nature of deformation in these models, and because the uppermost growth layer is not bonded, it is not surprising that there are places where growth and pregrowth particles are intermingled near the main unconformity. Particles of pregrowth strata within the growth strata may be the conceptual equivalent of slide blocks or breccia deposits that are shed off the emerging anticline. There is also the potential for growth strata particles to become involved in the fault zones.

The growth strata produced in these models display many features and subtleties of natural growth strata (Figure 14). With decreased particle size and stress-strain-rate tensor data, it is likely that in the near future we will gain insight into the evolution of strain and outcrop-scale deformation mechanisms responsible for folding the growth strata.

DISCUSSION

The main goal of this modeling is to investigate the influence of variable-thickness growth strata on the kinematics and mechanics of a fault-related fold. All of the results from these models must be taken in light of our assumptions on rock-mass strength and the validity of our subjective criteria for developing the base model (M1) upon which we carry out the modeling. We have taken an empirical approach to developing these models, because we are using a relatively new numerical technique, which is effectively a 2-D sandbox, to model a large and mechanically poorly understood rock mass. Because of the stochastic nature of these models and of natural rock systems, we make no effort here to develop any sort of mechanical theory of fault-related folding and growth strata; the systems are too complicated, even when the numerical model is made as simple as possible. Rather, the goal of this type of modeling, using the most realistic yet simplest model possible, is to generate basic rules of thumb or observations about how the models behave in different circumstances, with the aim of applying those rules to natural examples. The rules and observations can guide our field investigations and shape our hypotheses. They can also provide insight into a structure's development and the physical conditions that may have been responsible for, or at least present during, its formation.

FIGURE 12. Stress tensor results from the two extreme-end measurement circles (MC1 and MC5) of the end-member models M1 (no growth strata) and M5 (100% growth strata). Shown are graphs of both σ_x, σ_y and the principal stresses σ_1 and σ_2, and the angle θ between σ_1 and horizontal. For circle M1MC1, the principal and horizontal and vertical stresses are essentially coincident, and this is also shown by the very low θ angles. In model M5 and measurement circle MC1, there is an angular difference of about 30° between σ_1 and horizontal. The difference between this behavior and that at M1MC1 is likely the result of the increased load of the growth strata, which causes the material to better 'feel' the friction of the fault. In general, the results from MC5 from both models M1 and M5 are similar, with differences resulting from the increased overburden in M5. The hanging wall of model M1 transitions from being dominated by subhorizontal ($\theta < 45°$) principal stresses to subvertical ($\theta > 45°$) principal stresses, while the hanging wall in model M5 experiences similar flip-flops in principal-stress orientations, but in general is dominated by subhorizontal stresses.

Model M4

Backlimb Forelimb

a)

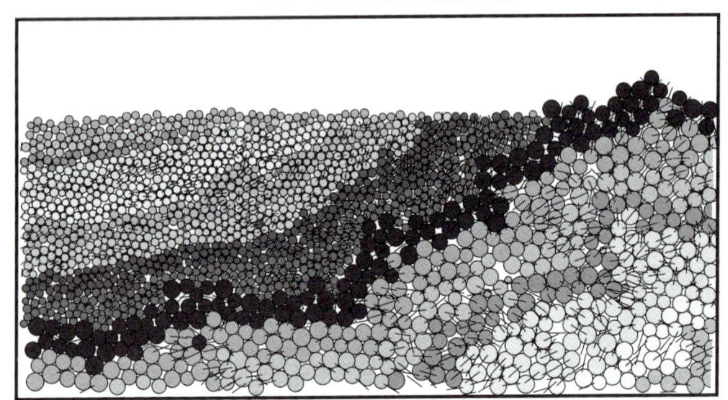

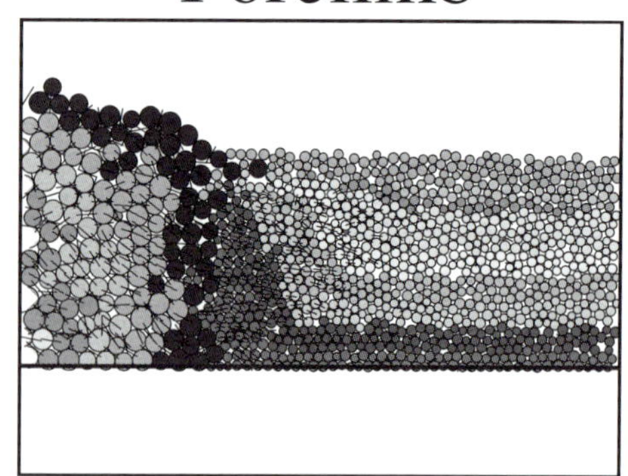

b)

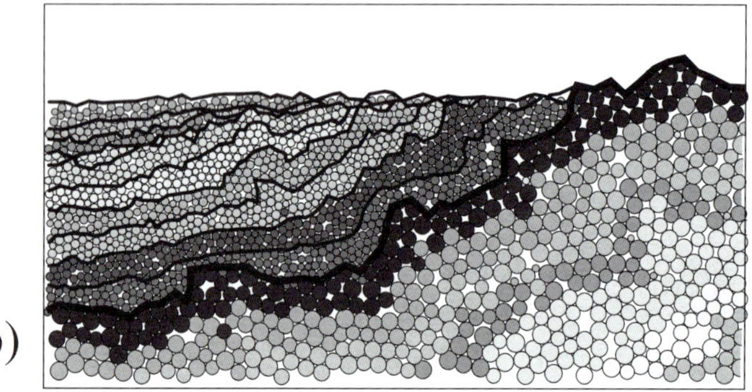

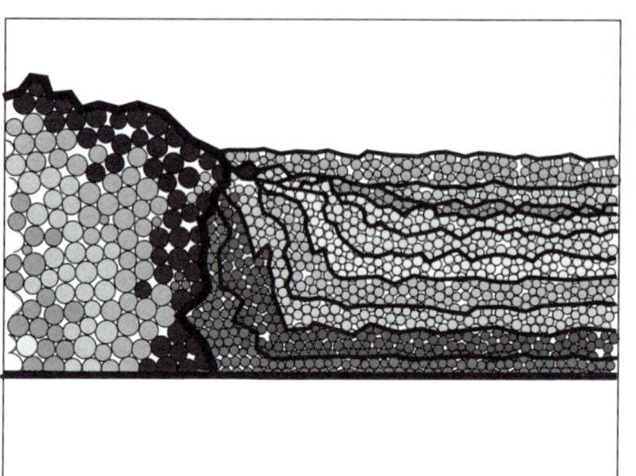

c)

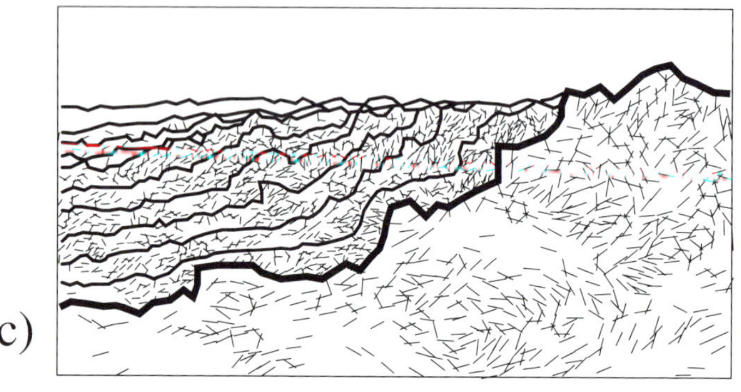

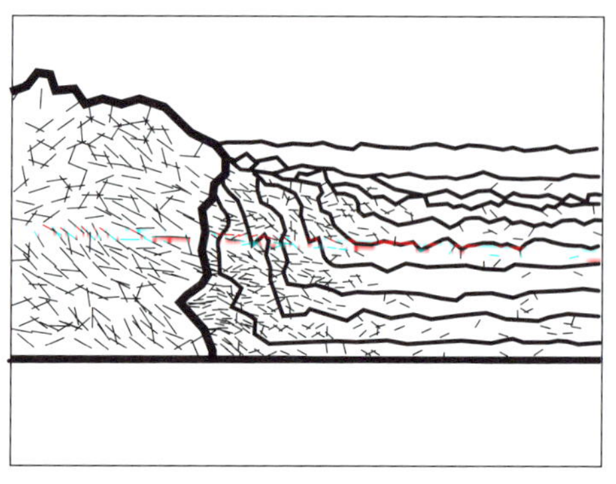

d)

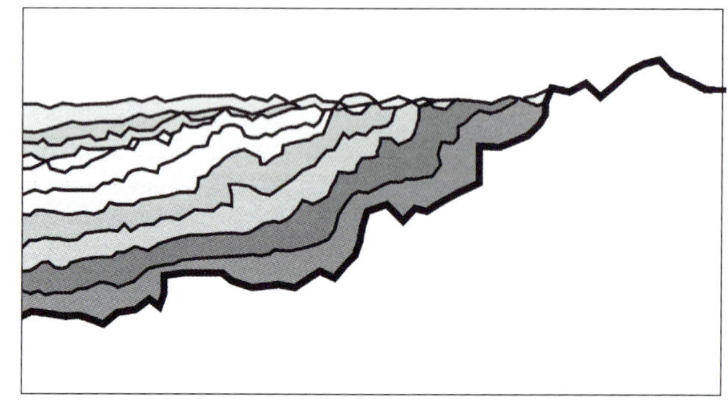

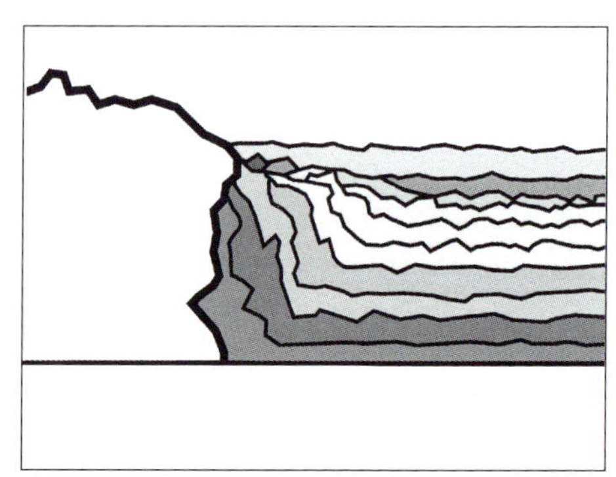

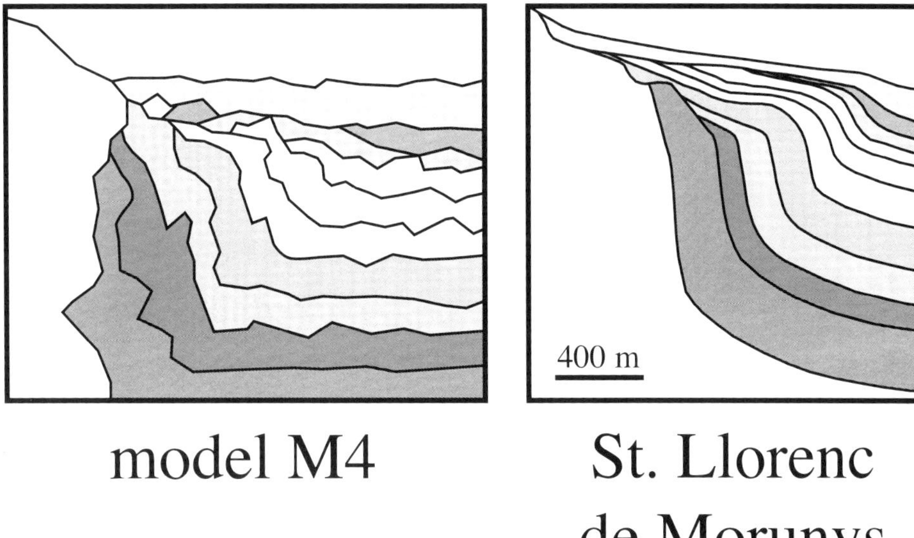

FIGURE 14. Comparison of forelimb growth strata from model M4 to a published cross section by Ford et al. (1997) of growth strata from Sant Llorenc de Morunys in the Spanish Pyrenees. This comparison is made to show similarities in the general architecture and scale of the two examples, even though the details of the kinematics of the underlying pregrowth strata are likely different.

The major conclusion we draw from this work is that increasing thickness of growth strata tends to force simple, emergent, fault-bend folds to become nonemergent, fault-propagation folds as a result of foreland buttressing. The hanging-wall strata of the no- or low-sedimentation-rate models (M1 and M2) conform to the underlying pregrowth footwalls on emergent faults. The upward-widening fault zones in higher sedimentation-rate models (M4 and M5) maintain the original ramp dip and lose displacement in the overturned pregrowth anticline and overlying forelimb growth strata. With a high sedimentation rate (M5), the fault-propagation fold continues to grow in an effectively self-similar manner; the fault does not flatten and produce footwall growth duplexes but instead maintains a steep trajectory and cuts through the overturned pregrowth and adjacent growth strata.

The notion that forelandward buttressing of fault-related folds affects the evolution of those folds is not new: It is a relatively intuitive concept. Chester et al. (1991), using unscaled analog and rock models, showed that near-field resistance to forelandward translation strongly controls the resultant fold shape. Using finite-difference numerical methods with outcrop-scale models, Strayer and Hudleston (1997a) showed very much the same thing. Our work and Barrier's (1998) show that growth strata can be very effective near-field buttresses to forelandward translation, and that fold shape is, in fact, *dependent* on the amount and nature of the growth strata. This dismisses the notions that the geometry of the growth package is strictly the result of the evolution of the underlying fold and that the presence of growth strata does not affect the mechanics of the underlying material. Instead, deformation and sedimentation are interdependent. Therefore, growth-strata geometries should not be used to derive rules about the evolution of fault propagation in general.

Two end-member mechanisms exist by which hanging-wall material can be deformed and transported up the fault ramp: (1) back thrusting and (2) flexural-slip or bedding-parallel shear. Back thrusting tends to be favored in massive units. However, spontaneous bedding-parallel shear has been seen in clay cake models (Kuenen and Desitter, 1938) and in outcrop-scale continuum numerical models using isotropic and homogeneous materials (Strayer and Hudleston, 1997a; Erickson et al., 2001, 2004)—both of which *may* have some anisotropy because of possible clay fabric and numerical mesh effects. Bedding-parallel slip tends to occur in well-layered or anisotropic materials (Erickson et al., 2001). From this, we might posit that small-scale structures—those with less likelihood of strong anisotropy—might be dominated by back thrusting, and that large, well-bedded Canadian Front Range–style structures should be dominated by flexural slip in the lower ramp. Erickson et al. (2001, 2004) used homogeneous and isotropic continuum models to show that a sharp lower fault bend produces back thrusts, whereas a rounded lower fault bend promotes bedding-parallel shear. In our particle models, we see a robust pattern of initial back thrusting followed by alternating back thrusting and bedding-parallel shear. In the extreme, at the last stages of model M5 (full sedimentation), it appears that brittle deformation at the lower ramp hinge ceases altogether, and

FIGURE 13. Detail of both back- and forelimb growth strata, showing (a) particles and fractures, (b) geometry of growth strata boundaries with respect to the particle configuration, (c) relationship of growth strata boundaries to fractures, and (d) solid-colored growth-strata boundaries. Note the extensive fractures in, and effectively limited to, the forelimb growth triangle, and the extension of back thrusting into the backlimb growth triangle.

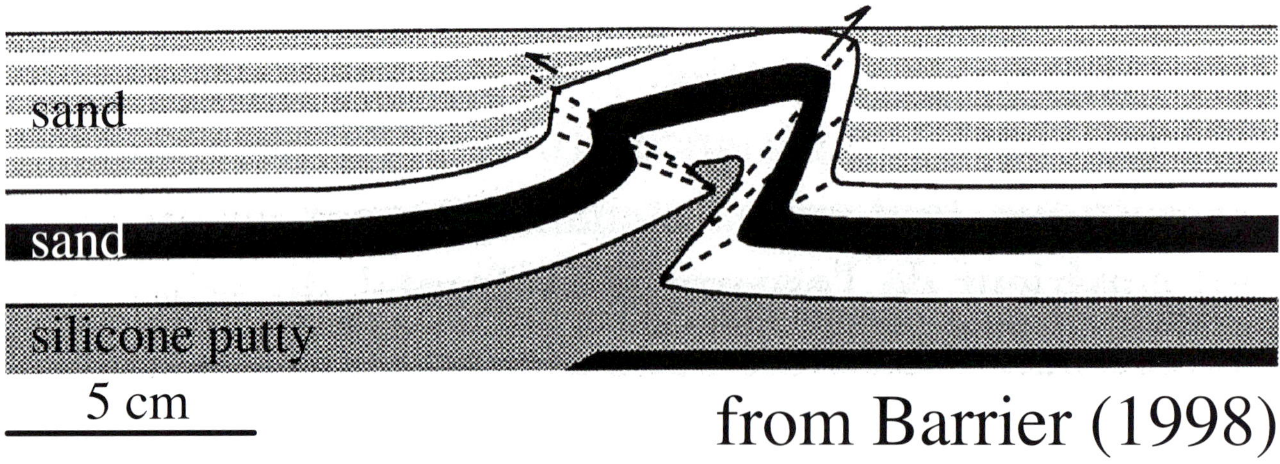

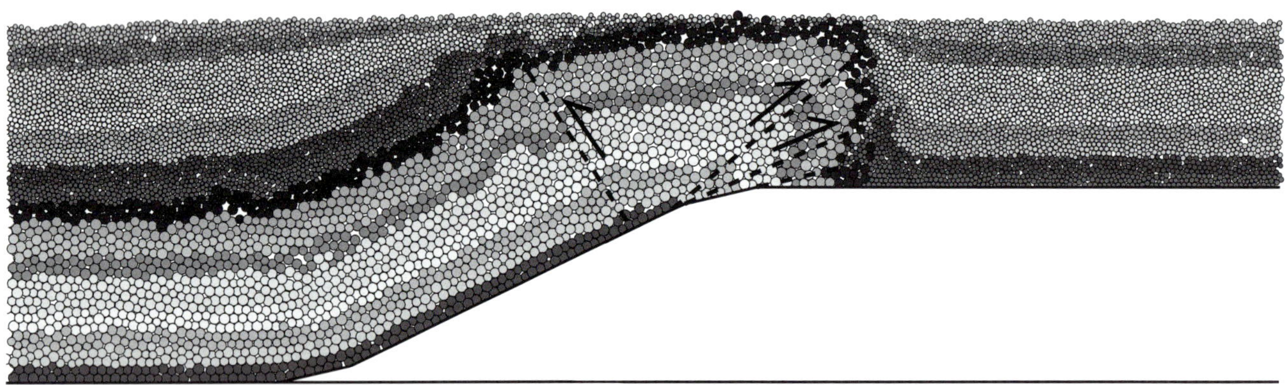

FIGURE 15. Comparison of an analog model of Barrier (1998) to model M5, showing very similar overall geometry, faulting, and growth package geometry. Barrier's (1998) models are composed of a basal silicone putty layer that underlies colored-sand pregrowth and growth strata.

the elasticity of individual bonded particles accommodates the deformation. The alternating nature of the back thrusting and bedding-parallel shear in these models may be the result of the compound, three-segment fault ramp, which has the effect of slightly rounding the fault bends. It may be that these models are at the transition between being dominated by either the sharp-bend/back-thrust or the rounded-bend/bedding-parallel slip deformation mechanism. Not surprisingly, the back thrusting plays a significant role in the final geometry of the backlimb growth package. Back-thrust zones extend into the growth strata and are often associated with small-scale folds in the growth triangle (Figure 13). Growth packages on both limbs of the fold experience both the strains associated with folding, be it kink-band migration or limb rotation, and those associated with faulting that originates from the loaded pregrowth beam. Barrier (1998), using analog materials (sand, silicone putty), also investigates the effect of growth strata on underlying fault-related fold evolution. The results of our study strongly resemble those of Barrier (1998), including outer-arc extension in low-sedimentation-rate models, the strong buttressing effect of the growth strata, fault-propagation folds with high sedimentation rate, the style of back thrusting, and the very similar growth packages (Figure 15). It is interesting to note that in Figure 15 there is one distinct back thrust that formed very early, and then there is effectively no other back-limb localization. Our models and Erickson et al.'s work (2001, 2004) suggest that the lack of faulting in Barrier's analog model may be the result of increased confining pressure caused by growth strata, thereby suppressing localized failure. In addition or alternatively, the lack of faulting may be because the initial back thrusting that occurs in response to motion over an initially angular fault bend is replaced by bedding-parallel shear, which is enhanced by the progressively more gently curved lower-ramp hinge that ultimately forms in the silicone putty.

Although certainly not the focus of this work, the realism and detail of the growth strata in our models are a fine bonus. Applied after every 250 m of shortening, the growth strata record the complete development of the fold-growth system, allowing for detailed testing of restoration methods and kinematic models. In light of the fact that we made no effort to engineer the distribution of the growth, except to limit thickness, our growth strata are remarkably similar to cross sections near St. Llorenc de Morunys (Ford et al., 1997) (Figure 14). Our growth strata resemble these cross sections specifically with regard to the erosional-depositional architecture and scale of individual growth packages and also to the concentration of strain in the overturned limbs. Deformation in our growth packages seems to derive from three sources. (1) Passive folding of the growth strata happens in response to deformation of the underlying pregrowth strata, which occurs by bedding-plane slip and small-scale faulting associated with neutral surface bending in the growth strata. (2) Faulting originates in the pregrowth strata and extends into the growth strata. (3) There is subhorizontal shortening of the overturned limb, which has the effect of thinning the bedding layers on that limb. This last mechanism is visible as a zone of subhorizontally oriented fractures in Figure 13, and may be equivalent to conjugate faulting in the steep limb of the St. Llorenc structures (Ford et al., 1997).

Because realistic growth packages result naturally from this type of model without any tweaking of the model, it will be difficult in future work to model precisely the growth stratigraphy of a specific individual natural or analog structure. Rather, it is more likely that we can match the essential relationships, such as deposition, erosion, and uplift rates, and deformation mechanisms and structural styles. Capturing the essential character rather than the minute detail of a structure will allow us to better understand how a specific class of structures forms.

There are no real surprises in the measurement-circle data. The various parts of the fold react as expected, given the deformation. That is, with uplift and faulting, we see such things as a porosity increase and a mean stress decrease, and also a sliding-fraction increase and shear-stress decrease. As material approaches the upper ramp hinge, we also see a clear transition from contractional to extensional deformation, which results from a switch in the orientation of σ_1 from subhorizontal to subvertical. We take this predictability as a good sign for our simple model, which can be used as the basis for future modeling of natural and analog examples of growth strata. The ability to extract mechanical data from within the interior of the model is standard in continuum finite-element or -difference techniques, and measurement circles allow similar data to be collected from the particle models. Because these numerical models have so many aspects in common with sandbox modeling, one of their most novel and useful components is access to mechanical data at a variety of scales, in damaged or faulted regions. Numerical models can then be used to understand the mechanics of sandbox models.

CONCLUSIONS

By a process of iteration between model building and model material testing, we developed and accepted a stable and well-behaved base model for fault-related folding, using the distinct-element code PFC2D. Model building, design, and final acceptance were governed by a set of simple subjective rules that guaranteed at least realistic geometry, deformation mechanisms, and strength values that do not exceed published laboratory values for sediments typical of fold-thrust belts. Given our particle diameters, we needed to reduce the values for Young's modulus, E, and unconfined compressive strength, UCS, by nearly two orders of magnitude to create a model that both looked realistic and behaved realistically. Deformation that occurred at the subparticle scale was partitioned into the weakened or softer individual elastic particles. This weakening of the rock mass is clearly in line with the well-accepted notion that the strength of a rock mass decreases as its scale increases.

Material property testing showed that the behavior of our initial assembly was pressure dependent, with diffuse or poor localization and nearly granular flow at surface conditions and well-localized failure following elastic loading and yield at approximately a 1-km depth. The assembly failed initially by discrete tensile (Mode I) failure of individual contact bonds that interacted and coalesced to form larger-scale faults and fault zones (Mode II), which has been shown to be the case for natural rocks.

The presence of thick growth strata had a profound effect on the kinematic and mechanical behavior of the pregrowth material. Acting as a buttress to forelandward translation, the increasing thicknesses of growth strata forced the fault-related fold to change from a fault-bend to a fault-propagation geometry. The presence of growth strata inhibited extension in the outer arc of the growing fold, suppressed failure by increasing mean stress at depth, and also affected the nature of deformation above the thrust ramp—back thrusting was common in all models, but bedding-parallel shear became more important with increased overburden.

Growth strata in the models showed a surprising amount of sedimentalogical and structural detail and complexity, given the simple displacement-based sediment application criteria and coarse discretization. Well-developed growth triangles, unconformities, and on- and offlaps formed. Growth strata recorded both kink-band migration *and* hinge rotation during the evolution of the fault-related fold. Deformation in the growth packages resulted from folding of the growth beds and also from faulting that extended upward from the pregrowth material. Although in this study there were no measurement circles in the accumulating growth strata, we will be implementing this in the future, using a system of nested measurement circles of differing diameters.

In summary, these models and the distinct-element method in general provide excellent tools with which to study shallow-level brittle deformation, simultaneous folding and faulting, and syntectonic erosional-depositional processes. Future work will focus on applying these models to both natural and analog examples of fault-related folding with growth strata. The goal will be to provide insight into their mechanical evolution, into deformation mechanisms in growth strata, and perhaps into the nature of seismicity in these important structures.

ACKNOWLEDGMENTS

The authors wish to thank Itasca Consulting Group of Minneapolis, Minnesota, especially T. Coetzee and David Potyondy, for the kind use of PFC2D, and for advice on using the code, and Peter Cundall for his suggestions to improve the model. The authors also wish to thank Mary Ford and Dave Waltham for their constructive reviews of the manuscript, volume editor Ken McClay for his organization of the Thrust Tectonics, 1999 conference, and April Harper for all her work behind the scenes. We also wish to thank the industrial associates of the Princeton 3-D Structure Project for their generous support of this research. Luther Strayer also thanks Peter Hudleston for his early and enthusiastic support of this modeling work during Strayer's time at the University of Minnesota.

REFERENCES CITED

Barrier, L., 1998, Relations tectonique-sedimentation en bordure de bassins compressifs intramontagneux: Memoire de DEA Dynamique de la Croute Continental, Universite de Rennes,16 p.

Barton, N., R. Lien, and J. Lunde, 1974, Engineering classifications of rock masses for the design of tunnel support: Rock Mechanics, v. 6, p. 183–236.

Bieniawski, Z. T., 1989, Engineering rock mass classifications: John Wiley and Sons, New York, 272 p.

Chester, J. S., J. M. Logan, and J. H. Spang, 1991, Influence of layering and boundary conditions on fault-bend and fault-propagation folding: Geological Society of America Bulletin, v. 103, p. 1059–1072.

Craig, R. F., 1992, Soil mechanics: 5th Edition, New York, Chapman & Hall, 427 p.

Cundall, P. A., 1988, Computer simulations of dense sphere assemblies, *in* M. Satake and J. T. Jenkins, eds., Micromechanics of granular materials: Amsterdam, Elsevier, p. 113–123.

Cundall, P. A., 1989, Numerical experiments on localization in frictional materials: Ingenieur-Archiv, v. 59, p. 148–159.

Cundall, P. A., and O. D. L. Strack, 1979, A discrete numerical models for granular assemblies: Geotechnique, v. 29, p. 47–65.

DeCelles, P. G., M .B. Gray, K. D. Ridgeway, R. B. Cole, D. A. Pivnik, N. Pequera, and P. Srivastava, 1991, Controls on synorogenic alluvial-fan architecture, Beartooth Conglomerate (Paleocene), Wyoming and Montana: Sedimentology, v. 38, p. 567–590.

Erickson, G., M. Ford, J. Suppe, L. Strayer, and S. Carena, 1998, Sequential restoration of folds using growth strata; an example from Sant Llorenc de Morunys, Southeast Pyrenees: Geological Society of America Abstracts with Programs, v. 30(7), p. 42.

Erickson, S. G., L. M. Strayer, and J. Suppe, 2001, Initiation and reactivation of faults during movement over a thrust-fault ramp: Numerical mechanical models. Journal of Structural Geology, v. 23, p. 11–23.

Erickson, S. G., L. M. Strayer, and J. Suppe, 2004, Numerical modeling of hinge-zone migration in fault-bend folds, *in* K. R. McClay, ed., Thrust tectonics and hydrocarbon systems: AAPG Memoir 82, p. 438–452.

Erslev, E. A., 1991, Trishear fault-propagation folding: Geology, v. 19(6), p. 617–620.

Ford, M., E. A. Williams, A. Artoni, J. Verges, and S. Hardy, 1997, Progressive evolution of a fault-related fold pair from growth strata geometries, Sant Llorenç de Morunys, SE Pyrenees: Journal of Structural Geology, v. 19(3–4), p. 413–441.

Hardy, S., and M. Ford, 1997, Numerical modeling of trishear fault propagation folding: Tectonics, v. 16(5), p. 841–854.

Hardy, S., and J. Poblet, 1994, Geometric and numerical model of progressive limb rotation in detachment folds: Geology, v. 22(4), p. 371–374.

Hardy, S., J. Poblet, K. McClay, and D. Waltham, 1996, Mathematical modelling of growth strata associated with fault-related fold structures, *in* P. G. Buchanan and D. A. Nieuwland, eds., Modern developments in structural interpretation, validation and modeling: Geological Society of London Special Publication 99, p. 265–282.

Hazard, J. F., S. C. Maxwell, and R. P. Young, 1998, Micho-mechanical modelling of acoustic emissions. Society of Petroleum Engineers, SPE 47320.

Hoek, E., and E. T. Brown, 1988, The Hoek – Brown failure criterion—a 1988 update: Proceedings of the 15th Canadian Rock Mechanics Symposium, University of Toronto, p. 31–38.

Horii, H., and S. Nemat-Nasser, 1985, Compression-induced

microcrack growth in brittle solids: axial splitting and shear failure: Journal of Geophysical Research, v. 90, p. 3105–3125.

Itasca Consulting Group, Inc., 1999a, PFC2D (Particle Flow Code in 2 Dimensions): Version 2.0, Minneapolis, Minnesota, U.S.A.

Itasca Consulting Group, Inc., 1999b, PFC3D (Particle Flow Code in 3 Dimensions): Version 2.0, Minneapolis, Minnesota, U.S.A.

Itasca Consulting Group, Inc., 2000, FLAC (Fast Lagrangian Analysis of Continua): Version 4.0, Minneapolis, Minnesota, U.S.A.

Jaeger, J. C., and N. G. W. Cook, 1969, Fundamentals of Rock Mechanics, Merhen & Co. Ltd. London, 513 p.

Kuenen, P. H., and L. U. de Sitter, 1938, Experimental investigation into the mechanisms of folding: Leidsche Geologische Mededeelingen, v. 10, p. 217–240.

Lawton, T. F., E. Roca, and J. Guimera, 1999, Kinematic-stratigraphic evolution of a growth syncline and its implications for the tectonic development of the proximal foreland basin, southeastern Ebro basin, Catalunya, Spain: Geological Society of America Bulletin, v. 111, p. 412–431.

Morgan, J. K., 1997, Studying Submarine Accretionary Prisms in a "Numerical Sandbox": Simulations using the Distinct Element Method: Eos, Transactions, American Geophysical Union, Fall Meeting Supplement 78, p. 78.

Morgan, J. K., and M. S. Boettcher, 1999, Numerical simulations of granular shear zones using the distinct element method: I. Shear zone kinematics and micromechanics of localization: Journal of Geophysical Research, v. 104, p. 2703–2719.

Morse, J., 1977, Deformation in ramp regions of overthrust faults: experiments with small-scale rock models: Twenty-ninth annual field conference, Wyoming Geological Association Guidebook, p. 457–470.

Morse, J., 1978, Deformation in ramp regions of thrust faults: Experiments with rock models: M. S. Thesis, Texas A & M University, College Station, Texas, 138 p.

Potyondy, D. O., and P. A. Cundall, 1999, Modeling of notch formation in the URL mine-by tunnel: Phase 4—Enhancements to the PFC model of rock: Itasca Consulting Group, Inc., Report to Atomic Energy of Canada Limited (AECL).

Reches, Z., and D. A. Lockner, 1994, Nucleation and growth of faults in brittle rocks: Journal of Geophysical Research B 99, v. 18, p. 159–18173.

Rouby, D., P. R. Cobbold, P. Szatmari, S. Demerician, D. Coelho, and J. A. Rici, 1993, Restoration in plan view of faulted Upper Cretaceous and Oligocene horizons and its bearing on the history of salt tectonics in the Campos Basin (Brazil): Tectonophysics v. 228, p. 435–445.

Rubin, A. M., 1992, Dike-induced faulting and graben subsidence in volcanic rift zones: Journal of Geophysical Research v. B 97, p. 1839–1858.

Saltzer, S. D., and D. D. Pollard, 1992, Distinct element modeling of structures formed in sedimentary overburden by extensional reactivation of basement normal faults: Tectonics, v. 11, p. 165–174.

Serafim, J. L., and J. P. Pereira, 1983, Considerations of the geomechanics classification system of Bieniawski: Proceedings of the International Symposium on Engineering Geology and Underground Construction LNEC, Lisbon, v. II.33–II, p. 42.

Schultz, R. A., 1996, Relative scale and the strength and deformabilty of rock masses: Journal of Structural Geology, v. 18, p. 1139–1149.

Storti, F., and J. Poblet, 1997, Growth stratal architectures associated to decollement folds and fault-propagation folds: inferences on fold kinematics, *in* S. Cloetingh, M. Fernandez, J. A. Munoz, W. Sassi, and F. Horvath, eds., Structural controls on sedimentary basin formation: Tectonophysics, v. 282, p. 353–373.

Strayer, L. M., and P. J. Hudleston, 1997a, Numerical modeling of fold initiation at thrust ramps: Journal of Structural Geology, v. 19, p. 551–566.

Strayer, L. M., and P. J. Hudleston, 1997b, Simultaneous folding and faulting and fold-thrust belt evolution: a distinct element model: Geological Society of America Abstracts with Programs v. 29(8), p. A44.

Suppe, J., G. T. Chou, and S. C. Hook, 1992, Rates of folding and faulting determined from growth strata, *in* K. R. McClay, ed., Thrust tectonics: London, Chapman & Hall, p. 105–121.

Suppe, J., F. Sabat, J. A. Munoz, J. Poblet, E. Roca, and J. Verges, 1997, Bed-by-bed growth by kink-band migration: Sant Llorenç de Morunys, eastern Pyrenees: Journal of Structural Geology, v. 19(3–4), p. 443–461.

Verges, J., D. W. Burbank, and A. Meigs, 1996, Unfolding; an inverse approach to fold kinematics: Geology, v. 24, p. 175–178.

Williams, E. A., M. Ford, J. Verges, and A. Artoni, 1998, Alluvial gravel sedimentation in a contractional growth fold setting, Sant Llorenc de Morunys, southeastern Pyrenees, *in* A. Mascle, C. Puigdefabregas, H. P. Luterbacher, and M. Fernandez, eds., Cenozoic foreland basins of Western Europe: Geological Society of London Special Publication 134, p. 69–106.

22

Erickson, S. G., L. M. Strayer, and J. Suppe, 2004, Numerical modeling of hinge-zone migration in fault-bend folds, *in* K. R. McClay, ed., Thrust tectonics and hydrocarbon systems: AAPG Memoir 82, p. 438–452.

Numerical Modeling of Hinge-zone Migration in Fault-bend Folds

S. Gregg Erickson[1]

Princeton 3D Structure Project, Department of Geosciences, Princeton University, Princeton, New Jersey, U.S.A.

Luther M. Strayer[2]

Princeton 3D Structure Project, Department of Geosciences, Princeton University, Princeton, New Jersey, U.S.A.

John Suppe

Princeton 3D Structure Project, Department of Geosciences, Princeton University, Princeton, New Jersey, U.S.A.

ABSTRACT

We use numerical models to investigate the deformation associated with folding by hinge-zone migration. The model geometry consists of a layer that is displaced over a basal fault with a flat-ramp trajectory. The models use an elastic-plastic Mohr-Coulomb material that allows deformation to be localized as shear bands. Bedding can be simulated as discrete interfaces of slip or as weak layers within the continuum. Two orientations of shear bands are dominant: (1) back thrusts that initiate along the axial surface of the fold and (2) layer-parallel shear bands. The style of deformation in the models is controlled by material properties, degree and type of bedding anisotropy, and geometry of the basal fault. In models with homogeneous, isotropic, elastic–perfectly plastic material properties, shear strain rate is localized in a band along the axial surface that is fixed above the fault bend. A uniform region of high shear strain develops above the ramp in material that has moved through the active axial surface. With cohesion softening, deformation is accentuated in the high-shear-strain-rate band. The active shear band is fixed to the material and moves up the ramp, but its base is pinned to the fault bend. As a result, the band rotates until it reaches an unfavorable orientation relative to the local stress field; then it is abandoned as a new shear band propagates up from the fault bend. This mechanism produces a series of abandoned back-thrust shear bands above the ramp. The introduction of bedding-parallel anisotropy or a rounded fault bend causes layer-parallel shear bands to develop. With increasing anisotropy, there is a transition from back thrusting to layer-parallel

[1]*Present address*: Science and Math, Sullivan County Community College, Loch Sheldrake, New York, U.S.A.
[2]*Present address*: Department of Geology, California State University, Hayward, California, U.S.A.

shear. The layer-parallel shear bands initiate at the fold's axial surface, and the propagating tips of the bands remain at the axial surface while inactive parts of the bands are carried up the ramp. The models provide insight into the development of back thrusting and bedding-parallel slip, two mechanisms of hinge-zone migration that are common features in both natural structures and analog models.

INTRODUCTION

Kinematic models of fault-bend folds (Suppe, 1983; Jamison, 1987) are based on the assumption of bedding-parallel shear, with no bedding-parallel shortening or extension of material as it moves over ramp hinges. For rock sequences in which bedding slip is the dominant deformation mechanism, this assumption is approximated, although rock layers between slip surfaces may undergo shortening and extension associated with neutral-surface bending. Bedding slip is dominant in strongly anisotropic sequences of rock (Ramsay, 1967). However, in isotropic rock sequences, bedding slip is inhibited, and deformation must be accommodated by some other mechanism. One such mechanism is back thrusting, in which reverse faults with a vergence opposite that of the main fault develop in the backlimb of the fault-bend fold above the ramp (Morse, 1977; Chester et al., 1991). However, even in materials with low anisotropy, bedding-parallel slip surfaces may develop. Layer-parallel faults develop spontaneously in homogeneous clay cakes (Figure 1a) (Kuenen and de Sitter, 1938) and commonly are observed in thick, competent, apparently isotropic units in natural folds. Ramberg (1961) has compared longitudinal strains (layer-parallel shortening and extension) with layer-parallel shear strains during bending of a viscous layer. According to his analysis, layer parallel shear strains are highest for folds in which layers are thick relative to their arc length. In buckle folds, layer-parallel longitudinal strains are greatest in inner and outer arcs of the hinges, whereas layer-parallel shear strains are greatest in the middle of the layer at the inflection points (Ramberg, 1961). Wiltschko (1979) treated the stresses in material moving over a ramp to be the same as those in a bending layer. In addition to being influenced by these stresses and strains associated with bending, a layer moving over a ramp is also likely to be influenced by the presence of the ramp, which leads to boundary conditions different from those of a buckling layer. Fault shape and friction on the fault also influence the state of stress in the hanging wall (Berger and Johnson, 1980, 1982; Kilsdonk and Fletcher, 1989).

The style of deformation that occurs during fold growth by hinge migration depends on the material properties and mechanical anisotropy as well as on boundary conditions. Morse (1977, 1978) and Chester et al. (1991), using unscaled rock models, showed that fold geometry is dependent on the layering of the model; materials with a high degree of bedding anisotropy have a kink-like fold geometry, whereas more isotropic materials produce rounded fold hinges. Anisotropy also produces sharper hinges in numerical models of buckle folding (Lan and Hudleston, 1996). In addition to influencing fold geometry, anisotropy also influences the state of stress and deformation style near fault bends (Chester and Fletcher, 1997). In the rock models, back thrusts develop above the ramp (Figure 1b) in thick isotropic units in which there is little or no slip on bedding surfaces. These back thrusts are interpreted to form at the lower fault bend and translate up the ramp with continued shortening (Morse 1977, 1978). Back thrusts also develop in natural folds (Jacobeen and Kanes, 1974; Jamison and Pope, 1996). As is the case in the rock models, the mechanical stratigraphy of mesoscopic folds formed at ramps influences the deformation style (Serra, 1977; Spang et al., 1981); back thrusts tend to develop in thick isotropic layers. On the other hand, most of the deformation in the Pine Mountain block in Tennessee is the result of bedding-plane slip, which indicates that the hanging-wall units are anisotropic (Wiltschko et al., 1985).

In this chapter, we use numerical models with dilatant, Mohr-Coulomb, elastic-plastic material properties to study the deformation associated with folding by hinge-zone migration. In previous numerical modeling of fault-bend folds, Erickson and Jamison (1995) used models with viscous and plastic material properties to simulate pressure solution and cataclasis during the development of fault-bend folds; plastic deformation is concentrated above the fault bends as material passes over them. In the numerical models of Strayer and Hudleston (1997), with material properties similar to those used in this study, back thrusting above the lower fault bend dominates. However, during periods in the displacement history in which there is no back thrusting, zones of bedding-parallel shear develop spontaneously. Here, we model the deformation of material in greater detail as it moves over a fault bend. Our goal is to simulate deformation features associated with fold growth by hinge-zone migration. In particular, we examine the competing processes of back thrusting and bedding-parallel slip, and the effects of material properties, anisotropy, heterogeneity, and fault geometry, on their development.

METHOD

We use FLAC (Fast Lagrangian Analysis of Continua) (Cundall and Board, 1988; Coetzee et al., 1995), a two-dimensional, explicit finite-difference code, which has been used to model geologic structures (Riley, 1996; Strayer and Hudleston, 1997). The explicit finite-difference method differs from finite-element methods in that grid points are numerically associated only with their immediate neighbors; there is no global stiffness matrix to invert. We use this method because it is much faster computationally than other methods, making it possible to use a fine grid to model deformation in detail. However, with the finite-difference method, time steps must be small enough to ensure numerical convergence (Cundall and Board, 1988). Coordinate positions are updated using displacements calculated from the previous time step, and the grid is displaced along with the material it represents. Because the dynamic equations of motion are included in the formulation, the numerical scheme is stable, even if unstable physical processes are modeled. The simulations are quasi-static; kinetic energy is both generated and dissipated (Cundall and Board, 1988).

The mechanical behavior of the material is elastic-plastic. For the constitutive rule, we use Mohr-Coulomb plasticity, which consists of a yield function and a plastic-flow rule. The yield function, f, which governs the onset of plastic behavior, is

$$f = \tau + \sigma \sin \phi - C \cos \phi, \tag{1}$$

where τ is the maximum shear stress, σ the mean stress, ϕ the internal friction angle, and C the cohesion. The plastic potential function, g, which governs plastic flow, is

$$g = \tau + \sigma \sin \psi - C \cos \psi, \tag{2}$$

where ψ is the dilation angle. The dilation angle is the plastic volume change divided by the plastic shear strain (Vermeer and de Borst, 1984) and is positive for an increase in volume with increasing plastic shear strain. The values of cohesion, internal friction angle, and dilation angle may harden or soften (increase or decrease) as a function of increasing plastic shear strain. Unequal friction and dilation angles result in nonassociated plasticity, in which the plastic-flow rule (equation 2) is not associated with the yield function (equation 1) (Vermeer and de Borst, 1984). For upper-crust geologic materials, in which plastic behavior is frictional and dilatant, values of the dilation angle generally are less than those of the internal friction angle (Cundall, 1990) and are between 10° and 20° (Vermeer and de Borst, 1984).

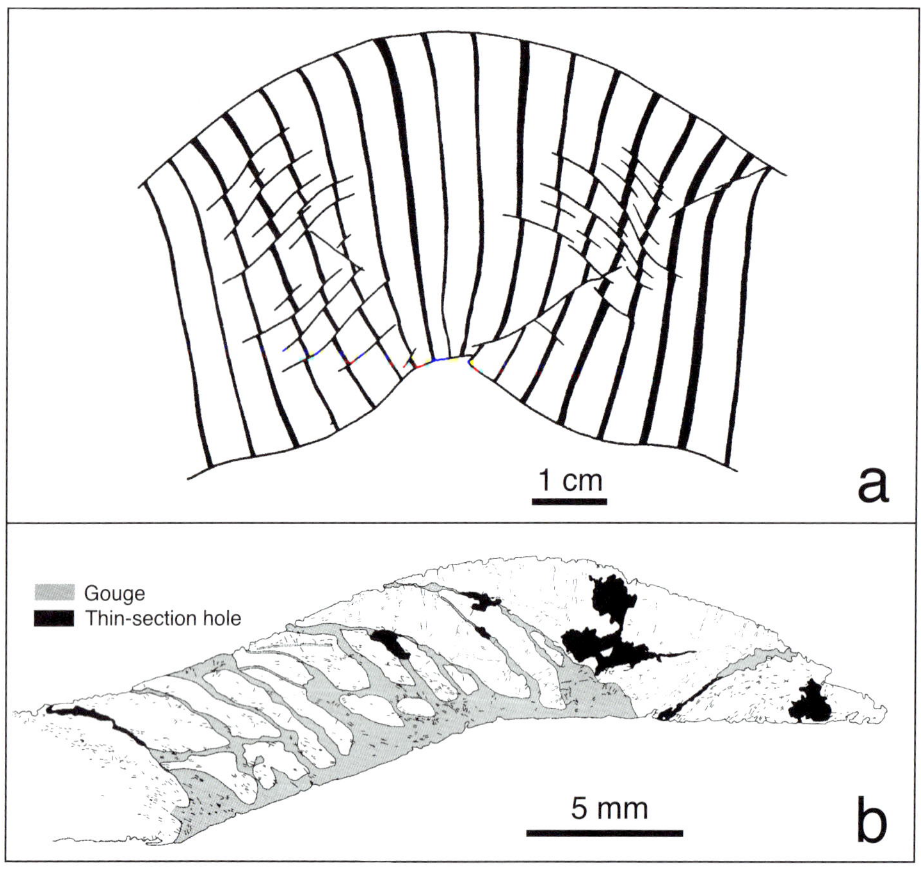

Figure 1. (a) Drawing of bedding-parallel slip surfaces developed spontaneously in a clay-cake model (after Kuenen and de Sitter, 1938, redrawn in Ramsay, 1967). (b) Drawing of back-thrust fault zones (gray) in a rock model (after Morse, 1978).

With nonassociated plasticity, localization of deformation is possible (Cundall, 1989; Hobbs and Ord, 1989). Because of bifurcation in the governing equations, strain may be accommodated either by distributed deformation or by initiation of shear bands. Localization may occur even without a strain-softening constitutive relation, provided that the dilation angle ψ is less than the internal friction angle ϕ (Rudnicki and Rice, 1975; Vardoulakis, 1980). Localization is grid dependent; for a shear band to develop, the grid must be fine relative to the shear-band thickness. The orientation of shear bands follows the relation

$$\theta = \pi/4 - (\phi + \psi)/4, \qquad (3)$$

where θ is the angle between the shear band and the maximum compressive stress. Localization of deformation into shear bands simulates faulting in the upper crust. However, the development of a shear band does not introduce any actual discontinuity into the numerical model; all continuous regions remain continuous throughout the simulation. In addition to the localized shear bands that may develop within the continuum, discrete interfaces, along which slip can occur, may be included. These interfaces, which represent faults and bedding surfaces, are assigned normal and shear stiffnesses and a coefficient of sliding friction. All such interfaces within the models are discontinuities created prior to the simulation. If the properties of the material are too strong, gaps may open along an interface. We choose material properties that do not produce gaps along interfaces.

The material properties we use for the hanging wall are based on the properties of Carrara Marble at a confining pressure of 100 MPa (Ord, 1991), from experiments of Edmond and Paterson (1972). To simplify the properties, we have averaged the values of cohesion, friction, and dilation over all the tabulated strains and used these as constant values. We use values of elastic shear modulus G = 2.1 GPa, elastic bulk modulus K = 2.1 GPa, cohesion C = 45.6 MPa, friction angle ϕ = 23.0°, and dilation angle ψ = 10.7°. Although it would be possible to model the experimental hardening and softening behaviors of cohesion, friction, and dilation in an attempt to be realistic, we have chosen to use a simple set of material properties. The only softening behavior we present is a simple linear softening relationship for cohesion, although tests with more complicated material behaviors give similar results.

The model geometry consists of a hanging-wall block with elastic-plastic material properties and a footwall block with elastic properties. The hanging wall is moved over the footwall by a constant horizontal velocity of 2×10^{-5} units per time step, applied at the left side of the hanging wall, for a total displacement of 5 units (Figure 2). The element size is 0.04 units, the model is 25 units long, and the hanging wall is 2 units thick; there are approximately 15,000 elements in each model. The limit to the number of elements and the smallest element size in a model is determined by computation time. Because there is no gravitational-body force in the models, the scale of the model is not important, except that the applied displacement per time step is slow enough to allow convergence. An overburden of 5 km is simulated by applying a normal traction of 132 MPa to the top of the model. The fault between the hanging wall and footwall blocks is a frictionless interface with a flat-ramp trajectory consisting of a flat with 0° dip and a ramp with 25° dip. The effects of friction on the fault will be addressed in a future paper. In most models, there is a sharp fault bend separating flat and ramp, although in one model we use a rounded fault bend. Because we are interested in deformation of material as it moves through a fold hinge, we only include a lower fault bend in the model and do not attempt to model the deformation at an upper fault bend. Our goal here is to study the detailed deformation in a migrating fold hinge zone. The advantage of zooming in on one part of a structure is that a fine mesh can be used to study the deformation in detail. Erickson and Jamison (1995) and Strayer and Hudleston (1997) have used numerical models to simulate fault-bend folds in their entirety.

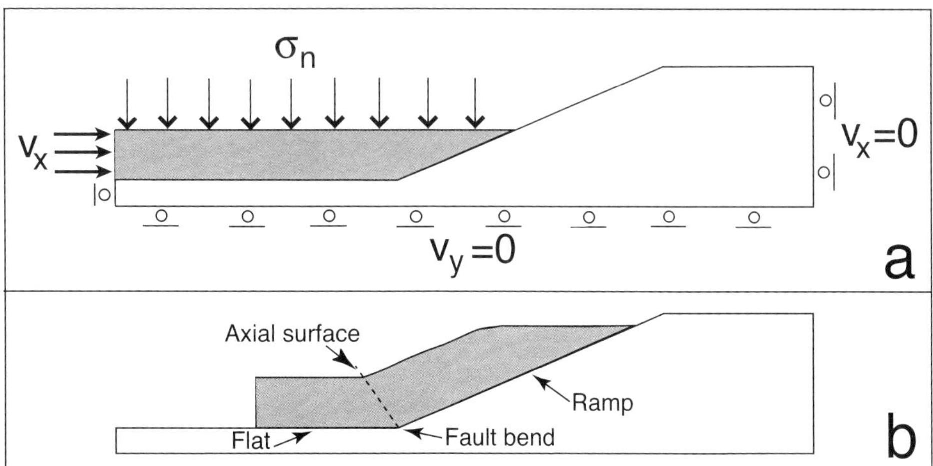

Figure 2. (a) Initial and (b) final model geometry, showing the boundary conditions. The hanging wall (gray) has Mohr-Coulomb elastic-plastic properties, the footwall (white) has elastic properties. V_x is the horizontal component of velocity, and V_y is the vertical component of velocity. σ_n is the normal component of stress.

RESULTS

The models presented here investigate the effects of material properties, anisotropy, heterogeneity, and fault geometry on the competing processes of back thrusting and bedding-parallel shear. Four basic model types, which are based on material properties of the hanging wall, are presented: (1) isotropic, homogeneous models, (2) initially isotropic, homogeneous models with cohesion softening, (3) anisotropic models, either with discrete interfaces to create bedding-parallel slip surfaces or with weak bedding layers to enhance the development of bedding-parallel shear bands, and (4) heterogeneous models, either isotropic or anisotropic. In addition, we investigate the effects of fault geometry on the results.

Isotropic, Homogeneous Models

For models in which the hanging wall has isotropic, homogeneous material properties, the hanging wall undergoes localized back thrusting as it passes over the fault bend. Active deformation is concentrated in a narrow band along the axial surface of the growing fold, which dips 59° toward the foreland for these material properties and ramp dip. Because stress is concentrated at the fault bend, shear strain rate decreases upward from it (Figure 3a). This zone of high shear strain rate along the axial surface is asymmetric with respect to the ramp geometry; it does not bisect the fault bend (for a 25° ramp dip, the bisector dips 77.5°). The asymmetry of the axial surface is associated with thickening of the layer on the dipping limb above the ramp. The orientation of the shear band is determined by the orientation of principal stresses and equation (3). There is a concave-downward curvature to the zone of high shear strain, as a result of the plunges of maximum principal

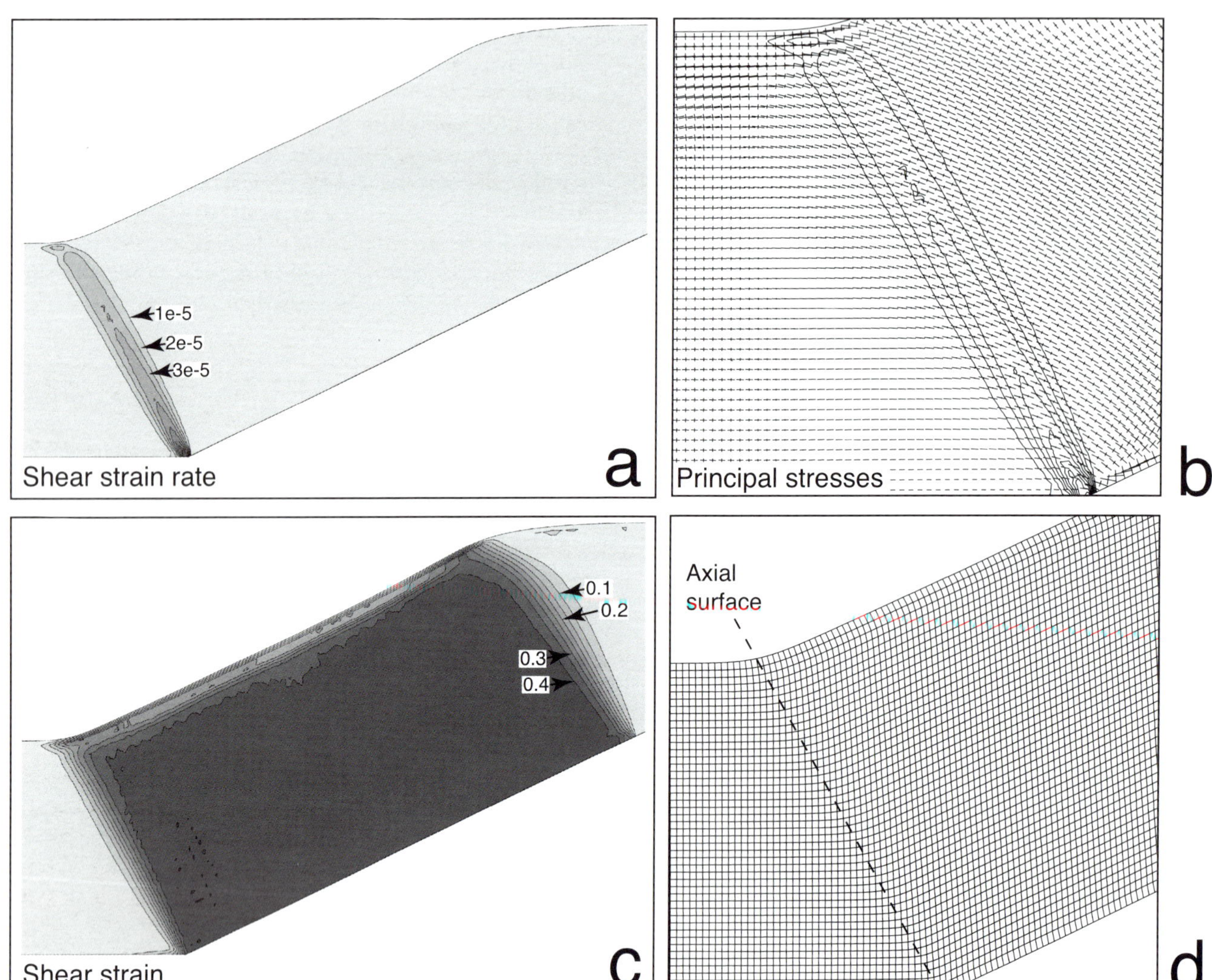

FIGURE 3. Model with an isotropic, homogeneous hanging wall, in which material moves through a zone of high shear strain rate above the fault bend to produce a uniform region of high shear strain. (a) Contours of shear strain rate (contour interval = 1×10^{-5} per time step). (b) Orientations and magnitudes of principal stresses across an axial-surface shear band. Contours of shear strain rate are also shown. (c) Contours of shear strain (contour interval = 0.05). (d) Deformed grid. Elements initially were square. "e-5" represents "$\times 10^{-5}$."

stresses becoming shallower with height in the layer. For given ramp dips and material properties, the magnitude and orientation of this zone of high shear strain rate remain fairly constant with time. This axial-surface shear band is different from shear bands that develop in uniaxial shortening, because, in addition to the shear strain parallel to the shear band, there is also strain associated with bending as the material moves over the fault bend. Principal-stress orientations within the shear band are not rotated with respect to those outside the band, and there appears to be no evidence for stress-state weakening— two features that have been observed in other studies of shear-band formation (Hobbs and Ord, 1989; Cundall, 1989, 1990; Ord, 1991). Instead, principal-stress orientations are unaffected by the formation of the shear band (Figure 3b). As material passes through the narrow band of high shear strain rate, plastic shear strain accumulates in a homogeneous zone within the dipping limb (Figure 3c). Material undergoes top-to-the-foreland shear strain as it moves through the axial surface (Figure 3d). The curvature of the fold above the fault bend decreases upward, becoming less kink-like with height above the fault. Differential stress is highest along the axial surface of the fold and along the ramp, whereas mean stress is lowest in the lower part of the layer above the flat and in the upper part of the layer above the ramp.

The orientation of the axial-surface band of high shear strain rate is controlled by the orientation of the principal stresses and the material properties (equation 3). The principal-stress orientations are determined, in turn, by the ramp dip. As the ramp dip is increased from 15° to 90°, the plunge of the maximum principal stress, σ_1, in the shear band decreases from 22° to 0° (Figure 4). However, for the range of realistic ramp dips, between 15° and 30°, changes in the ramp dip have little effect on either the orientation of σ_1 or the orientation of the axial-surface shear band that develops above the fault bend. For all angles of ramp dip, the dip of the axial-surface shear band is less than the dip of the fault-bend bisector. Although the shear band that forms along the axial surface is not a typical shear band, its

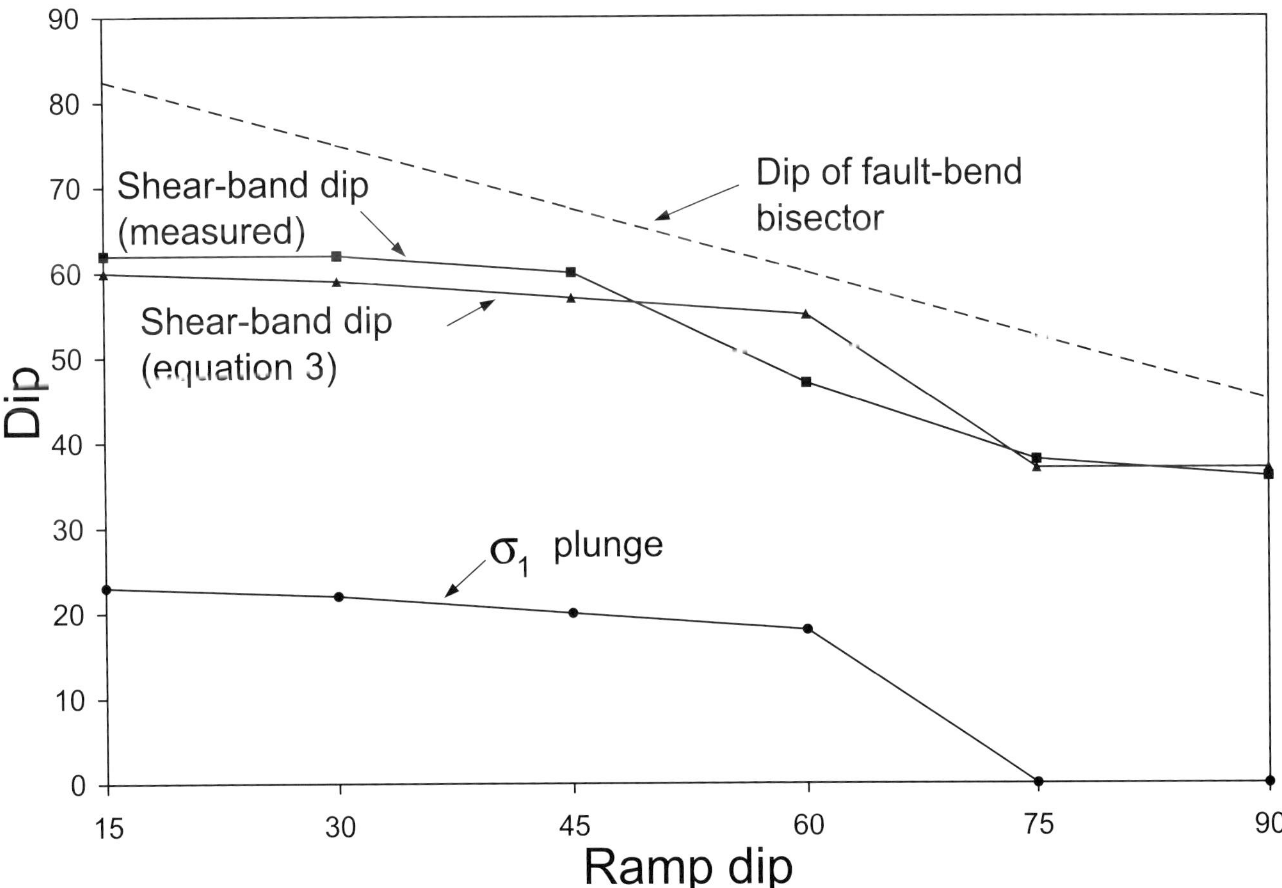

FIGURE 4. Graph of ramp dip versus axial-surface shear-band dip and plunge of maximum compressive principal stress σ_1. The shear-band dip, measured from the models, is compared with the shear-band dip determined by adding the plunge of the maximum compressive principal stress to the angle between the maximum compressive principal stress and the shear band, as determined by equation (3). Also shown is the dip of the fault-bend bisector, demonstrating the deviation of the axial surface from this bisector.

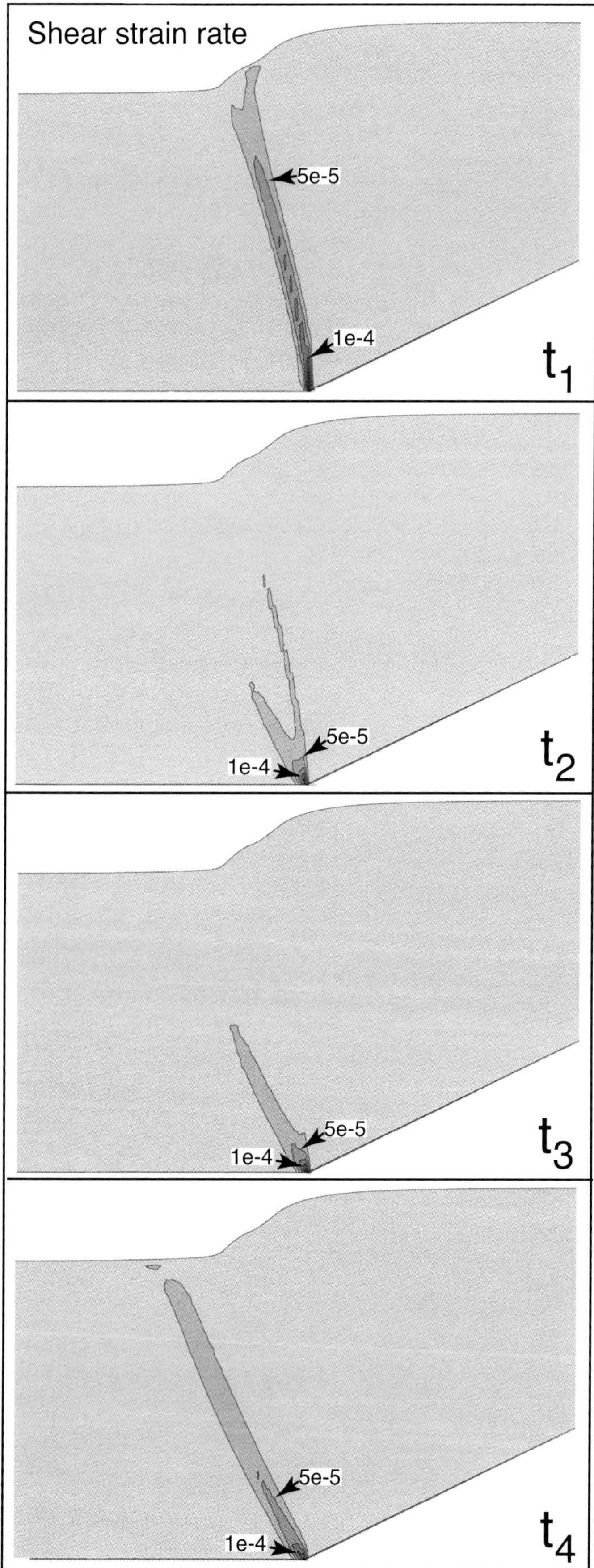

FIGURE 5. Contours of shear strain rate at four time steps (t_1-t_4) with an initially isotropic and homogeneous, cohesion-softening hanging wall, showing the abandonment of one shear band (t_1) and the initiation of a more favorably oriented one (t_4); t_2 and t_3 are transitional stages. Contour interval = 2.5×10^{-5} per time step.

orientation agrees reasonably well with that given by equation (3) for typical shear bands (Vardoulakis, 1980; Vermeer and de Borst, 1984; Hobbs and Ord, 1989). The predicted dip of the shear band, given by the plunge of σ_1 in the shear band and equation (3), agrees with the measured values (Figure 4). Equation (3) predicts the shear band's orientation with respect to the far-field stress. Because the ramp perturbs the stress field, we have measured the orientation of principal stresses within the shear band. This method of measurement is justified because the principal-stress orientations within shear bands have not rotated and because the far-field stress does not represent the local state of stress in which the shear band forms. The axial-surface shear-band orientation is not greatly affected by material properties. For instance, for a material with a friction angle of 45°, the results are similar to those with $\phi = 23°$, although there is more variability in the data and the agreement between the measured and theoretical values is not as good.

Cohesion-softening Models

Models in which the hanging wall initially has homogeneous and isotropic material properties but later undergoes cohesion softening result in some interesting variations. The cohesion-softening rule that we use is a simple function of plastic shear strain; the cohesion is linearly reduced to one-half its initial value at a plastic shear strain of 0.4, after which it remains constant with increasing plastic shear strain. This form of softening is not meant to model rock behavior exactly but is used because of its simplicity; tests of other softening curves give similar results. Shear strain rate is concentrated along the axial surface of the fold above the fault bend (Figure 5), as is the case without cohesion softening. However, because of the material memory introduced by the cohesion softening, deformation is concentrated in high-strain regions. Weakening of high-strain zones causes shear bands to become fixed in the material, and as material moves over the fault bend, the zone of high shear strain rate migrates and rotates in order to coincide with the shear band (Figure 5). For the material properties used here, shear bands initiate with dips of 59° toward the foreland and rotate to a 78° dip. Each shear band is abandoned when it rotates into an orientation that is unfavorable, with respect to the local stress field, for further slip. At this point, a new shear band initiates at the fault bend. For the material properties used in this model, shear bands rotate about 20° before they are abandoned; a more pronounced cohesion-softening rule produces larger angles of rotation before abandonment, because the weaker shear band remains favorably oriented for a longer time. The final result of this shear-band initiation and abandonment is a series of evenly spaced shear bands above the ramp (Figure 6a). These shear bands narrow upward, because strain is more localized in higher parts and more broadly distributed near the base of the hanging wall, as a result of the rotation of back-thrust shear bands. The shear bands separate lozenge-shaped regions of mildly deformed material, which are primarily transported through the fold-hinge zone by rigid-body translation (Figure 6b). Because there is shortening perpendicular to shear-band boundaries, there is a volume decrease in the shear bands of ~10%, even though the material's dilation angle is positive. This is another atypical aspect of the axial-surface shear bands and might represent pressure solution along axial surfaces of natural folds. Material properties and ramp dip determine the orientation of these back-thrust shear bands. The spacing of the shear bands is controlled in the models by the degree of cohesion softening. Greater cohesion softening (i.e., cohesion reduced to a lower value at a given shear strain) leads to a wider spacing of shear bands, because each shear band can rotate to a steeper orientation before the stress field becomes unfavorable for slip and the shear band is abandoned.

The development of back-thrust shear bands is closely related to the strong stress perturbation that develops at the sharp fault bend. To produce a more rounded fault bend, we replace the sharp fault bend by a circular arc connecting the flat and ramp. In this model, the lack of a sharp fault bend causes layer-parallel shear bands instead of back thrusts to develop (Figure 7a). The model in Figure 7 uses the same initial properties and cohesion-softening rule as that in Figure 6; the only difference is the geometry of the basal fault. In models with a rounded fault bend and without cohesion softening, the shear bands are less well developed and shear strain is more uniformly distributed throughout the layer, although it is highest in the center of the layer. In the cohesion-softening model, some shear bands step up or down relative to the grid and some initiate or terminate during displacement up the ramp. The shear bands are fairly evenly spaced but are concentrated in the central part of the layer, where shear strains are expected to be highest.

Grid Effects

It should be noted that there is a mesh dependence to the shear bands developed in the model shown in Figure 7. Whereas the back-thrust shear bands are several elements wide and cut across the grid (Figure 6b), the bedding-parallel shear bands are one element wide and lie parallel to the grid (Figure 7b, 8a). A model with

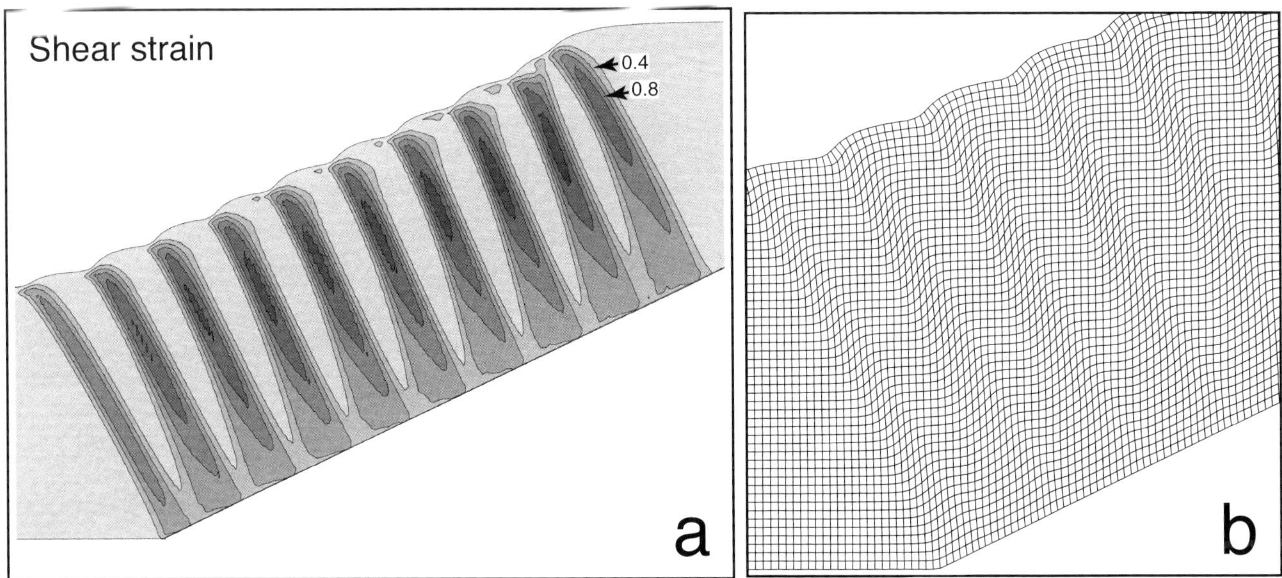

Figure 6. Model with an initially isotropic and homogeneous, cohesion-softening hanging wall, in which a series of back-thrust shear bands form above the fault bend and are carried up the ramp. (a) Contours of shear strain (contour interval = 0.2). (b) Deformed grid.

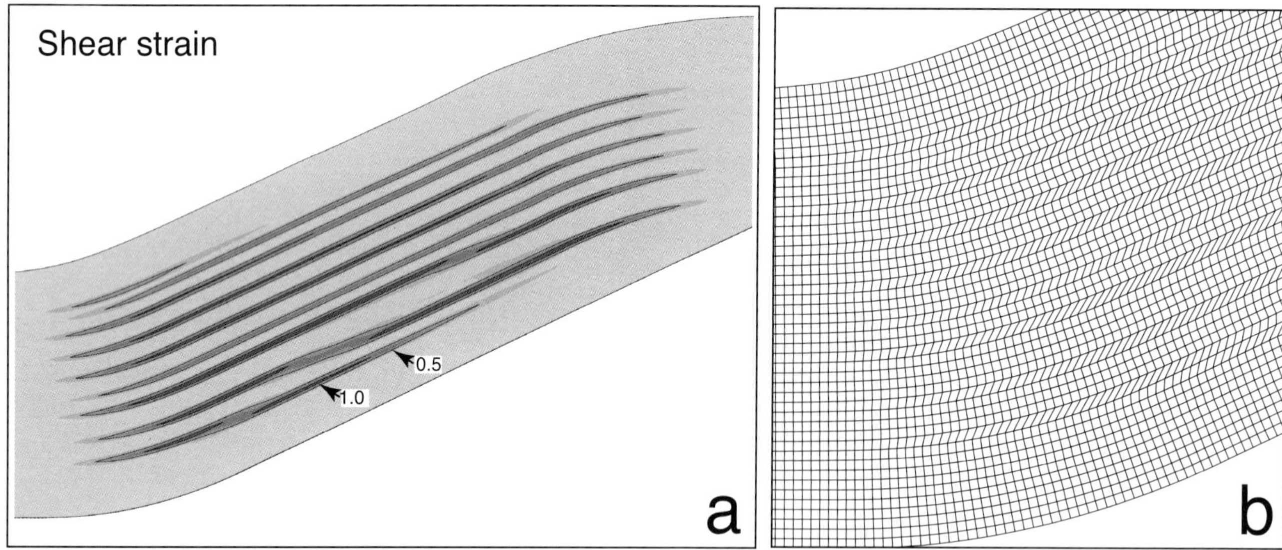

FIGURE 7. Model with an initially isotropic and homogeneous, cohesion-softening hanging wall and a rounded fault bend, in which layer-parallel shear bands form above the fault bend. (a) Contours of shear strain (contour interval = 0.5). (b) Deformed grid.

a rounded fault bend and a mesh in which columns and rows are oriented at 45° to layering produces a fairly uniform distribution of strain, with incipient back-thrust shear bands instead of layer-parallel shear bands (Figure 8b). It is impossible to produce a truly isotropic material with these models; the mesh introduces a slight anisotropy. However, because of depositional layering, sedimentary rocks are anisotropic. Thus, even though the material in the models is not isotropic, a material with layer-parallel anisotropy is appropriate for simulating sedimentary rocks. These models demonstrate that, with a rounded fault bend and the slight layer-parallel anisotropy introduced by the mesh, bedding-parallel shear bands develop, instead of back thrusts. The size of the mesh also influences the development of shear bands. With a coarser mesh, in which element sides are twice as long, layer-parallel shear bands develop, although these shear bands are fewer and more widely spaced than with a finer mesh (Figure 8c).

Anisotropic Models

There are two ways to introduce a bedding-parallel anisotropy into the models: (1) interfaces, along which discrete slip occurs, can be built into the model, or (2) material properties can be varied spatially to create weak layers. For the model with discrete interfaces, we use seven frictionless interfaces that divide the hanging wall into eight layers (Figure 9a). Each layer folds with its own neutral surface, and strain within each layer is dominated by extension in the outer arc and shortening in the inner arc as its passes over the fault bend. Fold curvature decreases upward, causing strain to be most concentrated in the hinges of the lowest layers. Shear sense on each interface is top to the foreland and accumulates as material moves through the fold axial surface above the fault bend (Figure 9b). Maximum slip on each interface is about one-half the thickness of layers between interfaces. In models with cohesion softening and a finer mesh, localized back thrusts develop in the lowest layer, where the fold hinge is sharp, whereas bedding-parallel shear bands develop in higher layers, where fold hinges are rounded. The fold that develops is symmetric, bisecting the angle between flat and ramp.

The second way to create bedding-parallel anisotropy is to vary material properties to simulate weak bedding layers. We use seven weak layers, each 1 element thick, with a cohesion of 22.8 MPa, which is one-half the value of the strong layers (Figure 10a). In this model, the highest shear strain rates are concentrated in the axial surface of the fold above the fault bend, but are also concentrated in the weak layers. The high shear strain rate is therefore concentrated in beads within the weak layers as they pass through the axial surface. As material passes over the fault bend, bands of high shear strain develop in the weak layers (Figure 10b), which experience top-to-the-foreland shear (Figure 10c). The beads of high shear strain rate (Figure 10a) above the fault bend are analogous to propagating crack tips, which remain stationary as material passes through them and into the dipping limb. Unlike the back-thrust shear bands, these bedding-parallel shear bands experience a volume increase of ~5%. This model with weak layers has a fold asymmetry that is similar to isotropic models. Effects of bending can be seen on several scales in this model. The highest shear strain within a bending layer is in the center of the layer, whereas the bedding-parallel longitudinal strain varies monotonically across the layer, causing inner-arc shortening and outer-arc extension. On the scale of the entire model, the highest shear strains within the weak layers in the hanging wall are in the layers near the center of the sequence. On a smaller scale, the highest shear strain within each strong layer is midway between the adjacent weak layers. This distribution of shear strain can also be seen in the deformed grid (Figure 10c); layer-parallel shear of the grid increases toward the center of each of the strong layers.

Heterogeneous Models

Rocks are not homogeneous materials, but rather are masses with spatially heterogeneous strength. Therefore, we study the behavior of a heterogeneous, isotropic material, which is created by assigning a random Gaussian distribution of cohesion to equidimensional zones that are 2 elements on a side. The length scale of the heterogeneity is kept larger than the grid size, to avoid mesh effects. The standard deviation of the Gaussian distribution controls the degree of heterogeneity. In addition to the heterogeneity, the cohesion of each element softens to one-half its initial value at a plastic shear strain of 0.4, after which it is constant. With initially equidimensional regions defining the heterogeneity in cohesion, back thrusting is dominant. The results are similar to those in the homogeneous cohesion-softening model, although the spacing and orientation of back-thrust shear bands are less systematic (Figure 11a). Shear bands initiate preferentially in regions that have initially low cohesion; scan lines along shear bands show that the average initial cohesion is less than the average initial cohesion of the entire model. Also, a shear band is abandoned sooner if it initiates in a region of higher (close to the average) cohesion. Variations in the lifetimes of shear bands lead to the irregular pattern of shear bands in Figure 11a. Shear bands that are abandoned early have lower dips than do the ones that last longer, because they rotate less before abandonment.

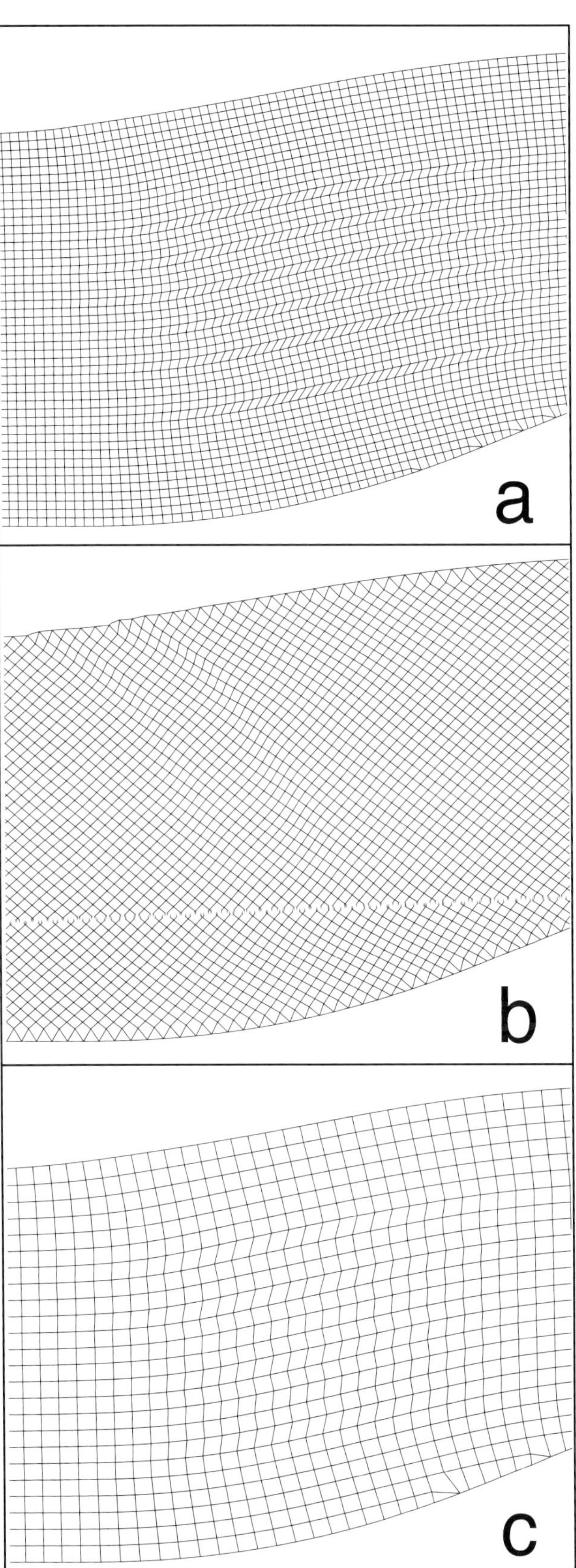

Figure 8. The effects of the grid in models with a rounded fault bend. All models have the same fault geometry. (a) The grid is parallel and perpendicular to the layer. Shear bands form parallel to the grid. (b) The grid is oriented 45° to the layer. The strain is fairly uniform and incipient back-thrust shear bands develop. (c) The grid is coarser; each element is twice as long on a side as elements in (a). Layer-parallel shear bands develop, but are fewer and more widely spaced than with a finer grid.

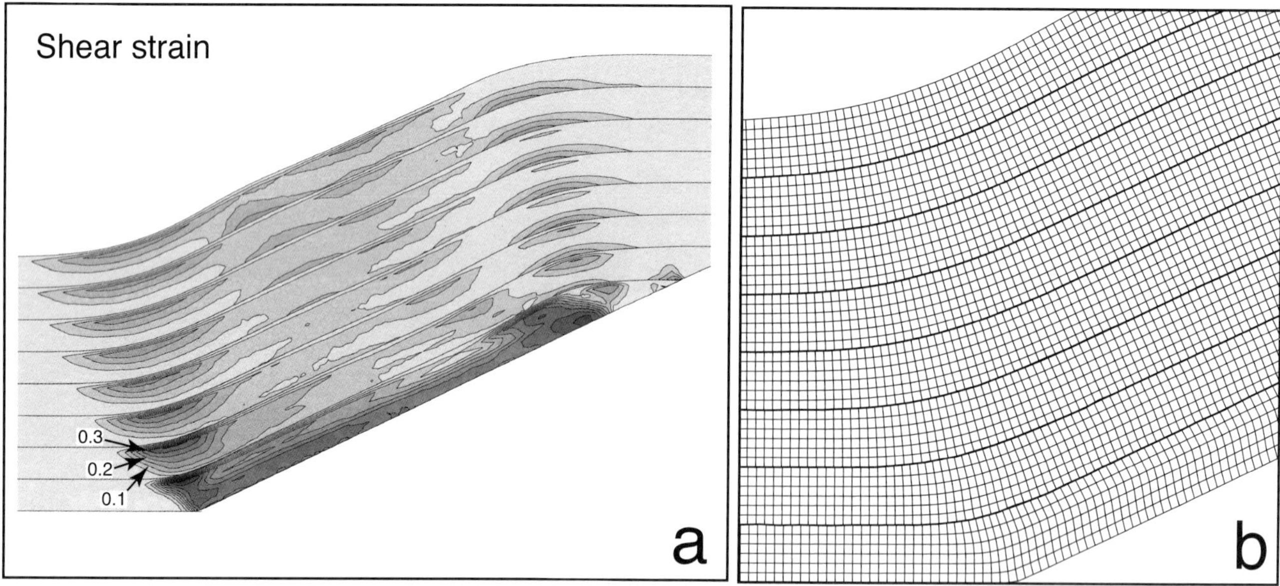

FIGURE 9. Model with an isotropic, homogeneous hanging wall divided by interfaces on which slip can occur. Slip on each interface is top to the foreland, and each layer folds with its own neutral surface. (a) Contours of shear strain (contour interval = 0.05). (b) Deformed grid.

In heterogeneous, anisotropic models, cohesion is randomly assigned to regions that have a thickness of 2 elements and a randomly determined length, parallel to bedding, of 2 to 40 elements. The degree of anisotropy depends on two factors: (1) the standard deviation of the Gaussian distribution for the value of cohesion assigned to each region and (2) the mean aspect ratio of regions with different cohesion. Standard deviations of 10%, 30%, and 50% of the mean cohesion (45.6 MPa) are used. The degree of anisotropy determines whether the deformation is controlled by bedding-parallel shear bands or by back thrusts. Comparison of Figure 11a with 11b demonstrates the effects of the mean aspect ratio of the heterogeneity; both models have a distribution of cohesion with the same mean and with a 50% standard deviation about the mean. Isotropic heterogeneity (Figure 11a) leads to the dominance of back thrusting, whereas anisotropic heterogeneity, with large aspect-ratio regions defining a bedding orientation, causes layer-parallel shear bands to develop (Figure 11b).

For a given aspect ratio to the heterogeneous regions and a given mean cohesion, the nature of deformation is determined by the standard deviation of the cohesion distribution. With a standard deviation of 50%, the deformation is dominated by layer-parallel shear bands (Figure 11b), whereas with a standard deviation of 10%, the deformation is dominated by back thrusting (Figure 11d). With a standard deviation of 30%, there is a combination of layer-parallel shear and back thrusting (Figure 11c), which represents a transition between the two end members. Even though the heterogeneity is 2 elements in height, localized bedding-parallel shear bands are 1 element wide. With a standard deviation of 50%, the distribution of bedding-parallel shear bands is fairly uniform across any given layer-perpendicular scan line of the hanging wall. Where the spacing is large between two weak layers, a shear band may localize midway between them, even if no weak layer is present. This phenomenon is similar to that discussed above, in which the highest shear strain is at the center of a layer; the center of a layer is the most likely location for a shear band to localize. Shear bands step up or down to connect regions of low strength. With a standard deviation of 30%, regions dominated by layer-parallel shear alternate with regions dominated by back thrusting, thereby indicating a spatial as well as temporal alternation between back thrusting and layer-parallel shear. Comparison of scan lines of the initial cohesions along back thrusts and those across regions dominated by layer-parallel shear bands indicates that back thrusts develop in zones of below-average cohesion and that layer-parallel shear bands develop in regions of above-average cohesion. There is no correlation between the standard deviation of cohesion along a scan and the development of back thrusts versus layer-parallel shear zones; that is, there is no statistical difference between the standard deviations on scan lines along back thrusts and those across bedding-parallel shear bands.

DISCUSSION

One of the goals of this numerical modeling was to investigate the competing processes of back thrusting and bedding-surface slip, which are manifested in the

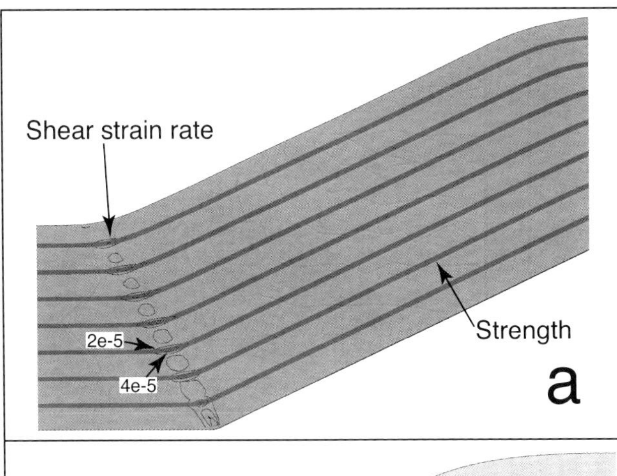

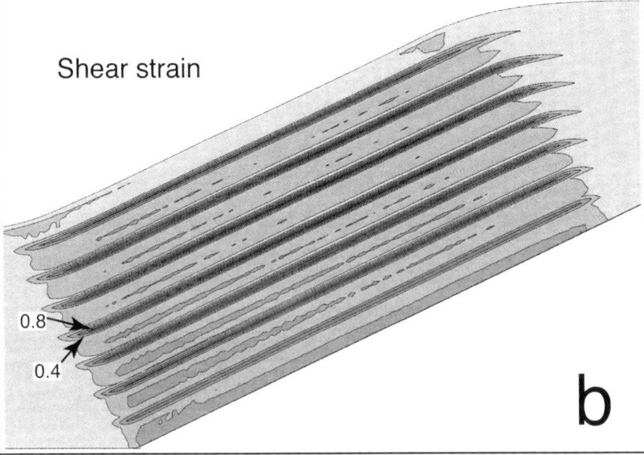

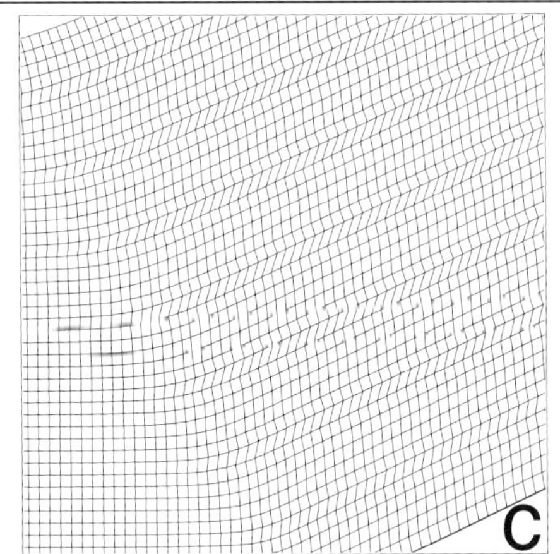

FIGURE 10. (a) Model with weak (low-cohesion) layers (dark gray) separating strong layers (light gray). Layer-parallel shear bands develop in the weak layers as they move over the fault bend. Contours of shear strain rate are also shown (contour interval = 2×10^{-5} per time step). (b) Contours of shear strain (contour interval = 0.2). (c) Deformed grid.

models as shear bands parallel to the axial surface of the fold and parallel to the layer, respectively. The models indicate that, for the material properties used here and a sharp fault bend, an isotropic, homogeneous material will undergo back thrusting. Cohesion softening causes the back-thrust zones to localize in a periodic fashion, if there is a constant-velocity boundary condition. According to Ramberg's (1961) analysis of buckle folds, layer-parallel shear strain is inhibited relative to longitudinal strains, unless the layer is thick compared with the arc length of the fold. In the models of our study, whether back thrusting or layer-parallel shear develops depends on the sharpness of the fault bend. With a sharp fault bend, the strong stress perturbation at the fault bend causes back thrusting to develop instead of layer-parallel shear. With a rounded fault bend, layer-parallel shear is enhanced relative to back thrusting. To produce layer-parallel shear with a sharp fault bend and the material properties used in this study, it is necessary to introduce a layer-parallel anisotropy into the models. With increasing anisotropy, there is a transition from back thrusting alone to combined back thrusting and layer-parallel shear to layer-parallel shear alone (Figure 11).

The back-thrust shear bands in the numerical models resemble the upward-narrowing cataclastic back-thrust zones in rock models (Figure 1b) (Morse, 1977, 1978; Chester et al., 1991). In our numerical models, the upward decrease in thickness results from rotation of the back-thrust zones, which in turn is caused by cohesion softening and locking of shear bands into highly strained material. The narrow parts of these shear bands have undergone the highest shear strains. The concave-downward geometry of the back-thrust shear bands in our models, as displayed by contours of shear strain rate, is also similar to that of back thrusts that develop in rock models (Morse, 1977, 1978; Chester et al., 1991) and in natural structures (Jacobeen and Kanes, 1974; Serra, 1977; Jamison and Pope, 1996). In our models, these back thrusts develop along the axial surface of the growing fold. The orientation of the axial surface does not bisect the fault bend, as is commonly assumed in kinematic models. The asymmetry of the axial surface is associated with thickening in the dipping limb. In rock models, back thrusts dip more shallowly than does the axial surface of the fold ($\sim 35^\circ$ dip for a 20° ramp dip) and more shallowly than the back-thrust shear bands developed in our models. The orientation of back-thrust shear bands appears to be fairly insensitive to ramp dip, at least for common ramp dips (Figure 4).

In our models, with some material properties and fault geometries, there is spontaneous development of layer-parallel shear bands. Strayer and Hudleston (1997) also observed this phenomenon in numerical models. Layer-parallel slip surfaces develop spontaneously during folding of clay cakes (Figure 1a) (Kuenen and de Sitter, 1938). In these clay-cake buckle folds, layer-parallel slip surfaces are best developed at the inflections of fold limbs and near the center of the layer, where the highest layer-parallel shear strains are predicted by Ramberg's (1961) theory. In our models, layer-parallel shear bands also are best developed at the centers of layers. In the clay cake models, there may be a layer-parallel anisotropy as a result of preferred orientation of clay particles. In our numerical models, there is also an anisotropy introduced by the grid orientation; for tests in which the grid is initially oriented at 45° to vertical, layer-parallel shear bands do not develop (Figure 8). Thus, it appears that layer-parallel shear bands do not develop in an isotropic material and that some anisotropy is required for their development. However, with appropriate material properties or geometries, layer-parallel shear bands may develop in material with low anisotropy. This phenomenon may explain observations of natural folds in which bedding-slip surfaces develop in the centers of thick competent layers that lack any apparent bedding anisotropy.

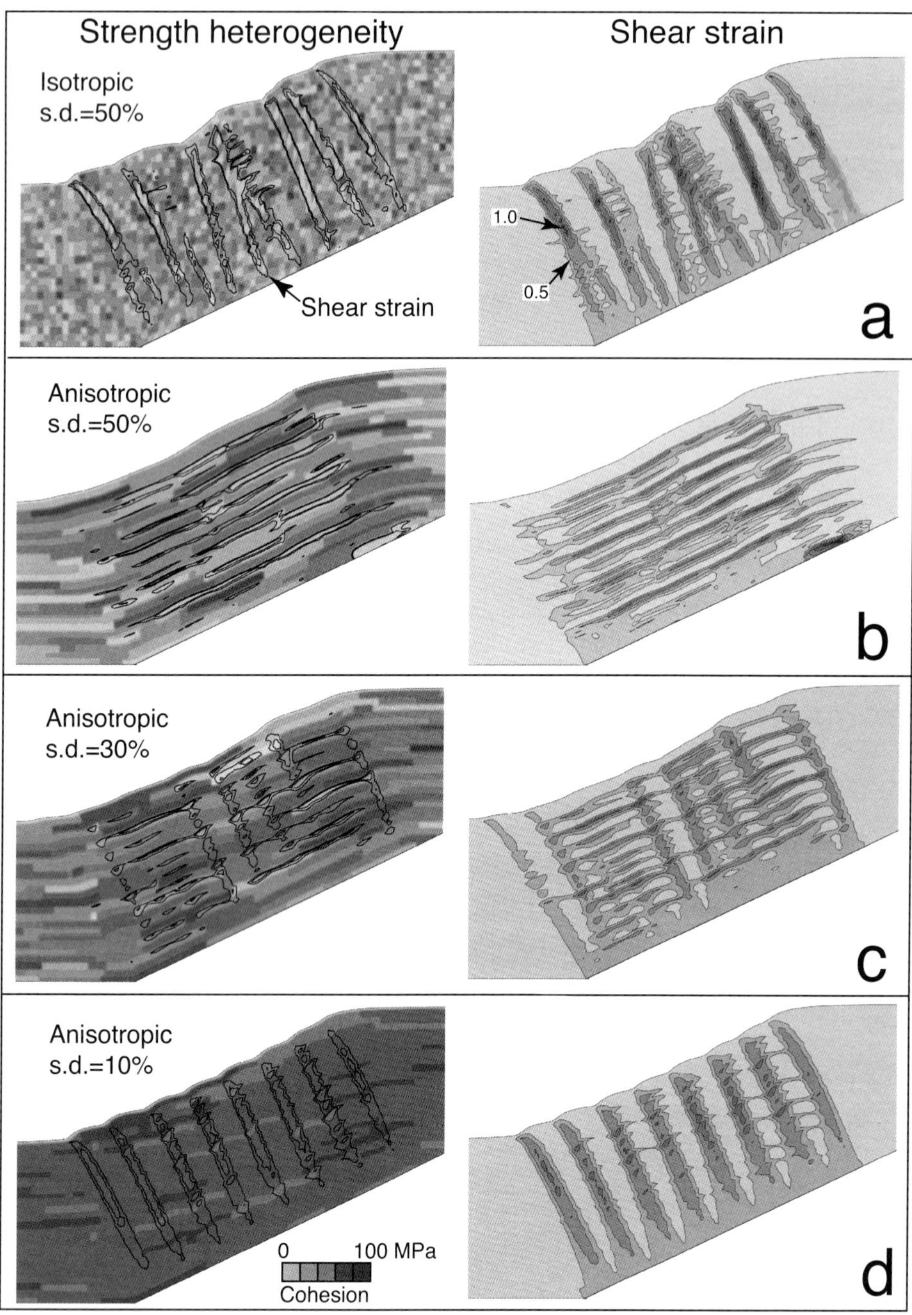

FIGURE 11. Models with heterogeneous and anisotropic strength. Models with isotropic heterogeneity produce back-thrust shear bands, whereas models with anistropic heterogeneity produce layer-parallel shear bands. (a) Isotropic, heterogeneous model with length scale of heterogeneities twice that of the grid size, and standard deviation of the Gaussian distribution of cohesion that is 50% of the mean cohesion. (b–d) Anisotropic, heterogeneous models, in which the length scale of heterogeneities perpendicular to the layer is twice that of the grid size, and that parallel to the layer is a random length between 2 and 40 times the grid size. Standard deviation of the Gaussian distribution of cohesion is 50% of the mean cohesion in (b), 30% of the mean cohesion in (c), 10% of the mean cohesion in (d). Contours of shear strain are also shown. Contour interval = 0.25.

As is the case in our models, deformation in natural fault-bend folds may be accommodated by back thrusting (Jacobeen and Kanes, 1974; Jamison and Pope, 1996) or by bedding-parallel slip (Wiltschko et al., 1985). The mechanical stratigraphy of mesoscopic folds formed at ramps influences the deformation style (Serra, 1977; Spang et al., 1981); back thrusts develop in thick isotropic layers. Jacobeen and Kanes (1974) and D'Alberto et al. (1995) describe back thrusts on the limbs of kilometer-scale fault-bend folds in the central Appalachians and southern Alps, respectively, indicating that they developed in isotropic hanging-wall sequences. On the other hand, most of the deformation in the Pine Mountain block in Tennessee is by slip on bedding surfaces, which indicates anisotropy of the hanging-wall units (Wiltschko et al., 1985). Along-strike changes in mechanical layering control the geometry and the deformation style of a fault-bend fold in the Canadian Rockies (Jamison and Pope, 1996). Thinning of the competent basal dolostone unit in the hanging wall causes a change from an open fault-bend fold to a complex structure with second-order folds and back thrusts.

Our models have important implications for hydrocarbon systems, because the style of deformation determines the nature of fractured reservoirs in thrust belts. The locations and orientations of faults and fractures within a thrust-belt structure are controlled by its mechanical stratigraphy and by the stress field during its development. Locations, densities, and orientations of fractures, in turn, determine the flow and storage of hydrocarbons. Whether deformation is by back thrusting or by bedding-surface slip determines the anisotropy of permeability. Knowledge of the timing and sequence of deformation, as determined by the models, also leads to a better understanding of reservoir potential. For example, fractures that develop during the early stages of movement over a thrust-fault ramp may later be sealed when material moves through the different stress states above the lower and upper fault bends (Erickson and Jamison, 1995).

CONCLUSIONS

Using numerical models, we have investigated the deformation associated with folding by hinge-zone migration. Material properties, degree and type of bedding anisotropy, and geometry of the basal fault together control the style of deformation in the models. Two orientations of shear bands are possible: (1) back thrusts that develop along the axial surface of the fold and (2) layer-parallel shear zones. In models with homogeneous and isotropic material properties, active deformation is localized as a band along the axial surface of the fold, and a uniform region of high shear strain develops in one limb as material moves through the active axial surface and up the ramp. With cohesion softening, deformation is further concentrated in axial-surface shear bands to produce a series of evenly spaced back thrusts above the ramp. With the introduction of layer-parallel anisotropy, layer-parallel shear bands develop in addition to the shear bands parallel to the axial surface. These layer-parallel shear bands initiate at the fold hinge, and the propagating tips of the zones remain at the hinge while inactive parts of the zones are carried into the limb. With an increasing degree of anisotropy, caused either by increasing the mean aspect ratio of heterogeneous regions or by increasing the standard deviation about the mean cohesion, there is a transition from back thrusts to layer-parallel shear bands.

ACKNOWLEDGMENTS

This chapter benefited from the constructive reviews of Dave Waltham and an anonymous reviewer. This research was funded by the members of the Princeton 3D Structure Project consortium; ARCO, Chevron, Elf, Intevep, Paradigm, Tarim, and Texaco.

REFERENCES CITED

Berger, P., and A. M. Johnson, 1980, First-order analysis of deformation of a thrust sheet moving over a ramp: Tectonophysics, v. 70, p. T9–T24.

Berger, P., and A. M. Johnson, 1982, Folding of passive layers and forms of minor structures near terminations of blind thrust faults— application to the central Appalachian blind thrust: Journal of Structural Geology, v. 4, p. 343–353.

Chester, J. S., and R. C. Fletcher, 1997, Stress distribution and failure in anistropic rock near a bend on a weak fault: Journal of Geophysical Research, v. 102, p. 693–708.

Chester, J. S., J. M. Logan, and J. H. Spang, 1991, Influence of layering and boundary conditions on fault-bend and fault-propagation folding: Geological Society of America Bulletin, v. 103, p. 1059–1072.

Coetzee, M. J., R. D. Hart, P. M. Varona, and P. A. Cundall, 1995, FLAC basics: An introduction to FLAC and a guide to its practical application in geotechnical engineering: Minneapolis, Itasca Consulting Group, 68 p.

Cundall, P. A., 1989, Numerical experiments on localization in frictional materials: Ingenieur-Archiv, v. 59, 148–159.

Cundall, P. A., 1990, Numerical modeling of jointed and faulted rock, *in* J. Rossmanith, ed., Mechanics of jointed and faulted rock: Rotterdam, A. A. Balkema, p. 11–18.

Cundall, P. A., and M. Board, 1988, A microcomputer program for modeling large-strain plasticity problems, *in* C. Swoboda, ed., Numerical Methods in Geomechanics: Proceedings of the 6th International Conference on Numerical Methods in Geomechanics, A. A. Balkema, Rotterdam, p. 2101–2108.

D'Alberto, L., A. Boz, and C. Doglioni, 1995, Structure of the Vette Feltrine (eastern southern Alps): Memorie di Scienze Geologiche, v. 47, p. 189–199.

Edmond, J. M., and M. S. Paterson, 1972, Volume changes during the deformation of rocks at high pressures: International Journal of Rock Mechanics and Mining Science, v. 8, p. 161–182.

Erickson, S. G., and W. R. Jamison, 1995, Viscous-plastic finite-element models of fault-bend folds: Journal of Structural Geology, v. 17, p. 561–573.

Hobbs, B. E., and A. Ord, 1989, Numerical simulation of shear band formation in a frictional-dilational material: Ingenieur-Archiv, v. 59, p. 209–220.

Jacobeen, F. Jr., and W. H. Kanes, 1974, Structure of the Broadtop syclinorium and its implications for Appalachian structural style: AAPG Bulletin, v. 58, p. 362–375.

Jamison, W. R., 1987, Geometric analysis of fold development in overthrust terranes: Journal of Structural Geology, v. 9, p. 207–219.

Jamison, W. R., and A. Pope, 1996, Geometry and evolution

of a fault-bend fold: Mount Bertha anticline: Geological Society of America Bulletin, v. 108, p. 208–224.

Kilsdonk, B. and R. C. Fletcher, 1989, An analytical model of hanging-wall and footwall deformation at ramps on normal and thrust faults: Tectonophysics, v. 163, p. 153–168.

Kuenen, Ph. H., and L. U. de Sitter, 1938, Experimental investigation into the mechanisms of folding: Leidsche Geologische Mededeelingen, v. 10, p. 217–240.

Lan, L., and P. Hudleston, 1996, Rock rheology and folds in single layers: Journal of Structural Geology, v. 18, p. 925–931.

Morse, J., 1977, Deformation in ramp regions of overthrust faults: Experiments with small-scale rock models: Twenty-ninth annual field conference, Wyoming Geological Association Guidebook, p. 457–470.

Morse, J., 1978, Deformation in ramp regions of thrust faults: Experiments with rock models: M.S. Thesis, Texas A&M University, College Station, Texas, 138 p.

Ord, A., 1991, Deformation of rock: A pressure-sensitive, dilatant material: Pure and Applied Geophysics, v. 137, p. 337–366.

Ramberg, H., 1961, Relationship between concentric longitudinal strain and concentric shearing stain during folding of homogeneous sheets of rocks: American Journal of Science, v. 259, p. 382–390.

Ramsay, J. G., 1967, Folding and fracturing of rocks: New York, McGraw–Hill, 568 p.

Riley, D. J., 1996, Applications of numerical modeling to thrust tectonics: Ph.D. thesis, University of Southampton, England, 180 p.

Rudnicki, J. W., and J. R. Rice, 1975, Conditions for the localization of deformation in pressure-sensitive dilatant materials: Journal of the Mechanics and Physics of Solids, v. 23, p. 371–394.

Serra, S., 1977, Styles of deformation in the ramp regions of overthrust faults: Twenty-ninth annual field conference: Wyoming Geological Association Guidebook, p. 487–498.

Spang, J. H., T. L. Wolcott, and S. Serra, 1981, Strain in the ramp regions of two minor thrusts, southern Canadian Rocky Mountains, *in* N. L. Carter, M. Friedman, J. M. Logan, and D. Stearns, eds., Mechanical behavior of crustal rocks: The Handin volume: American Geophysical Union Geophysical Monograph 24, p. 243–250.

Strayer, L. M., and P. J. Hudleston, 1997, Numerical modeling of fold initiation at thrust ramps: Journal of Structural Geology, v. 19, p. 551–566.

Suppe, J., 1983, Geometry and kinematics of fault-bend folding: American Journal of Science, v. 283, p. 684–721.

Vardoulakis, I., 1980, Shear band inclination and shear modulus of sand in biaxial tests: International Journal for Numerical and Analytical Methods in Geomechanics, v. 4, p. 103–119.

Vermeer, P. A., and R. de Borst, 1984, Non-associated plasticity for soils, concrete and rock: Heron, v. 29, p. 1–64.

Wiltschko, D. V., 1979, A mechanical model for thrust sheet deformation at a ramp: Journal of Geophysical Research, v. 84, p. 1091–1104.

Wiltschko, D. V., D. A. Medwedeff, and H. E. Millson, 1985, Distribution and mechanisms of strain within rocks on the northwest ramp of Pine Mountain block, southern Appalachian foreland: A field test of theory: Geological Society of America Bulletin, v. 96, p. 426–435.

23

Salvini, F., and F. Storti, 2004, Active-hinge-folding-related deformation and its role in hydrocarbon exploration and development—Insights from HCA modeling, *in* K. R. McClay, ed., Thrust tectonics and hydrocarbon systems: AAPG Memoir 82, p. 453–472.

Active-hinge-folding-related Deformation and its Role in Hydrocarbon Exploration and Development—Insights from HCA Modeling

Francesco Salvini
Dipartimento di Scienze Geologiche, Università "Roma Tre," Rome, Italy

Fabrizio Storti
Dipartimento di Scienze Geologiche, Università "Roma Tre," Rome, Italy

ABSTRACT

Thrust-related folding is an important and efficient mechanism for generating wide, deformed rock panels at shallow structural levels. In these environmental conditions, three basic forms of deformational features commonly occur: pressure-solution seams (stylolites), extensional fractures (joints and veins), and small-scale faults. Two end-member mechanisms of thrust-related folding can be identified: active-hinge folding and fixed-hinge folding. The spatial distribution of longitudinal deformational features (LDF) induced by fixed-hinge folding is confined within the axial-surface zone, and its intensity is roughly proportional to the fold interlimb angle. Conversely, the spatial distribution of LDF induced by active-hinge folding is arranged in well-defined rock volumes—the deformation panels, each of which is affected by a characteristic LDF intensity. Deformation panels result from the development and interference of deformation domains, which correspond to the rock volumes that underwent deformation during their rolling across the active axial surfaces. The geometry of deformation domains is therefore defined by the architecture of the axial surfaces and the amount of fault displacement. The spatial distribution of deformation panels bears important insights for predicting the distribution of secondary permeability in folded carbonate reservoirs where, in the past, the permeability in the anticlinal crests has been overestimated. The development of deformation panels and deformation domains related to longitudinal deformational elements has been numerically simulated by a hybrid cellular automata (HCA) modeling technique for fault-bend and décollement folding. Model results show that, in many cases, active-hinge anticlines have a crestal zone that is unaffected by folding-related, longitudinal deformational features, and that deformation panels develop along the corresponding limbs. Our numerical results compare favorably with the spatial distribution of deformation in field examples and in analog models. The numerical technique we developed provides an efficient tool for evaluating the spatial distribution of fracture porosity and permeability in folding-related fractured reservoirs.

INTRODUCTION

Folding of rock multilayers in the brittle field is a well-known fracture-generating mechanism (Price, 1966; Hancock, 1985). Understanding the relationships between fold kinematics and the spatial distribution and geometry of folding-related deformational features is of primary importance for hydrocarbon exploration and development (Narr and Currie, 1982; Laubach, 1997). Anticlinal crests commonly are assumed to be the best sites for locating exploration and development wells. This is because fixed-hinge folding models (De Sitter, 1956) are used to predict the spatial distribution of deformation as a function of layer curvature (Murray, 1968; Decker et al., 1989; Schultz-Ela and Yeh, 1992; Lemiszki et al., 1994; Lisle, 1994; Hennings and Olson, 1997; Fischer and Wilkerson, 2000). However, in some cases, fold limbs and their transition to the crest have been found to be more productive and deformed than the crests themselves (Murray, 1968; Apotria et al., 1996) (Figure 1). Such a variability may depend on the fact that thrust-related folds in the brittle field develop following evolutionary paths in between two end-member kinematics: fixed-hinge and active-hinge folding (Stewart and Alvarez, 1991). Fixed-hinge folding implies no material migration across axial surfaces during folding, while active-hinge folding implies material migration across axial surfaces (Suppe et al., 1992). Both kinematics can produce folds with similar shapes (Storti and Poblet, 1997), but very different deformation patterns are expected (Fischer et al., 1992; Salvini and Storti, 1997, 2001).

Computer modeling has been proved to provide a very useful tool to simulate fault-fold kinematics (Egan et al., 1999) and their impact on growth strata geometries (Hardy et al., 1996; Allmendinger, 1998). Finite-element models have been used to address the distribution of deformation in active-hinge folds (Erickson and Jamison, 1995; Strayer and Hudleston, 1997; Erickson et al., 1999). However, results from this technique are influenced by the architecture of the undeformed mesh, by the poor resolution of the mechanical stratigraphy, which is imposed by computational limitations, and by the intrinsic difficulty in accounting for interlayer slip and multiple faulting.

In this chapter we introduce an innovative numerical modeling technique to illustrate the influence of fold kinematics on the two-dimensional distribution of deformation intensity induced by active-hinge folding. The case of fixed-hinge décollement folding is also presented for comparison. We have developed the hybrid cellular automata (HCA) numerical approach for overcoming limitations imposed by the finite-element technique. Our modeling is purely kinematic, and no geometric constraints are required, apart from the fault shape. Model outputs are used here to predict the location and intensity of longitudinal deformational features associated with simple step fault-bend folding (Suppe, 1983) and décollement folding (Jamison, 1987). Our results show that, in many cases, fold limbs are preferred sites for intense deformation. Validation from natural examples and experimental works supports our results.

DEFORMATION PATTERNS AND FOLD KINEMATICS

The deformation fabric associated with cylindrical folding includes structural elements trending almost perpendicular and parallel to fold axial surfaces (Hancock, 1985) (Figure 2). Development of the former deformational features is not univocally related to folding, because it has been interpreted as being the result of either layer-parallel shortening that precedes folding (Kilsdonk and Wiltschko, 1988) or hinge-parallel stretching that postdates folding (Srivastava and Engelder, 1990). Development of structural elements trending almost parallel to the axial surfaces (longitudinal deformational features, or LDF) is commonly related to folding (Srivastava and Engelder, 1990; Cooper, 1992; Fischer et al., 1992; Lemiszki et al., 1994; Storti and Salvini, 1996). In this chapter we refer only to LDF; structural elements trending almost perpendicular to fold axial surfaces are not included in the modeling procedure.

Regardless of the fold kinematics and folding mechanisms, the LDF-generating zones approximately coincide with the axial-surface zones and have a finite width. In active-hinge folding, migration of material through the axial-surface zones allows the progressive widening of the deformed rock volumes. In contrast, fixed-hinge folds are characterized by the location of the same rock volumes within the axial-surface zones, which progressively undergo stronger and stronger deformation (Fischer et al., 1992; Salvini and Storti, 2001).

The differing time-space distribution of deformational features in fixed- and active-hinge folding is illustrated in the model folds of Figure 3 (Salvini and Storti, 2001). This figure compares the evolution of a fault-bend fold (Suppe, 1983) and a fixed-hinge fold (or buckle fold, Biot, 1961). The fault-bend fold develops by bending and rolling of rock layers around an active axial surface; in the fixed-hinge fold, rocks bend within the fixed axial-surface

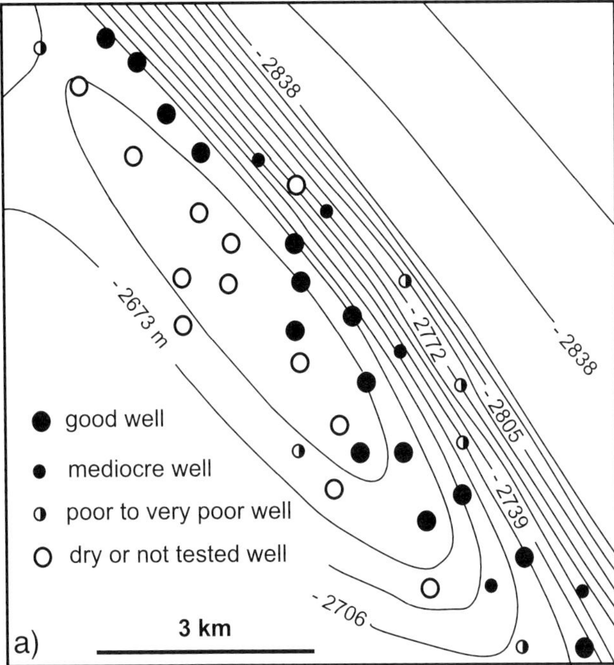

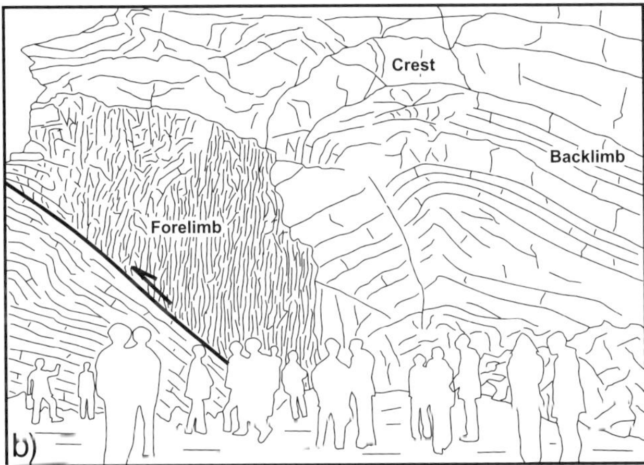

FIGURE 1. (a) Structural contour map of the Antelope Sanish Pool, McKenzie County, North Dakota (contours of top of Bakken Formation). Contour datum is sea level; contour interval is 16.5 m (after Murray, 1968). (b) Line drawing of an outcrop-scale thrust-related anticline, Hudson Valley Fold-and-thrust Belt, New York, U.S.A. (redrawn from the photo in Figure 1 of Woodward, 1999). Note the high deformation intensity in the forelimb, compared with the lower intensity in the crest and in the backlimb.

zone. In the early stages of folding (Figure 3c, d), deformed rocks lie close to the axial surfaces in both fold kinematics. The volumes of rocks affected by LDF have a comparable size, but deformation intensity is different: Rocks are weakly deformed in the buckle fold, whereas they are moderately deformed in the fault-bend fold. A characteristic feature of active-hinge folding is the asymmetric distribution of deformation with respect to the axial surface: Deformed rocks are confined to the limb above the ramp (Figure 3). With increasing contraction, the width of the deformed rock volume in the fault-bend fold increases by an amount corresponding to the displacement along the fault. Deformation intensity remains constant throughout the deformed zone. Conversely, the deformed rock volume in the buckle fold slightly increases, but remains localized in the fold-hinge zone. Deformation intensity is weak in the newly deformed sectors and increases to moderate and strong in the previously deformed sectors.

The late evolutionary stages of the two folds (Figure 3g, h) show the main differences between the two folding processes. The fault-bend fold has a wide deformed sector characterized by a homogeneous distribution of deformation intensity throughout the limb above the ramp. In contrast, the buckle fold is characterized by the presence of a heavily deformed narrow zone in the axial-surface sector, with deformation intensity that shows maximum values at the axial-surface pin and decreases toward the limbs. Where deformation is strong, its intensity exceeds that of the fault-bend fold. However, the volume of deformed rocks in the buckle fold is exceeded by the deformed rock volume in the fault-bend fold.

DEFORMATION DOMAINS AND DEFORMATION PANELS

Deformation patterns (Fischer and Jackson, 1999) in active-hinge folding can be conveniently described in

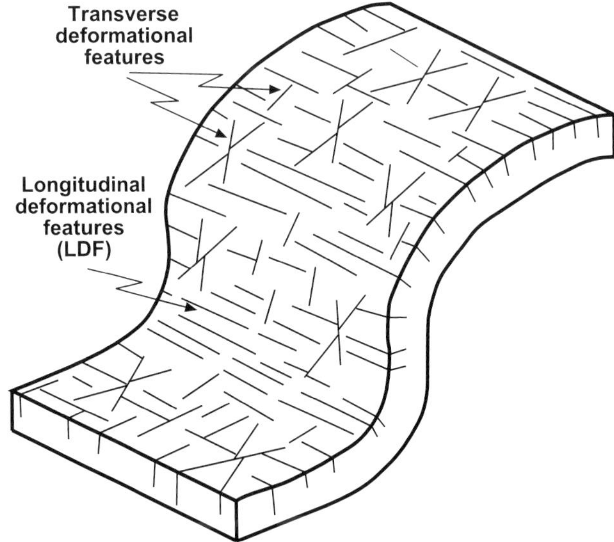

FIGURE 2. Geometry of folding-related deformational features commonly observed in thrust-related anticlines (after Cooper, 1992).

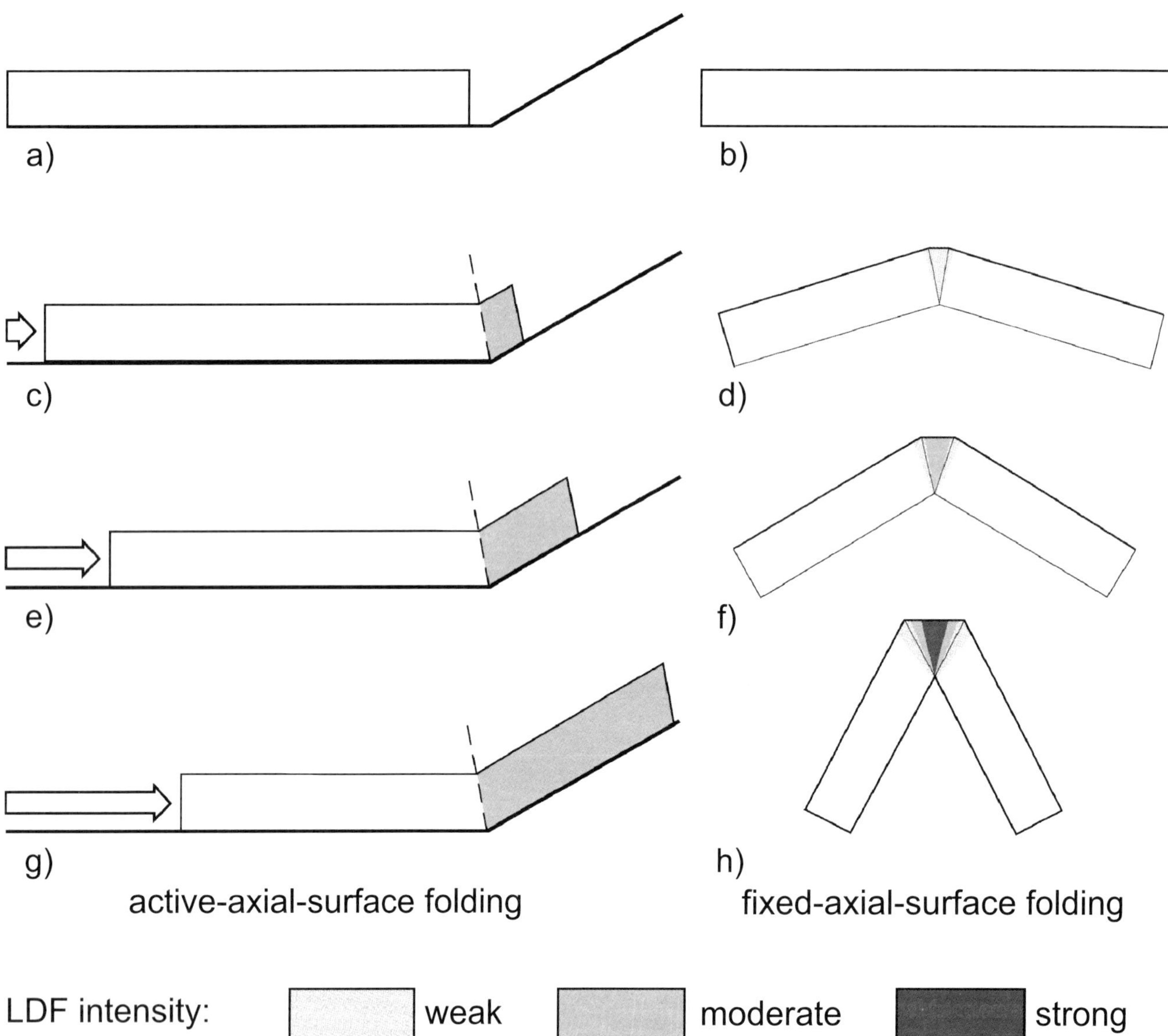

Figure 3. Comparison of the time-space evolution of longitudinal deformational features in an active axial-surface fold (a, c, e, g) and in a fixed axial-surface fold (b, d, f, h). Patterns indicate the location of deformed sectors and a qualitative estimate of deformation intensity.

terms of *deformation domains* and *deformation panels*. The first concept relates to the kinematics of the folding process, whereas the second concept describes the deformation fabric. A deformation domain is a rock volume that contains deformational features generated during a single bending event by rolling of layers about an active axial surface. This deformed rock volume extends from the generation zone (i.e., from the axial-surface zones) for a width that corresponds to the displacement of the rock panel passed through the active axial surface itself (Figure 4a). Following this definition, deformation domains are expected to be characterized by a homogeneous distribution of deformation intensity, apart from possible preexisting deformational features. Variations in the deformational features and intensity will depend on (1) variations in the rheology (i.e., on mechanical stratigraphy, Gross et al., 1995; Erickson, 1996; Fischer and Jackson, 1999) and/or on (2) variations in the preexisting deformation history. The 3-D geometry of a deformation domain can be predicted once the geometry of the generating active axial surface, the folding mechanism, the rock rheology, and the amount of fault displacement are known.

Deformation panels are rock volumes that have undergone the same deformation history (Figure 4a). Variations in the deformation style within each panel will depend only on corresponding variations in the rheology and boundary conditions (i.e., on overburden).

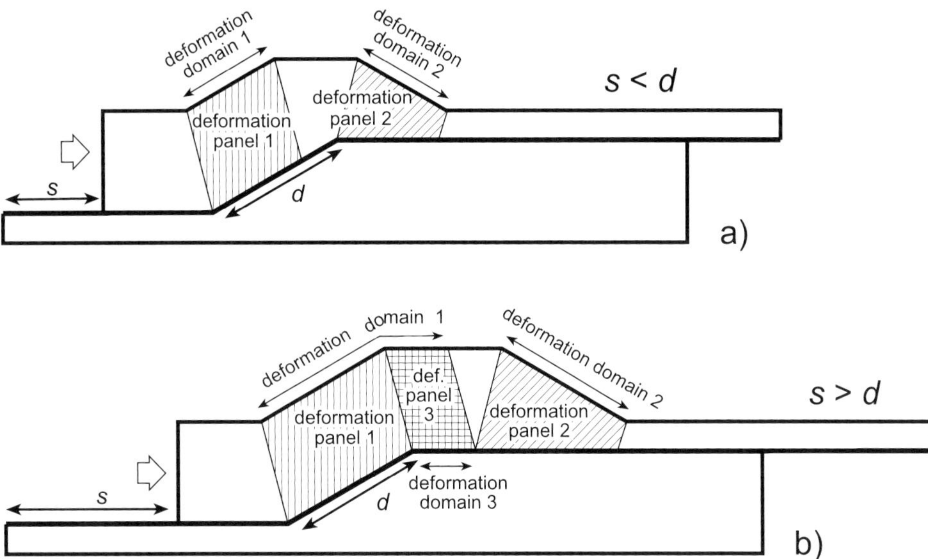

FIGURE 4. Evolution of deformation domains and deformation panels in a fault-bend anticline. (a) When fault displacement is less than the total length of the hanging-wall ramp, deformation domains (rock volumes that contain deformational features generated during a single bending event by rolling of layers about an active axial surface) and deformation panels (rock volumes that underwent the same deformation history) coincide. (b) When fault displacement exceeds the total length of the hanging-wall ramp, a new deformation panel develops in the crestal region (deformation panel 3) by refolding of material from the backlimb. The new panel still remains within deformation domain 1.

Deformation domains coincide with deformation panels until the former do not overprint to each other. When rocks belonging to a deformation domain roll about a second active axial surface and enter a new deformation domain, the overprinting between these two adjacent domains produces an additional deformation panel that includes deformational features related to both folding events. In the fault-bend anticline of Figure 4b, deformation domain 3 forms and interferes with deformation domain 1 when their relative distance, *d*, is shorter than the displacement, *s*. This overlapping produces three deformed rock volumes in the hanging wall that have different deformation histories (both in time and space): (a) rock belonging to deformation domain 1 and deformation panel 1; (b) rock belonging to deformation domain 2 and deformation panel 2, and (c) rock belonging to both deformation domains 1 and 3 and forming deformation panel 3.

It is worth noting that deformation domains may include deformation panels with different deformation histories, as is the case with deformation domain 1 in Figure 4b.

DEFORMATION STYLES

Deformation styles during folding depend on the interplay among three main factors (Fischer and Jackson, 1999): the folding mechanism (Price and Cosgrove, 1990), the rheology of the multilayer, that is, the mechanical stratigraphy (Woodward and Rutherford, 1989; Gross et al., 1995; Fischer and Jackson, 1999), and the stress conditions, including overburden and pore-fluid pressure (Jamison, 1992; Lemiszki et al., 1994). The first two factors are intimately related. A thick sequence of competent rocks is likely to undergo tangential-longitudinal deformation with tensile radial fractures in the outer arc region and pressure-solution cleavages in the inner arc region (Figure 5a) (Ramsay and Huber, 1987). Alternatively, it may deform by hinge-parallel shear, characterized by shear fracturing parallel to the active axial surfaces (Figure 5b) (Srivastava and Engelder, 1990; Cooper, 1992). The occurrence of one deformation mechanism rather then the other is determined by the stress conditions during folding. An increase of the confining stresses may inhibit and possibly suppress tangential-longitudinal deformation (Lemiszki et al., 1994).

A well-bedded rock multilayer consisting of alternating competent and incompetent beds is likely to deform by layer-parallel shear (or flexural-slip folding; Donath and Parker, 1964). A more complex deformation pattern is expected for flexural-slip folding, resulting from the concurrent action of tangential-longitudinal stresses in the competent layers, and layer-parallel shear stresses in the incompetent layers. The corresponding deformation may include formation of a pressure-solution cleavage system inclined to bedding, a tensile fracture system perpendicular to bedding, a system of en echelon tensile fractures inclined to bedding, and shear fracture systems inclined to bedding (Figure 5c) (Hedlund et al., 1994; Jamison, 1997; Fischer and Jackson, 1999; Gutièrrez-Alonso and Gross, 1999).

Depending on the distribution of the mechanical units within the deforming multilayer, different folding mechanisms can operate sequentially or coevally, thereby producing complex arrays of mesostructures (Fischer and Jackson, 1999). The change from layer-parallel to hinge-parallel strain maxima moving from anisotropic to isotropic rocks has been reproduced in finite-element models (Erickson et al., 1999).

THE HCA MODELING TECHNIQUE

The Numerical Multilayer

The forward-modeling algorithm we developed (hybrid cellular automata, or HCA) is a hybrid methodology between the cellular automata (CA) and the finite-element method (FEM) philosophies. From the CA method (Wolfram, 1986) it follows the principle of using large amounts of cells to simplify links. On the other hand, each point is linked by physical and geometric laws to the surrounding ones, thus following the FEM philosophy (Zienkiewicz and Taylor, 1991).

Layered rock units are simulated by a mesh of a very large number of semi-independent cells (Figure 6). Cells are related to the surrounding ones by three types of links that simulate the natural behavior of layered rocks: (1) intralayer relations, (2) interlayer relations, and (3) discontinuity relations. Intralayer relations link cells belonging to the same layer. They consist of rigid relationships derived from volume and surface preservation conditions, physical boundary conditions, and rock rheology. Interlayer relations regulate the relationships among adjacent layers. These relations take into account the weaker rheologies of interlayer material, physical-boundary conditions, and volume-preservation conditions, whereas partial freedom is given to surface variations. Discontinuity relations correspond to the presence of ruptures such as faults. No kinematic links exist across them, but physical boundary conditions and slip-induced stresses do. The relative position of each cell changes at each step, and these relations are recomputed only among adjacent cells. Preservation of volumes and surfaces is simulated by using large amounts of cells and preserving the average distance among adjacent cells.

Following the above-mentioned relations, the computation of the next position, P, of the i cell at time $t + \Delta t$, is

$$P(i, t + \Delta t) = P(i, t) + \Sigma_k\{\Sigma_j[F(j)\Delta t^2/D^2(i, k, t)]\} \quad (1)$$

where Δt is the step increment of time, k is the pointer to the surrounding cells, $F(j)$ are the links between i and k cells, and $D(i, k, t)$ is the distance at instant t between i and k cells.

The models run in a forward-modeling fashion, and a pace is selected that is slow enough to minimize the influence of nonadjacent cells. This reduces the links to first-order computations. Because the influence among cells is exponentially reduced by their relative distance, the pace is properly selected to limit the computation to closer cells, according to the CA philosophy. A multilayer of model beds with identical properties constitutes a mechanical unit. A multilayer of different mechanical units constitutes the mechanical stratigraphy.

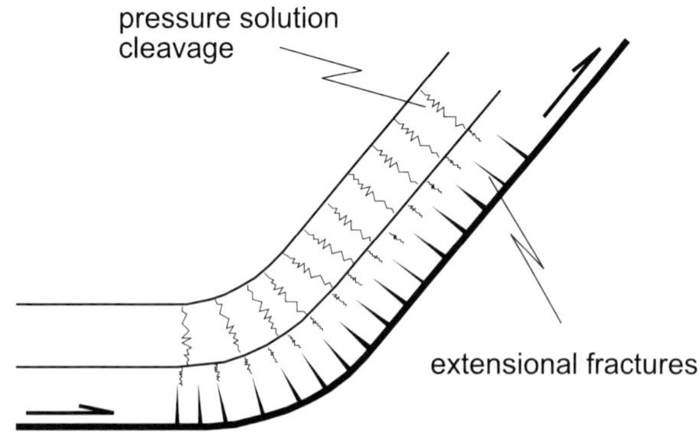

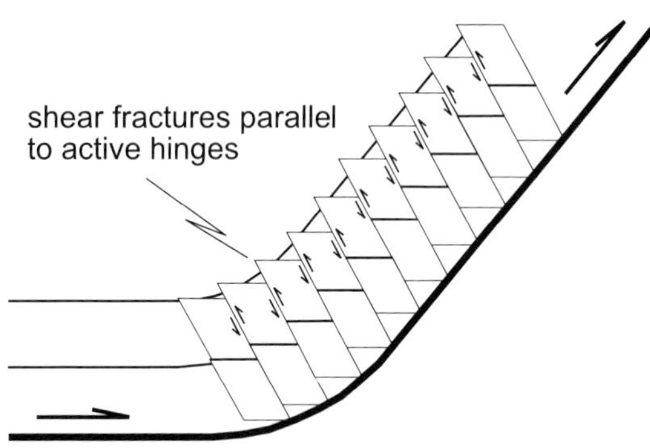

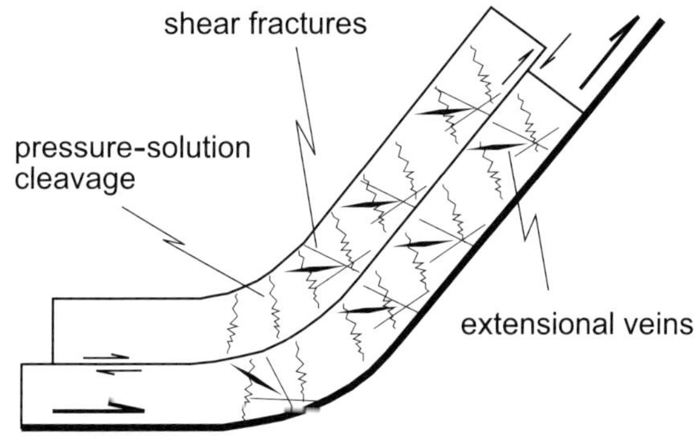

FIGURE 5. Different deformation styles occurring at active axial surfaces, depending on stress boundary conditions and rock rheology.

The physical parameters and types of links within the mechanical stratigraphy can be modified by the local conditions at any step during the run. For this reason, we call (1) a self-determining algorithm. In this way, faults are directly traced on the model section, and we can simulate fault propagation, multiple synchronous faulting, and synkinematic sedimentation. No other conditions are imposed to the model and the hanging wall/s "self-decide" their evolutionary pathway while obeying the physical boundary conditions.

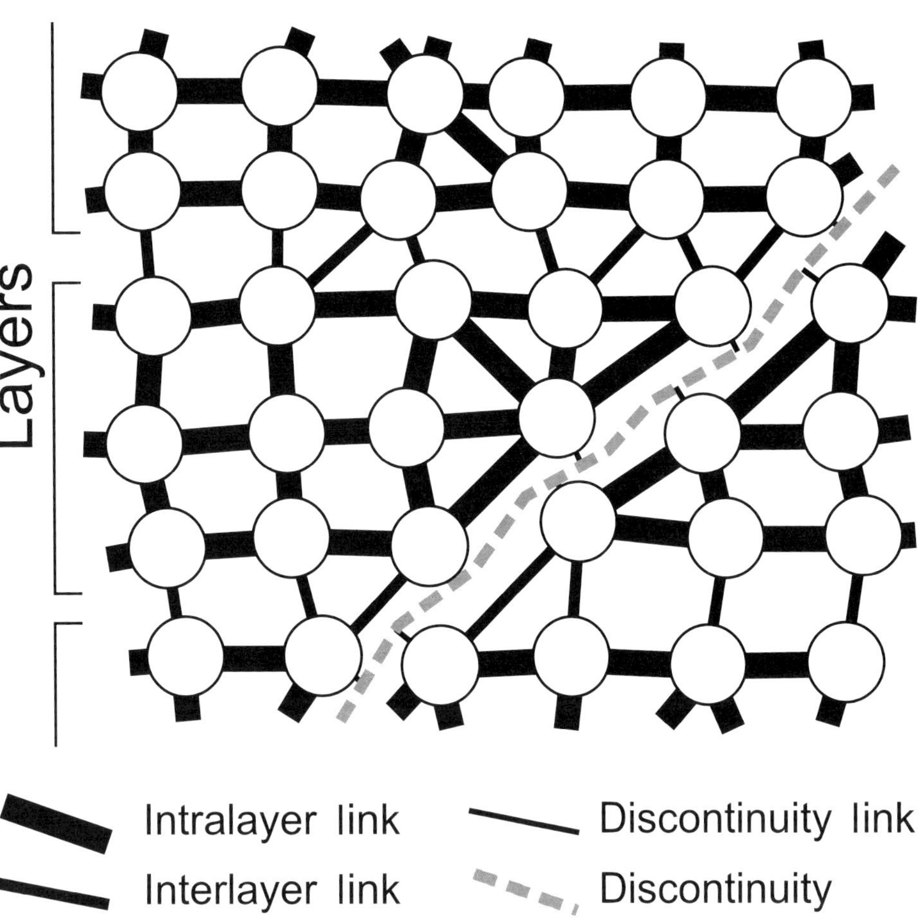

FIGURE 6. Conceptual sketch to show the different types of links that occur in a hybrid cellular automata multilayer. Cells are related to the surrounding ones by three types of links: (a) intralayer relations; (b) interlayer relations, and (c) discontinuity relations. Intralayer relations link cells belonging to the same layer, interlayer relations regulate the relationships among adjacent layers, and discontinuity relations correspond to the presence of ruptures such as faults. No kinematic links exist across them, but physical boundary conditions and slip-induced stresses occur.

Because of the large number of cells, typically on the order of hundreds of thousands for a medium-high-resolution experiment, results are independent of the location of each individual cell. If we draw the proper segments, the output may resemble typical FEM models and produce easy-to-comprehend results. The conceptual architecture of the modeling tool is independent of the fold kinematics and geometry, and also can simulate complex tectonic evolutionary paths, including out-of-sequence thrusting and refolding.

Computation of Deformation Intensity

The self-determining algorithm used in the modeling allows us to compute, at each step, the stress and (brittle) deformation conditions at each cell. Stresses at each step result from the combination of boundary conditions (including overburden) and stresses induced by the kinematics (Figure 7). Three kinematic effects were considered for computing the latter component: (1) torsion-induced fiber stress; (2) flexural slip- (i.e., interlayer slip-) induced stress; and (3) slip along the discontinuities. The fiber stress, σ_f, is computed as Turcotte and Schubert (1982) did it

$$\sigma_f = Eh/[2(1-\upsilon^2)]d^2\Theta/dt^2 \qquad (2)$$

where E and υ are the Young and Poisson modules of the rock, h is the average layer thickness, and Θ is the layer dip. Flexural-slip-induced stress relates to the rheology of the interlayer material and the amount of relative displacement among adjacent layers. It can be derived from general stress/strain equations (Turcotte and Schubert, 1982) and is computed as

$$\tau_f \equiv \sigma_f = E\varepsilon/(1+\upsilon) \qquad (3)$$

where τ_f and σ_f are the maximum shear and stress generated in the layer by interlayer slip, and ε represents the elastic deformation of the layer

$$\varepsilon = [(h^2 - har + {}^1/_2\, a^2r^2)/h - 1]^{1/2} \qquad (4)$$

where r is the observed flexural slip displacement and a is a factor (lower than 1) that takes into account the rheology and strength of the interlayer material. Similar equations are used to compute the deformation associated with the slip along active discontinuities. Failure conditions for each cell are computed by comparing, at each step, the desired stress conditions, with the cell strength S as derived by the Coulomb failure criterion

$$S = c + k\sigma_{eff} \qquad (5)$$

where c is the cohesion, k is the friction coefficient, and σ_{eff} is the effective stress on the considered surface (Storti et al., 1997). Results are then used to modify the

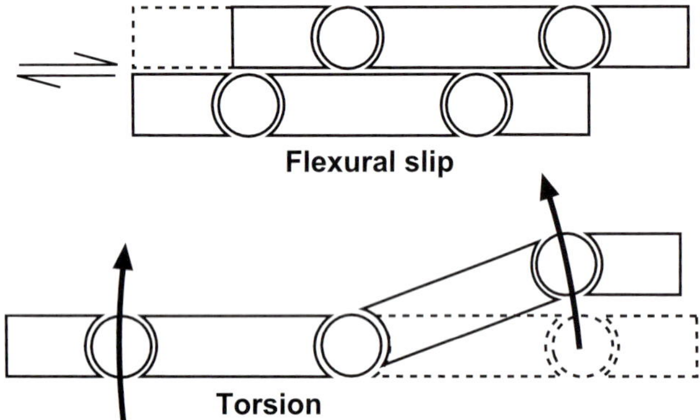

FIGURE 7. Cartoon showing the simulation of flexural slip and torsion by hybrid cellular automata. Flexural slip is computed by differential motion across selected interlayer links. Torsion is computed by differential rotation of adjacent, intralayer links.

cell rheology and accumulated to compute the deformation history. By varying the rheology, the elastic parameters, and the bedding geometry for each cell, it is possible to replicate the mechanical stratigraphy on the model's multilayer as well as along-strike variations.

MODELING THE DISTRIBUTION OF LDF INTENSITY IN SIMPLE STEP FAULT-BEND FOLDING

Fault-bend anticlines develop by the passive adjustment of hanging-wall rocks during their climbing up above nonplanar thrust surfaces (Rich, 1934). The evolution of a fault-bend anticline progresses through two main stages (Suppe, 1983). The first stage is characterized by the upward translation of the hanging wall ramp above the footwall ramp. Material rolls across two active axial surfaces, respectively pinned at the lower and upper inflection points of the fault (Figure 8a, b). A fixed axial surface limits the backlimb upward and another fixed axial surface limits the forelimb toward the foreland. A deformation domain/panel develops in the backlimb and another one in the forelimb, whereas the flat-lying crest remains virtually unaffected by LDF. In the second evolutionary stage, the hanging-wall ramp is translated forelandward along the upper décollement (Figure 8c, d). The previously fixed axial surface at the top of the backlimb reaches the upper ramp inflection point and becomes active, deforming material that migrates from the backlimb itself and enters the upper décollement. A new deformation domain/panel consequently develops in the inner side of the crest (Figure 8c). The former active axial surface at the upper inflection point becomes fixed and is passively translated forelandward, limiting the forelimb toward the hinterland (Suppe, 1983). The opposite dips of the two axial surfaces limiting the crest of the anticline causes the preservation of an inverted triangle adjacent to the forelimb made up by rocks virtually unaffected by LDF. The late-stage evolution of a fault-bend anticline is characterized by the progressive widening of the crestal deformation panel, whereas the width of deformation panels in the forelimb and in the backlimb remains constant (Figure 8d).

DEFORMATION IN A NATURAL FAULT-BEND ANTICLINE: THE GUADAGNOLO ANTICLINE

The Guadagnolo Anticline, in the central Apennines, Italy, developed in Pliocene times by fault-bend folding along a splay fault of the Olevano-Antrodoco thrust system (Salvini and Vittori, 1982; Bigi et al., 1990). The Olevano-Antrodoco thrust front characterizes the central Apennines with a north-south trend for a length of more than 80 km. It consists of the eastward overthrust, over the coeval Latium-Abruzzi carbonate platform, of a basin-to-platform transitional sedimentary succession that is Mesozoic to Late Cenozoic in age. The Guadagnolo Anticline lies 15 km west of the frontal thrust and involves the upper part of the transitional succession. The basal décollement is located at the base of Upper Cretaceous limestones and the upper décollement is at the base of upper Miocene clayish marls. Despite the complex structural evolution of the central Apennines, the Guadagnolo Anticline formed during a single tectonic event and is not significantly affected by later tectonism.

The frontal part of the Guadagnolo Anticline is shown in the cross section of Figure 9. The geometry of the anticline is relatively simple, and rocks are well exposed, thus providing the opportunity for testing the basic criteria of active-hinge folding. Most field analysis sites were located within the lower Miocene limestones that outcrop from the backlimb to the crest. The use of the same mechanical unit allowed us to avoid complications related to the influence of the mechanical stratigraphy (Gross et al., 1995). A comprehensive description of the three-dimensional architecture of deformation panels and domains and deformation styles in the Guadagnolo Anticline is beyond the scope of this chapter. Data presented here provide only a representative overview of the folding-related deformation pattern.

Pressure-solution cleavage is the best expressed and ubiquitously distributed deformation feature in the Guadagnolo Anticline. Solution seams typically are oblique to bedding and have a slightly curvilinear shape (Figure 10). Most cleavage surfaces are confined within bedding, that is, they are intrabed features. The difference between deformation intensities in the limbs and in the crest is best described by the spacing analysis of the

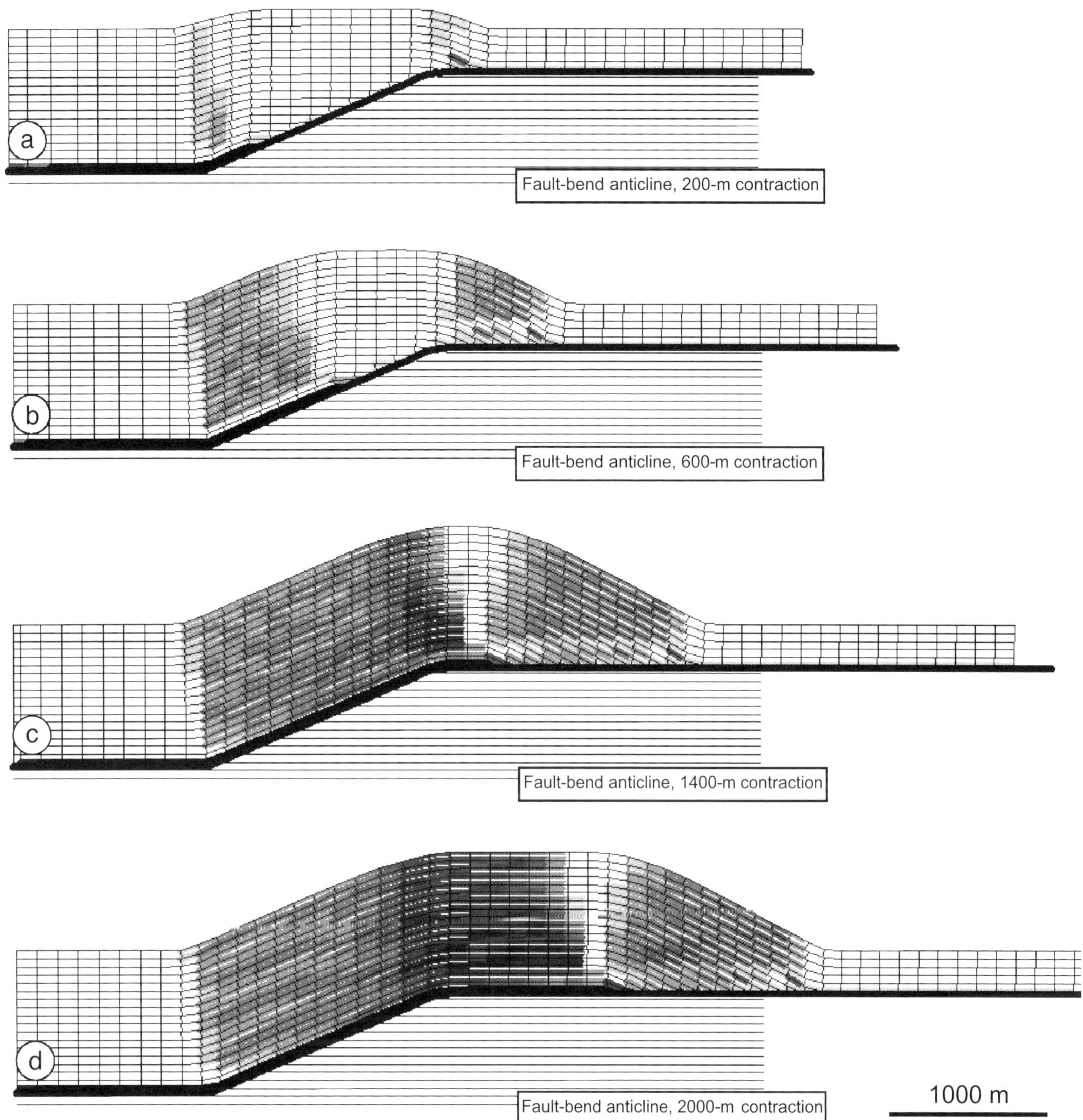

FIGURE 8. Hybrid cellular automata model of the evolution of deformation panels in a fault-bend anticline. Grey shade intensity is proportional to the intensity of deformation. (a, b) When fault displacement is significantly less than the length of the hanging-wall ramp, two deformation domains are clearly visible in the fold limbs, separated by a large, virtually undeformed crestal sector. (c, d) As the hanging-wall ramp entirely passes the upper ramp hinge, a new deformation domain develops in the inner crestal zone. Its interaction with the backlimb deformation domain produces a third, highly deformed panel. The width of the virtually undeformed crestal sector reduces to a narrow triangular zone at the transition to the forelimb.

pressure-solution longitudinal cleavage (Figure 9). Cleavage spacing in the backlimb is about 30–35 mm, as well as in the forelimb, where it lowers to about 20 mm close to the forelandward hinge. In contrast, the average cleavage spacing in the crest is varied, with an average of about 1200 mm. The first-order deformation pattern in the Guadagnolo Anticline consists of two distinctive deformation panels that coincide with corresponding deformation domains ($s \equiv d$), one in the backlimb and the other in the forelimb (Figure 11).

The panoramic view of the different sectors of the Guadagnolo Anticline illustrates the deformation pattern

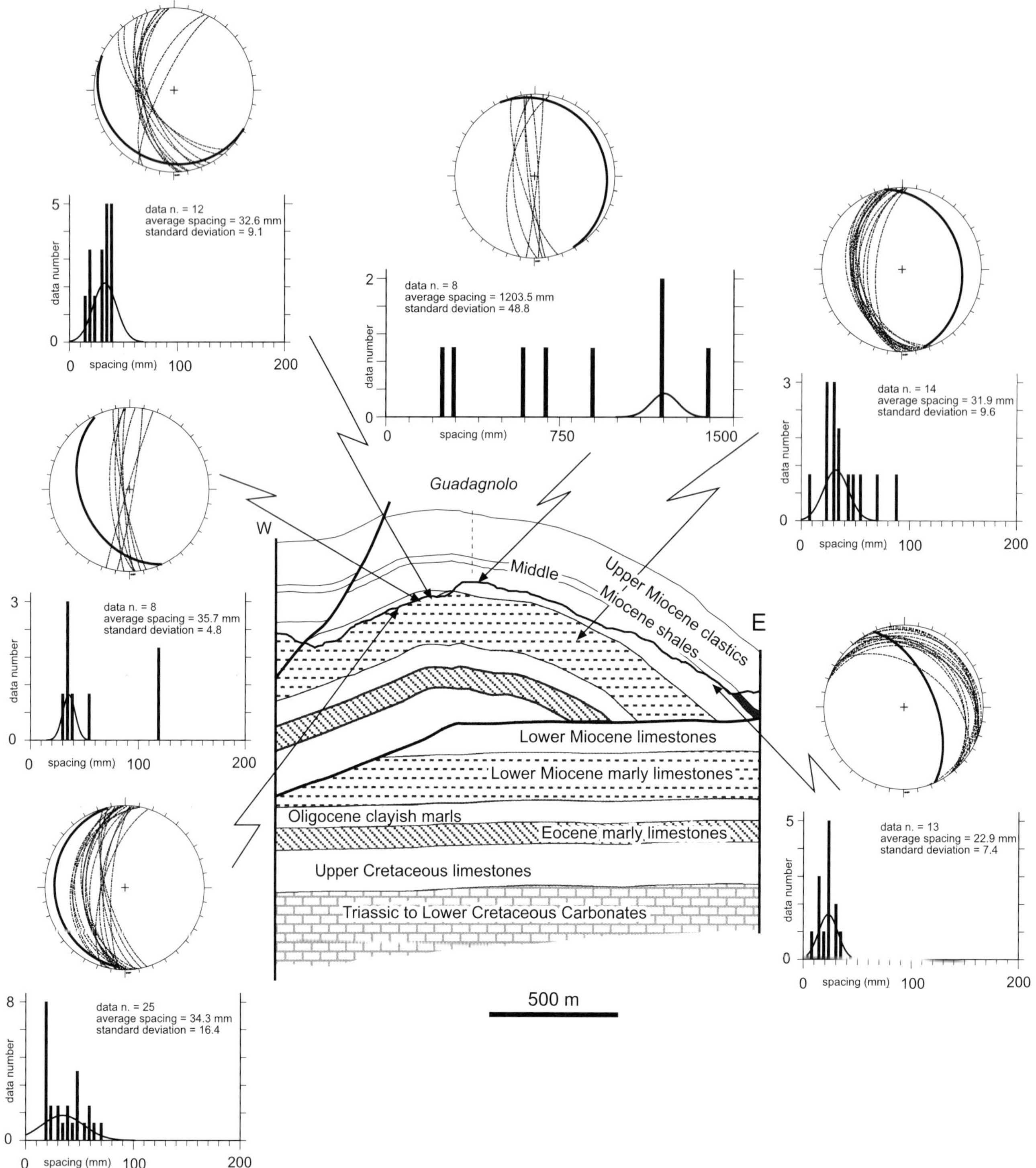

FIGURE 9. Mesoscopic deformation analysis of the Guadagnolo Anticline, 40 km east of Rome, central Apennines, Italy. Center shows a geological cross section of the structure. Results from representative field analysis sites are shown both as stereonets and histograms. Stereonets (Schmidt projection, lower hemisphere) indicate the attitude of pressure-solution cleavage (dashed circles) and bedding (unbroken, thick circles). Histograms show distribution of cleavage spacing. A Gaussian best-fit curve indicates the average values and standard deviation in each field site.

at a scale directly comparable with the resolution of the reflection seismic lines (Figure 11). Fracturing is highlighted by recent karst landscape evolution. The backlimb of the anticline appears to be moderately deformed. Bedding surfaces are clearly evident and gently dip toward the west. The crest is poorly deformed, as indicated by its massive appearance resulting from the minor karst evolution despite its morphological position.

FIGURE 10. Detail of the main folding-related cleavage set in the backlimb of the Guadagnolo Anticline. Note that cleavage surfaces are mostly confined within bedding and are oblique to bedding.

Deformation increases progressively entering the forelimb. The forelimb shows the greatest degree of damaging, that is, the greatest deformation intensity.

VALIDATION FROM ANALOG MODELS

Laboratory models provide dynamically scaled analogs that are useful for a direct comparison with numerical models. The evolution of deformation during laboratory simulations of fault-related folding, using rocks as analog materials, shows a well-defined spatial distribution of folding-related fracturing. Chester et al. (1991) simulated the evolution of thrust related anticlines developing by two different kinematics (Figure 12a, b). In both cases, anticlinal crests are much less fractured than the corresponding limbs. Fractures are organized in two well-defined deformation panels, in the backlimb and in the forelimb, respectively. Apart from their orientation, and depending largely on rock rheology and model scaling, note the geometric correspondence between the spatial distributions of fractures in these analog models and the distribution of deformation predicted in our numerical models (Figure 8). Similar considerations arise from Morse's (1977) experimental results, which show deformation of thin sandstone layers overlain by a thick layer of limestone (Figure 12c). Experimental conditions inhibited deformation in the thick limestone layer. In the sandstone multilayers, a well-defined deformation panel developed in the backlimb during translation of the hanging wall over the thrust ramp. The crestal region is much less deformed, and fracture intensity increases again in the forelimb.

In the experimental configuration designed by Patton et al. (1995), initially flat, homogeneous sand layers were translated over a rigid footwall ramp. The evolution of a fault-bend anticline progressed by the development of mainly extensional kink bands in the forelimb and compressional kink bands in the backlimb, while the crestal region remained almost undeformed (Figure 12d). Apart from the nature of these deformational features, which depends on the experimental technique and analog material, the deformation pattern is again similar to that obtained in the HCA models.

COMPARISON WITH FINITE-ELEMENT MODELS

Finite-element models of the deformation pattern induced by fault-bend folding generally show the spatial distribution of plastic strain. The motion of a homogeneous hanging wall above a rigid footwall (Patton et al., 1995) shows the localization of deformation at the fault-bend inflection points and the subsequent translation of highly strained regions along the ramp and upper flat (Figure 13). These highly strained regions correspond to the backlimb and forelimb deformation panels in our HCA models (Figure 8).

MODELING THE DISTRIBUTION OF DEFORMATION INTENSITY IN DÉCOLLEMENT FOLDING

Constant-limb-dip Décollement Folding

Décollement (detachment) folding is geometrically less constrained than either fault-bend or fault-propagation folding. This has led to the development of many geometric and kinematic solutions for décollement folds (De Sitter, 1956; Jamison, 1987; Dahlstrom, 1990; Epard and Groshong, 1995; Poblet and McClay, 1996; Homza and Wallace, 1997), with constant-limb-dip and constant-limb-length folding being two end-member models.

Constant-limb-dip décollement folding (Dahlstrom, 1990) is a self-similar kinematic model in which the limb dip remains constant during the evolution of the fold, as does the width of the anticlinal crest. Fold amplification occurs by rolling of material across the two active axial surfaces that limit the limbs outward (Poblet and McClay, 1996).

Hybrid cellular automata (HCA) models in Figure 14 illustrate the evolution of a constant-limb-dip décollement fold. The only geometric constraint is the preservation of the limb dips at the base of the model. During fold amplification, two deformation domains, or panels, develop in the backlimb and in the forelimb, respectively, and their lengths coincide with the length of the

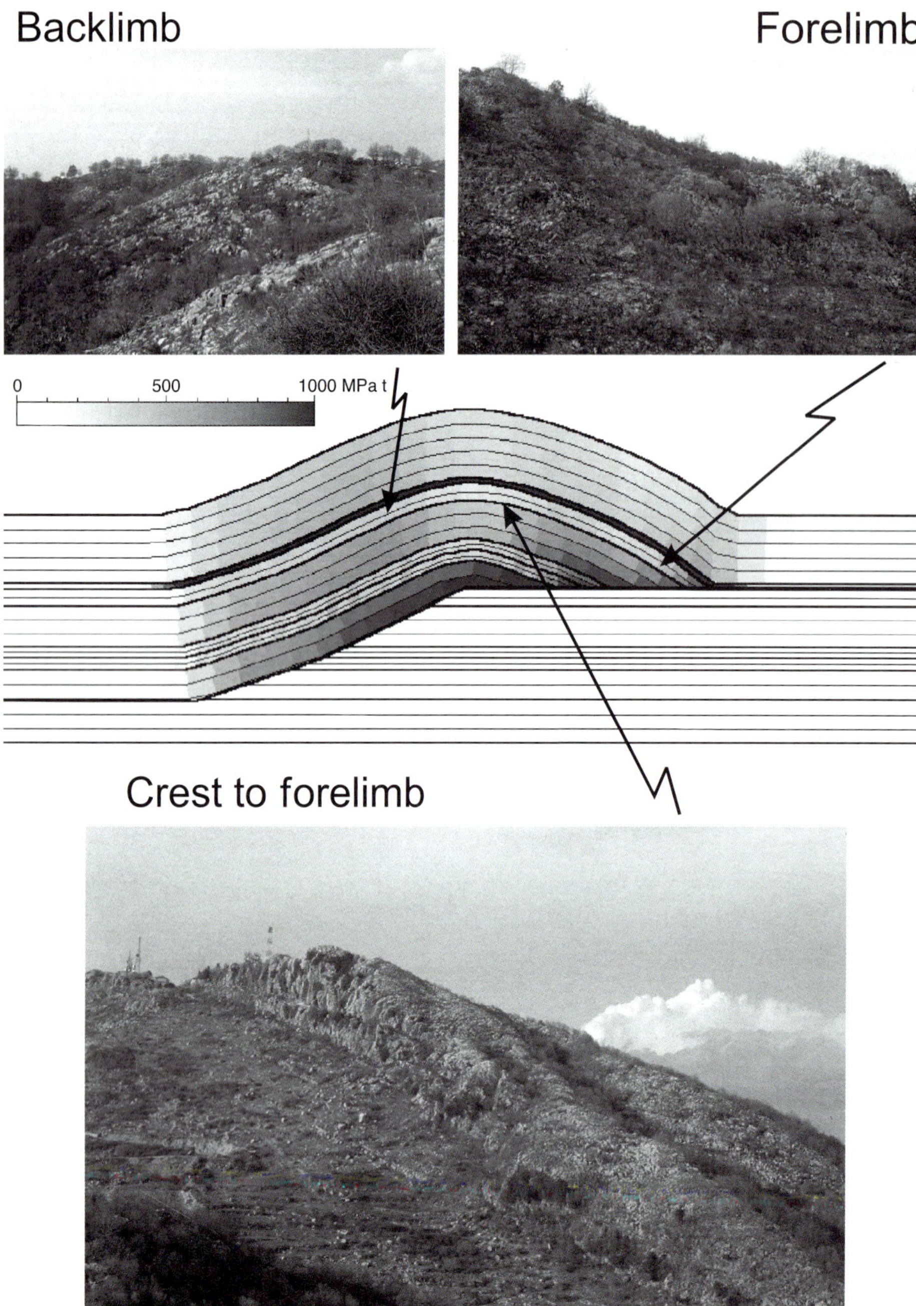

FIGURE 11. Karst landscape associated to the deformation panels in the Guadagnolo Anticline. Note the relatively undeformed crest and the increase of damaging moving towards the forelimb (lower right corner). An HCA model of the deformation-panel distribution for the Guadagnolo Anticline is shown for comparison.

by the progressive rotation of the limbs around fixed axial surfaces. Initial limb lengths remain unchanged during fold amplification (De Sitter, 1956). This is the only geometric constraint in the hybrid cellular automata models of Figure 15. The deformation pattern associated with constant-limb-length décollement folding is almost opposite that which characterizes constant-limb-dip-décollement folding (compare Figure 15 with Figure 14). Deformation is confined in the axial-surface regions of the structure, and its intensity increases progressively with increasing limb dip. Fold limbs are less affected by folding-related deformation. This fold kinematic model therefore produces small deformation panels.

corresponding limb. The model evolution in Figure 14 shows that the deformation pattern of constant-limb-dip décollement folding is roughly self-similar and consists of two homogeneously fractured limbs separated by an anticlinal crest that is virtually unaffected by folding-related deformation.

Constant-limb-length Décollement Folding

Constant-limb-length décollement anticlines represent a typical case of fixed-hinge folding. They develop

DEFORMATION IN A NATURAL DÉCOLLEMENT FOLD: THE MONKSHOOD ANTICLINE

A quantitative description of LDF intensity in a décollement fold has been provided by Jamison (1997) for the Monkshood Anticline, in the Canadian Foothills. Longitudinal extensional fractures (veins) are the most common deformational features in this anticline. Two main features characterize their spatial distribution. (1) Fold hinges have the lowest fracture densities (Figure 16);

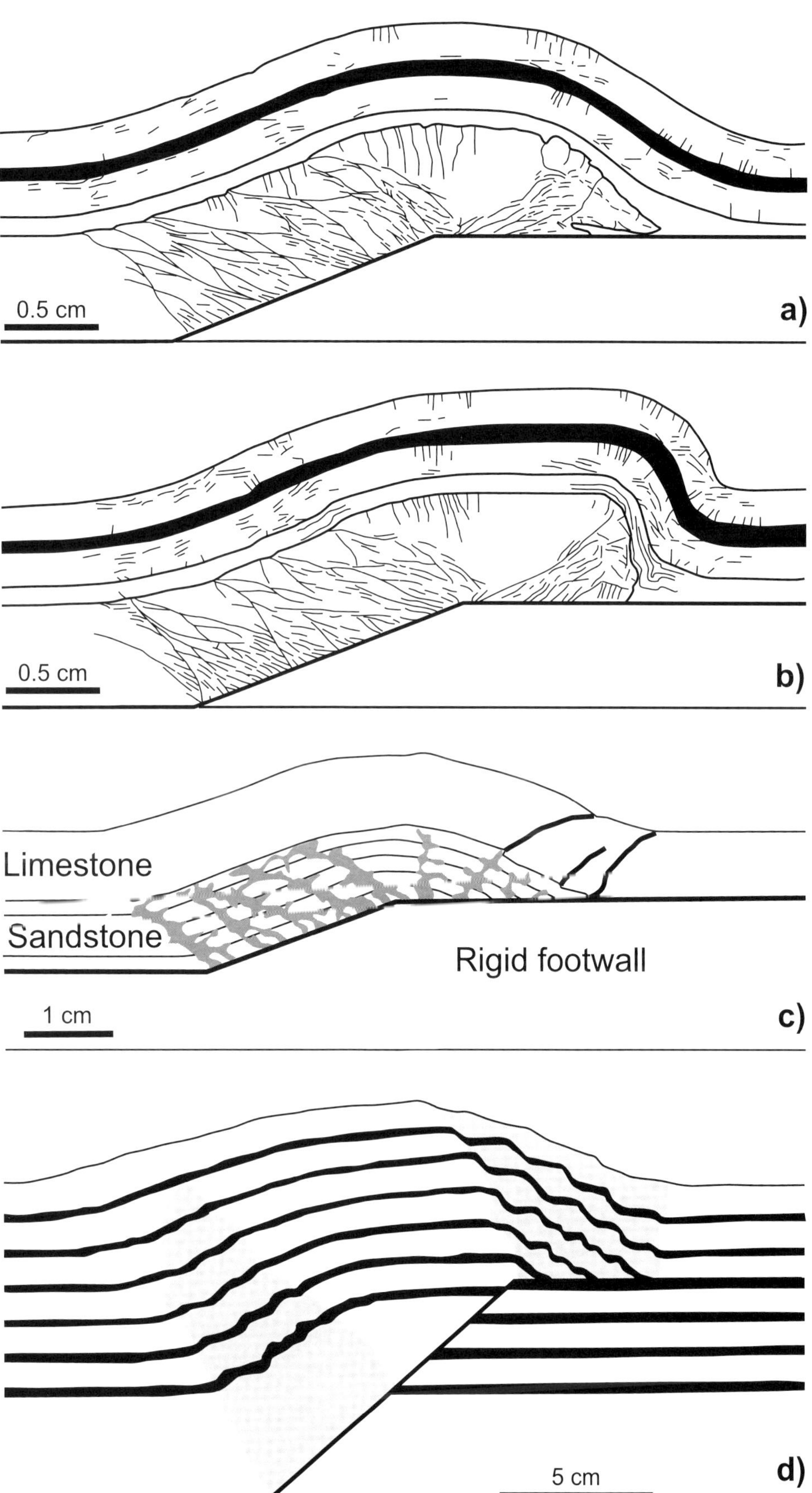

FIGURE 12. (a, b, c) Analog rock models showing the development of fault-bend anticlines and related deformational panels (a, b, after Chester et al., 1991; c after Morse, 1977). Note the two deformation panels developed in the limbs and the poorly fractured crests, as predicted by the numerical model in Fig. 8. (d) Analog sandbox model of a fault-bend anticline growing above a rigid footwall (after Patton et al., 1995). Two deformation panels (shaded areas) developed in the fold limbs, while the crest remains almost undeformed, apart from the transition to the forelimb. Compare with the model in Fig. 8.

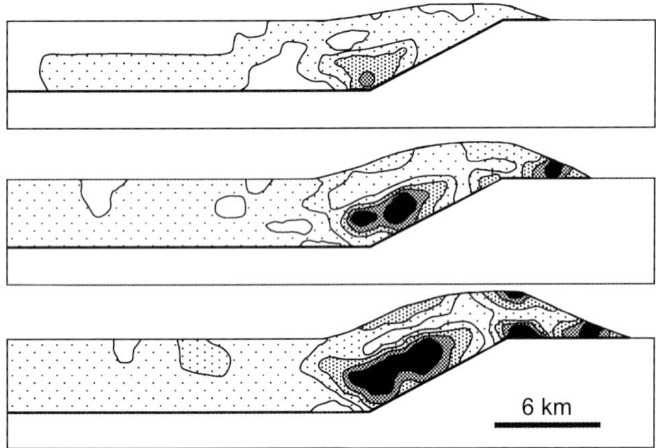

FIGURE 13. Finite-element model of the evolution of a fault-bend anticline by translating a homogeneous hanging wall above a rigid footwall (after Patton et al., 1995). The progressive development of two deformation panels in the fold limbs is clearly visible. Grey shades are proportional to the deformation intensity. Note the qualitative similarity with HCA models in Figure 8.

and (2) longitudinal fractures are inclined 5–15° toward the horizontal from the bedding normal. A possible explanation for (1) is the role of the regional tectonic stress to prevent fracturing in the hinge zones, whereas the occurrence of fracturing after some degree of folding can account for the fracture set (2) (Jamison, 1997). Our modeling results suggest that active-hinge folding may have occurred during the growth of the Monkshood Anticline. This fold kinematics can explain the presence of a slightly fractured to virtually unfractured crest (see Figure 14) and the development of inclined fractures, if we assume that they developed nearly parallel to the active axial surfaces during limb lengthening (e.g., Figure 5b).

DISCUSSION

For a fully constrained kinematic evolution of a thrust-related fold and associated LDF, the relationships among axial-surface activity, deformation domains, and deformation panels should be constrained. End-member fold kinematic models produce rather simple and diagnostic deformation panels, as previously illustrated in Figures 8, 14, and 15. Natural thrust-related folds develop according to a wide variety of kinematic paths where axial surfaces can be alternately active and fixed as the deformation proceeds (Beutner and Diegel, 1985; Stewart and Alvarez, 1991; Alonso and Teixel, 1992; Butler, 1992; Fischer et al., 1992; Edlund et al., 1994; Fisher and Anastasio, 1994; Vergés et al., 1996; Marrett and Bentham, 1997; Thorbjornsen and Dunne, 1997; Erslev and Mayborn, 1997; Homza and Wallace, 1997; Anastasio et al., 1997; Meigs, 1997; Tavarnelli, 1997; Poblet et al., 1998; Gutièrrez-Alonso and Gross, 1999). Moreover, fold kinematics may change with increasing contraction (Dixon and Liu, 1992; Storti et al., 1997), which introduces further complications in the final patterns of folding-related deformation panels. This is particularly true for structures located in tectonic wedges.

Thrust-related anticlines at the toe of thrust-fold belts are usually kinematically simpler, and simpler architectures of deformation panels are expected. In this case, HCA numerical model results can provide useful templates for predicting the architecture of deformation panels in subsurface folds, once their kinematics are inferred using, for instance, the analysis of growth-strata patterns (Suppe et al., 1992; Storti and Poblet, 1997).

The reverse approach, from the deformation pattern to fold geometry and kinematics, can also be used when the distribution of deformation panels can be inferred from field observations and subsurface data. Deformation domains can then be deduced by comparing the deformation style and intensity in the deformation panels. The geometry of deformation domains allows the identification of the active axial surfaces and the estimation of fault displacements.

IMPLICATIONS FOR HYDROCARBON EXPLORATION AND DEVELOPMENT

The presence of fracture systems can be a major factor for adequate permeability in carbonate reservoirs, such as the Val d'Agri oil field in the Southern Apennines. Fracturing and small-scale faulting have a primary effect on the deformation style of rocks in the foreland regions of thrust-fold belts (Wojtal, 1986; Wojtal and Mitra, 1986; Holl and Anastasio, 1992; Jamison, 1997). In many cases, overprinting between dilation and solution processes occurs on the same mechanical discontinuities (Geiser and Sansone, 1981; Wiltschko et al., 1985; Srivastava and Engelder, 1990), thereby favoring connectivity and permeability increase in hydrocarbon reservoirs. Recent work on carbonate reservoir rocks in the Ionian zone of Albania demonstrated that the main reservoir is not controlled by veins or joints, but that there is an important contribution related to stylolites (Van Geet et al., 1998). This is because the relevant stress conditions for hydrocarbon accumulation are not those related to the formation of the trap (when stylolites and joints developed) but instead are those that occur during hydrocarbon migration. The two stress fields can be significantly different, and stylolites can undergo dilation.

It follows that, when applied to hydrocarbon exploration and development, model outputs can be interpreted in terms of fracture intensity, and deformation panels can be regarded as fracture panels (Storti and

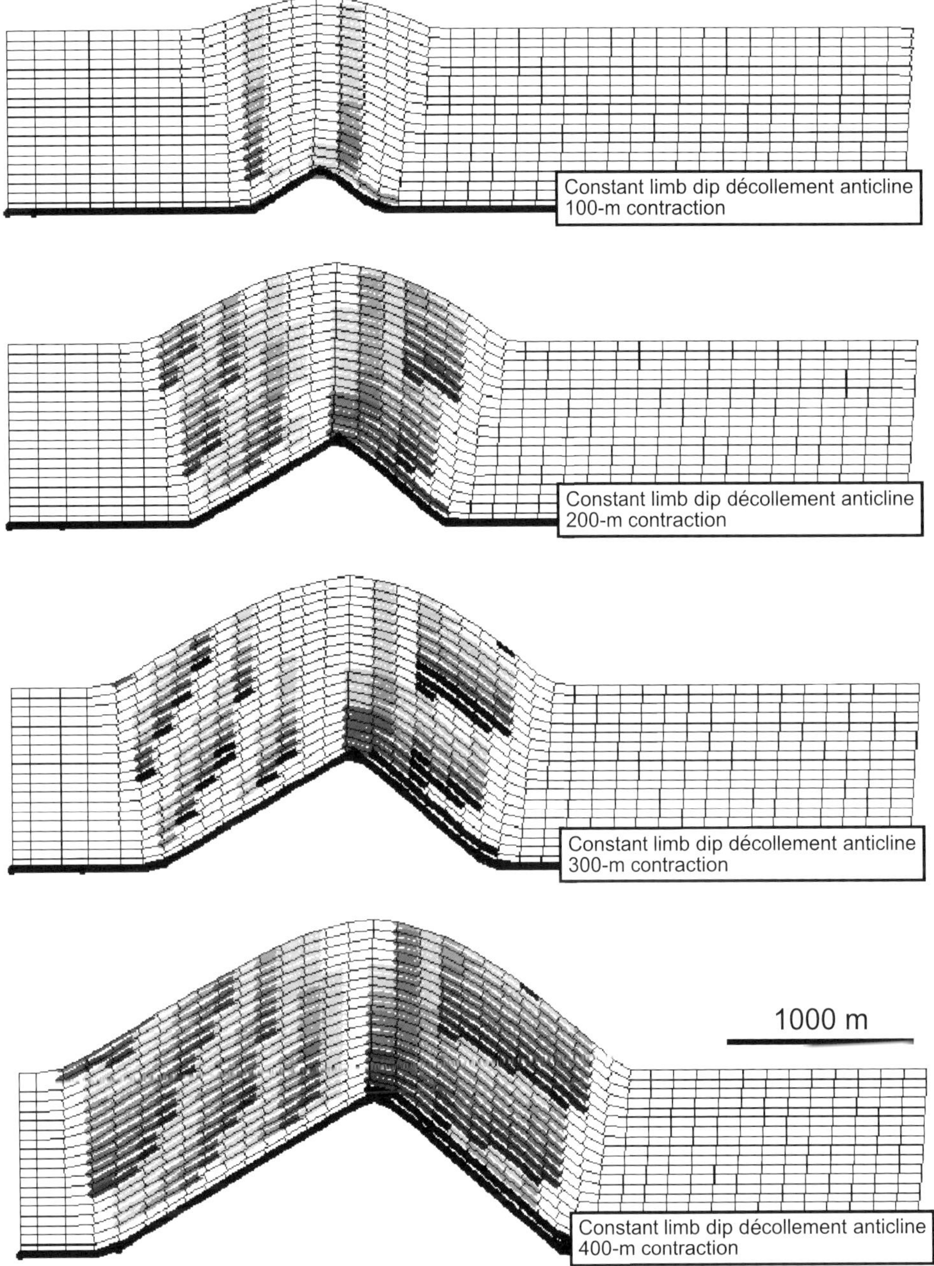

FIGURE 14. Hybrid cellular automata model of the evolution of deformation panels in a constant-limb-dip décollement anticline. Grey shade intensity is proportional to the amount of deformation. Two deformation panels develop in the fold limbs, as a result of the migration of material across the centrifugal migrating basal axial surfaces. A virtually undeformed, narrow panel is preserved in the anticlinal crest.

Salvini, 1996), if we use the term fracture in its broadest meaning, that is, to include all mesoscopic mechanical discontinuities produced by brittle deformation. Fracture panels describe the spatial distribution and frequency of longitudinal fractures (Mitra, 1987), that is, of fractures trending nearly parallel to fold axes and dipping perpendicularly or obliquely to bedding, (type 2 of Stearns, 1968). Longitudinal fracture intensity can be assumed to be proportional to the number of bending events (Sanderson, 1982) and to the angle and sense of bending (Storti and Salvini, 1996) that the material has suffered, at least until a fracture saturation intensity is reached (Becker and Gross, 1996).

The two-dimensional architecture of fracture panels predicted by our numerical models for different kinematics of active-hinge folding shows that anticlinal crests can be less fractured than the corresponding limbs. The evidence that fold limbs can provide the best fracture porosity and permeability conditions in reservoirs bears important implications for the location of exploration and development wells.

Some examples of ideal well locations for different kinematic mechanisms in models of thrust-related anticlines are shown in Figure 17. The most suitable well location sites in fault-bend anticlines are at the transition between the crest and the limbs (Figure 17a). For

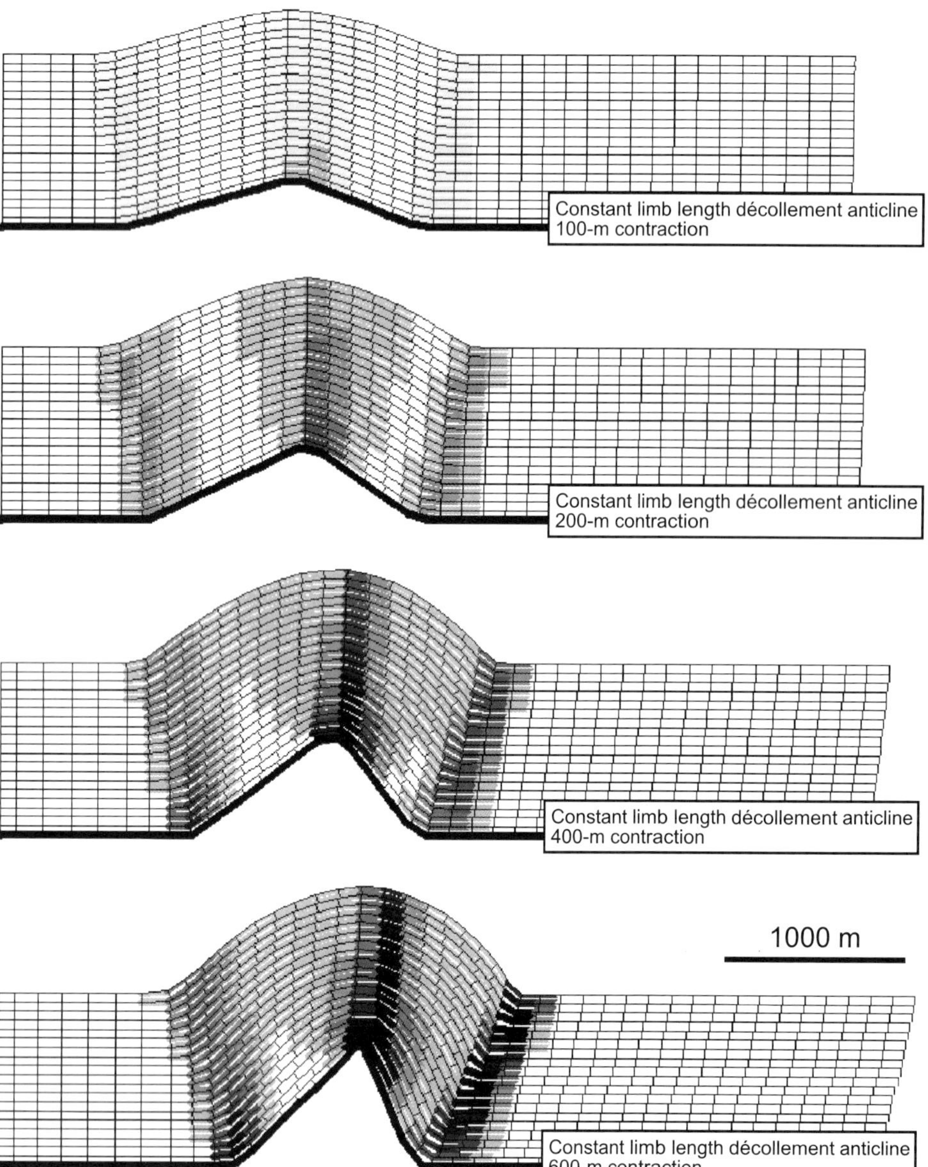

Figure 15. Hybrid cellular automata model of the evolution of deformation panels in a constant-limb-length décollement anticline. Grey shade intensity is proportional to the amount of deformation. Deformation panels develop at axial surfaces, separated by a less-deformed area in the limbs. The intensity of deformation increases with increasing contraction, but deformed material does not migrate away from the axial surfaces.

displacements greater than the ramp length, the fracture panel in the rear of the crest may represent the most productive sector of the reservoir. Constant-limb-dip décollement folding provides similar results (Figure 17b). The higher fracture intensity in the forelimb relates to its dip being steeper than that of the backlimb. In both cases, the crest-to-forelimb transitional region represents the best suitable drilling site. Figure 17c illustrates the case of a constant-limb-length décollement anticline that developed by fixed-hinge folding. In contrast to the previous examples, in this fold kinematic model, the crestal region represents the most suitable drilling site for hydrocarbon recovery (Figure 17c). The steeper dip of the forelimb results in a higher fracture density in the frontal portion of the crest.

CONCLUSIONS

The concepts of deformation domains and deformation panels associated with active-hinge folding impose important constraints on the presence and spatial distribution of deformational features in fault-related folding. Deformation-panel theory may be used to infer the kinematic mechanisms of folding, when the spatial distribution of deformed rock volumes is available.

We have developed an innovative mathematical tool, the hybrid cellular automata (HCA), to numerically predict the architecture and evolution of hanging-wall accommodation during fault-related folding. HCA modeling allows us also to predict the distribution of deformation panels produced during fault-related folding.

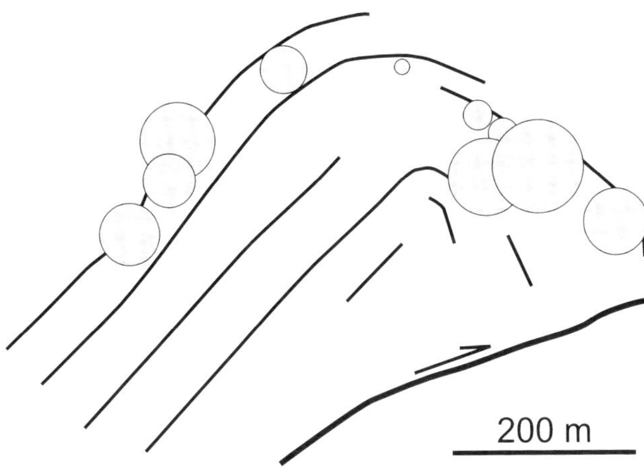

Figure 16. Deformation pattern in the Monkshood Anticline. The diameters of the gray circles are proportional to the density of fractures (after Jamison, 1997).

For application in hydrocarbon exploration and development, deformation panels can be interpreted in terms of fracture panels—that is, rock volumes affected by a characteristic fabric and intensity of longitudinal fractures.

We applied the HCA modeling technique to compressional fault-bend and décollement folding. It correctly simulated the development and spatial distribution of fracture panels in thrust-related anticlines, thus providing an effective tool for spatial evaluation of permeability in hydrocarbon reservoirs.

Model results illustrate the importance of understanding the kinematic link between folding mechanisms and fracture-panel development when we attempt to predict the architecture of fracture systems in reservoirs. These constraints are as follows:

1) Fracture systems develop within well-defined rock volumes, the fracture panels, whose location and dimensions depend on the presence of active axial surfaces, the amount of displacement, and the interactions between fracture domains;
2) Active axial surfaces can produce large volumes of homogeneously fractured rocks, which is opposite the case of fixed axial surfaces, in which fracturing is restricted to the axial zones; and
3) Active-hinge folding may justify the observed low productivity of crestal regions of some thrust-related folds. Conversely, it justifies higher productivity in the upper limb regions.

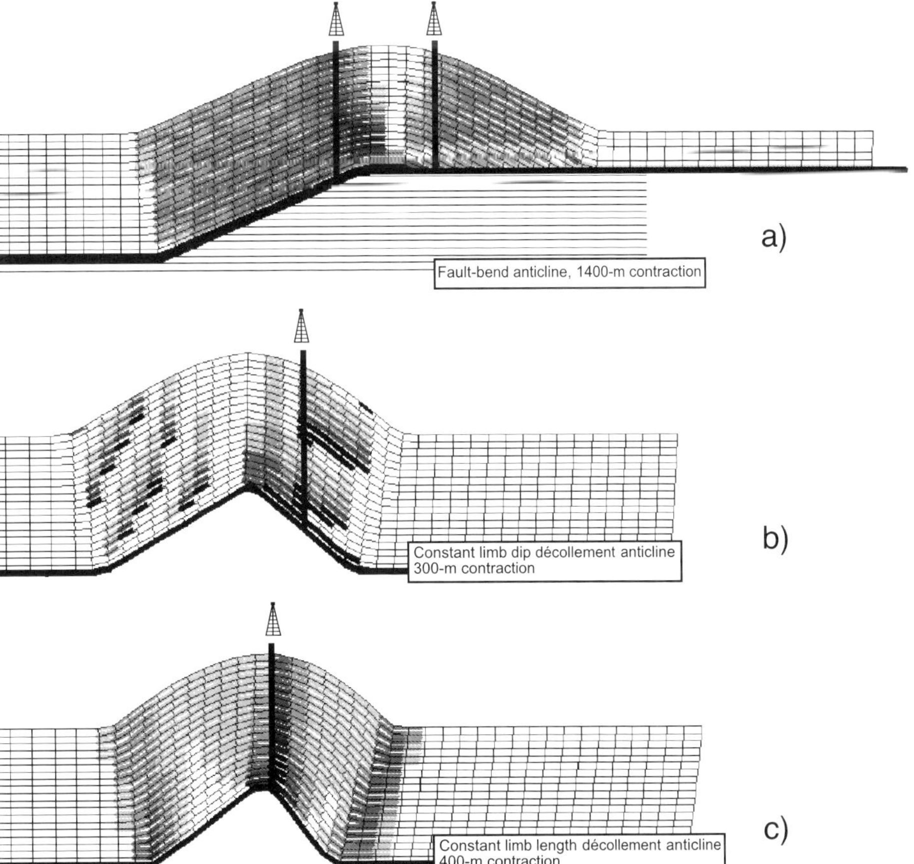

Figure 17. Conceptual sketch models of ideal well locations in (a) fault-bend anticlines, (b) constant-limb-dip décollement anticlines, and (c) a constant-limb-length anticline.

ACKNOWLEDGMENTS

We gratefully acknowledge K. McClay and J. A. Muñoz for useful discussions on the subject. Revision from M. Insley, K. McClay, and two anonymous reviewers allowed us to improve the paper. This work has been funded by MURST 60% research projects (grants to F. Salvini). Enterprise Oil Ltd. has being providing partial funding for testing and further developing the application of HCA to thrust-related folding and related fracture patterns.

REFERENCES CITED

Allmendinger, R. W., 1998, Inverse and forward numerical modeling of trishear fault-propagation folds: Tectonics, v. 17, p. 640–656.

Alonso, J. L., and A. Teixel, 1992, Forelimb deformation in some natural examples of fault-propagation folds, *in* K. R. McClay, ed., Thrust tectonics: London, Chapman & Hall, p. 175–180.

Anastasio, D. J., D. M. Fisher, T. A. Messina, and J. E. Holl, 1997, Kinematics of décollement folding in the Lost River Range, Idaho: Journal of Structural Geology, v. 19, p. 355–368.

Apotria, T. G., M. S. Wilkerson, and S. L. Knewtson, 1996, 3D geometry and controls on fracturing in a natural fault-bend fold: Rosario Field, Maracaibo Basin, Venezuela: AAPG Bulletin, v. 80, p. 1268.

Becker, A., and M. R. Gross, 1996, Mechanics for joint saturation in mechanically layered rocks: an example from southern Israel: Tectonophysics, v. 257, p. 223–237.

Beutner, E. C., and F. A. Diegel, 1985, Determination of fold kinematics from syntectonic fibers in pressure shadows, Martinsburg slate, New Jersey: American Journal of Science, v. 285, p. 16–50.

Bigi, G., A. Castellarin, R. Catalano, M. Coli, D. Cosentino, G. V. Dal Piaz, F. Lentini, M. Parotto, E. Patacca, A. Praturlon, F. Salvini, R. Sartori, P. Scandone, and G. B. Vai, 1990, Synthetic structural-kinematic map of Italy (1:2,000,000), *in* G. Bigi, D. Cosentino, M. Parotto, R. Sartori, and P. Scandone, eds., Structural model of Italy: Roma, Quaderni de "La Ricerca Scientifica" del CNR, n. 114.

Biot, M. A., 1961, Theory of folding of stratified viscoelastic media and its implication in tectonics and orogenesis: Geological Society of America Bulletin, v. 72, p. 1595–1632.

Butler, R. W. H., 1992, Evolution of alpine fold-thrust complexes: a linked kinematic approach, *in* S. Mitra and G. W. Fisher, eds., Structural geology of fold and thrust belts: Baltimore, Johns Hopkins University Press, p. 29–44.

Chester, J. S., J. M. Logan, and J. H. Spang, 1991, Influence of layering and boundary conditions on fault-bend and fault-propagation folding: Geological Society of America Bulletin, v. 103, p. 1059–1072.

Cooper, M., 1992, The analysis of fracture systems in subsurface thrust structures from the foothills of the Canadian Rockies, *in* K. R. McClay, ed., Thrust tectonics: London, Chapman & Hall, p. 391–405.

Dahlstrom, C. D. A., 1990, Geometric constraints derived from the law of conservation of volume and applied to evolutionary models for detachment folding: AAPG Bulletin, v. 74, p. 336–344.

De Sitter, L. U., 1956, Structural Geology: New York, McGraw-Hill, 552 p.

Decker, A. D., J. C. Close, and R. A. McBane, 1989, The use of remote sensing, curvature analysis, and coal petrology as indicators of higher coal reservoir permeability: International Coalbed Methane Symposium Proceedings, p. 325–340.

Dixon, J. M., and S. Liu, 1992, Centrifuge modelling of the propagation of thrust faults, *in* K. R. McClay, ed., Thrust tectonics: London, Chapman & Hall, p. 53–69.

Donath, F. A., and R. B. Parker, 1964, Folds and folding: Geological Society of America Bulletin, v. 75, p. 45–62.

Edlund, C. A., D. J. Anastasio, and D. M. Fisher, 1994, Kinematics of fault-related folding in a duplex, Lost River Range, Idaho, U.S.A.: Journal of Structural Geology, v. 16, p. 571–584.

Egan, S. S., S. Kane, T. S. Buddin, G. D. Williams, and D. Hodgetts, 1999, Computer modelling and visualisation of the structural deformation caused by movement along geological faults: Computers and Geosciences, v. 25, p. 283–297.

Epard, J. L., and R. H. Groshong Jr, 1995, Kinematic model of detachment folding including limb rotation, fixed hinges and layer-parallel strain: Tectonophysics, v. 247, p. 85–103.

Erickson, S. G., 1996, Influence of mechanical stratigraphy on folding vs faulting: Journal of Structural Geology, v. 18, p. 443–450.

Erickson, S. G., and W. R. Jamison, 1995, Viscous-plastic finite-element models of fault-bend folds: Journal of Structural Geology, v. 17, p. 561–573.

Erickson, G., L. Strayer, and J. Suppe, 1999, Mechanical modelling of deformation mechanisms in folds: Thrust Tectonics 99, Royal Holloway University of London, Abstracts, p. 188.

Erslev, E. A., and K. R. Mayborn, 1997, Multiple geometries and modes of fault-propagation folding in the Canadian thrust belt: Journal of Structural Geology, v. 19, p. 321–335.

Fischer, M. P., and P. B. Jackson, 1999, Stratigraphic controls on deformation patterns in fault-related folds: a detachment fold example from the Sierra Madre Oriental, northeast Mexico: Journal of Structural Geology, v. 21, p. 613–633.

Fischer, M. P., and M. S. Wilkerson, 2000, Predicting the orientation of joints from fold shape: Results of pseudo-three-dimensional modeling and curvature analysis: Geology, v. 28, p. 15–18.

Fischer, M. P., N. B. Woodward, and M. M. Mitchell, 1992, The kinematics of break-thrust folds: Journal of Structural Geology: v. 14, p. 451–460.

Fisher, D. M., and D. J. Anastasio, 1994, Kinematic analysis of a large scale leading edge fold, Lost River Range, Idaho: Journal of Structural Geology, v. 16, p. 337–354.

Geiser, P. A., and S. Sansone, 1981, Joints, microfractures and the formation of solution cleavage in limestone: Geology, v. 9, p. 280–285.

Gross, M. R., M. P. Fisher, T. Engelder, and R. J. Greenfield, 1995, Factors controlling joint spacing in interbedded sedimentary rocks: integrating numerical models with field observations from the Monterey Formation, USA, *in* M. S. Ameen, ed., Fractography: fracture topography as a tool in fracture mechanics and stress analysis: Geological Society of London Special Publication 92, p. 215–233.

Gutièrrez-Alonso, G., and M. R. Gross, 1999, Structures and mechanisms associated with development of a fold in the Cantabrian Zone thrust belt, NW Spain: Journal of Structural Geology, v. 21, p. 653–670.

Hancock, P. L., 1985, Brittle microtectonics: principles and practice: Journal of Structural Geology, v. 7, p. 437–457.

Hardy, S., J. Poblet, K. McClay, and D. Waltham, 1996, Mathematical modelling of growth strata associated with fault-related fold structures, *in*: P. G. Buchanan and D. A. Nieuwland, eds., Modern developments in structural interpretation, validation and modelling: Geological Society of London Special Publication 99, p. 265–282.

Hedlund, C. A., D. J. Anastasio, and D. M. Fisher, 1994, Kinematics of fault-related folding in a duplex, Lost River Range, Idaho, U.S.A.: Journal of Structural Geology, v. 16, p. 571–584.

Hennings, P. H., and J. E. Olson, 1997, The relationship between bed curvature and fracture occurrence in a fault-propagation fold: AAPG/SEPM Annual meeting abstracts, v. 6, p. 49.

Holl, J. E., and D. J. Anastasio, 1992, Deformation of a foreland carbonate thrust system, Sawtooth Range, Montana: Geological Society of America Bulletin, v. 104, p. 944–953.

Homza, T. X., and W. K. Wallace, 1997, Detachment folds with fixed hinges and variable detachment depth, northeastern Brooks Range, Alaska: Journal of Structural Geology, v. 19, p. 337–354.

Jamison, W. R., 1987, Geometric analysis of fold development in overthrust terranes: Journal of Structural Geology, v. 9, p. 207–219.

Jamison, W. R., 1992, Stress controls on fold thrust style, *in* K. R. McClay, ed., Thrust Tectonics: London, Chapman & Hall, p. 155–164.

Jamison, W. R., 1997, Quantitative evaluation of fractures on Monkshood Anticline, a detachment fold in the Foothills of western Canada: AAPG Bulletin, v. 81, p. 1110–1132.

Kilsdonk, B., and D. V. Wiltschko, 1988, Deformation mechanisms in the southeastern ramp region of the Pine Mountain block, Tennessee: Geological Society of America Bulletin, v. 100, p. 653–664.

Laubach, S. E., 1997, A method to detect natural fracture strike in sandstones: AAPG Bulletin, v. 81, p. 604–623.

Lemiszki, P. J., J. D. Landes, and R. D. Hatcher Jr., 1994, Controls on hinge-parallel extension fracturing in single-layer tangential-longitudinal strain fold: Journal of Geophysical Research, v. 99, p. 22,027–22,041.

Lisle, R. J., 1994, Detection of zones of abnormal strains in structures using Gaussian curvature analysis: American Association of Petroleum Geologists Bulletin, v. 78, p. 1811–1819.

Marrett, R., and P. A. Bentham, 1997, Geometric analysis of hybrid fault-propagation/detachment folds: Journal of Structural Geology, v. 19, p. 243–248.

Meigs, A. J., 1997, Sequential development of selected Pyrenean thrust faults: Journal of Structural Geology, v. 19, p. 481–502.

Mitra, S., 1987, Regional variations in deformation mechanisms and structural styles in the central Appalachian orogenic belt: Geological Society of America Bulletin, v. 70, p. 1087–1112.

Morse, J., 1977, Deformation in ramp regions of overthrust faults: experiments with small-scale rock models: 29th Annual Field Conference 1977 Wyoming Geological Association Guidebook, p. 457–470.

Murray, G. H. Jr., 1968, Quantitative fracture study—Sanish Pool, McKenzie County, North Dakota: AAPG Bulletin, v. 52, p. 57–65.

Narr, W., and J. B. Currie, 1982, Origin of fracture porosity – example from Altamont Field, Utah: American Association of Petroleum Geologists Bulletin, v. 66, p. 1231–1247.

Patton, T. L., S. Serra, R. J. Humphreys, and R. A. Nelson, 1995, Building conceptual structural models from multiple modelling sources: an example from thrust-ramp studies: Petroleum Geoscience, v. 1, p. 153–162.

Poblet, J., and K. McClay, 1996, Geometry and kinematics of single-layer detachment folds: AAPG Bulletin, v. 80, p. 1085–1109.

Poblet, J., J. A. Muñoz, A. Travé, and J. Serra-Kiel, 1998, Quantifying the kinematics of detachment folds using three-dimensional geometry: Application to the Mediano anticline (Pyrenees, Spain): Geological Society of America Bulletin, v. 110, p. 111–125.

Price, N. J., 1966, Fault and joint development in brittle and semi-brittle rocks: Oxford, Pergamon Press, 392 p.

Price, N. J., and J. W. Cosgrove, 1990, Analysis of geological structures: Cambridge, Cambridge University Press, p. 1–502.

Ramsay, J. G., and M. I. Huber, 1987, The techniques of modern structural geology: v. 2: Folds and fractures: London, Academic Press, 700 p.

Rich, J. L., 1934, Mechanics of low-angle overthrust faulting as illustrated by Cumberland thrust block, Virginia, Kentucky and Tennessee: AAPG Bulletin, v. 18, p. 1584–1596.

Salvini, F., and F. Storti, 1997, Spatial and temporal distribution of fractured rock panels from geometric and kinematic models of thrust-related folding [abs.]: AAPG Bulletin, v. 81, p. 1409.

Salvini, F., and F. Storti, 2001, The distribution of deformation in parallel fault-related folds with migrating axial surfaces: Comparison between fault-propagation and fault-bend folding: Journal of Structural Geology, v. 23, p. 25–32.

Salvini, F., and E. Vittori, 1982, Analisi strutturale della linea Olevano-Antrodoco-Posta (Ancona-Anzio Auct.): metodologia di studio delle deformazioni fragili e presentazione del tratto meridionale: Memorie della Società Geologica Italiana, v. 24, p. 337–355.

Sanderson, D. J., 1982, Models of strain variation in nappes and thrust sheets, a review: Tectonophysics, v. 88, p. 201–233.

Schultz-Ela, D. D., and J. Yeh, 1992, Predicting fracture permeability from bed curvature, *in* J. R. Tillerson and W. R. Wawersik, eds., Rock Mechanics: U. S. Symposium on Rock Mechanics Proceedings, v. 33, p. 579–589.

Srivastava, D. C., and T. Engelder, 1990, Crack-propagation sequence and pore-fluid conditions during fault-bend folding in the Appalachian Valley and Ridge, central Pennsylvania: Geological Society of America Bulletin, v. 102, p. 116–128.

Stearns, D. W., 1968, Certain aspects of fracture in naturally deformed rocks, *in* R. E. Rieker, ed., National Science Foundation advanced science seminar in rock mechanics, special report: Air Force Cambridge Research Laboratories, Bedford, Massachusetts, AD66993751, p. 97–118.

Stewart, K. G., and W. Alvarez, 1991, Mobile-hinge kinking in layered rocks and models: Journal of Structural Geology, v. 13, p. 243–259.

Storti, F., and J. Poblet, 1997, Growth stratal architectures associated to décollement folds and fault-propagation folds: inferences on fold kinematics: Tectonophysics, v. 282, p. 353–373.

Storti, F., and F. Salvini, 1996, Progressive rollover fault-propagation folding: a possible kinematic mechanism to generate regional-scale recumbent folds in shallow foreland belts: AAPG Bulletin, v. 80, p. 174–193.

Storti, F., F. Salvini, and K. McClay, 1997, Fault-related folding in sandbox analogue models of thrust wedges: Journal of Structural Geology, v. 19, p. 583–602.

Strayer, L. M., and P. J. Hudleston, 1997, Numerical modeling of fold initiation at thrust ramps: Journal of Structural Geology, v. 19, p. 551–566.

Suppe, J., 1983, Geometry and kinematics of fault-bend folding: American Journal of Science, v. 283, p. 684–721.

Suppe, J., G. T. Chou, and S. C. Hook, 1992, Rates of folding and faulting determined from growth strata, *in* K. R. McClay, ed., Thrust Tectonics: London, Chapman & Hall, p. 105–121.

Tavarnelli, E., 1997, Structural evolution of a foreland fold-and-thrust belt: the Umbria-Marche Apennines, Italy: Journal of Structural Geology, v. 19, p. 523–534.

Thorbjornsen, K. L., and W. M. Dunne, 1997, Origin of a thrust-related fold: geometric versus kinematic tests: Journal of Structural Geology, v. 19, p. 303–319.

Turcotte, D. L., and G. Schubert, 1982, Geodynamics—applications of continuum physics to geological problems: New York, John Wiley and Sons, 451 p.

Van Geet, M., R. Swennen, P. Muchez, and F. Roure, 1998, Diagenetic study of potential carbonate reservoirs in the Ionian Zone of Albania. Paper presented at the conference on Exploration and Production of Carbonate Plays on the Southern Margin of the Mediterranean Tethys, Geological Society of London, Petroleum Group, 16–18th June 1998.

Vergés, J., D. W. Burbank, and A. Meigs, 1996, Unfolding: an inverse approach to fold kinematics: Geology, v. 24, p. 175–178.

Wiltschko, D. V., D. A. Medwedeff, and H. E. Millson, 1985, Distribution and mechanisms of strain within rocks on the northwest ramp of Pine Mountain block, southern Appalachian foreland: a field test of theory: Geological Society of America Bulletin, v. 96, p. 426–435.

Wojtal, S., 1986, Deformation within foreland thrust sheets by populations of minor faults: Journal of Structural Geology, v. 8, p. 341–360.

Wojtal, S., and G. Mitra, 1986, Strain hardening and strain softening in fault zones from foreland thrusts: Geological Society of America Bulletin, v. 97, p. 674–687.

Wolfram, S., 1986, Theory and applications of cellular automata: Singapore, World Scientific.

Woodward, N. B., 1999, Competitive macroscopic deformation processes: Journal of Structural Geology, v. 21, p. 1209–1218.

Woodward, N. B., and E. Rutherford Jr., 1989, Structural lithic units in external orogenic zones: Tectonophysics, v. 158, p. 247–267.

Zienkiewicz, O. C., and R. L. Taylor, 1991, The finite element method: solid and fluid mechanics and nonlinearity: London, McGraw-Hill Book Company, 648 p.

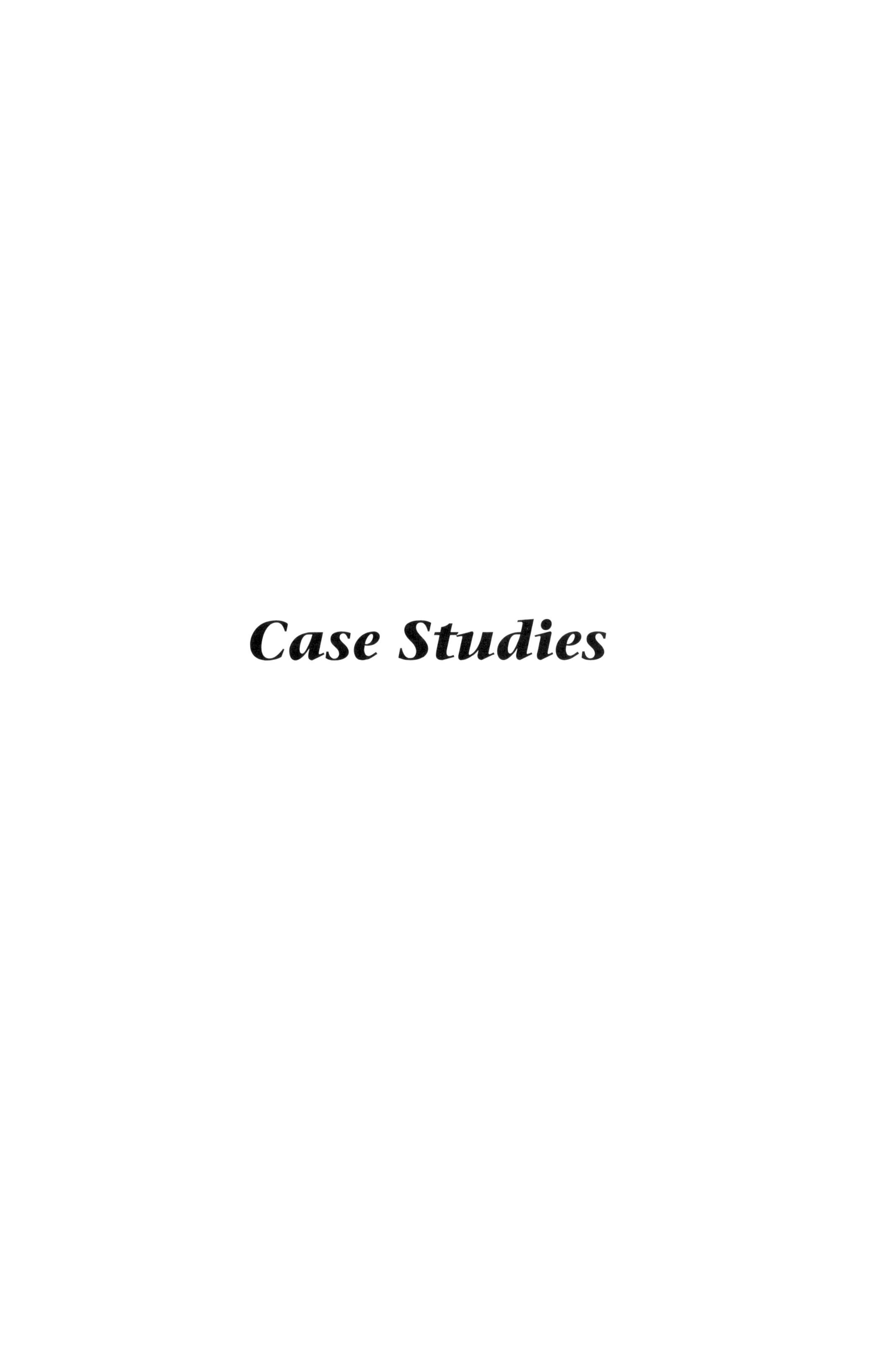

Case Studies

24

Roure, F., S. Nazaj, K. Mushka, I. Fili, J.-P. Cadet, and M. Bonneau, 2004, Kinematic evolution and petroleum systems — An appraisal of the Outer Albanides, *in* K. R. McClay, ed., Thrust tectonics and hydrocarbon systems: AAPG Memoir 82, p. 474–493.

Kinematic Evolution and Petroleum Systems—An Appraisal of the Outer Albanides

François Roure
Institut Français du Pétrole, Rueil-Malmaison, France

Shaqir Nazaj
Oil and Gas Institute, Fieri, Albania

Kristaq Mushka
Oil and Gas Institute, Fieri, Albania

Ilia Fili
Oil and Gas Institute, Fieri, Albania

Jean-Paul Cadet
University Pierre and Marie Curie, Paris, France

Michel Bonneau
University Pierre and Marie Curie, Paris, France

ABSTRACT

The lithostratigraphic column of the Outer Albanides records a long geodynamic evolution. It began with a Liassic rifted margin made up of tilted fault blocks, carbonate platforms, and euxinic basins (Posidonia Schist) and evolved into a Paleogene flexed foreland, part of which inverted during the Neogene. The complex Mesozoic paleogeography accounts for the distribution of potential décollement levels and lateral changes in structural styles.

Petroleum plays are numerous and result from the occurrence of various source rocks with contrasting burial histories and migration pathways. To document the respective timing of thrusting, petroleum generation, and trapping, that is, the critical timing of Albanian petroleum systems, we have reconstructed the kinematic and thermal evolution of two representative regional transects that cross the Peri-Adriatic Depression and the Kruja Zone, as well as the Ionian Basin, in the northern and southern segments of the Outer Albanides, respectively.

INTRODUCTION

The geology of Albania displays a typical foreland fold-and-thrust belt, including a flexed foreland domain in the west (Peri-Adriatic Depression and offshore), frontal thrusts and foothills structures involving the Mesozoic series of the former Tethyan passive margin (Kruja and Krasta Zones in the north, inverted Ionian Basin in the south), as well as an ophiolitic suture, well preserved in the hinterland (Mirdita Massif (Ophiolite), Figure 1).

Albania also constitutes a petroleum province, with oil being extracted from Neogene sandstone reservoirs in piggyback basins (Patos field), as well as from Upper Cretaceous and Paleogene carbonates in the Ionian allochthon (i.e., the Balshi and Delvina fields, Figure 1) (Albpetrol, 1993; Curi, 1993; Sedjini et al., 1994;

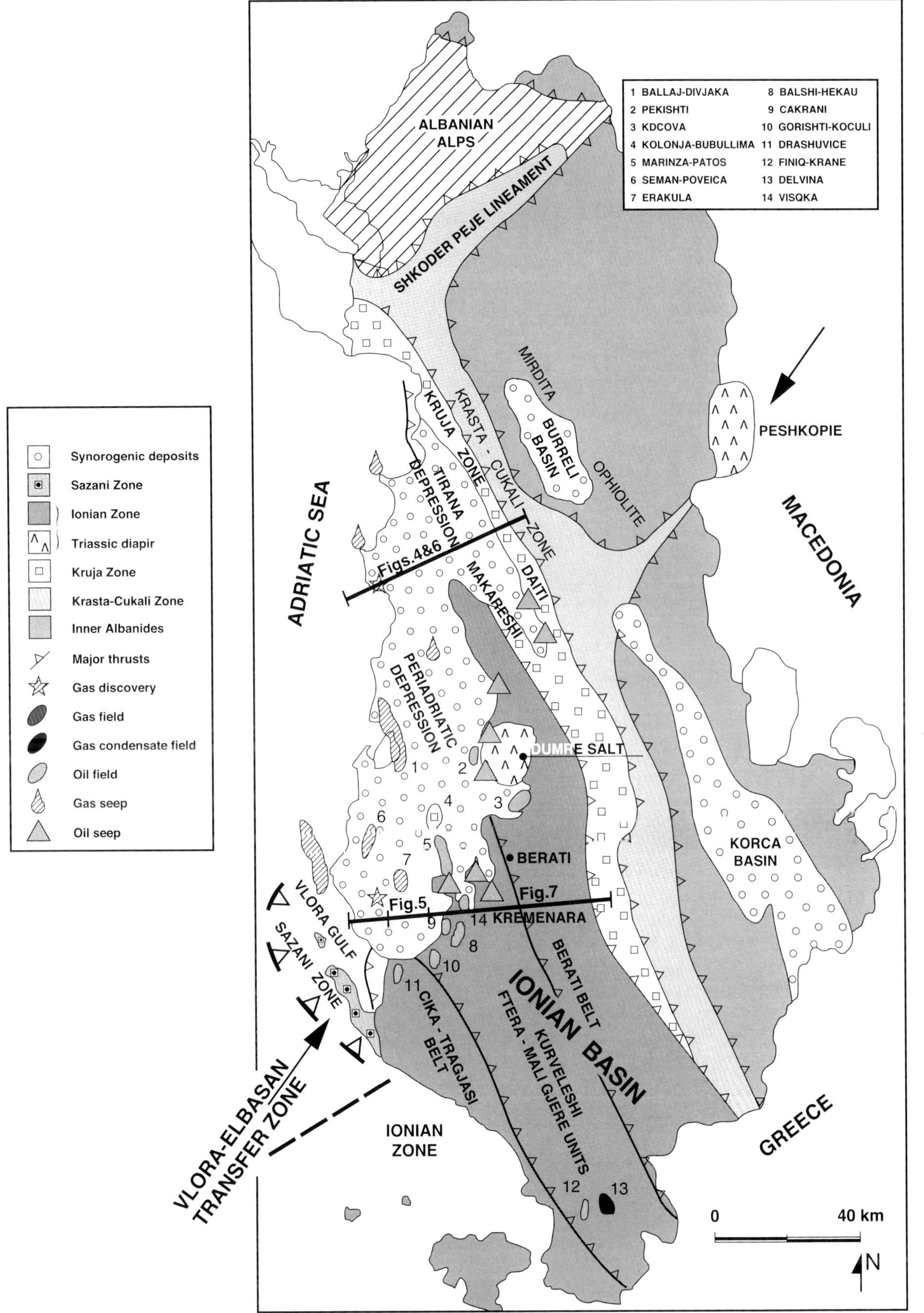

Figure 1. Structural map of the Outer Albanides, with the locations of the two modeled transects, wells, and outcrops referred to in the text (modified after Albpetrol, 1993).

Diamanti et al., 1995; Shteto et al., 1995; Valbona et al., 1995). In addition, biogenic gas is produced from the Pliocene series of the foredeep (Peri-Adriatic Depression).

Albania constitutes a key area for the study of petroleum systems in a foothills domain, where deformation is still currently active (Sorel et al., 1992; Tagari, 1993; Muço, 1994, 1998; Roure, 1999). Together with the Apennines and Romanian Carpathians, the Outer Albanides constitute one of the European foredeep basins in which the synorogenic series are the best preserved (Casero et al., 1991; Roure and Sassi, 1995). Because of these unusual conditions, it is possible to constrain the kinematics of the deformation and trace the burial and thermal evolution of potential source rocks and reservoirs.

GEOLOGIC BACKGROUND OF THE OUTER ALBANIDES

The Albanian foothills have been thrust westward over the Adriatic foreland and constitute a segment of the wide Circum-Mediterranean-Peri-Tethyan Thrust Belts, between the Dinarides in the north (former Yugoslavia) and the Hellenides in the south (Greece). The hinterland (i.e., the Mirdita ophiolite and basinal units of the Krasta-Cukali Zone, which are coeval with the Pindos Zone of Greece) displays a relatively simple linear geometry. However the Outer Albanides are dissected by a major transfer zone: the Vlora-Elbasan transfer zone (Figure 1).

Major Paleogeographic and Structural Units

The northern part of the Outer Albanides comprises two distinct domains (Figures 1 and 2; Table 1): (1) the Peri-Adriatic Depression in the west, where only the Neogene molasse of the foredeep crops out at the surface; and (2) the Kruja Zone or Platform, which comprises thrust anticlines of Mesozoic platform carbonates.

The southern part of the Outer Albanides is almost entirely made up of Mesozoic-Paleogene basinal units of the Ionian Zone, which is presently detached from its former substratum along an intra-Triassic evaporitic series (Figures 1 and 2). Only the frontal structure, which crops out along the Adriatic Sea, is made up of Cretaceous platform carbonates of the Sazani or Pre-Apulian Zone, which extends westward across the sea and connects directly with the autochthonous Apulian Platform exposed in Italy (Table 1) (Nikolaou, 1986; Paulucci et al., 1988; Flores et al., 1991; Veizaj and Frashëri, 1995).

The Kruja Zone

The Cretaceous platform carbonates of the Kruja Zone constitute a stack of duplexes, which have been partially identified by drilling, but which also crop out at the surface in the Makareshi and Daiti Anticlines (Figures 1, 3, and 4).

Outcrops and seismic profiles both suggest that these structures have been detached from their former pre-Cretaceous substratum along Aptian shales or conjectural Lower Cretaceous anhydritic horizons, which are inferred from drilling data in the Cretaceous platforms in Yugoslavia, because no Jurassic or Triassic series has ever been encountered in these shallow, thin-skinned tectonic units.

The Cretaceous and Paleocene-Eocene series of the Kruja Zone display shallow-water facies, with frequent hiatuses, evidence of emersion and erosion, and numerous bauxitic intervals (Heba, 1997) (Figure 2; Table 1).

Oligocene transgressive series are made up of deep-water turbidites (flysch). They attest to the progressive flexing of the Adriatic foreland and the involvement of the Kruja paleogeographic domain in the Albanides foredeep during Paleogene time.

Tortonian to Pliocene series postdate part of the deformation. However, they were deposited in piggyback basins, which indicates ongoing deformation; late-stage thrust-emplacement of the Outer Albanides was coeval with the deposition of these Neogene series.

The Peri-Adriatic Depression

The Peri-Adriatic Depression is characterized by a thick terrigenous synflexural (Oligocene flysch) and synkinematic (Neogene molasse) series, which is presently largely detached from its former Mesozoic carbonate subtratum. The total thickness of Cenozoic siliciclastic series imaged on seismic profiles frequently exceeds 7 km.

Although the pre-Oligocene substratum has never been reached by drilling, it is assumed to be basinal in type, likely resembling the Ionian Basin farther to the south or the Adriatic offshore in the west (Finetti et al., 1989; Frashëri et al., 1996). However, unlike in the Ionian Basin, no intra-Triassic décollement has ever been found beneath the Peri-Adriatic Depression. This may be explained by a major paleogeographic change that occurred in the vicinity of the Vlora-Elbasan transfer zone, with a basement fault limiting the northwestern extent of the Triassic evaporite (Figure 3).

The Ionian Basin

The Ionian Basin is located south of the Vlora-Elbasan transfer zone and extends across the Greek border into Epirus, where its stratigraphic and structural evolution have been widely documented (IGRS and IFP, 1966; Guzzetta, 1982; Karakitsios et al., 1988; Underhill, 1989; Waters, 1994). In Albania, the Ionian Basin is subdivided into three main tectonostratigraphic units, listed from west to east (Figure 1): the Cika-Tragjasi belt; the

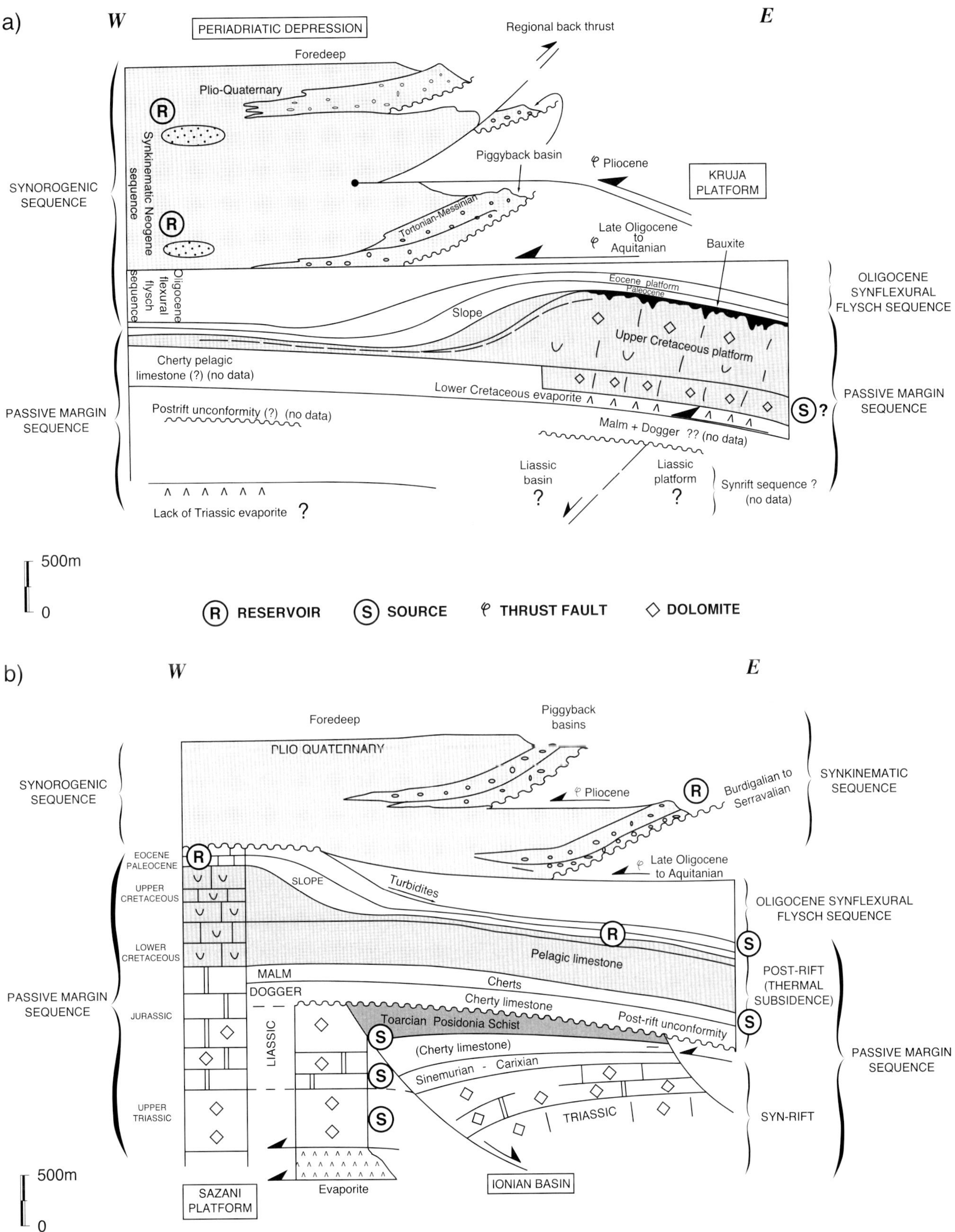

FIGURE 2. (a) Synthetic lithostratigraphic log of the Peri-Adriatic Depression (Neogene) and of the Kruja Zone (Cretaceous-Paleogene). (b) Synthetic lithostratigraphic log of the Ionian Basin (Mesozoic and Cenozoic).

Table 1. Simplified chronostratigraphic diagram showing the stratigraphy and tectonic events. The pre-Oligocene history records the evolution of the Tethyan rifting and subsequent development of the passive margin. Oligocene to Present history records the flexural evolution of the foreland and thrust emplacement of the foothills. The last column outlines the time intervals (stages) selected for the Thrustpack simulation. R = reservoir; S = source rock; Seal = seal horizon.

	SAZANI Z.	IONIAN Z.	PERI-ADRIATIC Z.	KRUJA Z.	Tectonic events	Thrustpack stages
Pliocene-Quaternary	erosion	erosion	molasse **R**	erosion	mud diapirs late shortening	**Stage 5** 5 Ma (B.P.)
Messinian Tortonian Langhian-Serravalian Burdigalian Aquitanian	turbidite? calcarenite	molasse **R** local emersion	turbidite **R** **S** deep water?	molasse emersion	Messinian evaporitic seal continuous shortening growth anticlines	5 Ma **Stage 4** 10 Ma **Stage 3** 15 Ma **Stage 2**
Oligocene	emersion	flysch **Seal**	flysch **S?**	flysch **Seal**	onset foreland flexure	20 Ma **Stage 1** 30 Ma
Eocene Paleocene	emersion bauxite	**R**		emersion bauxite	salt diapirs passive margin	30 Ma **Stage 0** 70 Ma
Upper Cretaceous Lower Cretaceous	**R ?** Platform	**R** carbonate turbidite **S**	no data	**S ? R ?** Platform **S ?**		
Upper Jurassic Middle Jurassic Lower Jurassic	Platform	basinal sequence Posidonia source **S - Seal**	no data	no data	synrift extension	
Triassic	**S** dolomite salt ?	**S** dolomite **R** basal salt	no salt?			
Hercynian basement	no data	no data	no data	no data		

Kurveleshi-Ftera-Mali Gjere units (Kurveleshi belt); and the Berati belt.

Unlike in other paleogeographic and structural domains, the Triassic evaporite constitutes the main décollement in the Ionian Basin, although Toarcian Posidonia Schist and Oligocene flysch are likely to provide secondary decoupling horizons. For instance, the basal décollement of the Mali Gjere unit is located in the Triassic evaporites in the north, where the Triassic series is exposed at the thrust front, but it is found in the Posidonia Schist farther to the south, where the Liassic is directly thrust on top of the Oligocene flysch series of the footwall (Figures 1 and 2) (Roure et al., 1995). Triassic salt constitutes huge diapirs, which are mostly unrooted and passively transported piggyback on the Neogene (Alpine) thrusts (Monopolis and Bruneton, 1982; Underhill, 1989; Bakiaj et al., 1990; Berberi et al., 1990; Velaj and Xhuli, 1995; Velaj et al., 1999).

As it is in Epirus, the Ionian Basin in Albania is characterized by large variations in thickness and facies within the Liassic horizons, which outline the past activity of former tilted blocks (with dolomite or platform carbonates developing at the crest of the blocks) and intervening euxinic basins filled with organic-rich Posidonia Schist (Danelian et al., 1986; Karakitsios et al., 1988; Dommergues et al., 2000).

Two intra-Dogger and intra-Malm cherty horizons have a wide regional extent (Danelian and Baudin, 1990). In addition, two Cretaceous phosphatic horizons (Turonian and lower Senonian) can be traced in the landscape, thanks to the more luxurious vegetation. Middle-Upper Jurassic and Lower Cretaceous series display a fairly constant thickness and constitute the basal portion of the postrift sequence. They are interpreted to record the episode of thermal subsidence of the Ionian Basin (Table 1).

Slumps and carbonate turbidites are common in the Upper Cretaceous and Paleogene strata, whereas fine-grained pelagic limestones (Scaglia) constitute most of the Eocene series.

Synorogenic Series and Agenda of the Deformation

Synorogenic deposits young westward. The first flysch is dated as Late Cretaceous-Eocene in the east (Krasta Zone), but no older than Oligocene in the Kruja Zone and Ionian Basin. Even younger synflexural deposits were still deposited during the Neogene in the Peri-Adriatic Depression and the Sazani Zone (Sorel et al., 1992; Muftari et al., 1995).

This progessive shift in the timing is also observed for the thrust-front's emplacement. However, most of the Ionian deformation and thrust emplacement of the Kruja anticlinal stack can be generally related to a continuum of deformation, extending from the late Oligocene until the Pliocene-Quaternary. Local unconformities recorded at the base of piggyback basins and the progressive tilt recorded by synkinematic series agree well with coeval active deformation (Tagari, 1993).

Structural Style

Although the present architecture of the Outer Albanides and its foreland largely results from Neogene Alpine tectonics, numerous extensional features inherited from the former Tethyan passive margin still contribute to the overall structural grain of the area. This is especially the case in the Delvina district, where the sole thrust of the Mali Gjere unit carries numerous hanging-wall Jurassic normal faults, thus resulting in a dominantly horst-and-graben architecture within the allochthon. Unfortunately, these early tectonic features are very difficult to trace on seismic profiles; therefore, they were not taken into consideration in our numerical models.

Potential Décollement Levels

Five décollement levels have a regional expression in the Outer Albanides and account for the thin-skinned fold-and-thrust architecture. They also account for the regional clockwise rotations of the allochthon observed with respect to the Adriatic foreland both in Epirus and the Outer Albanides (Waters, 1994; Speranza et al., 1995). They comprise:

1) The Triassic evaporite, which constitutes the major décollement level within the Ionian Basin (Roure et al., 1995; Velaj et al., 1999). This horizon is exposed here and there along major thrust fronts

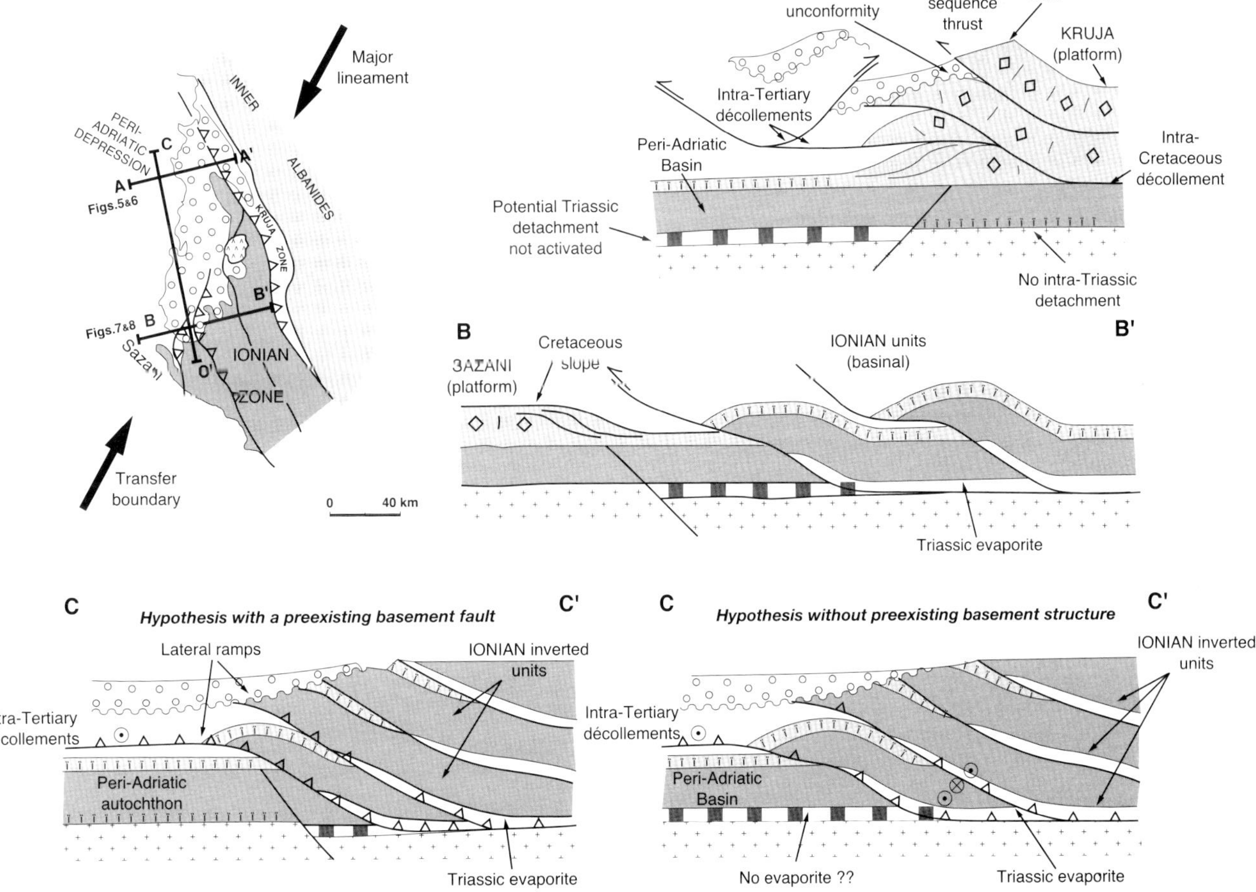

FIGURE 3. Schematic diagrams outlining the geodynamic significance of the Vlora-Elbasan transfer zone. Sections A–A′ and B–B′ outline the change observed in the structural style between the northern section, where the main décollement level is located within the Oligocene flysch, and the southern transect, where the carbonate reservoirs are detached from their substratum along Triassic evaporitic series. Section C–C′ images two distinct scenarios for the deep architecture of a north-south transect crossing the Vlora-Elbasan transfer zone. For many reasons discussed in the text, we favor the hypothesis with a preexisting basement fault (left solution) controlling the lateral extent of the Triassic salt.

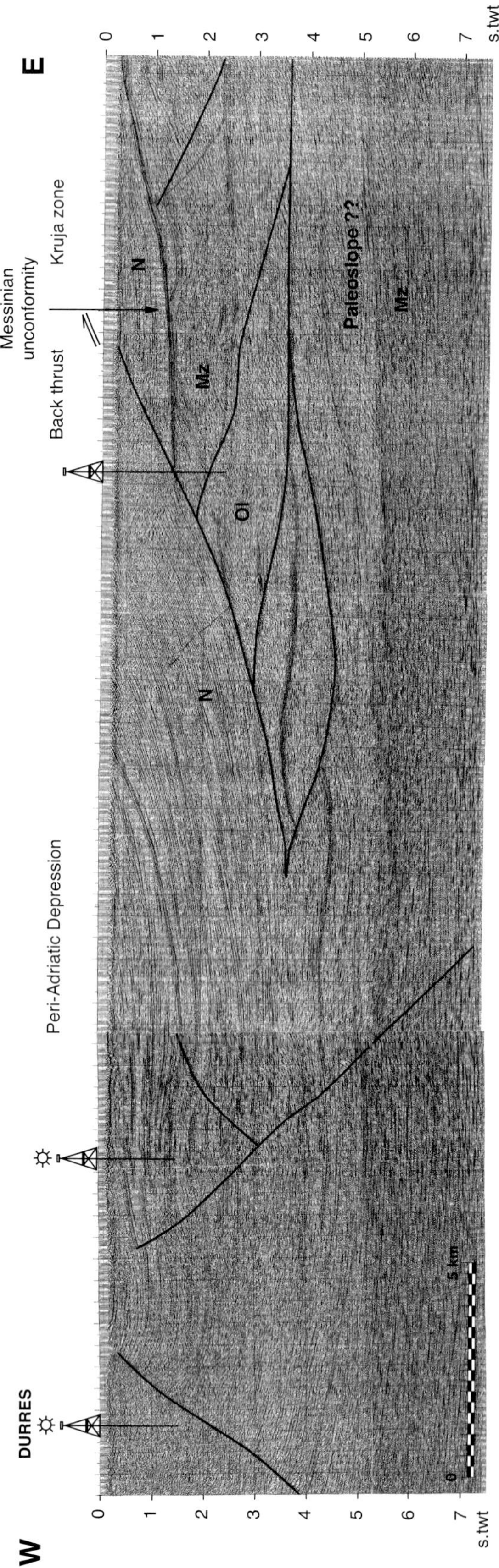

Figure 4. Seismic profile used for the construction of the northern transect (time section). Note the deep reflections imaged beneath the Kruja allochthon. Mz = Mesozoic; Ol = Oligocene flysch; N = Neogene.

in the Ionian Basin, but has also been perforated in the Dumre structure (Velaj et al., 1999)

2) The Toarcian Posidonia Schist, which constitutes a secondary décollement level in the Ionian Basin, when present (i.e., in the southern part of the Mali Gjere unit; Roure et al., 1995)
3) The Aptian-Albian shale, which is exposed in a number of Ionian structures, such as the Kremenara Anticline, and accounts for local disharmonic features and duplexing in the Delvina trend, in the footwall of the Mali Gjere unit (Figure 1) (Roure et al., 1995)
4) Conjectural intraplatform Cretaceous evaporites of the Kruja Zone, which do not crop out at the surface and have not yet been drilled but which could account for the decoupling horizon imaged on seismic profiles near Tirana (Figure 4)
5) The Oligocene flysch and pelitic interbeds of the Neogene molasse, which constitute major decoupling horizons in the Peri-Adriatic Depression. These Cenozoic décollement horizons are well imaged on seismic profiles and also merge to the surface west of the Kruja Zone, along the regional Peri-Adriatic back thrust (Figures 3, 4, 5, and 6).

Although clockwise rotations of as much as 20 or even 30° have been proposed by Speranza et al. (1995) for their so-called Albanian block, this is probably valid only for the Ionian allochthon, which becomes increasingly shortened and transported first southwestward above the Triassic evaporite across the Vlora-Elbasan transfer zone (Figure 3) and then farther south in Epirus (Waters, 1994).

Triangle Zones

A regional-scale triangle zone developed during Pliocene-Quaternary time in front of the Kruja duplexes, thus accounting for a major vertical offset of the basal Serravalian-Messinian unconformity (Figure 6) (Janopulli et al., 1995). At depth, this structure relates to a major facies transition between a platform domain in the east (Kruja Zone) and a basinal domain farther to the west (Peri-Adriatic domain), the intra-Cretaceous décollement level being abruptly interrupted in the vicinity of this former paleogeographic boundary.

During the Pliocene-Quaternary reactivation of the thrust system, the brittle Mesozoic carbonates of the Kruja Zone acted as an active backstop, inducing a progressive wedging of the siliciclastic infill of the Peri-Adriatic Depression.

The Vlora-Elbasan Transfer Zone

The Vlora-Elbasan transfer zone, which is well expressed on the regional geologic maps and satellite imagery (Industrial and Mining Research Institute, 1967; Roure et al., 1995), is probably located above a major paleogeographic boundary (Figures 1 and 3). In the south, the carbonate units of the Ionian Basin are detached along the Triassic evaporite. In the north, in contrast, the basal décollement level is located in the Oligocene flysch. There, the Mesozoic series of the Peri-Adriatic Depression are still attached to the autochthonous foreland, and only the Neogene series is actually accreted into the allochthonous wedge in this part of the belt.

The major lineament evident on the Landsat imagery is likely to connect with a deeper basement structure, unless it relates only to the lateral disappearance of the Triassic salt. On both sides of the Vlora-Elbasan transfer zone, the Mesozoic series display the same layered seismic pattern, which can be attributed to basinal units, making the Peri-Adriatic Depression and Ionian Basin part of a formerly single basin domain. Two major diapirs, the Dumre and Peshkopie structures (Figure 1), lie also along this lineament and are likely to have been triggered along a currently deeply buried basement fault. Such basement structure would also account for the middle-crust seismicity recorded parallel to this surface regional trend (Muço, 1994, 1998).

GEOPHYSICAL AND GEOLOGICAL DATA, CONSTRUCTION OF THE STRUCTURAL SECTIONS AND FORWARD KINEMATIC MODELING

Subsurface Data (Wells And Seismic) and Selection of the Two Modeled Transects

Ultimately, we selected two regional transects, one across the Peri-Adriatic domain and the other one across the Ionian Basin, for which the seismic coverage was mostly continuous and of sufficient quality, and for which there was a reasonable number of calibration wells (Figures 1, 4, and 5).

- The northern transect crossed the Peri-Adriatic Zone from the Adriatic Sea to the Daiti overthrust, that is, to the first outcrops of the Kruja Zone.
- The southern section crossed the northern part of the Ionian Basin, from the Vlora Gulf in the west, to the town of Berati in the east.

Calibration, Interpretation and Depth Conversion of the Seismic Profiles

Calibration wells were especially helpful in identifying the Neogene series, because the subsurface extension of the Burdigalian-Tortonian unconformity has been largely dissected by subsequent post-Messinian thrust reactivations at the front of the Ionian foothills (Figure 5). This is also true for the western extension of the carbonate duplexes of the Kruja Zone, which are interpreted to grade westward into look-alike structures involving solely the Oligocene flysch (Figure 4).

Interval velocities used for the depth conversion of the seismic profiles range from 2.5 km/s to 3.5 km/s in the Pliocene-Quaternary and Neogene series, respectively, and up to 6 km/s in the Mesozoic carbonates.

Cross-section Balancing

The two structural sections have been restored using the Locace software, with a line-length balance and constant thickness hypothesis.

Numerous iterations were necessary to improve the consistency between unfolded and present geometries, especially for the northern transect, where great uncertainties exist regarding the paleogeographic affinities of underthrust units. In the final interpretation, we favored the hypothesis of a Mesozoic paleoslope domain being still preserved in the autochthon beneath the presently allochthonous platform duplexes of the Kruja Zone (Figure 4). However, alternate explanations could also adequately account for the thick sedimentary wedge observed in the footwall of the Kruja duplexes. For instance, one could assume a lateral thickening of the Oligocene flysch series, a wider underthrusting of the Peri-Adriatic basinal domain (assumed to be of Ionian affinity), or the development of footwall duplexes within the basinal carbonate sequence.

Construction of Forward Kinematic Models Using the Thrustpack Software

Once the two structural sections were restored with Locace, their resulting initial geometries subsequently were used as the initial template in the Thrustpack forward kinematic modeling. These initial sections provided the requisite thickness values to simulate the former Mesozoic to Eocene passive-margin series, prior to their burial beneath the Oligocene flexural and Neogene synkinematic series, and also provided the spacing of future thrusts.

On the restored sections, two basement faults have been assumed to control the lateral extent of the Triassic salt basin. (1) A fault in the footwall of the Kruja allochthon in the northern transect (Figure 6) accounts for the lateral continuity between the intra-Cretaceous décollement in the east and the Oligocene décollement evident beneath the Peri-Adriatic Depression. (2) A fault near the surface trace of the Vlora-Elbasan transfer zone

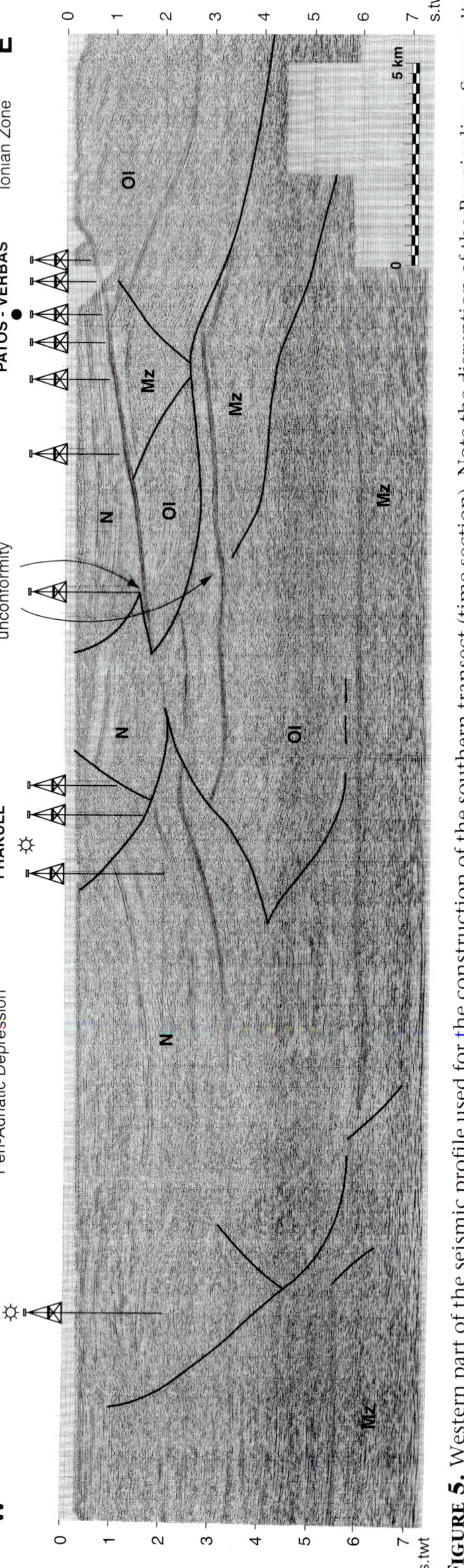

FIGURE 5. Western part of the seismic profile used for the construction of the southern transect (time section). Note the disruption of the Burgigalian-Serravalian unconformity, as a result of Pliocene-Quaternary fault reactivation. Mz = Mesozoic; Ol = Oligocene flysch; N = Neogene.

in the southern transect (Figure 7) accommodates the southeastward thickening observed within the Mesozoic series.

The following stratigraphic intervals have been used to define the successive evolutionary stages of the Thrustpack modeling (see also Table 1).

- ***Stage 0*** (from 70 to 30 Ma, Late Cretaceous to Eocene): The passive margin ended, with local emersions and bauxitic horizons in the Kruja Zone, but pelagic carbonates were still deposited in the Ionian Basin.
- ***Stage 1*** (from 30 to 20 Ma, Oligocene): The foreland flexure commenced, with deposition of the Oligocene flysch.
- ***Stage 2*** (from 20 to 15 Ma, latest Oligocene-early Miocene): The first deformation event occurred, followed by a progressive unconformity developing at the base of piggyback basins after Burdigalian time.
- ***Stage 3*** (from 15 to 10 Ma, middle Miocene): Foreland subsidence resumed in the Peri-Adriatic Depression, with progressive unconformities still developing at the base of the piggyback basins.
- ***Stage 4*** (from 10 to 5 Ma, late Miocene): The foreland underwent ongoing flexural subsidence, and progressive unconformities developed in the hinterland.
- ***Stage 5*** (from 5 Ma to Present, Pliocene-Quaternary): The foreland underwent ongoing flexural subsidence, and there was back thrusting and out-of-sequence thrusting near the former thrust fronts and emersion and karstification of the hinterland.

Incremental deformation was applied successively to the various thrust faults identified along each transect, with coeval activation of synorogenic sedimentation, erosion, and flexure, to account for the geometries and unconformities observed on seismic profiles and in structural sections (Figures 4 to 7). Using a trial-and-error procedure, it was possible to achieve a realistic kinematic model. The resulting geometries were reasonably similar to the modern seismic imagery and thus, were representative of the burial history of potential source-rock and reservoir intervals (Figures 8 and 9). However, because of the lack of supporting data, we did not involve the basement in the deformation, nor did we try to simulate any development of basement short-cuts, although this would be expected in the case of basin inversion reactivating formerly tilted blocks. Nevertheless, such complex structures have been described farther to the south, in the vicinity of Delvina, and at the front of the Mali Gjere unit (Roure et al., 1995; Dommergues et al., 2000), implying that our structural sections are simplistic and only partly representative of the regional structural style (i.e., they do not take into account

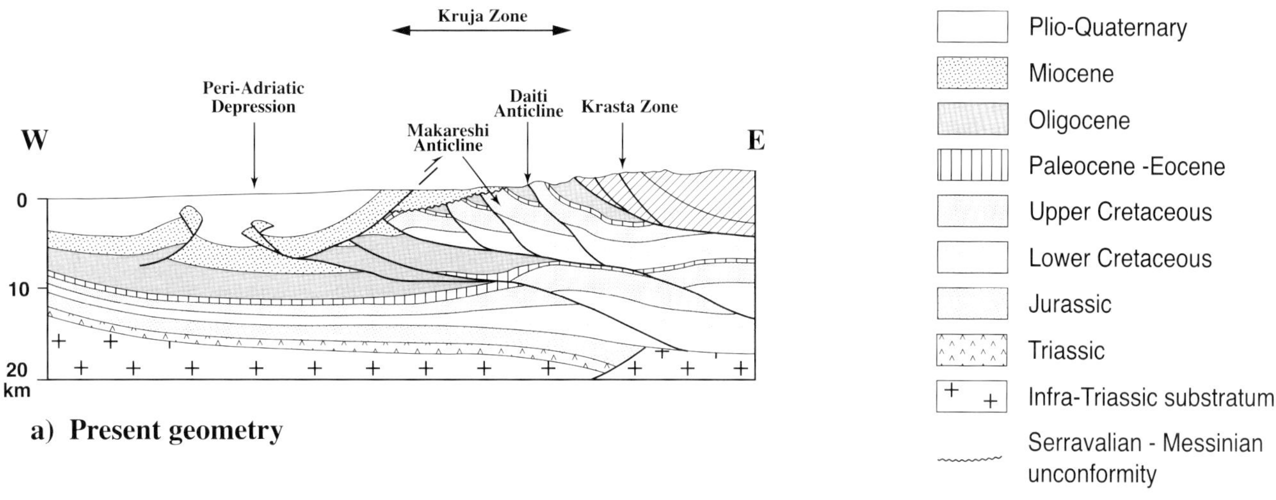

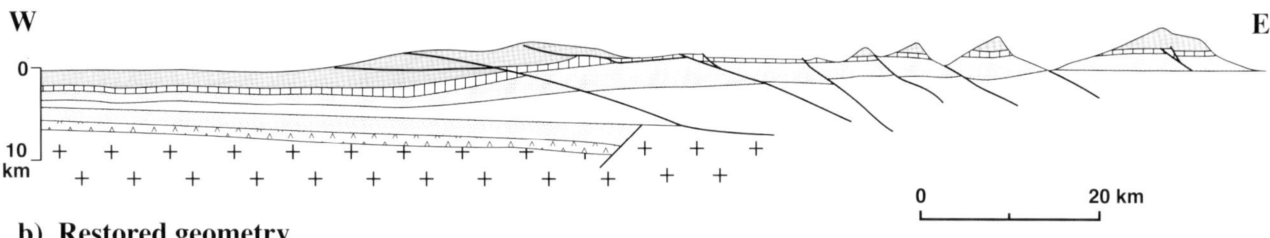

FIGURE 6. Northern cross section balanced with the Locace software. (a) Present geometry. (b) Restored geometry (Late Cretaceous–Paleogene stage).

possible lateral thickness variations within the Liassic series).

THERMAL AND GEOCHEMICAL DATA AND KEROGEN MATURATION MODELING

Distribution and Characteristics of the Potential Source Rocks

Numerous organic-rich horizons are evident within the synrift and passive-margin series, both in the presently allochthonous units of the Kruja Zone and in the Ionian Basin.

The Triassic Source Rocks

As in other parts of the Peri-Adriatic domain (i.e., in the Southern Alps, Apennines, and Sicily), the oldest potential source-rock intervals are located within Upper Triassic and Lower Liassic series (e.g., bituminous dolomite; Brosse et al., 1990; Koster et al., 1987).

Upper Triassic bituminous intervals have been drilled in the Sazani-1 well (Sazani or pre-Apulian Zone; Figure 1 and Table 1). In addition, 15-m-thick bituminous horizons crop out in the Cika unit of the Ionian Basin. In these Upper Triassic series, TOC values of as much as 5%, and R_o values between 0.7 and 0.9, indicate a rich, mature source rock. In more internal units, that is, in the Krasta-Cukali Zone, Upper Triassic argilite displays TOC values in the range of 1.5%.

The Jurassic Source Rocks

At least four organic-rich intervals have been described within the Ionian Jurassic series.

Organic-rich carbonates crop out in the Liassic series of the Cika and Kurveleshi units (Ionian Zone), with TOC values of between 15 and 24%, but they have not yet been drilled. These horizons are probably too thin (only a few centimeters thick) to account for an effective source-rock interval. Vitrinite reflectance values (R_o) are between 0.55 and 0.65 in surface outcrops, indicating the beginning of the oil window.

The Toarcian Posidonia Schist is well exposed in the Mali Gjere unit of the Ionian Basin, where it displays TOC values of 5% and R_o values of 0.55, indicating a still-immature source rock in surface conditions. Also known in Epirus (Danelian et al., 1986; Danelian and Baudin, 1990; Baudin et al., 1988, 1989, 1990; Jenkyns, 1988;

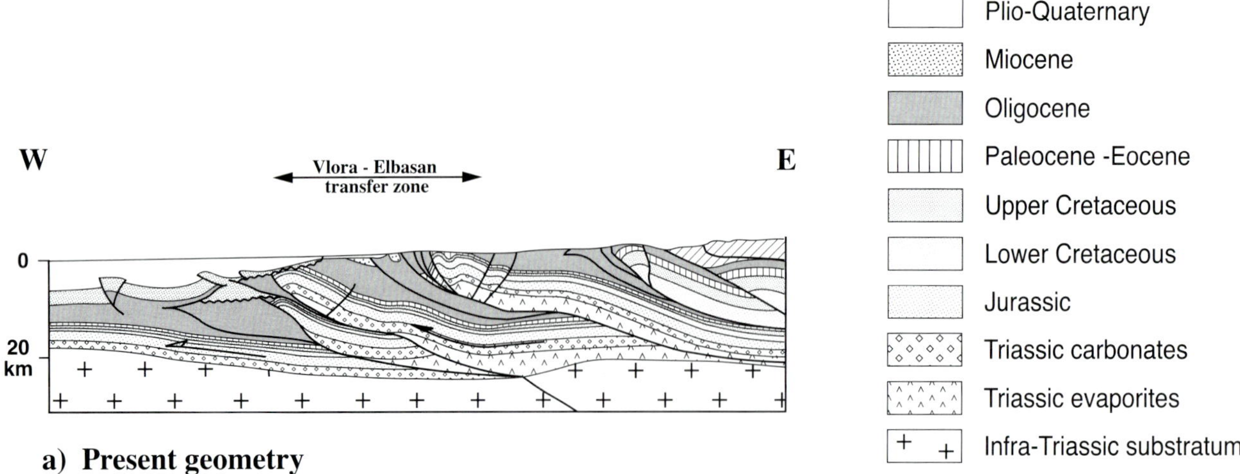

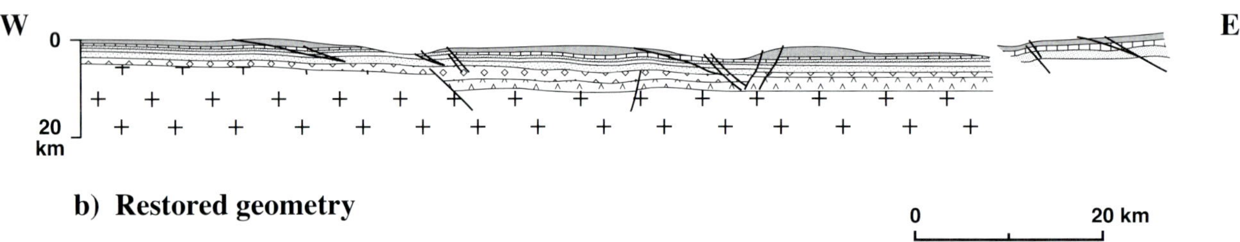

FIGURE 7. Southern cross section (Ionian transect), balanced with the Locace software. (a) Present geometry. (b) Restored geometry (late Cretaceous-Paleogene).

Baudin and Lachkar, 1990; Karakitsios, 1995; Karakitsios and Rigakis, 1996; Rigakis and Karakitsios, 1998), these Toarcian series can locally exceed 300 m in thickness, and they constitute the most prolific source-rock interval of the Ionian Basin (Palacas et al., 1985; Karakitsios et al., 1988).

Bituminous argilites have also been described in the Dogger of the Ionian Basin in the Kurveleshi unit, both in surface outcrops and wells, with TOC values of 5.25% and R_o values between 0.52 and 0.57 at the surface.

Organic-rich horizons also occur in the Upper Jurassic series, with TOC values of 1.5% and R_o values of 0.51 at surface.

The preservation of such low-maturity values at the surface in Liassic source rocks precludes their former burial as deep beneath the Oligocene to Neogene flexural sequence as is currently imaged for coeval series in the Peri-Adriatic Depression. This instead suggests that most Ionian structures started to develop as growth anticlines very early during the foothills evolution, most likely before or during the late Oligocene. This is also in agreement with the amounts of erosion locally detected beneath the early Miocene unconformity as the base of the piggyback basins.

Cretaceous Source Rocks

Euxinic episodes were numerous during the Cretaceous within the entire Peri-Adriatic basinal domains, with synchronous emersion episodes across adjacent intervening platforms (Van Graas et al., 1981, 1983; Schlanger et al., 1987; Jenkyns, 1991; Moldowan et al., 1992; Jerinic et al., 1994). Thus, bituminous shales and carbonates have been found at the boundary between Lower and Upper Cretaceous intervals in the Kurveleshi unit of the Ionian Basin, as well as in the Upper Cretaceous of the Cika unit of the same basin and within Late Cretaceous–Paleocene series of the Kruja Zone (Table 1). TOC values as high as 26% have been recorded in the Kurveleshi unit, and R_o values are between 0.48 and 0.53. In the Kruja Zone, TOC values barely reach 4%, and R_o values are lower than 0.5. In more internal units of the Krasta-Cukali Zone, Cretaceous shales, when not entirely oxidized, can reach TOC values of 1.5%.

Cenozoic Source Rocks

Apart from a few coal measures, the organic matter of the Cenozoic terrigenous series is quite dispersed in

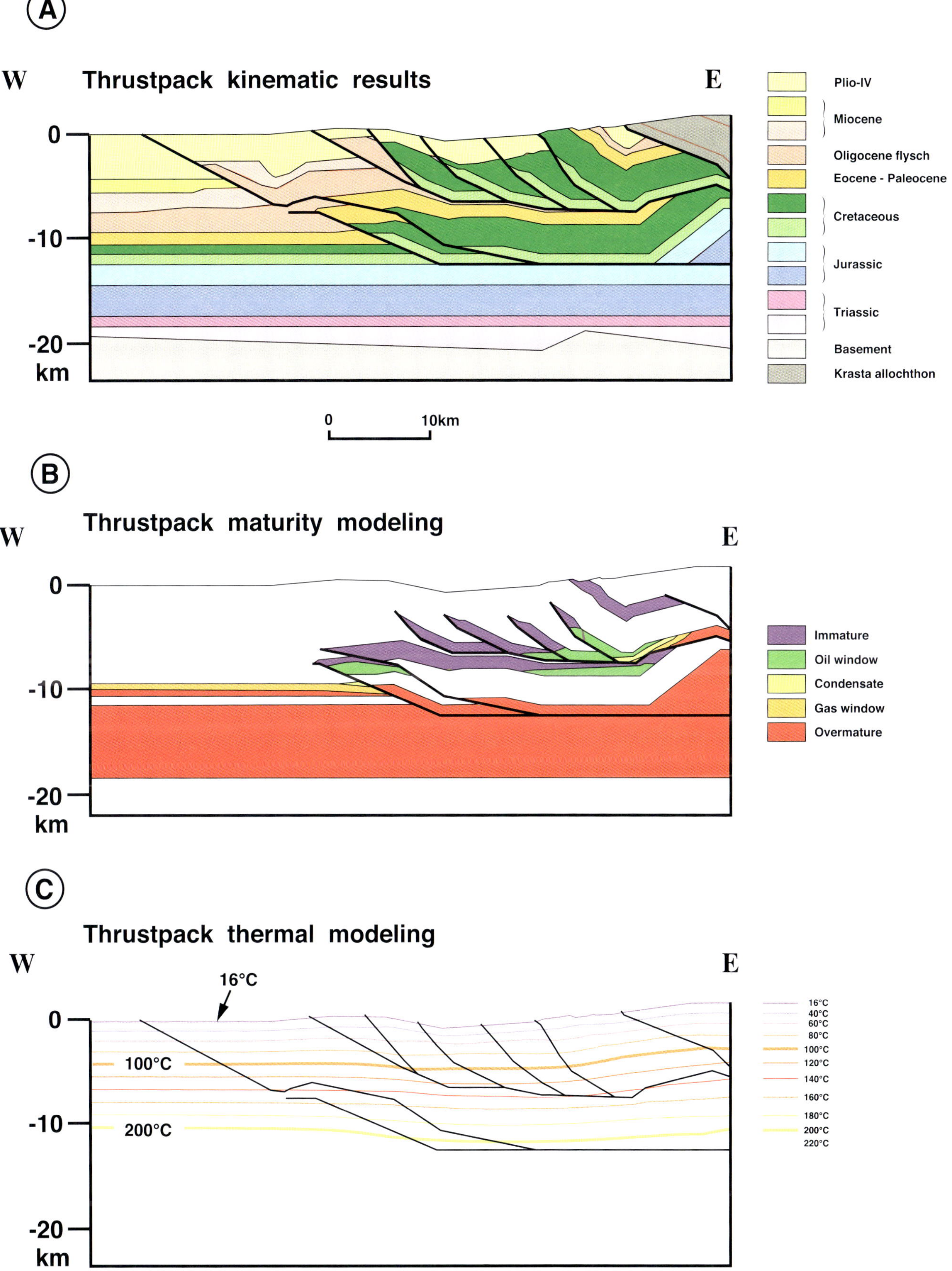

FIGURE 8. Results of the Thrustpack kinematic, thermal, and maturity modeling for the northern (Kruja) transect.

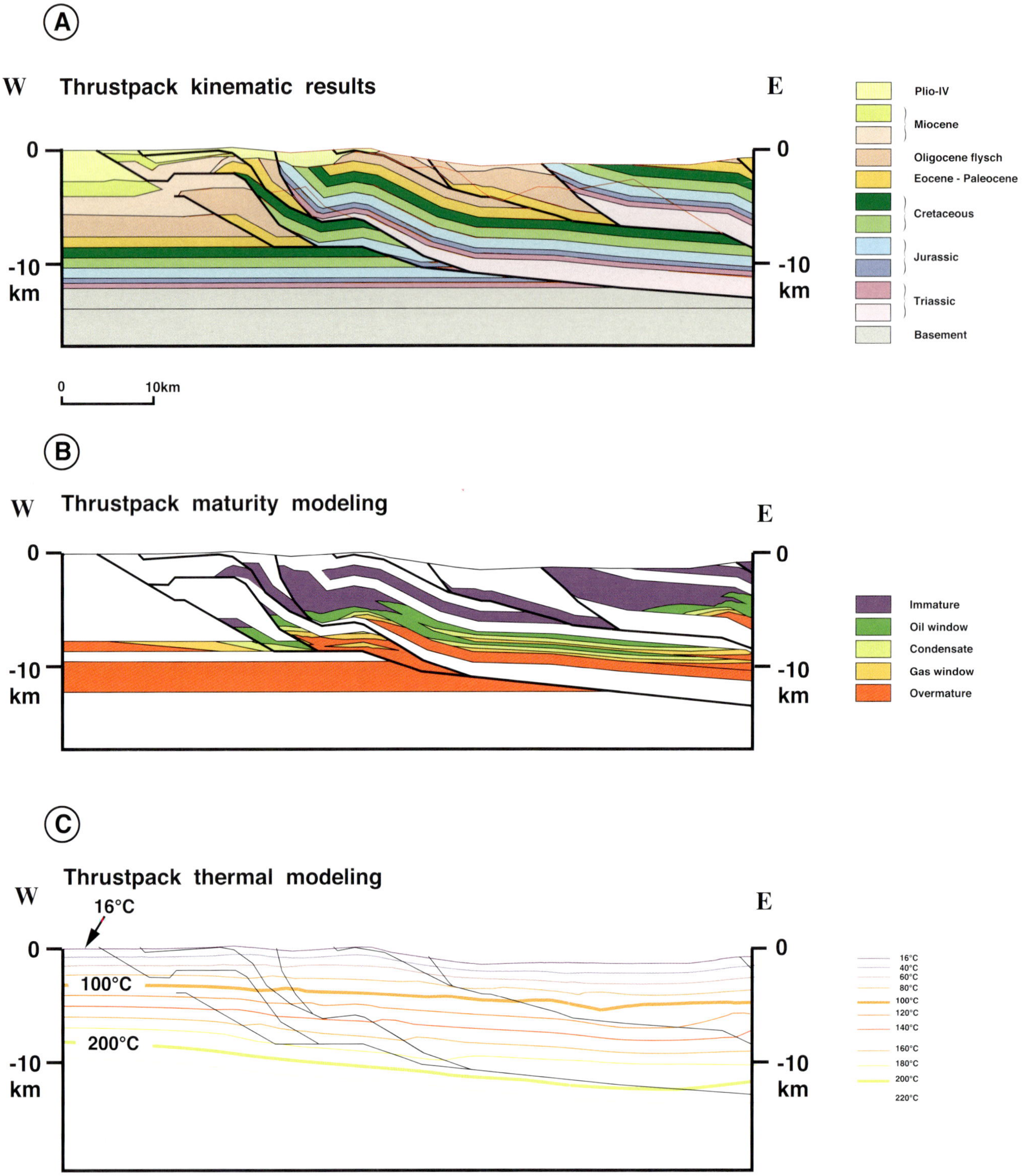

FIGURE 9. Results of the Thrustpack kinematic, thermal, and maturity modeling for the southern transect.

the Peri-Adriatic Depression (with low TOC values, averaging 0.3%) and displays a terrestrial signature with a type-III kerogen. This series is still mostly immature (R_O between 0.3 and 0.5) and is considered to be good only for generating biogenic gas (Valbona et al., 1995).

Bottom-hole Temperatures and Present Geothermal Gradients

Geothermal gradients are very low in the Outer Albanides, at least in the first 5 km below the earth surface for which numerous well data are available

(Frashëri et al., 1995). Average values of 20° C/km characterize the Peri-Adriatic Depression, whereas current geothermal gradients are in the range of 10° C/km in the Ionian Basin. Although the geothermal gradients in the Ionian Basin prior to the erosion of the Oligocene flysch were probably close to those still currently observed in the Peri-Adriatic Depression (having been protected from the high meteoric water flux by the Oligocene seal), these are extremely low values that require an explanation.

The rapid Neogene and Pliocene-Quaternary sedimentation created a blanketing effect in the Peri-Adriatic Depression. Thus, undercompacted sediments slow down thermal transfers and constitute an efficient barrier between the deep horizons and the earth's surface.

Strong karstification and fracturing of Ionian carbonates induced a major flow of meteoric water, which has resulted in a drastic cooling of the first 5 km below the earth's surface. In this case, it is not surprising to record local negative geothermal gradients in close proximity to active fluid conduits, such as faults or active aquifers.

Maturity Rates of the Organic Matter (T_{max} and R_o)

In the Upper Triassic series of the Cika unit, R_o values are between 0.7 and 0.9, thus indicating a mature source rock. Because they relate to surface samples, these values attest to an early maturation of the Triassic series, which eventually reached the oil window and started to expel hydrocarbons long before the onset of thrusting (the observed maturity rates indeed result from sedimentary burial, the maximum thickness of sediments probably being recorded prior to the late Oligocene-Aquitanian orogenic event—that is, during deposition of the lower Oligocene synflexural flysch).

Liassic carbonates, Toarcian Posidonia Schist, and other Mesozoic series were also sampled in surface outcrops of the Ionian Basin and Kruja Zone. However, all these samples display R_o values lower than 0.55 and thus, the rocks are immature (Roure et al., 1995). Nevertheless, these fossil maturity states only reflect the stage of maturation reached at the onset of compressional episodes (pre-Aquitanian stage). In contrast, the same source-rock intervals are likely to have reached the oil window locally in subthrust domains, during subsequent Neogene (Miocene and Pliocene-Quaternary) episodes of tectonic burial.

Mesozoic source-rock horizons of the Peri-Adriatic Depression are still deeply buried and attached to the autochthonous foreland. Therefore, their maturity has increased constantly during Neogene sedimentary and tectonic burial, with more than 5 km of Neogene series accounting for the present infill of the foredeep basin.

Again, samples collected within the Neogene molasse of the Peri-Adriatic Depression display R_o values of between 0.3 and 0.5 in both outcrops and exploration wells (at present depth, in the range of 2 or 3 km); thus, they are still immature. Nevertheless, because of the extremely wide range of burials recorded by these series, and despite the low geothermal gradients measured in the Peri-Adriatic Depression, part of the Neogene strata are probably already in the oil window. Thus, apart from the biogenic gas, Neogene coal measures are likely to contribute also to a local thermogenic source for hydrocarbons (Table 1).

Types and Distribution of Hydrocarbon Accumulations

Apart from biogenic gas, a few sandstone reservoirs are also producing gas condensate in the Peri-Adriatic Depression.

Although no commercial hydrocarbon accumulation has yet been found in the Kruja Zone, numerous surface oil seeps occur there along the Neogene unconformity (Shehu, 1995). The seeps attest to the occurrence of a currently active petroleum system in this area, and the Cretaceous series is the most likely to still be in the oil window in nearby underthrust synclines.

Major identified reserves of the Ionian Basin are located in the vicinity of the Vlora-Elbasan transfer zone (Albpetrol, 1993; Sedjini et al., 1994), where oil is extracted from both sandstone reservoirs of unconformable Neogene molasse and Cretaceous to Eocene fractured carbonates of the former Tethyan passive margin.

Structural History and Petroleum Systems of the Northern Transect (Kuja Zone and Peri-Adriatic Depression)

The simulation of the northern transect was initiated with the deposition of the Oligocene flexural sequence (flysch). Then, the initial thrust emplacement of the Kruja duplexes was simulated during the pre-Burdigalian compressional episode. Subsequently, the inner thrusts were modeled as having been active during the entire Miocene, together with a constant balance between erosion (in the hinterland) and sedimentation (in the foreland).

We assumed the western part of the platform domain was again involved in the Pliocene-Quaternary deformations, thus accounting for a progressive wedging of the terrigenous sequences of the Peri-Adriatic Depression, which became progressively thrust backward and eastward on top of the Kruja duplexes. We have also taken into account the possibility of an out-of-sequence reactivation of more-internal thrust fronts, such as the Daiti Thrust, because ongoing erosion could have changed the boundary conditions significantly within already accreted terranes, thus favoring such thrust reactivation (Hardy et al., 1998).

In the Thrustpack simulation, we have considered two distinct source-rock intervals: a type-II kerogen within the Lower Cretaceous series of the Kruja Platform, and a type-III kerogen in the Paleogene series (Figure 8).

Because rapid sedimentation rates and tectonic loading occurred during the Neogene, the Mesozoic source-rock horizons of the autochthon are entirely overmature, whereas coeval hydrocarbon products generated within the Peri-Adriatic Depression prior to the Pliocene-Quaternary are likely to be found updip in the regional foreland flexure, along the opposite (Italian) side of the Adriatic. In contrast, early entrapped hydrocarbons could have been preserved in relatively cold traps, within Late Cretaceous parautochthonous reservoirs, but these are deeply buried carbonates that would require open fractures or hydrothermal diagenesis to enhance their reservoir characteristics locally.

Also significant for exploration is apparent oil potential in the Cretaceous series of the Kruja Zone. It is indicated by the occurrence of low-maturity source rocks (still immature, or recording a fossil oil window). Therefore, Pliocene-Quaternary tectonism could have resulted locally in a new episode of petroleum generation, with out-of-sequence thrusts involving the Daiti and more-inner allochthonous units having contributed to increasing the tectonic burial above these potential kitchens (Figure 10).

Structural History and Petroleum Systems of the Southern Transect (Ionian Basin)

We selected the same structural agenda for the simulation of the Ionian transect, although we had to restrict the lateral extension of Neogene deposits to the frontal portion of the foothills, because no large piggyback basin occurs farther to the east.

As with the northern cross section, the Triassic autochthon displays an early maturation episode, coeval hydrocarbon products eventually having been trapped along the Sazani Platform, unless they migrated across the Adriatic up to the Puglia forebulge (i.e., the Aquila field; Mattavelli et al., 1991).

In contrast, more-external parts of the Ionian allochthon entered successively into the oil window at the time that foreland synclines became progressively underthrust beneath more-internal anticlinal structures (Figure 9). This induced a succession of expulsion and migration episodes, which contributed to a continuous infill of frontal prospects. However, the sealing efficiency of Neogene terrigenous cover constitutes the major exploration risk. Effectively, known Ionian petroleum occurrences clearly indicate that traps and migration pathways are most likely to be found within unconformable Neogene sequences rather than within the carbonates, in any location where the intervening Oligocene seal was removed by erosion during early Miocene time (Figure 11).

CONCLUSIONS

We did not take into consideration ductile shortening (pressure-solution) or layer-parallel thickening during our line-length restoration, with the resultant shortening estimates therefore being minimum values. Shortening of the Oligocene reference horizon reaches a value of 40% in the north (Kruja transect) and 35% in the south (Ionian transect) of the Outer Albanides (Figures 7 and 8). The total amount of shortening for this level is similar, in the range of 50–60 km along the two sections. However, these values become quite different when we look at the Mesozoic horizons, which have about zero shortening in the Triassic-Liassic series in the north, and approximately the same amount of shortening as for the Oligocene in underlying Mesozoic horizons in the southern transect. The deformed zone involving the Mesozoic carbonates is narrower in the north compared with the south. Such changes in the taper of the tectonic wedge relate to the regional shift observed in the basal décollement level across the Vlora-Elbasan transfer zone, the northern and southern sections comparing adequately with sand-box experiments involving either brittle (sand, northern transect) or ductile (silicone putty, southern transect) decoupling horizons (Figure 4) (Philippe, 1994; Letouzey et al., 1995; Roure et al., 1995). Assuming deformation was continuous in the Outer Albanides since the middle Oligocene, the resulting shortening values relate to an average convergence rate of 2 mm/yr since 25 Ma.

A heat-flow value of 30 mW/m^2 at the base of the sediments is sufficient to account for the maturity rates observed, as well as for the present distribution of the borehole temperatures (with an average surface temperature of 16°C).

Oil to source-rock correlations are still necessary to control or confirm some of the modeling results. However, these results are largely consistent with the present distribution of hydrocarbon accumulations in the Outer Albanides and provide helpful guidelines for the exploration.

In the Ionian Basin (southern transect, Figure 11), currently active kitchens are located in the underthrust synclines. As a result of increasing burial, hydrocarbon products generated during Pliocene-Quaternary time are likely to be more mature and lighter than hydrocarbons generated early in late Oligocene–early Miocene time. Although many carbonate anticlines have lost their early entrapped hydrocarbons as a result of extensive early

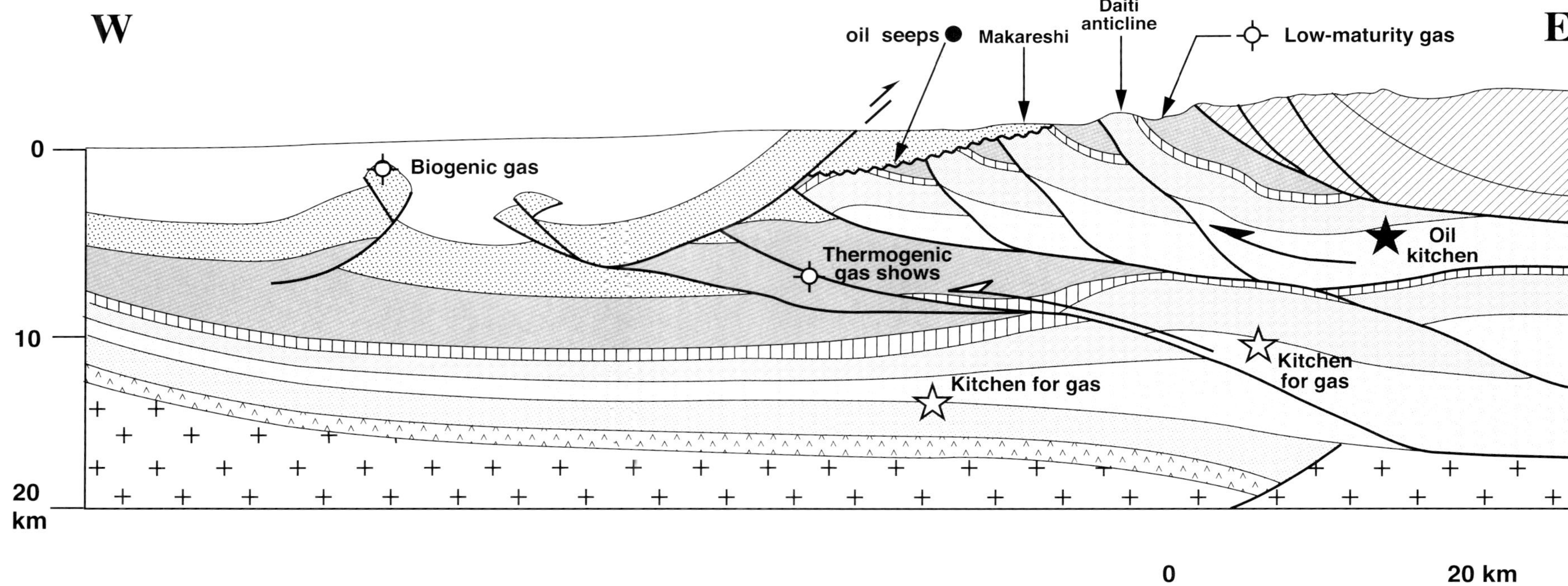

Figure 10. Migration pathways and petroleum systems along the northern transect. The lithostratigraphy is the same as in Figure 6.

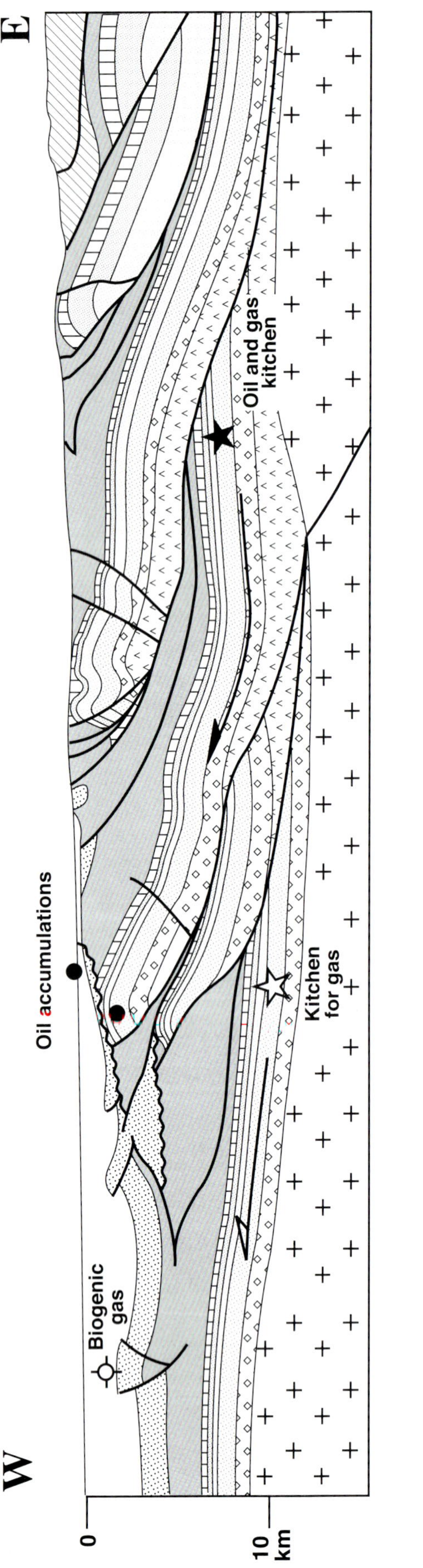

FIGURE 11. Migration pathways and petroleum systems along the southern transect. The lithostratigraphy is the same as in Figure 7.

Miocene erosion, distinct hydrocarbon phases are likely to have mixed in most of the productive carbonate fields.

The easiest migration pathway for the hydrocarbons would be lateral migration, with east-dipping Mesozoic series directly connecting the syncline kitchens with adjacent productive anticlines. However, an alternative hypothesis is to also consider vertical migration from the footwall, across or along the intervening thrust planes. In such a case, future exploration should consider not only the shallower Cretaceous-Eocene carbonate reservoirs, but also the deeper Triassic-Liassic dolomite, if very good seals can be documented in the intervening Posidonia Schist and other Jurassic and Cretaceous shaly intervals.

A large amount of hydrocarbons generated in the Peri-Adriatic Depression probably migrated across the Adriatic Basin, along the regional foreland flexure. However, late thermogenic gas and light oil are also likely to be entrapped locally in growth anticlines and in the vicinity of active mud diapirs of the Peri-Adriatic Depression itself, thus providing an additional, not-yet-documented target for exploration in Albania, in an area where only biogenic gas has been found to date.

Finally, the inner part of the Albanides has not been considered during this petroleum modeling study. However, the same type of simulations should be attempted farther to the east, at places where tectonic windows have been described and where the Oligocene flysch and Triassic evaporites are again exposed in the Peshkopie tectonic window at the rear of the ophiolites (Figure 1) (Collaku et al., 1990), if carbonate prospects can still be expected in the subsurface. Alternatively, T_{max} and/or R_o measurements could also provide a quick-look analysis on the maturity states reached by the Outer Albanian units beneath the ophiolites, and thereby provide an estimate of the maximum thickness ever reached by the currently eroded allochthonous series.

ACKNOWLEDGMENTS

This work was initiated in 1994 (OGI-IFP contract n°18002), and would not have been successfull without the strong support of numerous colleagues from OGI and Albpetrol. Thus, we greatly appreciated the collaboration of MM. Y. Sadiku, F. Fezgha, E. Seiti, L. Zaimi, and S. Dhrima. This work also benefited from the expertise of Mrs. C. Müller, MM. I. Mio, and E. Lafargue. MM. P. J. MacLaurin and I. Davison provided helpful comments on a former version of the manuscript.

REFERENCES CITED

Albpetrol, 1993, Petroleum exploration opportunities in Albania: 1st onshore licensing round in Albania: London, Western Geophysical publicity brochure.

Bakiaj, H., V. Misha, T. Velaj, L. Bandilli, L. Mehillka, and A. Berberi, 1990, Tectono-stratigraphic evolution of the Berati anticline and Dumre diapir, and its perspective for petroleum exploration (in Albanian): Nafta dhe Gazi, Fieri, p. 33–44.

Baudin, F., J. Dercourt, J. P. Herbin, and G. Lachkar, 1988, Le Lias supérieur de la zone ionienne (Grèce): une sédimentation riche en carbone organique: Paris, Comptes Rendus à l'Académie des Sciences, II, v. 307, p. 985–990.

Baudin, F., J. P. Herbin, J. P. Bassoullet, J. Dercourt, G. Lachkar, H. Manivit, and M. Renard, 1989, Distribution of organic matter during the Toarcian in the Mediterranean Tethys and Middle East, *in* A. Huc, ed., Deposition of organic facies: AAPG Studies in Geology no. 30, p. 73–91.

Baudin, F., J. P. Herbin, and M. Vandenbroucke, 1990, Mapping and geochemical characterization of the Toarcian organic matter in the Mediterranean Tethys and Middle East: Organic Geochemistry, v. 16, p. 677–687.

Baudin, F., and G. Lachkar, 1990, Géochimie organique et palynologie du Lias supérieur en zone ionienne (Grèce). Exemple d'une sédimentation anoxique conservée dans une paléo-marge en distension: Bulletin Société Géologique de France, p. 123–132.

Berberi, A., Z. Bega, K. Jano, and V. Stoli, 1990, Seismic imagery of the Dumre diapir, (in Albanian): Nafta dhe Gazi, v. 2, p. 77–93.

Brosse, E., A. Riva, S. Santucci, M. Bernon, J. P. Loreau, A. Frixa, and F. Laggoun-Defarge, F, 1990, Some sedimentological characters of the late Triassic Noto Formation, source rock in the Ragusa Basin, Sicily: Organic Geochemistry, v. 16, p. 715–734.

Casero, P., F. Roure, and R. Vially, 1991, Tectonic framework and petroleum potential of the southern Apennines, *in* A. M. Spencer, ed., Generation, accumulation and production of Europe's hydrocarbons: European Association of Petroleum Geoscientists Special Publication 1, p. 1–23.

Collaku, A., J. P. Cadet, V. Melo, A. Vranai, and M. Bonneau, 1990, Sur l'allochtonie des zones internes albanaises: mise en évidence de fenêtres à l'arrière de la nappe ophiolitique de la Mirdita (Albanie): Paris, Comptes Rendus Académie des Sciences, II, v. 311, p. 1251–1258.

Curi, F., 1993, Oil generation and accumulation in the Albanian Ionian basin, *in* A. M. Spencer, ed., Generation, accumulation and production of Europe's hydrocarbons: Springer, European Association of Petroleum Geoscientists Special Publication 3, p. 281–285.

Danelian, T., and F. Baudin, 1990, Découverte d'un horizon carbonaté, riche en matière organique, au sommet des radiolarites d'Epire (zone ionienne, Grèce): le Monte de Paliambela: Paris, Comptes Rendus Académie des Sciences, II, v. 311, p. 421–428.

Danelian, T., P. De Wever, and B. Vrielynck, 1986, Datations nouvelles fondées sur les faunes de Radiolaires de la série jurassique des Schistes à Posidonies (zone ionienne, Epire, Grèce): Revue de Paléobiologie, Genève, v. 5, no. 1, p. 37–41.

Diamanti, F., Y. Sadikaj, L. Zaimi, I. Tushe, M. Gjoka, I. Prifti, and B. Murataj, 1995, Hydrocarbon potential of Albania, in 1965–1995, 30 years: Oil and Gas Institute: Albpetrol ed., 300 p.

Dommergues, J. L., C. Meister, M. Bonneau, J. P. Cadet, and I. Fili, 2000, Les ammonites du Sinémurien supérieur et du Carixien inférieur à moyen du gisement de Lefterochori, près de Delvina, Est de Saranda (Albanie méridionale): Geobios, v. 33, no. 3, p. 329–358.

Finetti, I., G. Bricchi, A. DelBen, M. Pipan, and Z. Xuan, 1989, Geophysical study of the Adria plate, Memoria Societa Geologica Italiana, v. 40, p. 335–344.

Flores, G., M. Pieri, and G. Sestini, 1991, Geodynamic history and petroleum habitats of the southeast Adriatic region, *in* A. M. Spencer, ed., Generation, accumulation and production of Europe's hydrocarbons: Springer, European Association of Petroleum Geoscientists Special Publication 3, p. 389–398.

Frashëri, A., N. Kapedani, R. Lico, B. Canga, and E. Jareci, 1995, Geothermy of External Albanides, in 1965–1995, 30 years Oil and Gas Institute: Albpetrol ed., 300 p.

Frashëri, A., P. Nishani, S. Bushati, and A. Hyseni, 1996, Relationship between tectonic zones of the Albanides, based on results of geophysical studies, *in* P. Ziegler and F. Horvath, eds., Structure and prospects of Alpine basins and forelands: Mémoire Museum national d'Histoire naturelle, v. 170, p. 485–511.

Guzzetta, G., 1982, Thin-skinned style of deformation in Epirus, *in* HEAT Proceedings: Athens, v. I, p. 151–175.

Hardy, S, C. Duncan, J. Masek, and D. Brown, 1998, Minimum work, fault activity and the growth of critical wedges in fold and thrust belts: Basin Research, v. 10, p. 365–373.

Heba, G., 1997, Faciès, diagenèse et cycles sédimentaires des carbonates de la plate-forme de Kruja (Albanie) au Crétacé terminal, Master's Thesis, University of Lille and Paris XI-Orsay, 50 p.

I.G.R.S. and I.F.P., 1966, Etude géologique de l'Epire (Grèce nord-occidentale), Editions Technip, 306 p.

Industrial and Mining Research Institute, 1967, Geologic map of Albania, 1:200.000, Tirana.

Janopulli, V., Z. Bega, and C. Durmishi, 1995, Molasse backthrust phenomenon in Western Albania, in 1965–1995, 30 years: Oil and Gas Institute: Albpetrol ed., 300 p.

Jenkyns, H. C., 1988, The early Toarcian (Jurassic) anoxic event: Stratigraphic, sedimentary and geochemical evidence: American Journal of Science, v. 288, p. 101–151.

Jenkyns, H. C., 1991, Impact of Cretaceous sea level rise and anoxic events on the Mesozoic carbonate platform of Yugoslavia: AAPG Bulletin, v. 75, p. 1007–1017.

Jerinic, G., V. Jelaska, and A. Alajbeg, 1994, Upper Cretaceous organic-rich laminated limestones of the Adriatic carbonate platform, Island of Hvar, Croatia: AAPG Bulletin, v. 78, p. 1313–1321.

Karakitsios, V., 1995, The influence of preexisting structure and halokinesis on organic-matter preservation and thrust system evolution in the Ionian basin, northwest Greece: AAPG Bulletin, v. 79, p. 960–980.

Karakitsios, V., T. Danelian, and P. De Wever, 1988, Datations par les Radiolaires des Calcaires à filaments, Schistes à Posidonie supérieurs et Calcaires de Vigla (zone ionienne, Epire, Grèce) du Callovien au Tithonique terminal: Paris, Comptes Rendus Académie des Sciences, II, v. 306, p. 367–372.

Karakitsios, V., and N. Rigakis, 1996, New oil source rocks cut in Greek Ionian basin: Oil & Gas Journal, February 12, p. 56–59.

Koster, J., H. Wehner, and H. Hufnagel, 1987, Organic geochemistry and organic petrology of organic rich sediments within the "Hauptdolomit" Formation (Triassic, Norian) of the northern calcareous Alps: Organic Geochemistry, v. 13, p. 377–386.

Letouzey, J., B. Colletta, R. Vially, and J. C. Chermette, 1995, Evolution of salt-related structures in compressional settings, *in* M. P. A. Jackson, D. G. Roberts, and S. Snelson, eds., Salt tectonics: a global perspective: AAPG Memoir 65, p. 42–60.

Mattavelli, L., L. Novelli, and L. Anelli, 1991, Occurrence of hydrocarbons in the Adriatic Basin, *in* A. M. Spencer, ed., Generation, accumulation and production of Europe's hydrocarbons: Springer, European Association of Petroleum Geoscientists Special Publication 1, Oxford, p. 369–380.

Moldowan, J. M., P. Sundaraman, T. Salvatori, A. Ajalbeg, B. Gukic, C. Y. Lee, and G. Demaison, 1992, Source correlation and maturation assessment of select oils and rocks from the Central Adriatic basin (Italy and Yugoslavia), *in* M. Moldowan, P. Albrecht, and R. P. Philip, eds., Biological markers in sediments and petrol: New Jersey, Prentice-Hall, p. 370–401.

Monopolis, D., and A. Bruneton, 1982, Ionian Sea (Western Greece): its structural outline deduced from drilling and geophysical data: Tectonophysics, v. 83, p. 227–242.

Muço, B., 1994, Focal mechanism solutions for Albanian earthquakes for the years 1964–1988: Tectonophysics, v. 231, p. 311–323.

Muço, B., 1998, Catalogue of $ML \geq 3.0$ earthquakes in Albania from 1976 to 1995 and distribution of seismic energy released: Tectonophysics, v. 292, p. 311–319.

Muftari, S., S. Prillo, L. Hasanaj, P. Sadushi, and D. Shehu, 1995, Faunistic zones of Oligo-Miocene and Pliocene deposits of Ionian zone and Peri-Adriatic Depression of Albania, based on Foraminifera, in 1965–1995, 30 years: Oil and Gas Institute: Albpetrol ed., 300 p.

Nikolaou, C., 1986, Contribution to the knowledge of the Neogene, the geology and the Ionian and pre-Apulian limits in relation to the petroleum geology observations in Strophades, Zakynthos and Kephalynia islands: Ph.D. Thesis, University of Athens, 228 p.

Palacas, J. G., D. Monopolis, C. A. Nicolaou, and D. E. Anders, 1985, Geochemical correlation of surface and subsurface oils, western Greece: Organic Geochemistry, v. 10, p. 417–423.

Paulucci, G., L. Novelli, D. Bongiorni, and R. Cesaroni, 1988, Deep offshore exploration in the southern Adriatic Sea: Offshore Technology Conference, Houston, v. 5730, p. 413–421.

Philippe, Y., 1994, Transfer zone in the southern Jura thrust belt (Eastern France): geometry, development and comparison with analogue modeling experiments, *in* A. Mascle, ed., Hydrocarbon and petroleum geology of France: Springer, European Association of Petroleum Geoscientists Special Publication 4, p. 327–346.

Rigakis, N., and V. Karakitsios, 1998, The source rock horizons of the Ionian Basin (NW Greece): Marine and Petroleum Geology, v. 15, p. 593–617.

Roure, F., 1999, Stratégies de l'exploration dans les chaînes plissées et leurs avant-pays: Pétrole et Techniques, v. 418, p. 62–69.

Roure, F., E. Prenjasi, and Z. Xhafa, 1995, Petroleum geology of the Albanian thrust belt, Guide book to the Field Trip 7: AAPG International Conference, Nice, 50 p.

Roure, F., and W. Sassi, 1995, Kinematic of deformation and petroleum potential appraisal in Neogene foreland fold-and-thrust belt systems: Petroleum Geosciences, v. 1, p. 253–269.

Schlanger, S. O., M. A. Arthur, H. C. Jenkyns, and P. A. Scholle, 1987, The Cenomanian-Turonian oceanic anoxic event; I Stratigraphy and distribution of organic carbon-rich beds and the marine δC^{13} excursion, *in* Brooks and Fleet, eds., Marine petroleum source rocks: Geological Society of London Special Publication 26, p. 371–399.

Sedjini, B., P. Constantinescu, and T. Piperi, 1994, Petroleum exploration in Albania, *in* B. Popescu, ed., Hydrocarbons of Eastern Central Europe, p. 1– 28.

Shehu, H., 1995, The geological feature and oil gas bearing perspective of Kruja tectonic zone, in 1965–1995, 30 years: Oil and Gas Institute, Albpetrol ed., 300 p.

Shteto, T., S. Guri, E. Seiti, and Z. Xhafa, 1995, The structural model in the external Albanides and the hydrocarbon prospectivity, in 1965–1995, 30 years: Oil and Gas Institute, Albpetrol ed., 300 p.

Sorel, D., G. Bizon, S. Aliaj, and L. Hasani, 1992, Essai de synthèse sur l'âge et la durée des phases compressives des Hellénides externes (Grèce nord-occidentale et Albanie), depuis le Miocène: Bulletin Société Géologique de France, v. 163, p. 447–454.

Speranza, F., I. Islami, C. Kissel, and A. Hyseni, 1995, Paleomagnetic evidence for Cenozoic clockwise rotation of the external Albanides: Earth and Planetary Science Letters, v. 129, p. 121–134.

Tagari, D., 1993, Etude néotectonique et sismotectonique des Albanides: analyse des déformations et géodynamique du Langhien à l'Actuel, Ph.D. Thesis, University Paris XI, Orsay.

Underhill, J. R., 1989, Late Cenozoic deformation of the Hellenide foreland, Western Greece: Geological Society of America Bulletin, v. 101, p. 613–634.

Valbona, U., I. Tushe, H. Seiti, and S. Nazaj, 1995, The estimation of Adriatic-Ionian offshore hydrocarbon prospective (Albanian offshore): European Association of Geoscientists and Engineers, 57th Conference, Glasgow, Abstracts, P523.

Van Graas, G., T. C. Viets, J. W. de Leeuw, and P. A. Schenck, 1981, Origin of the organic matter in a Cretaceous blackshale deposit in the Central Apennines (Italy): Advances in Organic Geochemistry, p. 471–476.

Van Graas, G., T. C. Viets, J. W. de Leeuw, and P. A. Schenk, 1983, A study of the soluble and insoluble organic matter from the Livello Bonarelli, a Cretaceous blackshale deposit in the Central Apennines, Italy: Geochimica et Cosmochimica Acta, v. 47, p. 1051–1059.

Veizaj, V., and A. Frashëri, 1995, The relationships between Albanides orogeny and Apulian platform according to the gravity data, in 1965–1995, 30 years: Oil and Gas Institute, Albpetrol ed., 300 p.

Velaj, T., I. Davison, A. Serjani, and I. Alsop, 1999, Thrust tectonics and the role of evaporites in the Ionian zone of the Albanides: AAPG Bulletin, v. 83, p. 1408–1425.

Velaj, T., and C. Xhuli, 1995, The evaporite effect in the tectonic style of the internal Ionian subzone of the Albanides: European Association of Geoscientists and Engineers, 57th Conference, Glasgow, Abstracts, P574.

Waters, D. W., 1994, The tectonic evolution of Epirus, northwest Greece: Ph.D. Thesis, Cambridge University, U.K.

25

Hill, K. C., J. T. Keetley, R. D. Kendrick, and E. Sutriyono, 2004, Structure and hydrocarbon potential of the New Guinea Fold Belt, *in* K. R. McClay, ed., Thrust tectonics and hydrocarbon systems: AAPG Memoir 82, p. 494–514.

Structure and Hydrocarbon Potential of the New Guinea Fold Belt

Kevin C. Hill[1]
Victorian Basin Studies Centre, Earth Sciences, La Trobe University Melbourne, Australia

Jeffrey T. Keetley[1]
Victorian Basin Studies Centre, Earth Sciences, La Trobe University Melbourne, Australia

R. Dan Kendrick[1]
Victorian Basin Studies Centre, Earth Sciences, La Trobe University Melbourne, Australia

Edy Sutriyono[2]
Victorian Basin Studies Centre, Earth Sciences, La Trobe University Melbourne, Australia

ABSTRACT

Seismic, borehole, and field data acquired over the last 12 years demonstrate that the structural style in the Papuan Fold Belt comprises inverted basement extensional faults and asymmetric detachment folds that break through the overturned forelimb. The previous fault-bend fold model was flawed because of a lack of data and because of inappropriate analogs that ignored the mechanical stratigraphy.

In the oil province of Papua New Guinea, the deformation front has not yet impinged on strong Australian lithosphere, so a relatively low fold belt occupies its own foreland basin. The adjacent gas-condensate province in the western Papuan Fold Belt has just impinged on the strong lithosphere, developing a foreland basin and basement-cored anticlines. Along strike in the Irian Jaya Fold Belt, the deformation front has encountered the buttress of the strong Australian lithosphere, causing a 15-km-thick Paleozoic and Mesozoic sequence to be thrust to the surface along a previously extensional, basin-margin fault. Focusing deformation on one fault has created mountains 5 km high and an adjacent foreland basin. Structures in the mountains are breached, but the adjacent foreland basin has hydrocarbon potential. Farther west, in the "Bird's Neck" area, the poorly known Lengguru Fold Belt resembles the oil province in the Papuan Fold Belt, but Pleistocene extensional faulting may have caused breaching. In addition, deep burial by thick Tertiary carbonates has probably destroyed reservoir porosity, except in the north, where additional discoveries like the 30-tcf Tangguh gas province are possible.

[1]*Present address*: 3D-GEO, Earth Sciences, University of Melbourne, Victoria 3052, Australia.
[2]*Present address*: Universitas Sriwijaya, Palembang, Sumatra, Indonesia.

INTRODUCTION

As the leading edge of the northward-moving Australian plate, the island of New Guinea records the effects of Tertiary convergence with the Pacific and Philippine-Caroline plates (Figure 1) (Hall, 1997; Hill and Hall, 2003). This culminated in the late Miocene–early Pliocene New Guinea orogeny (Hill and Hall, 2003) forming the mobile belt and fold belt that traverse the island, adjacent to the stable Arafura/Fly Platform to the south.

The New Guinea Fold Belt and adjacent platform areas have been the sites of significant oil and gas discoveries, including the Kutubu-area and Gobe oil fields in Papua New Guinea (PNG) (Buchanan, 1996) and, most recently, the 30-tcf Tangguh gas province in Irian Jaya (Robertson, 1998). One of the keys to success and to future discoveries is unraveling the complex geologic structure of the fold belt. This task is made difficult by the paucity of data, which is the result of the jungle-covered, strongly karstified mountainous terrain.

More than ten years ago, on regional balanced and restored cross sections, Hill (1991) interpreted the structures of the Papuan Fold Belt as fault-bend folds. Since that time, substantial development drilling, improvements in the acquisition of seismic data (Hill, G. S. et al., 1996), further field mapping, structural modeling (Dixon, 1996), and an improved understanding of mechanical stratigraphy have better defined the structural style in PNG (Buchanan et al., 2000). Some of the new interpretations incorporating this knowledge are similar to the original PNG models presented in the 1960s (Smith, 1965).

In this chapter we review interpretations of the Papuan Fold Belt through time and then present revised regional balanced and restored cross sections incorporating the latest data, particularly seismic data. We then present new regional sections across the Central Ranges and the Lengguru Fold Belt of Irian Jaya for comparison, discuss why they are so different from those in PNG, and review their hydrocarbon potential. All sections were constructed, balanced, and restored using Geosec™.

STRATIGRAPHY

Stratigraphically, New Guinea is divided into two halves, particularly with respect to pre-Mesozoic beds. The Tasman Line in Australia (Scheibner, 1974) separates areas underlain by Proterozoic continental crust, to the west, from those underlain by Paleozoic continental crust, to the east— which is thought to overlie Cambrian oceanic crust. This line projects northward into New Guinea just on the Irian Jaya side of the border with PNG (Figure 1). In Irian Jaya, Proterozoic basement is overlain by thick sequences of Paleozoic sediments (Figure 2) similar to those along Australia's North West Shelf. In contrast, in PNG, the Paleozoic sediments have undergone greenschist-grade metamorphism and have been intruded by Triassic granites as part of the New England Orogeny of eastern Australia (Hill and Hall, 2003).

The Mesozoic and Tertiary sedimentary histories of PNG and Irian Jaya are similar (Figure 2). Triassic and Early Jurassic rifting led to crustal-scale extensional faults with major thickening of Jurassic facies and deposition of Jurassic source rocks in distal settings (Figure 2) (Pigram and Panggabean, 1984; Home et al., 1990). The Late Jurassic to Neocomian sandstone reservoirs were deposited during lowstands in the mudstone-dominated postrift sequence. Late Cretaceous rifting and Paleocene seafloor spreading in the Coral Sea and northern PNG (Pigram and Symonds, 1991) were associated with regional denudation in the area of the Papuan Fold Belt, but there was ongoing Cretaceous and Tertiary subsidence and deposition in Irian Jaya. Thick Tertiary limestones were deposited in Irian Jaya, whereas the Mesozoic beds in PNG were unconformably overlain by thinner Miocene limestone. All sequences were deformed during late Miocene–Pliocene orogenesis and, in some areas, were overlain by Pliocene molasse (Figure 2).

The known hydrocarbon reservoirs are the Mesozoic sandstones, particularly the Toro Sandstone in PNG. In the Salawati Basin and the Gulf of Papua, oil and gas are trapped in Miocene reefs (Figure 1). The main hydrocarbon source rocks are in distal Jurassic and possibly Cretaceous shales. In Irian Jaya, the Permian coal measures may also be good gas source rocks. Hydrocarbon generation is thought to have occurred mainly in the Pliocene, charging the traps, but it may also have occurred during the Late Cretaceous (Hill and Gleadow, 1989; O'Sullivan et al., 1995).

Importantly, from a structural point of view, the stratigraphic section in PNG consists of basement rock cut by extensional faults, overlain by a 2- to 5-km-thick Mesozoic clastic sequence, dominated by shales, but including a 100- to 300-m-thick, more competent Upper Jurassic-Neocomian Toro sandstone unit. The Mesozoic clastics are unconformably overlain by 1- to 2-km-thick, competent Miocene limestone (Figure 2). This combination of plastic shales and thick competent units leads to detachment geometries during compression. The main detachments are in the shales just above basement and those just below the Miocene limestone. In Irian Jaya, on the other hand, there is scope for more detachment horizons because of the thicker and more complete stratigraphic sequence.

In PNG, a fairly abrupt change to distal facies is recorded in the Miocene strata; this change generally coincides with a thickening of the Mesozoic strata (Hill et al., 1990, 2000; Hill, 1991). This was interpreted to be a long-lived crustal-scale fault system that controlled the shelf edge. The facies change can be observed throughout New Guinea and similarly interpreted.

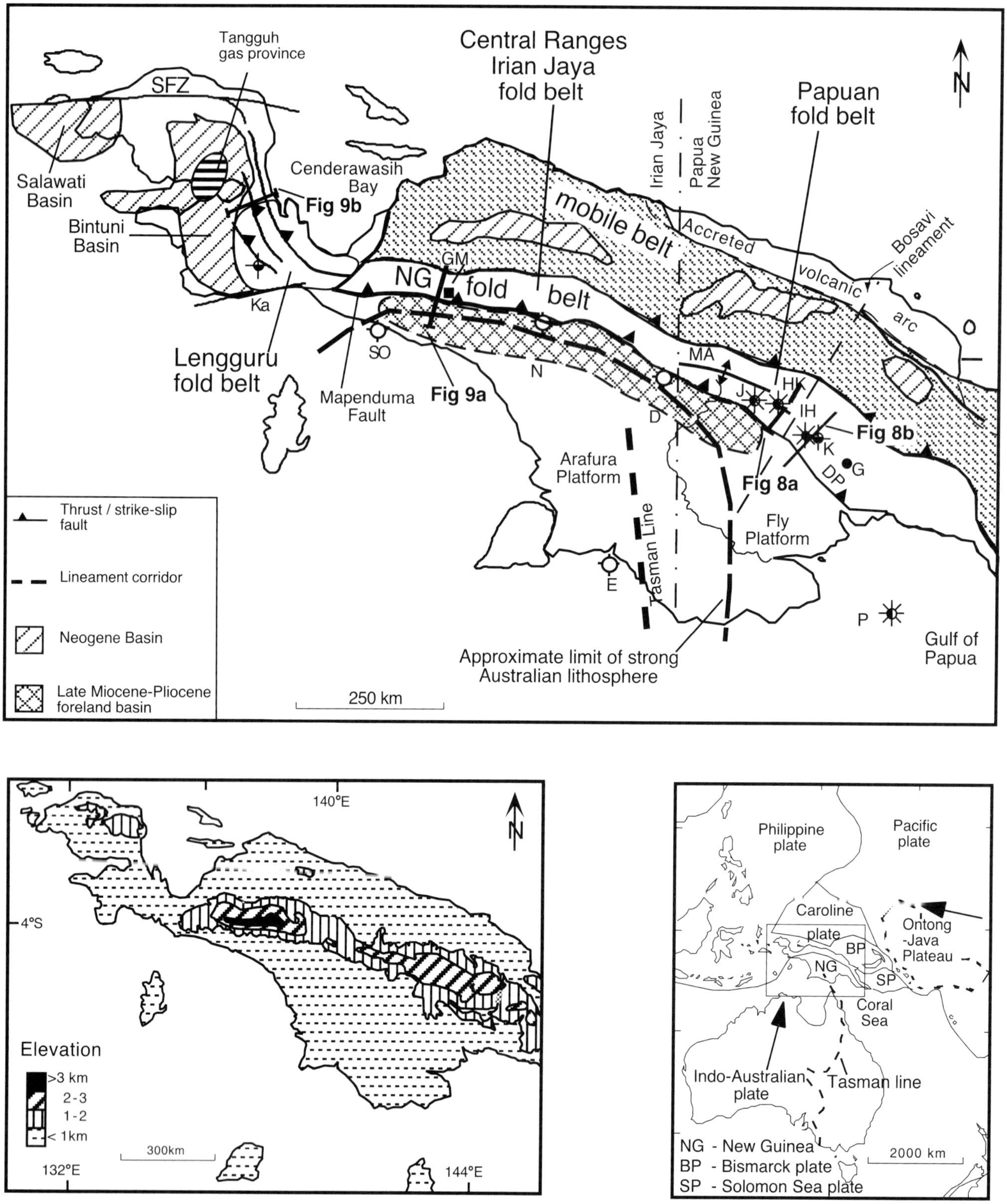

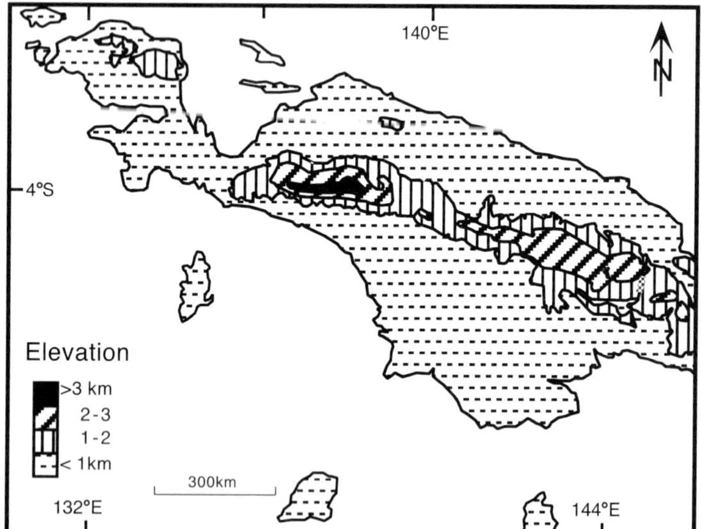

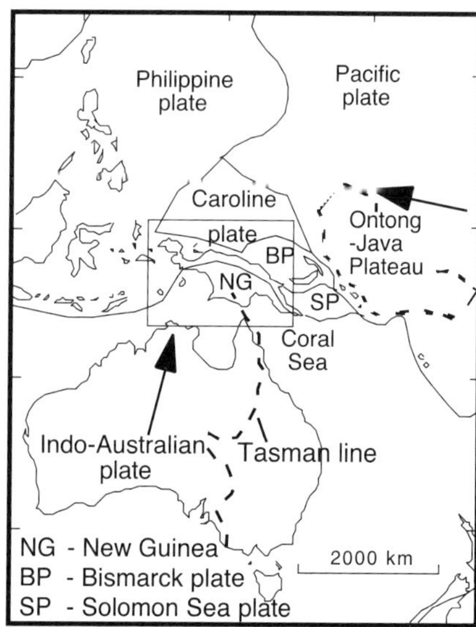

Figure 1. Location of the New Guinea Fold Belt, showing key wells/fields and approximate limits of some of the Neogene basins, after Quarles van Ufford (1996), Kendrick et al. (1995), Hill, K.C. et al. (1996). D = Digul-1; DP = Darai Plateau; E = E-2-X; G = Gobe-1; HK = Hides-Karius; IH = Iagifu-Hedinia; J = Juha; K = Kutubu; Ka = Kamakawala-1; MA = Muller Anticline; N = Noordwest-1; P = Pandora; SFZ = Sorong fault zone; SO = South Oeta. Below left is the topographic map of New Guinea adapted from Dow (1977), Dow et al. (1986), McDowell et al. (1996). Below right shows the plate tectonic setting of New Guinea and present relative plate motions.

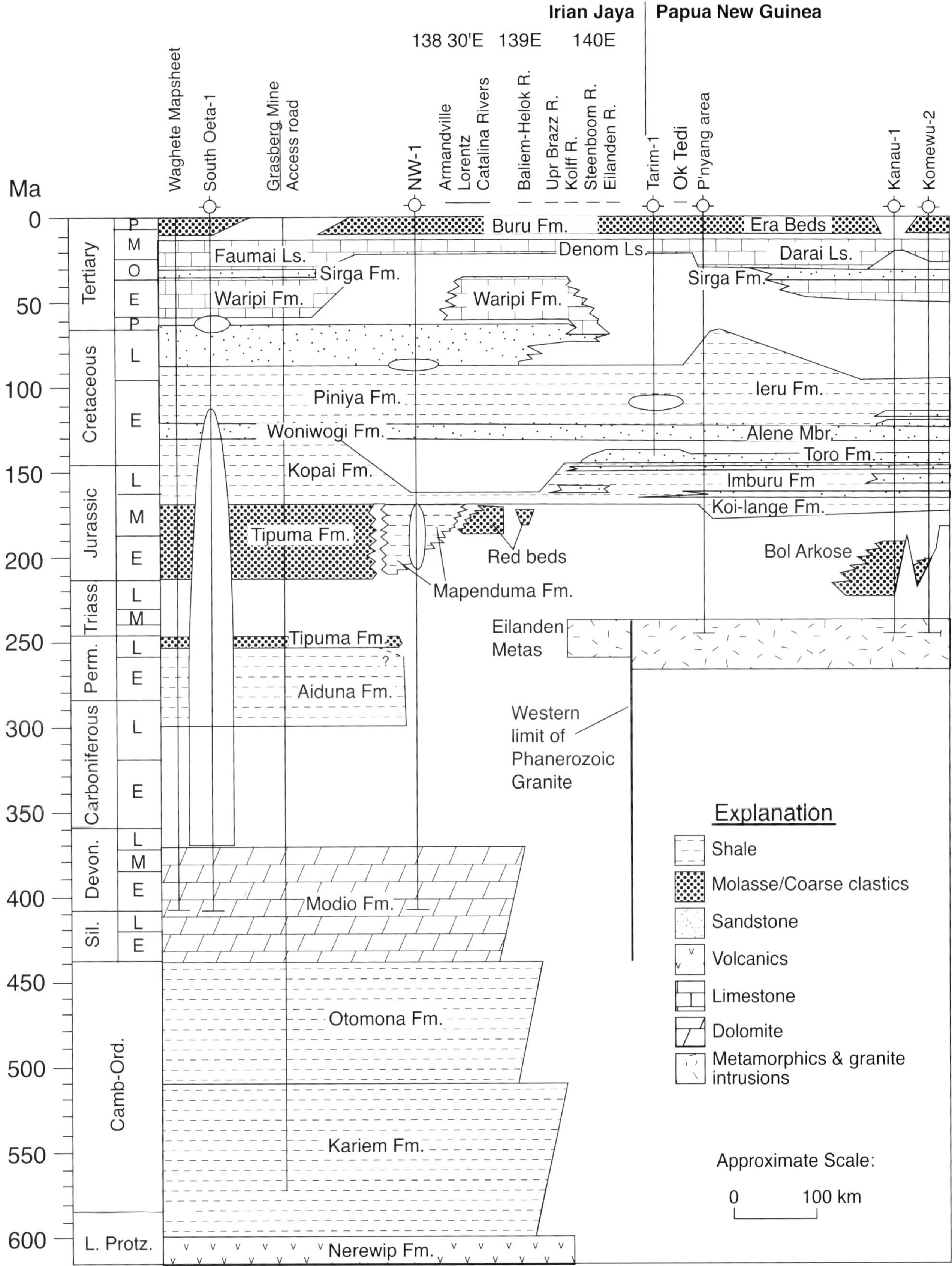

FIGURE 2. Generalized chronostratigraphy of the Central Ranges of Irian Jaya and western PNG, showing the main source and reservoir rocks. Adapted from Pigram and Panggabean (1989), Home et al. (1990), Granath and Hermeston (1993), and Valenti (1993).

PAPUA NEW GUINEA FOLD BELT

Both structural and tectonic models for the evolution of PNG have evolved through time, using the popular model of the day. This is largely because of the lack of data in the karstic jungle-covered, mountainous terrain of PNG, which requires the use of external models to interpret structure. However, the increased exploration associated with production of oil and gas in the last 10 years has resulted in a large amount of well data, in addition to field data with reliably dated samples (Hornafius and Denison, 1993) constraining the structural styles. Also, the development of robust, although expensive, seismic acquisition techniques (Hill, G.S. et al., 1996) has led to the collection of good-quality seismic data over key structures (Buchanan et al., 2000). Thus, we now have a better understanding of the structural styles in PNG, and we can use this understanding to predict further discoveries.

Structural Interpretation through Time

In the 1960s, interpretations were dominated by geosynclinal models. These involved vertical uplift of a hinterland basement complex and gravity sliding of the cover rocks to create the fold-and-thrust belts. The individual structures were interpreted in terms of asymmetric detachment folds, cored by ductile Jurassic strata, with occasional breakthrough thrust faults (Figure 3a) (Smith, 1965; Laws, 1971). Such an interpretation is likely to be restorable, with a relatively small amount of shortening.

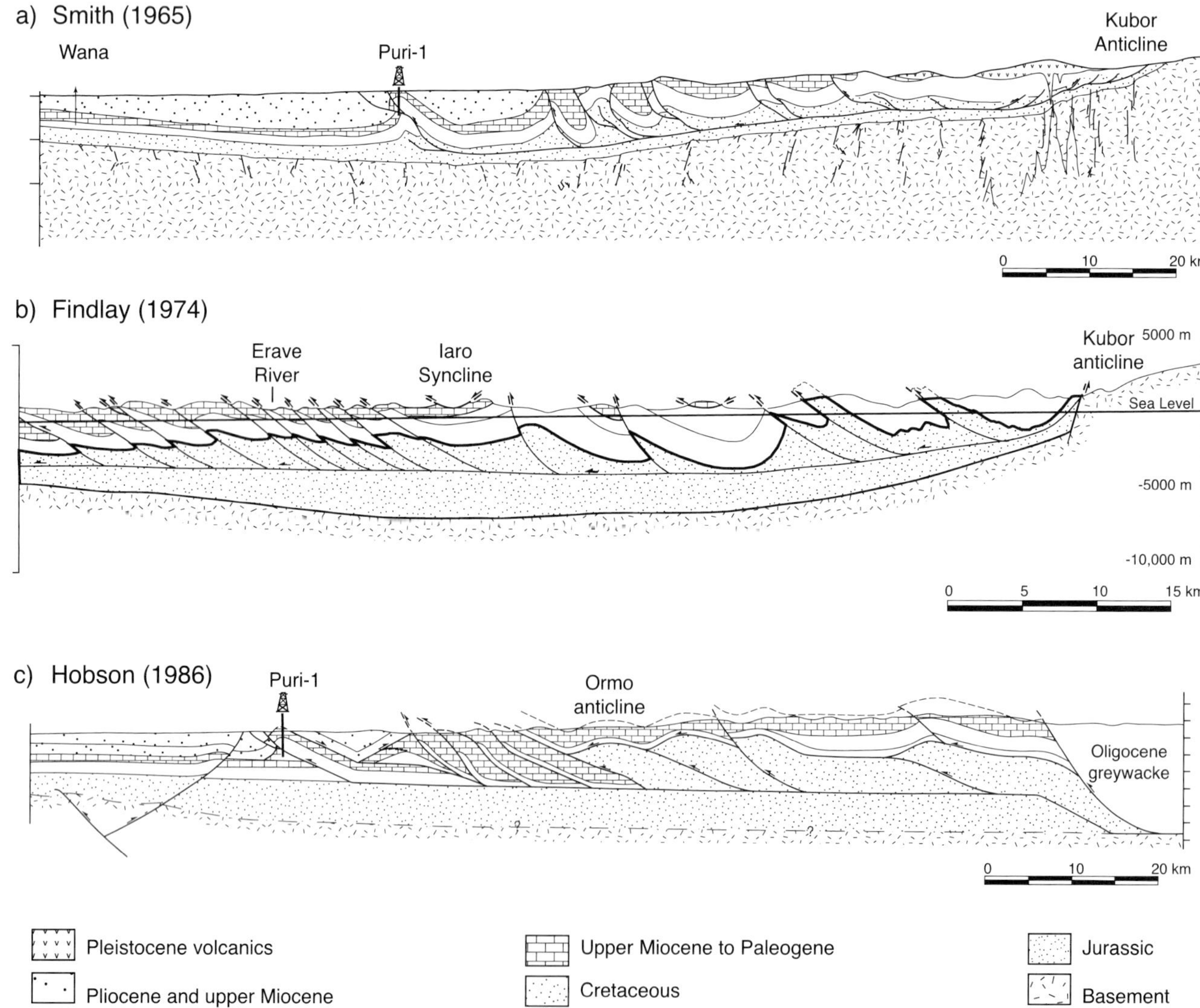

FIGURE 3. Evolution of regional structural interpretations across the eastern Papuan Fold Belt through time. (a) Gravity sliding and asymmetric detachment folds with thrusts through the overturned forelimb (Smith, 1965), (b) gravity sliding and imbricate thrusts (Findlay, 1974), and (c) fault-bend fold model (Hobson, 1986).

In the 1970s, interpretations evolved to incorporate the new plate-tectonic models— that is, arc-continent collision creating an uplifted basement complex with subsequent gravity sliding to build the fold-and-thrust belt. With the well-known mountain belts of North America as an analog, the individual structures were interpreted as imbricate thrusts with relatively little folding, which indicates more-competent Jurassic strata (Figure 3b) (Findlay, 1974; Jenkins, 1974). The section in Figure 3b cannot be restored as shown, because several faults have minimal offset of Jurassic strata but large offset of the Miocene limestone.

In the 1980s, the rigor of balanced cross sections was seemingly applied to the geometric analysis of fold-and-thrust structures, with the assumption that the fundamental structure was the through-going detachment and that associated thrust ramps resulted in fault-bend folds (Boyer and Elliot, 1982). Regional interpretations involved Neogene and Mesozoic duplexes detached above basement, but also suggested some basement inversion (Figure 3c) (Hobson, 1986; Hill, 1991). Most models for the tectonic evolution of New Guinea involved early and late Miocene arc-continent collisions that created an orogenic belt.

More-rapid progress was made in the 1990s, following the acquisition of considerably more field and borehole data as a result of the success of oil and gas exploration and new geophysical techniques. However, interpretations were still model driven. As occurred elsewhere in the world, inversion structures became popular, and, with some supporting evidence from magnetotelluric data, the fold belt was interpreted to have minimal shortening and to be the result of compressional reactivation of previous extensional faults in the basement (Figures 4, 5) (Bennet and Parish, 1993; Buchanan and Warburton, 1996). However, drilling of the Kutubu field wells demonstrated more-complex structures with steep forelimbs, asymmetric folds, and break-through thrust faults, as had been inferred in the 1960s (Lamerson, 1990; Eisenberg, 1993; Franklin and Livingston, 1996; Lingrey, 2000).

Hides Section

The Hides Anticline in the western part of the Papuan Fold Belt (Figure 4); (Grainge, 1993; Johnstone and Emmett, 2000) has a proven gas column of more than 1240 m in the Toro Sandstone reservoir. The ultimate height of the gas column is probably greater than 1800 m, with estimated reserves of more than 8 tcf gas and associated condensate. A small amount of gas is currently being produced to generate electricity for the nearby Porgera gold mine.

Early interpretations of the Hides Anticline show variations from fault-bend folds to basement inversion (Figure 5). The Karius-1 well, drilled in 1995, encountered a repetition of the Oligocene-Miocene limestone, with nearly vertical limestone in the bottom of the well, suggesting a strong folding component.

Reflection seismic data acquired in 1990 to the southwest of Hides and in 1995 over the Hides Anticline (Hill, G.S. et al., 1996) illustrate the structural style in the Hides area (Figures 4, 6, and 7). Line 90-02 traverses the narrow Cecilia Anticline, the frontal structure of the fold belt, which is interpreted as a thrust repeat of the Miocene limestone, as was encountered in the Cecilia-1 well. Although basement is difficult to pick, the line also indicates a relatively abrupt thickening of the Mesozoic sequence beneath the Cecilia Anticline, suggesting a large extensional fault cutting the basement (Figure 6). There also appears to be a significant low-amplitude anticline beneath the Cecilia Anticline, more than can be attributed to velocity pull-up, here interpreted to be the result of minor inversion of this extensional fault. To the northeast of the Cecilia Anticline, a larger, broad monocline is apparent, with a strong horizontal reflector at 2.3 s below the southwest-dipping limb. This reflector is interpreted to be a short footwall flat (Figure 6), with the Cecilia fault cutting down through basement to the northeast. This fault, too, originally may have been extensional.

Line BP95-01 (Figure 7) records good data below the Hides Anticline but noise below the Karius Anticline, where vertical to overturned beds were encountered at the surface (Hill, 1989) and in the Karius-1 well. The line shows the Jurassic sequence thickening to the southwest to the limit of good data, and a back thrust that intersects the surface.

Figure 8a shows a regional interpretation of the Hides Anticline, with the interpretations from the depth-converted seismic lines projected onto it. For alternative interpretations and more seismic data, see Johnstone and Emmett (2000) and Cole et al. (2000). The interpretation presented here suggests that preexisting extensional faults were important in controlling the development of the large surface anticlines. The Hides Anticline is interpreted to have developed by inversion of a basement fault, thereby generating an asymmetric fold in the cover with late-stage breakthrough thrusts in the forelimb and backlimb. The large anticline to the southwest shows the incipient growth of a similar structure. Here, though, a buckle fold has not yet developed, perhaps because of the thinner Jurassic sequence, illustrating the importance of mechanical stratigraphy. The overall shortening across the section is ~20%, although shortening is increasing to the northeast, where a duplex in the Miocene limestone is exposed.

Iagifu-Wage-Nembi Section

This section traverses the Iagifu-Hedinia and Kutubu Anticlines, to the southwest of Lake Kutubu. These

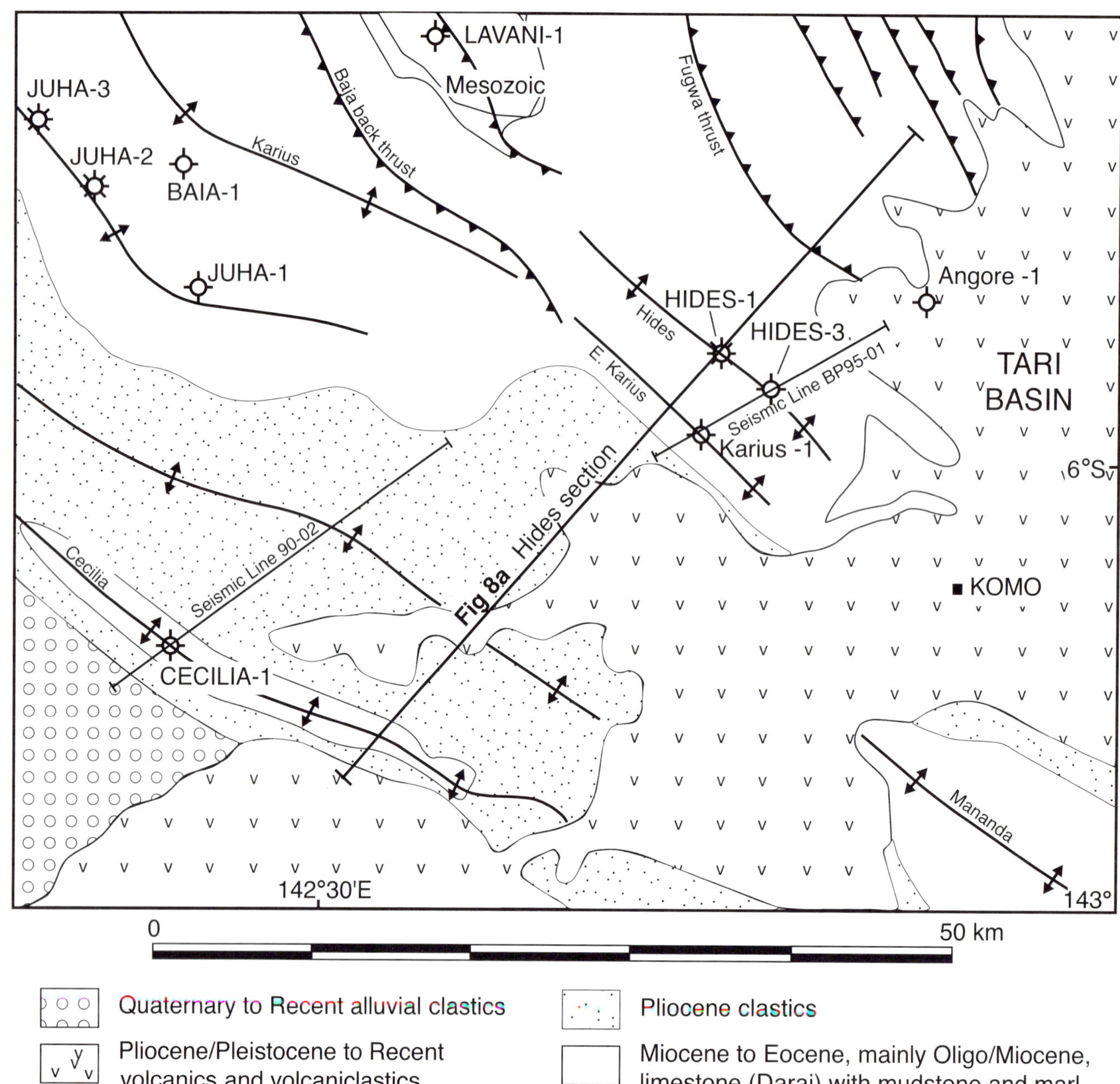

Figure 4. Location map of the Hides Anticline showing recent seismic lines.

anticlines contribute to the Kutubu area's oil production. Original-oil-in-place reserves were estimated to be 550 MMSTB (million standard bbl) (Wali, 1996).

Hill (1991) interpreted the same section line to be fault-bend folds of the entire sedimentary section to the southwest of Lake Kutubu, but with a Miocene limestone duplex overlying a Mesozoic duplex to the northeast. Overall shortening was estimated to be 54%. However, subsequent drilling of more than 30 development wells on the Iagifu-Hedinia structure, as well as new discoveries along strike, has revealed that folds with overturned forelimbs are common in the Mesozoic sequence (Franklin and Livingston, 1996; Lamerson, 1990; Eisenberg, 1993; Lingrey, 2000). In addition, Dixon (1996) presented analog models illustrating buckle folding in weaker sedimentary layers detached from thrusting of an adjacent competent layer.

These new data constraining the structural style are incorporated into the revised regional balanced and restored cross section shown in Figure 8b. The section incorporates the interpretation of Franklin and Livingston (1996) for the Iagifu-Hedinia Anticlines, inferring that the structure initially formed as a detachment fold with thrusts subsequently breaking through the steep forelimb and a minor thrust through the backlimb. The structures cropping out as far as 50 km northeast of Lake Kutubu are similarly interpreted to be to the result of initial folding of the Mesozoic section detached from

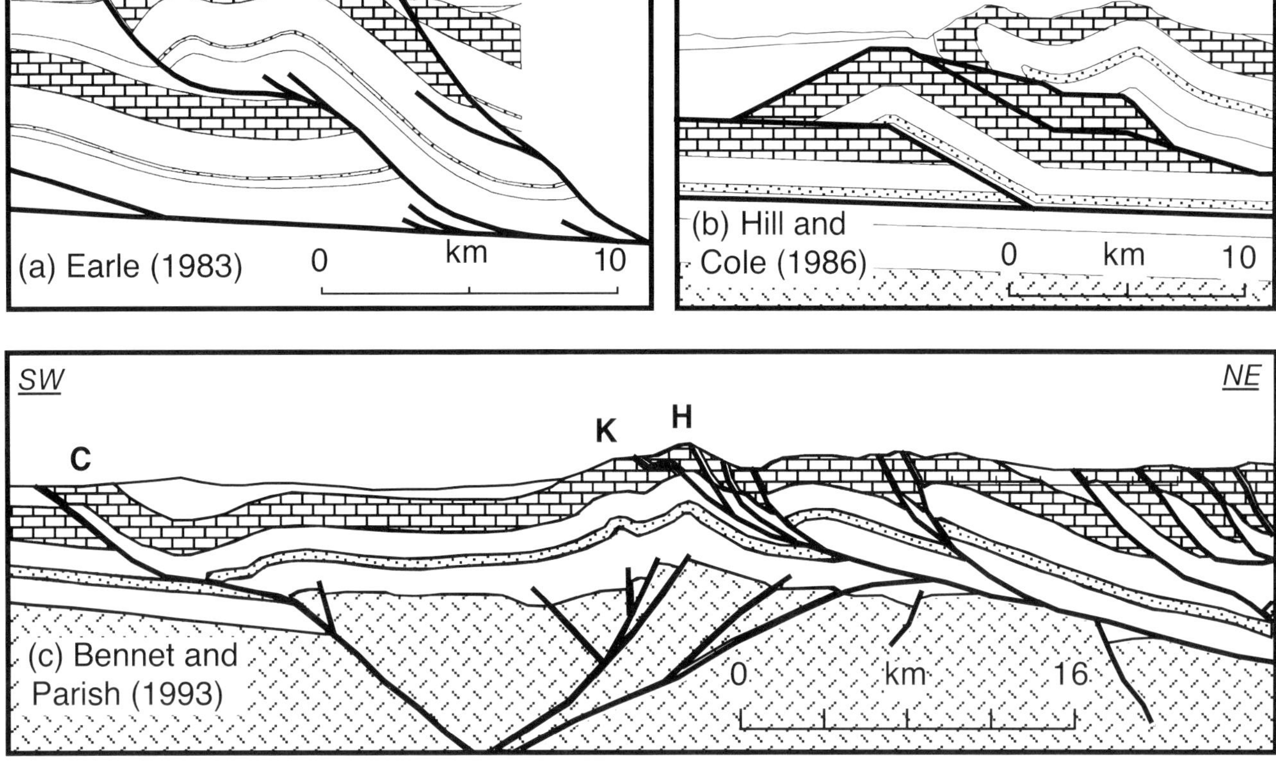

FIGURE 5. Structural interpretations across the Hides Anticline through time. K = Karius Anticline, H = Hides Anticline, KT = Karius Thrust, C = Cecilia Anticline. (a) simple fault-bend fold, (b) overturned fault-bend fold incorporating overturned dips on Karius Anticline, (c) inversion model incorporating interpretation of magnetotelluric data. See Figure 2 for stratigraphy.

thrusting of the Miocene limestone. To the northeast of the shelf-edge fault, deformation is more complex in the much thicker and more distal sedimentary section. A basement shortcut fault is interpreted at the shelf edge, following Cooper et al. (1989). In the distal Upper Cretaceous and Tertiary facies to the northeast, tight detachment faults are inferred by Jenkins et al. (1969), and along strike to the southeast by Hill et al. (1990). These are interpreted to overlie larger wavelength folds in the Jurassic section (Figure 8b). Overall the section shows ~40% shortening.

COMPARISON WITH IRIAN JAYA

In western New Guinea, the mountains are divided into the Irian Jaya Fold Belt and Lengguru Fold Belt (Figure 1). Both areas are represented by regional balanced and restored cross sections along the lines with the most available data (Figure 9a, b). However, both areas are less well known than the Papuan Fold Belt along strike, which is used as a model for some of the structures. In constructing the sections, it was found that the Papuan models were not generally applicable. This indicates that there are substantial and important differences between the areas that affect their structure and prospectivity. The main observed differences are as follows:

1) The Irian Jaya Fold Belt is nearly twice the length of the Papuan Fold Belt, which, in turn, is twice the length of the Lengguru Fold Belt.
2) The Irian Jaya Fold Belt reaches 5 km above sea level and is much higher than the Papuan Fold Belt. The Papuan Fold Belt is generally <3 km above sea level, whereas the Lengguru Fold Belt is low-lying, at ~1 km above sea level (Figure 1).
3) The Irian Jaya Fold Belt is generally narrower than the Papuan Fold Belt, whereas the Lengguru Fold Belt is of comparable width in the south, but tapers to nothing in the north. Coupled with the elevation data, this indicates that the Irian Jaya Fold Belt

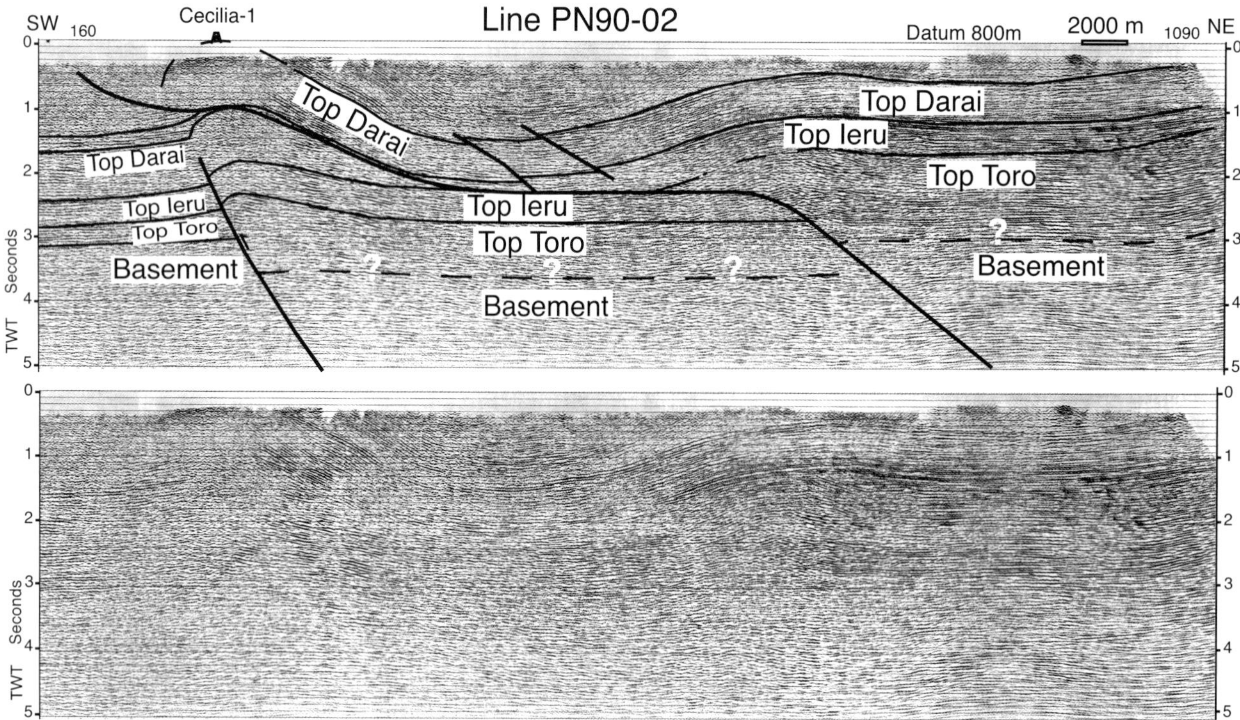

FIGURE 6. Interpretation of seismic line 90-02 (see Figure 4 for location). Note the asymmetric fold beneath the Cecilia structure, with an inferred faulted overturned forelimb, demonstrated from surface dips and the Cecilia well.

may have undergone comparable shortening but that the shortening was focused over a limited width. In contrast, the Lengguru Fold Belt has spread the shortening over a wide area. This suggests that the strength and discontinuities in the underlying lithosphere had a major influence on structural evolution.

4) The Irian Jaya Fold Belt trends uniformly east-west, only turning southeast near the Papuan border. The Papuan Fold Belt trends east-southeast, whereas the Lengguru Fold Belt is highly oblique and trends north-northwest. Again, this suggests an underlying structural control in the Irian Jaya Fold Belt.
5) Irian Jaya was far more influenced by late Proterozoic–Paleozoic rifting, as occurred along Australia's North West Shelf, which resulted in as much as 8 km of pre-Mesozoic sedimentary section, including late Proterozoic, Ordovician, and Permian detachment horizons, all of which are absent in PNG (Figure 2).
6) Thick Paleozoic strata commonly occur at the surface along major frontal thrusts in the Irian Jaya Fold Belt, indicating that the thrusts sole into deep detachments. In contrast, basement exposures are rare in the Papuan Fold Belt and occur only in the hinterland of the Lengguru Fold Belt, which implies that there are shallower, regional detachments above basement in the Papuan and Lengguru Fold Belts.
7) The east-west trend of the Irian Jaya Fold Belt suggests an underlying east-west crustal fabric, but it contains common east-southeast-trending oblique thrusts, indicating a direction of compression that is common with that of PNG.
8) Unlike the other two belts, the Lengguru Fold Belt has common normal faults cutting across the belt and substantial normal faults in the hinterland and throwing down to Cenderawasih Bay, which suggests post-orogenic collapse.
9) Like the Papuan Fold Belt, the Lengguru Fold Belt can be divided into structural provinces with different structural styles (Figures 8b, 9b), including large anticlinal structures similar to those of the oil fields in the Kutubu and Gobe areas and the gas-condensate field at Hides.

Irian Jaya Fold Belt

The section through the western Irian Jaya Fold Belt passes close to the Grasberg gold mine, one of the world's richest gold deposits, at an elevation of more than 5000 m. Mapping and prospecting around the mine has led to a significant amount of new data, although still much less than there is for PNG. Seismic-reflection data in the foreland (Phillips Petroleum, unpublished report, 1972; Umbach and Klepacki, 1994), combined with field, well, and stratigraphic data from the Grasberg and adjacent areas (Parris, 1994), were used to constrain the section's interpretation. This was in addition to the Timika (Rusmana et al., 1995), Hitalipa (Panggabean et al., 1995), and Rotanburg (Harahap

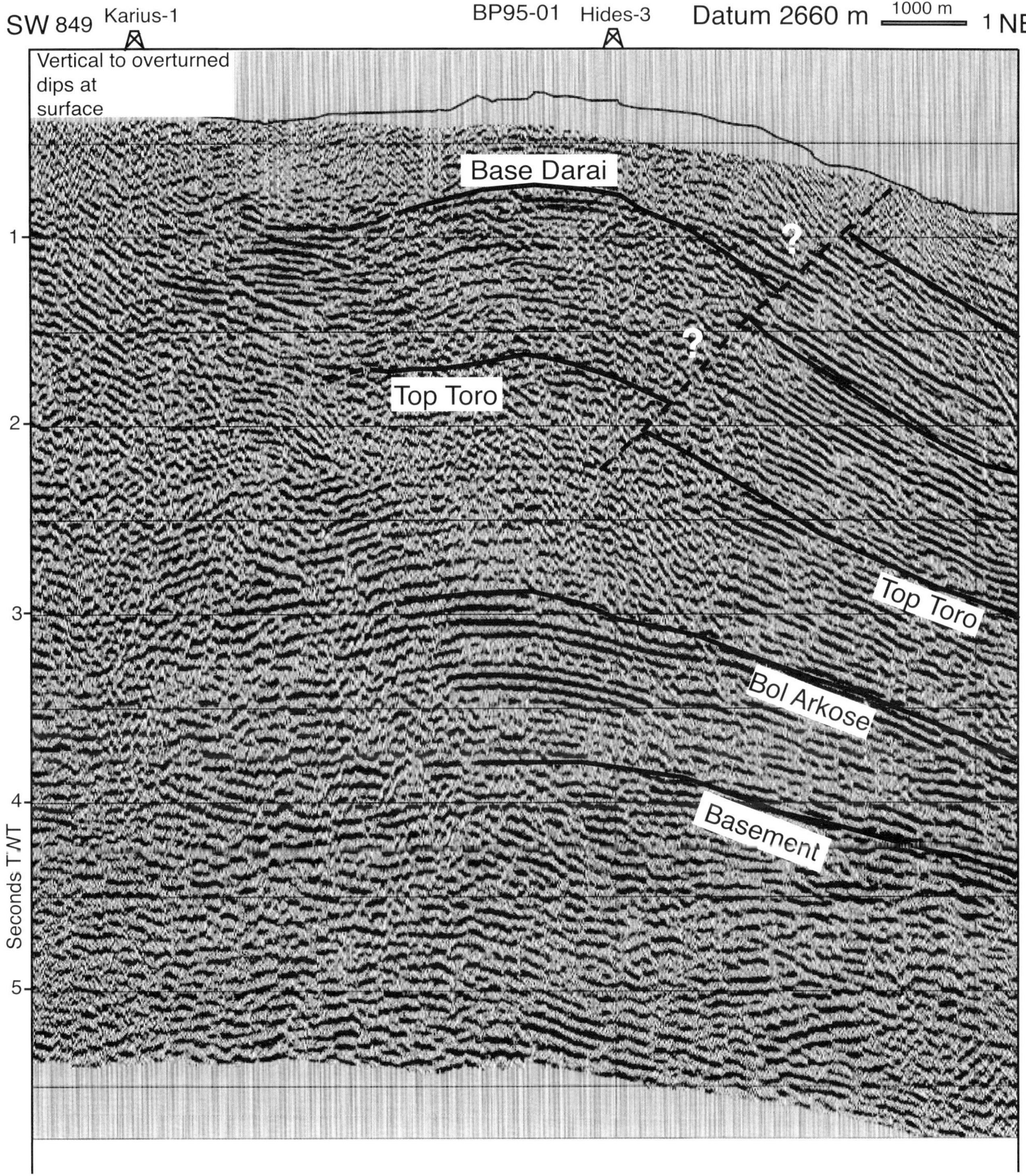

FIGURE 7. Interpretation of Seismic Line BP95-01 across the Hides and Karius Anticlines. See Figure 4 for location.

and Noya, 1995) 1:250,000 geologic maps and more recent mapping of Parris (1996a, b) and Parris and Warren (1996).

The dominant structure evident in the western fold belt is the Mapenduma Fault (Figure 9a). This is interpreted to be a crustal-scale fault, across which changes in facies and >8-km thickening of the pre-Tertiary section occur, as recorded from outcrop data along the mountain front and seismic data in the foreland (Phillips Petroleum, unpublished report, 1972; Umbach and Klepacki, 1994). To accommodate the significant thickening of the sedimentary section, the Mapenduma Fault is interpreted to have originally been a shelf-edge extensional fault with as much as 20 km of heave. However, it should be noted that the Mapenduma Fault could be steeper than is interpreted on the section shown in Figure 9a.

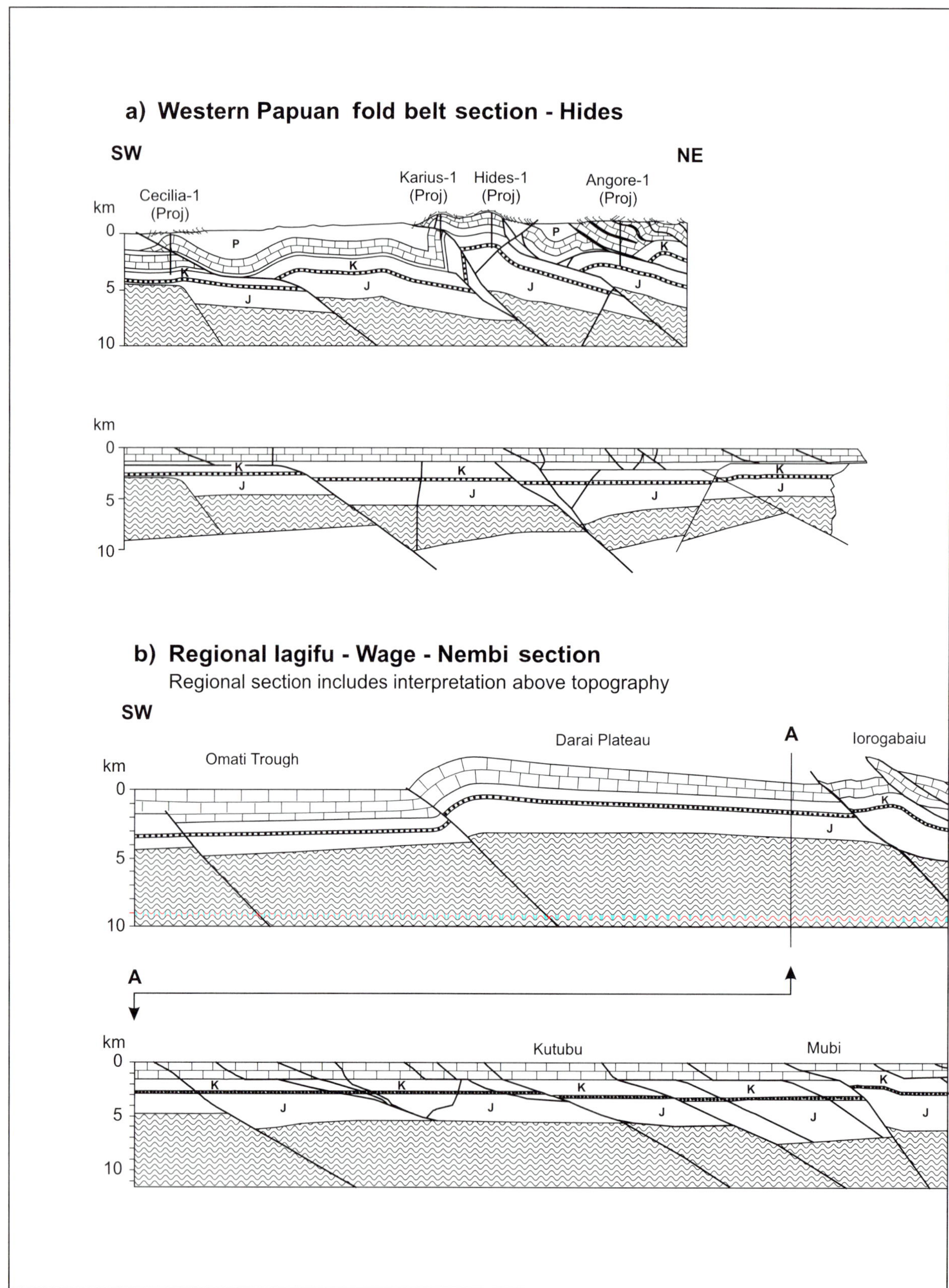

FIGURE 8. Balanced and restored sections across the Papuan Fold Belt. (a) Hides-Karius section. A structural interpretation based upon field data and the seismic lines in Figures 5 and 6. Note the asymmetric Cecilia and Karius folds with the overturned limbs cut by thrust faults. (b) Iagifu-Wage-Nembi section. A reinterpretation of the same section presented in Hill (1991) but incorporating folding in the Mesozoic as demonstrated by the several wells in the Kutubu area (see text). Note the typical elevation of 1–2 km.

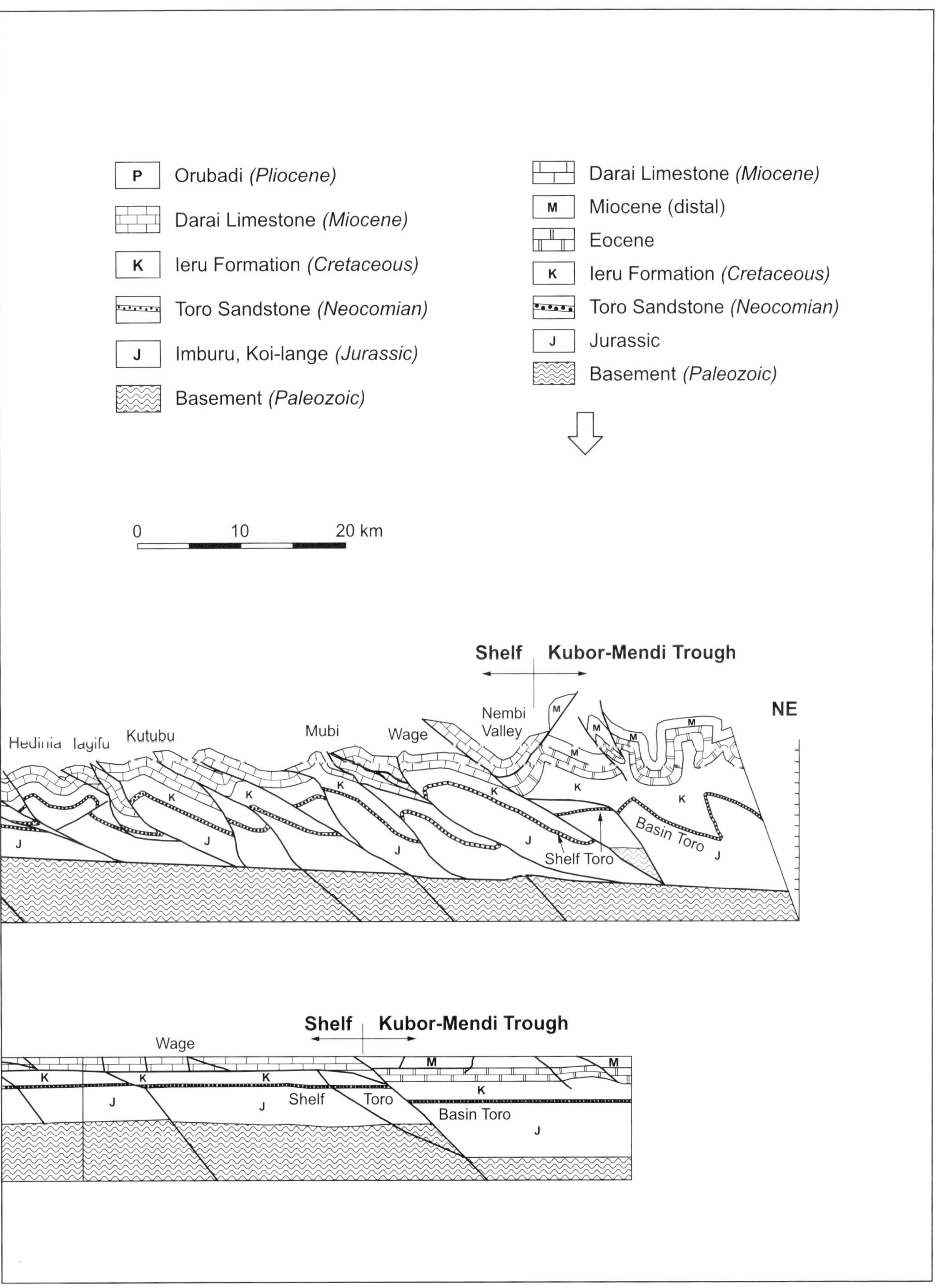

P Orubadi (Pliocene)
Darai Limestone (Miocene)
K Ieru Formation (Cretaceous)
Toro Sandstone (Neocomian)
J Imburu, Koi-lange (Jurassic)
Basement (Paleozoic)
Darai Limestone (Miocene)
M Miocene (distal)
Eocene
K Ieru Formation (Cretaceous)
Toro Sandstone (Neocomian)
J Jurassic
Basement (Paleozoic)
0 10 20 km
Shelf
Kubor-Mendi Trough
NE
Kutubu
Mubi
Wage
Nembi Valley
Shelf Toro
Basin Toro
Shelf
Kubor-Mendi Trough
Wage
Shelf Toro
Basin Toro

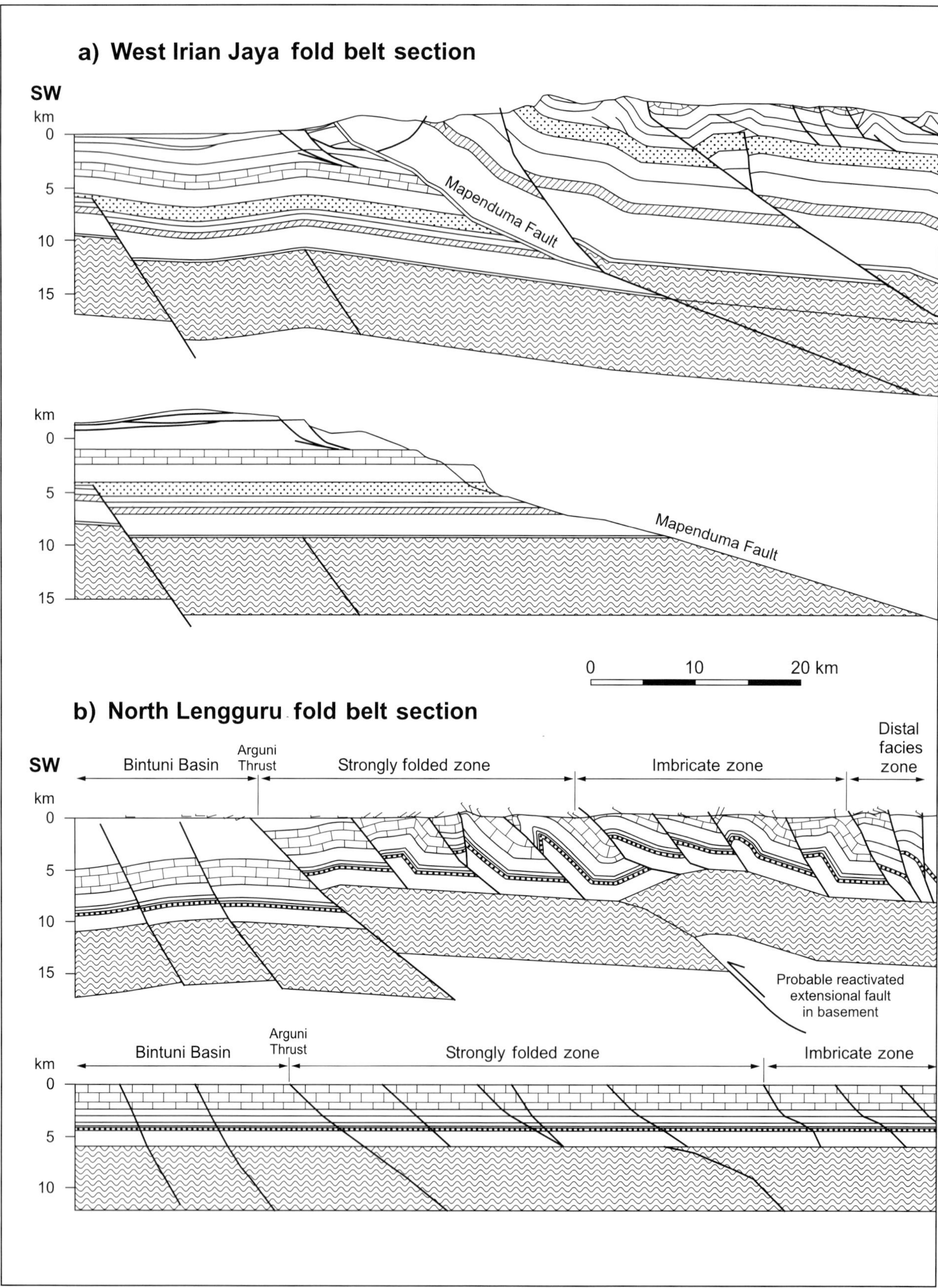

Figure 9. Balanced and restored sections across the Irian Jaya and Lengguru Fold Belts. (a) Section across the Irian Jaya Fold Belt (Figure 1). Note the 50 km of thrusting along the previously extensional Mapenduma Fault, building the 5-km-high mountains, and the older fold-and-thrust belt on the elevated plateau. (b) Section across the northern Lengguru Fold Belt. Note the generally low elevation and the late-stage extensional fault in the northeast.

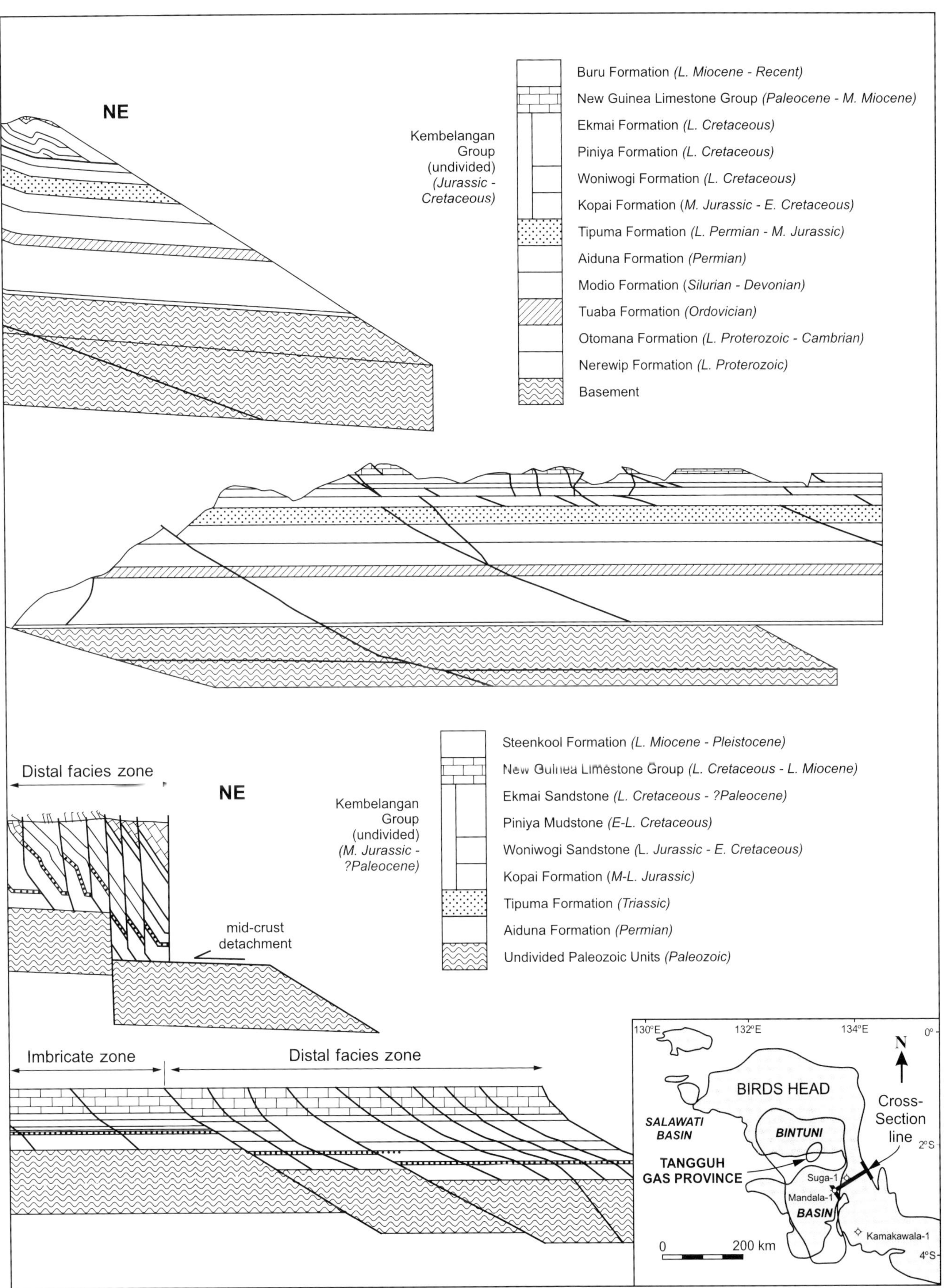
NE
Kembelangan Group (undivided) (Jurassic - Cretaceous)
Buru Formation (L. Miocene - Recent)
New Guinea Limestone Group (Paleocene - M. Miocene)
Ekmai Formation (L. Cretaceous)
Piniya Formation (L. Cretaceous)
Woniwogi Formation (L. Cretaceous)
Kopai Formation (M. Jurassic - E. Cretaceous)
Tipuma Formation (L. Permian - M. Jurassic)
Aiduna Formation (Permian)
Modio Formation (Silurian - Devonian)
Tuaba Formation (Ordovician)
Otomana Formation (L. Proterozoic - Cambrian)
Nerewip Formation (L. Proterozoic)
Basement
Distal facies zone
NE
Kembelangan Group (undivided) (M. Jurassic - ?Paleocene)
Steenkool Formation (L. Miocene - Pleistocene)
New Guinea Limestone Group (L. Cretaceous - L. Miocene)
Ekmai Sandstone (L. Cretaceous - ?Paleocene)
Piniya Mudstone (E-L. Cretaceous)
Woniwogi Sandstone (L. Jurassic - E. Cretaceous)
Kopai Formation (M-L. Jurassic)
Tipuma Formation (Triassic)
Aiduna Formation (Permian)
Undivided Paleozoic Units (Paleozoic)
mid-crust detachment
Imbricate zone
Distal facies zone
130°E
132°E
134°E
0°
N
BIRDS HEAD
SALAWATI BASIN
BINTUNI
Cross-Section line
2°S
TANGGUH GAS PROVINCE
Suga-1
Mandala-1
BASIN
Kamakawala-1
0
200 km
4°S

Minimal deformation of the Buru Formation molasse sediments is recorded south of the frontal, basement-involved thrust. The minor amount of tilting and folding observed is interpreted to be associated with downwarping of the basement at depth. The Mapenduma extensional fault is interpreted to have been reactivated so that it now accommodates ~52 km of shortening divided between detachments at the base of the Nerewip Formation and at a much deeper middle- to lower-crust level. If the Mapenduma Fault were steeper, the shortening would be reduced by 10–20 km.

An older, structurally higher thrust system is preserved in the hanging wall of the Mapenduma Fault. Thin-skinned thrusting within the Jurassic–Cretaceous Kembelangan Group and Tertiary New Guinea Limestone Group is largely the result of slip along detachments in the Cretaceous and Jurassic shales. Total shortening recorded in this system is approximately 17 km (Figure 9a).

The restored section shows two significant down-to-the-northeast basement extensional faults. The southern fault, southwest of the Mapenduma Fault, was responsible for a thickened Permian through Middle Jurassic section and it does not appear to have been inverted during compression. These extensional faults are spaced at ~35 km apart. To the north, minor, late-stage shortening is recorded on out-of-sequence basement-involved thrusts within the lower thrust sheet. Although the slip is only on the order of a few kilometers, portions of the northern fold belt are uplifted by ~2 km as a result. Significantly, these faults are interpreted to have provided a conduit for magmas and mineralizing fluids in the vicinity of the Grasberg Mine. Overall shortening across the section is ~45%.

Lengguru Fold Belt

The Lengguru Fold Belt in western Irian Jaya is a north-northwest-trending belt that tapers to the north, although this is partly the result of plunge beneath the adjacent Bintuni Basin (Figure 1). The fold belt is bounded to the east by Pleistocene high-grade metamorphic rocks in the Wandamen Peninsula (Hill, K.C. et al., 1996, 2002) and then Cenderawasih Bay. To the west is the Bintuni Basin, a 9-km-deep depocenter with >4 km of Pliocene-Pleistocene sediment. The northern margin of the basin, where the strata are thinner, contains the recently discovered Tangguh gas province, which has ~30 tcf of gas (Robertson, 1998). Data constraining the geologic interpretation of the cross section include the 1:250,000 geologic maps published by the Australian Bureau of Mineral Resources and the Indonesian Geological Research and Development Centre, a 1:250,000 SAR image, a regional offshore seismic profile across the southern part of the fold belt, and unpublished geologic reports by Mobil Indonesia (Mobil Oil Indonesia, Inc., unpublished report, 1991; Henage, 1993a, b).

The structure of the Lengguru Fold Belt commonly has been interpreted in terms of thin-skinned tectonics (Moffat et al., 1991; Robinson et al., 1990; Sulaeman et al., 1990). More recently, using schematic cross sections, Hobson et al. (1997) proposed both thin and thick-skinned interpretations of the southern part of the belt, the latter involving inversion of Triassic half grabens.

The Lengguru Fold Belt is highly arcuate and divisible into structural belts similar to those in the eastern Papuan Fold Belt (Hill, 1991). A key difference between the areas, however, is the presence of Pleistocene extensional faults in the Lengguru Fold Belt that both crosscut the belt and down-fault the northeastern part of the belt toward Cenderawasih Bay. We interpret this to be the result of Pleistocene orogenic collapse associated with the opening of oceanic crust in Cenderawasih Bay (Hill et al., 2002).

A second difference is the elevation of the fold belts and adjacent foreland basins. Seismic data from the offshore portion of the Lengguru Fold Belt show that the frontal anticline was once >1 km higher than it is at present, it was then denuded, and it has since subsided below sea level to be onlapped by Pleistocene sediments. This is consistent with regional subsidence of the Lengguru Fold Belt to its present low elevation; the subsidence may have been associated with Pleistocene extension.

A third and critical difference for hydrocarbon exploration involves the stratigraphic thicknesses, in that the Tertiary New Guinea Limestone is substantially thicker in the Lengguru Fold Belt than it is in the Papuan Fold Belt. In the northern Lengguru Fold Belt, the restored section indicates that the predeformation depth of the Woniwogi reservoir was ~4 km, approaching the limit for preservation of porosity. In the south, the predeformation depth was ~5 km, indicating significant porosity destruction.

On the northern Lengguru Fold Belt section (Figure 9b), a Permian detachment is inferred, such that the Woniwogi Sandstone reservoir is involved in the broad frontal anticlines. The anticlines are interpreted to have formed initially as fault-propagation folds that were similar to those observed on offshore seismic data to the southwest. Underlying, deep-seated thrust faults are interpreted to have elevated the basement. However, it should be noted that the stratigraphic section here is poorly constrained at depth. The basement thrusts could be reactivated extensional faults that may have been active in the Triassic/Jurassic, thereby leading to a thicker Mesozoic section, as is seen on the Hides section (Figure 8a).

In the northeast, many steeply dipping thrust faults are inferred that repeat the steeply dipping slabs of the distal facies section recorded on geologic maps. The change in structural style is interpreted to coincide with a regional shelf-edge fault, as is the case in PNG (Figure 8b). However, unlike in PNG, there is no mobile belt to the northeast. Instead, there are Pleistocene extensional faults, then an inferred Pleistocene metamorphic core complex adjacent to Cenderawasih Bay. We suggest here that the mobile belt was present originally but collapsed into Cenderawasih Bay during the Pleistocene. Overall shortening across the section is ~30%.

DISCUSSION

Analysis of interpretations of the Papuan Fold Belt through time demonstrates some major pitfalls in interpreting the structure of areas using insufficient data. It is interesting to note that, in some ways, the sections most like the current interpretations of detached structures are those presented in the 1960s with almost no supporting data. It is also worth noting that the advent of balanced and restored cross sections in the 1980s did not constrain the structural style, and in fact it may have been detrimental by (1) forcing application of a fault-bend fold model and (2) giving false confidence in the resulting interpretation.

It is clear that the primary constraint on structural interpretation of a geologic cross section has to be the observed structural styles, or the structural family, as Dahlstrom (1969) pointed out. In the case where there are insufficient data and analogs from elsewhere have to be applied, it is important to use areas with similar mechanical stratigraphy. In the PNG case, the thick, shale-dominated clastic section in the Mesozoic should be compared with a similar section elsewhere. Analogs with the thick competent limestones and quartzites of the Canadian Rockies (Hill, 1991) were inappropriate. The thrust repeats observed at surface in the competent Miocene limestone in PNG do resemble those in the Rockies, but the mechanical stratigraphy varies greatly with depth, so the interpretation cannot be extrapolated downward.

Deformation throughout the New Guinea Fold Belt appears to have occurred synchronously during the late Miocene–Pliocene, as is recorded on all sections by deformation of the Miocene limestones. This is supported by apatite fission track analyses in all areas that indicate late Miocene–Pliocene uplift, erosion, and cooling (Hill and Gleadow, 1989; O'Sullivan et al., 1995; Kendrick et al., 1995, 1997; Sutriyono et al., 1997). However, Kendrick et al. (1997) suggest that deformation and cooling in the Irian Jaya Fold Belt commenced in the middle Miocene, earlier than the inferred late Miocene commencement in the mobile belt of northeastern PNG (Hill and Raza, 1999). Hill and Raza suggested that the New Guinea orogenic belt was constructed ~12–4 Ma, by rapid oblique convergence of the Pacific/Caroline plate with New Guinea. It could be that the compression was slightly earlier in Irian Jaya, because the margin protruded farther north.

Although the timing was similar, the structural styles across New Guinea that resulted from the shortening were very different. The Kutubu area of the eastern Papuan Fold Belt and the Lengguru Fold Belt exhibit some similarities, both being broad, relatively low-lying thrust belts. They may prove to be more alike than shown here, once more is known about the latter. However, the thicker Tertiary section and the late-stage extensional faults in the Lengguru Fold Belt are important, particularly for hydrocarbon exploration.

The main change that occurs along the New Guinea Fold Belt is from the very high mountains in the narrow fold belt of Irian Jaya to the lower ranges in the broad fold belt of PNG. The frontal structure in the Irian Jaya Fold Belt exposes a 15-km-thick, complete thrust panel of Precambrian to Miocene strata, dipping consistently back to the northeast. In contrast, the basement is rarely exposed in PNG, and in the mountain front the dips are consistently toward the foreland to the southwest. Another important difference is the development of a 3- to 4-km-deep foreland basin adjacent to the Irian Jaya Fold Belt, which decreases to ~1.5 km adjacent to the Hides Anticline (Figures 8a, 9a). Southwest of the eastern Papuan Fold Belt, there is no foreland basin, and the top of the Miocene limestone is currently at sea level.

To account for the differences in structural style, we propose the following model. In the Triassic and Jurassic, rifting occurred along the northern margin of New Guinea, but it occurred farther south in PNG than in Irian Jaya, where the main rift-bounding fault was east-west at the current mountain front (Figure 10a). Subsequently, in PNG, but not in Irian Jaya, Coral Sea rifting occurred in the Late Cretaceous–Paleocene (Figure 10b), creating a broken, weak and hot lithosphere in PNG, as pointed out by Abers and Lyon-Caen (1990). Thus, when compression commenced in the middle Miocene, initially there was a low-lying fold belt throughout New Guinea as the weak lithosphere subsided beneath the mountains (Figure 10c). However, in the Pliocene, the deformation front in Irian Jaya impinged on the strong, cold Australian lithosphere south of the Mapenduma Fault, which acted as a buttress. This caused inversion of the fault, focusing the deformation in one place, building mountains 5 km high, and loading the strong lithosphere to create a foreland basin (Figure 10d). However, in PNG, the fold belt still overlies weak lithosphere and therefore is low-lying and occupies its own foreland basin.

The proposed limit of strong, cold Australian lithosphere is shown on Figure 1, which is the limit of the Arafura Platform. It can be seen that the mountain front is along the edge of the strong lithosphere in Irian Jaya and has yet to reach it in eastern PNG. In between, in

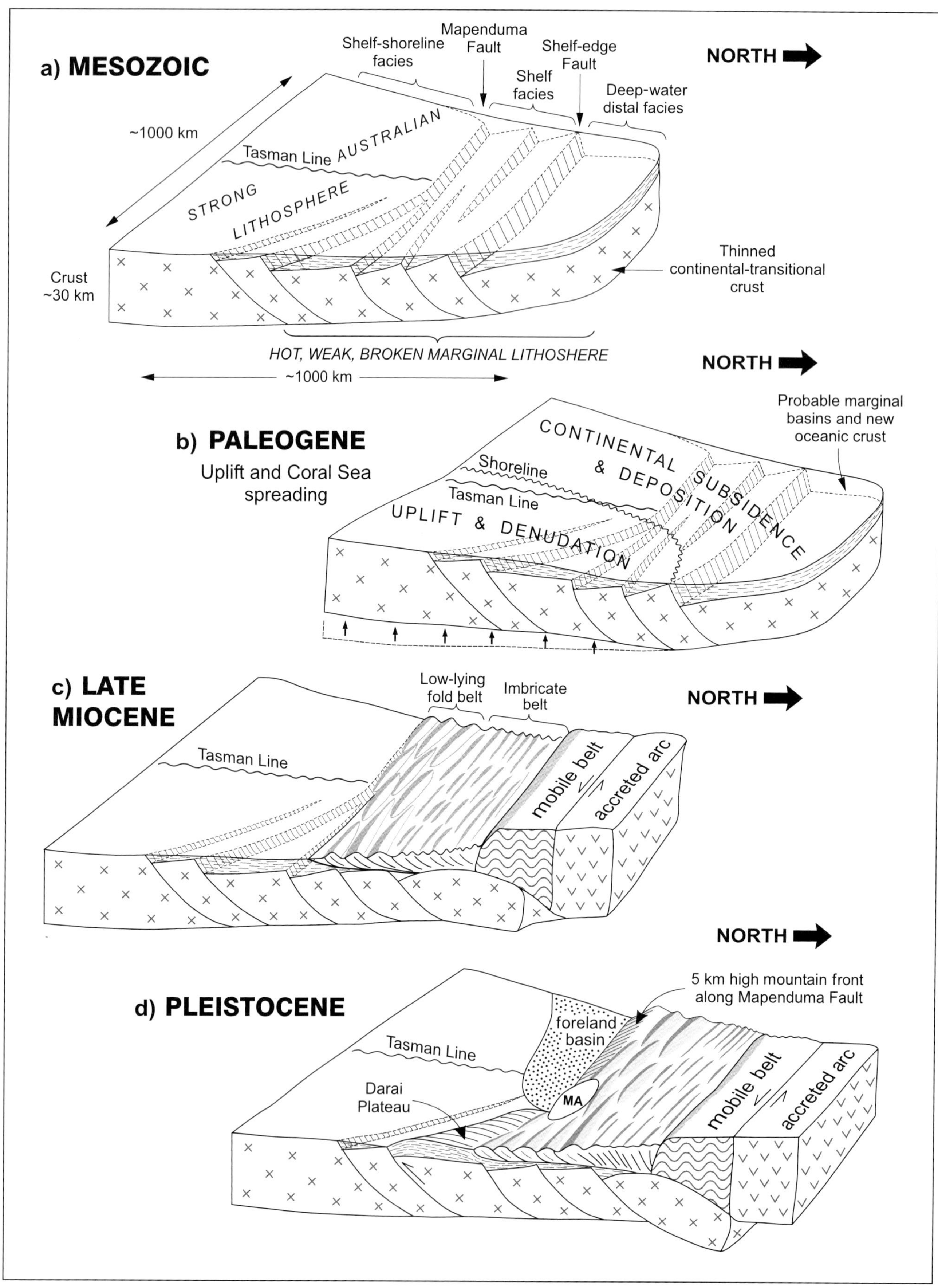
a) MESOZOIC
Shelf-shoreline facies
Mapenduma Fault
Shelf-edge Fault
Shelf facies
Deep-water distal facies
NORTH
~1000 km
Tasman Line
AUSTRALIAN
STRONG
LITHOSPHERE
Thinned continental-transitional crust
Crust ~30 km
HOT, WEAK, BROKEN MARGINAL LITHOSHERE
~1000 km
NORTH
b) PALEOGENE
Uplift and Coral Sea spreading
Probable marginal basins and new oceanic crust
CONTINENTAL SUBSIDENCE
& DEPOSITION
Shoreline
Tasman Line
UPLIFT & DENUDATION
c) LATE MIOCENE
Low-lying fold belt
Imbricate belt
NORTH
Tasman Line
mobile belt
accreted arc
NORTH
d) PLEISTOCENE
5 km high mountain front along Mapenduma Fault
foreland basin
Tasman Line
Darai Plateau
MA
mobile belt
accreted arc

western PNG, the mountain front is just impinging on the strong lithosphere, building the 150 km long, basement-cored Muller Anticline (MA on Figure 10d) (Hill 1991, 1989).

HYDROCARBON POTENTIAL

Lengguru Fold Belt

The regional cross section of the Lengguru Fold Belt illustrates that the main problems in terms of hydrocarbon prospectivity are deep burial of the reservoir and Pleistocene extensional faulting. In contrast to the Papuan Fold Belt, the Lengguru area was subject to continual subsidence in the Paleogene, which allowed deposition of thick New Guinea Limestone. Thus the Mesozoic sandstones typically were buried to a depth of 4–5 km, which destroyed porosity. This was illustrated by the Kamakawala well (Scotchmer and Livsey, 1991), which had oil shows and a low temperature gradient but a tight Woniwogi Sandstone reservoir. The extensional faulting in the Pleistocene is an additional risk, because traps may be breached.

The Lengguru areas with the most potential would be those with preserved highs that were not buried so deeply. These are most common at the northern end of the fold belt, where sediments lap onto the exposed Bird's Head basement high. The Tangguh gas province (Robertson, 1998) is a case in point, where the reservoir porosity is preserved and the area has been only very gently folded by the Lengguru deformation, the result being broad, gentle anticlines. This area has the additional advantage of being adjacent to the Bintuni Basin's Pliocene-Pleistocene depocenter, which supplies hydrocarbon charge.

Irian Jaya Fold Belt

Because much shortening on the Mapenduma Fault was focused in this part of the New Guinea Fold Belt, the thin-skinned fold belt has been raised to high elevations so that it is deeply dissected, thereby breaching most potential hydrocarbon traps. However, the adjacent foreland basin may have considerable potential for oil and gas discoveries. It may be akin to the Tangguh gas province, depending on local seals and the time of trap formation.

Papuan Fold Belt

The western Papuan Fold Belt, west of the Bosavi Lineament, is currently impinging on the strong Australian lithosphere, and therefore it has basement-cored anticlines and a shallow foreland basin. The consequent burial, that is deeper here than to the east (Hill and Gleadow, 1989), means that the area has mainly gas-condensate potential, as is the case in the producing Hides Anticline.

East of the Bosavi Lineament, where there is no foreland basin and less burial, there is proven oil potential in the Kutubu and Gobe areas. An outcome of the present structural interpretations, which apply the better constrained structural styles, is that there is less shortening than previously thought (Hill, 1991; Hobson, 1986) and thus more potential in the distal areas. If the Mesozoic structure is dominated by folding that drives thrusting of the overlying limestone, then it is likely to produce local culminations where the Mesozoic sandstone reservoirs are near the surface, as shown on the Iagifu cross section (Figure 8b).

ACKNOWLEDGMENTS

This research was funded by the Australian Geodynamics Cooperative Research Centre. We are grateful to Paradigm for making available Geosec™ and to the following organizations for support: the GRDC, Pertamina, Amoseas, Arco, BP, British Gas, Chevron JV, Esso, Mobil, Normandy, Oilsearch JV, P. T. Freeport, and Union Texas. AIDAB, the Australian Geodynamics Cooperative Research Centre and La Trobe University made available Ph.D. scholarships for Sutriyono, Keetley, and Kendrick. This paper is published with the permission of the Director of the Australian Geodynamics Cooperative Research Centre.

Figure 10. Schematic tectonic model for the evolution of the New Guinea Fold Belt. (a) Jurassic-Cretaceous rifting and sedimentation along the northern margin of the Australian craton, with different basement geology on either side of the Tasman Line (Figures 1, 2). Reservoir sandstones were deposited on the shelf, with source rocks in the basin facies. (b) Late Cretaceous–Paleocene uplift and denudation of the area adjacent to rifting in the Coral Sea. Jurassic sandstone reservoir porosity was reduced in areas that underwent continuous burial. (c) Early Miocene arc-continent collision generated a low-lying fold belt throughout New Guinea in the late Miocene, overlying the extended margin. (d) In the Pliocene-Pleistocene, the orogeny impinged on the strong Australian lithosphere in Irian Jaya, inverting the Mapenduma Fault to build 5-km-high mountains and breaching hydrocarbon traps. The mountains generated an adjacent foreland basin that may be prospective for hydrocarbons. In PNG, the low-lying fold belt migrated south, preserving hydrocarbons, with minor inversion of basement faults. MA = Muller Anticline.

REFERENCES CITED

Abers, G. A., and H. Lyon-Caen, 1990, Regional gravity anomalies, depth of the foreland basin and isostatic compensation of the New Guinea Highlands: Tectonics, 9, 1479–1493.

Bennet, J. K., and M. Parish, 1993, The use of magnetotelluric surveying as a risk reduction tool for hydrocarbon exploration in the Papuan Fold Belt, *in* G. J. Carman and Z. Carman, eds., Proceedings of the Second PNG Petroleum Convention, 1993, Port Moresby, 31st May–2nd June 1993, p. 335–350.

Boyer, S. E., and D. Elliot, 1982, Thrust systems: AAPG Bulletin, v. 66, p. 1196–1230.

Buchanan, P. G., 1996, Petroleum exploration, development and production in Papua New Guinea: Proceedings of the Third PNG Petroleum Convention, Port Moresby, September 1996, 590 p.

Buchanan, P. G., and J. Warburton, 1996, The influence of pre-existing basin architecture in the development of the Papuan Fold and Thrust Belt: implications for petroleum prospectivity, *in* P. G. Buchanan, ed., Petroleum exploration, development and production in Papua New Guinea: Proceedings of the Third PNG Petroleum Convention, Port Moresby, September 1996, p. 89–109.

Buchanan, P. G., A. M. Grainge, and R. C. N. Thornton, (eds.), 2000, Papua New Guinea's petroleum industry in the 21st century: Proceedings of the Fourth PNG Petroleum Convention, Port Moresby, 29th–31st May, 2000, p. 612.

Cole, J. P., M. Parish, and D. Schmidt, 2000, Sub-thrust plays in the Papuan Fold Belt: the next generation of exploration targets, *in* P. G. Buchanan, A. M. Grainge, and R. C. N. Thornton, eds., Papua New Guinea's petroleum industry in the 21st century: Proceedings of the Fourth PNG Petroleum Convention, p. 87–100.

Cooper, M. A., G. D. Williams, P. C. De Graciansky, R. W. Murphy, T. Needham, D. De Paor, R. Stonely, S. P. Todd, J. P. Turner, and P. A. Ziegler, 1989, Inversion tectonics— a discussion, *in* M. A. Cooper and G. D. Williams, eds., Inversion tectonics: Geological Society of London Special Publication 44, p. 335–347.

Dahlstrom, C. D. A., 1969, Balanced cross sections: Canadian Journal of Earth Sciences, v. 6, 743–757.

Dixon, J. M., 1996, Physical model investigation of the influence of early extensional (growth) faults on fold-thrust structures, with application to the Papuan fold and thrust belt., *in* P. G. Buchanan, ed., Petroleum exploration, development and production in Papua New Guinea: Proceedings of the Third PNG Petroleum Convention, Port Moresby, September 1996, p. 147–160.

Dow, D. B., 1977, A geological synthesis of Papua New Guinea: Australian Bureau of Mineral Resources, Geology and Geophysics Bulletin, v. 201, 41 p.

Dow, D. B., G. P. Robinson, U. Hartono, and N. Ratman, 1986, Geologic map of Irian Jaya, Indonesia: Geological Research and Development Centre, Indonesian Ministry of Mines and Energy, scale: 1:1,000,000.

Eisenberg, L. I., 1993, Hydrodynamic character of the Toro Sandstone, Iagifu/Hedinia area, Southern Highlands Province, Papua New Guinea, *in* G. J. Carman and Z. Carman, eds., Proceedings of the Second PNG Petroleum Convention, 1993, Port Moresby, 31st May–2nd June, 1993, p. 447–458.

Findlay, A. L., 1974, The structure of the foothills south of the Kubor Range, Papua New Guinea: Australian Petroleum Production and Exploration Association (APEA) Journal, v. 14, no. 1, p. 14–20.

Franklin, S. P., and J. E. Livingston, 1996, Development of an infill well program to maximise economic return from the Iagifu-Hedinia Field: Part I, Integrated structural, stratigraphic and reservoir attribute modelling as input to reservoir simulation and well targeting, *in* P. G. Buchanan, ed., Petroleum exploration, development and production in Papua New Guinea: Proceedings of the Third PNG Petroleum Convention, Port Moresby, September 1996, p. 573–590.

Grainge, A., 1993, Recent developments in prospect mapping in the Hides/Karius area of the Papuan Foldbelt. *in* G. J. Carman and Z. Carman, eds., Proceedings of the Second PNG Petroleum Convention, 1993, Port Moresby, 31st May–2nd June, 1993, p. 527–540.

Granath, J. W., and S. A. Hermeston, 1993, Relationship of the Toro Sandstone Formation and the Alene Sands of Papua to the Woniwogi formation of Irian Jaya: *in* G. J. Carman and Z. Carman, eds., Proceedings of the Second PNG Petroleum Convention, 1993, Port Moresby, 31st May–2nd June, 1993, p. 201–206.

Hall, R., 1997, Cenozoic tectonics of SE Asia and Australasia, *in* Petroleum systems of S.E. Asia and Australasia: Indonesian Petroleum Association conference, Jakarta, May 1997, p. 47–62.

Harahap, B. H., and Y. Noya, 1995, Geological map of the Rotanburg Quadrangle, Irian Jaya, scale 1:250,000: Geological Research and Development Centre, Bandung, Indonesia.

Henage, L. F., 1993a, Regional geological and geophysical interpretation, Babo, Lengguru West and Lengguru East PSCs: Mobil Exploration Bomberai Internal Report, 122 pages.

Henage, L. F., 1993b, Mesozoic and Tertiary tectonics of Irian Jaya: evidence for non-rotation of Kepala Burung: Proceedings of the 22nd Annual Convention, Indonesian Petroleum Association, Jakarta, 763–792.

Hill, G. S., S. J. Price, M. S. Foster, R. W. Stephenson, D. Ellis, and J. A. Lyslo, 1996, Seismic acquisition in the Papuan Fold Belt— a new approach, *in* P. G. Buchanan, ed., Petroleum exploration, development and production in Papua New Guinea: Proceedings of the Third PNG Petroleum Convention, Port Moresby, September 1996, p. 445–458.

Hill, K. C., 1989, The Muller Anticline, Papua New Guinea, basement-cored inverted extensional fault structures with opposite vergence: Tectonophysics, v. 158, p. 227–245.

Hill, K. C., 1991, Structure of the Papuan Fold Belt, Papua New Guinea: AAPG Bulletin, v. 75, p. 857–872.

Hill, K. C., and J. P. Cole, 1986, Report on the 1985 PPL 27

Geological Survey, Hides Anticline, Papua New Guinea: BP Australia Report to PNG Geological Survey; PNG Department of Minerals and Energy.

Hill, K. C., and A. J. W. Gleadow, 1989, Uplift and thermal history of the Papuan Fold Belt, Papua New Guinea: Apatite Fission Track Analysis: Australian Journal of Earth Science, v. 36, 515–539.

Hill, K. C., and R. Hall, 2003, Mesozoic-Tertiary evolution of Australia's New Guinea margin in a West Pacific context, *in* R. R. Hillis and R. D. Muller, eds., Evolution and Dynamics of the Australian plate: Geological Society of Australia Special Publication 22 and Geological Society of America Special Paper 372, p. 265–290.

Hill, K. C., and A. Raza, 1999, Arc-continent collision in Papua Guinea— constraints from fission track thermochronology: Tectonics, v. 18, p. 950–966.

Hill, K. C., D. Medd, and P. Darvall, 1990, Structure, stratigraphy, geochemistry and hydrocarbons in the Kagua-Kubor area, Papua New Guinea, *in* G. J. Carman and Z. Carman, eds., Petroleum exploration in Papua New Guinea: Proceedings of the First PNG Petroleum Convention, 1990, Port Moresby, p. 351–368.

Hill, K. C., R. J. Simpson, R. D. Kendrick, P. V. Crowhurst, P. B. O'Sullivan, and I. Saefudin, 1996, Hydrocarbons in New Guinea, controlled by basement fabric, Mesozoic extension and Tertiary Convergent margin tectonics, *in* P. G. Buchanan, ed., Petroleum exploration, development and production in Papua New Guinea, Proceedings of the Third PNG Petroleum Convention, Port Moresby, September 1996, p. 63–76.

Hill, K. C., M. S. Norvick, J. T. Keetley, and A. Adams, 2000, Structural and stratigraphic shelf-edge hydrocarbon plays in the Papuan Fold Belt, in P. G. Buchanan, A. M. Grainge, and R. C. N. Thornton, eds., Papua New Guinea's petroleum industry in the 21st century: Proceedings of the Fourth PNG Petroleum Convention, p. 67–85.

Hill, K. C., N. Hoffman, P. Lunt, and R. Paul, 2002, Structure and hydrocarbons in the Sareba block, "Bird's Neck," West Papua: Proceedings of the Indonesian Petroleum Association, p. 227–248.

Hobson, D. M., 1986, A thin skinned model for the Papuan thrust belt and some implications for hydrocarbon exploration. The Australian Petroleum Production and Exploration Association (APEA) Journal, v. 26, no. 1, 214–224.

Hobson, D. M., A. Adnun, and L. Samuel, 1997, The relationship between Late Tertiary basins, thrust belts and major transcurrent faults in Irian Jaya: implications for petroleum systems throughout New Guinea, *in* J. V. C. Howes and R. A. Noble, eds., Petroleum systems of S.E. Asia and Australasia: Indonesian Petroleum Association conference, Jakarta, May 1997, p. 261–284.

Home, P. C., D. G. Dalton, and J. Brannan, 1990, Geological evolution of Western Papuan Basin, *in* G. J. Carman and Z. Carman, eds., Petroleum exploration in Papua New Guinea: Proceedings of the First PNG Petroleum Convention, 1990, Port Moresby, p. 107–118.

Hornafius, S., and R. E. Denison, 1993, Structural interpretations based on strontium isotope dating of the Darai Limestone, Papuan Fold Belt, New Guinea, *in* G. J. Carman and Z. Carman, eds., Proceedings of the Second PNG Petroleum Convention, 1993, Port Moresby, 31st May–2nd June, 1993, p. 313–324.

Jenkins, D. A. L., 1974, Detachment tectonics in western Papua New Guinea: Geological Society of America Bulletin, v. 85, p. 533–548.

Jenkins, D. A. L., A. L. Findlay, and G. P. Robinson, 1969, Mendi geological survey, Permits 46, 27, Papua: BP Petroleum Development, Australia, Ltd., report (unpubl.), Geological Survey of Papua New Guinea Open File Report.

Johnstone, D. C., and J. K. Emmett, 2000, Petroleum geology of the Hides gas field, Southern Highlands, Papua New Guinea, *in* P. G. Buchanan, A. M. Grainge, and R. C. N. Thornton, eds., Papua New Guinea's petroleum industry in the 21st century: Proceedings of the Fourth PNG Petroleum Convention, p. 319–336.

Kendrick, R. D., K. C. Hill, K. Parris, I. Saefudin, and P. B. O'Sullivan, 1995, Timing and style of Neogene regional deformation in the Irian Jaya Fold Belt, Indonesia: Proceedings of the 24th Annual Convention, Indonesian Petroleum Association, Jakarta, p. 249–261.

Kendrick, R. D., K. C. Hill, P. B. O'Sullivan, K. Lumbanbatu, and I. Saefudin, 1997, Mesozoic to Recent thermal history and basement tectonics of the Irian Jaya Fold Belt and Arafura Platform, Irian Jaya, Indonesia, *in* J. V. C. Howes and R. A. Noble, eds., Petroleum systems of S.E. Asia and Australasia, p. 301–315.

Lamerson, P. R., 1990, Evolution of structural interpretations in Iagifu/Hedinia field, Papua New Guinea, *in* G. J.Carman and Z. Carman, eds., Petroleum exploration in Papua New Guinea: Proceedings of the First PNG Petroleum Convention, 1990, Port Moresby, February 1990, p. 283–300.

Laws, M. J., 1971, The use of the seismic refraction method in Papua: The Australian Petroleum Production and Exploration Association (APEA) Journal, v. 11, p. 107–114.

Lingrey, S., 2000, Structural interpretation and modelling of seismic data from the Moran and Paua Area, PNG Foldbelt, in P. G. Buchanan, A. M. Grainge, and R. C. N. Thornton, eds., Papua New Guinea's petroleum industry in the 21st century: Proceedings of the Fourth PNG Petroleum Convention, p. 385–396.

McDowell, F. W., T. P. McMahon, P. Q. Warren, and M. Cloos, 1996, Pliocene Cu-Au bearing igneous intrusions of the Gunung Bijih (Ertsberg) district, Irian Jaya, Indonesia: K-Ar geochronology: Journal of Geology, v. 104, p. 327–340.

Moffat, D. T., L. F. Henage, R. A. Brash, R. W. Tauer, and B. H. Harahap, 1991, Lengguru, Irian Jaya: Prospect selection using field mapping, balanced cross section, and gravity modeling: Proceedings of the 20th Annual Convention, Indonesian Petroleum Association, Jakarta, p. 85–106.

O'Sullivan, P. B., K. C. Hill, I. Saefudin, and R. D. Kendrick, 1995, Mesozoic and Cenozoic thermal history of sedimentary rocks in the Bintuni Basin, Irian Jaya, Indonesia: Proceedings of the 24th IPA Convention, p. 235–248.

Panggabean, H., Amiruddin, Kusnama, K. Sutisna, R. L. Situmorang, T. Turkandi, and B. Hermanto, 1995, Geological map of the Beoga Quadrangle, Irian Jaya. Scale 1:250,000: Geological Research and Development Centre, Bandung, Indonesia.

Parris, K. R., 1994, Preliminary Geological Data Record of the Timika (3211) 1:250,000 Sheet Area, Irian Jaya. PT Freeport Indonesia Report.

Parris, K. R., 1996a, Preliminary Geological Data Record of the Rotanburg (3312) 1:250,000 Sheet Area, Irian Jaya. PT Freeport Indonesia Report.

Parris, K. R., 1996b, Preliminary Geological Data Record of the Wamena (3311) 1:250,000 Sheet Area, Irian Jaya. PT Freeport Indonesia Report.

Parris, K. R., and P. Q. Warren, 1996, Preliminary geological data record of the Hitalipa (3212) 1:250,000 Sheet Area, Irian Jaya. PT Freeport Indonesia Report.

Pigram, C. J., and H. Panggabean, 1984, Rifting of the northern margin of the Australian continent and the origin of some microcontinents in eastern Indonesia: Tectonophysics, v. 107, p. 331–353.

Pigram, C. J., and H. Panggabean, 1989, Geology of the Waghete Sheet area, Irian Jaya, Explanatory notes and geological map: Geological Research and Development Centre, Bandung, Indonesia. Geological data record.

Pigram, C. J., and P. A. Symonds, 1991, A review of the timing of the major tectonics events in the New Guinea orogen: Journal of Southeast Asian Sciences, v. 6, p. 307–318.

Quarles van Ufford, A. I., 1996, Stratigraphy, structural geology, and tectonics of a young forearc-continent collision, western Central Range, Irian Jaya (western New Guinea), Indonesia: Ph.D. dissertation, University of Texas at Austin, 421 pages.

Robertson, J., 1998, Tangguh— Discovery of a major gas province in Irian Jaya, Indonesia. Abstract, *in* Indonesian Petroleum Association, Proceedings of the Gas Habitats of SE Asia and Australasia Conference: Technical Program Guide, p. 43.

Robinson, G. P., R. J. Ryburn, B. H. Harahap, S. L. Tobing, A. Achdan, G. M. Bladon, and P. E. Pieters, 1990, Geology of the Steenkool Sheet area, Irian Jaya: Geological Research and Development Centre, Bandung, Indonesia, 45 p.

Rusmana, E., K. Parris, U. Sukanta, and H. Samodra, 1995, Geological map of the Timika Quadrangle, Irian Jaya. Scale 1:250,000: Geological Research and Development Centre, Bandung, Indonesia.

Scheibner, E., 1974, Fossil fracture zones, segmentation and correlation problems in the Tasman Fold Belt System, *in* A. K. Denmead, G. W. Tweedale, and A. F. Wilson, eds., The Tasman Geosyncline: Brisbane, Geological Society of Australia, Queensland Division, p. 65–98.

Scotchmer, J., and A. R. Livsey, 1991, Mobil Lengguru West Exploration Inc. Kamakawala-1X well, Lengguru west block, Irian Jaya Petroleum Geochemistry. Unpublished report by P. T. Robertson, Utama Indonesia, 16 p.

Smith, J. G., 1965 v Orogenesis in western Papua and New Guinea: Tectonophysics, v. 2 (1), p. 1–27.

Sulaeman, A., A. Sjapawi, and S. Sosromihardjo, 1990, Frontier exploration on the Lengguru Fold Belt, Irian Jaya: Proceedings of the 20th Annual Convention, Indonesian Petroleum Association, Jakarta, p. 85–106.

Sutriyono, E., P. B. O'Sullivan, and K. C. Hill, 1997, Thermochronology and tectonics of the Birds Head region, Irian Jaya: Apatite fission track constraints, *in* J. V. C. Howes and R. A. Noble, eds., Petroleum systems of S.E. Asia and Australasia: Indonesian Petroleum Association conference, Jakarta, May 1997, p. 285–300.

Umbach, K. E., and D. Klepacki, 1994, A triangle zone along the active thrust front in southern Irian Jaya, Indonesia: Proceedings of the 23rd Annual Convention, Indonesian Petroleum Association, Jakarta, p. 305–321.

Valenti, G. L., 1993, P'nyang field: Discovery and geology of a gas giant in the western Papuan Fold Belt, Western Province, Papua New Guinea, *in* G. J. Carman and Z. Carman, eds., Proceedings of the Second PNG Petroleum Convention, 1993, Port Moresby, 31st May–2nd June, 1993, p. 413–430.

Wali, K. F., 1996, The Kutubu Project: Challenges of the next 150 million barrels, *in* P. G. Buchanan, ed., Petroleum exploration, development and production in Papua New Guinea: Proceedings of the Third PNG Petroleum Convention, Port Moresby, September 1996, p. 545–550.

26

Korsch, R. J., 2004, A Permian-Triassic retroforeland thrust system — The New England orogen and adjacent sedimentary basins, eastern Australia, *in* K. R. McClay, ed., Thrust tectonics and hydrocarbon systems: AAPG Memoir 82, p. 515–537.

A Permian-Triassic Retroforeland Thrust System—The New England Orogen and Adjacent Sedimentary Basins, Eastern Australia

R. J. Korsch
Australian Geodynamics Cooperative Research Centre, Geoscience Australia, Canberra, Australia

ABSTRACT

From the Late Devonian to the Triassic, eastern Australia was part of eastern Gondwana, where an active, convergent plate margin was influenced by a west-dipping subduction system. This system terminated as the result of global plate reorganization in the Middle to Late Triassic. The southern New England orogen changed from a prowedge (P) mode (terminology of Beaumont et al., 1999) in the Devonian-Carboniferous to an uplifted plug (P-U) mode in the Permian to Triassic. In contrast, the northern New England orogen was dominated by the retrowedge (P-U-R) mode in both time periods. This led to the development, in the Permian–Triassic, of a major west-directed retroforeland thrust belt in northern New England, with the formation of a thick foreland-basin phase in the adjacent Bowen Basin to the west. Thick-skinned and thin-skinned processes operated simultaneously. In the orogen, the thrusts are dominantly thick-skinned and planar, cutting deep into the crust. The eastern part of the Bowen Basin, however, is dominated by thin-skinned thrusting that, in places, propagated a considerable distance into the basin, with the formation of an imbricate thrust fan. The thick-skinned and thin-skinned thrusts are hard-linked into a single thrust system by a major middle-crust detachment surface. Contractional events at the plate margin associated with the formation of the thrust system were also responsible for the propagation of far-field stresses and the reactivation of older extensional faults as thrusts. These reactivated faults are well inboard of, not physically attached to, and soft-linked to, the retroforeland thrust system. Contraction in the Denison Trough in the western Bowen Basin produced a variety of geometries, including reactivation of Early Permian extensional faults as thrusts and the growth of fault-propagation folds. These inversion structures often house commercial quantities of hydrocarbons.

INTRODUCTION

In this chapter, I describe the development of a Permian-Triassic orogenic system in eastern Australia and its impact on the evolution of the adjacent petroleum-producing sedimentary basin system. A regional grid of industry seismic-reflection data, complemented by high-quality, regional deep seismic-reflection profiles across both the orogen and basins, documents the geometry of the basin system. The seismic data reveal the the system's shallow-to-deep structural linkages and its changes in vergence or orogenic style.

The easternmost tectonic unit in eastern Australia, the New England orogen (Figure 1), contains a geologic record dominated by subduction-related rocks and indicates that the orogen has been part of, or adjacent to, convergent plate margins of eastern Gondwana from the late Neoproterozoic until at least the end of the Early Cretaceous. Inboard of the orogen are two overlapping sedimentary basin systems, the Early Permian–Middle Triassic Bowen-Gunnedah-Sydney basin system, and the Early Jurassic–Early Cretaceous Clarence-Moreton-Surat-Eromanga basin system (Figure 1). The Bowen Basin and overlying Surat Basin contain more than 70 mainly small commercial hydrocarbon fields in Queensland, some of which occur in anticlinal structures whose formation are related to reactivation of older faults as thrusts (Elliott, 1993; Korsch et al., 1998). Farther inboard are the Cambrian to Carboniferous Lachlan and Thomson orogens (Figure 1), also components of the convergent plate margin system, but essentially cratonized by the Devonian.

The rock record prior to the Middle Devonian is fragmentary, but there are remnants of rocks as old as late Neoproterozoic (Watanabe et al., 1998; Bruce et al., 1999). During the Late Devonian to Late Carboniferous, plate convergence at the interface of eastern Gondwana and the Panthalassan (or proto-Pacific) Ocean related to a west-dipping subduction system produced, from west to east, a magmatic arc, a fore-arc basin, and an accretionary wedge (Leitch, 1975; Murray et al., 1987; Korsch et al., 1990a).

In the Devonian–Upper Carboniferous strata, the Connors Arc and the southern part of the Gogango Thrust Zone represent the magmatic arc. To the south, this component is not recognized on the surface. The fore-arc basin is represented by the Tamworth Belt in the south and by the area immediately east of the Gogango Thrust Zone in the north. The Tablelands Complex in the south and most of the area between the Gogango Thrust Zone and Gympie Province in the north represent the accretionary wedge. The Drummond Basin represents part of the back-arc basin (Figure 1).

During the Late Carboniferous, subduction ceased temporarily, or migrated well to the east, giving way to extension with some strike-slip faulting. This was followed by a new convergent-plate-margin system that operated from the Early Permian to the Cretaceous (Korsch et al., 1998). The fore-arc basin and accretionary wedge components of this youngest plate margin were later rifted off the Australian continental margin and are now located in continental slivers in the western Pacific Ocean, for example, in Lord Howe Rise, New Caledonia, and New Zealand. Only the magmatic arc component remained in eastern Australia. During the Early Permian to Cretaceous convergent-margin phase, sedimentation was punctuated by contractional events, defined by Korsch et al. (1998), which had different effects in different parts of the basin system. The main effect was that a westward-propagating thrust system, interpreted as a retroforeland thrust belt, developed in the New England orogen from the Late Permian to the Middle Triassic. Immediately to the west, the Bowen-Gunnedah-Sydney Basin changed from an Early Permian extensional system to a major foreland basin until the Middle Triassic.

In this chapter I will focus primarily on the Middle Permian to the Middle Triassic, when the New England orogen portion of the subduction system was a doubly vergent orogen, in the sense of Willett et al. (1993). During this period, a major retroforeland thrust belt (terminology of Beaumont et al., 1999) developed, leading to the formation of a significant foreland loading basin phase in the Bowen Basin west of and adjacent to the orogen. The associated contractional events played a significant role in the development of structures that became important in later hydrocarbon entrapment.

Throughout this chapter, I use the pro- and retro- terminology as defined by Willett et al. (1993). Beaumont et al. (1999) developed a conceptual model to describe components for the short-term operation of subduction-accretion systems at convergent plate margins (Figure 2). The basic geometric and mechanical components of a convergent plate margin are

1) the pro- (accretionary) wedge on the incoming side of the subduction zone (P),
2) a central uplifted plug (U),
3) the retrowedge in the retrolithosphere, inboard of the subduction zone (R), and
4) the conduit of the subduction zone (C).

At least eight conceptual modes, depending on which of the above four components are active at any time, have been recognized and described in detail by Beaumont et al. (1999).

CRUSTAL GEOMETRY OF THE NEW ENGLAND OROGEN

The New England orogen is conveniently divided by the Surat and Clarence-Moreton Basins into the southern New England orogen, mostly in New South Wales, and the northern New England orogen in Queensland (Figure 1). Deep seismic-reflection profiles, collected by Geoscience Australia (formerly the Australian Geological Survey Organization), have been important in defining the crustal architecture of the orogen (Korsch

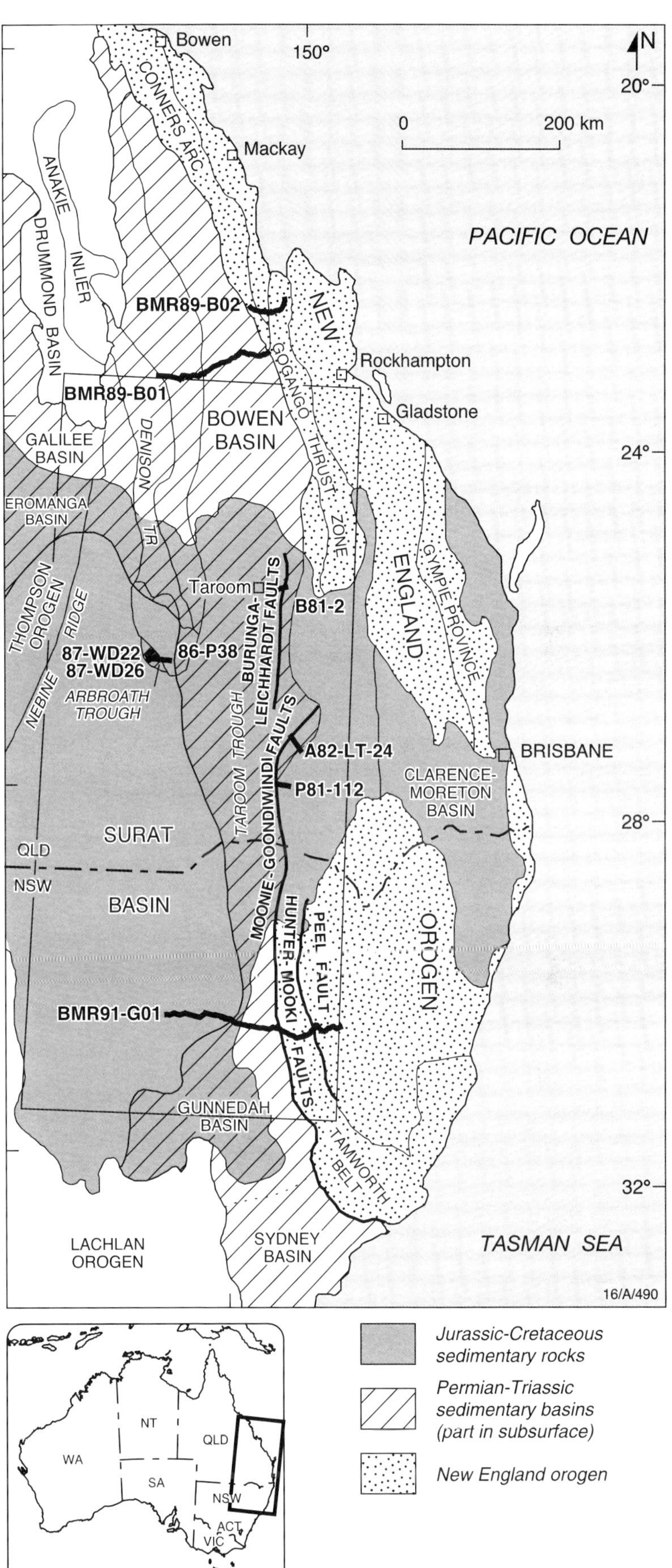

Figure 1. Map of eastern Australia, showing locations of the New England orogen, Bowen Basin, Gunnedah Basin, and Surat Basin. The box shows the location of Figure 8. Also shown are the locations of some of the seismic sections illustrated in the paper.

et al., 1990b, 1993a, 1997) and also allow the upper-crust foreland thrust belts to be examined in a full crustal context. Three deep seismic-reflection transects are oriented approximately east–west and run from the Bowen and Gunnedah Basins into the New England orogen. Here I will focus on data from the 1989 and 1991 surveys (Figure 1), as examples of the geometry of the northern and southern New England orogens, respectively.

RETROFORELAND THRUST SYSTEM IN NEW ENGLAND OROGEN

Southern New England Orogen

The eastern half of the deep seismic profile BMR91.G01 crossed the margin between the Gunnedah Basin in the west and the New England orogen in the east (Figure 3). In the New England orogen, east of the Peel Fault, there are predominantly west-dipping structures in the seismic section. The major boundary defining the western limit of the accretionary wedge dips moderately to the west, and I interpret the west-dipping structures east of it to represent original fabric that developed in the Late Devonian and the Carboniferous, during formation and growth of the accretionary wedge (Korsch et al., 1997). Some of these structures have been strongly reactivated during more recent contractional events. There is, as yet, no evidence for major deformation in the fore-arc basin during the Late Devonian to Late Carboniferous, and thus no evidence that there was a foreland fold-thrust belt component to the system at that time. This suggests that the backstop was the west-dipping structure separating the Lachlan from the New England orogen, shown in Figure 3.

Hence, during the Late Devonian to the Late Carboniferous, the subduction system in the southern New England orogen lacked a linked foreland fold-thrust belt and had an

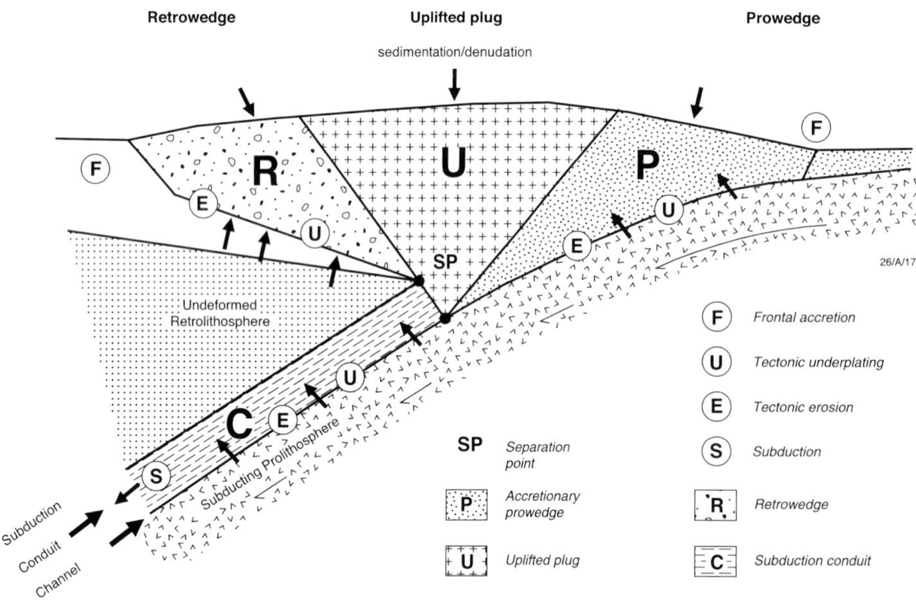

FIGURE 2. Conceptual model of subduction-accretion at a convergent margin, illustrating the P-U-R-C components, after Beaumont et al. (1999) and Waschbusch et al. (1999) (see text for details). The boundaries separate regions with different flow regimes, and hence are usually zones of high strain rate. Thus, although the boundaries are shown here as lines, they are, in reality, zones of finite width.

inland-dipping boundary between the prowedge and the inactive plug (Figure 3, lower-right inset). By comparison with the conceptual modes of Beaumont et al. (1999), the system appears to be in either the P or P-C mode (Figure 3; see also Waschbusch et al., 1999).

From the Late Carboniferous to the Early Permian, the southern New England orogen was dominated by extension. Late in the Early Permian or early in the Late Permian, there was a return to a subduction system. Inboard of the preserved magmatic arc, the Bowen-Gunnedah-Sydney basin system developed as a foreland basin phase in a back-arc setting until the Middle Triassic. South of about latitude 28°S, during the early Late Permian to the Middle Triassic, a relatively narrow, west-directed thrust belt developed in the Devonian-Carboniferous fore-arc rocks of the Tamworth Belt (Glen and Brown, 1993; Glen et al., 1993; Woodward, 1995). To the west, the eastern margin of the Gunnedah Basin is overthrust, at the Mooki Fault, by the Devonian-Carboniferous Tamworth Belt fore-arc basin rocks. Hence the eastern margin of the Gunnedah Basin is dominated by east-dipping structures and has only a relatively narrow and relatively thin foreland basin succession in the Gunnedah Basin (Korsch et al., 1993b; Totterdell and Krassay, 1995), compared with the area farther north in the Bowen Basin (see below).

The deep seismic profile BMR91.G01 (Korsch et al., 1993a, 1997) shows that the system has a Y-shaped geometry (Figure 3), which is typical of a doubly vergent orogen as described by Willett et al. (1993). Thus I interpret the west-dipping fault immediately to the east of the Peel Fault to represent the boundary between the prowedge and the uplifted plug, and the western limit of the Tamworth Belt, the Mooki Fault, to be the boundary between the active uplifted plug and the inactive retrowedge. By comparison

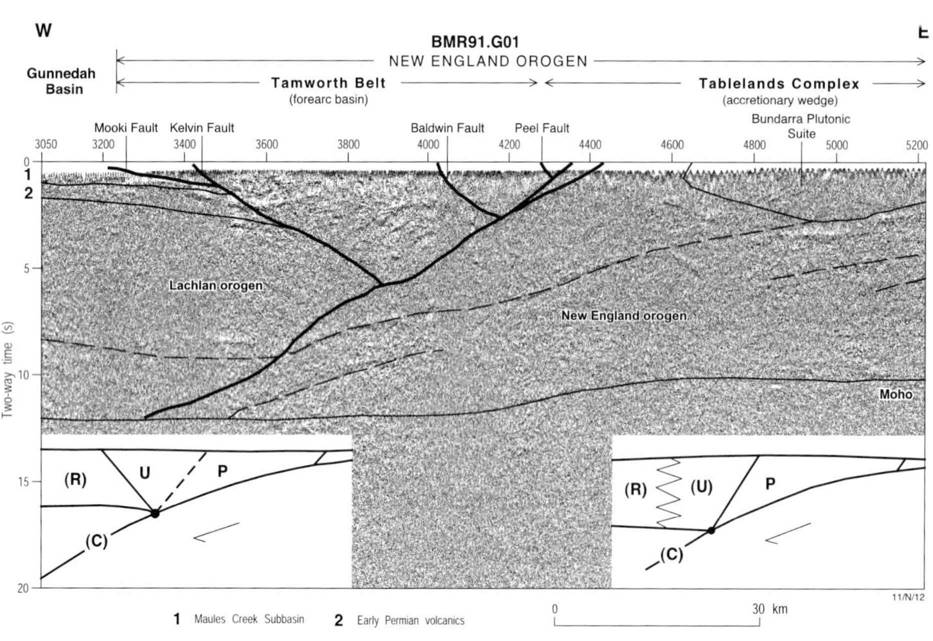

FIGURE 3. Unmigrated deep seismic profile BMR91.G01 across the eastern margin of the Gunnedah Basin and the Tamworth Belt and Tablelands Complex of the New England orogen. Vertical scale approximately equals horizontal scale, assuming an average crustal velocity of 6000 m/s. Insets show the geometries of the P (lower right) and P-U (lower left) subduction-accretion modes. For location see Figure 1.

with the conceptual modes of Beaumont et al. (1999), the system in the Permian-Triassic appears to be in either the P-U or P-U-C mode (Figure 3; see also Waschbusch et al., 1999).

Northern New England Orogen

Components of the convergent-plate-margin system recognized in the Upper Devonian to Upper Carboniferous rocks of the northern New England orogen by Korsch et al. (1990a) include the accretionary wedge, fore-arc basin, magmatic arc (including Connors Arc), and back-arc basin (Drummond Basin). The back-arc basin initiated as an extensional system in the Late Devonian, and then in the Carboniferous developed a foreland-basin phase west of a foreland fold-thrust belt (Johnson and Henderson, 1991; Henderson and Davis, 1993). Much of this fold-thrust belt is hidden beneath younger sedimentary rocks of the Bowen Basin, but thrusts in the pre-Bowen basement are seen in some seismic sections (Korsch and Totterdell, 1995). Because of the presence of a foreland fold-thrust belt and associated foreland-basin phase, the subduction system in the northern New England orogen, at least during the Carboniferous, appears to be in either the P-U-R or the P-U-R-C conceptual mode of Beaumont et al. (1999).

In the Late Permian to Middle Triassic, a major episode of west-directed thrusting has been documented in the northern New England orogen and in the eastern part of the Bowen Basin by Korsch et al. (1990b, 1997), Fergusson (1991), and Holcombe et al. (1997), with thin-skinned thrust models being proposed for this region. These models infer a west-dipping subduction zone located farther to the east.

In the region of the 1989 BMR seismic lines (Figure 1), the zone of west-directed thrusting is about four times wider than that seen in the southern New England orogen. I interpret this to be a very wide Middle Permian–Middle Triassic westward-propagating retroforeland thrust belt that produced massive loads in the crust, leading to a major retroforeland basin phase in the Bowen Basin (see below). The deep seismic lines BMR89.B01 (Figure 4) and BMR89.B02 (Figure 5) in the northern New England orogen show relatively planar reflections that dip moderately to the east and cut much deeper into the crust than those shown in Holcombe et al.'s (1997) interpretative cross section. These are interpreted to form part of a thick-skinned, crustal-scale imbricate thrust system.

The Marlborough Serpentinite, located about the eastern half of seismic line BMR89.B02 (Figure 1), is a thin flat sheet of upper Neoproterozoic sea floor (with a whole-rock Nd-Sm isotope age of 562 ± 22 Ma; Bruce et al., 1999). It now sits above deformed Lower Permian and older rocks of the Gogango Thrust Zone (Figure 5). The serpentinite was emplaced late in the sequence of Middle Permian–Middle Triassic thrusting, as a major out-of-sequence event, because it has been thrust over earlier, more steeply dipping planar thrusts in the northern New England orogen.

The seismic lines BMR89.B01 and BMR89.B02 (Figures 4 and 5) can be combined to form a composite section from the western Bowen Basin eastward into the northern New England orogen (Figure 6a). These seismic lines are dominated by a thrust system that dips to the east, suggesting that, during the Late Permian to the Middle Triassic, the New England orogen was thrust westward over the Thomson orogen (Korsch et al., 1990b; 1997). The deep seismic data show a much greater development of the west-directed foreland thrust belt in northern New England (Figure 6a) than in southern New England (Figure 6b).

Because of the presence of a major foreland fold-thrust belt and associated foreland-basin phase, the subduction system in the northern New England orogen, during the Late Permian to the Middle Triassic, appears to have been in either Beaumont et al.'s (1999) P-U-R or P-U-R-C conceptual mode (Figure 4).

Comparison of Southern and Northern New England Orogens

When compared with the conceptual modes of Beaumont et al. (1999), the geometry of the subduction systems for the northern and southern New England orogens show significant differences for the Late Devonian-Late Carboniferous and Late Permian–Middle Triassic periods. The northern New England orogen is dominated by the retrowedge mode, whereas the southern New England orogen changed from a prowedge mode to an uplifted plug mode (Figure 6).

In the southern New England orogen, the initial system during the Late Devonian–Late Carboniferous involved either a small input of material at the trench and a strongly negatively buoyant subducting plate or, more probably, a plate with significant subduction zone retreat (roll-back). This geometry changed, during the Late Permian–Middle Triassic, to a system in which there was some constriction in the subduction zone as a result of a buoyant subducting plate or an advance of the subduction zone, thereby producing an active, uplifted plug, with limited west-directed retrothrusting (for a full discussion of the geologic implications of the subduction zone geometries, refer to Beaumont et al., 1999).

In contrast, in the northern New England orogen, the subduction zone appears to have undergone severe constriction or choking during both time intervals, because both the uplifted plug and retrowedge were active, and significant amounts of retrodeformation took

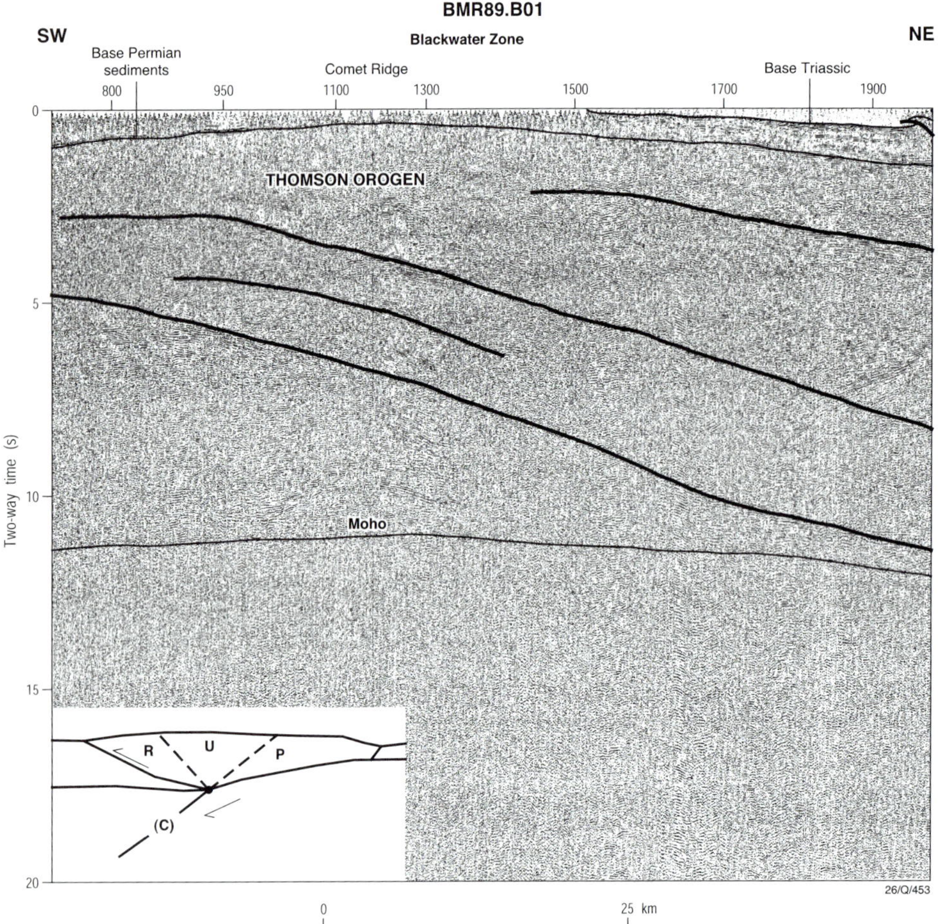

FIGURE 4. Unmigrated deep seismic profile BMR89.B01 across the Permian-Triassic Bowen Basin and Tertiary Duaringa Basin. Vertical scale approximately equals horizontal scale, assuming a crustal velocity of 6000 m/s. Inset shows the geometry of the P-U-R subduction-accretion mode. For location see Figure 1.

place. One or more of several geologic reasons are likely to have contributed to the choking of the subduction zone: docking and attempted subduction of terranes, a shallowly dipping subduction zone, magmatism that might weaken the retrocrust, attempted subduction of a thick, detached sedimentary succession or trench fill, and/or the growth of a large accretionary prowedge (see Beaumont et al., 1999, for further discussion on the geologic implications of the subduction-zone geometries).

RETROFORELAND BASIN PHASE IN BOWEN AND GUNNEDAH BASINS

The development of the early Late Permian to Middle Triassic retroforeland thrust belt in the New England orogen had a marked impact on the development of the Bowen, Gunnedah, and Sydney Basins. The event that initiated formation of these basins was extensional (Figure 7), and the continental crust was stretched to form the significant Early Permian East Australian Rift System, as defined by Korsch et al. (1998). This Early Permian extensional event was followed by a period with lower subsidence rates resulting from thermal relaxation of the lithosphere (Korsch et al., 1998). Basin-forming events from the early Late Permian to the Middle Triassic are associated with plate convergence and development of the retroforeland thrust belt in the New England orogen. Thus, the history of the basin system is intimately linked to tectonic events in the orogen (Fielding et al., 1997; Glen and Beckett, 1997; Korsch et al., 1998).

The Early Permian thermal-subsidence phase was interrupted in the early Late Permian by the onset of rapid subsidence in the Taroom Trough on the eastern side of the Bowen Basin, which is interpreted as a foreland loading phase because of thrusting and crustal thickening in the New England orogen to the east. Very thick successions were deposited for some of the individual foreland supersequences, with as much as 8 km of foreland deposits occurring on the eastern side of the Taroom Trough (e.g., Figure 8). The subsidence during the foreland-basin phase in the Bowen and Gunnedah Basins may have been a flexural response to loading of the retrolithosphere by supralithospheric loads, such as the accretionary-wedge-thickened orogenic crust and the retrothrust belt in the New England orogen. Alternatively, it may have been the result of dynamic platform tilting caused by viscous corner flow in the asthenospheric wedge above the subducting plate (for a full description refer to Mitrovica et al., 1989), or because of a combination of the two processes.

Geodynamic modeling of eight time slices between seismic horizons B45 and B90 (Figure 7) was carried out by Waschbusch et al. (1998) and P. Waschbusch, C. Beaumont, and R. J. Korsch (unpublished data) using

FIGURE 4. (cont.).

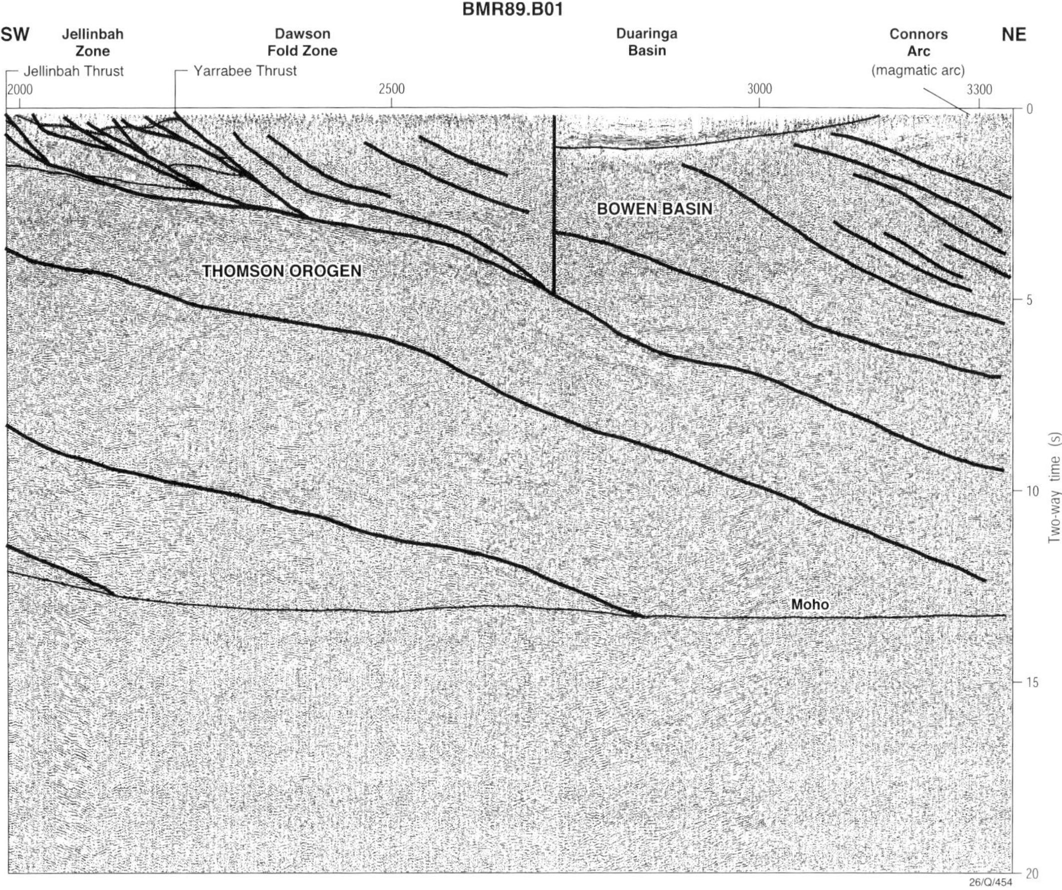

an elastic, planform, finite-element plate model to calculate the flexural response to orogenic and sedimentary loads. The models require the loads to be placed adjacent to the relatively undeformed, current eastern edge of the Bowen Basin, throughout the Late Permian to Middle Triassic period of retrothrusting in the northern New England orogen. This requirement is difficult to reconcile with geologic constraints for the earliest foreland phase, such as the initial distribution of stratigraphic units now correlated with the Bowen Basin (Fielding et al., 1997). The fine-grained turbiditic nature of the early foreland sediments and the lack of coarse-grained input suggest a depositional setting somewhat remote from the thrust front at that time. A second geologic constraint is the notion that foreland thrust belts usually evolve progressively from the internal (eastern New England) to external (western New England) positions as the width of the orogen increases. The dichotomy is removed if the early foreland phase of subsidence, in particular, was primarily a dynamic platform tilt, which eliminates the need for proximal supralithospheric loads. Mantle corner flow can create both the necessary near-field large amplitude–short wavelength subsidence and a far-field small tilt, such as is observed for the Springsure Shelf of the Galilee Basin, and which occurs to the west of the Bowen Basin (Figure 1). In later phases of the Bowen Basin's development, the two mechanisms may have acted together or supralithospheric loads may have dominated.

The positions of the crustal loads varied through time, although the loads in the northern New England orogen were usually an order of magnitude greater than those in the southern New England orogen, as is inferred from the sediment thicknesses in the foreland basin (e.g., Figure 8).

DEFORMATION IN THE BOWEN AND GUNNEDAH BASINS

Response to Thrusting along the Eastern Margin of the Basins

The current eastern margin of the Bowen and Gunnedah Basins has historically been regarded as a faulted margin that is defined by the Burunga-Leichhardt, Moonie-Goondiwindi, and Hunter-Mooki fault system, approximately along the 150°E meridian (Figure 1). This fault system consists of several discrete faults (Figure 8) separated by gaps interpreted to be displacement transfer zones. The geometries of these faults show considerable variation along strike. The faults are hard-linked to the

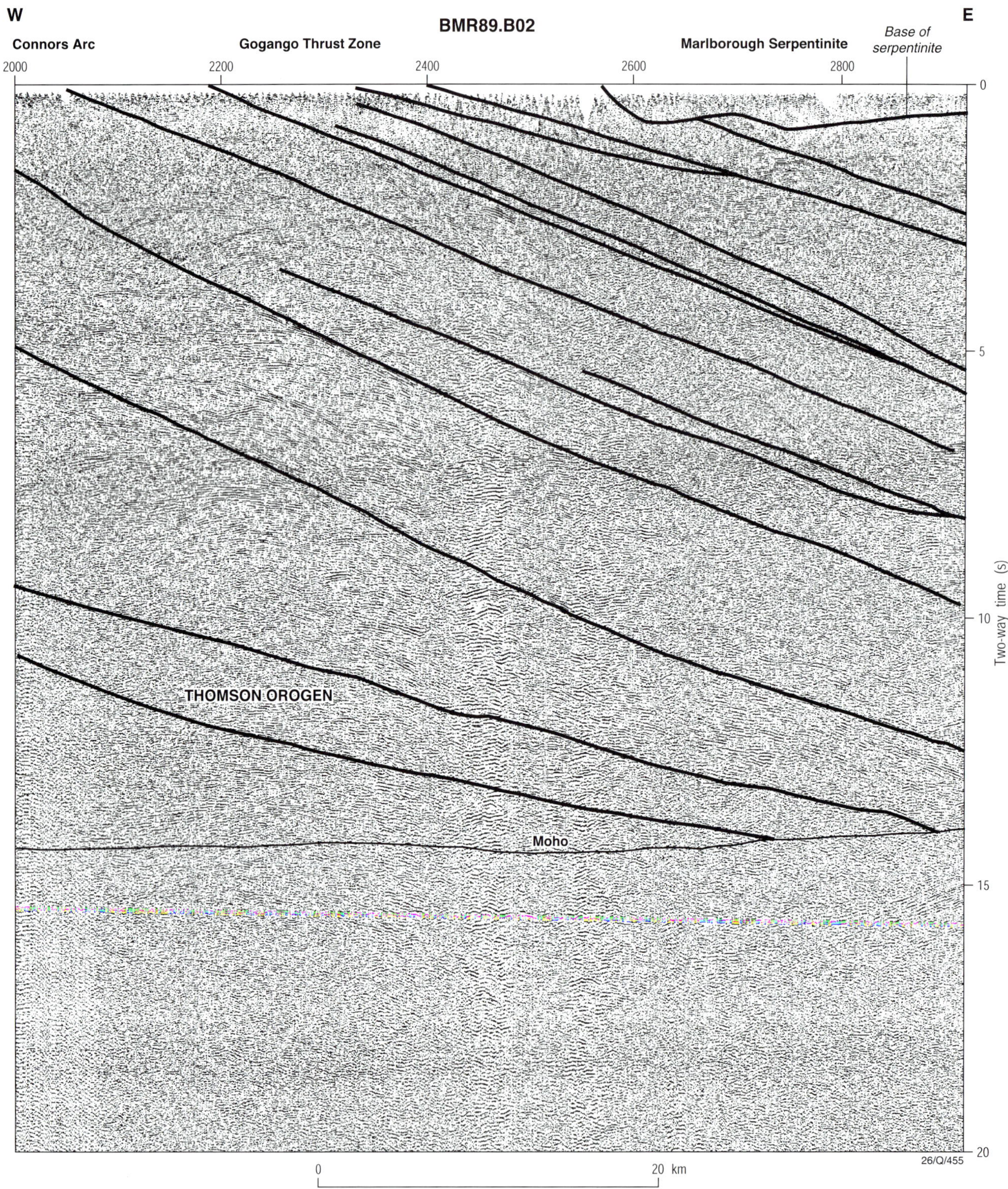

FIGURE 5. Unmigrated deep seismic profile BMR89.B02 across elements of the northern New England orogen, including the Connors Arc and Marlborough Serpentinite. Vertical scale approximately equals horizontal scale, assuming an average crustal velocity of 6000 m/s. For location see Figure 1.

thrusts in the retroforeland thrust belt in the New England orogen to the east (Korsch et al., 1997). In places, remnants of the Bowen Basin occur to the east of the faults, and hence they cannot be regarded as basin-bounding faults. The basin originally extended much farther to the east, and the present eastern margin is an erosional remnant located to the east of the thrust faults.

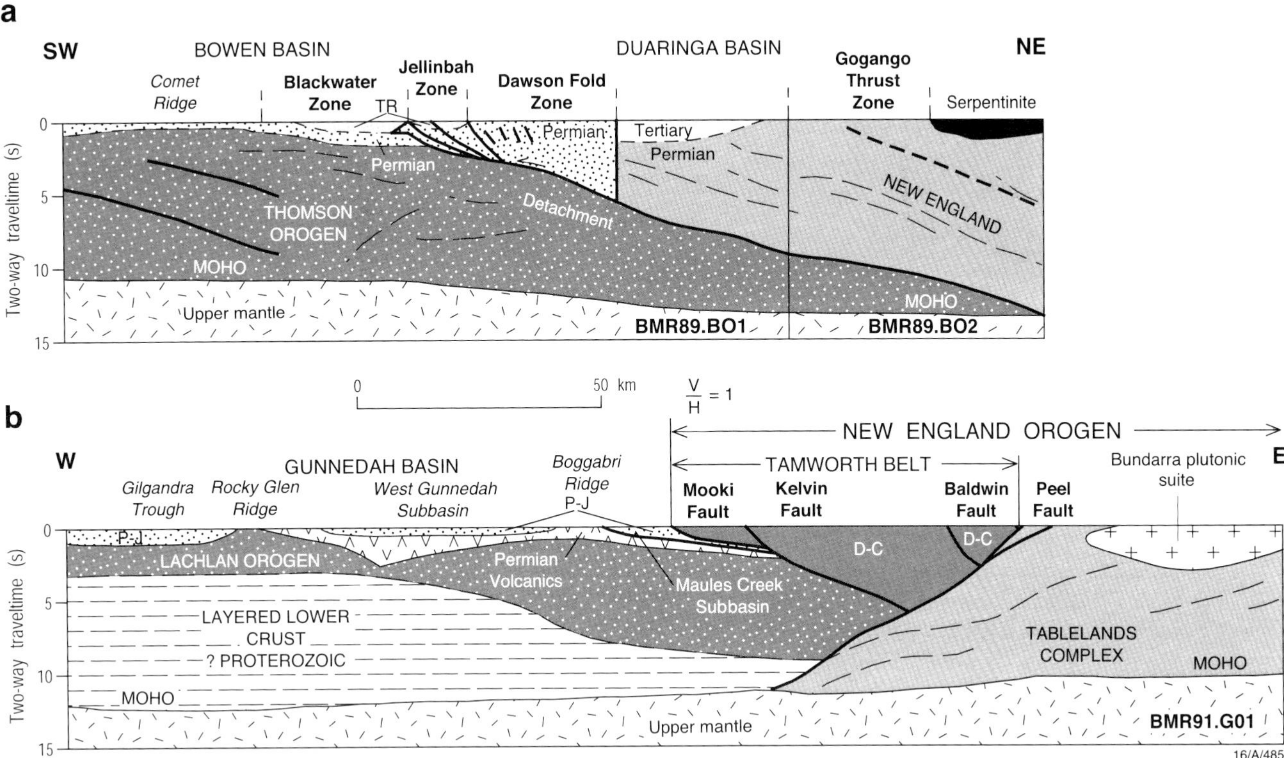

FIGURE 6. Diagrams of the crustal architectures in the vicinity of the two deep seismic surveys, showing the west-directed retroforeland thrust systems and the relationships between the New England orogen and the Thomson (a) and Lachlan (b) orogens. Note that (a) is a composite section consisting of seismic lines BMR89.B01 and BMR89.B02 (see Figure 1 for locations).

All of the thrusts along the eastern margin are new thrusts that are hard-linked to the growing thrust wedge in the New England orogen to the east that is propagating westward into the basin system. No evidence has been found to demonstrate that these faults are reactivated normal faults, as described from the Denison Trough in the western part of the Bowen Basin (see below).

Jellinbah Zone

At the northern limit of the study area, at approximately 23°30′S, in the Blackwater area (Figure 8), a series of low-angle thrusts forms an imbricate thrust fan that roots onto a major east-dipping detachment in the Jellinbah Zone. At the surface, the limits of the imbricate fan are defined by the Jellinbah and Yarrabee Thrusts (Figure 9).

Along the Jellinbah Thrust, the Upper Permian coal measures have overridden the Lower Triassic Rewan Group. The thrust soles onto the detachment at about 1.3 s two-way traveltime (TWT). The detachment flattens and becomes more diffuse in the ductile zone in the middle to lower crust, at about 7 s TWT beneath the Duaringa Basin (Figure 4). In the section dominated by the dipping detachment, the Blackwater, Jellinbah, and Dawson structural zones in the Bowen Basin, which were first recognized by Hobbs (1985), exhibit distinct structural styles both in their surface geology and seismic characteristics (Figure 9).

In the Blackwater Zone, to the west of the Jellinbah Thrust, the uppermost part of the seismic section is dominated by subhorizontal reflections that are interpreted to be mainly sedimentary layering (Figure 9). In the Jellinbah Zone, the seismic section is dominated by a series of mildly listric thrust faults forming an imbricate thrust fan that soles into a detachment fault dipping gently to the east. Ramp anticlines have developed in the hanging wall above some thrusts (Korsch et al., 1992). The boundary between the Upper Permian coal measures and overlying Lower Triassic sandstones shows minor displacement at several localities (Figure 9), indicating that the deformation, at least in this area, is Early to Middle Triassic or younger. The frontal splay below the Jellinbah Thrust (Figure 9) represents the westernmost hard-linked thrust of the retroforeland thrust belt.

The Dawson Fold Zone, to the east of the Yarrabee Thrust, consists of tightly folded Upper Permian Bowen Basin fill on the surface (Fergusson, 1991). In this zone, the deformation is so intense that indications of the original geometry of the basin before thrusting are all

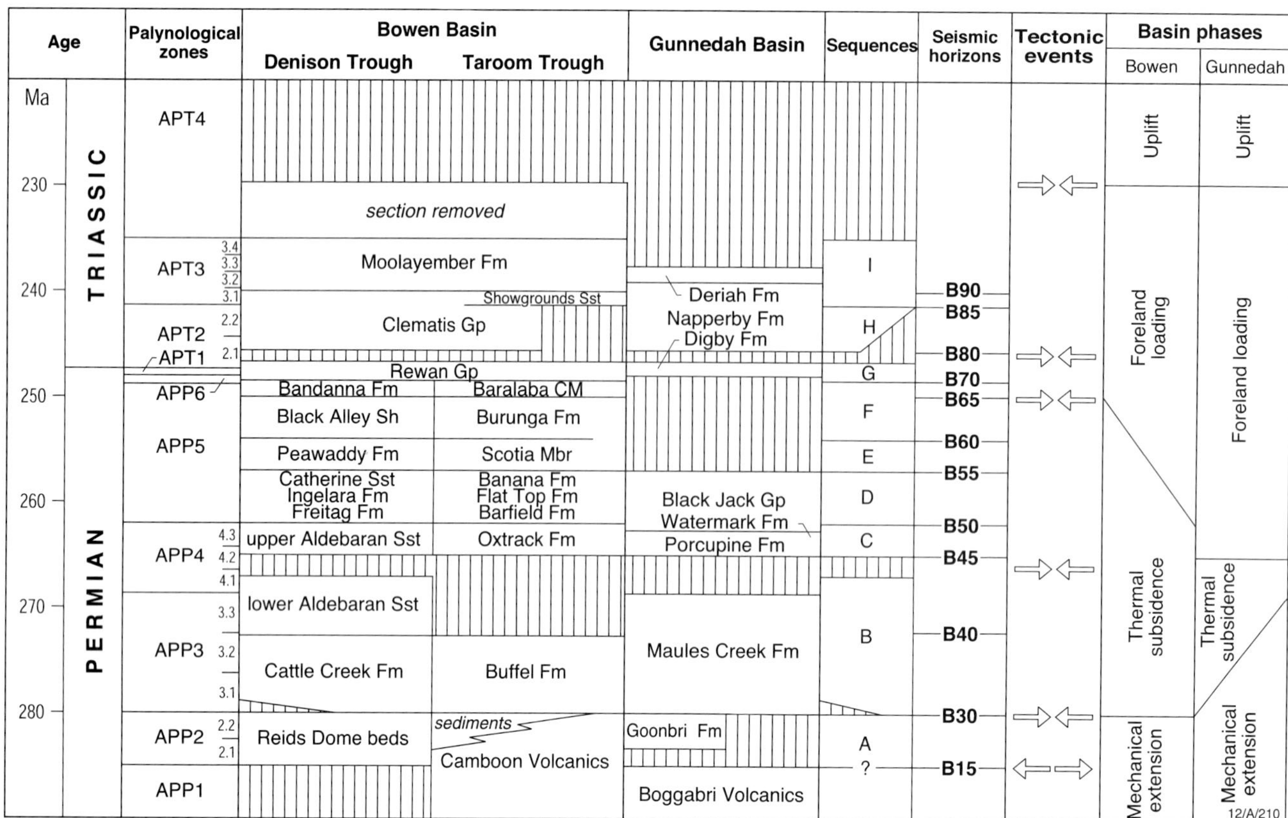

FIGURE 7. Permian and Triassic correlation chart of the Bowen and Gunnedah Basins, showing relationship between lithostratigraphy, sedimentary supersequences, and main tectonic events. Arrows indicate contractional (⇒⇐) and extensional (⇐⇒) events. For a description of the supersequences see Totterdell et al (1995).

but obliterated. Because of the steep dips, seismic reflectivity is low, but in the seismic section some reflections dip moderately to the east and are interpreted to be thrust faults above the east-dipping detachment (Figure 4). Therefore, the surface folds are interpreted to be ramp anticlines related to underlying thrusts. The base of the Dawson Fold Zone is the major detachment. Below the Blackwater Zone and the east-dipping detachment, strong, east-dipping reflections are interpreted to be major thrust faults in the basement (Figure 4).

In this area, the thrust front initiated in the New England orogen to the east and propagated a long distance into the basin—a characteristic not seen elsewhere in the basin system to the south. The imbricate thrust fan is a good example of westward-propagating thin-skinned deformation, which is hard-linked to the underlying detachment. At middle- to lower-crust levels, this detachment forms part of a series of planar, crustal-scale, thick-skinned faults. Thus the detachment forms the linkage between the thin-skinned and thick-skinned systems.

Burunga to Goondiwindi Faults

Farther south, the northernmost segment of the meridianal fault system is the Burunga Fault (Figure 8), a west-dipping back thrust associated with a basement duplex to the east (Totterdell and Korsch, 1992; Elliott, 1993). The Burunga Anticline, above the thrust, is a hanging-wall anticline inferred to have formed as an incipient fault-propagation fold (Figure 10). The Late Permian coal measures in it contain commercial quantities of coal-seam methane. Because of the offset of the regional, this fault must be planar to a considerable depth; that is, it is probably a deep-seated crustal fault. The timing of thrusting postdates deposition of the Middle Triassic Moolayember Formation but precedes deposition of the Surat Basin; that is, it occurred during the Goondiwindi Event (Korsch et al., 1998) in the Middle to Late Triassic. At its southern end, a gap, interpreted as a displacement transfer zone, separates the Burunga Fault from an east-dipping thrust, the Miles Fault (Figure 8). In contrast to the Burunga Fault, which is east-directed (Figure 10), the Miles Fault and all the other basin-margin faults to the south (from north to south: Leichhardt, Moonie, Tingan, Goondiwindi, and Mooki Faults) are west-directed (e.g., Figures 11 and 12).

Farther south, another segment, the Moonie Fault, consists predominantly of a low-angle thrust fault with a ramp-flat geometry, showing a typical fault-bend fold style (Figure 11). Frontal or shortcut thrusts with limited displacement commonly occur northwest of the

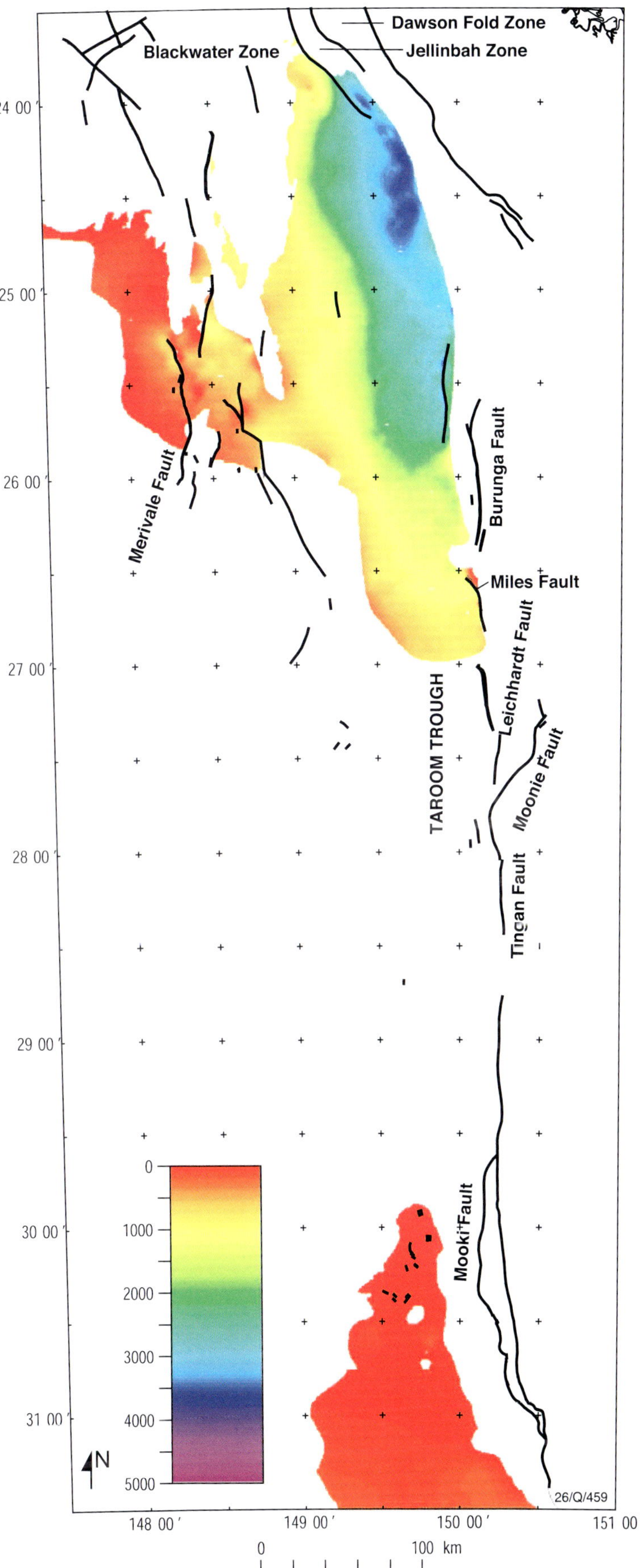

FIGURE 8. Isopach map of the B70-B80 interval, which is equivalent to Supersequence G (thicknesses in meters). The present map distribution, particularly in the north, is the result of later uplift and erosion of parts of the basin. Note that, on the basis of published palynological data, this supersequence is not present in the southernmost Bowen and northernmost Gunnedah Basins. The contour pattern reflects subsidence resulting from foreland loading in response to thrusting and mountain building at the western edge of the New England orogen. On the eastern side of the basin, the Taroom Trough extends north to about 25°S. Faults are mapped at the B70 horizon.

main thrust. The major movement on this thrust occurred after deposition ceased in the Bowen Basin but prior to commencement of deposition in the Surat Basin, that is, again, in Middle to Late Triassic time. The youngest Triassic unit, the Moolayember Formation, is preserved to the west of the Moonie Fault (immediately below S10 on Figure 11) but was uplifted in the hanging-wall block above the thrust and eroded off prior to deposition of the Surat Basin. The significant increase in the elevation of the regional (e.g., base of Bowen Basin, B30 horizon), to the east of the fault, indicates that at depth, below the section illustrated in Figure 11, there is an underlying planar crustal-scale fault. The entire Bowen Basin succession in the anticline above the thrust fault has been eroded, so that the Surat Basin succession sits directly on a basement of New England orogen rocks immediately southeast in the hanging wall of the fault, but with a remnant of the Bowen Basin preserved farther to the southeast (Figure 11). Farther to the south, this structure hosts the largest oil field in the basin system, with the reservoir being the basal unit of the Surat Basin.

The southern part of the Moonie Fault is also separated from the northern part of the Tingan Fault (Figure 12) by another displacement transfer zone. Here, significant uplift occurred prior to deposition of the Surat Basin, because the Bowen Basin succession has been completely removed from east of the Moonie Fault, which suggests that it is a planar, deep-seated crustal fault. Thus, there was major movement on the faults during the Middle to Late Triassic Goondiwindi Event, but also, additional reactivation occurred during the early Late Cretaceous Moonie Event of Korsch et al.

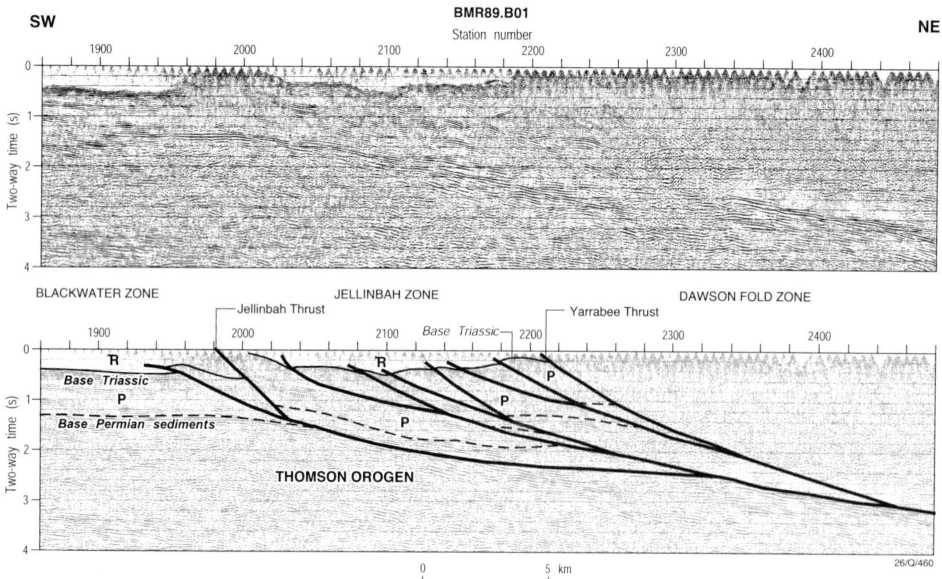

FIGURE 9. Enlargement of part of an unmigrated seismic profile BMR89.B01 across the Jellinbah Zone, between the Jellinbah and Yarrabee Thrusts, showing the imbricate thrust fan. The sole thrust, or detachment, forms a narrow discrete zone in the upper crust. The Upper Permian coal measures in the anticline at Station 2200 have been drilled to test the coals as a potential coal-seam methane exploration drainage target. Top of the coal measures is equivalent to the Base Triassic horizon. Vertical scale approximately equals horizontal scale, assuming an upper-crust velocity of 4000 m/s.

(1998). Seismic line P81-112 (Figure 12) suggests that there was significant reactivation of the thrusts, probably in the early Late Cretaceous after deposition of the Early Cretaceous succession. This is particularly true for the Tingan Fault, where there is about 455 ms (approximately 720 m) of uplift of the Surat succession east of the Tingan Fault relative to the same succession to the west of the Moonie Fault. The geometry of the two faults is inferred to be planar and deep-seated to maintain the elevation of the regional in the hanging wall, both pre- and postdeposition of the Surat Basin succession. Although the faults were reactivated at depth, here they appear not to have propagated upward into the Surat Basin succession. That is, there appears to be no brittle displacement of the Surat succession across the faults.

The geometry of the structures associated with the fault system near the eastern margin of the basin, that

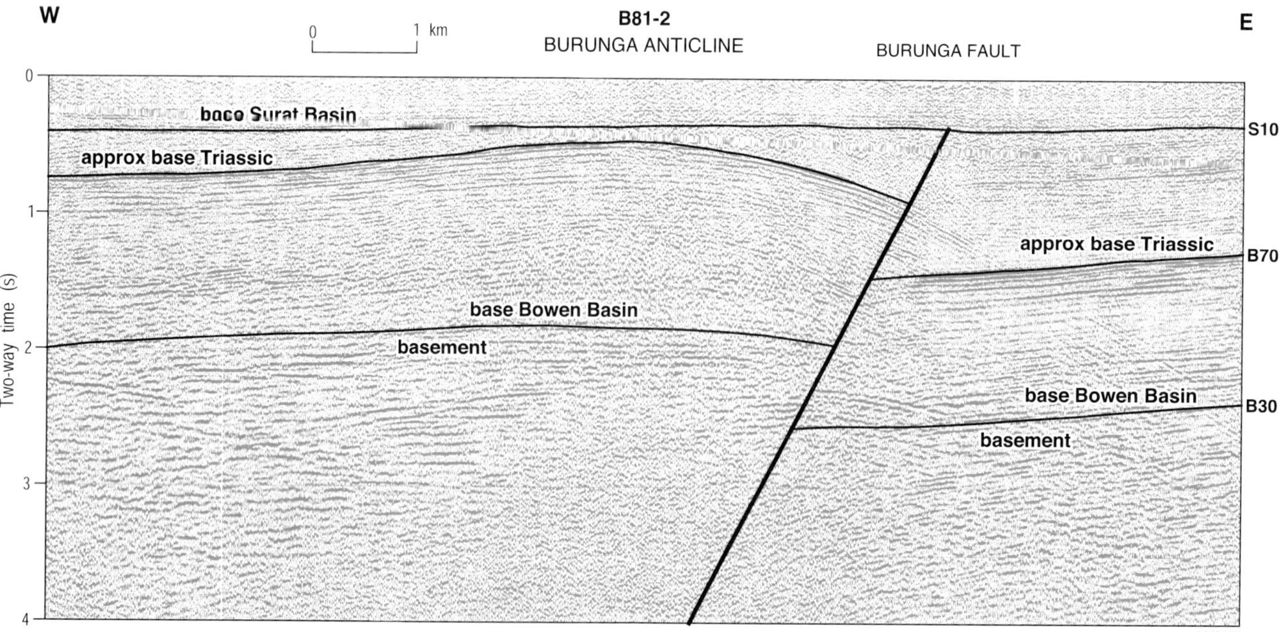

FIGURE 10. Seismic line B81-2, showing the geometry of the Burunga Fault, with the development of the Burunga Anticline as an incipient fault-propagation fold above the thrust. This anticline houses commercial quantities of coal-seam methane in the Upper Permian coal measures.

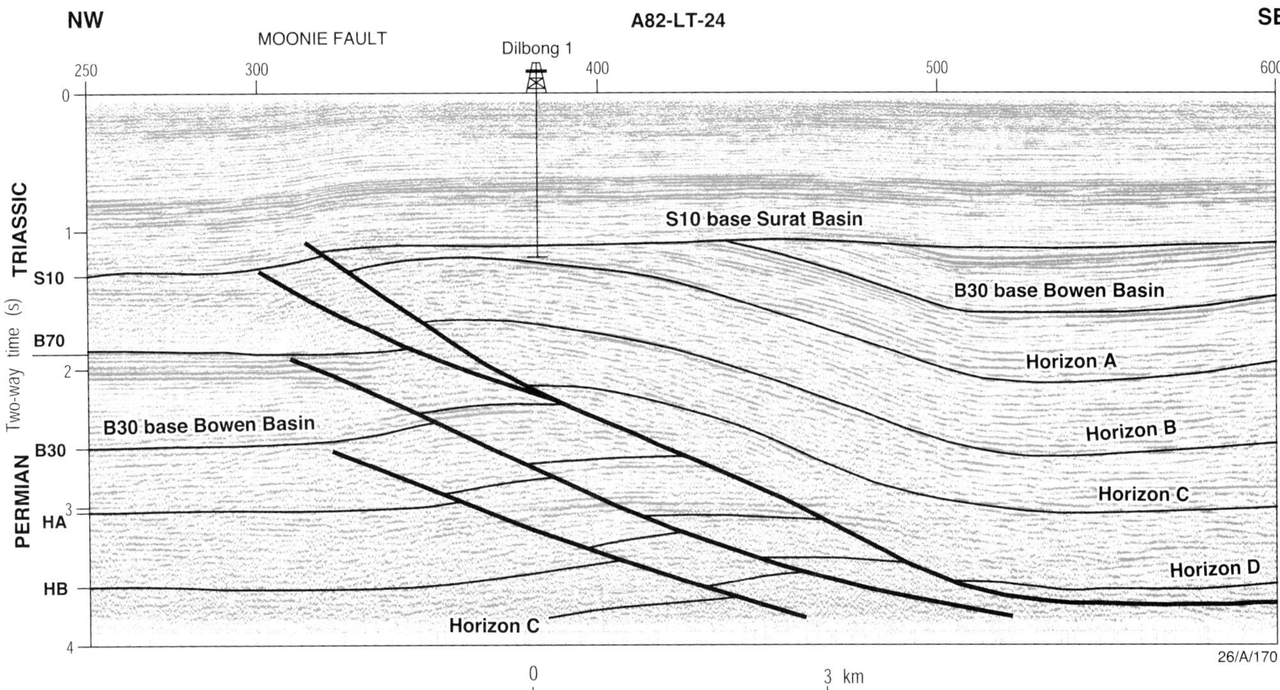

FIGURE 11. Seismic line A82-LT-24, showing typical geometry of the Moonie Fault to the north of Moonie oil field. The predominant movement is on a low-angle thrust fault with a ramp-flat geometry. There are also frontal (shortcut) thrusts and an out-of-sequence thrust, all with limited displacement. The section shows typical fault-bend fold geometry. Major movement on this thrust occurred after deposition ceased in the Bowen Basin but prior to commencement of deposition in the Surat Basin, that is, in Middle to Late Triassic time. The succession in the anticline above the thrust fault has been eroded so that the Surat succession sits directly on a basement of fore-arc rocks of the Tamworth Belt immediately east of the fault. Horizons labeled by a letter and number on this and the other seismic sections (e.g., B30, S10) refer to seismic horizons (Figure 7) described by Totterdell et al. (1995). Horizons A to D are within the fore-arc basin and are described by Wartenberg et al. (1999).

is, the Burunga to Goondiwindi Faults (Figures 10 to 12), and especially the elevation of the regional, infers that the faults are planar to considerable depths in the crust. This supports the interpretation of crustal-scale planar faults imaged in the deep seismic profiles (Figures 4 and 5).

Response to Retrothrusting in the Denison Trough, Western Bowen Basin

The development of the Early Permian East Australian Rift System (Korsch et al., 1998) is best exemplified by the geometry of the Denison Trough in the western Bowen Basin (Figure 13). During this mechanical extensional phase, the trough was actually a series of half grabens that, in places, coalesced. It was only during the subsequent thermal subsidence phase (Figure 7) that the troughlike geometry initially developed. The Denison Trough is located well to the west of the westernmost extent of the retrothrust belt, and hence the reactivated normal faults and new thrusts in the trough are not hard-linked to the thrust belt.

After the extensional faults and associated half grabens developed, several contractional events occurred between the late Early Permian and the Middle Triassic, as documented by Elliott (1993) and Korsch et al. (1998). The events are the responses to far-field stresses that resulted from contraction in the orogen and farther east. Of these events, the two most important for petroleum occurrences in the trough will be discussed here. These are the middle Permian Aldebaran Event and the Middle to Late Triassic Goondiwindi Event (as defined by Korsch et al., 1998).

The Aldebaran Event is important in eastern Australia, because it marks the initial development of the retroforeland thrust belt in the middle Permian. In the New England orogen, it is represented by a major middle Permian deformational event, which was the original definition of the Hunter-Bowen orogeny by early workers (Carey and Browne, 1938).

In the Bowen Basin, evidence for the middle Permian Aldebaran Event is best seen in seismic sections from the Denison Trough, where it is equivalent to the unconformity in the middle of the Aldebaran Sandstone. In places, many of the extensional faults bounding the

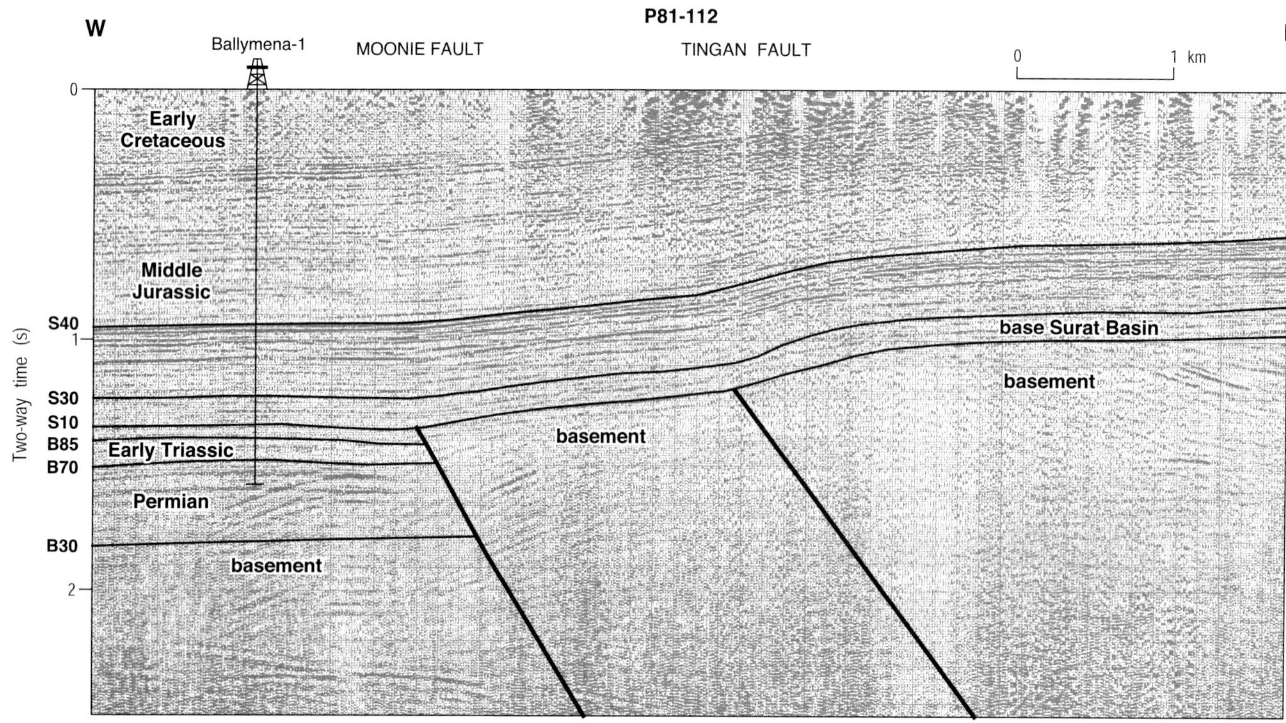

FIGURE 12. Seismic line P81-112, showing the geometry of the Surat Basin (from S10 to top of section) across the zone between the Moonie and Tingan Faults. This line shows significant reactivation of the thrusts, particularly the Tingan Fault, during the early Late Cretaceous Moonie Event, with about 455 ms (approximately 720 m) of uplift of the Surat Basin succession east of the Tingan Fault relative to the same succession west of the Moonie Fault. Although the faults were reactivated at depth, here they appear not to have propagated into the Surat Basin succession, that is, there appears to be no brittle displacement of the Surat Basin succession across the faults.

half grabens were reactivated as thrust faults during this event, leading to partial inversion of the sedimentary pile, with the lower Aldebaran Sandstone below the unconformity being uplifted and partially eroded. For example, in seismic section BMR78-02 (Figure 14), the lower Aldebaran Sandstone (B40–B45 interval) has been totally removed adjacent to the fault, indicating about 250 m of erosion. The amount of erosion decreases to zero several kilometers away from the fault. Elsewhere in the Denison Trough, the lower Aldebaran Sandstone has not been significantly eroded and forms one of the main gas reservoirs in the hanging-wall anticlines (Figure 13). For comparison, in the Taroom Trough to the east, the Aldebaran Event is represented by the significant time break between the Buffel and Oxtrack Formations (Figure 7).

The Middle to Late Triassic Goondiwindi Event deformed the youngest stratigraphic unit of the Bowen Basin, the Moolayember Formation (Figure 7), and older rocks, prior to initial deposition of the Surat Basin in the Early Jurassic. This event was complex and produced significantly different geometries in different parts of the basin. It was not a simple folding event, as proposed by Elliott (1993). In the Denison Trough, the Early Permian extensional faults were again reactivated as thrusts, although new thrusts also developed. These faults propagated upward into the younger part of the Denison Trough, with the shortening being accommodated by thrusting at depth and by folding in the upper levels of the pile. This produced fault-propagation folds. The amount of displacement on the thrusts decreased upward and was balanced by the fold shortening.

A good example of an extensional fault that was reactivated as a thrust is seen in seismic line AD91-03 from the northern Denison Trough (Figure 15). Here, there are varying displacements on the thrust, with the contraction being accommodated by thrusting at depth but by the development of the fault-propagation fold in the upper levels. The thrust was originally an extensional fault up to the B30 level and probably up to the B40 level. During the Middle Triassic contraction, the extensional fault inverted as a thrust and propagated upward through the sediment pile. The elevation of the B65 and other horizons above the regional suggests that the fault is a planar-dipping, regional-scale fault at depth. The amount of shortening on the fault decreases upward until it is effectively zero displacement just above the B70 horizon, with the shortening

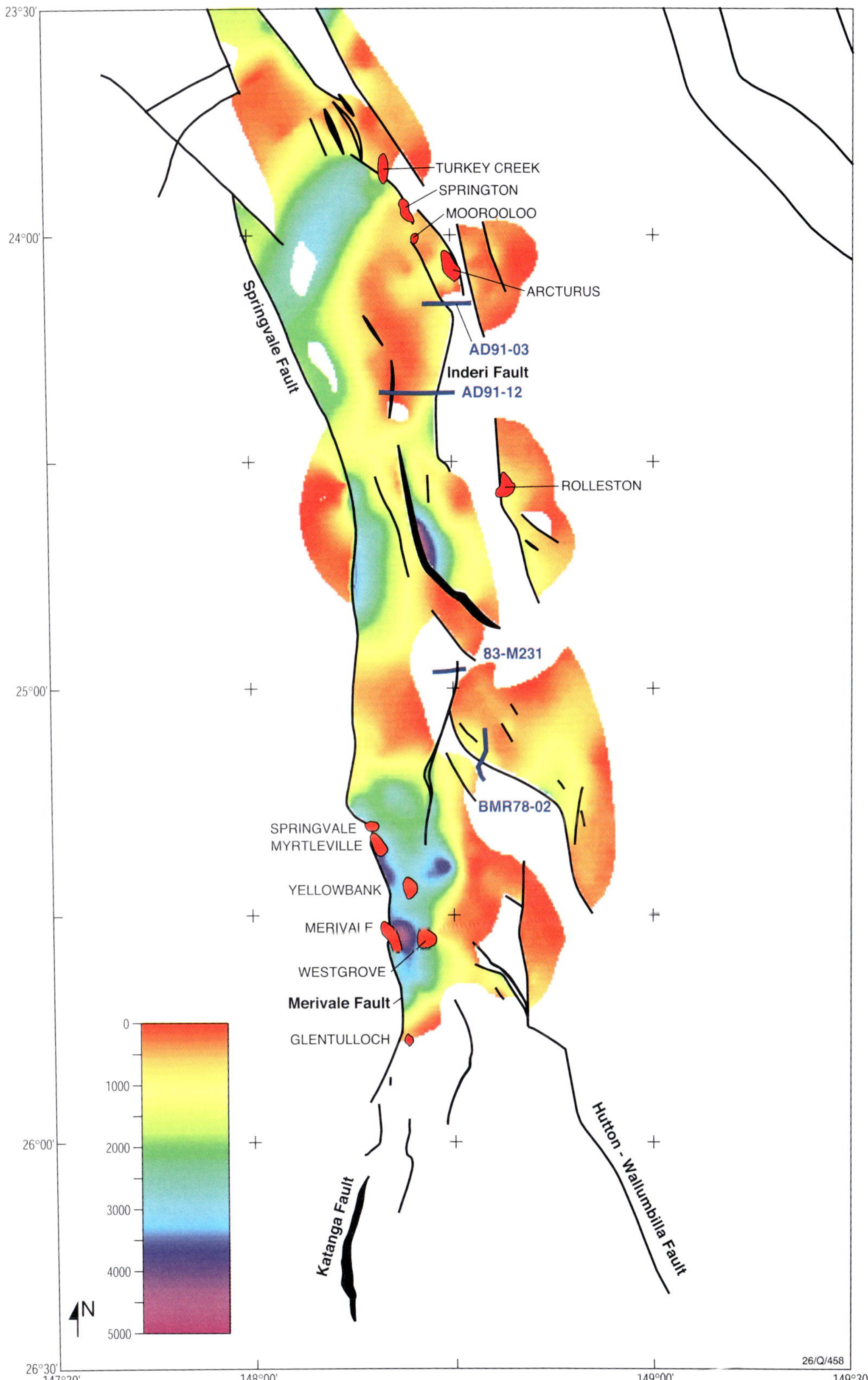

FIGURE 13. Isopach map of Denison Trough (in meters), showing the rift fill, B15 to B30 seismic horizons, equivalent to the Reids Dome beds on Figure 7, and the geometry of the extensional phase. The "trough" is actually a series of discrete half grabens, some of which coalesced to form the observed geometry. This is part of the Early Permian East Australian Rift System. Also shown are the locations of gas fields and the locations of figured seismic lines in the Denison Trough. Map area consists of the northwest part of Figure 8.

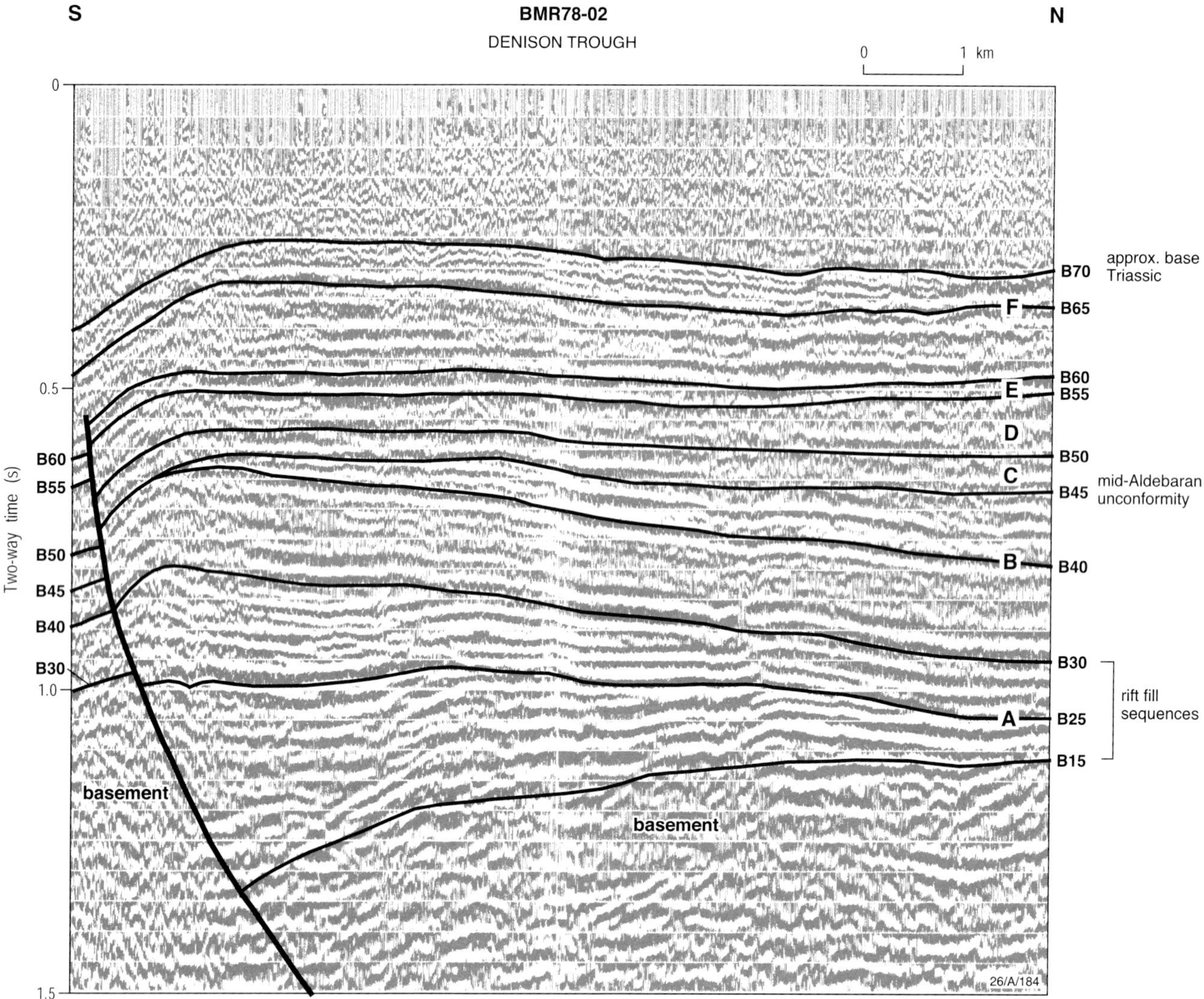

FIGURE 14. Portion of seismic line BMR78-02, showing the geometry of Supersequence A (rift fill) in the Denison Trough and the angular unconformity (B45) between Supersequences B and C that defines the Aldebaran Event. Note the inversion below the B45 horizon, where uplift adjacent to the fault has removed much of the sediment between the B40 and B45 horizons. There was also later reactivation in the Triassic, with the formation of a fault-propagation fold (B70 is close to the Permian-Triassic boundary). See Figure 13 for location of seismic line and Figure 7 for stratigraphic positions of the supersequences (A–F) and seismic horizons (B15–B70).

being accommodated by the growth of the anticline above the fault tip. The total amount of shortening in the section remains the same. However, at depth it is totally accommodated by thrusting, at the top of the section it is totally accommodated by folding, and in between, it is accommodated by a combination of thrusting and folding. Also, significant compaction of rift fill led to a much thicker succession immediately above the rift fill (B30–B40 interval) compared with that deposited on the footwall above basement where there was no compaction or where growth continued on the fault until B40 times. In this example, inversion has resulted in the B30 seismic horizon in the hanging wall being returned to the similar depth it occupies in the footwall. For comparison, in seismic section BMR78-02 (Figure 14), B30 shows significant uplift in the hanging wall compared with the footwall.

In some situations during contraction, an extensional fault does not invert and the strain is accommodated by the development of a back thrust, which cuts through basement. In other situations, an extensional fault inverts as a thrust, and a back thrust develops. This can be seen in seismic section AD91-12 from the northern Denison Trough (Figure 16a). In this case, the dip of the B30 horizon on either side of the pop-up reflects the regional dip.

Another response to contraction is the development of shortcut thrusts. In seismic line 87-WD22 from the

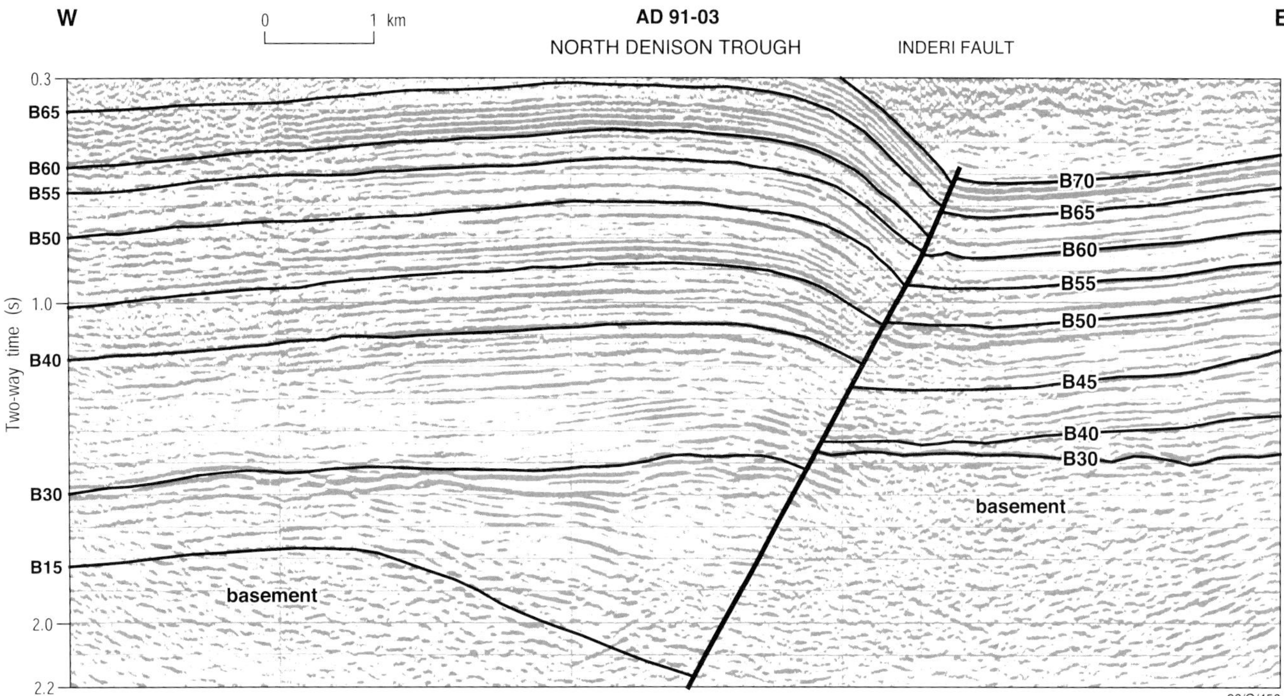

Figure 15. Seismic line AD91-03, showing the geometry of one half graben in the northern part of the Denison Trough. The rift fill, deposited during the active extensional phase, is bounded by the B15 and B30 seismic horizons and possibly by the B40 horizon. This section is a typical geometry for the half grabens within the Denison Trough, and also shows the later reactivation of the bounding fault, with inversion of the rift fill and formation of a hanging-wall anticline above the thrust. See Figure 13 for location of seismic line.

Arbroath Trough (Figure 16b; see Figure 1 for location), the extensional fault has been inverted as a thrust, and also a shortcut thrust has developed. Finally, a combination of inversion of the extensional fault and development of a new back thrust and a shortcut thrust also occurred, such as in seismic line 87-WD26, again from the Arbroath Trough (Figure 16c).

Thus, during the inversion of the half grabens, the shortening was accommodated in several ways. Commonly, inversion by thrusting on the original extensional fault led to the upper levels of the stratigraphic pile forming a fault-propagation fold above the thrust. Also, locking of the extensional fault led to development of a back thrust with the opposite polarity, or a combination of back thrusting and inversion of the extensional fault occurred. Shortcut thrusts also developed during the reactivation of extensional faults and inversion of half grabens.

Hayward and Graham (1989) proposed several theoretical models for the extrusion of half-graben synrift fill during contraction and consequent inversion of extensional basins. These models included deformation by internal strain, "bipolar" extrusion, back thrusting, and the development of shortcut thrusts. The Denison and Arbroath Troughs provide many excellent examples of these geometries in response to the Early Permian rifts during the Goondiwindi contractional event.

Along with reactivation of early extensional faults, there was new thrust development away from the reactivated faults. For example, there are no Early Permian half grabens close to seismic section 83-M231 from the southern Denison Trough (Figure 13). Here, the B30 horizon, the base of the thermal-subsidence phase, which is approximately equivalent to the base of the Cattle Creek Formation (Figure 7), is cut by a thrust fault. Also, the B70 horizon, approximately the Permian-Triassic boundary, is deformed as part of a fault-propagation fold above the thrust (Figure 17). If B30 is restored to a single horizontal surface, that is, if the effects of the Triassic contraction are removed, there is still a significant thrust fault in the basement, which consists of the Devonian–Early Carboniferous Drummond Basin. There is a classic angular unconformity between the Drummond and Bowen Basins, indicating that the Drummond Basin had been extensively deformed prior to initiation of the Bowen Basin. The deformation of the Bowen Basin succession occurred during the Middle to Late Triassic contractional event. When the base of the Permian succession is restored to a single horizontal surface, the ramp-flat geometry within the Drummond Basin is only partially removed, indicating that the initial footwall flat and footwall ramp of the basement thrust were present in the Drummond Basin prior to deposition of the Bowen Basin.

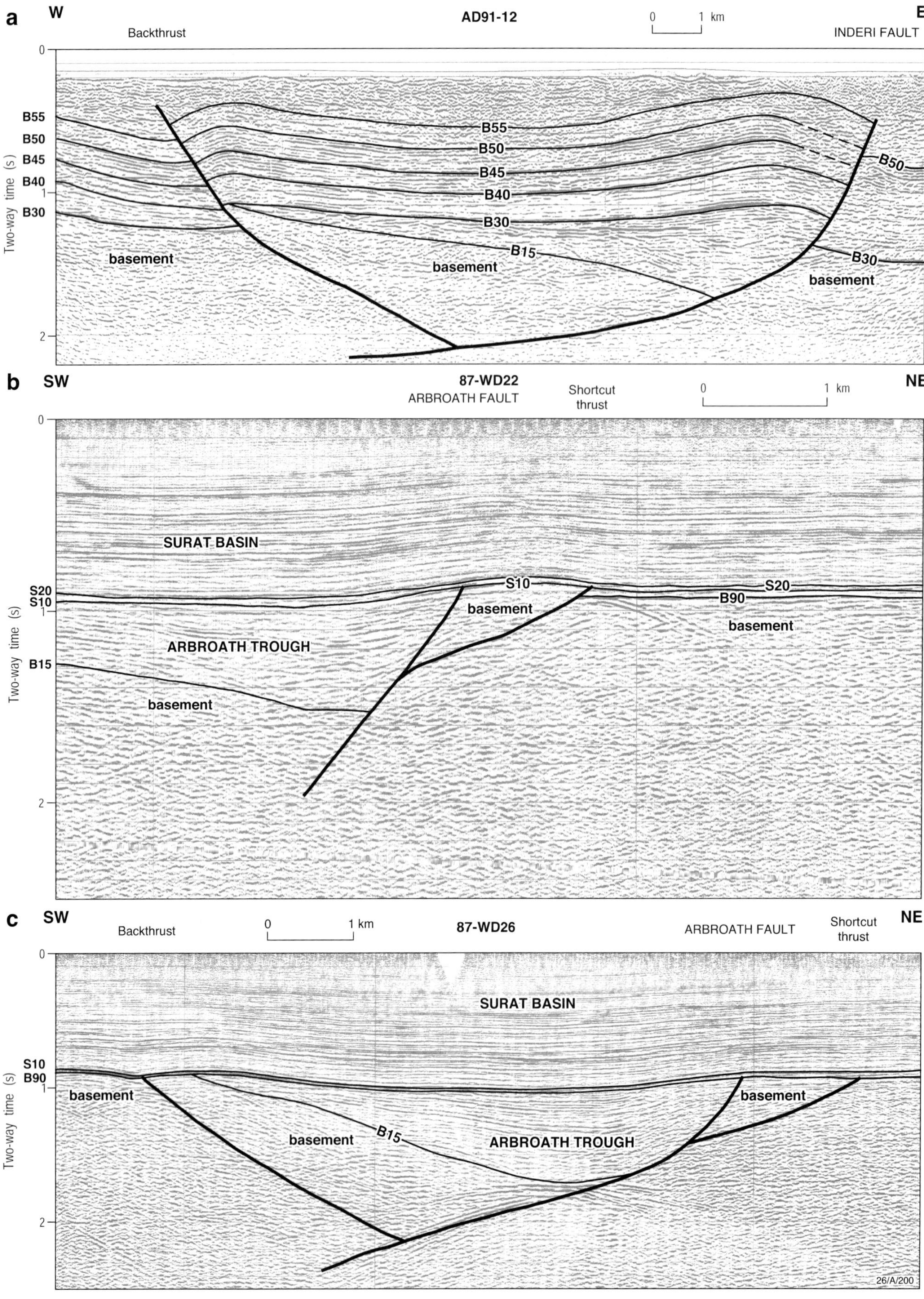
a
W
AD91-12
0 1 km
E
Backthrust
INDERI FAULT
Two-way time (s)
B55
B50
B45
B40
B30
basement
B15
B30
B50
b
SW
87-WD22
ARBROATH FAULT
Shortcut thrust
0 1 km
NE
SURAT BASIN
S20
S10
B90
ARBROATH TROUGH
B15
basement
c
SW
Backthrust
0 1 km
87-WD26
ARBROATH FAULT
Shortcut thrust
NE
SURAT BASIN
S10
B90
basement
B15
ARBROATH TROUGH
26/A/200

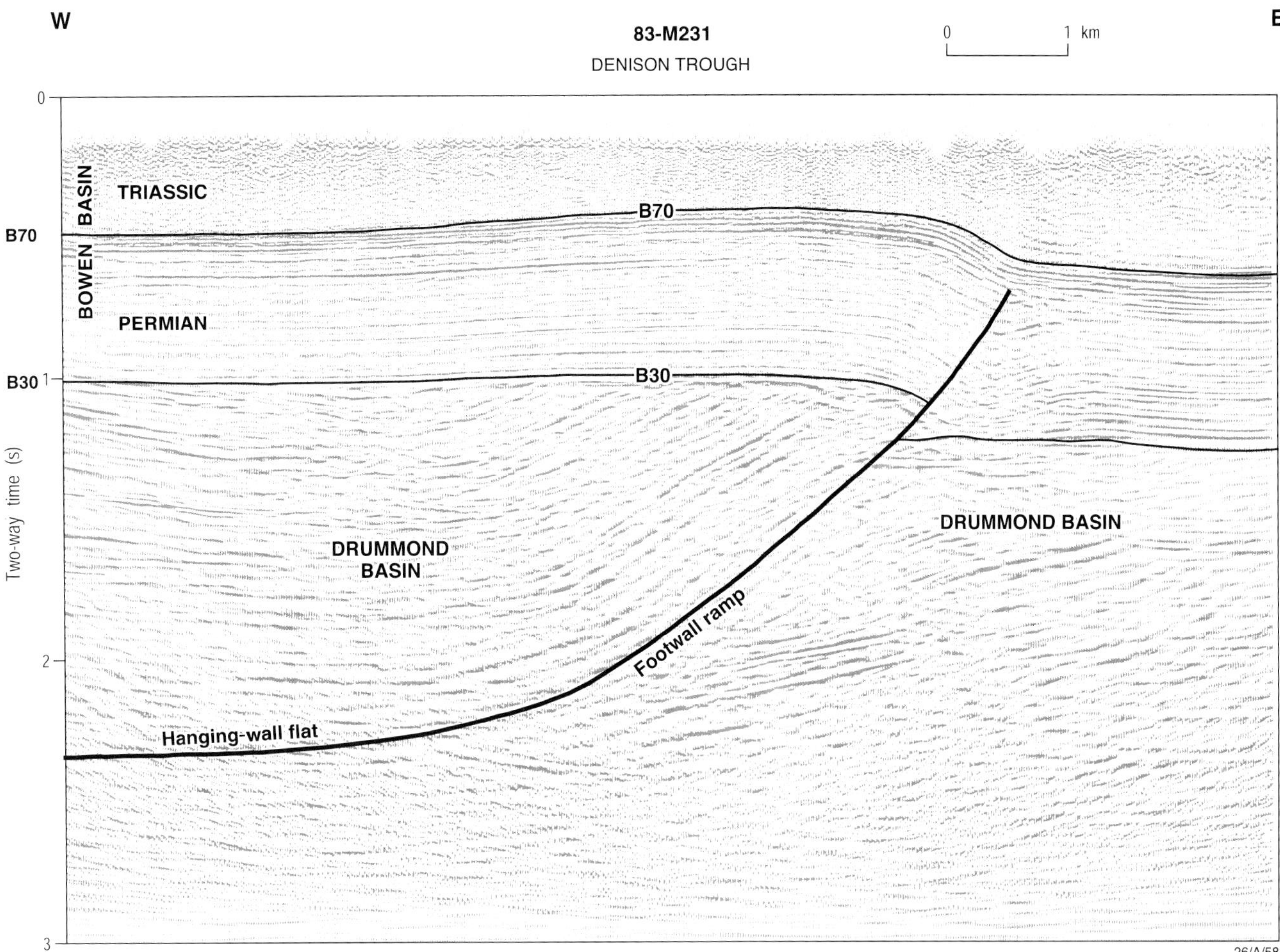

FIGURE 17. Seismic line 83-M231, from the southern Denison Trough, showing development of a fault-propagation fold in the Bowen Basin succession above a reactivated basement thrust fault. The three main seismic packages are defined by the different seismic reflection characteristics: 1, Basement, interpreted to be Drummond Basin, seen as well-defined reflections that lack extensive lateral continuity; 2, Permian rocks of the Bowen Basin, seen as strong, laterally continuous reflections (the very strong reflections at the top of the Permian are the Upper Permian coal measures succession), and, 3, Triassic rocks of the Bowen Basin, represented by poorly reflective to nonreflective units. Note the classic angular unconformity between the Drummond and Bowen Basins. See Figure 13 for location of seismic line.

That is, the deformation of the Drummond Basin was related to this thrust fault and occurred presumably during the Middle Carboniferous Kanimblan orogeny. It is interesting to note that the fault shows no sign of reactivation during the Early Permian extension, but that it was reactivated during the Middle to Late Triassic contraction, when the fault tip propagated upward into, but remained within, the Permian succession.

DISCUSSION

There is some debate over the importance of the contractional events in the history of the Bowen Basin. During the development of fault-propagation folds, the fault tips propagate upward through the sedimentary succession, thereby causing the displacement to

FIGURE 16. (a) Seismic line AD91-12, showing the geometry of one half graben in the northern part of the Denison Trough. The rift fill is bounded by the B15 and B30 seismic horizons. This section shows a typical geometry for the half grabens, and also shows Triassic reactivation of the bounding fault as a thrust, with inversion of the rift fill and formation of a fault-propagation fold above the thrust. In this case, the shortening is also accommodated by the development of a new back thrust. See Figure 13 for location. (b) Seismic line 87-WD22, from the Arbroath Trough, showing Middle to Late Triassic contraction being accommodated by the development of a new shortcut thrust, along with minor inversion on the initial extensional fault. See Figure 1 for location. (c) Seismic line 87-WD26, from the Arbroath Trough, showing Early to Middle Triassic contraction being accommodated by the development of a new back thrust, a new shortcut thrust, and some inversion on the initial extensional fault. See Figure 1 for location.

increase. In several of the seismic sections illustrated, the fault tips occur at different horizons within the succession, even though the deformation occurred during the same contractional event. This is to be expected in a fault system in which the fault displacements vary laterally along the faults. For example, in seismic line 83-M231, the fault tip is blind in the Upper Permian strata near the base of the coal measures (Figure 17). In line AD91-03, the fault tip is also blind just above the contact between the Upper Permian coal measures and the Lower Triassic Rewan Group (B70 horizon on Figure 15). In line B81-2 (Figure 10), however, the fault tip appears to have reached the surface or have been exposed by erosion prior to deposition of the Surat Basin. This interpretation differs from Elliott's (1993), which considered that all faults active during the Early Triassic Clematis Event (as defined by Korsch et al., 1998) extended up to, but stopped at, the top of the Rewan Group. The interpretations presented here, however, suggest that, in different seismic sections, the fault tips stop at different stratigraphic levels and that much of the deformation interpreted by Elliott to be Early Triassic is actually a different Middle to Late Triassic event. Thus, in terms of the tectonic development of the region, I consider the Middle to Late Triassic Goondiwindi Event to have been an order of magnitude greater in size than, and more important than, the Early Triassic Clematis Event. For example, in the Late Triassic, particularly along the eastern margin of the basin, there was uplift and erosion of at least 4 km of the stratigraphic succession. Harrington and Korsch (1985) speculated that the deformation associated with the Goondiwindi Event was caused by the arrival, docking, and suturing of the Gympie Terrane to the New England orogen. At that time, however, deformation was widespread across Australia (Etheridge and O'Brien, 1994; Davidson, 1995, 1997). This widespread deformation implies causation by a more global plate-tectonic event. This was a time of active global tectonism, with accretion and collision of terranes in Asia, extension prior to seafloor spreading in the Atlantic, and collision of Wrangellia with North America. There is also a very sharp bend in the North American apparent-polar-wander path at this time (the J1 cusp), which indicates a major change in plate relationships. There is also a less well-constrained cusp in the Australian apparent-polar-wander path over the interval 245–215 Ma (Klootwijk, 1996).

PETROLEUM IMPLICATIONS

The Bowen and Surat Basins in Queensland contain more than 70 mainly small commercial oil and gas fields, many of which are in anticlinal structures that formed in response to thrusting events in the New England orogen. Geographically, the Denison Trough, the eastern margin of the Taroom Trough in the vicinity of the Moonie Fault, and the western margin of the Taroom Trough south of the Denison Trough (Figure 1) are the most important hydrocarbon-producing areas (see Shaw et al., 1999, their Figure 2).

In Denison Trough, several commercial gas fields occur in Early to Late Permian sandstone reservoirs (Figure 13). These reservoirs are in fault-propagation folds that formed during the inversion of Early Permian extensional faults. During the inversion, growth of the folds produced the traps that now house the gas fields. Most of the fields occur very near the bounding faults of the Early Permian half grabens (Figure 13). One of the important reservoir horizons is the Aldebaran Sandstone (Figure 7). This unit is separated by a major unconformity, with the lower reservoir being eroded, in places near some inverted faults, by the middle Permian Aldebaran Event. The upper Aldebaran Sandstone reservoir had been folded into fault-propagation folds by the Middle to Late Triassic Goondiwindi Event (Figure 14).

The Moonie oil field, near the eastern margin of the Taroom Trough, is the largest oil field in the Bowen and Surat Basins, and contains more than 24 million barrels (recoverable), which represents more than half the oil produced from the basins to date. The reservoir is the Lower Jurassic Precipice Sandstone, the basal unit of the Surat Basin. The initial basement high is an eroded fault-bend fold that formed during the Middle to Late Triassic Goondiwindi Event, but that was later enhanced in the early Late Cretaceous contractional event (see Figure 11). Maximum paleotemperatures were reached in the Early Cretaceous, coincident with maximum burial. The Upper Permian source rocks in the eastern Bowen Basin entered the oil window at this time (Boreham et al., 1996), leading to the generation of large quantities of hydrocarbons that were potentially available for entrapment (Shaw et al., 1999). The Moonie oil field is adjacent to the Moonie Fault, but a possible migration pathway was up the displacement transfer zone between the Moonie and Tingan Faults to the south (Figure 8).

The Burunga Anticline, also near the eastern margin of the Bowen Basin, is an incipient fault-propagation fold above the west-dipping Burunga Fault (Figure 10). The Upper Permian Baralaba Coal Measures (Supersequence F, Figure 7) in this anticline contain commercial quantities of coal-seam methane, and substantial quantities of conventional gas occur in fractured volcaniclastic basement rocks.

Based on petroleum exploration results, the fault systems near the eastern margin of the Bowen Basin acted as major impediments to updip easterly migration, with most hydrocarbons apparently migrating up the fault plane and into overlying active aquifers of the Surat Basin before being dispersed and lost to the system. Shaw et al. (1999) suggested that the fault systems acted as impediments to migration onto the shallower shelf to the east. Therefore, those parts of the margin coincident with the displacement transfer zones between the fault systems represent potential areas for heightened exploration effort because of the monoclinal migration pathways that developed in the displacement transfer zones.

In the western Taroom Trough and overlying Surat Basin, reactivation of basement thrusts during both the Middle to Late Triassic Goondiwindi contractional event and early Late Cretaceous Moonie contractional event has produced several small traps that house oil and gas fields, particularly in the Lower Triassic Showgrounds Sandstone and the Lower Jurassic Precipice Sandstone. These are often combined stratigraphic-structural traps.

Shaw et al. (1999) developed an inventory of prospects and leads in their study of the undiscovered hydrocarbon resources in the Bowen and Surat Basins. They found that because of the late maturation of the source rocks over much of the basin, late structuring played a more significant role in trap integrity than did earlier structuring. Thus, the reactivation of earlier thrust and extensional structures is very important.

CONCLUSIONS

From the Late Devonian to the Middle to Late Triassic, eastern Australia was part of an active Gondwanan convergent plate margin influenced by an active west-dipping subduction system. The southern New England orogen changed from a prowedge mode in the Devonian-Carboniferous to an uplifted plug mode in the Permian-Triassic. In contrast, the northern New England orogen was dominated by the retrowedge mode in both time periods. This led to the development of a major west-directed retroforeland thrust belt in northern New England, with significant static crustal loading and the formation of a major foreland-basin phase in the adjacent Permian-Triassic Bowen Basin to the west.

There is evidence that both thick-skinned and thin-skinned processes operated simultaneously and that they are hard-linked into a single thrust system by a major middle-crust detachment surface. In the orogen, the thrusts are dominantly planar and cut deep into the crust. The eastern part of the Bowen Basin, however, is dominated by thin-skinned thrusting, which in places propagated a considerable distance into the basin, forming an imbricate thrust fan with hard linkages to the detachment surface.

Contractional events associated with the formation of the major retroforeland thrust system in the New England orogen have led to the propagation of far-field stresses and the reactivation of older faults well inboard of, and only soft-linked to, the retroforeland thrust system. In particular, contraction in the Denison Trough in the western Bowen Basin produced a variety of geometries, including reactivation of Early Permian extensional faults as thrusts and the growth of fault-propagation folds. Back thrusts and shortcut thrusts also developed. Small commercial hydrocarbon fields in the Bowen and Surat Basins in Queensland usually occur in traps that formed during one or more of the contractional events.

ACKNOWLEDGMENTS

This chapter presents some results of the Tectonic Framework of Eastern Australia project carried out within the Australian Geodynamics Cooperative Research Centre. Part of this project utilised data compiled in the earlier NGMA project Sedimentary Basins of Eastern Australia. I wish to thank: members of the NGMA Sedimentary Basins of Eastern Australia project from Geoscience Australia, Geological Survey of Queensland and Geological Survey of New South Wales, particularly Jennie Totterdell and Malcolm Nicoll; the Geoscience Australia Land Seismic Group, particularly Kevin Wake-Dyster and David Johnstone, for acquisition and processing of the Bowen and Gunnedah deep seismic data; and Paula Waschbusch and Chris Beaumont of the Canadian Institute of Advanced Research, Dalhousie University for their contribution to geodynamic modeling on the New England orogen and Bowen, Gunnedah and Surat Basins as part of the Tectonic Framework of Eastern Australia project. I thank Joe Mifsud, Information Management Branch, Geospatial Applications and Visualisation Group, Geoscience Australia, for drafting the figures, and Dick Glen and Kevin Hill for their constructive reviews. I also thank the AGCRC for funds to attend the Thrust Tectonics '99 Conference and Ken McClay for his encouragement to write this paper. This paper is published with the permission of the Chief Executive Officer, Geoscience Australia, and the Director, AGCRC.

REFERENCES CITED

Beaumont, C., S. Ellis, and A. Pfiffner, 1999, Dynamics of subduction-accretion at convergent margins: short-term modes, long-term deformation, and tectonic implications: Journal of Geophysical Research, v. 104, p. 17,573–17,601.

Boreham, C. J., R. J. Korsch, and D. C. Carmichael, 1996, The significance of mid-Cretaceous burial and uplift on the maturation and petroleum generation in the Bowen and Surat basins, eastern Australia: Geological Society of Australia, Abstracts, v. 43, p. 104–113.

Bruce, M. C., Y. Niu, and R. J. Holcombe, 1999, Remnants of a Neoproterozoic Oceanic Basin in the Northern New England orogen, Eastern Australia: Evidence for break-up of Rodinia?, *in* P. G. Flood, ed., New England orogen, eastern Australia: Department of Geology and Geophysics, Armidale, University of New England, p. 221.

Carey, S. W., and W. R. Browne, 1938, Review of the Carboniferous stratigraphy, tectonics and palaeogeography of New South Wales and Queensland: Journal and Proceedings of the Linnaean Society of New South Wales, v. 71, p. 591–614.

Davidson, J. K., 1995, Globally synchronous compressional pulses in extensional basins: implications for hydrocarbon exploration: Australian Petroleum Production and Exploration Association (APEA) Journal, v. 35, p. 169–188.

Davidson, J. K., 1997, Synchronous compressional pulses in extensional basins: Marine and Petroleum Geology, v. 14, p. 513–549.

Elliott, L. G., 1993, Post-Carboniferous tectonic evolution of eastern Australia: Australian Petroleum Production and Exploration Association (APEA) Journal, v. 33, p. 215–236.

Etheridge, M. A., and G. W. O'Brien, 1994, Structural and tectonic evolution of the Western Australian marginal basin system: Petroleum Exploration Society of Australia (PESA) Journal, v. 22, p. 45–63.

Fergusson, C. L., 1991, Thin-skinned thrusting in the northern New England orogen, central Queensland, Australia: Tectonics, v. 10, p. 797–806.

Fielding, C. R., C. J. Stephens, and R. J. Holcombe, 1997, Permian stratigraphy and palaeogeography of the eastern Bowen Basin, Gogango Overfolded Zone and Strathmuir Synclinorium in the Rockhampton-Mackay region, central Queensland, *in* P. M. Ashley and P. G. Flood, eds., Tectonics and metallogenesis of the New England orogen: Geological Society of Australia Special Publication 19, p. 80–95.

Glen, R. A., and J. Beckett, 1997, Structure and tectonics along the inner edge of a foreland basin: the Hunter Coalfield in the northern Sydney Basin, New South Wales: Australian Journal of Earth Sciences, v. 44, p. 853–877.

Glen, R. A., and R. E. Brown, 1993, A transect through a forearc basin: preliminary results from a transect across the Tamworth Belt at Manilla N.S.W., *in* P. G. Flood and J. C. Aitchison, eds., New England orogen, eastern Australia: Armidale, Department of Geology and Geophysics, University of New England, p. 105–111.

Glen, R. A., R. J. Korsch, and K. D. Wake-Dyster, 1993, A deep seismic cross section through the Tamworth Belt: preliminary interpretation of 4 seconds two-way time data, *in* P. G. Flood and J. C. Aitchison, eds., New England orogen, eastern Australia: Armidale, Department of Geology and Geophysics, University of New England, p. 101–104.

Harrington, H. J., and R. J. Korsch, 1985, Deformation associated with the accretion of the Gympie terrane in eastern Australia: Geological Society of Australia, Abstracts, v. 14, p. 104–108.

Hayward, A. B., and R. H. Graham, 1989, Some geometrical characteristics of inversion, *in* M. A. Cooper and G. D. Williams, eds., Inversion tectonics: Geological Society of London Special Publication 44, p. 17–39.

Henderson, R. A., and B. K. Davis, 1993, The Drummond and Bowen basins: sequential back-arc extensional elements of the Northern New England orogen, *in* P. G. Flood and J. C. Aitchison, eds., New England orogen, eastern Australia: Armidale, Department of Geology and Geophysics, University of New England, p. 609–616.

Hobbs, B. E., 1985, Interpretation and analysis of structure in the Bowen Basin: Geological Society of Australia, Abstracts, v. 17, p. 151.

Holcombe, R. J., C. J. Stephens, C. R. Fielding, D. Gust, T. A. Little, R. Sliwa, J. Kassan, J. McPhie, and A. Ewart, 1997, Tectonic evolution of the northern New England Fold Belt: the Permian-Triassic Hunter-Bowen event, *in* P. M. Ashley and P. G. Flood, eds., Tectonics and metallogenesis of the New England orogen: Geological Society of Australia Special Publication 19, p. 52–65.

Johnson, S. E., and R. A. Henderson, 1991, Tectonic development of the Drummond Basin, eastern Australia: backarc extension and inversion in a Late Palaeozoic active margin setting: Basin Research, v. 3, p. 197–213.

Klootwijk, C., 1996, Phanerozoic configurations of Greater Australia: Evolution of the North West Shelf. Part two: Palaeomagnetic and geologic constraints on reconstructions: Australian Geological Survey Organisation, Record, 1996/52, p. 1–85.

Korsch, R. J., and J. M. Totterdell, 1995, Structural events and deformational styles in the Bowen Basin, *in* I. W. Follington, J. W. Beeston, and L. H. Hamilton, eds., Bowen Basin Symposium 1995, Proceedings: Brisbane, Geological Society of Australia, Coal Geology Group, p. 27–35.

Korsch, R. J., H. J. Harrington, C. G. Murray, C. L. Fergusson, and P. G. Flood, 1990a, Tectonics of the New England orogen: Bureau of Mineral Resources, Australia, Bulletin, v. 232, p. 35–52.

Korsch, R. J., K. D. Wake-Dyster, and D. W. Johnstone, 1990b, Deep seismic profiling across the Bowen Basin, *in* J. W. Beeston, comp., Bowen Basin Symposium 1990, Proceedings: Brisbane, Geological Society of Australia, Queensland Division, p. 10–14.

Korsch, R. J., K. D. Wake-Dyster, and D. W. Johnstone, 1992, Seismic imaging of extensional and contractional structures in the Bowen and Surat basins, eastern Australia: Tectonophysics, v. 215, p. 273–294.

Korsch, R. J., K. D. Wake-Dyster, and D. W. Johnstone, 1993a, The Gunnedah Basin New England orogen deep seismic reflection profile: implications for New England tectonics, *in* P. G. Flood and J. C. Aitchison, eds., New England orogen, eastern Australia: Armidale, Department of Geology and Geophysics, University of New England, 85–100.

Korsch, R. J., K. D. Wake-Dyster, and D. W. Johnstone, 1993b, Deep seismic reflection profiling in the Gunnedah Basin: regional tectonics and basin responses, *in* G. J. Swarbrick and D. J. Morton, eds., Proceedings NSW Petroleum Symposium: Petroleum Exploration Society of Australia, New South Wales Branch, 30 p., (no pagination in volume).

Korsch, R. J., D. W. Johnstone, and K. D. Wake-Dyster, 1997, Crustal architecture of the New England orogen based on deep seismic reflection profiling, *in* P. M. Ashley and P. G. Flood, eds., Tectonics and metallogenesis of the New England orogen: Geological Society of Australia Special Publication 19, p. 29–51.

Korsch, R. J., C. J. Boreham, J. M. Totterdell, R. D. Shaw, and M. G. Nicoll, 1998, Development and petroleum resource evaluation of the Bowen, Gunnedah, and Surat Basins, Eastern Australia: Australian Petroleum Production and Exploration Association (APEA) Journal, v. 38, p. 99–237.

Leitch, E. C., 1975, Plate tectonic interpretation of the Paleozoic history of the New England Fold Belt: Geological Society of America Bulletin, v. 86, p. 141–144.

Mitrovica, J. X., C. Beaumont, and G. T. Jarvis, 1989, Tilting of continental interiors by the dynamical effects of subduction: Tectonics, v. 8, p. 1079–1094.

Murray, C. G., C. L. Fergusson, P. G. Flood, W. G. Whitaker, and R. J. Korsch, 1987, Plate tectonic model for the Carboniferous evolution of the New England Fold Belt: Australian Journal of Earth Sciences, v. 34, p. 213–236.

Shaw, R. D., R. J. Korsch, C. J. Boreham, J. M. Totterdell, C. Lelbach, and M. G. Nicoll, 1999, An evaluation of the undiscovered hydrocarbon resources of the Bowen and Surat basins, southern Queensland: AGSO Journal of Australian Geology and Geophysics, v. 17, no. 5/6, p. 43–65.

Totterdell, J. M., and R. J. Korsch, 1992, Structural configuration and tectonic development of the Bowen and Surat basins in the Taroom region: Bureau of Mineral Resources, Australia, Record 1991/102, p. 35–52.

Totterdell, J. M., and A. A. Krassay, 1995, Sequence evolution and structural history of the Bowen and Surat basins in northern New South Wales, *in* D. J. Morton, J. D. Alder, I. J. Grierson, and C. E. Thomas, eds., 1995 NSW Petroleum Symposium Proceedings: Petroleum Exploration Society of Australia, New South Wales Branch, 13 p. (no pagination in volume).

Totterdell, J. M., A. T. Brakel, A. T. Wells, and K. L. Hoffmann, 1995, Basin phases and sequence stratigraphy of the Bowen Basin, *in* I. W. Follington, J. W. Beeston, and L. H. Hamilton, eds., Bowen Basin Symposium 1995, Proceedings: Brisbane, Geological Society of Australia, Coal Geology Group, p. 247–256.

Wartenberg, W., R. J. Korsch, and A. Schäfer, 1999, Geometry of the Tamworth Belt in the New England orogen beneath the Surat Basin, southern Queensland, *in* P. G. Flood, ed., New England orogen, eastern Australia: Armidale, Department of Geology & Geophysics, University of New England, p. 211–219.

Waschbusch, P., C. Beaumont, and R. J. Korsch, 1998, Geodynamic modelling of aspects of the Bowen and Surat Basins in Eastern Australia: Geological Society of Australia Abstracts, v. 49, p. 457.

Waschbusch, P., C. Beaumont, and R. J. Korsch, 1999, Geodynamic modelling of aspects of the New England orogen and adjacent Bowen, Gunnedah and Surat basin, *in* P. G. Flood, ed., New England orogen, eastern Australia: Armidale, Department of Geology & Geophysics, University of New England, p. 203–210.

Watanabe, T., C. M. Fanning, and E. C. Leitch, 1998, Neoproterozoic Attunga Eclogite in the New England Fold Belt: Geological Society of Australia Abstracts, v. 49, p. 458.

Willett, S. D., C. Beaumont, and P. Fullsack, 1993, Mechanical model for the tectonics of doubly vergent compressional orogens: Geology, v. 21, p. 71–374.

Woodward, N. B., 1995, Thrust systems in the Tamworth Zone, southern New England orogen, New South Wales: Australian Journal of Earth Sciences, v. 42, p. 107–117.

27

Flottmann, T., M. Hand, D. Close, C. Edgoose, and I. Scrimgeour, 2004, Thrust tectonic styles of the intracratonic Alice Springs and Petermann Orogenies, Central Australia, *in* K. R. McClay, ed., Thrust tectonics and hydrocarbon systems: AAPG Memoir 82, p. 538–557.

Thrust Tectonic Styles of the Intracratonic Alice Springs and Petermann Orogenies, Central Australia

T. Flottmann

Santos Ltd., Brisbane, Queensland, Australia

M. Hand

Department of Geology and Geophysics, University of Adelaide, Adelaide, South Australia, Australia

D. Close

Northern Territory Geological Survey, Alice Springs, Northern Territory, Australia

C. Edgoose

Northern Territory Geological Survey, Alice Springs, Northern Territory, Australia

I. Scrimgeour

Northern Territory Geological Survey, Alice Springs, Northern Territory, Australia

ABSTRACT

The Petermann and Alice Springs orogens are late Neoproterozoic to early Cambrian and middle Paleozoic intracratonic fold-and-thrust belts that shaped the Amadeus Basin in Central Australia. Displacements during both orogenies were accommodated in the footwalls of crustal-scale fault systems. The deformational style is characterized by basement-involved triangle zones in which displacement is partitioned into two mechanical units. A lower succession is characterized by basinward thrusting and an upper succession with hinterland-directed displacement. The interleaving of basement and cover units within the basin-directed wedges is a reflection of the relatively low mechanical contrast between cover successions and basement. North-vergent shortening during the Petermann orogeny exceeded 100 km, the bulk of which was accommodated by dramatic thickening, which led to burial of the Amadeus Basin sequences deeper than 20 km. Despite the thickening, there was little foreland-basin development, suggesting that the lithosphere may have been too weak to form a long-wavelength flexure. Along the northern margin of the Amadeus Basin, the Alice Springs orogeny was associated with as much as 50 km of shortening that can be resolved into two phases: (1) south-directed, basinward overthrusting of a basement wedge and (2) underthrusting that led to the development of a large-displacement north-vergent passive back thrust within the northern Amadeus Basin. This back thrusting led to the formation of structural hydrocarbon traps that developed along the southern flank of a progressively narrowing and deepening foreland flexure. Our study provides a kinematic framework for basin-shaping deformation along the margins of the Amadeus Basin and suggests that future hydrocarbon exploration in the Central Australian basins should assess the implications of the style of basement-cover interaction, as highlighted by the structural styles of both the Petermann and Alice Springs orogenies.

INTRODUCTION

The late Neoproterozoic to middle Paleozoic intracratonic Amadeus Basin is one of Australia's major onshore gas and oil producers. Sedimentation and structuring in the Amadeus Basin are controlled by a system of compressional structures that formed during two intraplate orogenic events: the late Neoproterozoic to Early Cambrian (580–540 Ma) Petermann orogeny and the middle to late Paleozoic (400–300 Ma) Alice Springs orogeny. These orogenic events dissected a formerly continuous, continent-scale intracratonic basin (Centralian Superbasin, Walter et al., 1995), resulting in the formation of a series of structurally remnant basins that include Amadeus, Officer, Georgina, and Ngalia Basins (Figure 1). Viewed in this context, the Petermann and Alice Springs orogenies represent extreme examples of basin inversion. During both orogenies, basement lithologies were intercalated with cover sequences belonging to the Centralian Superbasin. In the Amadeus Basin, the southern margin formed during the Petermann orogeny as north-vergent structures were propagated into the foreland during exhumation of the Mesoproterozoic Musgrave Inlier (Figure 1). The northern margin of the Amadeus Basin formed during the Alice Springs orogeny as the Paleoproterozoic-to-Mesoproterozoic Arunta Inlier was exhumed, thereby resulting in separation of the Amadeus and Georgina Basins.

Despite differences in their ages and magnitudes of deformation, both the Petermann and Alice Springs orogenies show similarities in their styles of cover-basement interaction along the margins of the Amadeus Basin. During both events, basement lithologies of broadly similar composition were interleaved with the basal sequences of the Amadeus Basin in the footwall of crustal-scale fault systems (Forman, 1966; Marjoribanks, 1976). This interleaving generated foreland-propagating crustal wedges, above which the upper sequences of the Amadeus Basin were transported toward the hinterland in a back-thrust system. Similarly, along the southwestern margin of the basin, the basement sheets thrust northward during the Petermann orogeny, dipping at between 30 and 70° toward the foreland and producing features reminiscent of downward-facing nappe structures (Forman, 1966).

In this chapter we present results of a regional mapping campaign, which we use together with previously published data to construct balanced and restored cross sections that highlight development of the crustal wedges during intraplate orogeny in central Australia. The cross sections allow us to estimate the magnitude of intracratonic shortening for both the Petermann and Alice Springs orogenies in the vicinity of the Amadeus Basin margins.

The occurrence of hydrocarbons in the Amadeus Basin is structurally controlled (Schroder and Gorter, 1984; Do Rozario, 1991; Havord, 1991; Roe, 1991). To date, drilling has only tested the eastern and central parts of the basin (Figure 1). The hydrocarbon potential of the remote and poorly studied western part of the basin is currently being assessed. The thrust-tectonics models presented here provide a structural framework in which to consider future oil and gas exploration, particularly in the western part of the Amadeus Basin.

THE ALICE SPRINGS OROGENY

The middle Paleozoic Alice Springs orogeny (400–300 Ma) was a major intraplate contractional event in central Australia that resulted in exhumation of the Arunta Block from beneath a broad intracratonic basin. An intriguing feature of the Alice Springs orogeny is the presence of several "Alpine-style," basement-cored, downward-facing antiformal synclines along the northern margin of the Amadeus Basin (Figure 2) (Marjoribanks, 1976; Flottmann and Hand, 1999). In the past, these structures were interpreted to be the erosional remnants of the frontal parts of fold nappes, comparable to regional structures found in the Alps (Milnes, 1978; Merle and Guillier, 1989; Kurz et al., 1998). Here, we present the results of structural investigations in the western MacDonnell Ranges, at the northwestern margin of the Amadeus Basin. Based on Marjoribanks's (1974, 1976) excellent base map and our own mapping (Figure 3), we established balanced and sequentially

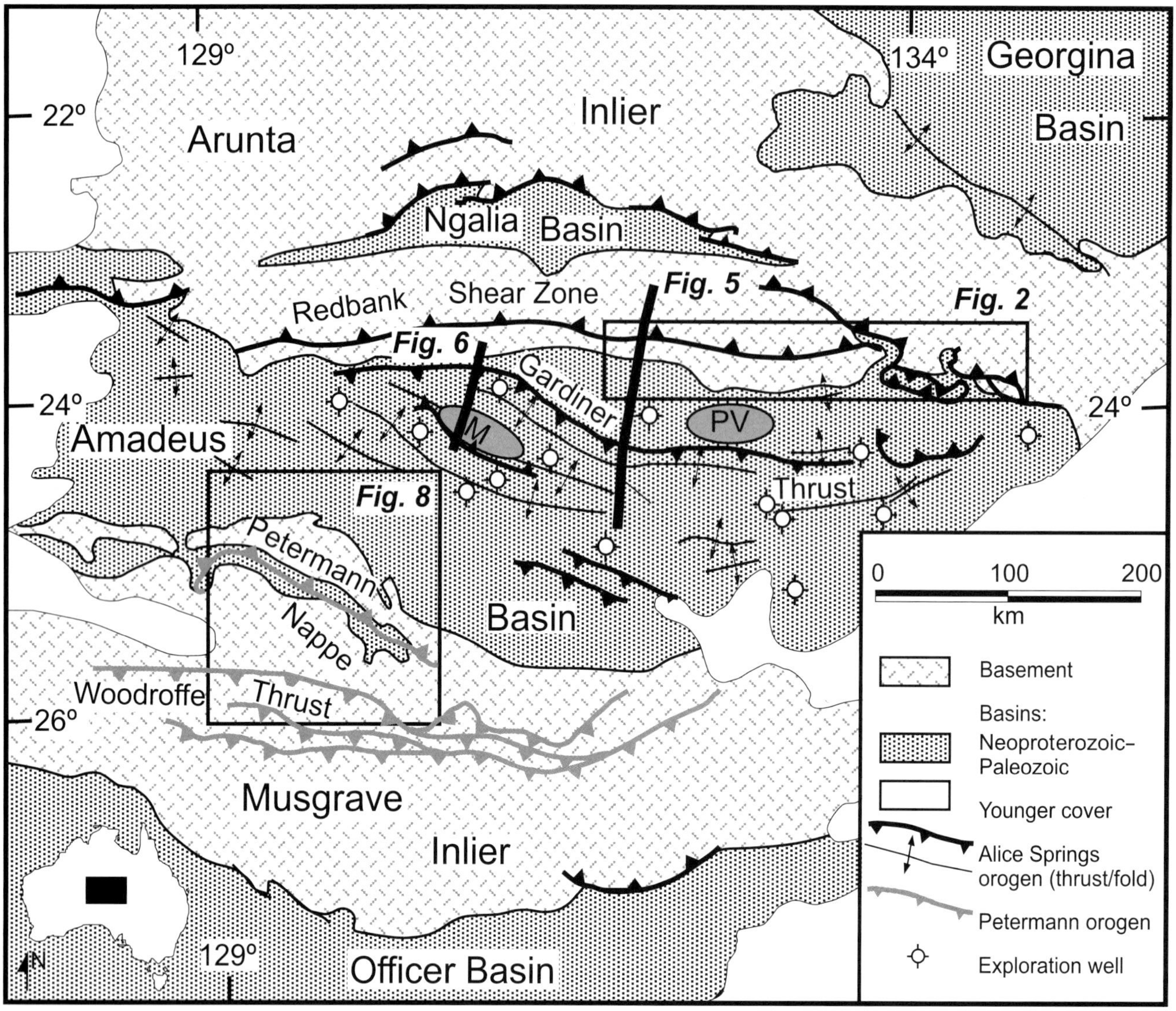

FIGURE 1. Regional map of Central Australia showing the Alice Springs and Petermann orogens and the structurally remnant basins that constituted part of a formerly continuous Central Superbasin (Walter et al., 1995). Boxes show the locations of inset maps of Figures 2 and 6 (north: Alice Springs orogeny, south: Petermann orogeny); also shown is line of cross section of Figure 5. M = Mereenie oil and gas field, PV = Palm Valley gas field.

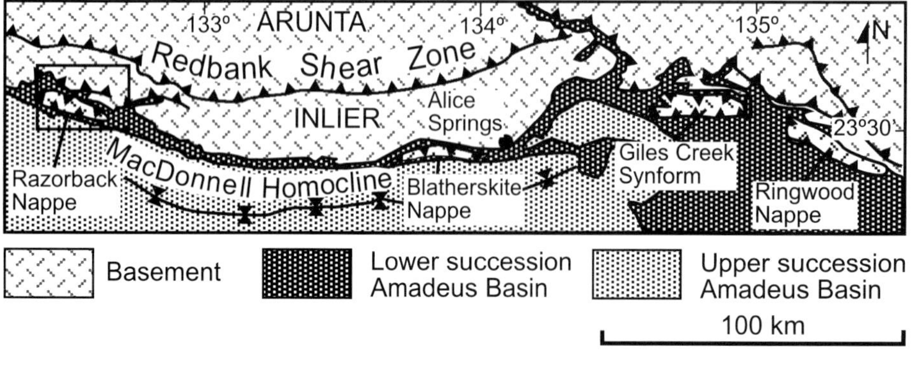

FIGURE 2. Basement-cored nappe structures at the northern margin of the Amadeus Basin. Inset shows location of Figure 3. The lower succession of the Amadeus Basin comprises the Heavitree Quartzite and the Bitter Springs Formation. The upper succession consists of all sequences between the Bitter Springs Formation and the foreland deposits of the Pertnjara Group.

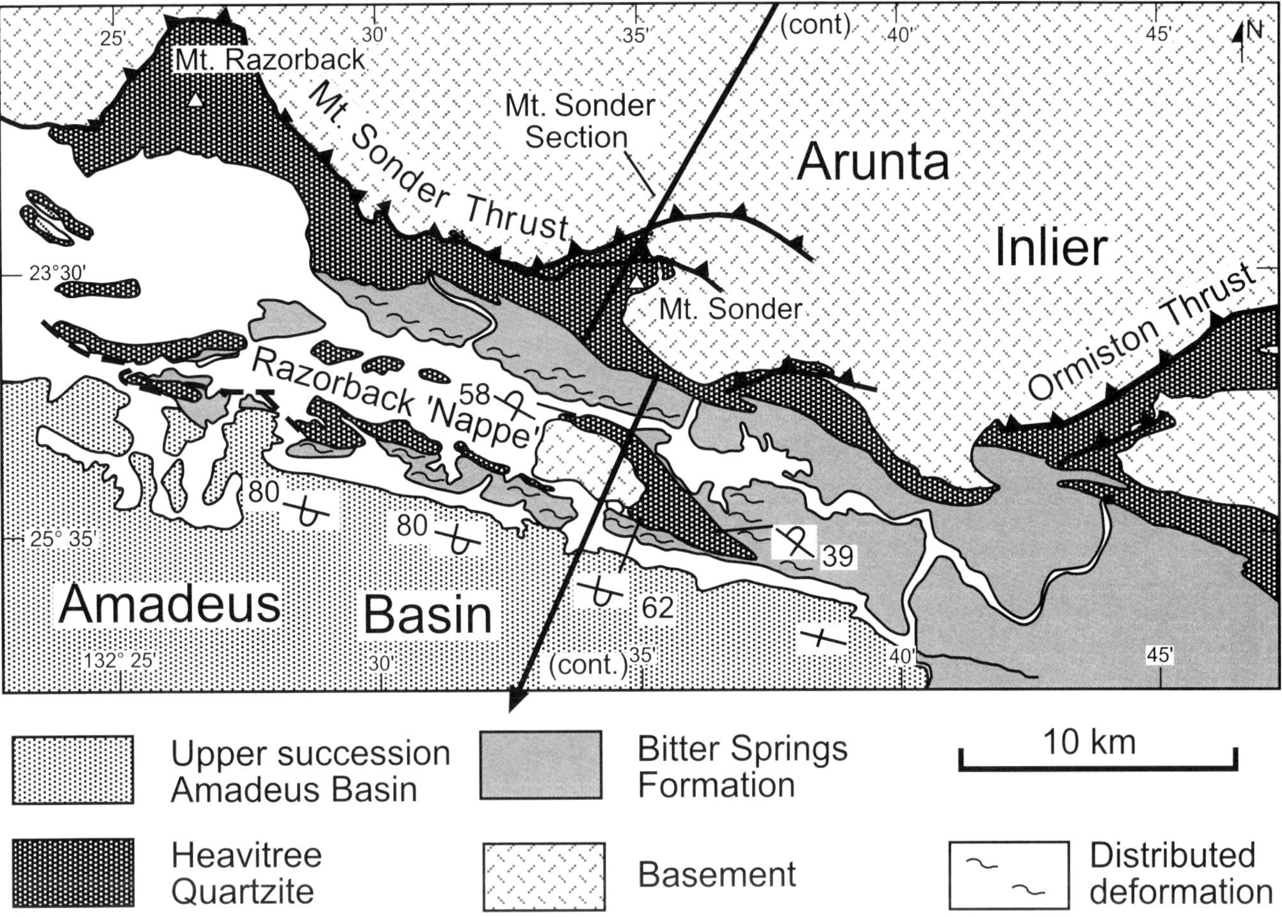

FIGURE 3. Map of the Razorback structure, with location of Mt. Sonder section of Figure 4. The Mt. Sonder and Ormiston Thrust Complexes are the basement region immediately east of Mt. Sonder.

restored cross sections across the western MacDonnell Ranges (Figure 4 a–d). These permit the formation of a kinematically and geometrically admissible model for the formation of basement-cored nappes and associated structures in the western MacDonnell Ranges.

The geologic framework of the western MacDonnell Ranges region (labeled MacDonnell Homocline in Figure 2) is characterized across strike by three distinct provinces: (1) a thick-skinned province in the Arunta Inlier, formed by Proterozoic crystalline basement, (2) a zone of cover-basement interaction, and (3) a thin-skinned zone in the northern Amadeus Basin. The Arunta Inlier to the north of the Amadeus Basin consists of Paleoproterozoic to Mesoproterozoic basement and contains a crustal-scale system of south-vergent thrusts that were operative beneath the now-eroded cover. These faults dip to the north at 30–45° into the lower crust, as indicated by seismic-reflection data (Goleby et al., 1989; Wright et al., 1991). At the boundary between the southern and central parts of the Arunta Inlier is the Redbank Shear Zone (Figure 2), which forms a major crustal discontinuity. This structure is interpreted to offset the Moho vertically by ~20 km. The horizontal Moho offset along the Redbank Shear Zone is estimated to be in the range of 40 km (Korsch et al., 1998). However, the Redbank Shear Zone loses displacement up section in the crust, and at the present surface level, the vertical offset along the Redbank Shear Zone is inferred to be less than ~3 km (Shaw et al., 1992).

The transition zone between the basement province to the north and the Amadeus Basin to the south is characterized by coupled basement-cover deformation. The most peculiar feature of this interface between thick-skinned and thin-skinned domains is the presence of large basement-cored nappe-style folds within an apparently simple stratigraphic envelope (Figure 2). The northern margin of the Amadeus Basin is characterized by large-scale strike ridges forming the MacDonnell Homocline (Figure 2), which has a strike length of more than 200 km. The homocline is characterized by steep southerly dips with local overturning that results in subvertical to steep northerly dips. The homocline is fronted by an asymmetric synclinal zone that merges with subhorizontal successions of the Amadeus Basin to the south. At the base of the homocline, north-dipping basement thrusts propagate into the synclinal region, and the entire structure is geometrically consistent with a large basement culmination that is fronted by a homocline representing the eroded remnant of the ~7-km-thick sedimentary carapace. The structural geometry of the basement culmination resembles in principle the core of a trishear fault-propagation fold (Erslev, 1991; Allmendinger, 1998). Displacement leading to the formation of the Trishear zone is taken up equally by the Redbank shear zone and the shear zones in its footwall (Figure 4).

The stratigraphy of the Amadeus Basin is divided into mechanically distinct components. The lower succession comprises the Neoproterozoic Heavitree Quartzite, which forms a mechanically competent stratigraphic unit. The overlying Bitter Springs Formation consists of carbonates and, as evidenced in drill holes, thick successions of evaporites (Wells et al., 1970; Lindsay, 1987) that are presumably dissolved at the current outcrop level. At depth the Bitter Springs Formation thus constitutes a mechanically weak layer along which overlying rocks can be detached. Formations overlying the Bitter Springs Formation are here referred to as the upper succession. The upper succession comprises Neoproterozoic through Devonian rocks. The Ordovician Pacoota Sandstone forms the reservoir of the main hydrocarbon-producing systems in the basin. Overlying the Pacoota Sandstone is the Horn Valley Siltstone, which forms a source-seal couplet for the Ordovician hydrocarbon system.

At the current level of exposure, thrusts associated with the basement culmination penetrate only the lower stratigraphic units of the Amadeus Basin succession. These thrusts truncate the Heavitree Quartzite but terminate in the upper parts of the overlying Bitter Springs Formation.

The Amadeus Basin province south of the zone of basement-cover interleaving is characterized by a thin-skinned (i.e., basement-detached) style of shortening, which produces structures typical of foreland fold-thrust belts (Bradshaw and Evans, 1988; Shaw et al., 1991b; Stewart et al., 1991). Apparently this thin-skinned deformation continued after deformation in the transition zone had largely ceased (Shaw et al., 1991b). Notably, thrusts that truncate sediments shed from the advancing basement block to the north are back thrusts, with top-to-the-north-directed displacement, such as the Gardiner Thrust (Figure 1), that produce a triangle-like zone in front of the homocline (Bradshaw and Evans, 1988). This region of thin-skinned deformation is part of a basinwide system of Alice Springs-aged deformation.

Key Structures of the Alice Springs Orogeny

Mt. Sonder Section

The northern part of the Mt. Sonder section is formed by basement, which is bounded to the south by the Mt. Sonder and Ormiston Thrust systems (Figure 3). The Mt. Sonder Thrust Complex forms a broad, basement-cored culmination with a structural envelope that dips northward in the north and southward in the south. Internally, the thrust system consists of six thrusts along which basement and the overlying Heavitree Quartzite are displaced southward (Figure 4a, b). Major thrusts are formed by quartzite mylonites in the 0.5- to 1.5-m thickness range (Flottmann and Hand, 1999). Basement exposures east of Mt. Sonder display an anastomosing network of steeply to moderately north-dipping shear zones that are 2–40 cm wide and have north-over-south-directed relative displacement. In suitably potassic rocks of the Mt. Sonder Thrust Complex, there is synkinematic muscovite growth, indicating that deformation occurred at lower greenschist facies conditions (~300–350°C).

The southern boundary of the Mt. Sonder Thrust Complex is formed by a lower boundary thrust that juxtaposes the Heavitree Quartzite over the Bitter Springs Formation. This lower boundary thrust effectively separates metamorphic structures (those containing new muscovite) from essentially unmetamorphosed Bitter Springs Formation to the south. Internally, the Bitter Springs Formation is irregularly folded, and dolomite and siliceous layers of this formation are crosscut by an irregular network of calcite and quartz veins with slickensides on numerous surfaces. These veins commonly are brecciated, suggesting elevated fluid pressures during deformation of the Bitter Springs Formation. Closer to the Razorback structure (see below), the Bitter Springs Formation displays a wide zone of distributed strain that is characterized by intense disruption, and in many domains little or none of the original lithologic layering is preserved (Flottmann and Hand, 1999). These highly deformed zones are structurally coplanar with the envelope of the Bitter Springs Formation.

The Razorback nappe structure (Marjoribanks, 1976; Teyssier, 1985) has a gentle (~5°) westerly plunge and consists of a northern limb of overturned Heavitree Quartzite, a core of Mesoproterozoic basement, and a southern limb of overturned Heavitree Quartzite (Figures 3 and 4). The overturned nature of the Heavitree Quartzite is indicated by abundant sedimentary way-up criteria (mainly trough crossbedding at centimeter to decimeter scale). These relationships make this structure a synformal anticline (see Hobbs et al., 1976), with a wavelength of about 2.5 km. South of the Razorback synformal anticline, the Bitter Springs Formation of the MacDonnell Homocline is subvertical to overturned toward the north (Figures 3 and 4), whereas farther east the dips are steeply to the south.

The cross section of Figure 4 explains the evolution of the structure while it attempts to maintain line-length balance throughout the deformational history for all successions of the Amadeus Basin. Area balance was also maintained, with the exception of the Bitter Springs Formation, where area, and hence volume balancing, are complicated by flow of evaporitic horizons (Lindsay, 1987). Detailed maps that form the basis for our interpretations are presented in Flottmann and Hand (1999).

We suggest that the Razorback structure was initiated as a wedge that formed during initial, south-directed thrusting along the Mt. Sonder Thrust (Figure 4b–d) in the footwall of the Redbank Shear Zone. This wedge incorporates basement in the core of a fold outlined by Heavitree Quartzite and the Bitter Springs Formation. The wavelength of the fold is broadly given by the across-strike width of the Razorback synformal anticline. The Ormiston Thrust forms a ramp in the basement and presumably in the Heavitree Quartzite, but it most likely forms a flat in the less-competent Bitter Springs Formation, which represents the main detachment horizon (Figure 4c). During southward displacement, the basement-cored wedge overrode the portion of the Heavitree Quartzite that now forms the thrust stack of Mt. Sonder. The current erosional surface intersects the downward-facing nose of the refolded Razorback wedge, with the thrust stack now located structurally above the remnant of the Razorback fold (Figure 4a). We suggest that the Bitter Springs Formation filled a triangle-shaped detachment horizon, which formed a detachment-wedge at the base of the upper succession of the Amadeus Basin. This is indicated by the intense, distributed deformation of the Bitter Springs Formation in the vicinity of the Razorback fold (Figure 3). Here, a diffuse movement horizon within the evaporitic Bitter Springs Formation may have accommodated distributed deformation in a regime of elevated pore pressures, as is suggested by the presence of abundant veined disruption of competent marl layers. Kinematic indicators in this zone of distributed deformation yield a contradictory sense of displacement. The overall displacement associated with the intense deformation of this zone cannot be resolved conclusively.

To maintain compatibility in the amount of shortening (and line-length balance) between the lower and upper Amadeus Basin successions, the shortening of the lower successions must also be accommodated in the upper succession. There is no evidence of south-vergent thrusts cutting the homocline. Therefore, we suggest that shortening of the upper succession was accommodated along a south-dipping back thrust(s). The combination of forward-propagating thrusts in the basement and the lower successions of the Amadeus Basin and of back thrusting in the upper succession leads to a triangle-zone type of displacement. The structures are similar to those described by Price (1986) for parts of the Rocky Mountains of western Canada and by Kraig et al. (1988) for the Rocky Mountain Fold-and-thrust Belt of the western U.S.

Back thrusting during formation of triangle zones is the favorable mode for accommodating regional shortening in the presence of a mechanically weak detachment horizon (Couzens and Wiltschko, 1996). In contrast, a forward-propagating scenario would be favored where competent rheologies dominate throughout the entire stratigraphic succession (Couzens and Wiltschko, 1996). Back thrusting is kinematically compatible with the observed kinematics of thrusts that displace the upper succession to the west and south of the area described here (Figure 1).

After the emplacement of the proto-Razorback structure, the northern part of the entire succession was tilted progressively southward and the Razorback structure reached its final, downward-facing position to form the synformal anticline exposed at the current erosional level. The switch from wedge emplacement to the large-scale tilting during the formation of the homocline requires that the locus of deformation shifted progressively toward the foreland in the south, thereby leading to a typical forelandward progression of deformation.

Overall, we explain the Razorback fold structure as a tectonic wedge that was subsequently tilted during the formation of the homocline. The sequence of deformational stages explains the formation of the downward-facing Razorback structure without invoking major shortening. It also involves all stratigraphic successions deposited in the Amadeus Basin prior to the onset of Alice Springs orogenic shortening. The overall structure of the enveloping surface of the homocline (including the tectonic wedge) resembles basement-involved structures described in the western United States (Spang and Evans, 1988; Cook, 1988; Erslev, 1991; Narr and Suppe, 1993; Schmidt et al., 1993; Chase et al., 1993; Erslev and Rogers, 1993).

Emplacement of the Razorback fold appears to have been prompted by deformation along the Redbank Shear Zone and faults in its footwall, which led to the formation of the MacDonnell Homocline in the western MacDonnell Ranges. At outcrop level, the actual displacement across the Redbank Shear Zone appears to have been in the order of 3 km (Shaw et al., 1992) during the Alice Springs orogeny. However, more than 20 km of vertical offset and 40 km of horizontal offset are seismically imaged at depth along the Redbank Shear Zone (Korsch et al., 1998). This suggests that most of the upper-crust displacement was accommodated by faults located in the footwall of the Redbank Shear Zone. These upper-crust fault and shear zones merge into one fault zone that forms the Redbank Shear Zone at middle- and lower-crust levels.

Shaw et al. (1992) inferred heating to ~350°C in the Ormiston Thrust Complex over the interval 390–370 Ma, which we suggest reflects the time of emplacement of the basement wedge and the associated back thrust that together create the crustal stacking reflected

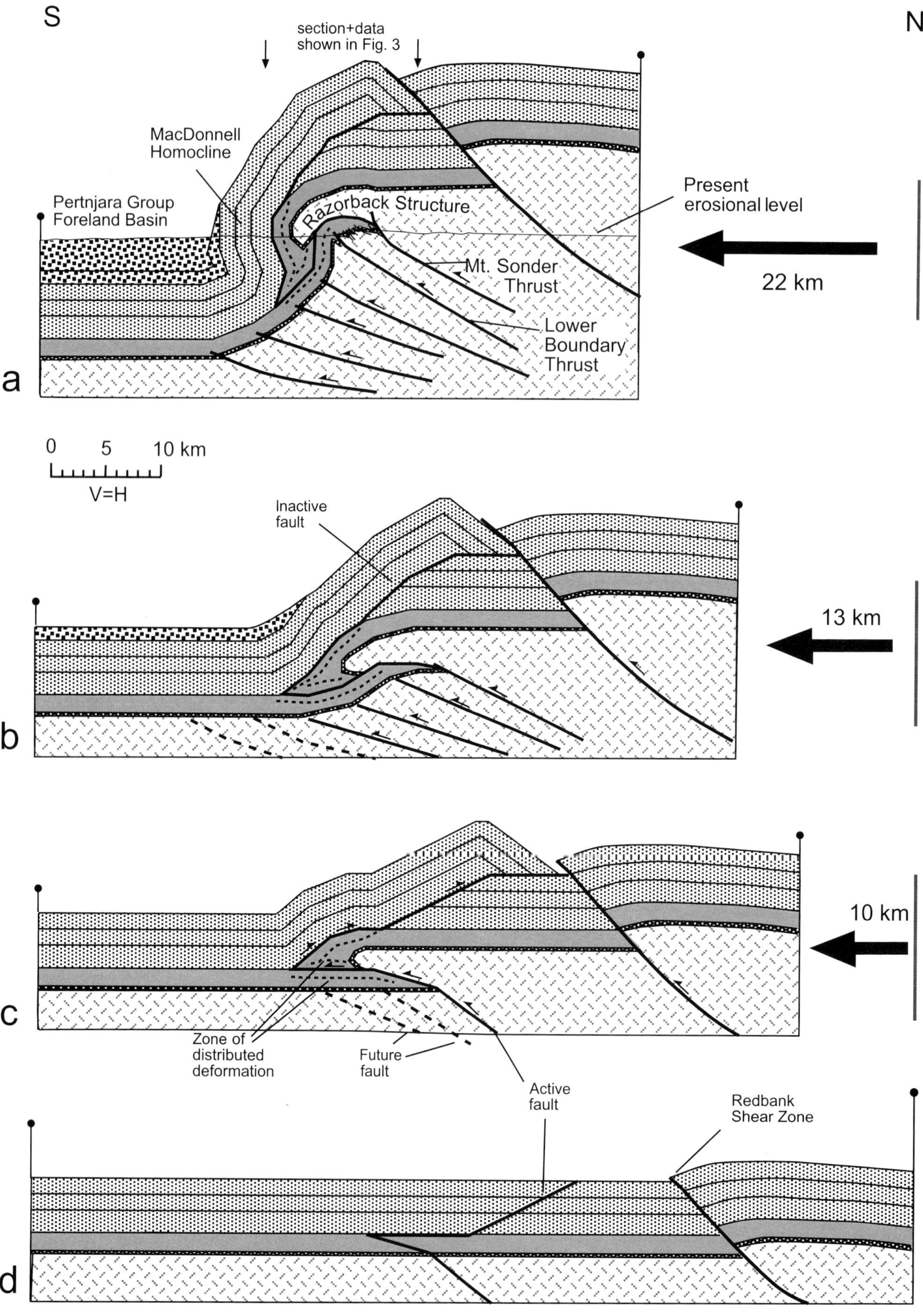
S
N
section+data
shown in Fig. 3
MacDonnell
Homocline
Pertnjara Group
Foreland Basin
Razorback Structure
Present
erosional level
Mt. Sonder
Thrust
Lower
Boundary
Thrust
22 km
a
0 5 10 km
V=H
Inactive
fault
13 km
b
10 km
c
Zone of
distributed
deformation
Future
fault
Active
fault
Redbank
Shear Zone
d

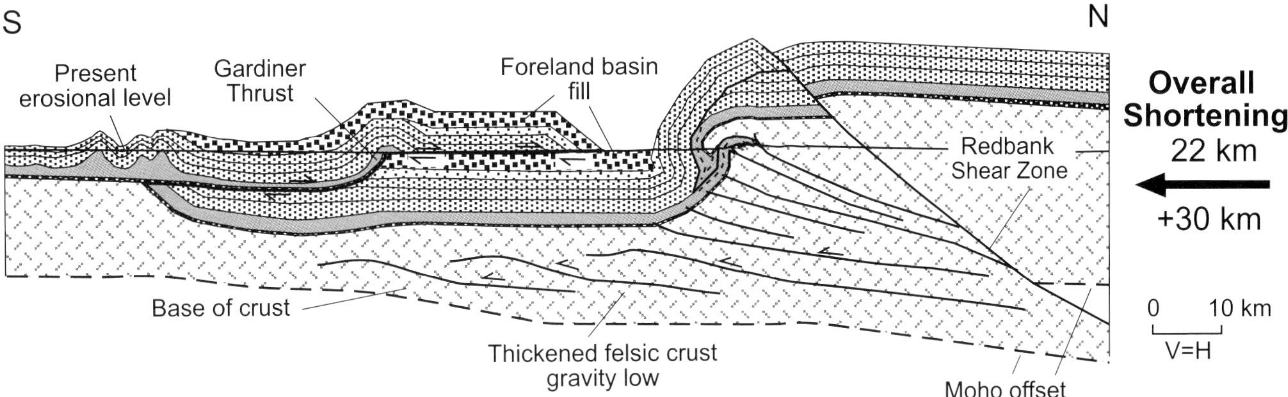

FIGURE 5. Displacement along Gardiner Thrust accommodates up to 30 km of shortening that accumulated mostly after deposition of foreland successions of Pertnjara Group (heavy stipple). Cross section modified from interpretation of regional seismic line of Stewart et al. (1993) and combined with the cross section presented here. Shortening along the Gardiner Thrust is explained as footwall underthrusting during continued shortening in the footwall of Redbank Shear Zone, leading to thickening of felsic crust and contributing to the pronounced gravity low in the northern Amadeus Basin (Mathur, 1976).

by this heating event. The deposition of the clastic members of the Pertnjara Group (Figures 4 and 5), which fills a narrow, deep foreland depression in front of the homocline, also commenced between 370 and 380 Ma and has been related to uplift of a source to the north (Jones, 1991). The northerly derived clastic members of the Pertnjara Group thus provide sedimentologic evidence for the erosion of the topography related to the emplacement of the tectonic wedge and the associated back thrusting proposed by our model.

The structural evolution suggested here might have been operative in one way or another along the entire western MacDonnell Homocline. In principle, the scenario we suggest here may also be applied to the northeastern Amadeus Basin, where deformation both in the hinterland and in the foreland appears more complex than in the western MacDonnell Ranges (Teyssier, 1985; Collins and Teyssier, 1989; Stewart et al., 1991; Dunlap et al., 1995). The preservation of structures similar to the Razorback structure at the present erosional level along the MacDonnell Homocline is directly related to the amount of initial horizontal displacement of the wedge(s). After the southward tilting of the homocline, the only parts of other possible wedges that will be intersected by the current erosional surface are those that were displaced southward beyond the principal tilt axis of the homocline (Flottmann and Hand, 1999).

Largely subsequent to the development of the MacDonnell Homocline, further shortening of as much as 30 km was accommodated along the Gardiner Thrust (Figure 5), which led to a duplication of the Amadeus Basin successions, including the foreland deposits of the Pertnjara Group. This indicates that a significant amount of the displacement along the Gardiner Thrust postdates the shortening that led to the emplacement of the Razorback structure. Shortening along the Gardiner Thrust has been explained by south-over-north-directed displacement that is apparently manifested in the hanging wall of the thrust (Shaw et al., 1991a). However, strata in the hanging wall to the south show a decrease in displacement and intensity of deformation. This suggests instead that the displacement leading to the formation of the Gardiner Thrust system was mainly accommodated by underthrusting in the footwall of that thrust system. This would explain why deformation intensity wanes toward the south. The footwall displacement thereby simply forms a forelandward progression of the displacement that led to the Razorback structure's emplacement. The combination of the displacement that led to the emplacement of the Razorback

FIGURE 4. Sequentially restored cross section showing the final deformational geometry along the Mt. Sonder section. Horizontal arrows on right indicate orogenic shortening in kilometers. Structures above present erosional level are interpreted. (a) Final structure and upward projection of strata in balanced cross section; note that topography was presumably progressively eroded to source sediments in foreland basin. (b) Possible preceding increment showing the initiation of southward tilt of the proto-Razorback structure. (c) Emplacement of the proto-Razorback structure in an intercutaneous wedge mainly confined to the Bitter Springs Formation. (d) Restored section. Notice the small displacement along the upper-crust part of the Redbank Shear Zone throughout the deformational history; displacement was mainly taken up in the footwall of Redbank shear zone. Legend as in Figure 3.

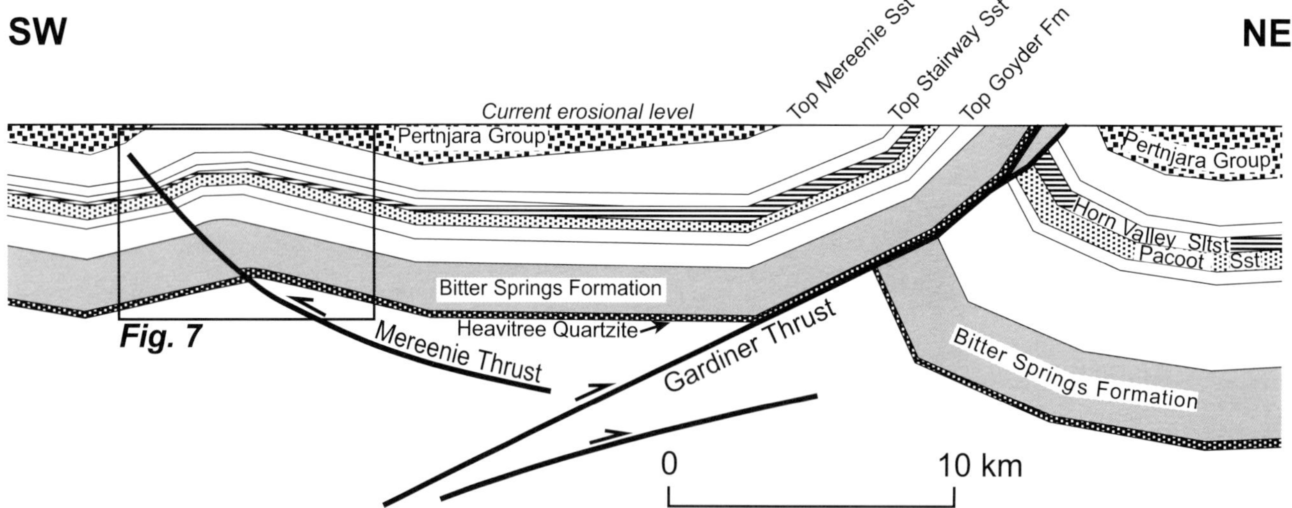

FIGURE 6. Schematic cross section through Mereenie oil and gas field, developed in the hanging wall of a north-dipping thrust. Note that displacement along Gardiner Thrust is unconstrained here (no seismic data). Box indicates location of seismic data given in Figure 7.

structure and the later out-of-sequence Gardiner Thrust together form a structural geometry that resembles thick-skinned triangle zones described by Zapata and Allmendinger (1996) from the Argentinian Precordillera. Overall, the shortening in the vicinity of the northern margin of the Amadeus Basin during the Alice Springs orogeny is in the range of 50 km.

Structural Setting of Major Oil and Gas Fields in the Amadeus Basin

The Amadeus Basin is one of Australia's major onshore oil and gas producers. All current hydrocarbon production in the Amadeus Basin is from Cambrian–Ordovician petroleum systems. The Cambrian–Ordovician Pacoota Sandstone is the major reservoir rock, and the overlying Horn Valley Siltstone forms the main source as well as a (stratigraphic) seal couplet (see Figures 6 and 7). However, hydrocarbon shows were also reported in the overlying Stairways Sandstone. Exploration drilling (Figure 1) has also revealed hydrocarbon shows in Neoproterozoic and Cambrian rocks (Schroder and Gorter, 1984). Established hydrocarbon occurrences are dominated by structural plays, because most of the currently productive fields are associated with late-phase thrust systems. This play is exemplified by the Mereenie Anticline, in which the Pacoota Sandstone contains 300 million barrels of oil in place. Oil and gas are produced from the Pacoota Sandstone. Potentially significant gas shows are also recorded from the Stairway Sandstone. The Mereenie field is located in a large, gently dipping anticline formed in the hanging wall of the northeast-dipping Mereenie Thrust. Thrust displacements are in the range of 50–150 m (Figures 6 and 7). The Mereenie Thrust forms an antithetic thrust to the Gardiner thrust system in the north (Figure 6). The Gardiner Thrust itself formed as a response to the previously described tectonics of the Alice Springs orogeny (see Figure 5). The Palm Valley gas field (Figure 1) is located in a large anticline in the footwall of the Gardiner Thrust. Fold-related natural fractures are the key contributors to permeability that allows economic gas flows from this field (Berry et al., 1996). There are no economic oil accumulations in the Palm Valley field.

Both producing fields in the Amadeus Basin are associated with structures that formed during late-stage thrusting of the Alice Springs orogeny. Practically all mapped structures of similar age that contain unbreached Pacoota through Stairways sandstones have been drilled without yielding economic hydrocarbon accumulations. Current exploration efforts focus on the hydrocarbon potential of folded older successions in the western part of the Amadeus Basin that were affected by the Alice Springs orogeny and, in part, also by the Petermann orogeny.

THE PETERMANN OROGENY

The late Neoproterozoic to Early Cambrian Petermann orogeny (580–540 Ma) led to the exhumation of the Musgrave Inlier that separates the Officer Basin to the south and the Amadeus Basin to the north (Figure 1). One of the key elements of this orogen is the Woodroffe Thrust, which is a south-dipping crustal-scale shear zone that is interpreted to offset the Moho by as much as 20 km (Lambeck and Burgess, 1992). In the hanging

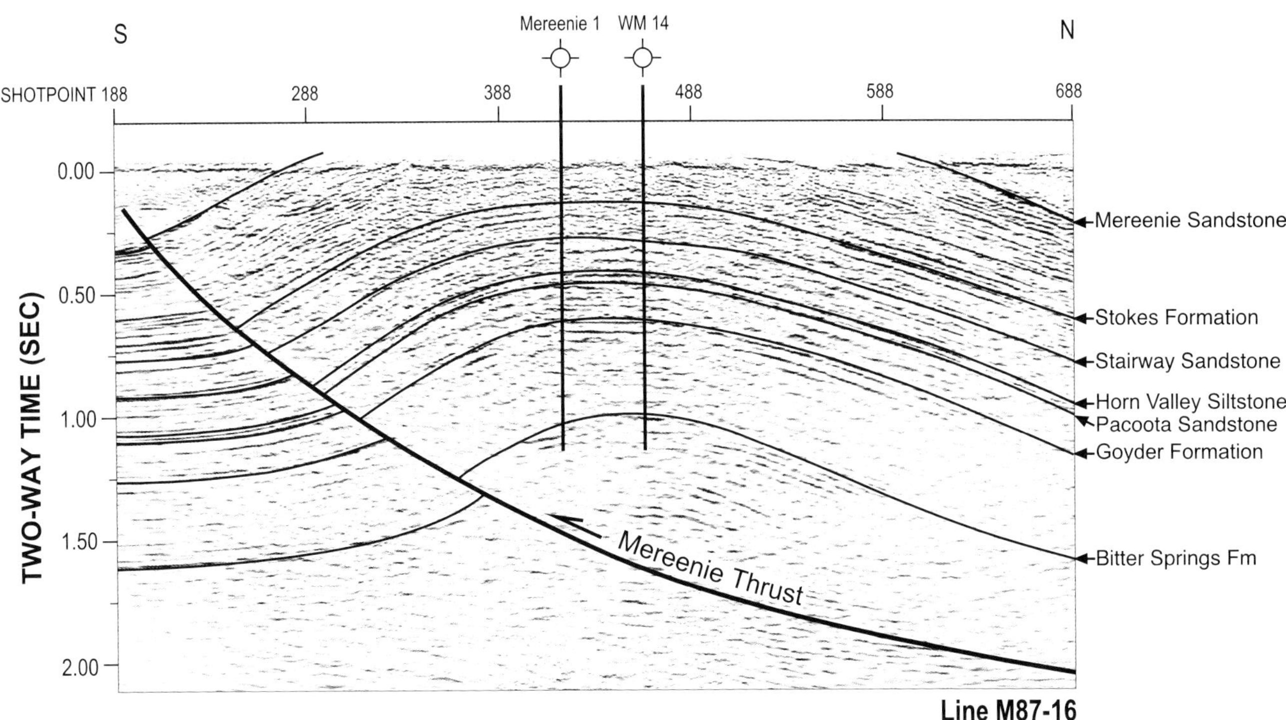

Figure 7. Seismic line m 87-16, showing an anticline in the hanging wall of Mereenie Thrust. Key formation tops are labeled. Wells are projected from an approximately 200-m distance. Heavitree Quartzite is not identifiable seismically.

wall of the Woodroffe Thrust, Mesoproterozoic granulites have been reworked by Early Cambrian subeclogite facies shear zones (Glikson et al., 1996; Camacho et al., 1997; Scrimgeour and Close, 1999), which represent the deepest exposed structures that formed during the Petermann orogeny. In the footwall of the Woodroffe Thrust is a crustal-scale thrust stack, referred to here as the Petermann Thrust Complex. This thrust complex involves (1) basement, which is defined here as including Mesoproterozoic granites and gneisses, and a cover sequence comprising the Mt. Harris Basalt and associated quartz-rich sequences belonging to the Bloods Range Beds, (2) the lower successions of the Amadeus Basin, which are the Dean Quartzite (equivalent to the Heavitree Quartzite) and the Pinyinna Beds (equivalent to Bitter Springs Formation).

There is a significant regional variation in deformational style and metamorphic grade within the Petermann Thrust Complex. At the northern limit of the complex, the units are essentially unmetamorphosed, and thrust displacements occur along discrete detachments. In the Mannanana Range (Figure 8), the coexistence of kyanite-pyrophyllite-quartz ± muscovite in the Dean Quartzite indicates that temperatures reached 400°C (Holland and Powell, 1998), which implies a minimum burial depth of ~8.5 km in that part of the wedge. In this region, strain is more homogeneously distributed, with recrystallization affecting most of the rock volume. The most deeply buried cover units occur in the Olia Chain and in the vicinity of Mt. Deering (Figure 8). In these areas, regional metamorphism reached mid-amphibolite facies, with the formation of coarse kyanite, and rare garnet-bearing assemblages in aluminous schists. Preliminary P-T data suggest that temperatures reached ~550–600°C (Scrimgeour and Close, 1999). At these temperatures, the presence of kyanite implies that the basal units of the Amadeus Basin were buried to a depth of at least 18 km during the Petermann orogeny. This burial estimate is consistent with a pressure of 6.3 kbar (630 MPa) obtained from a deformed and recrystallized 800-Ma mafic dyke in the Olia Chain (Scrimgeour and Close, 1999). In these more deeply buried parts of the thrust complex, structural fabrics are pervasively developed, and the structural style is more reminiscent of a basement terrain.

The degree of pervasive deformation in much of the thrust complex seriously hampers the construction of detailed restorable cross sections. To overcome this problem, we have adopted a stratigraphic template based on the sequences in the less internally deformed regions. Such a template allows us to construct a kinematically plausible cross section and forms the basis of our structural interpretation.

The lowermost unit in the southwestern Amadeus Basin is the middle Neoproterozoic (~750 Ma) Dean Quartzite. The least-deformed sections of this unit retain sedimentary structures and are 150–350 m thick. In many cases, the Dean Quartzite is structurally repeated,

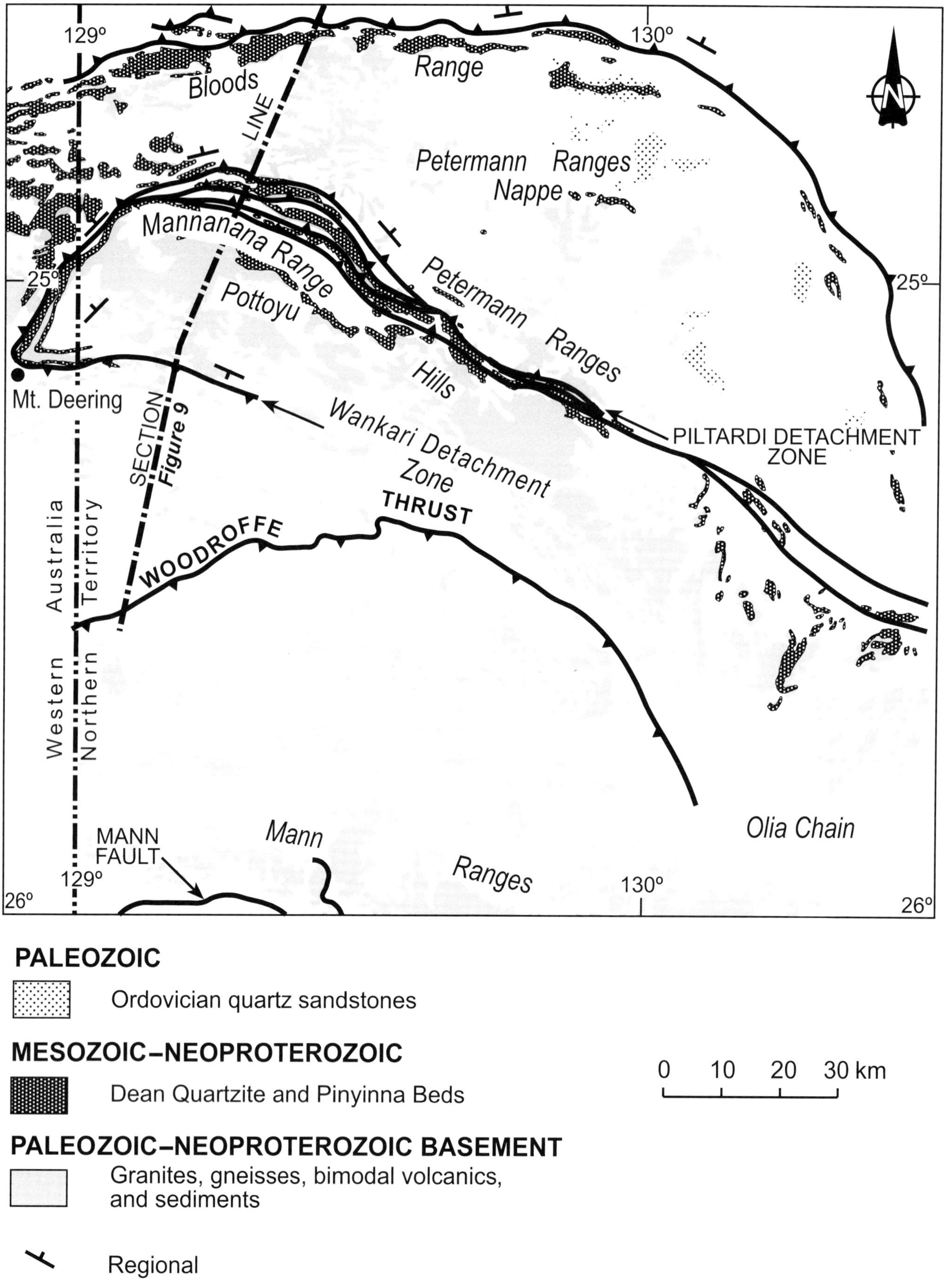

129° 130°
Bloods Range
LINE
Petermann Ranges Nappe
Mannanana Range
Petermann Ranges
Pottoyu Hills
25° 25°
Mt. Deering
SECTION Figure 9
Wankari Detachment Zone
PILTARDI DETACHMENT ZONE
WOODROFFE THRUST
Western Australia
Northern Territory
MANN FAULT
Mann Ranges
Olia Chain
129° 130°
26° 26°
N

PALEOZOIC

 Ordovician quartz sandstones

MESOZOIC–NEOPROTEROZOIC

Dean Quartzite and Pinyinna Beds

PALEOZOIC–NEOPROTEROZOIC BASEMENT

 Granites, gneisses, bimodal volcanics, and sediments

 Regional

and consequently, previous stratigraphic-thickness estimates of as much as 700 m (Wells et al., 1970) are overestimates. The Dean Quartzite is overlain by the Pinyinna Beds, which are a succession of metacarbonates, siltstones, and local evaporites. Locally, the Pinyinna Beds can be thicker than 2 km. Whether this reflects a true stratigraphic thickness is unclear, because the Pinyinna Beds typically are deeply eroded and form morphologic depressions that are covered by younger sediments, which makes detailed mapping and the identification of structural duplications impossible. The basal units of the Amadeus Basin were deposited on a basement consisting of (1) a late Mesoproterozoic (~1050 Ma) sequence of dominantly basaltic bimodal volcanics (Mount Harris Basalt) and associated quartzose sediments (Bloods Range Beds), and (2) ~1050- to 1600-Ma granites and gneisses, some of which were metamorphosed and deformed during the 1200–1100 Ma Musgravian orogeny. Together with the basement, the lower succession of the Amadeus Basin forms the intensely deformed core of the Petermann Thrust Complex. The upper succession of the southwestern Amadeus Basin is defined here as the sequences above the Pinyinna Beds that were deposited prior to the Petermann orogeny. The upper succession attains a thickness of about 3–4 km (Hand and Sandiford, 1999). Rocks of the upper succession do not crop out within the Petermann Thrust Complex; they lie to the north in the Amadeus Basin, where they dip northward at as much as 45°.

Key Structures of the Petermann Orogen

Our interpretation of the Petermann orogen is based on a regional remapping of the Petermann Ranges. In addition, we completed a regional traverse aimed at a structural and tectonic reinterpretation of the northern margin of the Petermann orogen. The preserved stratigraphic template provides key beds on which we base the structural analysis, which in turn allows minimum estimates to be made of orogenic shortening. Given the intense deformation in the structurally lowest part of the Petermann Thrust Complex, a full restoration and balancing of displacements is impossible. The regional cross sections were balanced at the scale of the remapping (1:100,000). We made no attempt to quantify the magnitude of internal strain in individual thrust packages, so the estimates of orogenic displacements are an absolute minimum. Across strike, six tectonic domains (Figure 8) can be distinguished on the basis of their structural and kinematic style, located on the northern margin of the Petermann orogen. From south to north, these are as follows.

South of the Woodroffe Thrust, fabrics that formed during the Musgravian orogeny (1200–1150 Ma) were reworked at temperatures in excess of 700°C during the Petermann orogeny (Scrimgeour and Close, 1999). In the immediate hanging wall of the Woodroffe Thrust, pressures are around 10 ± 1 kbar (1000 MPa), but farther south in the Mann Ranges (Figure 6), pressures reach as much as 13 ± 1 kbar (1300 MPa) (Scrimgeour and Close, 1999). The Petermann-aged structures are expressed as south-dipping mylonite zones as wide as several hundred meters and, in some regions, they are a pervasive structural fabric that largely obliterates earlier tectonic features. Displacement along these mylonites is south-over-north directed. The southward increase in pressure through the hanging wall of the Woodroffe Thrust suggests that the system of Petermann-aged high-strain zones produced a north-vergent imbricate structure. The Woodroffe Thrust itself is a mylonitic zone as thick as 1 km that dips 10–40° south with a north-directed displacement. The mylonitic fabrics are lower in grade than are the hanging wall and immediate footwall, which suggests that the zone was reactivated after a juxtaposition event(s).

The Wankari Detachment Zone is located in the footwall of the Woodroffe Thrust (Figure 6). The Wankari Detachment Zone dips south at 20–45°, and, over most of its length, it outcrops in basement. Deformational features include asymmetric shear ("s-c") structures and shear-banded mylonites that indicate a north- to northwest-directed sense of displacement. The displacement vector is indicated by mineral elongation and slip lineations that trend between 125° and 175°. At the western part of the Wankari Detachment Zone, thin slivers of Bloods Range beds and Dean Quartzite occur in the immediate footwall of the detachment but pinch out to the east (Figure 6). To the west, increasingly more Dean Quartzite is involved in the Wankari Detachment Zone, indicating that the entire structural package has a westerly plunge in this part of the orogen. The metamorphic assemblages within the Wankari Detachment Zone indicate that deformation occurred at ~640°C and 6.5 ± 1 kbar (650 MPa) (Scrimgeour and Close, 1999). This is approximately 4 kbar (400 MPa) less pressure than in the immediate hanging wall of the Woodroffe Thrust and implies a vertical offset of ~15 km across that structure.

Figure 8. Geological map of the northern Petermann orogen. The locations of the thrust faults south of the Wankari Detachment Zone are inferred from aeromagnetic data and discontinuities in regional metamorphic grade. Back thrust at the northern margin of the orogen is mapped in a few outcrops; the map trace of the back thrust is also inferred from aeromagnetic data. Note the scale difference compared with Figure 4.

The Pottoyu Hills domain is located in the footwall of the Wankari Detachment Zone and defines a broad, doubly plunging basement culmination composed of the Pottoyu Granite. The granite is dissected by numerous south-dipping high-strain zones, which are 1–10 m wide. The intensity and abundance of the high-strain zones decreases northward away from the Wankari Detachment Zone.

The Piltardi Detachment Zone bounds the northern flank of Pottoyu basement culmination. Internally, the detachment zone consists of four principal repetitions of cover sequence, which outcrop in strike ridges that dip, on average, 20–50° to the north. Individual ridges can contain basement and Bloods Range beds, and all contain Dean Quartzite. The overlying Pinyinna Beds are rarely exposed, but are interpreted to underlie morphologic depressions between ridges of quartzite. Dean Quartzite and, where involved, basement and Bloods Range beds, show kinematic indicators such as s-c fabrics, shear bands, and porphyroclasts consistent with south-over-north-directed displacement in the Piltardi Detachment Zone. The displacement vector is about 000–030, as indicated by mineral elongation and slip lineations. The Piltardi Detachment Zone narrows toward the west and is reduced to one major shear zone of high internal strain at the western end of the Pottoyu Culmination, where the Piltardi detachment merges with the Wankari detachment (Figure 8). In the eastern part of the orogen, the Piltardi Detachment Zone bifurcates into a principal system of southeast-trending detachments and a subsidiary east- to north-trending system (Figure 6). S-c mylonites and shear bands in micaceous units in the Dean Quartzite indicate north-over-south-directed thrusting (i.e., back thrusting).

North of the Piltardi Detachment Zone is a zone of flat-lying to gently north-dipping basement. Along the section in Figure 8, this basement is mainly formed by little-strained vesicular basaltic rocks comprising the Mt. Harris Basalt. Farther east, this basement domain is composed of granite that underlies the basalt.

The northern part of the Petermann orogen is characterized by both forward- and hinterlandward-propagating thrusts. The foreland-propagating thrust system duplicates the Dean Quartzite along thrusts with northward displacement. The northernmost parts of the Petermann Thrust Complex display north-over-south-directed displacements, which are manifest as south-vergent folds and small-scale south-vergent thrusts. North of this, the magnitude of shortening decreases dramatically, and there are few obvious Petermann-age structures within the broader extent of the Amadeus Basin.

Structural Interpretation

The lack of involvement of upper-succession rocks in the exposed core of the Petermann Thrust Complex, and the comparatively simple structure at the front of the complex, resemble the structural style of the basement wedges at the northern margin of the Amadeus Basin that formed during the Alice Springs orogeny. The key to a structural interpretation of the Petermann Thrust Complex is the interpretation of the Piltardi Detachment Zone, which has a northerly dip of 20–70°. The major structural repetitions in the Piltardi Detachment Zone were identified where the inverse stacking of key lithologies clearly indicates a thrust repetition (i.e., basement or Bloods Range Beds overlie Dean Quartzite or Pinyinna Beds). On the cross section in Figure 9, we have not discriminated among individual lithologies within the Piltardi Detachment Zone. Rather, we have grouped all lithologies present into the lower succession (which may also incorporate thin slivers of basement). There are four major thrust packages in the Piltardi Detachment Zone. Toward the western margin of the Pottoyu Dome, the individual detachments within the Piltardi Detachment Zone merge into the Wankari Detachment Zone. Projecting these relationships onto the section line of Figure 9a–e suggests that the splaying of the detachments and the associated thrust repetition occur above the Pottoyu Dome, and that the Piltardi Detachment Zone is a footwall structure linked to the Wankari Detachment.

The Piltardi Detachment is located along the northeastern flank of the Pottoyu Hills domain. In its present geometry, the detachment system dips northward, and consequently the north-directed transport results in an

Figure 9. Sequentially restored cross section showing the final deformational geometry of the northern Petermann orogen. Note that the internal strain has not been balanced. Horizontal arrows on left indicate orogenic shortening in kilometers. Structures above the present erosional level are interpreted. (a) Final structural geometry formed by (1) the Wankari Detachment Zone; mapped structures are below white line representing current erosional level, structures above white line are interpreted (2) Pottoyu Dome and (3) the north-dipping Piltardi Detachment Zone, with a fourfold repetition of the lower succession of the Amadeus Basin. Shortening of the upper succession of the Amadeus Basin is interpreted to have occurred by back thrusting. The total orogenic shortening is in excess of 100 km. (b) Preceding increment structures of the lower succession dip to the south or are subhorizontal. (c) Formation of stacked triangle zone. (d) Transfer of displacement from Woodroffe Thrust to Wankari Detachment Zone and southernmost triangle zone. (e) Restored section showing first increment of displacement along the Woodroffe Thrust.

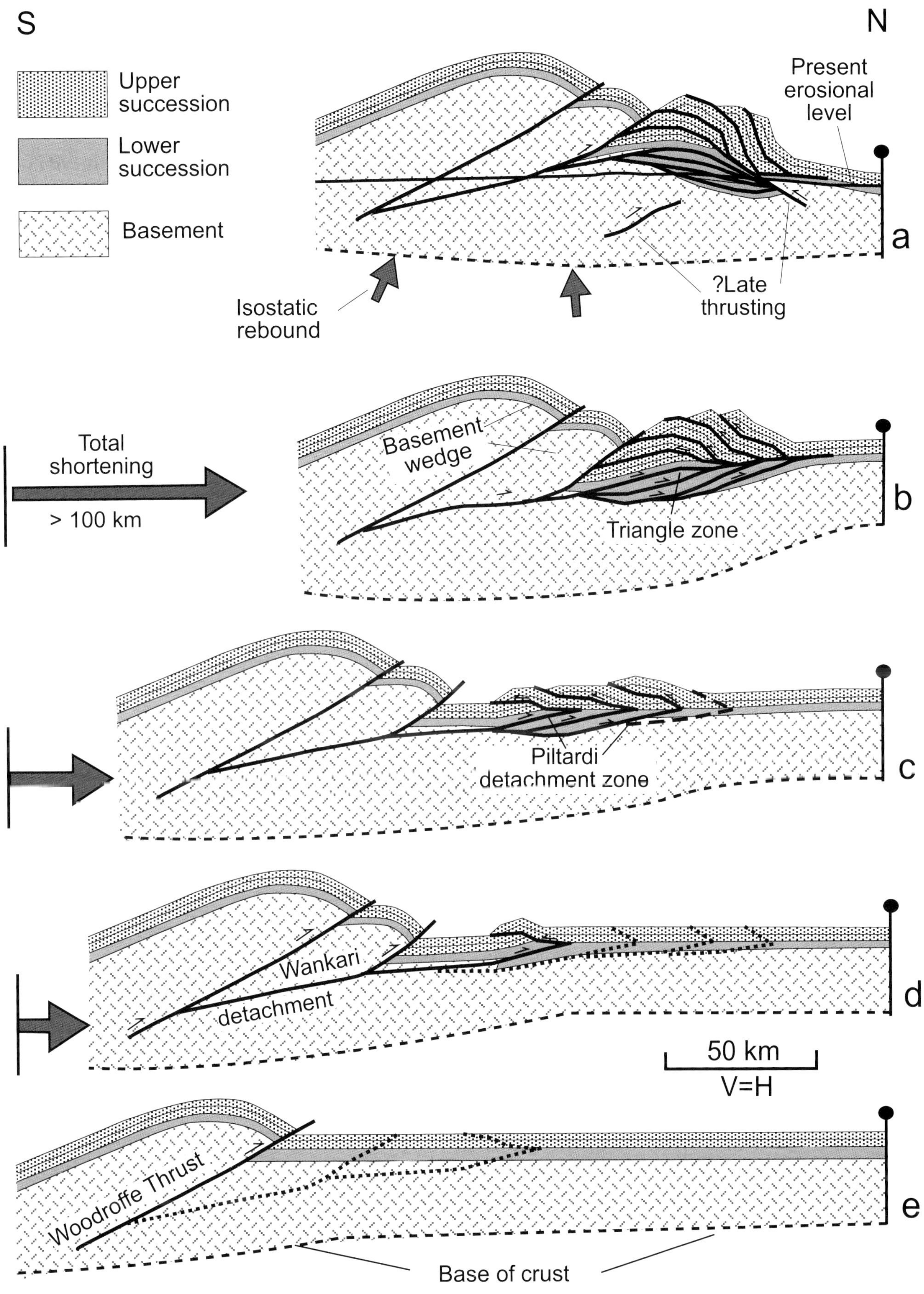
S
N
Upper succession
Lower succession
Basement
Present erosional level
a
Isostatic rebound
?Late thrusting
Total shortening
> 100 km
Basement wedge
b
Triangle zone
c
Piltardi detachment zone
Wankari detachment
d
50 km
V=H
Woodroffe Thrust
e
Base of crust

apparently normal sense of shear on the detachments. However, the presence of hanging-wall basement slabs overlying Dean Quartzite and Pinyinna Beds confirms that the structures are compressional. The northward dips are caused by doming within the Pottoyu Hills domain. We interpret this doming to reflect duplexing in deeper levels of the Pottoyu Granite. Up-plunge projection suggests that the total structural relief in the Pottoyu Hills domain is on the order of 3–4 km.

The Wankari and Piltardi Detachment Zones incorporate rocks of the lower succession of the Amadeus Basin and basement. However, as much as 3 km of sediment was present in the upper succession of the Amadeus Basin at the time of the Petermann orogeny (Hand and Sandiford, 1999). This upper sequence also must have been incorporated into the shortening history of the Petermann orogen. The northern margin of the Petermann Thrust Complex shows that the upper successions are displaced toward the south along ~45° north-dipping thrusts. In the footwall of this south-vergent thrust, the Dean Quartzite and underlying Mt. Harris Basalt are duplicated along south-dipping thrusts with north-directed displacement. These relationships suggest that the Petermann Thrust Complex has a structural style similar to that proposed for the basement wedge structures formed along the northern margin of the Amadeus Basin during the Alice Springs orogeny. We suggest that the upper succession likely was stacked along multiple back thrusts to form a crustal-scale triangle zone (Figure 9b) similar to those described by Stockmal et al. (1996) for the Canadian Rocky Mountains. There is geophysical evidence to suggest that the shortening of the upper sequence was accommodated by back thrusting at the front of the Petermann Thrust Complex. Aeromagnetic data show that a major fault system bounds the frontal edge of the complex, where it terminates against the upper succession of the Amadeus Basin. The geophysical style of this frontal zone is very similar to the Piltardi Detachment Zone, which suggests that it represents a major detachment system in which internal duplexing has thickened the zone locally to greater than 1 km.

Overall, deformation in the Petermann Thrust Complex presumably propagated toward the foreland in the north. The Woodroffe Thrust forms a crustal detachment similar to the Redbank Shear Zone in the Alice Springs orogen. In principle, the deformation that led to the formation of the Petermann Thrust Complex reflects a forelandward propagation of displacement in the footwall of the Woodroffe Thrust. Whether the triangle zone's formation strictly reflects an in-sequence propagation of the individual thrusts of the Wankari and Piltardi Detachment Zones is unclear. It appears feasible that most of the principal thrusts were active concomitantly (Figure 9c). Although the locus of deformation probably propagated northward, the relatively low-grade fabrics in the Woodroffe Thrust, compared with its immediate footwall, implies that it accommodated some out-of-sequence displacements. Our model suggests that the total minimum horizontal shortening on the northern side of the Petermann orogen exceeds 100 km (Figure 9a–e). The metamorphic character of the deeper parts of the Petermann Thrust Complex indicates that the cover sequence belonging to the Amadeus Basin was buried to at least 20 km. Exhumation of rocks in the core of the orogen is from a 20-km depth. The exhumation may have been the result of flexural rebound and/or it may have been assisted by further thrusting (Figure 9a). Together, they result in the overall doming of the Pottoyu Hills, which leads to the present northward tilt of the Piltardi Detachment Zone. The Petermann Thrust Complex is unconformably overlain by early Ordovician shallow-marine sediments, indicating that topography associated with the deformation had been largely removed by 475 Ma.

DISCUSSION

The Alice Springs and Petermann orogens are intracratonic fold-and-thrust belts that have, despite their different ages and different amounts of displacement, a common thrust-tectonic style. In both orogens, deformation in the footwalls of crustal-scale faults is characterized by significant horizontal displacements with a partitioning of deformation in a forward-propagating lower succession and an upper stratigraphic succession in which the same shortening is accommodated by back thrusting. The combination of forward and back thrusting, and the presence of a weak detachment horizon in the Bitter Springs Formation/Pinyinna Beds, led to the formation of deformation zones reminiscent of the triangle-zone terminations of fold-and-thrust belts of the western Canadian Cordillera. However, in the latter cases, the deformation is essentially thin-skinned, in that crystalline basement is not incorporated into the triangle zones (Hiebert and Spratt, 1996; Lebel et al., 1996). This highlights an important characteristic of the basement-cover interaction along the margins of the Amadeus Basin that are characterized by interleaving of basement and cover sequences. This is particularly the case in the Petermann Thrust Complex, where multiple basement slabs have been stacked in the Piltardi Detachment Zone. The basement-involved deformational style of the Amadeus Basin's margins resembles the interaction of the foreland and hinterland of a thrust belt (e.g., on the western side of the Idaho-Wyoming part of the Cordilleran fold belt (De Celles and Mitra, 1995).

From a mechanical viewpoint, the intercalation of multiple basement slabs and cover is intriguing. For example, in the Piltardi Detachment Zone in the

Petermann Thrust Complex, at least four basement-bearing slabs are interleaved with the relatively thick Pinyinna Beds, which elsewhere contain evaporites and argillaceous horizons. Intuitively, one would expect that the Pinyinna Beds would act as a detachment horizon, as appears to be the case at the leading edge of the triangle zones (Figure 4b and c), and would not be involved in stacking. The fact that the beds are stacked with the basement suggests that there was relatively little mechanical difference between the granitic basement and a cover sequence that would generally be regarded as incompetent, relative to basement.

The inference that the basement may have been relatively weak is consistent with the way in which the orogenic strains are partitioned. In both the Alice Springs and Petermann orogens, there is substantial shortening in the footwall of crustal-scale faults that cut the main load-bearing regions of the lithosphere. In a thick-skinned intraplate orogen, the crustal-scale faults normally would be expected to accommodate the bulk of the shortening (Brown, 1988; Korsch et al., 1998). However, during the Alice Springs orogeny, the Redbank Shear Zone at outcrop level apparently accommodated <3 km of shortening (Shaw et al., 1992). Instead ~ 50 km of shortening was transferred to the footwall (Figure 6). This implies that, despite containing fine-grained mylonites, the Redbank Shear Zone was not anomalously weak compared with its footwall. Similarly, although the Woodroffe Thrust accommodated significant shortening during the Petermann orogeny, as evidenced by the exhumation of >10 kbar (1000 MPa) rocks in the hanging wall, the magnitude of footwall shortening was far greater. Indeed, if the Woodroffe Thrust dips, on average, at 40° into the lower crust (Lambeck and Burgess, 1992), the metamorphic offset (Scrimgeour and Close, 1999) implies approximately 20–25 km of shortening. This compares with a minimum of 100 km of shortening in the footwall (Figure 9). In the deeper parts of the Petermann Thrust Complex, the shortening was accommodated by the development of pervasive structural fabrics. This implies that individual detachments played a less important role, relative to pervasive flow of material, and that the overall behavior was controlled by the bulk rheology rather than by individual faults. Along the northern Amadeus Basin margin, the level of denudation is less than in the Petermann Thrust Complex, and the structures in the footwall of the Redbank Shear Zone are relatively discrete. However, we see no reason why, at deeper levels, the footwall deformation is not also accommodated by diffuse flow. Hand and Sandiford (1999) put forward the notion that the crustal-scale faults in central Australia, such as the Woodroffe Thrust and Redbank Shear Zone, did not exert the first-order control on the distribution of intraplate deformation there. Based on the shifting patterns of deformation and pre-orogenic sedimentation, they suggested that instead, thermally induced lithospheric weakening may have exerted the first-order control on the distribution of the intraplate deformation.

Triangle zones associated with significant orogenic shortening, such as in the Canadian Rocky Mountains (Hiebert and Spratt 1996; Lebel et al., 1996; Stockmal et al., 1996), are still within the hydrocarbon window. This suggests that the shortening there did not result in deep burial of the sequences, which implies that the generated structural topography was eroded progressively (Jones, 1991; Stockmal et al., 1996) rather than undergoing long-term crustal thickening. This could occur if the triangle zone developed on relatively strong lithosphere, an inference that is consistent with the wide, shallow foreland basins (Leckie and Smith, 1992) that developed during the Canadian Rocky Mountain deformation. The structural-stratigraphic-metamorphic relationships in the Petermann Thrust Complex and the shape of the foreland basin to the Alice Springs orogeny point to a very different scenario and suggest that regionally weak lithosphere was involved in intraplate orogeny in central Australia.

In the Petermann Thrust Complex, the metamorphic assemblages that developed in the basal units of the Amadeus Basin (Dean Quartzite and Pinyinna Beds) indicate that the lowest sections of the lower succession were buried to ~20 km during the Petermann orogeny. At the northern edge of the thrust complex, the same sequences are covered by ~3 km of sediment belonging to the upper sequences of the Amadeus Basin (Hand and Sandiford, 1999). This implies that a significant proportion of the structural topography in Figure 9 was realized and was expressed as actual crustal thickening, as opposed to theoretical structural topography (Beaumont et al., 1992) in which erosion acts to constantly limit the extent of thickening. The structurally lowest exposures of the cover sequence occur immediately above the Pottoyu Granite in the vicinity of Mt. Deering and in the Olia Chain (Figure 8). Cover sequences in these areas are located approximately 60 km from the leading edge of the thrust complex. Allowing for shortening associated with the formation of the Pottoyu Hills Dome and for internal shortening, the distance from the leading edge of the thrust complex was probably originally on the order of 80 km. This implies differential burial of around 15–17 km over a distance of ~80 km. Conceivably, the deep burial of the structurally lowest cover sequences reflects a profound flexural response to loading that was generated by thickening in the Petermann Thrust Complex. If burial did arise as a consequence of a flexure, dip values of the base of the wedge were around 10°. This suggests that the lithosphere involved in the intraplate Petermann orogeny was comparatively weak (Bertotti et al., 1998), and it is consistent with thermomechanical modeling (Sandiford and Hand, 1998; Hand and Sandiford, 1999).

The inference that the Petermann orogen incorporated weak lithosphere may explain the apparent absence of a well-developed foreland basin to the north of the orogen. If the lithosphere is sufficiently weak, there will be a relatively short-wavelength, high-amplitude flexural response to loading (Watts, 1992; Bertotti et al., 1998). This is consistent with the presence of a narrow, elongate trough ~30 km wide and as much as 5 km deep (the Mount Currie subbasin), immediately in front of the Petermann Thrust Complex, that contains coarse clastics shed from the orogen. This basin overfilled, resulting in transport of synorogenic sediment northward for several hundred kilometers to depocenters in and beyond the northern Amadeus Basin (Lindsay and Korsch, 1991; Shaw et al., 1991a, b).

The evolving pattern of synorogenic sedimentation during the Alice Springs orogeny also suggests the involvement of weak lithosphere. In the Early Devonian, a shallow (<1.5-km deep) foreland basin extended southward for a distance of ~250 km (Shaw et al., 1991a, b). However, with continued shortening, the basin narrowed and deepened, to the extent that in the Early to Middle Carboniferous time, as much as 6 km of coarse-grained clastic sediment was accommodated in a trough approximately 70 km across (Jones, 1991). The final foreland basin shape is similar to basins such as the Appenines (Royden and Karner, 1984) and the Ebro (Zoetemeijer et al., 1990), which developed on relatively weak lithosphere with elastic thicknesses on the order of 5 km (Watts, 1992; Bertotti et al., 1998).

The importance of structure in controlling the distribution of hydrocarbons is not surprising, given that the Amadeus Basin was located in successively developed forelands. Despite this unusual tectonic setting, relatively little interaction appears to have occurred in the basin between structures that formed during the Petermann and Alice Springs orogenies. This is largely because of the rapid decrease in the intensity of deformation at the northern margin of the Petermann orogen, which we suggest may be the consequence of locally weak lithosphere within the Petermann orogen. This restricted distribution of foreland deformation contrasts with the Alice Springs orogeny, which produced basinwide deformation that is manifest as thin-skinned fold-thrust systems (Shaw et al., 1991a). The work presented here indicates that the Petermann Thrust Complex is similar to the foreland of the Canadian Cordillera in its kinematic style and magnitude of shortening. However, the inferred weak lithosphere involved in the Petermann orogeny was not conducive to developing a regionally extensive foreland-basin system. Therefore, the basin has limited potential for hydrocarbon plays specifically related to progradational sequences as described from other foreland basins (Barclay and Smith, 1992). The main hydrocarbon-producing reservoirs potentially resulting from the Petermann orogeny are successions such as the Arumbera and Cleland Sandstones, in the northern Amadeus Basin, that represent the dispersed overfill of the Petermann foreland deposits. The proximity of these successions to the northern margin of the Amadeus Basin makes them ideally suited to trap formation in response to structuring that resulted from the Alice Springs orogeny. However, it must be stressed that the stratigraphy, structure, and sedimentology of sequences in the western Amadeus Basin north of the Petermann orogen are very poorly known. In that area, regional-scale dome-and-basin patterns in the basin suggest that it is one of the few regions where the Petermann- and Alice Springs-aged fold systems have interacted. Compared with the Petermann orogeny, the Amadeus Basin's extensive structuring during the Alice Springs orogeny provides more scope for hydrocarbon trapping. The basic structures associated with the early phase of the Alice Springs orogeny strongly resemble structures of the Laramide orogeny in the western United States. Clearly, this early phase of structuring may form naturally fractured sweet spots with enhanced gas permeability in a proven natural-gas province.

ACKNOWLEDGMENTS

We thank the Central Lands Council for permission to access Aboriginal land in the Petermann Ranges. T. Flottmann and M. Hand are grateful for logistic support provided by the Northern Territory Geological Survey, with special thanks to Peter Crispe and Max Heggen. The chapter benefited from discussions with Fabrizio Storti, Russell Korsch, Mike Sandiford, and Dave Warner. Reviews by Rick Allmendinger, Tomas Zapata, Ken McClay, and an anonymous reviewer helped clarify and focus the chapter.

REFERENCES CITED

Allmendinger, R. W., 1998, Inverse and forward numerical modelling of trishear fault propagation folds: Tectonics, v. 17, p. 640–656.

Barclay, J. E., and D. G. Smith, 1992, Western Canada foreland basin oil and gas plays, *in* R. W. Macqueen and D. A. Leckie, eds., Foreland basins and fold belts: AAPG Memoir 55, p. 191–228.

Beaumont, C., P. Fullsack, and J. Hamilton, 1992, Erosional control of active compressional orogens, *in* K. R. McClay, ed., Thrust tectonics: London, Chapman & Hall, p. 1–18.

Berry, M. D., D. W. Stearns, and M. Friedman, 1996, The development of a fractured reservoir model for the Palm Valley gas field: Australian Petroleum Exploration Association Journal, v. 36, p. 82–103.

Bertotti, G., V. Picotti, and S. Cloetingh, 1998, Lithospheric weakening during "retroforeland" basin formation: Tectonic evoluation of the central South Alpine foredeep: Tectonics, v. 17, p. 131–142.

Bradshaw, J. D., and P. R. Evans, 1988, Paleozoic tectonics, Amadeus Basin, Central Australia: Australian Petroleum Exploration Association Journal, v. 28, p. 267–282.

Brown, W. G., 1988, Deformation style of Laramide uplifts in the Wyoming foreland, *in* J. C. Schmidt and W. J. Perry Jr., eds., Deformation style of Laramide uplifts in the Wyoming foreland: Geological Society of America Memoir 171, p. 1–23.

Camacho, A., W. Compston, M. McCulloch, and I. McDougall, 1997, Timing and exhumation of eclogite facies shear zones, Musgrave Block, Central Australia: Journal of Metamorphic Geology, v. 15, p. 735–751.

Chase, R. B., C. J. Schmidt, and P. W. Genovese, 1993, Influence of rock compositions and fabrics on the development of Rocky Mountain foreland folds, *in* C. J. Schmidt, R. B. Chase, and E. A. Erslev, eds., Laramide basement deformation in the Rocky Mountain foreland of the Western United States: Geological Society of America Special Paper 280, p. 45–72

Collins, W. J., and Teyssier, C., 1989, Crustal scale ductile fault systems in the Arunta Inlier, Central Australia: Tectonophysics, v. 158, p. 49–66.

Cook, D. G., 1988, Balancing basement cored folds of the Rocky Mountain foreland, *in* J. C. Schmidt and W. J. Perry Jr., eds., Interaction of the Rocky Mountain foreland and the Cordilleran thrust belt: Geological Society of America Memoir 171, p 53–64.

Couzens, B. A., and D. A. Wiltschko, 1996, The control of mechanical stratigraphy on the formation of triangle zones, *in* P. A. MacKay, T. E. Kubli, A. C. Newson, J. L. Varsek, R. G. Dechesne, and J. P. Reid, eds., Triangle zone geometry, terminology, and kinematics: Bulletin of Canadian Petroleum Geology, v. 44, p 139–152.

De Celles, P. G., and G. Mitra, 1995, History of the Sevier orogenic wedge in terms of critical taper models, northeast Utah and southwest Wyoming: Geological Society of America Bulletin, v. 107, p. 454–467.

Do Rozario, R. F., 1991, The Palm Valley Gas Field, Amadeus Basin, central Australia: Bureau of Mineral Resources Bulletin, v. 236, p. 477–492.

Dunlap, W. J., C. Teyssier, I. McDougall, and S. Baldwin, 1995, Thermal and structural evolution of the intracratonic Arltunga Nappe Complex, Central Australia: Tectonics, v. 14, p. 1182–1204.

Erslev, E. A., 1991, Trishear fault-propagation folding: Geology, v. 19, p. 617–620.

Erslev, E. A., and J. L. Rogers, 1993, Basement-cover geometry of Laramide fault-propagation folds, *in* C. J. Schmidt, R. B. Chase, and E. A. Erslev, eds., Laramide basement deformation in the Rocky Mountain foreland of the Western United States: Geological Society of America Special Paper 280, p. 125–146.

Flottmann, T., and M. Hand, 1999, Folded basement-cored tectonic wedges along the northern edge of the Amadeus Basin, Central Australia: evaluation of orogenic shortening: Journal of Structural Geology, v. 21, p. 399–412.

Forman, D. J., 1966, The geology of the south-western Amadeus Basin, central Australia. Australian Bureau of Mineral Resources Record, v. 87, p 1–54.

Glikson, A. Y., A. J., Stewart, C. G., Ballhaus, G. L.,Clarke, E. H. J. Feeken, J. H. Leven, J. W. Sheraton, and S. S. Sun, 1996, Geology of the western Musgrave Block, central Australia, with particular reference to the mafic-ultramafic Giles Complex: Australian Geological Survey Organisation Bulletin, v. 239, p. 206.

Goleby, B. R., R. D. Shaw, C. Wright, B. L. N. Kennett, and K. Lambeck, 1989, Geophysical evidence for 'thick-skinned' crustal deformation in central Australia: Nature, v. 337, p. 325–330.

Hand, M., and M. Sandiford, 1999, Intraplate deformation in central Australia, the link between subsidence and fault reactivation: Tectonophysics, v. 305, p. 121–140.

Havord, P. J., 1991, Mereenie oil and gas field, Amadeus Basin, Northern Territory: Bureau of Mineral Resources Bulletin, v. 236, p. 493–510.

Hiebert, S. N., and D. A. Spratt, 1996, Geometry of the thrust front near Pincher Creek, Alberta, *in* P. A. MacKay, T. E. Kubli, A. C. Newson, J. L. Varsek, R. G. Dechesne, and J. P. Reid, eds., Triangle zone geometry, terminology and kinematics: Bulletin of Canadian Petroleum Geology, v. 44, p. 195–201.

Hobbs, B. E., W. D. Means, and P. F. Williams, 1976, An outline of structural geology: New York, John Wiley and Sons, 467 p.

Holland, T. J. B., and R. Powell, 1998, An internally consistent thermodynamic dataset for phases of petrological interest: Journal of Metamorphic Geology, v. 16, p. 309–343.

Jones, B. G., 1991, Fluvial and lacustrine facies in the Middle to Late Devonian Pertnjara Group, Amadeus Basin, Northern Territory, and their relationship to tectonic events and climate, *in* R. J. Korsch and J. M. Kennard, eds., Geological and geophysical studies in the Amadeus Basin, central Australia: Australian Bureau of Mineral Resources, Geology and Geophysics Bulletin, v. 236, p. 333–348.

Korsch, R. J., B. R. Goleby, J. H. Leven, and B. J. Drummond, 1998, Crustal architecture of Central Australia based on deep sesismic profiling: Tectonophysics, v. 75, p. 57–69

Kraig, D. H., D. V. Wiltschko, and J. H. Spang, 1988, The interaction of the Moxa Arch (la Barge Platform) with the Cordilleran thrust belt, south of Snider Basin, southwestern Wyoming, *in* J. C. Schmidt and W. J. Perry Jr., eds., Deformation style of Laramide uplifts in the Wyoming foreland: Geological Society America Memoir 171, p. 395–410.

Kurz, W., F. Neubauer, J. Genser, and F. Dachs, 1998, Alpine geodynamic evolution of active and passive continental margin sequences in the Tauern Window (eastern Alps, Austria, Italy): a review: Geologische Rundschau, v. 87, p. 225–242.

Lambeck, K., and G. Burgess, 1992, Deep crustal structure of the Musgrave Block, central Australia: Results from teleseismic travel-time anomalies: Australian Journal Earth Sciences, v. 39, p. 1–19.

Lebel, D., W. Langenburg, and E. W. Mountjoy, 1996, Structure of the central Canadian Cordolleran thrust-and-fold belt, Athabasca-Brazeau area, Alberta: a large, complex intercutaneous wedge, *in* P. A. MacKay, T. E. Kubli, A. C. Newson, J. L. Varsek, R. G. Dechesne, and J. P. Reid, eds., Triangle zone geometry, terminology and kinematics: Bulletin of Canadian Petroleum Geology, v. 44, p. 282–298.

Leckie, D. A., and D. G. Smith, 1992, Regional setting, evolution, and depositional cycles of the western Canada foreland basin, *in* R. W. Macqueen and D. A. Leckie, eds., Foreland basins and fold belts: AAPG Memoir 55, p. 9–46.

Lindsay, J. F., 1987, Upper Proterozoic evaporites in the Amadeus Basin, central Australia and their role in basin tectonics: Geological Society of America Bulletin, v. 99, p. 852–865.

Lindsay, J. F., and R. J. Korsch, 1991, The evolution of the Amadeus Basin, central Australia, *in* R. J. Korsch and J. M. Kennard, eds., Geological and geophysical studies in the Amadeus Basin, central Australia: Australian Bureau of Mineral Resources, Geology and Geophysics Bulletin, v. 236, p. 7–32.

Marjoribanks, R. W., 1974, The structural and metamorphic geology of the Ormiston Region, Central Australia: Ph.D. thesis, University of Canberra.

Marjoribanks, R. W., 1976, Basement and cover relations on the northern margin of the Amadeus Basin, central Australia: Tectonophysics, v. 33, p. 15–32.

Mathur, H. J., 1976, Relationship of Bouguer anomaly to crustal structure in southwest and central Australia: Bureau of Mineral Resources Journal, v. 1, p. 277–286.

Merle, O., and B. Guillier, 1989, The building of the Central Swiss Alps: an experimental approach: Tectonophysics, v. 165, p. 41–56.

Milnes, A. G., 1978, Structural zones and continental collision, Central Alps: Tectonophysics, v. 47, p. 369–392.

Narr, W., and J. Suppe, 1993, Kinematics of basement involved compressive structures: American Journal of Science, v. 294, p. 302–360.

Price, R., 1986, The southeastern Canadian Cordillera: thrust faulting, tectonic wedging and delamination of the lithosphere: Journal of Structural Geology, v. 8, p. 239–254.

Roe, L. E., 1991, Petroleum Exploration in the Amadeus Basin, *in* R. J. Korsch and J. M. Kennard, eds., Geological and geophysical studies in the Amadeus Basin, central Australia: Australian Bureau of Mineral Resources, Geology and Geophysics Bulletin, v. 236, p. 463–476.

Royden, L., and G. D. Karner, 1984, Flexure of the lithosphere beneath the Apennine and Carpathian foredeep basins: Nature, v. 309, p. 142–144.

Sandiford, M., and M. Hand, 1998, Controls on the locus of intraplate deformation in central Australia: Earth and Planetary Science Letters, v. 162, p. 97–110.

Schmidt, C. J., P. W. Genovese, and R. B. Chase, 1993, Role of basement fabric and cover-rock lithology on the geometry and kinematics of twelve folds in the Rocky Mountain foreland, *in* C. J. Schmidt, R. B. Chase, and E. A. Erslev, eds., Laramide basement deformation in the Rocky Mountain foreland of the Western United States: Geological Society of America Special Paper 280, p. 1–44.

Schroder, R. J., and J. D. Gorter, 1984, A review of recent exploration and Hydrocarbon potential of the Amadeus Basin Northern Territory: Australian Petroleum Exploration Association Journal, v. 24, p. 19–41.

Scrimgeour, I., and D. Close, 1999, Regional high-pressure metamorphism during intracratonic deformation: the Petermann Orogeny, central Australia: Journal of Metamorphic Geology, v. 17, p. 557–572.

Shaw, R. D., R. J. Korsch, C. Wright, and B. R. Goleby, 1991a, Seismic interpretation and thrust tectonics of the Amadeus Basin, central Australia, along the BMR regional seismic line, *in* R. J. Korsch and J. M. Kennard, eds., Geological and geophysical studies in the Amadeus Basin, central Australia: Australian Bureau of Mineral Resources, Geology and Geophysics Bulletin, v. 236, p. 385–409.

Shaw, R. D., M. A. Etheridge, and K. Lambeck, 1991b, Development of the Late Proterozoic, intracratonic Amadeus Basin in central Australia: a key to understanding tectonic forces in plate tectonics: Tectonics, v. 10, p. 688–721.

Shaw, R. D., P. K. Zeitler, I. McDougall, and P. R. Tingate, 1992, The Palaeozoic history of an unusual intracratonic thrust belt in central Australia based on ^{40}Ar-^{39}Ar and fission track dating: Journal of the Geological Society, London, v. 149, p. 937–954.

Spang, J. H., and J. P. Evans, 1988, Geometrical and mechanical constraints on basement-involved thrusts in the Rocky Mountains foreland province, *in* J. C. Schmidt and W. J. Perry Jr., eds., Deformation style of Laramide uplifts in the Wyoming foreland: Geological Society of America Memoir 171, p. 41–52.

Stewart, A. J., R. Q. Oaks Jr., J. A. Deckelman, and R. D. Shaw, 1991, 'Mesothrust' versus 'megathrust' interpretations of the structure of the northeastern Amadeus Basin, central Australia, *in* Korsch, R. J. and J. M. Kennard, eds., Geological and geophysical studies in the Amadeus Basin, central Australia: Australian Bureau of Mineral Resources, Geology and Geophysics Bulletin, v. 236, p. 361–385.

Stewart, A. J., R. D. Shaw, and J. F. Lindsay, 1993, Cross sections of the Amadeus Basin, *in* J. F. Lindsay, ed., Geological Atlas of the Amadeus Basin: Australian Geological Survey Organisation, p. 4.

Stockmal, G. S., P. A. Mackay, D. C. Lawton, and D. Spratt, 1996, The Oldman River triangle zone: a complicated tectonic wedge delineated by new structural mapping and seismic interpretation, *in* P. A. MacKay, T. E. Kubli, A. C. Newson, J. L. Varsek, R. G. Dechesne, and J. P. Reid, eds., Triangle zone geometry, terminology and kinematics: Bulletin of Canadian Petroleum Geology, v. 44, p. 202–214

Teyssier, C., 1985, A crustal thrust system in an intracratonic tectonic environment: Journal of Structural Geology, v. 7, p. 689–700.

Walter, M. R., J. J. Veevers, C. R. Calver, and K. Grey,

1995, Neoproterozoic stratigraphy of the Centralian Superbasin, Australia: Precambrian Research, v. 73, p. 173–195.

Watts, A. B., 1992, The effective elastic thickness of the lithosphere and the evolution of foreland basins: Basin Research, v. 4, p. 169–178.

Wells, A. T., D. J. Forman, L. C. Ranford, and P. J. Cook, 1970, Geology of the Amadeus Basin, Central Australia: Bureau of Mineral Resources, Geology and Geophysics Bulletin, v. 100, p. 1–222.

Wright, C., B. R. Goleby, R. D. Shaw, C. D. N. Collins, R. J. Korsch, T. Barton, S. A. Greenhalgh, and S. Sugiharto, 1991, Seismic reflection and refraction profiling in central Australia: implications for understanding the evolution of the Amadeus Basin, *in* R. J. Korsch and J. M. Kennard, eds., Geological and geophysical studies in the Amadeus Basin, central Australia: Australian Bureau of Mineral Resources, Geology and Geophysics Bulletin, v. 236, p. 41–59.

Zapata, T. R., and R. W. Allmendinger., 1996, Thrust-front zone of the Precordillera, Argentina: a thick-skinned triangle zone: AAPG Bulletin, v. 80, p. 359–381.

Zoetemeijer, R., P. Desegaulx, S. Cloetingh, F. Roure, and L. Moretti, 1990, Lithospheric dynamics and tectonostratigraphic evolution of the Ebro Basin: Journal of Geophysical Research, v. 95, p. 2701–2711.

28

Turrini, C., and P. Rennison, 2004, Structural style from the Southern Apennines' hydrocarbon province—An integrated view, *in* K. R. McClay, ed., Thrust tectonics and hydrocarbon systems: AAPG Memoir 82, p. 558–578.

Structural Style from the Southern Apennines' Hydrocarbon Province—An Integrated View

C. Turrini[1]
Totalfina Italia, Milan, Italy

P. Rennison[2]
Totalfina Italia, Milan, Italy

ABSTRACT

The Campano-Lucano Arch is a major hydrocarbon province in the Southern Apennines of Italy. To provide new understanding of the thrust-belt architecture while constraining the derived interpretation of buried structure, we have analyzed data from seismic-reflection, wells, surface geology, 3-D computerized reconstructions, and sandbox simulations and integrated these data into a regional-scale structural model. Results show that refolding of already folded and thrusted units models the current structural setting, in which shallow and deep structures interact to form the recognized tectonic framework. In the region, recent folding and thrusting of the deeper autochthonous substratum reworked an early-phase vertical allochthonous assemblage, which ultimately behaved like a new pseudo-multilayered stratigraphy. Three-dimensional strain partitioning is moderately homogeneous across the thrust belt's internal domain, because shallow and deep structures show minor mechanical disharmony. Conversely, distribution of the deformation is strongly inhomogeneous across the thrust belt's external and marginal domains, both vertically and horizontally (along strike). Interference, through time and space, between the related noncylindrical, shallow (the "Apenninic" Thrust Belt) and deep (the "Apulian" Thrust Belt) contractional systems has developed a "dome and basin" structural pattern that characterizes the current tectonic architecture and that may localize, at depth, the main oil accumulations in the region.

INTRODUCTION

Since the middle 1980s, the Southern Apennines of Italy (Figure 1) have been proved to represent an important hydrocarbon province in the western Mediterranean area. As different oil and gas discoveries have been made throughout the region, a number of geologic models have attempted to fit the results progressively provided by the exploration activity (Mostardini and Merlini, 1986; Casero et al., 1988; D'Andrea et al., 1993;

[1]*Present affiliation*: Total, Paris, France.
[2]*Present affiliation*: Totalnorge A.F., Oslo, Norway.

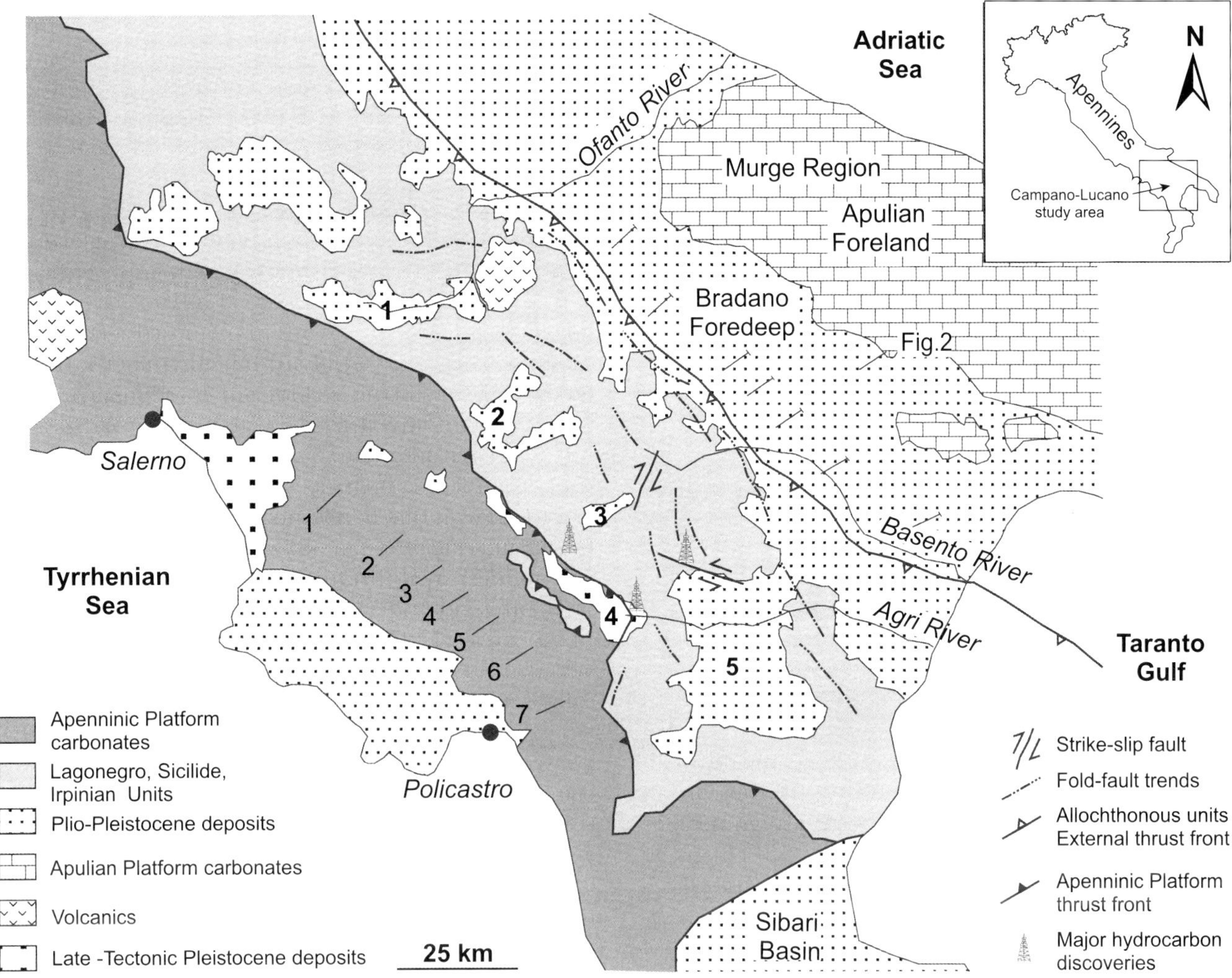

FIGURE 1. Regional surface geology framework. Locations are shown for the serial cross sections illustrated in Figures 5 and 8. 1 = Ofanto Basin, 2 = Potenza Basin, 3 = Anzi-Calvello Basin, 4 = Tanagro-Val D'agri Basin, 5 = Sant'Arcangelo Basin.

Finetti et al., 1996; Pasi et al., 1995; Lentini et al., 1996; Ferranti et al., 1997; Menardi Noguera and Rea, 1997; Bonini and Sani, 1998; Monaco et al., 1998). Despite their many important differences, all of the models describe the Campano-Lucano Arch as an arch-shaped, west-southwest to east-northeast displaced, fold-and-thrust belt in which contractional and extensional events have overprinted each other through geologic time to form the current mountain-chain structure. The related structural style indicates tectonism that is mainly thin-skinned (basement not involved in the external domains of the chain, to the northeast) to possibly thick-skinned (basement involved in the internal domains of the chain, to the southwest). An upper, allochthonous wedge of carbonate/clastic units (Apenninic Platform/Lagonegro/Sicilide/Irpinian units) is thrusting great distances onto the associated autochthonous deformed units of Apulian Platform carbonates (Figure 2). Late phases of compression and extension-related dynamics have controlled the recent activity within the orogen, and horizontal (transtensional, transpressional) movements have been reported throughout both its internal and external domains. Development of this tectonic system spans the period from the early Miocene (mainly compression only) to the late Pliocene-Pleistocene (extension and compression), during which multiple stages in the crustal-scale deformation led to a generic west-to-east migration of the associated chain-foredeep-foreland structural and sedimentary geometries (Figure 3).

Hydrocarbon distribution in the Campano-Lucano Arch directly relates to the geologic system mentioned above. In such a structural-stratigraphic context, the main exploration target is the autochthonous carbonate units of the Apulian Platform, which contain the largest and deepest (3000–5000 m below sea level) oil fields in the region. In addition, a number of shallow gas fields are concentrated in the Pliocene-Pleistocene foredeep area, corresponding with the advancing Southern Apennines' external front (Sella et al., 1988).

Although industry-driven fieldwork has recently produced detailed analyses of localized surface structures, a generic stratigraphy-based approach has been the key concept in past interpretations of the regional-scale geologic framework. This had the benefit of an extremely refined classification of the sedimentary formations but underestimated the geometric description and reconstruction of the structural families that characterize the region. As such, the many large-scale cross sections presented in the regional literature to date mainly addressed, with only a few exceptions, the possible relationships among the tectonic units deformed within the large-scale Southern Apennines' stack. Simultaneously, numerous and sometimes conflicting paleogeographic hypotheses regarding the pre-Apennines' setting have been proposed (D'Argenio, 1988).

This chapter presents and discusses the deformation geometries that model the Campano-Lucano Arch. Given such an objective, large simplifications have been made about the lithostratigraphy of the area as it has been reconstructed by many authors. Conversely, particular emphasis has been placed on the geometric analysis of the deformation features. We have considered the three-dimensional (3-D) reconstruction of the deep Apulian Platform carbonate structures to be the principal goal of this study. Nevertheless, we also discuss the related mechanics and kinematics, but they have only been approximated here, because they are part of ongoing analysis within Totalfina Italia.

To constrain the geometric reconstruction, we have integrated data from wells, surface geology, 3-D computerized reconstructions, and sandbox simulations into a methodology that aims to establish a regional and three-dimensional compatibility among the many and different deformational features. The structural geometries that characterize the allochthonous tectonic wedge have been defined mainly from surface geology (data from published literature, available geological maps, and an in-house field study). Seismic interpretation and well information in the area supplied the elements to link the shallow allochthonous units to the deeply buried autochthonous sequence. Integration of 2-D profile data into a 3-D model was performed to image the regional tectonic fabric of the target: the top Apulian Platform. Finally, sandbox modeling has been used to test and observe how far the mechanical tectonostratigraphy recognized in the region might have partitioned the development of structures within the two major (allochthonous and autochthonous) structural levels. It should be noted that, because many of the data presented are strictly confidential, names and locations of wells and seismic lines have been omitted, and the geologic cross sections' positions are only shown approximately.

THE MECHANICAL STRATIGRAPHY IN THE REGION

The mechanical stratigraphy recognized from the entire data set largely accounts for the wide range of deformation structures, mechanics, and kinematics in the study region. From southwest to northeast of the

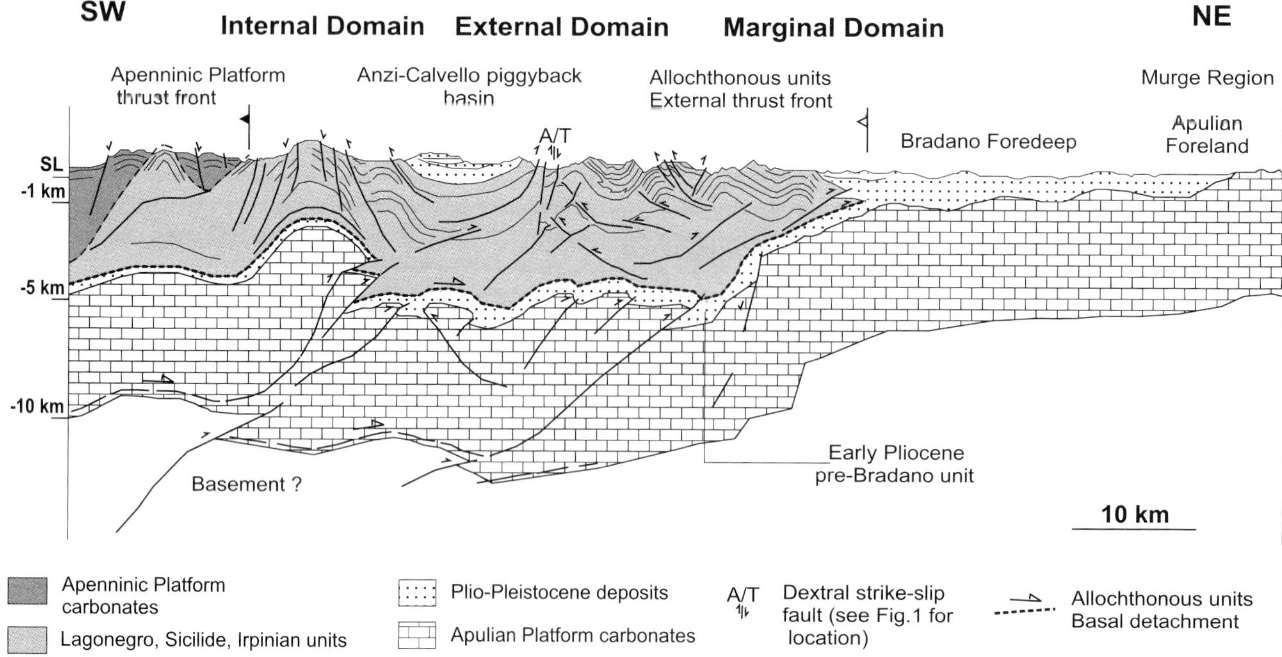

FIGURE 2. Structural cross section through the study area. See Figure 1 for location. Vertical scale is exaggerated approximately 2:1.

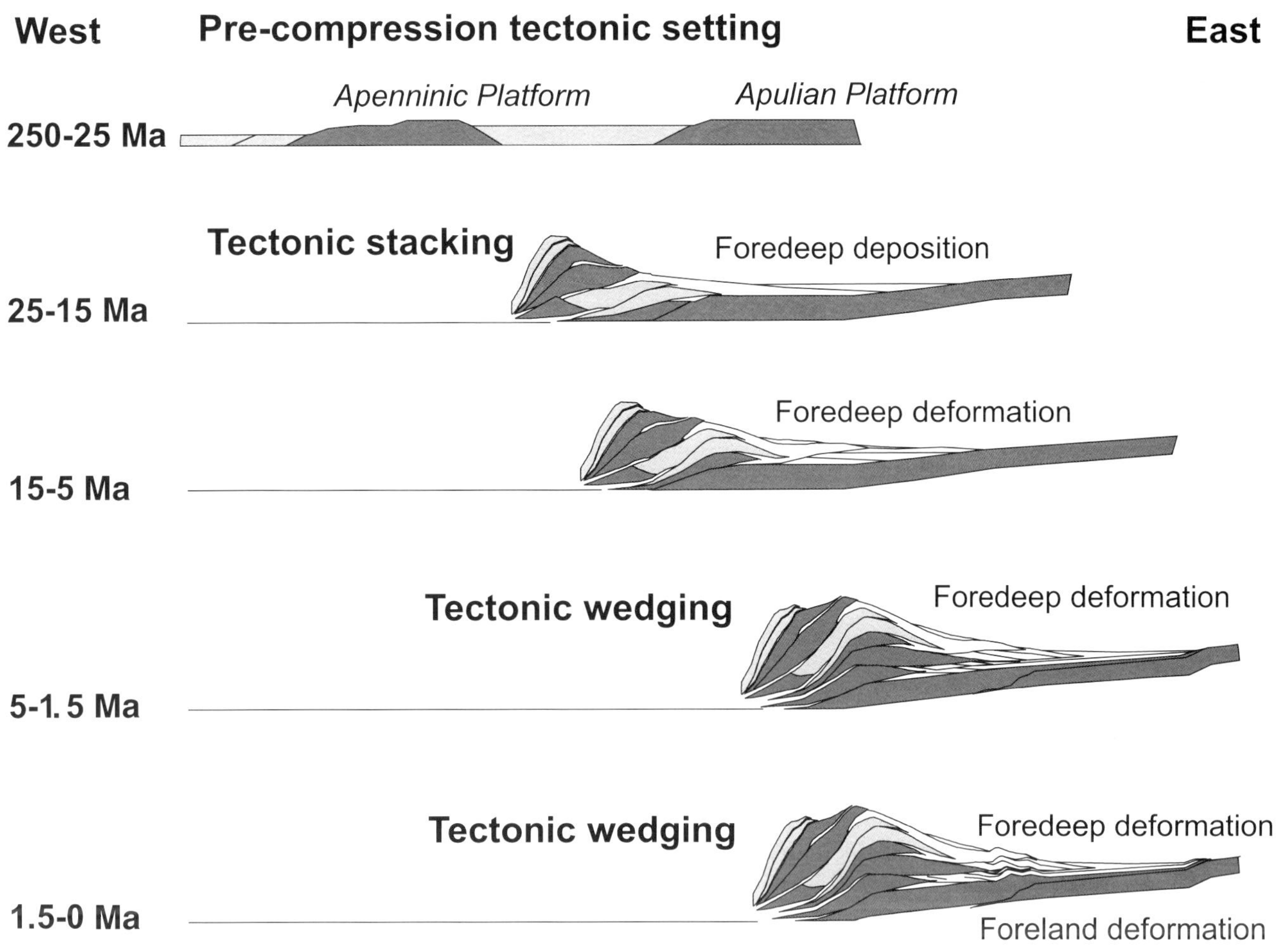

Figure 3. Southern Apennines simplified kinematics and west-to-east migration of the associated chain-foredeep-foreland structural and sedimentary geometries.

Campano-Lucano Arch (Figure 4; see also Figure 2), three principal tectonostratigraphic domains can be recognized: internal, external, and marginal domains, with respect to the regional architecture and the related southwest-to-northeast-displacement direction. The derived structural-lithologic assemblages suggest a vertical-strength profile that differs as we move across the chain from the internal to the marginal domains (see Figure 4), particularly when the thickness of the various units is considered.

The internal domain's strength profile (Figure 4a) suggests the existence of a series of stacked thrust slices confined above and below by massive carbonate units (the Apenninic Platform [partly eroded] and the Apulian Platform, respectively). The external domain's strength profile (Figure 4b) indicates a possible top (Sicilide and Irpinian unit or, locally, Pliocene piggyback clastics) to base (Apulian Platform) increase in mechanical competence and a lack of mechanical confinement above. The strength profile for the marginal domain (Figure 4c) is different from that of the external domain because of the drastically reduced thickness of the allochthonous unit (Irpinian-external flysch unit), which rapidly thins to the east–northeast toward the foreland area. Despite the observed differences, multiple weak horizons that are common to all three of the domains' stratigraphies (internal, external, and marginal) break the vertical mechanical continuity of the packages recognized. There is clear evidence for a major regional detachment surface (the allochthonous basal detachment) at the top of the uppermost autochthonous unit, the lower Pliocene clastics (see Figures 2 and 4), on which the allochthonous clastic units are decoupled and displaced. In terms of mechanical behavior, the basement's role in the region still remains uncertain and debatable among researchers. Depth to the magnetic (crystalline?) basement in the external and marginal domains, indicated by crustal-scale data (see Mostardini and Merlini, 1986, their cross sections 6 to 3), varies from 9 to 13 km and becomes shallower from north to south toward the Sant'Arcangelo Basin area. Although such data also suggest a generic northwest-to-southeast shallowing of the basement within the internal domain, its true presence and position at depth remains poorly constrained.

2-D STRUCTURAL GEOMETRIES IN THE ALLOCHTHONOUS UNITS

Surface geology indicates that the main structural geometries in the allochthonous units are ovrethrusts, thrusted folds, and ramp-flat thrust surfaces (Figure 5). Detailed field mapping across the region has shown that the overthrust surfaces between the different tectonic units have been recently folded and breached to model the current structural pattern (Figure 5)(Scandone, 1972; Carbone et al. 1991; Catalano, 1993; Catalano et al., 1993; Bonini and Sani, 1998). The related trends are generally oriented northwest to southeast (see fold-fault trends on Figure 1), yet an important curvature in the outcropping structures is evident near the Ofanto River in the northwestern sector of the study area, where both the deformation and the sedimentary geometries strike west–east. In addition to such a clear and progressive deviation from the regional northwest–southeast strike, other important interruptions along the fold-fault tectonic fabric are recognized at the surface. These correspond to local-scale Pliocene-Pleistocene sedimentary basin areas that are mildly deformed and generally passively transported on top of the tectonic stack as piggyback basins (see Figure 5, cross section 4; see also Figure 1, the Ofanto, Potenza, Anzi-Calvello Basins).

In the internal domain of the study area, the allochthonous unit is represented by the Apenninic Platform carbonates, which have deformed by thrusting to the northeast (Figure 5, cross sections 3–6; see Figure 1 for relative locations). Despite their large-scale tectonic significance as regionally thrusted units, their currently recognizable deformation geometries are mainly southwest-dipping (and northeast-dipping subordinately) normal faults. In the hinterland, these account for a regional collapse of the chain toward the Tyrrhenian coast (to the southwest). Directly at the footwall of the Apenninic Platform carbonate unit overthrust, the sediments of the Lagonegro unit are deformed and imbricated onto and into the external domain of the chain. Here, and across

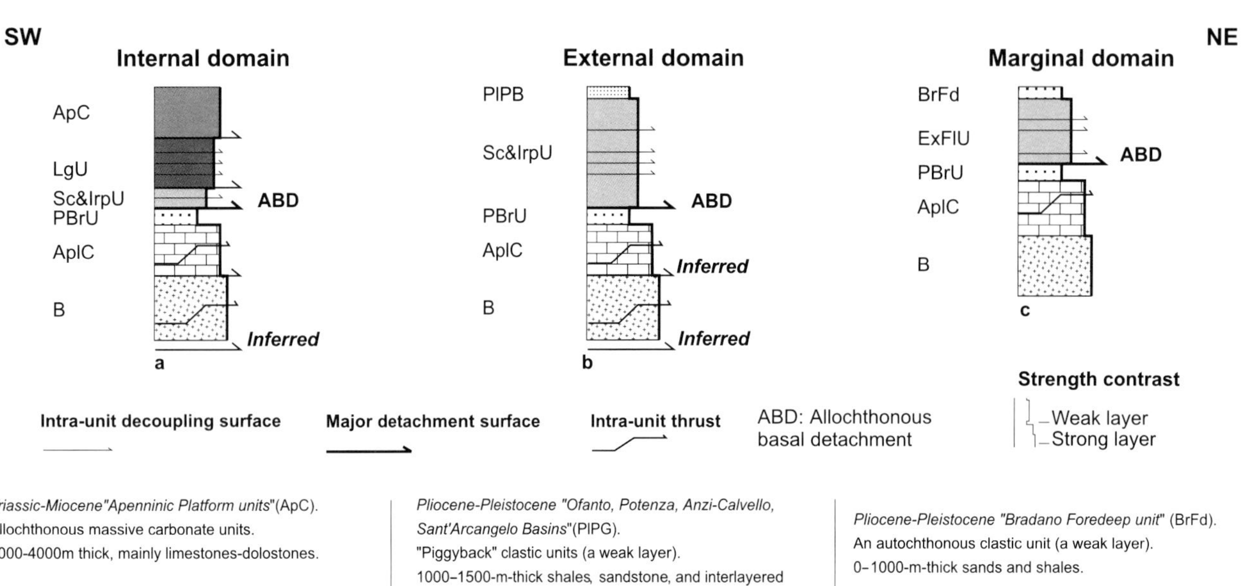

FIGURE 4. Mechanical tectonostratigraphy in the Campano-Lucano region. See Figure 2 for domain location.

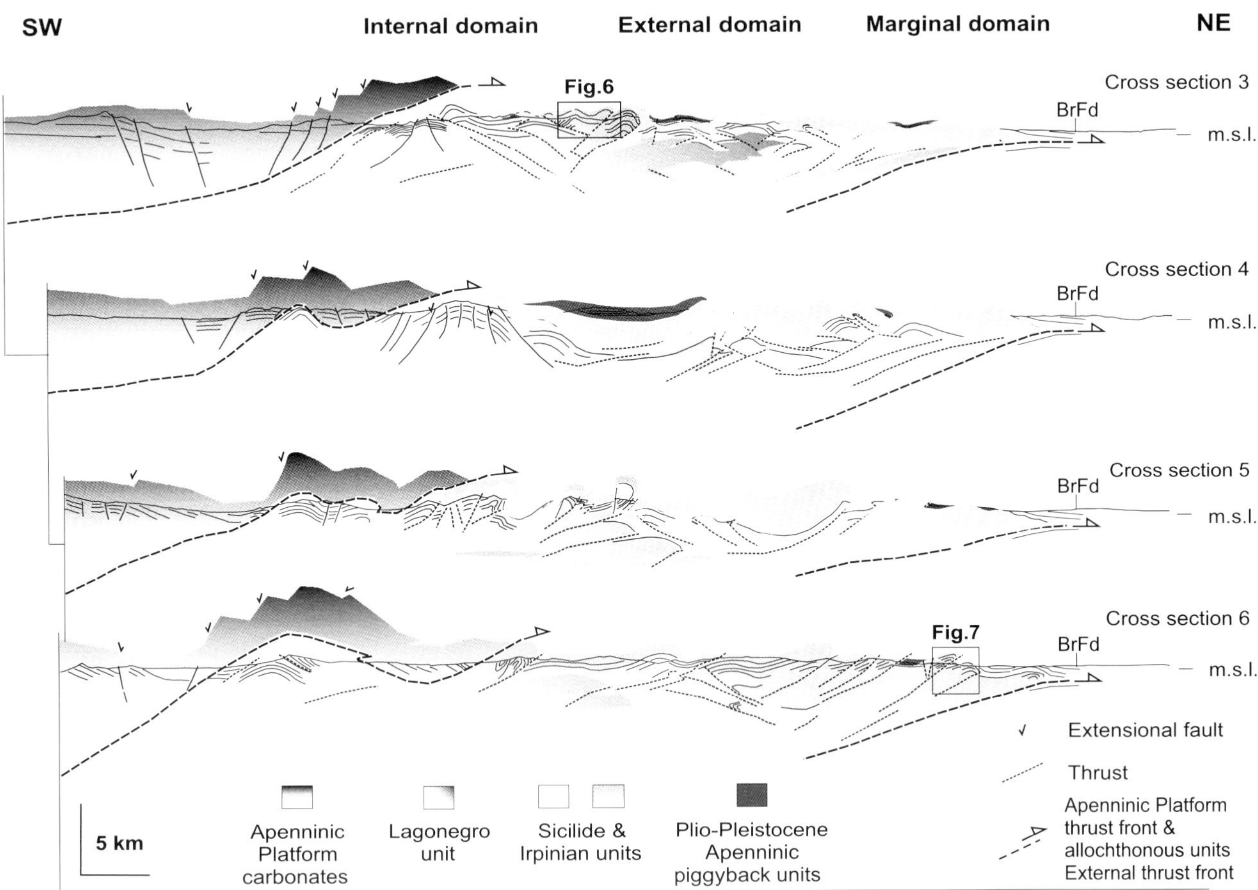

FIGURE 5. 2-D deformation geometries in the allochthonous units. Shallow structures in the external and marginal domains are simplified after Bonini and Sani (1998). Shallow structures in the outer internal domain are simplified after Scandone (1972) and Carbone et al. (1991). BrFd = Bradano Foredeep. See Figure 1 for location.

the marginal domain as well, the Sicilide and Irpinian sequences are structurally mainly folds and thrusts. Folds demonstrate different dimensions in structural wavelength (3–10 km), and they show transport directions toward both the hinterland (southwest) and the foreland (northeast) of the orogen. The derived pop-up geometries are both horizontally displaced and vertically extruded. As such, reverse-fault surfaces are crosscutting through the folded structures, while contractional stacking of the deformed units is widespread over all the region. Major strike-slip faults orient obliquely to the fold-fault regional pattern (see Figure 1). On seismic sections (Figures 6 and 7), the allochthonous units appear to be highly disturbed discontinuous packages that are deformed by minor folds and major thrusts. Folds, although they are broken and shattered, can be interpreted, and the related, particularly well-imaged, long-limb profiles can be useful for determining structures' asymmetry and vergence. Among the tectonic units described, the Lagonegro units provide some characteristically strong seismic reflections (Figure 6). However, below the Apenninic Platform units, the overall seismic response is practically masked, with local exceptions.

At the very front of the outcropping orogen (marginal domain, east-northeast part of the study area), the clastic allochthonous units (Irpinian external flysch) are wedging below the Pliocene-Pleistocene Bradano foredeep basin sediments (see Figure 5, northeast domains of each cross section). Additionally, both in seismic data (Figure 7) and in the field, we can clearly observe some tectonic imbrication within the allochthonous structures. In this structural situation, stacking of the deformed sediments can be interpreted thanks to the multiple thrusting surfaces and steepening of the sedimentary packages in the area (see Figure 6d).

2-D STRUCTURAL GEOMETRIES IN THE AUTOCHTHONOUS UNITS

Interpretation of the in-house available seismic grid and integration from some published data (Sella et al., 1988; Hippolyte et al., 1994) have resulted in the 2-D construction of the structural geometries at depth at the

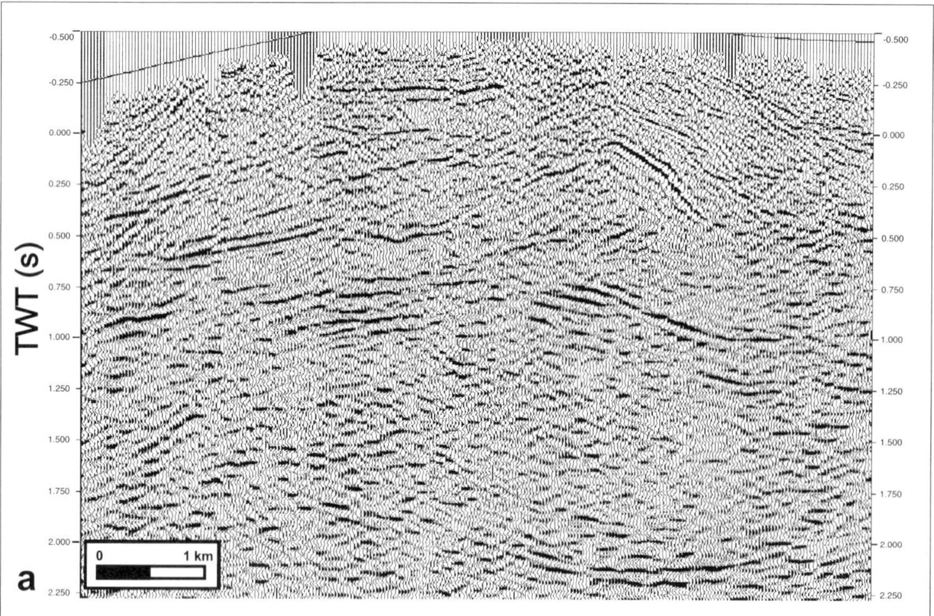

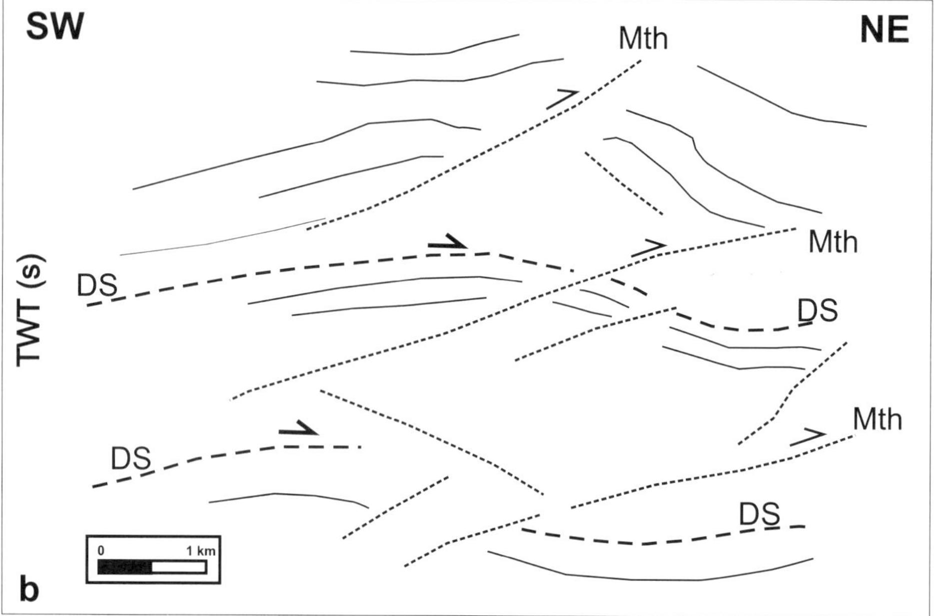

Figure 6. (a) Seismic expression of the Lagonegro Unit. (b) Tracing of Figure 6a. DS = Intra-unit decoupling surfaces. Mth = major thrusts. See Figure 5 for location.

top of the autochthon. Sensu strictu, the top of the autochthon is the top of the early Pliocene "Pre-Bradano" clastic unit (see Figures 2 and 4), as discussed above. This unit is thin to absent in the study area, so, for the purposes of this reconstruction, the top Apulian Platform carbonates have been regarded as the top of the autochthon. Heterogeneity of the seismic-velocity distribution has been considered to constrain depth conversion of the Apulian target in the area (see Figure 8 for referenced seismic-velocity values). Consequently, the derived geometries have been tied to the available well stratigraphies. Selected seismic images, although isolated, have also been used to validate the possible top-to-the-magnetic-basement depth in the region. This has constrained the eventual depth to the base of the Apulian unit and the thickness of the Apulian carbonates (2 s TWT = 5000 to 5500 m, and is surprisingly constant across the region).

In cross section (Figure 8; see Figure 1 for relative locations), the overall regional surface dips toward the southwest, below the allochthonous wedge (see also Figure 2). The inclination of this regional surface generally increases from the southeast to the northwest of the region (Figure 8, 4° in cross sections 5–7, 5–6° in cross sections 1–3), whereas local steepening of the top Apulian surface can be observed (Figure 8, cross section 3). The deformation geometries seem to relate to fault-bend folds in the internal domain of the study area, to faulted folds (i.e., folds that are cross-cut by thrusts) in the external domain, and occasionally, to compressionally deformed extensional terraces in the marginal domain (Figure 8, cross sections 5–6).

The folds in the internal domain show weakly rounded to gently flat hinge geometries, southwest-dipping, undeformed long-limb geometries, and faulted short-limb geometries (Figure 8). The faults are unequivocally southwest-dipping, and their inclination is approximately 20–40°.

Within the external domain, pop-up features appear that show weak structural asymmetry (Figure 8, cross sections 3–6). Nevertheless, they can be highly deformed, because tectonic imbrication can occur locally (Figure 9; see also Figure 8, cross section 6). Reverse faults dip 30–45° and are locally arranged in conjugate sets of southwest-dipping forethrust and northeast-dipping back-thrust

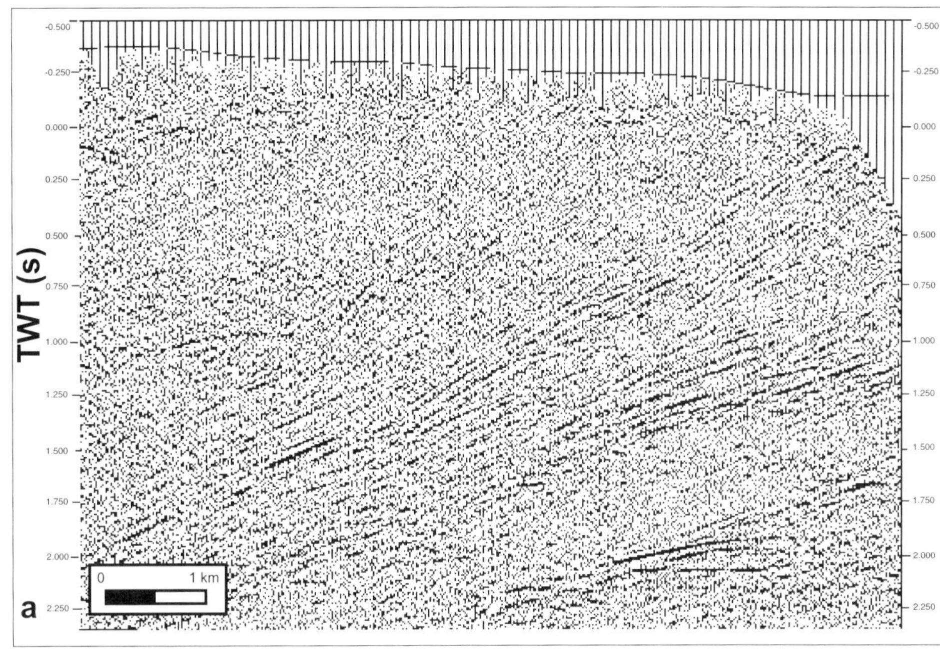

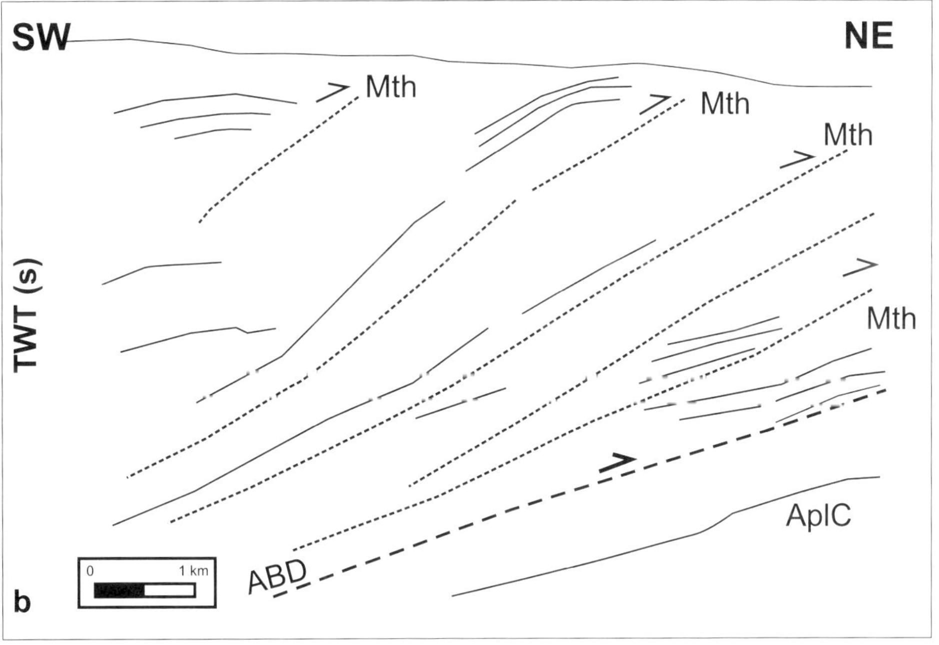

FIGURE 7. (a) Seismic expression of tectonic imbrication in the Irpinian external flysch units. (b) Tracing of Figure 7a. ABD = allochthonous basal detachment, Mth = major thrusts, AplC = top Apulian carbonates, See Figure 5 for location.

surfaces (Figure 8, cross sections 3, 5, and 6; Figure 9) with respect to the regional tectonic-transport direction. Detailed 3-D structural reconstructions have indicated that preexisting extensional faults (not presented in this paper) are locally inverted, and are sometimes cut by the newly formed compressional discontinuities and/or are passively transported by the related fold geometries (Figure 9).

The marginal domain is mainly deformed by normal faults that dip to the southwest and account for some extensional terrace development. In the southeast of this domain, shortcut faulting and initial folding of the top Apulian surface have been interpreted (Figure 8, cross sections 5–7) by the use of selected seismic images. Within the external-margin area, thrust trajectories show ramp geometries that flatten within or near the base of the estimated Mesozoic stratigraphic thickness, that is, above basement (Figure 8; see also Figure 2). Nevertheless, in the internal domain, a possible involvement of the pre-Mesozoic unit (basement?) could be invoked to honor the perceptible shallowing of the regional enveloping surface.

INSIGHTS FROM MECHANICAL SANDBOX MODELS

To support the deformation geometries described here, we compared the observed structures and some analog sandbox models drawn from the Totalfina Italia 1998 experiments catalog.

To approximate the vertically nonhomogeneous tectonostratigraphy interpreted in the study area (see above), mechanically strong layers of colored sand were alternated vertically with weak layers of glass microbeads, the latter being intended to localize disharmonic deformation and to model the major decoupling surfaces inferred from the regional tectonostratigraphic assemblage. For most of the experiments, a predeformation wedge-shaped initial configuration (topographic slope of 1–3°) was prepared to simulate an advanced stage in the Southern Apennines' thrust-belt evolution. This advanced stage was intended to represent the tectonic

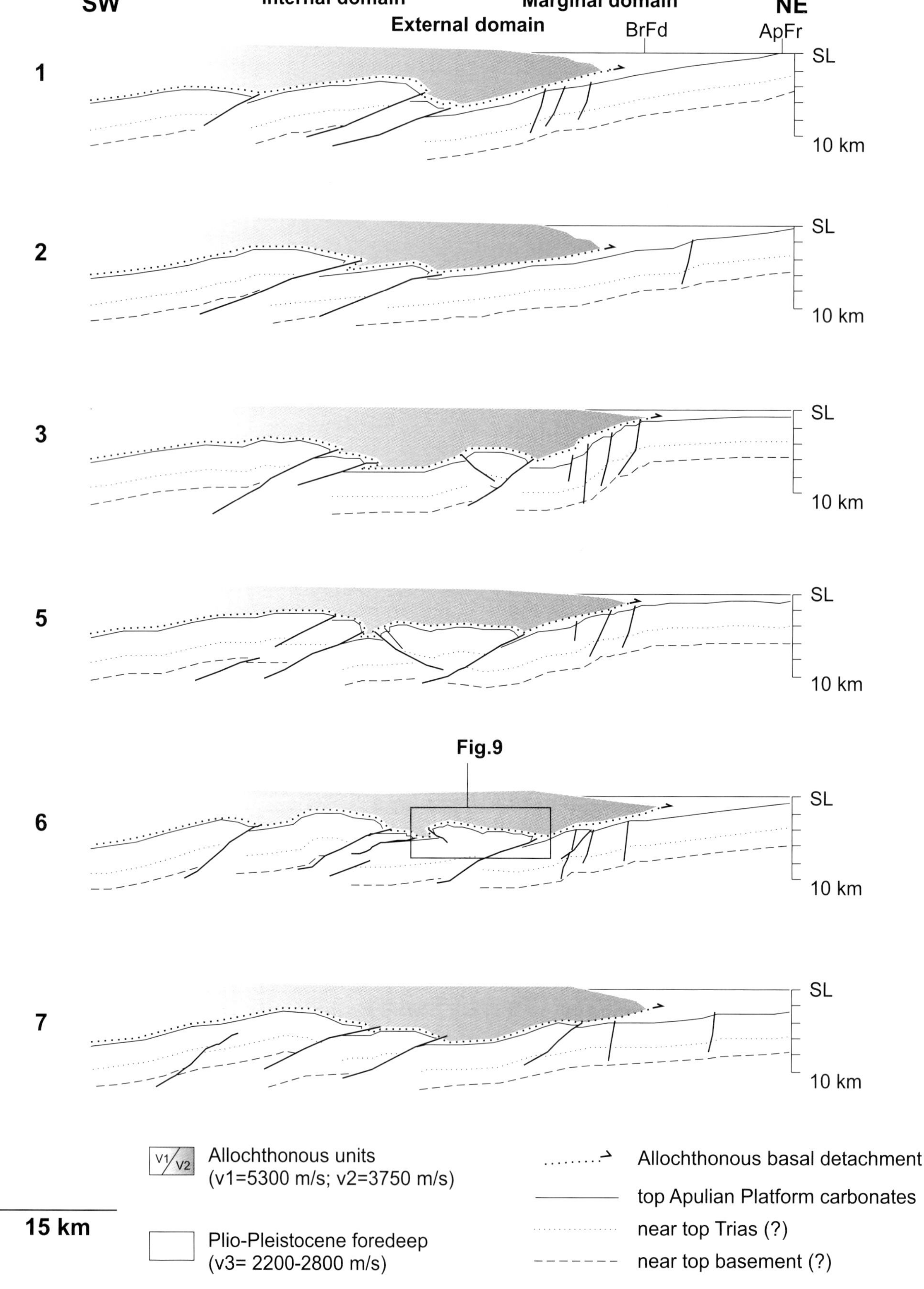

FIGURE 8. 2-D deformation geometries in the autochthonous units (Apulian Platform carbonates). v1, v2 = average seismic velocity for the allochthonous units, v3 = average seismic velocity for the Pliocene-Pleistocene sediments.

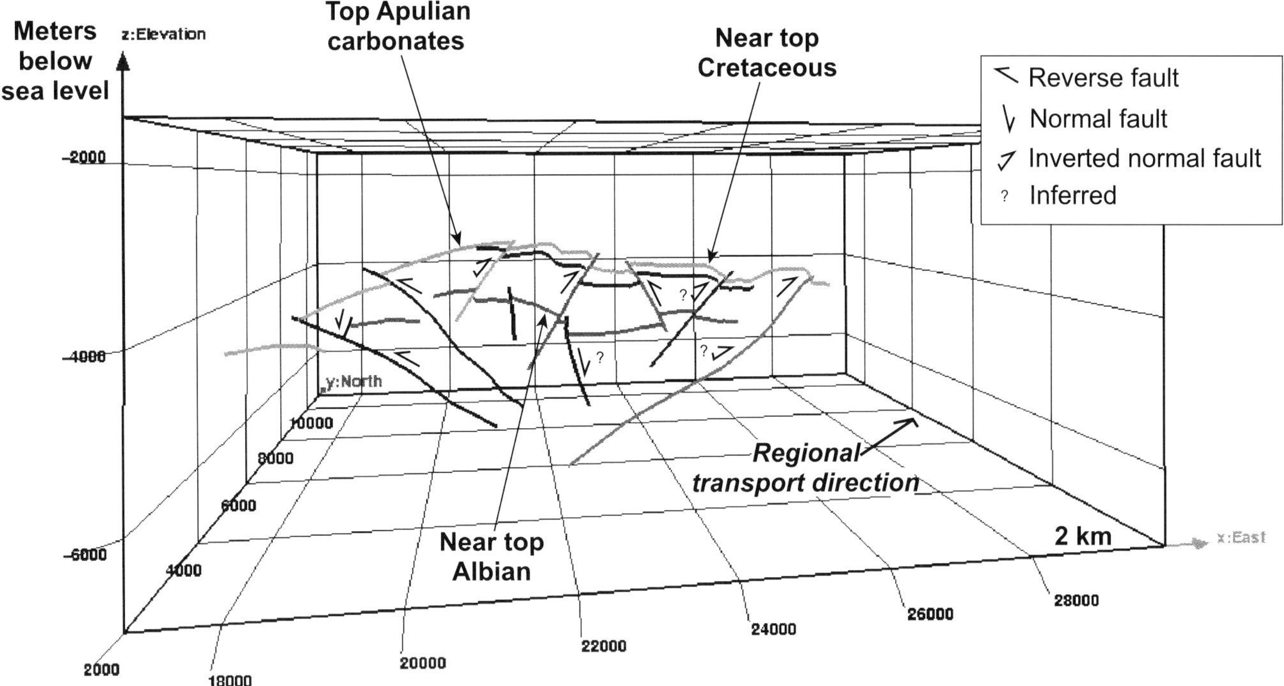

FIGURE 9. Fault pattern and imbrication in the Apulian Carbonates derived from detailed 3-D structure interpretation within the external domain. See Figure 8 for location. Note erosional truncation of the top Cretaceous horizon below the Apulian carbonate unconformity. Evidence for extensional faults within the structure is derived from interpretation of well stratigraphy and structural analysis.

framework that may have followed the emplacement of the allochthonous units onto the undeformed (or gently deformed) Apulian Platform carbonate foreland, and also anticipated the final tectonic stacking and imbrication in the region (see below). An "allochthonous" unit was constructed to overlie the "autochthonous" foreland from which it was mechanically separated by a weak decoupling layer. The "autochthonous" unit was modeled to be already flexured, draping over the underlying rigid basement. To replicate possible variations along strike in the foreland's rigid basement morphology, an undeformable oblique ramp was positioned on one side of the experimental apparatus (i.e., on the far side of the advancing backstop). Finally, deposition of new material (sand and microbeads) was performed during some of the experiments to simulate syntectonic sedimentation occurring at the front of the developing synthetic chain sector. Conversely, material was occasionally arbitrarily removed from the innermost part of the models (near the advancing backstop) to represent erosion of the growing structures at the top of the deforming tectonic stack.

Because a full description of the methodology and of the results from the sandbox models performed is outside the scope of this chapter (see Turrini et al., 2001 for complete discussion), we here present only the more interesting considerations derived from comparing the results of the mechanical experiments performed with the interpreted regional structures.

In particular, given the initial similarity established between the experimental and the natural boundary conditions (a similar mechanical tectonostratigraphy and an analog foreland autochthonous regional geometry), a comparative analysis between the selected sandbox models and 2-D structural geometries interpreted from the Campano-Lucano Arch suggests the following (Figures 10–12).

The tectonostratigraphy interpreted within the thrust belt and thus simulated by the models represents a key element in the deformation partitioning of the derived structures (Figures 10–12). Indeed, such a stratigraphy has likely enhanced mechanical decoupling between the observed major structural levels. Minor faulted folds and imbricates develop within the allochthonous units (i.e., the shallow structural level above the microbead layer in the model) as they detach from their substratum. Conversely, major anticlines deform the autochthonous units (i.e., the deep structural level below the microbead layer in the model) while fore thrusts and back thrusts (with respect to the "regional" transport direction) cut the derived geometries. As part of the described mechanics, the model's microbead layer

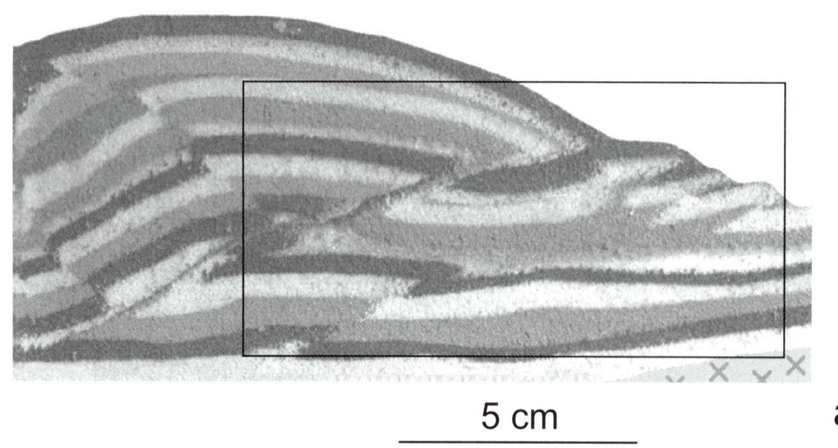

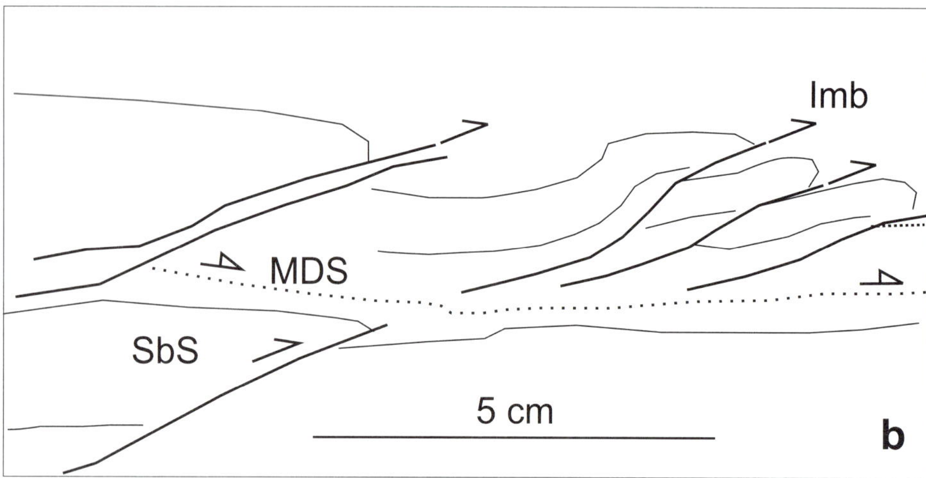

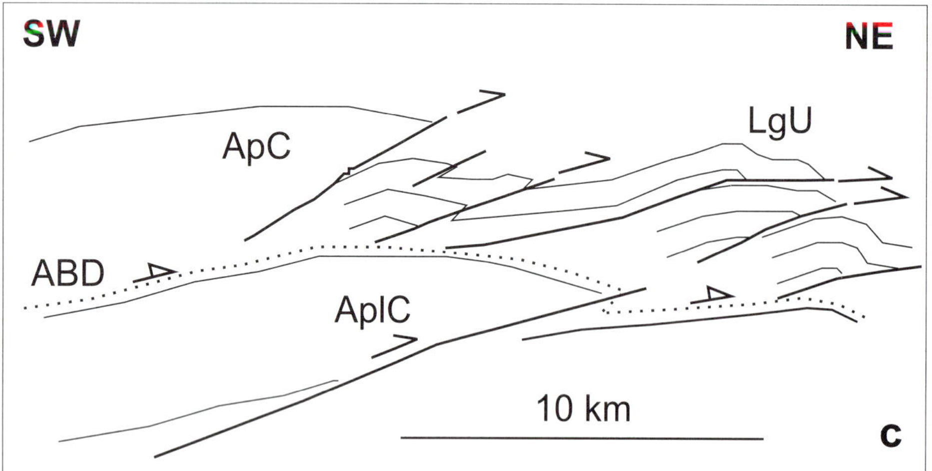

FIGURE 10. (a) Sandbox model from the Totalfina Italia 1998 experiments catalogue. (b) Tracing of Figure 10a. MDS = major decoupling surface (microbead layer). Imb = contractional imbricates. SbS = subthrust structure. (c) Structures from the Campano-Lucano internal domain (see box "A" in Figure 16 for location). ApC = Apenninic Platform carbonates, AplC = Apulian Platform carbonates, LgU = Lagonegro units, ABD = Allochthonous basal detachment.

and the autochthonous pre-Bradano Pliocene deposits represent the mechanically weak layer that allowed the observed decoupling to take place (Figures 10b and c and 11b and c).

Both in the model experiment and in the Campano-Lucano region (Figures 10 and 11), the mechanical decoupling between the shallow and the deep structural units seems to increase from the hinterland (left-hand

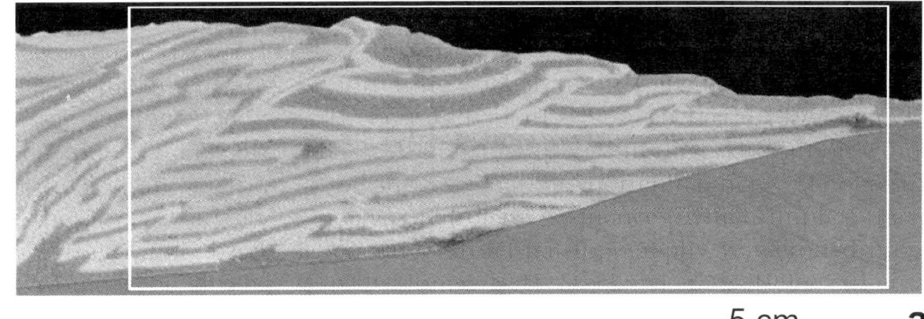

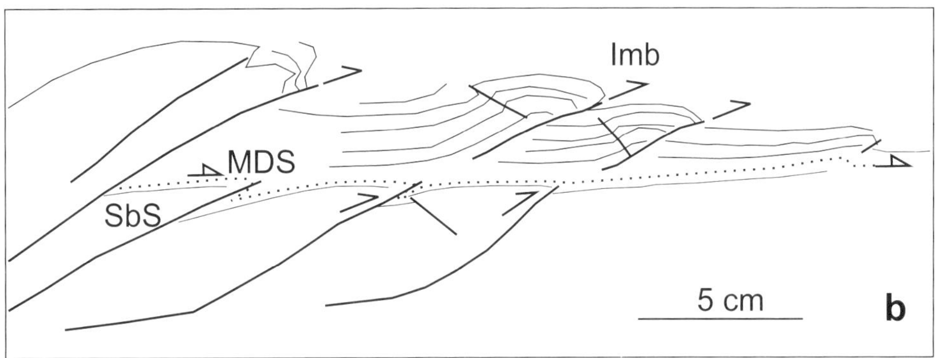

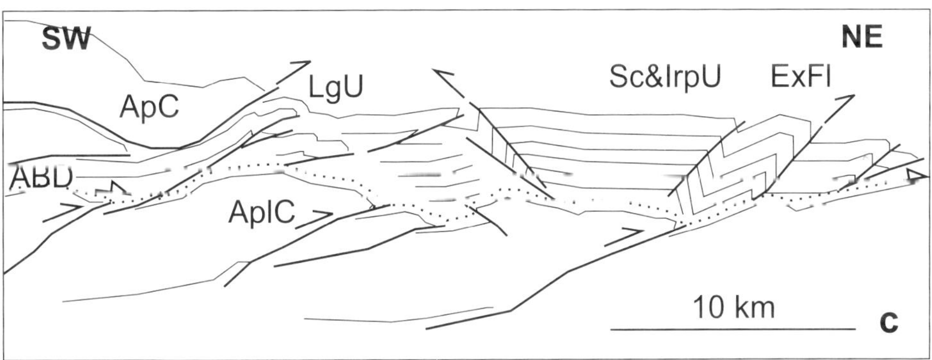

Figure 11. (a) Sandbox model from the Totalfina Italia 1998 experiments catalog. (b) Tracing of Figure 11a. MDS = major decouplng surface (microbead layer), Imb = contractional imbricates, SbS = subthrust structure. (c) Structures from the Campano-Lucano Thrust Belt (See box "B" in Figure 16 for location). ApC = Apenninic Platform carbonates, AplC = Apulian Platform carbonates, LgU = Lagonegro units, ScandIrpU = Sicilide and Irpinian unit, ExFl = external flysch unit, ABD = allochthonous basal detachment.

side in Figures 10 and 11) to the foreland (right-hand side in Figures 10 and 11) domain of the "thrust belt." This suggests a strong constraint from the "foreland" basement ramp morphology and position, with respect to the deformation propagation (Turrini et al., 2001).

Because of the decoupling mechanics interpreted between the shallow "allochthonous" structures and the deep "autochthonous" ones, subthrust anticlines that are not evident at surface can be observed in the internal part of the model (Figures 10a and b and 11a and b). Similar structures therefore can be validated or inferred within the deeper part of selected structural domains of the Southern Apennines chain.

The deformation events that took place during the model simulations (Figure 12) (see Turrini et al., 2001, their experiment F4 for a complete discussion of the model kinematics) roughly replicate the Pliocene-Pleistocene deformation progression recognized within the Calabro-Lucano region (see below). Thrusting of the deep structures occurs after the "allochthonous units" have been emplaced on top of the buried "autochthonous" ones. In such a context, development of the structures within the "autochthonous domain" (i.e., the Apulian carbonates in the Campano-Lucano region) would be out of sequence to the main thrust carrying the allochthon.

As a result of the above-described kinematics, the allochthonous basal detachment is breached (see Figures 10b and c and 11b and c), although this appears to be much more developed in the models than in "reality." We think that such a difference can result from three possible elements. (1) The sandbox simulations are

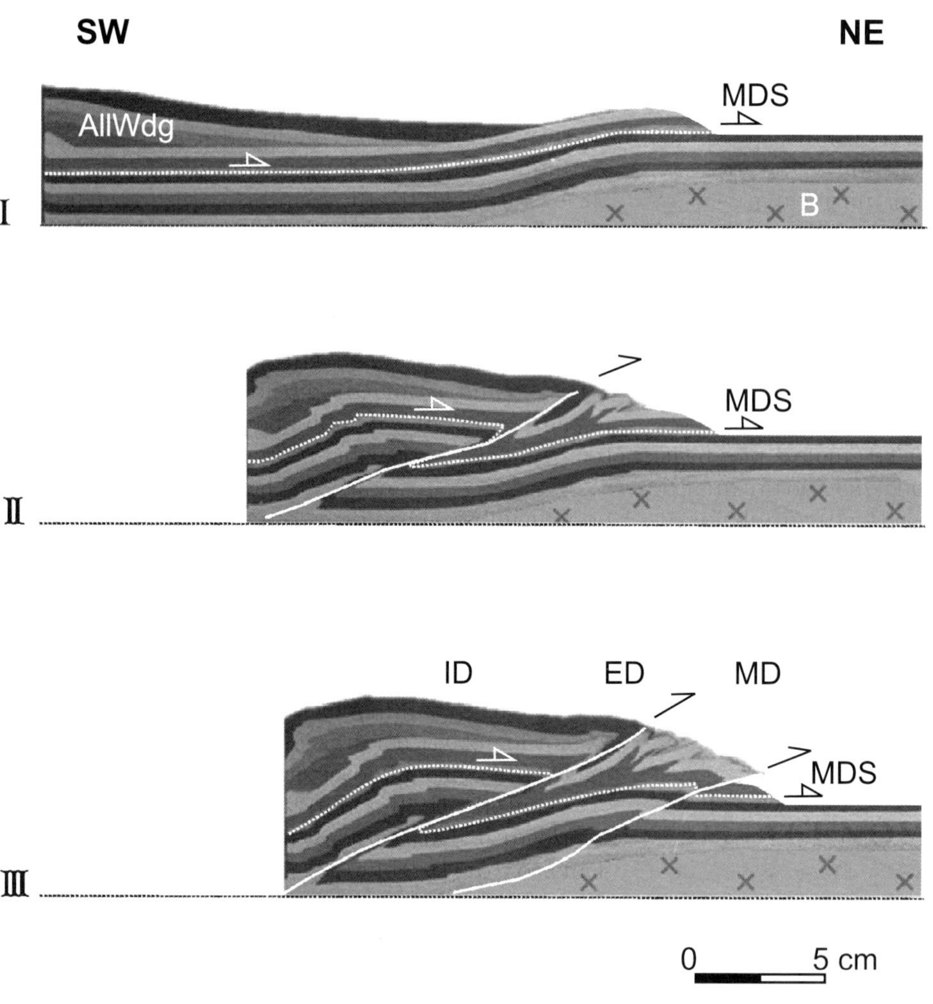

Figure 12. Sandbox model kinematics from the Totalfina Italia 1998 experiments catalog. The evolution of deformation is representative of the Campano-Lucano Pliocene-Pleistocene tectonics. Stage I = late-phase precontractional events. Note the allochthonous wedge (AllWdg) representing an advanced stage in the tectonic evolution of the Southern Apennines Thrust Belt (post-allochthonous emplacement). MDS = major decoupling surface (microbead layer, i.e., the allochthonous basal detachment), B = "basement." Stage II = "autochthonous" deformation and "allochthonous" imbrication. Note folding and breaching of the allochthonous basal detachment. Stage III = deformation of the external and marginal domain, ID = internal domain (i.e., the Apenninic Platform carbonates), ED = external domain (i.e., the Apulian Platform unit), MD = marginal domain (i.e., the Apulian Platform carbonates).

limited in their ability to replicate underthusting deformation mechanics (this would have accounted for part of the out-of-sequence thrusting of the deeper units). (2) "Synthetic" erosional and depositional rates may not be correctly scaled to the "natural" ones (this would have preserved the critical taper of the wedge, thus partly inhibiting out-of-sequence thrusting of the deep units). (3) The reconstruction of the precontractional framework (thickness of the original "autochthonous" and "allochthonous" units; dimension of the precompressional wedge; vertical position of the decoupling layers) is misleading (this has finally resulted in a "synthetic" thrust belt not properly similar, in terms of structure characteristics, to the "natural" one).

3-D STRUCTURAL GEOMETRIES IN THE AUTOCHTHONOUS UNITS

The 2-D depth cross sections interpreted in the study area have been integrated to provide a 3-D surface model of the top Apulian Platform carbonates. Despite some areas of poor-quality data, the resultant 3-D structural depth map (Figure 13) provides a clear picture of the local-scale and large-scale geometries and their orientations and distributions. Depth, in the internal and external domains, ranges from highs at approximately 2500 m below sea level (bsl) to lows as deep as 7000 m. From the northeast in the study area (the Murge region) the top autochthonous unit is a gently southwest-dipping homocline. Approximately aligned with the outcrop limit of the allochthonous unit, the homocline downsteps to the southwest, from the marginal domain into the external domain. The overall tectonic framework shows an anticlinal-synclinal northwest-southeast regional fabric that defines strong along-strike continuity in the internal domain. However, a dome-and-basin geometric pattern is seen in the external domain.

The internal domain is marked by an impressive and continuous northwest-southeast alignment of structures. The hinge of this internal domain is evident in Figure 13 (see contour lines at 3000–4000 m bsl). In the southern part of the region, northwest of the Sant'Arcangelo Basin, a clear bend causes this hinge to shift to

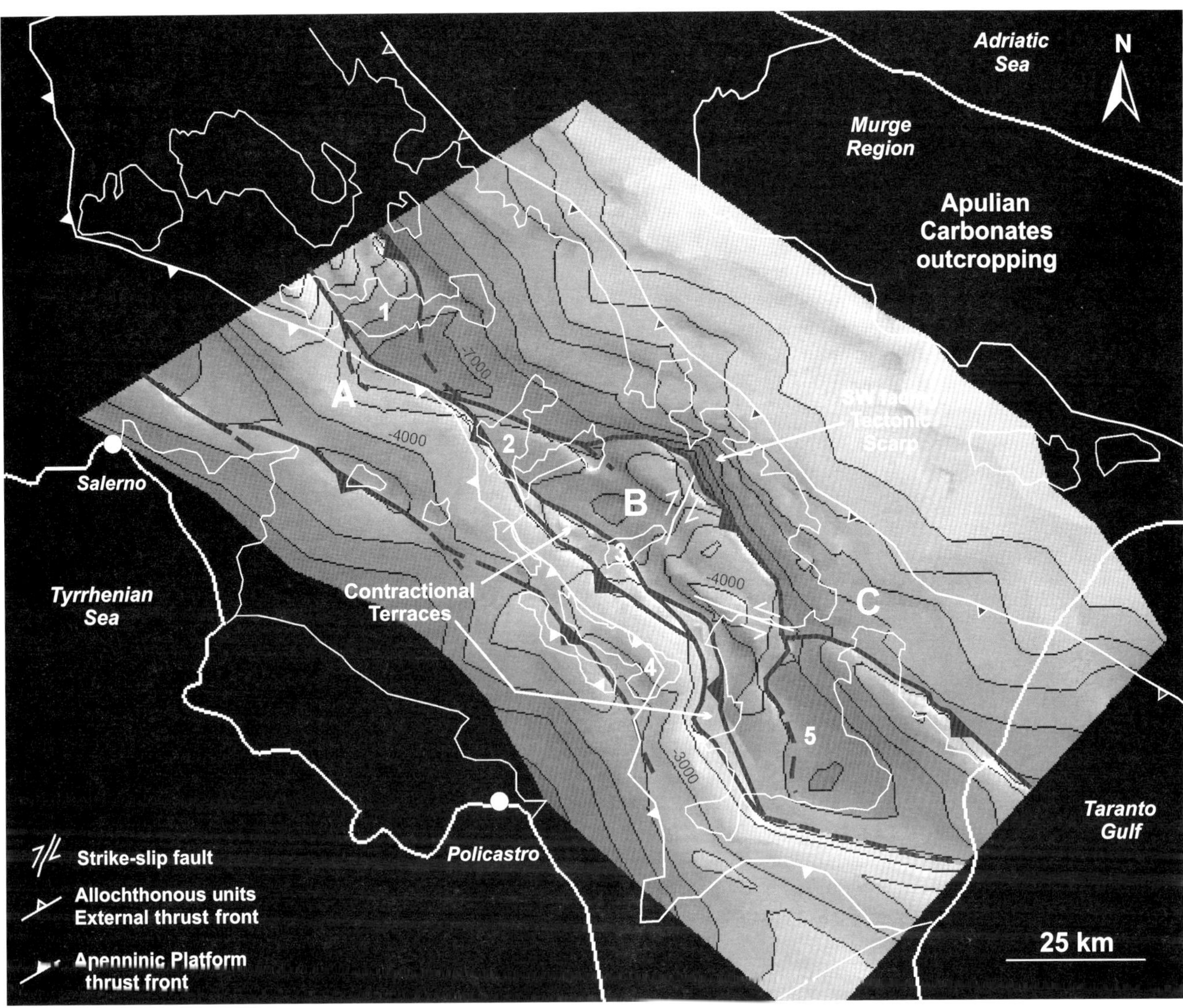

FIGURE 13. 3-D structure-depth map of the top Apulian Platform carbonates. Major thrust within the "Apulian" fold-and-thrust belt in bold black. Surface geology outlined in white. Surface illuminated from northeast. Contour lines every 1000 m. 1 = Ofanto Basin, 2 = Potenza Basin, 3 = Anzi-Calvello Basin, 4 = Tanagro-Val D'Agri Basin, 5 = Sant'Arcangelo Basin, A = Internal Domain, B = External Domain, C = Marginal Domain.

the southwest. The hinge demonstrates a regional plunge to the northwest, whereas significantly, the highest culmination corresponds with the geometric bend described above.

The external domain is by far the most areally extensive in the study region. This domain is a southwest-northeast depression between the marginal and the internal domain, with a series of anastomosing structural arches and depressions in longitudinal section, that is, in northwest-southeast section. Within the domain, the most outstanding geometric high is located to the northwest of the Sant'Arcangelo Basin and immediately to the northeast of the structural bend in the internal domain's hinge. The large-scale depressions that appear in the northwest (Ofanto Basin depression) and southeast (Sant'Arcangelo Basin depression) of the external domain (see Figure 13) underline the discontinuity of the structures within this domain, whereas smaller and narrower lows interfinger with northwest-to-southeast-trending highs.

The marginal domain lies to the east of the external domain. Here, the structure map (see Figure 13) shows a northwest-to-southeast-oriented terrace morphology causing the regional deepening to the southwest of the easternmost Apulian foreland. The dip of the terraces' enveloping surfaces varies as one moves from southeast to northwest (see also Figure 8), being steepest (cross section 3 of Figure 8) in the central sector of the domain, which corresponds to the outermost (northeast) limit of the external domain.

The structures that can be interpreted across the whole region primarily indicate folding and southwest-to-northeast thrusting of the Apulian carbonates onto the southwest-flexured Apulian foreland (Figure 14). Large-scale thrust surfaces (labeled "1" in Figure 14a, b, and c) separate the three recognized tectonic domains (see also Figure 13). Anticlinal features can also be defined ("2" in Figure 14b) and lateral/oblique ramps ("3" in Figure 14a, b, and c) can be clearly interpreted. Notably, and especially within the external tectonic domain, such ramps seem to be formed by the coalescence of minor en echelon structures ("4" in Figure 14b and c). Consequently, contractional relay zones can be defined accommodating a fault-fold structural transition along strike. So far, no major northeast-to-southwest-transecting discontinuities have been recognized or mapped in the area. Nevertheless, they can be speculated to control the major interruptions of the observed northwest-southeast structural grain while they trigger the lateral/oblique ramp development.

The external domain's major faults show a southwest dip, but conjugate sets of second-order southwest-dipping forethrust and northeast-dipping back-thrust surfaces (with respect to the regional tectonic-transport direction to the northeast; "5" and "6" in Figure 14c) control the vertical extrusion of the related pop-up structures (see also Figure 8, cross sections 3, 5, and 6, and Figure 9). Dip for the first-order internal domain's overthrust is constantly to the southwest, and discontinuous second-order thrust-related faulted terraces ("7" in Figure 14b and d) can be interpreted locally at its footwall (see also Figure 8, cross sections 3, 5, and 6).

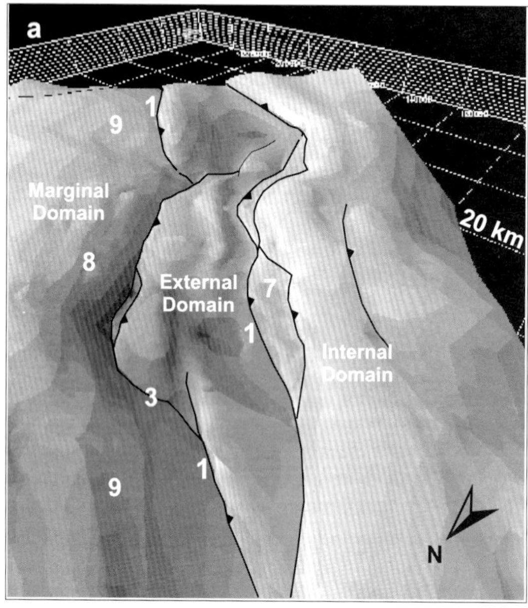

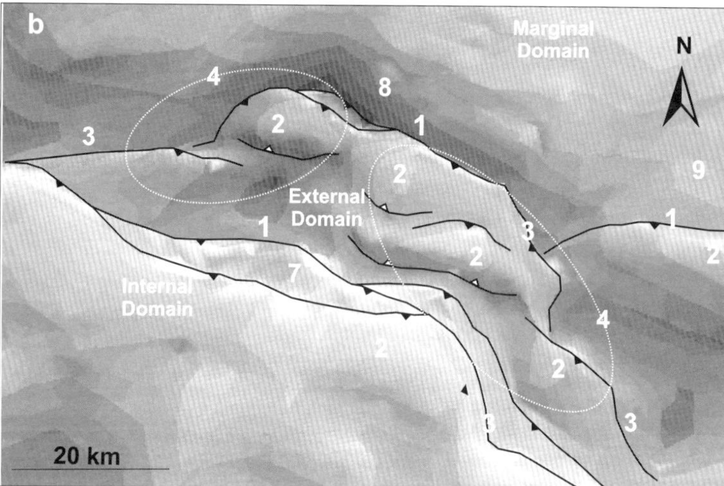

1 : Major thrust
2 : Thrust-related fold
3 : Lateral / oblique ramp
4 : En echelon structures
5 : Backthrust
6 : Forethrust
7 : Contractional terrace
8 : Steep marginal scarp
9 : Shallow marginal scarp
10: Normal fault

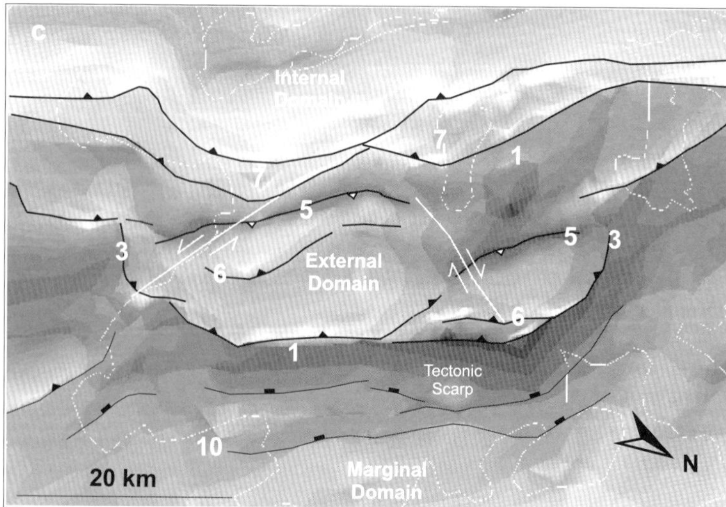

FIGURE 14. 3-D structures, with depth contours, at top Apulian carbonates. Details from Figure 13. (a) Low-angle perspective view looking south-southeast. (b) High-angle perspective view looking north. (c) High-angle perspective view looking northwest. Surface geology outlined in white. Note outcropping strike-slip faults in bold white. (a-b-c) Surface is illuminated from northeast.

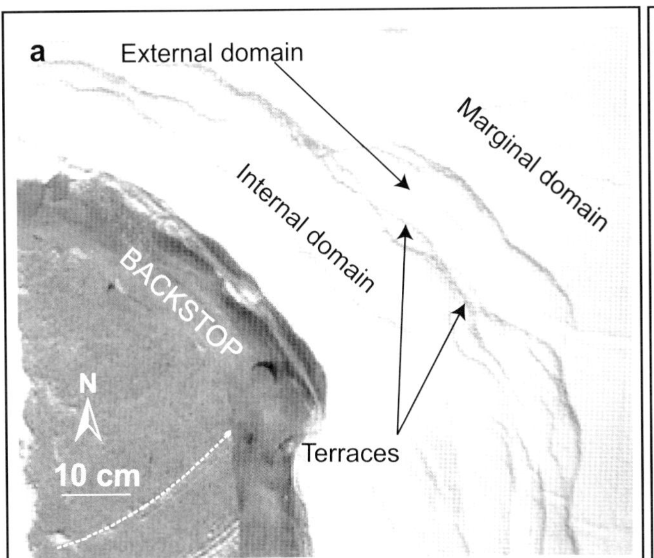

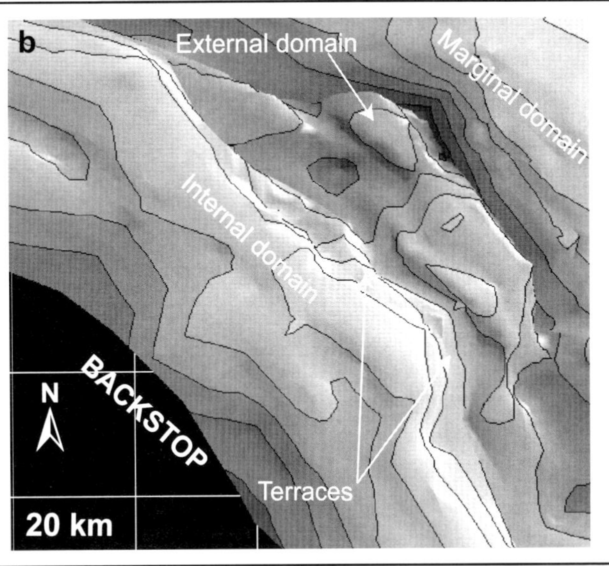

FIGURE 15. Comparison among analog and interpreted structures. (a) Map view of compression-induced sandbox model (model illuminated from southwest). (b) Contractional structures from the 3-D top Apulian surface in the Campano-Lucano Thrust Belt (3-D map illuminated from northeast). Note sand stripes at the rear of the backstop in Figure 15a, indicating anticlockwise rotation of the stress trajectory during the experiment.

STRUCTURAL STYLE AND HYDROCARBON DISTRIBUTION IN THE CAMPANO-LUCANO ARCH

Tectonic Regime, Deformation Mechanisms, and Structural Style

Totalfina Italia's proprietary borehole data from the Apulian Thrust Belt's external domain indicate that compression characterizes the active tectonic regime, at least to the northeast of the internal domain's hinge. This interpretation is supported by correlation between map views of compression-induced sandbox models (Figure 15a) and the 3-D regional structural model of the Apulian thrust-and-fold belt (Figure 15b). The same geometric comparison suggests that the arch-shaped thrust belt interpreted at depth indicates that folding and thrusting of the Apulian carbonates are the major deformation mechanisms in the region. Additionally, heterogeneity of the Apulian foreland's morphology (see above) and rotation in the regional compressional-stress regime (note displacement vector of the backstop in the sandbox model of Figure 15a) are likely to control the localization of oblique movements. Therefore, transpression-related mechanics can be inferred both within the shallow Apenninic structural level and the deep Apulian domains.

The interpreted structural style seems to fit generic thin-skinned tectonism within the external domains of the Campano-Lucano Arch, whereas preliminary balancing operations (not shown in this chapter) suggest thick-skinned tectonism within the internal domain. Irrespective of where and how far the deep basement involvement can be invoked, the 2-D and 3-D interpretations for the region indicate that refolding and breaching of previously folded and thrusted units model the current structural setting, as has already been mentioned in the existing literature (Casero et al., 1988; Carbone and Lentini, 1990; Finetti et al., 1996; Lentini et al., 1996). As such, recent folding and thrusting of the deep autochthonous unit (the Apulian Thrust Belt) deformed a preexisting early-phase vertical allochthonous assemblage (the Apenninic Thrust Belt) that was subsequently reworked as a pseudo-multilayered tectonostratigraphy. Decoupling of the allochthonous unit from the autochthonous one becomes evident from southeast to northwest and from the hinterland to the foreland of the chain (Figure 16; see also Figure 2). Three-dimensional strain partitioning seems moderately homogeneous across the thrust belt's internal domain, whereas shallow and deep structures show minor mechanical disharmony. Conversely, distribution of the deformation appears strongly inhomogeneous across the thrust belt's external and marginal domains where thrusting of the autochthonous Apulian unit enhances vertical extrusion, localized imbrication, and strike-slip mechanics in the overlying pseudostratigraphy (Figure 16; see also Figures 1 and 2).

Correlation of Deep to Shallow Geometries

The top Apulian carbonate's 3-D structure in the Campano-Lucano Arch clearly shows the tectonic architecture of the deeper (autochthonous) part of the present-day orogen (Figures 13 and 14a–c). The resultant 3-D model crucially images how deformation is distributed within this buried Apulian Thrust Belt and how and where the resulting geometries interact with shallower outcropping allochthonous features.

When we consider the correlations among the entire range of structures (see Figure 13), we can postulate an intriguing relationship between the deep and shallow deformation geometries. In general terms, the 3-D model's interpretation suggests that the thrusted folds at depth result in a corresponding, generic dome-and-basin interference pattern at the surface (Figure 13: compare the Pliocene-Pleistocene basin orientation and distribution with the underlying top Apulian Platform's morphology, which is shown by the structure contour map). In particular, the geometric modifications (i.e., along-strike variation in the dip of the foreland homocline) in the deep Apulian foreland (and the basement below?) appear to constrain the localization of deformation in the Apulian Thrust Belt, which, in turn, influences deformation in the shallower, allochthonous unit. The resulting geometric compatibility between subsurface and surface structures is indicated by the following and shown in Figures 13 and 14:

1) *subsurface*: the width of the central high in the Apulian Thrust Belt's external domain; *surface*: the

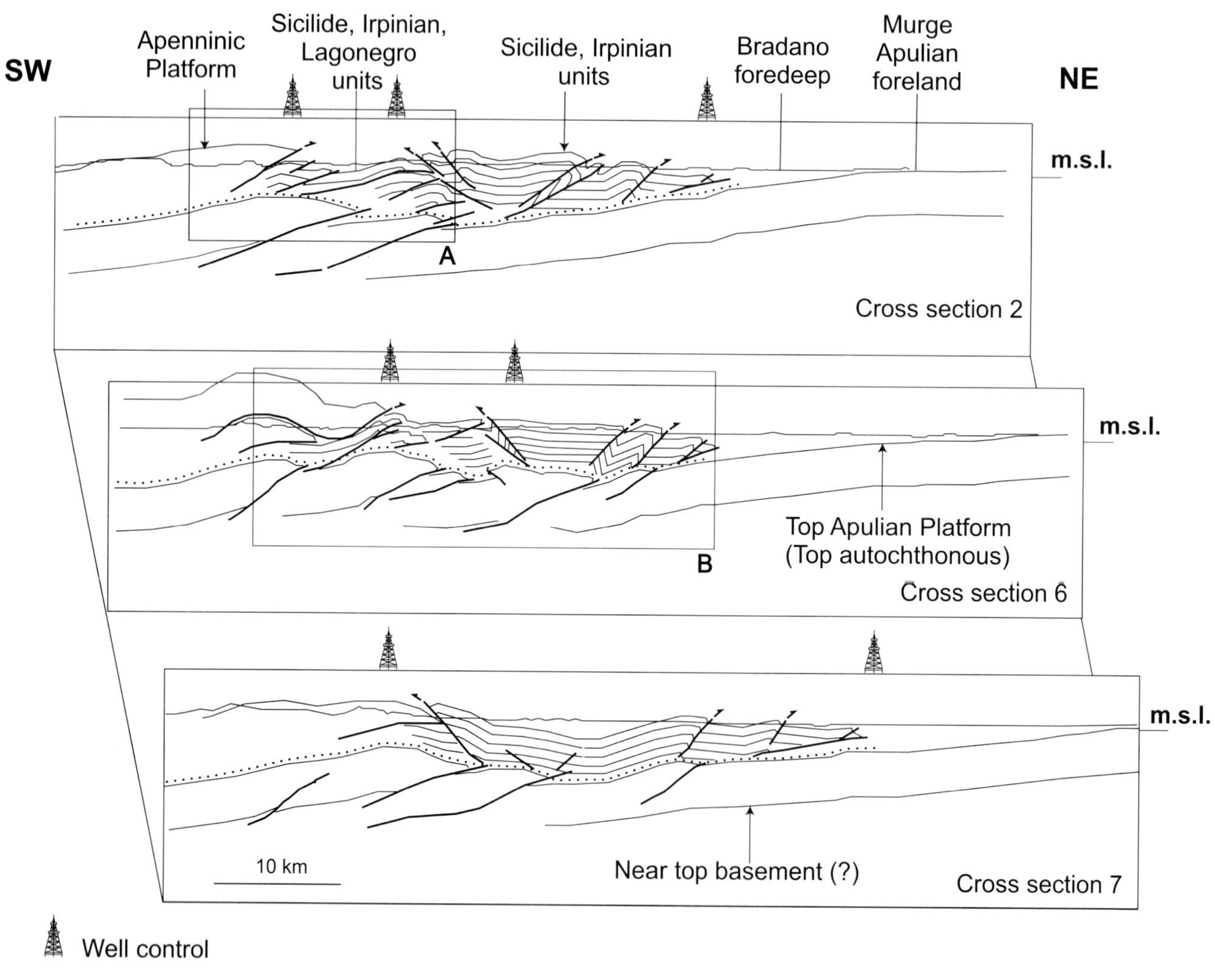

FIGURE 16. Perspective view of the deformation geometries in the Campano-Lucano Arch (cross section 7 is modified after Hippolyte et al., 1994). Note that "pseudostratigraphic" layering in the allochthonous units is not representative of the sedimentary geometries that can be observed in the structures. See Figure 1 for location.

morphology of the Italian coast south of Salerno, as far as Policastro;

2) *subsurface*: (a) the location of the northwest lateral ramp of the Apulian Thrust Belt's external domain (southeast of the Ofanto Basin); (b) the northern steep-to-shallow dip transition in the marginal domain's Apulian scarp; *surface*: the northwest termination of the Murge's outcropping Apulian foreland;
3) *subsurface*: the arch-shaped geometry of the central high in the Apulian Thrust Belt's external domain; *surface*: the arcuate distribution of folds and thrusts in the allochthonous external domain (see Figure 1);
4) *subsurface*: the fault-fold trend's orientation in the Apulian Thrust Belt's internal domain; *surface*: the northwest-southeast alignment of the Tanagro and Val D'Agri Pliocene-Pleistocene basins;
5) *subsurface*: (a) the southern steep-to-shallow dip transition in the marginal domain's Apulian scarp (north of the Sant'Arcangelo Basin); (b) the strong bend and tectonic imbrication in the Apulian Thrust Belt's internal domain, also coinciding with the highest culmination; *surface*: the major deviation in the structural trend of the Apenninic Platform (see also Figure 1);
6) *subsurface*: the structural orientation in the Apulian Thrust Belt's internal domain; *surface*: the boundary between an active, mainly extensional sector to the southwest and an active, compressional sector to the northeast (see previous discussion and Figure 5);
7) *subsurface*: the width and localization of culminations and depressions in the Apulian Thrust Belt's external domain; *surface*: the localization and orientation of the Pliocene-Pleistocene Ofanto, Potenza, Anzi-Calvello, and Sant'Arcangelo Basins; and
8) *subsurface*: the structural distribution in the Apulian Thrust Belt's external domain; *surface*: the strike-slip fault's localization and orientation east of the Anzi-Calvello Basin and north of the Sant'Arcangelo Basin (see also Figure 14c).

The last correlation is very relevant to understanding the link that may exist between deep and shallow structures, in terms of mechanics and kinematics. The ramp anticline in the Apulian Thrust Belt's external domain is asymmetric and unequivocally northeast-verging as it overthrusts the shallow-dipping southern segment of the Apulian foreland's marginal scarp ("9" in Figure 14a and b; see also Figure 8, cross section 7). Conversely, the ramp anticline develops symmetric folds and back-thrust faults ("5" and "6" in Figure 14d) as it attempts to ride over the steep, central segment of the Apulian foreland's marginal scarp ("8" in Figure 14a and b; see also Figure 8, cross section 6). This partitioning of deformation at the Apulian level causes the overlying tectonostratigraphy to deform by strike-slip faulting (Figure 14c). These faults intersect the northwest-to-southeast-outcropping thrusted folds; the faults locally control the formation of the Pliocene-Pleistocene Anzi-Calvello piggyback basin (Figures 13 and 14c).

Regional Structure and Hydrocarbon Distribution

Because the main topic of this work is to discuss the structural geometries in the Campano-Lucano sector of the Southern Apennines, only a simplified model of the structural kinematics is given.

Comparative analysis of the entire available data set indicates that two main tectonic events took place in the region. The first started in the early Pliocene, when an already folded and thrusted (Miocene tectonism) allochthonous unit was emplaced on top of the southwestern margin of the Apulian foreland (i.e., onto the Apulian Platform carbonates' internal domain). This tectonic event continued until both the internal and external Apulian Platform domains had been covered by the allochthonous nappe (late early Pliocene time). At this stage, the entire vertical tectonostratigraphic assemblage was being deformed while folding and thrusting in the deep Apulian Platform carbonates was taking place (middle to late Pliocene). Thus, refolding and imbrication of the overlying allochthonous first-phase tectonic layering (and the isolated Pliocene piggyback basins therein) occurred throughout the study area and likely continued until the late Pleistocene. It should be noted that the Pliocene-Pleistocene tectonism took place in the region after the Apulian Platform sediments had already been flexured below the advancing allochthonous unit and while the undeformed foreland (Murge region, to the northeast of the study area) bulged and eroded (Casero et al., 1988; Doglioni et al., 1996). In this context, normal faults related to the flexuring of the Apulian foreland contributed to the structural history by localizing and amplifying the deformation in the external domain. The tectonic history's description could become much more complex if selected local-scale examples from the in-house data set are also considered. In such situations, it is clear that important erosional events affected the Apulian Platform carbonates before the deposition of the autochthonous Pliocene clastic sediments on top of the platform and the subsequent emplacement of the allochthonous wedge (see the erosional truncation below the top Apulian carbonates, illustrated in Figure 9). It follows that part of the Mesozoic-Cenozoic sediments can be missing below the top Apulian unconformity surface. As well as these erosional events, some pre-Apenninic (pre-Miocene to late Cretaceous-Oligocene?) tectonic deformation events should also be considered, to honor the local-scale geometries encountered by the wells. This means that the carbonates may have been deformed previously, as is indicated by tilting of the Mesozoic-Cenozoic sequence prior to erosion at the end of Apulian time (although whether this was contractional or extensional remains unknown).

The hydrocarbon distribution in the region is one of gas-bearing plays in the foredeep Pliocene sediments and oil-bearing traps in the deep, deformed carbonates of the Apulian Platform foreland, the latter being considered to be the main exploration target in the area. The gas is mainly biogenic, while source, reservoir, and cap rocks are shale-sand sediments deposited to the northeast of the Southern Apennines' advancing front. Proven gas fields in the area are the Grottole Ferrandina, Pisticci and Accettura fields (Sella et al., 1988).

Oil density ranges from 17 to 30° API. Reservoir, source rocks, and cap rocks in the region are the Mesozoic-Cenozoic carbonates of the Apulian Platform, the associated Albian shaly limestone, and the middle-late Pliocene shale of the autochthonous pre-Bradano unit, respectively. Oil traps are faulted antiforms whose depths range from 3000 to 6000 m below sea level.

The two-stage Neogene tectonic evolution discussed above accounts for the hydrocarbon generation and expulsion in the region. The results from in-house geochemical modeling confirm that the generation of hydrocarbons in the region started as soon as a critical load/thickness on top of the Apulian carbonates had been reached. Such a critical load is easily achieved by emplacement of the allochthonous wedge. This means that hydrocarbon generation started soon after the early Pliocene first-phase tectonic event that controlled the emplacement of the Sicilide, Lagonegro, Irpinian, and Apenninic allochthons. This implies that expulsion and migration of the hydrocarbons only began after the late Pliocene,which is very recently. Considering the tectonic history discussed above, such hydrocarbon expulsion and migration therefore occurred during or soon after the final structuring of the deep Apulian Thrust Belt.

CONCLUSIONS

The Campano-Lucano Southern Apennines' chain sector is a refolded fold-and-thrust belt that developed from Miocene to Pliocene-Pleistocene times. As such, the early-phase vertical allochthonous assemblage was recently reworked as a pseudo-multilayered stratigraphy, by newer folding and thrusting of the deeper autochthonous substratum. The interpreted overall structural style fits generic thin-skinned tectonism (basement not involved) in the northeastern, external domains, and it may change to thick-skinned tectonism (basement involved) in the westernmost hinterland of the Southern Apennines' orogen.

The detailed top Apulian carbonate 3-D surface shows that an arch-shaped, buried thrust belt is currently deforming the deep substratum of the allochthon and that the resulting deformation geometries account for the majority of the outcropping structural features in the region (Figure 13).

The geometric heterogeneity (inherent structural heterogeneities?) in the underthrusting Apulian foreland (and the basement below?) seem to constrain the localization of deformation in the Apulian Thrust Belt, which, in turn, influences deformation in the shallower, allochthonous units. These units show geometric and kinematic compatibility to the related tectonized autochthonous units, although decoupling is evident (Figure 16).

The integrated methodology adopted here (well interpretation, surface-geology analysis, seismic interpretation, 2-D and 3-D structural mapping, and comparison with sandbox models) suggests that, despite the strike-slip structures occurring at the surface (Monaco et al., 1998; Paltrinieri et al., 1998), folding and thrusting of the Apulian carbonates are the major deformation mechanisms in the region (see Figure 15). Localized en echelon and lateral ramp structures in the deep Apulian units can, however, cause oblique movements and transpression-related mechanics in the shallow Apenninic structures (see also Catalano, 1993; Catalano et al., 1993).

The interpreted deformation history and the conceptual thermal-maturity model that it supports suggest that trap formation in the area occurred after hydrocarbon generation but synchronously with hydrocarbon expulsion. This conclusion implies that all the structural traps in the area should be filled with hydrocarbons, as is also indicated by fluid-migration simulation (Figure 17). This not being the case, a number of elements should be considered to explain the current hydrocarbon distribution. The most important among these is that source rocks are not always present in the interpreted structures. This suggests that migration occurs here only over a short distance. The definition of the pretrap geometric and sedimentary configurations in the Apulian carbonates is becoming as relevant as is the interpretation of a structural trap and should be regarded as the key factor for future exploration in the region. Such future exploration activity should rely both on detailed structural modeling and on advanced geophysical techniques.

ACKNOWLEDGMENTS

Totalfina Italia S.p.A. is kindly acknowledged for permission to present this work, and all the Totalfina Italia technical staff are thanked for their critical input. We thank Dr. Martin Insley, Dr. Jonathan Craig, Professor Ken McClay, and an anonymous reviewer for their constructive criticism and recommended revisions.

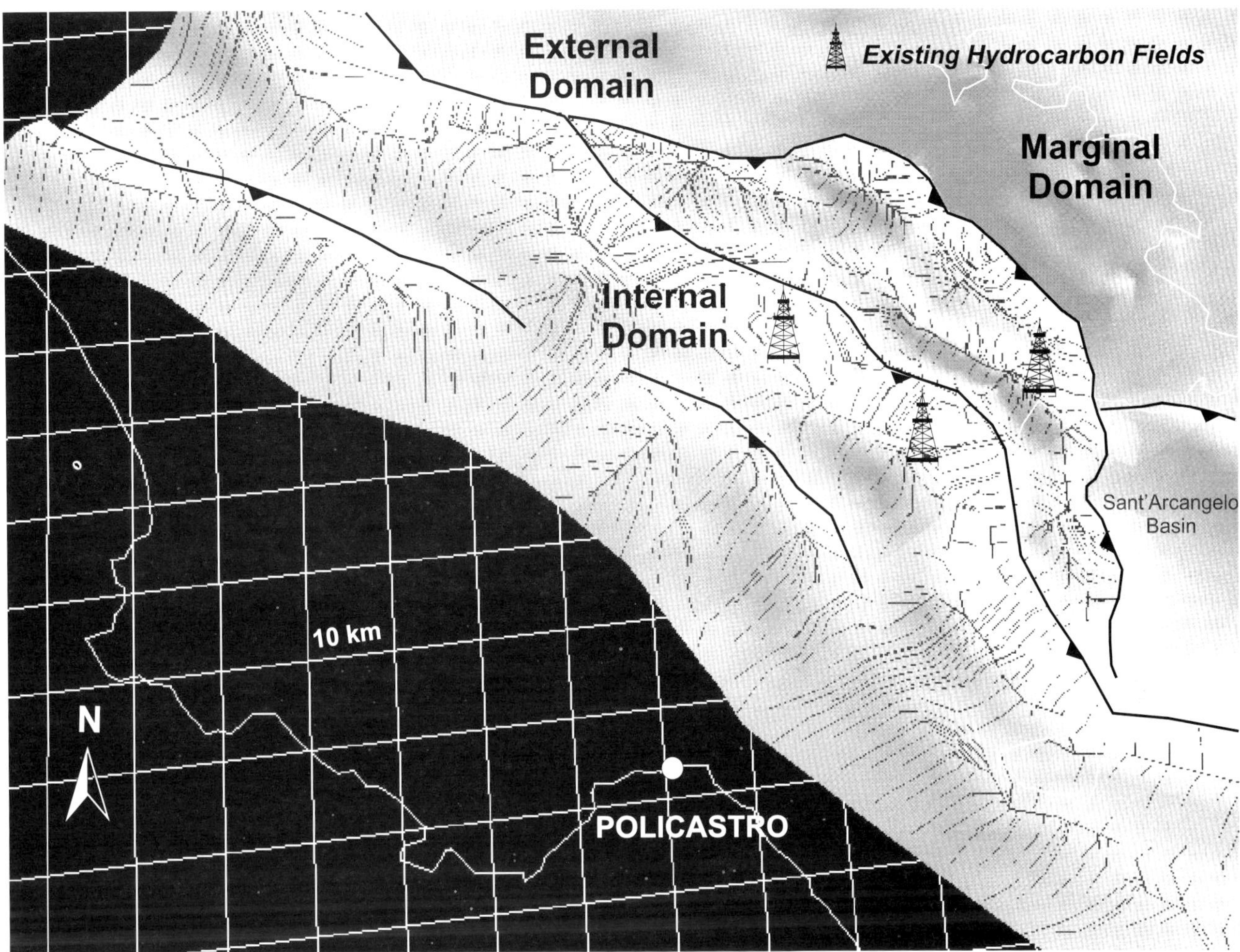

FIGURE 17. Simulated fluid migration at the top of the Apulian Platform carbonates of the Campano-Lucano region. Updip fluid migration path is mapped as a function of structural geometry. Migration paths follow the steepest geometric route, that is, are perpendicular to strike. Relationship to the existing hydrocarbon fields is shown. Depth surface is illuminated from northeast.

REFERENCES CITED

Bonini, M., and F. Sani, 1998, Structural evolution of the Lucanian Apennines between Potenza and the S. Arcangelo Basin: Atti del 79° Congresso nazionale della Società Geologica Italiana, Palermo, Sicily September 21–23, 1997.

Carbone, S., and F. Lentini, 1990, Migrazione neogenica del sistema Catena-Avampaese nell'Appennino Meridionale : problematiche paleogeografiche e strutturali: Rivista Italiana di Paleontologia e Stratigrafia, v. 96, p. 271–296.

Carbone, S., S. Catalano, S. Lazzari, F. Lentini, and C. Monaco, 1991, Presentazione della carta geologica del bacino del fiume Agri (Basilicata): Memorie della Società Geologica Italiana, v. 47, p. 129–143.

Casero P., F. Roure, L. Endignoux, I. Moretti, C. Muller, L. Sage, and R. Vially, 1988, Neogene geodynamic evolution of the Southern Apennines: Memorie della Società Geologica Italiana, v. 41, p. 109–120.

Catalano, S., 1993, Deformazioni polifasiche nel sistema a thrust dell'Appennino lucano: caso delle dorsali di stigliano ed accettura: Bollettino della Società Geologica Italiana, v. 112, p. 659–669.

Catalano, S., S. Carbone, and F. Lentini, 1993, Il Flysch di Gorgoglione nell'ambito dell'evoluzione dell'Appennino lucano: Giornale di Geologia, v. 55, no. 1, p. 165–178.

D'Andrea, S., R. Pasi, G. Bertozzi, and P. Dattilo, 1993, Geological model, advanced methods help unlock oil in Italy's Apennines: Oil & Gas Journal, v. 4, no. 8, p. 53–56.

D'Argenio, B., 1988, L'Appennino Campano-Lucano. Vecchi e nuovi modelli geologici tra gli anni sessanta e gli inizi degli anni ottanta: Memorie della Società Geologica Italiana, v. 41, p. 3–15.

Doglioni, C., M. Tropeano, F. Mongelli, and P. Pieri, 1996, Middle-late Pleistocene uplift of Puglia. An anomaly in the Apenninic foreland: Memorie della Società Geologica Italiana, v. 51, p. 101–117.

Ferranti, L., B. D'Argenio, E. Marsella, J. S. Oldow, G. Pappone, and M. Sacchi, 1997, Orogen-parallel extension during foreland imbrication; Southern Apennines fold-and-thrust belt, Italy: Geological Society of America Abstracts with Programs, v. 29, p. 318.

Finetti, I., F. Lentini, S. Carbone, S. Catalano, and A. Del Ben, 1996, Il sitema Appennino meridionale-Arco Calabro-Sicilia nel Mediterraneo centrale: studio geologico-geofisico: Bollettino della Società Geologica Italiana, v. 115, p. 529–559.

Hippolyte, J. C., J. Angelier, F. Roure, and F. Casero, 1994, Piggyback basin development and thrust belt evolution: structural and paleostress analyses of Plio-Quaternary basins in the Southern Apennines: Journal of Structural Geology, v. 16, no. 2, p. 159–173.

Lentini, F., S. Catalano, and S. Carbone, 1996, The External Thrust System in southern Italy : a target for petroleum exploration: Petroleum Geoscience, v. 2, p. 333–342.

Menardi Noguera, A., and G. Rea, 1997, The Campano-Lucano arc deep structure (Southern Apennines, Italy): Programme and abstracts from the 8th Workshop of the ILP Task Force: "Origin of sedimentary basins," Palermo, Sicily, June 7–13, 1997.

Monaco, C., L. Tortorici, and W. Paltrinieri, 1998, Structural Evolution of the Lucanian Apennines, southern Italy: Journal of Structural Geology, v. 20, no. 5, p. 617–638.

Mostardini, F., and S. Merlini, 1986, Appennino centro-meridionale : sezioni geologiche e proposta di modello strutturale: Memorie della Società Geologica Italiana, v. 35, p. 177–202.

Paltrinieri, W., C. Monaco, N. Steel, and L. Tortorici, 1998, Ruolo della tettonica trascorrente nell'evoluzione neogenica-quaternaria dell'Appennino Meridionale: indicazioni per lo sviluppo di trappole petrolifere: Atti del 79° Congresso nazionale della Società Geologica Italiana, Palermo, Sicily, September 21–23, 1997.

Pasi, R., P. Dattilo, G. Bertozzi, and T. Lakew, 1995, A novel approach to the exploration of the Southern Apennines, Italy: geological models and oil discoveries: AAPG Bulletin, v. 79, no. 8, p, 1241.

Sella, M., C. Turci, and A. Riva, 1988, Sintesi geopetrolifera della fossa bradanica (avanfossa della catena appenninica meridionale): Memorie della Società Geologica Italiana, v. 41, p. 87–107.

Scandone, P., 1972, Studi di geologia lucana: carta dei terreni della serie calcareo-silico-marnosa e note illustrative: Bollettino della Società Naturalistica di Napoli, v. 81, p. 255–300.

Turrini, C., A. Ravaglia, and C. Perotti, 2001, Compressional structures in a multilayered mechanical stratigraphy, *in* H. A. Koyi and N. S. Mancktelow, eds., Insights from sandbox modelling with 3-D variations in basal geometry and friction: Tectonic modeling, a volume in honour of Hans Ramberg: Geological Society of America Memoir 193, p. 153–178.

Cooper, M., C. Brealey, P. Fermor, R. Green, and M. Morrison, 2004, Structural models of subsurface thrust-related folds in the foothills of British Columbia—Case studies of sidetracked gas wells, *in* K. R. McClay, ed., Thrust tectonics and hydrocarbon systems: AAPG Memoir 82, p. 579–597.

Structural Models of Subsurface Thrust-related Folds in the Foothills of British Columbia—Case Studies of Sidetracked Gas Wells

Mark Cooper
EnCana Corporation, Calgary, Alberta, Canada

Chris Brealey
Shell Canada, Calgary, Alberta, Canada

Peter Fermor[1]
Shell Canada, Calgary, Alberta, Canada

Rick Green
Talisman Energy, Calgary, Alberta, Canada

Mike Morrison
Talisman Energy, Calgary, Alberta, Canada

ABSTRACT

The Foothills of the Canadian Rocky Mountains are a classic example of a thin-skinned fold-thrust belt. A regionally significant detachment in the shales of the Jurassic Fernie Formation separates the intensely folded and imbricated clastics of the Jurassic and Cretaceous from the thrust structures in the prospective dolomite reservoirs of the Triassic and upper Paleozoic.

The Monkman area lies within the foothills of northeastern British Columbia and contains a number of fields that produce sour gas from the platform carbonates of the Upper Triassic Pardonet and Baldonnel Formations. The current phase of drilling commenced in 1987 and has been remarkably successful, with numerous gas discoveries made at a 75% success rate. The area is topographically difficult and structurally complex, with a number of detachments that compartmentalize the deformation within distinct tectonostratigraphic units. The acquisition, processing, and interpretation of

[1]*Present address*: EnCana Corporation, Calgary, Alberta, Canada.

seismic data in this area is technically difficult and requires careful analysis to yield good results. The integration of the seismic, well, and surface-geology data is essential for the accurate definition of subsurface targets.

The traps are primarily fault-propagation and detachment folds that detach in the shale-dominated Lower Triassic. The structural style has been interpreted from stratigraphic data, VSP data, and detailed dip data provided by imaging tools. Fracturing is pervasive in the hinges and forelimbs of the folds, and it enhances the tight reservoir to yield typical flow rates of 40 million cubic feet of gas per day (mmcfg/d). The key to success in the play is to target these zones of the folds. However, this is not always successful initially, because the original wellbores often miss the structures. The case studies we present show how revised structural models that incorporate data from the original wellbores increase the chance of success in sidetracking a well into the preferred target area.

INTRODUCTION

The Foothills of the Canadian Rocky Mountains have spawned an extensive literature (Douglas, 1950, 1958; Bally et al., 1966; Dahlstrom 1969, 1970; Price and Mountjoy, 1970; and Thompson, 1979). Development of the fold-and-thrust belt commenced in the Late Jurassic and terminated in the Eocene, and is regarded conventionally as resulting from the accretion of a series of exotic terranes to the Pacific margin of North America (Monger et al., 1982). McMechan and Thompson (1989) give a good review of the fold-and-thrust belt.

The Foothills Fold-and-thrust Belt lies between the Front Ranges, a series of large thrust sheets of Paleozoic rocks to the west, and the largely undeformed Western Canada Sedimentary Basin to the east (Figures 1 and 2). The eastern boundary is characterized by a frontal monocline and/or a triangle zone (Jones, 1982), which becomes a less reliable indicator of the limit of deformation as the belt is traced northward. The intensity of the folding and the magnitude of the thrust displacements generally decrease to the north (Thompson, 1979).

The Monkman area is in the Foothills of northeastern British Columbia (Figures 1 and 2), which McMechan (1985) has described as a triangle zone with a low wedge taper. However, there are relatively few faults of any magnitude in the Jurassic-Cretaceous rocks at the surface (Barss and Montandon, 1981; Kilby and Wrightson, 1987) (Figures 2 and 14) and there are numerous folds. Beyond the well-developed frontal monocline (Figure 2), there extends a series of gentle, low-amplitude, large-wavelength structures. Little evidence exists at the surface for the development of any significant back thrusts within the frontal monocline, and only a few minor, west-vergent thrusts occur locally. The frontal monocline is primarily produced by thickening of the Late Jurassic–Early Cretaceous Nikanassin foreland basin megasequence from 850 m east of the monocline to 1800 m to the west (Cooper, 1996).

The reservoir units in the Monkman area are the dolomites of the Upper Triassic Pardonet and Baldonnel Formations (Figure 3). The Triassic is deformed by fault-propagation and detachment folds with amplitudes of as much as 1000 m. The geometric constraints provided by well and seismic data suggest that the detachment for the fault-propagation and detachment folds is within the shales of the Montney Formation (Figure 3). Seismic imaging of the Triassic structures is complicated by the presence of detachments within the numerous shale formations of the Lower Cretaceous section (McMechan, 1985) (Figure 3) and low-amplitude, short-wavelength folds developed above some of these detachments (Kilby and Wrightson, 1987). This shallow-level deformation also makes it difficult to extrapolate surface structure to depth with any confidence.

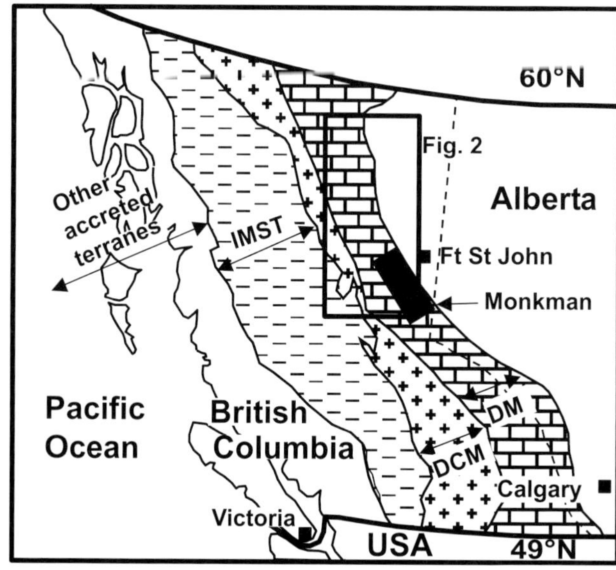

FIGURE 1. Map showing the relationships of the accreted terranes, the Rocky Mountains, and the Western Canada Sedimentary Basin to the Monkman area discussed in this chapter. IMST = Intermontane Superterrane; DCM = Displaced Cratonic Margin; DM = Deformed Cratonic Margin.

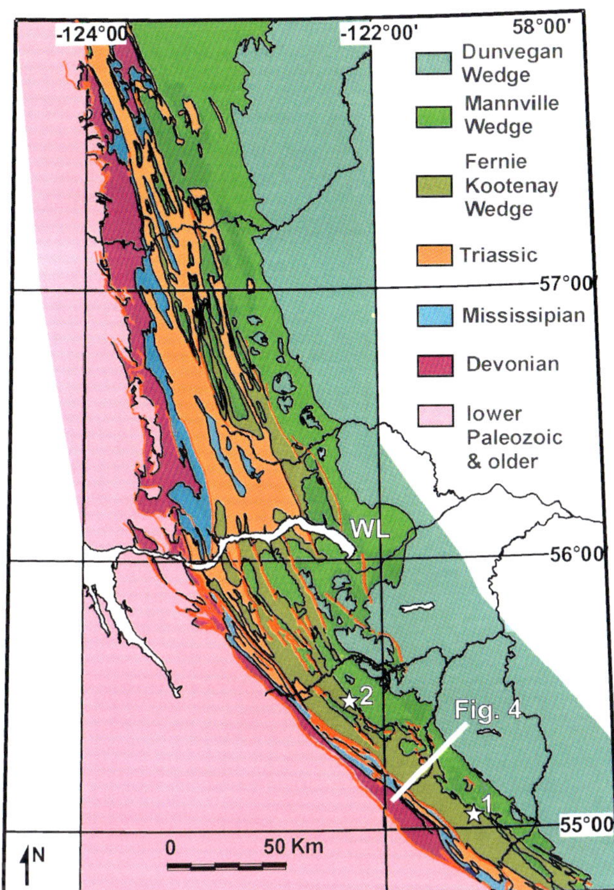

FIGURE 2. Surface geology map of the Foothills of northeastern British Columbia, showing major structural features and stratigraphic units. The locations of the two case study wells discussed are shown by white stars. WL = Williston Lake.

Exploration success in the area depends on locating the hinge zones of the fold structures at the level of the Triassic reservoirs. In this chapter, we describe two case studies of wells that used an integrated geologic and geophysical approach to locate the fold hinges, following initially unsuccessful wells.

STRATIGRAPHIC FRAMEWORK

Good reviews of the pre-Triassic stratigraphy of the area are provided by Bamber and MacQueen (1971, 1979). The Triassic rocks outcropping in the Front Ranges and Foothills of the Rocky Mountains in the Monkman area comprise a sequence of marginal-marine to deep-basinal siliciclastics and carbonates, with a few evaporites. The Monkman area was a significant Triassic depocenter, with as much as 1200 m of Triassic strata (Gibson and Barclay, 1989) that thins eastward onto the Canadian Shield as a result of onlap and erosional truncation.

Davies (1997a) provides an excellent summary of Triassic stratigraphy; hence only a brief outline is given here. The oldest formation in the Triassic is the Montney, a series of dolomitic shales and silts with occasional thin sands deposited in a basinal-marine setting (Davies, 1997a) (Figure 3). The Halfway Formation (Figure 3) contains a series of stacked shoreline sands. The Charlie Lake Formation (Figure 3) represents a mixture of hypersaline, restricted-marine, and nonmarine facies. The dominant lithologies are dolomitic and anhydritic shales, siltstones with dolomitic sands, thin dolomite and limestone beds, and dolomites with anhydrite nodules and anhydrite beds. The primary reservoir units in the Monkman area are the shallow subtidal bioclastic dolomites of the Baldonnel Formation and the uppermost Charlie Lake Formation. The carbonates of the Pardonet Formation (Figure 3) contribute locally to production, primarily from fracture networks. The lower Pardonet represents a deepening event following overall backstepping of shoaling-upward third-order sequences in the Baldonnel and Charlie Lake Formations (Davies, 1997b). The upper Pardonet is a regressive sequence that shallowed upward into the Bocock Formation (Davies, 1997b).

The stratigraphy of the post-Triassic formations (Figure 3) is conveniently reviewed in relation to the accretion history of the Intermontane and Insular Superterranes that make up the central core of the Canadian Cordillera. The accretion of these terranes controls the distribution and timing of deformation throughout the

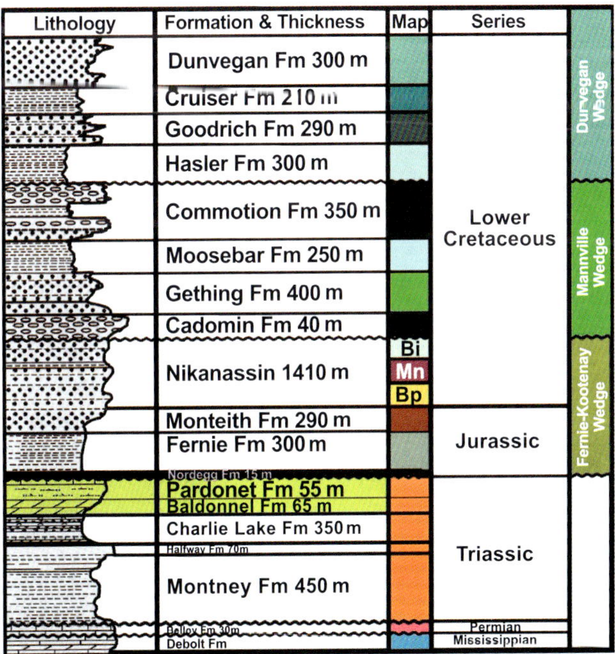

FIGURE 3. Generalized stratigraphic column for the Monkman area (Figure 1). The "Map" column in the figure shows the colors used on the detailed surface geology map (Figure 13) and the cross sections (Figures 10, 16, and 17). The subdivisions of the Nikanassin are: Bi = Bickford Fm; Mn = Monach Fm; Bp = Beattie Peaks Fm.

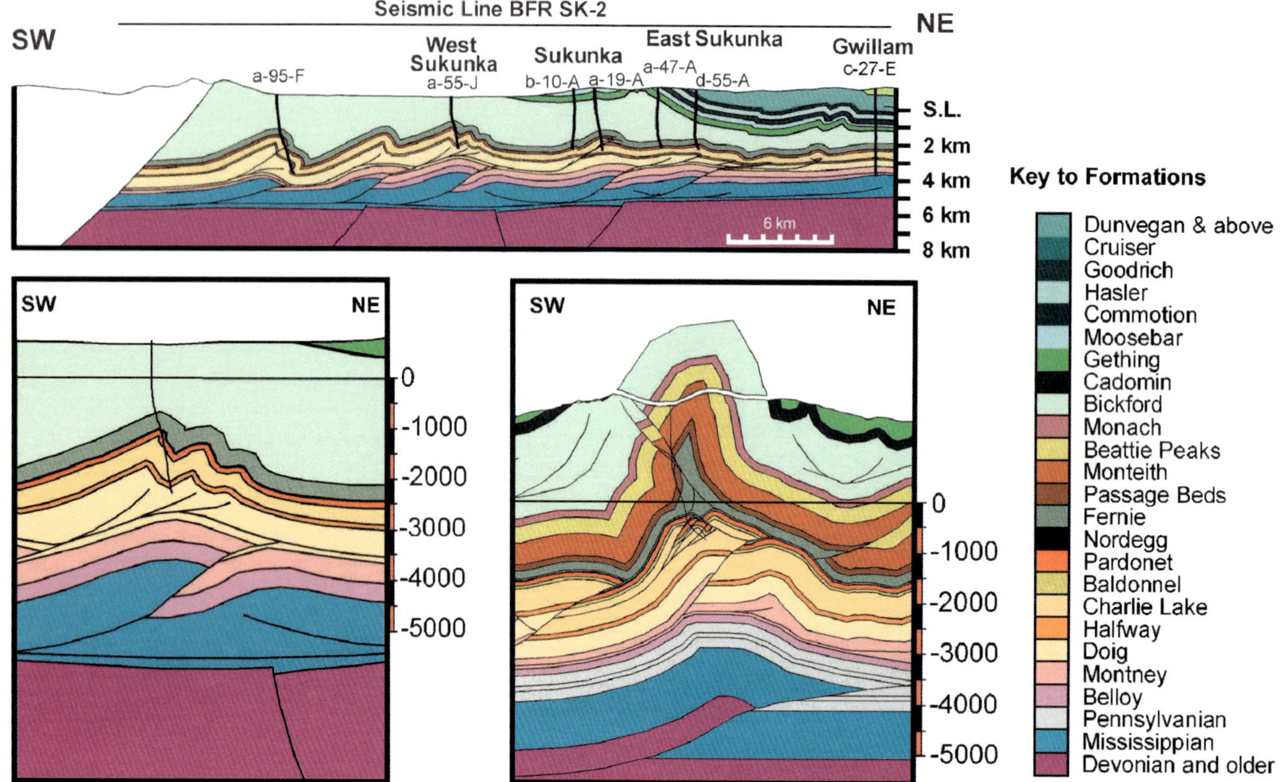

FIGURE 4. Regional cross section along the Sukunka River; for location see Figure 2. The detailed sections show the typical geometry of Triassic structures in the area.

Cordillera, including the Foothills (Monger and Price, 1979; Monger et al., 1982), and has also exercised control on the distribution and facies of the Jurassic and later sediments. The timing relationships between the accretion of the terranes and the foreland clastic wedges has been discussed by Eisbacher et al. (1974) and by Cant and Stockmal (1989), who recognize six clastic wedge sequences attributed to terrane-accretion events.

The oldest clastic wedge consists of the Fernie and Kootenay Groups (Cant and Stockmal, 1989); the Kootenay correlates with the Nikannassin in the Foothills of British Columbia (Stott, 1968). The encroaching Intermontane Superterrane produced a hiatus followed by rapid deepening of the protoforeland basin, in which the Nordegg and Fernie Formations (Figure 3) were deposited. The dominance of deep-water mudrocks in these units suggests that the sediment supply was inadequate to keep pace with the subsidence of the basin, which remained starved until, ultimately, it was infilled by the Nikanassin (Figure 3). The present-day isopach of the Fernie to Nikanassin section preserves the morphology of the early foreland basin. The Monkman area represents the thickest development of the Fernie-Kootenay wedge. To the north of Williston Lake (Figure 2), the sequence thins gradually and pinches out. The closest point of approach of the Intermontane Superterrane to the Foothills is to the west of the Monkman area (Figure 1), thus the depocenter of this foreland basin sequence may reasonably be expected to occur at this point where the load could have been the greatest.

The next clastic wedge (Mannville, Figure 3) recognized by Cant and Stockmal (1989) commenced with the conglomeratic debris of the Cadomin Formation (Figure 3) derived from the newly emergent and actively eroding orogenic hinterland to the west. The younger clastic wedges reflect cyclical regressive/transgressive sequences related to eustatic sea-level changes and to tectonic events in the orogen (Cant and Stockmal, 1989).

REGIONAL STRUCTURAL STYLE

The Monkman area lies in a narrow part of the Foothills (30–40 km wide) compared with the Alberta Foothills to the south (Figure 2). The Foothills widen again northward across a major lateral ramp at Williston Lake (Figure 2). Barss and Montandon (1981) describe the surface structure as being dominated by northwest–southeast-trending folds and southwest-dipping thrust faults in the Upper Jurassic to Cretaceous clastic strata (see also McMechan, 1994; Lingrey, 1996). The foothills belt is bounded to the west by the Front Range Fault, which carries (variously) Triassic and Paleozoic rocks in the hanging wall, and to the east by a prominent frontal monocline (Figures 2, 4). Barss and Montandon (1981)

Table 1. Reservoir unit lithologies.

Unit	*Description of lithology*
Pardonet P4	white to light grey, occasionally buff limestone, locally partially dolomitized, commonly fractured
Pardonet P3	dark to medium grey dolomites, dolomitic s&s, & dolomitic silts, commonly highly fractured
Pardonet P2	dark grey, flaggy, weathering dolomitic silts & shales
Pardonet P1	calcareous black shaly siltstone with a high organic & phosphatic content & local layers of bone fragments & pelecypod pearls
Baldonnel B4	pelloidal wackestones, as much as 50% fine s& or silt, some cleaner stringers of bioclastic dolomites toward the top
Baldonnel B3	pelloidal wackestones, as much as 50% fine s& or silt, some cleaner stringers of bioclastic dolomites toward the top
Baldonnel B2	bioclastic dolomites with good biomoldic porosities of 5–7%
Baldonnel B1	clean buff to cream bioclastic & pelloidal dolomites, porosities as much as 10%

also commented on the disharmony of the structures in the Upper Triassic carbonates and the underlying Mississippian and Devonian carbonates and noted detachments in the Fernie Formation, the anhydrites of the Triassic Charlie Lake Formation, and the Devonian Besa River Shales. McMechan (1985) recognized another detachment in the Lower Triassic Montney Formation shales.

Our observations, based on a compilation of surface geology mapping, analysis of seismic data, and analysis of well data, suggest additional detachment levels within shales of the Nikanassin, Moosebar, and Hasler Formations (Figure 3). These detachments subdivide the sedimentary sequence into a series of tectonostratigraphic packages that deform, to some extent independently and each with its own characteristic deformation style, and that generally exhibit different amounts of shortening.

Drilling results and seismic data indicate that the dominant tectonic style within the Triassic package is above the Montney detachment and fault-propagation folding, with the upper Charlie Lake and Baldonnel dolomites providing the rigid member that controls the periodicity of the structures (Figure 4). The folds can have amplitudes of as much as 1000 m and have a typical wavelength of 2 km. The elevation of the structures is a function of the amplitude of the Triassic structure and the position of the Triassic structures with respect to the deeper, broader-wavelength structures that appear to control the major regional structural highs and lows (Figure 4).

Several wells have penetrated the steeply dipping northeast limbs of the fault-propagation and detachment folds where fold morphologies are constrained by seismic data (Figure 4). Other wells have penetrated the backlimbs and hinges, thereby providing additional data on the typical fold geometries. The northeasterly dips of the forelimb strata commonly steepen with depth, and may even become vertical to overturned. Cooper (1991) gives a detailed analysis of the geometry of a typical fold structure in the area, based on resistivity image logging tools.

THE MONKMAN TRIASSIC PLAY

Play Elements and Exploration History

The southern limit of the Monkman Triassic play fairway is defined by the erosional truncation of the Pardonet and Baldonnel Formations' reservoirs (Table 1) (Barss and Montandon, 1981; Davies, 1997b) by the sealing Fernie Formation (approximately coincident with where the Foothills cross the Alberta/British Columbia boundary). The western limit is defined by the Front Range Thrust, which causes the reservoir formations to outcrop; the eastern limit is defined by the frontal monocline that marks the termination of significant deformation and hence of the fractured structural traps that characterize the fairway. The northern limit is the Williston Lake lateral ramp (Figure 2), across which a fundamental change in structural style is observed (Cooper, 1998).

The key elements of the play are summarized in Figure 5. Pervasive fracturing is of key importance, because the primary porosities in the reservoirs average only 4%. Heavily fractured reservoirs can produce at rates greater than 50 mmcfg/d, whereas wells with sparse fracturing are often subcommercial. Top and lateral seals for the traps are the deformed shales of the Fernie Formation. The source rocks may be the Jurassic Fernie and Nordegg Formations, but alternative possibilities are the shaly intervals within the Pardonet or deeper Triassic formations. It is not possible to type the gas, because it has virtually no ethane or higher hydrocarbons.

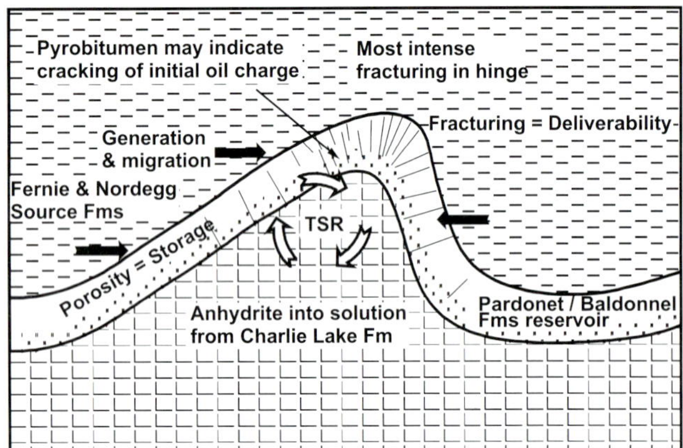

FIGURE 5. Generic structural-trap type in the Monkman area, illustrating the major play elements of the producing gas pools. TSR = thermochemical sulphate reduction.

The Monkman Triassic play has been explored intermittently since the late 1950s (Montgomery, 1994). The dominant company during the early phase of exploration, Triad Oil (now Talisman Energy), drilled three wells that they located using surface geology to test the Mississippian reservoirs that had been successfully explored in the Alberta Foothills. Drilling activity and success rates have fluctuated over the years (Figure 6a, b). The first commercial gas field, Sukunka, was discovered in 1965 by use of surface geology and seismic-reflection data (Barss and Montandon, 1981). The next commercial gas discovery, the Bullmoose field, was not drilled until 1975. During the intervening 10-year period, 12 wells were drilled, all of them dry and abandoned (D&A) except for two noncommercial gas discoveries. This reflects the difficulty of exploring the play with the poor-quality seismic data of the period. During the late 1970s, another 15 wells were drilled, yielding six commercial wells and hence sufficient reserves to allow infrastructure development. A series of dry and noncommercial wells in 1980 and 1981 coupled with a weak gas market effectively killed the play; from 1982 to 1987 only five wells were drilled. In 1988, a Shell discovery caused renewed interest in the play; in the next decade, 76 wells were drilled, of which 17 were D&A, 20 were noncommercial, and 39 were commercial gas wells (Figure 6b). Ten of the commercial wells yielded the high flow rates (>50 mmcfg/d) that make the play attractive. The success rate is attributed to an improvement in seismic acquisition, specialized seismic processing, and the application of rigorous principles of structural geology to the analysis of the seismic data, well data, and surface geology. This success has led to the development of many of the pools and an extensive pipeline infrastructure (Montgomery, 1994) (Figure 7).

Case Study 1: The Talisman Burlington Wolverine Well, b-30-C/093-P-03

The Talisman Burlington b-30-C/93-P-03 Wolverine well is located on a surface structural culmination northeast of the Murray River field (Figure 7). The well

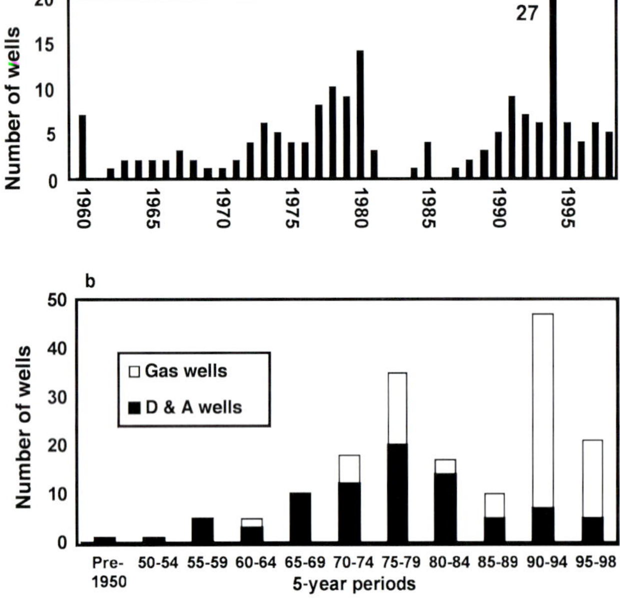

FIGURE 6. (a) Graph showing the number of wells drilled on the Monkman play trend since 1955. (b) D&A and successful wells drilled; including noncommercial gas tests.

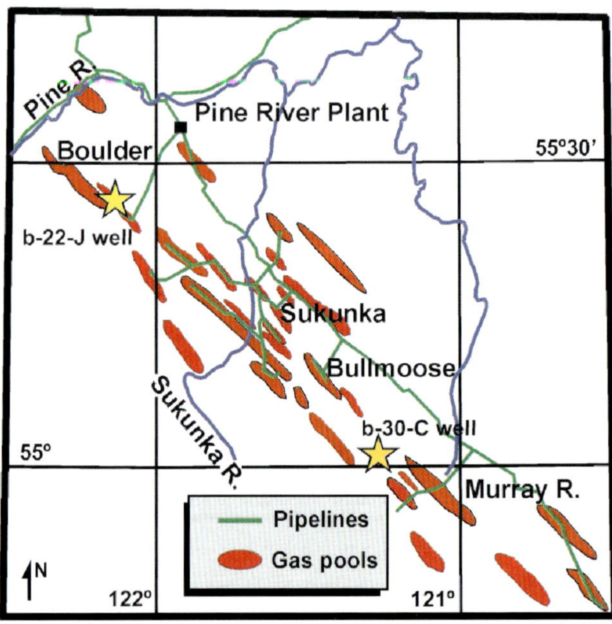

FIGURE 7. Map of the Monkman play trend, showing major fields and pipeline infrastructure; the two case study wells are located by yellow stars.

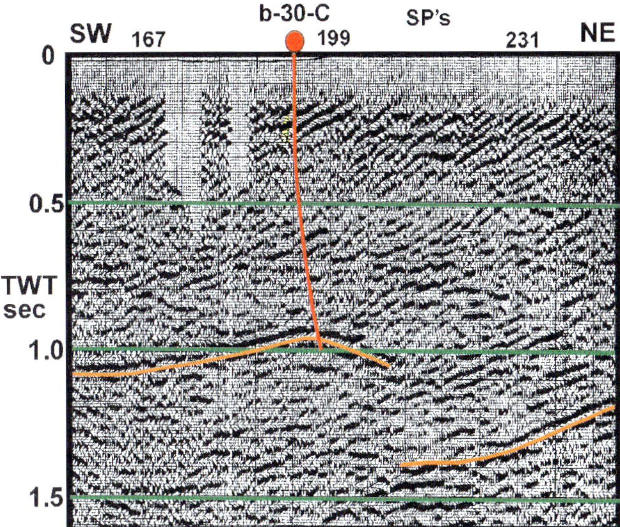

FIGURE 8. Seismic line through the b-30-C/93-P-03 well location; orange is the Top Triassic reflector.

was drilled in 1993 to test a structure identified on 2-D seismic data (Line BPC-31, shown in Figure 8). The predrill structural model, based on the seismic data and surface geology, predicted an east-vergent fault-propagation fold overlain by imbricated Fernie Formation and lower Nikanassin strata, with an upper detachment in the Nikanassin (Figure 9). The original hole was abandoned because of excessive torque and drag in the drill pipe in the stressed shales of the Fernie Formation. When the hole was logged, the dip data in the lower part of the hole indicated that the well had missed the hinge zone of the structure and was in the forelimb (Figure 10).

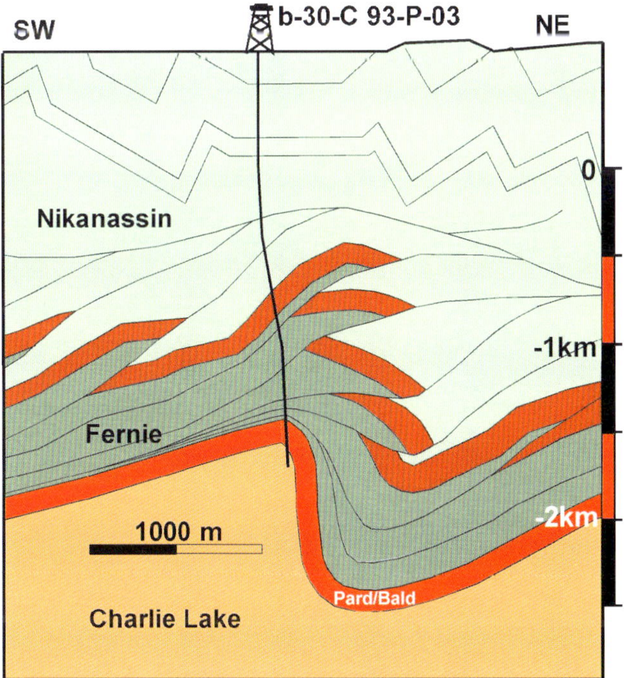

FIGURE 9. Predrill structural model of the b-30-C/93-P-03 well location, based on seismic data (Figure 8) and surface geology. For key to formation colors see Figure 3.

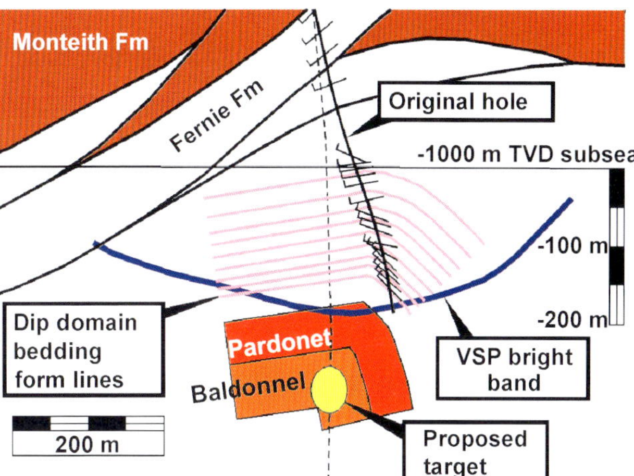

FIGURE 10. Proposed sidetrack target for the b-30-C/93-P-03 well, based on dip and VSP data.

The interpretation of seismic data to define an exact target location is at best only approximate. In areas of complex geology, the effects of refraction and anisotropy above the target often cannot be accounted for in seismic processing. The velocity field is never known accurately, and the quality of the seismic image depends on the choice of acquisition parameters and processing flows. Thus, it is not uncommon to drill to the proposed target location at depth without finding the target reservoir and to deepen the well until it is clear that the target has been missed or that the interpretation is incorrect. A VSP (vertical seismic profile), dipmeter data, and/or other logs can reduce the uncertainty in retargeting the structure for a sidetrack well.

Discussion of the VSP in the b-30-C Example

The VSP provides velocity information and also can look "away from the wellbore." It can only be used effectively in conjunction with dipmeter and other log information. The deviation of the wellbore and the location of the source, which in the Monkman area is generally restricted by local topography, are important. Some preliminary VSP modeling before acquiring the VSP helps us to determine which potential source location is best.

Ideally, for an acceptable 2-D migration, the VSP sources and wellbore should be coplanar in the dip direction (in this case, 225°/045°), but source locations were confined to the well-site access road because of the topography. The source location northeast of the well was approximately coplanar with regional dip and well

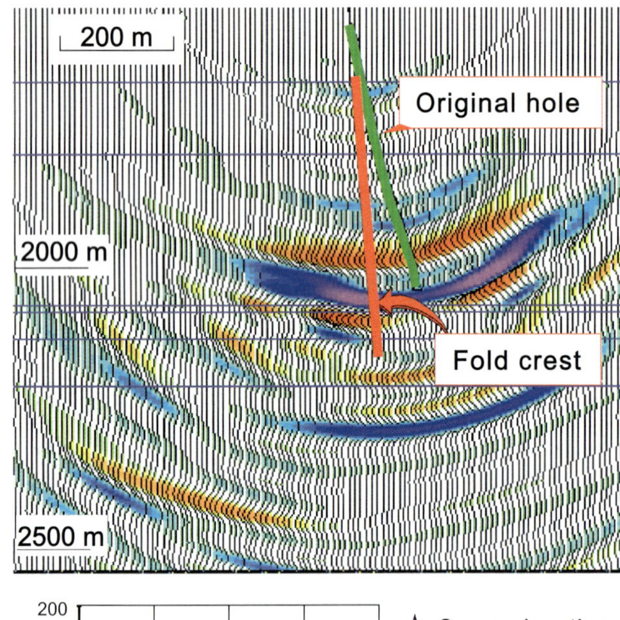

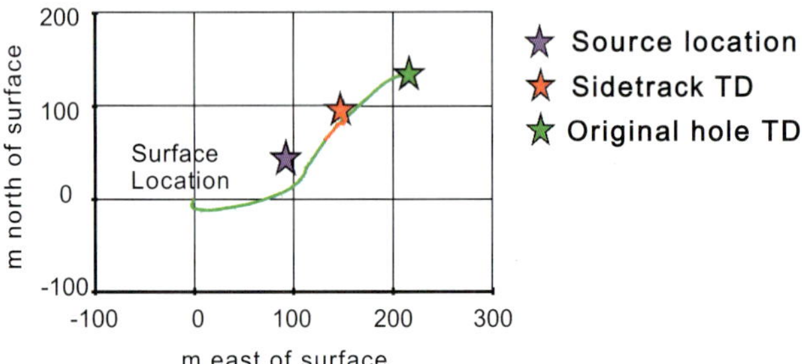

FIGURE 11. Migrated VSP data from the original b-30-C/93-P-03 well; see text for discussion.

deviation (Figure 11). This geometry permits the data to be processed employing the VSP Kirchhoff-Depth Migration algorithm in an attempt to image the position of the hinge zone using the method of Slawinski and Parkin (1996). The method relies on the fact that a point source, a point receiver, and a point scatterer in a homogeneous isotropic medium will yield an ellipse of possible locations for the position of the scatterer (Slawinski and Parkin, 1996). The point scatterer of the illuminating seismic energy is the hinge zone of the fold at the top of the Pardonet Formation, which is the preferred target for the well. Different combinations of source and receiver locations will yield different ellipses; in the migrated image of the VSP data, the intersection of many ellipses will create an area of high amplitudes indicating the likely position of the point scatterer (Slawinski and Parkin, 1996). Errors will be introduced by structural complexities distorting raypaths and by variations in velocity from the surface to the target.

The migrated VSP showed a prominent bright band of energy response, with a maximum 100 m to the west of the original hole's TD (Figures 10 and 11). The "smiling" geometry of the amplitude maximum results from the superposition of ellipses; it does not represent the geometry of the reflector (Slawinski and Parkin, 1996). The sidetrack well was designed using the dip and stratigraphic data from the original wellbore to create a new geometric model of the fold constrained by the VSP data (Figure 11). The sidetrack well encountered the top of the Pardonet Formation 30 m structurally higher than the model's prediction, which was remarkably close to actual structural geometry (Figure 12).

Case Study 1: Learnings

The sidetrack well confirmed the value of the VSP as a tool for refining the target for a sidetrack (Cooper et al., 1995; Slawinski and Parkin, 1996; Morrison et al., 1998). However, the results are most useful when integrated with other data, such as the dipmeter and stratigraphic data available in the original wellbore. The well was a technical success in that the target was tested, but it was a commercial failure because of low porosity in the reservoir. VSPs have now been used on several wells to more effectively target sidetracks from initially unsuccessful wellbores.

The result is more cost-effective testing of the chosen target with a sidetrack, when the original wellbore is mislocated with respect to the structure. The causes of the original wellbore being in the wrong position with respect to the targeted structure are varied but include the following.

1) Drilling is off the front of the structure. 2-D seismic data migration often tends to smear the top Triassic reflector farther to the northeast than is valid; this is because the migration algorithms are unable to correctly migrate the steep dips of the forelimb. In

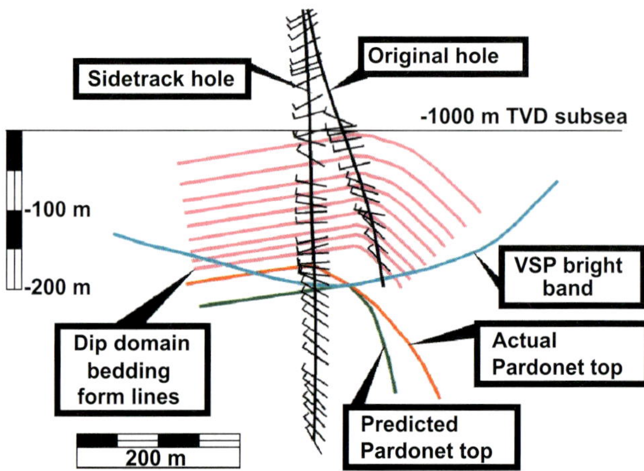

FIGURE 12. Pre- and post-sidetrack comparison of structure in the b-30-C/93-P-03 well, showing key well data, VSP data, and bedding form lines. Note the similarity of the predicted and actual positions for the top of the Pardonet Formation in the sidetrack well.

our experience in the Monkman area, the VSP method discussed by Slawinski and Parkin (1996), modified to use a single-offset VSP, is most effective in addressing this problem if the wellbore is within 150 m (and preferably within 60 m) of the targeted Triassic fold hinge. If the wellbore is much farther away, the information becomes too ambiguous to be reliable.

2) Drilling is on the backlimb, where fracturing and thus productivity are likely to be poor. In this situation in the Monkman area, the VSP is of limited value in designing the sidetrack well.
3) Difficult drilling conditions prevail in the Fernie Formation shales. The Fernie Formation shales are tectonically stressed and pervasively microfractured, particularly on the northeasterly dipping forelimbs of the propagation and detachment folds. Thus, wells that penetrate that part of the structure are prone to hole problems such as sloughing shales, deviation-control problems, and drillstrings that become stuck. This commonly results in the well having to be abandoned and a sidetrack well being required.
4) Variations occur in interval velocity to the top of the Pardonet Formation. The velocity of the strata from the surface to the top of the Pardonet Formation can vary from as low as 3900 m/s to as high as 4800 m/s. As a result, the top of the target can vary considerably from that predicted prior to drilling the well. Many of the wells are drilled directionally to a target that is a rectangular box in 3-D space; an increase in depth to target can affect the validity of the *x* and *y* coordinates of the target box. In this case, the velocity information from a VSP or a checkshot survey may be sufficient for us to decide on the next course of action.

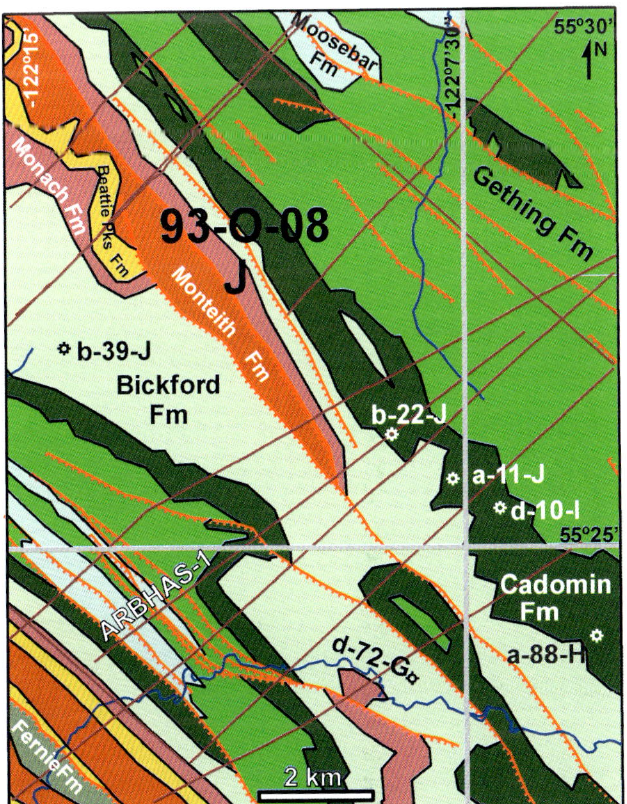

FIGURE 13. Surface geology of the Boulder area, showing well locations and 2-D seismic lines. For key to formation colors see Figure 3.

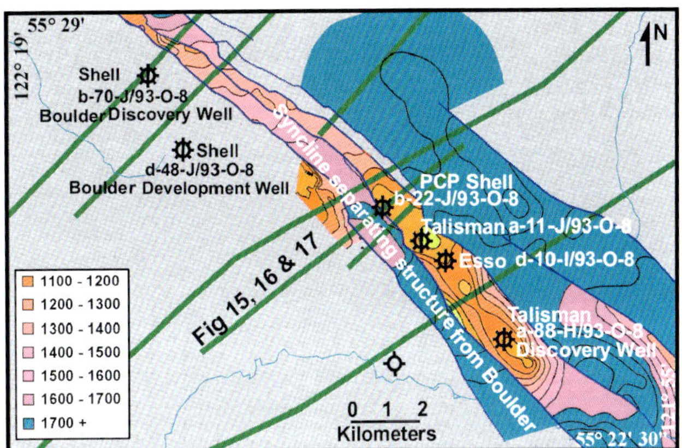

FIGURE 14. Predrill depth map on the top of the Triassic for the Boulder structure, showing key wells and seismic lines. The predrill depth map of the structure was based on 2-D seismic data in the area and on Shell's 3-D seismic survey of the Boulder area.

Case Study 2: The PCP Shell Boulder Well, b-22 J 93-O-08

The well is located on the eastern flank of a surface structural culmination (Figure 13) in the northern part of the established Monkman trend, southeast of the b-39-J 93-O-08 well in the Boulder field, which was discovered in 1988. The structure was first tested by the Talisman Ocelot a-88-H 93-O-08 well (Figures 13 and 14). The predrill structural model of the well used the surface geology and an arbitrary line extracted from a Shell 3-D seismic survey (Figure 15). The structure was interpreted to be a detachment fold in Triassic strata, with subvertical southwest and northeast limbs that detached in the Montney Formation shales (Figure 16). This interpretation was based on the high-amplitude Triassic reflectors on the crests of the two structures and the sudden shift downward in time of the Triassic high-amplitude reflector in the intervening syncline. Above the Fernie detachment, the Nikanassin Group is deformed into a disharmonic series of folds and minor thrusts. Of particular note is a major west-vergent back thrust, to the west of the well's surface location, that elevates the lower part of the Nikanassin Group to the surface in its hanging wall.

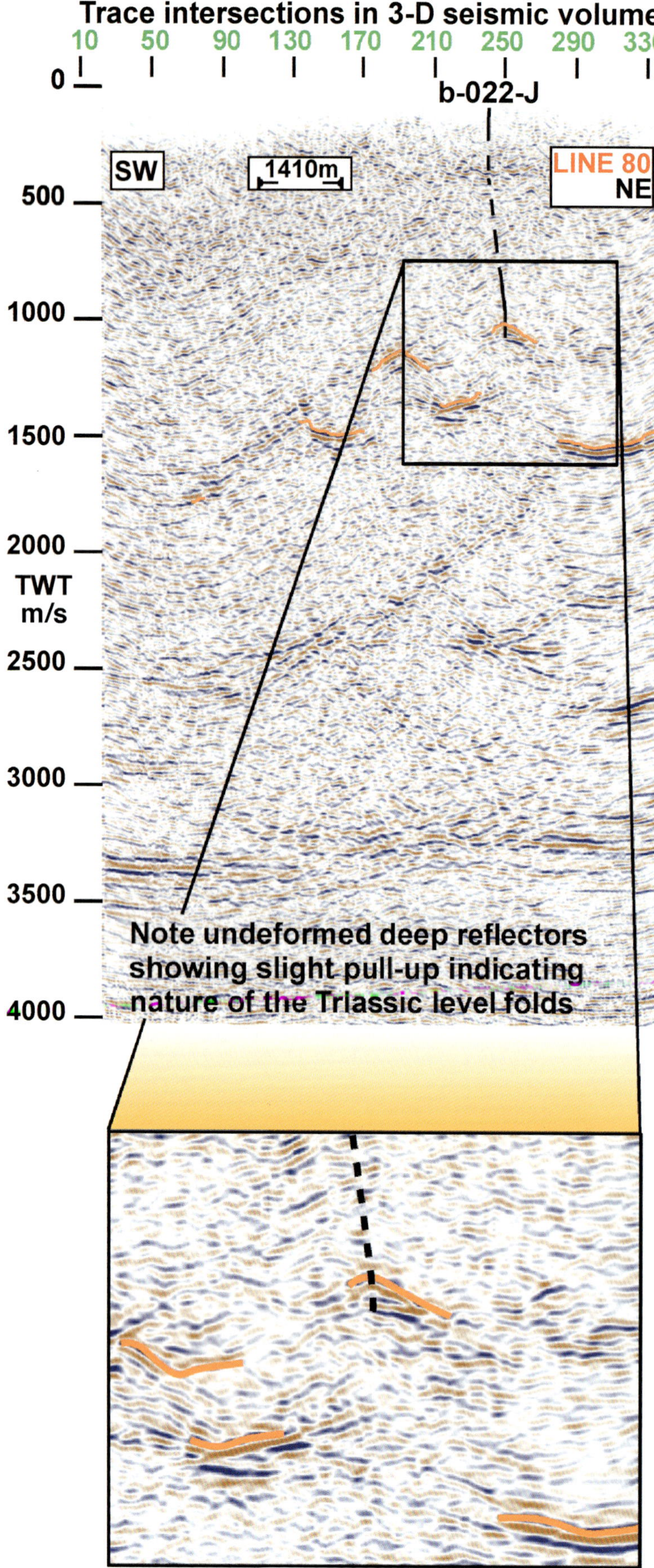

FIGURE 15. Predrilling interpretation of seismic Line 80 from the Shell 3-D seismic data volume. Top Triassic is shown in orange. For line location see Figure 14.

Postdrill Structural Model of the Original Wellbore

A stereogram of poles to bedding dips in the reservoir section of the original wellbore shows that the well penetrated the backlimb of the structure rather than the fold hinge predicted by the predrill interpretation (Figures 17 and 18). This positioning error may be the result of complex raypaths distorting the position of the structure on a time migration. The structural section is otherwise similar to the predrill model.

The original wellbore showed intense fracture development throughout the Baldonnel. The fractures were small and tight, with calculated hydraulic apertures rarely exceeding 0.001 mm. The fracturing thus appeared to be pervasive microfracturing, which does not, in general, yield the high flow rates characteristic of the trend. In the Pardonet, a few larger, discontinuously open fractures were developed, which suggested localized leaching. The logs indicated high resistivities throughout, suggesting that no water zone was present in the well. The porosity and net pay of the reservoir were significantly better than predicted and higher than is conventionally seen within the trend. It is possible that the higher porosity rendered the Baldonnel less brittle than normal, and hence, a lower density of effective fracturing developed. The well was completed open hole, and, following HCl acid stimulation, the well flowed 14.5 mmcf/d of gas at 7900 kPa. The production log of the wellbore during the flow test indicated that the majority of the gas was flowing into the wellbore from a small zone in the Baldonnel B1 sequence. This zone does not appear to be very heavily fractured on the FMI (Formation Micro-imager) log, but the rock does have a highly sheared appearance (Figure 19). It is not uncommon for highly fractured zones to contain relatively few measured fractures because the pervasive shattering makes it very difficult to confidently correlate an individual fracture across the image log. The conclusion is that the productive interval is pervasively microfractured.

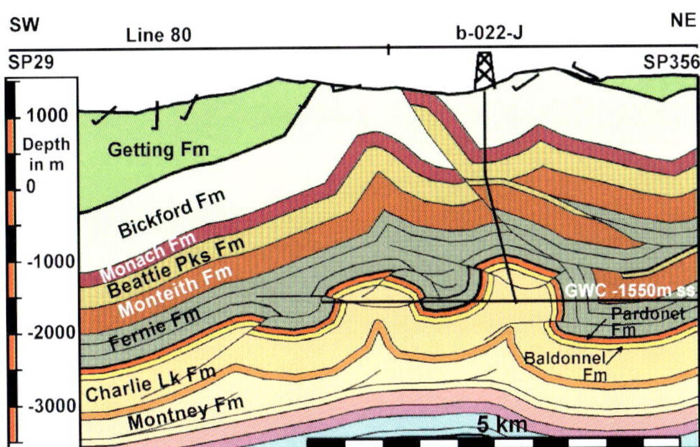

FIGURE 16. Predrilling structural cross section through the well, based on 3-D seismic and surface geology. For line location see Figure 14. For key to formation colors see Figure 3.

Planning the Sidetrack Well

The relatively disappointing flow rate of the well prompted the evaluation of a side-track well that was targeted at the hinge of the fold to intersect more fracture swarms and improve well deliverability. The 3-D seismic data suggested that a 500-m horizontal-displacement sidetrack would cross the crestal portion of the fold into the forelimb (Figure 20). The most efficient azimuth to reach the forelimb from the original well would have been to the northeast; however, this was not practicable because of the proximity of the eastern boundary of the gas spacing unit (Figure 21). The spacing-unit boundaries are designed to prevent unfair drainage of gas from lands adjacent to the unit on which the producing well is located. The revised depth map, which was based on remapping of the 3 D seismic data, the well data, and the average orientations of the fracture systems (Type 1, dipping 65° to 115° and Type 2, dipping 64° to 62°), suggested an azimuth for the sidetrack of 345° (Figure 21). This allowed a margin of error if the well turned

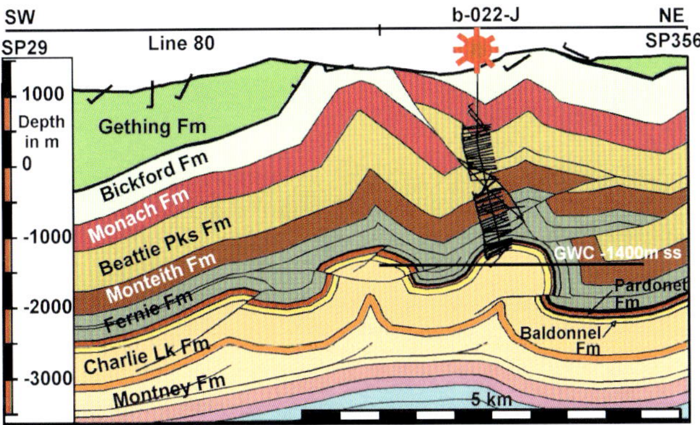

FIGURE 17. Postdrilling structural cross section through the well based on 3-D seismic, dip and stratigraphic data from the well and surface geology. For line location see Figure 14. For key to formation colors see Figure 3.

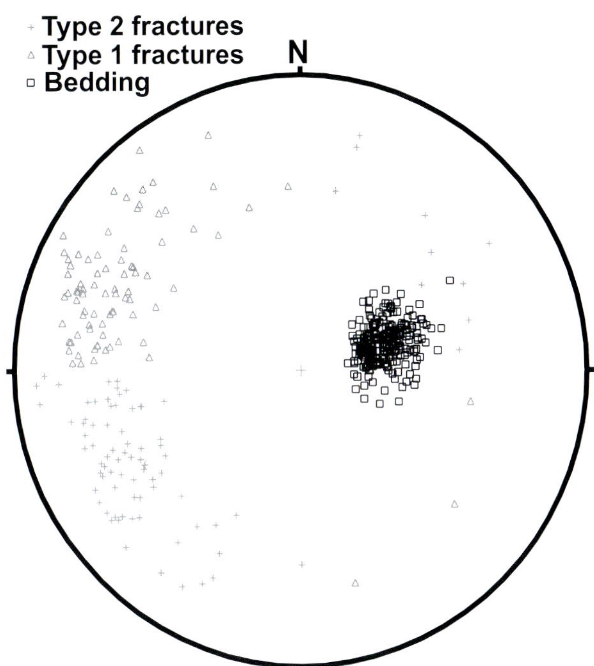

FIGURE 18. Stereographic projection of poles-to-bedding and fracture-dip data, derived from the FMI log run in the original wellbore.

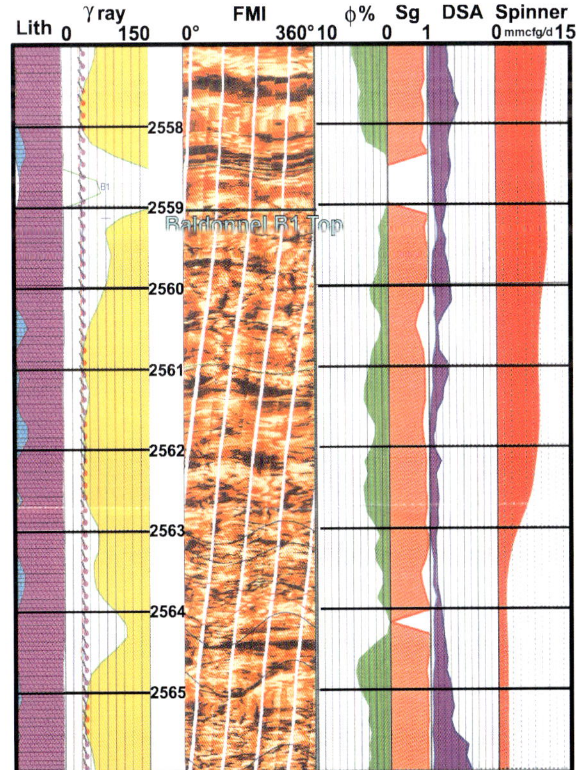

FIGURE 19. FMI resistivity image log and other log-derived data from the productive interval of the original wellbore. Lith = Calculated lithology; γ ray = Gamma-ray log; ϕ = calculated porosity; S_g = gas saturation; DSA = dipole shear anisotropy (color-filled curve separation indicates degree of anisotropy); Spinner = gas-production flow-rate log.

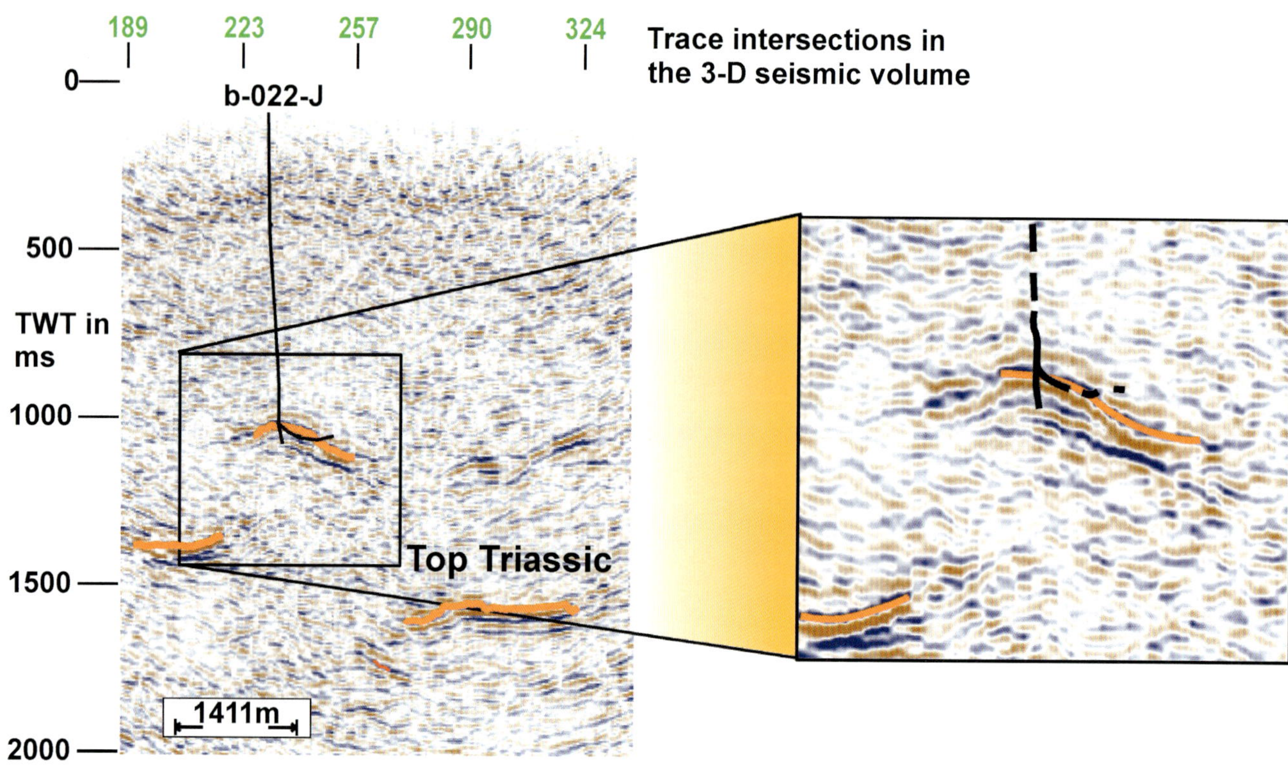

FIGURE 20. Pre-sidetrack interpretation of a seismic line extracted from the Shell 3-D seismic data volume; the line is oriented along the proposed sidetrack azimuth. Top Triassic is shown in orange. For line location see Figure 21.

naturally to the right once the drill bit had crossed into the forelimb. The dip data from the original well, combined with the new seismic map, were used to create a model along the chosen line of azimuth for the sidetrack (Figure 22). Three potential deviation-angle-well trajectories are shown on the model. A deviation angle of 80° from the vertical was chosen as the most suitable angle for achieving maximum wellbore exposure to the best reservoir section in the Baldonnel Formation (the

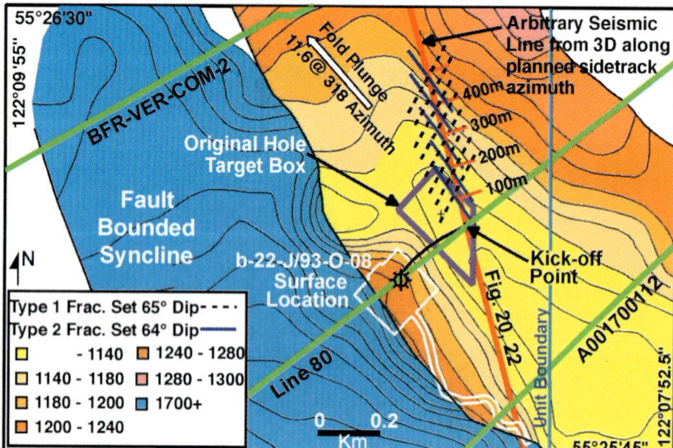

FIGURE 21. Pre-sidetrack depth map of the top of the Triassic on the Boulder structure, modified to account for the results of the original wellbore.

B1 and B2 units, Figure 22). However, the model had significant uncertainty in the distal portion because of the poorly constrained position of the fold hinge, based on the available data.

During the drilling of the sidetrack hole, stratigraphic position within the Baldonnel Formation was correlated to the original hole by use of an MWD (measurement-while-drilling) gamma-ray tool (Figure 23). The MWD gamma ray in true vertical depth yielded an acceptably undistorted log for correlation over most of the sidetrack hole. Where the wellbore was subparallel to the bedding, corrections for the assumed bed dip were applied, based on the pre-sidetrack structural model. The deviation angle of the sidetrack well was progressively increased to 95° once it became apparent that the hinge of the fold was flatter and broader than predicted.

Postdrill Structure of the Sidetrack Well

The sidetrack well just grazed the top of the Charlie Lake Anhydrite and then repenetrated the Baldonnel in reverse sequence, reaching a total depth of 3109 m in the Upper Baldonnel Formation. It achieved a total horizontal displacement of 670 m from the kickoff point (Figure 24). The dip of the beds began to change rapidly toward the end of the sidetrack well where the well crossed from the hinge into the forelimb of the structure (Figure 24). The sidetrack well cut obliquely across

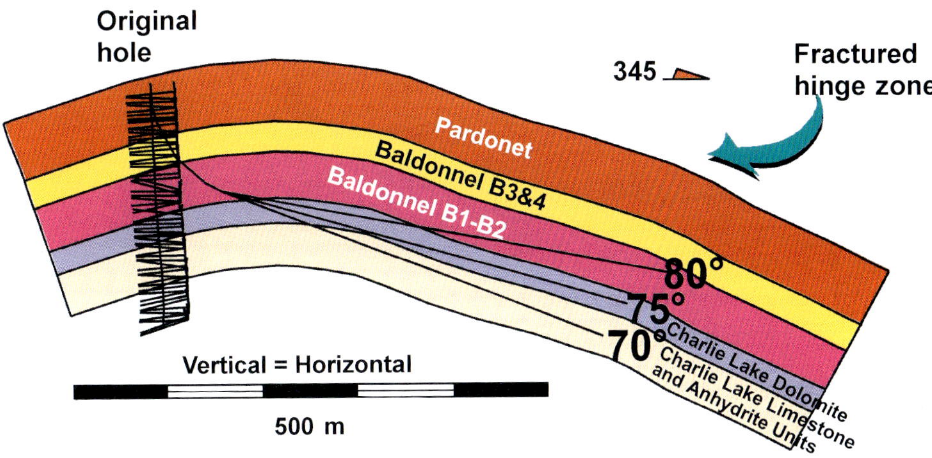

FIGURE 22. Pre-sidetrack structural cross section along the path of the proposed sidetrack well, based on 3-D seismic, dip, and stratigraphic data from the original well. For line location see Figure 21.

the fold at an acute angle to the fold axis (Figure 20), but, on a stereogram (Figure 25), the poles to bedding do not describe a great circle distribution characteristic of a cylindrical fold. Instead, the bedding poles are distributed around a small circle, which suggests that the fold is noncylindrical and has a variable plunge (Figure 25). The fracture data show a larger number of poles to Type 1 fractures. This may be the result of the orientation of the sidetrack, which is subperpendicular to the Type 1 fractures and subparallel to the Type 2 fractures (Figure 25). This will undersample the Type 2 fractures in the vicinity of the wellbore.

The structure has a relatively flat-topped crest and must have a sharp transition to the steeply dipping limb just tagged by the sidetrack (Figure 26). The 3-D seismic image was misleadingly positioned, and the sidetrack

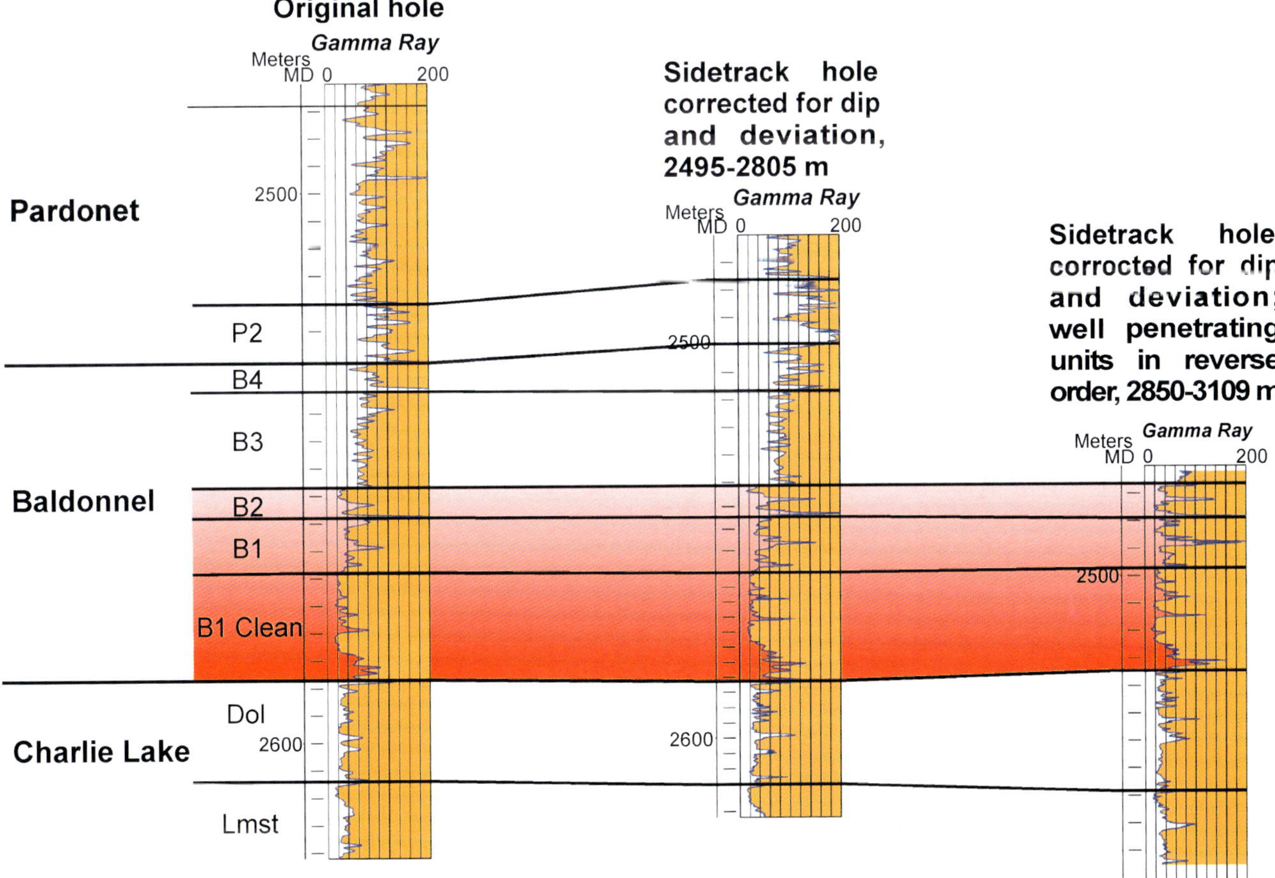

FIGURE 23. Correlation of gamma-ray logs from the original hole, the sidetrack hole, and the portion of the sidetrack hole penetrated in reverse stratigraphic order by the distal segment of the sidetrack hole. The gamma-ray logs for the sidetrack hole have been corrected for the distortions of log signature introduced by the wellbore deviation and the dip of the beds.

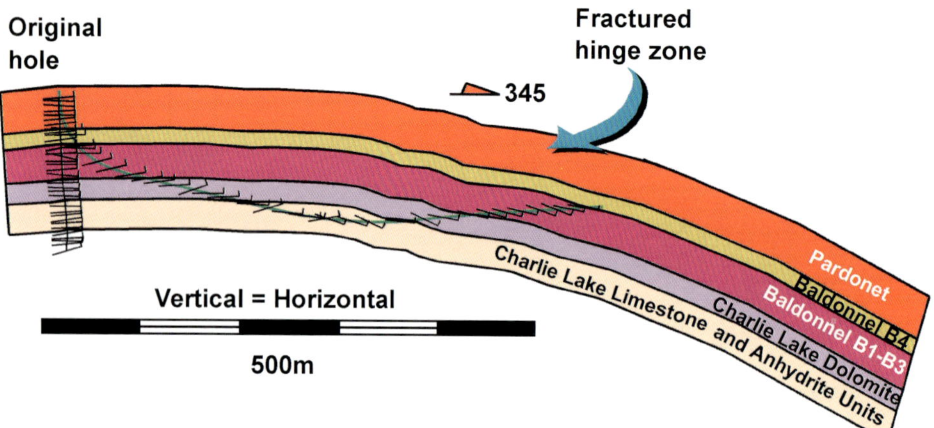

FIGURE 24. Post-sidetrack structural cross section along the path of the sidetrack well, based on 3-D seismic, dip, and stratigraphic data from the original and sidetrack wellbores. For line location see Figure 22.

well followed the west flank of the feature for much of its length before crossing into the plunging hinge in the last 200 m. When the sidetrack well is projected into a detailed northwest-southeast cross section, the elevation of the top B1 marker in the Baldonnel Formation drops 60 m from the original hole to the sidetrack hole (Figure 26). This is the result of the noncylindrical, variably plunging character of the folding. This 60-m drop occurs over a distance of about 600 m and implies that, in this area, the fold has an average plunge of 6° to the northwest.

Because of the small radius of the sidetrack's curvature, options for logging the well were limited to an image log and a gamma-ray log. Fracturing was pervasively developed throughout the sidetrack well, with several heavily fractured zones present (Figure 27). The well was stimulated with HCl acid and flowed at more than 30 mmcfg/d at pressures greater than 10,000 kPa; the well was put on production in October 1995.

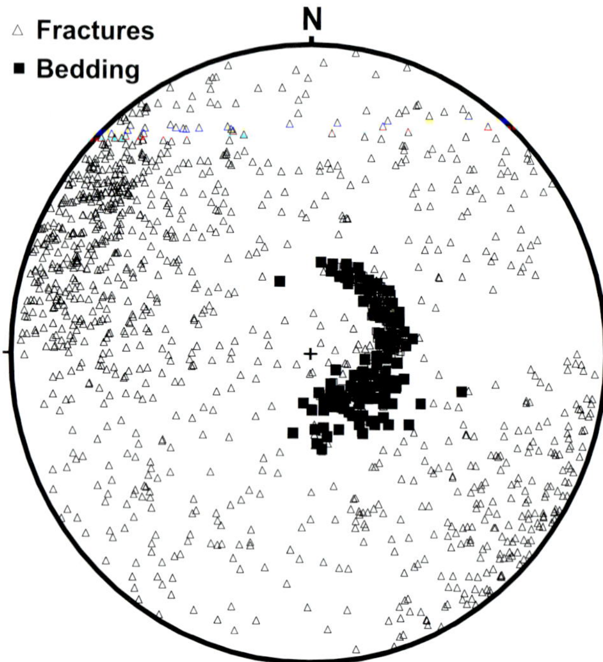

FIGURE 25. Stereographic projection of poles-to-bedding and fracture-dip data derived from the FMS log run in the sidetrack wellbore.

Production History of the Well

The production history of the well to May 31, 2000, shows that rates initially declined rapidly from 30 mmcfg/d, stabilized at 18 mmcfg/d for awhile, and then declined more slowly to a rate of 2.8 mmcfg/d (Figure 28). The well had produced 11.4 bcf of gas by May 2000, and the rate/cumulative graph implies an ultimate reserve of approximately 14 bcf from the well (Figure 29). However, in fractured reservoirs, typically the gradient will eventually flatten and a higher ultimate reserve is expected. This can be illustrated by an earlier analysis of ultimate reserves using the rate/cumulative plot by Cooper et al. (1997), which concluded that the ultimate reserves were 8 bcf, an estimate now exceeded by production from the well.

The discovery well on the structure was the Talisman Ocelot a-88-H 93-O-08 well, which by December 31,

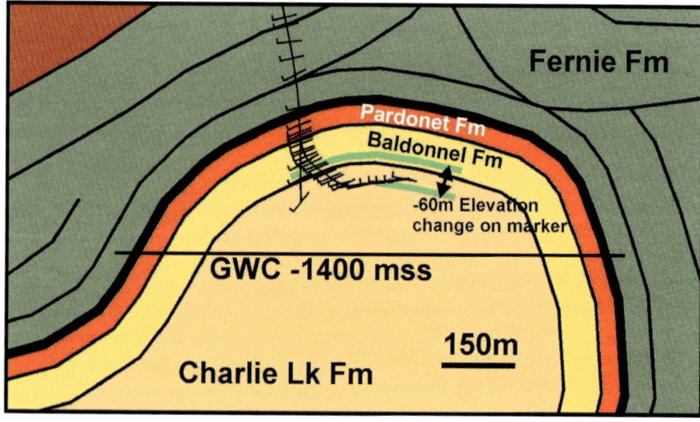

FIGURE 26. Post-sidetrack structural-dip-section model of the Boulder well, showing key well data and bedding form lines. Note the change in elevation of the Baldonnel Formation marker in the sidetrack well as a result of the plunge of the fold. MSS = meters subsea.

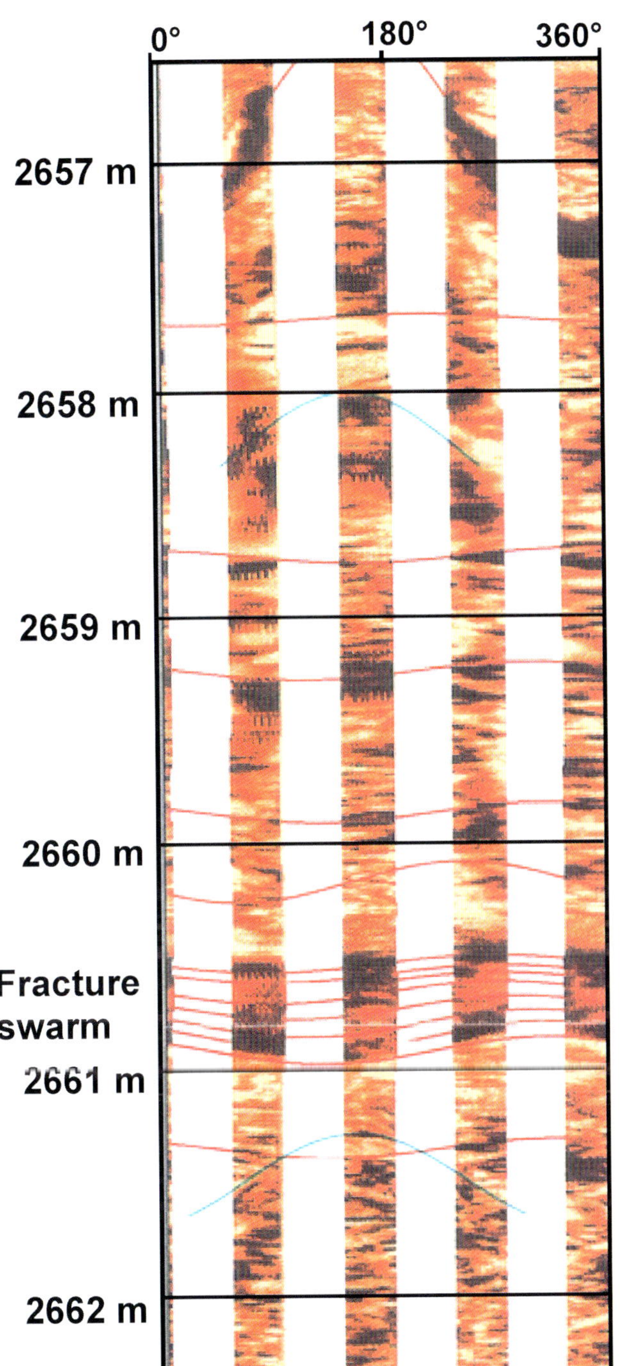

Figure 27. FMS resistivity image log from one of the inferred productive intervals of the sidetrack wellbore (2660.5-2661m measured depth). Red lines are Type 1 fractures and blue lines are Type 2 fractures.

1999, had produced 47 bcf with a reported rate of 10 mmcfg/d. The other two wells on the structure were both sidetracked twice in attempts to obtain commercial rates. In the case of the Esso Chevron Boulder d-10-I 93-O-08 well, commercial rates of gas were never obtained. The well was never tied in but was suspended as a possible water-disposal well. The Talisman Numac a-11-J 93-O-08 well produced only 2 bcf and is now shut in. All the wells were tied in to production facilities at the same time.

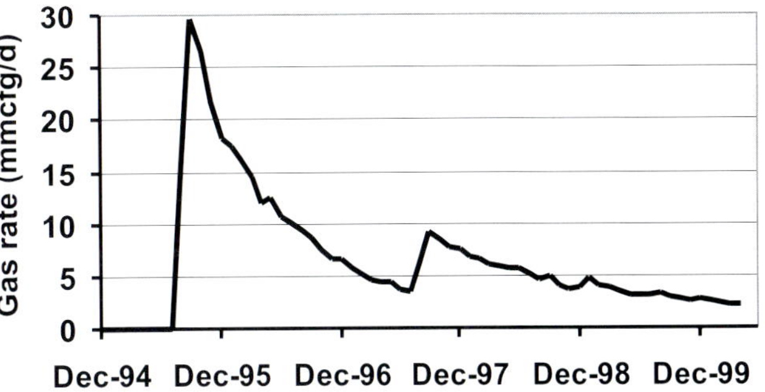

Figure 28. Production history profile of the b-22-J/093-O-08 well, from tie-in to May 31, 2000.

Case Study 2: Learnings and Business Impact

The predrill economics for the well were strongly positive, based on 47 bcf of gas reserves that could initially be produced at 40 mmcfg/d for a flat life of three years before the production rate would decline. Sidetracking the well resulted in incremental costs, and the production history shows that both the production rate and reserves did not meet predrill expectations. This well was a technical success in an area of complex structure with difficult drilling conditions. It has, however, been a commercially marginal venture.

The predrill model of well deliverability was too optimistic and assumed that, because the structure was tightly folded and highly elevated, it would be highly fractured and would yield flow rates similar to those from the best wells in the trend. This may be the result of the higher porosities in the Baldonnel Formation in the well making the rock less susceptible to fracturing. The flow rates of wells in the trend are variable and depend in large part on whether a major fracture swarm and/or fault is intersected by the wellbore. This must be considered as an irreducible risk in the play, because it cannot be predicted with any certainty.

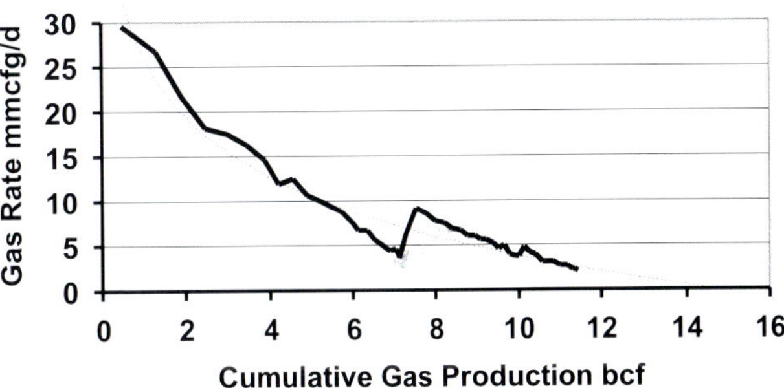

Figure 29. Cumulative gas production versus daily gas rate plot for the b-22-J/093-O-08 well, from tie-in to May 31, 2000. The dashed best-fit line predicts the ultimate recoverable reserves expected from the well.

Table 2. Sidetrack well operations in the Monkman trend, 1978–present.

Well Location	Year	VSP	Reason	Result	Status
c-27-I 93-I-15	1978	√	Unknown	4 mmcfg/d	SI gas
a-43-E 93-P-03	1979		Missed target	14 mmcfg/d	SI gas
d-78-K 93-P-05	1980		Missed target	Noncomm	SI gas
d-33-H 93-P-04	1981		Missed target	Lost hole	D&A
c-88-H 93-P-04	1990	√	Stuck in Fernie	10 mmcfg/d	Gas
d-33-H 93-P-04R	1991	√	Stuck in Fernie	10 mmcfg/d	SI gas
d-70-I 93-P-04	1991		Missed target	40 mmcfg/d	Gas
a-44-G 93-I-14	1992		Missed target	Noncomm	D&A
b-30-C 93-P-03	1992	√	Stuck in Fernie	Noncomm	SI gas
c-38-C 93-P-05	1992		Gas test rate	10 mmcfg/d	SI gas
b-02-E 93-P-05	1992	√	Stuck in Fernie	30 mmcfg/d	Gas
d-94-A 93-O-08	1993		Stuck in Fernie	40 mmcfg/d	Gas
d-02-C 93-P-05 R	1993	√	Gas production rate	5 mmcfg/d	Gas
c-87-C 93-P-05	1993	√	Missed target	Noncomm	SI gas
b-47-H 93-I-14	1994		Missed target	Noncomm	D&A
b-22-J 93-O-08	1994		Gas test rate	40 mmcfg/d	Gas
c-61-A 93-O-08	1994		Missed target	70 mmcfg/d	Gas
c-83-G 93-P-04	1994		Gas test rate	20 mmcfg/d	Gas
c-16-J 93-P-04	1994	√	Missed target	Dry hole	D&A
c-54-J 93-P-04*	1994	√	Missed target	20 mmcfg/d	Gas
d-10-I 93-O-08	1994		Missed target	Dry hole	D&A
c-22-E 93-I-15	1995	√	Backlimb test	Noncomm	D&A
a-11-J 93-O-08	1995		Poor reservoir	27 mmcfg/d	Gas
c-55-I 93-I-14	1995	√	Drilled structure to E	20 mmcfg/d	Gas
d-64-J 93-I-14	1995	√	Missed target	52 mmcfg/d	Gas
c-39-F 93-O-08 R	1997		New interpretation	50 mmcfg/d	Gas
b-81-B 93-P-05	1997		Poor reservoir	45 mmcfg/d	Gas
d-96-G 93-P-04	1997		Stuck in Fernie	Dry hole	D&A
a-05-G 93-P-04	1998		Oil mud correction	Dry hole	D&A
d-30-C 93-O-09 R	1998		Stuck in Fernie	Unknown	SI gas

"R" after the well location denotes a reentry of an existing well.
** denotes a multioffset VSP.*

Although the 3-D seismic data were believed to provide a reliable picture of the geometry of the structure, this was not the case, even when a well tie had been provided by the original wellbore. The seismic data suggested that the well was located on the fold crest, when the dip data clearly show that it was on the backlimb. 3-D seismic data in the area are prone to distortions of reflector geometries by complex raypaths from target to surface.

CONCLUSIONS

The two case studies presented in this paper illustrate how new technologies and careful, integrated analysis of all available data can significantly enhance the chance of successfully sidetracking a well to penetrate the target or to improve well deliverability. The successful design of a sidetrack well requires the construction of a new, revised model of the subsurface geometry of the fold, using the predrill information and new data from the drilling of the original wellbore. The most common reason for sidetracking a well in the play trend has been that the original well missed the target (Table 2). This has usually been caused by misinterpretation of the seismic imaging because of poor data quality, misleading time migration of the data, or limited seismic-velocity control. The original well provides additional constraining data for a revised structural model; formation tops, dip data from imaging logs such as the FMI, and detailed seismic-velocity data. In some situations, a VSP can be acquired in the original wellbore to confirm the position of the targeted fold hinge in the new structural model, such as in case study 1.

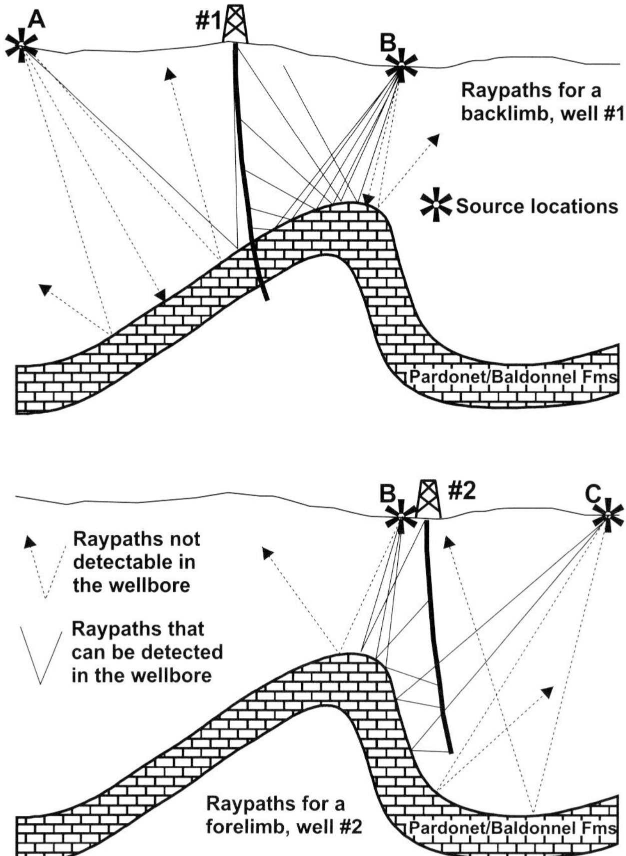

FIGURE 30. VSP raypaths for wells positioned on the backlimb and in front of the forelimb on a typical Monkman structure. Note how the well position on the structure strongly controls what portion of the structure can be imaged by the raypaths detectable in the wellbore geophones, even when the sources are located optimally.

In other instances, the VSP would be of limited value, as in case study 2, where the original wellbore penetrated the backlimb of the fold. In general, the VSP method described in case study 1 will only work when the original wellbore is on or beyond the forelimb of the fold. In this geometric configuration, the VSP raypaths can detect the hinge and forelimb of the fold; when the wellbore is on the backlimb of the fold, the raypaths can only image the backlimb (Figure 30).

Between 1978 and 1998 30 wells were sidetracked for a variety of reasons (Table 2). The results of the sidetrack operation have been mixed; in some cases, a noncommercial well has become marginally commercial, in others, the result has led to a high deliverability well, and in some cases, the sidetrack has failed. If we measure success as converting a dry or marginal well into a gas well capable of production at commercial rates, the success rate of sidetracking is 47%. This suggests that applying the integrated approach to sidetrack design outlined in this chapter enhances chance of success. Sidetracking has converted 14 of 30 originally unsuccessful wells into gas producers (Table 2) and significantly improved the drilling success rates in the play.

Shell Canada and Talisman have successfully horizontally sidetracked other wells in the Monkman area, and the technical success of these operations has stimulated Shell, Talisman, and others to conduct further horizontal drilling in the Foothills in British Columbia and in Alberta. Some of the sidetrack operations have been notable commercial successes, whereas others have been dismal commercial failures; most have been technically successful in testing the target. The well results show that the horizontal sidetracks will not necessarily increase the drainage area of a well, unless the well intersects a major fracture system that permits long-distance drainage of gas over a large area.

Some of the techniques described in this chapter have probably been applied in other thrust-and-fold belts, but we are not aware of any detailed case studies having been published, such as those presented here. Fold geometries similar to the Triassic detachment folds of the Monkman trend have been described from the Wyoming Overthrust Belt (Lamerson, 1982; West and Lewis, 1982) and the Sub-Andean Thrust Belt of southern Peru, Bolivia, and northern Argentina (Dunn et al., 1995; Belotti et al., 1995). Thus, this integrated approach could be applied profitably elsewhere to optimize the chance of commercial success in areas that have a complex structure.

ACKNOWLEDGMENTS

We wish to thank EnCana Corporation, Talisman Energy Inc., Burlington Resources, and Shell Canada for permission to publish this work. Thanks to J.C. Chameroy for preparation of the maps and PentaGraphix for drafting the figures. Paul Heffernan of Hef Petrophysical provided plots of the image log data for the b-22-C/93-P-03 well.

REFERENCES CITED

Bally, A. W., P. L. Gordy, and G. A. Stewart, 1966, Structure, seismic data and orogenic evolution of southern Canadian Rocky Mountains: Bulletin of Canadian Petroleum Geology, v. 14, p. 337–381.

Bamber, E. W., and R. W. Macqueen, 1971, Lower Carboniferous and Permian stratigraphy of the Monkman Pass and southern Pine Pass areas, northeastern British Columbia: Geological Survey of Canada, Paper 71–1A, p. 193–196.

Bamber, E. W., and R. W. Macqueen, 1979, Upper Carboniferous and Permian stratigraphy of the Monkman Pass and southern Pine Pass Area, northeastern British

Columbia: Geological Survey of Canada, Bulletin 301, 27 p.

Barss, D. L. and F. A. Montandon, 1981, Sukunka-Bullmoose gas fields: models for a developing trend in the southern Foothills of northeast British Columbia: Bulletin of Canadian Petroleum Geology, v. 29, p. 293–333.

Belotti, H. J., L. L. Saccavino, and G. A. Schachner, 1995, Structural styles and petroleum occurrence in the sub-Andean fold-and-thrust belt of northern Argentina, *in* A. J. Tankard, P. Suarez S., and H. J. Welsink, eds., Petroleum basins of South America, AAPG Memoir 62, p. 545–555.

Cant, D. J., and G. S. Stockmal, 1989, The Alberta foreland basin: relationship between stratigraphy and Cordilleran terrane-accretion events: Canadian Journal of Earth Sciences, v. 26, p. 1964–1975.

Cooper, M. A., 1991, The analysis of fracture systems in subsurface thrust structures from the Foothills of the Canadian Rockies, *in* K. R. McClay, ed., Thrust tectonics: London, Chapman & Hall, p. 391–406.

Cooper, M. A., 1996, Passive-roof duplexes, and pseudo-passive-roof duplexes at mountain fronts: a review: Bulletin of Canadian Petroleum Geology, v. 44, p. 410–421.

Cooper, M. A., 1998, Structural style variations in the BC Foothills: Canadian Society of Petroleum Geologists Annual Convention Abstracts, p. 277–278.

Cooper, M. A., R. Green, and M. Morrison, 1995, Analysis of well data to design side-tracks in structurally complex areas: case studies from the Monkman Triassic playtrend, NE British Columbia: Canadian Society of Petroleum Geologists Annual Convention Abstracts [unpaginated].

Cooper, M. A, C. Brealey, P. Fermor, and D. Bradish, 1997, From non-commercial to commercial; horizontal drilling in the fractured Triassic carbonates of the Boulder field, B.C. Foothills, Canada: AAPG/AMGP Research Symposium: Oil and Gas Exploration and Production in Fold and Thrust Belts, Abstracts, February 23–26, 1997, Veracruz, Mexico [unpaginated].

Dahlstrom, C. D. A., 1969, Balanced cross-sections: Canadian Journal of Earth Sciences, v. 6, p. 743–757.

Dahlstrom, C. D. A., 1970, Structural geology in the eastern margin of the Canadian Rocky Mountains: Bulletin of Canadian Petroleum Geology, v. 18, p. 332–406.

Davies, G. R., 1997a, The Triassic of the Western Canada Sedimentary Basin: tectonic and stratigraphic framework, paleogeography, paleoclimate and biota: Bulletin of Canadian Petroleum Geology, v. 45, p. 434–460.

Davies, G. R., 1997b, The Upper Triassic Baldonnel and Pardonet Formations, Western Canada Sedimentary Basin: Bulletin of Canadian Petroleum Geology, v. 45, p. 643–674.

Douglas, R. J. W., 1950, Callum Creek, Langford Creek and Gap map-areas, Alberta: Geological Survey of Canada Memoir 255, 124 p.

Douglas, R. J. W., 1958, Mount Head map-area, Alberta: Geological Survey of Canada Memoir 291, 241 p.

Dunn, J. F., K. G. Hartshorn, and P. W. Hartshorn, 1995, Structural styles and hydrocarbon potential of the sub-Andean thrust belt of southern Bolivia, *in* A. J. Tankard, P. Suarez S., and H. J. Welsink, eds., Petroleum basins of South America, AAPG Memoir 62, p. 523–543.

Eisbacher, G. H., M. A. Carrigy, and R. B. Campbell, 1974, Paleodrainage patterns and late-orogenic basins of the Canadian Cordillera, *in* W. R. Dickinson, ed., Tectonics and sedimentation: Society of Economic Paleontologists and Mineralogists Special Publication 22, p. 143–166.

Gibson D. W., and J. E. Barclay, 1989, Middle Absaroka Sequence; the Triassic stable craton, *in* B. D. Ricketts, Western Canada Sedimentary Basin: A case history: Calgary, Canadian Society of Petroleum Geologists, p. 219–231.

Jones, P. B., 1982, Oil and gas beneath east-dipping underthrust faults in the Alberta Foothills, *in* R. B. Powers, Studies of the Cordilleran Thrust Belt. Rocky Mountain Association of Geologists, p. 61–74.

Kilby, W. E., and C. B. Wrightson, 1987, Bullmoose mapping and compilation project (93P/3,4): British Columbia Ministry of Energy, Mines and Petroleum Resources, Geological Fieldwork 1986, Paper 1987-1, p. 373–378.

Lamerson, P. R., 1982, The Fossil Basin and it's relationship to the Absaroka thrust fault system, *in* R. B. Powers, ed., Geologic studies of the Cordilleran thrust belt: Denver, Rocky Mountain Association of Geologists, p. 279–340.

Lingrey, S.,1996, Structured patterns of imbrication in the Pine River area of northeastern British Columbia: Bulletin of Canadian Petroleum Geology, v. 44, p. 324–336.

McMechan, M. E., 1985, Low-taper triangle zone geometry: an interpretation for the Rocky Mountain Foothills, Pine Pass–Peace River area, British Columbia: Bulletin of Canadian Petroleum Geology, v. 33, p. 31–38.

McMechan, M. E., 1994, Geology and structure cross-section Dawson Creek, British Columbia: Geological Survey of Canada, Map 1858A, scale 1:250,000, 1 sheet.

McMechan, M. E., and R. I. Thompson, 1989, Structural style and history of the Rocky Mountain fold and thrust belt, *in* B. D. Ricketts, ed., Western Canada Sedimentary Basin: A case history: Calgary, Canadian Society of Petroleum Geologists, p. 47–72.

Monger, J. W. H., and R. A. Price, 1979, The geodynamic evolution of the Canadian Cordillera— progress and problems: Canadian Journal of Earth Sciences, v. 16, p. 770–791.

Monger, J. W. H., R. A. Price, and D. J. Tempelman-Kluit, 1982, Tectonic accretion and the origin of two major metamorphic and plutonic welts in the Canadian Cordillera: Geology, v. 10, p. 70–75.

Montgomery, S., 1994, Sukunka Bullmoose gas play: A frontier in Northeast British Columbia: Petroleum Frontiers, v. 11, pt. 3, 105 p.

Morrison, M., D. Mitchell, and R. Green, 1998, Analysis of well data to design side-tracks in structurally complex areas: case studies from the Monkman Triassic play-trend, NE British Columbia: Canadian Society of Petroleum Geologists Annual Convention Abstracts, p. 281.

Price, R. A., and E. W. Mountjoy, 1970, Geologic structure of the Canadian Rocky Mountains between Bow and Athabasca rivers, a progress report, *in* J. O. Wheeler, ed., Structure of the Southern Canadian Cordillera: Geological Association of Canada Special Publication 6, p. 7–25.

Slawinski, M. A., and J. M. Parkin, 1996, Migration of a multioffset VSP: a case study in NE British Columbia: Canadian Journal of Exploration Geophysics, v. 32, p. 104–112.

Stott, D. F., 1968, Lower Cretaceous Bullhead and Fort St. John Groups, between Smoky and Peace Rivers, Rocky Mountain Foothills, Alberta and British Columbia: Geological Survey of Canada, Bulletin 152, 273 p.

Thompson, R. I., 1979, A structural interpretation across part of the northern Rocky Mountains, British Columbia, Canada: Canadian Journal of Earth Sciences, v. 16, p. 1228–1241.

West, J., and H. Lewis, 1982, Structure and palinspastic reconstruction of the Absaroka thrust, Anschutz Field area, Utah and Wyoming, *in* R. B. Powers, ed., Geologic studies of the Cordilleran thrust belt: Denver, Rocky Mountain Association of Geologists, p. 633–640.

Restrepo-Pace, P. A., F. Colmenares, C. Higuera, and M. Mayorga, 2004, A Fold-and-thrust belt along the western flank of the Eastern Cordillera of Colombia— Style, kinematics, and timing constraints derived from seismic data and detailed surface mapping, *in* K. R. McClay, ed., Thrust tectonics and hydrocarbon systems: AAPG Memoir 82, p. 598–613.

A Fold-and-thrust belt along the western flank of the Eastern Cordillera of Colombia— Style, kinematics, and timing constraints derived from seismic data and detailed surface mapping

Pedro A. Restrepo-Pace
ConocoPhillips Indonesia Ltd., Jakarta, Indonesia

Fabio Colmenares
Geosearch Ltda., Bogota, Colombia

Camilo Higuera
Geosearch Ltda., Bogota, Colombia

Marcela Mayorga
Geosearch Ltda., Bogota, Colombia

ABSTRACT

The Eastern Cordillera of Colombia is considered to be the result of compression and uplift during the late Miocene–Pliocene Andean orogeny. Nevertheless, detailed mapping carried out on a portion of the fold belt exposed along the western flank of the Eastern Cordillera suggests that deformation began in late Paleocene–early Eocene time, in the form of a west-vergent, forward-propagating fold-and-thrust belt. Several structures related to this event remain concealed beneath the late Paleocene–late Eocene unconformity, and synkinematic sediments are preserved locally within some of these thrust sheets. The relief generated during this deformational episode constrained the depositional axis of late Eocene fluvial sediments to run along the present-day crestal zone of the Eastern Cordillera. Onlapping basal Miocene molassic sediments on tilted middle Miocene fluvial deposits suggest that a series of seismically delineated intercutaneous wedges located at the thrust front formed in late Miocene time. These structures mark the onset of the Andean deformational episode. Thrusting polarity reversed during the latter event, that is, toward the hinterland, and Paleogene structures were reactivated. Additionally, the late Paleocene–early Eocene unconformity surface was folded and incorporated into north-plunging folds, and Cretaceous extensional faults were inverted. As a result, the Eastern Cordillera developed its present pop-up, doubly vergent fold-belt geometry, flanked by thick molassic sequences that document the Miocene–Pliocene Andean orogeny. Regionally, the latest Campanian–Maastrichtian uplift of the Central Cordillera, Paleocene–early Eocene thrusting in the Middle Magdalena Valley and along the western flank of the Eastern Cordillera, and ultimately inversion and major uplift of the entire Eastern Cordillera in late Miocene–Pliocene time, attests to the eastward propagation of deformation from the Central toward the Eastern Cordillera. Kinematically, the driving element of deformation would have been the obliquely accreting western terranes of the Central and Western Cordilleras and the Panama arc. These transferred their easterly component of convergence to the Eastern Cordillera through a basal intracrustal detachment.

INTRODUCTION

Knowledge of the timing and kinematics of fold-belt structures and associated basins that developed around the Andean system of Colombia remains limited. This is primarily because of poorly determined Tertiary chronostratigraphy, the lack of detailed structural mapping, and the fact that the post-middle Miocene Andean orogeny is a formidable "orogenic veil" that makes understanding of pre-Andean deformation difficult. Poor correlations and the application of oversimplified structural models have resulted in implausible burial-history curves and hydrocarbon maturation and migration models for the fold-belt area.

Detailed surface and subsurface mapping carried out in a sector of the western fold belt of the Eastern Cordillera has brought new insight on the complex development of the Eastern Andean range and the Middle Magdalena Basin. In this chapter, we summarize the evidence for polyphase deformation at the western flank of the Eastern Cordillera of Colombia and along the eastern margin of Middle Magdalena Basin. The interpretations outlined here have significant implications with respect to trap formation, timing, and preservation, relative to maturation timing in the fold belt and adjacent Middle Magdalena Basin areas.

REGIONAL GEOLOGIC SETTING

The Andes of Colombia comprise three northeast–southwest- and north–south-trending ranges separated by narrow valleys (Figure 1a, b). The Andean domain is structurally separated from the Llanos foreland basin by the Borde Llanero thrust system (Figure 1a). Precambrian–Paleozoic-age polydeformed rocks constitute the basement of the Central and Eastern ranges. Oceanic and trench assemblage rocks accreted after Late Cretaceous time lie west of the Romeral suture zone. The Eastern Cordillera of Colombia contains a thick, shale-dominated succession of sediments deposited between Late Jurassic and Cretaceous time in a retroarc setting (Cooper et al., 1995). In the Bogota Basin and Cocuy Subbasin, these sediments are estimated to be more than 10 km thick (Campbell and Burgl, 1965; Irving, 1975). Crustal shortening estimates across the Eastern Cordillera are from 15 to 66% (Colletta et al., 1990; Dengo and Covey, 1993; Cooper et al., 1995; Roeder and Chamberlain, 1995). The calculated β stretching factor [sensu McKenzie (1978)] to account for the accommodation of these sediments ranges from 2.0 to 2.2 (Fabre, 1987). These figures are questionable, in view of the fact that detailed structural mapping has been carried out only in small areas (cf. Restrepo-Pace, 1989). The rocks that make up the Eastern Cordillera have been buckled and telescoped into an east-vergent fold-and-thrust belt along the eastern flank and a west-vergent fold-and thrust belt along the western flank of the range. Thus, in cross-sectional view, the Eastern Cordillera displays an asymmetric "pop-up" geometry flanked by the Llanos and Middle Magdalena Tertiary foreland basins (Figure 2a). Basement-cored anticlinoria are exposed in the hinterland areas of the range. These have been related to inversion of Jurassic–Early Cretaceous extensional faults, as indicated by thickness variations of correlative units across major faults (Julivert, 1970; Colletta et al., 1990; Cooper et al., 1995). On the western flank of the Eastern Cordillera, deep-seated inversion structures turn into thin-skinned structures at shallow-crust levels, detaching at the basal Cretaceous shales and gradually ramping their way up section, ultimately to involve the Tertiary succession at the thrust front. Also along this flank, high-frequency disharmonic folding or duplexing characterizes the deformation of the basal Cretaceous shales. The external portions of the thrust belt display fault-bend and fault-propagation folds. The Middle Magdalena Basin is an asymmetric, intermontane depression bordered by the Central and Eastern Cordilleras. The latter basin includes a Tertiary sedimentary wedge that thickens to the east below the fold belt. The

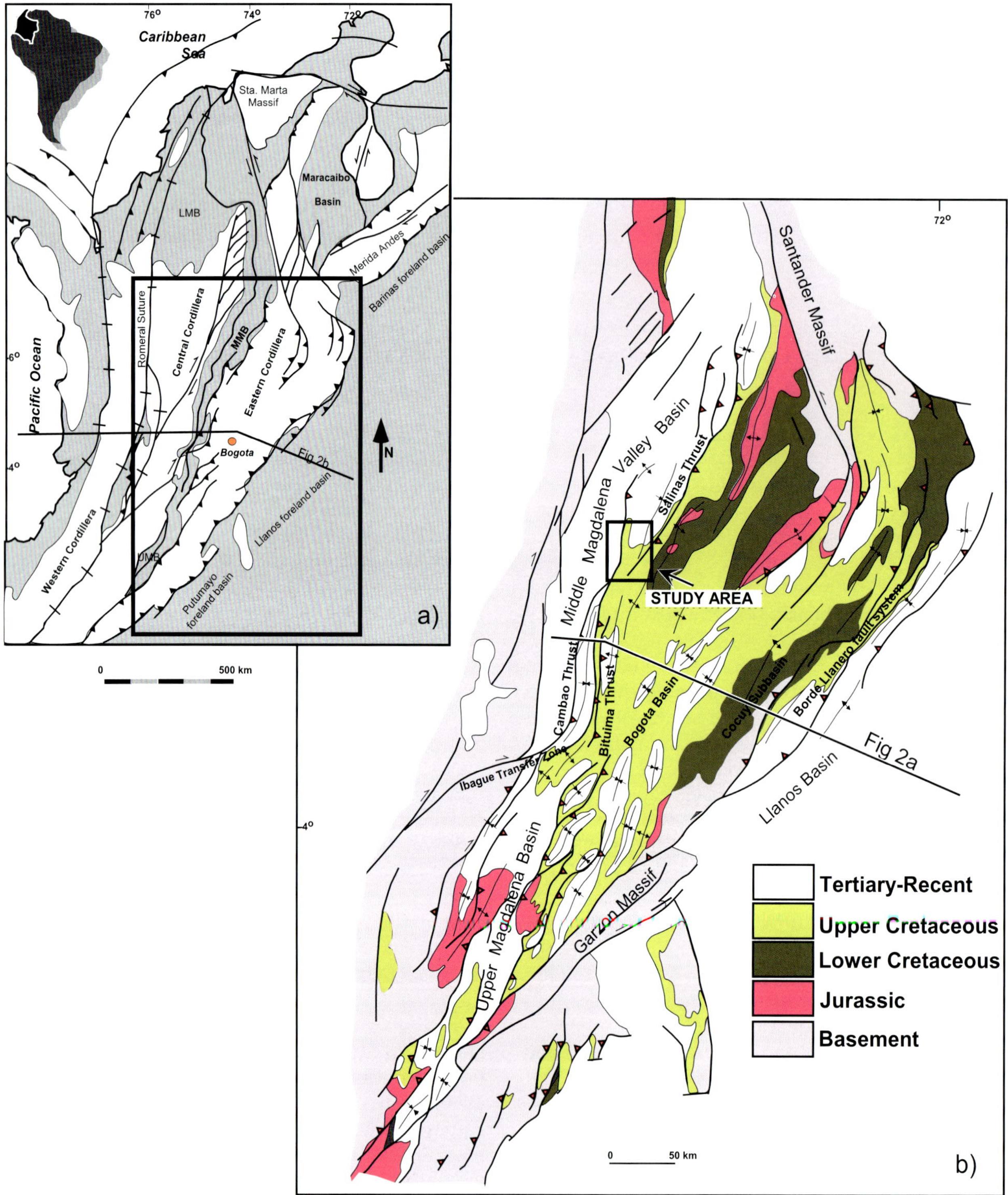

FIGURE 1. Index map showing the main structural features of the northern Andes and the location of the study area. UMB, MMB, and LMB correspond to the Upper, Middle, and Lower Magdalena Basins, respectively. The location of the regional cross sections of Figure 2 is also depicted.

base of the wedge is defined by an easterly dipping regional unconformity representing truncation of pre-Tertiary strata below it and progressive onlap of upper Eocene–Miocene sediments above it. This surface has been the main conduit for oil migration from the hydrocarbon-generating kitchen areas located within

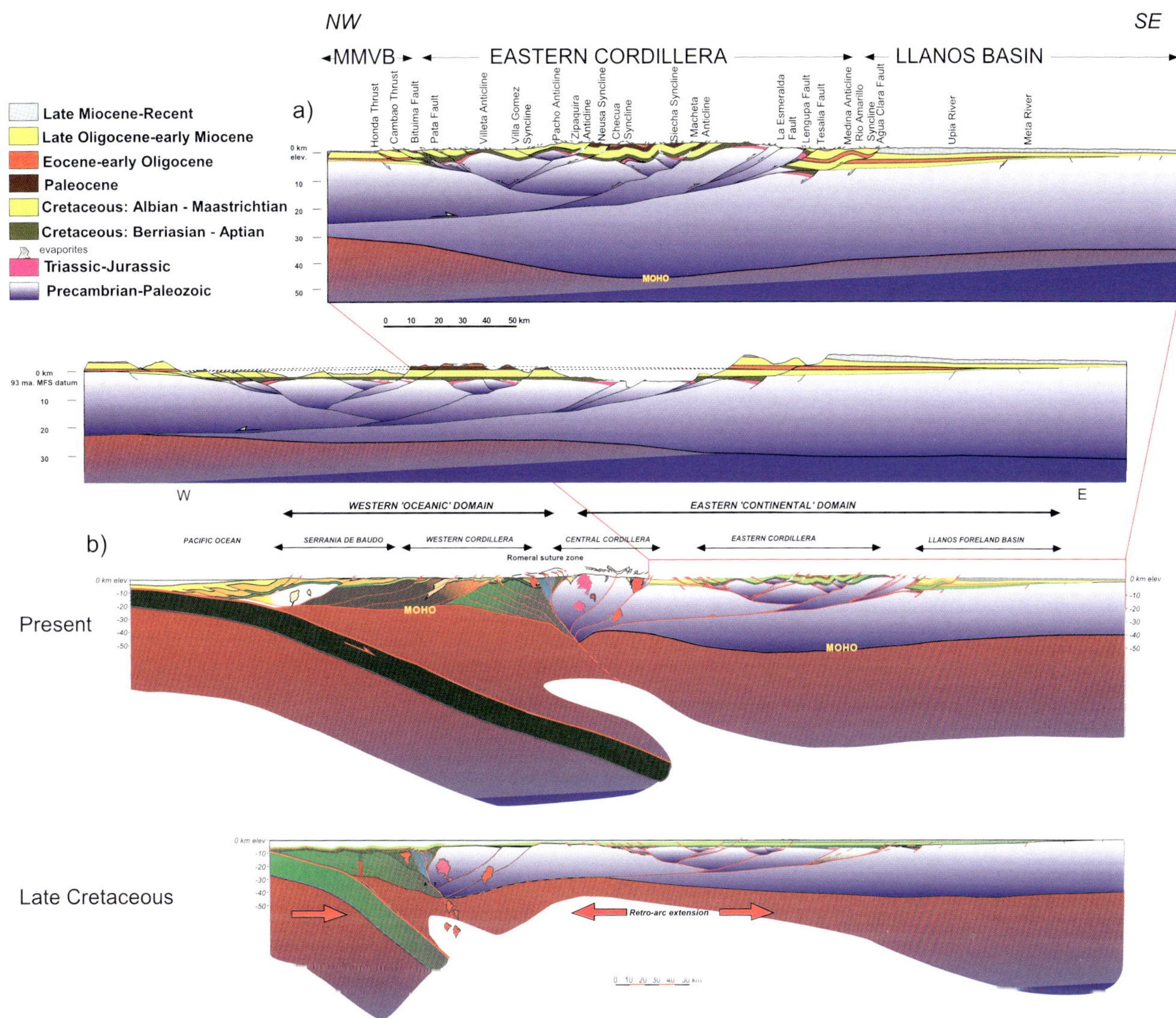

FIGURE 2. Crustal-scale cross section and restored section of the eastern Andes of Colombia. (a) Eastern Cordillera, (b) Colombian Andes. The restored section (a) (datum at the 93 Ma maximum flooding surface) is based on a delamination concept. Inversion during the Andean orogeny created the doubly vergent range. MMVB = Middle Magdalena Valley Basin.

and below the fold belt to traps in the basin. More than 3.5 BBOE reserves have been discovered in the Middle Magdalena basin since the early 1900s. This oil is generated by the Upper Cretaceous La Luna Formation and currently is trapped in fluvial sandstone reservoirs of the upper Eocene to Oligocene La Paz, Esmeraldas, and Mugrosa Formations.

A long-standing view is that thrusting and uplift in the Eastern Cordillera took place in late Miocene to Pliocene time (6–4 Ma) (Van der Hammen, 1957; Campbell and Burgl, 1965). However, authors invoke different processes to explain the data behind these ages. Dengo and Covey (1993) argue for a two-stage model in which basement-involved uplift in Pliocene–Holocene time was preceded by thin-skinned deformation in late Miocene to Pliocene time. Colletta et al. (1990) present a structural model that calls for a single-stage Miocene–Pliocene inversion event. Nevertheless, there are reported fission-track data and sediment-provenance studies that argue for a diachronous uplift of the eastern Andes (Fabre, 1987; Shagam et al., 1984; and a review by Kroonenberg et al., 1991). In the Santander Massif, uplift began in the early Eocene to early Oligocene (~55–32 Ma) (Fabre, 1987; p. 163). Uplift for the Garzon Massif began in the late Miocene (12–9 Ma), with possible initial uplift occurring as long ago as late Early Cretaceous time (~110 Ma) (Van der Wiel and Andriessen, 1991; Andriessen, 1995). Finally, uplift occurred in the late Miocene (<5 Ma) for the Cocuy and Bogota Basins (Hooghiemstra, 1984). These uplift ages most likely reflect the timing of structural inversion, which is responsible for the bulk of the uplift of the Cordillera.

STUDY AREA

This study focuses on an area of approximately 750 km^2 (which was part of a larger, 3800 km^2 field and seismic campaign) along the western flank of the Eastern Cordillera, between the Salinas and the Dos Hermanos Thrusts (Figure 1). The west-verging Dos Hermanos Thrust front places Cretaceous rocks over Tertiary rocks in the Middle Magdalena Basin. The rocks cropping out range in age from late Barremian to Holocene, with a prominent unconformity representing a hiatus between late Paleocene and early Eocene time. The Cretaceous assemblage consists predominantly of black marine shales, chert, and micrites, with minor limestone and sandstone beds. The Paleocene strata consist of shales and sandstones, with occasional coal seams marking the transition from marine- to continental-deposition conditions. Middle-upper Eocene to lower Miocene strata consist of a thick accumulation of alluvial/fluvial sediments that record increasing subsidence rates stratigraphically upward. Upper Miocene–Pliocene sediments consist of molassic coarse-grained clastics.

STRUCTURE AND MECHANICAL STRATIGRAPHY

The Salinas-Bituima-Cambao-Dos Hermanos thrust system exposed in the study area places Cretaceous rocks westward over the Tertiary sedimentary fill of the Middle Magdalena Basin (Figure 3). Thrust planes dip to the southeast between 20 and 35°; tectonic transport is to the west-northwest. The main detachment horizons of the fold belt are located within the Paja and Simiti Formation shales (Figure 4). Additional detachments are found within shales at the base of the Cretaceous succession, La Frontera Formation, at the base of the Hilo Formation, and within the Umir Formation. The shales of the Mugrosa and Colorado Tertiary formations act as local detachments in the frontal structures. The highly monotonous character of the Cretaceous succession requires strict paleontological control to determine structural repetitions. In outcrop, the shaly intervals of the Cretaceous succession exhibit tight buckle folds and duplexes with associated pressure-solution fabrics and penetrative foliation, and the cherty intervals are highly fractured and depict chevron-style folding (Figures 4 and 5). Tertiary strata have a more competent mechanical behavior and develop fewer disharmonies than do Cretaceous rocks, leading to the formation of geometrically simpler and broader folds. In the northern sector of the mapped area, the structures are wide and plunge to the north (e.g., La India syncline-anticline pair). Between the northeastern striking Laureles and the La Salinas Thrusts, the faults are more widely spaced and include larger sections of Cretaceous rocks. Northwest of the Laureles Thrust, faults strike to the north and are relatively closely spaced, forming part of an unroofed duplex that repeats the Simiti, Hilo, and Frontera units several times. The leading edge of the thrust front, defined to the north by the Dos Hermanos Fault, splits southward into the Cambao Fault, Caceres Fault, and, ultimately, the Honda thrust system (Figure 3). Toward the hinterland area of the thrust system east of the Salinas inversion fault, basal Cretaceous rocks are exposed in the south-plunging part of a basement-cored anticlinorium. The Cambao–Dos Hermanos thrust system strikes north-northeast to south-southwest, taking a west-northwest strike near the town of El Castillo. Seismic data indicate that this strike change is formed as the Dos Hermanos Thrust overrides a lateral ramping intercutaneous wedge that involves Cretaceous and Tertiary strata (Figure 6).

Balanced cross sections A to F (Figure 7) indicate that minimum structural contraction varies between 65 and 48%, decreasing toward the south. Shortening along the Dos Hermanos fault is gradually partitioned and relayed to the Cambao and Caceres Thrusts and also to the Honda leading thrust. South of the study area, the Honda Thrust gradually picks up 5–10% of the thrust-front shortening. A variance in contraction from section to section is attributed to the fact that the area has been subjected to multiple stages of shortening, thereby precluding the precise determination of the initial restored geometries. Additionally, the contribution of pressure dissolution to shortening has not been quantified.

Kinematics

Stratigraphic and crosscutting relationships derived from detailed mapping indicate that the fold belt developed through a series of discrete compressional events in Tertiary time (Figure 3). Cretaceous rocks faulted by the Dos Hermanos and the Laureles Thrusts, including the La Fiebre and La Alegria Thrusts and La Grilla Syncline, are all concealed beneath the pre-late Eocene unconformity. In the Betania-Cui Syncline, Paleocene Guaduas Formation rocks rest conformably over Cretaceous strata. These relationships suggest a post-Paleocene and pre-late Eocene thrusting event in the area. The unconformity at the base of the Tertiary succession is folded at La India Syncline, indicating a younger deformational episode affecting the region. Synkinematic deposits composed of locally derived sedimentary breccias containing angular fragments of Cretaceous rocks (Figure 8) that are related to the Paleogene event (La Paz Formation syntectonic sediments of Figure 5) are preserved in local piggyback basins in the northwestern part of the study area. These syntectonic

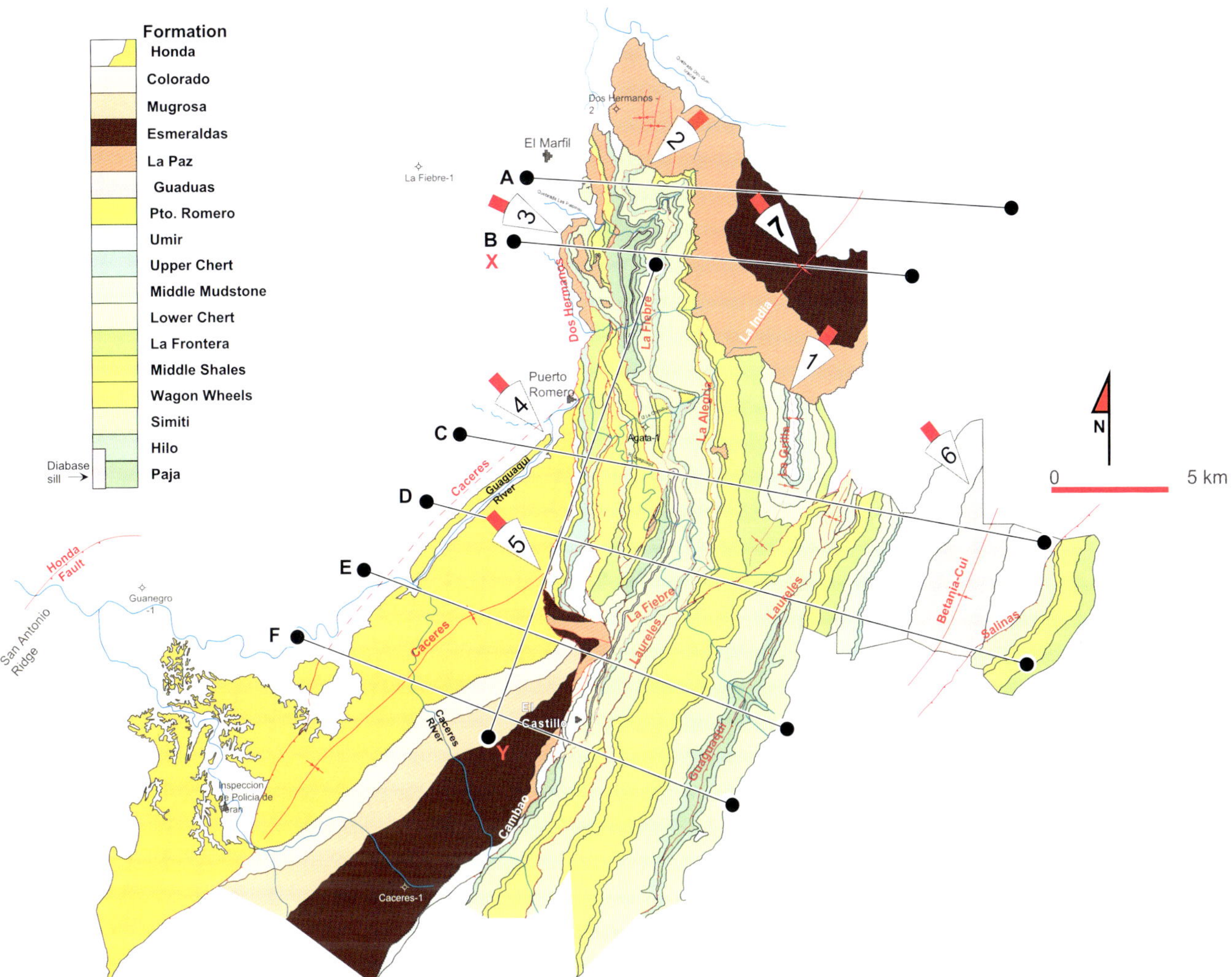

FIGURE 3. Geologic map of the study area. The arrows point to critical timing relationships: 1, 2 = La Grilla Syncline, La Fiebre and Alegria Thrusts are concealed below the basal unconformity of the upper Eocene La Paz rocks. This relationship constrains an upper age limit for Paleogene deformation. 3 = The preserved Paleogene synkinematic deposits, which are tilted and faulted by Andean-age (Miocene–Pliocene) thrusts. 4, 5 = The course of the Guaguaqui River shifted to the south by the late Miocene Caceres Thrust. Drainage in the Middle Magdalena valley is collected by the Magdalena River, which flows to the north. Hanging-wall rocks of the Dos Hermanos Thrust subsequently decapitate the Caceres Thrust and the Caceres Syncline. 6 = Conformable contact between Paleocene and older rock successions constrains the lower limit for Paleogene deformation. 7 = Characteristic large and open structures of Andean age have folded the basal upper Eocene unconformity and reactivated Paleogene structures.

sediments were subsequently faulted and tilted during late Miocene thrusting. Structures associated with the Caceres Thrust have a northeast-southwest trend and involve the lower Miocene rocks of the Honda group. The abandonment of northwesterly directed flow of the Guaguaqui River and adoption of its newly developed southwesterly directed course confirms the very young age of the Caceres Thrust (Figure 3). The fold axis associated with the seismically delineated frontal wedge structure strikes to the north. A constraint for the age of the intercutaneous wedge is derived from observing onlapping upper Miocene–Pliocene molassic sediments onto tilted middle(?) Miocene fluvial strata forming part of the forelimb of the structure (Figure 6). The frontal structural wedge is overridden and decapitated by the Dos Hermanos Thrust, indicating a reactivation of breached structures and a reversal of thrust propagation in the very late stages of shortening.

SOURCE ROCK MATURATION MODEL

Several Cretaceous intervals make up the source rock (the primary bituminous intervals are within the

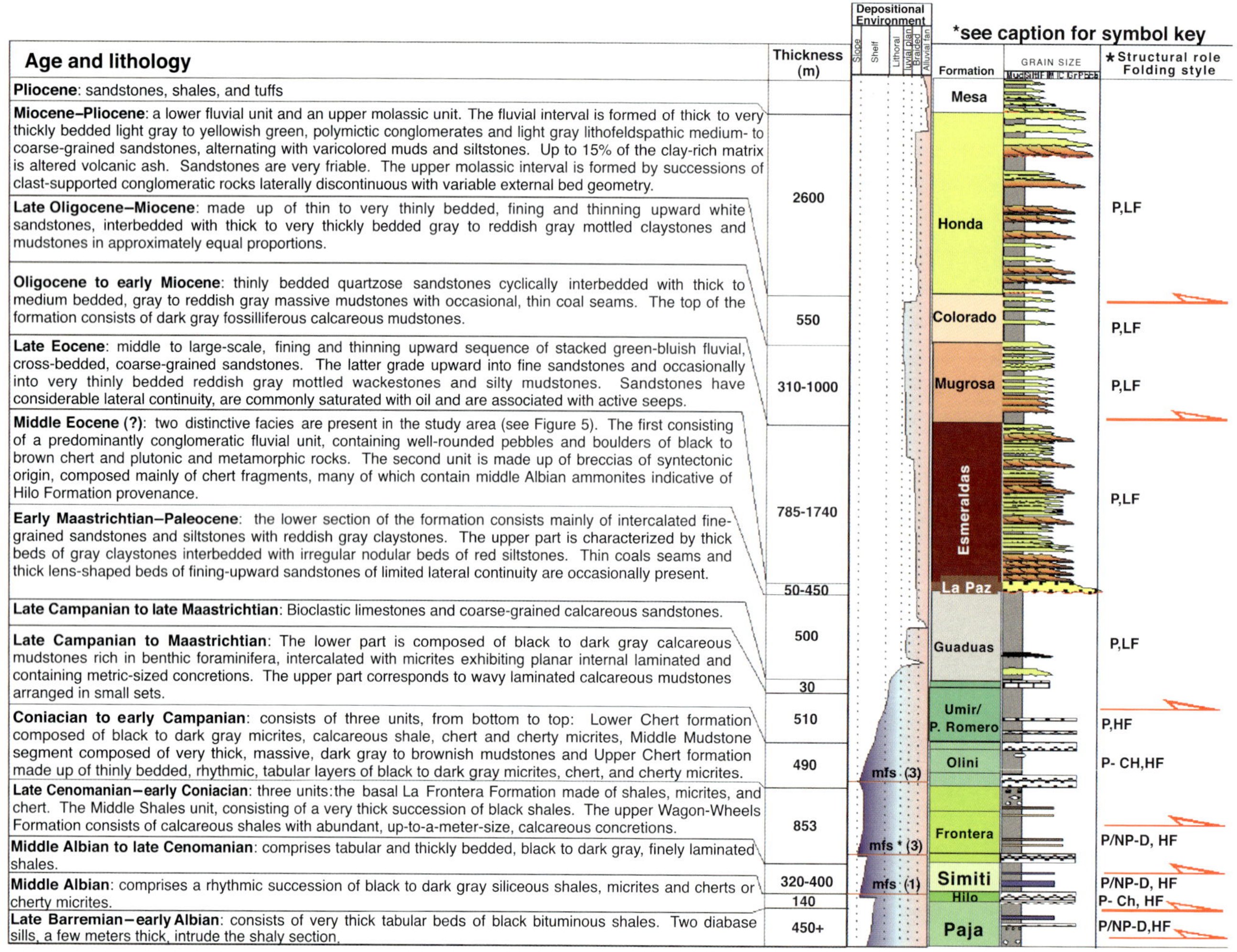

FIGURE 4. Summary of stratigraphy and mechanical stratigraphy. The right column indicates structural styles associated with particular formations: P = parallel folding, NP = nonparallel folding, D = disharmonic folding, HF = high-frequency folds, and LF = low-frequency folds. The arrows indicate the main detachment horizons. The stratigraphic nomenclature used here is according to Morales et al. (1956) and Caceres and Etayo-Serna (1969). La Frontera Formation is more widely known in industry as La Luna Formation.

Frontera Formation), reservoirs, and seals that are dispersed throughout the Tertiary clastic succession and constitute the main components of the petroleum system in the study area. A simplified version of cross section C of Figure 7 was forward modeled using Thrustpack software (Sassi and Rudkiewics, 2000, Thrustpack version 7.2, Institut Francais du Petrole). The forward structural model (Figure 9) depicts the onset of forward-propagating thrusting in late Eocene time and concurrent deposition of synkinematic and fluvial sediments of La Paz and Esmeraldas Formations. By the late Oligocene, accelerated shortening rates resulted in greater uplift in the hinterland area and deposition of the upper Honda Formation sediments in the foreland region. Additionally, at this time a frontal intercutaneous wedge formed, as constrained by subsurface data. Finally, continued shortening reactivated breached faults, indicating a reversal of thrust propagation. During the thrusting process, flexural-loading-related subsidence occurred along the eastern margin of the Middle Magdalena Basin.

Assuming that the Frontera Formation (equivalent of La Luna Formation) is the primary source rock in the system, type II, average TOC of 4%, and a steady-state basal heat flux of 60 mW/m^2, the hydrocarbon expulsion model predicts that oil generation began in late Oligocene time and continues today (Figure 10). This model is consistent with producing field data of the foreland region, indicating Frontera Formation type oils, trapped in Oligocene and Miocene strata that onlap the pre-Tertiary basement. The migration of oil from the fold belt's hydrocarbon-generating kitchen to the traps in the foreland region occurs along the basement-cover surface. The amount of oil expelled and lost prior to deposition of the Oligocene-Miocene reservoirs has not

FIGURE 5. Typical mechanical behavior and folding style of Cretaceous-age rocks.

been estimated. The intercutaneous wedge and similar traps that have been identified along the thrust front constitute primary targets for future exploration, because they remain largely untested.

REGIONAL IMPLICATIONS

The onset of Andean tectonism is relatively well documented in the Upper and Middle Magdalena Basins: Upper Maastrichtian, coarse alluvial clastics with a westerly provenance (Cimarrona and La Tabla Formations) (Gomez and Pedraza, 1994) signal the initial stages of uplift of the Central Cordillera. Schamel (1991), Butler and Schamel (1989), and others have documented pre-Oligocene thrusting from subsurface data in the Upper Magdalena Basin and pre-late Eocene thrusting in the northeastern sector of the Middle Magdalena Basin (Morales et al., 1956). Approximately 4000 m of westerly derived molassic sediments of Eocene age are exposed, mainly along the western flank of the Upper Magdalena Basin. These sedimentary rocks indicate thrusting and uplift of the Central Cordillera beginning in the middle(?) to late Eocene. North of the Ibague transfer zone (Figure 11) these Eocene-age sediments are not present along the western margin of the Magdalena Basin. Rather, they are encountered along the fold belt running on the eastern side of the Middle Magdalena Basin and in the study area.

The western boundary of the Middle Magdalena Basin corresponds to a sub-vertical north-south-striking basement-rooted fault, which contrasts to the moderately dipping thrust belt that runs along the western flank of the Upper Magdalena Basin. An eastward-tilted basement onto which Tertiary sediments onlap is also unique to the Middle Magdalena Basin. These contrasts reflect the importance of the Ibague transfer zone in partitioning the Paleogene deformation and compartmentalizing sedimentation along the Magdalena Basin as a whole. Our view is that the Ibague transfer zone acted as a Paleogene relay zone, by transferring the contraction in the eastern thrust belt of the Central Cordillera along the western sector of the Upper Magdalena Basin, to the western thrust belt of the Eastern Cordillera along the Middle Magdalena Basin (Figures 1, 11). A structural system similar to the Ibague transfer zone exists in the northern sector of the Middle Magdalena Basin near the Provincia field.

Uncertainty remains with regard to the amount and extent of relief generated by the Paleogene fold

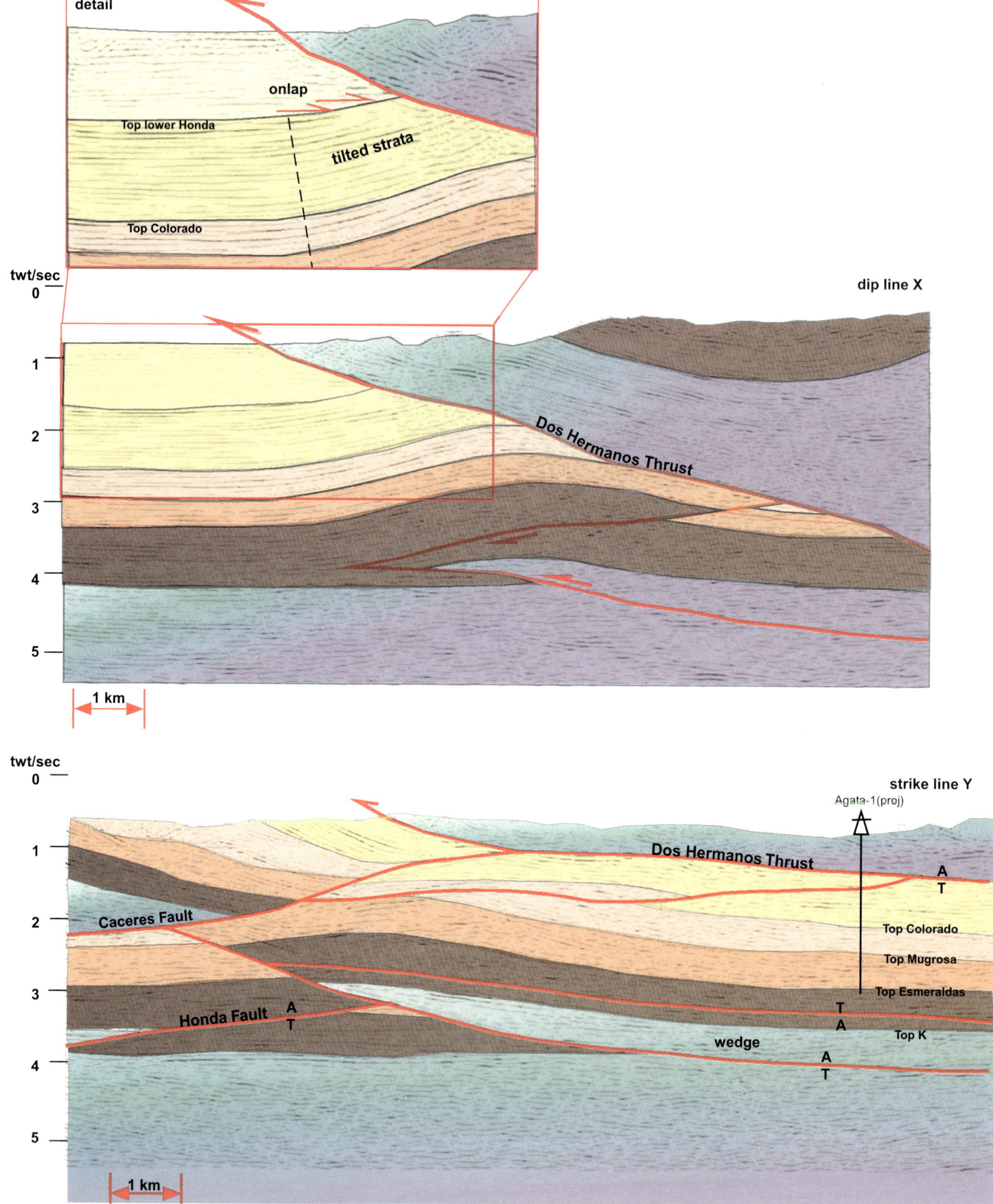

FIGURE 6. Dip and strike lines (location of seismic lines X and Y are on Figure 3), depicting subsurface structural geometry. Onlapping upper Miocene sediments on tilted middle Miocene strata indicate that the intercutaneous wedge formed in middle Miocene time. The strike line shows the lateral ramping of the wedge toward the south. Agata-1 well was used for subsurface stratigraphic calibration.

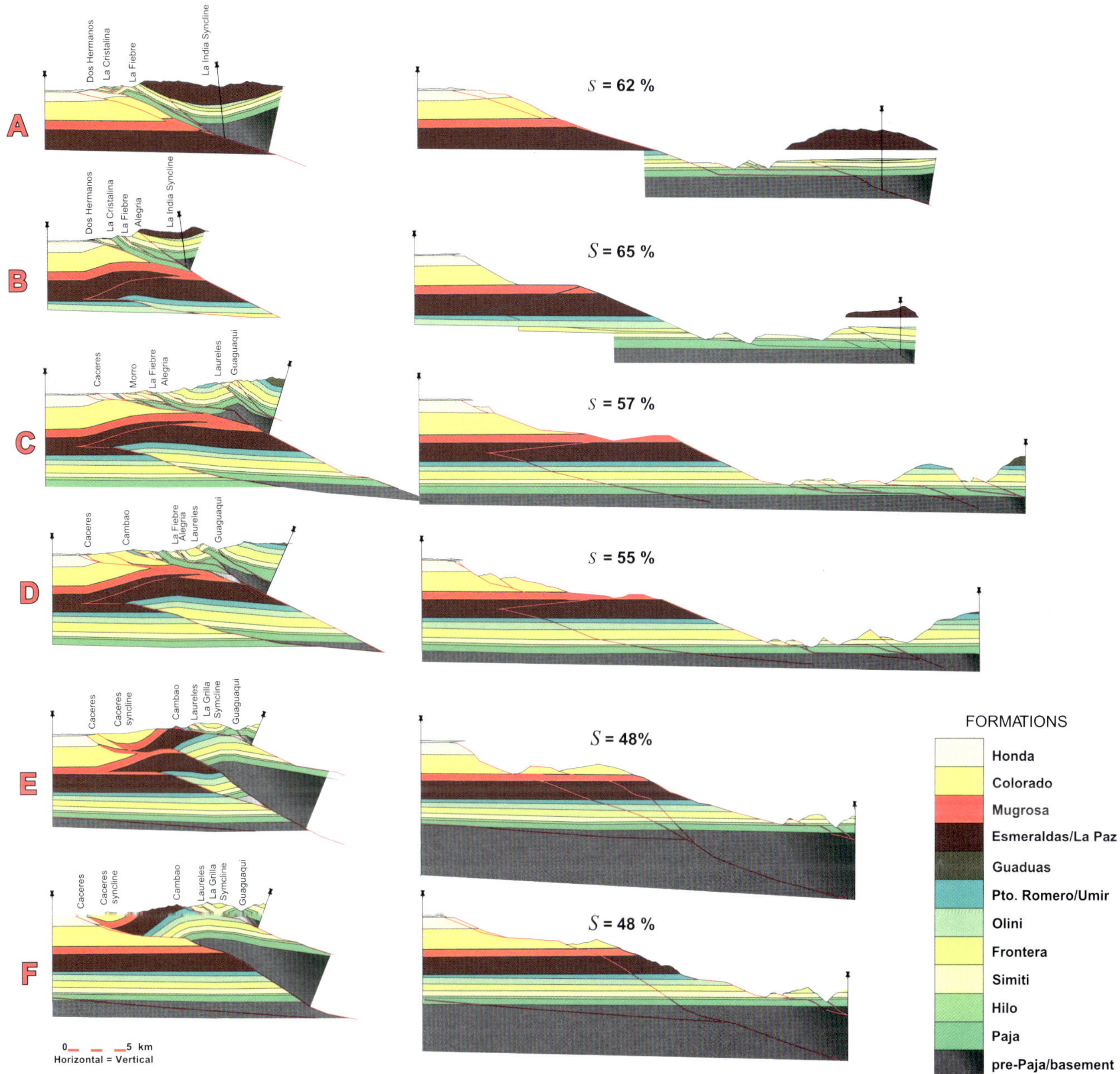

FIGURE 7. Balanced cross sections, from north to south in the study area. For section location, refer to map on Figure 3.

belt exposed in the study area. Regional stratigraphic studies suggest that the fold belt achieved sufficient elevation to allow it to act as a drainage divide, thereby constraining the late Eocene depocenter axis to run approximately along the axis of the Bogota Basin. Moreover, it may have caused the divergence of the Oligocene depocenter axis into two northeast-southwest systems (Villamil and Restrepo-Pace, 1997, 1998).

From the preceding discussion, we interpret a regional-scale, eastward propagation of deformation and uplift for the Colombian Andes. There was latest Campanian to Maastrichtian uplift of the Central Cordillera, Paleocene–early Eocene thrusting in the Middle Magdalena Basin and along the western margin of the Eastern Cordillera, and ultimately, inversion and major uplift of the entire Eastern Cordillera in late Miocene–Pliocene time (Figure 11). This interpretation is consistent with the easterly migration of the depocenter's axis during the Cenozoic, as Villamil and Restrepo-Pace (1997, 1998) have suggested. The driving element of deformation would have been the obliquely accreting western terranes of the Central and Western Cordilleras of Colombia. These would induce shortening in the Central and Eastern Andean region by transferring their easterly component of convergence through a basal crustal-scale detachment, as suggested by Dengo and Covey

a)

b)

c)

FIGURE 8. Photograph (a) illustrates the unconformable relationship between middle–upper Eocene syntectonic sediments of La Paz Formation and chert layers of the Hilo Formation. The remaining photographs depict two different facies of La Paz sediments: (b) syntectonic facies and (c) the coarse-grained fluvial sediments.

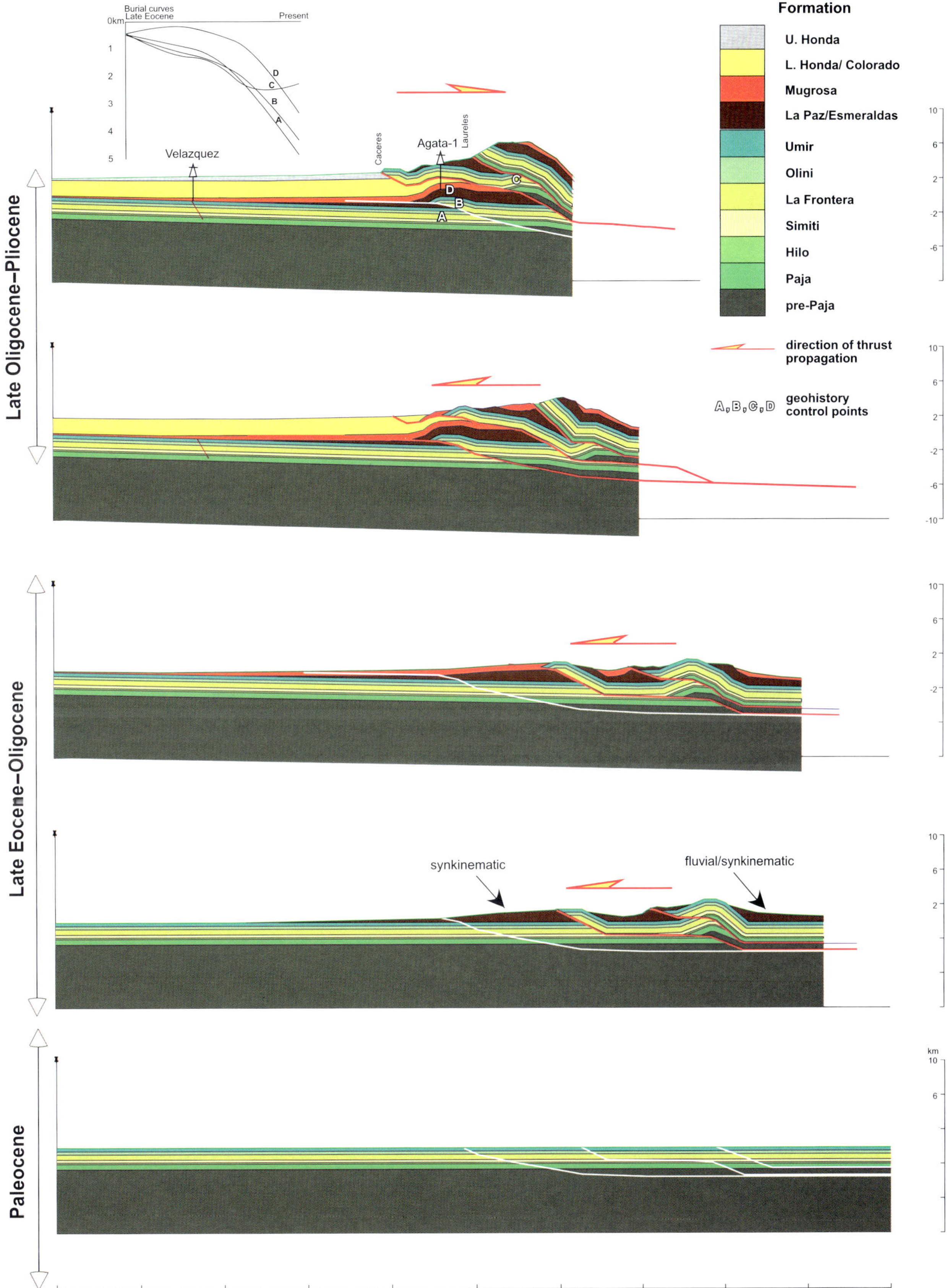

FIGURE 9. Forward structural model of section C, Figure 7, using IFP's Thrustpack 6.0 software. The top left inset qualitatively describes the burial history of several points within the fold-and-thrust belt. The red arrow indicates the direction of thrust propagation.

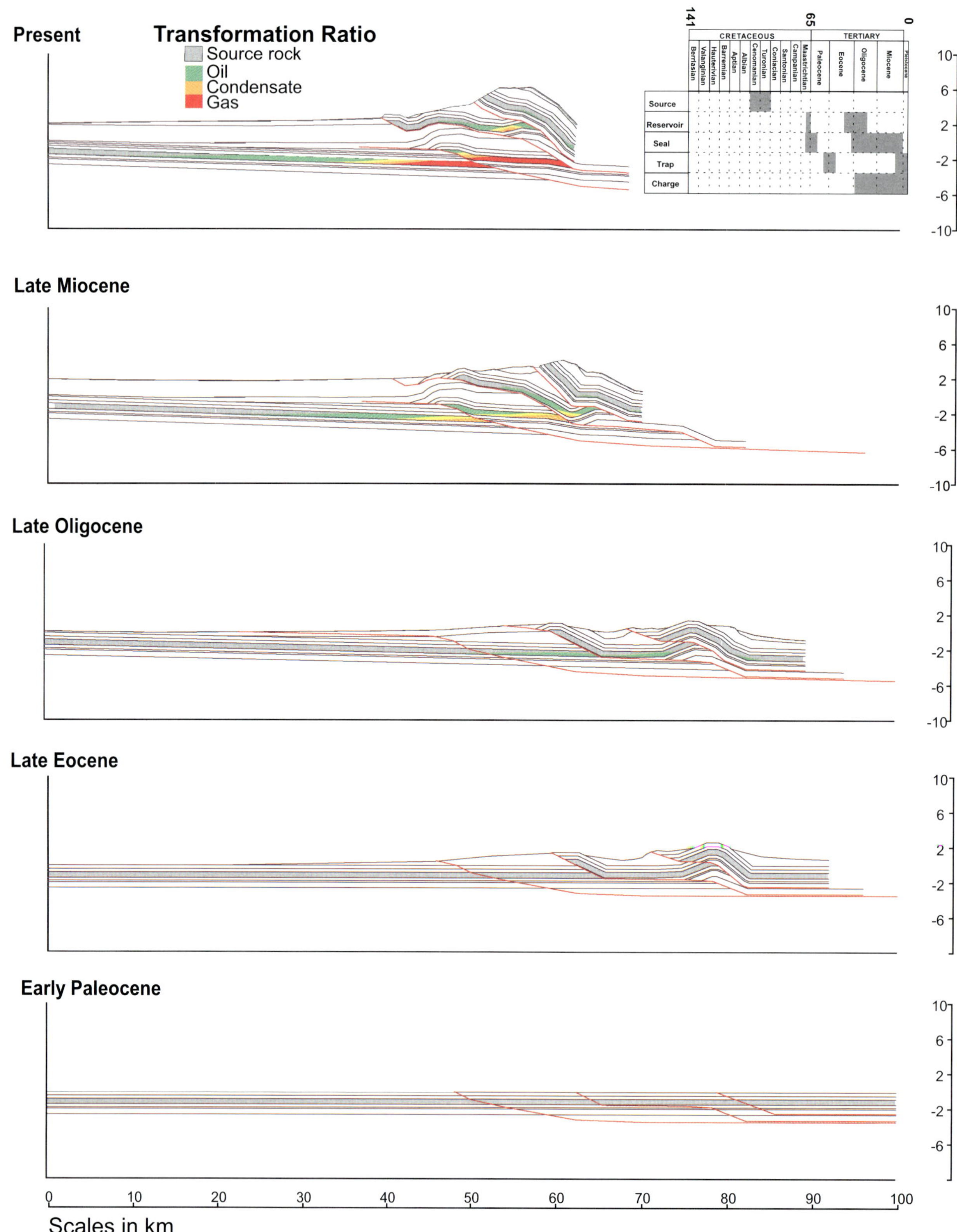

Figure 10. Hydrocarbon-generation model, based on the structural model of Figure 9. Hydrocarbon phases are shown as a function of the transformation ratio. The model predicts the onset of hydrocarbon generation occurring in late Oligocene time. The inset at the top right corner summarizes the main components of the petroleum system in the study area. Details of the model inputs may be requested from the authors.

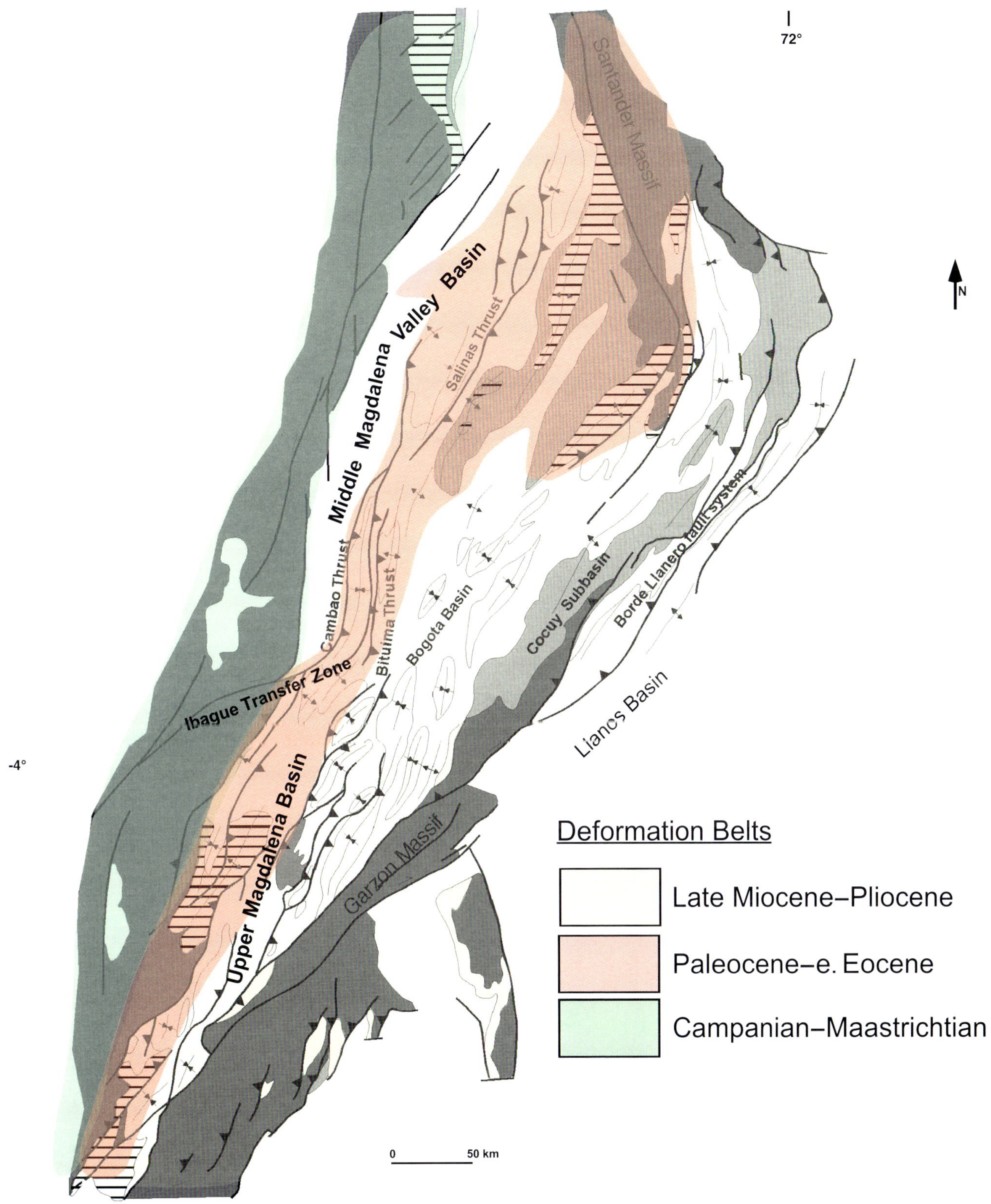

FIGURE 11. Diagram depicting the easterly migrating deformational episodes in the Colombian Andes. Boundaries are approximate and still poorly constrained. The Andean episode (late Miocene–Pliocene) affects the entire area, yet the boundary shows the sector most affected by this episode.

(1993), Cooper et al. (1995), Roeder and Chamberlain (1995), and our work here. An easterly transported crustal-scale wedge involving Middle Magdalena Basin basement rock may have acted as a structural backstop to produce the structures along the western flank of the Eastern Cordillera.

CONCLUSIONS

Detailed mapping along a sector of the western margin of the Eastern Cordillera of Colombia indicates that the fold belt exposed here resulted from a series of compressional events that began in late Paleocene–early Eocene time. Westerly derived Upper Cretaceous sediments exposed in the southern sector of the study area indicate uplift of the Central Cordillera at that time. The sequence of events and related evidence is outlined below.

1) Latest Campanian to Maastrichtian uplift of the Central Cordillera is signaled by the presence of extensive alluvial fans, that is, the Cimarrona Formation. These consist largely of coarse conglomerates and sandstones containing milky quartz pebbles derived from the crystalline rocks of the Central Cordillera.
2) Continued uplift of the Central Cordillera in Paleocene–Eocene time is signaled by the presence of thick sequences of alluvial sediments known as the Gualanday Formation (although they are mainly restricted to the Upper Magdalena Basin).
3) Late Paleocene–early Eocene thrusting occurred in a westerly directed, forward-propagating sequence along the western flank of the Eastern Cordillera. Synkinematic breccias containing fragments primarily of Cretaceous chert were deposited during this event. These sediments are preserved in piggyback basins within the main thrust sheets. The relief generated by this early Paleogene fold belt constrained the axis of upper Eocene sediments to run along the present-day crestal zone of the Eastern Cordillera (Villamil and Restrepo-Pace, 1997, 1998) and initiated the eastern tilting of the Middle Magdalena Basin basement by flexural loading.
4) Development of an east-dipping late Paleocene–early Eocene unconformity is recognized seismically throughout the Middle Magdalena Basin. Along the western flank of the Eastern Cordillera, this unconformity consists of a deep scouring surface that partially beveled the relief that had been created by Paleogene west-vergent, thin-skinned thrusting.
5) There was late Oligocene–early Miocene development of a series of intercutaneous wedges located at the thrust front. Onlapping upper Miocene molassic strata on folded middle Miocene fluvial deposits constrain the age of wedge formation. These structural traps constitute primary targets for future hydrocarbon exploration along the thrust front.
6) Late Miocene structural reactivation and exhumation of the hinterland area occurred in a break-back thrusting sequence. Major inversion of the Mesozoic extensional faults took place, with concomitant deposition of molassic sediments (the upper section of the Honda Group) in the Middle Magdalena Basin. Additionally, the Paleogene synkinematic deposits tilted and the late Paleocene–early Eocene unconformity folded.

ACKNOWLEDGMENTS

The authors wish to express gratitude to Charles Kluth for his reviews, which substantially improved this manuscript. Gratitude goes also to Peter Coney for our inspiration, for, as he would say, "a geologic map is a metaphor." This chapter is dedicated to Professor Dr. Otto Geyer, pioneer and mentor of Andean geology. We thank ConocoPhillips for allowing the release of the information contained in this study.

REFERENCES CITED

Andriessen, P. A. M., 1995, Fission-track analysis: principles, methodology, and implications for tectonothermal histories of sedimentary basins, orogenic belts, and continental margins: Geologie en Mijnbouw, v. 74, p. 1–12.

Butler, K., and S. Schamel, 1989, Structure along the eastern margin of the Central Cordillera, Upper Magdalena Valley, Colombia: Journal of South American Earth Sciences, v. 1, p. 109–120.

Caceres, C., and F. Etayo-Serna, 1969, Bosquejo geologico de la region del Tequendama. Opusculo guia de la excursion pre-congreso: Primer Congreso Colombiano de Geologia, Opusculo, p. 1–23.

Campbell, C. J., and H. Burgl, 1965, Section through the eastern Cordillera of Colombia, South America: Geological Society of America Bulletin, v. 76, p. 567–590.

Cooper, M. A., et al., 1995, Basin development and tectonic history of the Llanos basin, eastern Cordillera and middle Magdalena valley, Colombia: AAPG Bulletin, v. 79, p. 1421–1443.

Colletta, B., F. Hebrard, J. Letouzey, P. Werner, and J.-L. Rudkiewics, 1990, Tectonic style and crustal structure of the eastern Cordillera (Colombia) from a balanced cross-section, *in* J. Letouzey, ed., Petroleum and tectonics in mobile belts: Paris, Editions technip, p. 81–100.

Dengo, C. A., and M. C., Covey, 1993, Structure of the Eastern Cordillera of Colombia: Implications for trap styles and regional tectonics: AAPG Bulletin, v. 77, p. 1315–1337.

Fabre, A., 1987, Tectonique et Generation de hydrocarbures: un modele de l'evolution de la Cordillere Orientale de Colombie et du basin des Llanos pendant le Cretace et le Tertiaire: Archive de Science, Geneve, v. 40, Fasc. 2, p. 145–190.

Gomez, E., and P. E. Pedraza, 1994, El Maastrichtiano de la region Honda-Guaduas, limite N del Valle Superior del Magdalena: registro sedimentario de un delta dominado por rios trenzados, *in* F. Etayo-Serna, ed., Estudios geologicos del Valle Superior del Magdalena: Bogota, Universidad Nacional de Colombia, Chapter 3, p. 1–20.

Hooghiemstra, H., 1984, Vegetational and climatic history of the high plain of Bogota, Colombia: a continuous record of the last 3.5 million years: Disertaciones Botanicae, v. 79 , 368 p.

Irving, E. M., 1975, Structural evolution of the northernmost Andes, Colombia: USGS Professional Paper 846, 42 p.

Julivert, M., 1970, Cover and basement tectonics in the Cordillera Oriental of Colombia, South America, and a comparison with some other folded chains: Geological Society of America Bulletin, v. 81, p. 3623–3646.

Kroonenberg, S. B., J. G. M. Bakker, and A. M. Van der Wiel, 1991, Late Cenozoic uplift and paleogeography of the Colombian Andes: constraints on the development of high-Andean biota: Geologie en Mijnbouw 69, p. 279–290.

McKenzie, D. P., 1978. Some remarks on the development of Sedimentary Basins: Earth and Planetary Science Letters, v. 40, p. 25–32.

Morales, L. G., and the Colombian Petroleum Industry, 1956, General geology and oil occurrences of the Middle Magdalena Valley, Colombia, *in* L. G. Weeks, ed., Habitat of the middle and upper Magdalena basins, Colombia: Oil— A symposium: AAPG, p. 641–695.

Restrepo-Pace, P. A., 1989, Restauracion de la seccion geologica Caqueza-Puente Quetame: moderna interpretacion estructural del flanco este de la Cordillera Oriental: Universidad Nacional de Colombia, unpublished thesis, 57 p.

Roeder, D., and R. L. Chamberlain, 1995, Eastern Cordillera of Colombia: Jurassic-Neogene crustal evolution, *in* A. J. Tankard, R. Suarez-S., and H. J. Welsink, eds., Petroleum basins of South America: AAPG Memoir 62, p. 633–645.

Sassi, W., and J. L. Rudkiewics, 2000, Thrustpack 7.2: 2D Forward structural and hydrocarbon maturation modeling software: Paris, France, Insitut Francais du Petrole.

Schamel, S., 1991, Middle and Upper Magdalena Basins, *in* K. T. Biddle, ed., Active margin basins: AAPG Memoir 52, p. 283–301.

Shagam, R., B. P. Kohn, P. O. Banks, L. E. Dasch, R. Vargas, G. I. Rodriguez, and N. Pimentel, 1984, Tectonic implications of Cretaceous-Pliocene fission-track ages from rocks of the circum-Maracaibo Basin region, western Venezuela and Eastern Colombia: Geological Society of America Memoir 162, p. 385–412.

Van der Hammen, T., 1957, Estratigrafia palinologica de la Sabana de Bogota, Cordillera Oriental de Colombia: Boletin Geologico del Instituto Geologico Nacional, v. 5, no. 2, p. 189–203.

Van der Wiel, A. M., and P. A. M. Andriessen, 1991, Precambrian to recent thermotectonic history of the Garzon Massif (Eastern Cordillera of the Colombian Andes) as revealed by fission track analysis, *in* A. M. Van der Wiel, Uplift and volcanism of the SE Colombian Andes in relation to Neogene sedimentation of the Upper Magdalena Valley: Wageningen, Ph.D. thesis, 208 p.

Villamil, T., and P. A. Restrepo-Pace, 1997, Paleocene-Miocene Paleogeographic evolution of Colombia: Memorias del VI Simposio Bolivariano, Cartagena-Colombia, p. 275–287.

Villamil, T., and P. A. Restrepo-Pace, 1998, Stratigraphic and structural constraints on Campanian-Oligocene tectonostratigraphic development of northwestern South America: Abstracts of the AAPG International Conference and Exhibition, Rio de Janeiro, Brazil: AAPG Bulletin, v. 82, no. 10, p. 90.

31

Chambers, J. L. C., I. Carter, I. R. Cloke, J. Craig, S. J. Moss, and D. W. Paterson, 2004, Thin-skinned and thick-skinned inversion-related thrusting— A structural model for the Kutai Basin, Kalimantan, Indonesia, *in* K. R. McClay, ed., Thrust tectonics and hydrocarbon systems: AAPG Memoir 82, p. 614–634.

Thin-skinned and Thick-skinned Inversion-Related Thrusting— A Structural Model for the Kutai Basin, Kalimantan, Indonesia

John L. C. Chambers
Santos Ltd., Adelaide, Australia

Ian Carter
Lukoil Overseas Ltd., London, U.K.

Ian R. Cloke
ExxonMobil Production Company, Houston, Texas, U.S.A.

Jonathan Craig
Eni-Agip, Milan, Italy

Steve J. Moss
Apache Corporation, Perth, Australia

David W. Paterson
Consultant, Calgary, Canada

ABSTRACT

In the Kutai Basin, regional compression reactivated basement extensional faults, inverting the Paleogene depocenters as anticlines that are often flanked on one side by basement thrusts. Over most of the basin, the overlying Neogene section is detached near the top of an overpressured zone and deformed as a thin-skinned fold-thrust belt. Fieldwork and hydrocarbon exploration on the northern margin of the Kutai Basin, where the sedimentary section is relatively thin compared with the basin center, has provided data in an imaging window where both structural deformation styles can be observed. We contend that even in the deeper parts of the basin center, basement-involved inversion beneath the overpressure zone has influenced the shallower structures that are present in the Mahakam Delta depocenter. In the Mahakam Delta depocenter, the subsequent response to inversion of the Paleogene rift section was controlled in part by heterogeneity in the shallow section, including structures formed through syndepositional loading, delta progradation, normal faults, and marked facies changes. It is proposed that the structural model derived from study of the northern area is applicable to all of the Kutai Basin, including the basin depocenter, an area where only the thin-skinned deformation can be imaged because of the thickness of the Neogene section.

INTRODUCTION

Significant hydrocarbons occur in middle Miocene to Pliocene deltaic sandstone sequences in the Kutai Basin, East Kalimantan (Figure 1) (Paterson et al., 1997). Trap types in the basin include folds within the Samarinda Anticlinorium, folded and faulted structures along the coastline, and structural traps offshore of the

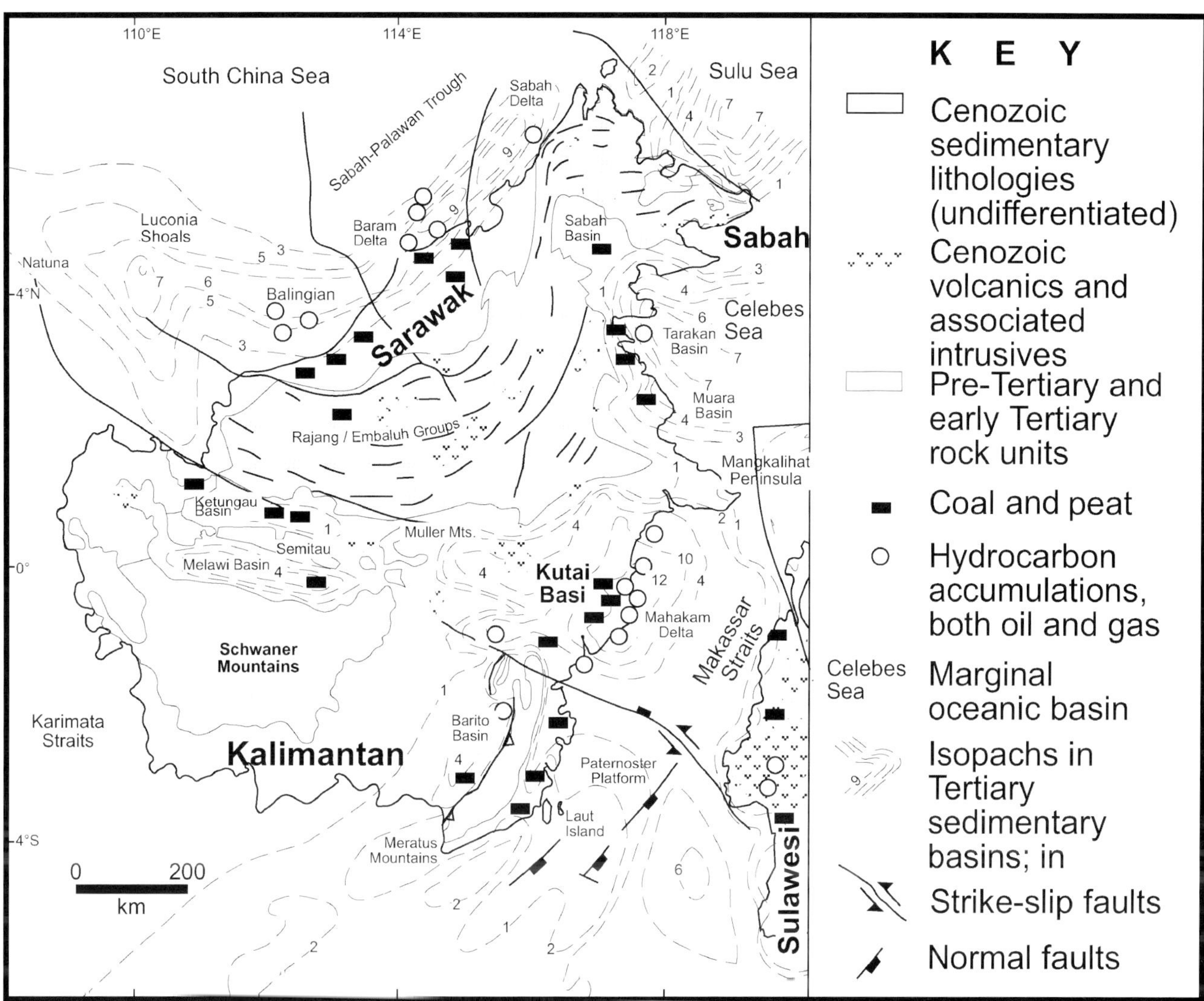

FIGURE 1. Simplified geologic and tectonic map of Borneo. Adapted from Wilson and Moss (1999).

present-day Mahakam Delta (Figure 2). A diverse range of previous explanations has been published regarding how some of these structures formed. These explanations have included regional gravity sliding of material off the rising Central Kalimantan Ranges (Ott, 1987); vertical shale diapirism, growth faulting and inversion through regional inversion (Biantoro et al., 1992); and a collision-related fold-and-thrust belt (Moriya and Nishidai, 1996). More recently, two structural models that have been proposed to explain how the onshore structures include folding and thrusting in response to basin inversion. Chambers and Daley (1995) proposed a direct cause-and-effect relationship between basement inversion and detachment thrusting and folding, using regional gravity data to infer an inverted rift sequence component. In contrast, Ferguson and McClay (1997) and McClay et al. (2000) concluded that basement inversion, inferred from scaled physical analog models, led indirectly to thin-skinned inversion of middle Miocene and younger delta-top basin fill that was already deformed by depositional loading and gravity-driven tectonism. The latter workers emphasized the influence that early syndepositional structures, such as growth faults, delta-toe thrust faults, and rollover anticlines, had on the observed geometries.

Rose and Hartono (1978) proposed a sediment thickness of more than 13 km in the Mahakam Delta depocenter, whereas Duval et al. (1992) proposed a maximum thickness of 12 km. Regional gravity modeling (Chambers and Daley, 1995) suggests that 15 km may be a more likely maximum figure for sediment thickness in parts of the basin. Structures in the shallow section cause distortions on seismic data, making the basement difficult to image through seismic data alone, especially in the basin's center. This makes it difficult to determine the relationship of deep structures to the shallow, detached fold-and-thrust structures. As part of a larger project, of which this chapter is only a small

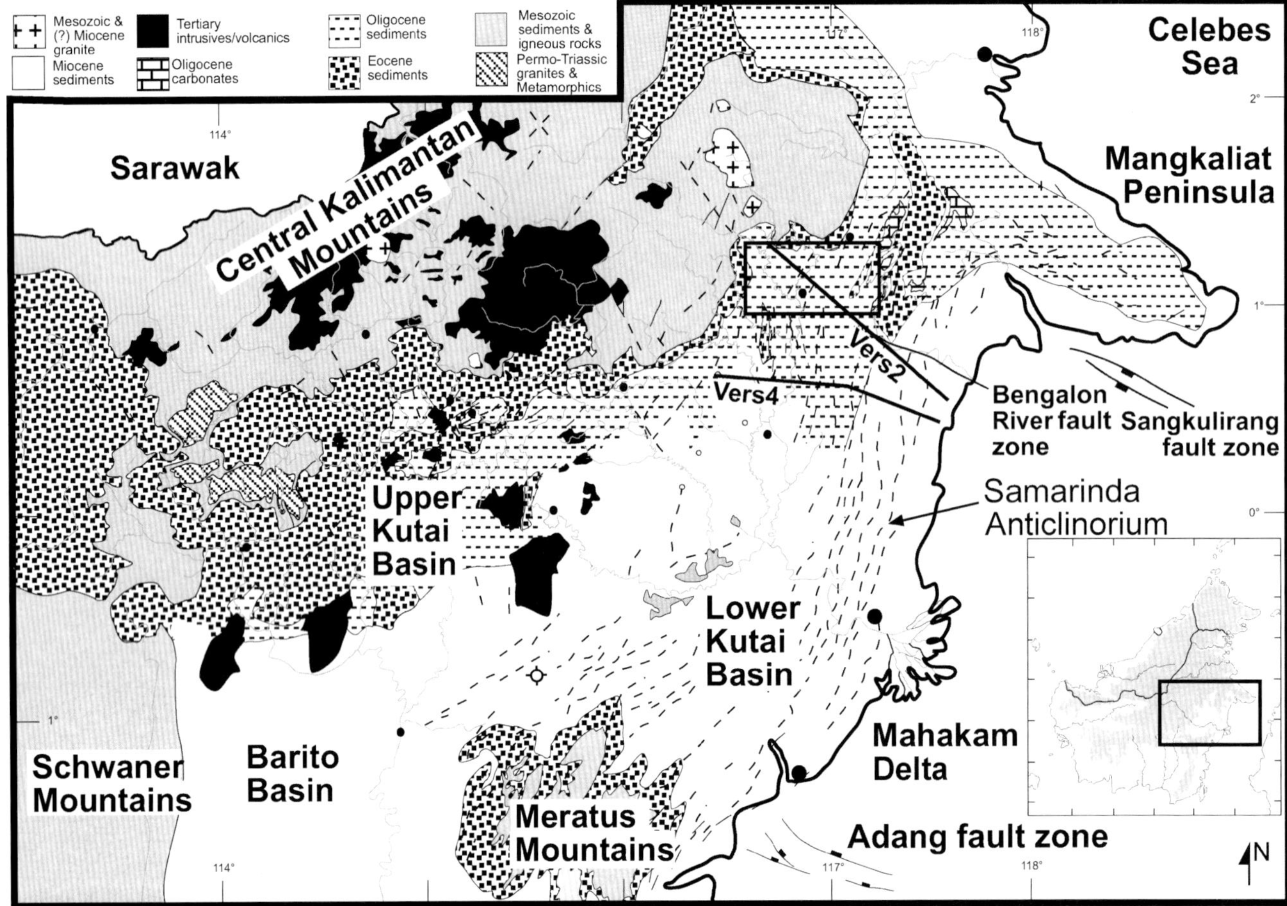

FIGURE 2. Geologic map of Kutai Basin. The location of the two regional cross sections shown in Figure 6 is shown. The location of Figure 5 is shown by the small inset box. From Moss and Chambers (1999).

facet, basement well intersections on the southern basin margin were projected into the basin center and used in combination with airborne gravity and aeromagnetic data, as well as a limited amount of deep, 10-s seismic data, to substantiate a 12- to 15-km depth to basement in the basin's center situated under the modern Mahakam Delta. A revised depth-to-basement map for the Kutai Basin is presented in Figure 4 of Moss and Chambers (1999).

This chapter examines the structural deformation on the northern margin of the basin (Figures 1 and 3), where the thin sedimentary section allows the entire Cenozoic section to be imaged more clearly with seismic data than is possible in other parts of the basin. We describe in detail the thrust structures formed through the reactivation of rift faults and examine their relationship to shallow detachment structures. We then present an aeromagnetic-data interpretation covering the Mahakam Delta's depocenter and test the validity of extrapolating the reactivated extensional rift-fault model from the northern study area to the Mahakam Delta's depocenter. Finally, we consider the relationship of the basement rift inversion to the shallower detachment folding and faulting in the Mahakam Delta's depocenter.

THE TERTIARY KUTAI BASIN

The island of Borneo represents the accreted southwestern tip of the eastern margin of Sundaland, with accretion occurring by the Late Cretaceous (see Metcalfe, 1999 for full discussion). The Meratus Mountains, which abut the southern edge of the Kutai Basin, contain a wide variety of tectonically intercalated sedimentary, metamorphic, and basic to ultrabasic igneous rocks formed during late Mesozoic accretion (Sikumbang, 1986, 1990). A similar mixture of cherts, deepwater clastics, layered gabbros, and basic and ultrabasic rocks occurs at numerous outcrops on the northern margin of the Kutai Basin (Moss and Chambers, 1999). Tertiary lacustrine, fluvial marginal-marine, or marine sedimentary rocks occur in the Kutai Basin over large tracts of eastern, central, and northern Borneo and unconformably overlie the above-described, variably deformed pre-Tertiary units (Figure 4). Earlier basement

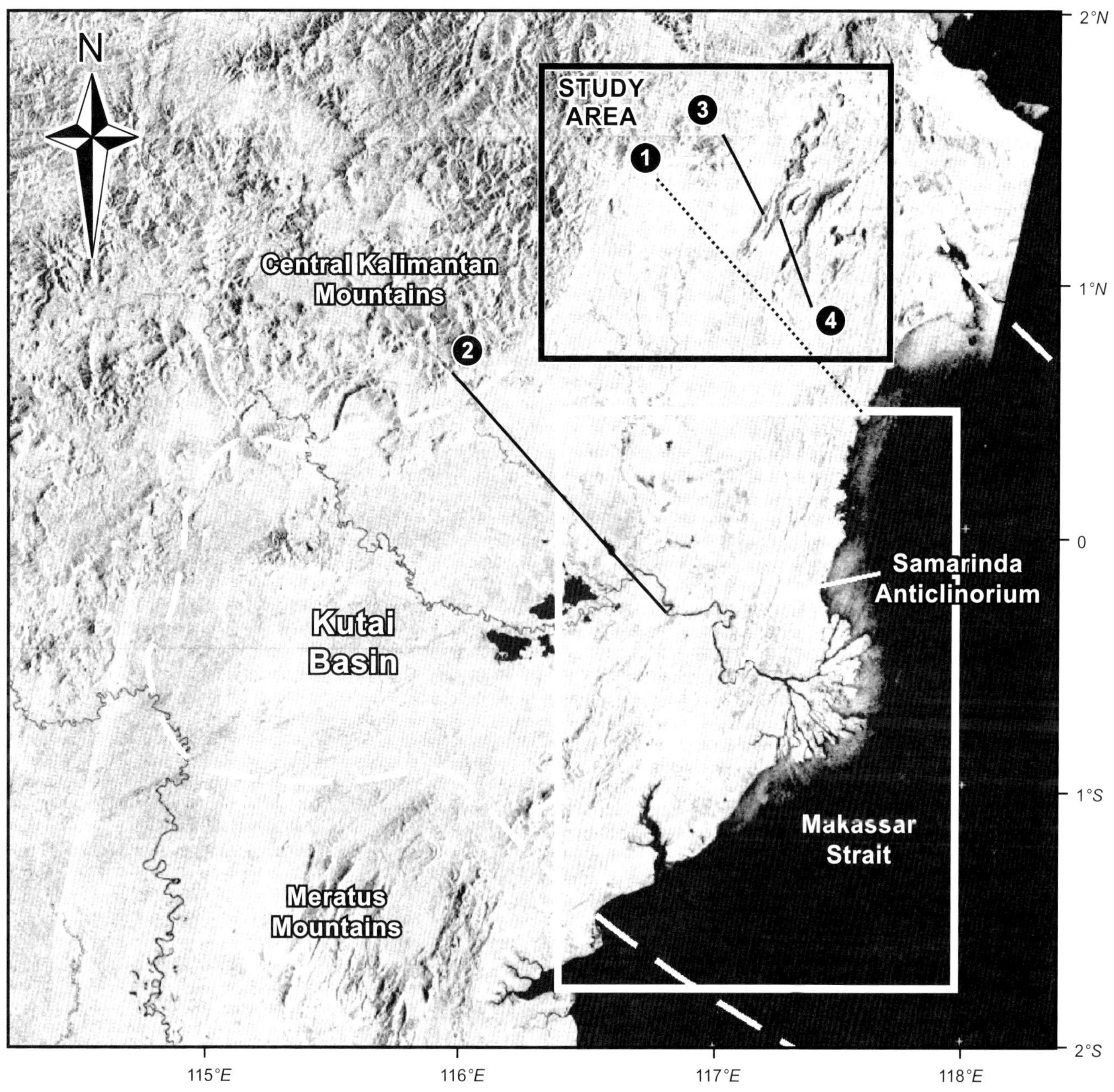

FIGURE 3. A satellite image of eastern Kalimantan, showing a northeast-southwest lineament in the metasediment basement outcrop to the north of the Kutai Basin that parallels the axis of the inversion anticlines found on the northern and southern basin margins. An orthogonal northwest-southeast lineament evident in the basement outcrop is more subtly expressed in the basin. A third major lineament direction is evident in the north-northeast to south-southwest orientation of the coastal fold-and-thrust belt, the Samarinda Anticlinorium. Study area (Figure 5) and Mahakam Delta depocenter map (Figures 11 and 12) locations outlined.

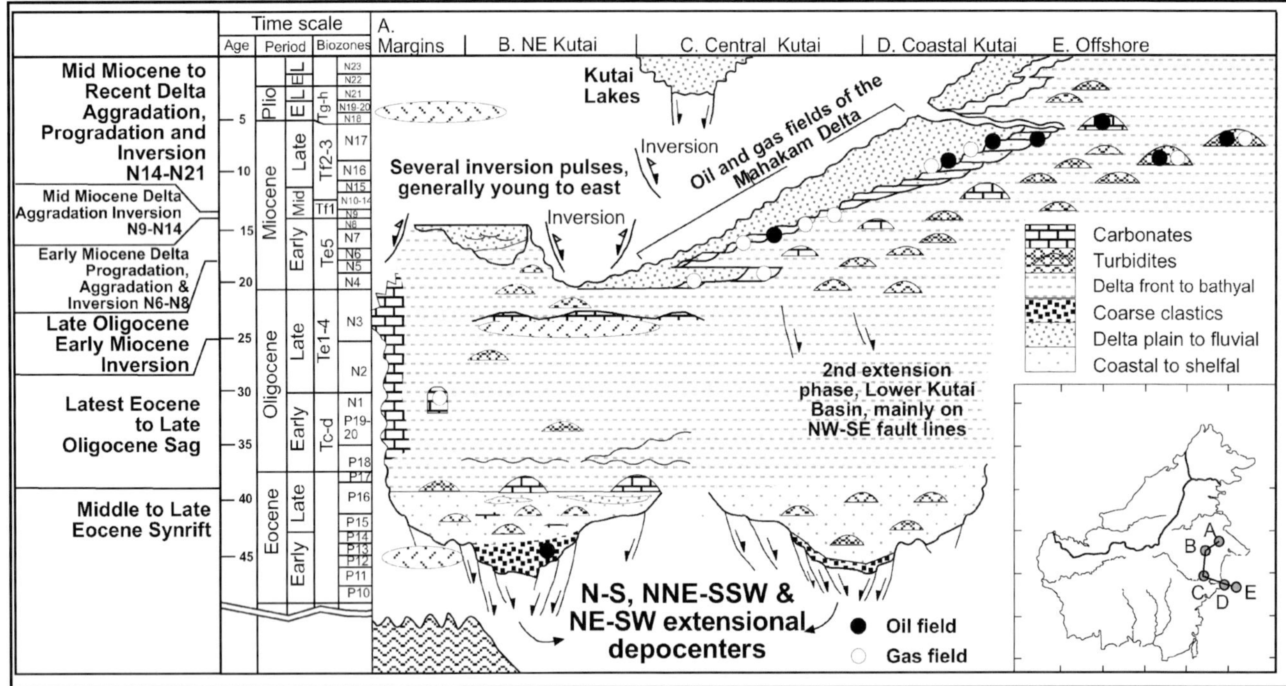

FIGURE 4. A two-dimensional stratigraphic section for the Kutai Basin. Adapted from Moss and Chambers (1999).

structures influenced basin evolution by providing loci for development of the initial basin rift faults (see descriptions in Cloke et al., 1997, 1999). The onset of Tertiary sedimentation was contemporaneous with, and subsequent to, a period of Paleogene extension and subsidence, which had begun by ~44 Ma Middle Eocene (Hutchinson, 1996; Moss et al., 1997; Moss and Finch, 1998; Moss and Chambers, 1999).

Deposition in the Kutai Basin can be divided into the following five phases (Moss and Chambers, 1999):

- Middle to late Eocene synrift phase
- Late Eocene to late Oligocene sag phase
- Late Oligocene second rift phase with coeval volcanic activity
- Early Miocene delta progradation–aggradation and early inversion phases
- Middle Miocene to present delta progradation–aggradation and main inversion phases

These depositional phases resulted in the accumulation of 9 to 15 km of siliciclastic, carbonate, and organic sediments in the Kutai Basin, punctuated by several phases of localized volcanism. Late Oligocene and younger uplift of sections of Borneo is related to reactivation of earlier-formed structures, including both basement lineaments and Eocene extensional faults in the Kutai Basin. In the Kutai Basin, fault reactivation caused inversion of rift depocenters and, we contend, contributed to the formation of the detached thrust-and-fold structures.

NORTHERN MARGIN OF THE KUTAI BASIN STUDY AREA

Field Description

As illustrated in Figure 5, four ~20-km-wavelength, open, asymmetric anticlines, with north- to north-northeast-trending fold axes, are located on the northern margin of the Kutai Basin. Major basement thrusts bound the western sides of two of these 30- to 70-km-long anticlines (Gongnyay, and Gergaji Anticlines in the east). All four anticlines have cores of middle Eocene sediments and have either linear eastern (Wahau Anticline) or western flanks (Gongnyay, Gergaji and Jeli Anticlines).

In the cores of the Gongnay and Gargaji Anticlines, parasitic chevron and box detachment folds occur and range in amplitude from decimeter to meter scale. Typically, these parasitic folds detach upon either carbonaceous shales or deltaic coals. For example, open chevron folds in interbedded synrift sandstones and shales in the core of the Gongnyay Anticline have approximately 5-m wavelengths and 3-m amplitudes. The regional fold hinge trend is 14–208° with a subvertical fold axial surface. Bulbous hinge zones (shale dominated) or limb thrusts (sandstone dominated) result from competence contrasts between sandstone and shale. Slickenfibers on fault planes of thrusts within fold limbs indicate west-northwest to east-southeast shortening. The 90° bed "cutoff" angles on these limb thrusts suggest that the thrusts developed in already-folded strata. At several fold

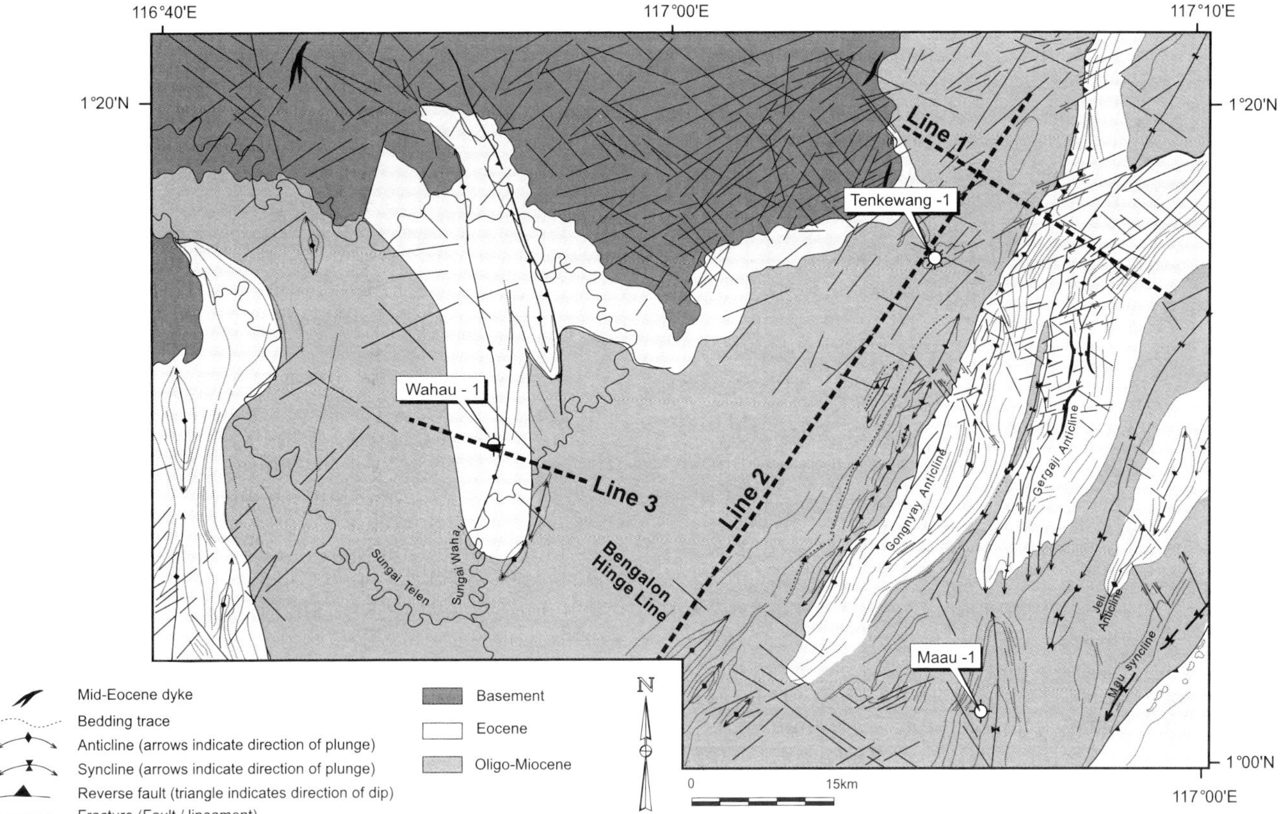

FIGURE 5. Geology map of the northern margin of Kutai Basin. Anticlinal features are interpreted to be inverted middle Eocene half grabens. Seismic line locations for Figures 7, 8, and 9 are shown. Location of the map is shown on Figure 2.

outcrops within heterolithic Eocene synrift sections, the small-scale thrusts that occur are oriented bedding-parallel to bedding-subparallel. These small bedding-parallel thrusts commonly form part of high-angle (~60°), west-facing centimeter- to meter-scale thrust duplexes observed on the limbs of chevron folds. The duplex structures formed prior to the formation of the small-scale chevron folds as minor, intrabed, low-angle duplexes, and with continued contraction these early-formed structures were rotated on steep fold limbs into their present position. Small-scale oblique slip thrusts also occur at outcrop scale and have been interpreted from remote-sensing data. Measurement of slickenfibers on thrust-fault surfaces, and removal of folding, again indicate west-northwest to east-southeast contraction. Measurements of slickenfibers on fault planes indicate that the shortening direction was originally oriented at 290°. This rotated through time, with demonstrably younger slickenfibers indicating a shortening direction oriented at 315°. Crude estimates of the shortening using line-length measurements suggest as much as 60% shortening on a local scale. On a regional scale, the degree of shortening is likely to be considerably lower (Ferinsyah et al., 2000), and this is supported by the results of section balancing (Figure 6).

Subsurface Description

Subsurface structure was determined from several seismic surveys shot over the area and from detailed mapping of key horizons. Interpretation of the seismic data was supported by systematic lithology and paleontologic logging and sampling of the seismic shot holes, using the same techniques described by Carter and Morley (1995) in the southern portion of the basin. Figure 7 is a northwest–southeast-orientated 2-D seismic profile across the Gongnyay and Gergaji Anticlines. The section is orthogonal to the structural grain. The profile, located at the northern end of the Gongnyay and Gergaji Anticlines where deformation is more severe, shows basement thrust faults. Interpretation of this line and numerous other seismic lines indicates that the basement thrusts are former extensional faults that were reactivated with a reverse sense of shear, which produced folds with the inversion of former half-graben fill. Vitrinite reflectance of surface samples, section restoration, and seismic velocity data suggest that 2500 m of section has been uplifted and eroded on the eastern features. The Tengkawang-1 well (shown in Figure 7) tested a small inversion anticline that abuts the basement to the west. Based on surface and subsurface vitrinite

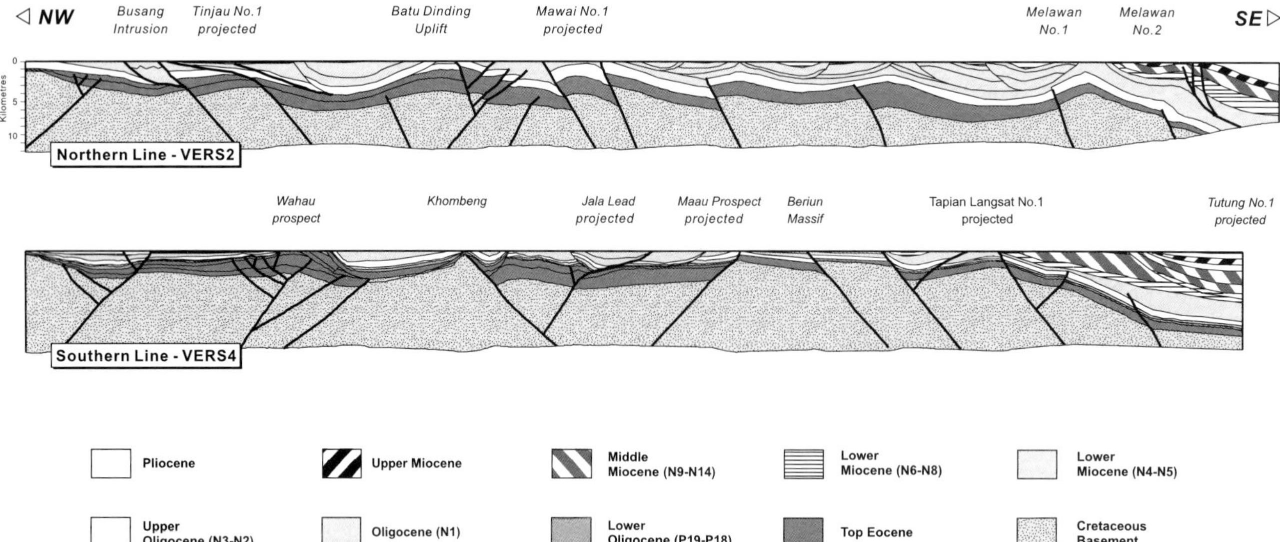

FIGURE 6. Two depth-converted regional cross sections constructed from several seismic lines, surface data, and well ties. These two profiles (and several others) were reconstructed using section-balancing software to calculate the amount of compression (see Ferinsyah et al., 2000).

reflectance data and velocity data, we estimate that this anticline has experienced less than 1000 m of uplift and erosion. The Tengkawang Anticline does not have a "surface footprint" because upper Oligocene shales at outcrop retain a minimal structural expression in the prevailing equatorial climate, whereas the Gongnyay and Gergaji Anticlines are cored by middle Eocene sandstones at outcrop and are visible on satellite images (Figure 5).

Figure 8 shows another west-northwest to east-southeast seismic line, this time through the east-vergent Wahau Anticline (located on Figure 5). Four west-dipping faults can be seen with several steeper and shorter antithetic faults in between. A prominent fold can be seen in the hanging wall of one the major west dipping faults. This fold was tested by the Wahau-1 well.

Figure 9 is a north-northeast- to south-southwest-oriented seismic profile that intersects the west-northwest to east-southeast profile of Figure 7. It extends southward across the Bengalon Lineament, a northwest-southeast late Oligocene to early Miocene extensional hinge line (shown on Figures 2 and 3). This lineament can also be identified on gravity data and satellite imagery as an important fault zone in the basin. The structures in this part of the basin lend themselves to field and SAR mapping because lower Miocene bathyal shales in the detached package are interbedded with more erosion-resistant turbidite sands that outcrop as topographic ridges. Note in Figure 9 the minimally deformed horst block between the hinge line and the inversion anticline to the north that was tested by the Tengkawang-1 well. We can see that folding and thrust faulting in the upper part of the section in the southwest end of the profile are detached from the inverted basement-rift section below them.

Well results, seismic-data interpretation, surface mapping, and the presence of mud volcanoes in the detached lower Miocene section south of the Bengalon fault zone's hinge line suggest extensive overpressure as a result of disequilibrium compaction, with overpressure being ubiquitous over most of the Lower Kutai Basin (Bates, 1996). The detachment thrusts are interpreted to sole out within the top of the overpressured shales that often reach fracture gradient within a few hundred meters of the top of the overpressure. Further, it seems geologically reasonable to infer that the basement thrusts flatten out as ceiling detachments in the base of the overlying overpressured shales, thereby uplifting and creating shortening in the overlying section. Seismic interpretation clearly shows that the detachment structures are not continuous with the deeper inversion structures. Basement and rift section lineations have a different orientation from that of the younger, detached-feature lineations. The deeper inversion structures and the detachment surfaces do, however, show similar amounts of shortening, which are estimated from section balancing to be 10–18% (Figure 6) (Ferinsyah et al., 2000). This observation is supported by studies in the middle Miocene and younger section (McClay et al., 2000) and by the differences in the orientation of lineation between basement and rift section to the younger detached features.

Figure 10 shows two isochron maps produced from detailed seismic mapping in the area shown in Figure 5. The Eocene isochron map (Figure 10a) shows predominant northeast-southwest fault orientation in the

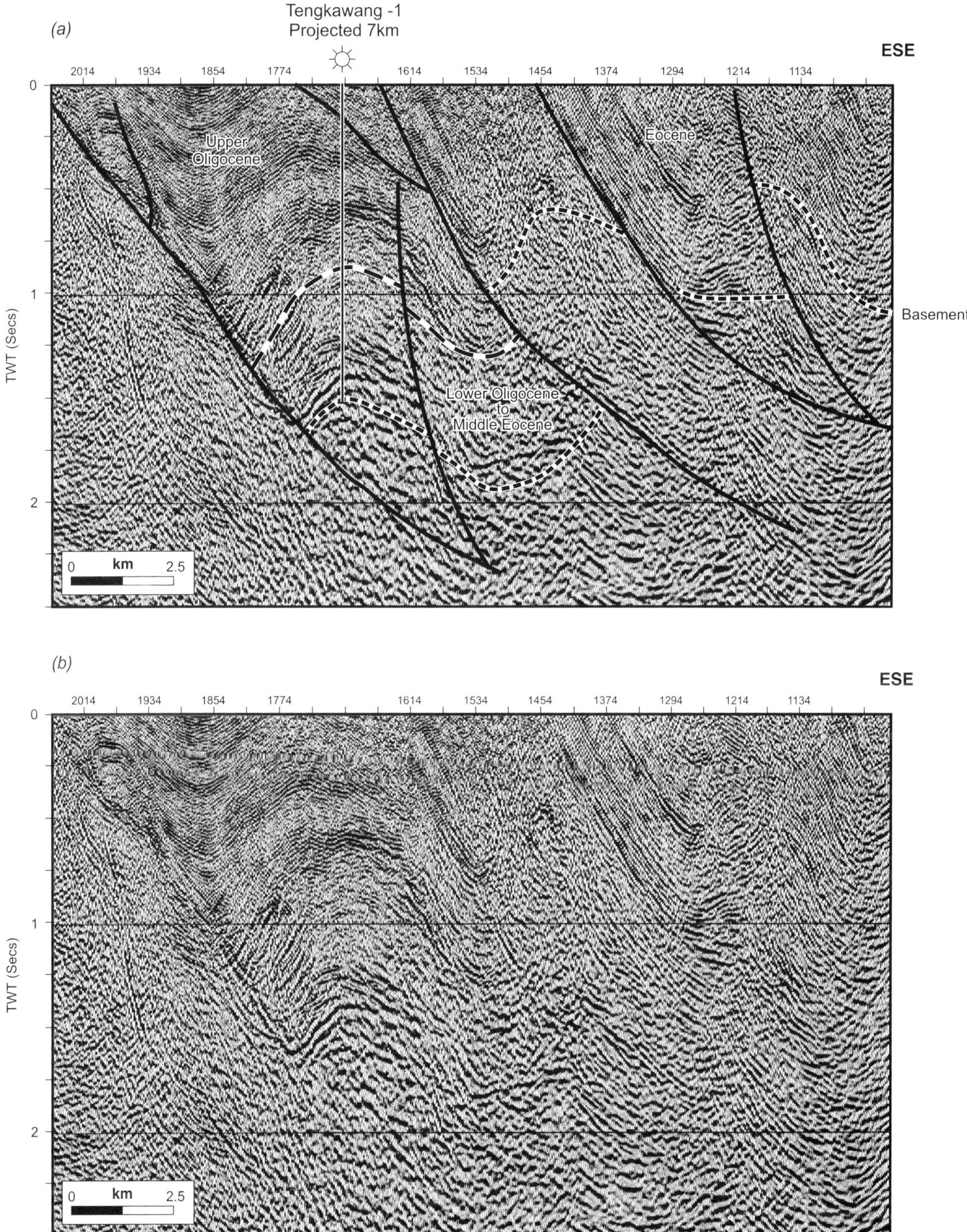

FIGURE 7. Interpreted (a) and uninterpreted (b) seismic line 1, a west-northwest–east-southeast-oriented seismic profile on the northern margin of the basin, showing west vergent basement thrust faults. The three inversion structures to the east become symmetric anticlines down-plunge to the south. The higher amplitude upper Oligocene events are conglomeratic turbidites that become volcanogenic, up-section. Vertical exaggeration is approximately 300%.

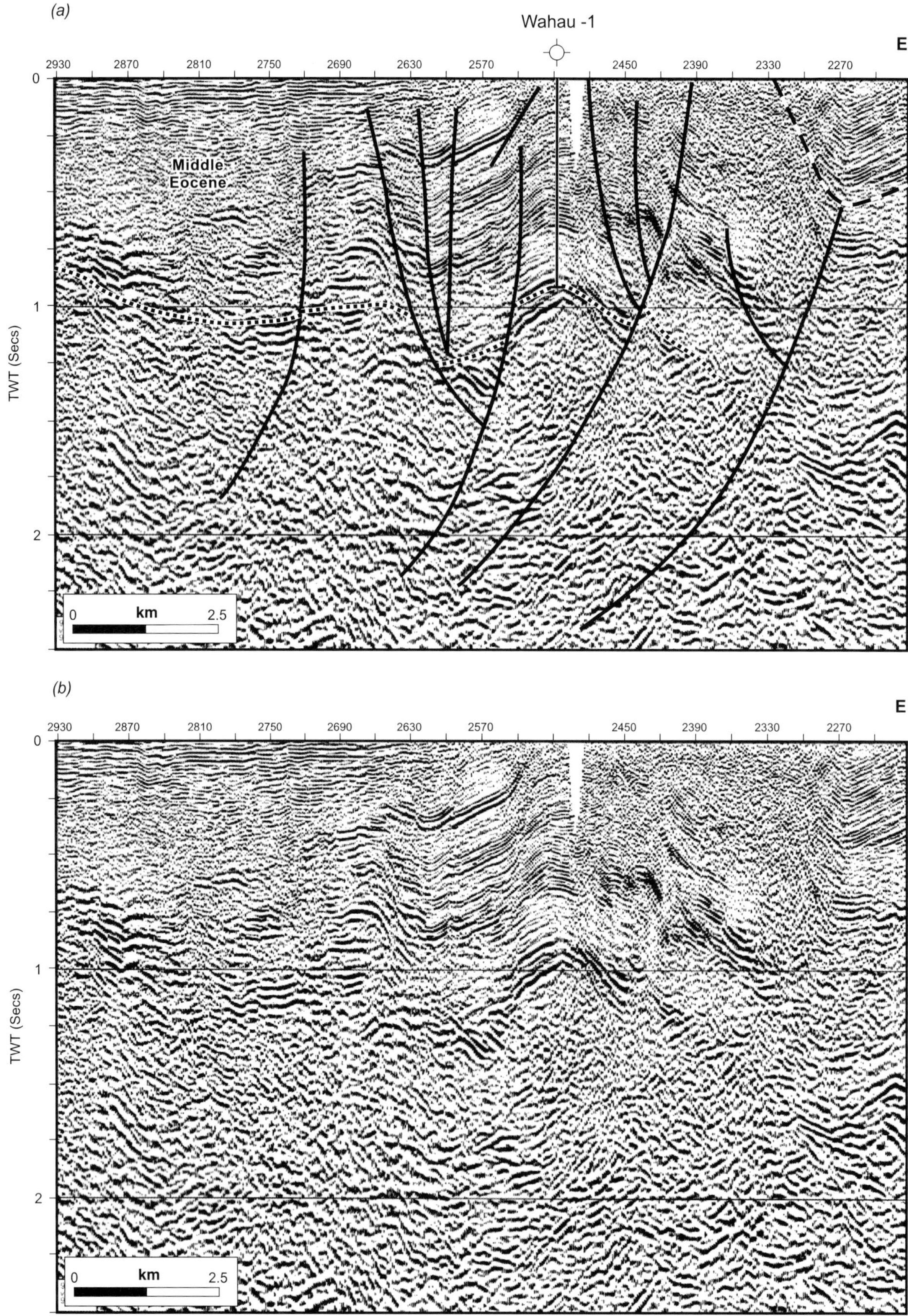

FIGURE 8. Interpreted (a) and uninterpreted (b) seismic line 3, an east-west seismic profile that ties with the Wahau-1 well. Vertical exaggeration is approximately 200%.

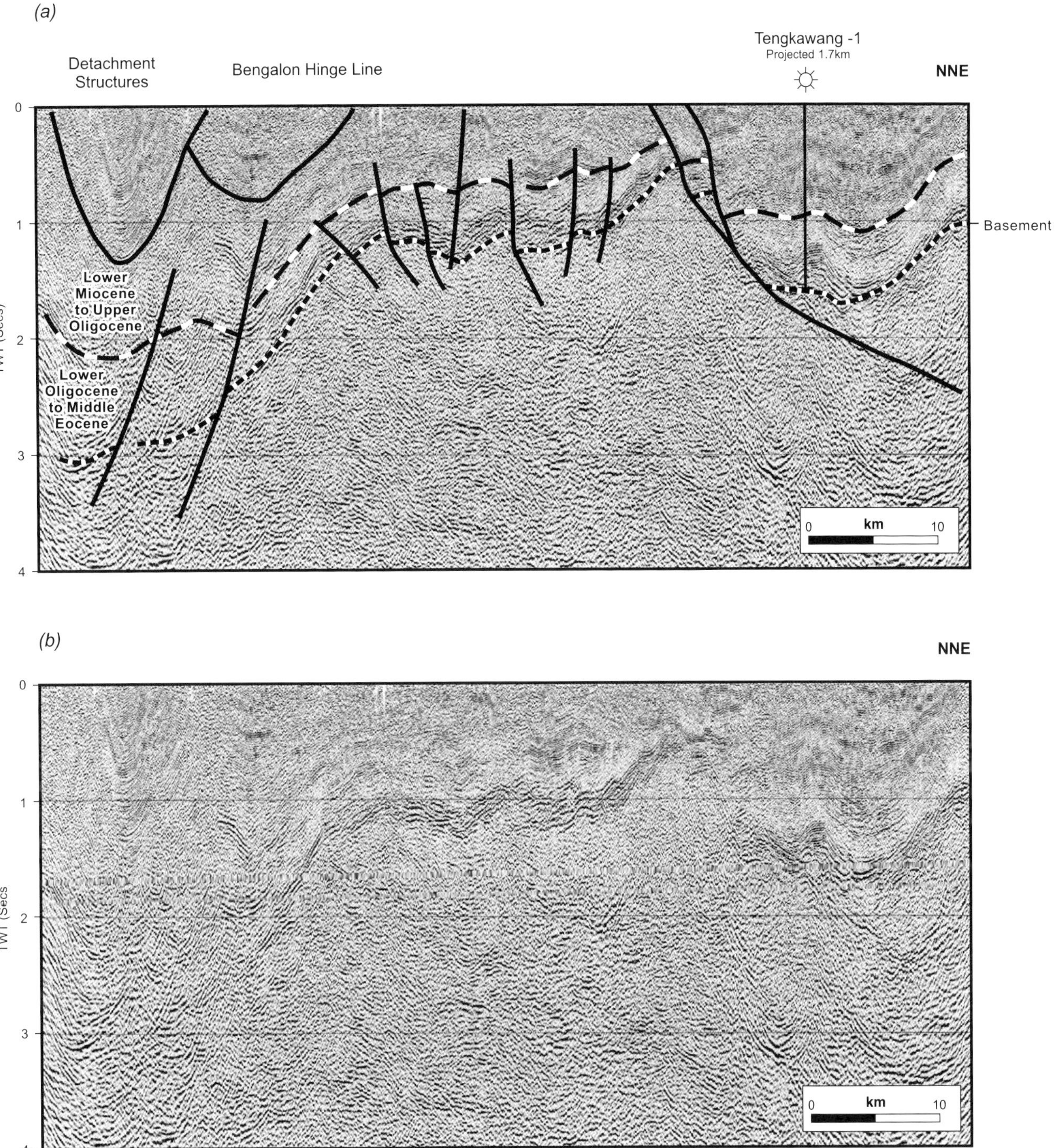

FIGURE 9. Interpreted (a) and uninterpreted (b) seismic line 2, a north-northeast- to south-southwest-oriented seismic profile on the northern margin of the basin that crosses the Bengalon lineament— interpreted to be a northwest–southeast flexural hinge line initiated during the late Oligocene. Basement and base of upper Oligocene horizon interpretations are shown. The high-amplitude events at approximately 1.0 s on the horst block are from limestones, which correlate to the top of a 300-m-thick upper Eocene limestone interval penetrated in the Tengkawang-1 well at 1.2 s (Figure 6). Reversed basement rift faults and shallow detachment thrust faults are interpreted. Vertical exaggeration is approximately 600%.

Gongnyay and Gergaji half grabens, matching the structural trend observed in the remote-sensing data. A north-south fault trend is evident near the Wahau half graben. Thickening of the Eocene section into the hanging wall of northeast-southwest faults can be seen. Interpretations made from the isochron map agree with field mapping, residualized gravity, and SAR data. Two depocenters corresponding to the Gongnyay and Gergaji

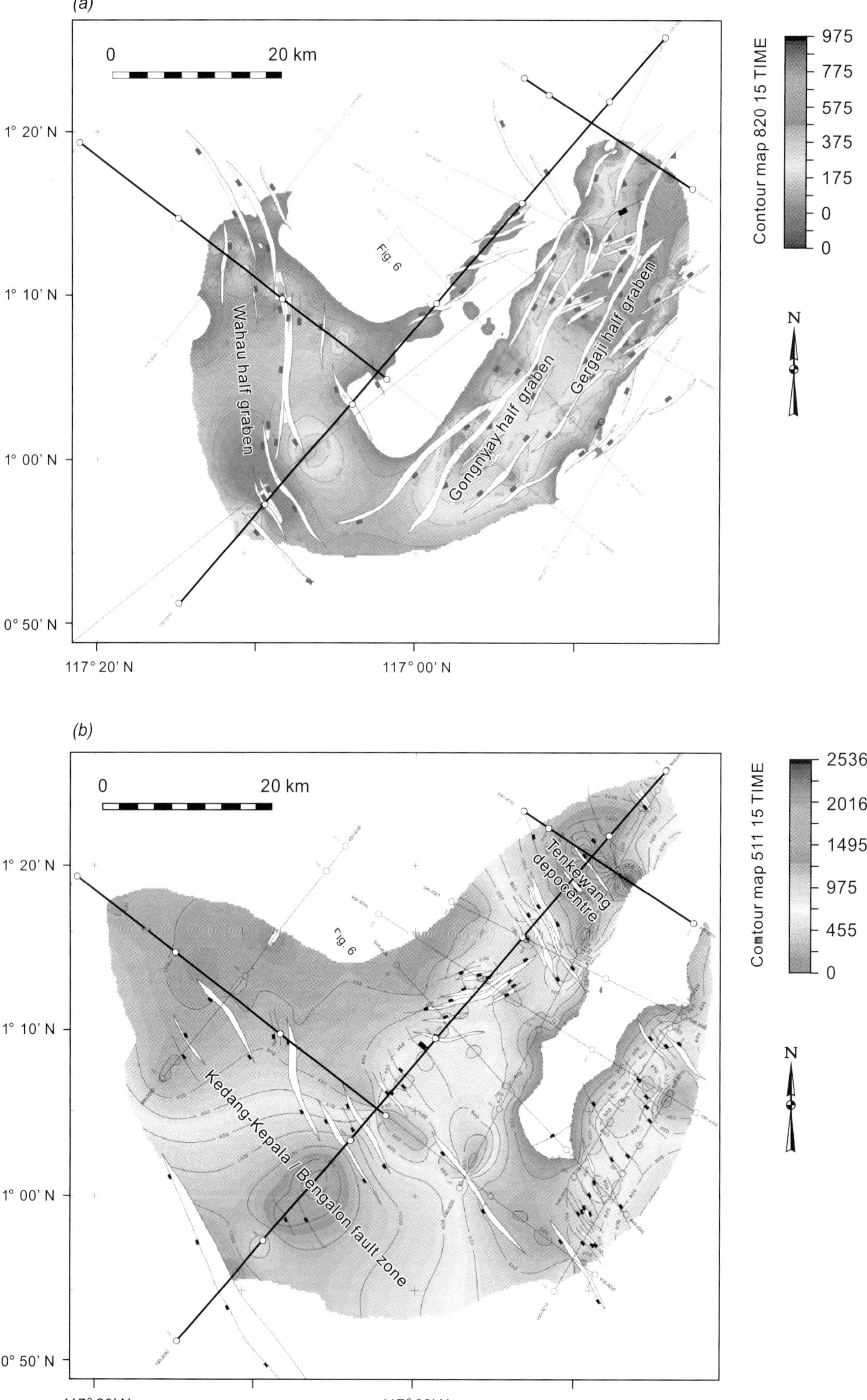

FIGURE 10. Isochron map for (a) the Eocene and (b) the Oligocene.

Anticlines shown on Figure 5 can be seen to trend north-northeast to south-southwest, whereas a third depocenter corresponds to the Wahau Anticline and trends almost north–south. The most significant aspect of the Eocene isochron map is that elliptical patterns of depositional thickness are apparent, with sequences thickening to the west in the Gongnyay and Gergaji half graben and to the east in the Wahau half graben (Figure 10a)— which is consistent with sediment deposition in a extensional half graben. The north-northeast–south-southwest-trending faults along this side of the basin appear to be arranged in a right-stepping en echelon pattern.

Figure 10b, an Oligocene isochron map, shows the northwest-southeast alignment of depocenters in the Oligocene. In the northern section of the map, we see a series of northwest–southeast-trending en echelon extensional faults, down-throwing to the north. In the southern section of the data, a similar set of parallel en echelon faults, but down-throwing to the south, are present. Rapid thickness changes are recorded across these faults, from 400 ms TWT over the footwall to 1.7 s TWT in the basin. Landsat TM data show a distinct surface lithologic boundary along the southern fault zone between karstified carbonates and shales (Moriya and Nishidai, 1996), both of which are of late Oligocene to early Miocene age (Moss and Chambers, 1999; Wilson et al. 1999), thereby confirming fault control on deposition.

Structural Synthesis

Pervasive east-northeast- to west-northwest-, northeast- to southwest-, and northwest–southeast-trending fractures are recognized from detailed field measurements, remote-sensing interpretation, and seismic mapping in the basement and the basin (Figure 5). Oblique extension (west-northwest to east-southeast) during the middle Eocene formed half grabens oriented north-south to northeast-southwest, with a dominant north-northeast to south-southwest orientation in the northern Kutai Basin. Half-graben polarity changes from locality to locality. For example, the Gongnyay and Gergaji half grabens were east-facing, whereas the Wahau half graben was west-facing. We derived the orientation of the half grabens mainly from seismic mapping in conjunction with surface data. The data described above and presented in other papers by Moss et al. (1997), Cloke et al. (1999), and Moss and Chambers (1999) are interpreted to show that middle Eocene north-northeast to south-southwest graben-bounding faults initially formed in conjunction with right-stepping en echelon arrays having northwest–southeast-trending transfer faults.

During the late Oligocene, a second rift phase modified the ancestral middle Eocene basin architecture and led to the present-day configuration of the Kutai Basin. Offset of middle Eocene carbonate reflectors and rapid syntectonic thickening of upper Oligocene sediments on seismic sections (see Figure 9) indicate late Oligocene extension on northwest–southeast-trending faults. This fault orientation is a clockwise rotation of 45° to 90° from the earlier middle Eocene rifts. The late Oligocene extensional faults are therefore orientated orthogonally to the Eocene extensional structures (Figure 10). Extension was accommodated over broad hinge zones (Cloke et al., 1999) comprising a diffuse zone of overlapping en echelon faults interpreted to follow preexisting crustal planes of weakness, possibly the northwest–southeast-trending transfer faults that separated middle Eocene graben systems (Cloke et al., 1997). The north-northeast to south-southwest middle Eocene rift faults were reactivated only locally by this late Oligocene extension.

Several lines of evidence indicate that, subsequent to the late Oligocene extension, faults were reactivated during the Miocene. The pattern of horizon offsets across faults often shows both normal and reverse shear, and this plus the presence of folds in the hanging wall and the thickening of the oldest Cenozoic sections (middle Eocene) toward faults are all taken to indicate that original rift-related extensional structures were reactivated, and this led to inversion of the half-graben fill. Northwest–southeast-trending late Oligocene extensional structures show little evidence for the reactivation seen on Eocene structures, although a minor amount of either sinistral or dextral strike-slip displacement is likely, given their configuration. Inversion of the Kutai Basin commenced at the end of the early Miocene, preceded by a short period of increased volcanic activity (Tanean et al., 1996). Most of the onshore section of the Kutai Basin has been inverted, although the precise amount of inversion varies with the orientation and hade of individual precursor half-graben faults (Cloke et al., 1997).

We suggest that detachment structures in the younger Oligocene-Miocene section immediately south-southwest of the Bengalon Hinge Zone (Figure 9) were directly caused by the basement inversion. We propose this to be the case along the basin margin, where the sedimentary section and the overpressured interval are relatively thin. The role of basement inversion in the deformation of the detached shallow section in the basin depocenter is discussed in the next section.

MAHAKAM DELTA DEPOCENTER

Figure 11 is a map of the present-day surface structures in the Mahakam Delta depocenter, from the

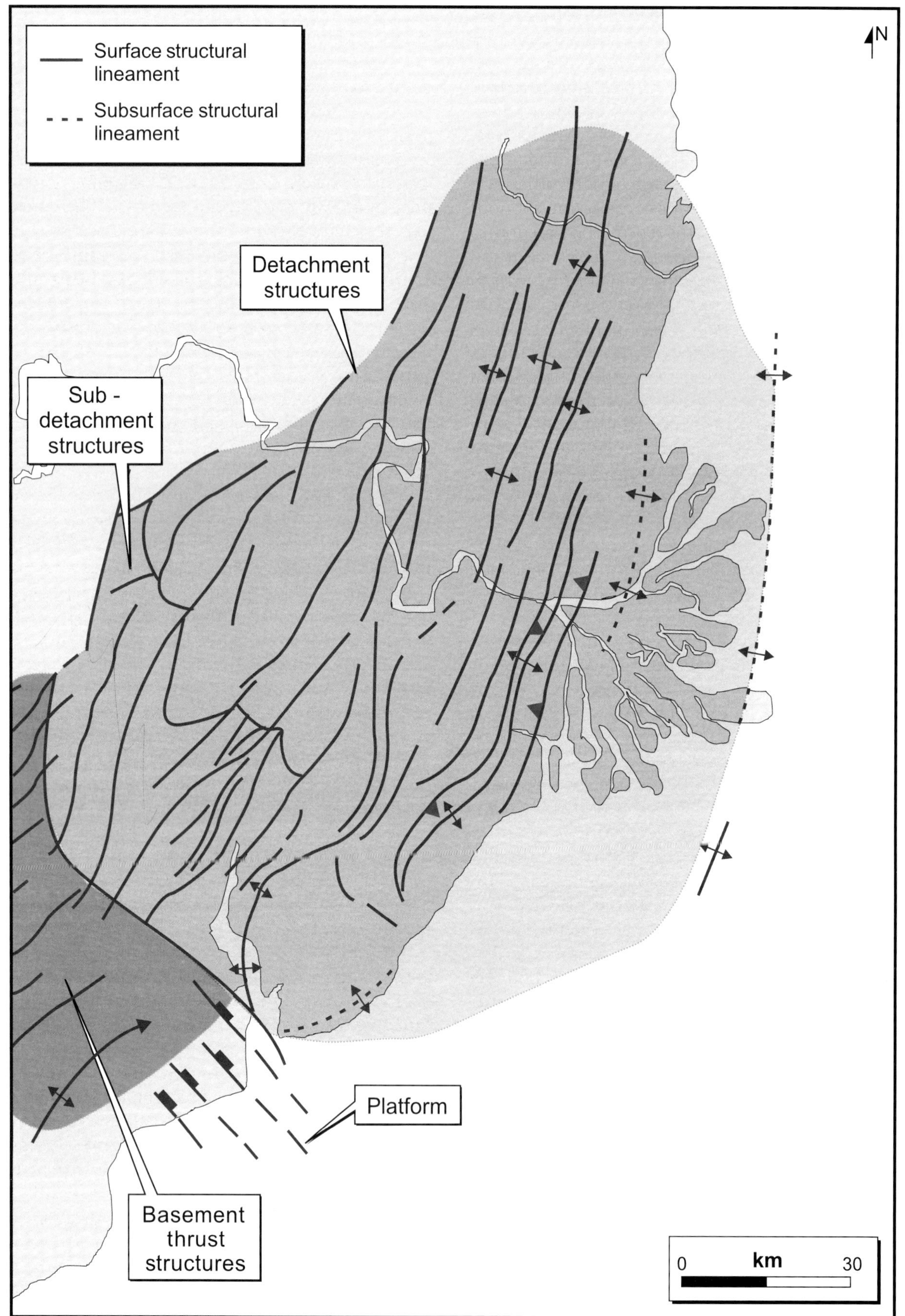

FIGURE 11. Mahakam Delta depocenter surface-structure map.

platform area south of the Adang fault zone on the southern margin of the basin, north to the Samarinda Anticlinorium. The onshore area structures can be classified into three structural zones on the basis of surface structural expression. First is a detached fold-and-thrust-belt area with north-northeast structural orientations, the Miocene outcrop of the Samarinda Anticlinorium, covering much of the onshore area. Second, in the south, there is an area with northeast–southwest-oriented surface structures, characterized by significant inversion and a Paleogene outcrop interpreted to be where the deeper structural grain has been unroofed. This second zone is separated from the Samarinda Anticlinorium by a transitional area where the 3000-m-thick normal-pressured section has been uplifted and eroded, exposing the previously overpressured lower Oligocene to lower Miocene bathyal shale section beneath the upper detachment surface. Third, a stable platform area with down-to-the-basin northwest-southeast faults is located south of the Adang fault zone.

Figure 12 shows magnetic basement lineaments in the Mahakam Delta depocenter, as originally interpreted in a proprietary 1970 regional aeromagnetic survey. The lineaments have orientations of east-northeast to west-southwest, northeast–southwest, and northwest–southeast, which is identical to the basement fracture orientations described on the northern basin margin (Figure 5) and different from the north-northeast azimuths of the overlying detached fold belt (Figures 1 and 11). The number of deep lineaments that can be imaged is limited because of the relatively low magnetic susceptibility of the metasediment basement, the thick Tertiary sedimentary section, and the equatorial latitude. Prominent magnetic lineaments include northwest-southeast lineaments associated with the Adang fault zone along the southern basin margin and also, farther to the west, a continuation of the Belayon lineament identified from gravity data (Chambers and Daley, 1995). Basement type underlying the Mahakam Delta depocenter is interpreted to be the same as that observed in outcrops on the northern margin of the basin only 150 km distant from the Mahakam depocenter and described earlier in this chapter. The regional Tertiary extensional and contractional directions are also likely to have been the same as those affecting the northern margin of the basin. It then seems reasonable to assume that basement and synrift deformation history in the Mahakam Delta depocenter will have been similar to that described for the northern margin of the basin.

Figure 13 shows a conceptual interpretation of the basement structure under the Mahakam Delta depocenter. This interpretation was created using the magnetic lineaments, northern margin analogs, and interpretations of basement influences on syninversion structure and stratigraphy. Northeast-southwest extensional faults that were active in the middle Eocene are shown as inverted in the onshore area, to explain the regional uplift (Grundy et al., 1996). Uplift increases from zero at the coastline to 1500 m over a distance of 60 km, and increases to greater than 3000 m in the area south of the Samarinda Fold Belt (transitional area, Figure 11). By analogy to the northern study area, thrust faults are shown to be 70 km long and 20 km apart, with polarity changing across northwest–southeast-oriented transfer zones that are inferred to have been reactivated in the late Oligocene to early Miocene as extensional fault zones. Northwest–southeast-oriented transfer zones are characterized by en echelon normal fault zones, as mapped along the Adang trend (Van de Weerd and Armin, 1992) and north-south faults (e.g., Raden Fault, Ferguson and McClay, 1997). Inversion is inferred to have been largely taken-up on the northeast–southwest-trending middle Eocene faults, although inversion structures do exist along the Adang fault trend (Courteney, 1991), where splays of individual faults are oriented obliquely to the prevailing interpreted northwest-southeast contractional stresses.

The Sebulu Block (Figure 13) is a less-faulted, stable basement high, as is indicated by the thinner sequence of transgressive N8 age marine shales and the lower values for Neogene uplift over the block, compared with those for the area immediately to the south (Kadar et al., 1996). The southeastern boundary of this block became a major middle Miocene hinge zone (see figures in Ferguson and McClay, 1997) that was controlled by deeper inversion. The Miocene section overlying the Sebulu basement block (Figure 12) is characterized by less sandstone and more carbonate than in the area immediately to the south. Carbonates outcrop in anticlines aligned to the north-northeast, with an axial length of as much as 60 km, about double the length of the delta-toe-thrust cored folds to the south (McClay et al., 2000). Similarly, the synclines are only 7 to 9 km apart, as opposed to the 10- to 12-km separations observed farther to the southeast in the main deltaic depositional areas. Both these observations suggest a different early depositional and structural genesis for the structures described in the area immediately to the south (cf. structures described in McClay et al., 2000). Stratigraphic thinning toward the western anticline is noted in the synclines from seismic sections, but there is no corresponding thinning to the east. On the western flanks of the synclines, the preserved shelf-edge clinoform units thicken markedly toward the east. The bases of these clinoforms units are not usually seen, because they are detached and in the footwall of the anticline. The traces of reverse faults in this area are aligned subparallel to the identified shelf edges, which suggests that reverse faulting is strongly controlled by facies. The same facies boundary is likely to have influenced the overpressure boundary's location as well,

FIGURE 12. Mahakam Delta depocenter basement lineaments map, as originally interpreted in a proprietary 1970 regional aeromagnetic survey.

as Bates (1996) proposed. It appears that aggradational carbonate-prone shelf edges acted as a locus of contractional deformation in this part of the basin. These folds and thrusts are interpreted to have formed without any pre-inversion folding.

Figure 14 shows an interpreted northwest-southeast seismic line through the western margin of the Wain Subbasin (see Figure 13 for location). This line was part of a limited local survey in which recording time was increased to 10 s. A series of three northwest-dipping, reactivated Eocene normal faults are interpreted on the seismic section. These faults are overlain by a 3- to 7-km-thick section of overpressured upper Oligocene and lower Miocene shale. A northwest-verging, shallow detachment fault is interpreted at the base of the normal-pressured lower to middle Miocene section. The shallow detachment fault is interpreted as a back thrust of a thickened wedge over what was a middle Miocene depositional hinge zone (see figures in Ferguson and McClay, 1997). The basement inversion in this part of the basin is interpreted to have commenced in the late Miocene. It resulted in the uplift and erosion

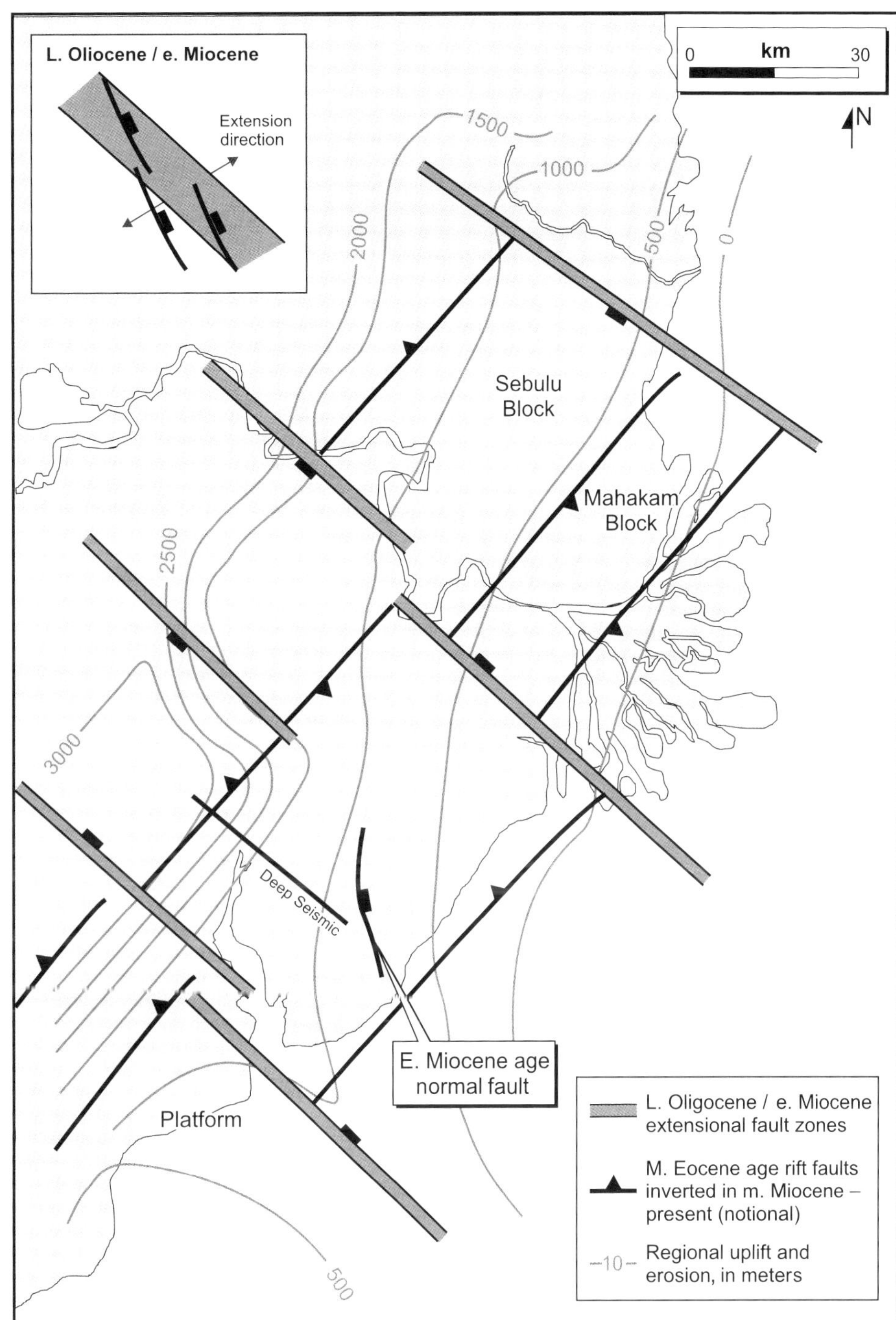

FIGURE 13. Mahakam Delta depocenter notional basement rift structure map.

of 1000 m of section at the southeastern end of the profile; this increases to 3000 m at the northwestern end.

DETACHED FOLD-AND-THRUST STRUCTURES

Ferguson and McClay (1997) and Chambers and Daley (1995) offer different explanations of how the shallower, Miocene-age detachment structures kinematically relate to the rift inversion. The Chambers and Daley model suggests that the shallow, shorter-wavelength fold belt was created directly by bed-length shortening above a detachment surface in the overpressured section, that is, the folds are syninversional. Scaled physical-analog modeling was used to test this model (Ferguson and McClay, 1997; McClay et al., 2000). The models revealed a broad, long-wavelength fold developing in the overpressured shales above inverted half grabens, as predicted by Chambers and

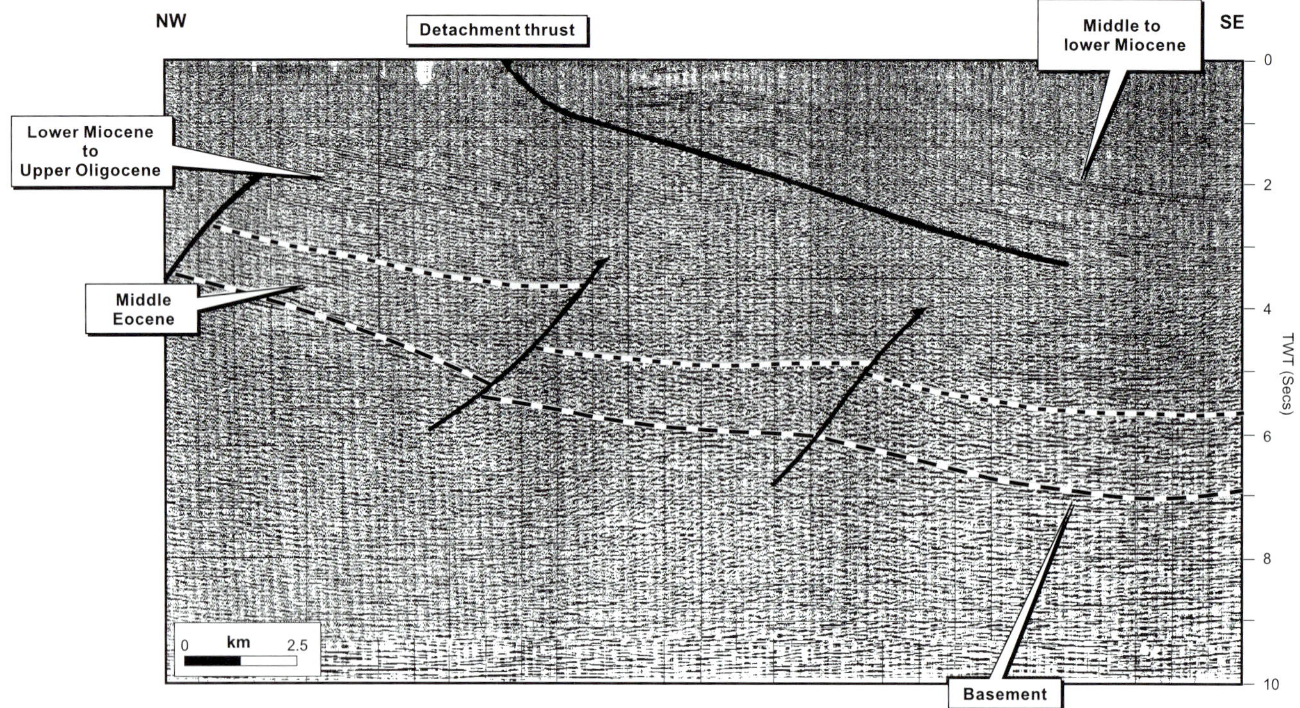

FIGURE 14. Deep seismic profile (see Figures 2 and 13 for location), showing shallow detachment structure above inverted-basement rift section. Vertical exaggeration is approximately 100%.

Daley, but no development of shorter-wavelength detachment folding in the shallow section above the overpressured interval. Instead, Ferguson and McClay proposed that the shallower structures observed in the field initiated as differential load features of delta toe thrusts that subsequently became contracted when inversion of the rift section occurred, that is, folding initiated before inversion. They deemed this latter model to be applicable to the portions of the basin that had major periods of shelf advance as a result of shelf-edge deltas and high deposition rates. The model that McClay et al. (2000) proposed for the formation of fold-and-thrust structures immediately inboard from the Mahakam Delta does not preclude basement-involved inversion below the overpressured shale zone.

Preexisting normal faults are another potential site of contractional deformation in the detached section with onset of rift inversion. Van de Weerd and Armin (1992) observed that the Neogene basin fill accumulated largely by the infill of accommodation-space-generated flexural subsidence. This appears to be the case from middle Miocene time onward, except on long-standing basement-related zones of weakness, such as the Adang fault zone. However, in the lower Miocene strata, normal faults are recognized that are oriented north–south and are interpreted here to be related to oblique extension along northwest-southeast late Oligocene extensional faults/hinge lines. Ferguson and McClay (1997) documented an example in which a fault is observed to retain a normal sense of displacement or to show reversal, depending on the fault's orientation relative to the direction of contraction.

DISCUSSION

Basement lineaments are interpreted to have controlled deposition of the middle and upper Miocene strata in a regional sense. The Mahakam Block (Figure 13) was the site of deltaic deposition from the middle to late Miocene (Van de Weerd and Armin, 1992), interpreted here to be to the result of basement subsidence along two northwest-southeast hinge zones. Periodically, fluvial-deltaic sediments were directed elsewhere, such as along the Adang fault trend (Van de Weerd and Armin, 1992).

Whereas the inverted rift component of the structural model is interpreted to be similar across the basin, we conclude that the shallower detachment structures have two different kinematic origins, depending on location (Figure 15). On the northern basin margin, where sediment is thinner and consequently overpressuring did not develop, field mapping supports the model of inversion leading directly to the formation of thrusts and folds. In the Mahakam Delta depocenter, where there is a thick sedimentary section and overpressuring has developed, the response to inversion of the rift section was controlled in part by heterogeneity in the shallow section, including structures formed through syndepositional loading, delta progradation, normal

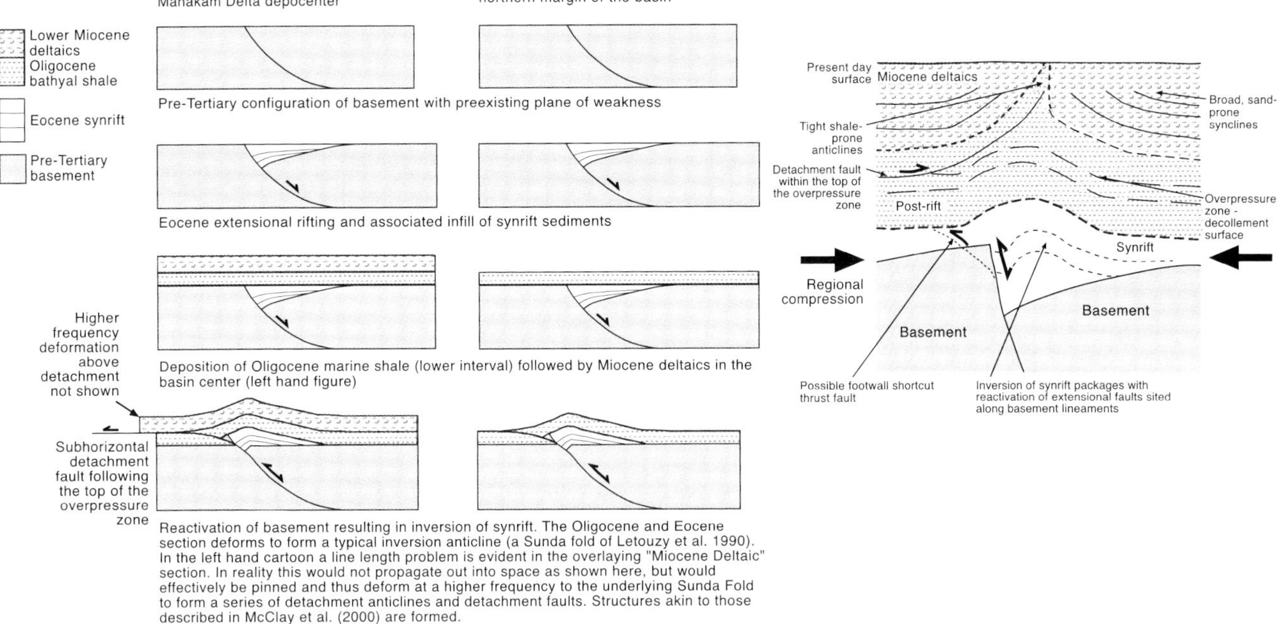

FIGURE 15. Summary model for the structure of the Kutai Basin.

faults, and marked facies changes. Hence, several contractional deformation styles are seen. We contend that basement-involved inversion beneath the overpressured zone has also influenced the shallower structures present in the Mahakam Delta depocenter (Figure 14).

Ferguson and McClay's (1997) progressively deformed delta-toe-thrust model appears to be a valid model for the Mahakam Delta area's detached fold-and-thrust structures. In the deep water offshore, upper Miocene and Pliocene deep-water fan sandstones that ponded between delta-toe-thrust structures have recently been the target of exploration success in the Kutai Basin. These structures formed by depositional loading soon after sedimentation and have not yet been buried by delta-top facies or affected by inversion. These structures therefore correspond to the early stages of the Ferguson and McClay (1997) structural model, prior to burial by delta-top facies and before application of regional contractional stresses. We have not studied the offshore area to determine whether this least-deformed portion of the basin has undergone inversion activity.

Our work recognizes a significant northwest-southeast extensional event of late Oligocene to early Miocene age that also appears to have influenced syninversional areas of deposition in the basin from middle Miocene time onward. These basement structures have directed sandstone sedimentation along the Adang fault zone (Van de Weerd and Armin, 1992), and there may be other similarly influenced areas in the Mahakam Delta's main producing trend. On a regional scale, the northwest-southeast structural grain influenced the location of the main deltaic depocenter and probably also influenced terminations of older lower and middle Miocene toe thrusts along strike, hence determining the length of the anticlines subsequently formed. The northwest-southeast structural grain is generally parallel to the compressive stresses; hence, structural response, which is likely to be strike-slip in character, is subtle and difficult to recognize in the dominant north-northeasterly detached fold-belt grain.

SIGNIFICANCE TO PETROLEUM EXPLORATION

Inversion anticlines are a major hydrocarbon trap type in Indonesian basins and are present in the onshore Kutai Basin. Typically, synrift sections within Indonesian Tertiary basins contain good-quality lacustrine source intervals, and adjacent and overlying coarse clastics provide reservoirs. The Tanjung field in the adjacent Barito Basin produces oil from Eocene sandstones in an inversion anticline (Kusama and Darwin, 1989) and is a typical example of an inversion-anticline play in Indonesia.

Only along the Kutai Basin margins, where postrift basin fill is of modest thickness, are the middle Eocene synrift sandstone reservoirs likely to have preserved sufficient porosity with burial. However, three wells drilled by LASMO Runtu Limited and its partners in the northern basin margins failed to discover economic hydrocarbon reserves. In the Kutai Basin, the Tengkawang-1 well had oil shows and found gas in subcommercial quantities in an inversion structure (Figure 6), and the Wahau-1 well (Figure 7) was a dry hole, although a significant middle Eocene source-rock section was encountered (Guritono and Chambers, 2000). Although these wells avoided the risk of overpressure and overmaturity, which had been found by earlier drilling campaigns testing anticlines located farther basinward to the south, reservoir quality was a problem with the Eocene reservoir objective in all three wells. The Tengkawang-1 well tested a small inversion anticline and encountered a basement of Cretaceous metasediments at 2512 m. Two additional wells, Wahau-1 and Maau-1 (Figure 5), targeted an inversion anticline trap and a carbonate buildup above an Eocene footwall (Figure 7), respectively. In all three wells, the low porosities and permeabilities encountered in the reservoir-target Eocene sandstones is ascribed to the mineralogic immaturity of the sandstones, which are interpreted to have derived from a nearby source. Hydrocarbon charge is also a risky element over most of the basin, because trap formation postdates initiation of hydrocarbon generation— there was no post-inversion subsidence and burial here, as there had been elsewhere in Indonesia. Trap breach is also a significant risk. MacGregor (1997) concludes that the potential for hydrocarbon preservation in these traps is low and that these types of traps represent very high-risk exploration targets.

To 1995, 28 tcf of gas (EUR) and 2.5 billion barrels of oil (EUR) had been discovered in the simple folds and four-way closures of the middle Miocene to Pliocene section located offshore and along the coastline of the Mahakam Delta. Paterson et al. (1997) attribute the lack of exploration success in the more-deformed fold-and-thrust belt, located to the west of the discoveries, as resulting primarily from the lack of local source kitchens in the adjacent synclines. Local source kitchens were needed because the large coastal anticlines acted as migration shadows for the structures farther inboard of the prolific offshore kitchen area. In addition, the inboard structural traps were exposed to more risk of breaching as they became progressively more contracted.

Our work recognizes a significant northwest-southeast, late Oligocene to early Miocene extensional event that also appears to have influenced syninversional areas of deposition in the basin from middle Miocene time onward. These basement structures have directed sandstone sedimentation along the Adang fault zone, and there may be other similarly influenced areas in the main Mahakam Delta's producing trend. On a regional scale, the northwest-southeast structural grain influenced the location of deltaic deposition and probably also influenced terminations of older, lower and middle Miocene toe thrusts along strike, thereby determining the length of the anticlines that formed subsequently. The northwest-southeast structural grain is generally parallel to the compressive stresses, and therefore, structural response, which is likely to be strike-slip in character, is subtle and difficult to recognize within the dominant, north-northeasterly detached fold-belt grain.

CONCLUSIONS

Interpretation of the structural deformation on the northern margin of the Kutai Basin, where the sedimentary section is relatively thin, has identified northeast–southwest- to north-northeast–south-southwest-oriented asymmetric anticlines. These are interpreted to be the result of inversion of preexisting middle Eocene half grabens that developed into inversion anticlines with the onset of middle Miocene compression and reverse reactivation of formerly normal faults. A second phase of rifting, during the late Oligocene, was accommodated over broad hinge zones consisting of segmented overlapping en echelon faults that follow the northwest–southeast-trending transfer faults that separated the middle Eocene graben systems. Using aeromagnetic data, we could reasonably extrapolate these same structural elements 150 km to the south to the Mahakam Delta's depocenter. In the Mahakam Delta's depocenter, the northwest-southeast extensional elements have not been fully recognized, because imaging of the basement rift section is difficult as a result of overprinting by a thick syninversional middle to upper Miocene section deformed into a detached north-northeast-trending fold-and-thrust belt.

Basement thrust faults flatten out and detach in the overlying overpressured shales, which results in a broad uplift and a shortening of the shallower section. The manner in which the shallow section deforms with the rift inversion depends on the thickness of the intervening overpressured shale section and the preexisting structural and facies elements of the shallow section. Preexisting shallow features, such as delta-toe-thrust-created folds and synsedimentary faults, become the locus of shallow-section contractional deformation when inversion commences. The shallower features detach in the top of the overpressured zone, likely where the fracture gradient is reached, often within a few hundred meters of the base of the normally pressured section. Along the basin margins, where sediment thickness is reduced and consequently overpressure did not develop, rift inversion is interpreted to directly cause the formation of folds and thrusts.

Basement structure influenced syninversion sedimentation from the middle Miocene onward. Inversion of northeast-southwest middle Eocene half grabens created depositional hinge lines at the boundary between inverted basin to the west and increased basinal dip to the east. Flexural subsidence that followed the northwest-southeast extensional fault zones restricted the main deltaic deposition to the area of the present-day Mahakam Delta and to periodic deltaic deposition along the Adang fault trend. This northwest-southeast structural element has not been fully recognized in the structural and stratigraphic interpretation of the Mahakam Delta's producing area, and hence, there may be merit in studying it further, given the large oil and gas reserves under production and development.

ACKNOWLEDGMENTS

The authors wish to thank the management of Pertamina and the Runtu PSC Partners (LASMO, Japex, OPIC, Ross Petroleum, and UGO) for permission to publish this paper. The SE Asia Consortium of Companies (then comprising ARCO, LASMO, Can Oxy, Exxon, Union Texas, and Mobil) is thanked for financial support for Steve Moss's involvement in this project.

REFERENCES CITED

Bates, J. A., 1996, Overpressuring in the Kutei Basin: Distribution, Origins and Implications for the Petroleum System: Proceedings of the Indonesian Petroleum Association, 25/1, 93–116.

Biantoro, E., B. P. Muritno, and J. M. B. Mamuaya, 1992, Inversion faults as the major structural control in the northern part of the Kutei Basin, East Kalimantan: Indonesian Petroleum Association, Proceedings of the 21st Annual Convention, I, p. 45–68.

Carter, I. S. and R. J. Morley, 1995, Utilizing outcrop and palaeontological data to determine a detailed sequence stratigraphy of the early Miocene sediments of the Kutei Basin, East Kalimantan, *in* C. A. Caughey et al., eds., Proceedings of the International Symposium of Sequence Stratigraphy in SE Asia, May 1995: Indonesian Petroleum Association, p. 345–361.

Chambers, J. L. C., and T. Daley, 1995, A tectonic model for the onshore Kutai Basin, East Kalimantan, based on an integrated geological and geophysical interpretation: Proceedings of the Indonesian Petroleum Association, 24/ I, p. 111–130.

Cloke, I. R., S. J. Moss, and J. Craig, 1997, The influence of basement reactivation on the extensional and inversional history of the Kutai Basin, Eastern Kalimantan: Journal of Geological Society, v. 154, p. 157–161.

Cloke, I. R., S. J. Moss, and J. Craig, 1999, Structural controls on the evolution of the Kutai Basin, East Kalimantan, Tectonics, Stratigraphy and Petroleum Systems of Borneo Workshop, Brunei 1997: Special Edition, Journal of Asian Earth Sciences, v. 17, p. 137–156.

Courteney, S., ed., 1991, Indonesia oil and gas fields atlas, Vol. 5: Kalimantan, Indonesian Petroleum Association, v. 27, 27 p.

Duval, B. C., G. Choppin de Janvry, and B. Loiret, 1992, Detailed geoscience reinterpretation of Indonesia's Mahakam delta scores: Oil & Gas Journal, August 10, 1992, p. 67–72.

Ferguson A., and K. McClay, 1997, Structural Modeling within the Sanga Sanga PSC, Kutei Basin, Kalimantan: Its Application to paleochannel orientation studies and timing of hydrocarbon entrapment *in* J. V. C. Howes and R. A. Noble, eds., Petroleum systems of SE Asia and Australia: Indonesian Petroleum Association, p. 727–744.

Ferinsyah, L. T., J. L. C. Chambers, S. H. Dewantohadi, M. Syaiful, T. Priantono, and D. N. Imanhardjo, 2000, Structural and stratigraphic framework of the Palaeogene in the northern Kutai Basin, East Kalimantan: Indonesian Petroleum Association, Proceedings of the 27th Convention, p. 443–455.

Grundy, R. J., D. W. Paterson, and F. H. Sidi, 1996, Uplift measurements in Tertiary sediments of the Kutei Basin, East Kalimantan, Indonesia, as it relates to VICO Indonesia's PSC and the surrounding area: Expanded abstracts from the Jakarta 1996 International Geophysical Conference, Society of Exploration Geophysicists, p. 81–85.

Guritono, E. E., and J. L. C. Chambers, 2000, North Runtu PSC: The first proven Eocene petroleum play in the Kutai Basin: Indonesian Petroleum Association, Proceedings of the 27th Convention, p. 361–380.

Hutchinson, C. S., 1996, The 'Rajang Accretionary Prism' and 'Lupar Line' problem of Borneo, *in* R. Hall and D. J. Blundell, eds., Tectonic evolution of SE Asia: Geological Society of London Special Publication 106, p. 247–261.

Kadar, A., D. W. Paterson, and Hudianto, 1996, Successful techniques and pitfalls in utilizing biostratigraphic data in structurally complex terrain: VICO Indonesia's Kutei basin experience: Proceedings of the Indonesian Petroleum Association, v. 25/1, p. 333–346

Kusama, I., and T. Darwin, 1989, The hydrocarbon potential of the lower Tanjung Formation Barito basin, SE Kalimantan: Indonesian Petroleum Association, Proceedings of the 18th Annual Convention, October 1989, p. 107–138.

Macgregor, D. S., 1997, The destruction and preservation of giant light oilfields, *in* J. V. C. Howes and R. A. Noble, eds., Petroleum Systems of SE Asia and Australia: Indonesian Petroleum Association, p. 843–852.

McClay, K., T. Dooley, A. Ferguson, and J. Poblet, 2000, Tectonic evolution of the Sanga Sanga Block, Mahakam Delta, Kalimantan, Indonesia: AAPG Bulletin, v. 84, p. 765–786.

Metcalfe, I., 1999, Gondwana dispersion and Asian accretion: An overview, *in* I. Metcalfe, ed., Gondwana dispersion and Asian accretion: Rotterdam, A.A. Balkema, p. 9–28.

Moriya, S., and T., Nishidai, 1996, Application of remote sensing data to structural analysis of East Kalimantan, Indonesia. 11th Thematic Conference and Workshops on Applied Remote Sensing, Las Vegas, Nevada.

Moss, S. J., and J. L. C. Chambers, 1999, Tertiary facies architecture in the Kutai Basin, Kalimantan, Indonesia: Journal of Asian Earth Sciences, v. 17, p. 157–181.

Moss, S. J., and E. M. Finch, 1998, New Nannofossil and Foraminifera determinations from east and west Kalimantan and their implications: Journal of Asian Earth Sciences, v. 15, p. 489–506.

Moss, S. J., J. Chambers, I. Cloke, A. Carter, D. Satria, J. R. Ali, and S. Baker, 1997, New observations on the sedimentary and tectonic evolution of the Tertiary Kutai Basin, East Kalimantan, *in* A. Fraser and S. Matthews, eds., Petroleum Geology of Southeast Asia: Geological Society of London Special Publication 126, p. 395–416.

Ott, H. L., 1987, The Kutei Basin—a unique structural history: Indonesia Petroleum Association, Proceedings 16th Annual Convention, I, p. 307–317

Paterson, D. W., A. Bachtier, J. A. Bates, J. A. Moon, and R. C. Surdam, 1997, Petroleum System of the Kutei Basin, Kalimantan, Indonesia, *in* J. V. C. Howes and R. A. Noble, eds., Petroleum systems of S.E. Asia and Australia: Indonesia Petroleum Association, p. 709–726.

Rose, R., and P. Hartono, 1978, Geological evolution of the Tertiary Kutei–Melawi Basin, Kalimantan, Indonesia: Indonesian Petroleum Association, Proceedings of the 7th annual convention, Jakarta, 1978, p. 225–252.

Sikumbang, N., 1986, Geology and tectonics of Pre-Tertiary Rocks in the Meratus Mountains, Southeast Kalimantan, Indonesia: PhD Thesis, University of London, 313 p.

Sikumbang, N., 1990, The geology and tectonics of the Meratus mountains, south Kalimantan, Indonesia: Geologi Indonesia: Journal of the Indonesian Association of Geologists, v. 13, no. 2, p. 1–31.

Tanean, H., D. W. Paterson, and M. Endharto, 1996, Source provenance interpretation of Kutei Basin sandstones and the implications for the tectono-stratigraphic evolution of Kalimantan: Proceedings of the Indonesian Petroleum Association, v. 25/1, p. 313–323.

Van de Weerd, A. A., and R. A. Armin, 1992, Origin and evolution of the Tertiary hydrocarbon bearing basins in Kalimantan (Borneo), Indonesia: AAPG Bulletin, v. 76, p. 1778–1803.

Wilson, M. E. J. and S. J. Moss, 1999, Cenozoic evolution of Borneo-Sulawesi: Palaegeography, Palaeoclimatology and Palaeooceanography, v. 145, no. 4, p. 303–337.

Wilson, M. E. J., J. L. C. Chambers, M. J. Evans, S. J. Moss, and D. S. Nus, 1999, Cenozoic carbonates in Borneo: case studies from northeast Kalimantan: Asia Journal of Earth Sciences, v. 17, p. 183–201.

32

Beauchamp, W., 2004, Superposed folding resulting from inversion of a synrift accommodation zone, Atlas Mountains, Morocco, *in* K. R. McClay, ed., Thrust tectonics and hydrocarbon systems: AAPG Memoir 82, p. 635–646.

Superposed Folding Resulting from Inversion of a Synrift Accommodation Zone, Atlas Mountains, Morocco

Weldon Beauchamp

Atlas Exploration and Production Company LLC, Dallas, Texas, U.S.A.

ABSTRACT

The conspicuous offset of the northern margin of the High Atlas Mountains is composed of several large superposed folds, one of which is known as the Ait Attab Syncline. The original northeast-trending syncline (F1) was folded by a second set of fold axes (F2) that trend to the northwest. The superposed folding was generated by one phase of compression, with thrusting of synrift rocks northwestward over a prior accommodation zone formed during rifting. This accommodation zone is expressed in the exposure of synrift rocks, the exposure of Paleozoic strata in the footwall, and a coincident offset of topography. Inversion was accomplished by the transport of synrift strata along reactivated normal faults and newly formed thrusts. The unique pattern of refolding is believed to be characteristic of inversion.

INTRODUCTION

2-D Models of Inversion

Many existing models of inversion are 2-D, accordion-style models that assume that both extension and subsequent compression are orthogonal to faulting. Rift systems characterized by long, straight faults bounding the rift system result from orthogonal extension (McClay and White, 1995). The inversion of rift systems where compression and extension are orthogonal produces structures that trend parallel to the preexisting rift structures.

However, many rift systems show patterns of en echelon normal faulting and segmented faults (McClay and White, 1995) that are the product of oblique extension during rifting. We can expect that compression relative to structures produced by oblique extension would produce structural geometries that are a result of 3-D strain. The study area in the High Atlas Mountains displays accommodation zones, en echelon normal faults, and pull-apart basins, indicating that extension was oblique during rifting. During inversion in the Tertiary, compression resulted in an unusual pattern of superposed folding that was influenced by preexisting extensional structural geometries.

Tectonic Setting of the Atlas Mountains

Tectonic inversion in the Atlas Mountains and elsewhere has shown that many intracontinental mountain belts are related to the uplift of preexisting intracontinental rift systems (Bally, 1984; Beauchamp et al., 1996). The High Atlas Mountains represent a major Mesozoic rift system (~2000 km in length) that was uplifted and inverted during the Cenozoic (Figure 1) (Beauchamp et al., 1996). The convergence of the African and Iberian plates in the Tertiary resulted in the inversion of Mesozoic synrift strata along preexisting synrift faults and also, by the transport of synrift strata, along newly formed low-angle thrusts.

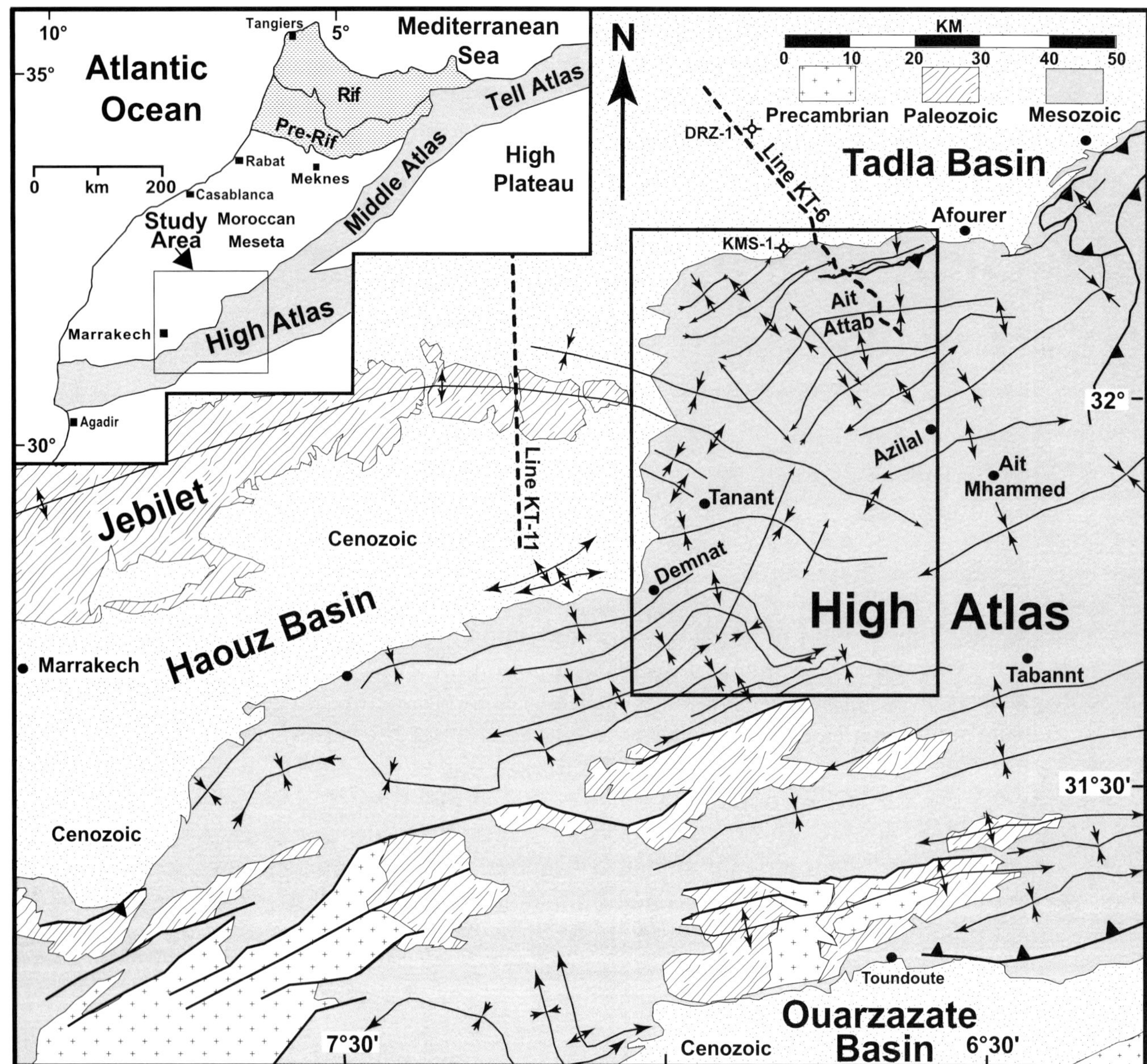

Figure 1. Location and simplified tectonic/structural map of the study area in the High Atlas Mountains, Morocco. Fold axes in the High Atlas are approximately parallel to the orogen and normal to the direction of thrusting, except where preexisting accommodation zones have influenced the regional stress field, resulting in polyphase deformation. Also, shown is the location of a more detailed geological map that displays superposed folding (Figure 3). Compiled from geological maps (Saadi et al., 1977; Rolley, 1978; Saadi et al., 1985; Jenny, 1988).

The Jebilet Accommodation Zone

The northern margin of the High Atlas Mountains contains a conspicuous offset (~90°) of the topography and exposure of the synrift Mesozoic rocks (Figure 1). Synrift strata thicken dramatically to the south and east, into the High Atlas Mountains. The present-day Tadla, Haouz, and Ouarzazate Basins were the shelf/platform margin areas of the paleo-Atlas rift (Figure 1). Synrift Jurassic-age strata are noticeably thinner or absent to the northwest in the Tadla and Haouz Basins (Jabour and Nakayama, 1988). Mesozoic strata are absent along the eastern margins of the Jebilet, where Paleozoic-age rocks crop out (Figure 1). Immediately to the south and east of the Haouz Basin and the exposed Paleozoic rocks of the Jebilet are 2–3 km of preserved synrift strata (Beauchamp et al., 1999). North of the Ait Attab Syncline (Figure 1), well data show that synrift Jurassic strata are absent and Triassic strata are condensed to a thickness

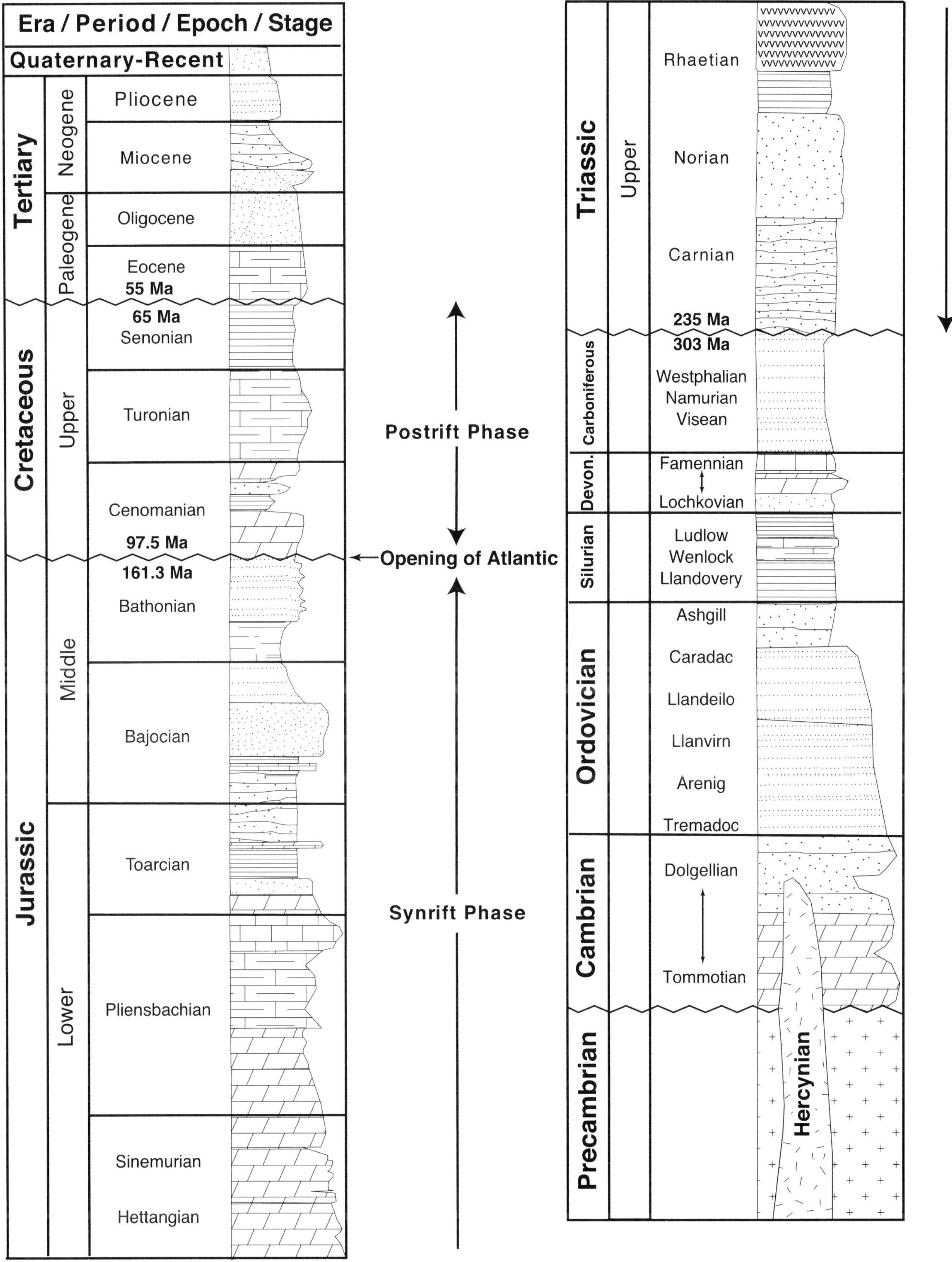

Figure 2. Generalized stratigraphic section of the High Atlas Mountains. Compiled from geological maps (Saadi et al., 1977; Rolley, 1978; Saadi et al., 1985; Jenny, 1988), measured sections (Rolley, 1978; Jabour and Nakayama, 1988; Jenny, 1988), and well data.

of less than 500 m (Jabour and Nakayama, 1988) (Figure 2). The Jebilet and topographic offset are related to a synrift accommodation zone. The accommodation zone affected the deposition of synrift strata during rifting, and later, it influenced the thrusting of synrift strata northward from the rift basin to the shelf margins. The geometry of this accommodation zone also controlled the thrusting and subsequent superposed folding (Figure 3).

The Jebilet extends eastward across the accommodation zone, where the anticlinorium plunges to the southeast (Figure 1). This anticline separates the Guettioua Syncline and the Ait Attab Syncline (Figure 3).

GEOPHYSICAL EVIDENCE OF AN ACCOMMODATION ZONE

The Paleozoic strata of the Jebilet are believed to be in the hanging wall of a thrust system that verges northward from the Atlas Mountains. This thrust may have formed as a footwall shortcut fault that transported Paleozoic rocks from the margin of the rift basin (Figure 4). Seismic line KT-11 extends across the Haouz Basin to the north across the Jebilet and onto the Moroccan Meseta. The condensed Mesozoic section found in the Haouz Basin was deposited on the margin of the rift basin. A thrust fault is interpreted that transports Paleozoic-age rocks in the hanging wall northward, to where they crop out in the Jebilet.

The Ait Attab Syncline is well imaged on seismic line KT-6 (Figures 1 and 5). The KMS-1 well north of the Ait Attab Syncline did not encounter rocks of Jurassic age, which indicates that the well was drilled on the rift-basin margin. Surface structural dips are in good agreement with those extracted from the seismic data, and a thrust is interpreted beneath the syncline. Thickening of the Middle and Lower Jurassic rocks across the syncline, from southeast to northwest, indicates that the synrift rocks were deposited in the hanging wall of an active normal fault that dipped southward into the paleo-Atlas rift basin. The synrift normal fault acted as a ramp during subsequent compression in the Tertiary, and the synrift rocks were transported northward along a footwall shortcut fault (Figure 6). This fault, interpreted on line KT-6 (Figure 5), is exposed east of the seismic line, where Jurassic-age rocks are thrust over rocks of Late Cretaceous age (Figure 3) (Beauchamp et al., 1999).

CENOZOIC REFOLDING PATTERNS

Ait Attab Area

Prominent large-scale refolded folds bound the accommodation zone (Figures 1 and 3). Although superimposed folding is a common structural occurrence in orogenic belts, the timing and sequence of folding in the Atlas Mountains display a unique pattern of folding. Folds trend approximately parallel to the margins of the Atlas mountain belt (Figure 1) on either side of the accommodation zone, indicating that the direction of transport (thrusting) was normal to the orogen.

Folding within the accommodation zone (Figure 3) displays two phases of folding activity (F1 and F2). The Ait Attab Syncline demonstrates an early phase of folding (F1) that was oriented northeasterly and that was refolded by a later oblique inversion phase (F2) oriented north-northwesterly. The western end of the Ait Attab Syncline is parallel to the Jebilet Anticline (Figure 1).

Jebel Guettioua-Sidal Area

Farther to the south, the Guettioua Syncline displays an early phase of folding (F1) that was oriented northwesterly followed by a later phase of folding (F2) that was oriented to the north-northeast. The Jebel Sidal Syncline was folded first (F1) along a northeasterly trend and folded later by a second phase (F2) that was north-northeasterly. The sequence and timing of folding in the area of study (Figure 3) indicate that the first phase of folding (F1) seen in the Ait Attab and Jebel Sidal structures was oriented to the northeast. These fold axes (F1) are normal to the first phase of folding (F1) of the Guettioua structure (to the northwest). The orientation of the second phases of folding illustrates a similar discrepancy. The second phase of folding (F2) of the Ait Attab Syncline (Figure 3) is normal to the second phase (F2) of folding of the Guettioua and Jebel Sidal structures.

Figure 3. Detailed geological map of the accommodation zone along the northern margin of the High Atlas (see Figure 1 for location). The first phase of folding (F1) of the Ait Attab and Jebel Sidal structures is normal to the first phases of folding in the Guettioua structure. Likewise, the second phase of folding (F2) of the Ait Attab and Guettioua synclinal structures are also normal to one another. The sequence and orientation of folding necessitates that folding was influenced by a preexisting accommodation zone during tectonic inversion The map was compiled from fieldwork, Landsat-TM images, and geological maps (Huvelin, 1972; Saadi et al., 1985; Jenny, 1988). Stereoplots of poles to bedding show the dip domains of folding.

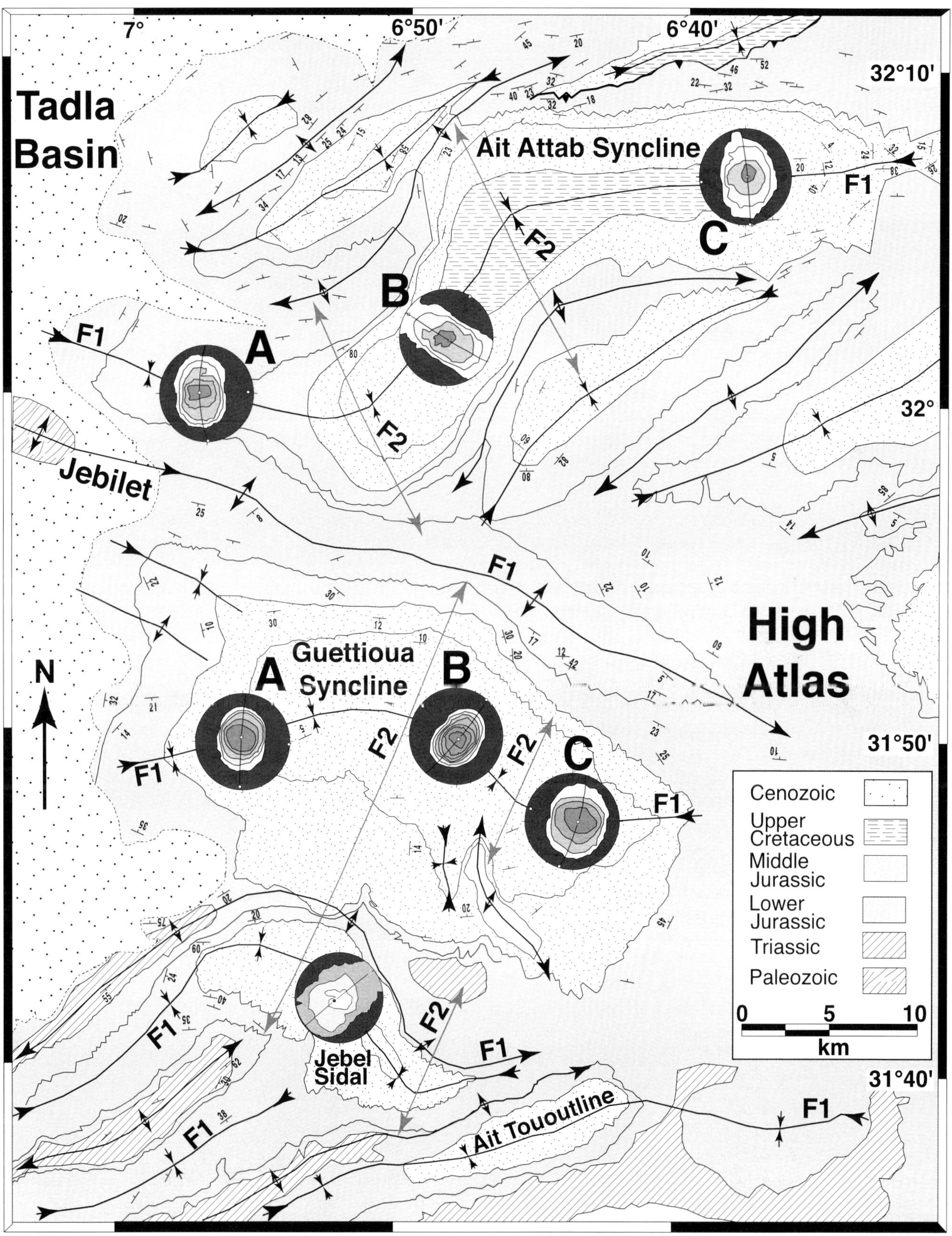
7°
6°50'
6°40'
32°10'
32°
31°50'
31°40'
Tadla Basin
Ait Attab Syncline
Jebilet
High Atlas
Guettioua Syncline
Jebel Sidal
Ait Tououtline
A
B
C
F1
F2
N
Cenozoic
Upper Cretaceous
Middle Jurassic
Lower Jurassic
Triassic
Paleozoic
0
5
10
km

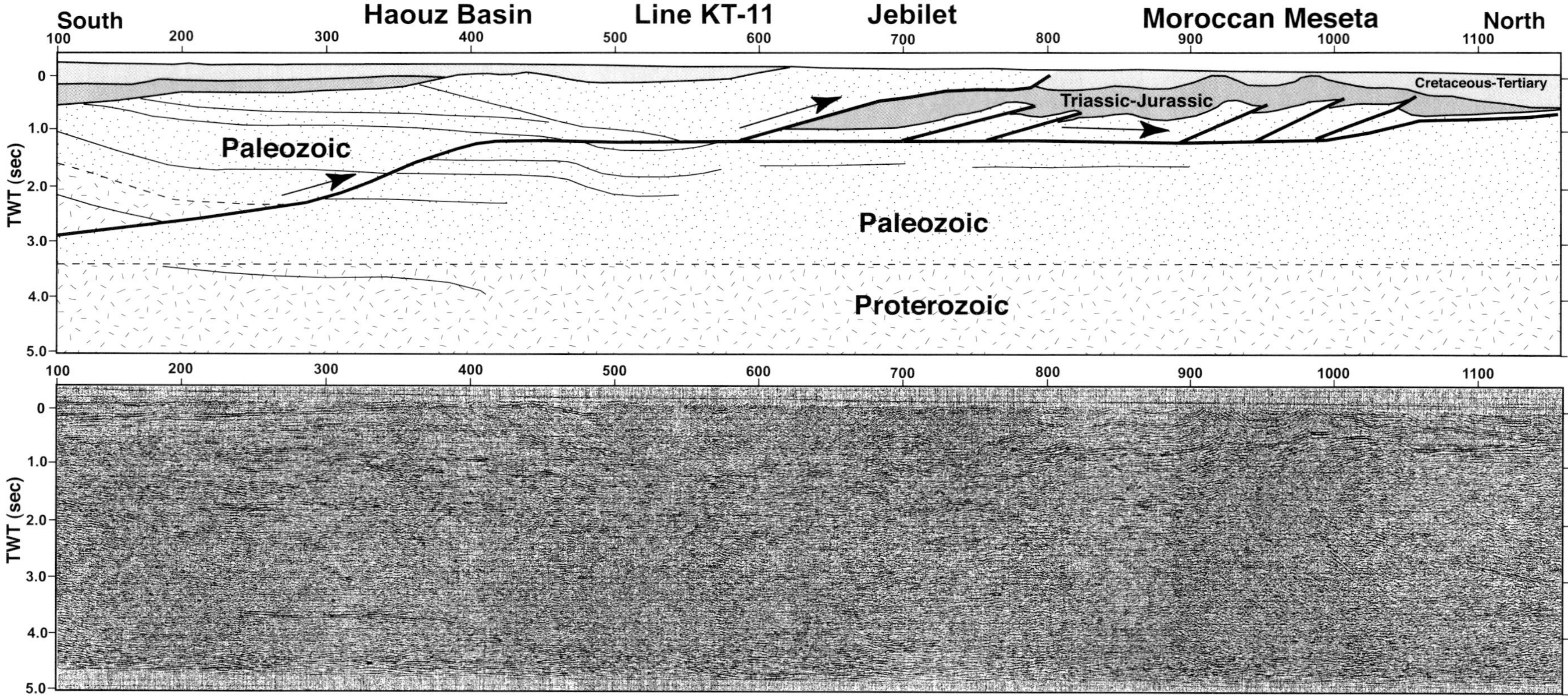

FIGURE 4. Seismic line KT-11 (see Figure 1 for location) extends northwards across the Haouz basin and the Paleozoic exposure in the Jebilet. Thin Mesozoic strata interpreted on part of this seismic line and in wells, indicate this region was on the margin of the rift basin. Shortening across the rift margin resulted in the transport of synrift rocks northwards up a preexisting normal fault, and then along a newly formed footwall short cut fault. Paleozoic rocks were transported to the north where they are exposed in the Jebilet.

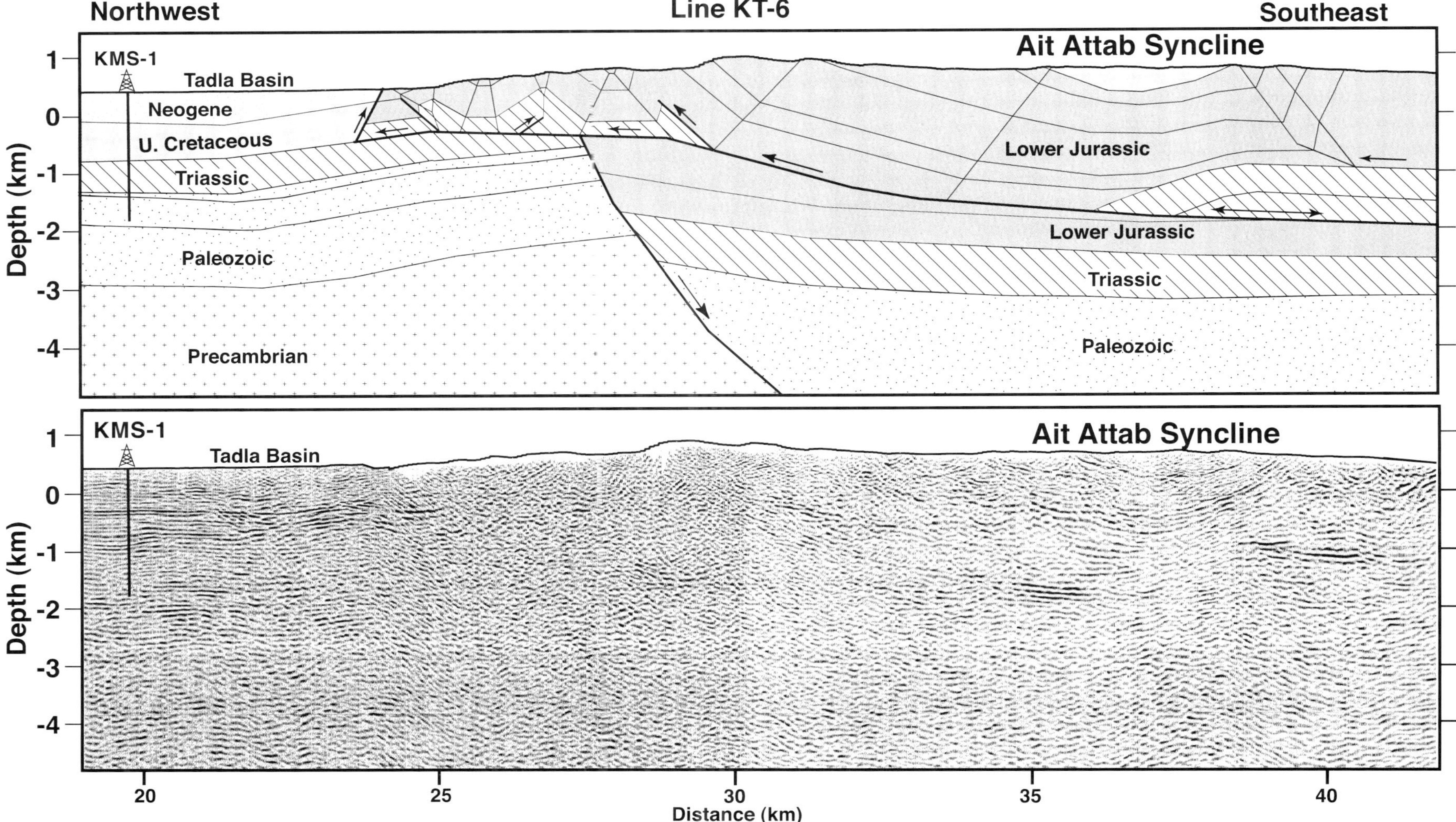

FIGURE 5. The southern part of seismic line KT-6 illustrates the absence of Jurassic age rocks in well KMS-1. Immediately to the south 2-3 km of Jurassic crops out in the Ait Attab Syncline (see Figures 1 and 3 for location). The Ait Attab Syncline shows a thickening of Jurassic age rocks northwards that are thought to have been deposited in a synrift half graben, and later thrust northward along a newly formed footwall shortcut fault similar to the fault interpreted on Figure 4.

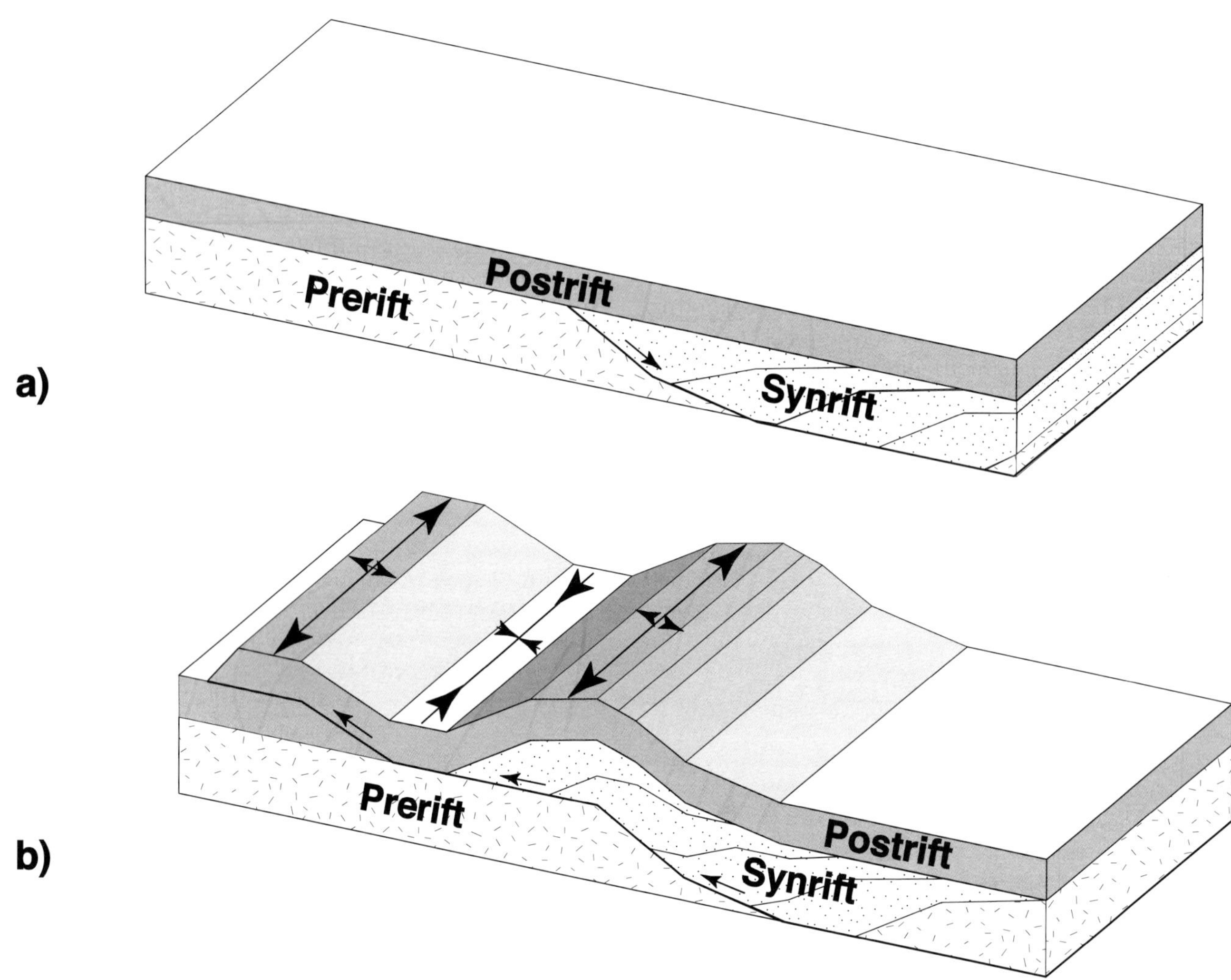

FIGURE 6. The shortening and inversion of the half graben may have occurred by the transport of strata from the hanging wall of the synrift half graben northward along low-angle thrust faults (a and b).

SUMMARY

Paradox of Different Orders of Superposition across the Accommodation Zone

The sequence of folding seen in the study area (Figure 3) implies that two separate, nonhomogeneous phases of deformation could not have generated the unique pattern of folding, without the presence of preexisting structural elements. Offsets in the basin margins result in 3-D strain across the accommodation zone. During compression and subsequent inversion across these accommodation zones, a 3-D strain is produced that is coupled to the preexisting accommodation zone. The structures formed by compression across these accommodation zones are caused by 3-D strain and result in unique refold patterns.

The Paleozoic exposure of the Jebilet Anticlinorium trends eastward and then plunges to the southeast (Figure 3). The expression of this anticline can be seen by the gentle dips in the Jurassic on both limbs of the fold. The trend of the Jebilet Anticline is not affected by the second phase of folding (F2) that refolds the Ait Attab and Guettioua Synclines. This unusual fold relationship was likely generated as a result of the fold forming parallel to the ramp of an accommodation zone, rather than forming over a preexisting normal fault.

Interpretation of the Structural Evolution of the Refolding Geometry: Accommodation Zone

Accommodation zones are common features of rift systems. The geometries of these extensional structures result from a change in strike of the rift-basin margin, produced by a ramp/relay fault system or transfer fault (McClay and White, 1995). The steplike accommodation of two parallel normal faults by a higher angle fault forming a ramp is a common feature in extensional rift basins (Rosendahl et al., 1986). En echelon normal faults have been found to form in rift systems when extension is oblique to the margin of the rift system (McClay and White, 1995). Transfer faults are also a common characteristic of extensional terrains and allow for the "transfer" of extensional slip between faults. These faults are analogous to lateral ramps in thrust tectonic terrains. Extensional transfer faults transport rotational strike and dip components during rifting, as do lateral ramps (Gibbs, 1984). These transfer faults occur as oblique or lateral accommodation zones and may involve a change in fault polarity. A transfer fault in cross section may have a high-angle flower geometry during rifting (Gibbs, 1987), in which case the fault may be normal to the basin margin, or it may be more oblique and involve a lower angle extensional ramp. Faults offset by these transfer faults can be parallel to the rift basin and have a planar, listric, or extensional fault-ramp geometry. The geometry of transfer faults often influences the development of younger folds upon inversion (Alonso, 1989). The transfer faults or ramps may be oblique or normal to the basin margin, and the orientation of the extensional/compressional ramp will result in differential movement in the hanging wall (in the sense of the transport direction) on the oblique ramp during inversion (Casas-Sainz, 1993).

Comparison with Other Studies of Refolding during Inversion

Superimposed folding is often associated with a change in the orientation of a regional stress field through time. Superimposed folds may occur by successive deformational events separated by long time intervals, multiple deformational phases in one orogenic cycle, continuous deformation in one orogenic cycle, and simultaneous folding from several directions in one orogenic phase (Ramsay and Huber, 1987).

The Cobar Basin of Australia is an inverted Paleozoic basin that exhibits superposed folding (Smith and Marshall, 1992). Superposed folding in the Cobar Basin is proposed to have occurred as a result of margin-normal shortening followed by progressive oblique deformational shear. The Davenport Province of Central Australia (Stewart, 1987) is another region where superposed folding of Proterozoic-age rocks resulted from two episodes of deformation that used preexisting synsedimentary normal and transfer faults that were reactivated in reverse and strike-slip senses, respectively. In this region, major sedimentary faults such as those associated with accommodation zones in a rift system, controlled the structural domains of each fold trend.

Steeply dipping faults offer greater resistance to reverse dip-slip movement during compression and may form buttresses where displacement is concentrated (Velasque et al., 1989). Folding can result from preexisting structures (e.g., folds, faults, or diapirs), which provide buttresses that concentrate strain. The geometry or configuration of preexisting structures will control or alter the stress field affecting folding. I propose that in the High Atlas Mountains, a single continuous phase of deformation across a preexisting structural feature, such as a extensional accommodation zone, resulted in the unique pattern of superposed folding. The sequence of superposed folding in the High Atlas Mountains of Morocco shows that these styles of folding are an important characteristic of inversion.

Exploration Potential

The unique style of folding found in the Atlas Mountains has useful applications for exploration in tectonic settings that involve the reactivation of structures created in previous tectonic phases. Unusual folding styles, such as those documented in the Ait Attab region of Morocco, may signal that significant shortening and tectonic burial could have placed potential source rocks in an active petroleum-generating setting. Hydrocarbon exploration may have been overlooked in Morocco because of the lack of well-developed foreland basins adjacent to the Atlas and Rif mountain belts. Source rocks that are immature on the paleomargins of the Atlas rift system and the present-day Moroccan Meseta may be in the oil window beneath thrust systems that have buried these rocks along the margins of the mountain belt.

DISCUSSION AND CONCLUSIONS

Laville (1981) attributed the superposed folding along the northern margin of the High Atlas Mountains to successive rotation of compressional phases, from the west-southwest–east-northeast to the north-northeast–south-southwest. Superposed folding in the eastern High Atlas Mountains was recognized by De Sitter (1960), and was attributed to separate tectonic phases and stress orientations.

However, we relate the unusual sequence and relationship of folding to preexisting structural control. Separate tectonic phases and stress orientations would have yielded a different pattern of folding than is found in this region of the High Atlas Mountains. Superposed folding resulting from two different stress orientations often yields two phases of folds (F1 and F2), each phase having parallel fold axes (Figure 7a and b). In the Atlas Mountains, as in the Davenport Province, oversteepened fold limbs (common in the Atlas) were probably caused by the second phase of folding. Both the examples from

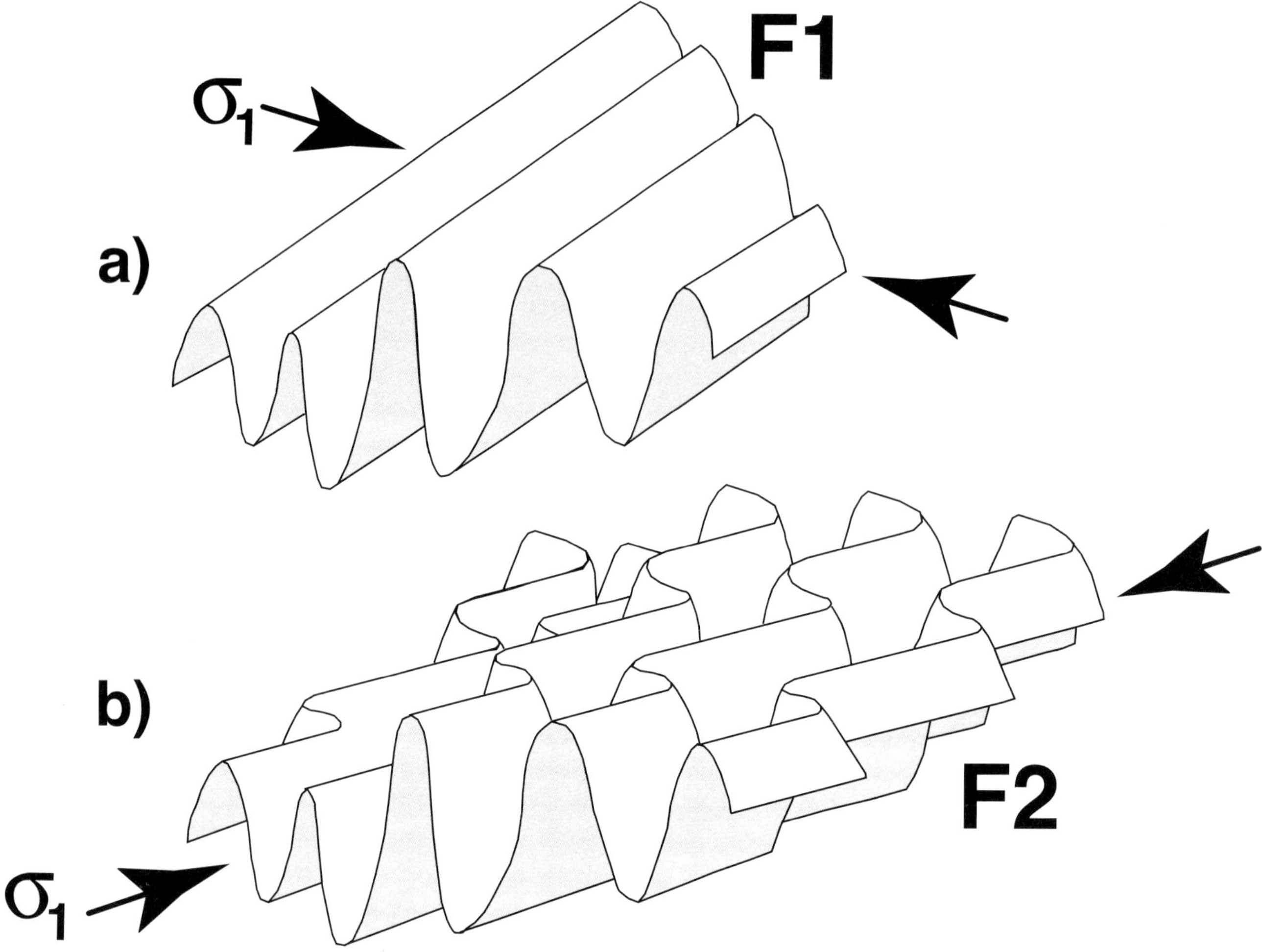

FIGURE 7. Schematic models show the formation of superimposed folding as a result of two different stress orientations (a and b). σ_1 = principal stress component; maximum principal stress.

Australia exhibit Type 1 and 2 interference patterns (Ramsay and Huber, 1987). The folding in the study area of the Atlas Mountains exhibits a Type 2 interference pattern (Ramsay and Huber, 1987), where the axial surfaces of the first folds are folded along with the limbs of the first folds. The superposed folding in the Atlas Mountains may have resulted from a synrift accommodation zone made up of a ramp/fault relay, or a transfer fault normal to the basin margin. This ramp transfers slip laterally between two major down-to-the-basin (southward-dipping) normal faults (Figure 8a). Folding in the area of this study yielded folds that are normal to one another in both phases (F1 and F2) (Figure 8b and c). Continued shortening across the accommodation zone may have created interference with the first F1 folds. The refolding of the F1 folds resulted in the F2 phase fold axes.

The presence of preexisting structural geometries, such as accommodation zones, fault ramps, fault relays, en echelon folding, and other features formed by rift processes, will have an effect on subsequent compressional stress fields generated by plate convergence and other tectonic processes. These structural geometries formed during rifting will affect the 3-D strain field, and that may result in superposed folding that is disharmonic. Superposed disharmonic folding may, in fact, be a unique characteristic of the inverted rift systems that result in intracontinental mountain belts.

ACKNOWLEDGMENTS

This study would not have been possible without the assistance of ONAREP (Office National de Recherches et d'Exploitations Petrolieres) and the Geological Survey of Morocco. Many thanks to my colleagues at Cornell that helped in this study and in the review of this manuscript. I would also like to thank Tarik Djebbar and C. Elders for their constructive review. Also, I would like to thank ARCO for assistance and expenses in the preparation of this manuscript.

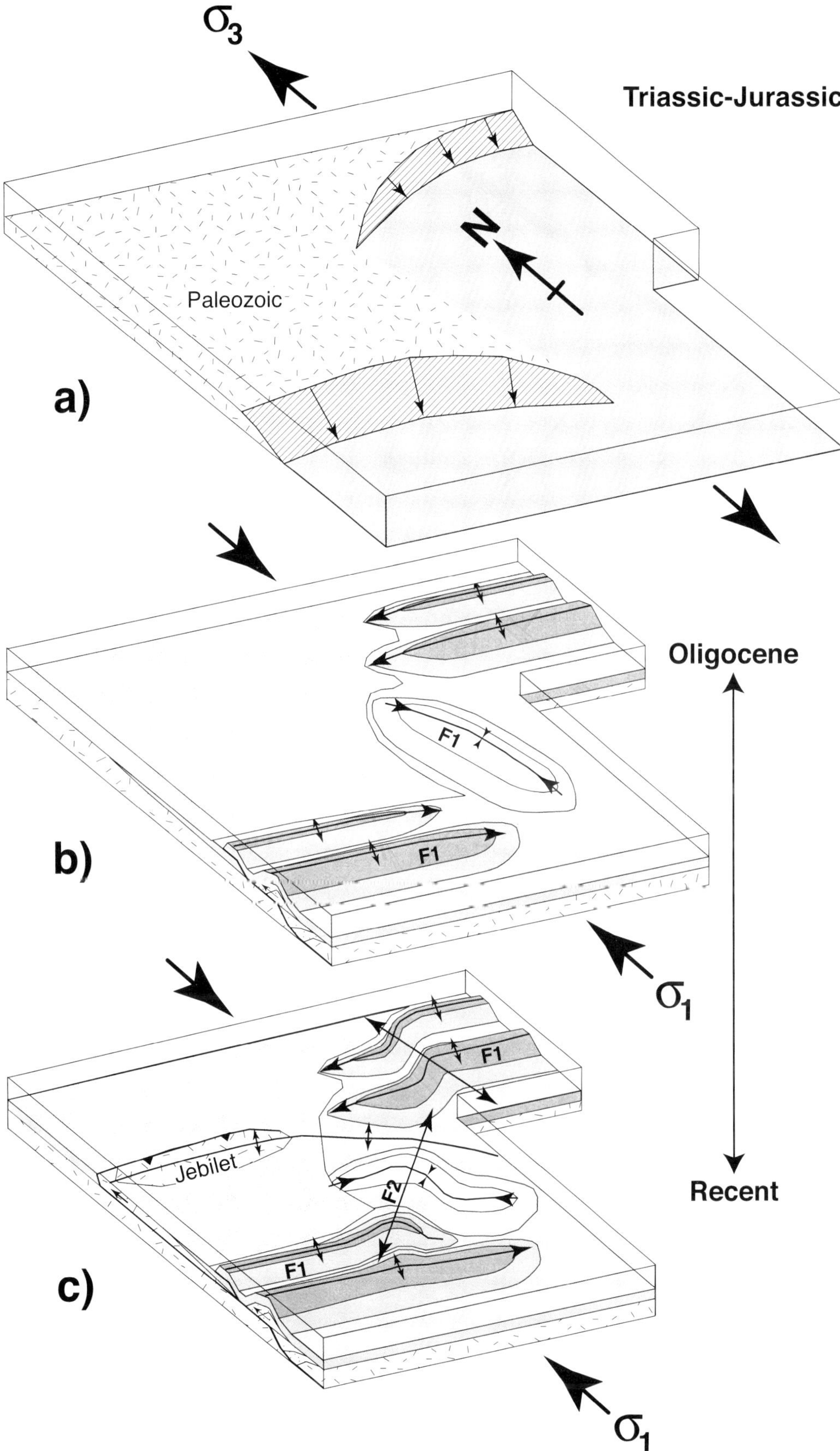

FIGURE 8. The inversion of a synrift accommodation zone (a) by margin-normal shortening created first folds (F1) that are oriented parallel to preexisting ramps and down-to-the-basin normal faults (b). Continued compression across the accommodation zone results in the interference of the first F1 folds and the formation of F2 folding (c). σ_1 = principal stress component; maximum principal stress.

REFERENCES CITED

Alonso, J. L., 1989, Fold reactivation involving angular uncomformable sequences: theoretical analysis and natural examples from the Cantabrian Zone (Northwest Spain): Tectonophysics, v. 170, p. 57–77.

Bally, A. W., 1984, Tectogenese et sismique reflexion: Bulletin de la Société Géologique de France, v. 7, p. 279–285.

Beauchamp, W., M. Barazangi, A. Demnati and M. El Alji, 1996, Intracontinental Rifting and Inversion: the Missour Basin and Atlas Mountains of Morocco: AAPG Bulletin, v. 80, no. 9, p. 1459–1482.

Beauchamp, W., R. W. Allmendinger, M. Barazangi, A. Demnati, and M. El Alji, 1999, Inversion tectonics and the evolution of the High Atlas Mountains, Morocco, based on a geological-geophysical transect: Tectonics, v. 18, no. 2, p. 163.

Casas-Sainz, A. M., 1993, Oblique tectonic inversion and basement thrusting in the Cameros Massif (Northern Spain): Paris, Geodinamica Acta, v. 6, no. 3, p. 202–216.

De-Sitter, L. U., 1960, Plissement Croise Dans Le Haut-Atlas: Geologie en Mijnbouw, v. 5, p. 277–282.

Gibbs, A. D., 1984, Structural evolution of extensional basin margins: London, Journal of the Geological Society, v. 141, p. 609–620.

Gibbs, A. D., 1987, Development of extension and mixed-mode sedimentary basins, *in* M. P. Coward, J. F. Dewey. and P. L. Hancock, eds., Continental extensional tectonics: Geological Society of London Special Publication 28, p. 19–33.

Huvelin, P., 1972, Carte Géologique et Gitologique des Jebilet, 1:200,000: Royaume du Maroc, Ministere de l'Energie et des Mines, 232C.

Jabour, H. and K. Nakayama, 1988, Basin modeling of Tadla Basin, Morocco, for hydrocarbon potential: AAPG Bulletin, v. 72, p. 1059–1073.

Jenny, J., 1988, Carte Géologique du Maroc au 1/100,000 Feuille Azillal (High Atlas central), Notes Mémoires du Service Géologique, Rabat, v. 339, 104 p.

Laville, E., 1981, Incidence des jeux sucessifs d'un accident synsédimentaire sur les structures plicatives du versant nord du Haut Atlas central (Maroc): Bulletin de la Société Géologique de France, no. 3, p. 329–337.

McClay, K. R., and M. J. White, 1995, Analogue modelling of orthogonal and oblique rifting: Marine and Petroleum Geology, v. 12, p. 137–151.

Ramsay, J. G., and M. I. Huber, 1987, The techniques of modern structural geology; Volume 2: Folds and fractures, Academic Press, London, p. 494.

Rolley, J.-P., 1978, Carte Geologique du Maroc au 1/100,000 Feuille Afourer (High Atlas central): Rabat, Notes Mémoires du Service Géologique, v. 247, 103 p.

Rosendahl, B. R., D. J. Reynolds, P. M. Lorber, C. F. Burgess, J. McGill, D. Scott, J. J. Lambiase, and S. J. Derksen, 1986, Structural expressions of rifting: lessons from Lake Tanganyika, Africa: Geological Society of London Special Publication 25, p. 25–43.

Saadi, M., E. A. Hilali, M. Bensaid, 1977, Carte Géologique du Maroc-Jebel Saghro-Dadés, 1:200,000: Royaume du Maroc, Ministere de l'Energie et des Mines, v. 161.

Saadi, M., M. Bensaid, M. Dahmani, 1985, Carte Géologique du Maroc-Azilal, 1:100,000, Royaume du Maroc, Ministere de l'Energie et des Mines, v. 339.

Smith, J. V., and B. Marshall, 1992, Patterns of folding and fold interference in oblique contraction of layered rocks of the inverted Cobar Basin, Australia: Tectonophysics, v. 215, p. 319–334.

Stewart, A. J., 1987, Fault reactivation and superposed folding in a Proterozoic sandstone-volcanic sequence, Davenport Province, central Australia: Journal of Structural Geology, v. 9, p. 441–455.

Velasque, P. C., L. Ducasse, J. Muller, R. Scholten, 1989, The influence of inherited extensional structures on the evolution of an intracratonic chain: the example of the Western Pyrenees: Tectonophysics, v. 162, p. 243–264.

33

Butler, R. W. H., S. Mazzoli, S. Corrado, M. De Donatis, D. Di Bucci, R. Gambini, G. Naso, C. Nicolai, D. Scrocca, P. Shiner, and V. Zucconi, 2004, Applying thick-skinned tectonic models to the Apennine thrust belt of Italy—Limitations and implications, *in* K. R. McClay, ed., Thrust tectonics and hydrocarbon systems: AAPG Memoir 82, p. 647–667.

Applying Thick-skinned Tectonic Models to the Apennine Thrust Belt of Italy—Limitations and Implications

R. W. H. Butler
School of Earth Sciences, University of Leeds, Leeds, U.K.

S. Mazzoli
Facoltà di Scienze Ambientali, Università di Urbino, Urbino (PS), Italy

S. Corrado
Università di Roma Tre, Rome, Italy

M. De Donatis
Facoltà di Scienze Ambientali, Università di Urbino, Urbino (PS), Italy

D. Di Bucci
Servizio Sismico Nazionale, Rome, Italy

R. Gambini
Shell Oil Italia E & P, S.P.A., Rome, Italy

G. Naso
Servizio Sismico Nazionale, Rome, Italy

C. Nicolai
Shell Oil Italia E & P, S.P.A., Rome, Italy

D. Scrocca
Shell Oil Italia E & P, S.P.A., Rome, Italy

P. Shiner
Shell Oil Italia E & P, S.P.A., Rome, Italy

V. Zucconi
Shell Oil Italia E & P, S.P.A., Rome, Italy

ABSTRACT

Fold-thrust belts are commonly interpreted as "thin-skinned" structures, developed above a detachment, with the underlying basement remaining undeformed. However, in many areas, particularly where compressional tectonism was preceded by rifting, models of basement fault reactivation may be more appropriate. The contrasts between thin-skinned and deep-rooting, inversion-dominated deformation in building fold-thrust complexes are investigated using a case history from the Italian Apennines. Three sectors were chosen to represent the marked lateral variations in structural style evident in the thrust belt. The outer portion of the Marche (in the north) is contrasted with a section through the Lucanian Apennines in the south and with the Molise district of the Central Apennines. The Marche structures are readily explained in terms of inversion, a model that is consistent with new deep seismic data onshore and conventional seismic from the nearby Adriatic Sea. The displacements implicit for the inversion model are a factor of five less than for existing thin-skinned interpretations. However, these styles are not applicable throughout the Apennines. Well data in the Southern Apennines of Lucania demonstrate large-scale thin-skinned thrusting, with 57 km of horizontal displacement since earliest Pliocene time. This includes 14 km of shortening that ramps up through the buried Apulian Platform carbonates. These deeper structures may be restored using ramp-dominated thrust geometries. The Molise sector shows broadly the same structural style as for Lucania: allochthonous shallow-water carbonates and pelagic basin units overlie the carbonates of the Apulian Platform, with the major difference being that here, the pelagic basin units are detached at the level of the Oligocene–lower Miocene Argille Varicolori. In this setting, the Apulian carbonates may be restored using only 5 km of displacement. The overlying allochthon probably has accommodated about 45 km of displacement since the earliest Pliocene. Therefore, the Apennines show differing structural styles with differing displacements along their length. Thick-skinned thrusting models may be applied to the Marche and to structures in the buried Apulian units.

INTRODUCTION

The Apennine Thrust Belt of Italy forms part of the Mediterranean-Alpine orogen of Neogene age. It has a foredeep that includes parts of the Po Plain in the north, part of the Adriatic Sea, and, onshore in southern Italy, the Bradanic Trough (Figure 1). The thrust belt itself is northeast-directed and, at outcrop, it is composed entirely of Mesozoic and Cenozoic sedimentary rocks. The outer parts of the thrust belt contain important accumulations of gas, whereas there are oil fields beneath more internal parts of the thrust belt (Pieri and Mattavelli, 1986). The absence of crystalline basement at outcrop in the outer part of the Apennines, together with the presence of evaporites toward the base of the sedimentary cover, has led to a proliferation of thin-skinned, detachment-dominated models of structural evolution (Figure 2) (Bally et al., 1986; Hill and Hayward, 1988). These models influence not only our current understanding of the regional tectonic evolution (Doglioni et al., 1998) but also our knowledge of the petroleum systems (Casero et al., 1991). The models imply that substantial displacements have carried allochthonous sheets over buried foreland successions, offering potential exploration targets (Roure et al., 1991). However, Coward et al. (1999) have questioned the validity of these models in parts of the mountain belt. These workers suggest that thrusts pass to depth into basement without significant duplication of cover-sedimentary successions. These types of thick-skinned interpretations (Coward, 1983; 1994) generally require far less orogenic shortening than do equivalent thin-skinned ones. They also restrict the propensity for subthrust structures and thus limit the range of structural traps for hydrocarbons in the subsurface. The purpose of this chapter is to investigate the opportunities to reinterpret Apennine structure in terms of thick-skinned thrust geometries. However, because the chain displays remarkable lateral variations along its length, we investigate an array of transects, each with its own structural styles. These offer different challenges to competing structural models. Thus our case history should have general relevance to other fold-thrust systems around the world.

THE APENNINE CHAIN: INTRODUCTORY NOTES

The Apennine Thrust Belt runs from northwest to southeast, forming the topographic backbone to Italy. The orogenic foreland is represented at outcrop by rocks of the Apulian and Gargano promontories (Figure 1)—collectively termed the Apulian unit. These strata, like those from the Apennine Thrust Belt, range in age from Late Triassic to Miocene. Collectively, the Apulian foreland and the deformed strata of the Apennines represent a telescoped continental margin with complex subbasins of differing ages. These complexities of pre-orogenic continental margin architectures are reflected in the Mesozoic-Eocene strata. These show substantial facies and thickness variations that relate to depositional highs and lows (Parotto and Praturlon, 1975). However, the continental margin saw very little siliciclastic input until the Oligocene, and so the Mesozoic sediments are dominantly carbonates, shales, and, in basinal sections, cherts. Deeper parts of the Apennine

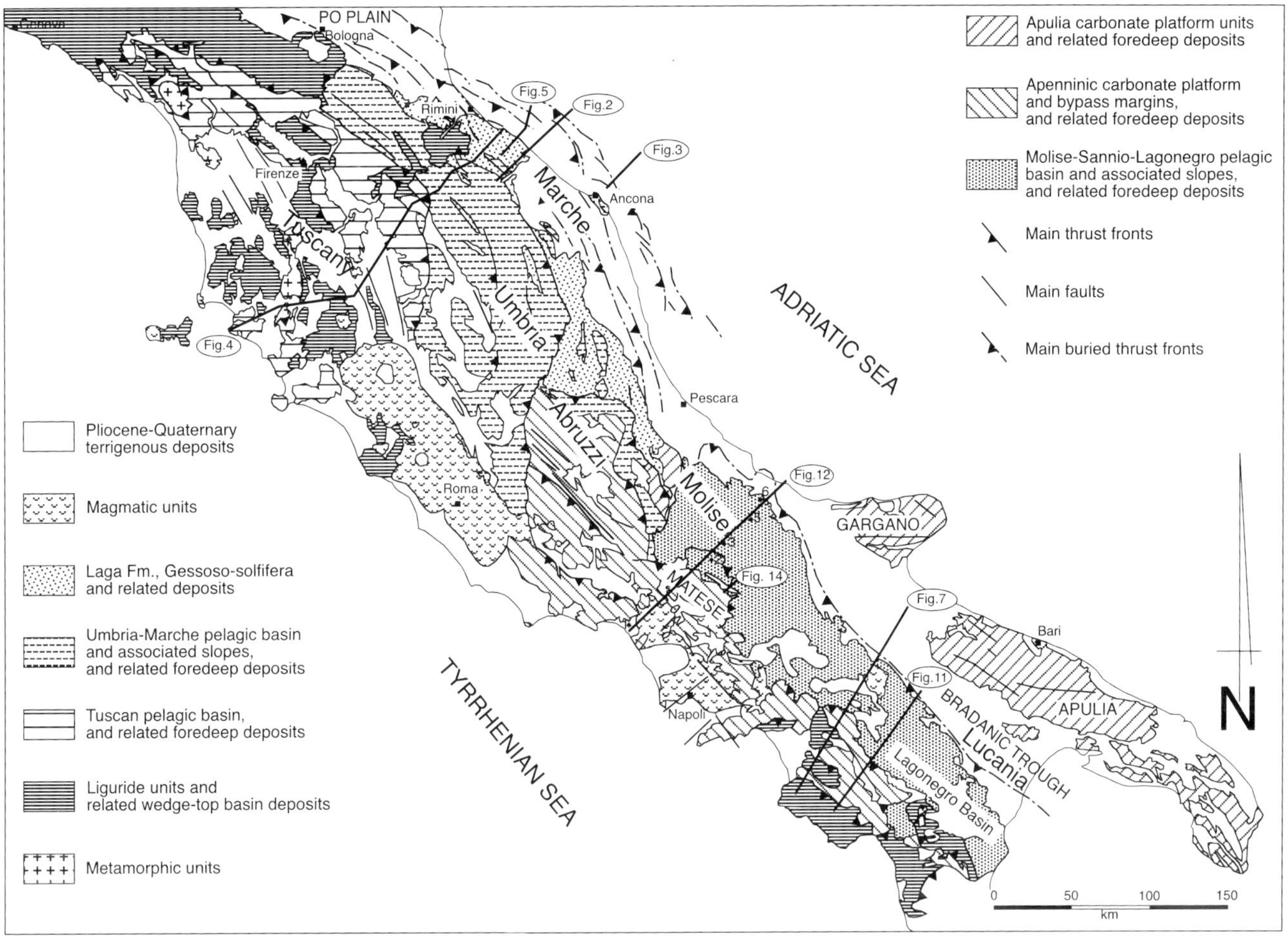

FIGURE 1. Geologic map of peninsular Italy, showing locations of cross sections.

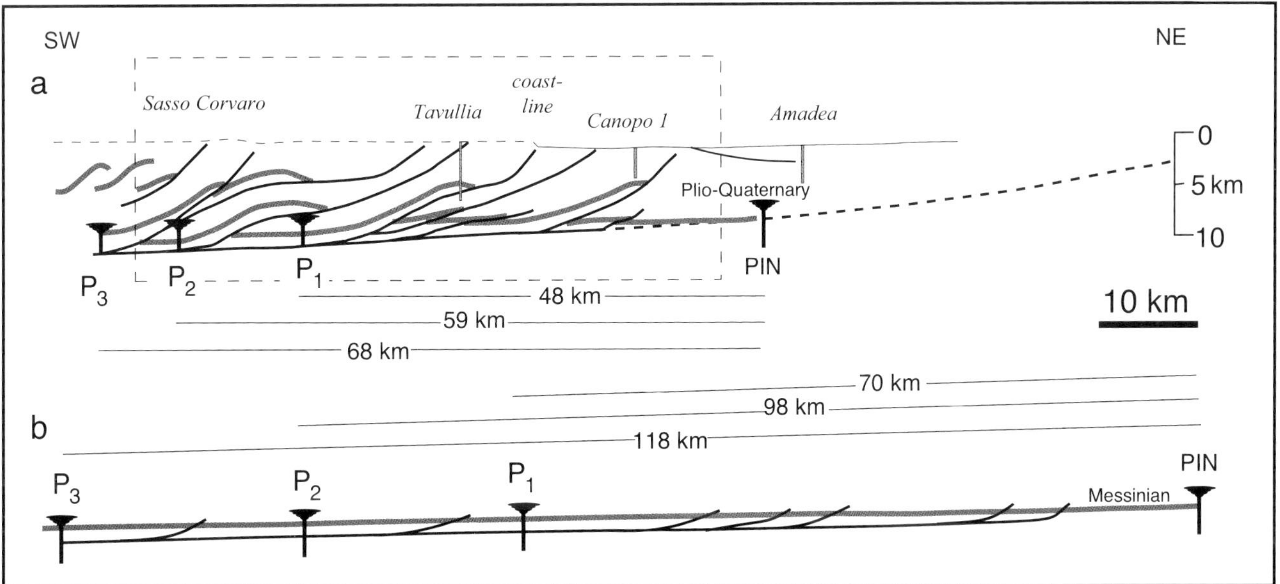

FIGURE 2. The thin-skinned interpretation of the northern Marche segment of the Apennine thrust belt (located on Figure 1), modified after Bally et al. (1986, their Figure 57). A variety of measurements is illustrated to permit comparison with the cross section of our study (boxed area, Figure 5).

geology are known from deep wells (Anelli et al., 1994) that penetrate Permian-Triassic continental siliciclastics (the Verrucano Group). These, in turn, overlie crystalline basement.

Understanding the depth to orogenic basement beneath the cover sequences of the Apennines has been a major problem. The Triassic strata include a thick wedge of evaporites (the Burano Formation, which is >300–500 m thick, according to Lavecchia et al., 1984) that mark the base of the carbonate-dominated continental-margin strata. The evaporites form an important structural decoupling horizon, therefore the underlying strata remain buried. Indeed, for many workers, the pre-evaporitic rocks are not involved in Apennine thrust tectonics until much farther back toward the orogenic hinterland (Bally et al., 1986). Certainly, the top to the magnetic basement lies at depths greater than 10 km below most of the thrust belt (Cassano et al., 1998). However, the Verrucano lies between the evaporites and the crystalline, magnetic basement. Deep seismic data of the CROP 3 transect show that this succession is as much as 6 km thick (Decandia et al., 1998).

Regardless of the controversies relating to the nature of the pre-evaporitic strata beneath the Apennines, these strata form a prerift succession for the complex platform and basin successions that characterize the Mesozoic sediments of the Italian peninsula. A plethora of elaborate paleogeographic terminology describes the various tectonostratigraphic units that now crop out in thrust sheets of the Apennines. It is not our intention to review this here; readers are referred to syntheses by Parotto and Praturlon (1975), Mostardini and Merlini (1986), Damiani et al. (1992) and Patacca et al. (1992). However, some comments are useful here.

The Mesozoic sediments include a platform-dominated succession of the Apulian foreland, deeper basinal successions of the Umbria-Marche (in the northern part of the chain; Santantonio, 1993) and Lagonegro (in the south; Scandone, 1972; Ciarapica et al., 1990), together with other, so-called "internal" platforms (i.e., originally more western platforms; Scandone, 1972; D'Argenio et al., 1973; Channell et al., 1979). The pre-orogenic setting of these various units has been controversial, with some authors restoring the more basinal successions (e.g., the Lagonegro) to the west of the more internal platforms (Marsella et al., 1992, 1995). Thus, there is a tradition of invoking great allochthoneity to some thrust sheets. Among these different units, the Apulian foreland and associated structures on the margin of the thrust belt have received the most attention for oil and gas exploration (Casero et al., 1991). An understanding of the structure of the thrust belt's outer parts is therefore central to ongoing petroleum development.

The variable nature of the Mesozoic strata along and across the Apennines leads to a great variety of thrust-and-fold architectures. While the Triassic evaporites are widely considered to have acted as a regional detachment, other levels within the Mesozoic and Cenozoic successions are also thought to have acted as thrust flats locally—particularly shales and marls within the otherwise carbonate-dominated units (Barchi et al., 1998). However, geoscientists have recently questioned the notion that some fold-thrust structures in the Apennines might not conform to the simple fault-bend relationships developed by compression acting on a layer-cake stratigraphy. In the Apennines of Umbria, Tavarnelli (1996) shows that folds initiated as buckles, which apparently nucleated at preexisting (Jurassic-age) normal faults. The thrusts that cut these folds have relatively little displacement, with the geometry essentially defined by "ramp-on-ramp" structures. However, Tavarnelli (1996) maintains that basement was not involved in these thrust systems and that there was regional detachment along the base of the sedimentary cover. Elsewhere, however, basement involvement has been postulated by several authors (Menichetti et al., 1991; Sage et al., 1991; Coward et al., 1999).

Much of the Apennine chain has been dissected by normal and strike-slip faults that locally postdate thrust structures. In the interior of the chain (e.g., Tuscany), these faults control Miocene-Pliocene basins (Carmignani et al., 1994; Decandia et al., 1998) and therefore are coeval with thrust structures that are active farther to the east. Indeed, the entire chain has been convincingly described as a paired tectonic belt, with extension in the orogenic hinterland balancing orogenic contraction on the forelandward side of the orogen (Lavecchia, 1988; Decandia et al., 1998). These larger considerations lie outside the scope of this chapter. However, extensional and strike-slip structures are important for the late-stage restructuring of, particularly, the Southern Apennines. Consequently, we will discuss these late structures where appropriate.

STRUCTURAL GEOMETRIES IN THE NORTHERN APENNINES

We start our investigation in the Northern Apennines and contrast the original thin-skinned interpretations of Bally et al. (1986) with the recent notions of Coward et al. (1999). Details of the local stratigraphy, geologic maps, and a summary of subsurface data are provided elsewhere (De Donatis et al., 1998; Coward et al., 1999).

Thin-skinned Interpretations

Existing thin-skinned models for the Northern Apennine thrusts involve detachment of cover units

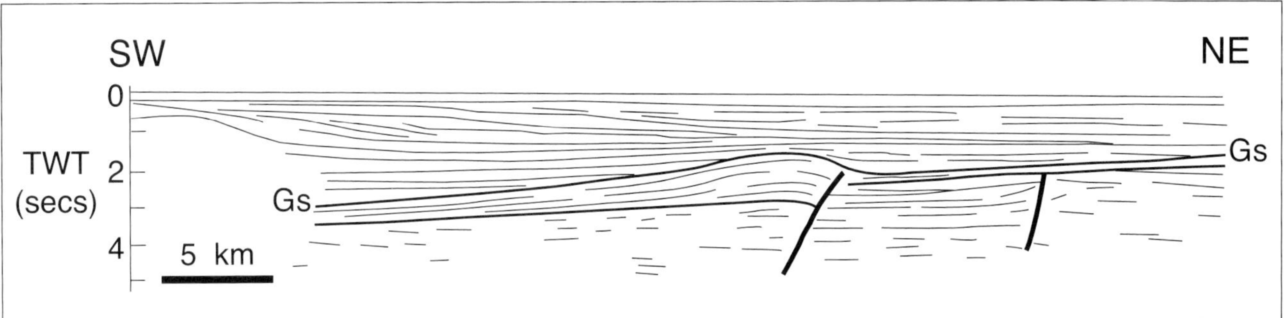

FIGURE 3. Line drawing of a marine seismic-reflection profile from the Adriatic Sea (located in Figure 1), modified after Argnani and Gamberi (1997).

from a panel of underthrust basement (Figure 2). The basement dips simply beneath the orogen, presumably to be subducted at depth beneath the interior of the chain. The cover units are shown as a layer-cake succession. In part, this is justifiable given the regional scale of Bally et al.'s cross sections (1986) (Figure 2). Jurassic strata in the Apennines currently crop out at elevations greater than 2 km above sea level. However, simple projection of the westward dip of strata from the foreland (the so-called "regional," as defined by Williams et al., 1989) would predict Jurassic rocks at a depth 13 km deeper. In Bally et al.'s model (1986; their Figure 54), the uplift above the regional is achieved by duplication of the stratigraphic section, at some sites by four times. Consequently, the displacement implicit in these sections is very great. More than 180 km of shortening has been inferred for regional sections through a present width of about 120 km across the Northern Apennines.

To illustrate the implications of Bally et al.'s (1986) model, Figure 2 has been prepared using only the frontal portion of one of their regional sections. A pin line (that of Bally et al., 1986) is inferred in the Adriatic where Messinian strata lap onto substrate. Seismic and well data confirm the integrity of this pin. Thrust structures ahead of a prominent antiformal structure (Sasso Corvaro, in Figure 2), restore to a width of 118 km from the pin line. The present distance is 68 km, which gives a bulk shortening of 50 km. Consideration of just the two main frontal structures, penetrated by the Tavullia and Canopo 1 wells, still yields a shortening of 22 km (70 km – 48 km = 22 km).

The structures illustrated on Figure 2 are post-Messinian in age. Thus, they developed between about 5 and 1 Ma. Time-averaged shortening rates may therefore be calculated. These are 12.5 mm/yr for the section as illustrated. Bally et al.'s (1986) regional sections through the frontal part of the thrust belt (footwall to the so-called Sibillini Thrust) show more than 100 km of post-Messinian shortening. For these sections, the time-averaged shortening rate exceeds 25 mm/yr.

Alternatives

Marine seismic data from the northern Adriatic (Argnani and Gamberi, 1997) provide excellent images of the Pliocene-Quaternary strata filling various subbasins in the foredeep. These strata are disconformable upon Messinian units that form particularly strong, characteristic reflectors. On Figure 3, the Pliocene sediments can be seen lapping onto a folded Messinian surface. There is clear evidence of early Pliocene folding. However, this fold structure has a more complex history. At depth, other reflectors show net extensional throws across faults. There is an increased stratigraphic thickness in the hanging wall to the fault. This structure has the geometry of an inverted normal fault.

The marine seismic data of Argnani and Gamberi (1997) do not provide good images of the top of basement beneath the Adriatic. Indeed, it is difficult to define the geometry of this horizon on seismic profiles throughout the Apennines. However, the acquisition of deep seismic-reflection profiles through the Apennines, as part of the CROP program, provides critical information (Barchi et al., 1998). Figure 4 illustrates the seismic model for the CROP 3 profile in the Northern Apennines (Morgante et al., 1998). Several features are displayed by this section. First, the crustal thickness varies markedly across Italy, with the greatest values beneath the highest topography of the Apennines. The crustal thickness is reduced in the western part of the section, underneath the Pliocene-Quaternary basins of Tuscany. Second, the Moho appears to be imbricated, with inter-fingering of high-velocity (mantle) and lower-velocity (crust) layers. This observation, adjacent to the thickened crustal units, strongly suggests crustal stacking beneath the Apennine chain. Note that the sections of Bally et al. (1986), constructed long before the acquisition of the CROP data, predict that basement involvement, and therefore crustal stacking, should occur only in the extreme southwestern portion of Figure 4.

Although the CROP data indicate that the Moho is imbricated, they are less clear for the internal structure

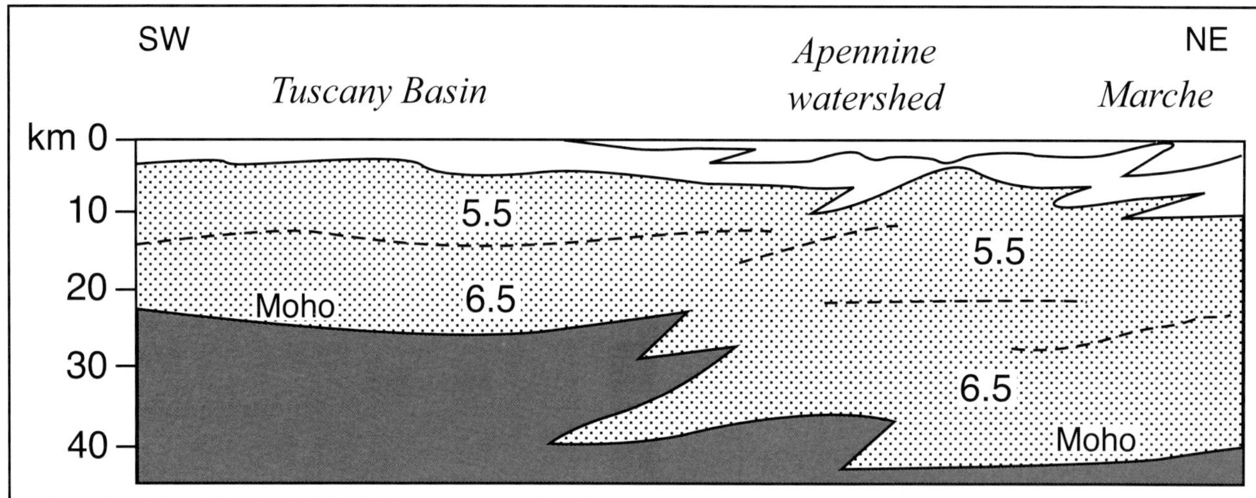

FIGURE 4. The seismic-velocity model of the CROP 3 line through the Apennines (located in Figure 1), modified after Morgante et al. (1998).

of the crust. Seismic-velocity models (Morgante et al., 1998) indicate that different crustal levels are imbricated (Figure 4). There is some indication that crystalline basement is imbricated into the Verrucano that underlies the Mesozoic cover (Decandia et al., 1998). Taken collectively, it is plausible that basement is stacked beneath the thrust belt.

The notion that the major anticlines of the outer part of the Northern Apennines are controlled by basement faults underpins a recent reinterpretation of the area by Coward et al. (1999). Using well and seismic reflection data in conjunction with geologic maps, they present three balanced cross sections. These show shortening values of less than 10 km over present distances of between 40 and 50 km. Note that as the present section lengths and positions of pin lines vary for these and Bally et al.'s (1986) sections, there is no virtue in giving the shortening in dimensionless units. The substantially lower values of bulk shortening of these sections compared with those of Bally et al. (1986) are emphasized when the shortening rates are compared. Coward et al. (1999) suggest that the average shortening rate was between 1.5 and 2.5 mm/yr, which is ten times slower than that of Bally et al. (1986).

Despite inferring that anticlines in the coastal areas of the Northern Apennines are underlain by uplifted basement blocks, Coward et al. (1999) also show important detachments within the cover sections. Thus, for these workers, individual folds do not balance within themselves. Thrusts at outcrop within Tertiary strata are shown passing onto extensive thrust flats. In part, this is reflected in the oblique, loose lines shown in their restorations, which imply substantial bed-parallel shear strains. There is little evidence for these strains in outcrop. Thus, the thrusts at high levels in the stratigraphy are likely to be accommodated within the larger anticlines upon which they occur.

Inversion and the Northern Apennines

We now examine the contention that the thrusts at outcrop in the frontal folds of the Northern Apennines merely accommodate shortening within these folds and do not indicate bed-parallel slip through the stratigraphy. This view is consistent with Coward et al.'s (1999) field observations and maps. Certainly, there is little evidence for the through-going thrusts that Bally et al. (1986) proposed. Indeed, published geologic maps (Servizio Geologico Nazionale, 1969) show fault patterns that are discontinuous and braided. Consequently, we believe that thrust faults are only weakly linked and have low displacements on cross sections.

To test our interpretation, we rebalanced Coward et al.'s cross section (1999, their Figure 10c) that shows the greatest thrusting in the shallow stratigraphic levels. In our new restoration (Figure 5), we have linked the steps on the top-of-basement horizon that can be inferred from seismic profiles with the folds in the overlying cover. This coincidence suggests that no significant detachment occurs along the basement-cover contact. Consequently, we have erected a loose line that pins basement and cover to the southwest of the last fold (Montefiore-Serrungarina Ridge on Figure 5). There is no requirement for the hypothetical bed-parallel shears invoked by Coward et al. (1999). The pin line lies offshore, beyond the most external thrust in the region.

To obtain a value for the shortening in the cover, we have used the top of the lower Liassic as a key horizon (in the sense of Geiser, 1988) and restored the remaining layers to achieve a compatible shortening (this method is outlined by Butler, 1992). In this way, the minor folds and faults, which are particularly evident in the younger stratigraphic units, are built into the restoration as a continuous layer-parallel-shortening strain. The post-Triassic, pre-Pliocene stratigraphy can

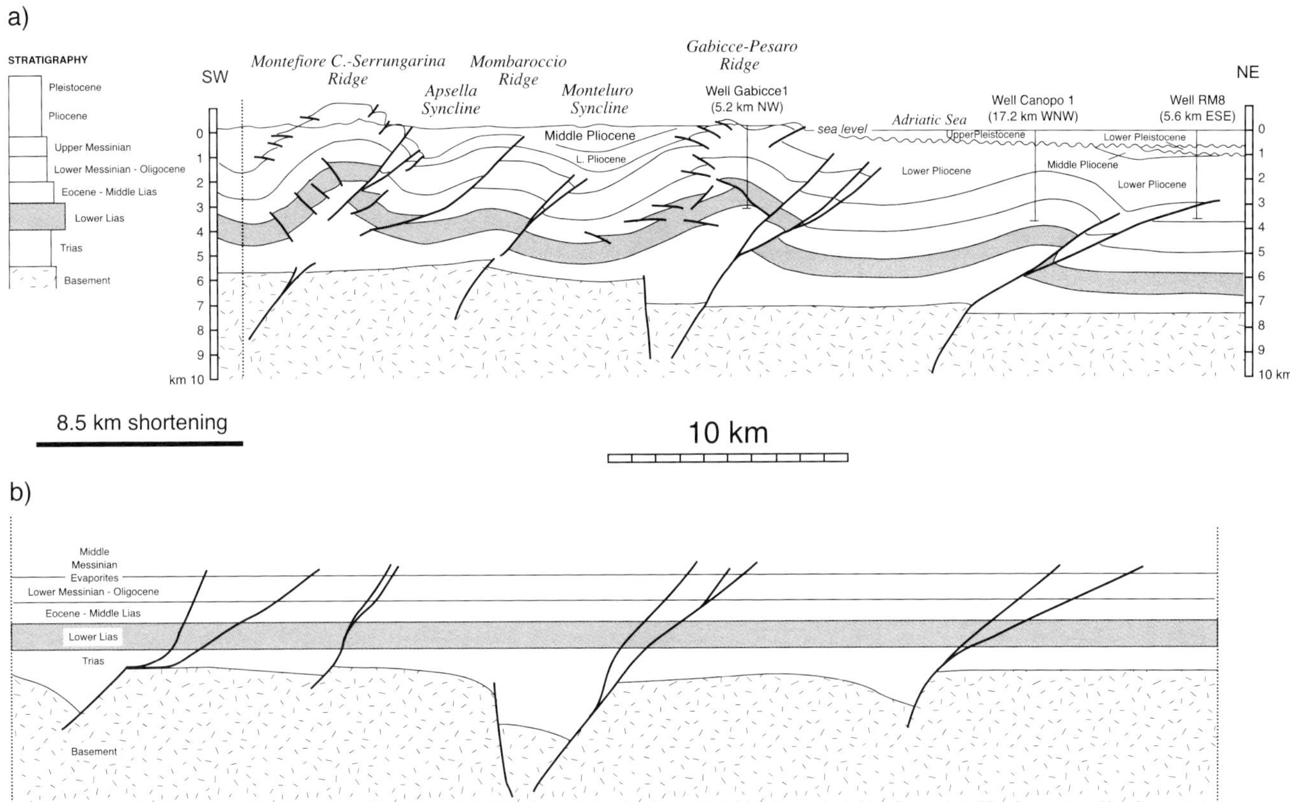

FIGURE 5. Balanced (a) and restored (b) geologic section across northern Marche (located in Figure 1). See text for discussion.

thus be restored to nearly a layer-cake geometry. Note that we have not restored the growth strata of Pliocene-Pleistocene age—this certainly would be a more complex challenge, and one that has little bearing on our aims here.

Clearly, if the top-of-basement level were returned to a subhorizontal, continuous elevation, only about 1 km of shortening could be accommodated on the faults. This compares with an estimate of shortening in the cover of 8.5 km (Figure 5). If basement is assumed to be uncoupled from the cover structures, then the discrepancy in shortening can be accommodated by slip in the detachment, thereby passing shortening onto basement structures at depth to the southwest. However, such a model would provide no explanation for the inferred excess area of Triassic rocks seen in the section. Although these strata include evaporites, diapirism is not considered important here, because the facies is dominated by anhydrites and silts rather than by more-buoyant halite. Furthermore, there is little evidence for diapirism elsewhere in the thrust belt or immediate foreland.

Retaining the integrity of the loose line, Figure 5a can balance only if the top-of-basement level has shortened by 8.5 km. This could be achieved by distributed layer-parallel shortening. However, the model we prefer here involves the reactivation of normal faults, as illustrated in Figure 5b. This model has the advantage of explaining the coincidence between steps along the top of the basement at depth and folds in the overlying cover. Furthermore, the excess area of Triassic rocks relative to the upper parts of the cover can be explained as stratigraphic variations. The Triassic strata represent fills of extensional basins, presumably of Triassic age. Some of the basement faults retain net extensional senses, indicating that thrust reactivation has not completely inverted the earlier displacements.

With our model, the top-of-basement level is broadly at the same level, even though we infer a complex history of normal faulting followed by contraction. At first sight, this may appear to be an implausible coincidence and to favor a detachment style of deformation. However, in examples of exposed basement-cover contacts in orogens (e.g., the western Alps; Gillcrist et al., 1987), reactivation of preexisting faults commonly returns the basement to its preextensional geometry. Indeed, increasing the reverse throws to exceed those of the earlier normal throws on inverted faults raises mechanical problems (Gillcrist et al., 1987). We conclude that the Marche thrust belt can be satisfactorily explained in terms of thick-skinned inversion, with no significant activation of thrust flats.

STRUCTURAL GEOMETRIES IN THE SOUTHERN APENNINES

Having reinterpreted aspects of Northern Apennine structure in terms of a thick-skinned, low-displacement structural style, we now turn our attentions to the Southern Apennines. In contrast to the north, the Southern Apennines of Lucania have relatively large amounts of subsurface data, particularly from wells. These data clearly demonstrate that the thrust belt at outcrop forms a displaced allochthon that has been carried onto by a footwall of foreland strata that are continuous with the Apulian Platform (Carbone et al., 1991). Clearly, these descriptions are inconsistent with the thick-skinned models of structural development presented above for the Northern Apennines.

The internal (western) part of the thrust belt consists of a separate, peritidal carbonate platform unit (the Apenninic or internal platform). Separating the two platforms (Apenninic and Apulian) are the thrusted pelagic basin successions of the Lagonegro units (Triassic to Paleogene in age). Stratigraphically overlying the Lagonegro are paleoforedeep deposits of Oligocene–early Miocene age; the whole complex is also capped by Miocene and Pliocene siliciclastic strata that were deposited in thrust top basins (Scandone, 1972; D'Argenio et al., 1973; Pescatore, 1978; Mostardini and Merlini, 1986; Cello et al., 1990; Marsella et al., 1995).

Unlike the Northern Apennines, where deep wells encounter no significant stratigraphic repetition, the Southern Apennines require substantial thrust detachments. Wells penetrating the Apenninic (internal) platform and Lagonegro units pass through into an additional buried platform succession (Mostardini and Merlini, 1986; Carbone et al., 1991; Marsella et al., 1995; Monaco et al., 1998). This deep platform represents the buried western continuation of the Apulian foreland, although it is cut by faults. The structure of the Apulian units that lie in the footwall to the allochthon has received considerable attention in recent years. This so-called "buried Apulian Belt" (Cello et al., 1989) consists of Mesozoic-Tertiary, 6- to 7-km-thick shallow-water carbonates stratigraphically overlain by upper Messinian and/or Pliocene terrigenous marine deposits. The youngest strata encountered in wells beneath the allochthon are Pliocene in age.

The detachment between the allochthon and the buried Apulian units is marked by a mélange zone several hundred meters thick. This unit consists predominantly of intensely deformed and overpressured deep-water mudstones and siltstones, with minor sandstones and limestones. As a result of overpressuring, gas shows are common throughout this interval (Figure 6). The log response of this unit is very characteristic, being marked by a uniform and moderate gamma-ray and a uniform and low sonic-log response. Lithologic and biostratigraphic information derived from cuttings returned to surface indicates the presence of common, reworked blocks of material derived from the overlying Lagonegro and Apenninic platform units within the mélange zone, particularly toward the top. Biostratigraphic data in this interval are generally poor, but what data there are indicate a Miocene–early Pliocene age for the bulk of the sediments within this interval (Mazzoli et al., 2001b). This unit is thought to represent the major décollement at the base of the allochthon and is interpreted to include Miocene-Pliocene foredeep deposits that were incorporated into the décollement zone as the advancing fold-and-thrust belt overrode its foreland basin. The exotic blocks of Lagonegro and Apenninic platform material in the mélange zone are interpreted to either be olistoliths derived from the exposed fold-and-thrust belt and deposited in the foreland basin, or to be horses that were tectonically accreted to the base of the allochthon by underplating of material derived from the subducted footwall. Both processes commonly are documented as contributing to mélange zone formation (Orange and Underwood, 1995).

Analysis of synorogenic deposits indicates that thrust accretion of the allochthonous units derived from the deformation of the Apenninic platform and the Lagonegro Basin's passive margin successions occurred mainly in early to late Miocene time, whereas deeper thrusting involving the Apulian Platform carbonates occurred mainly in late Pliocene to early Pleistocene time (Cello et al., 1990; Cello and Mazzoli, 1999).

Although the allochthon of the Southern Apennines clearly has been detached from its basement, the degree of thin-skinned thrusting in the buried Apulian Belt remains unclear. For many workers (e.g., Roure et al., 1991; Carbone et al., 1991; Casero et al., 1991), the buried structures show substantial imbrication. For instance, the buried Apulian Belt was interpreted by Casero et al. (1991) (Figure 7) as a duplex that roofed into the floor of the overlying allochthon. The shortening implicit in these models is additional to the slip on the basal detachment for the overlying allochthon and also increases the required slip for this detachment. Consequently, our task here is to examine the structure of the buried Apulian unit and to interpret these deeper structures as being relatively thick-skinned. We are aided in this venture by access to commercial seismic and well data.

Subsurface Data

Geoscientists' understanding of the structure of the Southern Apennines has consistently improved in recent years because of the availability of a large amount

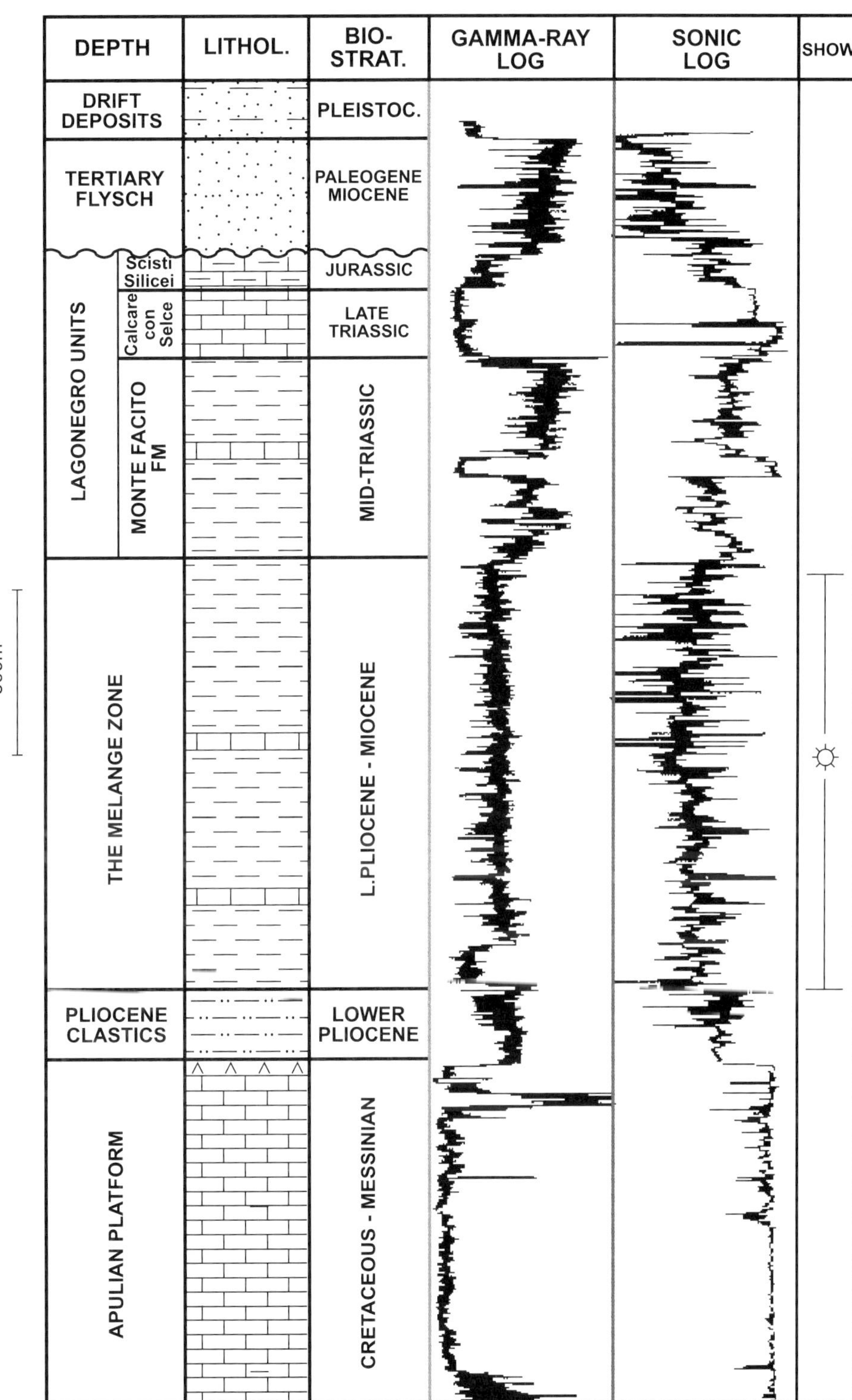

Figure 6. Typical hydrocarbon well in the Lucania area of the Southern Apennines, showing thick mélange zone at the base of the allochthonous Lagonegro and Tertiary flysch units. Biostratigraphic dating was by Patacca and Scandone in a 1999 unpublished Enterprise Oil internal report.

of subsurface data (both 2-D and, locally, 3-D seismic-reflection data and deep well logs) resulting from intensive hydrocarbon exploration, particularly in the Lucania area. The seismic response in this area is highly variable, ranging from moderate to poor, and is controlled by a number of factors commonly observed to limit seismic quality within fold-and-thrust belts: namely, extreme topographic variation, highly variable surface geology, complex structures commonly characterized by steep dips, strong lateral variations in velocity that result in "pull-up" phenomena, and the presence of marked velocity inversions, particularly at the base of the allochthon (general discussion of these issues is given by Tilander and Mitchel, 1995, and Burke and Knapp, 1995; Dell'Aversana et al., 2000, and La Bella et al., 1996, discuss specific seismic acquisition and processing issues in the Southern Apennines). Given this, the only seismic event that can be identified and mapped with a moderately high level of confidence, based on its seismic character alone, is the top Apulian Platform reflector. This event is seismically characterized by a high-continuity, high-amplitude, low-frequency singlet or doublet, arising either directly from the top Apulian carbonates or by interference between reflections arising from the strong positive reflection coefficients at the top of the platform and at the top of the Pliocene clastics (Figures 6, 8, and 9). Events within the allochthon locally show well-characterized seismic responses, but in general, mapping the structure above the top of the Apulian Platform is a highly iterative exercise that involves integration of seismic data with well interpretations and surface geology maps.

Figure 8 shows an interpretation of a typical seismic line from the Lucania area. A primary low-angle

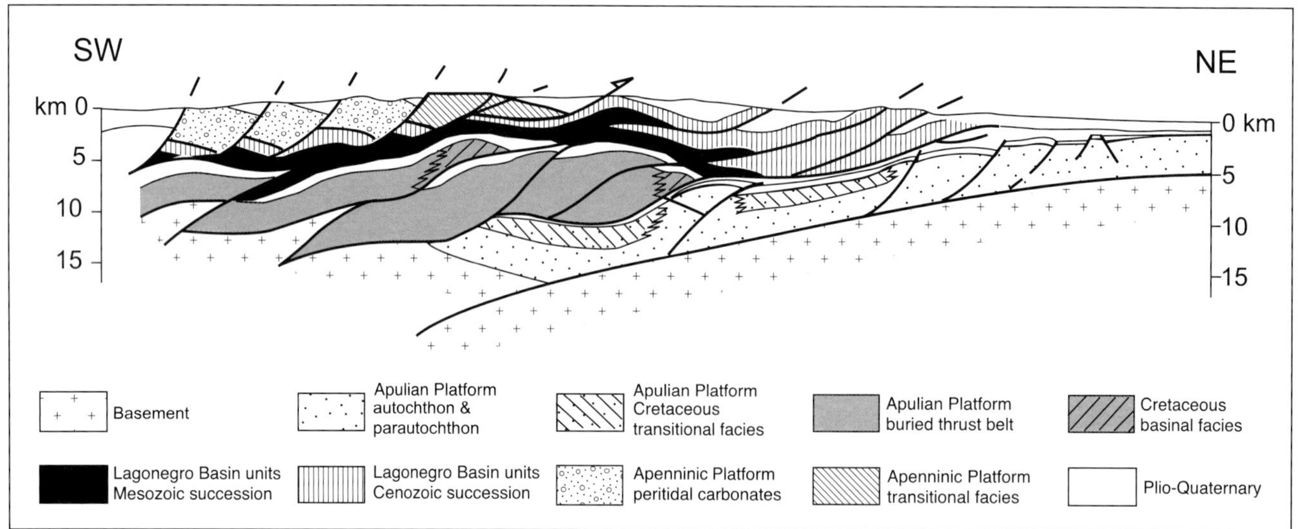

FIGURE 7. Interpreted geologic section across the Southern Apennines (after Casero et al., 1991; located in Figure 1).

detachment separates the Pliocene sequence, which stratigraphically overlies the Apulian carbonates in the footwall, from the allochthonous units in the hanging wall. This detachment also divides the section into two parts, each showing different styles of deformation. The allochthon is characterized by low-angle thrust surfaces, and the Apulian domain is characterized by steeply dipping reverse faults. A breaching thrust offsetting both the allochthon and the Apulian Platform is interpreted in the central part of the section. Within the allochthon, significant folding and imbrication of Mesozoic Lagonegro units is interpreted to occur in the southwestern part of the section and to represent the northern plunge of a structural culmination well exposed a few kilometers to the south (Mazzoli, 1992, 1995; Mazzoli et al., 2001a). To the northeast, the predominantly shaly, Upper Cretaceous–Paleogene part of the Lagonegro succession, unconformably overlain by Miocene synorogenic deposits (Gorgoglione Formation), is involved in imbricate sheets detached from their Triassic–Lower Cretaceous substratum.

The base of the Apulian Platform is also visible as a strong reflection in some areas and can sometimes be used to guide the interpretation of the top platform (Figure 9). This deep seismic event can be interpreted by reference to deep wells drilled in the Apulian foreland. For instance, the Puglia 1 well drilled 6112 m of shallow-water carbonates and evaporites before penetrating an unconformity underlain by approximately 1000 m of continental to transitional argillites and sandstones (Figure 10). The strong reflector often observed on seismic profiles below the top Apulian horizon both in the foredeep (Roure et al., 1991) and in the thrust belt is generated by the significant acoustic impedance contrast between the Upper Triassic dolomites and Lower Triassic–Upper Permian clastic deposits (Figure 10). This interpreted base Apulian Platform horizon is also clearly offset by thrusts on seismic data (Figure 9). Intraplatform events are sometimes visible on seismic sections. However, the data quality is generally too poor to allow a detailed interpretation of the internal structures of the carbonates.

Cross Section

The interpreted structure of the Southern Apennines in the Lucania area is shown in the cross section of Figure 11. We have not attempted to restore the structures in the allochthon (this has been done elsewhere by Mazzoli et al., 2001b). The majority of these formed early and are sealed by Miocene strata. Available stratigraphic data indicate that much of the shortening in the main allochthon predates its emplacement onto the Apulian foreland (Cello and Mazzoli, 1999). The Pliocene-Pleistocene-age Sant'Arcangelo Basin is only weakly deformed and seals thrust structures. Furthermore, the Serravallian-Tortonian-age Gorgoglione flysch strata effectively seal thrust structures within the Lagonegro units (Boiano, 1997; Pescatore et al., 1999). Our focus here is on the structure of the Apulian units at depth and on displacements on the basal detachment of the allochthon.

Using the ramp-dominated structural style for deformation in the Apulian Platform, we have performed a simple line-length restoration. Given the platform facies of the Apulian carbonates, we believe that this approach is valid, there being little propensity for these units to accommodate significant distortional strains. Our restoration yields a shortening of approximately 14 km (Figure 11). Well data allied to seismic data suggest that Pliocene strata that unconformably overlie

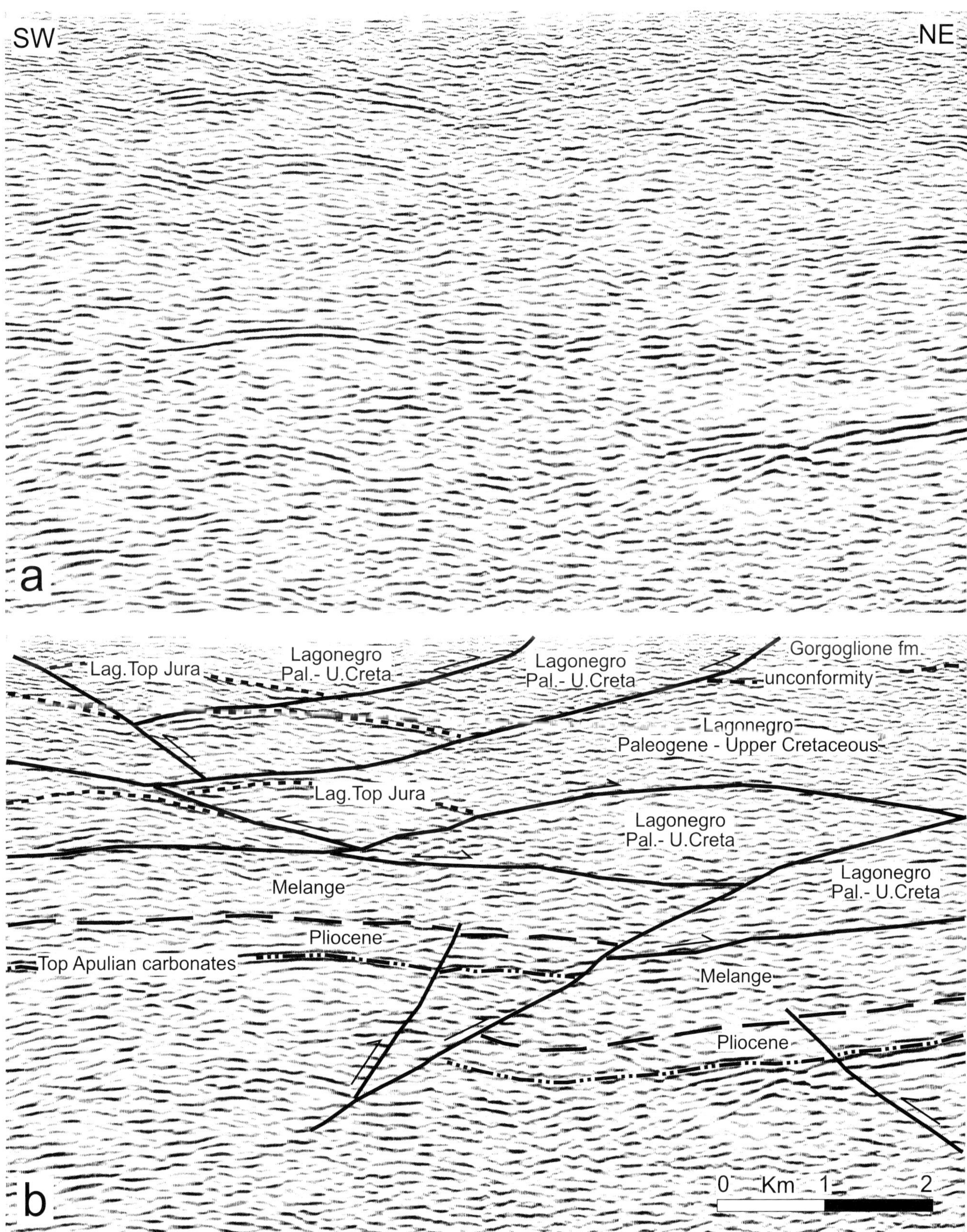

Figure 8. (a) Stack seismic line across central Lucania (precise location and depth in two-way time are confidential). (b) Interpretation, showing allochthonous thrust sheets overlying the Apulian Platform carbonates.

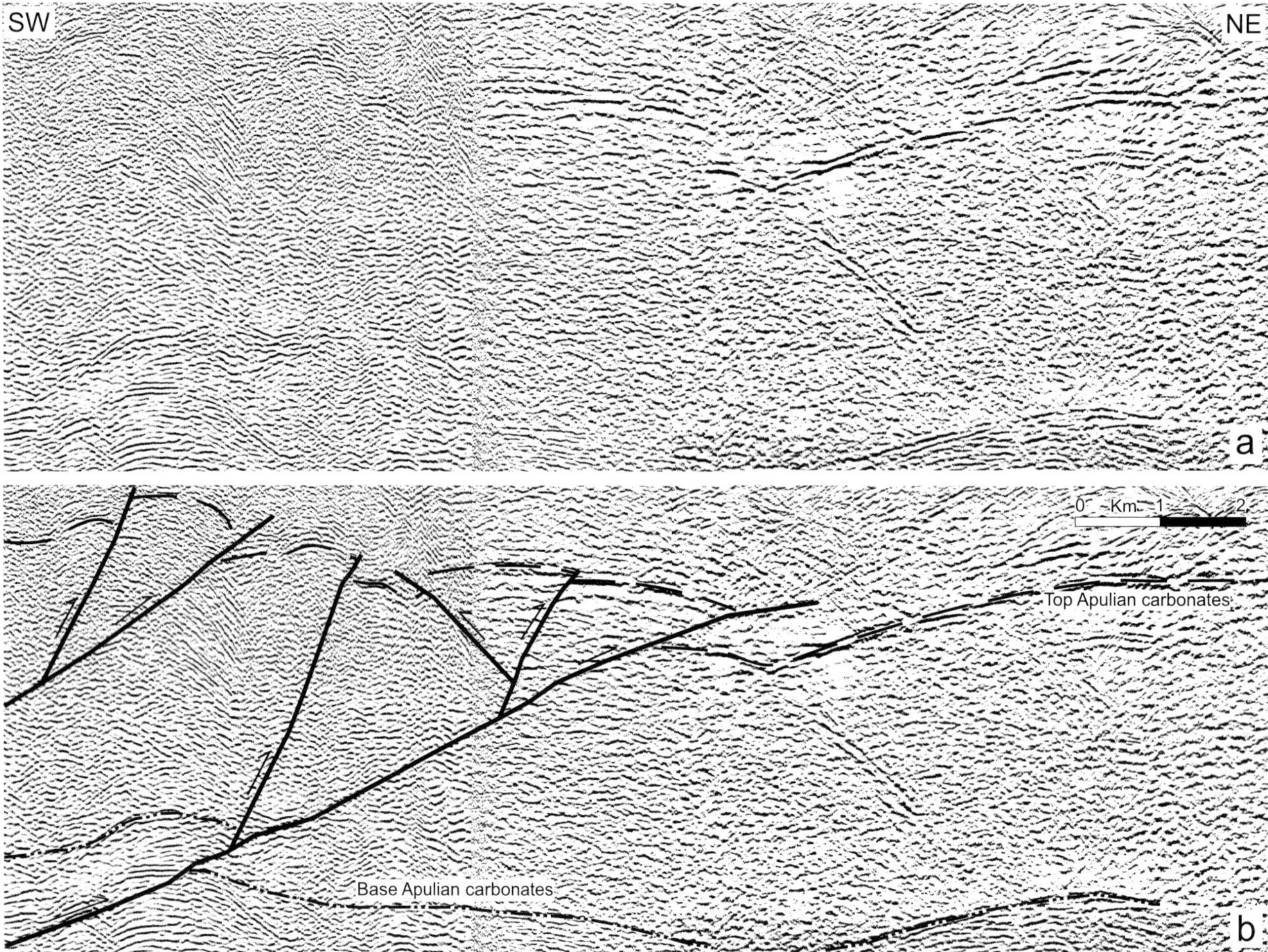

FIGURE 9. (a) Stack seismic line across southern Lucania (precise location and depth in two-way time are confidential). (b) Interpretation, showing reflectors ascribed to "top" and "base" Apulian Platform carbonates.

the Apulian carbonates are found, at depth, over a substantial distance to the west of the foreland. The restored length of Apulian carbonates that include a Pliocene cover at depth beneath the allochthon is 57 km. This value is, therefore, the minimum displacement on the base of the allochthon since earliest Pliocene times and includes the shortening within Apulia. The time-averaged displacement rate was therefore at least 12 mm/yr.

The thrust belt described above is dissected in the southwestern part of the section by steeply dipping, northwest–southeast-trending normal faults. Northeast of Tardiano (Figure 11), an array of young faults shows segmented geometries and cumulative throws greater than 1500 m, as defined by field geology and geomorphology. These link into north–south-trending, apparently left-lateral strike-slip faults. All of these faults, some of which are also seismically active (Valensise et al., 1993), are kinematically compatible with the northeast–southwest-oriented maximum extension direction associated with the middle Pleistocene–Holocene tectonic regime (Cello et al., 1982; Hyppolyte et al., 1994). In the more external part of the thrust belt, the Apulian carbonates, which dip gently to the southwest and also are offset by southwest-dipping normal faults, are stratigraphically overlain by a thick Pliocene foredeep succession.

STRUCTURAL GEOMETRIES IN THE CENTRAL APENNINES

The Central Apennines of Abruzzi (Figure 1) are characterized by various Tethyan continental-margin sedimentary successions (chiefly carbonate platforms, but with bypass margins and slopes locally) that range in thickness from 1000 to 5000 m (Damiani et al., 1992). These units are now telescoped into several thrust sheets (Accordi, 1966; Parotto and Praturlon, 1975; Bigi et al., 1990; Cosentino and Parotto, 1992). These structures are strongly influenced by the preexisting basin architecture and associated depositional environments that developed during the Mesozoic-Cenozoic along

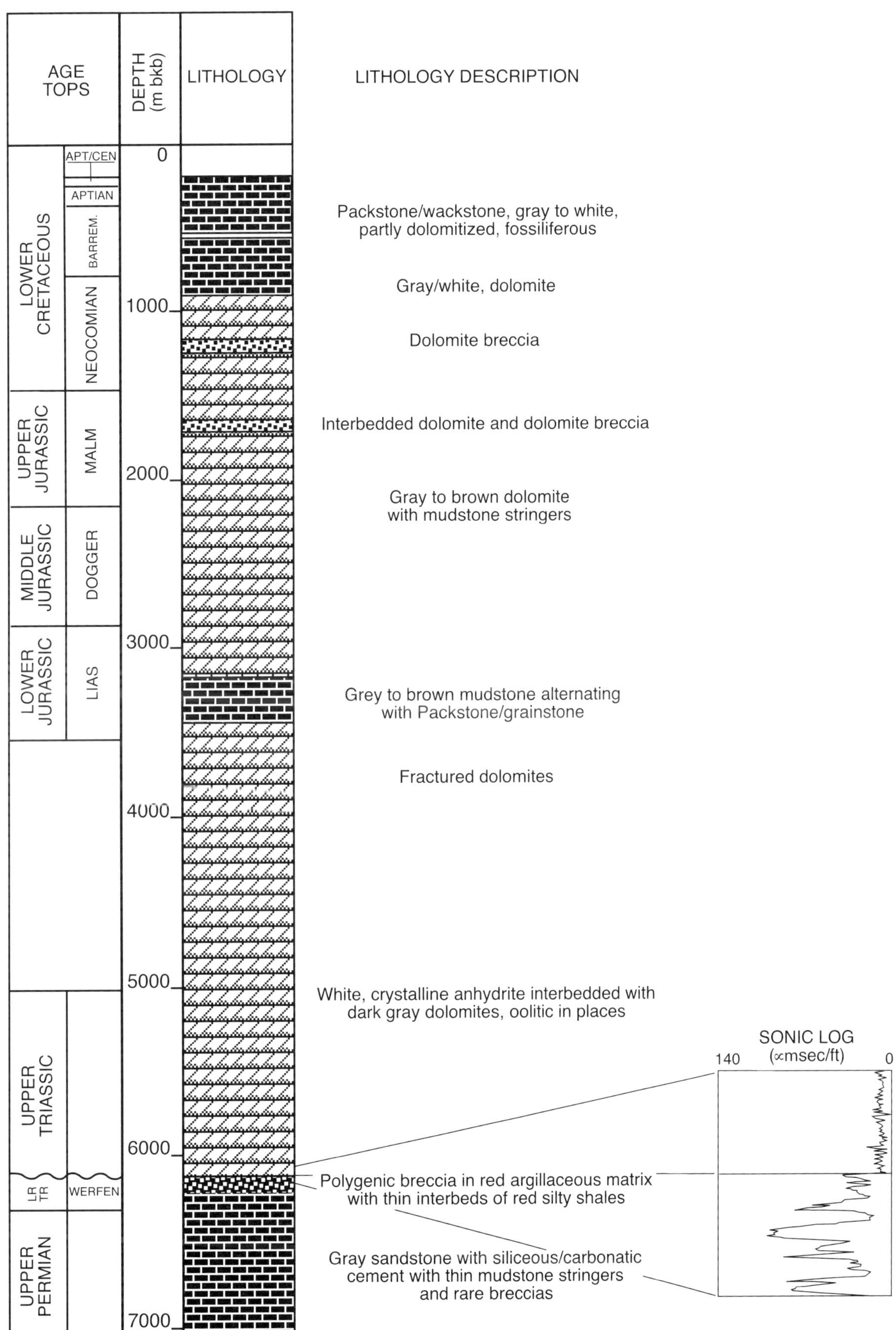

FIGURE 10. Simplified stratigraphic log of the Puglia 1 well drilled in the Apulian foreland. Inset shows detail of sonic log around the base of the shallow-water carbonate succession.

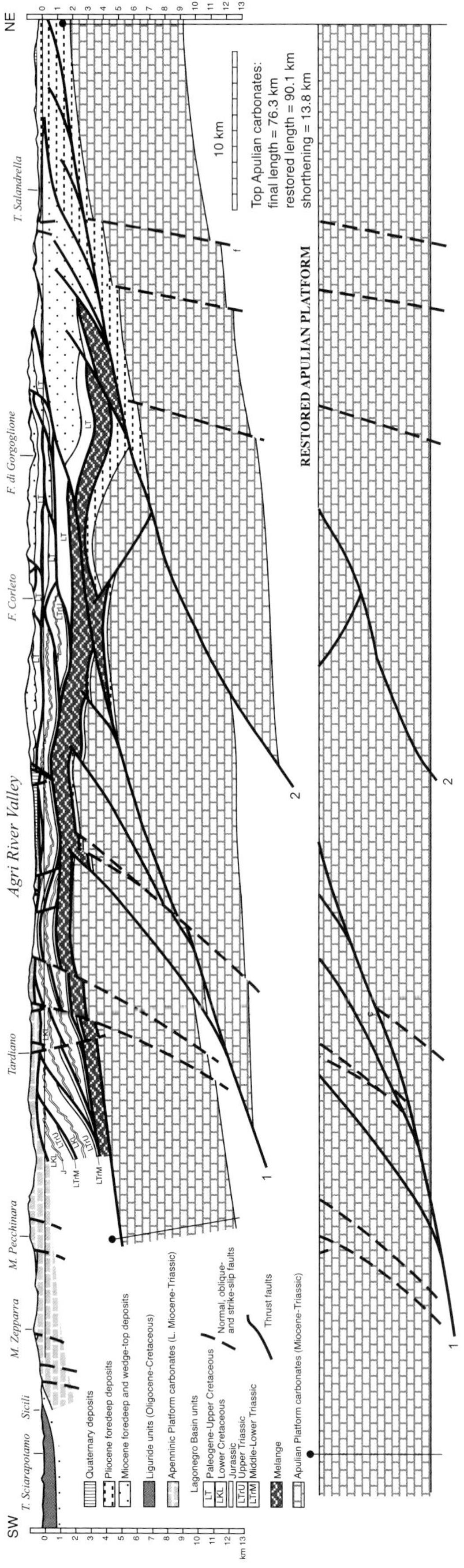

Figure 11. Geologic section across the Southern Apennines in the Lucania area (modified after Mazzoli et al., 2000; located in Figure 1). Restored Apulian Platform carbonates shown.

the southern continental margin of the Neotethys (Parotto, 1980). The region is particularly important, because it contains the most dramatic lateral change in tectonostratigraphic units in the Apennines. The Mesozoic rocks of the Abruzzi region are characterized by platform facies. However, to the southeast, the pre-orogenic strata are represented by the Sannio-Molise units, a succession of Late Cretaceous–Paleogene pelagic sediments that have been detached from their older substrate of Mesozoic cover and basement (Corrado et al., 1997). These units represent a good regional seal for one of the most productive reservoirs in Italy, which, as in the Southern Apennines, is represented by the buried and deformed Apulian carbonate platform (Zappaterra, 1990; Casero et al., 1991). Furthermore, various lithostratigraphic units in both the Apulian and Molise successions are considered to be potential source rocks (Zappaterra, 1990). West of the Sannio-Molise pelagic sequence is the platform unit of the Matese Mountains (part of the internal, Apenninic platform). It is worth noting that, according to some workers, the Sannio-Molise successions originally were transported from the southwest of the internal platform (Matese), implying multiple detachment and net horizontal displacements greater than 200 km. However, Corrado et al. (1998) have contested this view. Their vitrinite reflectance and illite crystallinity data demonstrated that the internal platform (Matese) has not been deeply buried by allochthonous thrust sheets. Therefore, the Sannio-Molise units must restore to the east of Matese, in broadly stratigraphic continuity. This view is supported by the Mesozoic geology of the eastern side of the Matese. From southwest to northeast, these strata show platform carbonates and associated bypass margin facies (Matese Mountains), slope facies (Montagnola di Frosolone), and pelagic basin facies (the Molise-Sannio units).

The structure of the southern Central Apennines in the northern Campania-Molise area is shown in cross section in Figure 12; it traverses the most promising hydrocarbon exploration area in the region. To construct the cross section, we integrated published surface geologic data (Servizio Geologico Nazionale, 1963, 1968, 1971; Bigi et al., 1990; Giordano et al., 1995; Scrocca et al., 1995; De Corso et al., 1998) and new geologic maps with available published (derived from the Italian Ministry for Industry database and Ippolito et al., 1973) and unpublished subsurface data (Figure 13). Contractional structures within the outcropping Apenninic carbonates consist of north- and northeast-vergent thrust ramps in the Matese Mountains and in the Montagnola di Frosolone area (Figure 12). An east–west-trending fault-propagation fold, showing a maximum shortening of a few hundred meters, occurs in the Matese Mountains (Scrocca et al., 1995; De Corso et al., 1998). The major frontal ramp to the Apenninic units is buried in front of the Montagnola di Frosolone, as shown in the detailed

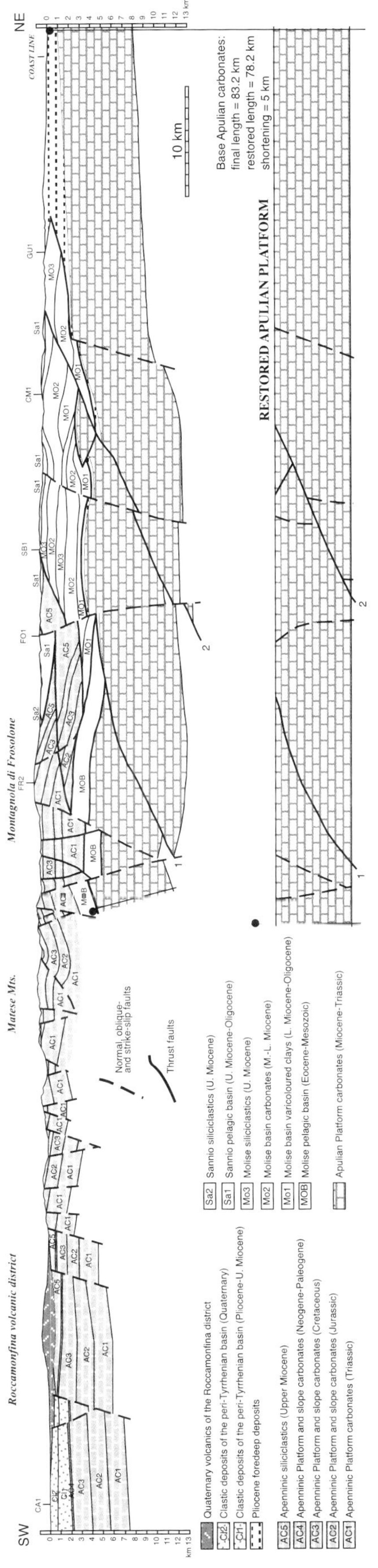

FIGURE 12. Geologic section across the Central Apennines in the northern Campania-Molise area (modified after Mazzoli et al., 2000; located in Figure 1). Restored Apulian Platform carbonates shown.

cross section of Figure 14 (also based on seismic interpretation). The frontal ramp is a triangular wedge made of Apenninic carbonates overthrusting Mesozoic to Eocene pelagic basin and slope deposits (Figure 14). The leading edge of the Apenninic carbonates is in turn tectonically overlain by a detached Oligocene–lower Miocene sequence (Sannio units). We interpret this as having been emplaced by a relatively low-displacement back thrust.

The Sannio-Molise units, detached at the level of the Oligocene–lower Miocene Argille Varicolori, crop out along the transect to the northeast of the Matese Mountains and Montagnola di Frosolone area, up to the front of the thrust belt near the present-day coastline. They are organized in a series of generally north- and northeast-vergent, thin thrust sheets overlying the buried Apulian carbonates. Locally, for example around the San Biase village area, transpressional strike-slip faults dissect the preexisting thrust stack, thereby causing a local uplift of the preexisting structures.

The Apulian carbonates, buried below the Sannio-Molise units, are deformed in at least two main ramp anticlines characterized by a total shortening of about 5 km. The restored section for Apulia shows the original distance of the platform, with its veneer of Pliocene strata that have been overthrust by the allochton of Sannio-Molise units. About 47 km of this restored section now lies in the footwall to the allochthon, making this the minimum displacement after earliest Pliocene time. This implies a time-averaged thrusting rate greater than 10 mm/yr. Note that there is little significance in this value being lower than that obtained for Lucania (Figure 11), because both are minimum values.

Younger (middle Pleistocene–Holocene), southwest- and northeast-dipping normal faults dissect the carbonates in the Montagnola di Frosolone area. East of the most external ramp anticline, the Apulian carbonates, gently dipping to the southwest, are also offset by southwest-dipping normal faults. In this external area, they are stratigraphically overlain by progressively younger and thicker clastics that fill the Pliocene foredeep.

DISCUSSION

All geologic cross sections illustrate structural interpretations that reflect combinations of data and some understanding of structural style. This dependence on structural models was recognized by early proponents of thin-skinned thrust system geometries (Dahlstrom, 1969; Elliott and Johnson, 1980). In the Apennines, the applicability of thrust geometries that involve substantial basement-cover detachments, such as those of Bally et al. (1986), have been highly influential. Our purpose has been to investigate how the same surface-geology data might be reinterpreted using

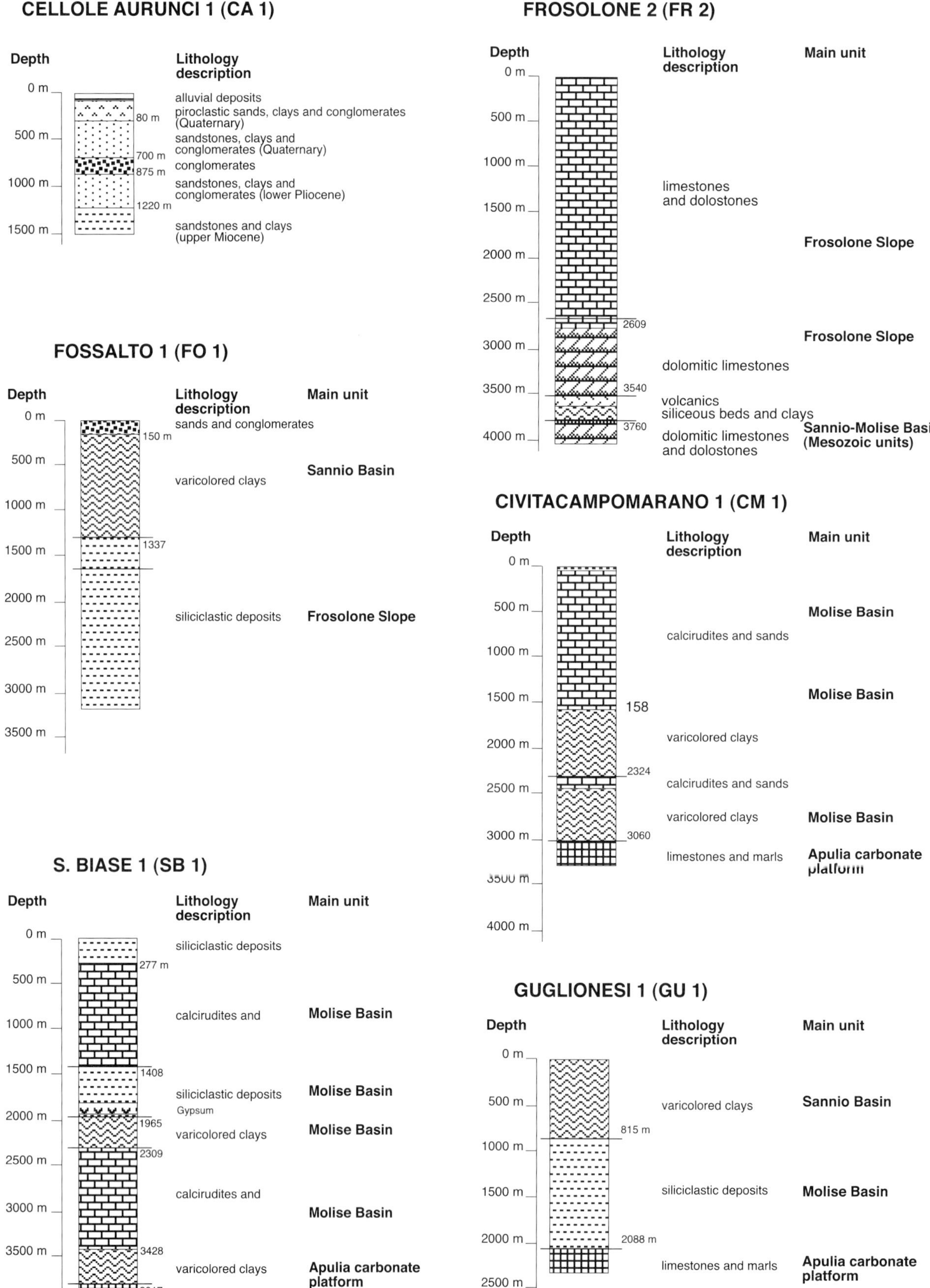

FIGURE 13. Simplified stratigrapic logs and structural interpretation of deep wells projected along the cross section of Figure 12.

an alternative structural model to build very different cross sections. We have been aided by new subsurface data. The different sectors of the Apennines in our case study offer different examinations of our working hypothesis of inversion tectonics.

In the Northern Apennines of the Marche, sections can be balanced with basement involvement and still honor available seismic, well, and outcrop data. In our study, shortening in the basement and cover can balance across individual structures without significant detachment by the reactivation of preexisting basement faults. This solution takes into account all available data. In particular, it accounts for (1) the kinematic link, apparent from the interpretation of available seismic data, between basement faults and thrust ramps in the sedimentary cover; (2) the much smaller net displacement of the top basement horizon with respect to thrust displacements in the sedimentary cover; and (3) the occurrence, below the major anticlines, of quite thick stratigraphic successions that can be best interpreted as tectonically extruded Triassic basin fills.

Thin-skinned thrusting models stand up much better in the Southern Apennines. The Lagonegro thrust complex is clearly detached from its original, pre-Mesozoic basement. This style of detachment also characterizes the thrust belt in the Central Apennines, as far north as the Molise district. There must be a major change in structural geometry, and perhaps in displacement, through the Abruzzi district that lies immediately north of Molise. These aspects are currently the topic of ongoing research.

Allochthonous structural models for the Lagonegro and Molise sheets go back to the work of Cello et al. (1989, 1990). Other workers have proposed that these allochthons themselves are made up of thin thrust sheets that showed vast displacements relative to each other (Scandone, 1972; Mostardini and Merlini, 1986; Carbone et al., 1991). Although the well data certainly require displacements greater than 45 km (since the early Pliocene) on the base of the allochthon, the presence of other detachments is far less secure. There is stratigraphic continuity between the internal Apenninic platform and flanking deposits that characterize the west-to-east transition in thrust sheet content in the Matese-Molise district. This argues that the allochthon is broadly intact and that there has not been a

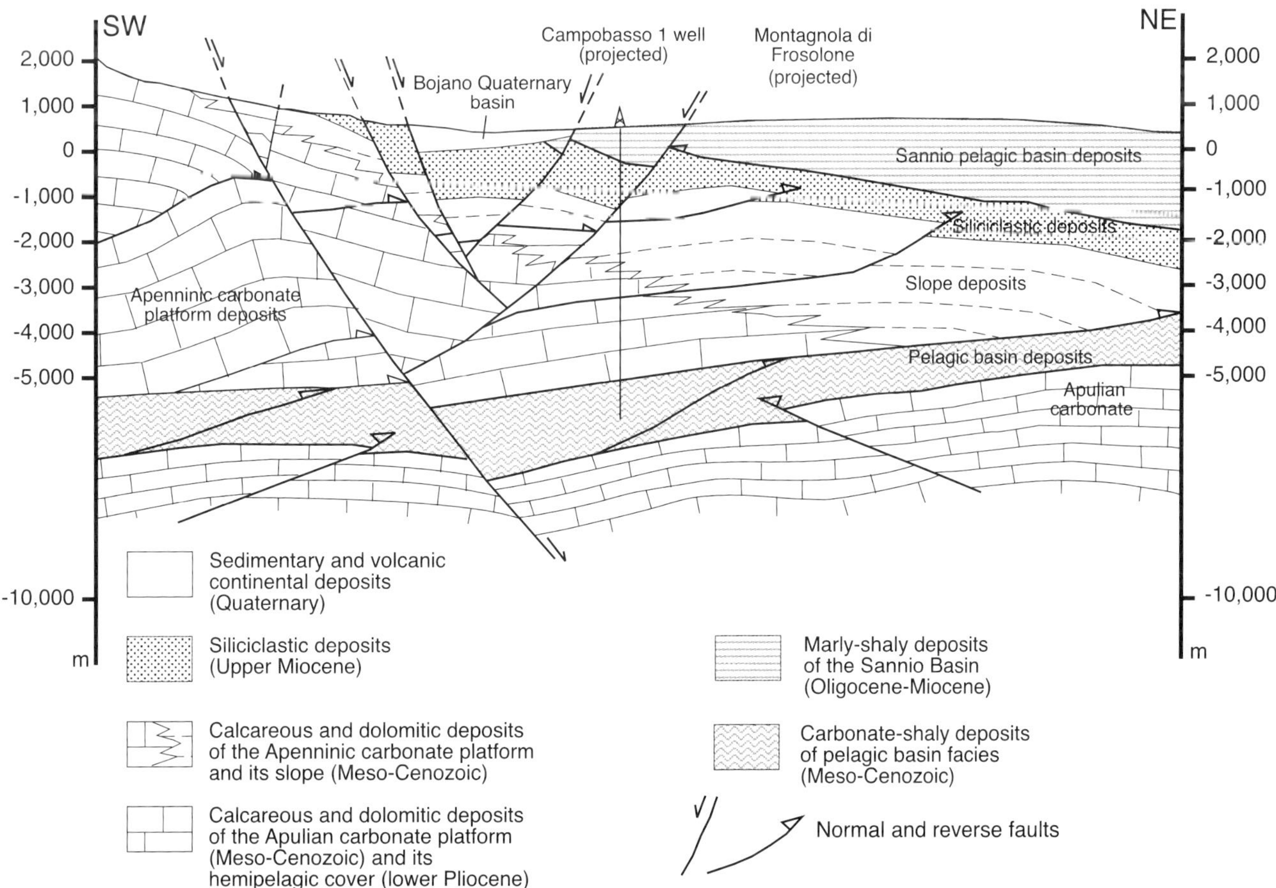

FIGURE 14. Detailed geologic section across the Matese Mountains–Campobasso area (located in Figure 1), based on field, seismic, and well data.

tectonic juxtaposition of originally disparate tectonostratigraphic units. In Lucania we cannot demonstrate the stratigraphic links so convincingly between the Apenninic platform and the Lagonegro basinal units. It is likely that the various synorogenic deposits (Oligocene to Miocene in age) will hold this key, provided the depositional facies are correlated. Such correlations are being established for some of these units that clearly tie together different thrust sheets as being part of the same small basin (e.g., the Gorgoglione Formation; Boiano, 1997).

In the Southern Apennines, evidence for faulting of the base Apulian horizon could be interpreted in two rather different ways. (1) A major detachment level may be present, located in the clastic Upper Permian–Lower Triassic deposits underlying the carbonates. (2) Alternatively, thick-skinned (crustal-ramp dominated) structures may occur in the deeper (and younger) part of the thrust belt underlying the allochthonous units. The geometry of the thrust system within the Apulian carbonates, together with the results of our structural modeling and balancing (Figure 11), favor the latter hypothesis—that of relatively low-displacement, deeply rooted thrust ramps. We should note that the buried Apulian Platform carbonates involved in the thick-skinned thrust belt do not form a classical duplex structure in the sense of Boyer and Elliott (1982). The overthrust detachment does not form a simple roof thrust. Rather, the deep-rooted thrusts in the Apulian Platform tend to cut upward, across the earlier detachment, causing widespread breaching (in the sense of Butler, 1987). Breaching thrusts can have different geometries. In some cases they are simply emergent, having cut up through the entire allochthon. Elsewhere, the breaching thrusts may transfer displacement onto zones of diffuse ductile strain. Alternatively, they may branch onto and reactivate higher thrusts in the allochthonous unit.

It is not known if the Apulian thrusts link downward onto a discrete floor thrust in the middle or lower crust. It is plausible, particularly given the low net displacements in this system, that the thrusts pass downward into a zone of more penetrative strain. These deep levels are not imaged on the available commercial seismic-reflection profiles.

Based on the foregoing discussion, we propose a "composite thin-thick-skinned" model for the structural evolution of the studied sectors of the Italian Apennines. It consists of a major allochthonous sheet of Mesozoic-Paleogene shallow-water carbonates and pelagic basin units, and of Miocene foredeep clastics thrust onto the Apulian Platform carbonates and underlying basement. The latter, deeper portions of the thrust belt show limited amounts of subsequent shortening on thick-skinned thrusts (5 km and 14 km for the Central and Southern Apennine examples, respectively) that developed mainly in late Pliocene to early Pleistocene time (Cello and Mazzoli, 1999). Structural style has varied over time. It seems likely that detachment-dominated, thin-skinned thrusting of passive-margin successions involved basement-cover decoupling and subduction of the most attenuated part of the Adriatic continental-margin lithosphere. This initiated during the Miocene. A transition to thick-skinned thrusting occurred later, with the involvement of the thicker lithosphere underlying the Apulian carbonate platform. This happened during the late Pliocene.

Hydrocarbon prospectivity in the Southern Apennines generally requires the presence of Apulian Platform carbonates to provide a reservoir beneath the allochthonous thrust pile of Lagonegro rocks (Casero et al., 1991). The mélange zone along the detachment forms an effective seal. Thus, the structure of the top Apulian surface is of critical importance in hydrocarbon potential. Thick-skinned, ramp-dominated structural models, such as the ones we have adopted here, make significantly different predictions for reservoir structure than do the thin-skinned models used elsewhere (Casero et al., 1991). Note, however, that to understand the complete structure, one must understand the later faults that dissect the whole chain and are associated with modern opening of the southern Tyrrhenian Sea.

An important distinction for us is that thick-skinned, ramp-dominated inversion models may be applied to explain the entire structure of the thrust belt in the eastern Marche area. These models predict that there is no repetition of Apulian units at depth. Therefore, hydrocarbon plays developed for the Southern Apennines are likely to have little relevance for the Marche.

CONCLUSIONS

The following conclusions can be drawn. In the Marche area, simple thick-skinned inversion models may be applied that satisfy available geologic data. They are more in keeping with modern seismic data in the area than some existing thin-skinned models are. Our sections require just 8.5 km of shortening (compared with 50 km for the equivalent transect using detachment-dominated, thin-skinned thrusting, per Bally et al., 1986). The rate of shortening for our model is just 1.5 mm/yr. The inversion model suggests that there is no significant repetition of stratigraphic units at depth—a prediction that may influence the choice of play when one is evaluating hydrocarbon potential.

In contrast to findings for the Northern Apennines, well data for the Southern and southern Central Apennines show that these regions contain an important allochthonous thrust sheet that overthrusts the Apulian carbonate platform. Whereas thick-skinned thrusting offers no explanation for the allochthon, it may be applied to predict the structure of the Apulian units in its footwall. Shortening on these buried Apulian structures is estimated to be 5 km for our Molise transect and 14 km for Lucania. These displacements pass onto and locally reimbricate the allochthon. Estimates for bulk displacements across these southern transects since Pliocene time may exceed 57 km (for Lucania), thereby giving a minimum time-averaged thrusting rate of 12 mm/yr.

The structural styles of the Southern (Lucania, Molise) Apennines differ from those of the Northern Apennines (Marche). Consequently, plays developed for hydrocarbon exploration in the Southern Apennines may be highly inappropriate for the northern parts of the chain. Future work is needed to establish the kinematic links between the allochthonous, thin-skinned thrusting styles and the inversion geometries, particularly along strike.

ACKNOWLEDGMENTS

We thank Enterprise Oil, Eni, Edison Gas, Fina and their partners in the Southern Apennines for granting permission to publish this paper.

REFERENCES CITED

Accordi, B., 1966, La componente traslativa nella tettonica dell'Appennino laziale abruzzese: Geologica Romana, v. 5, p. 355–406.

Anelli, L., M. Gorza, M. Pieri, and M. Riva, 1994, Subsurface well data in the Northern Apennines: Memorie della Società Geologica Italiana, v. 48, p. 461–471.

Argnani, A., and F. Gamberi, 1997, Stili strutturali al fronte delle catena appenninica nell'Adriatico centro-settentrionale: Studi Geologici Camerti, v. speciale 1995/1, p. 19–28.

Bally, A. W., L. Burbi, C. Cooper, and R. Ghelardoni, 1986, Balanced sections and seismic reflection profiles across the Central Apennines: Memorie della Società Geologica Italiana, v. 35, p. 257–310.

Barchi, R., G. Minelli, and G. Pialli, 1998, The CROP 03 profile: a synthesis of results on deep structures of the Northern Apennines: Memorie della Società Geologica Italiana, v. 52, p. 383–400.

Bigi, G., D. Cosentino, M. Parotto, R. Sartori, and P. Scandone, 1990, Structural Model of Italy 1:500,000: Quaderni de "La Ricerca Scientifica", C.N.R., v. 114, no. 3.

Boiano, U., 1997, Anatomy of a siliciclastic turbidite basin: the Gorgoglione Flysch, Upper Miocene, southern Italy: physical stratigraphy, sedimentology and sequence-stratigraphic framework: Sedimentary Geology, v. 107, p. 231–262.

Boyer, S. E., and D. Elliott, 1982, Thrust systems: AAPG Bulletin, v. 66, p. 1196–1230.

Burke, N. L., and S. Knapp, 1995, Imaging advances in the Colombian foothills: The Leading Edge, v. 14, p. 677–679.

Butler, R. W. H, 1987, Thrust sequences: Journal of the Geological Society of London, v. 144, p. 619–634.

Butler, R. W. H., 1992, Evolution of Alpine fold-thrust complexes: a linked kinematic approach, *in* S. Mitra and G. Fisher, eds., Structural geology of fold and thrust belts: Baltimore, Johns Hopkins University Press, p. 29–44.

Carbone, S., S. Catalano, S. Lazzari, F. Lentini, and C. Monaco, 1991, Presentazione della carta geologica del Bacino del Fiume Agri (Basilicata): Memorie della Società Geologica Italiana, v. 47, p. 129–143.

Carmignani., L., F. A. Decandia, P. L. Fantozzi, A. Lazzarotto, D. Liotta, and M. Meccheri, 1994, Tertiary extensional tectonics in Tuscany (Northern Apennines, Italy): Tectonophysics, v. 238, p. 295–315.

Casero, P., F. Roure, and R. Vially, 1991, Tectonic framework and petroleum potential of the Southern Apennines, *in* A. M. Spencer, ed., Generation, accumulation, and production of Europe's hydrocarbons: Special Publication of the European Association of Petroleum Geoscientists, v. 1, p. 381–387.

Cassano, E., L. Anelli, V. Cappelli, and P. La Torre, 1998, Interpretation of Northern-Apennine magnetic and gravity data in relation to profile CROP 03: Memorie della Societa Geologica Italiana, v. 52, p. 413–425.

Cello, G., and S. Mazzoli, 1999, Apennine tectonics in southern Italy: a review: Journal of Geodynamics, v. 27, p. 191–211.

Cello, G., I. Guerra, L. Tortorici, E. Turco, and R. Scarpa, 1982, Geometry of the neotectonic stress field in southern Italy: geological and seismological evidence: Journal of Structural Geology, v. 4, p. 385–393.

Cello, G., N. Martini, W. Paltrinieri, and L. Tortorici,1989, Structural styles in the frontal zones of the Southern Apennines, Italy: an example from the Molise district: Tectonics, v. 8, p. 753–768.

Cello, G., F. Lentini, and L. Tortorici, 1990, La struttura del settore calabro-lucano e suo significato nel quadro dell'evoluzione tettonica del sistema a thrust sudappenninico: Studi Geologici Camerti, v. speciale 1990/1, p. 27–34.

Channell, J. E. T., B. D'Argenio, and F. Horvath, 1979, Adria, the African promontory, in Mesozoic Mediterranean paleogeography: Earth Sciences Reviews, v. 15, p. 213–292.

Ciarapica, G., S. Cirilli, R. Martini, M. Panzanelli, R. Fratoni, G. Salvini-Bonnard, and L. Zaninetti, 1990, The Monte Facito Formation (Southern Apennines): Bollettino della Società Geologica Italiana, v. 59, p. 117–126.

Corrado, S., D. Di Bucci, G. Naso, and R. W. H. Butler, 1997, Thrusting and strike-slip tectonics in the Alto Molise region (Italy): implications for the Neogene–Quaternary Evolution of the Central Apennine Orogenic System: Journal of the Geological Society of London, v. 154, p. 679–688.

Corrado, S., D. Di Bucci, G. Naso, G. Giampaolo, and T. Adatte, 1998, Application of organic matter and clay minerals studies to the tectonic history of the Molise area, Central Apennines: Tectonophysics, v. 285, p. 167–181.

Cosentino, D., and M. Parotto, 1992, La struttura a falde della Sabina (Appennino Centrale): Studi Geol. Camerti, v. speciale 1991/2, p. 381–387.

Coward, M. P., 1983, Thrust tectonics, thin skinned or thick skinned, and the continuation of thrusts deep in the crust: Journal of Structural Geology, v. 5, p. 113–123.

Coward, M. P., 1994, Inversion tectonics, in: P.L. Hancock, ed., Continental Deformation: Oxford, Pergamon Press, p. 289–304.

Coward, M. P., M. De Donatis, S. Mazzoli, W. Paltrinieri, and F. C. Wezel, 1999, Frontal part of the Northern Apennines fold and thrust belt in the Romagna-Marche area (Italy): Shallow and deep structural styles: Tectonics, v. 18, p. 559–574.

Dahlstrom, C. D. A., 1969, Balanced cross-sections: Canadian Journal of Earth Sciences, v. 6, p. 743–757.

Damiani, A.V., M. Chiocchini, R. Colacicchi, G. Mariotti, M. Parotto, L. Passeri, and A. Praturlon, 1992, Elementi litostratigrafici per una sintesi delle facies carbonatiche meso-cenozoiche dell'Appennino Centrale: Studi Geologici Camerti, v. speciale 1991/2, p. 187–213.

D'Argenio, B., T. Pescatore, and P. Scandone, 1973, Schema geologico dell'Appennino meridionale: Atti dell'Accademia Nazionale dei Lincei, v. 183, p. 49–72.

Decandia, F. A., A. Lazzarotto, D. Liotta, L. Cernobori, and R. Nicolich, 1998, The CROP 03 traverse: insights on post-collisional evolution of Northern Apennines: Memorie della Societa Geologica Italiana, v. 52, p. 427–439.

De Corso, S., D. Scrocca, and M. Tozzi, 1998, Geologia dell'anticlinale del Matese e implicazioni per la tettonica dell'Appennino molisano: Bollettino della Società Geologica Italiana, v. 117, p. 419–441.

De Donatis, M., C. Invernizzi, A. Landuzzi, S. Mazzoli, and M. Potetti, 1998, CROP 03: Structure of the Montecalvo in Foglia-Adriatic Sea segment: Memorie della Società Geologica Italiana, v. 52, p. 617–630.

Dell'Aversana, P., E. Ceragioli, S. Morandi, and A. Zollo, 2000, A simultaneous acquisition test of high-density "global offset" seismic in complex geological settings: First Break, v. 18, p. 87–96.

Doglioni, C., F. Mongelli, and G. Pialli, 1998, Boudinage of the Alpine belt in the Apenninic back-arc: Tectonophysics, v. 52, p. 457–468.

Elliott, D., and M. R. W. Johnson, 1980, Structural evolution in the northern part of the Moine Thrust Zone, NW Scotland: Transactions of the Royal Society of Edinburgh: Earth Science, v. 71, p. 69–96.

Geiser, P. A., 1988, Mechanisms of thrust propagation: some examples and implications for the analysis of overthrust terranes: Journal of Structural Geology, v. 10, p. 829–845.

Gillcrist, R., M. P. Coward, and J.-L. Mugnier, 1987, Structural inversion and its controls: examples from the Alpine foreland and the French Alps: Geodinamica Acta, v. 1, p. 5–34.

Giordano, G., G. Naso and A. Trigari, 1995, Evoluzione tettonica di un settore particolare del margine tirrenico: l'area al confine tra Lazio e Campania. Prime considerazioni: Studi Geologici Camerti, v. speciale 1995/2, p. 269–278.

Hill, K. C., and A. B. Hayward, 1988, Structural constraints on the Tertiary plate tectonic evolution of Italy: Marine and Petroleum Geology, v. 5, p. 2–16.

Hyppolyte, J.-C., J. Angelier, and F. Roure, 1994, A major geodynamic change revealed by Quaternary stress patterns in the Southern Apennines (Italy): Tectonophysics, v. 230, p. 199–210.

Ippolito, F., F. Ortolani, and M. Russo, 1973, Struttura marginale tirrenica dell'Appennino campano: reinterpretazione dei dati di antiche ricerche di idrocarburi: Memorie della Società Geologica Italiana, v. 12, p. 227–250.

La Bella, G., L. Bertelli, and L. Savini, 1996, Monte Alpi 3d, a challenging 3D survey in the Appennine Range, Southern Italy: First Break, v. 14, p 285–294.

Lavecchia, G., G. Minelli, G. Pialli, G. Biella, P. Conversini, M. Demartin, A. Lozej, M. Maistrello, S. Scarascia, and I. Tabacco, 1984, Primi risutati del profilo sismico a rifrazione Perugia–Frontone: Bollettino della Società Geologica Italiana, v. 103, p. 447–466.

Lavecchia, G., 1988, The Tyrrhenian-Apennine system: structural setting and seismotectogenesis: Tectonophysics, v. 147, p. 263–296.

Marsella, E., G. Pappone, B. D'Argenio, G. Cippitelli, and A. W. Bally, 1992, L'origine interna dei terreni lagonegresi e l'assetto tettonico dell'Appenino meridionale: Rendiconti Accademia Scienze fisiche e matematiche della Società Nazionale di Scienze Lettere e Arti in Napoli, v. 59, p. 73–101.

Marsella, E., A. W. Bally, G. Cippitelli, B. D'Argenio, and G. Pappone, 1995, Tectonic history of the Lagonegro Domain and Southern Apennine thrust belt evolution: Tectonophysics, v. 252, p. 307–330.

Mazzoli, S., 1992, Structural analysis of the Mesozoic Lagonegro Units in SW Lucania (Southern Italian Apennines): Studi Geologici Camerti, v. 12, p. 117–146.

Mazzoli, S., 1995, Strain analysis in Jurassic argillites of the Monte Sirino area (Lagonegro Zone, Southern Apennines, Italy) and implications for deformation paths in pelitic rocks: Geologische Rundschau, v. 84, p. 781–793.

Mazzoli, S., S. Corrado, M. De Donatis, D. Scrocca, R. W. H. Butler, D. Di Bucci, G. Naso, C. Nicolai, and V. Zucconi, 2000, Time and space variability of "thin-skinned" and "thick-skinned" thrust tectonics in the Apennines (Italy): Rendiconti Lincei, Scienze Fisiche e Naturali, serie 9, v. 11, p. 5–39.

Mazzoli, S., V. Zampetti, and A. Zuppetta, 2001a, Very low-temperature, natural deformation of fine grained limestone: a case-study from the Lucania region, southern Apennines, Italy: Geodinamica Acta, v. 14, p. 213–230.

Mazzoli, S., S. Barkham, G. Cello, R. Gambini, L. Mattioni,

P. Shiner, and E. Tondi, 2001b, Reconstruction of continental margin architecture deformed by the contraction of the Lagonegro Basin, southern Apennines, Italy: Journal of the Geological Society, v. 158, p. 309–319.

Menichetti, M., A. J. De Feyter, and M. Corsi, 1991, CROP 03 – Il tratto Val Tiberina–Mare Adriatico: Sezione geologica e caratterizzazione tettonico-sedimentaria delle avanfosse della zona Umbria-Marchigiano-Romagnola: Studi Geologici Camerti, v. speciale 1991/1, p. 279–293.

Monaco, C., L. Tortorici, and W. Paltrinieri, W., 1998, Structural evolution of the Lucanian Apennines: Journal of Structural Geology, v. 20, p. 617–638.

Morgante, A., M. R. Barchi, S. D'Offizi, G. Minelli, and G. Pialli, 1998, The contribution of seismic modeling to the interpretation of the CROP-03 line: Memorie della Società Geologica Italiana, v. 52, p. 441–455.

Mostardini, F., and S. Merlini, 1986, Appennino centro-meridionale. Sezioni geologiche e proposta di modello strutturale: Memorie della Società Geologica Italiana, v. 35, p. 177–202.

Orange, D. L., and M. B. Underwood, 1995, Patterns of thermal maturity as diagnostic criteria for interpretation of melanges: Geology, v. 23, no. 12, p. 1144–1148.

Parotto, M., 1980, L'Apennin Central, *in* Consiglio Nazionale delle Ricerche, eds., Introduction à la géologie generale d'Italie: 26 Congrès International, Paris, p. 33–37.

Parotto, M., and A. Praturlon, 1975, Geological summary of Central Apennines: Quaderni de "La ricerca scientifica", Consiglio Nazionale delle Ricerche, v. 90, p. 257–311.

Patacca, E., P. Scandone, M. Bellatalla, N. Perilli, and U. Santini, 1992, La zona di giunzione tra l'arco appenninico settentrionale e l'arco appenninico meridionale nell'Abruzzo e nel Molise: Studi Geologici Camerti, v. speciale 1991/2, p. 417–441.

Pescatore, T., 1978, Evoluzione tettonica del Bacino Irpino (Italia meridionale) durante il Miocene. Bollettino della Società Geologica Italiana, v. 97, p. 783–805.

Pescatore, T., P. Renda, and M. Tramutola, 1999, Carta Geologica della Lucania Centrale (1:50,000): Societa del' Elaborazione della Carte, Florence.

Pieri, M., and L. Mattavelli, 1986, Geologic Framework of Italian Petroleum Resources: AAPG Bulletin, v. 70, p.103–130.

Roure, F., P. Casero, and R. Vially, 1991, Growth processes and melange formation in the Southern Apennines accretionary wedge: Earth and Planetary Science Letters, v. 102, p. 395–412.

Sage, L., A. Mosconi, I. Moretti, E. Riva, and F. Roure, 1991, Cross section balancing in the Central Apennines: An application of LOCACE: AAPG Bullettin, v. 75, p. 832–844.

Santantonio, M., 1993, Facies associations and evolution of pelagic carbonate platform/basin systems. Examples from the Italian Jurassic: Sedimentology, v. 40, p. 1039–1067.

Scandone, P., 1972, Studi di geologia lucana: carta dei terreni della serie calcareo-silico-marnosa e note illustrative: Bollettino della Società dei Naturalisti in Napoli, v. 81, p. 225–300.

Scrocca, D., M. Tozzi, and M. Parotto, 1995, Assetto strutturale del settore compreso tra il Matese, le Mainarde e l'Unità di Frosolone. Implicazioni per l'evoluzione neogenica del sistema di sovrascorrimenti nell'Appennino centro-meridionale: Studi Geologici Camerti, v. speciale 1995/2, p. 407–418.

Servizio Geologico Nazionale, 1963, Geological Map of Italy "Larino" 154 Sheet.

Servizio Geologico Nazionale, 1968, Geological Map of Italy "Gaeta" 171 Sheet.

Servizio Geologico Nazionale, 1969, Geological map of Italy "Pesaro" 109 Sheet.

Servizio Geologico Nazionale, 1971, Geological Map of Italy, "Isernia" 161 Sheet.

Tavarnelli, E., 1996, Tethyan heritage in the development of the Neogene Umbria-Marche fold-and-thrust belt, Italy: a 3D approach: Terra Nova, v. 8, p. 470–478.

Tilander, N. G. and R. G. Mitchel, 1995, Processing of seismic data from overthrust areas in Latin America: The Leading Edge, v. 14, p. 707–713.

Valensise, G., D. Pantosti, G. D'Addezio, R. F. Cinti, and L. Cucci, 1993, L'identificazione e la caratterizzazione di faglie sismogenetiche nell'Appennino centro-meridionale: nuovi risultati e ipotesi interpretative: Atti del 12° Convegno Gruppo Nazionale Geofisica della Terra Solida, p. 331–341.

Williams, G. D., C. M. Powell, and M. A. Cooper, 1989, Geometry and kinematics of inversion tectonics, *in* M. A. Cooper and G. D. Williams, eds., Inversion tectonics: The Geological Society of London Special Publication 44, p. 3–16.

Zappaterra, E., 1990, Carbonate paleogeographic sequences of the Periadriatic region: Bollettino della Società Geologica Italiana, v. 109, p. 5–20.